NMR IN STRUCTURAL BIOLOGY

WORLD SCIENTIFIC SERIES IN 20TH CENTURY CHEMISTRY

Consulting Editors: D. H. R. Barton (Texas A&M University)
F. A. Cotton (Texas A&M University)
Y. T. Lee (Academia Sinica, Taiwan)
A. H. Zewail (California Institute of Technology)

Published:

Vol. 1: Molecular Structure and Statistical Thermodynamics
— Selected Papers of Kenneth S. Pitzer
by Kenneth S. Pitzer

Vol. 2: Modern Alchemy
— Selected Papers of Glenn T. Seaborg
by Glenn T. Seaborg

Vol. 3: Femtochemistry: Ultrafast Dynamics of the Chemical Bond
by Ahmed H. Zewail

Vol. 4: Solid State Chemistry
— Selected Papers of C. N. R. Rao
by S. K. Joshi and R. A. Mashelkar

Vol. 5: NMR in Structural Biology
— A Collection of Papers by Kurt Wüthrich
by Kurt Wüthrich

Forthcoming:

Vol. 6: Reason and Imagination: Reflections on Research in Organic Chemistry
— Selected Papers of Derek H. R. Barton
by Derek H. R. Barton

Vol. 7: Frontier Orbitals and Reaction Paths
— Selected Papers of Kenichi Fukui
by Kenichi Fukui and Hiroshi Fujimoto

World Scientific Series in 20th Century Chemistry – Vol. 5

NMR IN STRUCTURAL BIOLOGY

A Collection of Papers by Kurt Wüthrich

Editor

Kurt Wüthrich

Eidgenössische Technische Hochschule Zürich
Switzerland

Published by

World Scientific Publishing Co. Pte. Ltd.

P O Box 128, Farrer Road, Singapore 912805

USA office: Suite 1B, 1060 Main Street, River Edge, NJ 07661

UK office: 57 Shelton Street, Covent Garden, London WC2H 9HE

Library of Congress Cataloging-in-Publication Data

NMR in structural biology : a collection of papers by Kurt Wüthrich /
editor, Kurt Wüthrich.
p. cm. -- (World Scientific series in 20th century chemistry ; vol. 5)
Includes bibliographical references and index.
ISBN 9810222424 (hc) -- ISBN 9810223846 (pbk)
1. Nuclear magnetic resonance spectroscopy. 2. Proteins--Analysis.
3. Nucleic acids--Analysis. I. Wüthrich, Kurt. II. Series.
QP519.9.N83N687 1995
574.19'245--dc20 95-20099
CIP

British Library Cataloguing-in-Publication Data

A catalogue record for this book is available from the British Library.

We are grateful to the following publishers for their permission to reproduce the articles found in this volume:

Academic Press (*Biochem. Biophys. Res. Comm.; J. Magn. Reson.; J. Mol. Biol.*)
American Association for the Advancement of Science (*Science*)
American Chemical Society (*Biochemistry; Inorg. Chem.; J. Am. Chem. Soc.*)
American Society for Biochemistry and Molecular Biology (*J. Biol. Chem.*)
Cell Press (*Cell*)
Cold Spring Harbor Laboratory Press (*DNA and Chromosomes; Cold Spring Harbor Symp. Quant. Biol.*)
Current Biology Ltd. (*Curr. Opinion Struct. Biol.*)
Elsevier Science Publishers B. V. (*Biochim. Biophys. Acta; FEBS Lett.; Progress in Bioorganic Chemistry and Molecular Biology*, ed. Yu. A. Ovchinnikov)
ESCOM Science Publishers B. V. (*J. Biomol. NMR*)
Federation of European Biological Societies (*Eur. J. Biochem.*)
MacMillan Magazines Ltd. (*Nature*)
National Academy Press (*Proc. Natl. Acad. Sci. USA*)
Oxford University Press (*EMBO J.; Nucl. Acids Res.*)
John Wiley and Sons, Inc. (*Biopolymers; Proteins*)

First published 1995
First reprint 1997

Copyright © 1995 by World Scientific Publishing Co. Pte. Ltd. and

All rights reserved. This book, or parts thereof, may not be reproduced in any form or by any means, electronic or mechanical, including photocopying, recording or any information storage and retrieval system now known or to be invented, without written permission from the Publisher.

For photocopying of material in this volume, please pay a copying fee through the Copyright Clearance Center, Inc., 222 Rosewood Drive, Danvers, MA 01923, USA. In this case permission to photocopy is not required from the publisher.

Printed in Singapore by Uto-Print.

Kurt Wüthrich

Kurt Wüthrich's background in physical inorganic chemistry was the stepping stone for a decade of research on the molecular architecture and electronic structure of metalloproteins. He then broadened the scope of his applications of magnetic resonance spectroscopy and computational techniques to other areas of structural biology, with applications in the fields of enzymology, molecular biology and biomedicine, as well as for studies of protein folding and the dynamics of biological macromolecules. From 1977 to 1984 his research group developed the now widely used NMR method for determination of atomic resolution three-dimensional protein structures in solution, which was subsequently extended to studies of receptor-bound drug molecules and the hydration of proteins and nucleic acids in solution. From 1985 to 1994, Wüthrich's laboratory solved more than 30 structures of proteins, protein–DNA and protein–drug complexes. The results of his work have been presented in two monographs, *NMR in Biological Research: Peptides and Proteins* (North Holland, 1976), and *NMR of Proteins and Nucleic Acids* (John Wiley, 1986), and in over 450 papers.

Wüthrich has held numerous endowed lectureships and visiting faculty appointments in the USA, Japan, India and Europe. He is a Foreign Associate of the US National Academy of Sciences, a Foreign Fellow of the Indian National Science Academy and a Member of the Deutsche Akademie der Naturfroscher Leopoldina. The numerous recognitions for his work include the Stein and Moore Award, the Louisa Gross Horwitz Prize, the Marcel Benoist Prize and the Prix Louis Jeantet de Medicine.

PREFACE

During the first three decades of my scientific career, I had the privilege of contributing to the amazing evolution of nuclear magnetic resonance (NMR) techniques from an analytical tool in chemistry to an indispensable method in structural biology as well as in medical diagnosis. In selecting for presentation in this volume 66 out of the close to 500 papers I have published during the period 1963–94, it is my hope that this collection will convey some of the excitement that filled our laboratories during these years.

In an earlier volume in this series, G.T. Seaborg expresses the following feeling: "As a nuclear chemist, I am perhaps unique as an editor in this new series of books because nuclear chemistry and nuclear physics are so closely intertwined." It would appear that the presently recounted development of NMR structure determination of biological macromolecules encompasses an even wider scope of interdisciplinarity. The work covered includes elements of inorganic, medical, organic and physical chemistry as well as biochemistry, molecular biology, cell biology, and experimental and theoretical physics. The interdisciplinary character of our activities is also reflected by our situation at the Eidgenössische Technische Hochschule (ETH): We share a building with the molecular biology institutes of the ETH and the University of Zürich, which was for many years the only building of the physics campus (ETH-Hönggerberg) that was not inhabited by physicists; I am a member of the biology department and an associate member of the physics department; my graduate students may have a background either in solid state physics, mathematical physics, chemistry or biology (see Appendix **A2**). Over the years, we had the opportunity to collaborate with colleagues from a variety of disciplines, mainly biochemistry, cell biology, chemistry and molecular biology. These contacts were essential for the success of many projects and greatly added to the quality of our scientific lives. Similarly to Professor Seaborg, I tend to think that I could turn out to be a unique borderline case as a contributor to this series. Conversely, I would not be overly surprised if similar volumes edited by esteemed friends and colleagues would bear out the notion that much of the success story of 20th century chemistry can be traced back to the encouragement given to individuals in letting them choose unconventional, interdisciplinary approaches in their quest for new frontiers.

The papers in this volume are organized under eight headings. The first part places the material covered in context with the role of structural biology in current biological, biomedical and chemical research, and briefly recounts the route from my University training to my major research themes. The main body of the volume consists of four chapters describing the foundations and selected practical applications of the NMR method for macromolecular structure determination. The remaining three parts are devoted to NMR studies of dynamic aspects of protein structures, hydration of proteins and nucleic acids in solution, and the protein folding problem, respectively. Within each part, the presentation of papers from the primary literature is preceded by introductory notes with additional references, which explain the relation of the contents of the individual papers with each other and with the general theme of the book.

An Appendix (p. 719) lists the academic appointments in my research group at the ETH Zürich from 1970–94, my Ph.D. students and the titles of their theses, the undergraduate students who performed their diploma theses in my laboratory, the visiting scientists and postdoctoral fellows, and the colleagues from other laboratories with whom we collaborated and coauthored publications over these years. The authors of the papers included in this volume are listed in the Author Index, (p. 732) and access to the contents is further facilitated by a Subject Index (p. 734).

This volume is primarily a tribute to the competent scientific contributions of the colleagues and students listed in the Appendix. I am deeply indebted to all of them for their generosity and enthusiasm. My research projects over the years have been supported primarily by the ETH Zürich, the Schweizerischer Nationalfonds, and the Kommission zur Förderung der wissenschaftlichen Forschung (KWF) jointly with Spectrospin AG, Fällanden, Switzerland. I want to thank Dr. Martin Billeter for the preparation of Table 2, and again jointly with Dr. Thomas Szyperski for reading drafts of the introductory texts. I also want to acknowledge the assistance in preparing this volume for publication by Mrs. R. Hug and Mr. R. Marani.

CONTENTS

Part III: Establishing Credibility for NMR Structures of Proteins

Part IV: Evolution of NMR Structure Determination with Biological Macromolecules 1985–94

Part V: Selected NMR Structure Determinations

Part VII: Hydration of Proteins and Nucleic Acids in Solution

Part VIII: Protein Folding

Appendices and Indices

I

NMR AND STRUCTURAL BIOLOGY

Introduction to papers 1 through 3

During the past three decades the use of physico-chemical techniques for the characterization of biological macromolecules has attracted the interest of a steadily increasing number of chemists and physicists, and the area of structural biology, in particular atomic resolution studies of the three-dimensional structure of biological macromolecules and their intermolecular interactions, has never before had as central a role in biological and biomedical research as today. Much of this popularity can be traced to the fact that although a virtually unlimited array of different polypeptide sequences can be generated with the use of DNA recombination and chemical synthesis, the relation of the resulting primary structures with corresponding biological functions can only be rationalized through knowledge of the three-dimensional structure. In addition to its key role in academic institutions, structure determination of proteins, nucleic acids and other classes of biomolecules has therefore become a discipline that is actively pursued also in profit-oriented organizations, especially in the major pharmaceutical companies.

X-ray diffraction with single crystals has from the initial structure determinations of myoglobin[1] and hemoglobin[2] until 1984 been unique in its potential for efficient determination of three-dimensional protein structures. Presently, it shares this role with nuclear magnetic resonance (NMR) spectroscopy in solution, as is clearly documented by annual surveys of new experimental macromolecular structures,[3] which are summarized in Table 1. The entries in the column "Other Methods" (Table 1) refer to two structures solved by X-ray fibre diffraction and by electron diffraction in crys-

---

[1] Kendrew, J.C. (1963) *Science 139*, 1259–1266. Myoglobin and the structure of proteins.

[2] Perutz, M.F. (1963) *Science 140*, 863–869. X-ray analysis of hemoglobin.

[3] Hendrickson, W.A. and Wüthrich, K., eds. (1994) *Macromolecular Structures 1994*. London: Current Biology.

Table 1. *New three-dimensional structures of biological macromolecules published during the period 1990–93.* Structures determined by X-ray crystallography, NMR, or by any other method are listed separately (from Ref. 3).

Year	X-rays (single crystals)	NMR (solution)	Other Methods
1990	109	23	2
1991	123	36	–
1992	168	61	–
1993	207	59	–

tals, respectively. These techniques can provide the desired information, but applications have mainly been focussed on highly complex systems so that the actual number of structures solved is still very small. Other experimental techniques that are traditionally associated with the structural biology of proteins and nucleic acids and have important roles in the empirical identification of different conformational states, *e.g.*, circular dichroism and other optical spectroscopies, or measurements based on studies of thermodynamic or hydrodynamic properties of macromolecules in solution, do not appear in Table 1 because they cannot provide sufficient data for *de novo* three-dimensional structure determinations. Considering that the number of structure determinations listed in Table 1 exceeds by far the total number of macromolecular structures solved up to 1989, we further learn that knowledge of three-dimensional structures is still very limited when compared to the available information on amino acid sequences in proteins, or nucleotide sequences in nucleic acids. Scarcity of data on three-dimensional structures is indeed still a major bottleneck in protein engineering and rational drug design.

Today, NMR solution structures of proteins and nucleic acids are an integral part of structural biology. However, in addition to structure determination, NMR spectroscopy can provide a wealth of information on dynamic properties of biological macromolecules and on associated conformational equilibria. Much of this information was in principle accessible long before NMR could be applied for *de novo* three-dimensional protein structure determination, but in conjunction with the availability of NMR-derived structures a more detailed characterization of dynamic phenomena can now be obtained. To provide a frame of reference for placing subsequent, more technical material on the different aspects of NMR with biomacromolecules into perspective, paper **1** is a "minireview" describing the state of NMR spectroscopy in structural biology in 1990.

During my student years from 1957–1964, NMR spectroscopy was just being introduced as an analytical tool in chemistry, molecular biology was not yet established as an independent discipline, and the initial three-dimensional protein crystal structures at atomic resolution were just emerging.[1,2] Clearly, my formal education could not possibly cover the areas of our current research. Starting in 1957 at the University of Bern, I studied inorganic chemistry with Prof. W. Feitknecht, physical chemistry with Prof. K. Huber, organic chemistry with Prof. R. Signer, biochemistry with Prof. H. Nitschmann, physics with Profs. J. Geiss, F.G. Houtermans and H. Schilt, and mathematics with Profs. H. Hadwiger, W. Nef and W. Scherrer. The emphasis was on the one hand on linear algebra, classical mechanics and chemical thermodynamics, on the other hand on physical chemistry of synthetic polymers and preparative biochemistry of proteins and nucleic acids. Only much later did I fully appreciate the extent to which this combination of undergraduate studies would provide an excellent foundation for my later scientific activities. In the spring of 1962, I enrolled as a student in the "Turn- und Sportlehrerkurs" at the University of Basel. In addition to about 25 weekly hours of intense physical exercise, these studies included premedical courses in human anatomy and physiology. Combined with experience gained from observations made in the pursuit of competitive sports, this provided an additional, important dimension to my education. In the fall of 1962 I also started activities to complement my training in chemistry. This included work on a Ph.D. thesis in inorganic chemistry with Prof. S. Fallab, and intensive courses in synthetic organic chemistry with Prof. C. Grob, in natural product chemistry with Prof. T. Reichstein, and in quantum chemistry with Prof. H. Labhardt. The subject of my Ph.D. thesis was the complex formation of copper ions with a variety of ligands, and studies on the catalytic activity of free and complexed copper ions toward autoxidation of aromatic amines.[4] In addition to the competent and amiable guidance by Prof. Fallab, two facets of my thesis project deserve special mention relative to my later scientific activities. Firstly, the project led to my initial practical experience with magnetic resonance spectroscopy, using a state-of-the-art electron paramagnetic resonance (EPR) spectrometer available in the physics institute at the University of Basel.[5] Secondly, although the actual experiments were performed with low molecular weight metal complexes, our intents were mainly focused on structure–function correlations in copper-containing metalloproteins. My formal University education was completed in March 1964, when I obtained my Ph.D. in chemistry and the "Eidgenössisches Turn- und Sportlehrerdiplom."

[4] Wüthrich, K. and Fallab, S. (1964) *Helv. Chim. Acta 47*, 1609–1616. Einfluss verschiedener Liganden auf den Mechanismus der Kupfer(II)-katalysierten Autoxydation von o-Phenylendiamin.

[5] Wüthrich, K., Loeliger, H. and Fallab, S. (1964) *Experientia 20,* 599–601. Elektronenspinresonanzmessungen zur Untersuchung von Kinetik und Mechanismus von Cu^{2+}-katalysierten Reaktionen.

After finishing my graduate studies, I obtained a postdoctoral fellowship from the *Stiftung für Stipendien auf dem Gebiete der Chemie* (Basel, Switzerland), which enabled me to spend another year in Basel concentrating on EPR studies of Cu^{2+} and VO^{2+} complexes in solution,[6] and from the spring of 1965 onward to join Prof. R.E. Connick at the University of California, Berkeley, for further investigations of vanadyl coumpounds, using nuclear spin relaxation measurements of ^{17}O, ^{2}H and ^{1}H in addition to EPR.[7] Paper **2** is representative of the research pursued during those years. In view of our recent studies on protein and nucleic acid hydration (see Part VII) it is of interest that the main emphasis of my activities in the laboratory of Prof. Connick was on studies of kinetic aspects of the hydration of metal ions and metal complexes.[7,8] Besides the experimental projects, the Berkeley period was devoted to intensive course work on the theory of nuclear spin relaxation, mostly with Prof. Connick, and on group theory and quantum mechanics, mostly with Prof. M. Tinkham in the physics department. In the summer of 1967 I spent three months as a visitor with Prof. H.H. Günthard at the physical chemistry laboratory of the ETH Zürich, where I participated in quantum-chemical studies on the electronic states in condensed aromatic systems.

In October 1967 I joined the biophysics department of Dr. R.G. Shulman at Bell Telephone Laboratories in Murray Hill, New Jersey. Shortly after my arrival, a super-

[6] Wüthrich, K. (1965) *Helv. Chim. Acta. 48,* 1012–1017. Elektronenspinresonanz-Untersuchungen von VO^{2+}-Komplexverbindungen in wässeriger Lösung II.

[7] Wüthrich, K. and Connick, R.E. (1967) *Inorg. Chem. 6,* 583–590. Nuclear magnetic resonance relaxation of oxygen-17 in aqueous solutions of vanadyl perchlorate and the rate of elimination of water molecules from the first coordination sphere.

[8] Connick, R.E. and Wüthrich, K. (1969) *J. Chem. Phys. 51,* 4506–4508. ^{17}O nuclear magnetic relaxation in aqueous solutions of diamagnetic metal ions.

conducting high resolution ^{1}H NMR spectrometer operating at 220 MHz was delivered to our department, and I was given responsibility for its maintenance and alotted plenty of instrument time for research on "protein structure and function." It turned out that with the exception of a high resolution NMR study of bicyclobutane,[9] all my projects at Bell Labs had to do with hemoproteins. Due to my background my interest was focused on the metal centers rather than the polypeptide chains. In an early, seminal paper by McDonald and Phillips,[10] I was intrigued by a description of NMR spectral differences between the reduced, diamagnetic and the oxidized, paramagnetic forms of cytochrome *c*; based on my work with Prof. Connick this seemed to present novel perspectives for studies of hemoproteins. Collaborations were then established, with Dr. J. Peisach on myoglobin, with Prof. E. Margoliash, who kindly supplied me with various cytochromes *c* and for many years with expert advice on this class of proteins, and with Dr. T. Yamane at Bell Labs, who prepared hemoglobin KW using blood sampled from my arm in the first aid station. Within a few months, numerous well separated resonance lines arising from electron–proton contact and pseudocontact coupling, and from interactions with the porphyrin ring current field were identified in the ^{1}H NMR spectra of all these proteins. This opened new avenues to information on conformation changes during the oxygenation of myoglobin and hemoglobin,[11] and on the electronic structure of the heme groups in different classes of hemoproteins.[12] Paper **3** is a typical example

[9] Wüthrich, K., Meiboom, S. and Snyder, L.C. (1970) *J. Chem. Phys. 52,* 230–233. Nuclear magnetic resonance spectroscopy of bicyclobutane.

[10] McDonald, C.C. and Phillips, W.D. (1967) *J. Am. Chem. Soc. 89,* 6332–6341. Manifestations of the tertiary structures of proteins in high-frequency nuclear magnetic resonance.

[11] Shulman, R.G., Ogawa, S., Wüthrich, K., Yamane, T., Peisach, J. and Blumberg, W.E. (1969) *Science 165,* 251–257. The absence of "heme–heme" interactions in hemoglobin.

[12] Wüthrich, K. (1970) *Structure and Bonding 8,* 53–121. Structural studies of hemes and hemoproteins by nuclear magnetic resonance spectroscopy.

of the work pursued during this period. The focus was on the metal center with the coordinatively linked ligands and its immediate environment in the protein. Within the folded protein, some elementary complexation reactions were performed. These resulted in the determination of an essential local detail of the three-dimensional cytochrome *c* structure, which had at the time not been accessible to X-ray crystal studies[13] but was independently inferred by comparative NMR studies of a group of evolutionarily related cytochromes *c*.[14] Paper **3** illustrates that I found a quite direct way from physical inorganic chemistry to the field of NMR with biological macromolecules. A group of colleagues with a similar background were among those who had a successful early start with NMR studies of proteins. Working with metalloproteins, or using paramagnetic metal ions as extrinsic chemical shift reagents had the advantage that nuclear spins located near the metal centers would experience large changes in chemical shift and/or relaxation times. Even with the limited sensitivity and spectral resolution of the NMR instrumentation available around 1970, these effects were quite readily accessible for precise measurements and detailed interpretation, which was in stark contrast to most other NMR spectral features of polypeptide or polynucleotide chains. For many years after I moved in 1970 from the Bell Telephone Laboratories to my present position at the ETH Zürich, our work involved projects with

[13] Dickerson, R.E. and Geis, I. (1969) *The structure and action of proteins,* p.63. New York: Harper & Row.

[14] McDonald, C.C., Phillips, W.D. and Vinograde, S.N. (1969) *Biochem. Biophys. Res. Commun. 36,* 442–449. Proton magnetic resonance evidence for methionine–iron coordination in mammalian-type ferrocytochrome *c*.

metalloproteins and metal-containing prosthetic groups,[15–19] and the unique spectral features of cytochromes *c* (see paper **3**) were an essential help in the early stages of the development of NMR as a method for three-dimensional protein structure determination (see Part II).

[15] Keller, R.M. and Wüthrich, K. (1972) *Biochim. Biophys. Acta 285,* 326–336. The electronic g-tensor in cytochrome b_5 : high resolution proton magnetic resonance studies.

[16] Wüthrich, K., Hochmann, J., Keller, R.M., Wagner, G., Brunori, M. and Giacometti, G. *J. Magn. Reson. 19,* 111–113 (1975). ^{1}H NMR relaxation in high spin ferrous hemoproteins.

[17] Senn, H., Keller, R.M. and Wüthrich, K. (1980) *Biochem. Biophys. Res. Comm. 92,* 1362–1369. Different chirality of the axial methionine in homologous cytochromes *c* determined by ^{1}H NMR and CD spectroscopy.

[18] Keller, R.M. and Wüthrich, K. (1981) in *Biological Magnetic Resonance* (L.J. Berliner and J. Reuben, eds.) Vol. 3, pp. 1–52. New York: Plenum Press. Multiple irradiation ^{1}H NMR experiments with hemoproteins.

[19] Senn, H. and Wüthrich, K. (1985) *Q. Rev. Biophys. 18,* 111–134. Amino acid sequence, haem iron coordination geometry and functional properties of mitochondrial and bacterial *c*-type cytochromes.

Minireview

THE JOURNAL OF BIOLOGICAL CHEMISTRY
Vol. 265, No. 36, Issue of December 25, pp. 22059-22062, 1990
© 1990 by The American Society for Biochemistry and Molecular Biology, Inc.
Printed in U.S.A.

Protein Structure Determination in Solution by NMR Spectroscopy*

Kurt Wüthrich

From the Institut für Molekularbiologie und Biophysik, Eidgenössische Technische Hochschule-Hönggerberg, CH-8093 Zürich, Switzerland

The introduction of nuclear magnetic resonance (NMR) spectroscopy as a second method for protein structure determination at atomic resolution, in addition to x-ray diffraction in single crystals, has already led to a significant increase in the number of known protein structures. The NMR method provides data that are in many ways complementary to those obtained from x-ray crystallography and thus promises to widen our view of protein molecules, giving a clearer insight into the relation between structure and function.

Biological Macromolecules and NMR Spectroscopy

The first nuclear magnetic resonance (NMR) experiments with biological macromolecules were reported more than 30 years ago (1), and with the advent of modern NMR techniques in the late 1960s and early 1970s, which included superconducting magnets, Fourier transform spectroscopy, and computer control of the instrumentation, NMR spectroscopy yielded an ever widening array of insights into the behavior of such molecules. Examples are studies of protein conformation changes, denaturation and internal mobility, pH titration of individual ionizable amino acid side chains in enzyme active sites, observation of hydrogen-bonded imino protons in tRNA, investigation of paramagnetic centers in metalloproteins, etc. (for surveys see Refs. 2 and 3). In addition, NMR in solution has become a technique for protein three-dimensional structure determination at atomic resolution (4), which is the subject of this review.

Survey of Protein Structure Determination by NMR

Fig. 1 presents an outline of the method (4, 5) that covers the preparation of the protein for the NMR experiments, the NMR measurements, the crucial problem of obtaining assignments of the NMR lines to individual atoms in the polypeptide chain, and two separate avenues for the structural interpretation of the NMR data.

Sample Preparation—The protein is usually dissolved in 0.5 ml of water, and the ionic strength, pH, and temperature may be adjusted so as to ensure near-physiological conditions (it is advantageous to work in the slightly acid pH range from 3 to 5 (4)). The protein concentration should be at least 1 mM, ideally 3–6 mM, so that 15–30 mg of a protein with molecular weight 10,000 should be available for a structure determination. Although this concentration is high relative to that of most proteins in their physiological milieu, it is not far from the *total* protein concentration in many body fluids. So far, structure determinations by NMR have been reported for proteins with molecular weights up to approximately 15,000. This upper size limit may perhaps be raised to about 30,000. For molecular sizes above approximately 12,000 the NMR study will have to include the preparation of protein enriched with ^{15}N and/or ^{13}C, which is best achieved with biosynthetic techniques.

NMR Measurements—Because of the large number of hydrogen atoms in a protein, a one-dimensional ^1H NMR spectrum is crowded with mutually overlapping lines. Therefore, two-dimensional (2D)[1] and three-dimensional (3D) NMR experiments are used. Fig. 2 shows a small region of a homonuclear 2D ^1H NMR spectrum. The NMR peaks are spread out along the two frequency axes ω_1 and ω_2, and they are therefore quite well separated. In the 3D spectrum of Fig. 3 the NMR peaks have been further spread out along a third frequency axis, which corresponds to the NMR frequencies of the ^{15}N spins in the ^{15}N-labeled protein. As a result, the NMR peaks of the 2D ^1H-^1H spectrum (Fig. 2) are distributed among several ^1H-^1H planes, typically 64 or 128. The ensuing further improved separation of the peaks is indispensable for work with larger proteins.

For an intuitive understanding of the information contained in the spectral region of Fig. 2 it may be helpful to imagine that the region from 3.75 to 4.15 ppm of the 1D ^1H NMR spectrum, which contains the resonance lines of α protons, is along ω_1, and the 1D ^1H NMR spectrum from 7.4 to 8.0 ppm, which contains resonance lines from amide protons and aromatic protons, is along ω_2. All the peaks seen in Fig. 2 are "cross-peaks" manifesting an interaction between a resonance line in the 1D spectrum along ω_1 and a line in the 1D spectrum along ω_2. For example, the peak Y11α-Q12N at the bottom of the figure correlates the α proton resonance of Tyr-11 along ω_1 with the backbone amide proton resonance of Gln-12 along ω_2. Depending on the experiment used, the cross-peaks manifest different types of interactions between the spins. The most important information needed for a *de novo* 3D structure determination can be obtained from nuclear Overhauser enhancement (NOE) spectroscopy, or NOESY. In a properly executed (see below) NOESY experiment (Fig. 2) a cross-peak between two hydrogen atoms is observed only if those two protons are separated by a shorter distance than approximately 5.0 Å. Since the NOE depends on the *through-space* distance, the locations of the two interacting protons in the primary structure may be far apart, as much as 100 residues or more (Fig. 4). In different 2D NMR experiments, which are primarily used to support obtaining the ^1H NMR assignments, the cross-peaks manifest *through-bond* relations between protons that are separated by not more than three covalent bonds, *i.e.* which are part of the same amino acid residue. Frequently used experiments are correlation spectroscopy (COSY) and total correlation spectroscopy (TOCSY) (4, 6).

In contrast to all other NMR parameters, ^1H-^1H distance measurements by NOE experiments can be directly related to the protein conformation. Recognizing that *NOE buildup experiments* (7), which enable one to eliminate derogatory effects from spin diffusion (8–10), yield reliable distance measurements in macromolecules was therefore one of three fundamental elements that constitute the foundations of the NMR method for protein structure determination. NOE

* This research was supported by the Schweizerischer Nationalfonds, Project 31.25174.88.

[1] The abbreviations used are: 1D(2D,3D), one-dimensional (...); NOE, nuclear Overhauser enhancement; NOESY, 2D NOE spectroscopy; COSY, 2D correlation spectroscopy; TOCSY, 2D total correlation spectroscopy; RMSD, root mean square deviation.

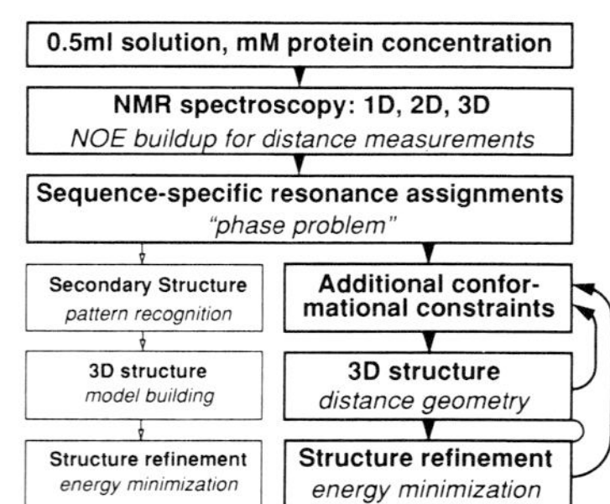

FIG. 1. **Diagram outlining the course of a protein structure determination by NMR.**

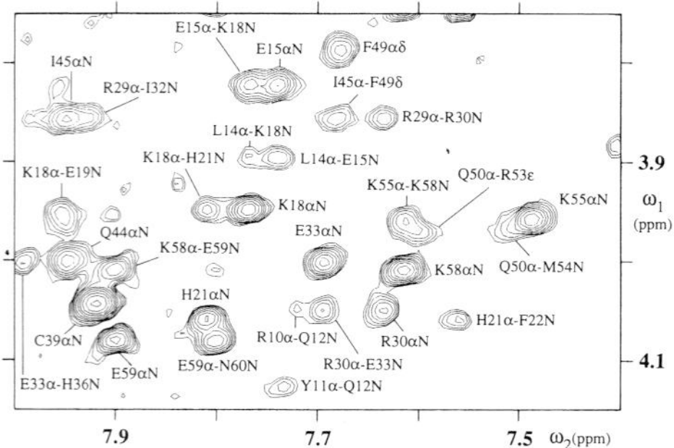

FIG. 2. **Contour plot of a small region from a ^{1}H soft-NOESY spectrum of the *Antennapedia* homeodomain.** To indicate the pairs of hydrogen atoms for which close proximity is evidenced by these data, the cross-peaks are identified either by the one-letter amino acid symbols of 2 residues, their sequence positions, and the proton types, or for intraresidual NOEs by the identification of 1 residue and two proton types. This spectral region was taken from a two-dimensional spectrum, but a similar presentation would be obtained by taking a part of one of the ^{1}H(ω_1)-^{1}H(ω_3) planes of the three-dimensional spectrum of Fig. 3 (reproduced from Ref. 22).

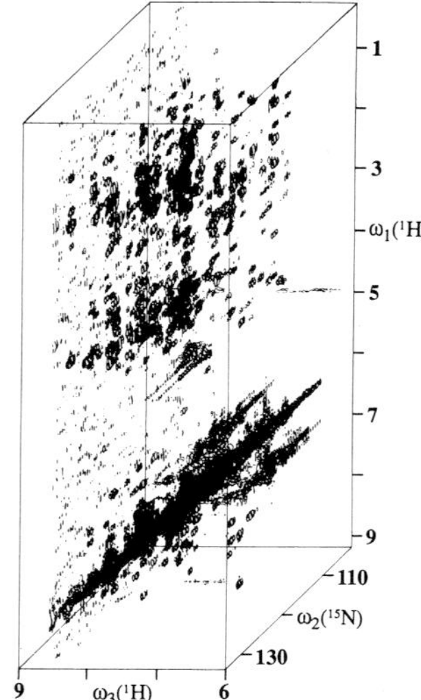

FIG. 3. **Three-dimensional ^{15}N-correlated ^{1}H-^{1}H NOESY spectrum of the N-terminal domain comprising residues 1–76 of the P22 c2 repressor.** The protein was uniformly labeled with ^{15}N to the extent of ≳95%.

buildup curves can be recorded with 1D experiments (7, 11) or with 2D and 3D NMR techniques (12–15).

Resonance Assignments—As proteins contain multiple units of the individual amino acids, spectral assignments are

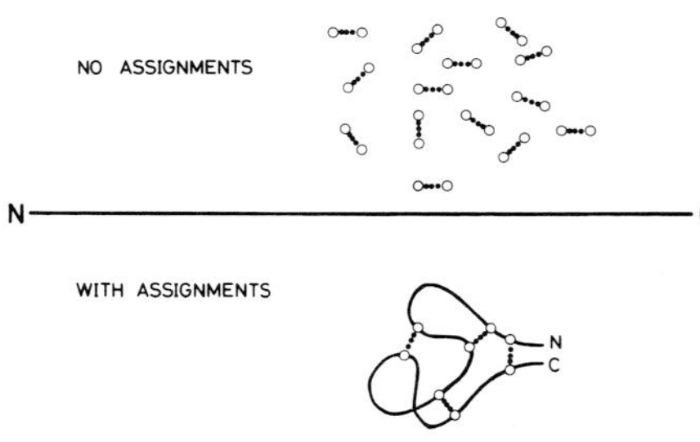

FIG. 4. **Scheme illustrating the information content of ^{1}H-^{1}H NOEs in a polypeptide chain (represented by the *horizontal line* in the center) with and without sequence-specific resonance assignments.** *Open circles* represent hydrogen atoms of the polypeptide, and *dotted lines* the short ^{1}H-^{1}H distances of less than 5 Å manifested by the NOEs (see text) (reproduced from Ref. 4).

nontrivial. The problem was solved with the *sequential assignment strategy* (4, 16–19), which can rely entirely on prior knowledge of the amino acid sequence and the use of homonuclear ^{1}H NMR experiments. For larger proteins it may profitably be supported with heteronuclear NMR experiments using isotope-labeled proteins (Fig. 3 (4, 20)). The importance of the resonance assignments is illustrated with Figs. 2–4. In the absence of sequence-specific resonance assignments each NOESY cross-peak (Fig. 3) merely indicates the presence of two nearby hydrogen atoms in the protein (*top* of Fig. 4). When resonance assignments have been made (Fig. 2), each cross-peak specifies an upper limit to the distance between two distinct locations along the polypeptide chain (*bottom* of Fig. 4). These distance constraints are the input needed for a structure determination (4, 5, 21). The sequential assignment strategy in its impact on the NMR structure determination method can be compared with the use of isomorphous heavy atom derivatives for solving the phase problem in protein crystallography (23, 24), and it is a second basic element of the method.

Structure Determination from NMR Data—As a by-product of the sequential resonance assignment procedure, the location of helical secondary structures, β-sheets and tight turns in the amino acid sequence can be identified (4, 17, 25, 26, 31; *lower left part* of Fig. 1). Therefore, one often finds preliminary reports on NMR studies of a protein that describe the resonance assignments and the secondary structure. The secondary structures so identified can be used as a starting point for interactive model building of the tertiary structure (*e.g.* Refs. 27–29), but this strategy has been little used as compared to computational structure determination (outlined in the *lower right* of Fig. 1).

The maximum possible number of conformational constraints must be collected as input for the calculation of the complete three-dimensional protein structure (30). Because the NMR data are entirely different from those obtained by x-ray diffraction, new techniques had to be developed for their structural interpretation (4, 32, 33), which is the third basic element of the method. The first structure calculations from NMR data used metric matrix distance geometry ((34–38) Fig. 5). Alternative techniques are a variable target function algorithm (39) and restrained molecular dynamics calculations (40). In present practice most structure determination protocols include either a structural interpretation of the NMR data with a variable target function algorithm supplemented by molecular mechanics energy minimization (*e.g.* Ref. 41) or an initial analysis with metric matrix distance geometry followed by molecular dynamics calculations (*e.g.* Ref. 42).

Each successful calculation with one of the aforementioned computational procedures yields a molecular structure that represents a good fit of the experimental data. To check that

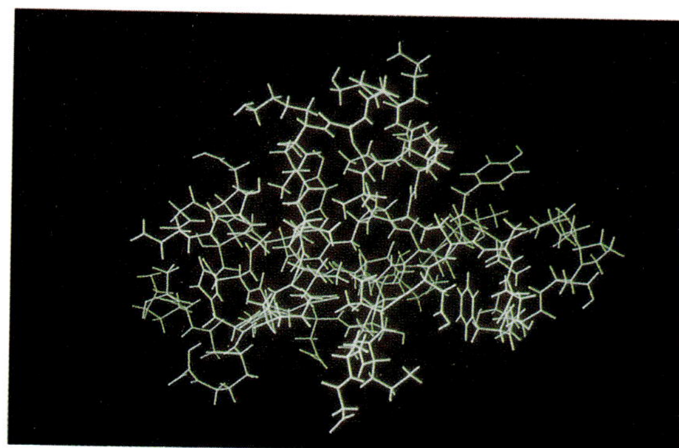

FIG. 5. **Three-dimensional structure of bull seminal proteinase inhibitor IIA (BUSI IIA) determined from NMR measurements in aqueous solution (38).** This protein consists of a polypeptide chain with 57 residues. All bonds connecting heavy atoms are shown.

the NMR data determine a *unique* three-dimensional structure, a group of conformers is compared that was obtained from a series of calculations using the same input but different, randomly chosen starting conditions (see below).

Since the first structure determination of a globular protein in solution was completed in 1984 (38) (Fig. 5), the method has been surveyed in a monograph (4), and special technical aspects were described in two volumes of *Methods in Enzymology* and in numerous reviews (5, 32, 43–53), so that ample reference material is available.

Protein Structures in Solution and in Single Crystals

There is already an impressive list of three-dimensional NMR structures of proteins that have never been crystallized, including, for example, the *Antennapedia* homeodomain from *Drosophila* (54), several zinc finger proteins (55, 56), epidermal growth factors (57, 58), and interleukin 8 (59).

Even when protein crystals are available it is often difficult to obtain isomorphous heavy atom derivatives suitable for solving the crystallographic phase problem (23, 24). A Patterson rotation search (24) using an independently solved NMR structure of the same or a closely related protein in solution may then be a viable alternative route to the phase determination (60, 61).

The availability of two methods for protein three-dimensional structure determination may have a useful mutual control function. For example, a crystal structure of rat metallothionein-2 (62) that was different from the NMR structure in solution (63, 64; Fig. 6) was subsequently found to need revision, and a new crystal structure of this metallothionein is nearly identical with the solution structure (65).

For several globular proteins a close similarity was observed between the molecular architectures in single crystals and in solution (*e.g.* Refs. 5, 38, 46, 61, 66, 67), including hydrogen-bonded secondary structures and the spatial arrangement of the interior amino acid side chains (61, 66). Protein surface areas, however, have been found to have, as a rule, significantly different structure and dynamic properties in crystals and in solution. The complementary information on the molecular surface obtained with the two methods is of special interest, since protein functions depend largely on the nature of molecular surface areas in direct contact with the substrates.

For polypeptides that do not form globular structures one may quite generally expect to find different conformations in crystals and in noncrystalline milieus. The polypeptide hormone glucagon is a typical example (34, 35).

Evaluating the Quality of a Protein Structure Determination by NMR in Solution

Figs. 5–7 illustrate different facets of protein structures calculated from NMR data. A structure calculation always uses the complete polypeptide chain with the amino acid side chains (Fig. 5) and possibly additional non-peptide components (Fig. 6) (otherwise the steric constraints would not be properly accounted for). A physically meaningful presentation of the solution structure consists of a superposition of a group of conformers calculated with different starting conditions from the same NMR data (4, 5, 32, 37). For clarity only the polypeptide backbone is usually drawn in such presentations (Fig. 7).

A critical assessment of a NMR structure determination can be based on the facts that nearly complete sequence-specific resonance assignments are indispensable as a basis for a structure determination (21), that the quality of a structure determination is improved if stereospecific assignments are obtained for the prochiral centers (68, 69), that at least 10 conformational constraints per residue should have been measured, and that each individual structure calculation must represent an acceptable fit of the experimental data, with small residual violations of the NMR and steric constraints. If the structure calculation is repeated with different starting conditions, a high quality structure determination generates a tight bundle of conformers, which corresponds to

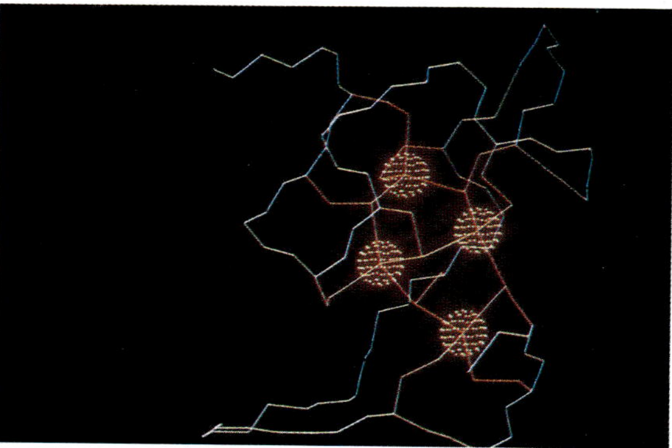

FIG. 6. **Three-dimensional structure of the α-domain from rat metallothionein-2 determined by NMR in solution (64).** The *purple line* represents the polypeptide backbone from residues 31–61, the *dotted spheres* the four metal ions, and the *orange lines* the 11 Cys side chains and the coordinative bonds between Cys sulfur atoms and the metal ions.

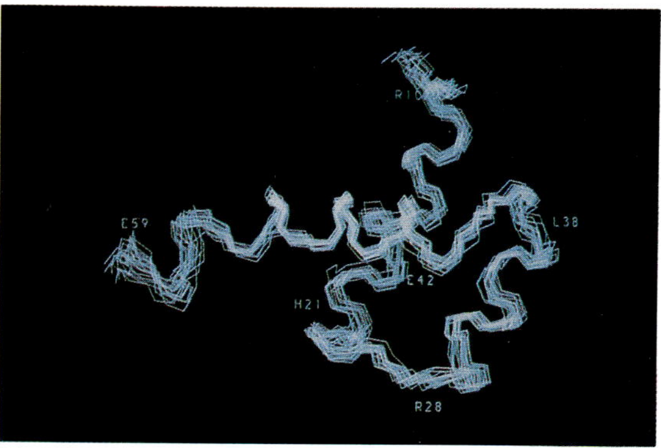

FIG. 7. **Three-dimensional structure of the *Antennapedia* homeodomain determined by NMR in solution.** Only the backbone of residues 7–59 of this 68-residue polypeptide is shown. A group of 19 conformers representing the solution structure have been superimposed for minimal RMSD (see text). The positions of selected residues are identified (reproduced from Ref. 54).

a small value for the average of the pairwise root mean square deviations (RMSDs). As an illustration, for the *Antennapedia* (*Antp*) homeodomain complete sequence-specific resonance assignments were obtained, with stereospecific assignments for 33 out of a total of 88 pairs of diastereotopic substituents. For the 53 residues shown in Fig. 7, 583 conformational constraints were collected. On average each of the 19 computed structures in Fig. 7 had eight residual NOE distance constraint violations larger than 0.2 Å, none of which exceeded 0.25 Å, and the average RMSD among the 19 conformers was 0.9 Å for the backbone atoms, 0.9 Å for the backbone plus interior "core" side chains, and 2.0 Å for the complete polypeptide structure 7–59. These numbers are representative of a high quality NMR structure determination, where the core of the molecule is comparable to a crystal structure refined at a crystallographic resolution of approximately 2.0 Å (66), but important parts of the protein surface may be dynamically disordered.

Conclusions and Outlook

NMR structure determination may be applied efficiently to small proteins with molecular weights up to about 12,000. For the architecture of the core of globular proteins the result of a high quality NMR structure determination is comparable to that achieved in a high resolution x-ray crystal structure. The onset of structural disorder toward the molecular surface is more pronounced in NMR protein structures than in the corresponding x-ray structures. One line of future research will undoubtedly focus on the functional significance of this observation.

Much effort is currently concentrated on improved experimentation, in particular with the use of stable isotopes, to extend NMR structure determination to larger proteins, maybe in the size range 15,000–30,000 (48–50, 52). The potential of the NMR method for studies of intermolecular interactions, for example, with the use of isotope-edited ^{1}H NMR spectroscopy (51, 70) and for structural and kinetic studies relating to the protein folding problem (*e.g.* Refs. 71–73) may be even more attractive with regard to obtaining new fundamental insights to be used as a platform for the design of functionally improved proteins.

Acknowledgments—I owe a great debt to a large number of colleagues who participated in this work and whose names appear in the references. I thank Dr. K. V. Chary and Dr. G. Otting for Fig. 3, and R. Marani for the careful processing of the manuscript.

REFERENCES

1. Saunders, M., Wishnia, A., and Kirkwood, J. G. (1957). *J. Am. Chem. Soc.* **79**, 3289–3290
2. Wüthrich, K. (1976) *NMR in Biological Research: Peptides and Proteins*, North-Holland Publishing Co., Amsterdam
3. Jardetzky, O., and Roberts, G. C. K. (1981) *NMR in Molecular Biology*, Academic Press, New York
4. Wüthrich, K. (1986) *NMR of Proteins and Nucleic Acids*, John Wiley & Sons, Inc., New York
5. Wüthrich, K. (1989) *Science* **243**, 45–50
6. Ernst, R. R., Bodenhausen, G., and Wokaun, A. (1987) *Principles of Nuclear Magnetic Resonance in One and Two Dimensions*, Clarendon Press, Oxford
7. Gordon, S. L., and Wüthrich, K. (1978) *J. Am. Chem. Soc.* **100**, 7094–7096
8. Solomon, I. (1955) *Physical. Rev.* **99**, 559–565
9. Hull, W. E., and Sykes, B. D. (1975) *J. Chem. Phys.* **63**, 867–880
10. Kalk, A., and Berendsen, H. J. C. (1976) *J. Magn. Reson.* **24**, 343–366
11. Wagner, G., and Wüthrich, K. (1979) *J. Magn. Reson.* **33**, 675–680
12. Kumar, A., Wagner, G., Ernst, R. R., and Wüthrich, K. (1981) *J. Am. Chem. Soc.* **103**, 3654–3658
13. Griesinger, C., Sørensen, O. W., and Ernst, R. R. (1987) *J. Magn. Reson.* **73**, 574–579
14. Fesik, S. W., Gampe, R. T., Jr., Zuiderweg, E. R. P., Kohlbrenner, W. E., and Weigl, D. (1989) *Biochem. Biophys. Res. Commun.* **159**, 842–847
15. Marion, D., Kay, L. E., Sparks, S. W., Torchia, D. A., and Bax, A. (1989) *J. Am. Chem. Soc.* **111**, 1515–1517
16. Dubs, A., Wagner, G., and Wüthrich, K. (1979) *Biochim. Biophys. Acta* **577**, 177–194
17. Billeter, M., Braun, W., and Wüthrich, K. (1982) *J. Mol. Biol.* **155**, 321–346
18. Wagner, G., and Wüthrich, K. (1982) *J. Mol. Biol.* **155**, 347–366
19. Wider, G., Lee, K. H., and Wüthrich, K. (1982) *J. Mol. Biol.* **155**, 367–388
20. Le Master, D. M., and Richards, F. M. (1988) *Biochemistry* **27**, 142–150
21. Wüthrich, K., Wider, G., Wagner, G., and Braun, W. (1982) *J. Mol. Biol.* **155**, 311–319
22. Billeter, M., Qian, Y. Q., Otting, G., Müller, M., Gehring, W. J., and Wüthrich, K. (1990) *J. Mol. Biol.* **214**, 183–197
23. Green, D. W., Ingram, V. M., and Perutz, M. F. (1954) *Proc. R. Soc. Edinb. Sect. A (Math. Phys. Sci.)* **225**, 287–307
24. Blundell, T. L., and Johnson, L. N. (1976) *Protein Crystallography*, Academic Press, New York
25. Wüthrich, K., Billeter, M., and Braun, W. (1984) *J. Mol. Biol.* **180**, 715–740
26. Zuiderweg, E. R. P., Kaptein, R., and Wüthrich, K. (1983) *Proc. Natl. Acad. Sci. U. S. A.* **80**, 5837–5841
27. Zuiderweg, E. R. P., Billeter, M., Boelens, R., Scheek, R. M., Wüthrich, K., and Kaptein, R. (1984) *FEBS Lett.* **174**, 243–247
28. Billeter, M., Engeli, M., and Wüthrich, K. (1985) *J. Mol. Graphics* **3**, 79–83, 97–98
29. Kaptein, R., Zuiderweg, E. R. P., Scheek, R. M., Boelens, R., and van Gunsteren, W. F. (1985) *J. Mol. Biol.* **182**, 179–182
30. Kline, A. D., Braun, W., and Wüthrich, K. (1988) *J. Mol. Biol.* **204**, 675–724
31. Pardi, A., Billeter, M., and Wüthrich, K. (1984) *J. Mol. Biol.* **180**, 741–751
32. Braun, W. (1987) *Q. Rev. Biophys.* **19**, 115–157
33. Crippen, G. M., and Havel, T. (1988) *Distance Geometry and Molecular Conformation*, John Wiley & Sons, Inc., New York
34. Braun, W., Bösch, C., Brown, L. R., Gō, N., and Wüthrich, K. (1981) *Biochim. Biophys. Acta* **667**, 377–396
35. Braun, W., Wider, G., Lee, K. H., and Wüthrich, K. (1983) *J. Mol. Biol.* **169**, 921–948
36. Havel, T. F., and Wüthrich, K. (1984) *Bull. Math. Biol.* **46**, 673–698
37. Havel, T. F., and Wüthrich, K. (1985) *J. Mol. Biol.* **182**, 281–294
38. Williamson, M. P., Havel, T. F., and Wüthrich, K. (1985) *J. Mol. Biol.* **182**, 295–315
39. Braun, W., and Gō, N. (1985) *J. Mol. Biol.* **186**, 611–626
40. Brünger, A. T., Clore, G. M., Gronenborn, A. M., and Karplus, M. (1986) *Proc. Natl. Acad. Sci. U. S. A.* **83**, 3801–3805
41. Widmer, H., Billeter, M., and Wüthrich, K. (1989) *Proteins* **6**, 357–371
42. Kraulis, P. J., Clore, G. M., Nilges, M., Jones, T. A., Pettersson, G., Knowles, J., and Gronenborn, A. M. (1989) *Biochemistry* **28**, 7241–7257
43. Kaptein, R., Boelens, R., Scheek, R. M., and van Gunsteren, W. F. (1988) *Biochemistry* **27**, 5389–5395
44. Wüthrich, K. (1989) *Acc. Chem. Res.* **22**, 36–44
45. Bax, A. (1989) *Annu. Rev. Biochem.* **58**, 223–256
46. Clore, G. M., and Gronenborn, A. M. (1989) *Crit. Rev. Biochem. Mol. Biol.* **24**, 479–564
47. Oppenheimer, N. J., and James, T. L. (eds) (1989) *Methods Enzymol.* Volumes 176 and 177
48. Le Master, D. M. (1990) *Q. Rev. Biophys.* **23**, 133–174
49. McIntosh, L. P., and Dahlquist, F. W. (1990) *Q. Rev. Biophys.* **23**, 1–38
50. Anglister, J. (1990) *Q. Rev. Biophys.* **23**, 175–203
51. Otting, G., and Wüthrich, K. (1990) *Q. Rev. Biophys.* **23**, 39–96
52. Fesik, S. W., and Zuiderweg, E. R. P. (1990) *Q. Rev. Biophys.* **23**, 97–131
53. Englander, S. W., and Wand, A. J. (1987) *Biochemistry* **26**, 5953–5958
54. Qian, Y. Q., Billeter, M., Otting, G., Müller, M., Gehring, W. J., and Wüthrich, K. (1989) *Cell* **59**, 573–580
55. Lee, M. S., Gippert, G. P., Soman, K. V., Case, D. A., and Wright, P. E. (1989) *Science* **245**, 635–637
56. Summers, M. F., South, T. L., Kim, B., and Hare, D. R. (1990) *Biochemistry* **29**, 329–340
57. Montelione, G. T., Wüthrich, K., Nice, E. C., Burgess, A. W., and Scheraga, H. A. (1987) *Proc. Natl. Acad. Sci. U. S. A.* **84**, 5226–5230
58. Cooke, R. M., Wilkinson, A. J., Baron, M., Pastore, A., Tappin, M. J., Campbell, I. D., Gregory, H., and Sheard, B. (1987) *Nature* **327**, 339–341
59. Clore, G. M., Appella, E., Yamada, M., Matsushima, K., and Gronenborn, A. M. (1990) *Biochemistry* **29**, 1689–1696
60. Brünger, A. T., Campbell, R. L., Clore, G. M., Gronenborn, A. M., Karplus, M., Petsko, G. A., and Teeter, M. M. (1987) *Science* **235**, 1049–1053
61. Braun, W., Epp, O., Wüthrich, K., and Huber, R. (1989) *J. Mol. Biol.* **206**, 669–676
62. Furey, W. F., Robbins, A. H., Clancy, L. L., Winge, D. R., Wang, B. C., and Stout, C. D. (1986) *Science* **231**, 704–710
63. Arseniev, A., Schultze, P., Wörgötter, E., Braun, W., Wagner, G., Vašák, M., Kägi, J. H. R., and Wüthrich, K. (1988) *J. Mol. Biol.* **201**, 637–657
64. Schultze, P., Wörgötter, E., Braun, W., Wagner, G., Vašák, M., Kägi, J. H. R., and Wüthrich, K. (1988) *J. Mol. Biol.* **203**, 251–268
65. Stout, C. D., McRee, D. E., Robbins, A. H., Collett, S. A., Williamson, M., and Xuong, X. H. (1989) *Abstracts of the International Chemistry Congress of Pacific Basin Societies, Honolulu, HI, December 18–22*
66. Billeter, M., Kline, A. D., Braun, W., Huber, R., and Wüthrich, K. (1989) *J. Mol. Biol.* **206**, 677–687
67. Clore, G. M., Gronenborn, A. M., James, M. N. G., Kjaer, M., McPhalen, C. A., and Poulsen, F. M. (1987) *Protein Eng.* **1**, 313–318
68. Wüthrich, K., Billeter, M., and Braun, W. (1983) *J. Mol. Biol.* **169**, 949–961
69. Güntert, P., Braun, W., Billeter, M., and Wüthrich, K. (1989) *J. Am. Chem. Soc.* **111**, 3997–4004
70. Griffey, R. H., and Redfield, A. G. (1987) *Q. Rev. Biophys.* **19**, 51–82
71. Roder, H., and Wüthrich, K. (1986) *Proteins* **1**, 34–42
72. Dyson, H. J., Rance, M., Houghten, R. A., Lerner, R. A., and Wright, P. E. (1988) *J. Mol. Biol.* **201**, 161–200
73. Oas, T. G., and Kim, P. S. (1988) *Nature* **336**, 42–48

[Reprinted from Inorganic Chemistry, **7**, 1377 (1968).]
Copyright 1968 by the American Chemical Society and reprinted by permission of the copyright owner.

CONTRIBUTION FROM THE INORGANIC MATERIALS RESEARCH DIVISION, LAWRENCE RADIATION LABORATORY, AND THE DEPARTMENT OF CHEMISTRY, UNIVERSITY OF CALIFORNIA, BERKELEY, CALIFORNIA 94720

Nuclear Magnetic Resonance Studies of the Coordination of Vanadyl Complexes in Solution and the Rate of Elimination of Coordinated Water Molecules

BY K. WÜTHRICH AND ROBERT E. CONNICK

Received March 15, 1968

The temperature dependence of the O^{17} nmr line width in O^{17}-enriched aqueous solutions of the vanadyl complexes with the chelating ligands ethylenediaminetetraacetate (EDTA), nitrilotriacetate (NTA), 2-picolyliminodiacetate (PIDA), iminodiacetate (IDA), 5-sulfosalicylic acid (SSA), and Tiron (TIR) has been measured. The dependence on temperature of the complex formation equilibria was obtained from esr studies, so that the concentrations of the various paramagnetic species present in solutions of VO^{2+} ions and one of the ligands were known over the whole temperature range studied. The data obtained for $VO(EDTA)^{2-}$ show that a possible exchange of the doubly bonded "vanadyl oxygen" would be too slow to be observed by the O^{17} nmr technique. The exchange of the water molecule in the axial position opposite the vanadyl oxygen in $VO(SSA)_2^{4-}$ and $VO(TIR)_2^{6-}$ contributes at most a very small line broadening which is consistent with a very short lifetime with respect to chemical exchange of the axial water. Large relaxation effects arise from the presence of the 1:1 complexes with IDA, SSA, and TIR which have equatorial positions available for coordination of water molecules. The influence of the ligands in adjoining positions on $\Delta H^{\ddagger}$, $\Delta S^{\ddagger}$, and the first-order rate constant k of the water exchange from the equatorial coordination sites and on the scalar coupling constant A/h of O^{17} in the equatorial positions has been studied. The data obtained from solutions of the complexes with the tetradentate ligands NTA and PIDA can be interpreted in terms of a pyramidal structure of these compounds. The vanadyl oxygen and the four equatorial positions would then be at the corners of a tetragonal pyramid, with V^{4+} somewhat above the plane of the base. In a similar pyramidal structure of the hydrated vanadyl ion one would expect only four waters to be tightly bound, which would be consistent with the experimental data. A comparison of the O^{17} relaxation data with chemical shift measurements and proton relaxation experiments reported by others indicates that, in addition to the effects arising from the chemical exchange from the equatorial positions, the nuclear resonance in the bulk water of VO^{2+} solutions is influenced by the exchange of loosely coordinated waters. This may correspond to weak coordination of water molecules in the axial position opposite the vanadyl oxygen and on the four faces of the pyramid formed by VO^{2+} and the four more tightly bound equatorial waters.

I. Introduction

In dilute aqueous solutions of vanadyl ions a single nuclear magnetic resonance of O^{17} or H^1 can be observed which corresponds to that of the bulk water modified by the exchange of O^{17} and protons in and out of the coordination spheres of V^{4+}. Measurement of the line width of that resonance in metal ion solutions is a convenient method for studying the rate of exchange of O^{17} and H^1 between the bulk water and the coordination spheres of the metal ions, as well as the interaction between the unpaired electrons of the metal ion and the nuclei of the coordinated water molecules.[1] Half the line width at half-height, $\delta\omega$, expressed in radians per second, is equal to the reciprocal of the apparent transverse relaxation time T_2 and is given by

$$\delta\omega = \frac{1}{T_2} = \frac{1}{T_{2H_2O}} + \frac{1}{T_{2p}} = \delta\omega_{H_2O} + \delta\omega_p \qquad (1)$$

(1) T. J. Swift and R. E. Connick, *J. Chem. Phys.*, **37**, 307 (1962); **41**, 2553 (1964).

T_{2H_2O} describes the relaxation of the nuclei in the bulk of the solution that would occur in the absence of paramagnetic ions, and T_{2p}, the relaxation effects arising from the presence of the paramagnetic ions.

The line width of the resonance observed in solutions containing the hydrated vanadyl ion may be influenced by the exchange of nuclei from four kinds of nonequivalent coordination sites (Figure 1), *i.e.*, the site of the doubly bonded "vanadyl oxygen," the four equatorial positions (I), the axial position opposite the vanadyl oxygen (II), and possibly additional coordination sites in a second coordination sphere (III). The observed line broadening arising from the presence of the VO^{2+} ions, $\delta\omega_p$, is then given by eq 2, where all the $\delta\omega_{pi}$ may be different.

$$\delta\omega_p = \sum_i \delta\omega_{pi} \qquad i = V{=}O, I, II, III$$

Experiments designed to distinguish between the exchange reactions from the different nonequivalent coordination sites have been described previously. Reuben and Fiat[2] measured the chemical shift of O^{17} in Dy^{3+} solutions which contained various amounts of $VOSO_4$. They found that four water molecules were tightly bound to the VO^{2+} ion at room temperature. Assuming that no appreciable relaxation effects arise from the exchange of the vanadyl oxygen, they concluded that only the exchange of the water molecules coordinated to the four equatorial positions leads to a marked broadening of the O^{17} resonance, while the extremely fast water exchange involving the axial position and possibly positions in a second coordination sphere leads to a small shift of the resonance. From studies of the O^{17} nmr relaxation in $VO(ClO_4)_2$ solutions[3] two exchange reactions could be distinguished. It was not possible, however, to assign the reactions with certainty to specific ones of the four different kinds of coordination sites. Analysis of the temperature dependence of the proton relaxation in $VOSO_4$ solutions also led to the conclusion that two different exchange reactions contribute to the observed data.[3,4] In the present paper it is shown how the different kinds of coordination sites can be studied separately in a series of vanadyl chelate complexes[5] and how the water exchange from one of the coordination sites can be influenced by the ligands coordinated to adjoining positions.

II. Theory

Nmr Relaxation Studies.—A thorough discussion of the transverse nuclear relaxation in dilute aqueous solutions of paramagnetic metal ions has been given by Swift and Connick.[1] They found that two relaxation mechanisms may contribute to the observed line-broadening $\delta\omega_p$, the "$\Delta\omega$ mechanism" involving relaxation through the change in precessional frequency

(2) J. Reuben and D. Fiat, *Inorg. Chem.*, **6**, 579 (1967).
(3) K. Wüthrich and R. E. Connick, *ibid.*, **6**, 583 (1967).
(4) R. K. Mazitov and A. I. Rivkind, *Dokl. Akad. Nauk SSSR*, **166**, 654 (1966).
(5) K. Wüthrich and R. E. Connick, paper presented at the 153rd National Meeting of the American Chemical Society, Miami Beach, Fla., April 1967.

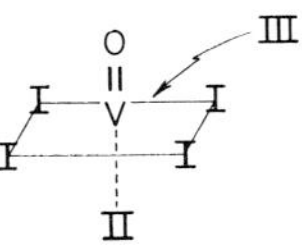

Figure 1.—Nonequivalent coordination sites of V^{4+} in vanadyl complexes.

which arises when the nuclei exchange between the bulk of the solution and the coordination sites of the metal ion,[6] and the "T_{2M} mechanism"[7] involving the fast relaxation of the coordinated nuclei. It has been shown that the $\Delta\omega$ mechanism is not of importance in solutions of vanadyl ions.[3] The effect of the exchange of nuclei between the ith kind of coordination sites of the vanadyl ion and the bulk water on the nuclear relaxation in the solution is then given by

$$\frac{1}{T_{2pi}} = \frac{P_{Mi}}{\tau_{Mi} + T_{2Mi}} \tag{3}$$

τ_{Mi} is the lifetime with respect to chemical exchange of a nucleus in the ith coordination site, T_{2Mi} is the transverse relaxation time of a nucleus in the ith coordination site, and the probability factor P_{Mi} is given closely by $n_i[VO^{2+}]/55.5$, where $[VO^{2+}]$ is the vanadyl ion concentration, and n_i is the number of water molecules in the coordination sites of type i. Two limiting cases may be distinguished, *i.e.*, where $1/T_{2pi}$ is controlled entirely by τ_{Mi} or by T_{2Mi}.[3]

The variation of τ_{Mi} with temperature will be that of rate constant (eq 6 of ref 3). The temperature dependence of T_{2Mi} is determined by the interactions between the coordinated nuclei and vanadyl ion. Three types of interaction might be of importance: scalar coupling between the nuclear spins and the unpaired electron, dipole–dipole coupling between the nuclei and the unpaired electron, and interaction of the nuclear quadrupole moment with the electric field in its vicinity. The limiting forms of equations applicable to the present systems are: scalar coupling[3,8]

$$\frac{1}{T_{2Mi}} = \frac{1}{3}S(S+1)\frac{A_i^2}{\hbar^2}\tau_{ci}; \quad \frac{1}{\tau_{ci}} = \frac{1}{T_{1e}} + \frac{1}{\tau_{Mi}} \tag{4}$$

dipole–dipole coupling[3,9]

$$\frac{1}{T_{2Mi}} = \frac{\gamma_I^2\gamma_S^2\hbar^2S(S+1)}{15d_i^6}\left[7\tau_{ci} + \frac{13\tau_{ci}}{1+\omega_s^2\tau_{ci}^2}\right];$$
$$\frac{1}{\tau_{ci}} = \frac{1}{\tau_r} + \frac{1}{\tau_{Mi}} \tag{5}$$

and quadrupole coupling[3,10,11]

$$\frac{1}{T_{2Mi}} = \frac{3(2I+3)}{40I^2(2I-1)}\left(1+\frac{\zeta_i^2}{3}\right)\left(\frac{eQq_i}{\hbar}\right)^2\tau_{ci};$$
$$\frac{1}{\tau_{ci}} = \frac{1}{\tau_r} + \frac{1}{\tau_{Mi}} \tag{6}$$

(6) H. M. McConnell and S. B. Berger, *J. Chem. Phys.*, **27**, 230 (1957).
(7) I. Solomon and N. Bloembergen, *ibid.*, **25**, 261 (1956).
(8) A. Abragam, "The Principles of Nuclear Magnetism," Oxford University Press, London, 1961.
(9) I. Solomon, *Phys. Rev.*, **99**, 559 (1955); see also ref 8.
(10) S. Meiboom, *J. Chem. Phys.*, **34**, 375 (1961).
(11) See ref 8, p 314.

Here A_i is the scalar coupling constant, d_i is the distance between the two dipoles, ω_s is the electronic Larmor frequency, τ_{ci} is the correlation time, τ_r is the correlation time for rotational tumbling, and T_{1e} is the longitudinal electronic relaxation time. The other symbols have their usual meaning.[3] In each case the temperature dependence of $1/T_{2Mi}$ is expected to arise from the correlation time only.

Chemical Shift Measurements.—The chemical shift in radians per second of the nuclear resonance in the bulk water relative to that in pure water is given by[1]

$$\Delta\omega_{H_2O} = -\sum_i \frac{P_{Mi}\Delta\omega_{Mi}}{\left(\dfrac{\tau_{Mi}}{T_{2Mi}} + 1\right)^2 + \Delta\omega_{Mi}^2\tau_{Mi}^2} \tag{7}$$

$\Delta\omega_{Mi}$ is the chemical shift relative to pure water of a nucleus in the ith position and is given by[12]

$$\Delta\omega_{Mi} = \omega S(S+1)\frac{\gamma_S}{\gamma_I}\frac{A_i}{3kT} \tag{8}$$

where ω is the Larmor frequency of the nucleus considered. When chemical exchange is fast compared to relaxation, *i.e.*, $1/\tau_{Mi}^2 \gg 1/(T_{2Mi}\tau_{Mi}) + \Delta\omega_{Mi}^2$, the contribution to the chemical shift of the bulk waters becomes

$$\Delta\omega_{H_2Oi} = -P_{Mi}\Delta\omega_{Mi} \tag{9}$$

Esr Studies of the Complex Formation.—In general, the complex formation of VO^{2+} with a chelating ligand will occur stepwise, as given by

$$VO(L)_{n-1}{}^{p+} + LH_q \rightleftharpoons VO(L)_n{}^{(p-q)+} + qH^+ \tag{10}$$

Under given conditions a solution may contain two, three, or more different paramagnetic species, which may contribute to the nuclear relaxation in the bulk water through water exchange from up to four kinds of nonequivalent coordination sites (see Figure 1). The observed line-broadening due to the presence of the VO^{2+} ions would then be given by

$$\delta\omega_p = \sum_i\sum_j \delta\omega_{pij} \quad (i = V{=}O, I, II, III; \; j = VO, VO(L), VO(L)_2, \text{etc.}) \tag{11}$$

For an interpretation of the nmr data of solutions of vanadyl complexes we therefore have to know the concentrations of the various species present.

Electron spin resonance measurements have been shown to be a convenient method to investigate the complex formation reactions in vanadyl ion solutions[13] and can easily be applied over the whole temperature range used for the nmr experiments. The esr signal of a vanadyl complex in solution is given by the spin Hamiltonian[14]

$$\mathcal{H}_S = g_0\beta HS_z + a\vec{S}\cdot\vec{I} \tag{12}$$

g_0 is the isotropic spectroscopic splitting factor, β the Bohr magneton, H the applied external magnetic field in the z direction, and a the isotropic hyperfine coupling constant of V^{51}. The eigenvalues of eq 12 to second order are given by[15]

$$H(m_I) = H_0 - am_I - \frac{a^2}{2H_0}(I(I+1) - m_I^2); \quad H_0 = \frac{h\nu}{g_0\beta} \tag{13}$$

where m_I are the eigenvalues of I_z and a is given in gauss. The signal consists of eight hyperfine components ($I = {}^7/_2$ for V^{51}). g_0 is only slightly different for different complexes, but the distances between the eight hyperfine components, which are essentially determined by the parameter a (see (13)), are greatly influenced by the ligands coordinated to VO^{2+}. Therefore different complexes which are present in the same solution can be distinguished.[13] This is illustrated in Figure 2 which shows the esr spectrum of a solution of VO^{2+} and Tiron at various temperatures. The positions of the eight hyperfine components of the signals corresponding to VO^{2+}, $VO(TIR)^{2-}$, and $VO(TIR)_2{}^{6-}$ are given at the bottom of Figure 2.[16] The lines corresponding to the signals of these three species are then easily identified in the high-field and low-field parts of the spectra recorded at various temperatures. It is seen that the relative intensities of the three signals vary greatly with temperature. $VO(TIR)_2{}^{6-}$ can hardly be detected in the spectrum at 25°, but it is the predominant species at 125°. These variations of the relative concentrations of the three complexes are fully reversible; *i.e.*, one gets the original spectrum back after cooling the solution from 125 to 25°.

For all of the ligands discussed in this paper, g_0 and a of the 1:1 and, where applicable, the 1:2 complexes with VO^{2+} are known,[13,16] and the line widths of the eight hyperfine components in solutions containing various concentrations of the complexes have been measured at different temperatures. Assuming that the shape of the resonance is Lorentzian, the esr signals of the vanadyl complexes have been reconstructed with eq 13 from these parameters on a CDC 6600 computer. From the calculated signals we computed for the various ligands the esr spectra of solutions containing variable relative concentrations of VO^{2+}, VO^{2+}-(L), and $VO^{2+}(L)_2$. The experimental spectra of the samples used for our experiments were then compared with the calculated spectra of corresponding solutions, and the relative concentrations of the different species were determined from the best fit. Since the total VO^{2+} concentration in the samples was known, we thus obtained the concentrations of VO^{2+}, $VO^{2+}(L)$, and $VO^{2+}(L)_2$.

These esr experiments further show that the lifetime in the first coordination sphere of the chelating ligands used for these studies is long compared to that of the coordinated water molecules. At temperatures above *ca.* 140° where the lifetimes of the water molecules in the first coordination sphere of some of the complexes are short compared to both the electronic relaxation

(12) N. Bloembergen, *J. Chem. Phys.*, **27**, 595 (1957).
(13) K. Wüthrich, *Helv. Chim. Acta*, **48**, 779 (1965).
(14) H. M. McConnell, *J. Chem. Phys.*, **25**, 709 (1956).
(15) R. N. Rogers and G. E. Pake, *ibid.*, **33**, 1107 (1960).
(16) K. Wüthrich, *Helv. Chim. Acta*, **48**, 1012 (1965).

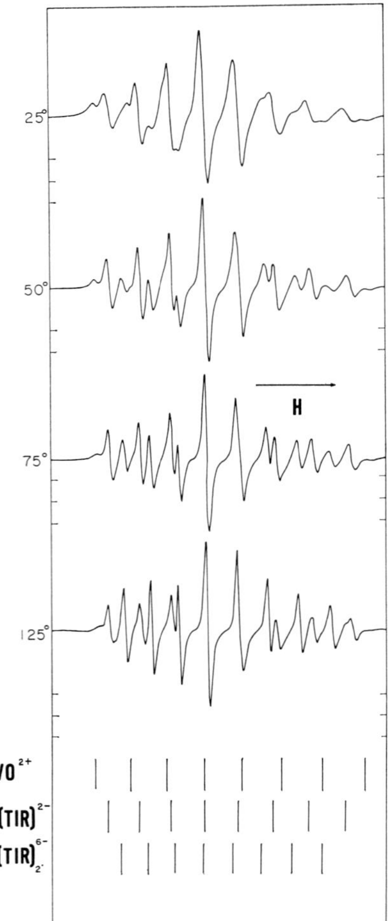

Figure 2.—Esr spectrum of an aqueous solution of 0.11 *M* $VO(ClO_4)_2$ and 0.26 *M* Tiron at various temperatures, pH (25°) 3.0. The line patterns at the bottom indicate the positions of the hyperfine components corresponding to the different complexes present. This solution was chosen to illustrate the sensitivity of the method for detecting several species and did not correspond to any of the nmr solutions.

time and the reciprocal of the difference between the *a* values corresponding to the various complexes present, one still observes the separate esr signals of the different chelate complexes.

Esr Relaxation Studies.—To calculate the scalar coupling constants from nmr relaxation measurements one has to know the longitudinal electronic relaxation time T_{1e} of the paramagnetic ion (see (4)). T_{1e} has been measured in $VO(ClO_4)_2$ solutions at room temperature,[17] but it is not known for any of the vanadyl complexes at high temperatures. As discussed previously[3] the experiments done by McCain and Myers[17] imply that it is a fairly good approximation to set $T_{1e} = T_{2e}$ for vanadyl ions under the experimental conditions of the nmr relaxation experiments at high temperatures. The value of T_{2e} used in this approximation was that corresponding to the average of the transverse relaxation times of the eight hyperfine components. The transverse relaxation times of the individual components were obtained from measurements of the separation of the positive and negative peaks in the first-derivative spectrum.

(17) D. C. McCain and R. J. Myers, *J. Phys. Chem.*, **71**, 192 (1967).

III. Experimental Section

Each experiment with one of the vanadyl complexes involved the following steps. (i) Known amounts of $VO(ClO_4)_2$ and one of the chelating ligands were dissolved in O^{17}-enriched water. (ii) Through addition of NaOH a certain pH value was established at room temperature. (iii) The O^{17} nmr spectrum was recorded at various temperatures. (iv) A corresponding solution in nonenriched water was analyzed by esr measurements for the species present at the temperatures where the nmr spectrum was recorded. From these esr spectra we also obtained the transverse electronic relaxation times.

The following chemicals were used: water enriched to 11% in O^{17} obtained from Oak Ridge National Laboratories, *ca.* 2.5 *M* $VO(ClO_4)_2$ solutions which were analyzed for paramagnetic impurities as described previously,[3] disodium ethylenediaminetetraacetate (EDTA) obtained from Fisher Scientific Co., nitriloacetic acid (NTA) and iminodiacetic acid (IDA) from Eastman Organic Chemicals, Tiron (TIR, 1,2-dihydroxybenzene-3,5-disulfonic acid, disodium salt) from Baker Chemical Co., 5-sulfosalicylic acid (SSA) from Merck Chemical Co., 2-picolyliminodiacetic acid (PIDA) obtained from Professor S. Fallab at the University of Basel, Basel, Switzerland, and NaOH from Allied Chemicals.

The $VO(ClO_4)_2$ concentration in the stock solution was determined through titration with $KMnO_4$. The samples for the nmr experiments were prepared by adding known volumes of the concentrated $VO(ClO_4)_2$ solution and weighed amounts of the ligands to a known volume of O^{17}-enriched water. By addition of very small amounts of concentrated NaOH or $HClO_4$ a certain pH value was then established at room temperature, using a combination glass–reference electrode especially designed for these experiments.[18] The solutions were finally degassed and then studied in sealed nmr tubes. The solutions of some of the complexes tend to decompose after standing for a few days. Therefore all of the experiments were done with freshly prepared solutions. It was checked with esr measurements that no irreversible changes occurred on heating the solutions up to the temperatures needed for the nmr relaxation studies.

All of the O^{17} nmr measurements were made at 8.134 Mc. The nmr spectrometer and the sample tubes used for the O^{17} relaxation studies were described previously.[3] For the O^{17} chemical shift measurements the solutions of the VO^{2+} complexes were studied in spherical bulbs and the spectra were compared to the resonance of pure water recorded in similar sample tubes. The esr measurements were done on a Varian V-4500 X-band spectrometer with 100-KHz field modulation, using a standard Varian V-4500 temperature controller for the experiments at high temperatures. The solutions were studied in sealed capillaries of *ca.* 1.5-mm outer diameter. The proton nmr experiments were done with a Varian A-60 spectrometer equipped with the standard Varian V-6031 variable-temperature probe.

IV. Results

The following VO^{2+} complexes have been studied: $VO(H_2O)_n^{2+}$ in $VO(ClO_4)_2$ solutions;[3] the 1:1 complexes, $VO^{2+}(L)$, with ethylenediaminetetraacetate (EDTA), nitrilotriacetate (NTA), 2-picolyliminodiacetate (PIDA), iminodiacetate (IDA), 5-sulfosalicylate (SSA), and Tiron (TIR, 1,2-dihydroxybenzene-3,5-disulfonic

(18) The combination glass–reference electrode was constructed for us by Gebr. Möller, Glasblässerei, Zürich, Switzerland.

TABLE I

ESR STUDIES OF THE COMPLEX FORMATION EQUILIBRIA AND AVERAGE TRANSVERSE ELECTRONIC RELAXATION TIMES

Complex studied	$[VO^{2+}]_{tot}$, M	$[L]_{tot}$, M	pH (at 25°)	T, °C	% of $[VO^{2+}]_{tot}$: $[VO^{2+}]$	$[VO^{2+}(L)]$	$[VO^{2+}(L)_2]$	$\overline{T}_{2e}$, sec
VO(IDA)	0.040	0.040	4.3	25	10	90	...	6.3×10^{-9}
				60	7	93	...	7.4×10^{-9}
				100	5	95	...	7.4×10^{-9}
				140	3	97	...	7.0×10^{-9}
$VO(SSA)^-$	0.010	0.011	3.8	25	$<4^a$	$>89, \leqslant 100$	$<7^a$	6.2×10^{-9}
				60	$<2^a$	$>94, \leqslant 100$	$<4^a$	9.3×10^{-9}
				100	$<2^a$	$>94, \leqslant 100$	$<4^a$	10.2×10^{-9}
				140	$<2^a$	$>94, \leqslant 100$	$<4^a$	10.0×10^{-9}
$VO(TIR)^{2-}$	0.010	0.020	4.0	25	20	$>75, \leqslant 80$	$<5^a$	6.0×10^{-9}
				60	15	$>80, \leqslant 85$	$<5^a$	8.3×10^{-9}
				100	10	85	5	9.7×10^{-9}
				140	7	86	7	9.7×10^{-9}

[a] The signal of that species was not observed. The numbers correspond to the limits of detection in the calculated spectra.

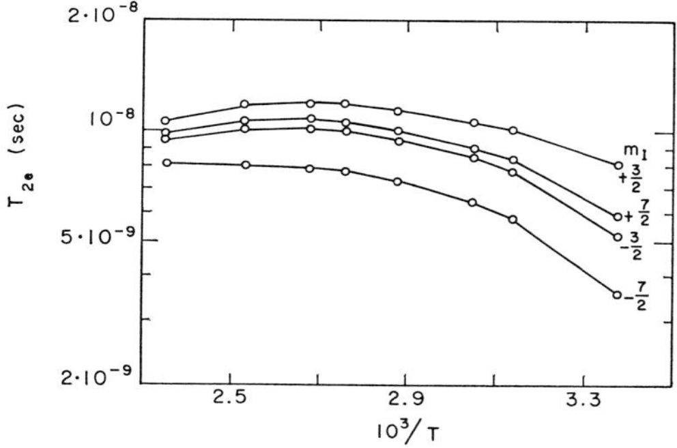

Figure 3.—Dependence on the reciprocal of temperature of log T_{2e} for four of the hyperfine components of the esr spectrum of a 0.01 M aqueous solution of $VO(SSA)^-$.

acid); and the 1:2 complexes $VO(SSA)_2^{4-}$ and VO-$(TIR)_2^{6-}$.

Esr Measurements.—Table I gives the results of the esr studies of three typical solutions used for the nmr studies of the 1:1 complexes of VO^{2+} with SSA, TIR, and IDA. For all the other complexes examined it was possible to prepare solutions in which only the esr signal of a single species was observed over the whole temperature region of interest. The temperature dependence of the transverse electronic relaxation times of the complexes used in our experiments follows qualitatively the behavior of T_{2e} of vanadyl acetylacetonate in toluene described by Wilson and Kivelson.[19] The data for $VO(SSA)^-$ are given in Figure 3, which shows a plot of log T_{2e} *vs.* $1/T$ for the four hyperfine components corresponding to $m_I = -^7/_2, -^3/_2, +^3/_2, +^7/_2$. It is seen that in the semilogarithmic plot, which is generally used to present the nmr relaxation data, T_{2e} appears to be essentially independent of temperature between *ca.* 60 and 150°. The values of $\overline{T}_{2e}$ in the last column of Table I correspond to the average of the transverse relaxation times of the eight hyperfine components.

O^{17} Nmr Relaxation Studies.—The experimental data are summarized in Figure 4, which shows plots of log T_{2p} *vs.* $1/T$ for some of the complexes, and in Table

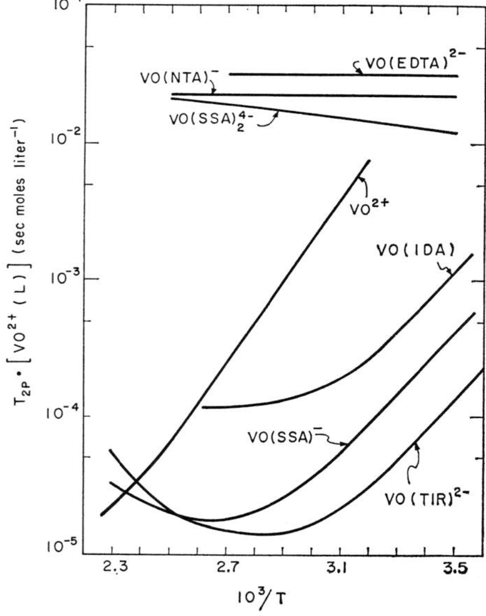

Figure 4.—Dependence on the reciprocal of temperature of log T_{2p} of O^{17} in aqueous solutions of various vanadyl chelate complexes.

II. It is seen that some of the chelate complexes give relaxation effects which are very small compared to those observed in $VO(ClO_4)_2$ solutions, while for other species T_{2p} is much shorter than that of $VO(ClO_4)_2$ in a part of the temperature region examined. A better understanding of the meaning of these data is obtained when one considers the schematic structures of the various complexes given in Figures 5 and 6.

Evidence has been found by others[20] that the coordination of protons and metal ions to EDTA occurs preferentially at the nitrogen atoms. Figure 5A appears then from model considerations to be the most likely structure of the VO^{2+}–EDTA complex. Figure 5A is also consistent with the stoichiometry of the complex formation reaction, if one assumes that the de-

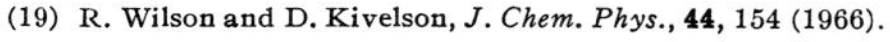
(19) R. Wilson and D. Kivelson, *J. Chem. Phys.*, **44**, 154 (1966).

(20) See R. J. Kula, Thesis, University of California, Riverside, Calif., 1964, wherein further references are given.

TABLE II

EXPERIMENTAL DATA ON THE O^{17} NMR RELAXATION IN AQUEOUS SOLUTIONS OF VANADYL COMPLEXES

Complex	Structure	Experimental observations (see Figure 4)	Parameters obtained[b]
VO^{2+} [a]	Figure 1 (I, II, III = H_2O)	Large $\delta\omega_p$, chemical exchange controlled in the major part of the temperature region 5–170°	$k = 5.0 \times 10^2$ sec^{-1} $\Delta H^{\ddagger} = 13.7$ kcal mol^{-1} $\Delta S^{\ddagger} = -0.6$ eu $A/h = 3.8 \times 10^6$ cps
		Small additional $\delta\omega_p$ at low temperatures, due to fast exchange of loosely bound waters (not given in Figure 4; see ref 3)	
$VO(EDTA)^{2-}$ $VO(EDTAH)^-$	Figure 5A	Very small $\delta\omega_p$ from 5 to 100°	k(vanadyl oxygen) < 20 sec^{-1}
$VO(SSA)_2^{4-}$	Figure 5B	Very small $\delta\omega_p$ from 5 to 130°	
$VO(Tiron)_2^{6-}$	Figure 5B	Very small $\delta\omega_p$ from 5 to 130°	
$VO(IDA)$ [c]	Figure 5C	Large $\delta\omega_p$, chemical exchange controlled from 5 to 50°; at higher temperatures T_{2M} controlled	$k = 1.2 \times 10^5$ sec^{-1} $\Delta H^{\ddagger} = 11.7$ kcal mole^{-1} $\Delta S^{\ddagger} = +3.9$ eu $A/h = 2.8 \times 10^6$ cps[d]
$VO(SSA)^-$ [c]	Figure 5D	Large $\delta\omega_p$, chemical exchange controlled from 5 to 50°; at higher temperatures T_{2M} controlled.	$k = 1.5 \times 10^5$ sec^{-1} $\Delta H^{\ddagger} = 10.8$ kcal mol^{-1} $\Delta S^{\ddagger} = +1.2$ eu $A/h = 4.5 \times 10^6$ cps[d]
$VO(Tiron)^{2-}$ [c]	Figure 5D	Large $\delta\omega_p$, chemical exchange controlled from 5 to 50°; at higher temperatures T_{2M} controlled	$k = 5.3 \times 10^5$ sec^{-1} $\Delta H^{\ddagger} = 11.8$ kcal mole^{-1} $\Delta S^{\ddagger} = +7.0$ eu $A/h = 4.9 \times 10^6$ cps[d]
$VO(NTA)^-$	Figure 6E *or* F	Very small $\delta\omega_p$ from 5 to 130°	
VO(PIDA)	Figure 6E *or* F	Very small $\delta\omega_p$ from 5 to 100°	
$VO(NTA)(OH)^{2-}$	Figure 6G	Very small $\delta\omega_p$ from 5 to 100°	
$VO(PIDA)(OH)^-$	Figure 6G	Very small $\delta\omega_p$ from 5 to 100°	

[a] Studied in $VO(ClO_4)_2$ solutions.[3] [b] k is the first-order rate constant at 25° for the loss from an equatorial position of the first coordination sphere of a particular one of the exchanging nuclei. $\Delta H^{\ddagger}$ and $\Delta S^{\ddagger}$ are the enthalpy and entropy of activation of the O^{17} exchange from the equatorial positions. A/h is the scalar coupling constant of O^{17} in the equatorial positions. Where only limits could be obtained for the concentration of $VO^{2+}(L)$ (Table I), the values of $[VO^{2+}(L)]$ used in the calculation of the reaction parameters were those which correspond to the average of the upper and the lower limits. [c] Solution composition is the same as in Table I. [d] Calculated using the approximation $T_{1e} = T_{2e}$.

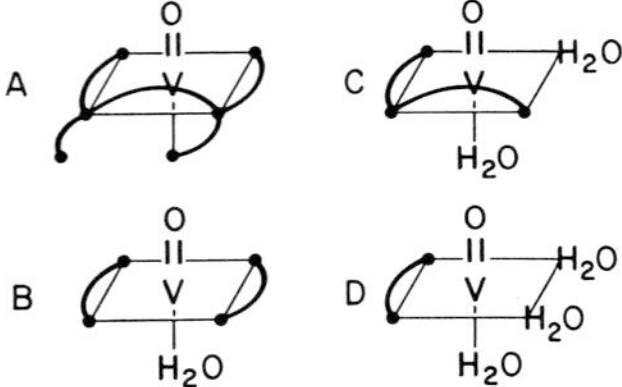

Figure 5.—Coordination of vanadyl complexes with chelating ligands.

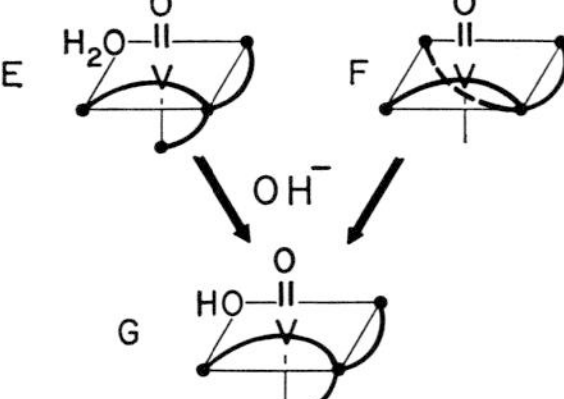

Figure 6.—Coordination of vanadyl complexes with tetradentate ligands.

protonation at pH *ca.* 3.0[21] involves the nonbonded carboxylic acid group. This would then explain that this deprotonation has no effect on the esr signal[13] and on the O^{17} relaxation in the bulk water as was found from a comparison of the O^{17} resonance in solutions of the protonated and the deprotonated EDTA complex. In the structure of Figure 5A only the vanadyl oxygen could possibly be involved in an O^{17} exchange between the first coordination sphere of the V^{4+} and the bulk water. The lack of any observable relaxation effects in $VO(EDTA)^{2-}$ solutions might be due to either very fast or very slow exchange of the vanadyl oxygen. We only consider the possibility of slow exchange and find the upper limit for the rate constant given in Table II. In the calculation of this limit we have used the datum that $T_{2p}[VO(EDTA)^{2-}] > ca.\ 3.0 \times 10^{-2}$ sec at 100° (Figure 4) and have assumed that the activation enthalpy for this reaction would be equal to or greater than that observed for the O^{17} exchange in $VO(ClO_4)_2$ solutions, *i.e.*, 13.7 kcal mol^{-1}.

X-Ray studies by Dodge, Templeton, and Zalkin[22] showed that the two bidentate ligands in the 1:2 complex of VO^{2+} with acetylacetonate are coordinated to the four equatorial positions of the first coordination sphere. Additional evidence for the correctness of the structures of the complexes with bidentate ligands given in Figure 5B and D has been obtained from esr experiments (see eq 5 and Table III of ref 16). From model

(21) G. Schwarzenbach and J. Sandera, *Helv. Chim. Acta*, **36**, 1089 (1953).

(22) R. P. Dodge, D. H. Templeton, and A. Zalkin, *J. Chem. Phys.*, **35**, 55 (1961).

considerations Figure 5C appears to be the most likely structure of the 1:1 complex with IDA. The experiments with $VO(SSA)_2^{4-}$ and $VO(TIR)_2^{6-}$ show that neither the exchange of the vanadyl oxygen nor water exchange from the axial position leads to appreciable broadening of the O^{17} resonance in the bulk water. On the other hand, large relaxation effects are observed in solutions of all of these complexes which have equatorial positions available for the coordination of water molecules (Figures 4 and 5). A comparison of these data with those obtained in $VO(ClO_4)_2$ solutions (Figure 4) implies that only the four equatorial coordination sites are involved in the water exchange which leads to the large $\delta\omega_p$ in $VO(ClO_4)_2$ solutions. The values of k, $\Delta H^{\ddagger}$, and $\Delta S^{\ddagger}$ of the water exchange from the four equatorial positions in the hydrated vanadyl ion (Table II) were taken from the data reported previously.[3]

The parameters given for the water exchange from the equatorial positions of the 1:1 complexes with IDA, SSA, and TIR were obtained from the curve-fitting processes given in Figures 7–9. It is easily shown that the observed large relaxation effects cannot arise from dipolar coupling or quadrupolar coupling but must be due to scalar coupling between the unpaired electron and the nuclear spin. Therefore the dependence on temperature of T_{2p} is, through eq 3, determined by the temperature dependence of τ_M (eq 6 of ref 3) and (4), where we assume that $T_{1e} = T_{2e}$ is independent of temperature between 60 and 150° (Table I). In $VO(ClO_4)_2$ solutions τ_M never became short compared to T_{2M} in the temperature region accessible for the nmr experiments, *i.e.*, *ca.* 5–170° (Figure 4). Since the exchange rate of the equatorial waters is greatly enhanced through the influence of the chelating ligands IDA, SSA, and TIR, it was now possible to observe the region where $\tau_M < T_{2M}$ and, in the case of $VO(TIR)^{2-}$, even the region where $\tau_M < T_{1e}$. The values found for A/h will be in error to the extent that T_{1e} differs from $\overline{T}_{2e}$.

Two types of coordination are possible for the complexes of VO^{2+} with the tetradentate ligands NTA and PIDA, which may occupy either the three equatorial positions and the axial position (Figure 6E) or the four equatorial positions (Figure 6F) of the first coordination sphere. Since one would expect the complex given in Figure 6E to give rise to relaxation effects comparable to those in VO(IDA) solutions the O^{17} relaxation experiments indicate that the structure of Figure 6F is more stable in aqueous solutions of these complexes. This structure seems, however, not to be stable in basic solutions, where a hydroxo complex is formed.[13,23] The most likely coordination for the latter seems to be that given in Figure 6G. No appreciable relaxation effects are observed in solutions of these hydroxo complexes, which is consistent with a slow rate of exchange of the OH^- group in the equatorial position.

Chemical Shift Measurements.—The shift of the O^{17} resonance in *ca.* 0.4 M solutions of $VO(ClO_4)_2$, $VO(NTA)^-$, $VO(TIR)_2^{6-}$, and $VO(SSA)_2^{4-}$ was studied at

(23) Th. Kaden and S. Fallab, *Chimia*, **20**, 51 (1966).

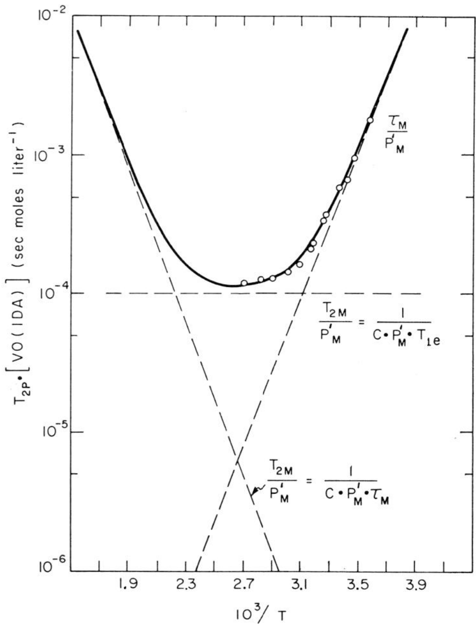

Figure 7.—Dependence on the reciprocal of temperature of log T_{2p} of O^{17} in solutions of VO(IDA) with the lines resulting from the curve fitting; $P_M' = P_M/[VO(IDA)]$.

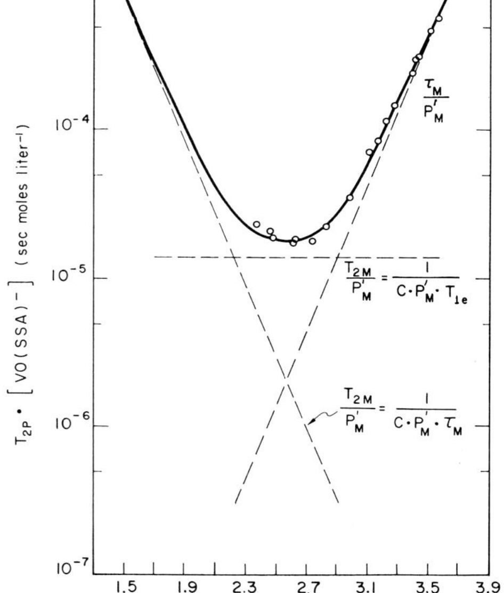

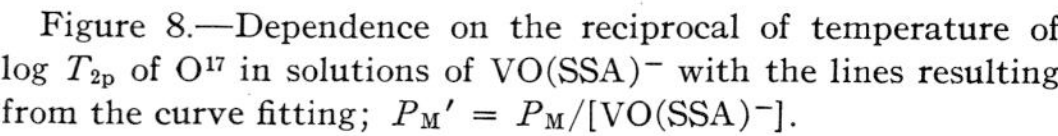

Figure 8.—Dependence on the reciprocal of temperature of log T_{2p} of O^{17} in solutions of $VO(SSA)^-$ with the lines resulting from the curve fitting; $P_M' = P_M/[VO(SSA)^-]$.

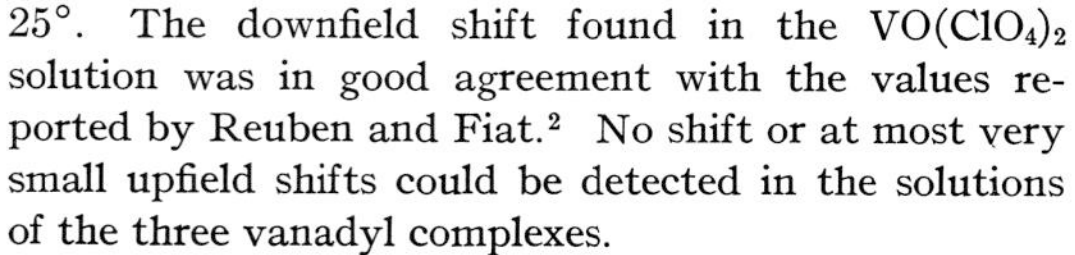

25°. The downfield shift found in the $VO(ClO_4)_2$ solution was in good agreement with the values reported by Reuben and Fiat.[2] No shift or at most very small upfield shifts could be detected in the solutions of the three vanadyl complexes.

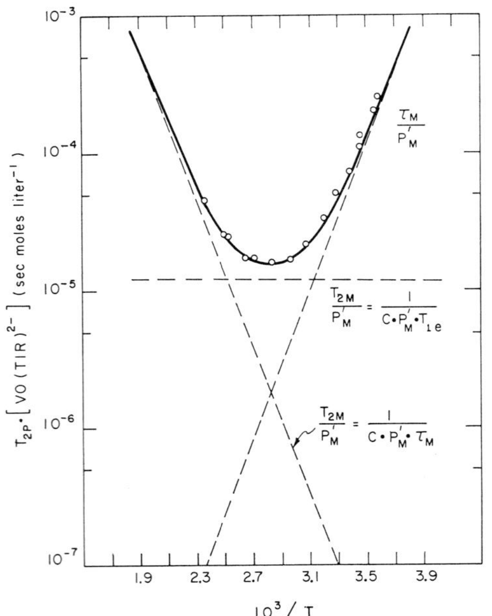

Figure 9.—Dependence on the reciprocal of temperature of log T_{2p} of O^{17} in solutions of $VO(TIR)^{2-}$ with the lines resulting from the curve fitting; $P_M' = P_M/[VO(TIR)^{2-}]$.

Proton Nmr Relaxation Studies.—The transverse proton relaxation was studied in the bulk water of solutions of $VO(EDTA)^{2-}$. At complex concentrations from 0.05 to 0.5 *M* no appreciable broadening of the resonance could be observed in the temperature range 5–100°.

V. Discussion

The experiments described in section IV have shown that in most of the solutions of vanadyl complexes studied only one of the many possible contributions to the relaxation of the bulk nuclei given in eq 11 is of importance. It was possible to distinguish between the effects on T_{2p} of the O^{17} exchange from the three nonequivalent kinds of positions in the first coordination sphere of V^{4+} (Figure 1) and to study quantitatively the influence on the rate of the water exchange from the equatorial coordination sites of the ligands coordinated to adjoining positions.

Exchange of the Vanadyl Oxygen.—From the results of previous investigations of vanadyl ion, it was to be expected that the vanadyl oxygen would exchange much more slowly than the oxygens of the water molecules in the other coordination sites of V^{4+}.[24] To our knowledge it has not been established that the vanadyl oxygen exchanges in aqueous solutions of vanadyl complexes. The reaction would certainly be too slow to be studied quantitatively by the O^{17} nmr relaxation technique. The upper limit for the rate constant of a possible vanadyl oxygen exchange given in Table II is probably far too high, because one would expect the enthalpy of activation of this reaction to be considerably greater than that used in the calculation of this limit, *i.e.*, $\Delta H^{\ddagger}$ of the water exchange from the equatorial positions.

Water Exchange from the Equatorial Positions.—The water exchange from the equatorial positions seems to be the only reaction which greatly influences the nuclear relaxation in the bulk of the solution. Its rate and its enthalpy of activation are such that all three limiting cases for the dependence on temperature of T_{2p} predicted by eq 3 and 4 can be observed for some of the complexes (Figures 8 and 9). The rate of the water exchange from one of the equatorial coordination sites can be changed by several orders of magnitude through the influence of the ligands coordinated to adjoining equatorial positions (Table II). A comparison of the water-exchange rates from the first coordination spheres of the hydrated VO^{2+} ion and of other hydrated doubly charged 3d metal ions led to the suggestion that the relatively slow exchange found for VO^{2+} is most likely due to large electrostatic contributions to the bonding of the water molecules.[3] The increase of the exchange rate brought about by the various chelating ligands might then be interpreted in terms of a simple electrostatic picture. Each of the ligands studied, *i.e.*, IDA, SSA, and TIR, occupies two equatorial positions of the first coordination sphere with negatively charged groups which might neutralize part of the high positive charge of V^{4+} effective in the bonding of the water molecules. One might even go further and explain the increase of the exchange rates when going from VO(IDA) to $VO(SSA)^-$ and $VO(TIR)^{2-}$ in terms of the increasing number of negatively charged sulfo groups which are not coordinated to the first coordination sphere of the metal ion. As one would expect, the negative charges of the bonding groups of the ligands would then have a much greater influence on the exchange rates than the charges localized on nonbonding groups. This partial neutralization of the high effective charge of V^{4+} through the chelating ligands presumably would also be reflected in the values found for $\Delta H^{\ddagger}$ (Table II) which appear to be somewhat smaller than $\Delta H^{\ddagger}$ of the water exchange from the hydrated VO^{2+} ion. The values of $\Delta S^{\ddagger}$ of the water exchange from the various complexes are close to 0, as is generally found for the elimination of water molecules from the first coordination sphere of metal ions.[1]

The influence of the ligands in adjoining coordination sites on the scalar coupling constant of the nuclei of waters coordinated to a paramagnetic metal ion has been discussed by Horrocks and Hutchison.[25] Following McConnell and Robertson,[26] they distinguished between two contributions to the observed coupling constant, $A = A_c + A_p$, where A_c is due to Fermi contact coupling and A_p to pseudo-contact interactions. From an analysis of the variations of the scalar coupling constant A of the water protons in a series of mixed

(24) J. Selbin, *Chem. Rev.*, **65**, 153 (1965).

(25) W. D. Horrocks, Jr., and J. R. Hutchison, *J. Chem. Phys.*, **46**, 1703 (1967).

(26) H. M. McConnell and R. E. Robertson, *ibid.*, **29**, 1361 (1958).

Co^{2+} complexes, they concluded that A_c might to a good approximation be constant for all of the complexes. The dependence of A on the ligands in adjoining positions observed in Co^{2+} complexes would therefore arise mainly from changes of the term A_p, which is closely related to the anisotropy Δg of the g tensor; *i.e.*, for a complex with axial symmetry $\Delta g = g_\perp - g_\parallel$. The data on VO^{2+} complexes (Table II) suggest that A_c of the nuclei of water molecules might in certain cases also be influenced by the ligands in mixed metal ion complexes. The anisotropy of the g tensor was found to be very small for the hydrated vanadyl ion,[17,19] and from estimates for a series of other vanadyl complexes Δg seems to be very little influenced by the ligands in the equatorial positions.[13] It then appears that the observed ligand influence on the O^{17} scalar coupling constant A/h in vanadyl complexes (Table II) is mainly due to changes of the contact interaction A_c.

The dependence on the ligands of the scalar coupling constants of V^{51} [13] and O^{17} of the coordinated waters indicates that, in addition to purely electrostatic effects, the bonding scheme in VO^{2+} complexes is influenced by the ligands in the equatorial positions. The general trend seems to be that the unpaired electron density at the V^{51} nucleus is decreased when VO^{2+} is coordinated to ligands which are known to form strongly "covalent" bonds in their metal ion complexes.[13,16] Since electrostatic interactions are probably important in all of the complexes studied so far, it has not yet been possible to determine how the water-exchange rates are influenced by the ligand effects indicated in the variations of the scalar coupling constants. Quite possibly these effects could cause the differences in the water-exchange rates from VO^{2+}, VO(IDA), $VO(SSA)^-$, and $VO(TIR)^{2-}$, rather than simple electrostatic interactions. This would not be very surprising, since others have shown that the rate of replacement of H_2O in Ni^{2+} complexes is not in all cases increased by the coordination of negatively charged groups to Ni^{2+}, while on the other hand some electrostatically neutral ligands seem to increase considerably the rate of the water replacement.[27]

Rate of Water Exchange from the Axial Position.— At most very small relaxation effects arise from the water exchange involving the axial position of VO^{2+} (Figure 1, II). Since these relaxation effects appear to be T_{2M} controlled even at low temperatures,[3] they could only come from fast exchange of the axial water molecule. We then have that according to the experimental data for hydrated vanadyl ion[4] the rates of the water exchange at 25° from the axial and the equatorial positions must differ by at least a factor of *ca.* 10^6.

Group theoretical considerations[28] show that different wave functions are involved in the bonding of the four equatorial and the axial ligands of the first coordination sphere of VO^{2+}. One would therefore anticipate that the water-exchange rates from these two kinds of coordination sites might be quite different. Furthermore there is some indication of a tendency for VO^{2+} complexes to form pyramidal molecules in which the V^{4+} would not be in the plane of the four equatorial ligands (Figure 10A), but rather somewhat above this plane (Figure 10B). X-Ray studies[22] have shown that in the solid state of the 1:2 complex of VO^{2+} with acetylacetonate, the vanadyl oxygen and the four equatorial positions are at the corners of a tetragonal pyramid with V^{4+} approximately at its center of gravity. Esr studies[29] imply that in solutions of vanadyl acetylacetonate a solvent molecule is coordinated to the axial position with a very short lifetime, which is at 25° only little longer than the electronic relaxation times of $VO(acac)_2$ in these solutions, *i.e.*, 1.0×10^{-8} to 1.0×10^{-9} sec. For steric reasons one would expect only weak coordination to the axial position in a structure of the type in Figure 10B. Therefore such a short lifetime seems comprehensible if one assumes that $VO(acac)_2$ maintains the pyramidal structure in solution. Further evidence for a pyramidal solution structure of vanadyl complexes comes from the O^{17} nmr experiments which appear to show that the tetradentate ligands NTA and PIDA are coordinated to the four equatorial positions of VO^{2+} (Figure 6F). From model considerations such a coordination of NTA and PIDA appears plausible in a pyramidal structure (Figure 10B), but it would for steric reasons be impossible in a planar structure (Figure 10A). In the pyramidal structure (Figure 10B) the V^{4+} would then be five-coordinated in the complexes with the tetradentate ligands, while for all of the other complexes (Figure 5) at most a weak coordination of a sixth group in the axial position of the first coordination sphere appears likely (Figure 11).

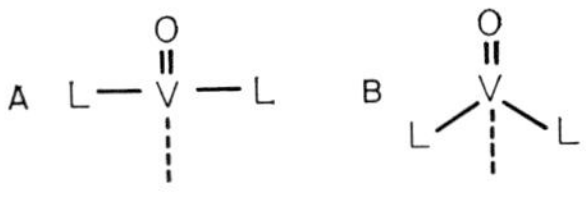

Figure 10.

Axial and Second-Coordination-Sphere Effects.— Two different effects have been found to arise from fast water exchange which could involve either the axial position (Figure 11, L_{ax}) or positions in the second coordination sphere above the triangular faces of the tetragonal pyramid (Figure 11, L_{II}). These are a small chemical shift of the O^{17} resonance in VO^{2+} solutions at room temperature[2] and small contributions to the line width of the O^{17} resonance observed at low temperatures.[3] In this section we discuss some experiments which indicate that two different kinds of interactions between the vanadyl ion and the loosely coordinated water molecules give rise to these two effects on the nuclear resonance in the bulk water.

The chemical shift of the resonance of the coordinated nuclei is given by (8), where A_i includes contributions from scalar and pseudo-scalar coupling between the un-

(27) D. W. Margerum and H. M. Rosen, *J. Am. Chem. Soc.*, **89**, 1088 (1967).

(28) C. J. Ballhausen and H. B. Gray, *Inorg. Chem.*, **1**, 111 (1962).

(29) F. A. Walker, R. L. Carlin, and P. H. Rieger, *J. Chem. Phys.*, **45**, 4181 (1966).

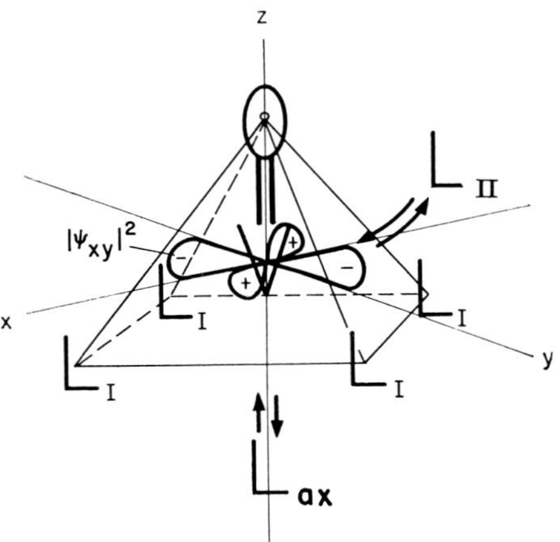

Figure 11.—Pyramidal coordination of the hydrated vanadyl ion. $|\psi_{xy}|^2$ outlines the areas of high unpaired electron density for the ground state of the molecule.

paired electron and the nuclear spin. Assuming that the shift of the O^{17} resonance in the bulk water (eq 7) is entirely due to fast water exchange from the axial position, Reuben and Fiat[2] found $A/h = 2.06 \times 10^6$ cps for O^{17} of the axial water. Alternatively, if one assumes that the chemical shift arises entirely from four second-coordination-sphere waters residing on the faces of the tetragonal pyramid (Figure 11), $A/h = 5.2 \times 10^5$ cps for O^{17} of these waters.

The enhanced nuclear relaxation could arise from dipolar coupling (eq 5) and quadrupolar coupling (eq 6) as well as scalar coupling (eq 4). To distinguish between the three possible relaxation mechanisms, we have studied the data on the proton nuclear relaxation in solutions of VO^{2+} ions, which have been reported by Hausser and Laukien,[30] and by Mazitov and Rivkind.[4] Mazitov and Rivkind's interpretation, which was unknown to us at the time we treated Hausser and Laukien's data,[5] is based on the same concepts as ours, although it is less detailed. Hausser and Laukien's data are given in Figures 12 and 13, together with our reinterpretation. In Figure 12, log T_{1p} is plotted *vs.* $1/T$. Only dipolar interactions are important for the enhancement of the longitudinal relaxation of the coordinated nuclei.[31] At high temperatures the resulting T_{1MI}, which is the longitudinal relaxation time of the protons of a water molecule coordinated to a position of type I, is long compared to τ_{MI}, but τ_{MI}, which was obtained from Figure 13, increases rapidly on lowering the temperature and controls the relaxation effects arising from exchange from the equatorial positions below *ca.* 25° (eq 3). Additional relaxation effects (T_{1II}/P_{MII}), which must come from fast water exchange from other positions, become important at lower temperatures. The temperature dependence of log T_{2p} (Figure 13) shows the same characteristic features as a similar plot of the O^{17} data.[3] The relaxation of the

(30) R. Hausser and G. Laukien, *Z. Physik*, **153**, 394 (1959).
(31) See ref 8, pp 309–311.

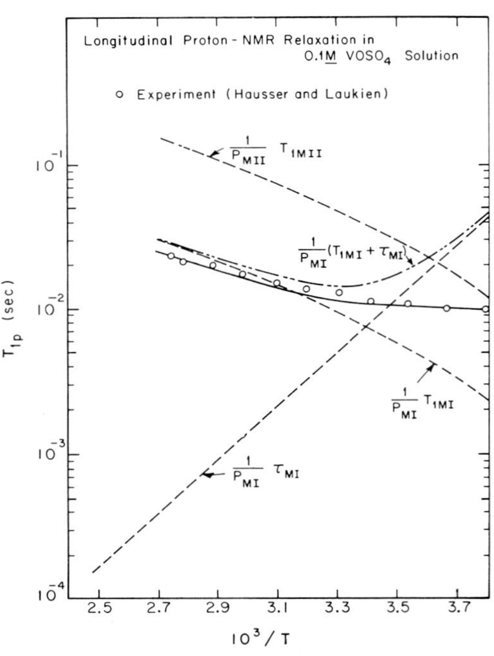

Figure 12.—Dependence on the reciprocal of temperature of log T_{1p} of protons in $VOSO_4$ solutions with the lines resulting from the curve fitting. Subscript I refers to the equatorial coordination sites; subscript II, to the positions which accommodate the labile water molecules. The experimental points which have been obtained from Figure 10 of ref 30 are corrected to 0.1 *M* VO^{2+}.

nuclei in the equatorial positions is due essentially entirely to scalar coupling (T_{2MIsc}/P_{MI}). At high and intermediate temperatures, T_{2p} is controlled by T_{2MI} and τ_{MI} (eq 3), while the relaxation due to the exchange of the labile waters in other positions (T_{2MII}/P_{MII}) becomes dominant below *ca.* 35°. The solid curves in Figures 12 and 13 correspond to the appropriate combinations (eq 2 and 3) of the individual contributions to T_{1p} and T_{2p} shown in the figures.

Protons have no quadrupole moment, and the longitudinal nuclear relaxation time T_{1p} in solutions of VO^{2+} is essentially unaffected by scalar coupling interactions (except for extremely short correlation times and large scalar coupling constants).[31] Therefore the contributions T_{1MI}/P_{MI}) and T_{1MII}/P_{MII} in Figure 12 must come from dipole–dipole coupling with the protons of the coordinated waters. From the curve-fitting process in Figure 13, it is seen that T_{2MII}/P_{MII} is to a good approximation equal to T_{1MII}/P_{MII}. This indicates that dipole–dipole coupling is also responsible for the transverse proton relaxation in the labile water molecules.

If we assume that the labile protons are parts of rapidly exchanging water molecules, we can use the proton data to help interpret the O^{17} data. The proton relaxation in environment II is related to the corresponding oxygen-17 relaxation in that dipole–dipole coupling will be present in both and the rotational and chemical exchange correlation times are presumably

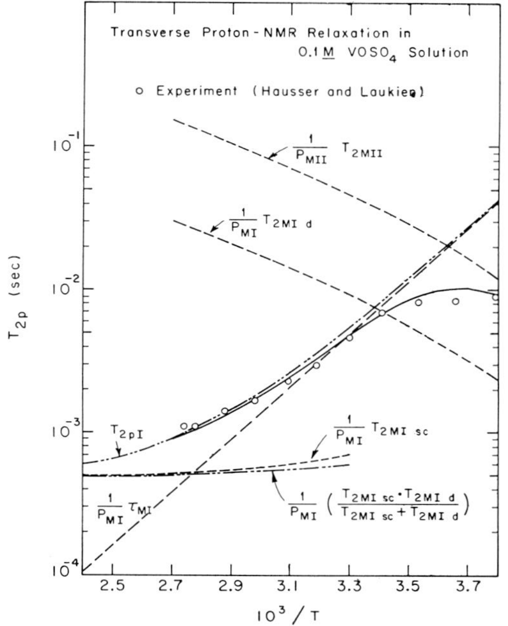

Figure 13.—Dependence on the reciprocal of temperature of log T_{2p} of protons in $VOSO_4$ solutions with the lines resulting from the curve fitting. Subscript I refers to the equatorial coordination sites, II to the positions which accommodate the labile water molecules, sc to scalar coupling, and d to dipolar coupling. The experimental points have been obtained from Figure 10 of ref 30 and are corrected to 0.1 *M* VO^{2+}. $T_{2MId}/P_{MI} = T_{1MI}/P_{MI}$ has been taken from Figure 12.

the same. The proton T_{1pII} data of Figure 12 were therefore used to estimate the dipole–dipole contribution to T_{2pII} of the oxygen-17 of aqueous vanadyl ion (eq 5).

The coordinates of the oxygens and protons of the water molecules were calculated from the vanadyl acetylacetonate structure parameters[22] assuming the equatorial water oxygens occupy the average positions of the acetylacetonate oxygens and the axial and second-coordination-sphere waters approach their neighboring oxygens to within the van der Waals distance of 2.80 Å. There are two serious sources of uncertainty: (a) the relative distances from the nuclei to the vanadium (d_i in eq 5 are not known precisely, and (b) it is a poor approximation in eq 5 to represent the unpaired electron by a point dipole at the vanadium nucleus. Using 2.61 and 3.14 Å for the vanadium–oxygen distances of the axial and second-coordination-sphere waters, respectively, and 3.31 and 3.83 Å for the vanadium–proton distances of the axial and second-coordination-sphere waters, respectively, one finds $1/T_{2pII}$ of oxygen-17 to be roughly half dipole–dipole relaxation arising from the axial and second-coordination-sphere waters. It should be noted that this calculation can be made without precise knowledge of the correlation times.

There remain the scalar and quadrupole couplings. The former could come from either the axial water oxygen or the four second-coordination-sphere water oxygens. If the observed scalar coupling is due solely to the axial water, the lifetime for chemical exchange of this water has to be less than 3×10^{-11} sec in order to give rise to no more than half of the observed $1/T_{2pII}$. Such a lifetime seems extremely short for the model chosen, in that the water oxygen approaches the vanadium to within 2.6 Å. Therefore it seems more likely that the scalar coupling is to be attributed primarily to the four second-coordination-sphere waters. Since the unpaired electron of the hydrated VO^{2+} ion is believed to be mainly in the d_{xy} atomic orbital[28] which lies above the plane of the first-coordination-sphere waters (Figure 11), it appears plausible that the four second-coordination-sphere waters on the faces of the tetragonal pyramid could have stronger scalar coupling than the axial water. This origin for the observed paramagnetic shift seems not impossible since scalar interactions of the same kind have been reported for O^{17} in the second-coordination-sphere waters of chromic ion.[32] Because similar electrons are involved (d_{xy} for VO^{2+} and d_{xy}, d_{xz}, and d_{yz} for Cr^{3+}), one might expect similar effects on the O^{17} resonance of water molecules residing on the faces of a pyramid formed by $VO(H_2O)_4{}^{2+}$ and on the faces of an octahedron formed by $Cr(H_2O)_6{}^{3+}$. Again, attributing at most half of $1/T_{2pII}$ of O^{17} to scalar coupling relaxation of the second-coordination-sphere waters sets an upper limit for their chemical exchange lifetime of 1.0×10^{-10} sec.

The quadrupole coupling components of O^{17} in the water molecule can be obtained from the data of Stevenson and Townes[33] for HDO^{17}. Assuming the correlation time is the same as for dipole–dipole coupling, the contribution of quadrupole coupling to the oxygen-17 relaxation should be severalfold that from dipole–dipole coupling—a value which is impossibly large. As will be discussed elsewhere, it seems likely that the quadrupole coupling is reduced by fast rotation of the water molecule around its twofold axis, and such an assumption removes the discrepancy.

The above model for the T_{2pII} relaxation of oxygen-17 on vanadyl ion can be fitted by a variety of combinations of correlation times. If one adopts the rotational lifetime for vanadyl ion found by McCain and Myers[17] by esr studies and corrects it to 25° ($\tau_r = 3.3 \times 10^{-11}$ sec), the exchange lifetime for the second-coordination-sphere waters is *ca.* 3×10^{-11} sec. In the calculation the exchange lifetime for the axial water was assumed to be long compared to τ_r. The scalar coupling from the second-coordination-sphere waters is roughly half the total quadrupole relaxation, which in turn is approximately equal to the total dipolar contribution.

The interpretation is not unique, but appears to be the most plausible one in the light of the present evidence. In any event, it is likely that the low-temperature transverse relaxation of oxygen-17 in a solution of vanadyl ion arises from two kinds of loosely bound

(32) M. Alei, Jr., *Inorg. Chem.*, **3**, 44 (1964).
(33) M. J. Stevenson and C. H. Townes, *Phys. Rev.*, **107**, 635 (1957).

water, *i.e.*, the axial water and water found in a second coordination sphere consisting of four positions on the faces of the pyramidal structure of the hydrated vanadyl ion (Figure 11). Furthermore there is evidence that $1/T_{2pII}$ contains appreciable contributions from dipole–dipole, scalar, and quadrupole coupling.

It would have been nice if the two kinds of loosely coordinated waters could have been studied separately in various chelate complexes, *e.g.*, by comparing the data on $VO(NTA)^-$ and VO(PIDA), where only the loose coordination on the faces of the pyramid could possibly be of importance (Figure 6F), with those on $VO(TIR)_2^{6-}$ and $VO(SSA)_2^{4-}$, where one would expect to observe the effects of both kinds of labile waters (Figure 5B). Since we are looking for extremely small effects on the nuclear resonance, such experiments have to be done with concentrated solutions of the vanadyl complexes, which seem not to be stable over an extended period of time. Furthermore the viscosity in these concentrated solutions is greatly enhanced compared to that of pure water. Therefore it seems beyond the limits of the method to measure the small enhancement of the relaxation of O^{17} which might arise from a possible exchange of labile waters in solutions of the above-mentioned vanadyl complexes. The small effects shown in Figure 4 for $VO(NTA)^-$, $VO(EDTA^{2-}$, and $VO(SSA)_2^{4-}$ are at about the limit of the experimental accuracy and cannot be taken to establish axial or second-coordination-sphere relaxation. Such relaxation may be appreciably diminished in these complexes through an increase in the rate of water exchange.

Chemical shift measurements appeared to give more reliable data. No measurable shift was observed in the solutions of complexes of the types in Figures 5B and 6F. This indicates that the hydration in the second coordination sphere of these complexes differs appreciably from that in the second coordination sphere of the hydrated vanadyl ion. It therefore appears rather unlikely that one might be able to deduce more information about the coordination of the labile waters in the hydrated vanadyl ion from further studies of vanadyl chelate complexes.

Proton Exchange from the Equatorial Positions.—From the curve-fitting process in Figure 13 we find for τ_{MI} at 25° a value of $n \times 1.15 \times 10^{-5}$ sec, where n is the number of water molecules coordinated to positions of type I. Since the proton exchange from any-one of the coordination sites cannot be slower than the corresponding water exchange, we can immediately exclude the possibility that τ_{MI} in Figure 13 corresponds to exchange from the axial position or from the second coordination sphere. Experiments with solutions of $VO(EDTA)^{2-}$ showed that the transverse proton nuclear relaxation in the bulk water is not affected by the presence of this complex. This is consistent with the assumption that a possible protonation of the vanadyl oxygen is not an important relaxation mechanism in nonacidified solutions. It follows that τ_{MI} in Figure 13 corresponds to the protons of the equatorial waters, as has been anticipated previously,[4] and the proton exchange from these positions can be characterized by $k(25°) = 2.2 \times 10^4$ sec^{-1}, $\Delta H^{\ddagger} = 7.8$ kcal mol^{-1}, $\Delta S^{\ddagger} = -13$ eu, and $A/h = 1.6 \times 10^6$ cps. As was pointed out previously,[3,34] these values clearly indicate that the proton exchange from the equatorial positions is not controlled by the rate of the water exchange (Table II) but by a hydrolysis mechanism.

Acknowledgment.—The authors wish to thank Dr. A. Bauder, who has written the program used for the analysis of the esr spectra, and Professor R. J. Myers, for many helpful discussions about the subject of this paper. This work was performed under the auspices of the United States Atomic Energy Commission.

(34) T. J. Swift, T. A. Stephenson, and G. R. Stein, *J. Am. Chem. Soc.*, **89**, 1611 (1967).

Reprinted from the PROCEEDINGS OF THE NATIONAL ACADEMY OF SCIENCES
Vol. 63, No. 4, pp. 1071–1078. August, 1969.

*HIGH-RESOLUTION PROTON NUCLEAR MAGNETIC RESONANCE SPECTROSCOPY OF CYTOCHROME c**

BY KURT WÜTHRICH†

BELL TELEPHONE LABORATORIES, INCORPORATED, MURRAY HILL, NEW JERSEY

Communicated by Robert E. Connick, April 28, 1969

Abstract.—In cytochrome *c* the axial positions of the heme iron are occupied by two amino acid residues, one of which is known from X-ray studies to be histidyl. Nuclear magnetic resonance spectroscopy provides strong evidence that the sixth ligand is a methionyl residue in both the ferric and ferrous oxidation states. It is further shown that in cyanoferricytochrome *c* cyanide ion replaces methionyl in the first coordination sphere of the heme iron. Additional data are obtained on the protein conformation and on the electronic structure of the heme group in ferricytochrome *c*. As in other heme proteins, the interactions with the polypeptide chain greatly affect the unpaired electron distribution in the heme group of cytochrome *c*. In particular, from a comparison of ferricytochrome *c* and cyanoferricytochrome c, the importance of the coordination of the sixth ligand is apparent.

Introduction.—Cytochrome *c*[1] is a protein of the respiratory chain which contains one heme group (Fig. 1) per molecule. The axial positions of the heme iron are occupied by two amino acid residues, one of which was found from X-ray studies to be the histidyl residue in position 18.[2] The present nuclear magnetic resonance (NMR) experiments yield information on the sixth ligand (Fig. 1). In the biological role of cytochrome *c*, interconversion between the ferric and ferrous oxidation states of the heme iron is an important factor. This paper presents a preliminary discussion of conformational changes arising from interconversion between the ferric and ferrous oxidation states, and of the electronic structure of the heme group in ferricytochrome *c*.

Nuclear magnetic resonance studies of cytochrome *c* have been described previously. Kowalsky[3] reported that hyperfine interactions with the heme iron give rise to large upfield and downfield shifts of three proton resonances of ferricytochrome *c*. More recently, McDonald and Phillips studied the denaturation of cytochrome *c*[4] and presented an interpretation of the NMR spectrum of ferrocytochrome *c*.[5]

Experimental.—Ferricytochrome *c* of Guanaco was obtained from Dr. E. Margoliash. For the NMR experiments, *ca.* 0.01 *M* solutions in 0.1 *M* deuterated phosphate buffer, pD 7.0, were prepared. Cyanoferricytochrome *c* was prepared by addition of KCN. Ferrocytochrome *c* was obtained by reduction of ferricytochrome *c* with ascorbic acid or lisodium-dithionite.

High-resolution proton NMR spectra were recorded on a Varian HR-220 spectrometer equipped with a standard Varian variable-temperature control unit. The temperature in the sample zone was determined from the chemical shifts of the resonances of ethylene glycol. Chemical shifts are expressed in parts per million (ppm) from internal DSS (sodium 2,2-dimethyl-2-silapentane-5 sulfonate), where shifts to low field are assigned negative values.

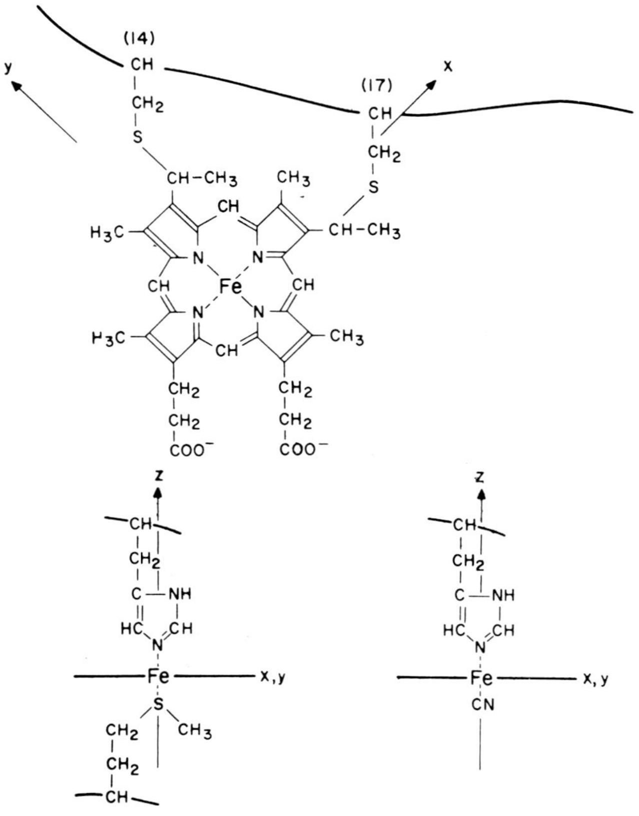

FIG. 1.—Heme group of cytochrome *c* and the axial ligands of the iron in cytochrome *c* and its complex with cyanide ion. In place of the methionyl residue, arginyl and lysyl residues have also been suggested as the sixth ligand.[1, 2]

Results and Discussion.—The proton NMR spectra at 220 Mc of ferricytochrome *c* and cyanoferricytochrome *c* are shown in Figure 2. Three parts of the spectra, which contain all the observed resonances, are reproduced with different vertical and horizontal scales. The spectrum of strongly overlapping resonances between 0 and -9 ppm comes from the bulk of the 650 protons of the polypeptide chain. The sharp lines between -4 and -6 ppm are the resonance of HDO and its side bands. The intensities of the resolved resonances observed in the regions 2 to 7 ppm and -10 to -35 ppm correspond to one to three protons. These resonances are shifted upfield or downfield by local magnetic fields arising both from aromatic ring currents[4, 6] and from the unpaired electron of the heme iron.[3, 7] The lines at -34.0, -31.4, and $+23.2$ ppm are the shifted resonances reported previously by Kowalsky.[3]

The resonances of amino acid residues located near the plane of an aromatic ring experience an upfield ring-current shift[4, 6] which may be as large as 2 ppm for protons located within a few angstroms above or below the plane of a phenylalanine ring.[9] Considerably larger shifts may result if the protons are located near several aromatic amino acid residues or near the plane of the heme group.[8] Resonances of aliphatic amino acid residues can thus be shifted to positions several ppm upfield from DSS and may be well resolved at 220 Mc. Ring-cur-

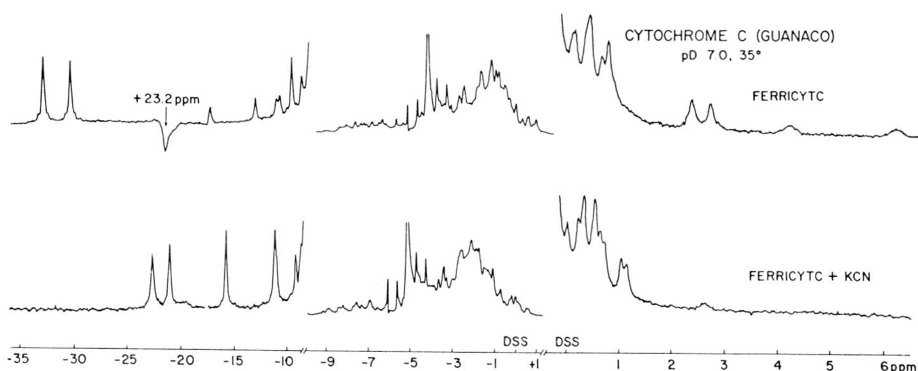

FIG. 2.—High-resolution proton NMR spectrum at 220 Mc of ferricytochrome *c* and cyanoferricytochrome *c*. No DSS was added to these samples. The sharp lines between -4 and -6 ppm correspond to the HDO resonance and its first and second spinning side bands. The vertical and horizontal scales are different for the three parts of the spectrum. The high-field line at $+23.2$ ppm is observed as an inversed resonance of the center band of the spectrum. (The HR-220 spectrometer operates with a 10 kc/sec field modulation. Usually one observes the first upfield side band. If large hyperfine shifts occur, parts of the center band and the different side bands of the spectrum overlap.)

rent shifts are very sensitive to the relative positions of the observed protons and the aromatic rings in the three-dimensional arrangement of the polypeptide chain and hence to conformational changes in the protein.[4, 6] On the other hand, in the absence of conformational changes, ring-current shifts are independent of temperature.[7]

In the NMR spectra of paramagnetic heme proteins, one observes hyperfine shifts in addition to the ring-current shifted resonances.[7] The unpaired electron of the iron in the low-spin ferric hemes (Fe^{3+}, $S = 1/2$) of ferricytochrome *c* and cyanoferricytochrome *c* is delocalized into the π-orbitals of the axial ligands and the porphyrin ring. Unpaired electron density is then transferred by spin polarization or hyperconjugation[10] from the carbon or sulfur atoms to the protons attached directly, or in methyl and methylene groups (Fig. 1). The resulting contact shifts of the heme proton resonances are proportional to the spin densities on the nearest ring carbon atoms.[10] It appears that for low-spin porphyrin iron (III) complexes contact shifts are large compared to pseudo-contact shifts.[3, 11] Furthermore, because of the very short electronic relaxation times of low-spin ferric hemes,[12] the line widths of the proton resonances are essentially unaffected by electron-proton interactions. Hence, information about the unpaired electron distribution in the π-orbitals of the heme group can be obtained from NMR studies.[7] For the following discussion it is of importance that hyperfine shifts are proportional to the reciprocal of temperature.

Ring-current shifts and hyperfine shifts can be distinguished from their temperature dependences.[7] In the spectra of Figure 2, one finds that all the resonances in the regions -10 to -35 ppm and 2 to 7 ppm are shifted by hyperfine interactions.[14] The temperature dependence of the ferricytochrome *c* spectrum between DSS and 3 ppm is shown in Figures 3 and 4. From Figure 4 it appears most likely that the resonances of intensities three and six protons observed at +100 cps and +40 cps are shifted by ring-current fields. Further evidence for the presence of ring-current shifted lines between 0 and 3 ppm comes from the observation that the total intensity of the resonances outside 0 to −9.5 ppm (Fig. 2) corresponds to a larger number of protons than are on the ligands bound to the iron (Fig. 1). From their temperature dependences all the other resonances upfield from DSS appear to be shifted by hyperfine interactions.

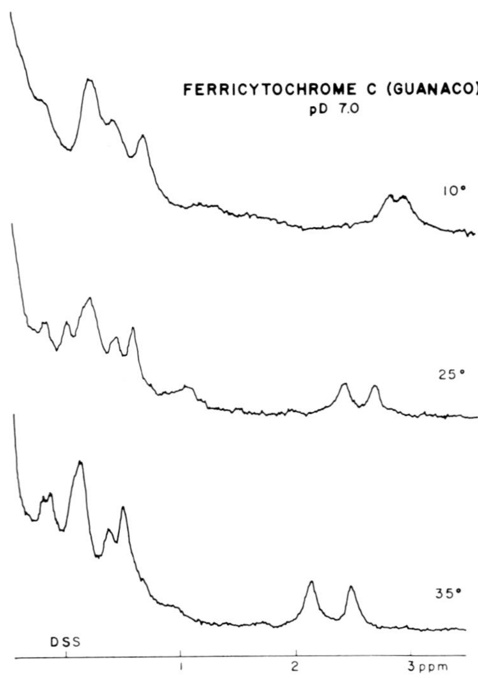

FIG. 3.—Dependence on temperature of the proton NMR spectrum at 220 Mc of ferricytochrome *c* between −1 and +3 ppm.

In the ferricytochrome *c* spectrum we then have the following hyperfine shifted resonances. At 35°C eight lines of intensity one proton are at −18.1, −13.8, −11.8, −11.5, 0.6, 1.0, 4.0, and 6.0 ppm (Figs. 2 and 3). Previous NMR investigations[7, 11] showed that in low-spin ferric hemes and heme proteins the resonances of the methyl and methylene groups are not usually split into single proton resonances. Therefore the observed one-proton lines come most likely from the six single protons of porphyrin *c* and the 2,4-imidazole protons of the axial histidyl residue (Fig. 1). Six methyl resonances are at −34.0, −31.4, −10.3, 2.1, and 2.5 ppm (Fig. 2) and at 0.2 ppm (Fig. 4).† Resonances of two, four, and five protons at −0.1, 0.5, and 23.2 ppm (Figs. 2 and 4) would account in intensity for all but two of the remaining ligand protons in the structure of Figure 1, which one would expect to experience sizable contact shifts. An additional methylene resonance might be between 0 and −9 ppm .

At least five of the six methyl resonances between −34 and 2.5 ppm come from methyl groups of porphyrin *c* (Fig. 1).‡ From the symmetry of the electronic wave functions of the heme group, and because no large negative spin densities would be expected on the carbon atoms of the porphyrin ring, it appears then extremely unlikely that any of the methyl or methylene resonances of the heme group could be shifted to +23 ppm. Furthermore no high-field resonances of intensity two or more protons have been observed above 5 ppm for any other low-spin ferric hemes[11] or heme proteins,[7, 12, 15] including cyanoferricytochrome *c* (Fig. 2). This implies that the resonance at 3.2 ppm comes from one of the axial ligands. From previous work the sixth ligand is known to be a hemochrome-forming aliphatic amino acid residue.[1, 2] Of these, only methionyl[16] could con-

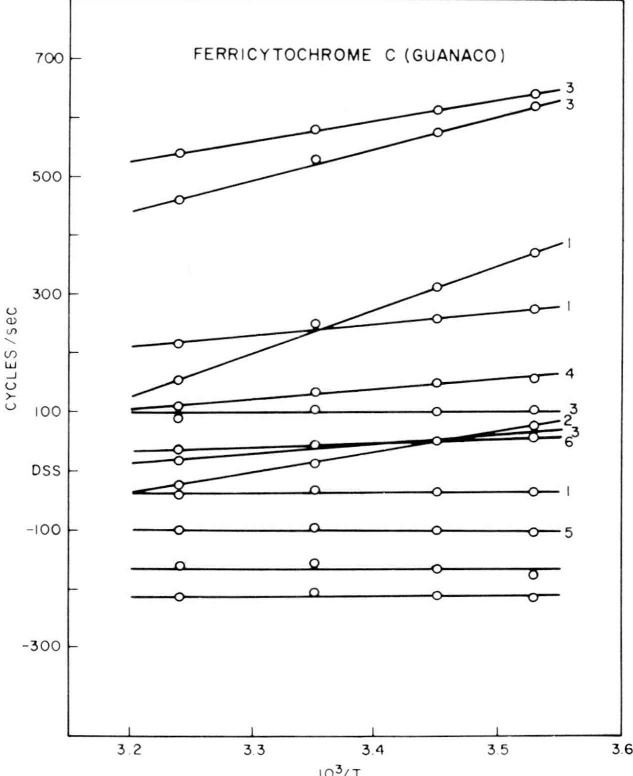

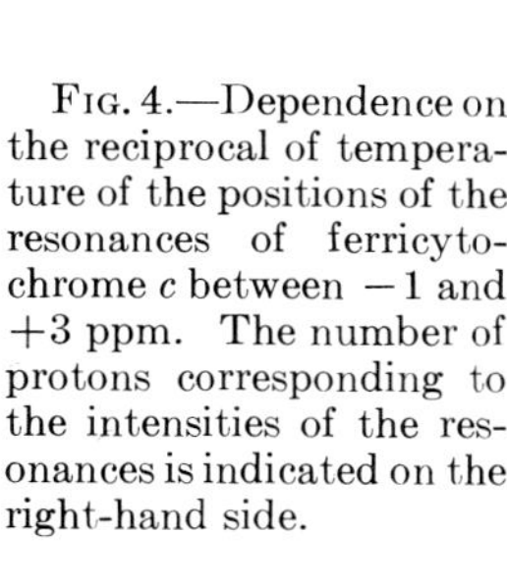
FIG. 4.—Dependence on the reciprocal of temperature of the positions of the resonances of ferricytochrome c between -1 and $+3$ ppm. The number of protons corresponding to the intensities of the resonances is indicated on the right-hand side.

ceivably give rise to a contact shifted resonance of five protons, i.e., if the resonances of the methyl and the methylene groups next to the sulfur (Fig. 1) were accidentally degenerate. The shape of the resonance at 23.2 ppm does indeed indicate that it consists of at least two overlapping lines. Thus the NMR spectrum implies that the sixth ligand of the heme iron in ferricytochrome c is methionyl.

Analysis of the cyanoferricytochrome c spectrum[14] indicates that the four ring methyls of the heme group (Fig. 1) are observed at -22.9, -21.1, -16.0, and -11.4 ppm (Fig. 2). Most of the other hyperfine shifted lines are at high field from DSS, but no resonance of intensity more than one proton is above 2 ppm. As in ferricytochrome c it appears that there are three ring-current shifted methyl resonances between 0 and 1 ppm. The NMR spectrum between -10 and -35 ppm can be used for studies of the reaction of ferricytochrome c with cyanide ion. It was found that the cyanoferricytochrome c spectrum in Figure 2 corresponds to a 1:1 complex.[14]

In Figure 5 the high-field regions of the NMR spectra of ferricytochrome c, cyanoferricytochrome c, and ferrocytochrome c are compared. The latter, which is diamagnetic (Fe^{2+}, $S = 0$), contains ring-current shifted resonances of intensity three protons at 3.3, 0.7, 0.6, and 0.6 ppm, a resonance of two protons at -0.1 ppm, and resonances of one proton at 3.7, 2.7, 1.9, 0.2, and -0.2 ppm

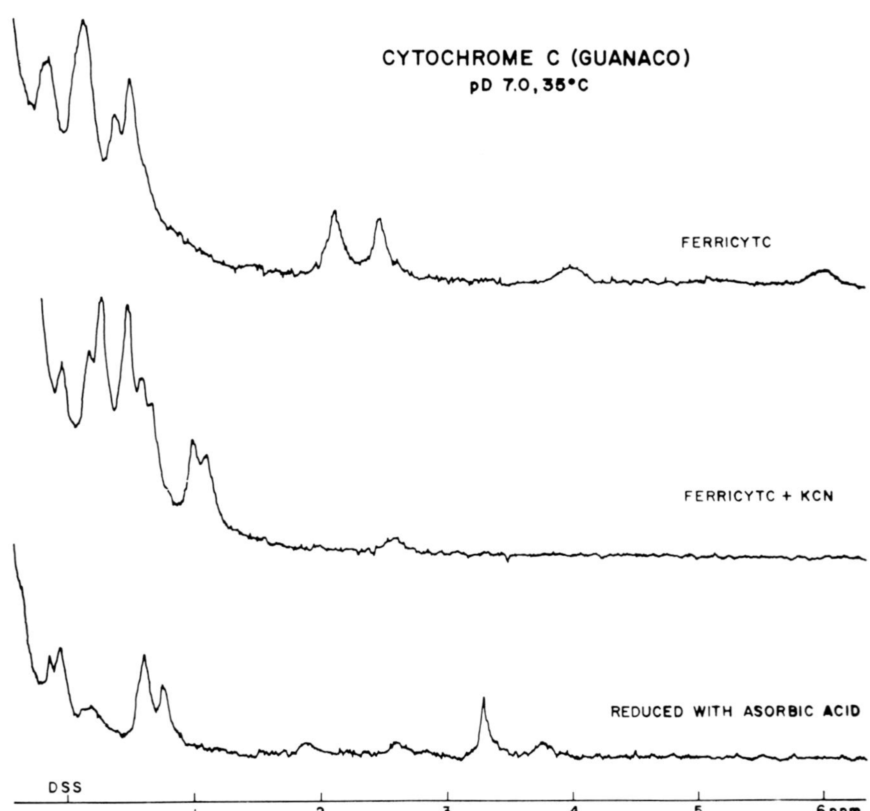

FIG. 5.—Proton NMR spectra between −1 and +3 ppm of ferricytochrome *c*, cyanoferricytochrome *c*, and ferrocytochrome *c*.

McDonald and Phillips[5] suggested that the lines between 1.9 and 3.7 ppm come from the methyl group and three protons of the γ- and β-methylenes of an axial methionyl residue (Fig. 1) which would experience the strong ring-current field of the porphyrin ring. As discussed above, other explanations for the unusual positions of these resonances could be found. However, the following experiment implies that the assignment to the methionyl protons is correct.

A solution of cyanoferricytochrome *c* was reduced with dithionite. Figure 6 shows the resulting changes in the NMR spectrum. As judged from the disappearance of the hyperfine shifted resonances, the reduction was very fast at 9°C. On the other hand, the four resonances between 1.9 and 3.7 ppm of the ferrocytochrome *c* spectrum appeared very slowly. After 50 minutes, the reaction was not complete, as is seen from a comparison of the last two spectra of Figure 6. These observations agree with the following reaction mechanism proposed by George and Schejter[17]:

$$\text{cyanoferricytochrome } c + e \xrightarrow{-\text{fast}} \text{cyanoferrocytochrome } c \quad (1)$$

$$\text{cyanoferrocytochrome } c \xrightarrow{\text{slow}} \text{ferrocytochrome } c + \text{CN}^-. \quad (2)$$

After the dissociation of the unstable cyanoferrocytochrome *c* (2), the axial amino acid residue which was displaced by cyanide ion (Fig. 1) goes back into its place in the native protein. The data in Figure 6 show that the ferrocytochrome *c* resonances between 1.9 and 3.7 ppm come from this axial ligand which then has to be methionyl, since this is the only hemochrome-forming amino acid residue[1, 16] that contains a methyl group.

In addition to the identification of the sixth ligand, the NMR spectra yield data on the protein conformation and on the electronic structure of the heme group in ferricytochrome *c*. For example, a comparison of the spectra in Figure 5 shows that the ring-current shifted methyl resonances at 0.7, 0.6, and 0.6 ppm in ferrocytochrome *c* are at different positions in both ferricytochrome *c* and cyanoferricytochrome *c*. Furthermore, it is seen from Figure 6 that two methyl resonances at 0.6 and 0.7 ppm are in identical positions in cyanoferrocytochrome *c* and ferrocytochrome *c*, whereas one methyl resonance moves from 0.6 to 0.7 ppm upon dissociation of the cyanide complex. This shows that at least minor conformational changes occur upon both interconversions between ferric and ferrous oxidation states and complex formation with cyanide ion. Extension of the analysis to the entire spectrum of the polypeptide chain will lead to a more detailed description of these conformational changes.

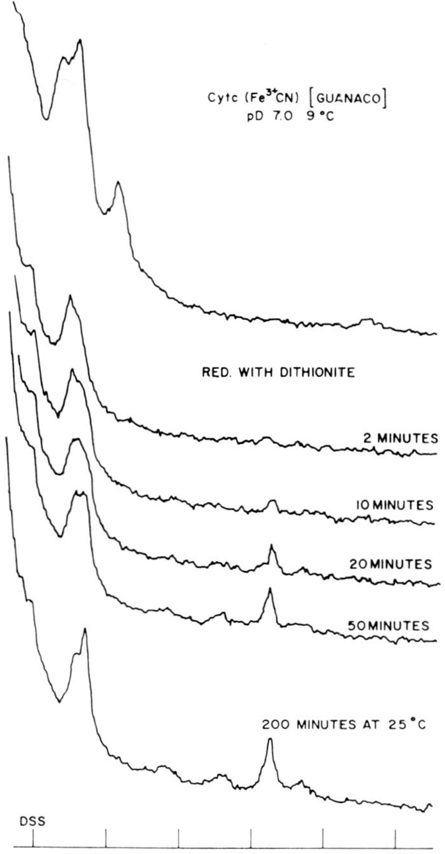

FIG. 6.—Proton NMR spectra between −1 and +3 ppm of cyanoferricytochrome *c* and of the reaction products observed at various times after reduction of cyanoferricytochrome *c* with dithionite.

From Figure 2 the unpaired electron distribution in the heme group of ferricytochrome *c* differs greatly from that found in cyanoporphyrin iron (III) complexes.[11] Thus, as was found in other heme proteins,[7, 15] the polypeptide-heme interactions have a strong influence on the electronic structure of the heme group. In particular, the comparison of ferricytochrome *c* and cyanoferricytochrome *c* implies that the coordination of the sixth ligand to the heme iron is an important factor. The proton resonances at −34.0 and −31.4 ppm‡ of ferricytochrome *c* correspond most likely to two ring methyls of the heme group.[14] The two remaining ring methyl resonances are then much closer to −3.5 ppm, which is the resonance position of the ring methyls in diamagnetic porphyrins.[8] This shows that there are large positive unpaired electron densities on ring carbon atoms of two pyrrole rings of porphyrin *c*, and small positive or negative spin densities on the two other pyrroles. This is interesting because in its biological role ferricytochrome *c* takes up an electron. Since X-ray studies have shown that only one edge of porphyrin *c* is exposed to the solvent,[2] a possible path for the electron transfer would seem to be through this edge. Negative or small positive electron density at one of the exposed pyrrole rings might contribute toward a favorable free energy for the electron uptake.

It is a pleasure to thank Dr. W. A. Eaton for interesting discussions, Dr. E. Margoliash for providing the samples of cytochrome *c*, and Dr. R. G. Shulman for the hospitality extended to me at his laboratory.

* Presented in part at the fourth Middle Atlantic Regional Meeting of the American Chemical Society, Washington, D.C., February 13–15, 1969.

† Present address: Laboratorium für Molekularbiologie, ETH, Universitätsstr. 6, Zürich, Switzerland.

‡ In the methyl resonances at −34.0 and −31.4 ppm one observes a narrow and a broader component, which have approximately equal intensities. This arises most likely because the dipole-dipole coupling between the methyl protons is modulated by two different rotational motions—a very fast motion about the C—C bond from the porphyrin ring to the methyl group, and the slower rotation of the entire cytochrome *c* molecule. This effect was predicted from theoretical calculations by A. Redfield.[13] To our knowledge it has not been observed previously in a protein NMR spectrum.

[1] For a review on cytochrome *c*, see Margoliash, E., and A. Schejter, *Advan. Protein Chem.*, **21**, 113 (1966).

[2] Dickerson, R. E., M. L. Kopka, J. Weinzierl, J. Varnum, D. Eisenberg, and E. Margoliash, *J. Biol. Chem.*, **242**, 3015 (1967).

[3] Kowalsky, A., *Biochemistry*, **4**, 2382 (1965).

[4] McDonald, C. C., and W. D. Phillips, *J. Am. Chem. Soc.*, **89**, 6332 (1967).

[5] McDonald, C. C., and W. D. Phillips, in *Proceedings of the Third International Conference on Magnetic Resonance in Biology*, Warrenton, Virginia, October 1968.

[6] Kowalsky, A., *J. Biol. Chem.*, **237**, 1807 (1962).

[7] Wüthrich, K., R. G. Shulman, and J. Peisach, these PROCEEDINGS, **60**, 373 (1968).

[8] Becker, E. D., and R. B. Bradley, *J. Chem. Phys.*, **31**, 1413 (1959).

[9] Johnson, C. E., and F. A. Bovey, *J. Chem. Phys.*, **29**, 1012 (1958).

[10] Carrington, A., and A. D. McLachlan, *Introduction to Magnetic Resonance* (New York: Harper and Row, 1967), pp. 81–85.

[11] Wüthrich, K., R. G. Shulman, B. J. Wyluda, and W. S. Caughey, these PROCEEDINGS, **62**, 636 (1969).

[12] Wüthrich, K., R. G. Shulman, and T. Yamane, manuscript in preparation.

[13] Redfield, A., private communication. See also Andrew, E. R., and R. Bersohn, *J. Chem. Phys.*, **18**, 159 (1950).

[14] Wüthrich, K., in *Structure-Function Correlations in Hemoproteins and Membranes*, ed. B. Chance, C. P. Lee, and T. Yonetani (New York: Academic Press, 1969).

[15] Wüthrich, K., R. G. Shulman, and T. Yamane, these PROCEEDINGS, **61**, 1199 (1968).

[16] Harbury, H. A., J. R. Cronin, M. W. Fanger, T. P. Hettinger, A. J. Murphy, Y. P. Myer, and S. N. Vinogradov, these PROCEEDINGS, **54**, 1658 (1965).

[17] George, P., and A. Schejter, *J. Biol. Chem.*, **239**, 1504 (1964).

II

THE AVENUE TO THREE-DIMENSIONAL PROTEIN STRUCTURE DETERMINATION IN SOLUTION

Introduction to papers 4 through 25

The development of the NMR method for protein structure determination outlined in Fig. 1 of paper **1** involved four principal factors: (*i*) identification of NMR parameters that are experimentally accessible under the conditions of the spin physics in macromolecules in solution, and that can be related in a straightforward way to molecular conformation (papers **4–7**); (*ii*) multi-dimensional NMR (papers **8–13**); (*iii*) an efficient technique for sequence-specific assignment of the many hundred to several thousand NMR lines in a protein (papers **14–19**); (*iv*) suitable techniques for the structural interpretation of the NMR data (papers **20–25**). The 22 papers in this section, which are from the period 1976–85, have been arranged so as to address these four themes in the above order. The following comments on the role of multi-dimensional NMR may be helpful for a better appreciation of the time course of events: Two-dimensional (2D) NMR is not a fundamental element of the method in the sense that all parameters used as input for a structure determination can be obtained using 1D NMR techniques. Conversely, 2D NMR measurements, and subsequently 3D and 4D NMR experiments were indispensible for obtaining the degree of spectral resolution and the high efficiency of data collection that make macromolecular structure determination by NMR a practical approach. During the period 1976–80 I worked with a group of about 20 students and postdoctoral fellows on the aforementioned subjects (*i*), (*iii*) and (*iv*). In addition to developing the sequential resonance assignment strategy and algorithms for structure calculation from NMR data, we used 1D NMR experiments for partial structure determination of globular proteins and for a structure determination of the C-terminal decapeptide segment of the intact polypeptide hormone glucagon incorporated into lipid micelles of molecular weight about 20,000 (paper **21**). In parallel, in a joint project with Prof. R.R. Ernst from the Laboratory of Physical Chemistry at the ETH Zürich, 2D NMR techniques were developed to the stage where they could be applied for studies of biological macromolecules. From [illegible] this project, which in addition to the intense personal involvement of Rich-[illegible] and myself included 1 to 1.5 postdoctoral positions at a time, was hardly no-ti[illegible] the other members of my research group, although all the experimental work was performed in my laboratory. This situation changed once Kuniaki Nagayama had demonstrated the practical application of correlation spectroscopy for spin system identification in a protein (paper **14**), and Anil Kumar made the best possible use of two weeks of instrument time allotted to him during the Christmas break 1979 by recording the first 2D nuclear Overhauser enhancement spectra (paper **11**). During the two-year period 1980–81, my entire group started to use 2D experiments in daily practice, and thus experience from more than a decade of NMR with proteins was joined with this new technique.

(i) NMR parameters yielding conformational constraints. Nuclear Overhauser effects are due to dipolar (through-space) interactions between different nuclei and are correlated with the inverse sixth power of the internuclear distance. The possibility of measuring inter-atomic distances between pairs of hydrogen atoms (or, in practice, of establishing upper bounds on ^{1}H–^{1}H distances over the range of about 2.0–5.0 Å) *via* observation of NOEs represents the physical basis for macromolecular structure determination by NMR. NOEs can be observed in double-irradiation 1D NMR experiments as the fractional change in intensity of one NMR line when another resonance is irradiated (paper **4**), or as cross peaks in 2D NOE spectroscopy (NOESY) or higher-dimensional spectra (paper **1**). Our work started to focus on NOE measurements two decades after the establishment of the theoretical foundations for this experiment.[1] However, in part because of limitations imposed by the available instrumentation,

[1] Solomon, I. (1955) *Phys. Rev. 99,* 559–565. Relaxation processes in a system of two spins.

NOEs had not been widely used, as is spelled out in the following quotation from the first monograph on the subject:[2] "The NOE has found limited use over the past two decades in the study of chemical kinetics and, somewhat more recently, in the assignment of NMR spectra. Lately, interest in the NOE has grown enormously following the realization that detailed qualitative and quantitative information on molecular configuration and conformation can be obtained from it. The uniqueness of this approach to problems in molecular structure, together with the increasing availability of NMR spectrometers sufficiently sophisticated for NOE studies and sufficiently simple in operation to be used on a routine basis, have increased and will continue to increase the applications of the method manyfold. However, the numerous existing books on NMR written primarily for chemists barely mention the NOE and do not provide the background in nuclear relaxation theory necessary to understand it. Aside from original research papers, only a few references of a highly theoretical nature are extant. The difficulty and rigor of these references has been a source of frequent misunderstandings and has surely limited the growth of the field." Although NOE measurements had previously been used for studies of cyclic oligopeptides[3,4] and observation of NOEs in proteins had also been reported,[5–7] it was not clear in 1976 whether with the inevitable presence of spin diffusion[1] the desired distance information could be

[2] Noggle, J.H. and Schirmer, R.E. (1971) *The nuclear Overhauser effect, chemical applications.* New York: Academic Press.

[3] Gibbons, W.A., Crepaux, D., Delayre, J., Dunand, J.J., Hajdukovic, G. and Wyssbrod, H.R. (1975) in *Peptides: Chemistry, Structure, Biology* (Walter, R. and Meienhofer, J., eds.) pp. 127–137. Ann Arbor: Ann Arbor Science Press. The study of peptides by INDOR, difference NMR and time-resolved double resonance techniques.

[4] Glickson, J.D., Rowan, R., Pitner, T.P., Dadok, J., Bothner-By, A.A. and Walter, R. (1976) *Biochemistry 15,* 1111–1119. ^{1}H NMR double resonance study of oxytocin in aqueous solution.

[5] Redfield, A.G. and Gupta, R.K., (1971) *Cold Spring Harbor Symp. Quant. Biol. 36,* 405–411. Pulsed NMR study of the structure of cytochrome *c*.

obtained in macromolecules.[8] The results of the experiments of Regula Keller described in paper **4** then demonstrated that selective intramolecular ^1H–^1H NOEs can be obtained in a protein with molecular weight 12,000, and that the conditions for NOE distance measurements are actually more favorable in macromolecules than in low molecular weight compounds. Based on this work and his own experience with NOE studies of small molecules, Sidney Gordon, who had come from the Georgia Institute of Technology to spend a sabbatical in Zürich, developed the transient NOE experiments described in paper **5**, which together with the truncated-driven NOEs (TOE) used in paper **6** [9] established the recording of NOE build-up curves as a basis for ^1H–^1H distance measurements in macromolecules. NOE build-up curves corresponding to those in paper **5** were subsequently recorded with 2D NOESY (see below), and the same principles govern the use of 3D and 4D NOESY experiments for ^1H–^1H distance measurements in present-day practice.

Scalar spin–spin coupling constants can provide important supplementary conformational constraints to the input for structure calculations. Paper **7** describes an approach for determination of conformational mobility about a single bond based on measurements of two vicinal coupling constants that are both related to the dihedral angle about this bond.

---

[6] Balaram, P., Bothner-By, A.A. and Dadok, J. (1972) *J. Am. Chem. Soc. 94,* 4015–4017. Negative nuclear Overhauser effects as probes of macromolecular structure.

[7] Campbell, I.D., Dobson, C.M. and Williams, R.J.P. (1974) *J. Chem. Soc., Chem. Comm. 894,* 888–889. Intramolecular nuclear Overhauser effects in proton magnetic resonance spectra of proteins.

[8] Kalk, A. and Berendsen, H.J.C. (1976) *J. Magn. Reson. 24,* 343–366. Proton magnetic relaxation and spin diffusion in proteins.

[9] Wagner, G. and Wüthrich, K. (1979) *J. Magn. Reson. 33,* 675–680. Truncated driven nuclear Overhauser effect (TOE): a new technique for studies of selective ^1H–^1H Overhauser effects in the presence of spin diffusion.

(*ii*) *2D NMR.* The paper **9** presents not only the first 2D NMR spectrum of a protein, but also the first 2D spectra recorded with a data size that made the technique useful for practical applications beyond observation of the classical test compounds, such as CH_3I or CH_3CH_2OH. Papers **10** and **11** introduced two key experiments for structural studies of proteins, spin-echo-correlated spectroscopy (SECSY) and NOESY. SECSY was designed as an alternative to correlated spectroscopy (COSY) as previously described in the classical 1976 paper by the Ernst group,[10] to overcome limitations imposed by the computer capacities available to us in 1977. The paper **12** established the practice of measuring NOE build-up curves for ^1H–^1H distance measurements with NOESY. In these early, absolute-value 2D NMR experiments, as well as in most of our 1D NMR studies with biological macromolecules, resolution enhancement with the sine bell window (paper **8**) was the standard way of obtaining the desired results. The continued wide-spread use of this routine is a fitting tribute to the beautiful spectroscopy performed by the late Antonio DeMarco during his all-too-short life span. A further decisive improvement of the spectral resolution was achieved with the practical implementation of high resolution phase-sensitive 2D NMR (paper **13**). A comprehensive coverage of 2D NMR during the decade 1976–85 is presented in a monograph by Ernst *et al.*,[11] and applications with biomolecules are described more extensively in a monograph by the editor.[12]

(*iii*) *Sequence-specific resonance assignments.* The pivotal role of resonance assignments in the course of a protein structure determination is illustrated in paper **1**.

[10] Aue, W.P., Bartholdi, E. and Ernst, R.R. (1976) *J. Chem. Phys. 64,* 2229–2246. Two-dimensional spectroscopy: application to nuclear magnetic resonance.

[11] Ernst, R.R., Bodenhausen, G. and Wokaun, A. (1987) *Principles of nuclear magnetic resonance in one and two dimensions*. Oxford: Clarendon.

[12] Wüthrich, K. (1986) *NMR of proteins and nucleic acids.* New York: Wiley.

In the present section, paper **14** describes the first systematic application of 2D *J*-resolved spectroscopy and SECSY for the identification of the spin systems of the individual amino acid residues in a protein. However, the decisive breakthrough that opened the way to NMR structure determination of proteins came years earlier with the successful "sequential NOE walk" by Andreas Dubs and Gerhard Wagner for obtaining sequential assignments of the β-sheet in BPTI (paper **6**). This approach was based on detailed inspection of the BPTI crystal structure, which had to take the interplay of secondary and tertiary structural features into account, and on 1D NMR techniques used previously by Regula Keller for a similar sequential walk around the heme group in hemoproteins (paper **4**), which had demonstrated the feasibility of this type of experiment in molecules with the size of proteins. The four papers **15–18**, which were published back to back in 1982, provide a comprehensive account of the sequential assignment strategy with fully integrated use of 2D NMR and demonstrate that this approach is generally applicable, irrespective of the secondary and tertiary structure types present in the protein studied. The last entry in this section, paper **19**, describes an early application of isotope labeling with an NMR-active isotope to resolve chemical shift degeneracies in the ^{1}H NMR spectra. The observation of an "E.COSY-type" multiplet fine structure in Fig. 6 of paper **19** is of special interest, as it led to a novel approach for measurements of scalar coupling constants that has in the meantime been exploited with a variety of homonuclear and heteronuclear experiments.

(*iv*) *Structure calculation from NMR data.* The first contribution to this section, paper **20**, relates to a unique feature of NMR structure determination: regular secondary structures can be identified at an early stage of a project, before the calculation of the complete structure is actually started (Fig. 1 of paper **1**). This is based on the following: The sequential NOEs are dependent on the local backbone conformation and thus indicative of regular secondary structure elements (paper **16**), and corresponding information on the local backbone conformation comes from $^3J_{HN\alpha}$ coupling constants.[13] These are features that have also been used in studies of small peptides.[14,15] Furthermore, when identifying sequential NOEs in proteins, all other NOEs between amide protons, H^α and H^β must also be analyzed, which includes medium-range NOEs that are unique for helices and tight turns (paper **20**). Using additional long-range NOE connectivities and amide proton exchange data, β-sheet structures can similarly be identified.[12] Identification of the regular secondary structures is not a necessary step in a protein structure determination (Fig. 1 of paper **1**), since the conformational constraints used are in any case part of the input for the calculation of the complete structure. However, it may be a valuable result in itself in situations where a structure determination cannot immediately be completed, which explains why reports on NMR structure determinations of proteins are often preceded by papers describing the secondary structure (*e.g.,* paper **26**, ref. 16 and 17).

---

[13] Pardi, A., Billeter, M. and Wüthrich, K. (1984) *J. Mol. Biol. 180,* 741–751. Calibration of the angular dependence of the amide proton–C^α proton coupling constants, $^3J_{HN\alpha}$, in a globular protein: use of $^3J_{HN\alpha}$ for identification of helical secondary structure.

[14] Leach, S.J., Némethy, G. and Scheraga, H.A. (1977) *Biochem. Biophys. Res. Comm. 75,* 207–215. Use of proton nuclear Overhauser effects for the determination of the conformations of amino acid residues in oligopeptides.

[15] Kuo, M. and Gibbons, W.A. (1979) *J. Biol. Chem. 254,* 6278–6287. Total assignments, including four aromatic residues, and sequence confirmation of the decapeptide tyrocidine A using difference double resonance. Qualitative nuclear Overhauser effect criteria for β turn and antiparallel β-pleated sheet conformations.

[16] Zuiderweg, E.R.P., Kaptein, R. and Wüthrich, K. (1983) *Proc. Natl. Acad. Sci. USA 80,* 5837–5841. Secondary structure of the *lac* repressor DNA-binding domain by two-dimensional ^1H nuclear magnetic resonance in solution.

[17] Williamson, M.P., Marion, D. and Wüthrich, K. (1984) *J. Mol. Biol. 173,* 341–359. Secondary structure in the solution conformation of the proteinase inhibitor IIA from bull seminal plasma by nuclear magnetic resonance.

The calculation of complete three-dimensional NMR structures was initially performed with metric matrix distance geometry. We started from software that was kindly provided to us by Dr. G.M. Crippen, who had pioneered the use of distance geometry for studies on protein folding.[18] The papers **21** and **22** describe a software package written by Werner Braun and its practical application with the polypeptide hormone glucagon. This program took account of the complete, detailed covalent structure of the amino acid residues and their chiral properties in addition to the NMR constraints, which is indispensible for NMR structure determinations, but it was limited for use with polypeptide segments of up to about ten residues. It was therefore left to Timothy Havel and Michael Williamson to compute the first globular protein structure from NMR data, using the program DISGEO[19] (paper **25**), and in paper **24** we demonstrated that the NMR data that can be collected with proteins are indeed sufficient for experimental determination of novel polypeptide folds. The paper **23**, finally, describes a treatment of NOEs with groups of two or more hydrogen atoms that could not be individually assigned, including pairs of diastereotopic substituents, which is essential for the preparation of the input for NMR structure calculations.

[18] Crippen, G.M. (1979) *Int. J. Peptide Protein Res. 13,* 320–326. Distance constraints on macromolecular conformation.

[19] Havel, T.F. and Wüthrich, K. (1984) *Bull. Math. Biol. 46,* 673–698. A distance geometry program for determining the structures of small proteins and other macromolecules from nuclear magnetic resonance measurements of intramolecular ^1H–^1H proximities in solution.

Reprinted from *Biochimica et Biophysica Acta*, Vol. 533, pp. 195–208 (1978)
Copyright © 1978, with kind permission from
Elsevier Science B. V. Amsterdam, The Netherlands.

BBA 37859

ASSIGNMENT OF THE HEME *c* RESONANCES IN THE 360 MHz ^{1}H NMR SPECTRA OF CYTOCHROME *c*

REGULA M. KELLER and KURT WÜTHRICH

Institut für Molekularbiologie und Biophysik, Eidgenössische Technische Hochshule, CH-8093 Zürich-Hönggerberg (Switzerland)

(Received August 2nd, 1977)

Summary

In the 360 MHz ^{1}H NMR spectra of horse heart ferrocytochrome *c* recorded after suitable digital resolution enhancement, the resonances of all the heme *c* protons with the exception of those of the propionic acid side chains were observed as well resolved lines. From spin decoupling and nuclear Overhauser effects in homonuclear double resonance experiments, all these resonances were assigned to their respective positions in heme *c*. With saturation transfer experiments in solutions of partially reduced cytochrome *c*, individual assignments were further obtained for the six heme *c* methyl resonances in ferricytochrome *c*. The present experiments add individual assignments to the earlier identifications of the heme *c* ring methyl and meso-proton resonances, and show that the earlier identifications of the thioether bridge methyl resonances must be revised. These data provide a basis for more detailed descriptions of the electronic structure of heme *c* and its possible relations with the pathway of the electron transfer in and out of the cytochrome *c* molecule. Furthermore, the pseudocontact shifts of the thioether bridge methyl resonances could be related to the electronic *g*-tensor measured by EPR in ferricytochrome *c* single crystals at low temperature. From this it will now be possible without chemical modification of the protein, to compare in detail the solution conformations near the heme *c* in reduced and oxidized cytochrome *c* and thus hopefully to obtain additional insights into the mechanism of the biological redox reaction of this protein.

Introduction

Wide interest is directed both to the functional properties of electron transferring systems which include cytochromes *c* [1—4] and to the evolutionary history of these systems [5,6]. It is therefore not surprising that cytochromes *c* are among the most extensively studied proteins. In addition to single crystal

X-ray studies of oxidized and reduced cytochromes *c* [7—12], a variety of spectroscopic techniques have also been used to characterize the structural properties and their relations with the biological roles of the protein [1,2,6,13—20]. Yet, in spite of the wealth of data obtained so far, many questions relating to mechanistic aspects of cytochrome *c* functions remain to be further investigated. The present high resolution ^1H NMR investigation is mainly directed to one of these questions, i.e. the nature of the intermolecular electron transfer in the biological redox reaction of cytochrome *c*.

Early ^1H NMR studies of cytochromes *c* had been mainly concerned with a limited number of well resolved resonances in extreme high and low field positions [6,13,21—23]. These experiments resulted in a qualitative structural characterization of the heme group and the heme crevice. More recently, with the use of modern high field NMR techniques, detailed interpretations were extended to other regions of the ^1H NMR spectra, in particular the resonances of aromatic amino acid residues [24—28]. In this paper, high resolution ^1H NMR at 360 MHz was used to obtain individual assignments of numerous resonance lines of heme *c* in reduced and oxidized cytochrome *c*. These data provide a basis for quantitative studies of the relations between cytochrome *c* functions and the electronic structure of heme *c* and the protein conformation near the heme crevice.

Materials and Methods

Horse heart ferricytochrome *c* "Type VI" was obtained from Sigma. For the NMR experiments, the protein was dissolved in 0.05 M deuterated phosphate buffer, p^2H 7.1. The p^2H values correspond to pH meter readings without correction for isotope effects. Solutions of partially or fully reduced cytochrome *c* were obtained by addition of solid $Na_2S_2O_4$ to ferricytochrome *c* solutions. Protein concentrations were determined spectrophotometrically [1]. Sodium 3-trimethylsilyl-[2,2,3,3-$^2H_4$]propionate was used as an internal reference for calibration of the chemical shifts.

High resolution Fourier transform ^1H NMR spectra were recorded on a Bruker 360 MHz spectrometer. In some experiments the spectral resolution was improved by multiplication of the free induction decay with a phase shifted sine bell window [29,30]. Double irradiation techniques were employed for the identification of individual spin systems through homonuclear ^1H-^1H spin decoupling, studies of through space interactions between neighboring protons by measurement of the nuclear Overhauser enhancement [31], and identification of corresponding resonance lines in oxidized and reduced cytochrome *c* by saturation transfer experiments in solutions of the partially reduced protein [22]. When the observed lines were well resolved, the double resonance irradiation was applied continuously while the spectrum was recorded. For observation in crowded spectral regions, free induction decays were sampled alternatingly with and without double resonance irradiation. The difference between the two accumulated free induction decays was then Fourier transformed, yielding the difference spectrum in the frequency domain.

Results

From the structure of heme *c* (Fig. 1) one expects that the four ring methyl groups 1, 3, 5 and 8 and the four meso-protons α to δ each give rise to a singlet resonance in the ^{1}H NMR spectrum of cytochrome *c*. The spin systems of the thioether bridges 2 and 4 are of the A_3X type [32]. In previous studies, the four ring methyl resonances had been identified in ferricytochrome *c* and the corresponding chemical shifts in ferrocytochrome *c* were determined by saturation transfer studies [21,22]. In the reduced protein the four meso-proton singlets could readily be identified from their chemical shifts between 9 and 10 ppm [13,23]. In the following, these earlier data will first be completed by the identification of the thioether bridge A_3X spin systems in ferrocytochrome *c* through spin decoupling. Next, individual assignments of the resonances of the heme *c* substituents 1—5, 8 and α to δ in reduced cytochrome *c* will be discussed on the basis of nuclear Overhauser enhancements [31]. Finally, saturation transfer in solutions of partially oxidized cytochrome *c* will be used to obtain individual assignments for the heme *c* ring methyl and the thioether bridge methyl resonances in ferricytochrome *c*.

The 360 MHz ^{1}H NMR spectra of fully oxidized, fully reduced and half reduced cytochrome *c* (Fig. 2) show the typical features which had previously been observed in the spectra recorded at lower field strength [13,21]. While earlier investigations had mainly concentrated on the well resolved lines at extreme high and low fields, some of which are outside the spectral region shown in Fig. 2 [13,21], the resolution obtained at 360 MHz makes it possible to recognize individual resonances also in the more crowded central spectral

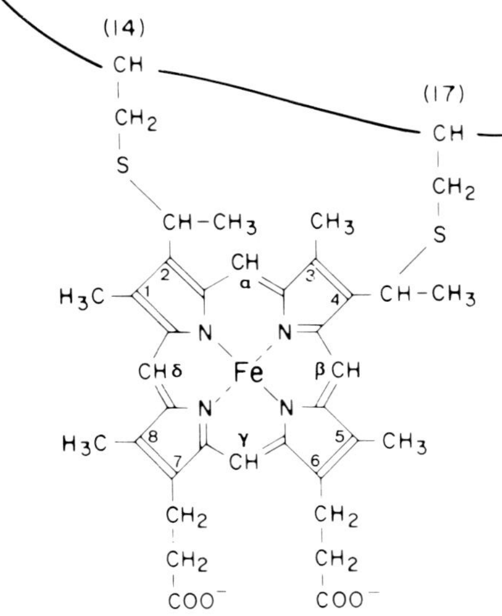

Fig. 1. Structure of heme *c*. The β positions of the four pyrrole rings are numbered 1—8 and the four meso positions α to δ. The covalent links of the heme group with the polypeptide chain are also shown. In this representation the axial histidine 18 would be below and the axial methionine 80 above the heme plane.

198

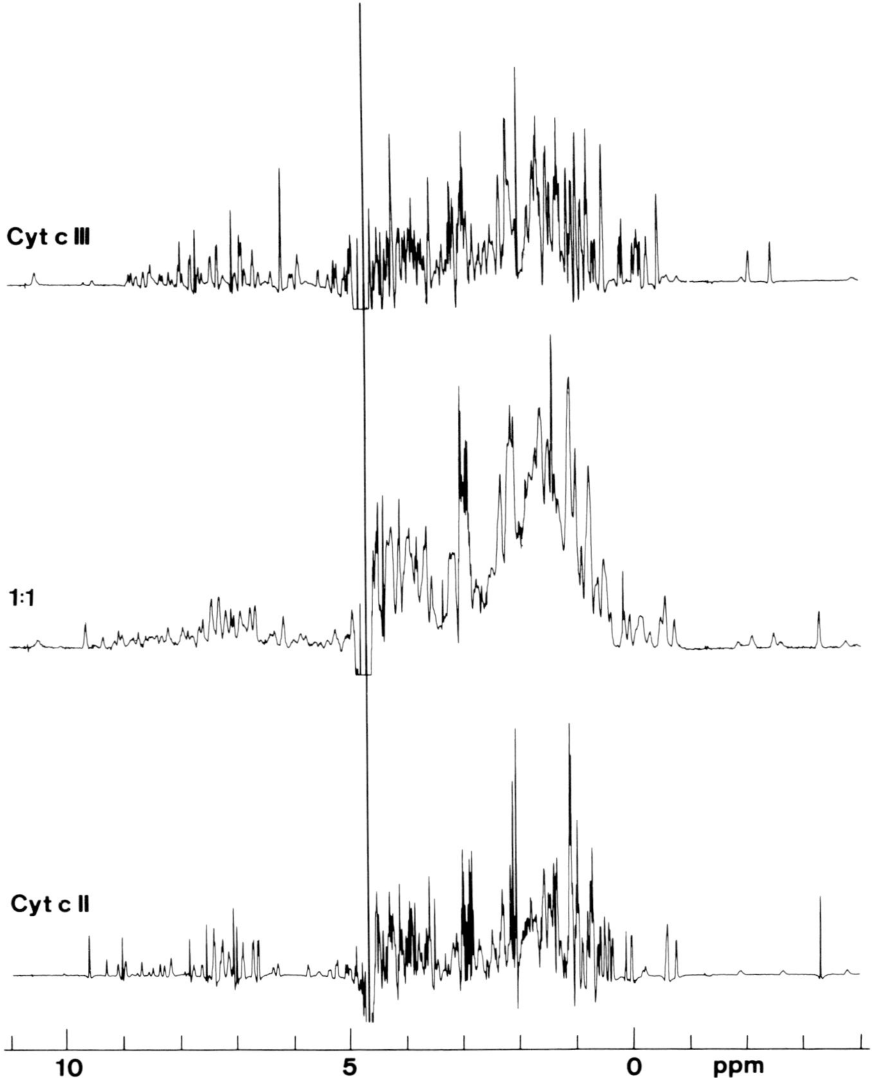

Fig. 2. ^{1}H NMR spectra at 360 MHz of 0.003 M solutions of horse heart cytochrome *c* in 0.05 M deuterated phosphate buffer, p^2H 7.1, temperature 35°C. From top to bottom the spectra correspond to oxidised cytochrome *c*, a 1 : 1 mixture of the two oxidation states, and reduced cytochrome *c*, respectively. In all three spectra the resolution was improved by multiplication of the free induction decay with a phase shifted sine bell window, $\sin\left(\frac{\pi t}{t_s} + \frac{\pi}{32}\right)$, with t_s equal to the acquisition time.

regions. Fig. 2 shows, however, that mutual overlap of resonances occurs throughout the spectrum, so that much care had to be employed in the interpretation of double resonance phenomena. In the partially oxidized protein the spectral resolution is further limited by the simultaneous presence of the spectra of the two species and by exchange broadening of the lines.

The identification of the A_3X spin system of one of the thioether bridges is illustrated in Fig. 3, which shows the regions from 2.4 to 2.8 ppm and 6.2

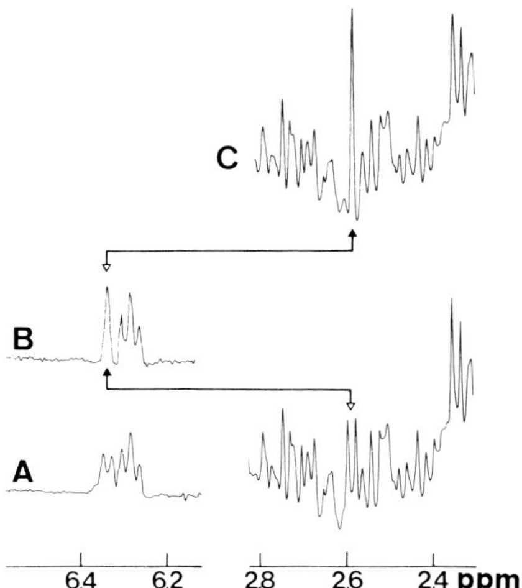

Fig. 3. 360 MHz ^{1}H NMR of a 0.005 M solution of ferrocytochrome *c* in 0.05 M deuterated phosphate buffer, p^{2}H 7.1, temperature 52°C. The spectral resolution was improved by the technique described in Fig. 2. (A) Spectral regions from 2.4 to 2.8 ppm and 6.1 to 6.5 ppm. (B) Spin decoupling at 6.34 ppm (filled arrow) upon double resonance irradiation at 2.59 ppm (empty arrow). (C) Spin decoupling at 2.59 ppm (filled arrow) upon double resonance irradiation at 6.34 ppm (empty arrow).

to 6.4 ppm of the ferrocytochrome *c* spectrum (Fig. 2) on an expanded scale. The doublet and quartet fine structures of the resonances in the undecoupled spectrum and the spin decoupling can be readily observed. Similar experiments revealed that two resonances at 1.53 and 5.24 ppm, which were assigned to a thioether bridge from nuclear Overhauser enhancement studies (see below) are connected by spin-spin coupling. However, because of overlap with other resonances, the multiplet structures could not unambiquously be determined for this pair of resonance lines.

Measurements of nuclear Overhauser enhancements were of prime importance for the present study. The structure of heme *c* (Fig. 1) indicates that considerable contributions to the overall nuclear spin relaxation of the meso-protons in ferrocytochrome *c* must come from dipole-dipole coupling with other protons of heme *c*. As a consequence, one would expect nuclear Overhauser enhancements of the meso-proton resonances upon double resonance irradiation of the neighboring heme *c* protons [31]. Indeed, as listed in Table 1, irradiation at each of the previously established [22] heme *c* ring methyl resonance positions and at the four positions corresponding to the A_3X spin systems of the two thioether bridges (Fig. 3) produced a negative nuclear Overhauser enhancement for one of the meso-proton lines. Three of these experiments are illustrated in Fig. 4, where it is seen that the nuclear Overhauser enhancements were as large as −50%. From inspection of the structure of heme *c* (Fig. 1), the nuclear Overhauser enhancements observed on the meso-protons resulted in the following assignments (Table I and Fig. 4): Meso-proton δ at 9.04 ppm, since it was affected by irradiation of two ring methyls; mesoproton γ at 9.62 ppm where no nuclear Overhauser enhancement resulted from irradiation of any of the substituents 1—5 or 8; the pair of meso-protons α and β

TABLE I

^{1}H RESONANCE ASSIGNMENTS FOR HEME *c* IN HORSE HEART CYTOCHROME *c*

Nuclear Overhauser enhancements observed upon homonuclear double irradiation of heme *c* resonance lines, on which the resonance assignments were based, are also indicated.

Assignment	Ferrocytochrome *c*		Ferricytochrome *c* Chemical shift (ppm from TSP *) at 35°C
	Chemical shift (ppm from TSP *) at 52°C	Irradiation causes negative nuclear Overhauser enhancements on the following heme *c* resonances	
Meso-proton α	9.32	Ring methyl 3, methine and methyl of thioether 2	
Meso-proton β	9.59	Ring methyl 5, methine and methyl of thioether 4	
Meso-proton γ	9.62		
Meso-proton δ	9.04	Ring methyls 1 and 8	
Ring methyl 1	3.52	Meso δ	7.4
Ring methyl 3	3.87	Meso α	31.2
Ring methyl 5	3.61	Meso β	10.5
Ring methyl 8	2.19	Meso δ	33.9
Thioether bridge 2: methine	5.24	Meso α	
methyl	1.53	Ring methyl 1, Meso α	−2.1
Thioether bridge 4: methine	6.34	Meso β	
methyl	2.59	Ring methyl 3, Meso β	3.1

* TSP, sodium 3-trimethylsilyl-[2,2,3,3-$^{2}H_4$]propionate.

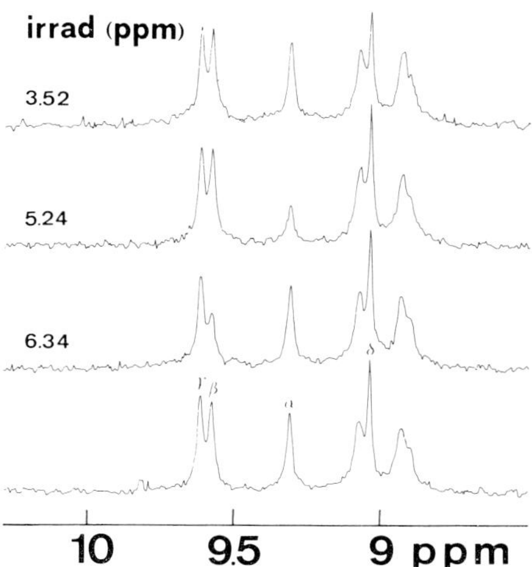

Fig. 4. Nuclear Overhauser enhancements observed in the spectral region from 8.5 to 10 ppm of the 360 MHz ^{1}H NMR spectrum of the ferrocytochrome *c* solution described in Fig. 3. No digital filtering was employed in these experiments. The bottom trace was obtained without double resonance irradiation. The four heme meso-proton resonances are indicated by the greek letters α to δ. The additional lines correspond to slowly exchanging amide protons. The upper three traces show the effects of double resonance irradiation at the positions indicated on the left. These correspond, from top to bottom, to the resonance positions of the heme ring methyl group 1, the methine proton of the thioether bridge 2, and the methine proton of the thioether bridge 4.

at 9.32 and 9.59 ppm; the pair of heme ring methyls 1 and 8 at 2.19 and 3.52 ppm, since irradiation at either one of these positions produced a nuclear Overhauser enhancement on the same meso-proton; the pair of ring methyls 3 and 5 at 3.61 and 3.87 ppm.

The assignments obtained from observation of the nuclear Overhauser enhancements on the meso-proton resonances (Fig. 4) could then be confirmed by the observation of nuclear Overhauser enhancements on the heme *c* substituents 1—5 and 8 upon irradiation of the meso-protons (Table I). These experiments are illustrated in Fig. 5. Since the observed resonances were in more crowded spectral regions than those of the meso-protons (Fig. 2), difference spectra were recorded. As one would expect [31], the nuclear Overhauser enhancements on the methyl resonances were markedly smaller than those on the meso-protons; from the difference spectra they were estimated to be approx. −15%. While the nearest neighbor relations between meso-proton and heme *c* methyl groups are unambiguously confirmed by Fig. 5, the appearance of additional weaker lines also indicate the limited selectivity of the nuclear Overhauser enhancement experiments. Thus, the pulse width was such that irradiation at 9.6 ppm affected also the meso-proton at 9.3 ppm, irradiation at 9.3 ppm affected the meso-proton at 9.0 ppm, and vice versa (Fig. 5). Furthermore, through-space interactions of the heme *c* protons with nearby protons of the polypeptide chain caused the appearance of additional lines in the difference spectra of Fig. 5, which will be discussed in detail in a forthcoming paper.

Since one methyl group of each of the two pairs of ring methyl 1 and 8, and

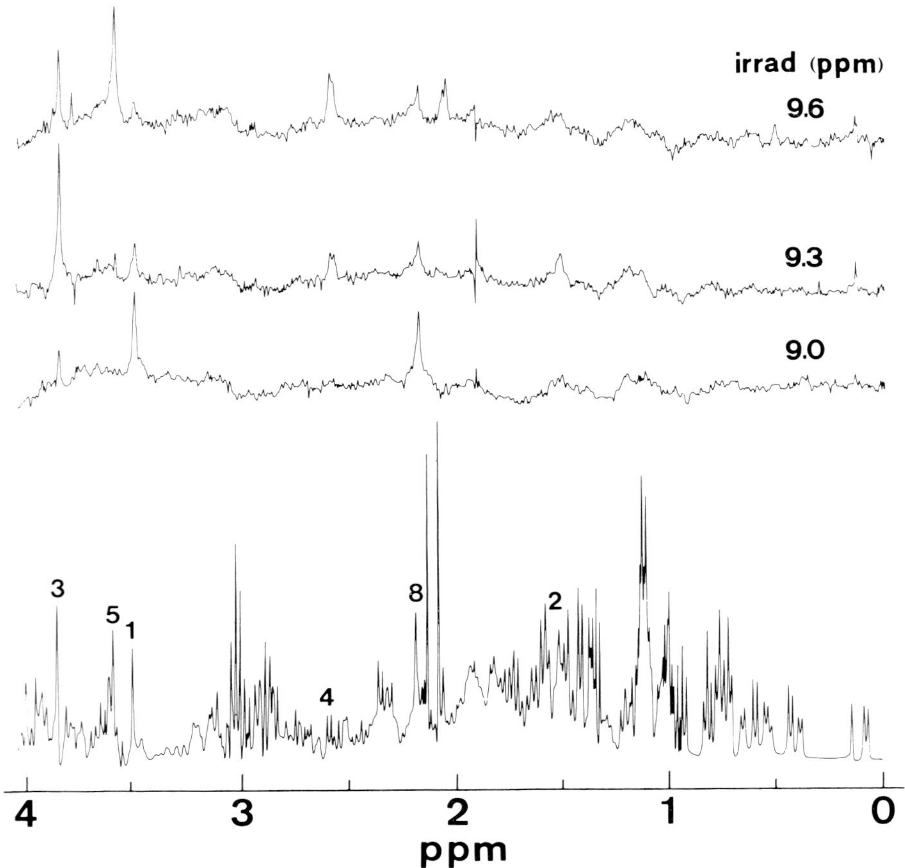

Fig. 5. Nuclear Overhauser enhancements observed in the spectral region from 0 to 4 ppm of the 360 MHz ^{1}H NMR spectrum of the ferrocytochrome *c* solution described in Fig. 3. The bottom trace was obtained without double resonance irradiation after digital resolution enhancement with the technique described in Fig. 2. The singlet resonances of the four heme ring methyls are numbered 1, 3, 5 and 8, the doublets of the two thioether methyls 2 and 4 (see Fig. 1). The upper three traces correspond to the differences between spectra obtained without double resonance irradiation and with double resonance irradiation at the positions indicated in the figure. These correspond, from top to bottom, to the positions of the meso-protons β, α, and δ. These spectra were recorded without digital filtering. Negative Overhauser enhancements appear as positive peaks in the difference spectra.

3 and 5 is next to a thioether bridge (Fig. 1), the nuclear Overhauser enhancements observed on the ring methyl resonances upon irradiation of the thioether methyls finally provided individual assignments for all the six heme substituents 1—5 and 8, and the four meso-protons (Table I).

In previous work, Redfield and Gupta [22] used saturation transfer experiments in solutions of partially oxidized cytochrome *c* to identify corresponding resonances in the spectra of the reduced and the oxidized protein. This technique was adopted here. The earlier correspondences [22] could be confirmed, resulting in the individual assignments of the four heme *c* ring methyl resonances and the thioether methyl resonance 2 in ferricytochrome *c* (Fig. 6, Table I). For the assignments of the thioether methyl resonances 4, which are

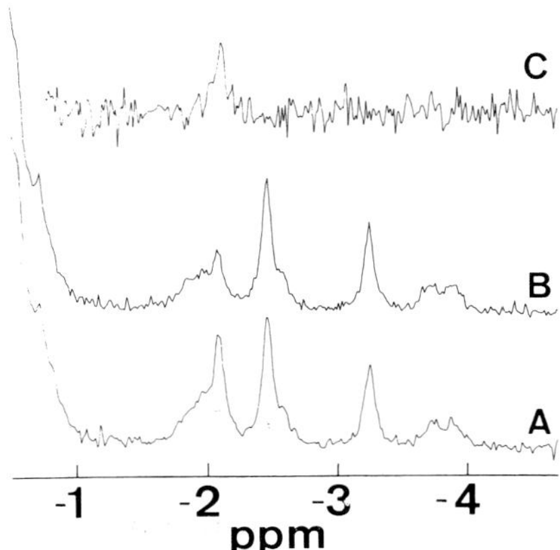

Fig. 6. Correspondence between the thioether methyl resonances 2 in ferro- and ferricytochrome *c* by saturation transfer in the 360 MHz ^{1}H NMR spectrum of a ^2H_2O solution of partially reduced cytochrome *c*, p^{2}H 7.1, temperature 35°C, relative concentrations of oxidised and reduced protein 2 to 1. (A) Spectral region from —4 to —1 ppm obtained with double resonance irradiation at —4.8 ppm, where there is no resonance nearby. (B) Corresponding spectrum obtained with double resonance irradiation at 1.5 ppm, i.e. the position of the thioether methyl resonance 2 in reduced cytochrome *c*. (C) Difference spectrum A — B.

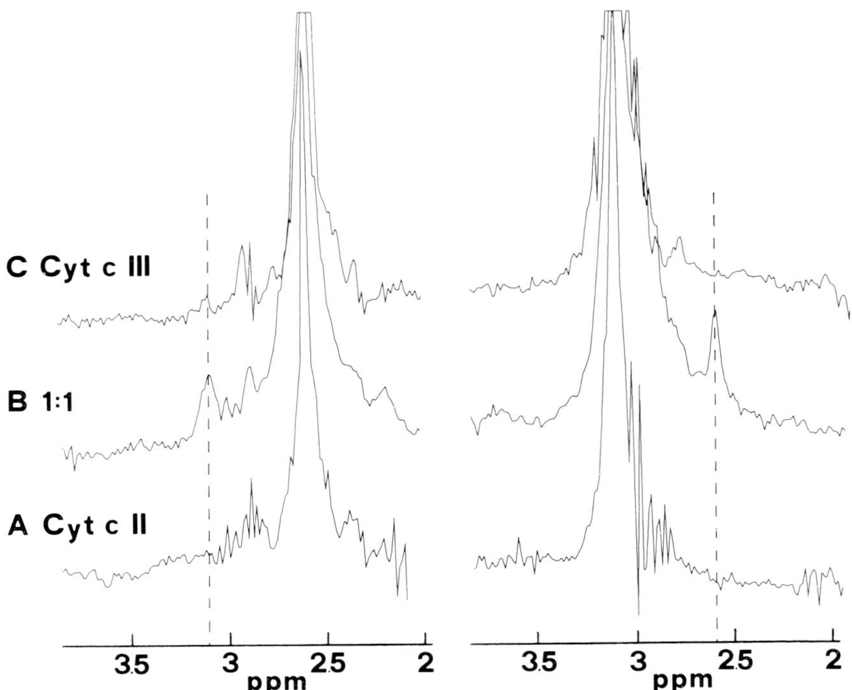

Fig. 7. Correspondence between the thioether methyl resonances 4 in ferro- and ferricytochrome *c* by saturation transfer in the 360 MHz ^{1}H NMR spectrum. (B) Partially reduced cytochrome *c* solution. The left trace corresponds to the difference between the spectra obtained on the one hand with double resonance irradiation at —1.3 ppm, where there is no resonance nearby, and on the other hand at 2.59 ppm, i.e. the position of the thioether methyl 4 in reduced cytochrome *c*. The conditions of this experiment were otherwise identical to those for Fig. 6C. Saturation transfer produces a positive peak at 3.1 ppm, which is tentatively assigned to the thioether methyl position in ferricytochrome *c*. (The intense peak at 2.6 ppm results from the irradiation at this position). The right trace corresponds to the difference between spectra obtained with double resonance irradiation at —1.3 ppm, and at 3.1 ppm, respectively. The appearance of a peak at 2.6 ppm confirms the saturation transfer with a line at 3.1 ppm. A and C, as a control, identical experiments to those of trace B were done with fully oxidised and fully reduced cytochrome *c*. It is seen that the pair of saturation transfer-linked resonance positions at 2.6 and 3.1 ppm are missing in both these solutions, thus confirming the interpretation of the results in trace B.

located in crowded spectral regions both in ferri- and ferrocytochrome *c*, observation of the saturation transfer was checked by various control experiments. First, optimal conditions for the observation of saturation transfer in a difference spectrum were established using the resonances of the thioether methyl line 2 (Fig. 6). Using these conditions with irradiation at 2.59 ppm, i.e. the chemical shift of the thioether methyl resonance 4 in reduced cytochrome *c*, the spectra of the fully reduced, the partially oxidized and the fully oxidized protein were scanned for double resonance effects. As is shown in Fig. 7, a line appeared at 3.1 ppm in the difference spectrum obtained from the partially oxidized protein, while there is no line at this position in either ferri- or ferrocytochrome *c*. A corresponding experiment with irradiation at 3.1 ppm (Fig. 7) confirmed that there was saturation transfer between the thioether methyl doublet at 2.59 ppm in reduced cytochrome *c* (Fig. 3) and a resonance at 3.1 ppm in the oxidized protein, which was thus assigned to the thioether bridge 4 (Table I). This resonance assignment is different from that suggested earlier [22]. Since the resonance in ferricytochrome *c* which had in the earlier studies [22] been found to be connected by saturation transfer with the methyl line at −2.46 ppm in ferricytochrome *c* does not coincide with either of the thioether bridge spin systems identified by the spin decoupling experiments (Fig. 3) and the nuclear Overhauser enhancement studies, these resonances have to correspond to a methyl group of the polypeptide chain.

Discussion

While earlier studies [13,21—23] had resulted in the identification of the ^{1}H NMR lines of different types of heme *c* protons in cytochrome *c*, i.e. in particular the group of four ring methyls, the four meso-protons and the two thioether methyls, the experiments described in the preceding section now provided individual assignments of these resonances (Table I). Furthermore it was found that the earlier identification of the thioether methyl lines [22] had to be revised, which has important consequences on the assessment of the pseudocontact shifts in ferricytochrome *c*, as will be discussed in more detail below.

While spin decoupling and saturation transfer experiments were important for some of the resonance assignments in Table I, studies of nuclear Overhauser enhancements played an essential role for most of the results obtained. In all the experiments, negative nuclear Overhauser enhancements were observed (Figs. 4 and 5). This is in agreement with earlier experimental and theoretical investigations [33,34]. These had shown that while positive nuclear Overhauser enhancements of up to 50% are to be expected from proton-proton dipolar coupling under extreme motional narrowing conditions [35], negative nuclear Overhauser enhancements of up to −100% can arise from proton-proton dipolar coupling in macromolecular species such as protein molecules. In the present investigation, the negative nuclear Overhauser enhancements were combined with the effective proton-proton distances estimated from the covalent structure of heme *c* to obtain assignments of the heme *c* resonances. It is obvious that besides the nuclear Overhauser enhancements arising from dipolar coupling between protons of the heme group, additional effects are to be expected from interactions with protons of the polypeptide chain. This is born out by numerous observations made during the present study, such as the appearance of additional lines in the spectra of Fig. 5. Systematic applications of nuclear Overhauser enhancement experiments for studies of static and dynamic aspects of the heme crevice in cytochrome *c* are currently in progress in our laboratory. Overall, the present experiments with cytochrome *c* support the earlier suggestions [33,34] that nuclear Overhauser enhancement measurements are a valuable addition to the NMR techniques available for resonance assignments in macromolecular systems and in particular for studies of certain aspects of protein conformation.

Comparison of the chemical shifts of heme *c* in ferrocytochrome *c* (Table I) with the corresponding chemical shifts in isolated diamagnetic metalloporphyrins [13] shows that the three ring methyl resonances 1, 3 and 5 are only little influenced by the protein environment. Ring methyl 8, however, is shifted upfield by approx. 1.4 ppm. This observation coincides with the prediction based on calculations using the single crystal X-ray data [32], which showed that the ring methyl protons 8 are exposed to the ring current field of the indole ring of Trp-59. Comparison with predicted ring current shifts had previously been used for a tentative assignment of the highest field ring methyl resonance to position 8 [23]. Comparison with ring current calculations supports also the assigments of the thioether bridge resonances in Table I. While the chemical shifts of the substituent in position 4 coincide closely with those observed for the thioether bridges in isolated heme *c* [36], the high field shifts of the thioether bridge 2 protons can readily be explained by the proximity of this heme substituent to the aromatic ring of Phe-82. It should be pointed out, however, that the ring current shifts caused by Phe-82 are markedly smaller than what had previously been suggested in an attempt [23] to fit the predictions from ring current calculations with the earlier assignments of the thioether methyl resonances [22].

From the early observations it had readily been recognized that the qualitative patterns of the hyperfine shifts of the heme *c* proton resonances were determined by Fermi contact interactions [13,21]. Consideration of the inherent symmetry properties of the molecular orbitals in heme groups then led to the suggestion, that the unpaired electron spin density distribution in the heme group of ferricytochrome *c* showed a pronounced C_2 symmetry [13,37]. This is now confirmed by the individual resonance assignments in Table I: Large spin densities of the order of 2% of an unpaired electron [32] are localized on the porphyrin ring carbon atoms 3 and 8, whereas much smaller spin densities of less than 0.5% of an electron are on the ring carbons 1 and 5. The pronounced asymmetry of the spin density distribution among the four pyrrole rings of heme *c* in ferricytochrome *c* appears to be an invariant of evolution [6,38—40], indicating that this structural feature might be an essential factor for the biological function of cytochrome *c*. This was further supported by the observation that chemical modifications of cytochrome *c* which led to a change of the electron spin density distribution in the heme, also abolished the function of cytochrome *c* in mitochondria [41]. As a possible role in the biological redox reactions of cytochrome *c*, it was suggested that the pronounced asymmetric spin density distribution in heme *c* of ferricytochrome *c* might facilitate direct electron transfer through the exposed edge of heme *c* [6,21]. Such a direct electron transfer mechanism has also been suggested from other evidence [4]. With the individual resonance assignments of Table I, the regions of high and low spin density on the periphery of heme *c* are now known, so that future discussion of the electron transfer mechanism can at least from this point of view be based on unambiguous experimental evidence. It should be pointed out in particular that the high spin density on the ring carbon atom 3 is located on the solvent exposed edge of cytochrome *c* [7—12].

In recent years, the question as to whether or not the biological redox reaction of cytochrome *c* involved major changes of the polypeptide conformation held an important position in the discussions on the electron transfer mechanism, especially since the interpretations of the X-ray data on the two oxidation states of the protein were highly ambiguous [7—12]. Even though to-day the X-ray data imply that there are no major conformation changes between reduced and oxidized cytochromes *c* [4,11,12], it seems important to extend comparative studies of the molecular conformations in the two oxidation states of cytochrome *c* also to the protein in solution. As was extensively discussed earlier [32,42—44], ^{1}H NMR studies would be ideally suited for such comparative conformational studies, provided a reliable basis could be obtained for a quantitative interpretation of the pseudocontact shifts in ferricytochrome *c*. From a preliminary analysis the new assignments of the thioether methyl resonances (Table I) seem to provide this information. Since the location of the thioether methyls relative to the heme iron is quite well defined by the covalent structure of heme *c* (Fig. 1) and at most very small contributions to the hyperfine shifts of these protons are expected to come from contact interactions [32], these resonances are suitable natural probes for the calibration of the local dipolar magnetic field of the unpaired electron in ferricytochrome *c* *. Thus the orientation of the principal axes of the paramagnetic susceptibility tensor $\overline{\chi}$ in the heme *c* plane could be derived from the NMR hyperfine shifts, which were taken as the chemical shift differences between ferro- and ferricytochrome *c* (Table I). It was found that the thioether bridge methyl NMR data were compatible with that orientation of the electronic *g*-tensor [15] which had originally been proposed from the low temperature single crystal EPR data [14]. According to this interpretation the *x*-axis of the electronic *g*-tensor in the solution structure of ferricytochrome *c* is directed near the nitrogen atom of the pyrrole ring which carries the substituents 1 and 2 (Fig. 1), with the highest spin density in the d_{yz} orbital. These results are compatible with the large contact shifts of ring methyls 3 and 8 (Table I) and with predictions based on theoretical model calculations [37]. On the basis of this qualitative assessment of the paramagnetism in ferricytochrome *c*, a comparative investigation of the conformation of the heme crevice in reduced and oxidized cytochrome *c* using the EPR data on the electronic *g*-tensor and the atomic coordinates of the X-ray structure [32] is currently under way in our

* It may be pointed out that the incompatibility of the earlier assignments of the thioether bridge methyl resonances [22] with the single crystal EPR data [14] has held up our analysis of pseudocontact shifts in ferricytochrome *c* for several years. In this context it was also unfortunate that the ^{1}H NMR data on cytochrome *c*-557 seemed to support the earlier identification of the thioether methyl resonances [38].

laboratory. It will be of particular interest to combine the results of this study with the data obtained from the above-mentioned nuclear Overhauser enhancement experiments with protons of the polypeptide chain.

As an additional consequence of the new assignments of the thioether methyl resonances, the ^{1}H NMR data no longer imply that the steric arrangement of one of the thioether bridges in mammalian type ferricytochromes *c* must necessarily be different from that in the ferricytochromes c_2 [39] and *c*-552 [40], where only one pseudocontact-shifted methyl resonance had been observed at around −2 ppm.

Acknowledgement

Financial support by the Swiss National Science Foundation (project Nr. 3.0040.76) is gratefully acknowledged.

References

1 Margoliash, E. and Schejter, A. (1966) Adv. Protein Chem. 21, 113—286
2 Lemberg, R. and Barrett, J. (1973) Cytochromes, Academic Press, London
3 Takano, T., Kallai, O.B., Swanson, R. and Dickerson, R.E. (1973) J. Biol. Chem. 248, 5234—5255
4 Salemme, F.R., Kraut, J. and Kamen, M.D. (1973) J. Biol. Chem. 248, 7701—7716
5 Dickerson, R.E., Timkovich, R. and Almassy, R.J. (1976) J. Mol. Biol. 100, 473—491
6 Wüthrich, K. (1971) in Probes of Structure and Function of Macromolecules and Membranes: Probes of Enzymes and Hemoproteins (Chance, B., Yonetani, T. and Mildvan, A.S., eds.), Vol. II, pp. 465—486, Academic Press, New York
7 Dickerson, R.E., Takano, T., Eisenberg, D., Kallai, O.B., Samson, L., Cooper, A. and Margoliash, E. (1971) J. Biol. Chem. 246, 1511—1535
8 Salemme, F.R., Freer, S.T., Xuong, Ng.H., Alden, R.A. and Kraut, J. (1973) J. Biol. Chem. 248, 3910—3921
9 Tamaka, N., Yamane, T., Tsukihara, T., Ashida, T. and Kakudo, M. (1975) J. Biochem. (Tokyo) 77, 147—162
10 Timkovich, R. and Dickerson, R.E. (1976) J. Biol. Chem. 251, 4033—4046
11 Swanson, R., Trus, B.L., Mandel, N., Mandel, G., Kallai, O.B. and Dickerson, R.E. (1977) J. Biol. 252, 759—775
11 Takano, T., Trus, B.L., Mandel, N., Mandel, G., Kallai, O.B., Swanson, R. and Dickerson, R.E. (1977) J. Biol. Chem. 252, 776—785
13 Wüthrich, K. (1970) Struct. Bonding 8, 53—121
14 Mailer, C. and Taylor, C.P.S. (1972) Can. J. Biochem. 50, 1048—1055
15 Taylor, C.P.S. (1977) Biochim. Biophys. Acta 491, 137—149
16 Brautigan, D.L., Feinberg, B.A., Hoffman, B.M., Margoliash, E., Peisach, J. and Blumberg, W.E. (1977) J. Biol. Chem. 252, 574—582
17 Lang, G.H. and Yonetani, T. (1968) J. Chem. Phys. 49, 944—950
18 Spiro, T.G. and Strekas, T.C. (1972) Proc. Natl. Acad. Sci. U.S. 69, 2622—2626
19 Brunner, H. (1973) Biochem. Biophys. Res. Commun. 51, 888—894
20 Vickery, L., Nozawa, T. and Sauer, K. (1976) J. Am. Chem. Soc. 98, 351—357
21 Wüthrich, K. (1969) Proc. Natl. Acad. Sci. U.S. 63, 1071—1078
22 Redfield, A.G. and Gupta, R.K. (1971) Cold Spring Harbor Symp. Quant. Biol. 36, 405—411
23 McDonald, C.C. and Phillips, W.D. (1973) Biochemistry 12, 3170—3186
24 Dobson, C.M., Moore, G.R. and Williams, R.J.P. (1975) FEBS Lett. 51, 60—65
25 Moore, G.R. and Williams, R.J.P. (1975) FEBS Lett. 53, 334—338
26 Campbell, I.D., Dobson, C.M., Moore, G.R., Perkins, S.J. and Williams, R.J.P. (1976) FEBS Lett. 70, 96—100
27 Keller, R.M., Wüthrich, K. and Pecht, I. (1976) FEBS Lett. 70, 180—184
28 Keller, R.M. and Wüthrich, K. (1976) Biochim. Biophys. Acta 491, 416—422
29 De Marco, A. and Wüthrich, K. (1976) J. Magn. Resonance 24, 201—204
30 Wagner, G. (1977) Ph.D. Thesis, ETH Zürich
31 Noggle, J.H. and Schirmer, R.E. (1971) The Nuclear Overhauser Effect, Academic Press, New York.
32 Wüthrich, K. (1976) NMR in Biological Research: Peptides and Proteins, North-Holland Publ. Co., Amsterdam

208

33 Balaram, P., Bothner-By, A.A. and Dadok, J. (1972) J. Am. Chem. Soc. 94, 4015—4017
34 Campbell, I.D., Dobson, C.M. and Williams, R.J.P. (1974) J.C. Soc. Chem. Commun. 1974, 888—889
35 Abragam, A. (1962) in The Principles of Nuclear Magnetism, p. 313, Clarendon Press, Oxford
36 Slama, J.T., Smith, H.W., Wilson, C.G. and Rapoport, H. (1975) J. Am. Chem. Soc. 97, 6556—6562
37 Shulman, R.G., Glarum, S.H. and Karplus, M. (1971) J. Mol. Biol. 57, 93—115
38 Keller, R.M., Pettigrew, G.W. and Wüthrich, K. (1973) FEBS Lett. 36, 151—156
39 Smith, G.M. and Kamen, M.D. (1974) Proc. Natl. Acad. Sci. U.S. 71, 4303—4306
40 Keller, R.M., Wüthrich, K. and Schejter, A. (1977) Biochim. Biophys. Acta 491, 409—415
41 Wüthrich, K., Aviram, I. and Schejter, A. (1971) Biochim. Biophys. Acta 253, 98—103
42 Wüthrich, K., Keller, R.M. and Baumann, R. (1973) in Dynamic Aspects of Conformation Changes in Biological Macromolecules (Sadron, C., ed.), pp. 151—163, Reidel Publ. Co., Dordrecht, The Netherlands
43 Wüthrich, K. and Keller, R.M. (1973) in Symposial Papers of the 4th International Biophysics Congress, Moscow and Pushchino, pp. 722—735
44 Wüthrich, K. (1975) in Metalloprotein Studies Utilizing Paramagnetic Effects of the Metal Ions as Probes (Kotani, M. and Tasaki, A., eds.), pp. 151—179, The Taniguchi Foundation

[Reprinted from the Journal of the American Chemical Society, **100**, 7094 (1978).]
Copyright © 1978 by the American Chemical Society and reprinted by permission of the copyright owner

Transient Proton–Proton Overhauser Effects in Horse Ferrocytochrome *c*

Sir:

The nuclear Overhauser effect (NOE) is the fractional change in intensity of one NMR resonance when another resonance is irradiated, and has long been a valuable tool for structural studies of small molecules.[1] More recently theoretical aspects of using NOE's for investigations of biological macromolecules at high frequencies were discussed,[2] and

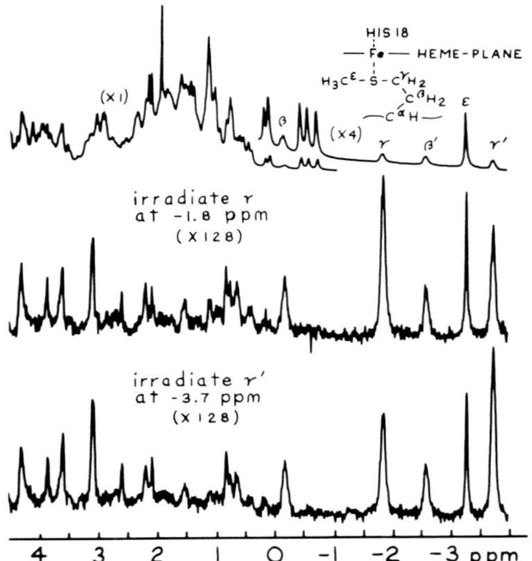

Figure 1. 360-MHz Fourier transform ¹H NMR and steady-state NOE difference spectra of a 0.008 M solution of horse heart ferrocytochrome *c* in 0.05 M deuterated phosphate buffer, pD 6.8, $T = 49$ °C. These and the spectra in Figures 2 and 3 were recorded on a Bruker HX-360 spectrometer. The chemical shifts are referenced to internal TSP.[15] The NOE difference spectra were obtained by subtracting spectra with NOE's from reference spectra. The spectra with NOE's were obtained by applying a 2-s low-power saturating pulse at the frequency indicated above each trace, followed immediately by a 90° observation pulse. The reference spectra were obtained by offsetting the saturation pulse to −5 ppm. Each spectrum was the result of 2000 accumulations. The NOE and reference spectrum FID's[16] were accumulated alternately in order to minimize drifts.

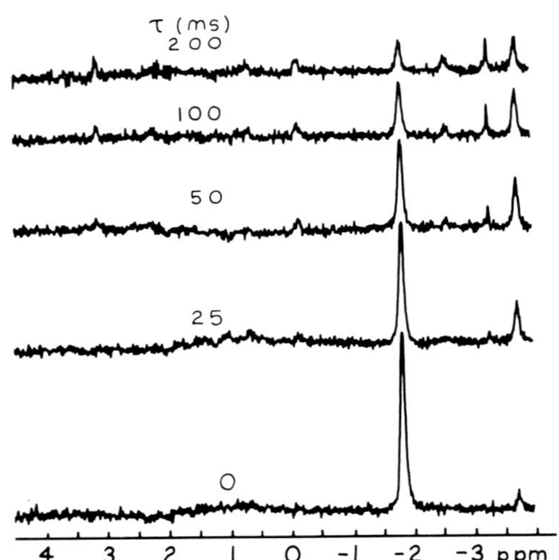

Figure 2. Transient NOE difference spectra resulting from inversion of the Met 80 γ-proton resonance. Values of the delay time τ in milliseconds are given above each spectrum. The difference spectra here and in Figure 3 were obtained by subtracting spectra with transient NOE's from reference spectra. Spectra with transient NOE's were obtained by applying a 15-ms inversion pulse at the γ resonance frequency (−1.8 ppm) followed, after a delay time τ, by a 90° observation pulse. The reference spectra were obtained by offsetting the inversion pulse to −5 ppm. Each spectrum was the result of 1000 accumulations. The NOE and reference spectrum FID's were accumulated alternately in order to minimize instrumental drifts.

several applications of ¹H–¹H NOE experiments for improving the spectral resolution and assigning individual resonances in the ¹H NMR spectra of proteins, as well as for characterization

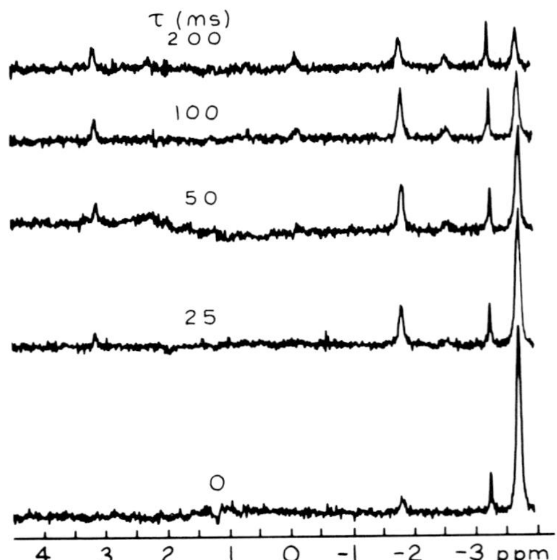

Figure 3. Transient NOE difference spectra resulting from inversion of the Met 80 γ′ proton resonance. See Figure 2 for experimental details.

of local structures in proteins, were described.[3–8] Theoretical considerations[3,5] indicate a more favorable situation for NOE studies of macromolecules at high fields than for the more conventional experiments with small molecules under the conditions of extreme motional narrowing. This is because the limiting magnitude of the NOE in macromolecules is larger by a factor 2, and the NOE's are almost completely determined by intramolecular dipole–dipole interactions. On the other hand it was also pointed out that spin diffusion will be of considerable importance in proteins, causing the NOE's to be less specific and hence less useful.[2,9] Here, we describe some experiments with ferrocytochrome *c* to demonstrate that transient NOE experiments[10,11] provide a means for obtaining specific NOE's even in the presence of strong spin diffusion. Ferrocytochrome *c* was particularly suited for this study. Despite the molecular weight of 12 500, the ¹H NMR spectrum contains numerous well-resolved lines[12,13] (Figure 1), which could be used to unambiguously outline spin diffusion pathways in this protein.

The high-field portion from 4 to −4 ppm of the 360-MHz ¹H NMR spectrum of horse heart ferrocytochrome *c* is shown in the top trace of Figure 1, and different NOE experiments are shown in Figures 1–3. Experimental details are given in the figure captions. We will focus our attention on the high-field shifted resonances of the heme iron ligand Met 80, which are identified in Figure 1 by the greek letters β, β′, γ, γ′, and ε.[12,13]

The lower traces of Figure 1 show two steady-state NOE difference spectra[7] obtained by irradiation of the Met 80 γ and γ′ peaks. NOE's are displayed directly, with negative Overhauser enhancements appearing as positive peaks. The steady-state NOE difference spectra contain a large number of lines which, except for the Met 80 resonances, have not yet been individually assigned. Except for the γ and γ′ peaks, the two NOE difference spectra are essentially identical, even though, as will be shown later by the transient NOE studies, the Met 80 γ and γ′ protons have substantially different local environments. This is a consequence of cross relaxation.

Transient NOE's observed at variable delay times τ after irradiation of the Met 80 γ and γ′ resonances with a selective inversion pulse are shown in Figures 2 and 3, respectively. For increasing τ, the intensity of the pulsed line decreases, while the intensities of the other lines build up by spin diffusion at characteristic rates. After reaching a maximum, the lines decay to zero via spin-lattice relaxation. The initial buildup rates of the NOE's depend only on the cross relaxation coefficients between the irradiated spin and the observed nuclei, and are thus directly related to the inverse of the sixth power of the proton-proton distances in the three-dimensional structure of the protein.

The following are some details to be observed in the transient NOE's of Figures 2 and 3. At $\tau = 0$, the line corresponding to the geminal methylene proton with respect to the pulsed nucleus has already emerged because of spin diffusion during the 15 ms of the pulse duration. In Figure 3 the additional appearance of the methyl signal ϵ in the $\tau = 0$ trace is a trivial consequence of the limited selectivity of the inversion pulse applied to the line γ'. In both Figures 2 and 3 a line at 3.1 ppm grows at about the same rate as the β-methylene signals of Met 80. From two decoupled NOE difference spectra, where, respectively, the β or β' line was irradiated for spin pumping prior to spin decoupling during data acquisition,[7,8] this resonance was independently assigned to the α proton of Met 80. It is seen that the α-proton resonance grows faster when the γ' peak is pulsed than when the γ peak is pulsed. In Figure 3 the α-proton resonance has already appeared at $\tau = 25$ ms, whereas it has not emerged until 50 ms in Figure 2.

Corresponding transient NOE's were obtained after application of selective inversion pulses to the resonances β and β' (Figure 1). It was found that the α-proton line of Met 80 grows faster after irradiation of resonance β' than after irradiation of β. In an additional experiment the resonances γ and γ' were found to grow faster than β and β' after pulse inversion of the Met 80 methyl line ϵ.

From these experiments the increased information content of transient NOE studies in macromolecules (Figures 2 and 3) as compared to the more conventional steady-state experiments (Figure 1) is readily apparent. While the steady-state NOE was able to distinguish between the β- and γ-methylene protons, the transient NOE's further distinguished between β and β' and γ and γ', respectively, of the axial Met 80. These assignments agree with those generally accepted,[14] which were originally suggested from ring-current calculations based on the X-ray structure.[13] The transient NOE's provided further information on static and dynamic aspects of the spatial arrangement of Met 80 in the protein. The different growth rates of the α-proton line (Figures 2 and 3) clearly show that the protons β' and γ' are located more closely to the α proton than the protons β and γ. Since they are at higher field (Figure 1), the β' and γ' protons must also be closest to the heme ring plane. That the different local environments of the individual β- and γ-methylene protons are manifested in the transient NOE's further shows that the rotational mobility about the single bonds in the side chain of Met 80 is severely limited in ferrocytochrome *c*. Finally, experiments of the type of Figures 2 and 3 provide a convincing demonstration of spin diffusion pathways in proteins. Overall, the present experiments imply that spin diffusion in macromolecules, rather than leading necessarily to less specific and hence less useful NOE's,[2,9] may through suitable use of the two-dimensional frequency-time space of transient NOE experiments lead to novel insights into the molecular structures which might not be available otherwise.

Acknowledgments. Financial support by the Roche Research Foundation for Scientific Exchange and Biomedical Collaboration with Switzerland (fellowship to S. L. Gordon) and the Swiss National Science Foundation (project 3.0046.76) is gratefully acknowledged.

References and Notes

(1) Noggle, J. H.; Schirmer, R. E. "The Nuclear Overhauser Effect", Academic Press: New York, 1971.
(2) Kalk, A.; Berendsen, H. J. C. *J. Magn. Reson.* **1976,** *24,* 343–366.
(3) Balaram, P.; Bothner-By, A. A.; Dadok, J. *J. Am. Chem. Soc.* **1972,** *94,* 4015–4017.
(4) Campbell, I. D.; Dobson, C. M.; Williams, R. J. P. *J. Chem. Soc., Chem. Commun.* **1974,** 888–889.
(5) Glickson, J. D.; Gordon, S. L.; Pittner, T. P.; Agresti, D. G.; Walter, R. *Biochemistry.* **1976,** *15,* 5721–5729.
(6) Keller, R. M.; Wüthrich, K. *Biochim. Biophys. Acta.* **1978,** *533,* 195–208.
(7) Richarz, R.; Wüthrich, K. *J. Magn. Reson.* **1978,** *30,* 147–150.
(8) Wüthrich, K.; Wagner, G.; Richarz, R.; Perkins, S. J. *Biochemistry.* **1978,** *17,* 2253–2263.
(9) Hull, W. E.; Sykes, B. D. *J. Chem. Phys.* **1975,** 867–880.
(10) Solomon, I. *Phys. Rev.* **1955,** *99,* 559–565.
(11) Experiments of the type discussed in this communication were alluded to in the closing paragraph of ref 2. Transient NOE experiments on small molecules have been carried out frequently, for example, Freeman, R.; Hill, H. D. W.; Tomlinson, B. L. *J. Chem. Phys.* **1974,** *61,* 4466–4473.
(12) Wüthrich, K. *Proc. Natl. Acad. Sci. U.S.A.* **1969,** *63,* 1071–1078.
(13) McDonald, C. C.; Phillips, W. D. *Biochemistry.* **1973,** *12,* 3170–3186.
(14) Cookson, D. J.; Moore, G. R.; Pitt, R. C.; Williams, R. J. P.; Campbell, I. D.; Ambler, R. P.; Bruschi, M.; Le Gall, J. *Eur. J. Biochem.* **1978,** *83,* 261–275.
(15) Sodium 2,2,3,3-tetradeuterio-3-trimethylsilylpropionate.
(16) Free induction decays.
(17) School of Chemistry, Georgia Institute of Technology, Atlanta, Georgia 30332.

Sidney L. Gordon,*[17] Kurt Wüthrich*
Institut für Molekularbiologie und Biophysik
Eidgenössische Technische Hochschule
CH-8093 Zürich-Hönggerberg, Switzerland
Received May 17, 1978

Reprinted from *Biochimica et Biophysica Acta*, Vol. 577, pp. 177–194 (1979)
Copyright © 1979, with kind permission from
Elsevier Science B. V. Amsterdam, The Netherlands.

BBA 38125

INDIVIDUAL ASSIGNMENTS OF AMIDE PROTON RESONANCES IN THE PROTON NMR SPECTRUM OF THE BASIC PANCREATIC TRYPSIN INHIBITOR

ANDREAS DUBS, GERHARD WAGNER and KURT WÜTHRICH

Institut für Molekularbiologie und Biophysik, Eidgenössische Technische Hochschule, 8093 Zürich-Hönggerberg (Switzerland)

(Received July 24th, 1978)

Key words: NMR, Nuclear Overhauser effect; Trypsin inhibitor; Amide proton Conformation

Summary

Studies of proton-proton nuclear Overhauser effects were used to obtain individual assignments of 17 amide proton resonances in the 360 MHz proton nuclear magnetic resonance spectrum of the basic pancreatic trypsin inhibitor. First, optimizing the conditions for obtaining selective nuclear Overhauser effects in the presence of spin diffusion in macromolecules is discussed. Truncated driven nuclear Overhauser experiments were used to assing the amide proton resonances of the β-sheet in the inhibitor. It is suggested that these techniques could serve quite generally to obtain individual resonance assignments in β-sheet secondary structures of proteins. Combination of nuclear Overhauser studies with spin decoupling further resulted in individual assignments of the γ-methyl resonances of the two isoleucines and numerous C^{α} and C^{β} protons.

Introduction

The basic pancreatic trypsin inhibitor is a small globular protein with a highly refined single crystal X-ray structure [1], which has recently been much used as a model compound for theoretical and experimental studies of fundamental aspects of protein conformation [2—8]. Particular emphasis was on the investigation of internal mobility in proteins [2—4,9—12]. For experimental studies in this field, high resolution NMR spectroscopy is a powerful method [13]. In the inhibitor, important structural information has come from ^{1}H NMR studies of the labile protons (Refs. 14—21, and Wüthrich and coworkers, unpublished results), whereby the analysis of the experimental data was greatly aided by the identification of the resonances. The present paper describes the techniques used to obtain individual assignments of numerous amide proton lines.

Early experiments showed that the ^1H NMR spectrum of the inhibitor contains numerous resolved resonance lines which correspond to labile protons [22,23]. Previously, individual assignments for five of these resonances were obtained with the use of lanthanide shift reagents in studies of chemical modifications of the inhibitor [24,25]. Here, individual assignments for these five and twelve additional lines were independently obtained from studies of proton-proton nuclear Overhauser effects [26,27]. Much care was exercised in the selection of the experimental conditions, so that specific Overhauser effects were obtained in spite of spin diffusion [28]. Technical details of these measurements are described in Materials and Methods. Interpretations of individual experiments and the strategy used to obtain individual assignments for numerous polypeptide backbone proton resonances in the inhibitor are presented in the Results section.

Materials and Methods

The basic pancreatic trypsin inhibitor (Trasylol®, Bayer Leverkusen, F.R.G.) was obtained from the Farbenfabriken Bayer AG. For the NMR studies, the lyophilized protein was dissolved either in ^2H_2O or in H_2O. In some experiments, praseodymium(III) was added as a NMR shift reagent as described previously [29]. While 0.005 M inhibitor solutions were used for all the other experiments, nuclear Overhauser difference spectra were obtained with 0.02 M solutions in order to obtain a satisfactory signal-to-noise ratio within reasonable periods of time.

^1H NMR spectra were recorded on a Bruker HXS-360 spectrometer. Sample tubes with 10 mm outer diameter were used for measurements of nuclear Overhauser effects, standard 5-mm tubes for all the other experiments. Nuclear Overhauser difference spectra [30] were obtained by subtracting spectra with Overhauser effects from reference spectra. Spin decoupled Overhauser difference spectra [30] were obtained with the application of a selective decoupling field during data acquisition. The free induction decays with and without Overhauser effects were accumulated alternately in order to minimize instrumental drifts. Accumulation times of approximately 90 min were used to obtain difference spectra with a satisfactory signal-to-noise ratio.

Selective proton-proton Overhauser effects in the presence of spin diffusion. The nuclear Overhauser effect is the fractional change in intensity of one NMR line when another resonance is irradiated. It has long been a valuable tool for structural studies of small molecules, where the conventional 'steady-state' Overhauser effects are simply related to the distance between irradiated and observed nuclei [26,27]. When working with macromolecules at high frequencies, however, spin diffusion is of considerable importance [28,31,32], which makes steady-state Overhauser effects less specific. For the basic pancreatic trypsin inhibitor, this is born out by the data in Fig. 1. It is seen that the steady-state Overhauser difference spectrum obtained with selective presatura-

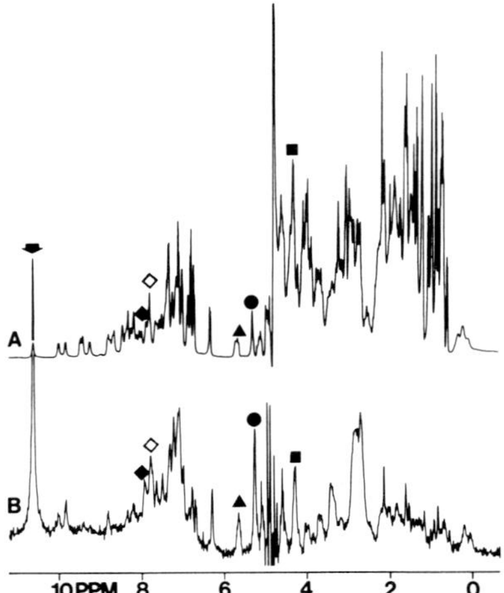

Fig. 1. 360 MHz ^{1}H NMR spectra of the inhibitor recorded in 0.02 M solution in ^2H_2O in 10-mm sample tubes, p^{2}H 4.5, $T = 15°$C. (A) Normal Fourier transform spectrum. (b) Steady-state nuclear Overhauser difference spectrum obtained with presaturation of the amide proton line at 10.6 ppm, as indicated by the arrow. ♦ and ◇, identify two amide proton lines and ▲, ● and ■, the three α-proton lines which will be further discussed below.

tion of a well resolved one-proton resonance contains numerous lines in different spectral regions, so that an unambiguous identification of protons located near the irradiated amide proton is not possible.

Theory shows that in contrast to the steady-state Overhauser effects, the initial build-up rates of nuclear Overhauser effects are simply related to the inverse sixth power of the distance between the observed and the presaturated proton even in the presence of spin diffusion [26—28]. It is hence of great practical interest to develop techniques where upon selective irradiation of individual resonance lines, spectral features can be observed which are in a simple way related to the initial build-up rates of the Overhauser effects. In the following two fundamentally different experiments are distinguished. In studies of 'transient nuclear Overhauser effects', no radio frequency field is applied during the build-up of the Overhauser effects. 'Radio frequency-driven nuclear Overhauser effects' or simply 'driven Overhauser effects' result when a radio frequency field is applied during the build-up process.

Transient nuclear Overhauser effects. Transient Overhauser difference spectra were obtained with the pulse sequence (1):

$$(-180°(\omega_A) - \tau_1 - \text{Observation Pulse} - \tau_2 - 180°(\omega_{\text{off-res.}}) - \tau_1$$
$$- \text{Observation Pulse} - \tau_2 -)_n \qquad (1)$$

The experiment is initiated by a selective 180° pulse of short duration, typically 10 ms, at the resonance frequency of a spin A. This is followed, after a delay time τ_1 during which the nuclear Overhauser effects are built up in the absence of a radio frequency field, by an observation pulse. After the data acquisition, the spin system is allowed to recover during a waiting period τ_2, where τ_2 was typically of the order of 2 s in the experiments with the inhibitor.

The second half of the pulse sequence is identical, except that the 180° pulse at $\omega_{off\text{-}res.}$ is applied in an empty region of the spectrum, so that no Overhauser effects are produced. The two free induction decays obtained in one cycle are stored in two different sections of the computer memory, and after accumulation of n cycles they are subtracted to get Overhauser difference spectra. When the experiment (1) is repeated with different delay times τ_1, the time course of the magnetizations of the irradiated nucleus A and nuclei located near A in the protein structure can be recorded (Fig. 2) [32].

Fig. 2 illustrates that in the sequence of transient Overhauser difference spectra recorded at different times τ_1 after the 180° pulse, the intensity of the pulsed line decreases, while the intensities of other lines build up by spin diffusion. If the observation had been continued for sufficiently long times τ_1, the pulsed line would have further decreased to zero, while the other lines would first have passed through an intensity maximum before decaying to zero by spin relaxation [32]. Comparison of Figs. 1 and 2 clearly illustrates the improved selectivity of the transient experiment. Between 5 and 10 ppm only the three lines marked with ♦, ◇ and ● appear in Fig. 2. Since the initial build-up rates of the resonance intensities are simply related to the inverse sixth power of the distance between the pulsed proton and the observed proton [26,27], transient Overhauser experiments reliably manifest nearest-neighbor relations between individual protons in proteins [32].

While transient Overhauser difference spectra are a particularly straight-forward technique for measurements of initial build-up rates, the continued practical applications with proteins showed that a reasonable compromise between high selectivity of the presaturation pulse and workable signal-to-noise ratio was difficult to obtain when working in crowded spectral regions. Thus, except for the experiment of Fig. 2, where the presaturation pulse was applied to the well separated lowest field amide proton resonance, transient Overhauser experiments were only of limited use for the present investigation, and more information was obtained from truncated driven Overhauser experiments.

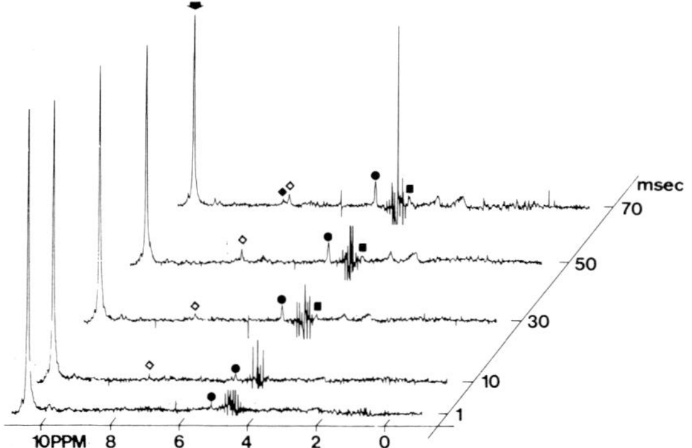

Fig. 2. Transient nuclear Overhauser difference spectra at 360 MHz of the inhibitor solution in Fig. 1, obtained after a 10 ms inversion pulse was applied to the amide proton line at 10.6 ppm (arrow). The delay time τ_1 (Eqn. 1) is indicated on the right. The same resonances are labelled by ♦, ◇, ● and ■ as in Fig. 1.

Driven nuclear Overhauser effects. The experiment (2) was used to obtain driven Overhauser difference spectra.

$$(-t_1(\omega_A) - \text{Observation Pulse} - \tau_2 - t_1(\omega_{\text{off-res.}}) - \text{Observation Pulse} - \tau_2 -)_n \tag{2}$$

The nuclear Overhauser effects are built up during the period of time t_1, while a selective low power radio frequency field is applied to an individual resonance A. This is followed immediately by the observation pulse. After data acquisition the system is allowed to recover during a waiting time τ_2, where τ_2 was typically of the order of 2 s in the experiments with the inhibitor. If t_1 is sufficiently long, the driven nuclear Overhauser experiment provides a steady-state Overhauser effect [27]. It was found that with sufficiently low power for obtaining selective irradiation of individual lines, steady-state Overhauser effects for the inhibitor resulted with $t_1 \gtrsim 2$ s (Fig. 1). On the other hand, highly selective Overhauser difference spectra are obtained when the driven Overhauser effects are 'truncated' after a short time t_1. From a series of truncated driven Overhauser difference spectra recorded with different preirradiation times t_1, the build-up rates of the Overhauser effects for individual nuclei are obtained (Fig. 3). Fig. 3 shows that the truncated driven Overhauser difference spectrum recorded with a pulse length of 150 ms was essentially identical to the transient Overhauser difference spectrum obtained after a delay time of 70 ms (Fig. 2).

A more reliable basis for the analysis of driven nuclear Overhauser effects was obtained by comparison of the experimental data with model calculations

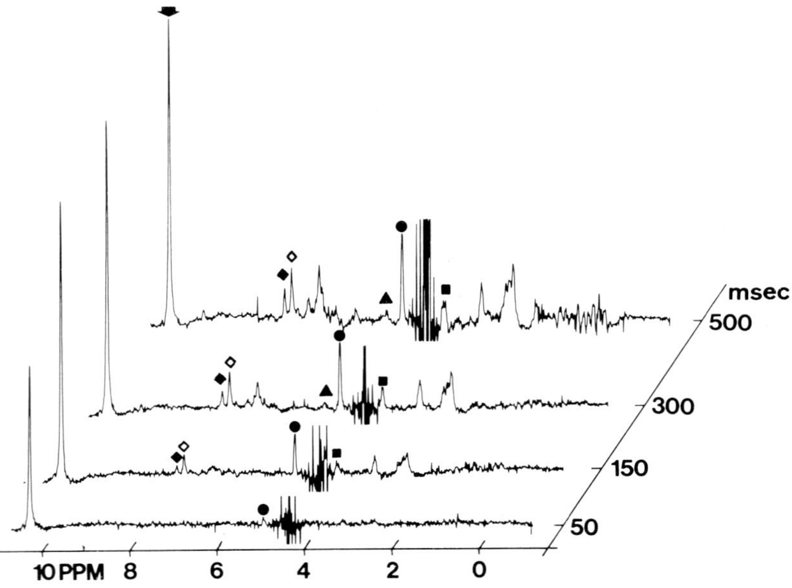

Fig. 3. Driven Overhauser difference spectra of the inhibitor solution in Fig. 1 obtained with irradiation of the amide proton resonance at 10.6 ppm (arrow). The irradiation time t_1 (Eqn. 2) is indicated on the right. The same resonances are labelled by ◆, ◇, ▲, ● and ■ as in Fig. 1.

for a three-spin system [33]. In Fig. 4 the experimental and calculated Overhauser effects are shown for the α-protons of Phe 22, Tyr 23 and Cys 30 in an experiment where the irradiation frequency ω_A (Eqn. 2) was on the amide proton of Tyr 23. The interatomic distances used in this calculation were taken from the refined crystal structure [1]. The algebraic form of the function plotted in Fig. 4A is given in the Appendix. Fig. 4A shows that the computed relative build-up rates are 1.0, 0.27 and 0.04 for the α-protons of Phe 22, Tyr 23 and Cys 30, respectively, which are at distances of 2.3, 2.6 and 4.6 Å from the irradiated proton. The agreement with the corresponding experimental data in Fig. 4B, which gave relative build-up rates for the three protons of 1.0, 0.38 and 0.07, is quite satisfactory.

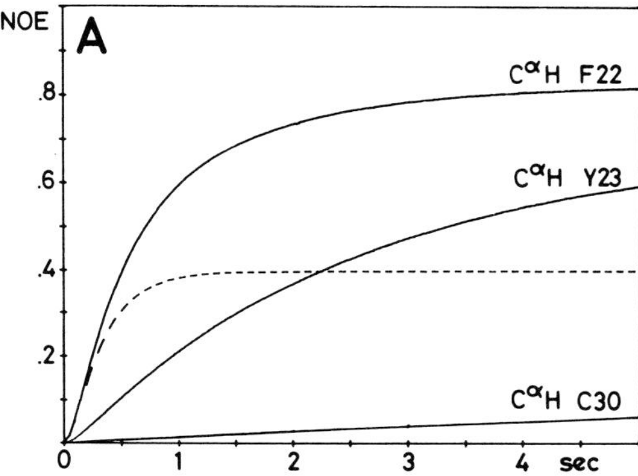

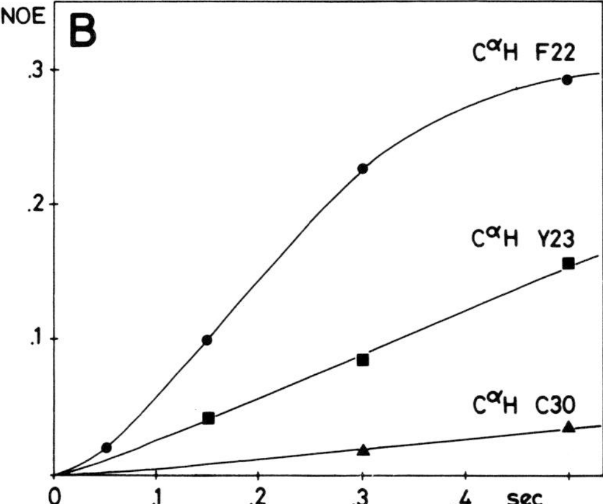

Fig. 4. Plot of the resonance intensities of the α-protons of Phe 22, Tyr 23 and Cys 30 vs. t_1 in the driven Overhauser experiment of Fig. 3, where the irradiation frequency ω_A was on the amide proton of Tyr 23. (A) Calculated with the three spin approximation, where the nearest-by vicinal β-proton was taken as the third proton (———). The interatomic distances used in the calculation were taken from the X-ray structure [1]. - - - - - -, the result of a calculation for the α-proton of Phe 22, where besides the β-methylene proton the influence of five additional protons was considered (see text). (B) Experimentally observed driven Overhauser effects from Fig. 3. Note that the scale extendes only over 0.5 s, as compared to 5 s in (A). The non-linearity of the initial time course of the resonance intensities is a consequence of the non-instantaneous saturation of the irradiated proton, which was also considered in the theoretical calculation.

The description of the situation in a protein by a three-spin system is of course a rather crude approximation, in which the influence of the bulk of the protons is accounted for by the location in space of a single proton relative to the irradiated proton and the observed proton. Yet, since the initial build-up rate of the Overhauser effect on the observed proton depends only on the inverse sixth power of the distance from the irradiated proton and is essentially unaffected by the presence of additional nuclei [26,28,32,33], this simple model provides an adequate description of the initial stages of a driven Overhauser experiment. On the other hand it cannot account for the observed steady-state Overhauser effects, which are largely affected by the presence of additional protons. This is readily seen from Fig. 4. The solid curves in Fig. 4A were computed for a three-spin system consisting of the irradiated amide proton of Tyr 23, the observed α-proton and the nearest-by vicinal β-methylene proton. They largely overestimate the steady-state Overhauser effects. If the influence of the bulk protons in the protein is accounted for more realistically by inclusion of the effects of additional neighboring protons on the relaxation time of the observed nucleus, which enters as a parameter into the three-spin model calculation (see Appendix), a more realistic value for the steady-state Overhauser effects is obtained (Fig. 4a, dashed line).

Results

The resonance assignments described in this paper are compiled in Table I. These results depended critically on certain properties of the amide protons in the protein, certain features of the spatial polypeptide structure and the combined use of proton-proton nuclear Overhauser and spin decoupling experiments. These general aspects are in the following briefly considered, followed by the description of individual resonance assignments in the β-sheet and α-helix secondary structures contained in the inhibitor [1].

Amide protons in the basic pancreatic trypsin inhibitor

In freshly prepared solutions of the inhibitor in ^2H_2O, numerous ^1H NMR lines of relatively slowly exchanging labile protons are observed between 6.5 and 11.0 ppm [22,23]. From comparison with model peptides [34] these resonances must correspond to amide protons (Refs. 22, 23 and 35, and Wüthrich and coworkers, unpublished results). In all, the inhibitor contains 53 backbone amide protons and eight amide protons of asparagine and glutamine side chains. Depending on the location in the protein structure, largely different exchange rates with the solvent prevail for the individual protons; quite generally, protons on the protein surface exchange too rapidly to be seen in ^2H_2O solution of the protein. As is discussed in detail elsewhere (Ref. 35, and Richarz, R., Sehr, P., Wagner, G. and Wüthrich, K., unpublished results), the amide proton exchange rates are further strongly dependent on pH and temperature. At the pH minimum and ambient temperature, 33 amide proton resonances were identified in a freshly prepared ^2H_2O solution of the inhibitor. These have been numbered in the order of decreasing chemical shifts (Refs. 19 and 35). Under the experimental conditions of Fig. 5, twelve of these protons

TABLE I

PROTON SPIN SYSTEMS IN THE BASIC PANCREATIC TRYPSIN INHIBITOR IN WHICH AMIDE PROTONS WERE INDIVIDUALLY ASSIGNED

Temperature: 36°C, p^2H 4.5. Spin systems where the amide proton could be observed in ^2H_2O are numbered according to Fig. 5, those where they were, because of fast exchange, observed only in H_2O are indicated by a star. Chemical shifts (δ) are in ppm relative to internal sodium-2,2,3,3-tetradeutero-3-trimethylsilyl-propionate. Spin-spin coupling constants (J) are in Hz.

Spin system	δ (NH)	δ ($C^{\alpha}H$)	δ ($C^{\beta}H$)	$^3J_{HN\alpha}$	$^3J_{\alpha\beta}$	Assignment *
1	10.55	4.31	{2.72 3.45	7.0		Tyr 23
2	9.94	5.12	2.79	11.0	{13.0 2.5	Phe 45
3	9.79	5.25	{2.87 2.81	9.0	{3.0 3.0	Phe 22 (Phe 22)
4	9.37	4.89		7.0		Phe 33 (Phe 33)
5	9.39	4.89	2.69 2.52	10.5		Tyr 35
6	9.18	5.70	2.69	9.5	{6.5 6.5	Tyr 21 (Tyr 21)
7	8.77	4.87				Gln 31 (Arg 20)
9	8.61					Gly 36
10	8.58			≈4.0		Met 52
*	8.40	5.60	{2.67 3.69		{3.5 14.0	Cys 30
11	8.39	4.70	1.62	9.0		Arg 20 (Gln 31)
15	8.25	4.31		8.5		Cys 55
16	8.12	4.25	1.86			Ile 18
18	8.07	5.28	4.00	9.0		Thr 32
19	7.98					Asn 43 $N^{\delta}H^1$
22	7.78	4.41	2.69			Asn 24
24	7.78					Asn 43 $N^{\delta}H^2$

* The experiments leading to these assignments are described in the text. The previous assignments by Marinetti et al. [24,25] are included in parentheses for comparison.

were already completely exchanged, i.e. the proton 8, 12, 14, 17, 18, 20, 21, 23 and 29—33.

Fig. 5 shows that certain amide proton lines overlap mutually, so that selective irradiation in nuclear Overhauser or spin decoupling experiments is difficult. It was therefore important for certain experiments that, because of the largely different exchange rates of different protons, the total number of amide proton lines is selectively reduced with time after the sample preparation, resulting in improved resolution of the remaining lines [22,23]. On the other hand additional resolved lines, besides the 33 resonances observed in ^2H_2O, can be seen in H_2O solutions of the inhibitor (Ref. 35, and Wüthrich, K. and Wagner, G., unpublished results).

Useful features of the protein conformation

In the interpretation of the nuclear Overhauser experiments it was assumed that the refined single crystal atomic coordinates [1] are conserved in the globular solution conformation of the inhibitor. Evidence in support of this assumption was presented elsewhere [29,35—37]. It is then of interest that entirely different proton-proton nearest-neighbor relations occur in the β-sheet

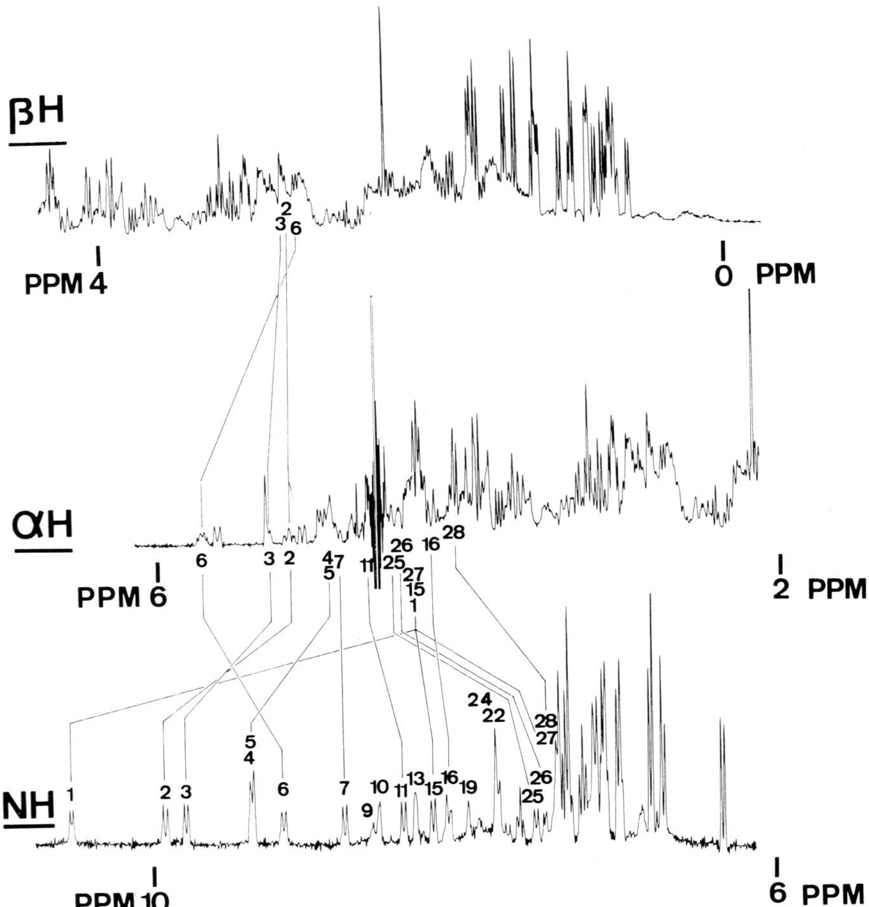

Fig. 5. 360 MHz ^{1}H NMR spectrum of the inhibitor in $^{2}H_2O$ solution at 45°C, $p^{2}H$ 4.5. The spectrum was recorded 90 min after sample preparation. The spectral resolution was improved by multiplication of the free induction decay first with an increasing exponential with time constant 4 s^{-1} and then with a shifted sine bell [19]. From top to bottom the spectral regions of the aliphatic side chain resonances, the α-proton resonances, and the aromatic and amide proton resonances are presented. The resonances of the labile protons are numbered in the order of the chemical shifts, as proposed previously (Refs. 19 and 35). Certain numbers are missing because these protons were already fully exchanged when this spectrum was recorded. Some of the correspondence between resonances of the same spin systems, which were evidenced by spin decoupling and are listed in Table 1, are indicated by the lines connecting the different spectral regions.

and in the α-helix of the inhibitor [1]. This is illustrated in Fig. 6 for all the amide protons involved in hydrogen bonds of the β-sheet or the α-helix. The sixth power of the distances between the amide protons and the neighboring protons was plotted, since the initial build-up rates for nuclear Overhauser effects are simply related to this quantity [26,28,32].

In the β-sheet there are, with the exception of residues 16 and 27, always α-protons among the nearest-by protons (Fig. 6). More precisely, the nearest neighbor of the amide proton of residue n is almost exclusively the α-proton of residue $(n-1)$. The second-nearest α-proton is that of the residue n, which can

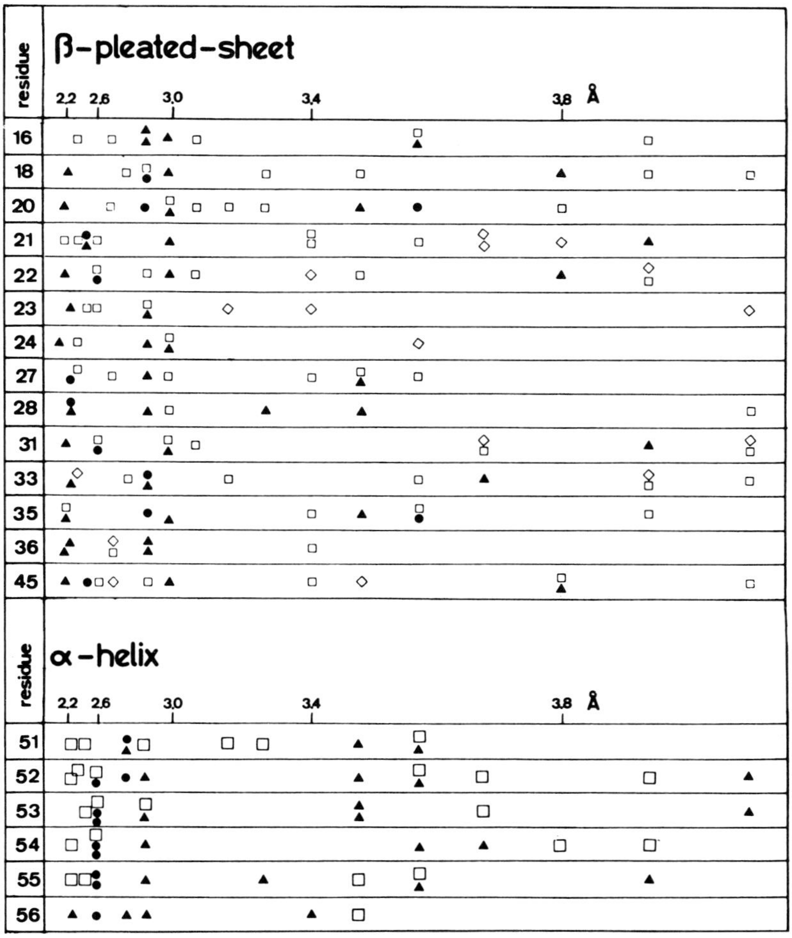

Fig. 6. Representation of the sixth power of the distances between the hydrogen-bonded peptide protons of the β-sheet and the α-helix in the inhibitor (indicated in the first column) and those α-protons (▲), β-protons (□), aromatic protons (◇) and other hydrogen-bonded amide protons (●) which are located within 4.0 Å from the respective peptide proton. The distances which correspond to the plotted sixth power values are also indicated.

easily be identified by spin decoupling. The next nearest α-proton is in no case closer than 3.0 Å and can therefore by nuclear Overhauser studies be distinguished from the nearest-by α-proton, which is typically at a distance of 2.2 Å (Fig. 6). Hence, once a backbone resonance in the β-sheet is assigned one can obtain additional assignments in a sequential way: from the assigned amide proton n, the neighboring α-proton of residue $(n-1)$ is identified by nuclear Overhauser experiments; the amide proton $(n-1)$ is then identified by spin decoupling and is next used to find the α-proton $(n-2)$ by nuclear Overhauser studies, etc.

In the α-helix the nearest neighbors of the peptide protons are mostly β-protons (Fig. 6). Thus a sequential assignment as in the β-sheet was not possible.

Selective nuclear Overhauser effects

In spite of the strong spin diffusion in the inhibitor (Fig. 1), both transient (Fig. 2) and truncated driven Overhauser experiments (Fig. 3) can give reliable information on intramolecular proton-proton distances [26,28,32,33]. However, while the two experiments were readily feasible when the lowest field amide proton 1 at 10.6 ppm was irradiated (Figs. 2 and 3), it proved much more difficult, if not impossible, to obtain selective transient Overhauser effects when irradiating in more crowded spectra regions (Fig. 5). Therefore, truncated-driven Overhauser experiments were used in the following.

From the above considerations on the protein structure it is evident that individual assignments of backbone resonances will largely depend on the determination of interatomic distances. While strictly speaking only the initial build-up rates of the Overhauser effects are simply related to the proton-proton distances [26—28,32], it is of considerable practical importance that the resonance intensities in truncated driven nuclear Overhauser difference spectra are simply related to the build-up rates when the preirradiation time t_1 is shorter than approximately 400 ms (Fig. 4). Therefore, rather than recording the time course of the driven Overhauser effect in each individual case, an optimal value for t_1 was determined from Fig. 4 and the relative intensities of individual lines (in percent of the corresponding resonance intensity in the unperturbed Fourier transform NMR spectrum) in a single truncated Overhauser experiment were used to compare relative distances from the irradiated nucleus. In all these experiments, t_1 was 400 ms.

Proton-proton spin decoupling

Spin decoupling data were obtained either with the use of conventional difference spectra [38] or with spin decoupled nuclear Overhauser difference spectra [30]. Spin decoupling experiments served two purposes. Firstly, once the amide proton or α-proton line was identified by nuclear Overhauser experiments in the sequential procedure used for the β-sheet, they provided resonance assignments within the spin systems of the individual amino acid residues. The correlations between amide, α- and β-protons, which were thus established, are listed in Table I and in part also shown in Fig. 5. Secondly, numerous α- and β-proton lines had previously independently been assigned [29,35,36,38] with different techniques. Spin decouplings with these resonances then provided double checks for some of the resonance assignments described in this paper.

Distinction between amide protons of the α-helix and the β-sheet

Three criteria were used for this global distinction. The first one is a direct consequence of the structural features presented in Fig. 6: when upon irradiation of an amide proton the strongest Overhauser effects were exclusively on β-protons, it was concluded that the amide proton in question is not in the β-sheet and hence could quite likely be located in the α-helix. Secondly, the vicinal amide proton-α proton spin-spin coupling constants $^3J_{NH\alpha}$ are expected to be of the order of 7—11 Hz in the β-sheet and, with the exception of Cys 55, smaller than 5 Hz in the α-helix [13]. Thirdly, potential lanthanide binding sites in the native inhibitor are all far away from the β-sheet, whereas some are at a short distance from the α-helix [29]. Hence addition of praseodymium(III) as a shift reagent [29] should not affect any of the amide proton lines in the β-sheet, but might well cause shifts for amide protons in the α-helix.

On the basis of these three criteria the most slowly exchanging amide protons could all be located in the β-sheet. Indication for a location in the α-helix was obtained for three resonances, i.e. line 10 and one line of each of the two groups 14 and 15, and 26–28.

Individual resonance assignments in the β-sheet

The antiparallel β-sheet in the inhibitor is formed by the residues 16–36. It is strongly twisted and extends through the full length of the molecule [1]. The above-mentioned sequential procedure for resonance assignments was started from the amide proton 6 at 9.25 ppm (Fig. 5), which was assigned to Tyr 21. The corresponding α-proton is at 5.70 ppm (Fig. 5). This assignment of the Tyr 21 amide proton is identical to that obtained previously from a different approach [24,25]. It was here evidenced by nuclear Overhauser experiments with the previously assigned [10] aromatic protons of Tyr 21, and confirmed by the entity of the experiments described in the following.

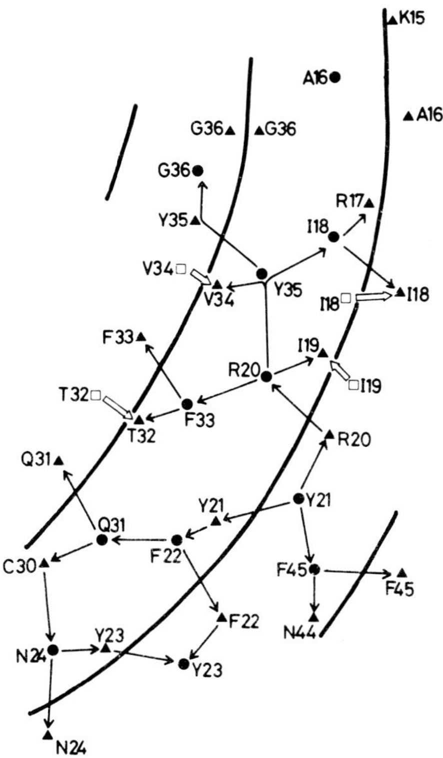

Fig. 7. Planar projection of the central part of the β-sheet in the inhibitor. The amino acids are denoted using the IUPAC one-letter code. The distances between amide protons (●), α-protons (▲) and β-protons (□) are projections of the actual distances in the three-dimensional structure. Three pathways for sequential resonance assignments, all starting from Y 21, are indicated by the arrows (see text). The open arrows indicate the locations where the resonance assignments were double checked by spin decoupling with previously assigned side chain resonances [29,36,38].

Starting from the amide proton and the α-proton of Tyr 21, further assignments were obtained by sequential use of Overhauser and spin decoupling measurements along the three pathways outlined in Fig. 7. The first pathway goes from Tyr 21 to the amide proton of Phe 22, from where it branches out into two directions, i.e. via the α-proton of Phe 22 to Tyr 23 and via Gln 31, Cys 30 and Asn 24 to Tyr 23. From the α-proton of Tyr 21 the amide resonance 3 at 9.79 ppm was assigned to Phe 22 by an Overhauser experiment. From there the α-proton resonance of Phe 22 was identified by spin decoupling at 5.25 ppm. Having identified the α-proton 22, the amide resonance 1 at 10.6 ppm was assigned to Tyr 23 (Figs. 2—4), and then the α-proton signal of Tyr 23 was identified at 4.30 ppm by spin decoupling. The second branch started with the Overhauser effect from the amide proton resonance of Phe 22 to the amide resonance 7 at 8.77 ppm, which was thus assigned to Gln 31 in the other peptide strand. From spin decoupling the α-proton of Gln 31 was then localized at 4.87 ppm. Overhauser effects obtained with irradiation of the amide proton resonance of Gln 31 assigned the α-proton signal at 5.60 ppm to Cys 30. The amide proton of Cys 30 exchanges rapidly, but could be observed in H_2O solution at 8.40 ppm. The two β-proton resonances of Cys 30 are largely inequivalent. The β-proton in gauche configuration to the α-proton was localized at 2.67 ppm and that in trans configuration at 3.69 ppm. From the α-proton resonance of Cys 30 the amide resonance 22 at 7.78 ppm was assigned to Asn 24 by an Overhauser experiment and subsequently the α-proton resonance of Asn 24 was found at 4.41 ppm by spin decoupling. Overhauser effects observed with irradiation of the amide proton resonance of Asn 24 resulted in the assignment of the α-proton resonance of Tyr 23, confirming the assignment via the first branch. The two singlet-like resonances 19 and 24 at 7.98 and 7.78 ppm were identified as the N^{δ} amide protons of Asn 43 by Overhauser effects obtained with irradiation of the amide proton of Tyr 23 (see Figs. 2 and 3). In the crystal structure the proton of signal 19 forms a hydrogen bond to the carbonyl oxygen of Glu 7, and that of resonance 24 is in a hydrogen bond to the carbonyl oxygen of Tyr 23 [1].

The second assignment pathway goes from the amide proton of Tyr 21 to Phe 45 and Asn 44. The amide proton of Phe 45 at 9.94 ppm was identified by Overhauser effects. Spin decouplings showed that the α- and β-proton signals of Phe 45 are at 5.12 ppm and 2.79 ppm, respectively. Overhauser effects obtained with irradiation of the amide proton of Phe 45 identified the α-proton resonance of Asn 44 at 4.88 ppm.

The third assignment pathway goes from the amide proton of Tyr 21 to Arg 20 and is branching at the amide proton of Arg 20 into three directions. The first branch leads to Phe 33 and Thr 32, the second branch to Tyr 35, Val 34, Gly 36, Ile 18 and Arg 17, and the third branch to Ile 19. From the amide proton of Tyr 21 the α-proton resonance of Arg 20 was localized at 4.70 ppm by an Overhauser experiment. By spin decoupling the amide proton resonance of Arg 20 was found to be line 11 at 8.39 ppm. Next, the amide resonance 4 at 9.37 ppm was identified as that of Phe 33 from Overhauser measurements. The α-proton signal of Phe 33 was detected at 4.89 ppm with spin decoupling. From the amide proton of Phe 33 the sharp α-proton resonance at 5.28 ppm, which overlaps with the α-proton resonance of Phe 22, was assigned to Thr 32 by an Overhauser experiment. The small value of approx. 2.0 Hz for the spin-spin coupling constant $^3J_{\alpha\beta}$ of Thr 32 was previously described [36]. The α-proton of Thr 32 is coupled to the amide proton resonance 18 at 8.07 ppm, which exchanges too fast to be seen in Fig. 5.

The second branch of the third pathway starts with the assignment of the amide proton line 5 at 9.39 ppm to Tyr 35 by the Overhauser technique. By spin decoupling in nuclear Overhauser difference spectra [30], the α-proton and the two β-protons of Tyr 35 were then localized at 4.89, 2.69 and 2.52 ppm. Overhauser effects between the α-proton resonance of Tyr 35 and the amide proton resonance 9 at 8.61 led to the assignment of the latter to Gly 36. In Fig. 5 the triplet-type fine structure of the amide proton resonance 9 can be recognized. From nuclear Overhauser effects obtained with irradiation of the amide proton of Tyr 35, the α-proton of Val 34 was identified at 3.92 ppm. The latter assignment was confirmed by spin decoupling with the previously assigned β-proton of this residue at 1.95 ppm [36,38]. From Overhauser effects between the amide proton of Tyr 35 and the amide resonance 16 at 8.12 ppm, the latter was assigned to Ile 18. The α-proton resonance of Ile 18 was then detected at 4.25 ppm by spin decoupling.

The third branch of the third pathway consists of a single step, i.e. the Overhauser effect built up from the amide proton of Arg 20, resulted in the identification of the α-proton resonance of Ile 19 at 4.30 ppm.

In previous work, the resonances of the C^{β}, $C^{\gamma 1}$, $C^{\gamma 2}$ and C^{δ} protons of the two isoleucines 18 and 19 in the inhibitor were identified but not individually assigned [36,38]. Spin decoupling with the two lines which were previously assigned to the C^{β} protons of the two isoleucines now provided on the one hand a double check of the assignments of the backbone protons (Fig. 7), and on the other hand resulted in the individual assignments of the β-methine and the γ-methyl resonances of Ile 18 and Ile 19 (Table I). For Ile 18, the β-proton line is at 1.87 ppm and the γ-methyl at 0.97 ppm, for Ile 19 the corresponding chemical shifts are 1.96 ppm and 0.73 ppm.

Individual resonance assignments in the α-helix

Evidence for individual assignments was found for two among the three resonance lines for which a location in the α-helix appeared likely. When the amide proton resonance 15 at 8.25 ppm was irradiated, the previously identified [36] spin system of the side chain of Thr 54 appeared in the nuclear Overhauser difference spectrum. From inspection of the atomic coordinates and comparison of the expected coupling constants with the experiment, the amide proton resonance 15 was thus assigned to Cys 55. Irradiation of the amide proton resonance 10 at 8.58 ppm produced Overhauser effects on two spin systems which would be compatible with the location of this amide proton between the side chains of a cysteinyl and a methionyl residue. Line 10 was therefore tentatively assigned to the amide proton of Met 52.

Discussion

The main purpose of this paper was to describe the experiments used to obtain individual assignments for 18 amide proton resonances in the inhibitor.

These data complement earlier work which resulted in the individual assignments of all the aromatic resonances (Refs. 10,19—21,29,35, and Wagner, G., Tschesche, H. and Wüthrich, K., unpublished results) and most of the methyl resonances of the aliphatic side chains [29,36,38].

Individual assignments of proton resonances of the polypeptide backbone are for a fundamental reason even more difficult to obtain than those for amino acid side chains. While each type of amino acid side chain occurs usually in a quite small number in a protein, the backbone protons present a large number of resonances with very similar spectral properties. Therefore, comparison of different homologous or chemically modified proteins may often result in unambiguous assignments of side chain resonances [13,36], but will not usually be a suitable method for assignments of backbone proton lines.

It would appear that the sequential use of Overhauser and spin decoupling experiments should be a generally applicable technique for identifying nearest-neighbor relations in antiparallel β-sheets. The perspective drawing of the

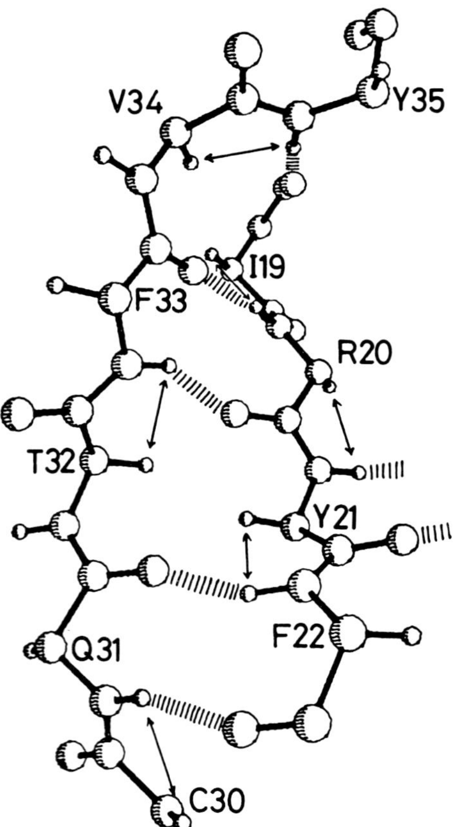

Fig. 8. Perspective computer drawing of the central part of the β-sheet in the inhibitor. This figure illustrates the close distances between the amide protons of residues n and the α-protons of residues $(n - 1)$, which was essential for the sequential resonance assignments by alternative use of spin decoupling and nuclear Overhauser effect measurements. Intramolecular hydrogen bonds are indicated by IIIIIII.

twisted β-sheet in the inhibitor (Fig. 8) clearly illustrates two structural features which were essential for the use of the spectroscopic techniques described here for resonance assignments. First, within each peptide strand of the β-sheet, the crucial property was the close proximity of the amide proton with the α-proton of the preceding residue (indicated by arrows in Fig. 8). Second, the arrangement of the hydrogen bonds brings always one amide proton of the first strand into close proximity with an amide proton of the antiparallel second strand, which allowed to extend the sequential assignment procedure also across the β-sheet. In certain cases this provided new additional assignments, in other cases important double checks on assignments made within each individual peptide strand. It may well be that close examination of other regular polypeptide structures could open similar avenues.

Since spin decoupling can in many cases be used to identify the spin systems of entire amino acid residues [13], the techniques described here could in principle probably also be applied to establish the amino acid sequences in β-sheet regions of proteins. In view of the other methods available for peptide sequencing, such applications would possibly be of rather limited practical interest.

Using lanthanide shift reagents with a chemically modified inhibitor, assignments were previously obtained for five amide proton resonances in the inhibitor [24,25]. From Table I it is seen that three of these earlier assignments could be confirmed by the present experiments. On the other hand, the resonances 7 and 11 (Fig. 5) are now found to correspond to Gln 31 and Arg 20, respectively, rather than to Arg 20 and Gln 31, as suggested previously [24,25]. From an examination of the assumptions used for the analysis of the lanthanide shifts [24,25], it would appear that this technique could not really differentiate between these two protons. Hence, one should probably argue that there is satisfactory agreement between the two sets of data, since the same two lines were assigned to the pair of amide protons of Arg 20 and Gln 31.

Structural interpretations of the NMR properties of the amide protons in the inhibitor, which depend to a large measure also on the assignments in Table I, are described elsewhere (Refs. 14—18, and Wüthrich and co-workers, unpublished results). Therefore, only two outstanding observations shall be listed here. One is that all the most slowly exchanging labile protons of the inhibitor are in the β-sheet. In particular, all the amide protons in the peptide fragment 20—24 exchange very slowly (Richarz, R., Sehr, P., Wagner, G. and Wüthrich, K., unpublished results). This fragment occupies a quite central position in the molecule. On the one side it has the second strand of the β-sheet as a near neighbor, on the opposite side the peptide fragments 6—9 and 43—46. Secondly, it is quite striking that the six resonances at lowest field (Fig. 5) correspond to those six aromatic residues which are preserved in homologous inhibitors [19,20] and which appear from this and other observations to be essential for the architecture of the molecule [14—20].

Appendix

This appendix describes the algebraic functions which were used to compute the time course of the driven nuclear Overhauser effects in Fig. 4A. A three-spin system was considered, including the irradiated spin, the observed spin and a third spin which represents the bulk of the protons in the protein.

The time dependence of the magnetisation of the non-irradiated spins i in a driven Overhauser experiment is determined by Eqn. 3 [28].

$$\frac{dM_i}{dt} = -\rho_i M_i - \sum_{j \neq i} \sigma_{ij} M_j \tag{3}$$

M_i is the difference between the actual magnetisation M_{zi} of spin i and its equilibrium magnetisation M_i^0, and correspondingly $M_j = M_{zj} - M_j^0 \cdot \rho_i$ and σ_{ij} describe the spin lattice relaxation and the spin diffusion.

$$\rho_i = \frac{\hbar^2\gamma^4}{10} \sum_{j \neq i} \frac{1}{r_{ij}^6} \left[\tau_c + \frac{3\tau_c}{1 + (\omega\tau_c)^2} + \frac{6\tau_c}{1 + 4(\omega\tau_c)^2} \right] \tag{4}$$

$$\sigma_{ij} = \frac{\hbar^2\gamma^4}{10} \frac{1}{r_{ij}^6} \left[\frac{6\tau_c}{1 + 4(\omega\tau_c)^2} - \tau_c \right] \tag{5}$$

ω is the Larmor frequency, r_{ij} the distance between spins i and j, $2\pi\hbar$ the Planck constant, γ the gyromagnetic ratio and τ_c the effective rotational correlation time. In the three-spin system, we denote as spin 1 the irradiated nucleus, as spin 2 the observed nucleus and as spin 3 the nucleus which represents the influence of additional spins upon the driven Overhauser effect. The two coupled differential Eqn. 3 for the spins 2 and 3 were solved with the assumption that the time course of the magnetization of the irradiated spin 1 can be described by

$$M_1 = \begin{cases} 0\,, & \text{for } t < 0 \\ M_1^0(1 - e^{-ct})\,, & \text{for } t \geqslant 0\,, \end{cases} \tag{6}$$

where M_1^0 is the equilibrium magnetization of spin 1 and c is a constant which determines the rate of saturation of spin 1. The expression (6) for M_1 was selected empirically on the basis of the time course of the saturation observed with the experimental conditions used [33]. For the magnetization M_2 of the observed spin one thus obtains

$$\begin{aligned} \frac{M_2}{M_1^0} = {} & \frac{\sigma_{21}}{(b-a)}(e^{-at} - e^{-bt}) + \frac{(\sigma_{21}\rho_3 - \sigma_{23}\sigma_{31})}{(a-b)ab}(b\,e^{-at} - a\,e^{-bt} + (a-b)) \\ & + \frac{\sigma_{21}\rho_3 - \sigma_{23}\sigma_{31}}{(a-b)(b-c)(c-a)}((b-c)\,e^{-at} + (c-a)\,e^{-bt} + (a-b)\,e^{-ct}) \\ & - \frac{\sigma_{21}}{(a-b)(b-c)(c-a)}(c(a-b)\,e^{-ct} + a(b-c)\,e^{-at} + b(c-a)\,e^{-bt} \end{aligned} \tag{7}$$

with

$$a = \tfrac{1}{2}[(\rho_2 + \rho_3) - \sqrt{(\rho_2 - \rho_3)^2 + 4\sigma_{23}^2}] \tag{8}$$

and

$$b = \tfrac{1}{2}[(\rho_2 + \rho_3) + \sqrt{(\rho_2 - \rho_3)^2 + 4\sigma_{23}^2}] \tag{9}$$

The t_1 dependence of the driven Overhauser effects in Fig. 4A was computed with Eqn. 7.

References

1 Deisenhofer, J. and Steigemann, W. (1975) Acta Cryst. B31, 238—250
2 Gelin, B.R. and Karplus, M. (1975) Proc. Natl. Acad. Sci. U.S. 72, 2002—2006
3 Hetzel, R., Wüthrich, K., Deisenhofer, J. and Huber, R. (1976) Biophys. Struct. Mech. 2, 159—180
4 McCammon, J.A., Gelin, B.R. and Karplus, M. (1977) Nature 267, 585—590
5 Levitt, M. and Warshal, A. (1975) Nature 253, 694—698
6 Levitt, M. (1976) J. Mol. Biol. 104, 59—107
7 Chothia, C. (1976) J. Mol. Biol. 105, 1—14
8 Creighton, T.E., Dykes, D.F. and Sheppard, R.C. (1978) J. Mol. Biol. 119, 507—518
9 Wüthrich, K. and Wagner, G. (1975) FEBS Lett. 50, 265—268
10 Snyder, G.H., Rowan III, R., Karplus, S. and Sykes, B.D. (1975) Biochemistry 14, 3765—3777
11 Wagner, G., De Marco, A. and Wüthrich, K. (1975) J. Magn. Resonance 20, 565—569
12 Wagner, G., De Marco, A. and Wüthrich, K. (1976) Biophys. Struct. Mech. 2, 139—158
13 Wüthrich, K. (1976) NMR in Biological Research: Peptides and Proteins, North-Holland, Amsterdam
14 Wagner, G. and Wüthrich, K. (1978) Nature 275, 247—248
15 Wüthrich, K. and Wagner, G. (1978) Proceedings Int. Symp. on Biomolecular Structure, Conformation, Function and Evolution, Madras, India, in press
16 Wüthrich, K. and Wagner, G. (1978) Trends Biochem. Sci. 3, 227—230
17 Wüthrich, K., Wagner, G. and Bundi, A. (1978) In Nuclear Magnetic Resonance Spectroscopy in Molecular Biology, Proceedings of the 11th Jerusalem Symposium on Quantum Chemistry and Biochemistry (Pullman, B., ed.), pp. 201—210, Reidel, Dordrecht, Holland
18 Wüthrich, K., Wagner, G. and Richarz, R. (1978) Proceedings of the 12th FEBS Meeting in Dresden, in press
19 Wagner, G., Wüthrich, K. and Tschesche, H. (1978) Eur. J. Biochem. 86, 67—76
20 Wagner, G., Wüthrich, K. and Tschesche, H. (1978) Eur. J. Biochem. 89, 367—377
21 Brown, L.R., De Marco, A., Richarz, R., Wagner, G. and Wüthrich, K. (1978) Eur. J. Biochem. 88, 87—95
22 Masson, A. and Wüthrich, K. (1973) FEBS Lett. 31, 114—118
23 Karplus, S., Snyder, G.H. and Sykes, B.D. (1973) Biochemistry 12, 1323—1329
24 Marinetti, T.D., Snyder, G.H. and Sykes, B.D. (1976) Biochemistry 15, 4600—4608
25 Marinetti, T.D., Snyder, G.H. and Sykes, B.D. (1977) Biochemistry 16, 647—653
26 Solomon, I. (1955) Phys. Rev. 99, 559—565
27 Noggle, J.H. and Schirmer, R.E. (1971) The Nuclear Overhauser Effect, Chemical Applications, Academic Press, New York
28 Kalk, A. and Berendsen, H.J.C. (1976) J. Magn. Resonance 24, 343—366
29 Perkins, S.J. and Wüthrich, K. (1978) Biochim. Biophys. Acta 536, 406—420
30 Richarz, R. and Wüthrich, K. (1978) J. Magn. Resonance 30, 147—150
31 Sykes, B.D., Hull, W.E. and Snyder, G.H. (1978) Biophys. J. 21, 137—146
32 Gordon, S.L. and Wüthrich, K. (1978) J. Am. Chem. Soc. 100, 7094—7096
33 Wagner, G. and Wüthrich, K. (1978) J. Magn. Resonance, in press
34 Englander, S.W., Downer, N.W. and Teitelbaum, H. (1972) Annu. Rev. Biochem. 41, 903—924
35 Wagner, G. (1977) Ph.D. Thesis 5992, E.T.H. Zürich
36 Wüthrich, K., Wagner, G., Richarz, R. and Perkins, S.J. (1978) Biochemistry 17, 2253—2263
37 Perkins, S.J. and Wüthrich, K. (1978) Biochim. Biophys. Acta 576, 409—423
38 De Marco, A., Tschesche, H., Wagner, G. and Wüthrich, K. (1977) Biophys. Struct. Mech. 3, 303—315

Eur. J. Biochem. *115*, 653 – 657 (1981)
© FEBS 1981

Structural Interpretation of Vicinal Proton-Proton Coupling Constants $^3J_{H^\alpha H^\beta}$ in the Basic Pancreatic Trypsin Inhibitor Measured by Two-Dimensional *J*-Resolved NMR Spectroscopy

Kuniaki NAGAYAMA and Kurt WÜTHRICH

Institut für Molekularbiologie und Biophysik, Eidgenössische Technische Hochschule, Zürich-Hönggerberg

(Received October 31, 1980)

Two-dimensional *J*-resolved ^{1}H NMR spectroscopy was used to measure the vicinal spin-spin coupling constants $^3J_{H^\alpha H^\beta}$ for numerous, previously individually assigned amino acid residues in the basic pancreatic trypsin inhibitor at various temperatures between 30 and 85 °C. An analysis of this data is proposed which enables one to compare the spatial arrangements of individual amino acid side chains in solution and in single crystals of the protein, and which also provides information on the mobility of the side chains in the solution conformation. As a rule, the amino acid side chains in the interior of the protein were found to be locked into unique spatial orientations, with the mobility restricted to rapid rotational fluctuations about this unique value for the dihedral angle χ^1. In most, but not all, instances the data for the interior amino acids indicate identical average conformations for the amino acid side chains in single crystals and in solution. For residues on the protein surface structural rearrangements between crystal and solution appear to be common, and the mobility in the solution conformation may include rapid averaging between two or several distinct, preferentially populated values of χ^1, analogous to the *gauche-trans-gauche* isomerization in isolated amino acids.

The basic pancreatic trypsin inhibitor from bovine organs is a small protein consisting of one polypeptide chain with 58 amino acid residues and with a molecular weight of 6500. The inhibitor was extensively studied by X-ray methods in single crystals [1], spectroscopic techniques in solution [2] and theoretical calculations [3, 4], and is at present a much used model compound for investigations on various fundamental aspects of protein conformations. This paper describes a comparison of the spatial arrangement of amino acid side chains in the crystal structure and in solution. The crystal data were taken from the refined structure by Deisenhofer and Steigemann [1]. The data on the solution conformation were obtained from measurements of the vicinal proton-proton coupling constants $^3J_{H^\alpha H^\beta}$ in two-dimensional *J*-resolved ^{1}H NMR spectra recorded at 360 MHz.

METHODS AND RESULTS

The proton spin systems of numerous amino acid residues in the inhibitor were recently assigned with two-dimensional NMR techniques [5, 6]. For the assigned residues the vicinal spin-spin coupling constants $^3J_{H^\alpha H^\beta}$ were now measured at various temperatures between 30 °C and 85 °C with the use of two-dimensional *J*-resolved ^{1}H NMR spectra at 360 MHz. Selected C$^\alpha$proton multiplets are shown in Fig. 1 and the results are compiled in Table 1.

The two-dimensional *J*-resolved experiment [7, 8] and the experimental procedures for obtaining the spectra used in this paper [5] were described recently and a brief survey of the experimental conditions is contained in the caption to Fig. 1. Compared to conventional, one-dimensional ^{1}H NMR spectra, the resolution in these two-dimensional spectra is greatly improved [7 – 9]. In the δ/J representation [10] the chemical shifts and the spin-spin coupling constants are displayed along two perpendicular axes. Cross sections perpendicular to the chemical shift axis then provide a particularly suitable presentation of the multiplet structures, which allows accurate measurements of the spin-spin coupling constants, *J*. Since overall views of 360-MHz ^{1}H two-dimensional *J*-resolved spectra of the inhibitor were previously presented [5, 8, 9] only cross sections of selected multiplets are shown here (Fig. 1).

Fig. 1 contains, in addition to the multiplet structures, the chemical shifts where the cross sections were taken, the resonance assignments and indications of how the parameters of Table 1 were extracted from the spectrum. For Tyr-21 and Asn-43, which are well separated from other resonances in the spectrum [5], the multiplet fine structure can readily be recognized. For many of the other residues some overlap with nearby resonances occurs; i.e. since the line widths are large compared to the digital resolution, 'tails' of neighbouring multiplets may appear in the cross sections. The spectral analysis was in these cases based on inspection of several cross sections on each side of the one shown in Fig. 1. In this context it must be emphasized that Fig. 1 contains only a small selection of the multiplets between 4.0 ppm and 4.9 ppm [5, 9] so that the appearance of spurious tails cannot be judged from Fig. 1 alone. Some additional comments concern the following amino acid residues. For Tyr-35 and Asn-44 the C$^\alpha$H chemical shifts are too close for the resonances to be resolved. Very similar spin-spin couplings seemed to prevail for the two residues. For Ser-47 the small coupling observed in the C$^\alpha$H multiplet was also seen in the C$^\beta$H resonances (Fig. 1), which further manifest the small

Abbreviations. The inhibitor, basic pancreatic trypsin inhibitor; NMR, nuclear magnetic resonance; ppm, part per million; δ, chemical shifts in ppm; *J*, spin-spin coupling constant in hertz.

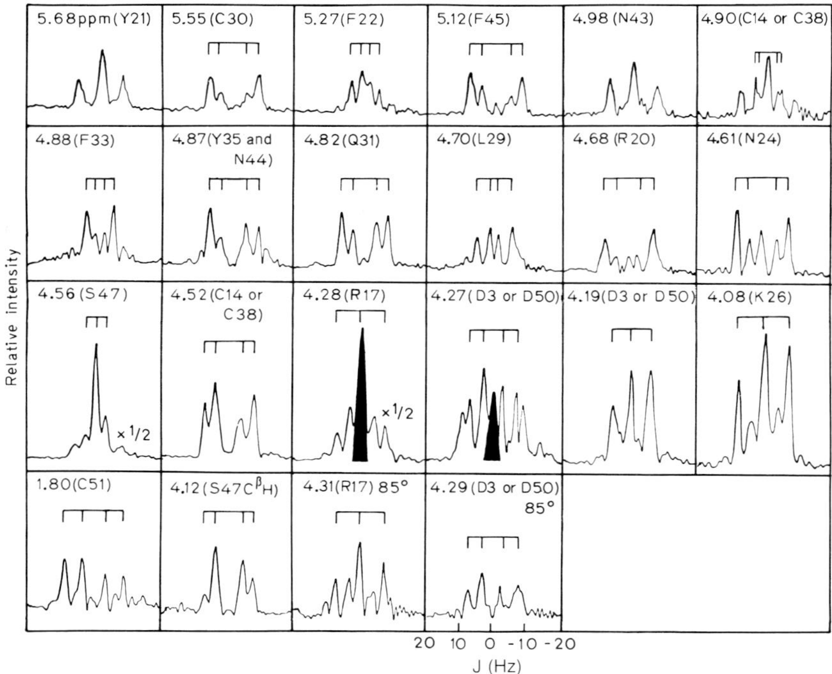

Fig. 1. *Cross-section representation of C^α proton multiplets in the two-dimensional* J*-resolved* ^1H *NMR spectrum at 360 MHz of a 0.01-M solution of the basic pancreatic trypsin inhibitor in* ^2H_2O, p^2H *7.0, 68 °C.* The labile protons had been replaced with deuterium. The frequency domain spectrum was developed on a data matrix of size 4096 × 124. The digital resolution was 0.61 Hz in both dimensions. Three-dimensional views of the entire spectrum in the δ/J representation and cross sections through all the methyl resonances and the C^α proton multiplets of Ala, Val, Ile and Thr were presented previously [8–10]. The singlet line of the residual solvent protons is near 4.30 ppm (shaded peak in two cross sections). The chemical shifts where the cross sections parallel to the J axis have been taken are indicated by the numbers in the upper left corner of each multiplet representation, the resonance assignments are indicated in parentheses. For reasons discussed in the text, a cross section through one of the C^β proton multiplets of Ser-47, and cross sections through the C^α proton multiplets of Arg-17 and Asp-3 or Asp-50 in the spectrum at 85 °C were added in the bottom row of the figure

geminal coupling $^2J_{H^\beta H^\beta}$ in this amino acid side chain [5,11]. For Arg-17 the cross section at 68 °C contained a strong central peak from the residual water protons. The identification of the Arg multiplet was confirmed at 85 °C, where the water line was shifted to higher field [11] (Fig. 1). The multiplet at 4.27 ppm, which was assigned to Asp-3 or Asp-50, was at 68 °C overlapped with a tail from the water line and a doublet of Gly-36 [5]. The identification of the multiplet components was confirmed at 85 °C, where both the water line and Gly-36 are shifted away. No data are listed for Tyr-23 (Table 1) since neither the $C^\alpha H$ not the $C^\beta H$ resonances were sufficiently well resolved to identify the multiplet structures unambiguously.

The χ^1 angles listed in Table 1 are from Deisenhofer and Steigemann [1]. Corresponding values for $^3J_{H^\alpha H^{\beta 2}}$ and $^3J_{H^\alpha H^{\beta 3}}$ were calculated using the angular dependence of the vicinal coupling constants proposed by De Marco et al. [12].

DISCUSSION

The use of the well known angular dependence of vicinal spin-spin coupling constants [15] for studies of the conformations of amino acids and peptides has been extensively discussed [11,16]. For proteins the structural information from spin-spin couplings was more limited since only a few experimental data were available. The first part of this discussion describes some fundamental considerations on the interpretation of vicinal spin-spin couplings in proteins, which will then be applied to the data on the inhibitor (Table 1).

The correlations between the dihedral angles which characterize the polypeptide conformation [13] and the vicinal spin-spin coupling constants depend critically on the internal mobility of the spatial structures. One limiting situation is a fully rigid structure. This was assumed in the interpretation of the data on ferrichrome peptides, which yielded the angular dependence of vicinal proton-proton couplings in amino acid side chains used here [12]. On the other hand a fully flexible molecule is assumed in interpretations of spin-spin couplings by a population analysis of rotamer configurations [16,17]. In the present treatment for a protein we propose to distinguish between three different situations. Firstly, we consider the hypothetical case of a rigid protein. Secondly, flexibility of the protein structure is taken into consideration with the assumption that the angles χ^1 fluctuate rapidly about a unique position corresponding to a single minimum of the conformational energy. Thirdly, a population analysis of multiple energy minima assumes rapid exchange among several distinct χ^1 values. In contrast to the analogous treatment for amino acids and small peptides [16,17], these distinct states will not in general coincide with the classical rotamer configurations *gauche*, *gauche* and *trans*, since in globular

Table 1. *Vicinal proton-proton coupling constants* $^3J_{H^\alpha H^{\beta 2}}$ *and* $^3J_{H^\alpha H^{\beta 3}}$ *for the basic pancreatic trypsin inhibitor in aqueous solution and in single crystals*

^{1}H NMR spectra were measured at 68 °C, p^{2}H 7.0. Chemical shifts, δ, are given $\pm$ 0.01 ppm, spin-spin coupling constants, J, $\pm$ 0.5 Hz. X-ray crystallography data were taken from [1] with angles χ^1 in degrees according to the IUPAC-IUB conventions [12]. All those residues are included for which proton resonance assignments have so far been obtained [5,6,14] and the proton-proton couplings determined in the two-dimensional spectra. First, residues located in the β-sheet are listed, then the additional individually assigned residues, and finally those where assignments to pairs of residues of the same type have been obtained. Gly and Ala have not been included in this table. The resonances of H$^{\beta 2}$ and H$^{\beta 3}$ were not individually assigned

Amino acid residue	^{1}H NMR			X-ray crystal structure		
	δ(H$^\alpha$)	δ(H$^{\beta 2}$/H$^{\beta 3}$)	$^3J_{H^\alpha H^{\beta 2}}$/$^3J_{H^\alpha H^{\beta 3}}$	χ^1	$^3J_{H^\alpha H^{\beta 2}}$	$^3J_{H^\alpha H^{\beta 3}}$
	ppm		Hz	deg.	Hz	
Arg-17	4.28	1.59/1.59	7.5/7.5	− 62	13.0	3.0
Ile-18	4.20	1.89	11.0	− 65	13.0	
Ile-19	4.32	1.95	11.5	− 57	13.0	
Arg-20	4.68		11.5/4.0	− 64	13.0	3.0
Tyr-21	5.68	2.72/2.72	7.0/7.0	− 72	12.5	2.0
Phe-22	5.27	2.92/2.81	5.5/3.0	76	5.5	2.0
Asn-24	4.61	2.88/2.15	12.0/3.5	173	4.5	13.0
Lys-26	4.08	1.88/1.88	8.0/8.0	− 88	10.5	2.0
Leu-29	4.70	1.68/1.41	6.5/4.0	44	2.0	5.5
Cys-30	5.55	3.61/2.67	12.0/3.0	− 68	12.5	2.5
Gln-31	4.82	2.15/1.71	11.0/3.5	− 67	12.5	2.5
Thr-32	5.25	4.01	2.5	53	2.5	
Phe-33	4.88	3.11/2.96	5.5/2.5	74	5.5	2.0
Val-34	3.96	1.93	11.0	172		12.5
Tyr-35	4.87	2.64/2.50	11.5/3.5	161	6.0	11.8
Asn-43	4.98	3.28/3.28	7.5/7.5	− 161	12.0	2.0
Asn-44	4.87	2.74/2.48	11.5/3.5	176	4.0	13.0
Phe-45	5.12	3.39/2.79	12.0/3.5	− 56	13.0	4.0
Thr-11	4.51	4.05	8.0	− 77	12.0	
Ser-47	4.56	4.12/3.85	3.5/3.5	76	5.5	2.0
(Cys-51)	1.80	3.17/2.86	12.5/5.0	175	4.0	13.0
Thr-54	4.06	3.98	10.0	− 60	13.0	
Asp-3 }	{ 4.19	{ 2.74/2.74	{ 6.0/6.0	− 86	11.0	1.5
Asp-50 }	{ 4.27	{ 2.86/2.70	{ 10.5/4.5	− 97	9.0	2.0
Cys-14 }	{ 4.52	{ 3.47/2.79	{ 12.0/3.5	− 73	12.5	2.0
Cys-38 }	{ 4.90	{ 3.77/3.14	{ 6.5/1.5	74	5.5	2.0

proteins the spatial arrangement of the amino acid side chains appears to be largely determined also by interactions with neighbouring fragments of the polypeptide chain, so that interactions within the side chain may not be the dominant factor [3,18].

Even when the dynamic situation is clearly defined there are in general four different values of χ^1 which correspond to a measured coupling constant $^3J_{H^\alpha H^\beta}$ [11,12,15–17]. On the other hand a set angle χ^1 is unambiguously correlated with a corresponding value for $^3J_{H^\alpha H^\beta}$. In Table 1 we therefore compare the measured coupling constants with those calculated from the atomic coordinates of the crystal structure of the inhibitor. For the amino acids with a β-methylene group and hence two correlated coupling constants, $^3J_{H^\alpha H^{\beta 2}}$ and $^3J_{H^\alpha H^{\beta 3}}$, this provides a basis for distinguishing between different limiting dynamic situations. Fig. 2 may be helpful to clarify this point further. The solid curves represents the correlation between $^3J_{H^\alpha H^{\beta 2}}$ and $^3J_{H^\alpha H^{\beta 3}}$ in a rigid structure with dihedral angles χ^1_0. The dotted curve was computed from the same Karplus-type angular dependence of the coupling constants [12], but with the assumption of rapid fluctuations of χ^1 over a range of $\pm 30°$ about χ^1_0, with equal population for all values of χ^1 within this range. Obviously for different types of fluctuations, e.g. for a harmonic fluctuation about χ^1_0, the dotted curve can correspond to a situation where considerably larger amplitudes than $\pm 30°$ may occur. From the location of the experimental points ($^3J_{H^\alpha H^{\beta 2}}$, $^3J_{H^\alpha H^{\beta 3}}$) in the correlation diagram (Fig. 2) the following qualitative conclusions can hence be drawn. (a) Data points located on the curves or in the narrow band between the solid and the dotted curve are compatible with a situation where the amino acid residues are locked in unique positions χ^1_0, with the internal mobility about χ^1 restricted to rapid fluctuations about this position. (b) Data points located in the area bounded by the peripheral branches of the dotted curve and outside the interior branches of the narrow band between solid and dotted curve indicate rapid averaging between two or several distinct values of χ^1, corresponding to multiple populated minima of the conformational energy. (c) Data points located on those parts of the curves, or the narrow band between solid and broken curve, which are inside the area bounded by the peripheral branches of the broken curve may correspond to either of the dynamic situations (a) and (b). (d) Data points outside the area bounded by the solid curve would not be compatible with the Karplus-type curves used here [12]. A possible explanation might be

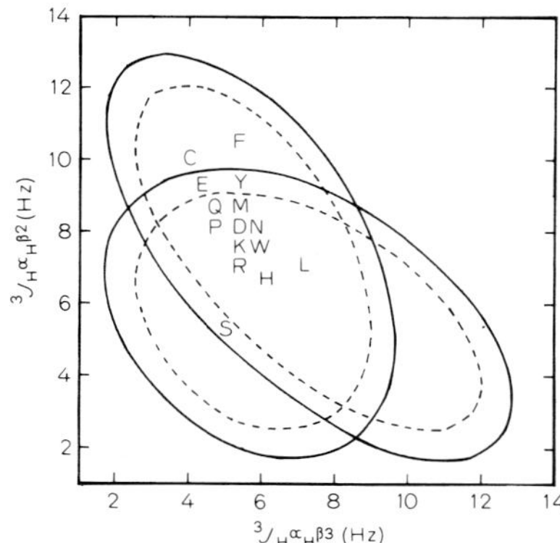

Fig. 2. *Correlation diagram for the vicinal coupling constants* ${}^3J_{H^\alpha H^{\beta 2}}$ *and* ${}^3J_{H^\alpha H^{\beta 3}}$ *in amino acid residues.* The solid line represents the correlation for a rigid molecule with fixed dihedral angles χ_0^1 and was computed with the Karplus-type curve proposed by De Marco et al. [12]. The broken line represents the correlation for a flexible molecule, where it was assumed that the time-dependent variations of χ^1 extended over the range $\chi_0^1 \pm 30^\circ$ and that within this range each value of χ^1 was equally populated (see text). Experimental points, identified by the IUPAC-IUB one-letter symbols, are given for the 15 common amino acid residues Xaa which contain a β-methylene group. These data were measured in the synthetic model peptides H-Gly-Gly-Xaa-Ala-OH [19]. Since the measured coupling constants had not been stereospecifically assigned, all the data were arbitrarily plotted in the upper left triangle of the correlation diagram

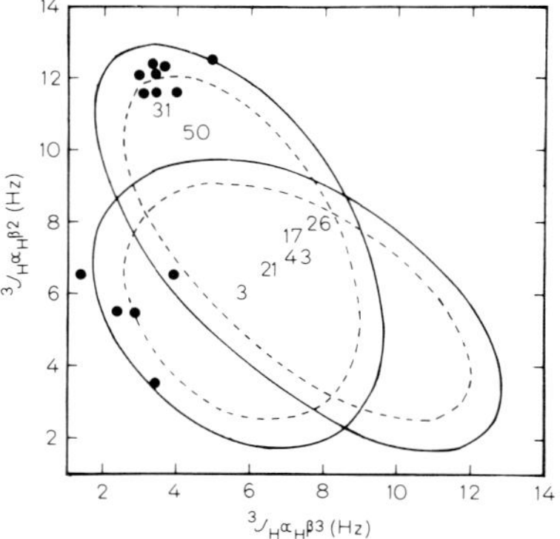

Fig. 3. *Locations in the correlation diagram of Fig. 2 of the experimental values for* (${}^3J_{H^\alpha H^{\beta 2}}$, ${}^3J_{H^\alpha H^{\beta 3}}$) *of the basic pancreatic trypsin inhibitor in* ${}^2H_2O$ *solution at* p^2H *7.0 and 68 °C.* Since the measured coupling constants had not been stereospecifically assigned, all the data were arbitrarily plotted in the upper left triangle of the correlation digram. Data points located on the two curves or in the narrow band between the curves are indicated by filled circles, the other ones by the position of the amino acid in the sequence

that unusual values for the coupling constants result because of steric strain leading to distortion of bond angles etc.

In Fig. 2 the data points (${}^3J_{H^\alpha H^{\beta 2}}$, ${}^3J_{H^\alpha H^{\beta 3}}$) measured in synthetic model peptides H-Gly-Gly-Xaa-Ala-OH [19] are indicated for the 15 common amino acid residues Xaa which contain a β-methylene group. It is seen that for all 15 amino acids the correlation between ${}^3J_{H^\alpha H^{\beta 2}}$ and ${}^3J_{H^\alpha H^{\beta 3}}$ is compatible with rapid averaging between two or more distinct values of χ^1. This agrees with the original interpretation of this data in terms of a population analysis of the *gauche*, *gauche* and *trans* rotamer configurations [19].

Fig. 3 shows the location in the correlation diagram of Fig. 2 of the presently available experimental values (${}^3J_{H^\alpha H^{\beta 2}}$, ${}^3J_{H^\alpha H^{\beta 3}}$) for the inhibitor (Table 1). None of the experimental points are outside the area bounded by the solid curve, which indicates that the 'ferrichrome curve' [12] is adequate for the interpretation of the data on the inhibitor. Within the experimental accuracy of ± 0.5 Hz, 13 of the 20 points lie on the solid or the dotted curve, or in the area between the two curves. Seven points are located in the inside areas bounded by the dotted curve. These correspond to the amino acid residues Asp-3, Arg-17, Tyr-21, Lys-26, Gln-31, Asn-43 and Asp-50. With the exception of one residue, Asn-43, qualitatively similar plots to that of Fig. 3, which corresponds to data recorded at 68 °C and p^2H 7.0, are obtained from the data at 30 °C and 85 °C, and at 68 °C and p^2H 4.2. For Asn-43 the spin-spin couplings (${}^3J_{H^\alpha H^{\beta 2}}$ and ${}^3J_{H^\alpha H^{\beta 3}}$) vary from 11.5 Hz and 3.5 Hz at 30 °C to 7.5 Hz and 7.5 Hz at 85 °C. Lys-26, where ${}^3J_{H^\alpha H^{\beta 2}}$ and ${}^3J_{H^\alpha H^{\beta 3}}$ are actually within 0.5 Hz of the dotted curve (Fig. 3), is located in a region of the correlation diagram where the differentiation between the different dynamic situations is rather ambiguous. Since identical chemical shifts were observed for the two β-methylene protons (Table 1) it was concluded that rapid rotational averaging between several states prevailed in this case.

Obviously, direct comparison of the spin-spin coupling constants calculated from the crystal structure with those measured in solution (Table 1) is sensible only when the latter fall within the narrow band between the solid and the dotted curve in Fig. 2. Among the 13 amino acid residues in the inhibitor which satisfy this condition, 10 give close agreement with the crystal data, i.g. Arg-20, Phe-22, Asn-24, Cys-30, Phe-33, Asn-44, Phe-45, Cys-51 and (Cys-14/Cys-38). The side chains of these residues thus appear to be locked into unique orientations χ_0^1 both in the crystal and in solution, whereby mobility about the angles χ_0^1 is clearly indicated in the spin-spin coupling constants, which are generally smaller than those predicted for a rigid structure from the 'ferrichrome curve' [12] (Fig. 3) and the X-ray data (Table 1). In the crystal structure these side chains are located in interior parts of the protein. The present data thus provide additional direct evidence that extended interior regions of the spatial structure of the inhibitor are closely similar in solution and in single crystals [1, 6, 20]. For Tyr-35 the apparent discrepancy between crystal and solution data might possibly result from an experimental error, since the multiplets of Tyr-35 and Asn-44 overlap in the two-dimensional spectrum (Fig. 1). The side chain of Ser-47 is located on the surface of the molecule in the crystal structure. It is therefore not unexpected that a different spatial arrangement should prevail in solution. On the other hand it is rather interesting that the NMR data indicate that this side chain is locked in an energy minimum centered about a unique dihedral angle χ_0^1, even though it is located on the protein surface. Leu-29 appears in the position (6.5 Hz, 4.0 Hz) in Fig. 3, which could correspond either to a unique angle χ_0^1 or averaging two or more distinct, preferentially populated values of χ_0^1. Since the spin-spin couplings were only slightly influenced by variation of temperature between 30 °C and 85 °C, the latter explanation seems rather unlikely. The data in Table 1 would then indicate a small variation of χ^1 for Leu-29 when going from the crystal to aqueous solution. The close agreement between crystal and solution structure thus found for numerous amino acid side chains depends on the assumption that the stereospecific assignments of the larger and smaller values of the spin-spin coupling constants in solution correspond to those in the crystal structure (Table 1). In view of the close fits obtained, reversed assignments appear unlikely. On this basis the data set of Table 1 thus provides further individual assignments of $^3J_{H^\alpha H^{\beta 2}}$ and $^3J_{H^\alpha H^{\beta 3}}$ for numerous amino acid residues.

Inspection of the crystal structure [1] shows that five of the seven residues observed in the interior areas in Fig. 3 are located on the protein surface, with the side chains pointing away from the protein. These are Asp-3, Arg-17, Lys-26, Gln-31 and Asp-50, for which population of several distinct values of χ^1 is therefore not unexpected. The observed spin-spin couplings might possibly correspond to averages over the classical *gauche-trans-gauche* rotamer configurations. From the location of Tyr-21 in the X-ray structure it would appear that this side chain should be locked into two, or possibily several, distinct orientations of χ^1 by interactions with neighbouring side chains rather than by intra-chain forces favoring the *gauche-trans-gauche* rotamers. Asn-43 presents the most intriguing case in this group, since it is clearly locked into a unique orientation χ_0^1 at 30 °C, with spin-spin couplings of 11.5 and 3.5 Hz, while averaging between several distinct rotamers is indicated at temperatures between 55 °C and 85 °C. In the crystal structure, the side chain of Asn-43 is involved in three hydrogen bonds with the backbone carbonyl oxygens of Glu-7 and Tyr-23 and the amide hydrogen atom of Tyr-23 [1]. The NMR observations thus indicate that while at low temperatures the side chain of Asn-43 is probably locked into a unique hydrogen-bonded conformation, a rapid equilibrium between different states prevails at higher temperatures, whereby this exchange probably involves opening and formation of different types of internal hydrogen bonds.

For Val, Ile and Thr, where only a single coupling constant $^3J_{H^\alpha H^\beta}$ is available, a similar analysis is not possible. However, in those cases where $^3J_{H^\alpha H^\beta}$ is near one of the extreme values [12] it appears that a qualitative comparison with the crystal data is warranted. This situation prevails for Val-34, Ile-18, Ile-19 and Thr-32, where it appears that the side chains are locked in a unique orientation χ_0^1, with the internal mobility restricted to rotational fluctuations about χ_0^1. This conclusion is further supported by the small temperature variations of $^3J_{H^\alpha H^\beta}$, which are $\leq |0.5\ \text{Hz}|$ for all four residues. For Thr-11 and Thr-54 the data are not as clearcut, and a definite statement on the conformation of these two side chains in the solution structure of the inhibitor seems unwarranted at this point.

Overall the present qualitative considerations imply that measurements of vicinal spin-spin coupling constants can provide valuable information on static and dynamic aspects of the spatial arrangements of individual amino acid side chains in globular proteins. Provided that individual resonance assignments have been obtained independently, an interesting basis for comparison of the protein structures in single crystals and in solution can thus be obtained. The present procedures appear particularly promising for studies of surface residues in globular proteins, which are on the one hand most likely to undergo changes in both static and dynamic aspects of conformation when going from the crystal into solution [21, 22] and among which the groups involved in the specific functions of globular proteins are often to be found [23].

We thank Prof. R. R. Ernst for continued advice on all aspects of two-dimensional NMR spectroscopy, Dr G. Wagner for interesting discussions on the conformation of the inhibitor and the Farbenfabriken Bayer AG. (Leverkusen) for a gift of Trasylol®. Financial support by a special grant of the Eidgenössische Technische Hochschule Zürich and the Swiss National Science Foundation (project 3.528.79) is gratefully acknowledged.

REFERENCES

1. Deisenhofer, J. & Steigemann, W. (1975) *Acta Crystallogr. B31*, 238–250.
2. Wagner, G. & Wüthrich, K. (1979) *J. Mol. Biol. 134*, 75–94.
3. Hetzel, R., Wüthrich, K., Deisenhofer, J. & Huber, R. (1976) *Biophys. Struct. Mech. 2*, 159–180.
4. McCammon, J. A., Gelin, B. R. & Karplus, M. (1977) *Nature (Lond.) 267*, 585–590.
5. Nagayama, K. & Wüthrich, K. (1981) *Eur. J. Biochem. 114*, 365–374.
6. Wagern, G., Antil Kumar & Wüthrich, K. (1981) *Eur. J. Biochem. 114*, 375–384.
7. Aue, W. P., Karhan, J. & Ernst, R. R. (1976) *J. Chem. Phys. 54*, 4226–4227.
8. Nagayama, K., Wüthrich, K., Bachmann, P. & Ernst, R. R. (1977) *Biochem. Biophys. Res. Commun. 78*, 99–105.
9. Wüthrich, K., Nagayama, K. & Ernst, R. R. (1979) *Trends Biochem. Sci. 4*, N178–N181.
10. Nagayama, K., Bachmann, P., Wüthrich, K. & Ernst, R. R. (1978) *J. Magn. Reson. 31*, 133–148.
11. Wüthrich, K. (1976) *NMR in Biological Research: Peptides and Proteins*, North-Holland, Amsterdam.
12. De Marco, A., Llinás, M. & Wüthrich, K. (1978) *Biopolymers. 17*, 617–636.
13. IUPAC-IUB Commission on Biochemical Nomenclature (1970) *Eur. J. Biochem. 17*, 193–201.
14. Wüthrich, K. & Wagner, G. (1979) *J. Mol. Biol. 130*, 1–18.
15. Karplus, M. (1959) *J. Chem. Phys. 30*, 11–15.
16. Bystrov, V. F. (1976) *Progr. NMR Spectrosc. 10*, 41–81.
17. Pachler, K. G. R. (1964) *Spectrochim. Acta, 20*, 581–587.
18. Gelin, B. R. & Karplus, M. (1979) *Biochemistry, 18*, 1256–1268.
19. Bundi, A. & Wüthrich, K. (1979) *Biopolymers, 8*, 285–297.
20. Perkins, S. J. & Wüthrich, K. (1979) *Biochim. Biophys. Acta, 576*, 409–423.
21. Brown, L. R., De Marco, A., Wagner, G. & Wüthrich, K. (1976) *Eur. J. Biochem. 62*, 103–107.
22. Brown, L. R., De Marco, A., Richarz, R., Wagner, G. & Wüthrich, K. (1978) *Eur. J. Biochem. 88*, 87–95.
23. Schulz, G. E. & Schirmer, R. H. (1979) *Principles of Protein Structure*, Springer-Verlag, Berlin, Heidelberg, New York.

K. Wüthrich, Institut für Molekularbiologie und Biophysik der Eidgenössischen Technischen Hochschule Zürich, Zürich-Hönggerberg, CH-8093 Zürich, Switzerland

K. Nagayama, Department of Physics, Faculty of Science, University of Tokyo, Hongo, Bunkyo-ku, Tokyo, Japan 113

JOURNAL OF MAGNETIC RESONANCE **24**, 201–204 (1976)

Digital Filtering with a Sinusoidal Window Function: An Alternative Technique for Resolution Enhancement in FT NMR

ANTONIO DE MARCO AND KURT WÜTHRICH

Institut für Molekularbiologie und Biophysik, Eidgenössische Technische Hochschule, 8093 Zürich–Hönggerberg, Switzerland

Received February 13, 1976

As an alternative to convolution difference techniques for resolution enhancement in the ^{1}H NMR spectra of proteins, digital filtering of the FID with a sinusoidal window function is suggested. As an illustration, this "sine bell routine" was applied to the high-field region of the ^{1}H NMR spectrum at 360 MHz of the basic pancreatic trypsin inhibitor.

Even with the highest currently available magnetic fields, the spectral resolution in the ^{1}H NMR spectra of proteins and other macromolecules is limited by the mutual overlap of the resonance lines of individual groups of protons (Fig. 1A). In FT NMR,

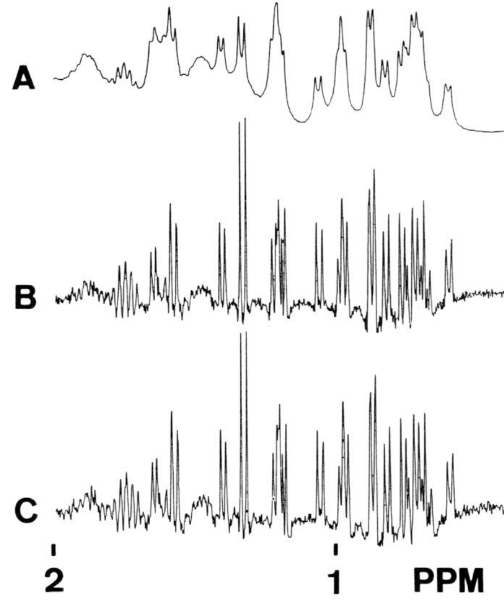

FIG. 1. Aliphatic region of the FT ^{1}H NMR spectrum at 360 MHz of the basic pancreatic trypsin inhibitor, 0.01 M solution in D_2O, 225 scans, 4000 Hz spectral width, 2 sec acquisition time, 16K of memory in the time domain. (A) Normal spectrum. The natural linewidth $\Delta\omega_{1/2}$ is approximately 20 rad sec^{-1}. (B) Same spectrum as (A) after digital filtering of the FID with the sine bell routine. (C) Same spectrum as (A) after application of the convolution difference routine with $\tau_1 = \infty$, $\tau_2 = 0.31$ sec, $K = 1$.

Copyright © 1976 by Academic Press, Inc.
All rights of reproduction in any form reserved.
Printed in Great Britain

an artificial resolution enhancement can be obtained by suitable manipulation of the FID (*1*); several resolution-enhancement routines were proposed in the past (*2–4*). For work with proteins, convolution difference techniques (Fig. 1C) have so far generally been preferred (*5–9*). The present paper suggests digital filtering of the FID with a sinusoidal window function as an alternative technique, which yields a comparable resolution enhancement with comparable or lesser side effects of lineshape distortion and reduced signal:noise ratio (Fig. 1B). This "sine bell" routine seems attractive because of the simplicity of its practical applications.

In the sine bell routine, the free induction decay is multiplied by a sinusoidal function with zero phase and a period of twice the acquisition time AT. For this operation, we used the "Hanning Window" from the Lab-1180 General Signal Averaging Package, Nicolet Instrument Corporation. The lineshape after this procedure is given by

$$L(\mathrm{SIN}) \propto \int_0^{AT} \exp(-t/T_2^*)\cos(\Delta\omega t)\sin(\pi t/AT)\,dt, \qquad [1]$$

where T_2^* is the characteristic time for the decay of the transverse magnetization and $\Delta\omega$ the difference between the frequency considered and the resonance frequency. It is readily apparent that analogous to the convolution difference method, the broad components are drastically reduced compared to the sharp ones. Moreover, there should be essentially no truncation effects since the free induction decay is forced to be zero at time AT. Neglecting terms in $\exp(-AT/T_2^*)$, lineshape function [1] becomes

$$L(\mathrm{SIN}) \propto \left(\frac{\pi T_2^{*2}}{AT}\right)\frac{1+(\pi T_2^*/AT)^2-\Delta\omega^2 T_2^{*2}}{[1+(\pi T_2^*/AT)^2-\Delta\omega^2 T_2^{*2}]^2+4\Delta\omega^2 T_2^{*2}}. \qquad [2]$$

In the following, lineshape function [2] is compared with the Lorentzian line, with the lineshape obtained in convolution difference spectra. This comparison will mainly focus on the linewidth and the signal intensity I_0 at the resonance frequency, and the line distortion will also be considered. For practical reasons we take the width $\Delta\omega_{1/2}^0$ of the distorted lines obtained from the sine bell routine and the convolution difference routine as one-half of the half-width $\Delta\omega^0$ at the intensity zero (Fig. 2); the distortion D is defined as $|d/I_0|$ (Fig. 2). The resolution enhancement RE with respect to the Lorentzian is defined as $1/\Delta\omega_{1/2}^0 \cdot T_2^*$. Comparing the sine bell lineshape with the Lorentzian, we have from Eq. [2],

$$RE(\mathrm{SIN}) = \frac{2}{[1+(\pi T_2^*/AT)^2]^{1/2}}, \qquad [3]$$

$$\frac{I_0(\mathrm{SIN})}{I_0(\mathrm{LOR})} = \frac{\pi T^*/AT}{1+(\pi T_2^*/AT)^2}, \qquad [4]$$

$$D(\mathrm{SIN}) = \frac{1}{4}\frac{1+(\pi T_2^*/AT)^2}{1+[1+(\pi T_2^*/AT)^2]^{1/2}}. \qquad [5]$$

As shown in Fig. 3, RE(SIN) reaches a plateau close to 2 for relatively small values of AT/T_2^*, where the distortion is already close to its minimum value of $\frac{1}{8}$, but the relative intensity is still in a favorable range.

In the convolution difference method (*2*), the three parameters T_A, T_B, and K, where

$1/T_A = (1/T_2^*) + (1/\tau_1)$ and $1/T_B = (1/T_2^*) + (1/\tau_2)$, can be adjusted to obtain a suitable compromise of optimal resolution enhancement with acceptable signal intensity and line distortion. To obtain a meaningful comparison with the sine bell routine, we selected T_A and K so that the RE(CD) was at a maximum i.e., $T_A = T_2^*$ and $K = 1$, and then

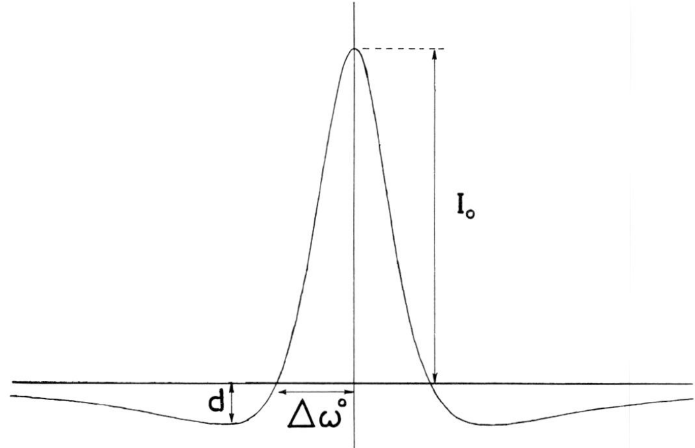

FIG. 2. Definition of the signal intensity I, the linewidth $\Delta\omega^0$, and the distortion $D = |I_0/d|$ for the resonance lines obtained with the resolution-enhancement routines.

adjusted τ_2 so that $I_0(\text{CD}) = I_0(\text{SIN})$. Assuming that $T_2^* = 0.05$ sec, which is quite typical for ^{1}H NMR spectra of proteins, we found that for an acquisition time $AT = 1$ sec, $RE(\text{SIN})/RE(\text{CD}) = 1.08$; with $AT = 2$ sec, $RE(\text{SIN})/RE(\text{CD}) = 1.04$. In both examples, the line distortions for the two routines were approximately the same and equal to ≈ 0.12. As an illustration, the two routines were applied to the aliphatic region of the basic pancreatic trypsin inhibitor, which is a "miniprotein" with molecular weight 6500 (*8*) (Fig. 1).

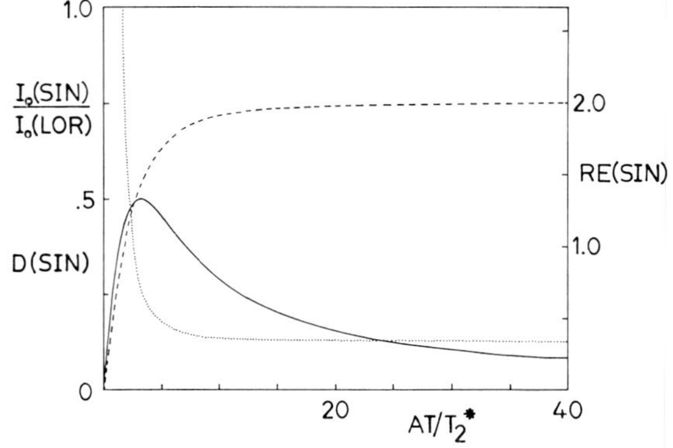

FIG. 3. Plots versus AT/T_2^* of the lineshape parameters resulting from the application of the sine bell routine. ——, intensity at centerband I_0(SIN) relative to the intensity in the Lorentzian lineshape I_0(LOR); -----, resolution-enhancement factor RE(SIN) relative to the Lorentzian lineshape; ....., distortion D(SIN) relative to the intensity at centerband.

The foregoing considerations show that for work with proteins the sine bell routine [1], which does not require selection of any arbitrary parameters besides the experimental acquisition time AT, yields a somewhat improved resolution enhancement as compared to the convolution difference routine applied with optimal choice of the parameters T_A, T_B, and K. For the sake of completeness it may be added that the sine bell routine can also be used in a form which includes an arbitrary parameter $K_S \geqslant 0$ corresponding essentially to the parameter $K \leqslant 1$ in the convolution difference method. The resulting lineshape is given by

$$L(\text{SIN}) \propto \int_0^{AT} \exp(-t/T_2^*) \cos(\Delta\omega t)[K_S + \sin(\pi t/AT)]\, dt. \qquad [6]$$

When this appears desirable, e.g., when working with very dilute solutions or with rare nuclei, K_S can be adjusted to obtain a suitable compromise of acceptable signal: noise ratio and resolution enhancement. On the other hand, since the characteristics of the sine bell window tend to suppress truncation effects, no undue line distortions result when the FID is deconvoluted by multiplication with $e^{t/\tau}$ prior to the application of the sine bell routine. *RE* greater than 2 can thus in practice be obtained by successive applications of the two routines.

The availability of convoluted spectra (Figs. 1B and 1C) in addition to the normal spectrum (Fig. 1A) is of considerable practical interest for studies of macromolecular systems. Since the broad components of the spectrum are eliminated in the altered data (Eq. [1]), the sharp lines are strongly emphasized. As a consequence the multiplet patterns of the narrow components are more readily recognized; spin decoupling thus becomes a more reliable technique for identification of the constituent spin systems, e.g., of individual amino acid residues in proteins (*5–8*). The convoluted spectra are also more readily amenable for a systematic determination of chemical shifts and spin–spin coupling constants. In certain cases it may further be of interest to determine the numbers of nuclei giving rise to broad and narrow resonances, respectively. In the normal spectrum real broad resonances cannot readily be distinguished from broad lines arising as a consequence of mutual overlap of many sharp lines. Comparison with the altered spectra, however, can in general provide this information.

ACKNOWLEDGMENT

Financial support by the Schweizerischer Nationalfonds (Project 3.1510.73) and by the Italian N.R.C. (stipend to A.DeN.) is gratefully acknowledged.

REFERENCES

1. R. R. Ernst, *Advan. Magn. Resonance* **2,** 1 (1966).
2. I. D. Campbell, C. M. Dobson, R. J. P. Williams, and A. V. Xavier, *J. Magn. Resonance* **11,** 172 (1973).
3. W. B. Moniz, C. F. Poranski, Jr., and S. A. Sojka, *J. Magn. Resonance* **13,** 110 (1974).
4. N. N. Semendyaev, *Stud. Biophys.* **47,** 151 (1974).
5. I. D. Campbell, S. Lindskog, and A. I. White, *J. Mol. Biol.* **90,** 469 (1974).
6. I. D. Campbell, C. M. Dobson, G. Jeminét, and R. J. P. Williams, *FEBS Lett.* **49,** 115 (1974).
7. L. R. Brown, A. De Marco, G. Wagner, and K. Wüthrich, *Eur. J. Biochem.* **62,** 103 (1976).
8. G. Wagner, A. DeMarco, and K. Wüthrich, *J. Magn. Resonance* **20,** 565 (1975).
9. E. Oldfield, R. S. Norton, and A. Allerhand, *J. Biol. Chem.* **250,** 6381 (1975).

Vol. 78, No. 1, 1977 BIOCHEMICAL AND BIOPHYSICAL RESEARCH COMMUNICATIONS

TWO-DIMENSIONAL J-RESOLVED ^{1}H n.m.r. SPECTROSCOPY FOR STUDIES OF BIOLOGICAL MACROMOLECULES

K. Nagayama, K. Wüthrich, P. Bachmann and R.R. Ernst, Institut für Molekularbiologie und Biophysik and Laboratorium für Physikalische Chemie, Eidgenössische Technische Hochschule, CH-8093 Zürich, Switzerland

Received July 14, 1977

SUMMARY: The technique of two-dimensional J-resolved ^{1}H n.m.r. spectroscopy has been extended to handle the very wide spectra of proteins and other macromolecules at 360 MHz. The potential of the method to resolve and assign individual spin multiplets in the complex spectra encountered in structural studies of biopolymers is illustrated with some experiments with amino acids and with a protein, the basic pancreatic trypsin inhibitor.

The potential of high resolution nuclear magnetic resonance (n.m.r.) for structural studies in biological systems depends critically on the ability to resolve and assign individual spin multiplets in the inherently complex spectra of biological macromolecules (1). In addition to using high polarizing magnetic fields, suitable digital filtering techniques (2-4) and the use of spin echo experiments (5) have recently resulted in markedly improved spectral resolution for conventional ^{1}H n.m.r. spectra of proteins. Here, a new concept is applied which promises to greatly enhance the resolution in the ^{1}H n.m.r. spectra of biopolymers and to facilitate assignment of individual spectral components, i.e. two-dimensional J-resolved ^{1}H n.m.r. spectroscopy (6-9) at high field.

The present paper describes for the first time two-dimensional high field ^{1}H n.m.r. experiments at 360 MHz. With the versatile software written to handle the large data matrices required for studies of macromolecules, two-dimensional ^{1}H n.m.r. can now be applied to greatly simplify the interpretation of the spectra of biopolymers. The initial experiments with aqueous solutions of amino acids and proteins clearly illustrate the potential of the method for resolving and assigning individual spin multiplets in complex spectra of biological materials, where extensive overlap of resonance lines is typically observed in conventional one-dimensional spectra.

MATERIALS AND METHODS: The basic method of 2-dimensional (2 D) J-resolved n.m.r. has been described in Ref. (6). Its principles shall briefly be summarized in this section. The technique is an example of the general class of 2 D resolved spectroscopy techniques (6-9); in this case, the multiplet splitting is utilized to obtain a 2 D spread of the proton magnetic resonance spectrum.

Copyright © 1977 by Academic Press, Inc.
All rights of reproduction in any form reserved.

ISSN 0006-291X

The technique requires the performance of N two-pulse spin echo experiments of the type $90^o - k\tau - 180^o - k\tau$, with k = 0, 1, ...(N-1). Due to the refocusing effect of the 180° pulse, an echo appears with its maximum at time $k\tau$ after the 180° pulse. Of the second half of each echo, M equidistant sample values are recorded, defining the M by N data matrix ($S_{k\ell}$) k = 0,...(N-1); ℓ = 0,...(M-1). Each of the N echo traces (corresponding to a row of the data matrix) is then Fourier-transformed to separate the contributions originating from the various resonance lines.

It is well-known that for weak spin-spin coupling the echo amplitude is independent of the chemical shift. The echo amplitude and correspondingly the separated signal amplitudes are thus modulated exclusively by the multiplet splittings. This modulation is utilized to differentiate between the resonance lines with different multiplet splittings by means of a second Fourier transformation, this time transforming the columns of the data matrix. This produces, finally, a 2 D spectrum, $S(\omega_1,\omega_2)$. The coordinates of each signal peak in the 2 D spectrum are given in the ω_2-direction by the resonance frequency of the conventional one-dimensional spectrum and in the ω_1-direction by the corresponding multiplet splitting. The resonance intensities are equal to those in the one-dimensional spectrum.

Applications of 2 D spectroscopy to protein n.m.r. require the handling of particularly large data matrices. In the present configuration, our equipment readily handles up to 10^6 data points in the time domain. The spectra in Figs. 2 and 3 were computed from a data matrix with N = 128 and M = 8192, that in Fig. 4 with N = 64 and M = 8192. Additional experimental details are given in the figure captions.

RESULTS: To explain some principal features of the high field 2 D ^1H n.m.r. spectra, we first consider the resonances observed in a D_2O solution of an equimolar mixture of the five amino acids alanine, isoleucine, methionine, tyrosine and histidine. For this sample the resonances of the aromatic protons of His and Tyr are already well resolved in the region from 6 to 8 ppm of the conventional 360 MHz spectrum (Fig. 1 A). In the 2 D spectrum, since the two singlet resonances of His are not modulated by a J coupling, these lines are not affected by the 2 D spread in the ω_1-direction and hence appear with their respective chemical shifts on the $\omega_1 = 0$ line (Fig. 2). For the doublets of Tyr, the modulation results, after Fourier transformation, in displacements in the ω_1-direction by $\pm \pi J_{AB}$ rad/sec relative to the $\omega_1 = 0$ line.

In the high field spectral region, Met gives rise to a singlet methyl resonance at 2.1 ppm (Fig. 1 A) which is located on the $\omega_1 = 0$ line in the 2 D spectrum (Fig. 2). The two component lines of the methyl doublet resonance of Ala at 1.2 ppm are displaced in the ω_1-direction, analogous to the doublets of Tyr. Ile gives rise to a doublet and a triplet methyl resonance which partially overlap at 0.9 ppm in the conventional spectrum (Fig. 1 A). In the 2 D spectrum these two multiplets are fully resolved and complete assignments of the two multiplets are readily obtained. The appearance of the doublet corresponds to

Vol. 78, No. 1, 1977 BIOCHEMICAL AND BIOPHYSICAL RESEARCH COMMUNICATIONS

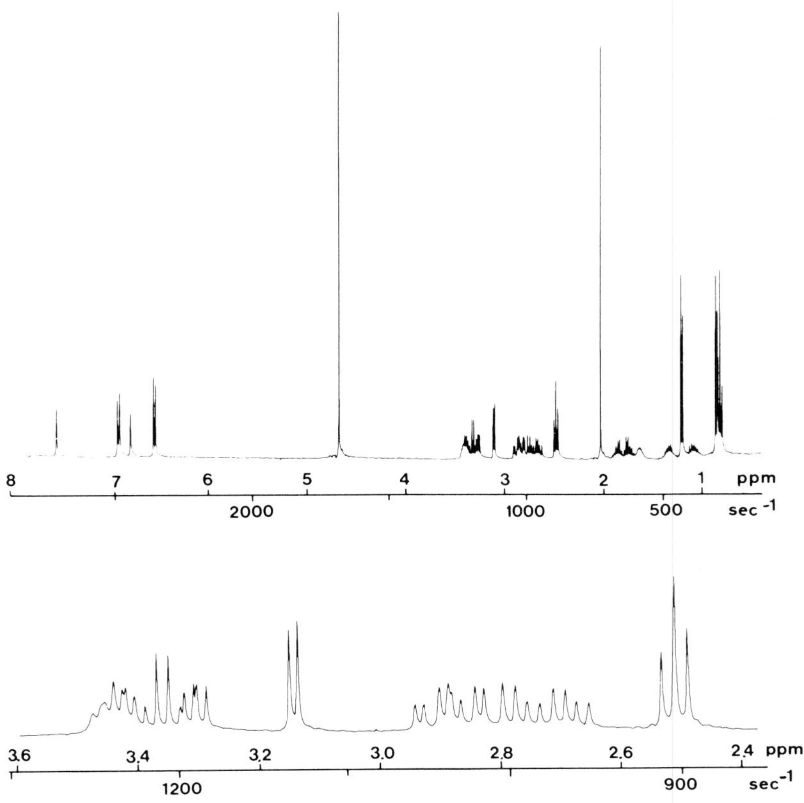

Fig. 1 A. Fourier transform ^{1}H n.m.r. spectrum at 360 MHz of a D_2O solution containing 0.1-M of each of the five amino acids Ala, Ile, Met, Tyr and His, pD = 10.5, T = 25°. Chemical shifts relative to internal 2,2-dimethyl-2-silapentane-5-sulfonate (DSS) are indicated both in Hertz (sec^{-1}) and in parts per million (ppm). The spectrum was computed from 8192 data points. B. Expanded representation of the spectral region from 2.4 to 3.6 ppm.

that described above for Tyr and Ala; for the triplet, the center line is at $\omega_1 = 0$ and the peripheral lines are displaced by $\pm 2\pi J$ rad/sec in the ω_1 direction. It is essential to recognize that, as a result of the spread in a second dimension, all lines belonging to a particular multiplet will be alined on a straight line which forms an angle of 45° with the ω_2-axis and intersects the $\omega_1 = 0$ line at the chemical shift of the proton considered. This is not well visible from Fig. 2 as the scales of the two axes differ by two orders of

Vol. 78, No. 1, 1977 BIOCHEMICAL AND BIOPHYSICAL RESEARCH COMMUNICATIONS

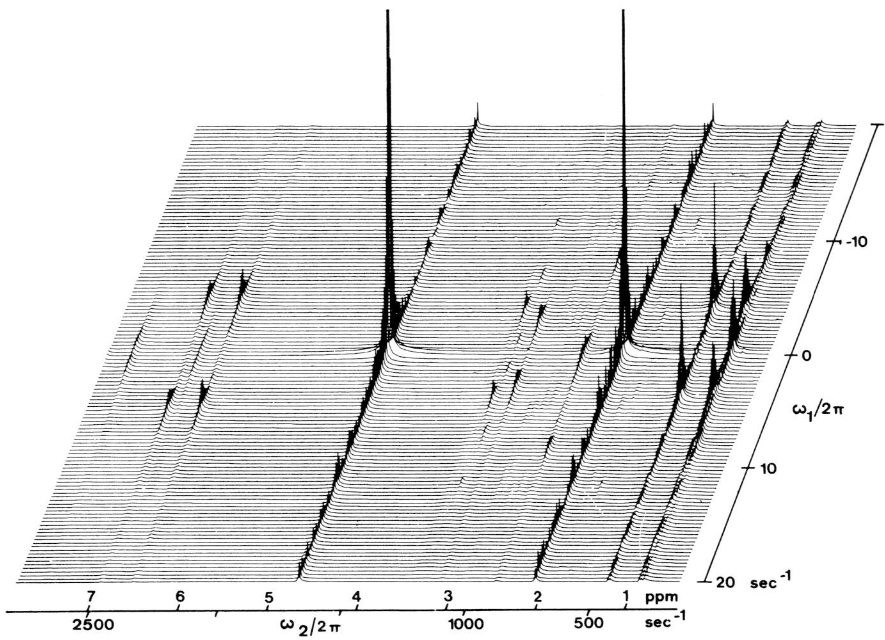

Fig. 2 Two-dimensional J-resolved 360 MHz ^1H n.m.r. spectrum of the amino acid mixture of Fig. 1. An absolute value spectrum is shown. The spectrum was computed from 128 x 8192 data points corresponding to 128 single echoes represented by 8192 sample values. The two frequency axes are calibrated in Hz, for the ω_2-axis the corresponding ppm scale relative to DSS is also indicated.

magnitude. The expanded spectra of Figs. 3 and 4 provide a somewhat better demonstration of this basic feature of 2 D J-resolved spectra.

A nice illustration of the power of 2 D NMR to resolve and assign individual spin multiplets is provided by the spectral region from 2.4 to 3.6 ppm in Fig. 1 A, which is represented on an expanded scale in Fig. 1 B. This region contains 10 spin multiplets, some of which mutually overlap in the one-dimensional spectrum (Fig. 1 B). In the expanded representation of the corresponding region of the 2 D spectrum in Fig. 3, all the resonance lines are well separated and the spin multiplets can readily be assigned. In the order of increasing chemical shift in this spectral region, we first have a two-proton triplet of the γ-methylene protons of Met (1). Next, there are four one-proton multiplets of the β-methylene protons of Tyr and His, each of which corresponds to a doub-

Vol. 78, No. 1, 1977 BIOCHEMICAL AND BIOPHYSICAL RESEARCH COMMUNICATIONS

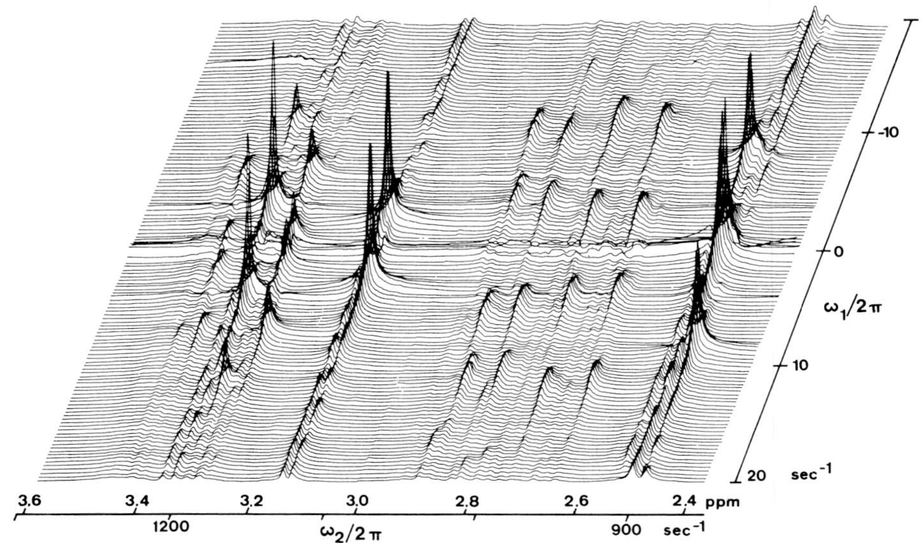

Fig. 3 Expanded representation of the spectral region from 2.4 to 3.6 ppm in the two-dimensional spectrum of Fig. 2, corresponding to the 1-dimensional expanded spectrum of Fig. 1 B.

let of doublets. The α-proton doublet resonance of Ile was already well resolved in the conventional spectrum. The remaining four α-proton resonances are a doublet of doublets for Met, a quartet for Ala, a doublet of doublets for Tyr and a doublet of doublets consisting of rather broad lines for His.

Fig. 4 shows the high field region from 0.5 to 1.7 ppm of the 1 D and 2 D spectra of the basic pancreatic trypsin inhibitor (BPTI). This protein consists of one polypeptide chain with 58 amino acid residues, including 6 Ala, 1 Val, 2 Leu, 2 Ile and 3 Thr, which give rise to 19 methyl doublet and triplet resonances at high field. These methyl resonances had previously been assigned to the different types of aliphatic amino acid side chains (10). In the 2 D spectrum, the components of the individual multiplets, which are partially overlapped in the 1 D spectrum, can readily be assigned. Between 1.61 and 1.04 ppm there are eight doublet resonances 2 - 9 corresponding to Ala and Thr. It is worth noting that the narrow lines of resonance 5, which corresponds to the C-terminal Ala 58 (10), are very prominent after digital resolution enhancement in both the 1 D and 2 D spectrum. Between 1.0 and 0.71 ppm, there are the eight doublets 10 - 17 of Val, Leu and Ile. In the presentation of Fig. 4, the pairs

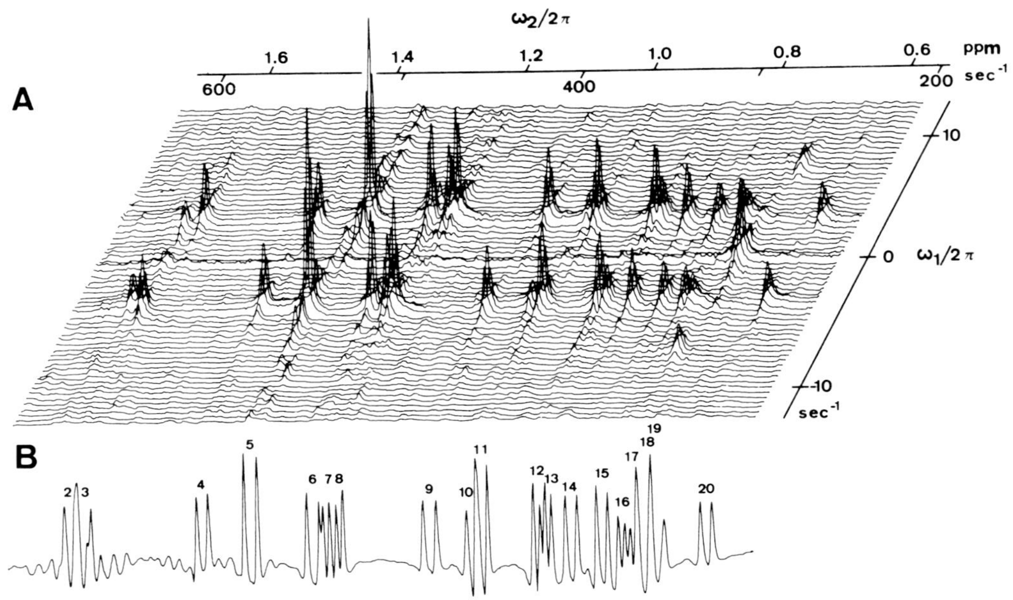

Fig. 4 A. Two-dimensional J-resolved 360 MHz ^{1}H n.m.r. spectrum of a 0.01-M solution of the basic pancreatic trypsin inhibitor in D_2O, pD = 4.5, T = 60°. The figure shows the expanded region from 0.5 to 1.7 ppm of a spectrum computed from 64 x 8192 data points. Prior to the Fourier transformations the FID's were multiplied with an increasing exponential to reduce the line widths (1,2). B. One-dimensional ^{1}H n.m.r. spectrum obtained after digital filtering with the sine bell routine (4). The individual methyl resonances are indicated by the numbers 2 - 20 (10).

of neighboring doublets 10 and 11, 12 and 13, and 16 and 17 appear partially overlapped, but can be well separated in a more expanded plot. The two triplet resonances of Ile, 18 and 19, have nearly identical chemical shifts (10) and appear strongly overlapped in Fig. 4, while the doublet 20 at the high field end, which comes from Ala or Thr (10) is well separated at 0.59 ppm. It is apparent also in this figure that the individual component lines of the multiplets can readily be assigned and the coupling constants J obtained by inspection of the 2 D spectrum.

DISCUSSION: Two-dimensional J-resolved ^{1}H n.m.r. spectroscopy is a very general technique to unravel complicated n.m.r. spectra. Conceptually, it appeared particularly suited for the simplification of the analysis of high field ^{1}H n.m.r.

Vol. 78, No. 1, 1977 BIOCHEMICAL AND BIOPHYSICAL RESEARCH COMMUNICATIONS

spectra of proteins. This is to a large extent born out by the results presented in this paper. These initial experiments indicate that the utility of n.m.r. for studies of biopolymers will be greatly enhanced by the use of 2 D techniques.

The following features of 2 D J-resolved n.m.r. spectra appear to be of particular relevance for biological applications: (i) As mentioned in Section 3, multiplet lines in a 2 D spectrum are alined along straight lines with 45° slopes relative to the ω_2 axis. It is, therefore, possible to project the 2 D spectrum along this distinguished direction to eliminate the multiplet splitting and, effectively, to obtain completely homonuclear decoupled ^{1}H n.m.r. spectra (6). Similarly, spectra with reduced multiplet separations, corresponding in appearance to the residual couplings obtained in off-resonance $\{^{1}H\}^{13}C$ n.m.r. (1), may be obtained. (ii) 2 D experiments, in general, are time-consuming. It should be emphasized, however, that the additional time required reflects itself in an enhanced sensitivity. Indeed, within a given performance time it is possible to obtain a 2 D spectrum with almost the same sensitivity as that achievable for a one-dimensional spectrum (8). (iii) Since the appearance of simple features in the 2 D J-resolved n.m.r. spectra depends on the weak coupling assumption, it is essential to combine the use of this technique with the application of the highest available magnetic field strength. For strong coupling, new characteristic lines will appear in addition to the higher order features already observed in the conventional spectra (7), which can complicate the analysis of the 2 D spectra.

ACKNOWLEDGEMENTS: Research grants of the ETH and the Swiss National Science Foundation are gratefully acknowledged.

REFERENCES:

1. Wüthrich, K. (1976) NMR in Biological Research: Peptides and Proteins. North Holland, Amsterdam.
2. Ernst, R.R. (1966) Advan.Magn.Reson. 2, 1-135.
3. Campbell, I.D., Dobson, C.M., Williams, R.J.P. and Xavier, A.V. (1973) J. Magn.Reson. 13, 172-181.
4. De Marco, A. and Wüthrich, K. (1976) J.Magn.Reson. 24, 201-204.
5. Campbell, I.D., Dobson, C.M., Williams, R.J.P. and Wright,P.E.(1975) FEBS Lett. 57, 96-99.
6. Aue, W.P., Karhan, J. and Ernst, R.R. (1976) J.Chem.Phys. 64, 4226-4227.
7. Aue, W.P., Bartholdi, E. and Ernst, R.R. (1976) J.Chem.Phys. 64, 2229-2246.
8. Kumar, A., Aue, W.P., Bachmann, P., Karhan, J., Müller, L. and Ernst, R.R. (1976) Proc. XIXth Congrès Ampère, Heidelberg, 473-478.
9. Ernst, R.R., Aue, W.P., Bachmann, P., Karhan, J., Kumar, A. and Müller, L. Ampère Summer School IV (Pula, Yugoslavia, 1976) (in press).
10. De Marco, A., Tschesche, H., Wagner, G. and Wüthrich, K. (1977) Biophys. Structure and Mechanism (in press).

Vol. 90, No. 1, 1979 BIOCHEMICAL AND BIOPHYSICAL RESEARCH COMMUNICATIONS
September 12, 1979 Pages 305-311

TWO-DIMENSIONAL SPIN ECHO CORRELATED SPECTROSCOPY (SECSY) FOR ^{1}H NMR STUDIES OF BIOLOGICAL MACROMOLECULES

K. Nagayama*, K. Wüthrich* and R.R. Ernst**, *Institut für Molekularbiologie und Biophysik and **Laboratorium für Physikalische Chemie, Eidgenössische Technische Hochschule, CH-8093 Zürich, Switzerland

Received July 28,1979

SUMMARY: The principle and the experimental realization of spin echo correlated spectroscopy (SECSY) are described and its use for studies of proteins is illustrated with ^{1}H n.m.r. spectra of the basic pancreatic trypsin inhibitor. This technique yields two-dimensional homonuclear correlated spectra which manifest connectivities between spin coupled nuclei and thus provide first level assignments of individual spin systems in biopolymers. Compared to previously described two-dimensional correlated n.m.r. techniques, which were applied exclusively to small molecules, SECSY uses a smaller data size both in the time domain and the frequency domain, which makes it particularly suitable for studies of macromolecules.

One of the fundamental steps in the analysis of complex n.m.r. spectra of biological macromolecules is the identification of the spin systems of individual structural fragments by investigation of connectivities between different components of the spin systems (1). Even though conventional n.m.r. methods include experiments for delineating these connectivities and very high magnetic fields are now available, n.m.r. studies of biopolymer conformation still depend critically on the ability to resolve and assign numerous resonance multiplets in the crowded ^{1}H n.m.r. spectra of the macromolecules. Recently we have demonstrated that greatly improved resolution can be obtained in two-dimensional (2D) J-resolved ^{1}H n.m.r. spectra of proteins at high field (2,3), and that selective spin decoupling in the 2D J-resolved spectra allows to use the improved spectral resolution for the identification of entire spin systems (4,5). Here we describe how numerous individual spin systems in complex spectra can be identified at once by the application of an experimental scheme which provides homonuclear 2D correlated ^{1}H n.m.r. spectra.

It was shown previously that 2D correlated n.m.r. spectra provide a direct manifestation of the connectivities between spin-coupled resonances (6). These techniques have, however, so far been used exclusively for studies of small molecules. Extension for the use with macromolecular systems requires special provisions for handling the large data matrices in the time domain and frequency domain. Spin echo correlated spectroscopy (SECSY) uses a

0006-291X/79/170305-07$01.00/0
Copyright © 1979 by Academic Press, Inc.
All rights of reproduction in any form reserved.

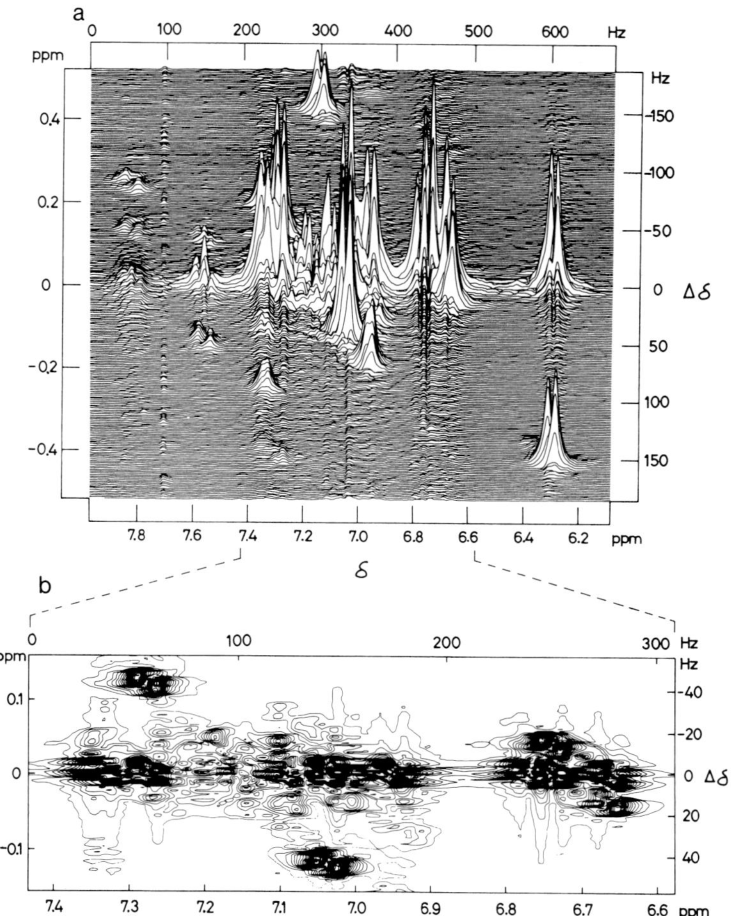

Fig. 1 Aromatic region from 6-8 ppm of the spin echo correlated (SECSY) spectrum recorded in a 0.01 M solution of BPTI in ^2H_2O, p^2H=4.4, T=70° C. (a) Three-dimensional presentation of the entire aromatic region. The chemical shift δ on the horizontal axis corresponds to that in conventional 1-dimensional spectra. Δδ on the vertical axis indicates the difference frequencies of coupled nuclei resulting in a pair of cross peaks at ±Δδ. For convenience both axes have also been calibrated in Hz. (b) Contour plot of the spectral region from 6.6 to 7.4 ppm. In this presentation contour lines connect points of equal signal amplitude.

smaller data size than the previously described experiments (6) and yields a novel presentation of the frequency domain spectra which is more suitable for work with large, complex systems. In the following we first demonstrate the use of SECSY with a protein and describe the analysis of the 2D correlated spectra thus obtained. The principles of the technique will then be briefly discussed and compared with previously proposed experimental schemes for recording 2D n.m.r. spectra.

APPLICATION TO THE ASSIGNMENT OF SPIN SYSTEMS IN A PROTEIN: Fig. 1 shows the aromatic region of the spin echo correlated ^1H n.m.r. spectrum of the basic pancreatic trypsin inhibitor (BPTI), a small globular protein with molecular weight 6'500. BPTI contains eight aromatic amino acid residues, i.e. four tyrosines and four phenylalanines, with a total of 36 aromatic protons. The spin systems of the eight aromatic rings were previously identified (7).

To gain access to the information contained in Fig. 1 it may be useful to first consider the schematic SECSY spectrum of an AX spin system in Fig. 2c'. The one-dimensional four-line AX-spectrum is shown at the bottom of Fig. 2c'. A four-line pattern corresponding to the one-dimensional spectrum is contained on the $\omega_1'=0$ line in the 2D spectrum as well. In addition, each of the four component lines appears a second time shifted parallel to the ω_1'-axis by $+\frac{1}{2}J$ or $-\frac{1}{2}J$, so that for each of the two doublets, these two additional peaks define a line which forms an angle of 135° with the ω_2'-axis. This results in two four-peak patterns centered on the $\omega_1'=0$ line. Two identical four-peak patterns appear also in positions resulting from shifts of $+\frac{1}{2}\delta_{AX}$ and $-\frac{1}{2}\delta_{AX}$ along the ω_1'-axis. The centers of these two shifted four-peak patterns again lie on a line which forms an angle of 135° with the ω_2'-axis. Fig. 2c' shows that a SECSY spectrum manifests simultaneously the chemical shifts, the spin-spin coupling constants and the connectivities between coupled resonances.

For the analysis of Fig. 1 we shall concentrate on the connectivities between coupled resonances. From Fig. 2c' we know that the cross-peaks indicating coupling between two resonances A and X are displaced from the $\Delta\delta=0$ line by $\Delta\delta = \pm\frac{1}{2}(\delta_A-\delta_X)$ (note that ω_2' and ω_1' in Fig. 2c' have been substituted by δ and $\Delta\delta$ in Fig. 1). Inspection of Fig. 1a shows that there is a cross peak pattern at δ=6.30 ppm and $\Delta\delta$=152 Hz and a corresponding pattern at δ=7.14 ppm and $\Delta\delta$=-152 Hz. The two patterns, which lie on a 135° line, correspond to the AA'XX' spin system of Tyr-23 (7). Moving from the periphery towards the $\Delta\delta$=0 line we then have a peak at $\Delta\delta$=86 Hz and δ=7.34 ppm and the correlated multiplet at $\Delta\delta$=-86 Hz and δ=7.82 ppm. At δ=7.82 ppm there is also a peak with $\Delta\delta$=-40 Hz which is correlated with a line at δ=7.56 ppm and $\Delta\delta$=40 Hz. These peaks are the AA'MXX' spin system of Phe-45 (7). Here it is particularly striking that the connectivities of the 3,5-ring proton reson-

Vol. 90, No. 1, 1979 BIOCHEMICAL AND BIOPHYSICAL RESEARCH COMMUNICATIONS

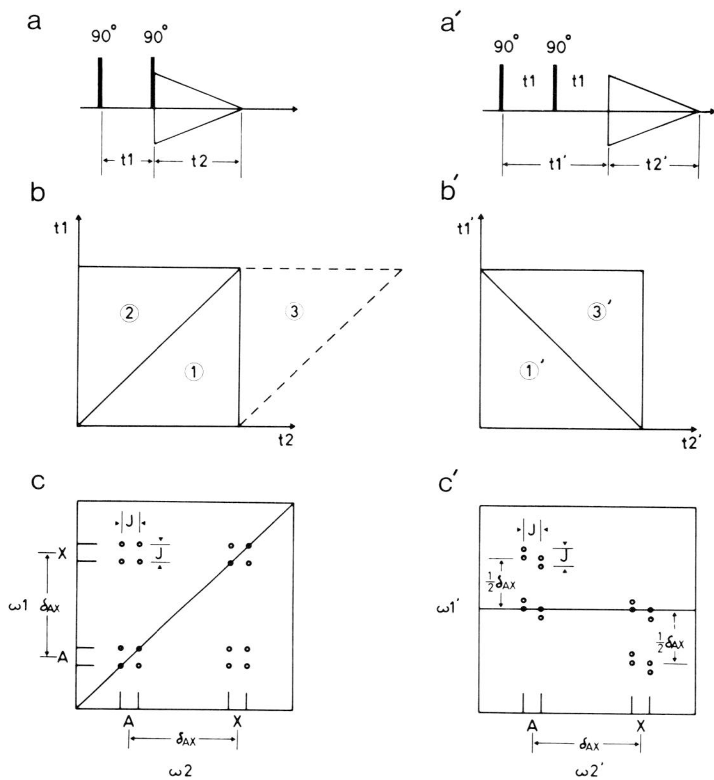

Fig. 2 Comparison of 2D correlated (a,b,c) and 2D spin echo correlated spectroscopy (SECSY) (a',b',c'). The experimental schemes are indicated by the diagrams a and a'. The relations between the recorded data sets $s(t_1,t_2)$ and $s'(t_1',t_2')$ are visualized in the diagrams b and b', where (1) and (1'), and (3) and (3') contain identical data values. The resulting 2D spectra are shown in the diagrams c and c'. They are related by a conformal mapping.

ance at 7.82 ppm with the 4-proton line at 7.56 and the 2,6-proton peak at 7.34 ppm are separately manifested. Next we have a doublet peak at $\Delta\delta$=68 Hz and δ=6.95 ppm which is correlated with a triplet at δ=7.33 ppm and $\Delta\delta$=-68 Hz. These peaks correspond to the two two-proton multiplets of the AA'MM'X spin system of Phe-4.

To analyse the crowded spectral region near the $\Delta\delta$=0 line the contour plot of Fig. 1b provides a more suitable presentation. Two correlated doublets are readily seen at δ=7.04 and 7.28 ppm, with $\Delta\delta$=43 Hz and -43 Hz, respectively. These correspond to the AA'XX' spin system of Tyr-10 (7). Similarly the correlation between the two doublets of the AA'XX' spin system of Tyr-21

(7) at δ=6.67 and 6.74 ppm, with $\Delta\delta$=13 Hz and -13 Hz, respectively, is clearly evidenced.

PRINCIPLES OF 2D SPIN ECHO CORRELATED SPECTROSCOPY: The basic scheme of the experiment is shown in Fig. 2a'. A first 90° pulse creates transverse magnetization which freely precesses for a time $t_1'/2$. Then, a second 90° pulse is applied which initiates the refocusing of the transverse magnetization components. At the peak of the echo at $t=t_1'$, the data acquisition is started, recording the signal $s'(t_1',t_2')$ as a function of t_2'. In accordance with the general 2D spectroscopy principle (6), the experiment is repeated for a set of equidistant t_1' values. A two-dimensional Fourier transformation of $s'(t_1',t_2')$, finally, produces the desired frequency domain spectrum.

SECSY forms an intermediate case between the previously described 2D correlated spectroscopy (6) (Fig. 2a) and 2D J-resolved spectroscopy (2,3,8), where the second 90° pulse of Fig. 2a' is replaced by a 180° pulse. We will at first explain SECSY starting from 2D J-resolved spectroscopy and afterwards show the connection with the experiment of Fig. 2a.

Let us consider for simplicity a weakly coupled AX spin system (Fig. 2). The 180° pulse in 2D J-resolved spectroscopy not only induces refocusing but simultaneously interchanges the multiplet components such that the defocusing before and the refocusing after the 180° pulse proceed with frequencies different by the coupling constant J. The apparent precession frequency during t_1' is therefore just $\pm$ J/2, leading to the well-known spread of the multiplet peaks by $\pm$ J/2 in the ω_1'-direction. These peaks are also visible in a SECSY spectrum. Replacing now the 180° pulse by a 90° pulse, there will be complete mixing of all four transverse magnetization components of the AX spin system. At the echo peak at $t=t_1'$, the resulting precession angle corresponds then to an average precession frequency $\omega_1'=\frac{1}{2}(\omega_i-\omega_k)$ where ω_i and ω_k are the precession frequencies before and after the 90° pulse, respectively. This leads then to the four ω_1' frequencies associated with each of the four resonance frequencies along the ω_2' axis in Fig. 2c'. The possible ω_1' in an AX system are 0, $\pm$ J/2, $\pm\frac{1}{2}(\delta_A-\delta_X)$ and $\pm[\frac{1}{2}(\delta_A-\delta_X)\pm J/2]$. These frequencies obviously contain both spin coupling and connectivity information as demonstrated in a previous section.

Fig. 2 demonstrates in addition the close relation between SECSY and the previously described (6) correlated spectroscopy. The two techniques

Vol. 90, No. 1, 1979 BIOCHEMICAL AND BIOPHYSICAL RESEARCH COMMUNICATIONS

differ merely in the time at which acquisition starts (Figs. 2, a and a'), This leads to two related data sets as shown in Figs. 2, b and b', where (1) and (1'), and (3) and (3') contain identical data values. Correspondingly, there is also a close relation between the two resulting spectra. Both contain all possible 16 cross peaks and it is possible to convert one spectrum into the other by a simple conformal mapping. However, the major advantage of the SECSY experiment is that it involves, in the ω_1'-domain, exclusively difference frequencies and requires therefore in many cases less t_1'-values and consequently less experiments and less data storage than conventional 2D correlated spectroscopy.

When using the simple technique indicated in Fig. 2a', an additional set of peaks appears in the 2D spectrum which form a mirror image spectrum. These undesired peaks can be eliminated by the following modified pulse sequence with phase alternation: $[90^\circ_x—90^\circ_x$—acquisition(+);$90^\circ_x—90^\circ_y$—acquisition(-); $90^\circ_x—90^\circ_{-y}$—acquisition(-);$90^\circ_x—90^\circ_{-x}$—acquisition(+)$]_n$.

EXPERIMENTAL: The basic pancreatic trypsin inhibitor (BPTI, Trasylol®) was obtained from the Farbenfabriken Bayer A.G.. ^{1}H n.m.r. spectra were recorded on a Bruker HX-360 spectrometer, using the software which was developed for 2D J-resolved spectroscopy (2,3). The 2D correlated spectrum in Fig. 1 was calculated from a 256x4096 time domain data matrix. The digital resolution is 1.46 Hz on both the ω_1' and ω_2' frequency axes. The accumulation time was approximately 6 hr.

CONCLUSIONS: Fig. 1 demonstrates that 2D spin echo correlated spectroscopy is capable of assigning coupled spin systems even in very complex spectra like those of medium size proteins. From a single SECSY spectrum of BPTI four aromatic spin systems could be completely assigned by inspection. A more careful analysis can reveal many more relations in the coupling network.

2D spin echo correlated spectroscopy can be considered as an alternative to selective decoupling experiments (4,5). It has the advantage to provide all coupling information at once. Many selective decoupling experiments would be necessary to gather an equivalent amount of information. Although spin coupling information is also contained in a SECSY spectrum (Fig. 1c'), due to limited digital resolution, it can usually not be exploited. A complete description of the spin systems in terms of connectivity and accurate spin coupling constants can be obtained if, in addition, a 2D J-resolved spectrum is also recorded. 2D spin echo correlated spectroscopy is of particular advantage when the coupled nuclei exhibit chemical

Vol. 90, No. 1, 1979 BIOCHEMICAL AND BIOPHYSICAL RESEARCH COMMUNICATIONS

shifts in the same spectral region, such that the shift differences $\Delta\delta$ cover a limited range only. However, the technique is not limited to this situation and with a sufficiently large data matrix it is also possible to correlate nuclei with widely separated chemical shifts.

REFERENCES:

1. Wüthrich, K. (1976) NMR in Biological Research: Peptides and Proteins. North Holland. Amsterdam.
2. Nagayama, K., Wüthrich, K., Bachmann, P. and Ernst, R.R. (1977) Biochem. Biophys. Res. Commun. 78, 99-105.
3. Nagayama, K., Bachmann, P., Wüthrich, K. and Ernst, R.R. (1978) J. Magn. Reson. 31, 133-148.
4. Nagayama, K., Bachmann, P., Ernst, R.R. and Wüthrich, K. (1979) Biochem. Biophys. Res. Commun. 86, 218-225.
5. Nagayama, K. (1979) (to be submitted).
6. Aue, W.P., Bartholdi, E. and Ernst, R.R. (1976) J. Chem. Phys. 64, 2229-2246.
7. Wagner, G., De Marco, A. and Wüthrich, K. (1976) Biophys. Struct. Mechanism 2, 139-158.
8. Aue, W.P., Karhan, J. and Ernst, R.R. (1976) J. Chem. Phys. 64, 4226-4227.

Vol. 95, No. 1, 1980 BIOCHEMICAL AND BIOPHYSICAL RESEARCH COMMUNICATIONS
July 16, 1980 Pages 1-6

A TWO-DIMENSIONAL NUCLEAR OVERHAUSER ENHANCEMENT (2D NOE) EXPERIMENT FOR THE ELUCIDATION OF COMPLETE PROTON-PROTON CROSS-RELAXATION NETWORKS IN BIOLOGICAL MACROMOLECULES

Anil Kumar*, R.R. Ernst and K. Wüthrich, Institut für Molekularbiologie und

Biophysik and Laboratorium für Physikalische Chemie, Eidgenössische Technische Hochschule, CH-8093 Zürich, Switzerland.

Received March 19,1980

SUMMARY: The recently developed technique of two-dimensional (2D) cross-relaxation spectroscopy is utilized for systematic measurements of selective nuclear Overhauser enhancements (NOE) in the high resolution ^{1}H nuclear magnetic resonance (NMR) spectra of biological macromolecules in solution. Compared to conventional one-dimensional NOE studies, the 2D NOE experiment has the principal advantage that it avoids detrimental effects arising from the limited selectivity of preirradiation in crowded spectral regions. Furthermore, it yields with a single instrument setting a complete network of NOE's between all the protons in the macromolecule. The resulting information on intramolecular proton-proton distances provides a new avenue for studies of the spatial structures of biopolymers.

Recent technological developments resulted in important improvements of the spectral resolution (1,2) and the possibilities to delineate J-coupling connectivities (3-5) in the ^{1}H nuclear magnetic resonance (NMR) spectra of macromolecules. At this stage the utility of high resolution NMR for studies of biopolymer conformations depends further critically on procedures to correlate NMR parameters with spatial macromolecular structures (6). Selective ^{1}H-^{1}H nuclear Overhauser effects (NOE) manifest the distances between different fragments of a polymer chain, which can be directly related to the molecular conformation (7-9). This paper describes the first use of two-dimensional (2D) cross-relaxation spectroscopy (10-12) for measurements of intramolecular NOE's and presents data obtained with a small globular protein, the basic pancreatic trypsin inhibitor (BPTI).

The NOE is the fractional change in intensity of one NMR line when another resonance is perturbed and has long been a valuable tool for structural studies of small molecules (13). In macromolecules at high magnetic fields, however, spin diffusion can become quite efficient (7,14,15), causing the conventional steady-state NOE's (13) to be less specific and hence less useful. Theory

* On leave from the Department of Physics, Indian Institute of Science, Bangalore, India.

0006-291X/80/130001-06$01.00/0
Copyright © 1980 by Academic Press, Inc.
All rights of reproduction in any form reserved.

shows that, in contrast, the initial build-up rates of NOE's are simply related to the inverse sixth power of the distance between the observed and the presaturated proton (7-9,13-16). One-dimensional experiments for measurements of NOE build-up rates have been developed (7,9). Their practical use is limited, however, since they are rather time-consuming and because of the poor selectivity for preirradiation of individual resonance lines in crowded regions of the ^{1}H NMR spectra. These practical difficulties can be largely overcome with the use of the 2D NOE experiment described in the following.

PRINCIPLES OF THE 2D NOE TECHNIQUE: The 2D NOE experiment uses a recently developed 2D NMR method for investigations of cross-relaxation and chemical exchange processes (10,11). As shown in the scheme of Fig. 1 the experiment consists of a sequence of three non-selective 90° pulses. During the evolution time between the first and the second pulse, t_1, the various magnetization components are frequency-labelled. During the mixing period between the second and the third pulse, τ_m, cross-relaxation leads to exchange of magnetization between nearby protons through mutual dipolar interactions. The interval τ_m is kept fixed and the signal recorded immediately after the third pulse as a function of t_2. In accordance with the general 2D spectroscopy principle (17) the experiment is repeated for a set of equidistant t_1 values. A two-dimensional Fourier transformation of the data matrix $s(t_1,t_2)$ then produces the desired frequency domain spectrum.

The general features of a 2D NOE spectrum are outlined in the lower part of Fig. 1. Magnetization components which do not exchange with other components during the mixing time τ_m have the same frequencies during t_1 and t_2. Hence the corresponding peaks in the ω_1-ω_2 spectrum lie on the diagonal which disects the two frequency axes. Exchange of magnetization between two components due to dipolar coupling during the mixing period is manifested by cross-peaks between the coupled components. In the scheme of Fig. 1 peak A is dipole-dipole coupled with peak C, and peak B with D and E. Two sets of cross-peaks appear in symmetrical locations with respect to the diagonal peaks. The fundamental aspects of this experiment have recently been worked out by Macura and Ernst (12).

EXPERIMENTAL: Application of the 2D NOE technique for studies of biological macromolecules requires the handling of particularly large data matrices. For this a slightly modified version of the software previously developed for recording 2D J-resolved ^{1}H NMR spectra of macromolecules (1,2) was used. 2D NOE spectra were recorded on a Bruker HX-360 spectrometer equipped with an Aspect 2000 data system. The basic pancreatic trypsin inhibitor (BPTI, Trasylol®) was obtained from the Farbenfabriken Bayer AG. Further experimental details are given in the figure caption 2.

Vol. 95, No. 1, 1980 BIOCHEMICAL AND BIOPHYSICAL RESEARCH COMMUNICATIONS

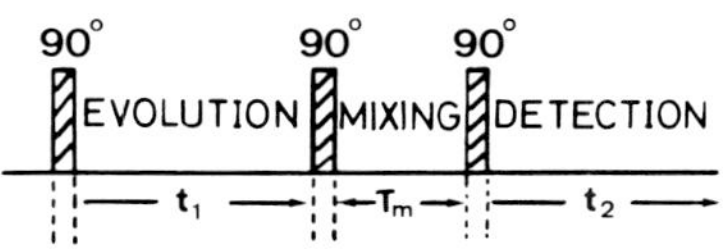

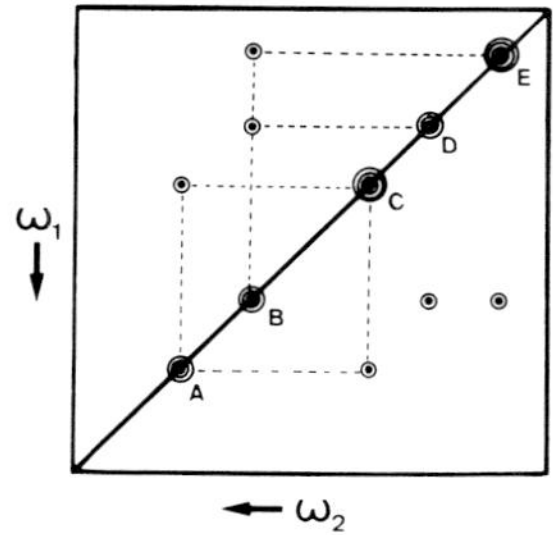

Fig. 1 The top trace shows a scheme for the 2D NOE experiment, which consists of a sequence of three 90° pulses. The pulses are separated by the evolution period, t_1, and the mixing period, τ_m, respectively. Immediately after the third pulse, the signal $s(t_1,t_2)$ is recorded. The bottom trace shows a contour plot of a schematic 2D NOE spectrum. Among the five resonance lines A-E, NOE's are manifested by the cross-peaks between A and C, B and D, and B and E.

RESULTS: A proton 2D NOE spectrum of the globular protein BPTI is shown in Fig. 2. Experimental details are given in the figure caption. The spectral range in each of the two dimensions, ω_1 and ω_2, extends from 1 to 10 ppm, so that with the exception of the lowest field amide proton line at 10.55 ppm all the resonances of BPTI are contained in Fig. 2. The distribution of signal intensity of the chemical shift axis seen previously in conventional one-dimensional ^1H NMR spectra of BPTI (6,19) is faithfully manifested on the diagonal from the lower left to the upper right corner, as one would have expected from the scheme in Fig. 1. Bands of intense signals extend from the water line at 4.85 ppm parallel to both the ω_1 and ω_2 axes. This is mainly due to the dispersion mode signal contained in the absolute value plot shown here. The informative feature of the spectrum are the cross-peaks seen throughout the spectral plane. On the basis of previously established resonance assignments in BPTI (19) and from comparison with earlier one-dimensional NOE studies (20-22) many of these cross-peaks have been identified as selective NOE's between individual groups of protons which manifest intramolecular proton-proton distances.

To illustrate how a 2D NOE spectrum is analyzed, some previously established (20-22) NOE connectivities are indicated by the broken lines in Fig. 2. The backbone amide proton of Glu 31 is connected with the α-proton of Cys 30. The

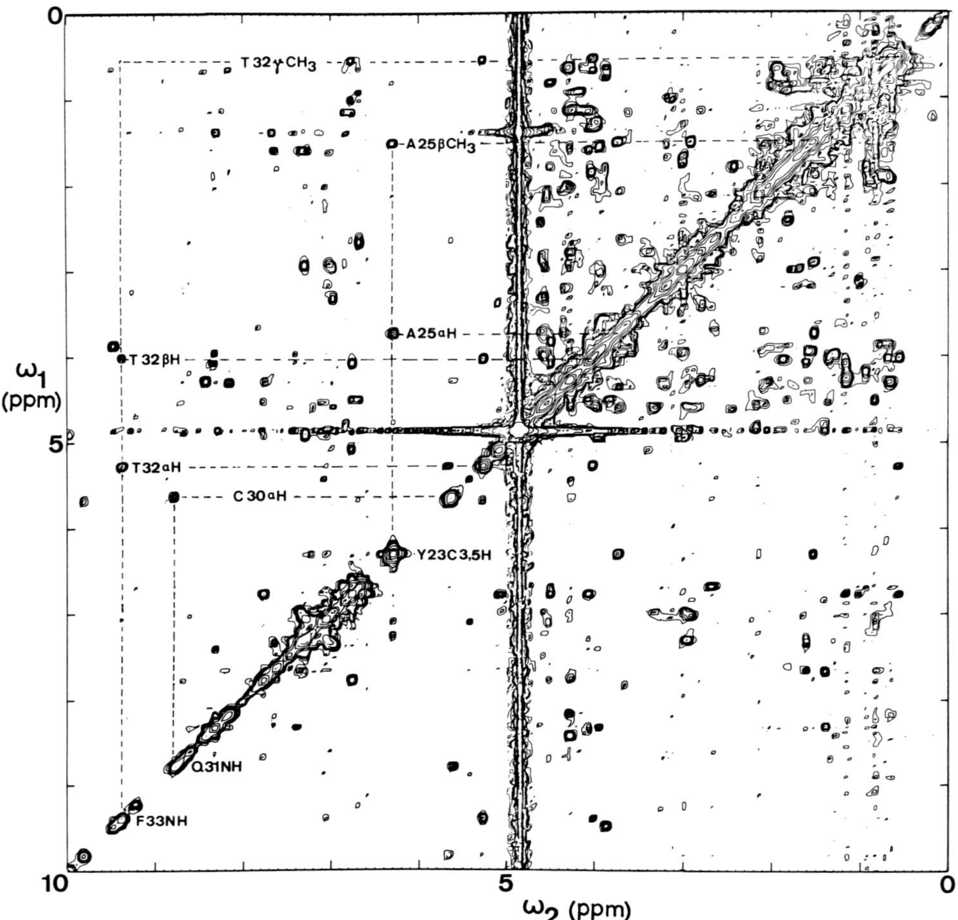

Fig. 2 Contour plot of a proton 2D NOE spectrum at 360 MHz of the basic pancreatic trypsin inhibitor. The protein concentration was 0.02 M, solvent ^2H_2O, $p^2H=3.8$, T=18°C. The spectral width was 4000 Hz. 512 data points were used in each dimension. 56 transients were accumulated for each value of t_1. The mixing time τ_m was 100 msec. The total accumulation time was 18 hours. The absolute value spectrum obtained after digital filtering in both dimensions with a shifted sine bell (18) is shown. Cross-relaxation connectivities for selected amino acid residues are indicated by the broken lines (see text). Connected peaks are identified by the one-letter symbol for amino acids (A=alanine, T=threonine, C=cysteine, Q=glutamine, F=phenylalanine, Y=tyrosine), the position in the amino acid sequence and the type of protons observed.

connectivity is manifested by a second cross-peak in a symmetrical location with respect to the diagonal (see also Fig. 1). NOE's between the 3,5 ring protons of Tyr 23 and the α- and methyl protons of Ala 25 (21) give rise to strong cross-peaks. The NOE's between the amide proton of Phe 33 and Thr 32 are manifested by strong cross-peaks for the α- and β-protons of Thr 32, and a weak peak for the methyl protons. All the NOE's previously observed by one-dimensional experiments have been seen in the 2D NOE spectra and at the present early stage of the spectral analysis numerous new connectivities have already been established. These will be further discussed elsewhere.

Fundamentally, the 2D NOE spectrum manifests for all possible combinations of protons in the macromolecular structure the phenomena seen in a conventional transient NOE experiment (7) for selected individual protons. After a mixing time, τ_m, of 100 ms the strongest cross-peaks are due to direct cross-relaxation between protons located at distances smaller than ca. 3.0 Å (22). With a shorter mixing time τ_m one would further emphasize the very short proton-proton distances (7,9,22). With the use of longer mixing times the intensities of the cross-peaks due to direct dipolar coupling between nearby protons would decrease on account of new peaks arising from spin diffusion. Hence, by recording 2D NOE spectra with a short mixing time we obtain a map of all the short proton-proton distances in a macromolecular structure. These can readily be correlated with the spatial molecular structure. Further studies are needed to explore how the information on indirect cross-relaxation pathways, "spin diffusion", obtained with longer mixing times can be employed for studies of static and/or dynamic aspects of protein conformations.

CONCLUSIONS: Compared to conventional one-dimensional experiments for measurements of selective proton-proton NOE's in macromolecules the 2D NOE experiment used in this paper has the following principal advantages: (i) With a single instrument setting it provides a complete set of NOE's between all closely spaced groups of protons in the macromolecular structure. (ii) It avoids adverse effects arising due to non-selective preirradiation of nearby resonances in crowded spectral regions. (iii) A complete set of NOE's is obtained with a significant saving in time. Since the NOE's measured with a suitably chosen mixing time τ_m (Fig. 1) can be directly related with interatomic distances in spatial molecular structures (7-9,22) the easy availability of NOE data from 2D NOE experiments promises to open new avenues for studies of biopolymer conformations.

ACKNOWLEDGEMENTS: This project is financed by a research grant of the ETH, Zürich. We would like to thank Drs. K. Nagayama and E. Bartholdi for help with the software modifications and Drs. L.R. Brown and G. Wagner for helpful discussions on the interpretation of the spectral data.

Vol. 95, No. 1, 1980 BIOCHEMICAL AND BIOPHYSICAL RESEARCH COMMUNICATIONS

REFERENCES:

1. Nagayama, K., Wüthrich, K., Bachmann, P. and Ernst, R.R. (1977), Biochem. Biophys. Res. Commun. 78, 99-105.
2. Nagayama, K., Bachmann, P., Wüthrich, K. and Ernst, R.R. (1978), J. Magn. Reson. 31, 133-148.
3. Nagayama, K., Bachmann, P., Ernst, R.R. and Wüthrich, K. (1979), Biochem. Biophys. Res. Commun. 86, 218-225.
4. Nagayama, K., Wüthrich, K. and Ernst, R.R. (1979) Biochem. Biophys. Res. Commun. 90, 305-311.
5. Nagayama, K., Kumar, Anil, Wüthrich, K. and Ernst, R.R. (1980), J. Magn. Reson. (in press).
6. Wüthrich, K. (1976) NMR in Biological Research: Peptides and Proteins. North Holland, Amsterdam.
7. Gordon, S.L. and Wüthrich, K. (1978) J. Amer. Chem. Soc. 100,7094-7096.
8. Bothner-By, A.A. (1979), in "Magnetic Resonance Studies in Biology" (R.G.Shulman, Ed.) Academic Press, New York, 177-219.
9. Wagner, G. and Wüthrich, K. (1979), J. Magn. Reson. 33, 675-680.
10. Meier, B.H. and Ernst, R.R. (1979), J. Amer. Chem. Soc. 101, 6441-6442.
11. Jeener, J., Meier, B.H., Bachmann, P. and Ernst, R.R. (1979), J. Chem. Phys. 71, 4546-4553.
12. Macura, S. and Ernst, R.R. (1980) Molec. Phys. (submitted).
13. Noggle, J.H. and Schirmer, R.E. (1971), "The Nuclear Overhauser Effect", Academic Press, New York.
14. Kalk, A. and Berendsen, H.J.C. (1976), J. Magn. Reson. 24, 343-366.
15. Hull, W.E. and Sykes, B.D. (1975), J. Chem. Phys., 867-880.
16. Bothner-By, A.A. and Noggle, J.H. (1979), J. Amer. Chem. Soc. 101, 5152-5155.
17. Aue, W.P., Bartholdi, E. and Ernst, R.R. (1976), J. Chem. Phys. 64, 2229-2246.
18. Wagner, G., Wüthrich, K. and Tschesche, H. (1978), Eur. J. Biochem. 86, 67-76.
19. Wüthrich, K. and Wagner, G. (1979), J. Mol. Biol. 130, 1-18.
20. Richarz, R. and Wüthrich, K. (1978), J. Magn. Reson. 30, 147-150.
21. Wüthrich, K., Wagner, G. Richarz, R. and Perkins, S.J. (1978), Biochemistry 17, 2253-2263.
22. Dubs, A., Wagner, G. and Wüthrich, K. (1979), Biochim. Biophys. Acta 577, 177-194.

3654 *J. Am. Chem. Soc.* **1981**, *103*, 3654–3658

Buildup Rates of the Nuclear Overhauser Effect Measured by Two-Dimensional Proton Magnetic Resonance Spectroscopy: Implications for Studies of Protein Conformation

Anil Kumar,[†,‡,1] Gerhard Wagner,[†] Richard R. Ernst,[‡] and Kurt Wüthrich[†,*]

Contribution from the Institut für Molekularbiologie und Biophysik and Laboratorium für Physikalische Chemie, Eidgenössische Technische Hochschule, 8093-Zürich, Switzerland.
Received January 19, 1981

Abstract: It is demonstrated, by means of experiments with the basic pancreatic trypsin inhibitor, that the buildup rates of the nuclear Overhauser effect can be measured by two-dimensional NMR spectroscopy. Qualitative correlations between the buildup rates of first-order Overhauser effects, which arise from direct dipole–dipole coupling between closely spaced protons, and the proton–proton distances in the protein conformation are established. Second-order Overhauser effects due to spin diffusion by cross-relaxation between more distant protons are also identified. On the basis of these observations, potentialities and limitations of two-dimensional nuclear Overhauser enhancement spectroscopy for studies of the conformations of biological macromolecules are discussed and suggestions made for improved experimental procedures. For quantitative measurements of Overhauser effects, the use of phase-sensitive spectra and of techniques for selective suppression of *J* cross-peaks in data sets recorded with very short mixing times appears particularly important.

The nuclear Overhauser effect (NOE)[2] is the fractional change in intensity by cross-relaxation of one NMR line when another resonance is perturbed. In the presence of fast motional processes, the steady-state NOE is a sensitive measure of the distance between observed and perturbed nuclei, and it has long been a valuable tool for structural studies of small molecules.[3] In macromolecules at high magnetic fields, however, spin diffusion can become quite efficient,[4,5] causing the conventional steady-state NOE's[3] to be less specific and hence less useful. In contrast, the initial buildup rates of NOE's in macromolecular systems are simply related to the inverse sixth power of the distance between the observed and the presaturated proton.[3-7] One-dimensional experiments for measurements of NOE buildup rates in macromolecules have been developed.[5,6,8] Their practical use is limited, however, since they require long accumulation times and because of the poor selectivity for preirradiation of individual resonance lines in crowded regions of the one-dimensional ^1H NMR spectra. Two-dimensional nuclear Overhauser enhancement spectroscopy (NOESY) is a more powerful and more efficient method for studies of selective NOE's between neighboring protons in the spatial structures of biological macromolecules. With a single instrument setting, NOESY can provide a complete set of proton–proton Overhauser effects in a protein.[9] For biological work it is further of particular interest that NOESY spectra can be recorded essentially as easily in H_2O solution as in deuterated solvents.[10] Overall, with the use of two-dimensional (2D) spectroscopy, measurements of proton–proton NOE's should become a generally applicable approach for the determination of the molecular conformations of biological macromolecules.[11-14] So far important information on protein conformations was obtained from semiquantitative interpretations of NOESY data.[10,12,14,15] Futher refinement of the spatial structures will depend on more quantitative interpretations of the NOE's. As a first step in this direction, the present paper reports on measurements of τ_m-dependent features in NOESY spectra of the basic pancreatic trypsin inhibitor (BPTI), a small globular protein.

Methods and Experimental Procedures

Two-dimensional nuclear Overhauser enhancement spectroscopy (NOESY)[9] uses a recently developed 2D NMR method for investigations of cross-relaxation and chemical exchange processes.[16] As shown in the experimental scheme of Figure 1A, the experiment consists of a sequence of three nonselective 90° pulses. During the evolution period between the first and the second pulse, t_1, the various magnetization components are frequency-labeled to mark their origin. During the mixing period between the second and the third pulse, τ_m, selective homonuclear NOE's between different protons are building up by cross-relaxation through mutual dipolar interactions. The signal is recorded immediately after the third pulse as a function of t_2. In accordance with the general 2D spectroscopy principle,[17] the experiment for fixed τ_m is repeated for a set of equidistant t_1 values. A two-dimensional Fourier transformation of the data matrix $\mathbf{s}(t_1,t_2;\tau_m)$ then produces the desired frequency domain spectrum, $S(\omega_1,\omega_2;\tau_m)$ (Figure 1B). In the NOESY spectrum one observes peaks on the diagonal which disects the two frequency axes (Figure 1B). These peaks correspond to magnetization components which do not exchange with other components during the mixing time τ_m, and hence have the same frequencies during t_1 and t_2. Off-diagonal cross-peaks

[†] Institut für Molekularbiologie und Biophysik.
[‡] Laboratorium für Physikalische Chemie.

(1) On leave from the Department of Physics, Indian Institute of Science, Bangalore, India.

(2) Abbreviations used: NMR, nuclear magnetic resonance; NOE, nuclear Overhauser enhancement; NOESY, two-dimensional nuclear Overhauser enhancement spectroscopy; 2D, two-dimensional; δ, chemical shift; ppm, parts per million; *J*, spin–spin coupling constant in Hz; FID, free induction decay; BPTI, basic pancreatic trypsin inhibitor (Trasylol, Bayer A.G., Leverkusen).

(3) Noggle, J. H.; Schirmer, R. E. "The Nuclear Overhauser Effect"; Academic Press: New York, 1971.

(4) Kalk, A.; Berendsen, H. J. C. *J. Magn. Reson.* **1976**, *24*, 343–366. Hull, W. E.; Sykes, B. D. *J. Chem. Phys.* **1975**, *63*, 867–880.

(5) Gordon, S. L.; Wüthrich, K. *J. Am. Chem. Soc.* **1978**, *100*, 7094–7096.

(6) Wagner, G.; Wüthrich, K. *J. Magn. Reson.* **1979**, 33, 675–680.

(7) Bothner-By, A. A.; Noggle, J. A. *J. Am. Chem. Soc.* **1979**, *101*, 5152–5155.

(8) Johnston, P. D.; Redfield, A. G. *Nucleic Acid Res.* **1978**, *5*, 3913–3927. Krishna, N. R.; Agresti, D. G.; Glickson, J. D.; Walter, R. *Biophys. J.* **1978**, *24*, 791–814.

(9) Kumar, A.; Ernst, R. R.; Wüthrich, K. *Biochem. Biophys. Res. Commun.* **1980**, *95*, 1–6.

(10) Kumar, A.; Wagner, G.; Ernst, R. R.; Wüthrich, K. *Biochem. Biophys. Res. Commun.* **1980**, *96*, 1156–1163.

(11) Bothner-By, A. A. In "Magnetic Resonance Studies in Biology"; Shulman, R. G., Ed.; Academic Press: New York, 1979; pp 177–219.

(12) Wagner, G.; Kumar, A.; Wüthrich, K. *Eur. J. Biochem.* **1981**, *114*, 375–384.

(13) Wüthrich, K.; Bösch, C.; Brown, L. R. *Biochem. Biophys. Res. Commun.* **1980**, 95, 1504–1509.

(14) Braun, W.; Bösch, C.; Brown, L. R.; Go, N.; Wüthrich, K. *Biochim. Biophys. Acta* **1981**, *667*, 377–396.

(15) Bösch, C.; Kumar, A.; Baumann, R.; Ernst, R. R.; Wüthrich, K. *J. Magn. Reson.* **1981**, *42*, 159–163.

(16) (a) Jeener, J.; Meier, B. H.; Bachmann, P.; Ernst, R. R. *J. Chem. Phys.* **1979**, *71*, 4546–4553. (b) Meier, B. H.; Ernst, R. R. *J. Am. Chem. Soc.* **1979**, *101*, 6441–6447. (c) Macura, S.; Ernst, R. R. *Mol. Phys.* **1980**, *41*, 95–117.

(17) Aue, W. P.; Bartholdi, E.; Ernst, R. R. *J. Chem. Phys.* **1976**, *64*, 2229–2246.

Reprinted from the Journal of the American Chemical Society, 1981, *103*, 3654.
Copyright © 1981 by the American Chemical Society and reprinted by permission of the copyright owner.

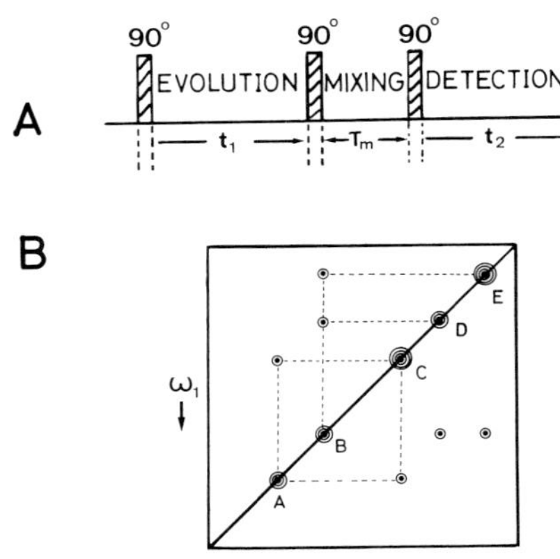

Figure 1. (A) Experimental scheme for two-dimensional nuclear Overhasuer enhancement spectroscopy (NOESY), which consists of a sequence of three nonselective 90° pulses. The pulses are separated by the evolution period, t_1, and the mixing period, τ_m, respectively. Immediately after the third pulse, the signal $s(t_1,t_2;\tau_m)$ is recorded. (B) Contour plot of a schematic NOESY spectrum. Among the five resonance lines A–E, NOE's are manifested by the cross-peaks between A and C, B and D, and B and E.

manifest exchange of magnetization. The magnetization transfer is due to dipole–dipole cross-relaxation during the mixing time. In the schematic NOESY spectrum of Figure 1B, peak A is dipole–dipole coupled with peak C, and peak B with D and E. Two sets of cross-peaks appear in symmetrical locations with respect to the diagonal peaks. Similar phenomena occur also in the presence of chemical exchange.[16a,b]

Fundamentally the physical situation in a NOESY experiment is similar to that in one-dimensional transient NOE experiments.[5] In both techniques the exchange of magnetization is initiated by a short radio-frequency pulse, and no radiofrequency field is applied while the NOE's are building up. In analogy to the transient NOE experiment,[5] the time development of NOE's in NOESY can be investigated when different two-dimensional spectra are recorded with different mixing times τ_m. For each τ_m value a complete set of NOE's is obtained. For work with macromolecules, NOESY is therefore much more efficient than one-dimensional NOE experiments.[9] Another important advantage is that NOESY uses *nonselective* pulses (Figure 1A) to obtain selective NOE's, whereas the selectivity of one-dimensional transient NOE experiments depends critically on the selectivity of the presaturation pulse.[5]

For the experiments in this paper we used a 0.02 M solution of the basic pancreatic trypsin inhibitor (BPTI) in ^2H_2O, p^{2}H 4.6. BPTI (Trasylol®) was a gift from the Farbenfabriken Bayer A.G., Leverkusen, F.R.G.; 360-MHz ^{1}H NMR spectra were recorded on a Bruker HX 360 spectrometer equipped with an Aspect 2000 data system. The previously developed software for handling of the large data matrices obtained in 2D experiments with biological macromolecules[18,19] was used. NOESY spectra were recorded using quadrature detection in both dimensions, with the carrier frequency at one end of the spectrum. Transverse components at the beginning of the mixing period and the axial peaks at $\omega_1 = 0$ were cancelled by addition of groups of 16 experiments with different phases. These phase cycles will be described in detail elsewhere. Absolute value spectra are presented either as stacked plots providing a three-dimensional view of the spectra, contour plots, or by means of cross sectins.[19,20] Further improvement of the 2D spectra can be obtained by triangular multiplication,[21] a technique based on the symmetry of the cross-peak pattern with respect to the diagonal peaks.

Results

Figure 2A shows a NOESY spectrum of BPTI which was recorded with a mixing time of 300 ms. On the diagonal from the upper right to the lower left corner of the figure one recognizes a pattern of peaks which corresponds to that in the normal, one-dimensional ^{1}H NMR spectrum of BPTI.[22] Many of these resonances were previously individually assigned.[22,23] At 4.8 ppm there are a horizontal band and a vertical band of spurious noise which correspond to "tails" of the strong diagonal peak of the residual water protons. All other off-diagonal peaks manifest NOE's between resonances with corresponding frequencies ω_1 and ω_2 (see Figure 1). Obviously, a large number of selective NOE's can be observed in this spectrum. Many of these have previously been assigned to specific pairs of protons in the protein.[9,10,12,22,24,25]

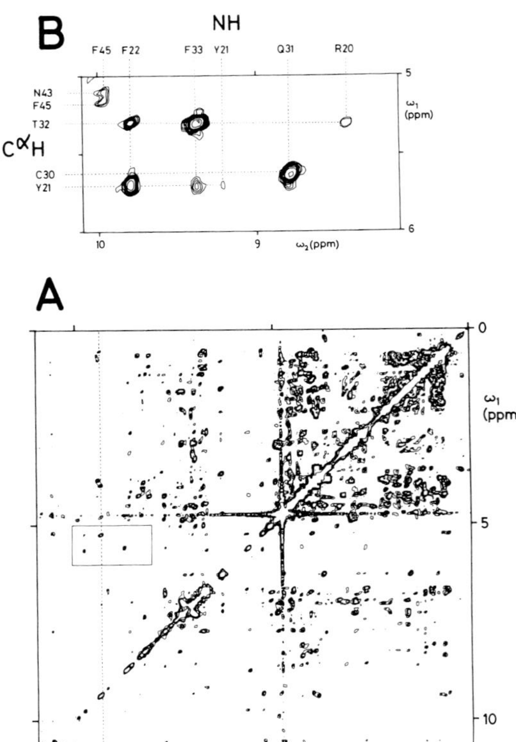

Figure 2. (A) Contour plot of a proton NOESY spectrum at 360 MHz of a 0.02 M solution of the basic pancreatic trypsin inhibitor (BPTI) in ^2H_2O, T = 24 °C, p^{2}H 4.6. The mixing time, τ_m, was 300 ms. The spectral width was 4000 Hz. The data set consisted of 512 points in both dimensions, 48 free induction decays were accumulated for each value of t_1, and the total accumulation time was 18 h. Before Fourier transformation the free induction decays were multiplied with a phase-shifted sine bell, sin $[\pi(t + t_0)/t_s]$, where t_s was the experimental acquisition time and t_0/t_s was 1/64.[30] After Fourier transformation the spectrum was further improved by triangular multiplication.[21] An absolute value plot is shown. The vertical and horizontal spikes at 4.8 ppm are due to the resonance of the residual solvent protons. The dotted, vertical line at 9.39 ppm inidicates where the cross sections of Figure 4 were taken. (B) Plot on an expanded scale of the spectral region from 5.0 to 6.0 ppm in ω_1 and 8.1 to 10.1 ppm in ω_2, which is indicated by a solid rectangle in spectrum A. Lower contour levels were plotted than in A, so that additional peaks appear in spectrum B. The cross peaks are identified by the IUPAC-IUB one-letter symbols for amino acids and the position in the amino acid sequence of the C$^\alpha$ and amide protons.

(18) Nagayama, K.; Wüthrich, K.; Bachmann, P.; Ernst, R. R. *Biochem. Biophys. Res. Commun.* **1977**, *78*, 99–105.
(19) Nagayama, K.; Bachmann, P.; Wüthrich, K.; Ernst, R. R. *J. Magn. Reson.* **1978**, *31*, 133–148.
(20) Nagayama, K.; Kumar, A.; Wüthrich, K; Ernst, R. R. *J. Magn. Reson.* **1980**, *40*, 321–334.
(21) Baumann, R.; Kumar, A.; Ernst, R. R.; Wüthrich, K. *J. Magn. Reson.*, in press.

(22) Dubs, A.; Wagner, G.; Wüthrich, K. *Biochim. Biophys. Acta* **1979**, *577*, 177–194.
(23) Wüthrich, K.; Wagner, G. *J. Mol. Biol.* **1979**, *130*, 1–18.
(24) Richarz, R.; Wüthrich, K. *J. Magn. Reson.* **1978**, *30*, 147–150.
(25) Wüthrich, K.; Wagner, G.; Richarz, R.; Perkins, S. J. *Biochemistry* **1978**, *17*, 2253–2263.

3656 J. Am. Chem. Soc., Vol. 103, No. 13, 1981

They include NOE's between covalently linked protons which are also connected by J-coupling, e.g., along amino acid side chains,[9,10,24,25] between protons in neighboring amino acid residues in the amino acid sequence (e.g., in the β-sheet secondary structure of BPTI[9,10,12,22]), and between different amino acid side chains which are closely spaced in the tertiary protein structure.[9,24,25] It is readily apparent also from the locations of the peaks that there are NOE's between all the different types of protons in the protein, e.g., amide protons with chemical shifts from 8 to 11 ppm,[23] on the one hand, and backbone C^α protons from 4 to 6 ppm and aliphatic side chain protons from 0.5 to 4 ppm, on the other hand, or between aromatic protons from 6 to 8 ppm and aliphatic side chain protons, etc. In Figure 2B the previously established[12,22] individual assignments for a small number of NOESY peaks between backbone amide and C^α protons have been indicated.

When the spectrum of Figure 2A is compared with previously published NOESY spectra of BPTI, which were recorded with a mixing time of 100 ms,[9,12] it is readily seen that the total number of observable peaks and the relative peak intensities are different throughout the entire spectrum. To investigate the dependence of the spectral features on τ_m in more detail, six NOESY spectra with different mixing times in the range from 20 to 300 ms were recorded. Figures 3 and 4 show the time development of two spectral regions outlined in Figure 2A.

Figure 3 presents three-dimensional views of the spectral region of Figure 2B in the NOESY spectra recorded with different mixing times. The correspondence of the 300-ms trace with Figure 2B can readily be checked. Inspection of the series of six spectra reveals two classes of magnetization transfer exemplified by cross-peaks with different time variations. The first class includes the strong peak between Phe-33 NH and Thr-32 C^αH (Figure 2B). The peak intensity increases at short mixing times, goes through a maximum between 100 and 200 ms, and decreases again. These incoherent magnetization transfers are due to dipole–dipole coupling between the protons modulated by the molecular motions, and are the NOE's of interest. The second class is exemplified by the peaks which are shadowed in the spectra recorded with τ_m = 20 and 30 ms, e.g., the peak between NH and C^αH of Phe-45. At the short mixing times the intensities of these peaks vary in the form of damped oscillations, which disappear when τ_m gets longer. These oscillations are due to coherent transfer of magnetization by J-coupling between protons, as was recently shown in different systems,[26] and will be further elaborated in the Discussion. Either after the coherent transfer has decayed or even earlier, these peaks reappear with incoherent transfer of magnetization, since the J-coupled protons usually also show NOE's (see the 300-ms spectrum of Figure 3).

Figure 4 shows the τ_m dependence for a cross section through the amide proton resonance of Phe-33 and parallel to ω_1 (Figure 2A). The spectra in Figures 3 and 4 both contain the two cross-peaks from Phe-33 NH to Thr-32 C^αH and to Tyr-21 C^αH. The two classes of magnetization transfer discussed in Figure 3 can also be observed here (a J-coupled cross peak is again shadowed). In addition, the decrease of the diagonal peak with increasing τ_m can be followed. For two methyl resonances, Tyr-32 $C^\gamma H_3$ and Ile-19 $C^\gamma H_3$, the NOE buildup starts after a lag time of 60 to 100 ms, indicating second-order NOE's through sequential cross-relaxation via protons located between the two protons.[4-7]

Figure 5 shows plots of the peak heights vs. τ_m for peaks in three different cross sections through the spectrum of Figure 2A. Except for the cross-peaks with J-coupling, which have not been included in this presentation, the classes of time behavior discussed above during the description of Figure 3 and 4 can be observed. Distances between the corresponding protons computed from the X-ray structure of BPTI[27] are also indicated in this figure. A qualitative correlation between proton–proton distances and NOE buildup rates is clearly evidenced, with faster rates for shorter

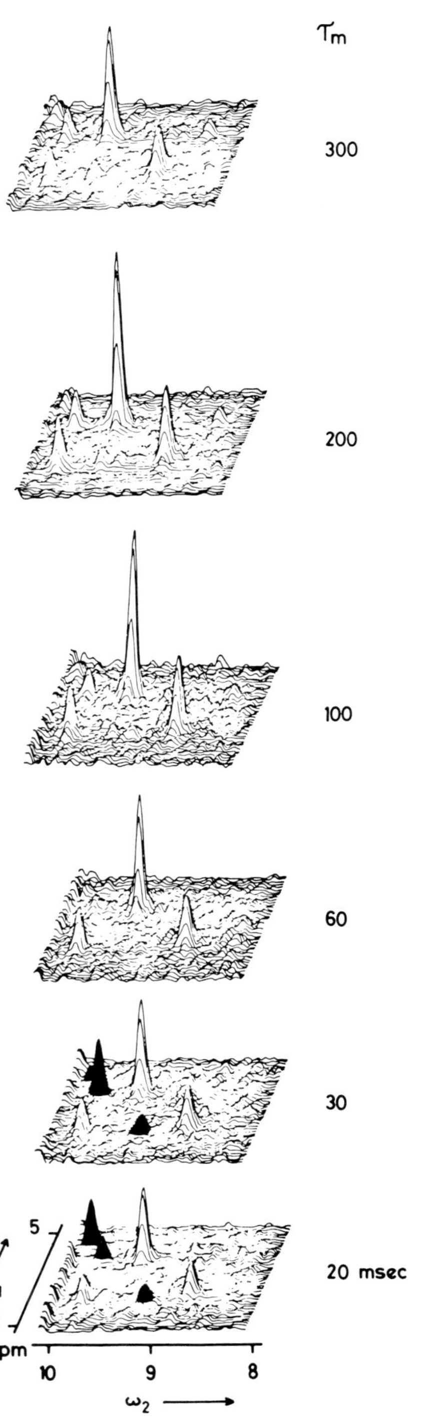

Figure 3. Stacked plots providing a three-dimensional view of the spectral region of Figure 2B in a series of NOESY spectra of BPTI recorded with different mixing times, τ_m, as indicated in the figure. Peaks manifesting NOE connectivities may be identified by comparing the spectrum recorded with τ_m = 300 ms with Figure 2B. The shadowed peaks in the spectra with τ_m = 20 and 30 ms are due to J couplings (see text).

(26) Macura, S.; Huang, Y.; Suter, D.; Ernst, R. R. *J. Am. Chem. Soc.*, in press.

(27) Deisenhofer, J.; Steigemann, W. *Acta Crystallogr., Sect. B* **1975**, *31*, 238–350.

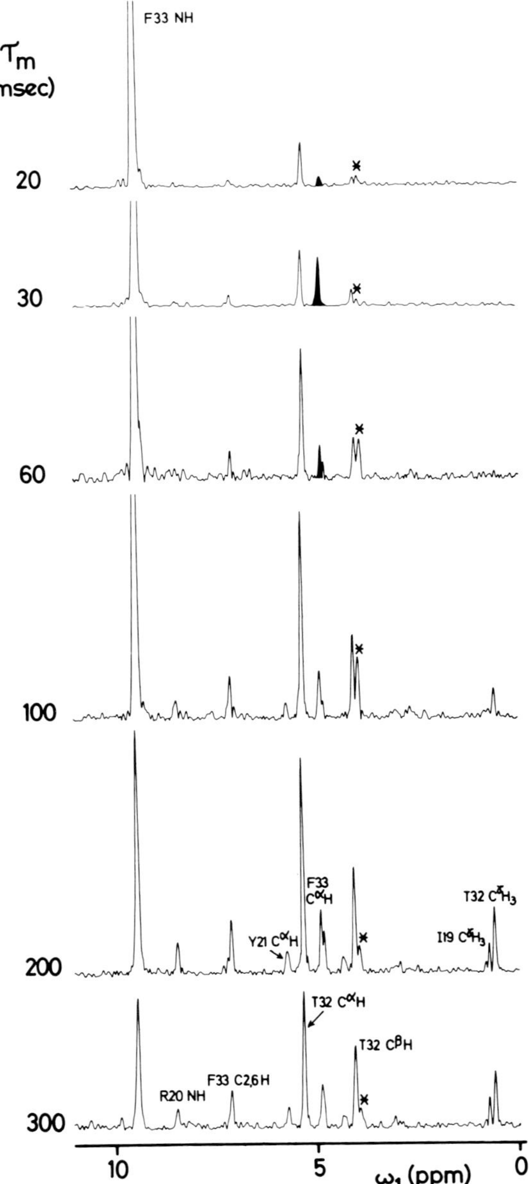

Figure 4. Cross sections parallel to ω_1 and through the diagonal amide proton resonance of Phe-33 at 9.39 ppm (dotted line in Figure 2) from the same series of NOESY spectra of BPTI as shown in Figure 3. Peaks which manifest first-order NOE's are identified in the 300-ms spectrum by the IUPAC-IUB one-letter symbols for amino acids, the position in the amino acid sequence, and the type of protons observed. Additional resonance assignments in the 200-ms spectrum include Tyr-21 C$^{\alpha}$H, Ile-19 C$^{\gamma}$H$_3$, and Thr-32 C$^{\gamma}$H$_3$ which show second-order NOE's (see text and Figure 5) and Phe-33 C$^{\alpha}$H, which appears as a strong peak at short mixing times (shadowed in the spectra with τ_m = 20, 30, and 60 ms) due to *J* coupling with the amide proton of Phe-33 (see text). The peak identified by * corresponds to the tail of a resonance centered in an adjoining cross section.

distances. Furthermore, the two methyl groups which show secondary NOE's are at a distance where first-order NOE's in proteins are hardly effective.[14] Structural analysis of the curves in this figure is discussed in the following section.

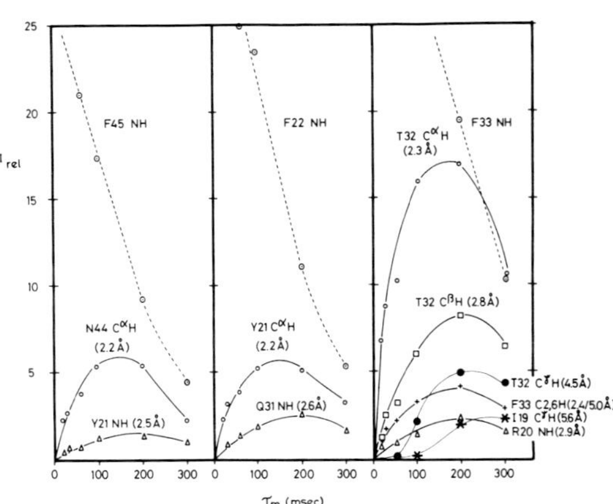

Figure 5. Dependence of the relative peak heights on the mixing time τ_m of cross-peaks in the NOESY spectrum of BPTI (Figures 2–4). The data were measured in cross sections through the diagonal peaks of the amide protons of Phe-45 at 9.98 ppm, Phe-22 at 9.82 ppm, and Phe-33 at 9.39 ppm (Figure 4). The broken lines show the decay of the intensity of the diagonal peaks and are identified by the assignment of these peaks. The time course of the NOE's for protons located near the proton corresponding to the diagonal peak is indicated by the solid curves which connect the experimental points. Resonance assignments and, in parentheses, the proton–proton distances in the X-ray structure[27] are added for each curve, whereby an average proton position was assumed for methyl groups. Two cases of second-order NOE's are clearly evidenced in the diagram on the right, i.e., for Thr-32 C$^{\gamma}$H$_3$ and Ile-19 C$^{\gamma}$H$_3$ (see also Figure 4).

Discussion

The nuclear Overhauser effect is one of the few quantities which can be utilized to deduce quantitative information on intramolecular proton–proton distances in molecules in solution. For work with macromolecules, measurements of NOE's must be combined with the high spectral resolution and sensitivity achived only at high frequencies. It is readily apparent from theory[3,4,6] and was borne out by one-dimensional NMR experiments[5–8] that the NOE buildup rates must be studied to obtain the distance information needed for the determination of macromolecular conformations.[12–14] Comparison of the data in Figures 3–5 with the data presented in ref 5 shows that fundamentally the time course of the magnetization caused by cross-relaxation is identical in one-dimensional transient NOE experiments and NOESY. An essential advantage of NOESY, which was already discussed in the introduction to the present paper, is that the problem of limited selectivity of irradiation of individual lines in crowded spectra, which is inherent in one-dimensional techniques, is solved in 2D NMR by the frequency labeling during t_1 and subsequent Fourier analysis. Many aspects of the spectral analysis, however, are closely similar for the two experiments, and some consequences for work with NOESY are discussed in the following.

The τ_m dependence of the peak intensities in NOESY (Figures 3 and 4) is characterized by magnetization buildup in the early phases of the experiment, transition through a maximum, and subsequent decay of the magnetization (Figure 5). The crucial point for the analysis of these curves is that only the initial magnetization buildup can reliably be used for measurements of internuclear distances. For short mixing times τ_m (i.e., in practice $\tau_m \lesssim 100$ ms when working with proteins of molecular weights $\lesssim$20 000), the cross-peak intensities at any given τ_m value are to a good approximation determined by the cross-relaxation rate and by the rate of decay of the magnetization (leakage relaxation) (Figure 5)[16a,c]. From the cross-relaxation rate the proton–proton distance can be calculated. The peak intensity and the τ_m value for the maximum magnetization as well as the time course of the magnetization decay (Figure 5) are further influenced by spin diffusion to additional nuclei.

From Figures 3–5 and the above comments it is clear that variation of the mixing time is indispensable for an unambiguous interpretation and in particular for a quantitative analysis of NOESY spectra. A single 2D spectrum for fixed τ_m value permits at most a qualitative assignment of cross-peaks to pairs of nuclei between which magnetization is transferred during the mixing time. For a more quantitative analysis a set of NOESY spectra for well-selected τ_m values must be performed. At short mixing times, additional cross-peaks occur due to J-coupling.[26] Since the NOE's at short values of τ_m are of most direct interest, it becomes important to distinguish the coherent transfer of magnetization due to J-coupling from the incoherent transfer of magnetization in the NOE cross-peaks.

For a detailed discussion of J cross-peaks, we refer to ref 26. Here we describe only the major features. A number of different pathways exist which can lead to coherent magnetization transfer. They cause characteristic oscillation frequencies by which the J cross-peak intensities vary with the mixing time τ_m. Transfer via zero quantum coherence involves the difference frequency $|\Omega_k - \Omega_l|$, where Ω_k and Ω_l are the Larmor frequencies of the two involved nuclei. Transfer via single quantum coherence, on the other hand, is oscillatory with the characteristic resonance frequencies of the two nuclei, while transfer through double quantum coherence involves the sum frequency $(\Omega_k + \Omega_l)$. All those frequencies are normally quite high and cause the J cross-peak intensities to vary rapidly with the mixing time. J peak intensities are therfore very difficult to predict and suppression or identification of such coherent transfers is imperative for successful, qualitative or quantitative, analysis of NOESY spectra.

Besides the possbilities to identify J cross-peaks by using 2D correlated spectra, such as COSY, SECSY, and FOCSY,[20] two principal possibilities have been proposed to suppress coherent transfer in the NOESY experiment.[26] By suitable phase-shifted pulse sequences, it is possible to selectively compensate J cross-peaks arising from single and double quantum coherence. This possibility utilizes the different behavior of coherence components of different order when a phase shift is introduced.[28] It is, however, not possible to suppress those J cross-peaks which involve zero quantum coherence because these components behave identically with the NOE cross-peaks under phase shifts. A second possibility for the elimination of coherent transfer is by slight random variation of the mixing time τ_m.[26] Random variation of the mixing time within the sequence of t_1 values required for a NOESY spectrum leads to smearing of all J cross-peaks in the ω_1 direction, and random variation of τ_m during accumulation of several transients for a fixed value of t_1 causes the J cross-peaks to be averaged out. The NOE cross-peak intensities, on the other hand, are slowly varying functions of τ_m and are therefore not markedly affected by a slight variation of τ_m.

It is desirable to determine the cross-relaxation rates from the initial buildup rates of the NOE cross-peak intensities since the intensities for longer mixing times τ_m depend also on the leakage mechanisms, like spin diffusion to further nuclei and spin–lattice relaxation, which tend to reduce the transferred magnetization. To obtain the cross-relaxation rates from the initial buildup rates, however, absolute intensity measurements are necessary. Peak intensities are for this purpose unsuitable because they are affected by the peak widths in both frequency dimensions. Broad peaks appear less intense than sharper peaks of the same integrated intensity. This feature is further accentuated by the use of digital resolution enhancement techniques.[29] In the present experiments, the phase shifted sine bell technique[30] has been applied. Therefore, the absolute buildup rates appear to be too low for broad lines, e.g., the amide protons peaks, and quantitative agreement with the proton–proton distances indicated in Figure 5 is rather poor in these cases. For the future, measurement of accurate absolute intensities in NOESY spectra is therefore an important task which may be solved by measuring integrated intensities in pure absorption mode NOESY spectra.

In spite of the respectable time consumption of NOESY experiments for several τ_m values, such studies appear to be extremely promising for investigations of the three-dimensional structure of biopolymers in solution. For the time being there will certainly remain cases where it will be worthwhile to combine NOESY experiments with selective one-dimensional frequency driven NOE experiments.[6] Especially when the resonances are well resolved, the one-dimensional NOE experiments still allow more straightforward intensity measurements and thus more accurate distance estimations and will thus in favorable cases remain useful experiments to complement the two-dimensional techniques.

Acknowledgment. Financial support by a special grant of the ETH Zürich and by the Swiss National Science Foundation (Project No. 3.528.79) is gratefully acknowledged.

(28) Wokaun, A.; Ernst, R. R. *Chem. Phys. Lett.* **1977**, *52*, 407–409.

(29) Ernst, R. R. *Adv. Magn. Reson.* **1966**, *2*, 1.

(30) Wagner, G.; Wüthrich, K.; Tschesche, H. *Eur. J. Biochem.* **1978**, *86*, 67–76.

Vol. 113, No. 3, 1983 BIOCHEMICAL AND BIOPHYSICAL RESEARCH COMMUNICATIONS
June 29, 1983 Pages 967-974

APPLICATION OF PHASE SENSITIVE TWO-DIMENSIONAL CORRELATED SPECTROSCOPY (COSY) FOR MEASUREMENTS OF ^1H-^1H SPIN-SPIN COUPLING CONSTANTS IN PROTEINS

D. Marion* and K. Wüthrich

Institut für Molekularbiologie und Biophysik,
Eidgenössische Technische Hochschule, CH-8093 Zürich, Switzerland

Received May 18, 1983

SUMMARY: Two-dimensional correlated spectroscopy (COSY) is used for measurements of proton-proton spin-spin coupling constants in protein ^1H NMR spectra. High digital resolution along the frequency axis ω_2 is achieved by placing the carrier frequency in the center of the spectrum, using quadrature detection in both dimensions and presenting the spectrum in the phase sensitive mode. Compared to other techniques for studies of spin-spin coupling constants, COSY provides greatly improved spectral resolution. This is illustrated by experiments with H_2O solutions of the small globular protein BUSI IIA (bull seminal inhibitor IIA).

Recently we proposed a general strategy for spatial structure determinations in non-crystalline polypeptides and proteins by high resolution nuclear magnetic resonance (NMR)(1). These procedures rely on sequence-specific assignments for the ^1H NMR lines, which have by now been completed for several small proteins (2-7). The initial steps of the determination of the three-dimensional structure make use of intramolecular distance constraints between distinct groups of protons, which are obtained by nuclear Overhauser enhancement experiments (8-10). Subsequently, a maximum number of additional experimental parameters should be collected and checked for consistency with the protein conformation determined from the ^1H-^1H distance constraints. In this context spin-spin coupling constants between vicinal hydrogen atoms are of particular interest, since the correlations with the molecular conformation have been extensively investigated (11-15). The present paper describes the use of two-dimensional (2D) correlated spectroscopy (COSY) for measurements of spin-spin coupling constants in crowded regions of protein ^1H NMR spectra.

FUNDAMENTAL CONSIDERATIONS: A schematic diagram of a COSY spectrum containing the resonances of two AX spin systems is shown in Fig. 1 (16-18). We distin-

* Present address: Centre de Biophysique Moléculaire C.N.R.S.,
45045 Orléans Cedex, France

0006-291X/83 $1.50

Copyright © 1983 by Academic Press, Inc.
All rights of reproduction in any form reserved.

guish two types of signals. The "diagonal peaks" are on the diagonal indicated by the broken line and the "cross peaks" lie at positions (ω_1^A, ω_2^X) and (ω_1^X, ω_2^A), where ω^A and ω^X are the chemical shifts of the coupled nuclei. All signals in this example are split into four fine structure components by spin-spin coupling, where the separation of the lines along ω_1 and along ω_2 corresponds to the coupling constant J_{AX}. In a phase-sensitive presentation the cross peaks are in absorption, with alternating sign of the individual fine structure components (Fig. 1), and the diagonal peaks are in dispersion. The presently proposed application of COSY for studies of spin-spin coupling constants in proteins relies on measurement of the separation of fine structure components along ω_2 in the cross peaks.

In small molecules the fine structure of COSY cross peaks has been resolved even in absolute value presentations of the spectra (16,18). However, in absolute value COSY spectra of proteins recorded at high field, which extend over a wide frequency range in both dimensions ω_1 and ω_2 and often contain rather broad lines, the spin-spin coupling fine structure has not usually been exploited. These experiments used relatively low digital resolution (e.g.2-7), both to obtain a workable signal-to-noise ratio with reasonable aquisition times and because of the limited data storage capacities. In the present experiments more efficient use of the storage capacity is obtained by

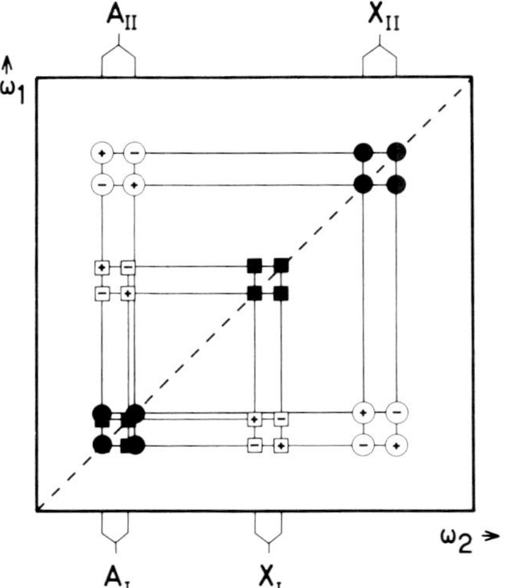

Fig. 1 Schematic diagram of a phase-sensitive COSY spectrum containing two AX spin systems. The diagonal peaks, which have a dispersion shape, are indicated by filled symbols. The cross peaks, which have an absorption line shape and alternate in sign, are indicated by open circles and open squares, respectively, with the sign given by + and -.

Vol. 113, No. 3, 1983 BIOCHEMICAL AND BIOPHYSICAL RESEARCH COMMUNICATIONS

placing the carrier frequency in the center of the spectrum, which required a modification of the data accumulation routine. Further it is important that high digital resolution is needed only along one frequency axis (Fig. 1). ω_2 is a natural choice for the high resolution axis both in view of sensitivity of the experiment and of minimal perturbations, for example by the additional irradiation needed for solvent suppression in H_2O solutions (19). Because of the intrinsically broad NMR lines in macromolecules, phase sensitive spectra had to be obtained for better separation of the fine structure components (In absolute value spectra the lines are broadened by the admixture of dispersion mode). Details of the experimental procedures used are given in the last section below.

Compared to conventional 1D NMR and to 2D J-resolved NMR (20-22), which have hitherto been used for studies of spin-spin couplings in proteins (12,13,21,22), COSY can provide greatly improved spectral resolution. This is illustrated by the resonances A_I and A_{II} in Fig. 1. In spite of the near coincidence of the chemical shifts for A_I and A_{II} the cross peaks with X_I and X_{II}, respectively, are well resolved in the ω_1-ω_2 plane. The resonance positions in a 1D spectrum or a 2D J-resolved spectrum can be represented by the diagonal or the non-diagonal fine structure components, respectively, of the diagonal peaks in Fig. 1. The figure shows that the resonances A_I and A_{II} would be overlapped in both of these experiments. In COSY cross peaks such overlap is expected only when both chemical shifts in the two spin systems coincide, which is relatively rare even in the crowded spectra of polypeptides and small proteins.

RESULTS AND DISCUSSION: The following experiments have been recorded with the small globular protein BUSI IIA (bull seminal inhibitor IIA), for which essentially complete sequence-specific resonance assignments are available (7). Fig. 2 shows corresponding regions of a low resolution absolute value COSY spectrum and a high resolution phase sensitive COSY spectrum, which contain the amide proton-C^{α}proton cross peaks of ca. 40 amino acid residues. It is readily apparent that the spectrum 2B contains much sharper peaks and that for each resonance of the low resolution spectrum a fine structure pattern can be observed at the higher resolution. Even though only the positive fine structure components are included in Fig. 2B it is readily apparent that more intricate cross peak patterns appear than in the scheme of Fig. 1. These correspond to the increased complexity of the amino acid spin systems (13) as compared to the AX case. For Cys 57 the NH-C^{α}H cross peak is shown on an enlarged scale (Fig. 3). Along ω_2 only the spin-spin coupling $^3J_{HN\alpha}$ is mani-

Vol. 113, No. 3, 1983 BIOCHEMICAL AND BIOPHYSICAL RESEARCH COMMUNICATIONS

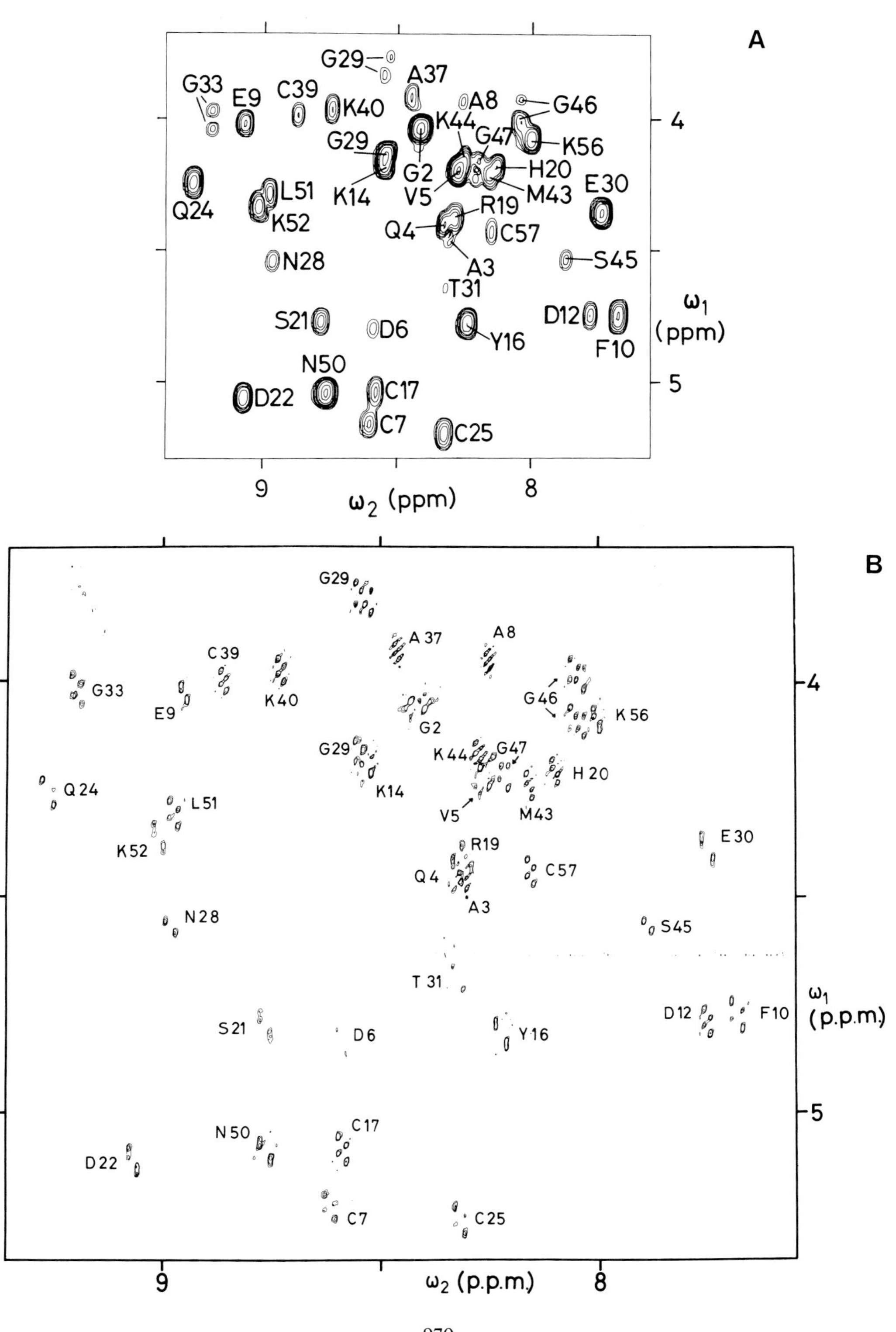

Vol. 113, No. 3, 1983 BIOCHEMICAL AND BIOPHYSICAL RESEARCH COMMUNICATIONS

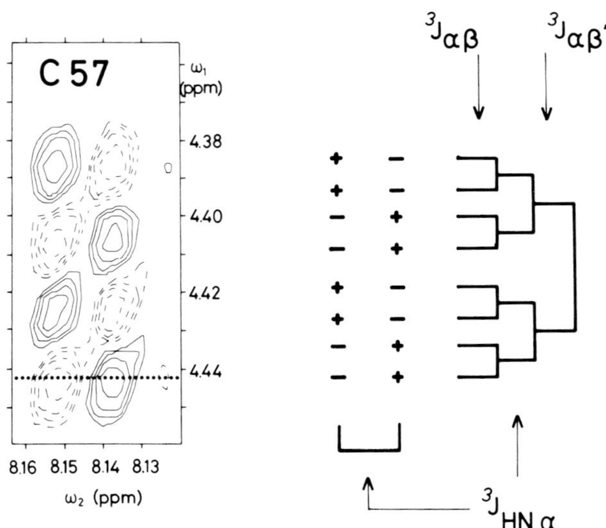

Fig. 3 Analysis of the fine structure of the NH-C$^{\alpha}$H cross peak of Cys 57 in the phase sensitive COSY spectrum of BUSI IIA in Fig. 2B. On the left this cross peak is shown on an expanded scale. Positive peaks are plotted with solid lines, negative peaks with broken lines. The dotted horizontal line near $\omega_1 = 4.44$ppm indicates where the cross section of Cys 57 in Fig. 4 was taken. The scheme on the right indicates the theoretical fine structure of this cross peak. Along ω_2 only one line separation by the spin-spin coupling constant $^3J_{HN\alpha}$ is expected. Along ω_1 all vicinal coupling constants in the fragment NH-C$^{\alpha}$H-C$^{\beta}$H$_2$ are manifested, which gives rise to two doublets of doublets (13). Comparison with the experimental spectrum on the left shows that closely spaced fine structure components of equal sign are not resolved, whereas the peak to peak distance between components with different sign is enhanced by the mutual overlap of the absorption lines (see also caption to Fig. 4).

fested, whereas all the vicinal couplings in the fragment -HN-C$^{\alpha}$H-C$^{\beta}$H$_2$- are represented in the fine structure along ω_1. More details on the spectral analysis are given in Fig. 3 and the caption to this figure. A suitable procedure for measurements of $^3J_{HN\alpha}$ in the direction of the high resolution axis ω_2 is by evaluation of the peak separation in suitably selected (Fig. 3) cross sections (21). As an illustration the cross sections through the NH-C$^{\alpha}$H cross peaks of all 6 cysteinyl residues in BUSI IIA are shown in Fig. 4.

Fig. 2 Contour plots of the spectral region ($\omega_1 = 3.7$-5.3ppm, $\omega_2 = 7.5$-9.4ppm) of two ^{1}H COSY spectra recorded in a 0.016M solution of BUSI IIA (bull seminal inhibitor IIA)(23) in a mixed solvent of 90% H_2O and 10% ^2H_2O, pH 4.9, T = 45°C. This spectral region contains most of the amide proton-C$^{\alpha}$proton cross peaks (7). A. Absolute value spectrum at 500 MHz. The digital resolution was 4.8 Hz/point in both directions. B. Absorption mode spectrum, digital resolution 2.4 Hz/point along ω_1 and 0.4 Hz/point along ω_2. Only the positive fine structure components are displayed.

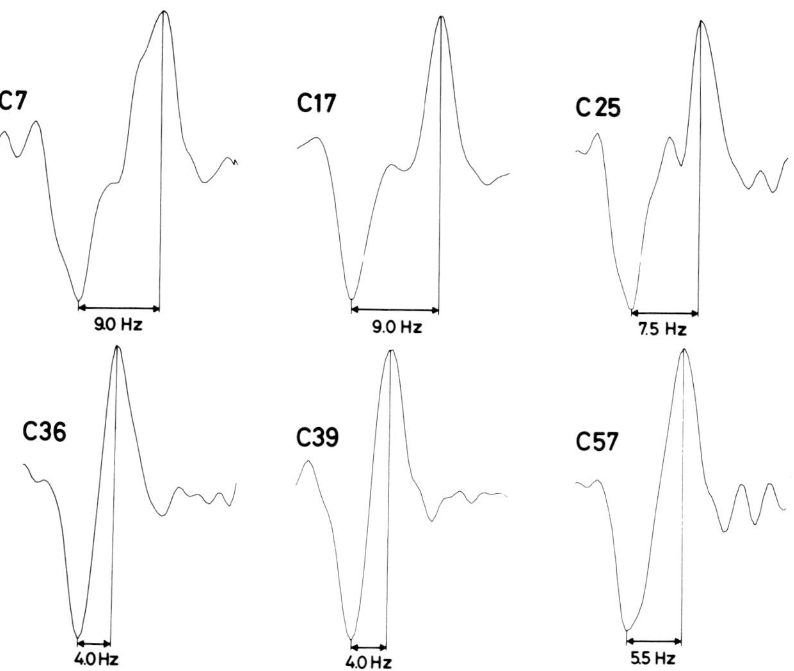

Fig. 4 Cross sections parallel to ω_2 (see Fig. 3) through the NH-C^{α}H cross peaks of the 6 cysteinyl residues of BUSI IIA (23) in the phase sensitive COSY spectrum of Fig. 2B. The separation of the positive and negative peak is indicated for each multiplet. For the larger couplings (upper row) this peak separation corresponds to $^3J_{HN\alpha}$ (Fig. 3). For the smaller couplings (lower row), where the line width is of the same order as the coupling constant, the peak separation represents an upper limit for $^3J_{HN\alpha}$ and a more accurate evaluation of the coupling constants must rely on comparison with spectrum simulations (D.Marion and K.Wüthrich, to be published).

From analysis of the complete NH-C^{α}H region of the COSY spectrum in Fig. 2B the coupling constants $^3J_{HN\alpha}$ were determined for all 57 residues in BUSI IIA except for the prolines and glycines. This data will be presented elsewhere in conjunction with a secondary structure determination in this protein (M. Williamson, D. Marion and K. Wüthrich, in preparation). In principle, similar experiments can be used for measurements of spin-spin couplings for example in the amino acid side chains of proteins and in other macromolecules. However, the spectral analysis may in certain cases be more involved when strong coupling has to be taken into account.

MATERIALS AND METHODS: BUSI IIA was obtained as a gift from Dr. P. Štrop and the NMR sample was prepared as described elsewhere (7).

The phase-sensitive COSY spectrum was recorded on a Bruker HX 360 spectrometer equipped with an Aspect 2000 computer containing a pulser board and a Control

Vol. 113, No. 3, 1983 BIOCHEMICAL AND BIOPHYSICAL RESEARCH COMMUNICATIONS

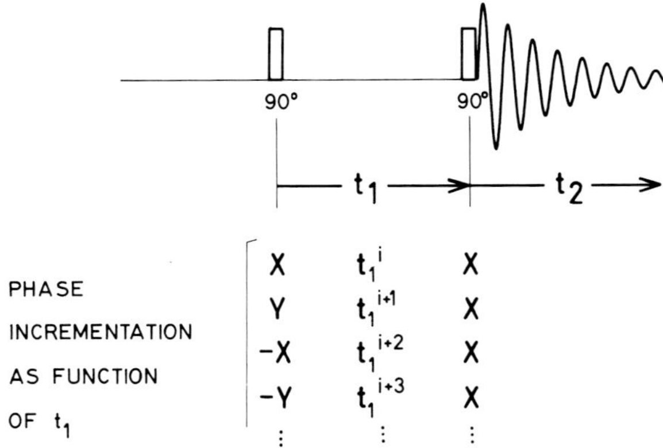

Fig. 5 Experimental scheme used to record COSY spectra of proteins in H_2O solution with four-quadrant phase sensitive display. The basic COSY experiment consists of two 90° pulses, which are separated by the evolution period t_1 (16-18). Immediately after the second pulse the signal is recorded during the observation period t_2. To obtain a 2D spectrum the measurement is repeated for a set of equidistant t_1 values ($t_1^i, t_1^{i+1}, t_1^{i+2}, \ldots$). To end up with a phase sensitive display in four quadrants the phase of the first pulse is incremented by 90° in parallel with each incrementation of t_1, whereas the phase of the second pulse is kept constant. This scheme may be amended with the common routines for suppression of artefacts, for example cancellation of axial peaks (17) and quadrature image suppression (24).

Data Corporation disk drive with a storage capacity of 20 000 kilowords. The pulse sequence of Fig. 5 was used. The phase incrementation of the first pulse amounts to an apparent shift of the rotating frame frequency by $\omega_1/2$ before the analog-to-digital conversion. A comparable two-channel detection has been used to obtain a four-quadrant phase sensitive display of NOESY (2D nuclear Overhauser enhancement spectroscopy) spectra (25). However, the implementation of the experimental scheme of Fig. 5 is quite different in ref. 25 and in the present work, since two spectrometers using a fundamentally different quadrature detection mode were employed. To suppress the resonance of H_2O a continuous irradiation of the solvent line was applied at all times except during t_2 (19). Special care had to be taken in the selection of the carrier frequency near the center of the spectrum, so that the quadrature image of the perturbations arising from the solvent irradiation were outside of the spectral region of interest. 1000 measurements were made, with t_1 values from 1μs to 200ms (The corresponding frequency range along ω_1 of 2400 Hz is less than the total range covered by the spectrum; hence part of the data were folded over). 4096 points were recorded in t_2, with a spectral width along ω_2 of 3600 Hz. To attain the digital resolution indicated in Fig. 2 the time domain data matrix was expanded by "zero filling" to 2048 points in t_1 and 16 384 points in t_2, and to improve the spectral resolution it was multiplied with a phase shifted sine bell in both directions. Because of the limitations on storage space only a part of the data were retained after Fourier transformation in t_2. The spectrum was recorded during ca. 40h, the Fourier transformation for the NH-C^{α}H region required ca. 10h.

The absolute value COSY spectrum of Fig. 2A was recorded on a Bruker WM 500 spectrometer. Quadrature detection was used in ω_2, the carrier frequency was at the low field end of the spectrum. 512 measurements with t_1 values from

Vol. 113, No. 3, 1983 BIOCHEMICAL AND BIOPHYSICAL RESEARCH COMMUNICATIONS

0.3 to 51ms were made. To end up with the digital resolution indicated in the Fig. caption 2, the time domain data matrix was expanded by "zero filling" to 2048 points in t_1 and 4096 points in t_2. Similar filter functions were applied as for the phase-sensitive spectrum. The spectrum was recorded in ca. 17h, the Fourier transformation for the entire spectrum required ca. 3h.

ACKNOWLEDGEMENTS: Financial support was obtained from the European Molecular Biology Organisation (EMBO long term fellowship to D. Marion) and the Schweizerischer Nationalfonds (project 3.528.79). We thank Dr. P. Štrop, Institute of Organic Chemistry and Biochemistry CSAV, Prague, CSSR for a gift of BUSI IIA.

REFERENCES:

1. Wüthrich, K., Wider, G., Wagner, G. and Braun, W. (1982) J.Mol. Biol. 155, 311-319.
2. Wagner, G. and Wüthrich, K. (1982) J. Mol. Biol. 155, 347-366.
3. Wider, G., Lee, K.H. and Wüthrich, K. (1982) J. Mol. Biol. 155, 367-388.
4. Arseniev, A.S., Wider, G., Joubert, F.J. and Wüthrich, K. (1982) J. Mol. Biol. 159, 323-351.
5. Keller, R.M., Baumann, R., Hunziker-Kwik, E.H., Joubert, F.J. and Wüthrich, K. (1983) J. Mol. Biol. 163, 623-646.
6. Hosur, R.V., Wider, G. and Wüthrich, K. (1983) Eur. J. Biochem. 130, 497-508.
7. Štrop, P., Wider, G. and Wüthrich, K. (1983) J. Mol. Biol. in press.
8. Wagner, G. and Wüthrich, K. (1979) J. Magn. Reson. 33, 675-680.
9. Roques B.P., Rao, R. and Marion, D. (1980) Biochemie 62, 753-773.
10. Anil Kumar, Wagner, G., Ernst, R.R. and Wüthrich, K. (1981) J. Am Chem. Soc. 103, 3654-3658.
11. Karplus, M. (1959) J. Phys. Chem. 30, 11-15.
12. Bystrov, V.F. (1976) Progr. in NMR spectrosc. 10, 41-81.
13. Wüthrich, K. (1976) NMR in Biological Research:Peptides and Proteins. North-Holland Publishing Company,Amsterdam.
14. DeMarco, A., Llinás, M. and Wüthrich, K. (1978) Biopolymers 17, 617-636.
15. DeMarco, A., Llinás, M. and Wüthrich, K. (1978) Biopolymers 17, 637-630.
16. Aue, W.P., Bartholdi, E. and Ernst, R.R. (1976) J. Chem. Phys. 64, 2229-2246.
17. Nagayama, K., Anil Kumar, Wüthrich, K. and Ernst, R.R. (1980) J. Magn. Reson. 40, 321-334.
18. Bax, A. and Freeman, R. (1981) J. Magn. Reson. 44, 542-561.
19. Wider, G., Hosur, R.V. and Wüthrich, K. (1983) J. Magn. Reson.,in press.
20. Aue, W.P., Karhan, J. and Ernst, R.R. (1976) J. Chem. Phys. 64, 4226-4227.
21. Nagayama, K., Bachmann, P., Wüthrich, K. & Ernst, R.R. (1978) J. Magn. Reson. 31, 133-148.
22. Nagayama, K. and Wüthrich, K. (1981) Eur. J. Biochem. 115,653-657.
23. Čechová, D., Jonáková, V., Sedlaková, E. and Mach, O. (1979) Hoppe-Seyler's Z. Physiol. Chem. 360, 1753-1758.
24. Hoult, D.J. and Richards, R.E. (1975) Proc. Roy. Soc. London A 344, 311-320.
25. States, D.J., Haberkorn, R.A. and Ruben, D.J. (1982) J. Magn. Reson. 48, 286-292.

Eur. J. Biochem. *114*, 365–374 (1981)
© FEBS 1981

Systematic Application of Two-Dimensional ^{1}H Nuclear-Magnetic-Resonance Techniques for Studies of Proteins

1. Combined Use of Spin-Echo-Correlated Spectroscopy and *J*-Resolved Spectroscopy for the Identification of Complete Spin Systems of Non-labile Protons in Amino-Acid Residues

Kuniaki NAGAYAMA and Kurt WÜTHRICH

Institut für Molekularbiologie und Biophysik, Eidgenössische Technische Hochschule, Zürich-Hönggerberg

(Received October 3, 1980)

This and the following paper describe the practical application of recently developed, two-dimensional nuclear magnetic resonance techniques for studies of proteins. In the present report spin-echo-correlated spectroscopy and two-dimensional *J*-resolved spectroscopy are used to identify complete spin systems of non-labile, aliphatic protons in the basic pancreatic trypsin inhibitor. Overall, 41 out of the 58 aliphatic spin systems in this protein were identified; for the first time the spin systems of all the glycyl residues in a protein have been identified in the ^{1}H NMR spectrum. Combined with the following paper, the present data yield new individual assignments for numerous amino acid residues and provide a new avenue, based on accurate measurements of spin-spin coupling constants in the two-dimensional *J*-resolved spectra, for studying changes of static and dynamic aspects of protein conformation between single crystals and solution, or between different conditions of solvent and temperature.

In several recent papers fundamental aspects of the use of two-dimensional (2D) ^{1}H nuclear magnetic resonance (NMR) spectroscopy [1–3] for studies of biological macromolecules have been described [3–12]. This included the development of the software capable of handling the large data matrices encountered in work with biopolymers [4,5], realization of 2D *J*-resolved spectroscopy [4,6], 2D correlated spectroscopy [7], spin-echo correlated spectroscopy [8,9], foldover-corrected correlated spectroscopy [9] and 2D nuclear Overhauser enhancement (NOE) spectroscopy [7,10] with proteins, and development of data handling routines for suitable presentation of the spectra [5,11,12]. Compared to conventional, one-dimensional NMR, 2D experiments can be much more efficient in that they provide, with a single instrument setting, complete information on all the *J* couplings in a macromolecule [1,7–9] or a complete set of NOE's between all nearby protons in a three-dimensional protein structure [7,10]. Furthermore 2D experiments avoid adverse effects which arise in the one-dimensional experiments due to the limited selectivity of radio-frequency irradiation of individual lines in crowded spectral regions [7–10], and 2D *J*-resolved spectroscopy affords greatly improved spectral resolution [3–6]. Overall the introduction of 2D NMR promises to make high-resolution NMR studies of biological macromolecules much more efficient and greatly to enhance the potential of nuclear magnetic resonance for many-parameter characterizations of biopolymer conformations in solution. This and the following paper [13] describe a systematic approach for the use of 2D NMR experiments for studies of proteins.

Abbreviations. The inhibitor, basic pancreatic trypsin inhibitor; reduced inhibitor, modified inhibitor obtained by reduction of the disulfide bond 14-38 and protection of the cysteinyl side chains by carboxamidomethylation; δ, chemical shift; *J*, spin-spin coupling constant; NOE, nuclear Overhauser effect; 2D, two-dimensional. The term SECSY, for spin-echo-correlated spectroscopy, was used in previous publications.

To obtain a truly many-parameter characterization of biopolymer conformations the largest possible number of resonance lines in the NMR spectra must be resolved, grouped into spin systems connected by *J* couplings and assigned to specific residues in the amino acid sequence [14]. The strategy outlined in this and the following paper foresees that on the one hand spin-echo-correlated or conventional 2D correlated spectroscopy are used in combination with 2D *J*-resolved spectroscopy to identify complete spin systems of the non-labile hydrogen atoms of individual amino acid residues. On the other hand 2D correlated spectroscopy and 2D NOE experiments are combined to obtain assignments of polypeptide backbone resonances to specified locations in the amino acid sequence. Since the two data sets thus obtained overlap at the α-proton positions, numerous complete spin systems can thus be assigned to specific locations in the amino acid sequence without reference to information obtained from X-ray crystallography or other techniques.

This paper describes the use of 2D spin-echo-correlated and 2D *J*-resolved spectroscopy for the identification of numerous spin systems of non-labile protons in the basic pancreatic trypsin inhibitor, a small globular protein of 58 amino acid residues and a molecular weight of 6500. The inhibitor is ideally suited for the demonstration of the practical applications of 2D NMR methods. A considerable number of resonance lines was previously assigned with conventional experiments [15] and these will be used to check on the reliability of the new techniques.

MATERIALS AND METHODS

Two-Dimensional J*-Resolved Spectroscopy*

For those used to working with conventional, one-dimensional NMR techniques, 2D *J*-resolved spectra are probably the most easily visualized of all the different types of 2D NMR spectra [16]. Compared to the corresponding one-dimensional spectrum each multiplet in a 2D *J*-resolved spectrum has been rotated to a certain extent about its center, e.g. by 90° in the δ/J representation [5]. Multiplets which overlap in the one-dimensional spectrum are therefore separated in a 2D *J*-resolved spectrum whenever they posses different chemical shifts, and the individual multiplet components are readily identified from inspection of the 2D *J*-resolved spectrum. Since the chemical shifts, δ, and the spin-spin coupling constants, J, are displayed in different dimensions, J couplings can be recorded with higher digital resolution than is usually practicable in one-dimensional spectroscopy. For work with proteins 2D *J*-resolved spectroscopy is an attractive experiment for determination of the multiplicity of individual resonances in crowded spectral regions and for accurate measurements of the spin-spin coupling constants, J.

The 2D *J*-resolved experiment uses the following pulse sequence [1,4,5]: $[90° - t_1/2 - 180° - t_1/2 - t_2]_n$. The experiment is initiated by a non-selective 90° pulse. In the middle of the 'evolution period', t_1, a refocusing 180° pulse is applied. The data acquisition as a function of t_2 starts at the time of the spin echo. In accordance with the general 2D spectroscopy principle [1], the experiment is repeated for a set of equidistant t_1 values. To attain a workable signal-to-noise ratio, n signals are accumulated for each value of t_1. A two-dimensional Fourier transformation of the data matrix $s(t_1,t_2)$ then produces the desired frequency domain spectrum $S(\omega_1,\omega_2)$. Through a spectrum tilt [5] the δ/J representation of the spectrum can then be obtained and for weakly coupled spin systems the individual multiplets are best recorded in the form of cross-sections through their chemical shift positions and parallel to the J axis.

Spin-Echo-Correlated Spectroscopy

This method yields 2D homonuclear correlated spectra which manifest connectivities between *J*-coupled nuclei. Compared to normal 2D correlated spectroscopy [1], it uses a smaller data size, which makes it particularly suitable for studies of macromolecules [8,9]. When used instead of the conventional one-dimensional spin-decoupling experiments, it is a powerful method for the identification of *J*-coupled spin systems. With a single instrument setting it can provide a complete set of all the J connectivities in a protein and further avoids detrimental effects arising from limited selectivity of irradiation in crowded spectral regions [8,9]. Compared to one-dimensional spectra the resolution in a 2D spin-echo-correlated spectrum is greatly improved since the correlation peaks are spread out in two dimensions.

The 2D spin-echo-correlated method uses the following pulse sequence [8,9]: $[90° - t_1/2 - 90° - t_1/2 - t_2]_n$. Compared to the 2D *J*-resolved experiment, the 180° pulse in the middle of the evolution period is substituted by a 90° pulse. In the frequency domain spectrum obtained after Fourier transformation in two dimensions, peaks corresponding to the normal, one-dimensional spectrum are seen on a straight line through the center of the spectrum and parallel to the chemical shift axis. The second axis is perpendicular to the chemical shift axis and indicates the chemical shift differences $\Delta\delta$, between *J*-coupled protons. Correlation peaks occur at the chemical shifts of the *J*-coupled protons and are displaced from the center of the spectrum by $\pm 0.5\,\Delta\delta$ along the $\Delta\delta$ axis. As a consequence a straight line connecting the two correlation peaks which indicate *J*-coupling between two protons forms an angle of 135° with the chemical shift axis [8]. Table 1 shows schemes of correlation peaks for weakly coupled spin systems corresponding to those of the common amino acid residues [14]. The simplest patterns are those of glycine and alanine which contain only two groups of non-equivalent protons. In the more complex spin systems the spin-echo-correlated spectra manifest the sequential J connectivities between all the individual spins.

Table 1. *Connectivity diagrams for 2D spin-echo-correlated spectra of weakly coupled spin systems of the non-labile, aliphatic protons in the common amino acid residues*

The spin systems of Leu, Arg, Lys and Pro are not shown, since identifications of the complete J connectivities for the residues were not obtained in the present paper. 'Diagonal peaks' are represented by filled circles, correlation peaks between *J*-coupled spins by filled triangles. The manifestations of spin-spin coupling fine structure in the diagonal as well as the cross peaks [8,9] are not shown

Spin system	Amino acid residues	Spin-echo-correlated connectivities
AX A_3X	Gly Ala	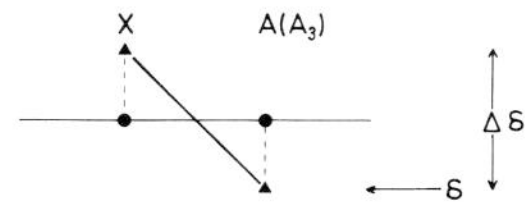 X A(A_3) (axes: δ, $\Delta\delta$)
AMX	Asn, Asp, Cys, His, Phe, Ser, Trp, Tyr	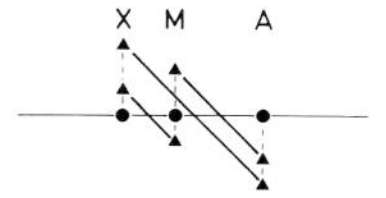 X M A
A_3MX	Thr	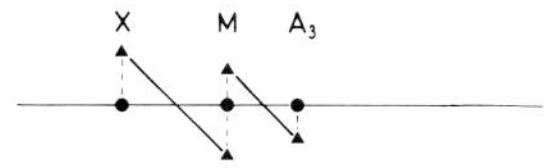 X M A_3
A_3B_3MX	Val	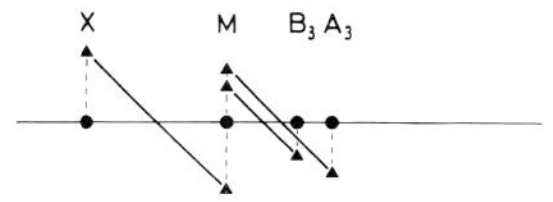 X M B_3 A_3
AMPTX	Gln, Glu, Met	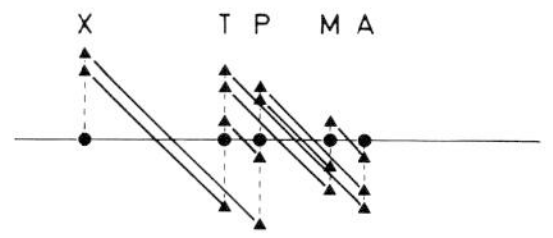 X T P M A
A_3B_3MPTX	Ile	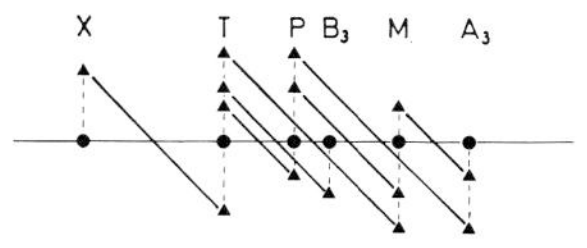 X T P B_3 M A_3

NMR Sample Preparation and Equipment

Basic pancreatic trypsin inhibitor (Trasylol, Bayer Leverkusen) was obtained as a gift from the Farbenfabriken Bayer AG. The reduced inhibitor was prepared by selective reduction of the disulfide bond 14-38 with borohydride as a reducing agent and iodoacetamide as a protective group [17]. The material used in the present experiments was previously extensively characterized [18,19]. For the NMR experiments 0.01 – 0.015 M solutions of the proteins in 100% ^2H_2O obtained from Stohler Chemicals were prepared. The p^2H of the solutions was adjusted by the addition of minute amounts of ^2HCl or NaO^2H. The p^2H values correspond to pH meter readings without correction for isotope effects [20,21]. Prior to the NMR measurements the labile protons were exchanged with deuterium by heating the protein solution in ^2H_2O followed by lyophilization and redissolving in ^2H_2O.

^1H NMR spectra were recorded on a Bruker HXS-360 spectrometer equipped with an Aspect 2000 data system and a standard Bruker temperature control unit. The software used to handle the large data matrices obtained in 2D experiments with biological macromolecules was described previously [4,5]. In the present experiments the size in the time domain was 1024×256 for spin-echo-correlated spectra and 4096×64 for *J*-resolved spectra. After two-dimensional Fourier transformation the frequency domain spectra were developed on data matrices of size 1024×512 for spin-echo-correlated spectra and 4096×128 for *J*-resolved spectra. The digital resolution in both dimensions was 2.44 Hz in the spin-echo-correlated spectra and 0.61 Hz in the *J*-resolved spectra. The spectral resolution was improved by multiplication of the free induction decays either with an increasing exponential and a cosine function [4] or with a phase-shifted sine bell [22]. For the spin-echo-correlated spectra the total accumulation time was 4 – 6 h, depending on the experiment, for *J*-resolved spectra it was 12 – 16 h. All the measurements were made in tubes with 5-mm outer diameter. Chemical shifts are quoted relative to the internal standard sodium 3-trimethylsilyl-(2,2,3,3-$^2H_4$)propionate, with correction for the p^2H dependence of the chemical shift of this reference compound [23].

RESULTS AND DISCUSSION

Fig. 1 presents a three-dimensional view of the 2D *J*-resolved spectrum of the inhibitor. The region of 0.5 – 6.0 ppm includes the resonances of 288 protons, i.e. all the non-labile protons with the exception of those of the aromatic rings. The 2D *J*-resolved spectrum contains well resolved multiplets for a large number of resonances; 15 of these multiplets are displayed in the cross-sections of Fig. 2. Additional cross-sections through *J*-resolved spectra of the inhibitor, including those of all the methyl groups, were previously presented [3,5,6]. A three-dimensional view of the 2D spin-echo-correlated spectrum of the inhibitor which was recorded under the same conditions as the 2D *J*-resolved spectrum of Fig. 1 is shown in Fig. 3. A large number of correlation peaks indicating proton-proton *J* couplings is readily apparent. However, for a detailed spectral analysis a countour plot presentation is more suitable. Fig. 4 shows a contour plot of the spectrum in Fig. 3, where the connectivities for a group of spin systems have been indicated. Connectivities for additional spin systems are displayed in a contour plot of a spin-echo-correlated spectrum of the reduced inhibitor (Fig. 5) and in three contour plots of selected regions of similar spectra of the native inhibitor recorded under various different conditions of p^2H and temperature (Fig. 6 – 8). Table 2 presents a list of the chemical shifts for spin systems of non-labile aliphatic protons which were identified on the basis of the 2D spin-echo-correlated

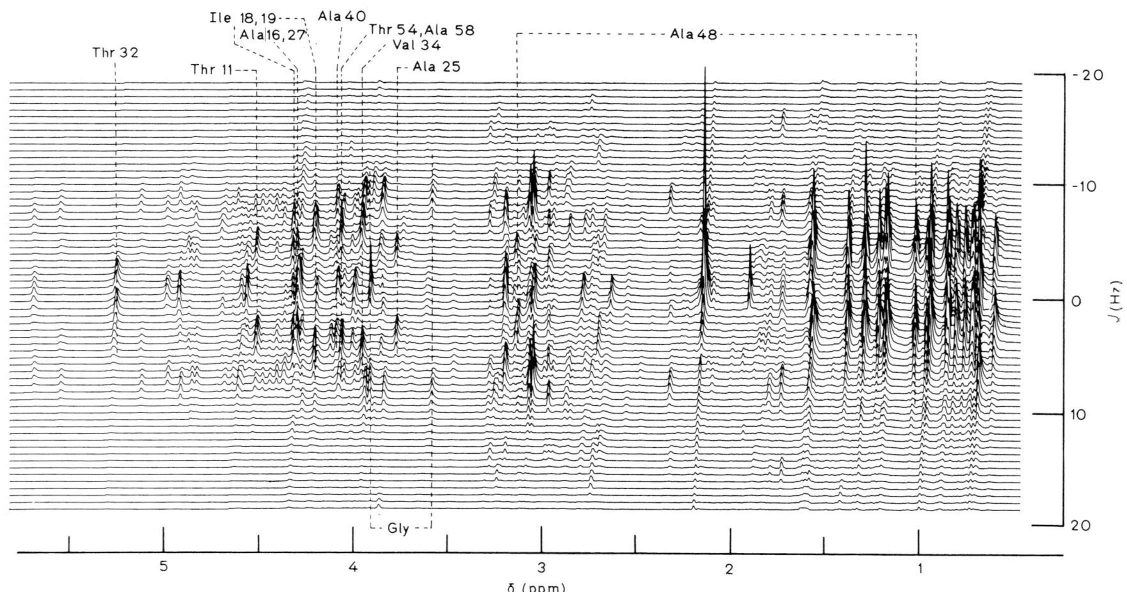

Fig. 1. *Spectral region of 0.5 – 6.0 ppm of the 2D J-resolved ^1H NMR spectrum at 360 MHz of a 0.01-M solution of the basic pancreatic trypsin inhibitor in ^2H_2O, p^2H 7.0, 68 °C.* The labile protons had been replaced with deuterium prior to this experiment and the protein was lyophilized repeatedly from ^2H_2O. The singlet line of the residual water protons is at 4.35 ppm. The spectrum is in the δ/J representation [5]. The stacked plots in the *J* direction provide a three-dimensional view of the spectrum. The broken lines indicate the chemical shifts where the cross-sections in Fig. 2 were taken. Resonance assignments for the cross-sections are also given

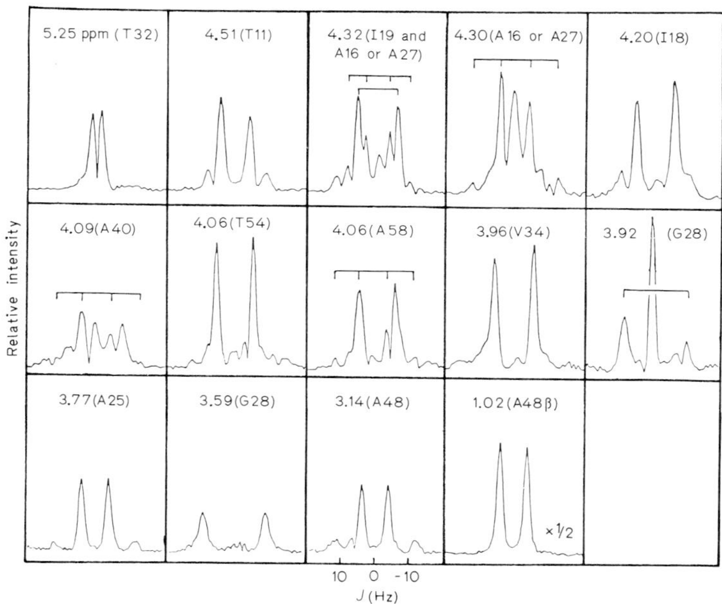

Fig. 2. *Cross-section representation of selected multiplets in the 360-MHz 2D* J*-resolved spectrum of the basic pancreatic trypsin inhibitor in Fig. 1.* The chemical shifts where the cross-sections parallel to the J axis have been taken are indicated by the numbers in the upper left corner of each multiplet representation, the resonance assignments are indicated in parentheses. Note that because of the different conditions of temperature and p^2H, all the chemical shifts do not coincide with those in Table 2

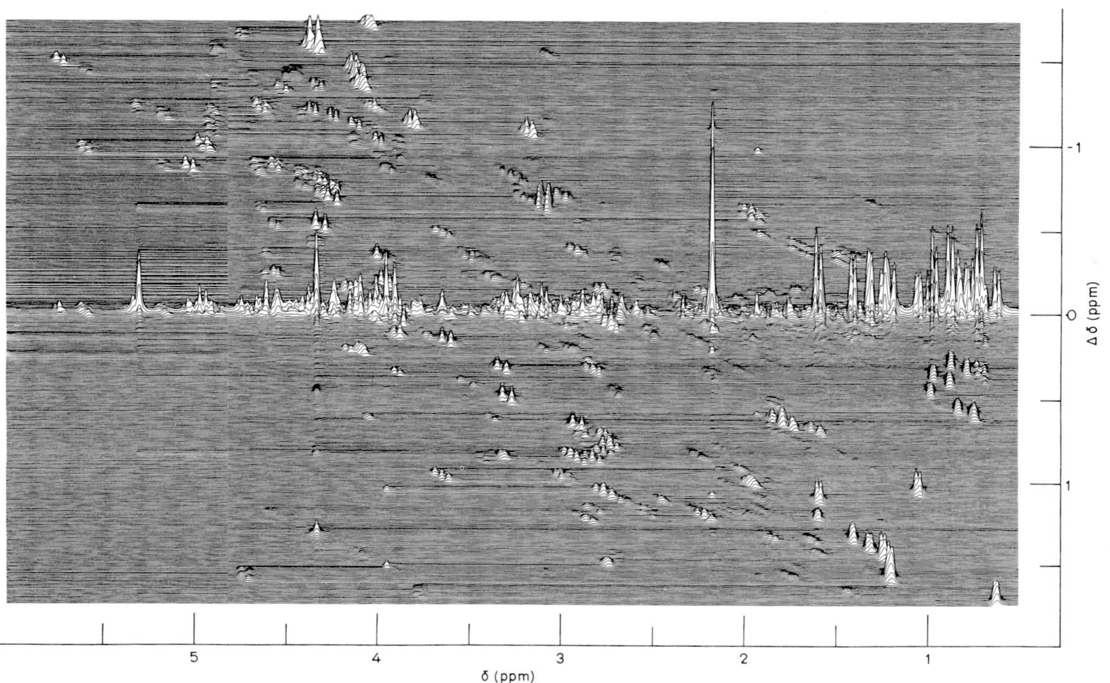

Fig. 3. *Three-dimensional presentation of the spectral region 0.5–6.0 ppm of a 360-MHz spin-echo-correlated* ^1H *NMR spectrum recorded in the same solution of the inhibitor used in Fig. 1 at 68 °C.* The chemical shift, δ, on the horizontal axis corresponds to that in conventional, one-dimensional spectra. $\Delta\delta$ on the vertical axis stands for the difference frequencies between correlated nuclei. Cross-peaks between J-coupled nuclei are at $\pm$ 0.5 $\Delta\delta$. The solvent singlet resonance is at 4.35 ppm. The apparent discontinuity at 4.8 ppm resulted since the spectrum had, for practical reasons, to be plotted in fragments extending over approximately 4 ppm

369

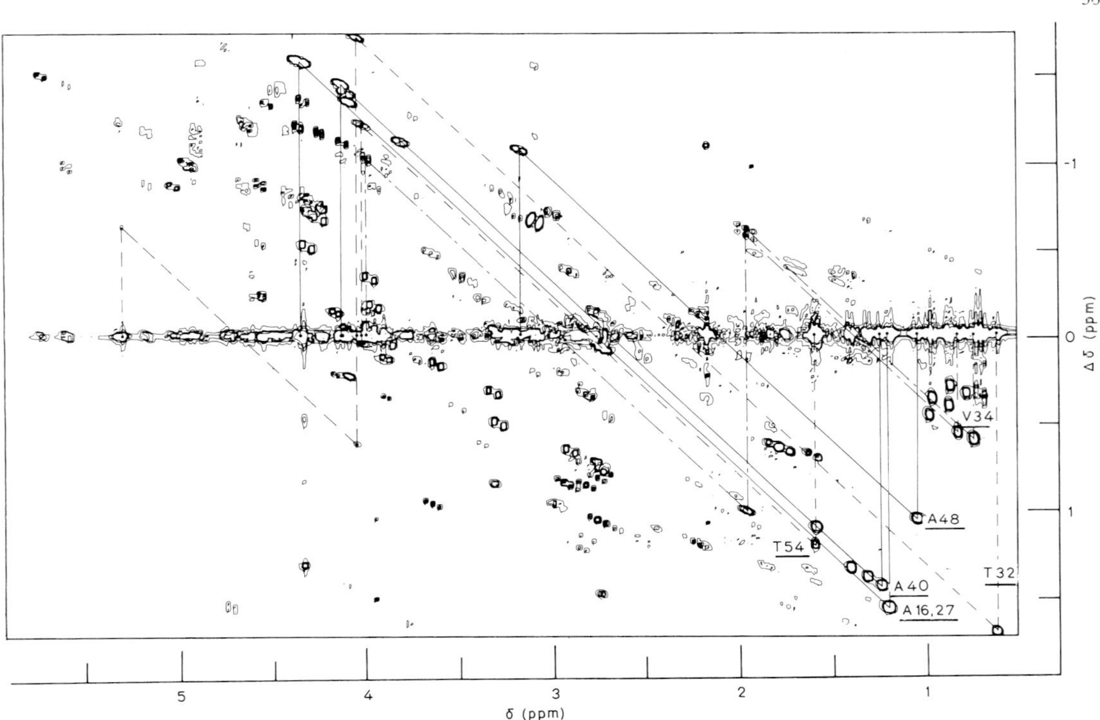

Fig. 4. *Contour plot of the spin-echo-correlated spectrum of the inhibitor in Fig. 3.* Connectivities between the individual components of the following spin systems are indicated: (——) alanines 16, 27, 40, 48; (– – – –) threonines 32, 54; (– · – · –) Val-34

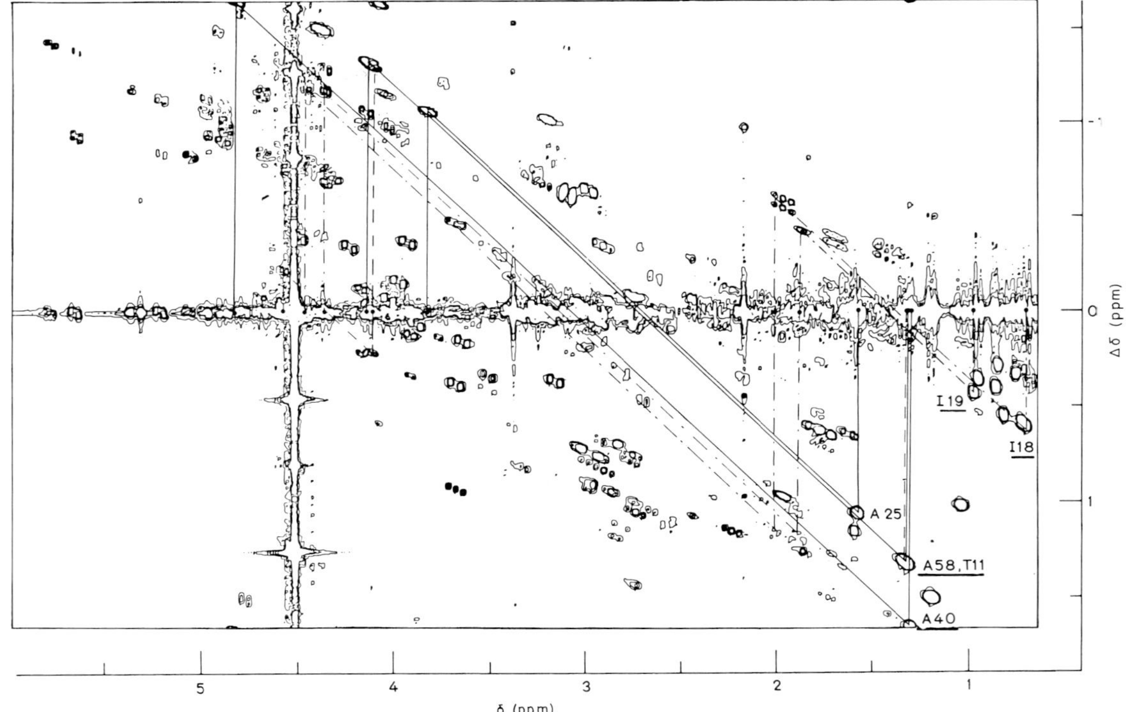

Fig. 5. *Contour plot of the region 0.7 – 6.0 ppm of a spin-echo-correlated spectrum of the reduced inhibitor.* A 0.015-M solution of the protein in ^2H_2O was measured at p^{2}H 4.1, 55 °C. The strong vertical noise band at about 4.4 ppm is at the chemical shift of the residual water protons. Connectivities are indicated for the following spin systems: (——) alanines 25, 40, 58; (– – – –) Thr-11; (– · – · –) isoleucines 18, 19 (only C^{α}-H-C^{β}-H-C^{γ}-H_3)

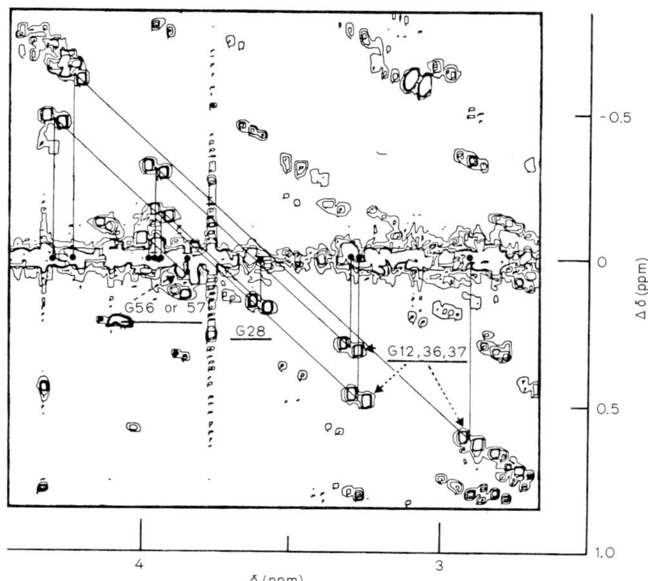

Fig. 6. *Contour plot of the region 2.7–4.5 ppm of a spin-echo-correlated spectrum recorded in a 0.01-M solution of the inhibitor in ^2H_2O, p^2H 4.2, 68 °C.* The solid lines indicate the connectivities for five glycyl spin systems

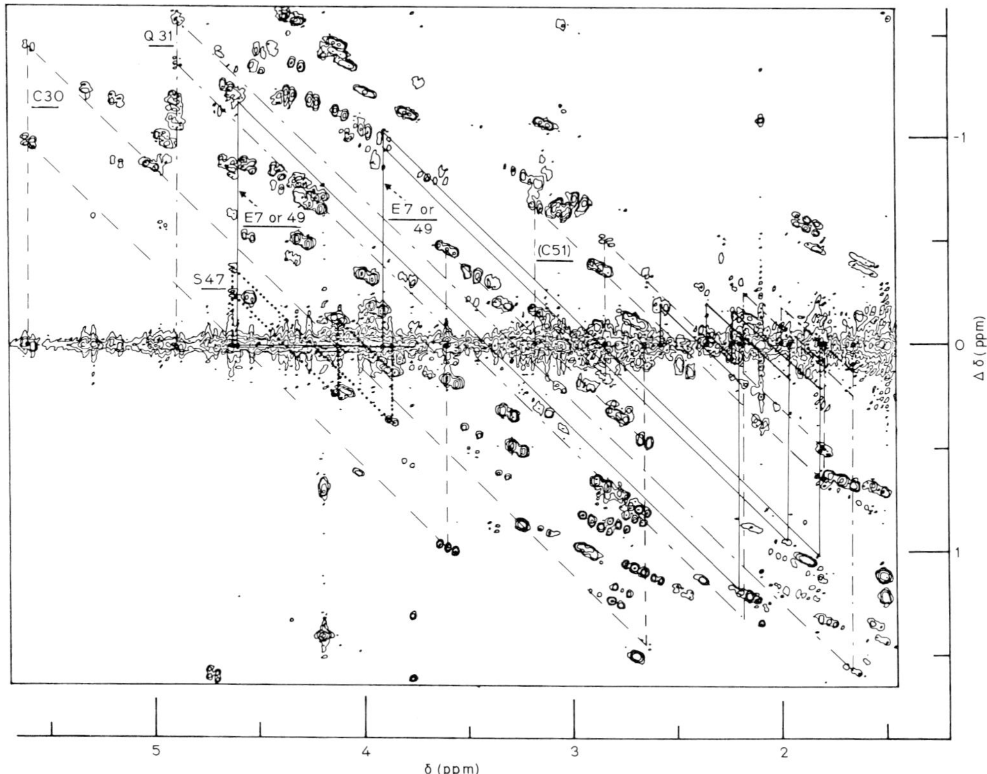

Fig. 7. *Contour plot of the region 1.5–5.7 ppm of a spin-echo-correlated spectrum recorded in a 0.01-M solution of the inhibitor in ^2H_2O, p^2H 4.2, 85 °C.* Connectivities are indicated for the following spin systems: (—) Glu-7, Glu-49; (----) Cys-30, (Cys-51); (-·-·-) Gln-31; (· · · · ·) Ser-47

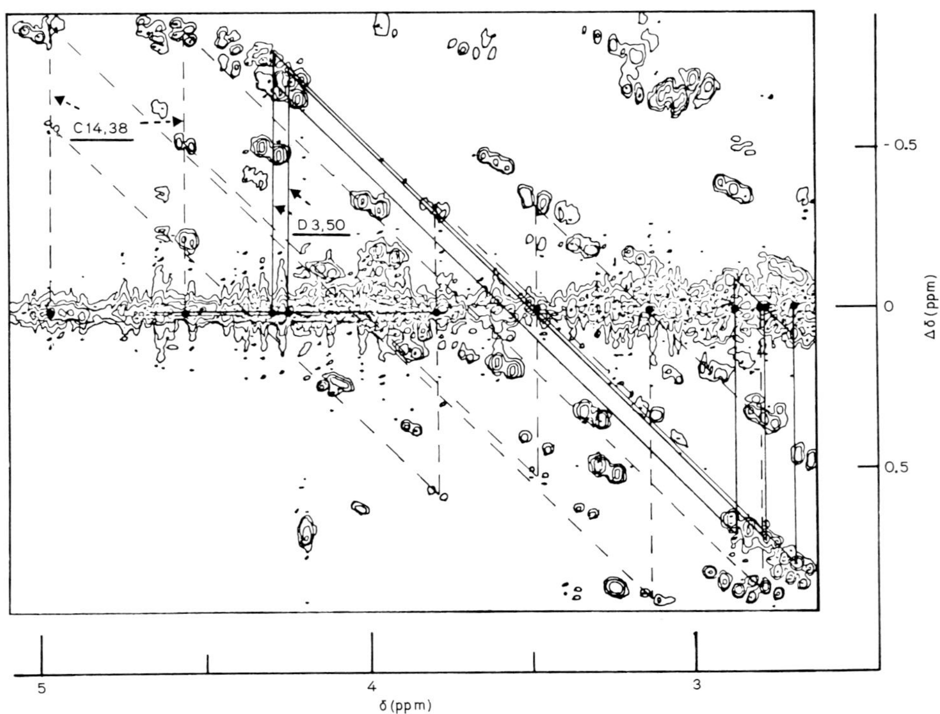

Fig. 8. *Contour plot of the region 2.4–5.4 ppm of the same spectrum as in Fig. 7.* The connectivities are indicated for the following spin systems: (——) Asp-3, Asp-50; (– – – –) Cys-14, Cys-38

and 2D *J*-resolved spectra and previous work with conventional one-dimensional NMR experiments [24–26].

Strategy Used for the Resonance Assignments

Three essential steps may be distinguished in the procedures used to obtain the resonance assignments listed in Table 2. (a) Identification of the components of distinct spin systems using the *J* connectivities manifested in the 2D spin-echo-correlated spectra. (b) Use of 2D *J*-resolved spectra to determine spin multiplicities and measure spin-spin coupling constants for distinct resonances. (c) Comparison of 2D *J*-resolved and spin-echo-correlated spectra recorded at different p^{2}H and different temperature for native and reduced inhibitor.

For some amino acid residues the identification with step (a) is unambiguous, since they have unique connectivity diagrams in the spin-echo-correlated spectra. From Table 1 it is seen that these include Thr, Val and Ile, and further unique spin systems would be those of Leu, Lys and the aromatic rings of Phe, Tyr, His and Trp. For all the other spin systems additional information is required to substantiate assignment to a particular type of amino acid residue. In the present investigation this additional information was mainly obtained from determination of the multiplet structures and/or accurate measurements of spin-spin coupling constants in the 2D *J*-resolved spectra, from the p^{2}H dependence of the chemical shifts of amino acid residues with ionizeable groups [14], from comparison of native and reduced inhibitor, and from previous sequential assignments of backbone C-α proton resonances [25]. In many instances combination of the new data from 2D spectroscopy with the information available from previous studies resulted in individual assignments of complete spin systems (Table 2).

The foregoing considerations show clearly that the information obtained from spin echo correlated spectroscopy and 2D *J*-resolved spectroscopy is complementary. The importance of the combined use of the two techniques is further emphasized since spin-echo-correlated experiments with proteins use, for practical reasons [9], generally a rather low digital resolution. In the present study the resolution was 2.44 Hz/point for spin-echo-correlated spectra and 0.61 Hz/point for 2D *J*-resolved spectra. Therefore accurate chemical shifts were obtained from the location in the *J*-resolved spectrum of the individual multiplets connected by cross peaks in the spin-echo-correlated spectra.

When inspecting the spin-echo-correlated spectra one must consider that the multiplet structure of the resonances is also manifested in the correlation peaks [8,9]. Therefore, while the peripheral protons, e.g. on C-α and the methyl groups, usually give prominent correlation peaks, the peaks from the connecting methylene or methine protons are, as a consequence of both high multiplicity and splitting into several correlation peaks, often rather diffuse and have small peak heights. The complexity and hence the diffuse appearance of the correlation peaks is further enhanced in strongly coupled spin systems. As a consequence of these spectral features the two correlation peaks which manifest *J* connectivity between two groups of protons (Table 1) may appear with markedly different peak heights, and connectivities between non-peripheral groups of protons are as a rule rather ambiguous. In the present investigation these difficulties prevented reliable identification of the complete spin systems of Pro, Leu, Arg and Lys.

Table 2. *Survey of 41 spin systems of non-labile, aliphatic protons in the inhibitor, which have so far been identified with the use of one-dimensional and two-dimensional NMR*

The 2D NMR experiments described in this paper cannot *a priori* yield individual assignments. However, in many instances earlier one-dimensional experiments provided assignments for certain components of spin systems, e.g. only the α-proton [25] or only the peripheral protons of an amino acid side chain [24]. In many of these cases the present 2D experiments resulted in the complete identification of the individually assigned spin systems. Furthermore some new individual assignments resulted from comparison of 2D spectra of the native and modified inhibitor, and from studies of the p²H dependence (see text). For the residues marked with *, individual assignments were obtained from combination of the present data with the sequential assignments described in the following paper [13]. The spin systems which have been individually assigned are listed first, in the order of the amino acid sequence, then all the other spin system which have so far been assigned only to groups of two or more residues. Chemical shifts are given (± 0.01) relative to internal sodium 3-trimethyl-silyl-(2,2,3,3-$^2H_4$)propionate at 24 °C and p²H 4.6. In the references column, 0 stands for 'this paper'. In all the spin systems listed the *J* connectivities were observed by the 2D NMR studies in this paper. However, in this column reference to the present work is included only where new resonance identifications were obtained

Resonance assignment	δ for protons on			References
	C-α	C-β	other	
	ppm			
Thr-11	4.53	4.04	1.39 (C-γ)	0, 24
Ile-18	4.19	1.87	1.07 (C-γ1)	24, 25
			0.97 (C-γ2)	
			0.68 (C-δ)	
Ile-19	4.30	1.96	1.46 (C-γ1)	24, 25
			0.73 (C-γ2)	
			0.68 (C-δ)	
Tyr-21	5.70	2.70		25
		2.70		
Phe-22	5.29	2.92		25
		2.81		
Tyr-23	4.31	3.45		25
		2.72		
Asn-24*	4.60	2.85		0
		2.18		
Ala-25	3.76	1.57		24
Gly-28	3.91			0
	3.61			
Cys-30	5.63	3.69		25
		2.67		
Gln-31	4.82	2.15	2.02 (C-γ)	0, 25
		1.72	1.83 (C-γ)	
Thr-32	5.28	4.04	0.59 (C-γ)	24, 25
Phe-33	4.86	3.11		0, 25
		2.98		
Val-34	3.92	1.95	0.81 (C-γ)	24, 25
			0.71 (C-γ)	
Tyr-35	4.89	2.69		25
		2.52		
Gly-36*	4.35			0
	3.25			
Ala-40	4.09	1.21		24
Asn-43*	5.09	3.39		0
		3.31		
Asn-44*	4.94	2.79		0
		2.51		
Phe-45	5.12	3.42		0, 25
		2.79		
Ser-47	4.51	4.08		0
		3.84		
Ala-48	3.14	1.06		24

Table 2 (Continued)

Resonance	δ for protons on			References
	C-α	C-β	other	
	ppm			
Met-52	4.17	2.07	2.70 (C-γ)	0
		2.00	2.70 (C-γ)	
Thr-54	4.07	3.95	1.61 (C-γ)	0, 24
Ala-58	4.01	1.31		24
Asp-3	4.30	2.89		0
		2.73		
Asp-50	4.26	2.79		0
		2.79		
Glu-4	4.55	2.19	2.64 (C-γ)	0
		2.19	2.64 (C-γ)	
Glu-49	3.86	2.03	2.41 (C-γ)	0
		1.87	2.28 (C-γ)	
Gly-12	4.21			0
	2.92			
Gly-37	3.89			0
	3.25			
Cys-14	4.96	3.80		0
		3.15		
Cys-38	4.55	3.48		0
		2.79		
Ala-16	4.30	1.19		24
Ala-27	4.28	1.19		24
Gly-56 or	3.96			0
Gly-57	3.80			
A_2X	4.93	2.97		0
		2.97		
Phe-4 AMX	4.62	3.33		0
Cys-5		2.97		
Tyr-10 AMX	4.62	2.26		0
Cys-51		2.03		
Cys-55 AMX	4.33	2.88		0
		2.74		
AMX[a]	1.68[a]	3.18[a]		0
		2.89[a]		

[a] On the basis of ring current calculations using the same parameters as in [27] this AMX spin system was tentatively assigned to Cys-51, whereby the highest field line would correspond to the C-α proton.

For a protein even the 2D NMR spectra contain crowded regions where individual peaks may overlap (Fig. 1 and 3). However, the chemical shifts of the individual lines usually have different p²H and temperature dependences so that possible ambiguities in the resonance assignments can be resolved by variation of these parameters. In the present study the native inhibitor was measured at p²H 3.1 and 68 °C, at p²H 4.2 and 30 °C, 68 °C and 85 °C, and at p²H 7.0 and 68 °C. The reduced inhibitor was studied at p²H 4.1 and 30 °C, 41 °C and 55 °C. Between these different spectra each of the spin systems listed in Table 2 could be completely resolved.

Identification of the Spin Systems for the Different Amino Acid Types

In the amino acid sequence of the inhibitor [28],

R¹P D F C⁵ L E P P Y¹⁰ T G P C K¹⁵ A R I I R²⁰ Y F Y N A²⁵ K A G L C³⁰-
Q T F V Y³⁵ G G C R A⁴⁰ K R N N F⁴⁵ K S A E D⁵⁰ C M R T C⁵⁵ G G A⁵⁸,

there are six glycines. Fixe AX spin systems of glycine could be identified from the connectivity patterns (Table 1) in the spin-echo-correlated spectrum (Fig. 6) and the multiplet fine structures seen in the *J*-resolved spectra (Fig. 2) includes the cross-sections for one glycine.

The inhibitor contains six alanines. The alanine spin systems are unambiguously manifested by the spin-echo-correlated connectivity diagrams (Table 1) shown in Fig. 4 and 5 and the multiplet structures of the C-α proton resonances in the *J*-resolved spectra (Fig. 2). These results coincide with the previously published assignments of the alanyl A_3X spin systems [24]. In agreement with the earlier work five alanines have nearly identical NMR parameters in native and reduced inhibitor, while Ala-40 is markedly different in the two proteins [18]. Therefore the connectivities for Ala-40 are shown for the inhibitor (Fig. 4) as well as the reduced inhibitor (Fig. 5).

The unique connectivity pattern (Table 1) of the single Val-34 in the inhibitor is indicated in Fig. 4. For the two isoleucines the connectivities between the γ-methyl doublets, C-β and C-α protons are indicated in Fig. 5; they coincide with the previously established assignments [26]. The connectivity peaks between the C-β proton and the previously identified [24] C-γ H_2 could not be unambiguously identified. The unique spin-echo-correlated connectivities for threonine (Table 1) are indicated in Fig. 4 and 5. For Thr-32 the previous assignment of the complete spin system [26] was thus confirmed. For Thr-11 and Thr-54 the previous assignments for the C-β H and C-γ H_3 were complemented with the identification of the C-α H resonances. For Val, Ile and Thr the identification of the C-α H resonances by the spin-echo-correlated connectivity diagrams (Fig. 4 and 5) was further supported by analysis of the multiplet structures in the *J*-resolved spectra (Fig. 2).

The inhibitor contains 20 AMX spin systems of 1 Ser, 2 Asp, 3 Asn, 6 Cys, 4 Phe and 4 Tyr. 17 AMX spin systems and 3 A_2X system were identified from the spin-echo-correlated connectivity patterns (Table 1) and the multiplicities in the *J*-resolved spectra observed in the experiments recorded at various p^2H values and temperatures. 12 AMX spin systems were further assigned to particular types of amino acids. This includes Asp-3 and Asp-50, which were identified from the p^2H dependence of the chemical shifts between p^2H 3.1 and 7.0 [21]. The single Ser-47 was distinguished from all the other AMX spin systems by the outstandingly small geminal coupling constant [14], i.e. $^3J_{\beta\beta} = 12.4$ Hz, measured in the *J*-resolved spectra. Cys-14 and Cys-38 were identified on the basis of the outstanding chemical shift differences between native and reduced inhibitor. The connectivities for these five residues are indicated in Fig. 7 and 8. The assignments of Tyr-21, Phe-22, Tyr-23, Cys-30 (Fig. 7), Phe-33, Tyr-35 and Phe-45 were obtained from combination of the 2D NMR data with previous sequential assignments of the C-α H resonances for these residues [25]. Of the remaining eight AMX spin systems one was tentatively assigned to Cys-51 (Fig. 7) on the basis of ring current calculations.

The inhibitor contains four five-spin systems of Glu-7, Glu-49, Gln-31 and Met-52. These could be identified from the spin-echo-correlated connectivity diagrams (Table 1) (Fig. 7). The two spin systems of Glu were then assigned from the p^2H dependence of the chemical shifts of the C-γ H_2 resonances [21]. One five-spin system was assigned to the single Gln-31 on the basis of the coincidence of the C-α H resonance with the previous assignment for this proton [25]. The fourth five-spin system thus had to come from Met-52.

Of the remaining spin systems in the inhibitor, those of the aromatic rings were previously completely assigned [29]. For the spin systems of the 2 Leu, 4 Pro, 4 Lys and 5 Arg complete connectivities could not at present be obtained. Obviously the spin-echo-correlated spectra (Fig. 3–5) contain additional correlation peaks besides those used for the assignments in Table 2, e.g. from the δ-methyl groups of Leu and Ile and in the C-α region. However, for the reasons discussed in the preceding section, the connectivities between non-peripheral protons in the long amino acid side chains could not yet be reliably outlined and therefore no data on these residues are included in Table 2.

Resonance Assignments to Individual Amino Acid Residues

In Table 2 individual assignments are listed for 25 amino acid residues. Among these are Ile-18, Ile-19, Tyr-21, Phe-22, Tyr-23, Ala-25, Cys-30, Thr-32, Val-34, Tyr-35, Ala-40, Ala-48 and Ala-58, which were taken from earlier work [24–26]. For Thr-11, Gln-31, Phe-33, Phe-45 and Thr-54 individual components of the spin systems were previously assigned [24–26] and combined with this earlier data the 2D NMR experiments resulted in the individual assignments for the complete spin systems. For the single Ser, identification of the spin system yielded also the assignment to position 47. For the single Met-52, assignment of the five-spin system resulted since all the other five-spin systems in the inhibitor could be assigned to different residues. The assignment of Gly-28 was based mainly on comparison of native and reduced inhibitor and is further discussed below. Asn-24, Gly-36, Asn-43 and Asn-44 were assigned from the combination of the present data with the sequential assignments described in the following paper [13].

A tentative assignment for Cys-51 is included in Table 2 since ring current calculations using the previously described parameters [27] indicated that the C-α resonance of this residue should be in an unusual high field position at about 1.7 ppm. The *J*-resolved spectra showed that the highest field component at 1.68 ppm of the AMX spin system assigned to Cys-51 was a doublet of doublets with spin-spin couplings of 4.9 and 12.8 Hz, which would be compatible with the assignment to a C-α resonance of Cys.

Inspection of Fig. 4 and 5 shows that overall the spectra of native and reduced inhibitor are closely similar, which manifests that the two proteins adopt nearly identical average spatial structures [18]. For example, the correlation peaks of the methyl groups in the lower right part of the spectrum are in identical locations for the two proteins, with the exception of Ala-40 (Fig. 4 and 5 [18]) and a small shift of Thr-11. On the other hand a small number of peaks are in largely different positions in the two proteins. This includes the spin systems of Cys-14 and Cys-38 (Fig. 8), which were assigned on the basis of this comparison, and three of the five identified glycine spin systems (Fig. 6). The following additional information on the glycines was available. One of Gly-56 and Gly-57 has an A_2 spin system [30], which explains that only five AX spin systems were observed in the spin-echo-correlated spectra. While four of the AX spin systems were independent of p^2H, the fifth system was shifted by 0.04 ppm between p^2H 3.1 and 4.2 at 68 °C. This system was therefore assigned to the second one of Gly-56 and Gly-57. Since the three glycines at positions 12, 36 and 37 are near the modification site in the reduced inhibitor, the single AX spin system which was not affected by the cleavage of the disulfide bond 14-38 was assigned to Gly-28 (Table 2).

CONCLUSIONS

It was the main purpose of this paper to illustrate the use of two recently developed 2D NMR techniques for the identification of complete proton spin systems of individual amino acid residues in proteins. Even though a considerable number of assignments in the inhibitor had already been made by conventional NMR experiments [26], partial or complete new assignments were obtained for 26 aliphatic spin systems (Table 2). In the following paper [13] the information on distinct spin systems in Table 2 is combined with new sequential assignments of polypeptide backbone resonances to obtain new assignments for individual amino acid residues. From this and the following paper [13] it is quite obvious that 2D NMR techniques provide a more powerful and more efficient approach for future studies of proteins than conventional, one-dimensional NMR experiments.

Information on the conformation of the inhibitor derived from the new spectroscopic data presented in this paper will be described elsewhere. Suffice it here to indicate that for nearly all the identified spin systems in Table 2, accurate values for the vicinal spin-spin coupling constants $^3J_{\alpha\beta}$ were obtained from the *J*-resolved spectra. This opens a new avenue for comparison of the spatial structures of the protein in single crystals and in solution, and for studies of subtle changes of conformation with temperature, p^2H or other external parameters (Nagayama, K. and Wüthrich, K., unpublished results). Furthermore, to the best of our knowledge this is the first time that the spin systems of the glycyl residues were identified in the ^1H NMR spectrum of a protein. Since a lot of evidence is available that glycyl residues are located in unique and essential positions in spatial protein structures, particularly also with regard to internal mobility of the molecules [31], the prospects for future structural studies using the glycine resonances appear quite exciting.

We would like to thank Prof. R. R. Ernst for continued advice on all aspects of 2D NMR, Dr G. Wagner for many long discussions on the resonance assignments in the inhibitor, Mr G. Wider for help with some experiments and with the preparation of the illustrations and the Farbenfabriken Bayer AG. (Leverkusen) for a gift of Trasylol. Financial support by a special grant of the ETH Zürich and the Swiss National Science Foundation (project 3.528.79) is gratefully acknowledged.

REFERENCES

1. Aue, W. P., Bartholdi, E. & Ernst, R. R. (1976) *J. Chem. Phys. 64*, 2229–2246.
2. Freeman, R. & Morris, G. A. (1979) *Bull. Magn. Reson. 1*, 5–26.
3. Wüthrich, K., Nagayama, K. & Ernst, R. R. (1979) *Trends Biochem. Sciences, 4*, N178–N181.
4. Nagayama, K., Wüthrich, K., Bachmann, P. & Ernst, R. R. (1977) *Biochem. Biophys. Res. Commun. 78*, 99–105.
5. Nagayama, K., Bachmann, P., Wüthrich, K. & Ernst, R. R. (1978) *J. Magn. Reson. 31*, 133–148.
6. Nagayama, K., Bachmann, P., Ernst, R. R. & Wüthrich, K. (1979) *Biochem. Biophys. Res. Commun. 86*, 218–225.
7. Kumar, A., Wagner, G., Ernst, R. R. & Wüthrich, K. (1980) *Biochem. Biophys. Res. Commun. 96*, 1156–1163.
8. Nagayama, K., Wüthrich, K. & Ernst, R. R. (1979) *Biochem. Biophys. Res. Commun. 90*, 305–311.
9. Nagayama, K., Kumar, A., Wüthrich, K. & Ernst, R. R. (1980) *J. Magn. Reson, 40*, 321–334.
10. Kumar, A., Ernst, R. R. & Wüthrich, K. (1980) *Biochem. Biophys. Res. Commun. 95*, 1–6.
11. Wider, G., Baumann, R., Nagayama, K., Ernst, R. R. & Wüthrich, K. (1980) *J. Magn. Reson* in the press.
12. Baumann, R., Kumar, A., Ernst, R. R. & Wüthrich, K. (1981) *J. Magn. Reson.* in the press.
13. Wagner, G., Kumar, A. & Wüthrich, K. (1981) *Eur. J. Biochem. 114*, 375–384.
14. Wüthrich, K. (1976) *NMR in Biological Research: Peptides and Proteins*, North-Holland, Amsterdam.
15. Wüthrich, K. & Wagner, G. (1979) *J. Mol. Biol. 130*, 1–18.
16. Ernst, R. R., Aue, W. P., Bachmann, P., Höhener, A., Linder, M., Meier, B., Müller, L., Wokaun, A., Nagayama, K. & Wüthrich, K. (1978) *Proceedings of the XXth Congress Ampere*, Tallinn, pp. 15–18.
17. Jering, H. & Tschesche, H. (1976) *Eur. J. Biochem. 61*, 443–452.
18. Wagner, G., Tschesche, H. & Wüthrich, K. (1979) *Eur. J. Biochem. 95*, 239–248.
19. Wagner, G., Kalb (Gilboa), A. J. & Wüthrich, K. (1979) *Eur. J. Biochem. 95*, 249–253.
20. Kalinichenko, P. (1976) *Stud. Biophys. 58*, 235–240.
21. Bundi, A. & Wüthrich, K. (1979) *Biopolymers, 18*, 285–297.
22. Wagner, G., Wüthrich, K. & Tschesche, H. (1978) *Eur. J. Biochem. 86*, 67–76.
23. De Marco, A. (1977) *J. Magn. Reson. 26*, 527–528.
24. Wüthrich, K., Wagner, G., Richarz, R. & Perkins, S. J. (1978) *Biochemistry, 17*, 2253–2263.
25. Dubs, A., Wagner, G. & Wüthrich, K. (1979) *Biochim. Biophys. Acta, 577*, 177–194.
26. Wüthrich, K. & Wagner, G. (1979) *J. Mol. Biol. 130*, 1–18.
27. Perkins, S. J. & Wüthrich, K. (1979) *Biochim. Biophys. Acta, 576*, 409–423.
28. Kassell, B. & Laskowski, M. (1965) *Biochem. Biophys. Res. Commun. 20*, 463–468.
29. Wagner, G., De Marco, A. & Wüthrich, K. (1976) *Biophys. Structure Mechanism, 2*, 139–158.
30. Perkins, S. J. & Wüthrich, K. (1978) *Biochim. Biophys. Acta, 536*, 406–420.
31. Schulz, G. E. & Schirmer, R. H. (1979) *Principles of Protein Structure*, Springer-Verlag, New York.

K. Nagayama, Department of Physics, Faculty of Science, University of Tokyo, Bunkyo-ku, Hongo, Tokyo, Japan

K. Wüthrich, Institut für Molekularbiologie und Biophysik der Eidgenössischen Technischen Hochschule Zürich, Zürich-Hönggerberg, CH-8093 Zürich, Switzerland

J. Mol. Biol. (1982) **155**, 311–319

Sequential Resonance Assignments as a Basis for Determination of Spatial Protein Structures by High Resolution Proton Nuclear Magnetic Resonance

Kurt Wüthrich, Gerhard Wider
Gerhard Wagner and Werner Braun

Institut für Molekularbiologie und Biophysik
Eidgenössische Technische Hochschule
ETH-Hönggerberg, CH-8093 Zürich, Switzerland

(Received 17 August 1981)

A general scheme is proposed for the determination of spatial protein structures by proton nuclear magnetic resonance. The scheme relies on experimental observation by two-dimensional nuclear magnetic resonance techniques of complete through-bond and through-space proton–proton connectivity maps. These are used to obtain sequential resonance assignments for the individual residues in the amino acid sequence and to characterize the spatial polypeptide structure by a tight network of semi-quantitative, intramolecular distance constraints.

1. Introduction

For over 20 years, single-crystal X-ray studies have set the standards for characterization of protein conformations (Schulz & Schirmer, 1979; Richardson, 1981). The highly sophisticated structural information obtained has long called for complementary techniques to obtain corresponding data for polypeptide chains in solution and in other non-crystalline environments. The potential of nuclear magnetic resonance for such studies has been indicated by various early experiments (see e.g. Roberts & Jardetzky, 1970; Wüthrich, 1976). However, as in n.m.r.† structure determinations for small molecules, the realization of the full potentialities of the method for structural studies of proteins depends on the identification of the individual resonance lines, and lack of reliable, generally applicable ways for obtaining individual assignments in polypeptide chains has so far been a severely limiting factor. With the use of high magnetic fields and two-dimensional spectroscopy techniques (Aue *et al.*, 1976; Freeman & Morris, 1979; Wüthrich *et al.*, 1979), individual assignments for almost all the resonance lines in protein ^{1}H n.m.r. spectra can now be obtained efficiently, as is described in detail in the following three papers (Billeter *et al.*, 1982; Wagner & Wüthrich, 1982;

† Abbreviations used: n.m.r., nuclear magnetic resonance; COSY, 2-dimensional correlated spectroscopy; SECSY, 2-dimensional spin echo correlated spectroscopy; NOE, nuclear Overhauser enhancement; NOESY, 2-dimensional NOE spectroscopy.

0022-2836/82/070311-09 $02.00/0 © 1982 Academic Press Inc. (London) Ltd.

Wider *et al.*, 1982). Of course, this provides a more solid basis for "conventional" applications of n.m.r.; for example, for delineation of active centers in enzymes, comparison of crystal and solution conformations from ring current calculations of the chemical shifts, or studies of internal mobility in globular proteins (see e.g. Roberts & Jardetzky, 1970; Wüthrich, 1976). Even more excitingly, experience from work with smaller sets of assigned resonances (Braun *et al.*, 1981; Crippen *et al.*, 1981) implies that with nearly complete individual assignments of the resonance lines in a protein ^{1}H n.m.r. spectrum, the three-dimensional structure could be determined entirely from the known amino acid sequence and a suitable set of n.m.r. data. This paper outlines a general scheme for ^{1}H n.m.r. determination of spatial protein structures, which relies on sequential individual resonance assignments.

2. Fundamental Considerations

During almost 25 years of n.m.r. studies with proteins, a variety of different ^{1}H n.m.r. manifestations of the spatial structures of polypeptide chains were described (for a survey, see e.g. Wüthrich, 1976). The dispersion of the chemical shifts that arises from interactions between nearby groups of atoms in interior parts of globular proteins (Fig. 1) is perhaps the most readily apparent, conformation-dependent n.m.r. spectral feature (McDonald & Phillips, 1967). Studies based on empirical comparison of chemical shifts yielded a wealth of interesting data on differences between protein conformations in different solvent media, internal flexibility of globular proteins, protein folding and various aspects of protein functions. However, one tackles a considerably more difficult problem when trying to determine the conformation of a polypeptide chain from the known amino acid sequence and n.m.r. data. While the dispersion of the chemical shifts (Fig. 1) obviously presents a criterion to discriminate between individual amino acid residues, and thus provides a basis for obtaining a many-parameter characterization of the spatial structure by n.m.r., it also leads to very complex, crowded protein ^{1}H n.m.r. spectra (compare traces A and B in Fig. 1). Furthermore, theoretical understanding of chemical shifts is too limited to allow determination of spatial structures based on measurements of this parameter. Therefore, a promising scheme for ^{1}H n.m.r. determination of protein conformations must include techniques able to resolve and assign the complex ^{1}H n.m.r. spectra, and to measure spectral parameters that can be directly correlated with the spatial structure.

The strategy for protein structure determination outlined in Figure 2 relies on very recent advances in n.m.r. techniques. Two-dimensional n.m.r. provides well-resolved protein ^{1}H n.m.r. spectra, and two different types of two-dimensional experiments are used to map two kinds of connectivities between the protons in the polypeptide chain. These are through-bond, scalar "J-connectivities" between hydrogen atoms linked *via* two or three bonds in the covalent polypeptide structure and through-space, dipolar "NOE-connectivities" between hydrogen atoms located at short distances in the spatial structure. J-connectivity maps of proteins have

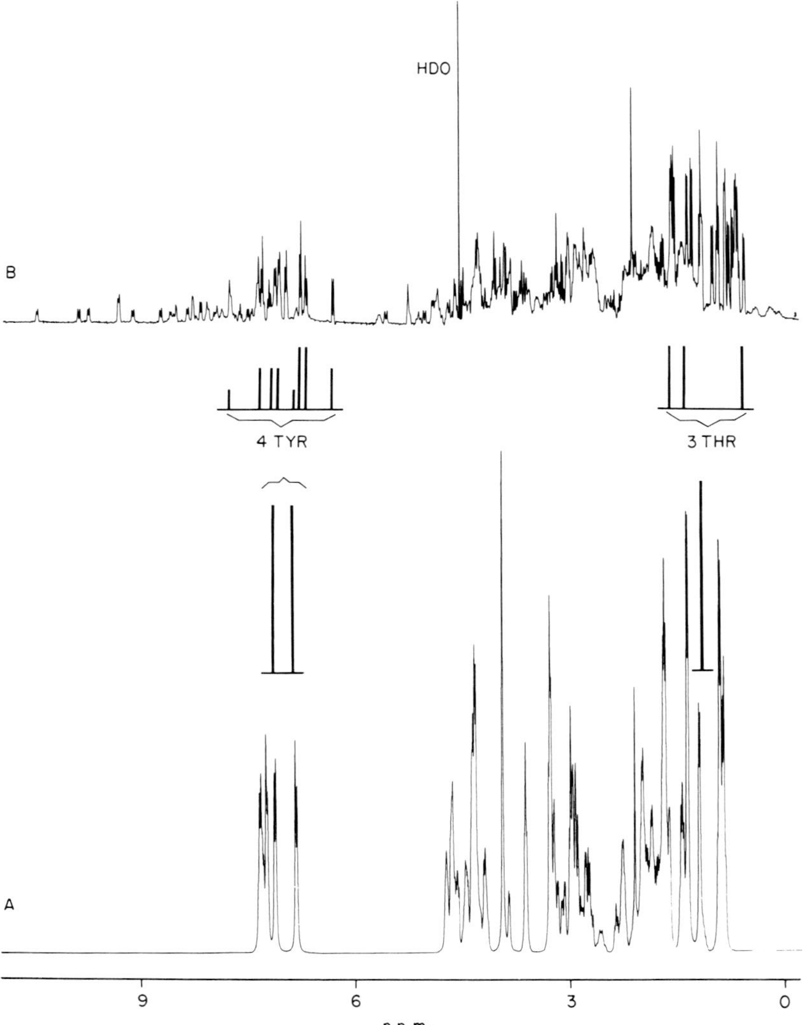

FIG. 1. Illustration of conformation-dependent ^{1}H n.m.r. chemical shifts in a globular protein. The lower trace (A) represents a hypothetical random coil spectrum for the polypeptide chain of the basic pancreatic trypsin inhibitor (BPTI) in ^2H_2O solution, which was computed from the random coil parameters presented by Bundi & Wüthrich (1979*a*). Trace B is the 360 MHz ^{1}H n.m.r. spectrum of a freshly prepared solution of BPTI in ^2H_2O (Wüthrich & Wagner, 1979). The stick diagrams below and above the letters 4 TYR and 3 THR in the centre of the Figure indicate chemical shifts and resonance intensities for the γ-methyl groups of the 3 threonine residues in positions 11, 32 and 54 and the aromatic protons of the 4 tyrosine residues in positions 10, 21, 23 and 35 in the spectra A and B, respectively. They show that in the random coil polypeptide, the peripheral side-chain hydrogen atoms of identical residues at different locations in the amino acid sequence have identical chemical shifts, whereas in the globular protein, one observes a dispersion of the chemical shifts due to the different microenvironments of the individual residues (McDonald & Phillips, 1967; Wüthrich, 1976). p.p.m., parts per million.

been recorded with two-dimensional correlated spectroscopy or spin echo correlated spectroscopy (Aue *et al.*, 1976; Nagayama *et al.*, 1979,1980; Anil Kumar *et al.*, 1980*b*; Wagner *et al.*, 1981). A COSY or SECSY spectrum provides, with a single instrument setting, a complete map of all ^1H–^1H J-connectivities in a polypeptide chain. From this, the spin systems of the different types of amino acid residues (Wüthrich, 1976) can be identified. NOE connectivity maps have been

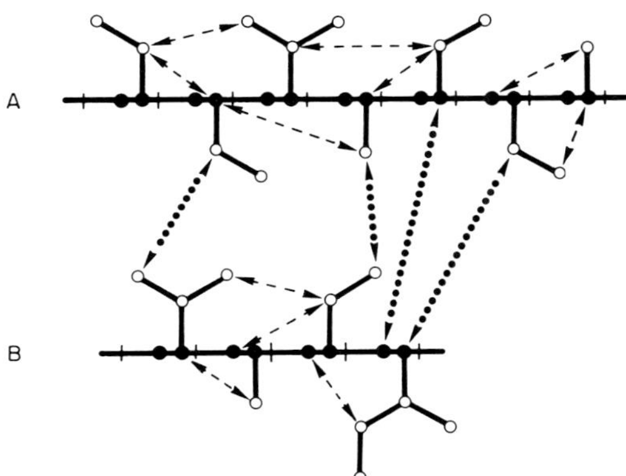

FIG. 2. Schematic representation of the proposed strategy for determination of spatial protein structures by ^1H n.m.r. At the top of the Figure, the horizontal line intersected by vertical dashes represents a polypeptide chain. In the larger scale drawings of segments A and B, short vertical dashes separate neighbouring residues, filled circles represent CH and NH groups of the backbone, and open circles represent CH groups of the amino acid side-chains. The first step in the structure determination is the sequential assignment of the resonances to the individual amino acid residues (Dubs *et al.*, 1979; Wagner *et al.*, 1981; Billeter *et al.*, 1982; Wagner & Wüthrich, 1982; Wider *et al.*, 1982). From distance constraints between distinct hydrogen atoms in the amino acid sequence, which are manifested in NOE connectivity maps, the spatial structure is then evaluated (Braun *et al.*, 1981). For example, distance constraints between backbone amide, C^α and C^β hydrogen atoms of nearby residues in the amino acid sequence (◀– – – –▶ in A and B) characterize the secondary structure, and long-range distance constraints (with respect to the locations of the residues in the amino acid sequence) between backbone and amino acid side-chain hydrogen atoms (◀●●●●▶) define the supersecondary and tertiary structures.

obtained by two-dimensional nuclear Overhauser enhancement spectroscopy (Jeener *et al.*, 1979; Anil Kumar *et al.*, 1980*a*,*b*; Wagner *et al.*, 1981). As the spectral manifestations of NOE connectivities develop slowly during a period of several hundred milliseconds, different NOESY spectra can be obtained by variation of one of the instrument settings, the so-called "mixing time". Semi-quantitative measurements of non-bonding distances in the range 2 to 5 Å can thus be obtained

(Anil Kumar *et al.*, 1981)†. These distance constraints mapped in the NOESY spectra can then be correlated directly with the protein conformation (Braun *et al.*, 1981).

3. Sequential Resonance Assignments

Individual assignments of a large proportion of the hundreds or even thousands of resonance lines (Fig. 1, trace B) to particular residues in the amino acid sequence represent the core of the presently described scheme for ^{1}H n.m.r. determination of protein conformations. This can be appreciated readily if one considers that intramolecular distance constraints (arrows in Fig. 2) or other potentially suitable parameters for structure determination can be correlated with the protein conformation only when they have been assigned to particular hydrogen atoms in the polypeptide chain.

Unambiguous assignments for a small number of resonances have been obtained in previous work with selective modification or substitution of individual residues (e.g. Snyder *et al.*, 1975). This will undoubtedly be a suitable approach also for future work that relies on the observation of a small number of resonance lines, but it appears unrealistic to use selective modification of individual residues by chemical or biological techniques for assignments of the bulk of a protein ^{1}H n.m.r. spectrum.

Most of the resonance assignments described so far have relied in part or entirely on reference to the crystal structure of the protein, or on observations linked with the functional properties. While this may have been an appropriate procedure in

† In principle, it should be possible to relate proton–proton distances shorter than ~5·0 Å quantitatively with suitably executed NOE measurements (see e.g. Noggle & Schirmer, 1971; Gordon & Wüthrich, 1978; Wagner & Wüthrich, 1979; Bothner-By & Noggle, 1979; Poulsen *et al.*, 1980; Anil Kumar *et al.*, 1981). In practice, however, a variety of different factors tend to mask the quantitative distance information contained in NOE data. Thus, distance measurements with the use of NOE experiments can be complicated by fundamental physical properties of the systems studied. These may include spin relaxation by other mechanisms than proton–proton dipolar coupling, for example by ^{14}N–^{1}H dipolar coupling or by paramagnetic impurities (Wüthrich, 1976), flexibility of the molecular structure (Braun *et al.*, 1981), and adverse effects of spin diffusion by cross relaxation (Kalk & Berendsen, 1976; Gordon & Wüthrich, 1978; Bothner-By & Noggle, 1979; Wagner & Wüthrich, 1979; Anil Kumar *et al.*, 1981). Practical experience (Dubs *et al.*, 1979; Wagner *et al.*, 1981; Braun *et al.*, 1981; Wagner & Wüthrich, 1982; Wider *et al.*, 1982) has shown that these fundamental complications do not, in general, preclude the use of NOE experiments for obtaining the semi-quantitative distance constraints needed for the presently discussed sequential resonance assignments and spatial structure determinations in globular proteins. It remains to be seen to what extent these effects could, in specified systems, be properly accounted for in order to obtain quantitative distance measurements from NOE studies. In addition, there are also various technical complications involved in the quantitative evaluation of NOE peak intensities in the NOESY spectra. One of these problems has been solved recently, i.e. the complete separation of the contributions from incoherent and coherent magnetization transfer to the peak intensities in NOESY spectra recorded with short mixing times (Macura *et al.*, 1981,1982). Other problems currently under investigation include, for example, the influence of a variety of filtering functions on the relative intensities of different peaks in NOESY spectra and the use of absolute value spectra *versus* absorption mode spectra for quantitative NOE measurements. This short and incomplete list of complications that can arise in the quantitative analysis of NOESY spectra illustrates that there is a lot of room for further advances. It should also explain why we are so pleased that the presently described approach for structure determination relies exclusively on qualitative analysis of the peak intensities in the NOESY spectra.

some of the studies concerned, it is not acceptable for the present scheme, which should provide a basis for meaningful, detailed comparisons of protein conformations in single crystals and non-crystalline environments, and for structural interpretations of functional properties of proteins in non-crystalline environments.

A straightforward concept for obtaining individual resonance assignments in a polypeptide chain is to identify the resonance lines of the individual residues step by step along the amino acid sequence. In practice, combined use of J-connectivity maps obtained from COSY or SECSY experiments and NOE connectivity maps obtained from NOESY experiments with short mixing times presents an efficient, reliable and generally applicable method for obtaining sequential resonance assignments. This is described in detail in the following three papers (Billeter *et al.*, 1982; Wagner & Wüthrich, 1982; Wider *et al.*, 1982). In this procedure, the connectivities between neighbouring residues are established *via* the amide protons. It is therefore important that COSY, SECSY and NOESY spectra can be recorded with comparable ease in ^2H_2O and H_2O solutions of proteins (Anil Kumar *et al.*, 1980*b*; Wagner *et al.*, 1981), because some or all of those critical amide protons may be exchanged too rapidly with the solvent to be observed in ^2H_2O (Wüthrich, 1976). Since the sequential assignments depend on ^1H–^1H distance constraints between neighbouring residues, one obtains simultaneously a first, qualitative outline of the secondary structure, which can be specified by those same distances (Dubs *et al.*, 1979; Wagner *et al.*, 1981; Billeter *et al.*, 1982).

Sequential assignments can, in principle, also be obtained with combined use of homonuclear and heteronuclear J-connectivities between ^1H, ^{13}C and ^{15}N (Llinás *et al.*, 1977; Okhanov *et al.*, 1980). At present, these experiments appear to be too laborious to be of practical use even for small proteins. However, any approach that depends entirely on through-bond J-connectivities would be appealing, and it is very possible that with future improvements in n.m.r. methodology, these experiments will become a viable alternative to those used at present.

4. Determination of Spatial Polypeptide Structures *via* Intramolecular Distance Constraints

As an alternative to the more familiar listings of atomic co-ordinates or dihedral angles (Schulz & Schirmer, 1979), spatial protein structures may also be characterized by a list of intramolecular distances between specified pairs of atoms (Crippen, 1979; Havel *et al.*, 1979), provided the chirality of the structure is independently specified (Cohen & Sternberg, 1980; Braun *et al.*, 1981). An extension of Crippen's distance geometry algorithm for use with the distance constraints obtained from NOE connectivity maps was described recently (Braun *et al.*, 1981) and applied to determine spatial structures for short segments of the polypeptide chains of lipid-bound glucagon (Braun *et al.*, 1981) and melittin (Brown *et al.*, 1982). While these structure determinations were limited by the small number of resonance assignments available at the time, they indicated a general feature that should have important consequences also when working with complete sets of assigned resonances. The implication is that it will be more effective to obtain a large number of relatively inaccurate distance constraints; for example, stating that the distances separating particular pairs of hydrogen atoms are $\lesssim 5 \cdot 0$ Å, rather than a small number of accurate interatomic distances. This further supports the suggestion that the NOE connectivity maps in NOESY spectra present suitable data sets for spatial structure determination, since they provide numerous semi-quantitative distance constraints (Anil Kumar *et al.*, 1980*a*,*b*;1981).

On the basis of the sequential resonance assignments, the NOE connectivities in the NOESY spectra recorded with different mixing times (Anil Kumar *et al.*, 1981) can be correlated with distance constraints between distinct hydrogen atoms in the amino acid sequence†. The distance constraints can then be grouped into different classes, depending on the hierarchic level of structure (Schulz & Schirmer, 1979) with which they are correlated (Fig. 2). Thus, distance constraints between backbone amide, C^α and C^β hydrogen atoms in residues that are closely spaced in the amino acid sequence would be used to characterize the secondary structure. Distance constraints among backbone and C^β hydrogen atoms of residues that are further apart in the sequence (e.g. between residues located in segments A and B in Fig. 2) provide the most reliable data on supersecondary structures, such as the combination of extended polypeptide segments into β-sheets (Wagner *et al.*, 1981) or the packing of α-helices. Finally, the connectivities with peripheral side-chain hydrogen atoms would be used to orient the amino acid side-chains, and thus obtain a characterization of the tertiary structure. It remains to be seen to what extent the structural analysis of the experimental distance constraints will be done by global evaluation of the data for the entire protein; for example, with the use of a distance geometry algorithm (Braun *et al.*, 1981), or by more intuitive methods, such as interactive use of a computer graphics system (Billeter, 1980).

5. Concluding Remarks

The general scheme for structure determination outlined in Figure 2 and sections 3 and 4 above should be applicable to obtain essentially complete descriptions of the conformations of small proteins, and also for partial structure elucidations in high molecular weight species. For example, it is conceivable that the resonances of more flexible polypeptide segments in large proteins could, on the basis of the different relaxation times (Wüthrich, 1976), be separated from those of the more rigid molecular regions, and thus become accessible for detailed studies. Similarly, the resonances of polypeptide chains bound to ordered lipid structures (Braun *et al.*, 1981; Brown *et al.*, 1982; Wider *et al.*, 1982) or to high molecular weight nucleic acids might be singled out for detailed investigation.

It is an important asset of the presently proposed strategy for protein structure determination by n.m.r. that it relies on qualitative or, at most, semi-quantitative interpretation of NOE connectivity maps†. The available experience indicates that a tight network of qualitative distance constraints, providing the data needed for a structure determination, can quite generally be obtained for the interior regions of

† See footnote to p. 315.

globular proteins and for polypeptide segments that are otherwise restricted in their local mobility; for example, by binding to an ordered lipid surface or to other macromolecules (Braun *et al.*, 1981). Additional structure refinements might then be obtained for these molecular regions by quantitating certain NOE measurements (Noggle & Schirmer, 1971; Kalk & Berendsen, 1976; Gordon & Wüthrich, 1978; Wagner & Wüthrich, 1979; Anil Kumar *et al.*, 1981; Macura *et al.*, 1981,1982). Furthermore, theoretical methods, for example energy refinement (Levitt, 1974), might profitably be employed to further improve the structures. In contrast, meaningful distance constraints from NOE-connectivities cannot generally be expected for highly flexible surface areas (Noggle & Schirmer, 1971), for example the chain termini and peripheral fragments of long amino acid side-chains. Therefore, in the final stages of the structure determination, different n.m.r. experiments should be used for obtaining additional information needed to characterize the protein surface. These may include measurements of spin–spin coupling constants by J-resolved two-dimensional spectroscopy (Nagayama *et al.*, 1977; Nagayama & Wüthrich, 1981), studies of pH titration connectivities between different groups (Brown *et al.*, 1976,1978; Bundi & Wüthrich, 1979*b*), use of paramagnetic shift and relaxation reagents (Dwek, 1973) and many other experiments that have hitherto profitably been used in n.m.r. studies of various aspects of protein structure and functions (see e.g. Dwek, 1973; Wüthrich, 1976). Combined with the individual resonance assignments and with the data on the core of the protein structure obtained by the new techniques outlined in this paper, such "conventional" n.m.r. experiments should be able to provide additional precise structural data, and thus contribute towards a rather complete determination of protein conformations in non-crystalline environments.

The development of two-dimensional n.m.r. techniques for studies of biopolymers, which have a crucial role in the realization of the presently outlined scheme for protein structure determination, is a joint project with Professor R. R. Ernst and is financed by a special grant of the Eidgenössische Technische Hochschule, Zürich. We thank Professor Ernst for continued advice on all aspects of two-dimensional n.m.r. Financial support by the Schweizerischer Nationalfonds (project 3.528.79) is gratefully acknowledged.

REFERENCES

Anil Kumar, Ernst, R. R. & Wüthrich, K. (1980*a*). *Biochem. Biophys. Res. Commun.* **95**, 1–6.

Anil Kumar, Wagner, G., Ernst, R. R. & Wüthrich, K. (1980*b*). *Biochem. Biophys. Res. Commun.* **96**, 1156–1163.

Anil Kumar, Wagner, G., Ernst, R. R. & Wüthrich, K. (1981). *J. Amer. Chem. Soc.* **103**, 3654–3658.

Aue, W. P., Bartholdi, E. & Ernst, R. R. (1976). *J. Chem. Phys.* **64**, 2229–2246.

Billeter, M. (1980). Diploma thesis, ETH-Zürich.

Billeter, M., Braun, W. & Wüthrich, K. (1982). *J.Mol. Biol.* **155**, 321–346.

Bothner-By, A. A. & Noggle, J. A. (1979). *J. Amer. Chem. Soc.* **101**, 5152–5155.

Braun, W., Bösch, C., Brown, L. R., Gō, N. & Wüthrich, K. (1981). *Biochim. Biophys. Acta*, **667**, 377–396.

Brown, L. R., De Marco, A., Wagner, G. & Wüthrich, K. (1976). *Eur. J. Biochem.* **62**, 103–107.

Brown, L. R., De Marco, A., Richarz, R., Wagner, G. & Wüthrich, K. (1978). *Eur. J. Biochem.* **88**, 87–95.
Brown, L. R., Braun, W., Anil Kumar & Wüthrich, K. (1982). *Biophys. J.* **37**, Nr. 1.
Bundi, A. & Wüthrich, K. (1979*a*). *Biopolymers*, **18**, 285–298.
Bundi, A. & Wüthrich, K. (1979*b*). *Biopolymers*, **18**, 299–312.
Cohen, F. E. & Sternberg, M. J. E. (1980). *J. Mol. Biol.* **138**, 321–333.
Crippen, G. M. (1979). *Int. J. Pept. Protein Res.* **13**, 320–326.
Crippen, G. M., Oppenheimer, N. J. & Connolly, M. L. (1981). *Int. J. Pept. Protein Res.* **17**, 156–169.
Dubs, A., Wagner, G. & Wüthrich, K. (1979). *Biochim. Biophys. Acta*, **577**, 177–194.
Dwek, R. A. (1973). *Nuclear Magnetic Resonance (NMR) in Biochemistry*, Clarendon Press, Oxford.
Freeman, R. & Morris, G. A. (1979). *Bull. Magn. Reson.* **1**, 5–26.
Gordon, S. L. & Wüthrich, K. (1978). *J. Amer. Chem. Soc.* **100**, 7094–7096.
Havel, T. F., Crippen, G. M. & Kuntz, I. D. (1979). *Biopolymers*, **18**, 73–81.
Jeener, J., Meier, B. H., Bachmann, P. & Ernst, R. R. (1979). *J. Chem. Phys.* **71**, 4546–4553.
Kalk, A. & Berendsen, H. J. C. (1976). *J. Magn. Reson.* **24**, 343–366.
Levitt, M. (1974). *J. Mol. Biol.* **82**, 393–420.
Llinás, M., Wilson, D. M. & Klein, M. P. (1977). *J. Amer. Chem. Soc.* **99**, 6846–6850.
Macura, S., Huang, Y., Suter, D. & Ernst, R. R. (1981). *J. Magn. Reson.* **43**, 259–281.
Macura, S., Wüthrich, K. & Ernst, R. R. (1982). *J. Magn. Reson.* **46**.
McDonald, C. C. & Phillips, W. D. (1967). *J. Amer. Chem. Soc.* **89**, 6332–6344.
Nagayama, K. & Wüthrich, K. (1981). *Eur. J. Biochem.* **114**, 365–374.
Nagayama, K., Wüthrich, K., Bachmann, P. & Ernst, R. R. (1977). *Biochem. Biophys. Res. Commun.* **78**, 99–105.
Nagayama, K., Wüthrich, K. & Ernst, R. R. (1979). *Biochem. Biophys. Res. Commun.* **90**, 305–311.
Nagayama, K., Anil Kumar, Wüthrich, K. & Ernst, R. R. (1980). *J. Magn. Reson.* **40**, 321–334.
Noggle, J. H. & Schirmer, R. E. (1971). *The Nuclear Overhauser Effect*, Academic Press, New York.
Okhanov, V. V., Afanas'ev, V. A. & Bystrov, V. F. (1980). *J. Magn. Reson.* **40**, 191–195.
Poulsen, F. M., Hoch, J. C. & Dobson, C. M. (1980). *Biochemistry*, **19**, 2597–2607.
Richardson, J. S. (1981). *Advan. Protein Chem.* **34**, 167–339.
Roberts, G. C. K. & Jardetzky, O. (1970). *Advan. Protein Chem.* **24**, 447–545.
Schulz, G. E. & Schirmer, R. H. (1979). *Principles of Protein Structure*, Springer, New York.
Snyder, G. H., Rowan III, R., Karplus, S. & Sykes, B. D. (1975). *Biochemistry*, **14**, 3765–3777.
Wagner, G. & Wüthrich, K. (1979). *J. Magn. Reson.* **33**, 675–680.
Wagner, G. & Wüthrich, K. (1982). *J. Mol. Biol.* **155**, 347–366.
Wagner, G., Anil Kumar & Wüthrich, K. (1981). *Eur. J. Biochem.* **114**, 375–384.
Wider, G., Lee, K. H. & Wüthrich, K. (1982). *J. Mol. Biol.* **155**, 367–388.
Wüthrich, K. (1976). *NMR in Biological Research: Peptides and Proteins*, North-Holland Publishing Company, Amsterdam.
Wüthrich, K. & Wagner, G. (1979). *J. Mol. Biol.* **130**, 1–18.
Wüthrich, K., Nagayama, K. & Ernst, R. R. (1979). *Trends Biochem. Sci.* **4**, N178–N181.

Edited by S. Brenner

J. Mol. Biol. (1982) **155**, 321–346

Sequential Resonance Assignments in Protein ¹H Nuclear Magnetic Resonance Spectra

Computation of Sterically Allowed Proton–Proton Distances and Statistical Analysis of Proton–Proton Distances in Single Crystal Protein Conformations

MARTIN BILLETER, WERNER BRAUN AND KURT WÜTHRICH

Institut für Molekularbiologie und Biophysik
Eidgenössische Technische Hochschule
8093 Zürich-Hönggerberg
Switzerland

(Received 17 August 1981)

Two different, theoretical studies of intramolecular proton–proton distances in polypeptide chains are described. Firstly, the distances between amide, C^{α} and C^{β} protons of neighbouring residues in the amino acid sequence, which correspond to the sterically allowed values for the dihedral angles ϕ_i, ψ_i and χ_i^1, were computed. Secondly, the frequency with which short distances occur between amide, C^{α} and C^{β} protons of neighbouring and distant residues in the amino acid sequence were statistically evaluated in a representative sample of globular protein crystal structures. Both approaches imply that semi-quantitative measurements of short, non-bonding proton–proton distances, e.g. by nuclear Overhauser experiments, should present a reliable and generally applicable method for sequential, individual resonance assignments in protein ¹H nuclear magnetic resonance spectra. Similar calculations imply that corresponding distance measurements can be used for resonance assignments in the side-chains of the aromatic amino acid residues, asparagine and glutamine, where the complete spin systems cannot usually be identified from through-bond spin–spin coupling connectivities.

1. Introduction

The preceding paper (Wüthrich *et al.*, 1982) described a general scheme for the use of high resolution ¹H nuclear magnetic resonance to determine spatial protein structures in non-crystalline environments. The first crucial step in this scheme is the assignment of the individual resonance lines in the complex protein ¹H n.m.r.† spectra to particular hydrogen atoms in the amino acid sequence. Our strategy for achieving this primary level in structural studies of polypeptide chains exploits first the fact that the complete spin systems of most amino acid residues can be identified from through-bond proton–proton J-couplings (Wüthrich, 1976;

† Abbreviation used: n.m.r., nuclear magnetic resonance; NOE, nuclear Overhauser enhancement; NOESY, nuclear Overhauser enhancement spectroscopy; RMSD, root-mean-square distance.

0022–2836/82/070321–26 $02.00/0 © 1982 Academic Press Inc. (London) Ltd.

Nagayama & Wüthrich, 1981). Secondly, neighbouring residues in the amino acid sequence are identified by nuclear Overhauser enhancement studies of non-bonding, through-space distances between hydrogen atoms located in different residues. Delineation of such non-bonding connectivities has been used to obtain individual resonance assignments along the extended polypeptide chains of β-sheet secondary structures in a globular protein (Dubs *et al.*, 1979; Wagner *et al.*, 1981). The present theoretical study investigates whether measurements of through-space proton–proton distances combined with the J-coupling connectivities could provide a generally applicable procedure for sequential resonance assignments in proteins, irrespective of the secondary structures adopted by the polypeptide chain.

At present, our preferred technique for investigations of through-space proton–proton connectivities in biological macromolecules is two-dimensional nuclear Overhauser spectroscopy (Anil Kumar *et al.*, 1980,1981). At the current state of development this experiment is able to provide semi-quantitative information on proton–proton distances in the range 2 to 5 Å (see the footnote in the preceding paper, Wüthrich *et al.*, 1982, for further discussion of this statement). By careful selection of the experimental parameters, in particular the "mixing time" during which the NOEs are built up (Anil Kumar *et al.*, 1981), the range of distances manifested in a NOESY spectrum can be further confined. The upper limit for observable distances between hydrogen atoms attached to the non-terminal regions of the polypeptide backbone in a small globular protein may then, for example, be $\lesssim$ 2·6 Å, or $\lesssim$ 3·0 Å or $\lesssim$ 3·5 Å. The present theoretical study on proton–proton distances in polypeptide chains covers the entire range from 2 to 5 Å, but for discussions of the practical implications the results obtained with an upper distance limit of $\lesssim$ 3·0 Å will be used throughout this paper. From past experience (Dubs *et al.*, 1979; Wagner *et al.*, 1981) and model considerations (Braun *et al.*, 1981), this is a realistic distance limit for the NOESY experiments used to obtain sequential resonance assignments in small and medium-size proteins. Numerical data for different experimental conditions, which correspond to different upper distance limits, can readily be extracted from Table 2 and Figures 2 to 10. We may further generalize by adding that practical applications of the results of this theoretical study are not necessarily restricted to NOE experiments, but could extend to other methods that are able to delineate short proton–proton distances in proteins.

The first part of the paper investigates whether there are proton–proton distances between neighbouring amino acid residues in a polypeptide chain that are sufficiently short to be manifested in NOE experiments. For this, the ranges of sterically allowed (Ramachandran & Sasisekharan, 1968) distances between particular types of protons in neighbouring residues, for example the amide, C^{α} and C^{β} protons (Fig. 1), are computed. From these calculations, which are closely related to previous work in connection with NOE studies of oligopeptides (Leach *et al.*, 1977; Kuo & Gibbons, 1979; Roques *et al.*, 1980), suitable proton–proton combinations for use in sequential resonance assignments by NOEs can then be singled out.

The data presented in the second part of this paper provide a basis for estimating the reliability of sequential resonance assignments obtained from proton–proton

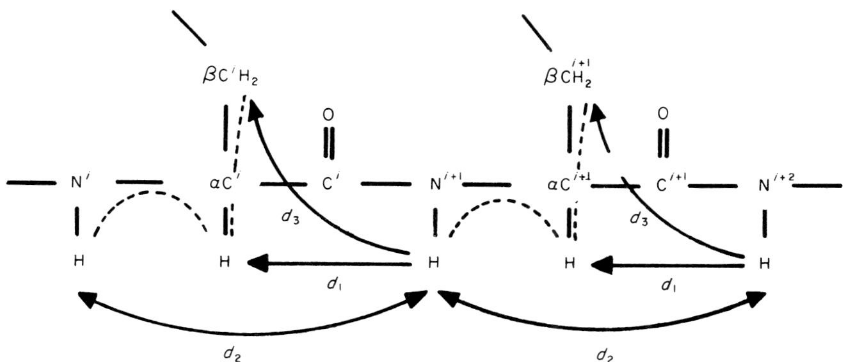

FIG. 1. Definition of the through-space proton–proton distances d_1, d_2 and d_3 between protons of neighbouring amino acid residues in polypeptide chains. The broken lines indicate through-bond connectivities by J-coupling between hydrogen atoms of the same residue.

NOE connectivities. As a consequence of the spatial arrangement of the polypeptide chains in globular proteins, short distances of 2 to 5 Å may prevail between protons in neighbouring residues, as well as between residues that are far apart in the amino acid sequence. *A priori*, a NOE experiment cannot distinguish between these two situations. Hence, the reliability of sequential resonance assignments by NOE connectivities will depend on the relative frequencies with which short proton–proton distances occur between neighbouring and non-neighbouring residues in the amino acid sequence. This information is obtained from statistical analyses of ^{1}H–^{1}H distances in a group of protein structures determined by high resolution X-ray crystallography and completed by the addition of hydrogen atoms.

In addition to the calculations that bear on sequential resonance assignments, potential uses of NOEs to aid in the identification of complete spin systems for amino acid side-chains are also explored, in particular for the aromatic amino acids.

2. Methods

(a) *Notation for proton–proton distances used for sequential resonance assignments*

To identify distances between individual atoms in a polypeptide chain, the notation $d(A_i, B_j)$ is used, where A and B define specific atom types and i and j indicate the position in the amino acid sequence of the residue to which these atoms belong. With this notation, the proton–proton distances that play essential roles for sequential resonance assignments in ^{1}H n.m.r. spectra of proteins (Dubs *et al.*, 1979; Wagner *et al.*, 1981) can be written as follows:

$$d_1(i,j) = d(C^\alpha H_i, NH_j). \qquad (1a)$$

Equation (1a) applies when any of the 19 common L-amino acid residues is in position i. When position i is occupied by glycine, d_1 is taken to be the shorter of the 2 amide proton–C^α proton distances (eqn (1b)):

$$d_1(i,j) = \min\{d(C^\alpha H_i^1, NH_j), d(C^\alpha H_i^2, NH_j)\}. \qquad (1b)$$

The amide proton–amide proton distance d_2 is defined for all the common amino acid residues except L-proline by:

$$d_2(i,j) = d(NH_i, NH_j). \qquad (2)$$

The amide proton–C^{β} proton distance d_3 is defined for all the common amino acid residues except glycine. With any of the 15 amino acids with β-methylene groups in position i, we have:

$$d_3(i,j) = \min \{d(C^{\beta}H_i^1, NH_j), d(C^{\beta}H_i^2, NH_j)\}. \quad (3a)$$

For valine, isoleucine and threonine, equation (3a) is reduced to:

$$d_3(i,j) = d(C^{\beta}H_i, NH_j). \quad (3b)$$

For alanine, the 3 β-methyl proton positions are represented by a single location on the extension of the C^{α}–C^{β} bond and at a distance of 0·36 Å from C^{β}, i.e. $\overline{C^{\beta}H_i}$. In view of the rapid rotation of the methyl group (Wüthrich, 1976), this appears to be a reasonable approximation for the present purposes:

$$d_3(i,j) = d(\overline{C^{\beta}H_i}, NH_j). \quad (3c)$$

The distances between amide protons, C^{α} protons and C^{β} protons of neighbouring residues in the amino acid sequence are of particular importance for sequential resonance assignments in the ^{1}H n.m.r. spectra as well as for characterization of the secondary structure (Fig. 1). For convenience, the following abbreviated notation will therefore be used:

$$d_1 = d_1(i, i+1), \quad (4)$$

$$d_2 = d_2(i, i+1), \quad (5)$$

$$d_3 = d_3(i, i+1). \quad (6)$$

In the Appendix, precise relations between d_1, d_2 and d_3 and the well-known dihedral angles ϕ_i, ψ_i and χ_i^1 (Ramachandran & Sasisekharan, 1968) are described.

(b) *Protein structures used for statistical studies of intramolecular proton–proton distances*

From the data files of the Protein Data Bank in Cambridge (Bernstein *et al.*, 1977), 19 structures determined at a resolution of $\leq 2{\cdot}0$ Å were selected (Table 1). Table 1 contains also 2 quantities used for the selection of the proteins. These are N_{err}, the number of residues with root-mean-square distances from standard ECEPP geometries (Momany *et al.*, 1975) exceeding 0·3 Å (see below) and N_{miss}, the number of non-terminal residues for which complete sets of co-ordinates were not available. Criteria for the selection of the proteins were that N_{err} and N_{miss} should be less than 5% of the total number of residues in the protein, N_{tot} (Table 1). For all the proteins in Table 1, N_{err} and N_{miss} are actually less than 3% of N_{tot}, and in most cases they are equal to zero. The structures that satisfied these criteria were then screened to assure that the major protein structure types (Richardson, 1981) were represented to a comparable extent. Furthermore, when data on several homologous proteins were available, only one was retained.

It turned out that, with the exception of carboxypeptidase A, all the protein structures used for the statistical analysis (Table 1) had been refined. In the computations, polypeptide chains in oligomeric proteins that are related by simple symmetry operations were counted only once, and N and C-terminal residues for which no or only incomplete co-ordinates are available were not counted.

(c) *Attachment of hydrogen atoms to the X-ray protein structures*

In a standard geometry for amino acid residues, for example ECEPP (Momany *et al.*, 1975), the co-ordinates for the non-hydrogen atoms would uniquely define the spatial locations of the amide protons, C^{α} protons and, with the exception of alanine (see eqn (3c)), the C^{β} protons. In the experimental protein structures, however, deviations from standard geometry abound. Therefore, the following root-mean-square fitting procedure was used to attach the hydrogen atoms. First, the atom to which a proton was to be attached, i.e. N_i, C_i^{α}

TABLE 1

Crystallographic protein data files used for statistical analysis of intramolecular proton–proton distances

Protein	Reference	Code†	N_{tot}‡	Resolution (Å)	N_{err}‡	N_{miss}‡
Actinidin (*Actinidia chinensis*)	Baker & Dodson (1980)	2ACT	218	1·7	0	1
Parvalbumin B (*Cyprinos carpio*)	Moews & Kretsinger (1975)	1CPV	108	1·85	0	0
Carbonic anhydrase C (*Homo sapiens*)	Kannan *et al.* (1972)	1CAC	256	2·0	8	0
Carboxypeptidase A (*Bos taurus*)	Quiocho & Lipscomb (1971)	1CPA	306	2·0	1	0
Concanavalin A (*Canavalia ensiformis*)	Reeke *et al.* (1975)	2CNA	237	2·0	7	0
Ferricytochrome b_5 (*Bos taurus*)	Mathews *et al.* (1972)	2B5C	85	2·0	0	0
Ferrocytochrome *c* (*Thunnus alalunga*)	Takano & Dickerson (1980)	4CYT	103	1·5	0	0
Haemoglobin (*Chironomus thummi thummi*)	Steigemann & Weber (1979)	1ECD	136	1·4	0	1
Flavodoxin (*Clostridium* MP)	Smith *et al.* (1977)	4FXN	138	1·8	0	0
High-potential iron protein (*Chromatium vinosum*)	Carter *et al.* (1974)	1HIP	85	2·0	0	3
Immunoglobulin Fab (*Homo sapiens*)	Saul *et al.* (1978)	2FAB	426	2·0	3	1
Insulin (*Sus scrofa*)	Dodson *et al.* (1979)	1INS	51	1·5	0	0
Lactate dehydrogenase (*Squalus acanthus*)	White *et al.* (1976)	4LDH	329	2·0	2	0
Lysozyme (*Gallus gallus*)	Diamond (1974)	2LYZ	129	2·0	3	0
Plastocyanin (*Populus nigra* variant *italica*)	Colman *et al.* (1978)	1PCY	99	1·6	0	0
Prealbumin (*Homo sapiens*)	Blake *et al.* (1978)	2PAB	113	1·8	4	1
Ribonuclease S (*Bos taurus*)	Fletterick & Wyckoff (1975)	1RNS	124	2·0	3	1
Trypsin (*Bos taurus*)	Fehlhammer & Bode (1975)	1PTN	223	1·5	2	0
Basic pancreatic trypsin inhibitor (*Bos taurus*)	Deisenhofer & Steigemann (1975)	3PTI	58	1·5	0	0
19 Proteins			3224			

† File identification code of the protein data bank.

‡ N_{tot}, N_{err} and N_{miss} are, respectively, the number of amino acid residues used for the present analysis, the number of residues with RMSD values relative to the ECEPP standard geometry larger than 0·3 Å (see the text) and the number of non-terminal residues for which not all the atom positions are listed in the data files.

or C_i^β, was placed at the origin of the co-ordinate system. Next, the atoms that have fixed positions relative to this proton were fitted to the standard ECEPP co-ordinates by a rigid-body rotation, such that the root-mean-square distance (RMSD);

$$\mathrm{RMSD} = \left(\frac{1}{N}\sum_{\nu=1}^{N}(\mathbf{x}_\nu^0 - \mathrm{R}\mathbf{x}_\nu)^2\right)^{\frac{1}{2}}, \tag{7}$$

between experimental co-ordinates and standard geometry for residue i was minimized. The hydrogen atom was then attached in the location corresponding to the ECEPP standard geometry. (In eqn (7) N is the number of atoms used in the fitting procedure, $\mathbf{x}_\nu^0$ terms are ECEPP co-ordinates of the atoms used in the fitting procedure, $\mathbf{x}_\nu$ terms are the co-ordinates in the X-ray structure, and R is the rotation matrix.)

The above procedure ensures that the bond lengths to the attached protons have the standard values, and that deviations of the bond angles from the standard values are distributed over all the angles involving the bond to the proton. As mentioned above and in Table 1, the RMSDs (eqn (7)) were also used as a criterion for the selection of the protein data files to be used for the statistical analysis; i.e. with very few exceptions, the RMSDs for the amino acid residues of the proteins in Table 1 are smaller than 0·3 Å.

Computer programs for attaching the protons and analysing the proton–proton distances were written in Fortran IV and run on a DEC-10 computer. The best-fit rotation R value (eqn (7)) was computed using an algorithm proposed by McLachlan (1979). To locate the amide proton NH_i, the fitting procedure involved C_{i-1}^α, C'_{i-1}, O_{i-1} and C_i^α. To locate $C^\alpha H_i$, the atoms N_i, C'_i and, except for Gly, C_i^β were used, and $C^\beta H_i$ was located based on best fits for C_i^α and C_i^γ (O_i^γ and S_i^γ in the case of serine and cysteine, respectively). For isoleucine, valine and threonine, the second γ-atom was also considered. Visual inspection of the structures on an Evans and Sutherland picture system (MPS) or on a graphics terminal with the use of the program XRAY (Feldmann, 1976) were useful for checking the attachment of protons.

3. Results and Discussion

In the early stages of this investigation the distances between all the different types of protons in neighbouring residues i and $(i+1)$ of a polypeptide chain were calculated as functions of the dihedral angles ϕ, ψ and χ^j, and the frequencies with which short distances between neighbouring and non-neighbouring residues occurred in protein crystal structures were evaluated. These preliminary data showed that sequential ^{1}H resonance assignments *via* NOE measurements should be based primarily on the distances d_1, d_2 and d_3 between C^α and amide protons, different amide protons, and C^β and amide protons, respectively (Fig. 1). Independently of the polypeptide backbone conformation, at least one of these three proton–proton distances is sufficiently short that a strong NOE between the corresponding protons should be observed (unless, of course, it is quenched, e.g. by local flexibility of the protein or by paramagnetic centres (Noggle & Schirmer, 1971)). Furthermore, the statistical analysis of proton–proton distances in X-ray structures of proteins revealed that short distances d_1 (i,j), $d_2(i,j)$ and d_3 (i,j) (eqns (1) to (3)) between amide, C^α and C^β protons located in non-neighbouring residues occurred only very rarely. In the following, we therefore concentrate on the presentation of the quantitative data on d_1, d_2 and d_3 (Fig. 1) and the corresponding distances between non-neighbouring residues.

(a) *Calculations of the stereochemically allowed distances between amide, C^{α} and C^{β} protons in neighbouring residues of a polypeptide chain*

The functions $d_1(\psi_i)$, $d_2(\phi_i, \psi_i)$ and $d_3(\chi_i^1, \psi_i)$ calculated with equations (A7), (A16) and (A18) from the Appendix are shown in Figures 2 to 4. Details of the calculations are given in the Appendix.

Since the average of the slightly different ECEPP parameters for the different amino acids (Momany *et al.*, 1975) was used, a common $d_1(\psi_i)$ curve was obtained for the 19 L-amino acids (Fig. 2). A similar curve was presented by Leach *et al.* (1977). With the assumption that the upper distance limit for through-space connectivities to be used for sequential resonance assignments is $\lesssim 3{\cdot}0$ Å (see Introduction), one has that d_1 is a suitable parameter for the entire sterically allowed regions B and C in the ϕ_i–ψ_i plane (Fig. 3). A particularly small d_1 value of $\sim 2{\cdot}2$ Å prevails for β-structures. This was exploited in the first sequential assignments obtained (Dubs *et al.*, 1979, Wagner *et al.*, 1981). In the region A, which includes the α-helix, d_1 is, however, $> 3{\cdot}0$ Å. For glycine in position i, d_1 is $\leq 3{\cdot}1$ Å in the entire sterically accessible region.

Figure 3 shows the well-known ϕ_i–ψ_i plane with the sterically allowed regions A, B and C (Ramachandran & Sasisekharan, 1968) and contour lines for fixed values of $d_2(\phi_i, \psi_i)$ between 1·5 and 4·5 Å. These contour lines coincide closely with data presented by Kuo & Gibbons (1979). It is seen that d_2 is $\leq 3{\cdot}0$ Å for the entire region A and $\sim$half of the region C, while most of the region B with β_p and β lies

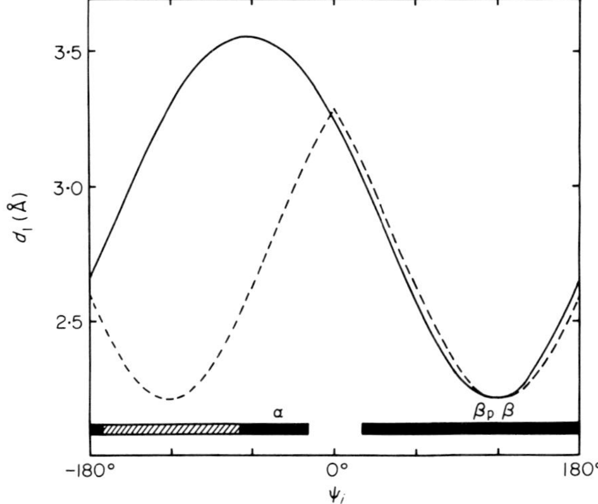

FIG. 2. Plot of the distance d_1 between NH_{i+1} and $C^{\alpha}H_i$ (see Fig. 1) *versus* the dihedral angle ψ_i. (——) 19 common L-amino acids in position i. (– – – –) Gly; at each value of ψ_i, the smaller of the 2 distances between NH_{i+1} and the two α-protons of Gly in position i is displayed. In the range from 0 to +180°, small differences between the 2 curves arise as a consequence of small deviations between the ECEPP parameters for Gly and for the other residues (Momany *et al.*, 1975). The range of sterically allowed ψ_i values for the common L-amino acids is indicated by the black bar at the bottom of the Figure. For glycine, the ψ_i values indicated by the hatched bar are also allowed. The letters α, β_p and β identify the ψ_i values for the α-helix, the parallel β-structure and the antiparallel β-structure. The curves were computed with eqn (A7).

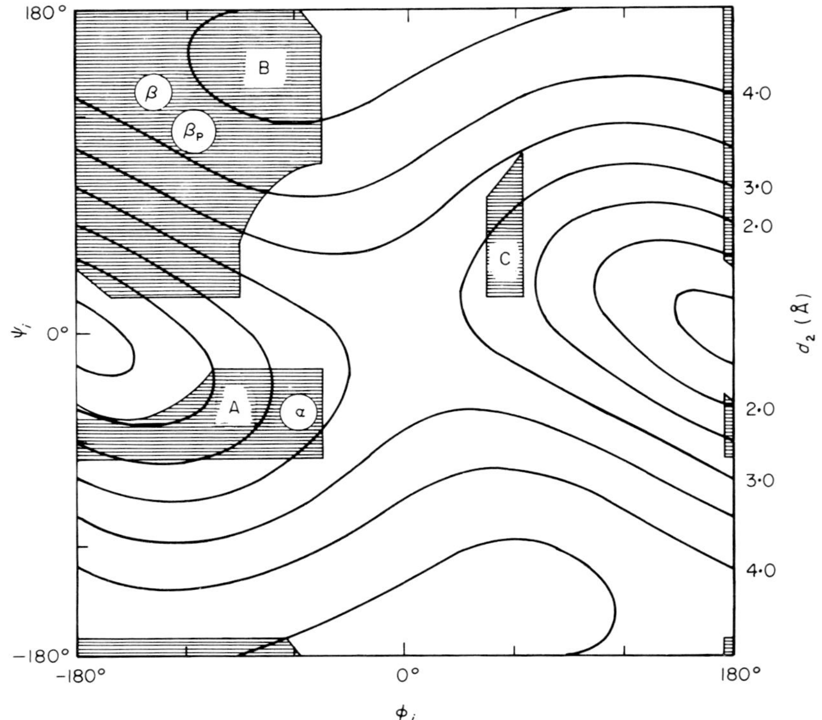

FIG. 3. Contour lines for fixed values of the distance d_2 between NH_{i+1} and NH_i (see Fig. 1) in the ϕ_i–ψ_i plane. The contour lines were computed with eqn (A16), using the average of the ECEPP parameters for the common amino acid residues (except, of course, Pro). The d_2 values for the individual contour lines are given on the right. The areas in the ϕ_i–ψ_i plane that are sterically allowed for Ala (Ramachandran & Sasisekharan, 1968; Mandel *et al.*, 1977) are hatched and labelled A, B and C. The ψ_i–ϕ_i combinations for the regular α-helix and parallel and antiparallel β-structure are indicated by α, β_p and β.

outside the 3·0 Å contour line. The curves in Figure 4 show that d_3 can adopt values smaller than 3·0 Å for nearly the entire region A and for a small part of region B, excluding β_p and β. Among the different types of amino acids in position i (eqns (3a) to (3c)), proline presents the most favourable case for sequential assignments *via* d_3. The curve for alanine indicates that this residue does not present a particularly favourable case for assignments *via* d_3. However, in view of the simple geometry used for alanine (eqn (3c)), this result should be used only as a general guideline.

Overall, inspection of Figures 2 to 4 shows that, with the assumption of an upper distance limit of $\lesssim$ 3·0 Å, none of the three distances d_1, d_2 and d_3 would provide a universally applicable criterion for sequential assignment *via* NOEs. However, since all three distances depend on ψ_i, their variations with conformation are correlated, and a more reliable criterion for sequential assignments is obtained when one considers the combination of any two of these distances. Figure 5 shows that the image of the ϕ_i–ψ_i plane in the d_1–d_2 plane extends from $d_1 \approx$ 2·2 Å to

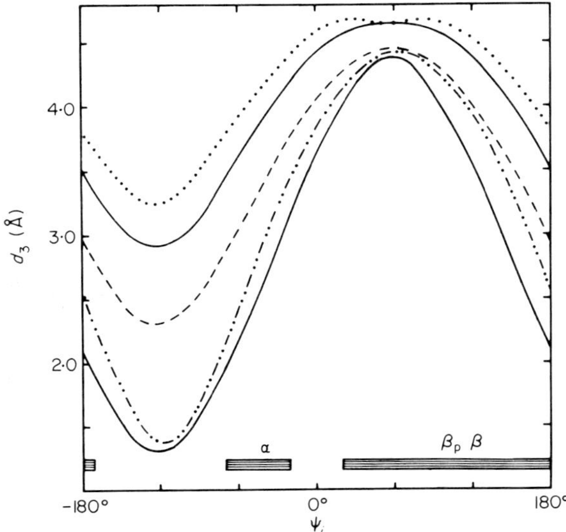

FIG. 4. Dependence of the distance d_3 (Fig. 1) on the dihedral angle ψ_i. The 5 curves, which were computed with eqn (A18), provide the data for the 19 common L-amino acid residues as follows: (1) for 14 amino acids with β-methylene groups (all except Pro), χ_i^1 was varied from $-180°$ to $+180°$ for each value of ψ_i. The d_3 values thus obtained lie between the 2 continuous curves. (2) For Val, Ile and Thr, a similar treatment showed that the d_3 values vary between the lower continuous curve and the dotted curve. (3) For alanine, the broken curve was obtained with the use of an average location for the 3 methyl protons on the extended C^{α}–C^{β} bond at 0·36 Å from C^{β}. (4) The curve for Pro (— · · — · · —) was computed for a planar Pro ring Momany *et al.*, 1975), without flexibility about χ_i^1. The shaded bar near the bottom of the Figure indicates the sterically allowed values for ψ_i. α, β_p and β identify the ψ_i values for the α-helix and the parallel and antiparallel β-structures.

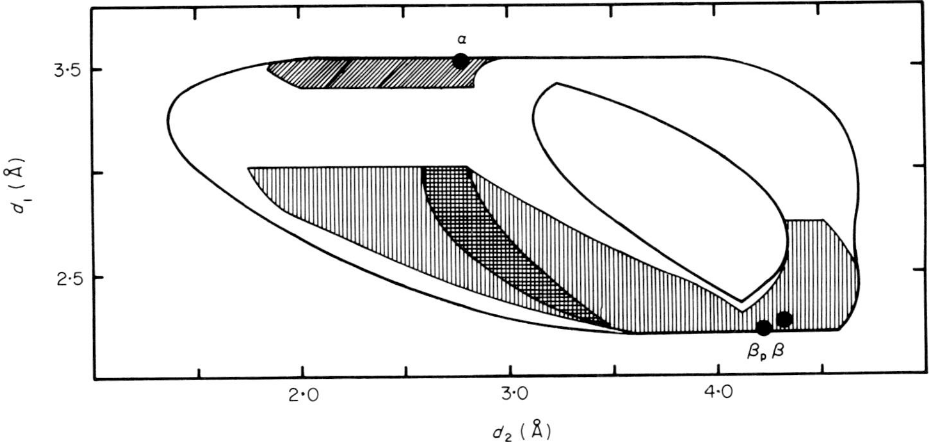

FIG. 5. Image of the ϕ_i–ψ_i plane in the d_1–d_2 plane. The area between the heavy continuous curves corresponds to the entire ϕ_i–ψ_i plane. The sterically allowed regions A, B and C (Fig. 3) are shown as follows: A /////; B |||||; C ≡. The data in this Figure apply for 18 of the common L-amino acid residues, i.e. all except Gly (see Fig. 6) and Pro. The filled circles indicate the d_1–d_2 combinations for the α-helix and the parallel and antiparallel β-structures.

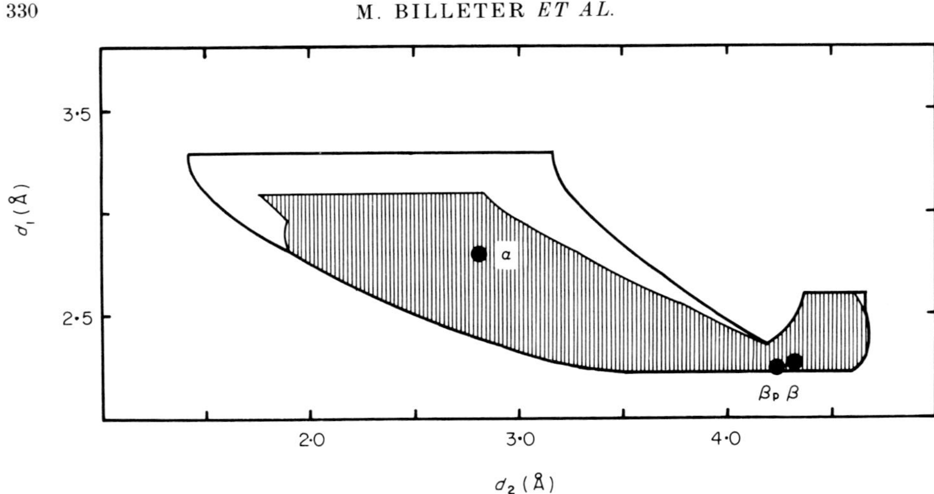

FIG. 6. Image of the ϕ_i–ψ_i plane in the d_1—d_2 plane for Gly, where for each combination of ϕ_i and ψ_i the smaller of the 2 distances between NH_{i+1} and the two α-protons of Gly in position i was taken (see Fig. 2 and eqn (1b)). The sterically allowed area (Ramachandran & Sasisekharan, 1968; Mandel *et al.*, 1977) is shaded. The filled circles indicate the d_1–d_2 combinations for the α-helix and the parallel and antiparallel β-structures.

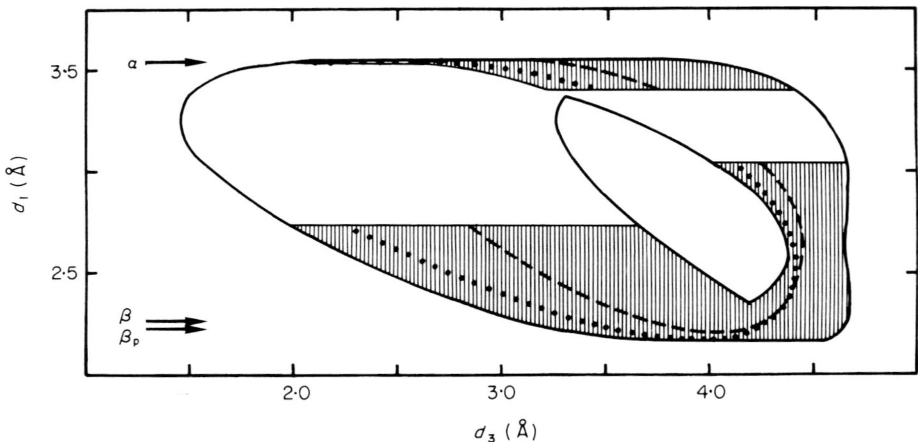

FIG. 7. Image of the ψ_i–χ_i^1 plane in the d_1–d_3 plane. The area between the heavy continuous curves corresponds to the entire ψ_i–χ_i^1 plane, the hatched area to those regions of the ψ_i–χ_i^1 plane that contain the sterically allowed values for ψ_i (see Figs 2 to 4) and all values from $-180°$ to $+180°$ for χ_i^1 (see Fig. 4). The hatched area was computed for Val, Ile and Thr, but it applies to a good approximation for all the common amino acids except Ala, Pro and Gly (see the text). The allowed (d_1–d_3) values for Ala and Pro are indicated by the broken curve and the dotted curve, respectively (see Fig. 4 for the geometries used for Ala and Pro). α, β and β_p indicate the d_1 values for the α-helix and the parallel and antiparallel β-structures.

$\approx 3{\cdot}6$ Å and from $d_2 \approx 1{\cdot}4$ Å to $\approx 4{\cdot}7$ Å. The shape is reminiscent of a cross-section through a plum, with a hole near the location of the stone. This hole is a consequence of the two-dimensional mapping and is not related to the steric restrictions of allowed ϕ_i–ψ_i values. The sterically allowed regions A, B and C

(Fig. 3) correspond to the shaded areas. It is seen that in the entire d_1–d_2 plane, all sterically accessible combinations of d_1 and d_2 values include at least one distance smaller than 3·0 Å. For glycine (Fig. 6), even d_1 is $< 3{\cdot}0$ Å in the region of the α-helix. Figure 7 shows an image of the ψ_i–χ_i^1 plane in the d_1–d_3 plane. Similar to Figure 5, this image resembles the outline of a plum, with a hole near where the stone would be located. The shaded area corresponds to the sterically allowed regions for valine, isoleucine and threonine (see Fig. 4) and extends only very slightly beyond the corresponding area for the 14 amino acids with $C^{\beta}H_2$ groups (see Fig. 4; eqns (3a) and (3b); Tables A1 and A2 of the Appendix). With the special geometries chosen for alanine (eqn (3c)) and proline (legend to Fig. 4), the two curves inside the shaded area were obtained. Figure 7 shows that, with the exception of a region extending from $d_1 \approx 3{\cdot}3$ to $\approx 3{\cdot}5$ Å and $d_3 = 3{\cdot}0$ to $\approx 4{\cdot}4$ Å, all the sterically allowed combinations of d_1 and d_3 include at least one distance shorter than 3·0 Å. A similar plot in the d_2–d_3 plane (not shown) would illustrate that there are large areas where both d_2 and d_3 are $>3{\cdot}0$ Å (see Figs 3 and 4).

(b) *Individual statistics of the proton–proton distances* $d_1(i, j)$, $d_2(i, j)$ *and* $d_3(i, j)$ *for* $j-i = 1$ *and* $j-i \neq 0, 1$ *in globular proteins*

In the Methods section we described the selection of 19 proteins with 3224 amino acid residues for the statistical studies (Table 1) and the attachment of hydrogen atoms to the X-ray structures. Table 2 lists how many times the three distances $d_1(i,j)$, $d_2(i,j)$ and $d_3(i,j)$ were shorter than the limits given in the first column, both for $j-i = 1$ and $j-i \neq 0, 1$. The last column in Table 2 lists the percentage of the distances smaller than this limit that occur between neighbouring residues with $j-i = 1$. This quantity presents a measure for the reliability of sequential resonance assignments based on measurements of a single one of the three distances in Table 2, since n.m.r. can, on the basis of spin–spin couplings and chemical shifts, distinguish between amide, C^{α} and C^{β} protons and, furthermore, can identify groups of protons belonging to the same residue (Wüthrich, 1976; Nagayama & Wüthrich, 1981).

The numbers in Table 2 show that short distances for $d_1(i,j)$, $d_2(i,j)$ and $d_3(i,j)$ arise mainly from neighbouring residues. For example, 88% of the distances $d_1(i,j) \leq 3{\cdot}0$ Å and 76% of the distances $d_3(i,j) \leq 3{\cdot}0$ Å are between neighbouring residues with $j-i = 1$. For $d_2(i,j)$, which is, in contrast to the other two distances, symmetrical with respect to the direction of the polypeptide chain (Fig. 1), the significance of the numbers in Table 2 is that, for example, 88% of the distances $\leq 3{\cdot}0$ Å are between neighbouring residues with either $j-i = 1$ or $j-i = -1$. While the results of the stereochemical considerations in Figures 2 to 4 showed that none of the three distances d_1, d_2 or d_3 could provide a universal criterion for sequential resonance assignments in the entire sterically allowed conformation space, Table 2 thus demonstrates that whenever a short distance $d_1(i, j)$, $d_2(i, j)$ or $d_3(i, j)$ is observed, it is a useful and quite reliable quantity for obtaining sequential assignments.

Besides the percentage of nearest-neighbour connectivities for a given distance, the dependence of this quantity on the distance limit is of great practical interest.

TABLE 2

Statistics of the distances $d_1(i, j)$, $d_2(i, j)$ *and* $d_3(i, j)$ *in the 19 protein structures of Table 1*

	d(Å)	n† $j-i=1$	n† $j-i \neq 0, 1$	% $j-i=1$
$d_1(i, j)$	$\leq 2{\cdot}4$	1194	23	98
	$\leq 2{\cdot}6$	1491	68	96
	$\leq 2{\cdot}8$	1615	137	92
	$\leq 3{\cdot}0$	1722	237	88
	$\leq 3{\cdot}2$	1895	435	81
	$\leq 3{\cdot}4$	2206	750	75
	$\leq 3{\cdot}6$	3068	1219	72
$d_2(i, j)$‡	$\leq 2{\cdot}4$	289	18	94
	$\leq 2{\cdot}6$	550	45	92
	$\leq 2{\cdot}8$	965	101	91
	$\leq 3{\cdot}0$	1231	160	88
	$\leq 3{\cdot}2$	1343	232	85
	$\leq 3{\cdot}4$	1420	317	82
	$\leq 3{\cdot}6$	1459	415	78
$d_3(i, j)$	$\leq 2{\cdot}4$	261	69	79
	$\leq 2{\cdot}6$	484	128	79
	$\leq 2{\cdot}8$	740	207	78
	$\leq 3{\cdot}0$	1005	310	76
	$\leq 3{\cdot}2$	1231	457	73
	$\leq 3{\cdot}4$	1478	621	70
	$\leq 3{\cdot}6$	1690	876	66

† n, The frequency with which the value for d indicated in the first column occurs between the 3224 amino acid residues of the proteins in Table 1.

‡ Because of the symmetry of $d_2(i, j)$ with respect to the chain direction (Fig. 1), only d_2 values with $j > i$ were counted.

Table 2 shows that with the use of any of the three parameters d_1, d_2 or d_3, the reliability of sequential assignments decreases with increasing values of the applicable upper distance limit. However, the percentage of connectivities between neighbouring residues is well above 70% for all distance limits between 2·4 Å and 3·4 Å. We conclude that a good experimental technique for sequential resonance assignments should be able to delineate short proton–proton distances up to a limiting value anywhere in the range from ~2·6 Å to 3·4 Å, but that sequential assignments will not have to rely on accurate distance measurements in this range.

The statistical data of Table 2 further support the conclusion that emanated from Figures 2 to 4, that $d_1(i, j)$ is the most useful among the three individual distances. In particular, when using an experiment capable of delineating proton–proton distances smaller than ~ 2·5 Å, the reliability of sequential assignments *via* d_1 is better than 96%. Hence, resonance assignments in the extended polypeptide chains of β-structures may quite generally be obtained on the basis of d_1 (Fig. 2). This was recognized several years ago from inspection of mechanical protein models and used for resonance assignments in β-sheets (Dubs *et al.*, 1979; Wagner *et al.*,

1981). For conformations in the sterically allowed region A (Fig. 3), d_2 provides more reliable evidence for sequential connectivities than either d_1 or d_3. However, information on the direction of the connectivity must be obtained independently from, and in addition to, the measurements of d_2.

(c) *Joint statistics of the proton–proton distance pairs* ($d_1(i, j)$, $d_2(i, j)$), ($d_1(i, j)$, $d_3(i, j)$) *and* ($d_3(i, j)$, $d_2(i, j)$) *for* $j-i = 1$ *and* $j-i \neq 0, 1$ *in globular proteins*

Joint statistics of pairs of proton–proton distances were obtained for the groups of 19 proteins in Table 1. The results are presented in Figures 8 to 10. For a detailed comparison of the individual statistics of Table 2 and the joint statistics of Figures 8 to 10, consult the footnote†.

Compared to the individual statistics, the joint statistics provide a further improved discrimination between sequential nearest-neighbour and long-range connectivities *via* proton–proton distances. In addition, in contrast to the individual statistics for $d_2(i, j)$, the joint statistics with $d_2(i, j)$ include also the direction of the connectivities. We first consider the data for ($d_1(i, j)$, $d_2(i, j)$). In Figure 8(a) and (b), those areas of the d_1–d_2 plane that are heavily populated with (d_1, d_2) combinations between neighbouring residues with $j-i = 1$ are outlined. These regions contain only very few short distances between non-neighbouring residues. In the upper left outlined area, with d_1 from 2·6 to 3·6 Å and d_2 from 2·0 to 3·4 Å, 99% of the d_1-d_2 combinations correspond to neighbouring residues with $j-i = 1$. In the lower right area, with d_1 from 2·0 to 2·8 Å and d_2 from 3·6 to 4·8 Å, this percentage is 96%. It is further seen that, with the exception of the small region with d_1 from 3·4 to 3·6 Å and d_2 from 3·0 to 3·4 Å, the outlined areas contain at least one distance smaller than 3·0 Å. The joint statistics thus confirm the prediction from stereochemical considerations (Fig. 5) that the combination of d_1 and d_2 provides a quite universal criterion for sequential resonance assignments, irrespective of the polypeptide conformation.

† To compare the numerical results of the joint statistics in Figs 8 to 10 with the data of the individual statistics in Table 2, one must in principle sum over the numbers in all the rows, or columns, respectively, up to the distance value corresponding to the limit given in Table 2. We first consider the statistics for neighbouring residues with $j-i = 1$. In Fig. 8(a), the sum of the values in the columns for $d_2 \leq 3·0$ Å is 1231 and in Fig. 9(a) the corresponding quantity for $d_3 \leq 3·0$ Å is 1005. These numbers are identical to the corresponding numbers in Table 2. For all other comparisons, one has to consider that the combinations of 2 types of distances may not be defined for all the amino acids for which the individual distances are defined (Fig. 1). In Fig. 8(a), the sum of the rows up to $d_1 \leq 3·0$ Å is 1629, which is smaller than the corresponding value in Table 2. The difference of 93 arises because, in contrast to the individual statistics for d_1, (d_1, d_2) is not defined with proline or the N terminus in position i. Similarly, when summing rows in Fig. 10(a), (d_3, d_2) is, in contrast to d_3, not defined with proline or the N terminus in position i. When summing rows in Fig. 9(a) or columns in Fig. 10(a), one has to consider that, in contrast to d_1 and d_2, (d_1, d_3) and (d_3, d_2) are not defined with Gly in position i.

The statistics for distances with $j-i \neq 0, 1$ in Table 2 and Figs 8(b), 9(b) and 10(b) can be compared on the same basis, with the exception of d_2 in Figs 8(b) and 10(b). Here, one needs to consider further that, in contrast to $d_2(i, j)$, ($d_1(i, j)$, $d_2(i, j)$) and ($d_3(i, j)$, $d_2(i, j)$) are not invariant with respect to the direction of the polypeptide chain (Fig. 1). Therefore, while the individual statistics for $d_2(i, j)$ covered only the values for $j-i > 1$, the joint statistics include $|j-i| > 1$, which is equal to twice the value for $j-i > 1$, and the values for $j-i = -1$, which are for symmetry reasons identical to those for $j-i = 1$ (Table 2). For example in Fig. 8(b), the sum of the numbers in the rows up to $d_2 \leq 3·0$ Å is 1551, which is equal to $2 \times 160 + 1231$, i.e. twice the values for $j-i \neq 0, 1$ plus the value for $j-i = 1$ in Table 2.

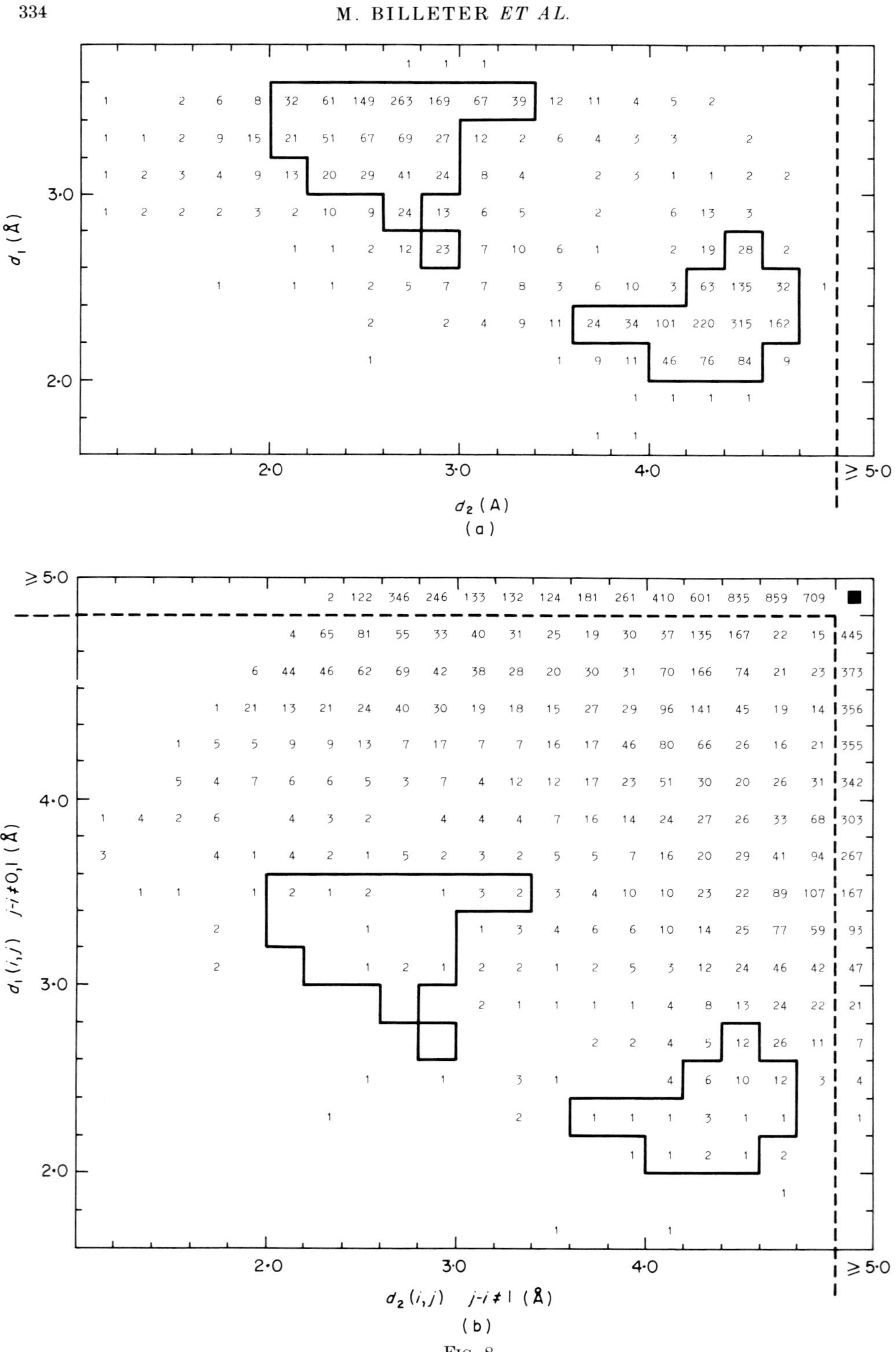
334
M. BILLETER *ET AL.*
3·0
2·0
d_1 (Å)
2·0
3·0
4·0
≥ 5·0
d_2 (Å)
(a)
≥ 5·0
4·0
3·0
2·0
d_1 (i,j) $j-i \neq 0,1$ (Å)
2·0
3·0
4·0
≥ 5·0
d_2 (i,j) $j-i \neq 1$ (Å)
(b)

FIG. 8.

Comparison of Figures 5 and 8(a) shows that the heavily populated regions of the d_1–d_2 plane fall mostly within the sterically allowed regions, with pronounced peaks near the d_1–d_2 values for the α-helix and the β-structures. The main discrepancy between Figures 5 and 8(a) is the sizeable population of the region from $d_1 \approx 3{\cdot}0$ to 3·4 Å and $d_2 \approx 2{\cdot}0$ to 3·0 Å. This area of the d_1–d_2 plane corresponds to the bridge region between the sterically allowed areas A and B in the ϕ_i–ψ_i plane (Fig. 3), which is known to be populated to some extent in globular proteins (Richardson, 1981). To further check on the coincidence between stereochemical prediction and statistics in globular proteins, Figure 8(c) shows the distribution of (d_1, d_2) values for the special case of glycine in position i. Again, the populated region coincides closely with the sterically allowed region in Figure 6. In particular, the cutoff at d_1 values of $\sim 3{\cdot}1$ Å is confirmed by the statistical analysis.

The joint statistics for $(d_1(i,j),\ d_3(i,j))$ in Figure 9 show very similar characteristics to those for $(d_1(i,j), d_2(i,j))$ in Figure 8. Most important, there is an almost complete distinction between neighbouring residues $(j-i = 1)$ and non-neighbouring residues $(j-i \neq 0, 1)$. In both outlined areas, 95% of the d_1–d_3 combinations correspond to neighbouring residues in the sequence. The heavily populated regions coincide closely with the stereochemical predictions in Figure 7,

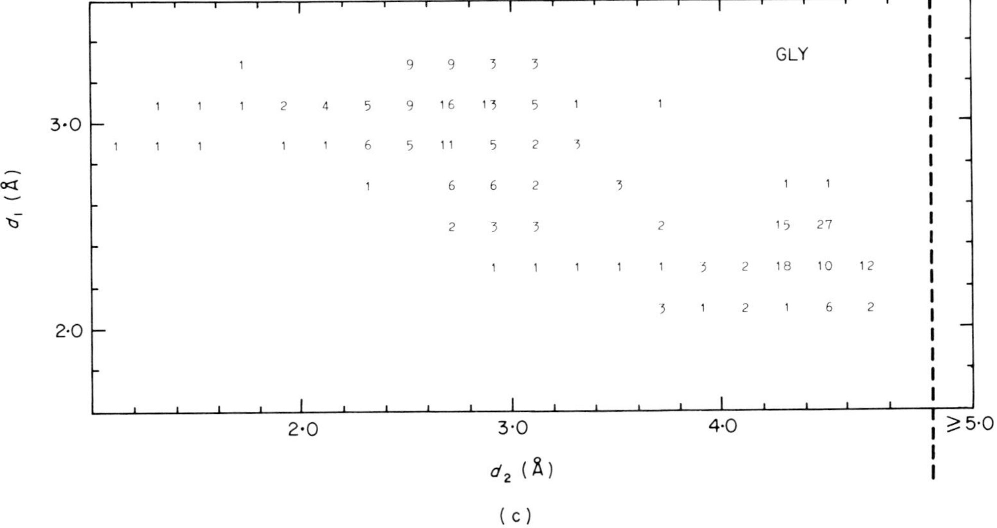

FIG. 8. Presentation in the d_1–d_2 plane of joint statistics for the ^{1}H–^{1}H distances $d_1(i,j)$ and $d_2(i,j)$. The numbers indicate how many of the $d_1(i,j)$–$d_2(i,j)$ combinations between the 3224 amino acid residues in the 19 proteins of Table 1 fall within a square of side length 0·2 Å centred about the location of the number. (Note that $(d_1(i,j), d_2(i,j))$ is not defined when proline is in either of the positions i or j.)† (a) Statistics for (d_1, d_2) between neighbouring residues with $j-i = 1$. Regions where the frequency of d_1–d_2 combinations is ≥ 20 per square of area 0·04 Å^2 are bounded with a thick line. This plot should be compared with the stereochemical predictions in Fig. 5. (b) Statistics for $(d_1(i,j), d_2(i,j))$ with $j-i \neq 0; 1$. The thick lines indicate those areas that are heavily populated for $j-i = 1$. (■) A number $>10^5$. (c) Statistics for (d_1, d_2) for neighbouring residues with $j-i = 1$ and glycine in position i. This plot should be compared with the stereochemical predictions in Fig. 6.

† See footnote to p. 333.

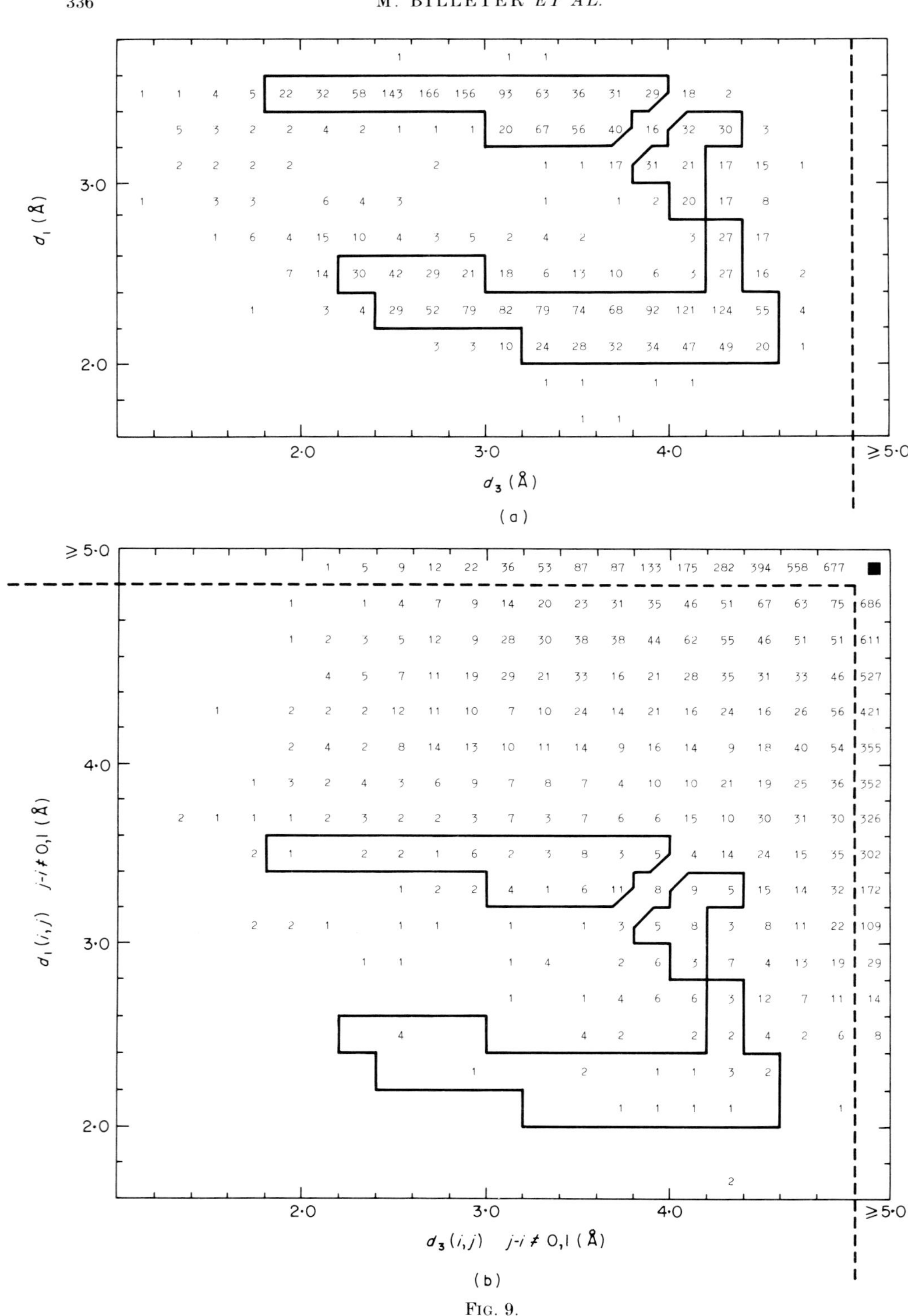

3·0
2·0
d_1 (Å)
2·0
3·0
4·0
≥5·0
d_3 (Å)
(a)
≥5·0
4·0
3·0
2·0
$d_1(i,j)$ j-i ≠ 0,1 (Å)
2·0
3·0
4·0
≥5·0
$d_3(i,j)$ j-i ≠ 0,1 (Å)
(b)

FIG. 9.

except that d_1 values from 3·0 to 3·4 Å are sizeably populated near $d_3 = 4{\cdot}0$ Å. Again, this corresponds to ψ_i values from $-20°$ to $+20°$ in the ϕ_i–ψ_i plane (Fig. 3), which are known to occur in globular proteins (Richardson, 1981).

Figure 10(a) and (b) shows the statistics for $(d_3(i,j), d_2(i,j))$ with $j-i = 1$ and $j-i \neq 0, 1$, respectively. In agreement with the stereochemical predictions, Figure 10(a) shows that the d_3–d_2 combinations between neighbouring residues cover large areas of the d_3–d_2 plane where both distances d_3 and d_2 are between 3·0 and 4·8 Å. Nonetheless, a potentially useful criterion for sequential assignments emerges from these data. If both distances $d_3(i,j)$ and $d_2(i,j)$ are between 2·0 and 3·0 Å, the connectivity is, with a reliability of 92%, between neighbouring residues with $j-i = 1$.

(d) *Calculations of the distances between $C^{\beta}H_2$ and the ring protons in aromatic amino acids*

For most of the common amino acids, the spin system of the entire side-chain can be identified on the basis of through-bond J-connectivities (Wüthrich, 1976; Nagayama & Wüthrich, 1981). The exceptions are the four aromatic residues, where the spin–spin coupling constants between $C^{\beta}H_2$ and the ring protons are usually too small to provide reliable connectivities (Wider *et al.*, 1981), methionine, where the methyl group usually cannot be connected *via* J-coupling and asparagine and glutamine, where the amide protons cannot usually be connected *via* J-coupling. Here we investigate the molecular geometries of the aromatic residues, asparagine and glutamine, with regard to the use of NOE measurements to aid in the assignments.

Figure 11 shows plots *versus* χ^2 of the shortest $C^{\beta}H_2$–ring proton distances in the four aromatic residues. Since there is evidence that the orientation of the ring planes in globular proteins is determined primarily by non-bonding interactions with the environment (Gelin & Karplus, 1975; Hetzel *et al.*, 1976), no discrimination of χ^2 on the basis of local interactions within the side-chain was considered. The Figure shows that for tyrosine, phenylalanine and tryptophan, at least one $C^{\beta}H_2$–ring proton distance is smaller than 3·0 Å over the entire range of χ^2 values from $-180°$ to $180°$. For histidine, this holds only for the range from $\chi^2 = -100°$ to $+100°$, whereas outside of this range there is a steep rise of the shortest distance. We conclude that delineation of short ^1H–^1H distances, for example by NOEs, should reliably establish the connectivities to the rings of

Fig. 9. Presentation in the d_1–d_3 plane of joint statistics for the ^1H–^1H distances $d_1(i,j)$ and $d_3(i,j)$. The numbers indicate how many of the $d_1(i,j)$–$d_3(i,j)$ combinations between the 3224 amino acid residues in the 19 proteins of Table 1 fall within a square of side length 0·2 Å centred about the location of the number. (Note that $(d_1(i,j), d_3(i,j))$ is not defined with either glycine at position i or proline at position j.)† (a) Statistics for (d_1, d_3) between neighbouring residues with $j-i = 1$. Regions where the frequency of d_1–d_3 combinations is ≥ 20 per square of area 0·04 Å^2 are bounded with a thick line. This plot should be compared with the stereochemical preditions in Fig. 7. (b) Statistics for $(d_1(i,j), d_3(i,j))$ with $j-i \neq 0, 1$. The thick lines indicate the areas that are heavily populated for $j-i = 1$. (■) A number $>10^5$.

† See footnote to p. 333.

tyrosine, phenylalanine and tryptophan, and that this approach should also work for some but not all histidine residues.

The geometry at the periphery of an asparagine or glutamine side-chain closely matches that of glycine $C^{\alpha}H_2$ and the amide group formed with the residue following glycine. Therefore, a plot of the shortest distance between $C^{\beta}H_2$ in asparagine, or $C^{\gamma}H_2$ in glutamine, respectively, and the side-chain amide protons in these residues coincides almost quantitatively with the curve for d_1 of glycine in Figure 2, when ψ_i is substituted by χ_i^2, or χ_i^3, respectively. Hence, NOE measurements should also provide the connectivities to the peripheral amide groups.

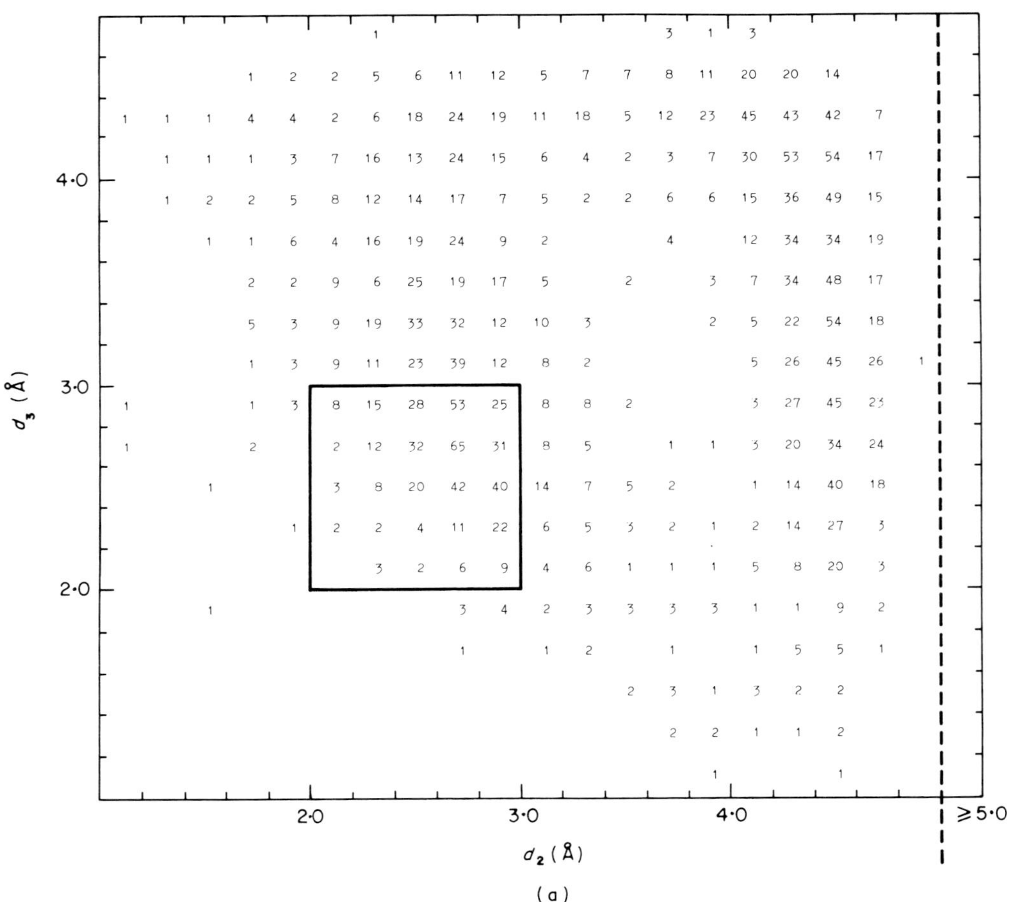

FIG. 10. Presentation in the d_3–d_2 plane of joint statistics for the ${}^1H$–${}^1H$ distances $d_3(i, j)$ and $d_2(i, j)$. The numbers indicate how many of the $d_3(i, j)$–$d_2(i, j)$ combinations between the 3224 amino acid residues in the 19 proteins of Table 1 fall within a square of side length 0·2 Å centred about the location of the number. (Note that $(d_3(i, j), d_2(i, j))$ is not defined with glycine at position i or proline at positions

4. Conclusions

The theoretical studies of intramolecular proton–proton distances presented in this paper imply that most individual ^{1}H n.m.r. assignments in polypeptides that cannot be established *via* through-bond J-connectivities might be obtained by measurements of through-space ^{1}H–^{1}H distances, for example with NOE experiments. Besides some connectivities in the amino acid side-chains of the aromatic residues, methionine, asparagine and glutamine, this concerns primarily connectivities between neighbouring residues in the amino acid sequence. The most important conclusion is that the present model studies would predict that sequential resonance assignments can be obtained with high reliability from

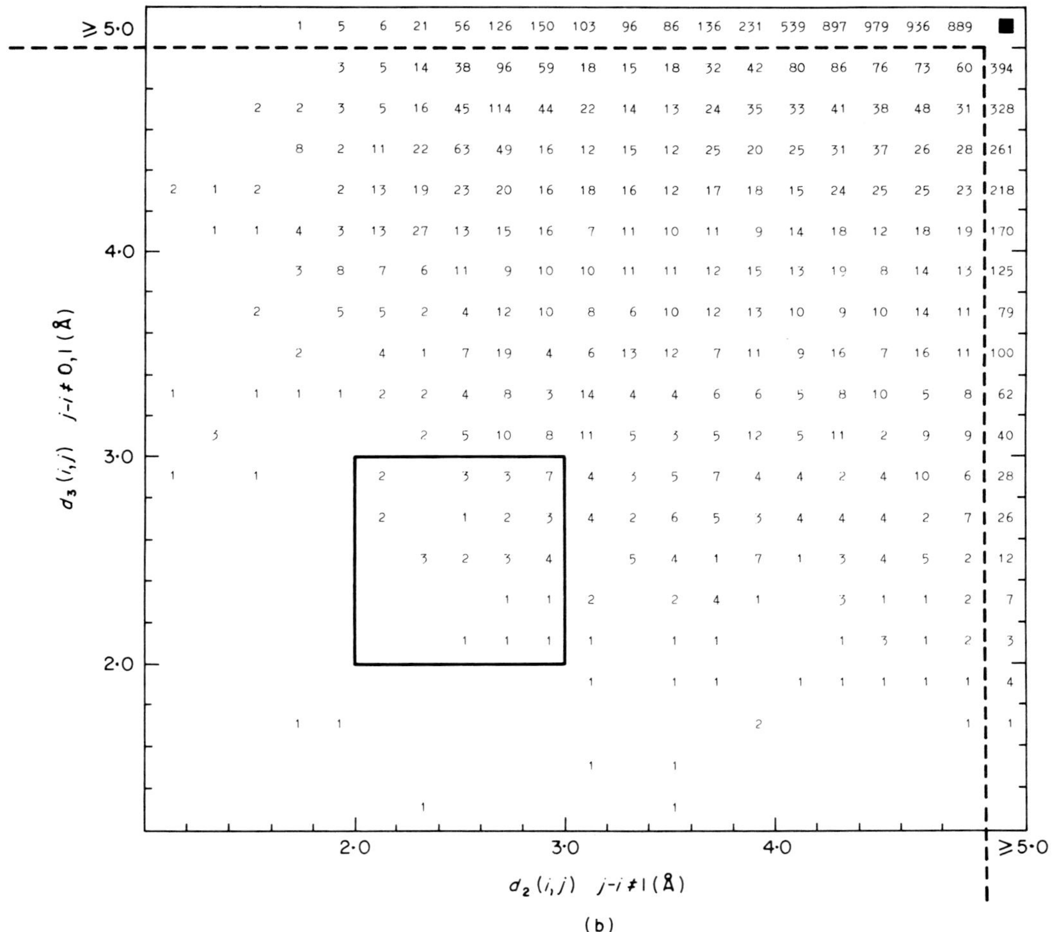

i or *j*.)† Thick lines bound the square within which 2·0 Å $\leq d_3(i,j)$, $d_2(i,j)$ $\leq$ 3·0 Å. (a) Statistics for (d_3, d_2) between neighbouring residues, with $j-i = 1$. (b) Statistics for $(d_3(i,j), d_2(i,j))$, with $j-i \neq 0, 1$. (■) A number $> 10^5$.

† See footnote to p. 333.

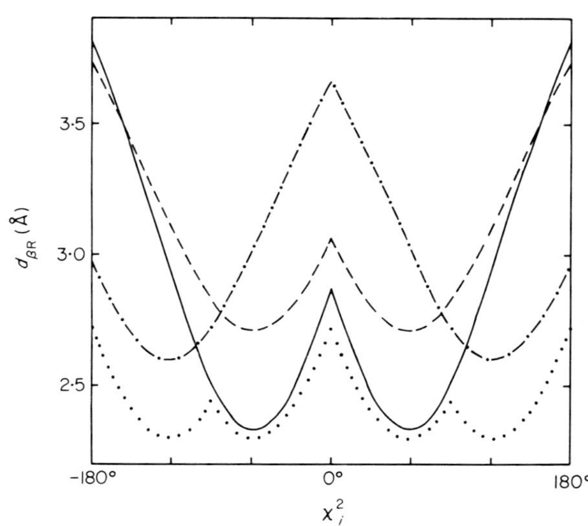

FIG. 11. Plots *versus* χ^2 of the shortest distances between the C^β protons and any of the ring protons, $d_{\beta R}$, in the 4 aromatic amino acids. The curves were calculated using the ECEPP parameters (Momany *et al.*, 1975) and an algebraic expression corresponding to eqn (A7). (·····) Minimum distance between $C^\beta H_2$ and $C^\delta H$ in Phe and Tyr; (– – – – –) minimum distance between $C^\beta H_2$ and $C^\delta H$ in His; (– · –) and (————) minimum distances between $C^\beta H_2$ and $C^\delta H$, and $C^\epsilon H$, respectively, in Trp.

combinations of semi-quantitative measurements of any two of the three distances $d_1(i,j)$ between amide protons and C^α protons, $d_2(i,j)$ between different amide protons and $d_3(i,j)$ between amide protons and C^β protons.

The following considerations may be helpful to further assess the practical significance of the results obtained here. (1) The reliabilities for sequential assignments *via* d_1, d_2, d_3 or a combination of any two of these distances, which can be calculated from the data in Table 2 and Figures 8 to 10, rely on the assumption that the experimental technique used to delineate the distance constraints would unambiguously manifest all the distances up to a preselected limit. For NOE experiments, one will have to allow for some uncertainty in the upper limit for ^1H–^1H distances manifested in the spectra. Furthermore, limited spectral resolution or other factors may adversely affect certain measurements. As an illustration, we may mention that quite often superior spectral resolution is obtained for the C^β protons than for the C^α protons, which can make assignments *via* d_3 (i,j) more useful, in spite of the lower inherent reliability. Therefore, while the calculated reliabilities for sequential assignments represent a result of fundamental importance, the numbers obtained should be considered as general guidelines rather than as accurate predictions. (2) In the present paper, only the backbone amide protons, C^α protons and C^β protons of the polypeptide chain were considered (Fig. 1). In reality, because there are 20 different amino acids there is much more variability in the covalent structure. By n.m.r. experiments, a large proportion of the amino acid side-chains can usually be identified (Nagayama & Wüthrich, 1981), which provides additional, independent checks of the sequential assignments and thus further improves their reliability (see the following two papers): Wagner &

Wüthrich, 1982; Wider *et al.*, 1982). (3) With the criteria presented in this paper, the sequential assignments come to a stop at each prolyl residue. This may again be used profitably to check the sequential assignments against the amino acid sequence. Furthermore, as will be discussed in detail in a forthcoming paper on distance criteria for n.m.r. determination of polypeptide secondary structures and demonstrated in the following paper on sequential assignments in basic pancreatic trypsin inhibitor (Wagner & Wüthrich, 1982), the distances between $C^{\delta}H_2$ of proline in position $(i+1)$ and the amide, C^{α} and C^{β} protons of residue i can, in favourable cases, be used to establish the missing sequential connectivities.

Practical applications of the criteria for sequential resonance assignments presented in this paper have been described for the β-sheet region of basic pancreatic trypsin inhibitor (Dubs *et al.*, 1979; Wagner *et al.*, 1981) and are to be found in the two following papers, which describe assignments for the complete polypeptide chains in basic pancreatic trypsin inhibitor (Wagner & Wüthrich, 1982) and in micelle-bound glucagon (Wider *et al.*, 1982).

APPENDIX

Dependence of the short-range distances d_1, d_2 and d_3 on the dihedral angles ϕ_i, ψ_i and χ_i^1

The short-range distances d_1, d_2 and d_3 (eqns (4) to (6)) between NH, $C^{\alpha}H$ and $C^{\beta}H$ of neighbouring amino acid residues can be used in the place of the well-known dihedral angles ϕ_i, ψ_i and χ_i^1 (Ramachandran & Sasisekharan, 1968) to characterize the local conformations along a polypeptide chain. Here, precise relations between d_1, d_2, d_3 and ϕ_i, ψ_i, χ_i^1 will be computed with the use of the ECEPP parameters for amino acid residues (Momany *et al.*, 1975).

If **a** and **n** denote the difference vectors from C_i^{α} to $C^{\alpha}H_i$ and NH_{i+1}, respectively, for the dihedral angle $\psi_i = 0$, then $d_1(\psi_i)$ is given by:

$$d_1(\psi_i) = |\mathbf{a} - T(\mathbf{e}_i, \psi_i)\mathbf{n}|, \tag{A1}$$

where $\mathbf{e}_i$ is the unit vector along the C_i^{α}–C_i' bond; $T(\mathbf{e}, \psi)\mathbf{x}$ describes a rotation of a vector **x** by the angle ψ about an axis defined by the unit vector **e**, i.e.:

$$T(\mathbf{e}, \psi)\mathbf{x} = (\mathbf{e}\cdot\mathbf{x})\mathbf{e} + \sin\psi\,(\mathbf{e}\wedge\mathbf{x}) - \cos\psi\,\mathbf{e}\wedge(\mathbf{e}\wedge\mathbf{x}), \tag{A2}$$

where · denotes the scalar product and $\wedge$ the vector product. A convenient co-ordinate system is given uniquely by centering C_i^{α}, setting C_i' on the positive x axis and placing $C^{\alpha}H_i$ in the x–y plane, with $y > 0$. In this co-ordinate system the vector **a** is written $\hat{\mathbf{a}}$. The unit length is 1·00 Å. Since $\hat{\mathbf{a}}$ coincides with the covalent bond C_i^{α}–$C^{\alpha}H_i$, which has a length of 1·00 Å (Momany *et al.*, 1975), it becomes:

$$\hat{\mathbf{a}} = \begin{pmatrix} \cos\tau(C^{\alpha}H_i, C_i^{\alpha}, C_i') \\ \sin\tau(C^{\alpha}H_i, C_i^{\alpha}, C_i') \\ 0 \end{pmatrix}, \tag{A3}$$

where $\tau(C^{\alpha}H_i, C^{\alpha}_i, C'_i)$ is the angle formed by the two bonds C^{α}_i–$C^{\alpha}H_i$ and C^{α}_i–C'_i. The vector **n** is placed in the x–y plane, with $y > 0$, through a rotation by an angle ψ^0 about an axis defined by:

$$\hat{\mathbf{e}}_i = \begin{pmatrix} 1 \\ 0 \\ 0 \end{pmatrix}. \qquad (A4)$$

After this rotation, the co-ordinates of the amide proton are:

$$\hat{\mathbf{n}}^0 = T(\hat{\mathbf{e}}_i, \psi^0)\hat{\mathbf{n}} = \begin{pmatrix} 1{\cdot}58 \\ 2{\cdot}06 \\ 0 \end{pmatrix}. \qquad (A5) \qquad (A5)$$

We then have that:

$$T(\hat{\mathbf{e}}_i, \psi_i)\hat{\mathbf{n}} = T(\hat{\mathbf{e}}_i, \psi_i - \psi^0)\hat{\mathbf{n}}^0. \qquad (A6)$$

$T(\hat{\mathbf{e}}_i, \psi_i)\hat{\mathbf{n}}$ in equation (A6) is now evaluated with the use of equation (A2), and the resulting expression inserted into equation (A1). After computing the modulus of the vector on the right-hand side of equation (A1) as the square root of the scalar product of this vector with itself, one obtains:

$$d_1(\psi_i) = (a_1 + b_1 \cos(\psi_i - \psi^0))^{\frac{1}{2}}, \qquad (A7)$$

where a_1 and b_1 are given by:

$$a_1 = (\hat{\mathbf{a}} \cdot \hat{\mathbf{a}}) + (\hat{\mathbf{n}}^0 \cdot \hat{\mathbf{n}}^0) - 2(\hat{\mathbf{e}}_i \cdot \hat{\mathbf{a}})(\hat{\mathbf{e}}_i \cdot \hat{\mathbf{n}}^0), \qquad (A8)$$

$$b_1 = 2((\hat{\mathbf{e}}_i \cdot \hat{\mathbf{a}})(\hat{\mathbf{e}}_i \cdot \hat{\mathbf{n}}^0) - (\hat{\mathbf{a}} \cdot \hat{\mathbf{n}}^0)). \qquad (A9)$$

Values for $\tau(C^{\alpha}H_i, C^{\alpha}_i, C'_i)$, a_1, b_1 and ψ^0 for the different amino acids were calculated using the ECEPP parameters (Momany *et al.*, 1975) and are listed in Table A1.

TABLE A1

Parameters used to compute the distance $d_1(\psi_i)$ *(Fig. 1) with equation (A7)*

Amino acid†	$\tau(C^{\alpha}H, C^{\alpha}, C')$(deg.)	a_1(Å^2)	b_1(Å^2)	ψ^0(deg.)
19 amino acids (all except Gly)	109·5	8·77	−3·88	117·0
Gly	109·7	8·78	−3·88	±121·4
18 amino acids (all except Gly, Pro)	109·7	8·78	−3·88	116·9
Pro	106·5	8·61	−3·95	119·8
Val, Ile, Thr	108·1	8·70	−3·91	117·3
Ala	109·0	8·74	−3·89	117·5

The parameters were calculated with equations (A8) and (A9) using the ECEPP geometry for the common amino acid residues (Momany *et al.*, 1975).

† Where a group of amino acids is listed, the average of the ECEPP parameters for these amino acids was used.

In analogy to equation (A1), the angular dependence of the amide proton–amide proton distance d_2 (Fig. 1) is calculated with:

$$d_2(\phi_i, \psi_i) = |T(\mathbf{e}_i^1, \phi_i)\mathbf{m} - T(\mathbf{e}_i^2, \psi_i)\mathbf{n}|, \tag{A10}$$

where **m** and **n** are the difference vectors from C_i^α to NH_i and NH_{i+1}, respectively, when $\phi_i = 0$ and $\psi_i = 0$, and $\mathbf{e}_i^1$ and $\mathbf{e}_i^2$ are unit vectors along the N_i–C_i^α bond and the C_i^α–C_i' bond, respectively. In a convenient co-ordinate system for calculating $d_2(\phi_i, \psi_i)$, C_i^α is centered, C_i' is on the positive x axis and N_i is placed in the x–y plane, with $y > 0$. In this system, the co-ordinates of NH_i at $\phi_i = 0$ are then given by:

$$\hat{\mathbf{m}} = T(\hat{\mathbf{c}}, \tau(N_i, C_i^\alpha, C_i'))\begin{pmatrix} 1{\cdot}88 \\ 0{\cdot}91 \\ 0 \end{pmatrix} \tag{A11}$$

with

$$\hat{\mathbf{c}} = \begin{pmatrix} 0 \\ 0 \\ 1 \end{pmatrix}. \tag{A12}$$

In analogy to equation (A5), the co-ordinates for NH_{i+1} are:

$$\hat{\mathbf{n}}^0 = \hat{\mathbf{n}} = \begin{pmatrix} 1{\cdot}58 \\ 2{\cdot}06 \\ 0 \end{pmatrix}. \tag{A13}$$

The equality $\hat{\mathbf{n}}^0 = \hat{\mathbf{n}}$ arises since, in this frame, NH_i and NH_{i+1} are in the *cis* arrangement and hence, according to the IUPAC/IUB convention (1970), $\psi^0 = 0$. Furthermore, in this co-ordinate system, the $\mathbf{e}_i^1$ and $\mathbf{e}_i^2$ terms of equation (A10) are:

$$\hat{\mathbf{e}}_i^1 = \begin{pmatrix} -\cos\tau(N_i, C_i^\alpha, C_i') \\ -\sin\tau(N_i, C_i^\alpha, C_i') \\ 0 \end{pmatrix} \tag{A14}$$

and

$$\hat{\mathbf{e}}_i^2 = \begin{pmatrix} 1 \\ 0 \\ 0 \end{pmatrix}. \tag{A15}$$

One then obtains:

$$d_2(\phi_i, \psi_i) = (a_2 + b_2\cos\phi_i + c_2\cos\psi_i + e_2\cos\phi_i\cos\psi_i + f_2\sin\phi_i\sin\psi_i)^{\frac{1}{2}}. \tag{A16}$$

With the use of the average of the ECEPP parameters for the 20 common amino acid residues (Momany *et al.*, 1975), the following numerical values were computed with the five equations corresponding to equations (A8) and (A9) for $d_1(\psi_i)$:

$$\begin{aligned} a_2 &= 13{\cdot}06\ \text{Å}^2 & b_2 &= 2{\cdot}69\ \text{Å}^2 & c_2 &= -7{\cdot}26\ \text{Å}^2 \\ e_2 &= 1{\cdot}26\ \text{Å}^2 & f_2 &= -3{\cdot}73\ \text{Å}^2. \end{aligned} \tag{A17}$$

For calculating $d_3(\psi_i, \chi_i^1)$, the formalism developed for d_2 (eqns (A10) to (A16)) was used, with N_i, NH_i and ϕ_i replaced by C_i^β, $C^\beta H_i$ and χ_i^1. The resulting equation

(A18) is analogous to equation (A16), except that χ^{10} and ψ^0 do not vanish for all the different amino acid residues:

$$d_3(\chi_i^1, \psi_i) = (a_3 + b_3 \cos(\chi_i^1 - \chi^{10}) + c_3 \cos(\psi_i - \psi^0) + e_3 \cos(\chi_i^1 - \chi^{10}) \cos(\psi_i - \psi^0) + f_3 \sin(\chi_i^1 - \chi^{10}) \sin(\psi_i - \psi^0))^{\frac{1}{2}}. \quad \text{(A18)}$$

For amino acids containing a $C^\beta H_2$ group, d_3 was calculated as defined in equation (3a). Proline was treated as a rigid, planar ring structure. For valine, isoleucine and threonine, d_3 was computed according to equation (3b), and for alanine d_3 was obtained according to equation (3c). Numerical values for the parameters in equation (A18), which were calculated for the different types of amino acids from their ECEPP geometries (Momany *et al.*, 1975), are listed in Table A2. As is further

TABLE A2

Parameters used to compute the distance $d_3(\chi_i^1, \psi_i)$ *(Fig. 1) with equation (A18)*

Amino acid†	a_3(Å^2)	b_3(Å^2)	c_3(Å^2)	e_3(Å^2)	f_3(Å^2)	χ^{10}(deg.)	ψ^0(deg.)
17 amino acids (all except Gly, Ala, Pro)	13·42	−2·97	−7·22	−1·50	4·16	n.u.‡	−122·1
14 amino acids (all with $C^\beta H_2$, except Pro)	13·45	−3·03	−7·27	−1·50	4·24	n.u.‡	−122·0
Val, Ile, Thr	13·37	−2·73	−7·01	−1·55	3·89	n.u.‡	−122·7
Ala	12·58	0	−7·21	0	0	n.u.‡	−121·8
Pro	13·64	−2·95	−7·15	−1·64	4·19	{0·3, 117·6}	−117·7

The parameters were calculated using the ECEPP geometry for the common amino acid residues (Momany *et al.*, 1975).

† Where a group of amino acids is listed, the average of the ECEPP parameters for these amino acids was used.

‡ Since for these amino acids only the upper and lower limits for d_3, which result from variation of χ_i^1, were computed (see the text), no specific value for χ^{10} was needed. For the alanine geometry defined in eqn (3c), χ^{10} is not defined.

discussed in Results, the dependence of d_3 on χ_i^1 was not considered explicitly. Instead, for each value of ψ_i, the upper and lower limits for d_3 were determined when χ_i^1 was allowed to vary between −180° and +180°.

Important input for the present paper came from the experimental studies carried out by Dr G. Wagner, G. Wider and Dr K. H. Lee. In particular, the fundamental ideas for the joint statistics were developed in discussions with Dr Wagner. We are also indebted to Mrs J. Richardson for stimulating discussions on various aspects of protein conformations, and thank Mrs E. H. Hunziker and Mrs E. Huber for the careful preparation of the illustrations and the manuscript. Financial support was obtained from the Schweizerischer Nationalfonds (project 3.528.79) and through special grants of the Eidgenössische Technische Hochschule (ETH) Zürich. One of the authors (M.B.) was the recipient of a fellowship of the Schweizerische Kommission für Molekularbiologie. Use of the facilities at the Zentrum für Interaktives Rechnen of the ETH is gratefully acknowledged.

REFERENCES

Anil Kumar, Ernst, R. R. & Wüthrich, K. (1980). *Biochem. Biophys. Res. Commun.* **95**, 1–6.

Anil Kumar, Wagner, G., Ernst, R. R. & Wüthrich, K. (1981). *J. Amer. Chem. Soc.* **103**, 3654–3658.

Baker, E. N. & Dodson, E. J. (1980). *Acta Crystallogr. sect. A*, **36**, 559–572.

Bernstein, F C., Koetzle, T. F., Williams, G. J. B., Meyer, E. F. Jr, Brice, M. D., Rodgers, J. R., Kennard, O., Shimanouchi, T. & Tasumi, M. (1977). *J. Mol. Biol.* **112**, 535–542.

Blake, C. C. F., Geisow, M. J., Oatley, S. J., Rerat, B. & Rerat, C. (1978). *J. Mol. Biol.* **121**, 339–356.

Braun, W., Bösch, C., Brown, L. R., Gō, N. & Wüthrich, K. (1981). *Biochim. Biophys. Acta*, **667**, 377–396.

Carter, C. W. Jr, Kraut, J., Freer, S. T., Xuong, N.-H., Alden, R. A. & Bartsch, R. G. (1974). *J. Biol. Chem.* **249**, 4212–4225.

Colman, P. M., Freeman, H. C., Guss, J. M., Murata, M., Norris, V. A., Ramshaw, J. A. M. & Venkatappa, M. P. (1978). *Nature (London)*, **272**, 319–324.

Deisenhofer, J. & Steigemann, W. (1975). *Acta Crystallogr. sect. B*, **31**, 238–250.

Diamond, R. (1974). *J. Mol. Biol.* **82**, 371–391.

Dodson, E. J., Dodson, G. G., Hodgkin, D. C. & Reynolds, C. D. (1979). *Can. J. Biochem.* **57**, 469–479.

Dubs, A., Wagner, G. & Wüthrich, K. (1979). *Biochim. Biophys. Acta*, **577**, 177–194.

Fehlhammer, H. & Bode, W. (1975). *J. Mol. Biol.* **98**, 683–692.

Feldmann, R. J. (1976). *Annu. Rev. Biophys. Bioeng.* **5**, 477–510.

Fletterick, R. J. & Wyckoff, H. W. (1975). *Acta Crystallogr. sect. A*, **31**, 698–700.

Gelin, B. R. & Karplus, M. (1975). *Proc. Nat. Acad. Sci., U.S.A.* **72**, 2002–2006.

Hetzel, R., Wüthrich, K., Deisenhofer, J. & Huber, R. (1976). *Biophys. Struct. Mech.* **2**, 159–180.

IUPAC-IUB Commission on Biochemical Nomenclature (1970). *J. Mol. Biol.* **52**, 1–17.

Kannan, K. K., Liljas, A., Warra, I., Bergsten, P.-C., Lovgren, S., Strandberg, B., Bengtsson, U., Carlbom, U., Fridborg, K., Jarup, L. & Petef, M. (1972). *Cold Spring Harbor Symp. Quant. Biol.* **36**, 221–231.

Kuo, M. & Gibbons, W. A. (1979). *Peptides: Structure and Biological Function, Proc. Sixth Amer. Peptide Symp.* (Gross, E. & Meienhofer, J., eds), pp. 229–232, Pierce Chem. Co., Rockford.

Leach, S. J., Némethy, G. & Scheraga, H. A. (1977). *Biochem. Biophys. Res. Commun.* **75**, 207–215.

Mandel, N., Mandel, G., Trus, B. L., Rosenberg, J., Carlson, G. & Dickerson, R. E. (1977). *J. Biol. Chem.* **252**, 4619–4636.

Mathews, F. S., Argos, P. & Levine, M. (1972). *Cold Spring Harbor Symp. Quant. Biol.* **36**, 387–395.

McLachlan, A. D. (1979). *J. Mol. Biol.* **128**, 49–79.

Moews, P. C. & Kretsinger, R. H. (1975). *J. Mol. Biol.* **91**, 201–228.

Momany, F. A., McGuire, R. F., Burgess, A. W. & Scheraga, H. A. (1975). *J. Phys. Chem.* **79**, 2361–2381.

Nagayama, K. & Wüthrich, K. (1981). *Eur. J. Biochem.* **114**, 365–374.

Noggle, J. H. & Schirmer, R. E. (1971). *The Nuclear Overhauser Effect*, Academic Press, New York.

Quiocho, F. A. & Lipscomb, W. N. (1971). *Advan. Protein Chem.* **25**, 1–78.

Ramachandran, G. N. & Sasisekharan, V. (1968). *Advan. Protein Chem.* **23**, 283–437.

Reeke, G. N. Jr, Becker, J. W. & Edelman, G. M. (1975). *J. Biol. Chem.* **250**, 1525–1547.

Richardson, J. S. (1981). *Advan. Protein Chem.* **34**, 167–339.

Roques, B. P., Rao, R. & Marion, D. (1980). *Use of Nuclear Overhauser Effect in the Study of Peptides and Proteins*, Imprimerie Declume, Z. I. Lons-Perrigny.

Saul, F. A., Amzel, L. M. & Poljak, R. J. (1978). *J. Biol. Chem.* **253**, 585–597.

Smith, W. W., Burnett, R. M., Darling, G. D. & Ludwig, M. L. (1977). *J. Mol. Biol.* **117**, 195–225.
Steigemann, W. & Weber, E. (1979). *J. Mol. Biol.* **127**, 309–338.
Takano, T. & Dickerson, R. E. (1980). *Proc. Nat. Acad. Sci., U.S.A.* **77**, 6371–6375.
Wagner, G. & Wüthrich, K. (1982). *J. Mol. Biol.* **155**, 347–366.
Wagner, G., Anil Kumar & Wüthrich, K. (1981). *Eur. J. Biochem.* **114**, 375–384.
White, J. L., Hackert, M. L., Buehner, M., Adams, M. J., Ford, G. C., Lentz, P. J. Jr, Smiley, I. E., Steindel, S. J. & Rossmann, M. G. (1976). *J. Mol. Biol.* **102**, 759–779.
Wider, G., Baumann, R., Nagayama, K., Ernst, R. R. & Wüthrich, K. (1981). *J. Magn. Reson.* **42**, 73–87.
Wider, G., Lee, K. H. & Wüthrich, K. (1982). *J. Mol. Biol.* **155**, 367–388.
Wüthrich, K. (1976). *NMR in Biological Research: Peptides and Proteins*, North-Holland Publishing Company, Amsterdam.
Wüthrich, K., Wider, G., Wagner, G. & Braun, W. (1981). *J. Mol. Biol.* **155**, 311–319.

Edited by S. Brenner

J. Mol. Biol. (1982) **155**, 347–366

Sequential Resonance Assignments in Protein ^{1}H Nuclear Magnetic Resonance Spectra

Basic Pancreatic Trypsin Inhibitor

Gerhard Wagner and Kurt Wüthrich

Institut für Molekularbiologie und Biophysik
Eidgenössische Technische Hochschule
ETH-Hönggerberg, CH-8093 Zürich
Switzerland

(Received 17 August 1981)

The assignment of the ^{1}H nuclear magnetic resonance spectrum of the basic pancreatic trypsin inhibitor with the use of two-dimensional ^{1}H nuclear magnetic resonance techniques at 500 MHz is described. The assignments are based entirely on the known amino acid sequence and the nuclear magnetic resonance data. Individual resonance assignments were obtained for all backbone and C^{β} protons, with the exception of those of Arg1, Pro2, Pro13 and the amide proton of Gly37. The side-chain resonance assignments are complete, with the exception of Pro2 and Pro13, the N^{δ} protons of Asn44 and the peripheral protons of the lysine residues and all but two of the arginine residues.

1. Introduction

The basic pancreatic trypsin inhibitor is a small globular protein of 58 amino acid residues, M_r 6500, which inhibits the function of trypsin and other proteases by formation of inert complexes with the enzymes (Tschesche, 1974). A refined crystal structure at 1·5 Å resolution has been described (Deisenhofer & Steigemann, 1975) and, during the last decade, BPTI† was the subject of a large number of experimental and theoretical studies on static and dynamic aspects of protein conformations and on protein folding. This paper describes individual ^{1}H nuclear magnetic resonance assignments for BPTI, which provide a basis for determination of the solution conformation and hence a meaningful, detailed comparison of the spatial structures of this protein in single crystals and in solution.

BPTI was previously extensively investigated by n.m.r. In particular, individual assignments obtained by a variety of different experiments (some of these depended also on reference to the crystal structure; for a survey of the original references and data, see Wüthrich & Wagner, 1979) for the eight aromatic rings, the

† Abbreviations used: BPTI, basic pancreatic trypsin inhibitor; n.m.r., nuclear magnetic resonance; p.p.m., parts per million; COSY, 2-dimensional correlated spectroscopy; NOESY, nuclear Overhauser enhancement spectroscopy; SECSY, spin echo correlated spectroscopy.

0022–2836/82/070347–20 $02.00/0

© 1982 Academic Press Inc. (London) Ltd.

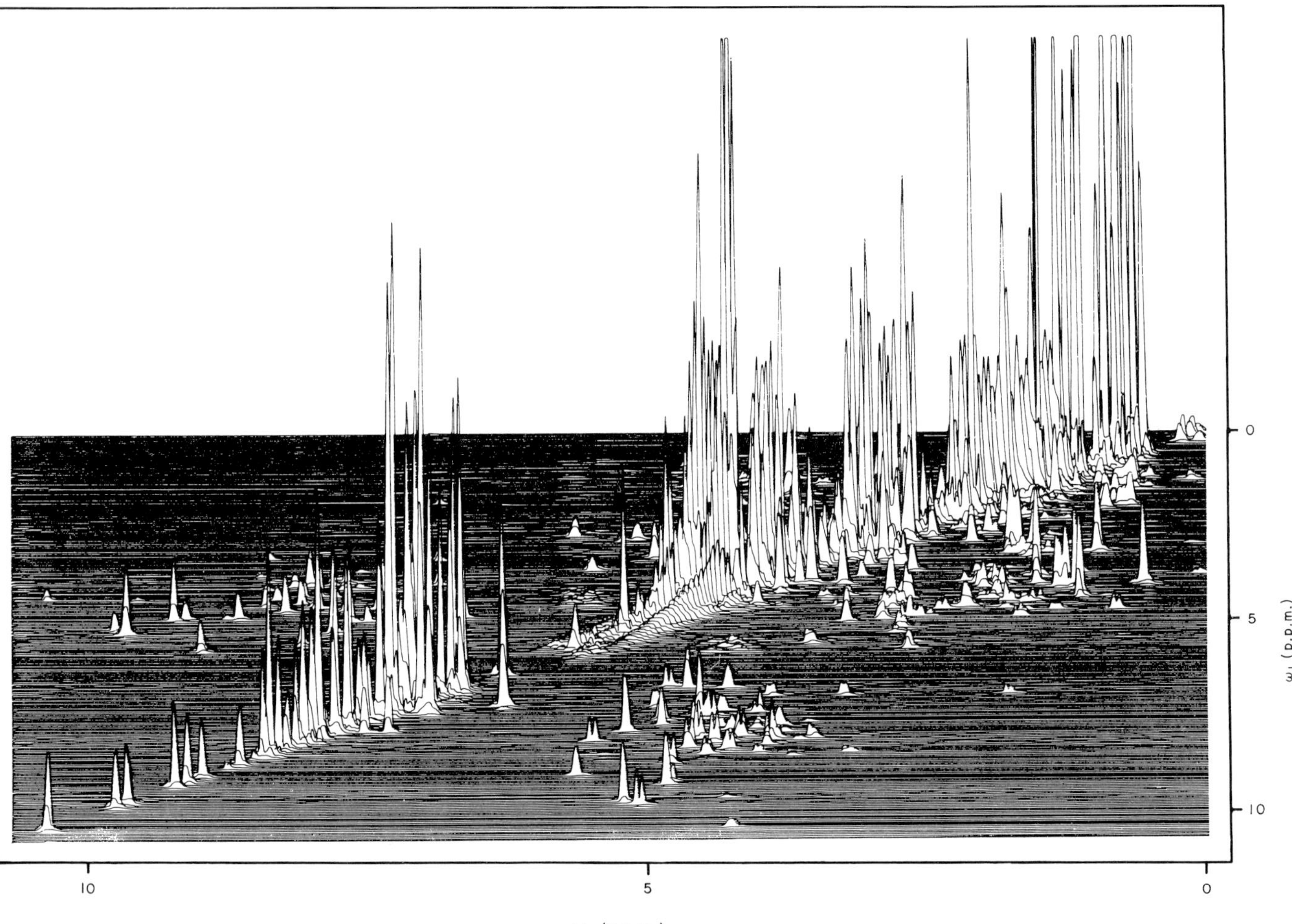

FIG. 1. Stacked-plot representation of a symmetrized (Baumann *et al.*, 1981), absolute value 500 MHz ^{1}H COSY spectrum of a 0·02 M solution of BPTI in a mixed solvent of 90% H_2O and 10% ^2H_2O, pH 4·6, $t = 80$°C. The spectrum was recorded in ~24 h, the digital resolution is 5·3 Hz/point. The stacked plots afford a "3-dimensional" view of the spectrum. The 2 perpendicular frequency axes ω_1 and ω_2 are calibrated with the chemical shifts. Peaks corresponding to the one-dimensional spectrum are displayed on the diagonal from the upper right to the lower left corner, where some of the highest peaks have been truncated. "Cross peaks" manifesting J connectivities between distinct lines on the diagonal are located in pairs, symmetrical with respect to the diagonal. Between 4·0 and 5·2 p.p.m., a band of artifactual peaks parallel to and in front of the diagonal spectrum are seen. These are a consequence of the water irradiation and the symmetrization of the spectrum.

20 methyl groups and selected backbone amide protons provided a basis for locating internal motions manifested in aromatic ring flips (Wüthrich & Wagner, 1975), amide proton exchange rates (Richarz *et al.*, 1979) and ^{13}C relaxation parameters (Richarz *et al.*, 1980) in specific regions of the molecule (Wagner & Wüthrich, 1979*a*; Wüthrich *et al.*, 1980). More recently, sequential resonance assignments with the use of two-dimensional n.m.r. experiments resulted in complete resonance assignments for the polypeptide segments residues 16 to 36 and 43 to 45 in the β-sheet of BPTI (Wagner *et al.*, 1981). Here, we describe sequential, individual resonance assignments for the other regions of the polypeptide chain.

2. Materials and Methods

^{1}H–^{1}H J-connectivity maps were obtained with 2-dimensional correlated spectroscopy. COSY spectra were recorded with a sequence of 2 non-selective 90° pulses (Aue *et al.*, 1976):

$$(90^\circ - t_1 - 90^\circ - t_2)_n.$$

The first 90° pulse creates transverse magnetization. During the evolution period, t_1, the various magnetization components precess with their characteristic precession frequency in the x–y plane of the rotating frame and are thus frequency labelled. The second 90° pulse causes transfer of magnetization components among those transitions that belong to the same J-coupled spin systems. The free-induction decay is recorded immediately after the second 90° pulse as a function of t_2. The experiment is repeated for a set of equidistant t_1 values. To obtain an adequate signal to noise ratio, n transients are accumulated for each value of t_1. At the end of each recording, the system was allowed to reach equilibrium during a fixed relaxation delay of 1 to 1·5 s.

Two-dimensional Fourier transformation of the data matrix $s(t_1, t_2)$ produced the desired frequency domain spectrum $S(\omega_1, \omega_2)$. As an illustration, Figure 1 shows a stacked-plot representation of a COSY spectrum of BPTI. The 2 perpendicular axes are calibrated with the chemical shifts, which increase from right to left and from the upper to the lower end, respectively. In this representation, peaks corresponding to the unidimensional spectrum appear on the diagonal from the upper right to the lower left of the ω_1–ω_2 plane. J-connectivities between individual lines are manifested by pairs of cross peaks in symmetrical locations with respect to the diagonal peak. A COSY spectrum can, with a single instrument setting, provide a complete map of all proton–proton J-connectivities in the polypeptide chain, i.e. a map of through-bond connectivities between hydrogen atoms that are normally separated by not more than 3 chemical bonds in the covalent structure.

Through-space ^{1}H–^{1}H connectivity maps were obtained with 2-dimensional nuclear Overhauser enhancement spectroscopy. NOESY spectra were recorded with a sequence of 3 non-selective 90° pulses (Jeener *et al.*, 1979; Anil Kumar *et al.*, 1980*a*):

$$(90^\circ - t_1 - 90^\circ - \tau_m - 90^\circ - t_2)_n.$$

After frequency labelling of the various magnetization components during t_1, cross-relaxation leads to incoherent magnetization exchange during the mixing time τ_m. The signal is recorded immediately after the third pulse as a function of t_2. Otherwise, the recording of the data is analogous to the procedures used for the COSY spectra, and the final appearance of the 2 spectra is very similar. In the place of the J-cross peaks in COSY, the NOESY spectra contain NOE cross peaks, which manifest spatial proximity between individual hydrogen atoms. In Fig. 2, a NOESY spectrum is presented as a contour plot. While aesthetically a contour plot may be less appealing than the 3-dimensional view of the spectrum in Fig. 1, it is a much more useful presentation for spectrum analysis, and all further spectra in this and the following paper are shown in this form. To illustrate the

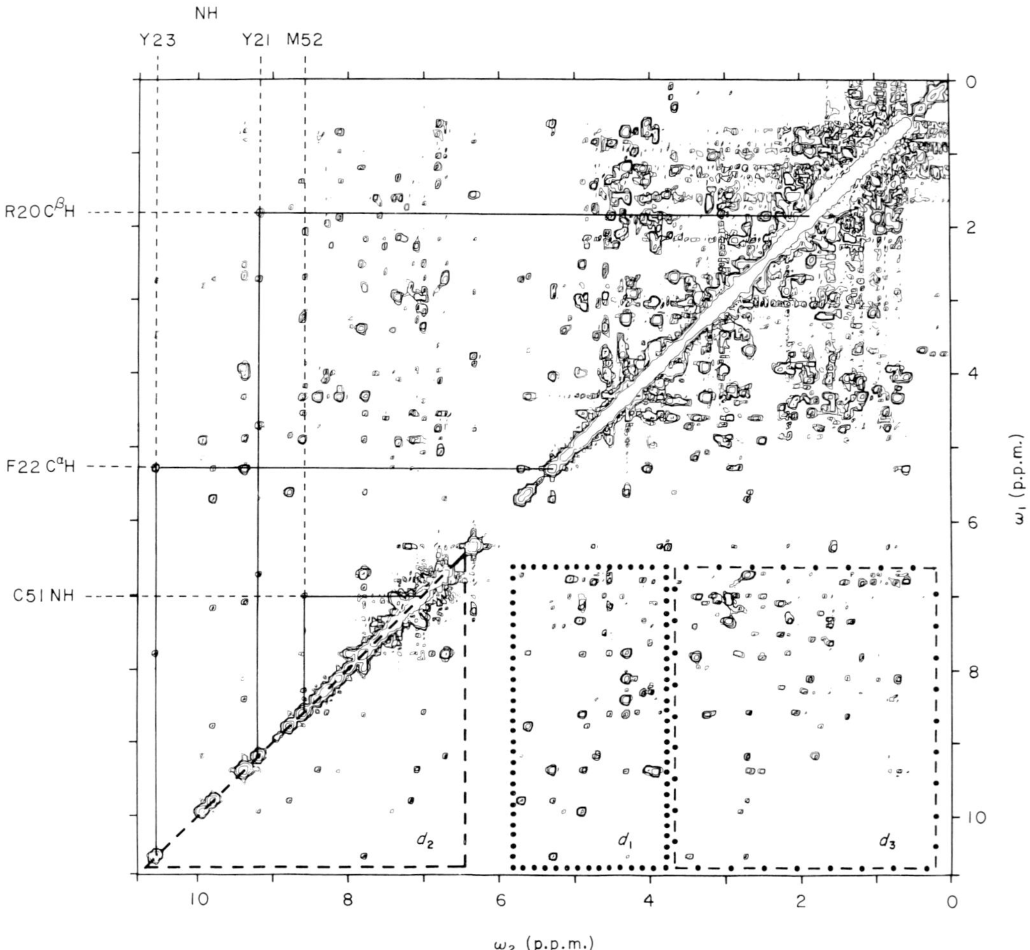

FIG. 2. Contour plot of a symmetrized, absolute value 500 MHz ^{1}H NOESY spectrum of a 0·02 M solution of BPTI in ^2H_2O, p^{2}H 4·6, $t = 36$°C. The digital resolution is 5·3 Hz/point. The spectrum was recorded in ~ 6 h, immediately after dissolving the protein in ^2H_2O, so that in addition to the non-labile protons the resonances of ~ 30 backbone amide protons are seen between 7 and 10·6 p.p.m. A 3-dimensional view of this spectrum would be closely similar to that of Fig. 1. The 2 frequency axes ω_1 and ω_2 are again calibrated with the chemical shifts, and peaks corresponding to the resonance positions in the normal, one-dimensional spectrum are on the diagonal from the upper right to the lower left corner. In the lower right triangle, 3 spectral regions of interest for sequential resonance assignments are outlined, i.e. the regions where NOE connectivities between different amide protons (— — —, d_2), between amide protons and C$^{\alpha}$ protons (· · · ·, d_1) and between amide protons and C$^{\beta}$ protons (—·—·—, d_3) are usually observed. In the upper left triangle, the analysis of such spectra is illustrated with 1 of each of these 3 types of connectivities. The continuous lines indicate the connectivities between cross peaks and diagonal peaks and the broken lines lead to the assignments of the connected resonances, which are indicated on the periphery of the Figure.

interpretation of such spectra, the connectivities for selected cross peaks have been indicated, and the connected protons identified in the upper left triangle of the spectrum in Fig. 2. In the lower right triangle, those spectral regions are outlined where the NOE cross peaks needed for sequential resonance assignments (Billeter *et al.*, 1982) are usually observed, i.e. those between different amide protons (d_2), between amide and C^α protons (d_1), and between amide and C^β protons (d_3). Obviously, in the COSY spectrum the J-connectivities between NH_i and $C^\alpha H_i$ are usually manifested in the spectral region that corresponds to d_1 in Fig. 2.

In contrast to the J-connectivities in COSY, a NOESY experiment cannot, with a single instrument setting, provide complete information on all NOE connectivities in a protein. The intensities of the NOE cross peaks in NOESY spectra vary over a period of several hundred ms (Anil Kumar *et al.*, 1981) and, for each individual peak, the initial build-up rate is proportional to the reciprocal of the sixth power of the distance between the 2 connected groups of protons (Solomon, 1955; Noggle & Schirmer, 1971). When long mixing times are used, this quantitative relation between internuclear distance and NOE intensity may be masked by spin diffusion and other relaxation effects, which jeopardize quantitative analyses of the data (Kalk & Berendsen, 1976; Gordon & Wüthrich, 1978; Wagner & Wüthrich, 1979*b*). For sequential resonance assignments, however, we need at most semi-quantitative distance information, i.e. the mixing time should be chosen just long enough that NOEs manifesting distances of $\lesssim 3{\cdot}0$ Å between different protons of the polypeptide backbone (Billeter *et al.*, 1982) are manifested with a workable signal to noise ratio. Detailed studies of the NOE buildup rates in BPTI (Dubs *et al.*, 1979; Wagner & Wüthrich, 1979*b*; Anil Kumar *et al.*, 1981) showed that, at frequencies between 360 and 500 MHz, a mixing time of 100 ms, which is the value used in all NOESY experiments reported in this paper, is adequate for this purpose.

Two-dimensional ^{1}H n.m.r. spectra at 500 MHz were recorded on a Bruker WM500 spectrometer. The spectra in Figs 1 to 3 and 5 to 8 were obtained from 512 measurements with t_1 values from 0 to 48 ms, the spectrum in Fig. 9 from 256 measurements with t_1 values from 0 to 28 ms. Quadrature detection was used for detection of the individual free induction decays, with the carrier frequency at the low field end of the spectrum. To eliminate experimental artifacts, groups of 16 recordings with different phases were added for each value of t_1 (Nagayama *et al.*, 1979,1980). For measurements in H_2O, the solvent resonance was suppressed by selective, continuous irradiation at all times except during data acquisition (t_2) (Anil Kumar *et al.*, 1980*b*). Usually, 2048 data points were used to store the data for each value of t_1 for all spectra, except that of Fig. 9, where 1024 data points were used. Before Fourier transformation, the time domain data matrix was multiplied in the t_1 direction with a phase-shifted sine bell, $\sin(\pi(t+t_0)/t_s)$, and in the t_2 direction with a phase-shifted sine-squared bell, $\sin^2(\pi(t+t_0)/t_s)$. The length of the window functions, t_s, was adjusted for the bells to reach zero at the last experimental data point in the t_1 or t_2 direction, respectively. The phase shifts, t_0/t_s, were $\frac{1}{32}$ and $\frac{1}{64}$ in the t_1 and t_2 directions, respectively. Furthermore, to end up with a 1024×1024 point data matrix in the frequency domain, which corresponds to a digital resolution of 5·3 Hz/point in Figs 1 to 3 and 5 to 8 and 4·3 Hz/point in Fig. 9, the time domain matrix was expanded to 2048 points in t_1 and 4096 points in t_2 by "zero-filling". The spectra in Figs 1 and 2 were further improved by symmetrization (Baumann *et al.*, 1981). All spectra are shown in the absolute value presentation.

Basic pancreatic trypsin inhibitor (Trasylol®, Bayer Leverkusen) was obtained from the Farbenfabriken Bayer AG. Four different samples were used for the present studies. All 4 contained 0·02 M-BPTI and the pH was adjusted by addition of minute amounts of HCl and NaOH, whereby in the ^2H_2O solutions the pH meter readings were used without correction for isotope effects (Kalinichenko, 1976; Bundi & Wüthrich, 1979*a*). The solvent for the first sample was a mixture of 90% H_2O and 10% ^2H_2O (pH 4·6), so that all the backbone amide proton resonances were present in the spectrum. The second and the third sample used ^2H_2O as a solvent, with p^{2}H 4·6 and 3·5, respectively, so that at low temperatures ~30 slowly exchanging amide protons could be observed (Wüthrich & Wagner, 1979). In the fourth sample, all the labile protons had been replaced by ^2H by heating a ^2H_2O solution of BPTI (p^2H 4·6), to 85°C for 10 min. In the ^2H_2O solutions, the concentration of residual solvent protons was minimized by repeated lyophilization from ^2H_2O. Chemical shifts are quoted relative to internal sodium 3-trimethylsilyl-[2,2,3,3-$^2H_4$]propionate.

3. Results and Discussion

(a) *General considerations on the experimental realization of sequential resonance assignments*

The fundamental elements of sequential resonance assignments in polypeptide chains are given on the one hand by the experience gained in the first investigations of this type (Dubs *et al.*, 1979; Wagner *et al.*, 1981) and the results of the theoretical study in the preceding paper (Billeter *et al.*, 1982), and on the other hand by the potentialities of the two-dimensional n.m.r. experiments COSY and NOESY described in Materials and Methods. Hence, from COSY one determines the J-connectivities between the amide, C^{α} and C^{β} protons of the individual residues. From NOESY spectra recorded with suitable mixing times, one then obtains connectivities between neighbouring residues. Depending on whether this connectivity is based on the observation that one or two of the distances d_1 (from NH_{i+1} to $C^{\alpha}H_i$), d_2 (from NH_{i+1} to either NH_{i+2} or NH_i) or d_3 (from NH_{i+1} to $C^{\beta}H_i$) are $\lesssim 3{\cdot}0$ Å, its reliability varies from ~ 70% to 99% (Billeter *et al.*, 1982). In addition to these fundamental considerations, the following are helpful practical aspects.

(1) In the $NH_i-C^{\alpha}H_i$ region of the COSY spectrum of a protein in H_2O (Fig. 3), each residue gives rise to one peak, with the exception of glycine, which may give one or two peaks, proline, which is not represented in this region, and possibly the residues in the N-terminal dipeptide segment, where the NH exchange with the solvent may be too fast for the cross peaks to be observed (Bundi & Wüthrich, 1979*b*). The COSY spectrum thus provides a first "fingerprint" of the protein.

(2) As far as possible, the complete spin systems of the individual amino acid residues (Wüthrich, 1976) should be identified at the outset of the investigation. All the COSY peaks in the NH_i–$C^{\alpha}H_i$ region that belong to an identified spin system can then be assigned either to a specific amino acid type or to a small selection of amino acid types (but of course not to the sequence positions as in Fig. 3, where the final result of the present paper is indicated). This can greatly aid the subsequent sequential resonance assignments. Firstly, inspection of amino acid sequences (Dayhoff, 1972) shows that, once the sequential assignments in the early stages of the analysis extend over two to four identified amino acid types, one can locate the assigned peptide segment in the sequence. Secondly, for all subsequent sequential assignments leading to COSY peaks of specific amino acid types, the amino acid sequence provides an immediate check on the sequential assignment. This is particularly helpful for resolving ambiguities in the sequential assignments that may arise when two or several residues have identical NH and/or $C^{\alpha}H$ chemical shifts. Thirdly, when the spin system of the amino acid residues is known and a sequential connectivity *via* either d_1, d_2 or d_3 has been established, the locations of

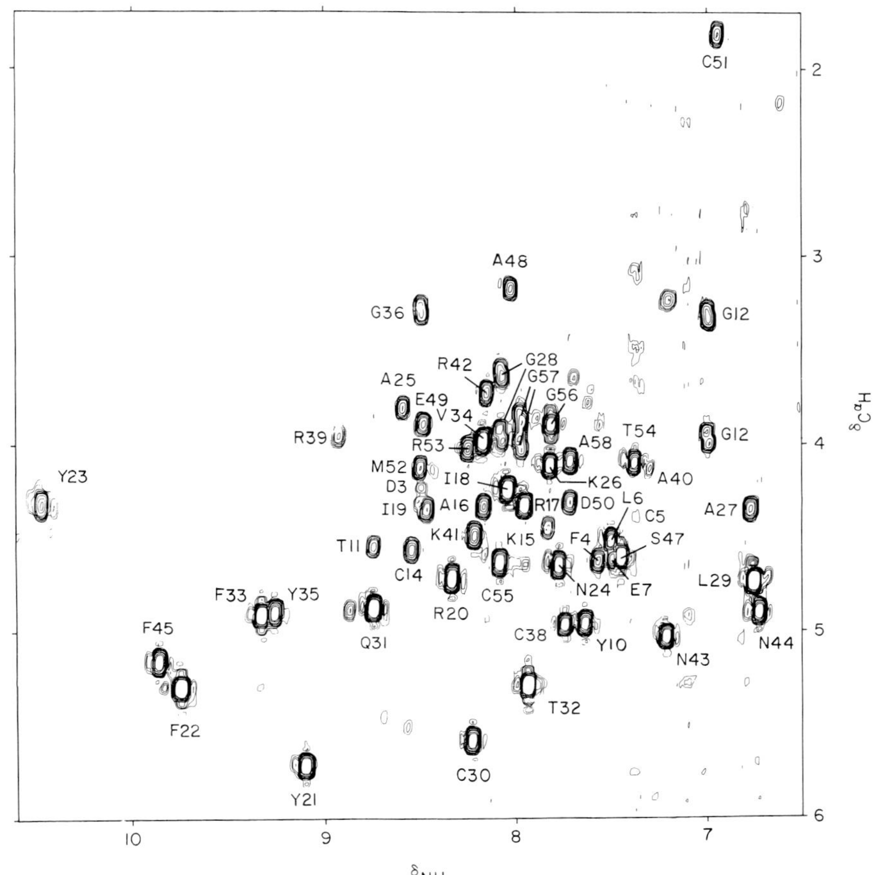

FIG. 3. Spectral region (1·7 to 6·0 p.p.m.) × (6·6 to 10·6 p.p.m.) of a 500 MHz ¹H COSY spectrum of BPTI in H_2O recorded at 68°C in the sample described in the legend to Fig. 1. The spectrum was recorded in ~ 17 h, the digital resolution is 5·3 Hz/point. The letters and numbers indicate the resonance assignments of the C^αH–NH J-connectivities for all the residues in the amino acid sequence of BPTI, except Arg1, where the N-terminal amino protons exchange probably too rapidly to be observed, the 4 proline residues, and Gly37. The cross peak of Lys46 is not visible in this spectrum, since at 68°C it coincides in the ω_1 direction with the water signal and is therefore bleached out by the solvent irradiation (see Table 1).

the NOESY peaks manifesting the other two connectivities is known *a priori*, and hence these can be studied more efficiently.

(3) In many proteins, a considerable number of interior amide protons exchange slowly with the solvent. It may be advantageous to start the sequential assignments in the simplified spectra obtained when such proteins are dissolved in ^2H_2O (Wüthrich, 1976). When the interior amide protons are pre-exchanged with ^2H, a simplified spectrum may also be obtained in H_2O solution.

(4) In the advanced stages of the sequential assignments, the protein fingerprint in the NH_i–$C^\alpha H_i$ region of the COSY spectrum provides a quite natural check for

each successive assignment, in that each peak must be assigned once and only once in the entire analysis (Fig. 3).

(5) As soon as all but one residue of a specific amino acid type are assigned, identification of the side-chain spin system is sufficient to locate the remaining residue in the amino acid sequence. Quite sizeable peptide segments may thus be assigned in the final stages of the experiment. However, with regard to the determination of the secondary structure, it is nevertheless advantageous to further investigate the sequential connectivities for these residues (W. Braun, M. Billeter, & K. Wüthrich, unpublished results).

(6) The complex spin systems of the long amino acid side-chains are usually difficult to identify. However, once the chemical shifts of the C^{α} and C^{β} protons are known from the sequential assignments, one has often a much better chance to extend the assignments to the peripheral groups of hydrogen atoms.

(b) *Strategy used to assign the ^{1}H n.m.r. spectrum of BPTI*

A survey of the experimental evidence accumulated to assign the ^{1}H n.m.r. spectrum of BPTI is presented in Figure 4, and the chemical shifts of the assigned resonances are listed in Table 1. Initial sequential assignments were obtained in the simplified ^{1}H n.m.r. spectra recorded in freshly prepared $^{2}H_2O$ solutions of the protein (Dubs *et al.*, 1979). These were then extended by studies in H_2O, as indicated by the stars in Figure 4 (Wagner *et al.*, 1981). All these earlier assignments were based on single NOE connectivities between the individual residues, i.e. d_1 for the peptide segments 16 to 25, 29 to 36 and 43 to 45, and

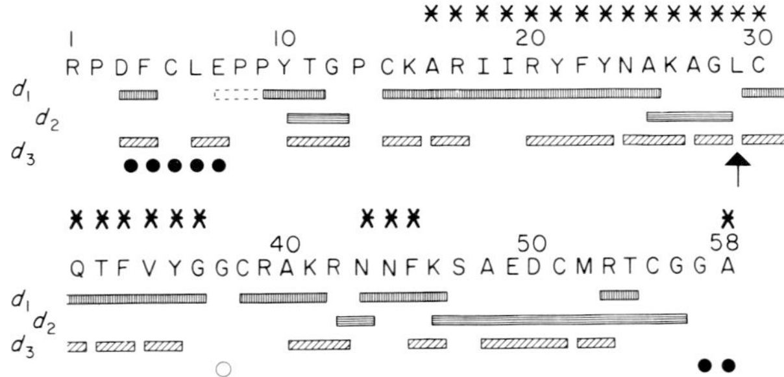

FIG. 4. Amino acid sequence of BPTI and survey of the experimental evidence by which the individual resonance assignments were obtained. (vertically hatched bar) Sequential assignments *via* d_1 (NOE from NH_{i+1} to $C^{\alpha}H_i$); (horizontally hatched bar) sequential assignments *via* d_2 (NOE from NH_{i+1} to NH_i); (diagonally hatched bar) sequential assignments *via* d_3 (NOE from NH_{i+1} to $C^{\beta}H_i$); (dashed bar) sequential assignments *via* NOEs from Pro $C^{\delta}H_{i+1}$ to $C^{\alpha}H_i$. (○, ●) The individual assignment relied primarily on the identification of the complete spin system of the amino acid residue without or with the amide proton, respectively. Once all but one residue of a given type has been assigned, this information is obviously sufficient for the individual assignment in the amino acid sequence even when neither of the connectivities d_1, d_2 or d_3 could be established. The arrow indicates a location where all the resonances were assigned, but the connectivity between 2 neighbouring residues was not established. Stars (*) above the sequence identify the residues for which complete resonance assignments had been obtained in previous work (Dubs *et al.*, 1979; De Marco *et al.*, 1977; Anil Kumar *et al.*, 1980*b*; Wagner *et al.*, 1981).

TABLE 1

Chemical shifts, δ†, of the assigned ^{1}H n.m.r. lines of BPTI, pH 4·6, t = 68°C

Amino acid residue	δ (±0·01 p.p.m.)†‡			
	NH	$C^{\alpha}H$	$C^{\beta}H$	Others
Arg1				
Pro2				
Asp3	8·49	4·23	2·79, 2·79	
Phe4	7·57	4·63	2·99, 3·34	$C^{\delta}H_2$ 7.01
				$C^{\epsilon}H_2$ 7·40
				$C^{\zeta}H$ 7·06
Cys5	7·37	4·39	2·75, 2·91	
Leu6	7·50	4·52	1·88, 1·88	$C^{\gamma}H$ 1.71
				$C^{\delta}H_3$ 0·88, 0·98
Glu7	7·50	4·61	2·18, 2·27	$C^{\gamma}H_2$ 2·58, 2·58
Pro8		4·67	1·88, 2·45	$C^{\gamma}H_2$ 2·12
				$C^{\delta}H_2$ 3·70, 3·98
Pro9		3·77	0·14, 0·27	$C^{\gamma}H_2$ 0·19, 1·30
				$C^{\delta}H_2$ 2·97, 3·18
Tyr10	7·64	4·96	2·95, 2·95	$C^{\delta}H_2$ 7·34
				$C^{\epsilon}H_2$ 7·10
Thr11	8·74	4·55	4·07	$C^{\gamma}H_3$ 1·40
Gly12	7·00	3·32, 3·97		
Pro13				
Cys14	8·54	4·56	2·81, 3·48	
Lys15	7·83	4·45	1·59, 2·08	$C^{\gamma}H_2$ 1·31, 1·41
Ala16	8·17	4·33	1·18	
Arg17	7·95	4·33	1·63, 1·63	
Ile18	8·04	4·24	1·90	$C^{\gamma}H_2$ 1·07
				$C^{\gamma}H_3$ 0·97
				$C^{\delta}H_3$ 0·73
Ile19	8·46	4·35	1·96	$C^{\gamma}H_2$ 1·46
				$C^{\gamma}H_3$ 0·74
				$C^{\delta}H_3$ 0·73
Arg20	8·33	4·72	0·86, 1·83	$C^{\gamma}H_2$ 1·37, 1·72
				$C^{\delta}H_2$ 3·08, 3·48
				$N^{\delta}H$ 7·37
Tyr21	9·10	5·71	2·72, 2·72	$C^{\delta}H_2$ 6·74
				$C^{\epsilon}H_2$ 6·79
Phe22	9·74	5·30	2·85, 2·96	$C^{\delta}H_2$ 7·16, 7·27
				$C^{\epsilon}H_2$ 6·98, 7·07
				$C^{\zeta}H$ 7·31
Tyr23	10·46	4·32	2·74, 3·49	$C^{\delta}H_2$ 7·19
				$C^{\epsilon}H_2$ 6·35
Asn24	7·78	4·65	2·18, 2·86	$N^{\delta}H_2$ 6·96, 7·79
Ala25	8·59	3·81	1·58	
Lys26	7·82	4·11	1·89	
Ala27	6·77	4·35	1·20	
Gly28	8·07	3·64, 3·95	—	
Leu29	6·76	4·73	1·45, 1·69	$C^{\gamma}H$ 1·44
				$C^{\delta}H_3$ 0·78, 0·87
Cys30	8·23	5·58	2·71, 3·63	
Gln31	8·74	4·87	1·77, 2·16	$C^{\gamma}H_2$ 1·83, 2·02
				$N^{\epsilon}H_2$ 7·22, 7·30
Thr32	7·94	5·24	4·04	$C^{\gamma}H_3$ 0·62
Phe33	9·32	4·91	2·98, 3·13	$C^{\delta}H_2$ 7·13
				$C^{\epsilon}H_2$ 7·20
				$C^{\zeta}H$ 7·33

TABLE 1 (*continued*)

Amino acid residue	δ(±0·01 p.p.m.)†‡ NH	$C^{\alpha}H$	$C^{\beta}H$	Others
Val34	8·17	3·98	1·96	$C^{\gamma}H_3$ 0·73, 0·83
Tyr35	9·26	4·90	2·52, 2·64	$C^{\delta}H_2$ (6·77, 7·76)§ $C^{\epsilon}H_2$ 6·83
Gly36	8·49	3·28, 4·33		
Gly37	n.o.	2·92, 4·24		
Cys38	7·74	4·97	3·15, 3·80	
Arg39	8·92	3·95	2·26, 2·26	
Ala40	7·30	4·13	1·23	
Lys41	8·22	4·49	1·67, 2·26	$C^{\gamma}H_2$ 1·32, 1·49
Arg42	8·15	3·73	0·53, 1·18	$C^{\gamma}H_2$ 1·22, 1·48 $C^{\delta}H_2$ 2·76, 2·86 $N^{\epsilon}H$ 6·80
Asn43	7·22	5·03	3·28, 3·34	$N^{\delta}H_2$ 7·77, 7·97
Asn44	6·73	4·90	2·54, 2·80	
Phe45	9·85	5·16	2·80, 3·41	$C^{\delta}H_2$ 7·39 $C^{\epsilon}H_2$ 7·87 $C^{\zeta}H$ 7·62
Lys46	9·71	4·37	0·95, 2·01	
Ser47	7·46	4·60	3·90, 4·16	
Ala48	8·03	3·17	1·05	
Glu49	8·48	3·89	1·77, 2·02	$C^{\gamma}H_2$ 2·23, 2·36
Asp50	7·72	4·31	2·73, 2·73	
Cys51	6·95	1·81	2·90, 3·15	
Met52	8·50	4·12	2·00, 2·06	$C^{\gamma}H_2$ 2·70, 2·70 $C^{\epsilon}H_3$ 2·16
Arg53	8·25	4·02	1·61	
Thr54	7·38	4·10	3·95	$C^{\gamma}H_3$ 1·59
Cys55	8·09	4·63	1·98, 2·23	
Gly56	7·82	3·90, 3·90		
Gly57	7·98	3·87, 4·01		
Ala58	7·72	4·09	1·32	

† The chemical shifts, δ, are relative to internal sodium 3-trimethylsilyl-[2,2,3,3-$^2H_4$]propionate.

‡ Where no numbers are given in the columns for NH, $C^{\alpha}H$ and $C^{\beta}H$ and where more peripheral side-chain hydrogen atoms are not listed in the last column, no individual resonance assignments were obtained (see the text).

§ At 68°C, the C^{δ} proton resonances of Tyr35 are broadened by the ring flipping (Wagner *et al.*, 1976). Therefore, the chemical shifts observed at 36°C are listed. n.o., not observed.

d_2 for the segment 25 to 29. To further improve the reliability (Billeter *et al.*, 1982) of these earlier results, a second connectivity was now established in most cases (Fig. 4). The sequential assignments were then extended to the other regions of the polypeptide chain. As far as possible, we tried to add onto the previously assigned peptide segments, so that the amino acid sequence could be consulted continuously to check on the sequential assignments. Figure 4 shows that, to a considerable extent, the new assignments relied on two connectivities, mainly either on d_1 and d_3, or on d_2 and d_3. In Figures 5 to 8, each step of the new sequential assignments is documented by one connectivity, usually the one that was established first.

In Figure 4, an important contribution to the spectral assignments is explicitly shown only for a few residues, i.e. the identification of the amino acid side-chain spin systems. For the technical aspects, we refer to the descriptions by Nagayama & Wüthrich (1981) of the spin system identifications in BPTI using spin echo correlated spectroscopy (SECSY) and by Wider *et al.* (1982, following paper) of the spin system identifications in glucagon using COSY. In BPTI, 41 side-chains were identified at the outset of the study. With regard to using the amino acid sequence (Fig. 4) as a check of the sequential assignments in BPTI, the following groups of different spin systems were then available: five Gly, five Ala, Ala58, three Thr, Val34, two Asp, two Glu, Ser47, 17 ABX and A_2X spin systems, two five-spin systems (Met and Gln), two Ile, and 12 "long side-chains" (Leu, Lys, Arg) (Nagayama & Wüthrich, 1981). Since many of these spin systems were already assigned to specific locations in the β-sheet region (starred residues in Fig. 4), the remaining spin systems provided quite stringent criteria to check on the new sequential assignments.

In the final stages of the spectrum analysis, a sizeable proportion of the resonances in the side-chains of leucine, proline, lysine and arginine could be assigned, starting from the sequentially assigned $C^{\alpha}H$ and $C^{\beta}H$ lines. Furthermore, as documented in Figure 9, connectivities with the peripheral protons in the aromatic side-chains of asparagine and glutamine were established *via* NOE measurements (Billeter *et al.*, 1982).

(c) *Sequential resonance assignments in the C-terminal α-helix region*

The first step in the assignments from the previously identified tripeptide, segment 43 to 45, to the C terminus was *via* a d_1 connectivity between Phe45 $C^{\alpha}H$ and Lys46 NH (Fig. 5). As previously shown (Wagner *et al.*, 1981), d_1 connectivities are readily documented in "combined COSY–NOESY connectivity diagrams". These diagrams make use of the fact that the information in a COSY or NOESY spectrum is contained redundantly in the two triangles separated by the diagonal peaks (Figs 1 and 2). When the upper left triangle of NOESY and the lower right triangle of COSY are added, the combined plot manifests the through-space NOE connectivities between NH_{i+1} and $C^{\alpha}H_i$, as well as the through-bond J-connectivities between $C^{\alpha}H_i$ and NH_i. A record of sequential assignments *via* d_1 then consists of a spiral-like connectivity pattern. In Figure 5, where only those regions are displayed that contain the $C^{\alpha}H$–NH cross peaks (Fig. 2), a horizontal line leads from the position of the Lys46 $C^{\alpha}H$–NH COSY peak to the virtual diagonal position of the amide proton resonance. There is a single NOESY cross peak with the Lys46 NH chemical shift, which is therefore assigned to $C^{\alpha}H$ of Phe45 (in Fig. 5, a vertical line connects this peak with the diagonal position of Lys46 NH). Continuing on, a horizontal line leads to the virtual diagonal position of $C^{\alpha}H$ of Phe45, from where a vertical line connects with the previously assigned $C^{\alpha}H$–NH COSY peak of Phe45.

The following assignments were *via* d_2 connectivities between amide protons of neighbouring residues, as documented in the low-field region of the NOESY

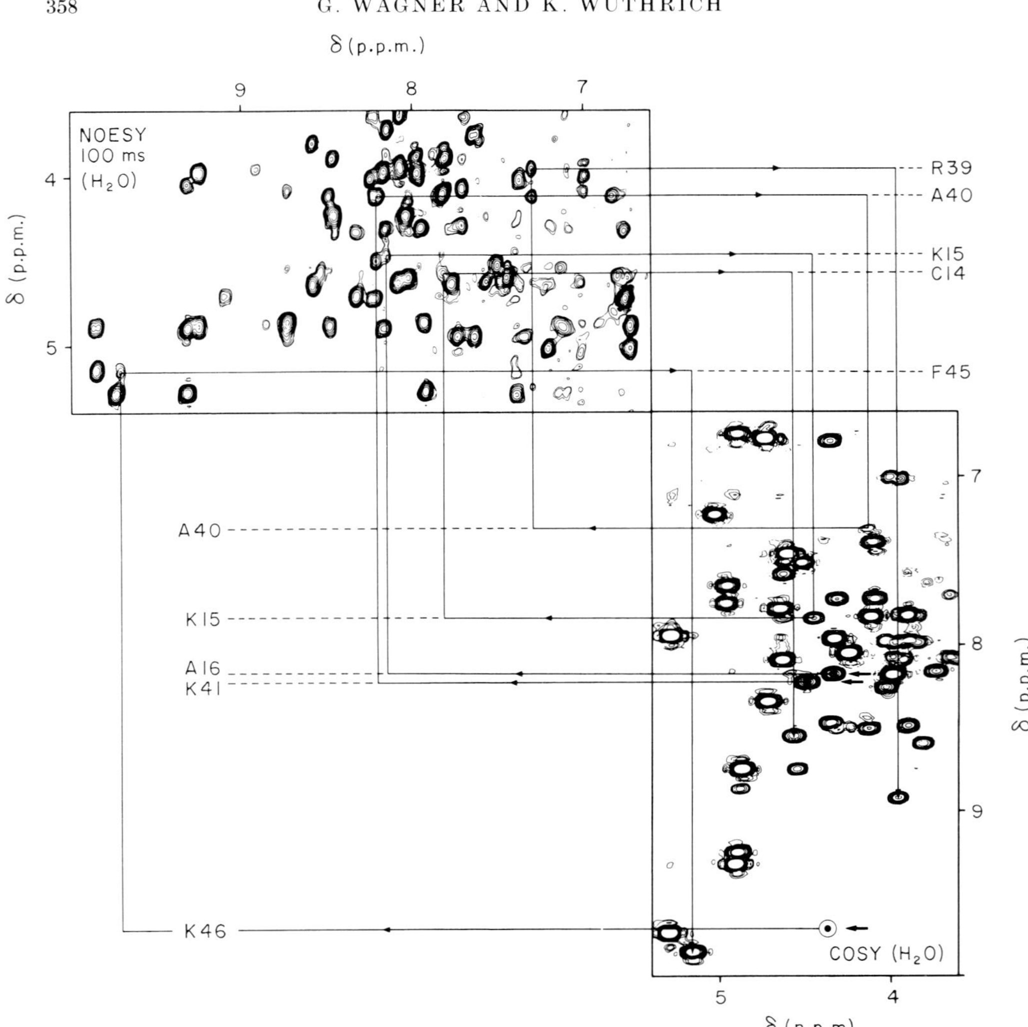

FIG. 5. Combined COSY–NOESY connectivity diagram for sequential resonance assignments *via* NOEs between NH and the $C^{\alpha}H$ of the preceding residue (d_1) (Wagner *et al.*, 1981). In the upper left the region (3·6 to 5·4 p.p.m.) × (6·6 to 10·0 p.p.m.) from the NOESY spectrum of Fig. 6 is shown. In the lower right, the corresponding region from the COSY spectrum in Fig. 3 is shown, which was recorded from the same sample and under identical conditions. The straight lines and arrows indicate the sequential resonance assignments obtained for the segments 46 to 45, 41 to 39 and 16 to 14. Starting points are indicated by the arrows in the COSY spectrum. At 68°C, the NH–$C^{\alpha}H$ cross peak for Lys46 in the COSY spectrum was bleached out by the irradiation of the solvent, since the $C^{\alpha}H$ resonance coincides with that of H_2O. It was, however, observed at different temperatures.

spectrum in Figure 6. As was discussed in detail in the preceding paper (Billeter *et al.*, 1982), d_2 connectivities are symmetrical with respect to the direction of the polypeptide chain, and hence an NH–NH NOE connectivity may equally well be with either of the two neighbouring residues in the sequence. However, when the NH chemical shift of one of these residue is known independently, the direction of

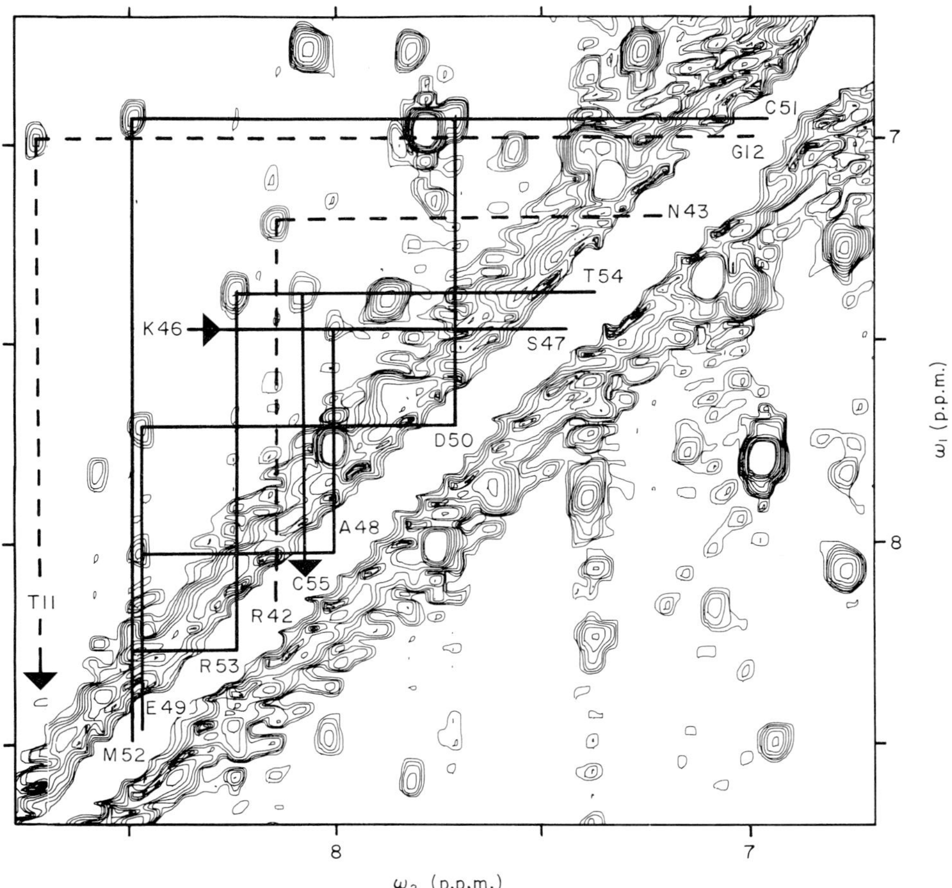

FIG. 6. Contour plot of the spectral region from 6·6 to 8·7 p.p.m. of a NOESY spectrum recorded at 68°C from the same sample and under identical conditions as the COSY spectrum of Fig. 3. This spectral region contains the diagonal peaks of most of the backbone amide protons and the cross peaks manifest NOEs between different amide protons. The solid lines indicate the sequential assignments for the polypeptide segment 46 to 55, which were obtained from NH–NH NOEs (d_2). The arrows indicate the start and the end of this sequence. The NOE cross peak between the amide proton resonances of Lys46 and Ser47 is not shown, since at (ω_1 = 7·47 p.p.m., ω_2 = 9·71 p.p.m.) it is outside of this spectral region. Connectivities between the amide protons of Thr11 and Gly12, and of Arg42 and Asn43 are indicated by broken lines.

the sequential connectivity is determined. Here, the first connectivity to be established was from Lys46 to Ser47. Since Phe45 NH was known (Fig. 5), the only NH–NH NOESY peak involving Lys46 had to be with Ser47. Because of the symmetry of d_2-connectivities, a second cross peak linking NH of Ser47 with NH of residue Ala48 might be present, depending on the local conformation (Billeter *et al.*, 1982). Since the amide proton of Ala48 might be to higher or lower field, this second cross peak could be either vertically above or horizontally to the left of the diagonal peak of Ser47 NH. Figure 5 shows that it is on the left, at ω_2 = 8·03 p.p.m. The second cross peak with NH of Ala48 is further to lower field at 8·48 p.p.m., leading to Glu49. The next two connectivities lead upfield, first to Asp50 NH at 7·72 p.p.m. and then to Cys51 NH at 6·95 p.p.m. Continuing on, each residue up to Thr54 has two NH–NH connectivities. The d_2 connectivity between Cys55 and Gly56 was established in a ^2H_2O solution of BPTI at p^{2}H 3·5 and 24°C. This cross peak was not present in the NOESY spectrum recorded in H_2O at 68°C and pH 4·6, and this connection is therefore not shown in Figure 6.

For most of these assignments, the d_3 connectivities between NH_{i+1} and $C^{\beta}H_i$ could also be established (Fig. 4). Furthermore, almost every sequential assignment in this peptide segment could be checked unambiguously against the amino acid sequence, since all but two side-chain spin systems had been identified (Nagayama & Wüthrich, 1981). It may also be added that the simultaneous occurrence of d_2 and d_3 connectivities is a quite strong indication that the helical structure from residues 47 to 55 observed in the crystal structure is preserved in solution (Billeter *et al.*, 1982).

(d) *Sequential assignments in the central and N-terminal regions*

At the N-terminal end of the assigned peptide segment 43 to 56, a d_2 connectivity leads to the amide proton of Arg42 (Fig. 6). Next, a sequence of three residues was obtained *via* d_1 (Fig. 5), with alanine in the central position and residues with long side-chains in the peripheral positions. Among the as yet unassigned segments, only Arg39-Ala40-Lys41 could be fitted to these data. Additional connectivities *via* d_3 to Arg42 and, at different temperature (56°C) or pH (3·5 at 68°C), *via* d_1 to Cys38 could then be established (Fig. 4).

At the N-terminal end of the assigned segment residues 16 to 36, d_1 connectivities could be established from Ala16 $C^{\alpha}H$ to Cys14 NH (Fig. 5). The connectivity from Gly12 to Thr11 *via* d_2 (Fig. 6) was unambiguous, since there is only one dipeptide segment Thr-Gly in BPTI (Fig. 4). In addition, the d_3 connectivity between these two residues could also be established and, as described in Figure 7, two successive d_1 connectivities provided the assignments for Tyr10 and for Pro9 $C^{\alpha}H$. The assignment of the proline C^{α} proton is consistent with the observation that there is no NH–C^{α}H J-peak at 3·77 p.p.m. in the COSY spectrum of Figure 3.

The spin system of Pro9 could now be completely characterized in the COSY spectrum (Fig. 8). It is characterized by extreme high-field positions of the two C^{β} protons and one C^{γ} proton (Table 1). On the basis of NOEs with Pro9 $C^{\delta}H$ (Billeter *et al.*, 1982), the C^{α} proton of Pro8 was assigned and subsequently linked to the C^{β}, C^{γ} and C^{δ} protons *via* J-connectivities in the COSY spectrum (Fig. 8). The assignment of Pro8 $C^{\alpha}H$ coincided again with the observation that there is no NH–C^{α}H J-peak at 4·67 p.p.m. in the COSY spectrum of Figure 3. Finally, NOE connectivities with Pro8 $C^{\delta}H_2$ led to $C^{\alpha}H$ of Glu7, which was also assigned independently from the identification of the side-chain spin system.

At this stage, only the residues 1 to 6, 13, 37 and 57 were not yet assigned. Each residue type in the sequence 3 to 6 thus occurred only once among the unassigned residues (Fig. 4) and was therefore unambiguously assigned from the identification of the side-chain spin system. In addition, sequential connectivities could be

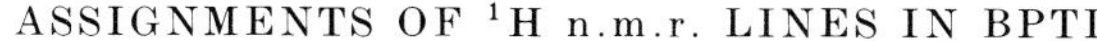

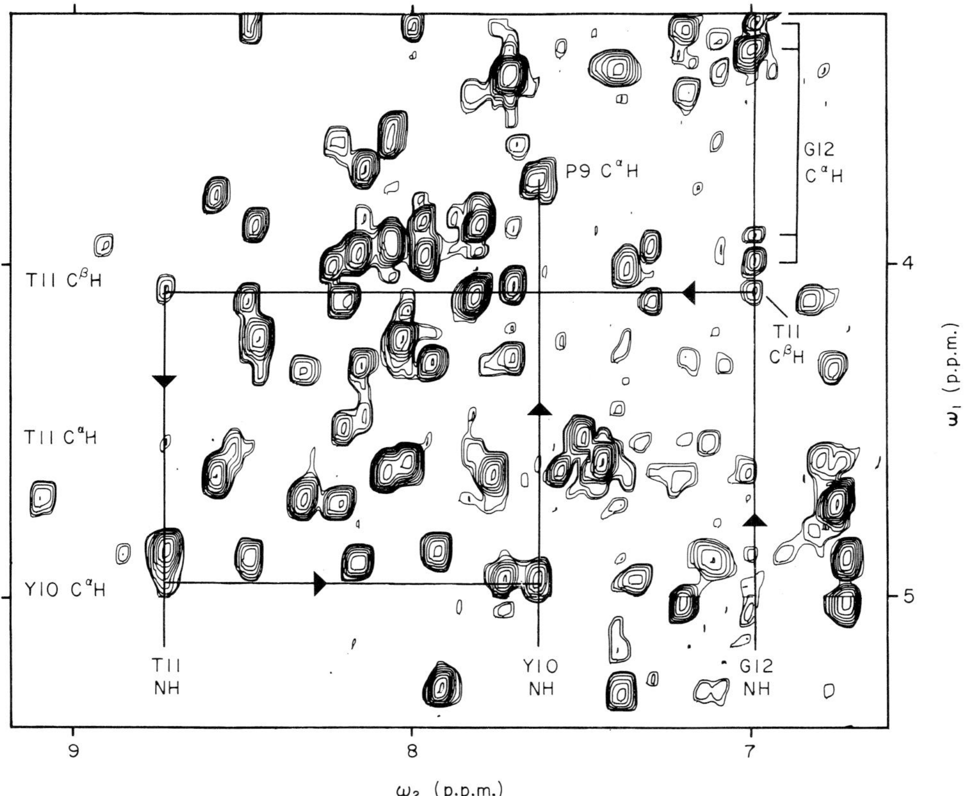

FIG. 7. Contour plot of the spectral region (3·2 to 5·4 p.p.m.) × (6·6 to 9·2 p.p.m.) of the same NOESY spectrum as in Fig. 6. The continuous lines indicate the sequential resonance assignments for the polypeptide segment 9 to 12, which were obtained from NOEs between the amide proton of Gly12 and the C^{β} proton of Thr11 (d_3), the amide proton of Thr11 and the C^{α} proton of Tyr10 (d_1), and the amide proton of Tyr10 and the C^{α} proton of Pro9 (d_1). The arrows indicate the direction of the sequence of assignments from residues 12 to 9.

established between Asp3 and Phe4 *via* d_1 and d_3, and between Leu6 and Glu7 *via* d_3. The $C^{\alpha}H_2$ resonances of Gly12 and Gly37 were previously assigned as a group from comparison of native BPTI with a chemical modification of the protein, where the disulfide bond 14–38 had been reduced (Nagayama & Wüthrich, 1981). Since Gly12 was identified (Figs 6 and 7), there remained only one AX spin system for $C^{\alpha}H_2$ of Gly37 (Table 1). The COSY cross peak with the amide proton of Gly37 was not observed, however, and no NOE connectivity with either Gly36 or Cys38 could be established. The C-terminal alanine was identified previously from the pH titration shift (De Marco *et al.*, 1977). The assignment of Gly57 relied on the identification of the spin system (Nagayama & Wüthrich, 1981), which was at this point the only glycine spin system that had not been assigned individually.

Returning at this point to the COSY spectrum of Figure 3, we find that with the assignments outlined in Figure 4 the $C^{\alpha}H$–NH cross peaks of 52

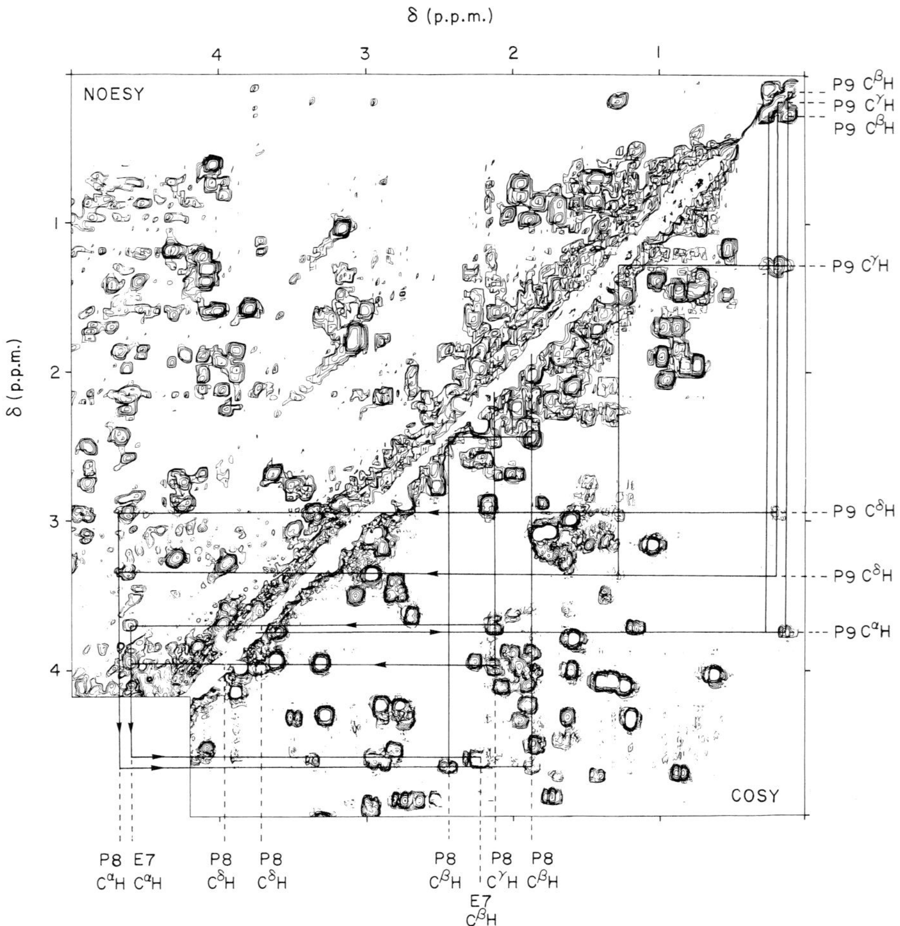

FIG. 8. Resonance assignments for the peptide segment Glu7-Pro8-Pro9 in BPTI. To document these data, the region (0·0 to 4·2 p.p.m.) × (0·0 to 5·0 p.p.m.) from the NOESY spectrum of BPTI in H_2O solution at 68°C (same spectrum as in Fig. 6) in the upper left triangle was combined with the corresponding region from a COSY spectrum (same as in Fig. 3) recorded from the same sample under identical conditions. Connectivities between peaks within COSY are indicated by continuous lines, connectivities between COSY and NOESY peaks by continuous lines with arrows. Starting from the previously assigned C^{α} proton (Fig. 7), the J-connectivities between the individual protons of Pro9 are indicated in the COSY spectrum, and the peaks identified on the right-hand margin. At the chemical shifts of each of the C^{δ} protons of Pro9, a cross peak was observed in the NOESY spectrum, which was assigned to $C^{\alpha}H$ of Pro8 (Billeter *et al.*, 1982). A vertical line leads through these 2 NOESY cross peaks to the diagonal position for $C^{\alpha}H$ of Pro8. Back in the COSY spectrum, the J-connectivities within the spin system of Pro8 are indicated, and the peaks identified at the bottom of the Figure. Only 1 connectivity with $C^{\gamma}H_2$ was observed, possibly because the 2 protons have identical chemical shifts. At the chemical shifts of each of the C^{δ} protons of Pro8, a strong peak was observed in the NOESY spectrum and assigned to $C^{\alpha}H$ of Glu7. Also indicated in the COSY spectrum is the previously identified (Nagayama & Wüthrich, 1981) $C^{\alpha}H$–$C^{\beta}H$ connectivity of Glu7, which provides an independent, final check for the assignments in this Figure.

residues were identified. There are a number of as yet unassigned, relatively weak peaks in the spectral region shown in Figure 3. Four of these could be assigned to the J-couplings between the labile side-chain protons of Arg20 and Arg42 with the C^δ methylene protons (see Table 1).

(e) *Resonance assignments in the long amino acid side-chains*

Starting with the sequentially assigned C^α and C^β protons, several spin systems of long amino acid side-chains could be completely or in part assigned *via* J-connectivities in the COSY and SECSY spectra. New complete assignments include Leu6, Pro8, Pro9, Ile18, Ile19, Leu29, Arg20 and Arg42.

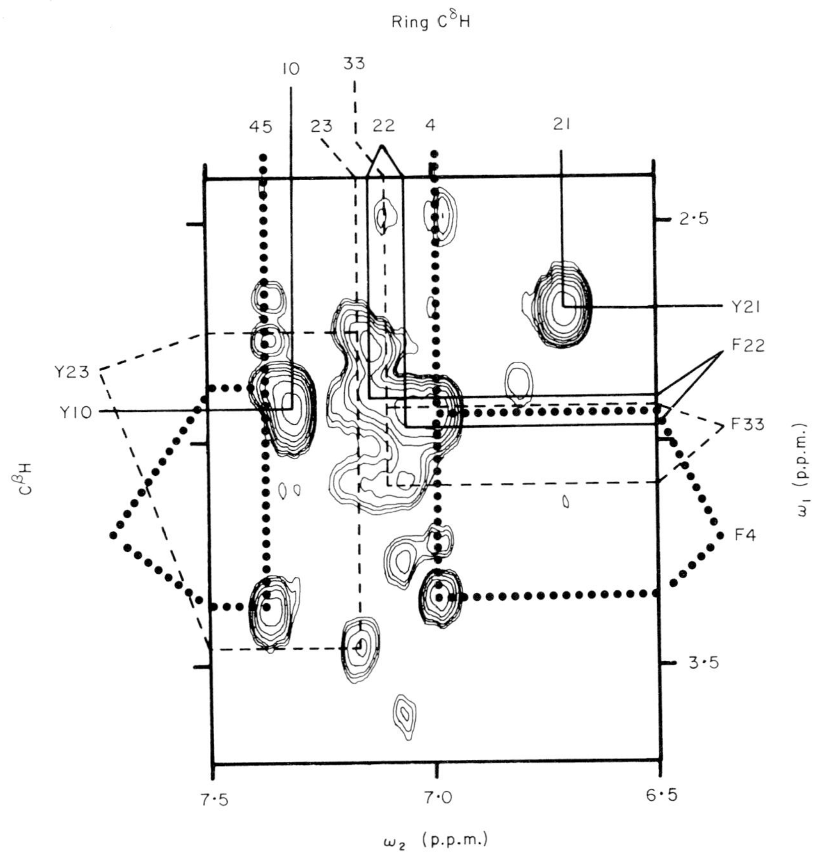

FIG. 9. Spectral region (2·4 to 3·7 p.p.m.) × (6·5 to 7·5 p.p.m.) of a 500 MHz ^{1}H NOESY spectrum of a 0·02 M solution of BPTI in ^2H_2O at 68°C, p^{2}H 4·6. All the labile protons had been replaced by ^{2}H. The spectrum was recorded in ~6 h, the digital resolution is 4·3 Hz/point. The spectrum shows NOE connectivities between the C^β protons and the aromatic C^δ protons of 7 of the 8 tyrosine and phenylalanine side-chains in BPTI. Tyr35 cannot be observed at this temperature, since the C^δ–H resonances are broadened due to low frequency ring flips (Wagner *et al.*, 1976). At lower temperatures, where sharp resonances are observed for the C^δ protons of Tyr35, the connectivity could be established readily.

Following the model calculations in the preceding paper (Billeter *et al.*, 1982), NOE connectivities were used to link the spin systems of the aromatic rings with the corresponding $C^{\alpha}H$–$C^{\beta}H_2$ fragments. The experimental evidence is presented in the NOESY spectrum of Figure 9. For the side-chains of Phe4, Tyr10, Tyr21, Tyr23, Tyr35 (not shown in Fig. 9, since it had to be measured at a different temperature (Wagner *et al.*, 1976)) and Phe45, unambiguous assignments were thus obtained. In all instances, these assignments coincide with those previously proposed on the basis of chemical modifications (Snyder *et al.*, 1975) or comparison of homologous proteins (Wagner *et al.*, 1978*a*,*b*). For Phe22 and Phe33, cross peaks in Figure 9 are compatible with the assignments obtained from comparison of homologous proteins, but the peaks are not sufficiently well-resolved to present clear-cut evidence for these assignments.

From similar studies of NOE peaks in the NOESY spectra recorded in H_2O, individual assignments were obtained for the side-chain amide protons of Asn24, Gln31 and Asn43 (Table 1).

4. Conclusions

Almost complete assignments of the ^1H n.m.r. spectrum of BPTI are now available (Table 1). Among the backbone and C^{β} protons, only those of Arg1, Pro2, Pro13 and the NH of Gly37 have not been assigned. It is unclear why the NH–$C^{\alpha}H$ COSY cross peaks of Gly37 were not observed (Table 1; Fig. 3). Possibly, the NH resonance is broadened by a dynamic rate process. For Arg1 and the two proline residues, it appears unlikely that assignments could be obtained along the lines followed in the present paper. The NH signal of Asp3 overlaps with four other lines and the $C^{\alpha}H$ signal overlaps with that of Ile18 (Fig. 3). Thus, NOE cross peaks between Asp3 and Pro2 might be hidden by overlap with other peaks. With Pro2 not assigned, one could mainly hope to achieve an assignment for Arg1 *via* complete identification of all the arginine side-chain spin systems or *via* pH titration of the N-terminal amino group. For Pro13, no NOE, either with NH of Cys14 or with Gly12, was observed that could be assigned unambiguously to a connectivity with $C^{\alpha}H$, $C^{\beta}H_2$ or $C^{\delta}H_2$ of proline. The connectivity between Gly28 and Leu29 (arrow in Fig. 4) could not be determined unambiguously, since the NH resonances of Ala27 and Leu29 have identical chemical shifts (Fig. 3; Table 1). The NOE cross peaks between the amide proton of Gly28 and those of Ala27 and Leu29 are therefore at the same position, so that the two d_2 connectivities could not both be established unambiguously (Wagner *et al.*, 1981). Of the spin systems of non-labile side-chain protons, all have been completely assigned except for those of the lysine residues, all but two arginine residues, and Pro2 and Pro13. For all the labile protons not listed in Table 1, it is most likely that they were not observed because of rapid exchange with the solvent (Wüthrich, 1976; Bundi & Wüthrich, 1979*a*,*b*).

The chemical shifts for all assigned resonances listed in Table 1 were measured at 68°C and pH 4·6. The Table is therefore more complete than previously published data sets, and also complementary, since the previous listings were for 36°C and pH 4·6 (Wüthrich & Wagner, 1979) and for 24°C and pH 4·6 (Nagayama & Wüthrich, 1981; Wagner *et al.*, 1981), respectively.

The principal project for further use of the presently described resonance assignments is the determination of the three-dimensional structure of BPTI in solution, as outlined in one of the preceding papers (Wüthrich *et al.*, 1982). There is already a lot of evidence that the core of the crystal structure is preserved in solution, both from the presently observed distribution of d_1, d_2 and d_3-connectivities along the amino acid sequence (Fig. 4) (Billeter *et al.*, 1982), as well as from earlier observations, e.g. NOEs between protons in the different strands of the β-sheet (Wagner *et al.*, 1981) or the coincidence with the present assignments of numerous previous assignments obtained with reference to the crystal structure (e.g. Wüthrich *et al.*, 1978; Perkins & Wüthrich, 1979). If further detailed checks confirm that there is close coincidence between extensive interior regions of the molecule in single crystals and in solution, BPTI will be a unique vehicle to further investigate the correlations between NOE intensities and proton–proton distances in proteins, which will be of crucial importance in view of further refined protein structure determinations by n.m.r. (Wüthrich *et al.*, 1982). Similarly, correlations between protein conformation and other n.m.r. parameters, for example chemical shifts and relaxation times, could then be investigated on a reliable basis. Finally, and perhaps most intriguing in view of biological interests, the resonance assignments provide a basis to extend the studies of internal flexibility of BPTI (Wagner & Wüthrich, 1979*a*; Wüthrich *et al.*, 1980) to cover essentially the entire molecular structure.

We thank Dr W. Braun, M. Billeter and G. Wider for stimulating discussions, and Mrs E. Huber and Mrs E. H. Hunziker for the careful preparation of the manuscript and the illustrations. Financial support was obtained from the Schweizerischer Nationalfonds (project 3.528.79) and by a special grant from Eidgenössische Technische Hochschule for the purchase of the 500 MHz spectrometer.

REFERENCES

Anil Kumar, Ernst, R. R. & Wüthrich, K. (1980*a*). *Biochem. Biophys. Res. Commun.* **95**, 1–6.

Anil Kumar, Wagner, G., Ernst, R. R. & Wüthrich, K. (1980*b*). *Biochem. Biophys. Res. Commun.* **96**, 1156–1163.

Anil Kumar, Wagner, G., Ernst, R. R. & Wüthrich, K. (1981). *J. Amer. Chem. Soc.* **103**, 3654–3658.

Aue, W. P., Bartholdi, E. & Ernst, R. R. (1976). *J. Chem. Phys.* **64**, 2229–2246.

Baumann, R., Wider, G., Ernst, R. R. & Wüthrich, K. (1981). *J. Magn. Reson.* **44**, 402–406.

Billeter, M., Braun, W. & Wüthrich, K. (1982). *J. Mol. Biol.* **155**, 321–346.

Bundi, A. & Wüthrich, K. (1979*a*). *Biopolymers*, **18**, 285–298.

Bundi, A. & Wüthrich, K. (1979*b*). *Biopolymers*, **18**, 299–312.

Dayhoff, M. O. (1972). Editor of *Atlas of Protein Sequence and Structure*, National Biomed. Res. Found., Washington.

Deisenhofer, J. & Steigemann, W. (1975). *Acta Crystallogr. sect. B*, **31**, 238–250.

De Marco, A., Wagner, G. & Wüthrich, K. (1977). *Biophys. Struct. Mech.* **3**, 303–315.

Dubs, A., Wagner, G. & Wüthrich, K. (1979). *Biochim. Biophys. Acta*, **577**, 177–194.

Gordon, S. L. & Wüthrich, K. (1978). *J. Amer. Chem. Soc.* **100**, 7094–7096.

Jeener, J., Meier, B. H., Bachmann, P. & Ernst, R. R. (1979). *J. Chem. Phys.* **71**, 4546–4553.

Kalinichenko, P. (1976). *Stud. Biophys.* **58**, 235–240.

Kalk, A. & Berendsen, H. J. C. (1976). *J. Magn. Reson.* **24**, 343–366.

Nagayama, K. & Wüthrich, K. (1981). *Eur. J. Biochem.* **114**, 365–374.

Nagayama, K., Wüthrich, K. & Ernst, R. R. (1979). *Biochem. Biophys. Res. Commun.* **90**, 305–311.
Nagayama, K., Anil Kumar, Wüthrich, K. & Ernst, R, R. (1980). *J. Magn. Res.* **40**, 321–334.
Noggle, J. H. & Schirmer, R. E. (1971). *The Nuclear Overhauser Effect*, Academic Press, New York.
Perkins, S. J. & Wüthrich, K. (1979). *Biochim. Biophys. Acta*, **576**, 409–423.
Richarz, R., Sehr, P., Wagner, G. & Wüthrich, K. (1979). *J. Mol. Biol.* **130**, 19–30.
Richarz, R., Nagayama, K. & Wüthrich, K. (1980). *Biochemistry*, **19**, 5189–5196.
Snyder, G. H., Rowan III, R., Karplus, S. & Sykes, B. D. (1975). *Biochemistry*, **14**, 3765–3777.
Solomon, I. (1955). *Phys. Rev.* **99**, 559–565.
Tschesche, H., (1974). *Angew. Chemie, Int. Ed. Engl.* **13**, 10–28.
Wagner, G. & Wüthrich, K. (1979*a*). *J. Mol. Biol.* **134**, 75–94.
Wagner, G. & Wüthrich, K. (1979*b*). *J. Magn. Reson.* **33**, 675–680.
Wagner, G., De Marco, A. & Wüthrich, K. (1976). *Biophys. Struct. Mech.*, **2**, 139–158.
Wagner, G., Wüthrich, K. & Tschesche, H. (1978*a*). *Eur. J. Biochem.* **86**, 67–76.
Wagner, G., Wüthrich, K. & Tschesche, H. (1978*b*). *Eur. J. Biochem.* **89**, 367–377.
Wagner, G., Anil Kumar & Wüthrich, K. (1981). *Eur. J. Biochem.* **114**, 375–384.
Wider, G., Lee, H. K. & Wüthrich, K. (1982). *J. Mol. Biol.* **155**, 367–388.
Wüthrich, K. (1976). *NMR in Biological Research: Peptides and Proteins*, North-Holland Publishing Company, Amsterdam.
Wüthrich, K. & Wagner, G. (1975). *FEBS Letters*, **50**, 265–268.
Wüthrich, K. & Wagner, G. (1979). *J. Mol. Biol.* **130**, 1–18.
Wüthrich, K., Wagner, G., Richarz, R. & Perkins, S. J. (1978). *Biochemistry*, **17**, 2253–2263.
Wüthrich, K., Wagner, G., Richarz, R. & Braun, W. (1980). *Biophys. J.* **10**, 549–560.
Wüthrich, K., Wider, G., Wagner, G. & Braun, W. (1982). *J. Mol. Biol.* **155**, 311–319.

Edited by S. Brenner

J. Mol. Biol. (1982) **155**, 367–388

Sequential Resonance Assignments in Protein ^{1}H Nuclear Magnetic Resonance Spectra

Glucagon Bound to Perdeuterated Dodecylphosphocholine Micelles

GERHARD WIDER, KONG HUNG LEE AND KURT WÜTHRICH

Institut für Molekularbiologie und Biophysik
Eidgenössische Technische Hochschule
ETH-Hönggerberg, CH-8093 Zürich, Switzerland

(Received 17 August 1981)

The assignment of the ^{1}H nuclear magnetic resonance spectrum of glucagon bound to perdeuterated dodecylphosphocholine micelles with the use of two-dimensional ^{1}H nuclear magnetic resonance techniques at 360 MHz is described. Sequential resonance assignments were obtained for all backbone and C^{β} protons except the N-terminal amino group and the amide proton of Ser2. The assignments of the non-labile amino acid side-chain protons are complete except for the γ-methylene protons of Gln20 and Gln24. These assignments provide a basis for the determination of the three-dimensional structure of lipid-bound glucagon.

1. Introduction

Glucagon is a hormone that consists of a linear polypeptide chain of 29 amino acid residues, M_r 3500. The primary target organ for glucagon is the plasma membrane of liver and other cells. Specific binding to a plasma membrane receptor site mediates activation of glycogenolysis (Pohl *et al.*, 1969). Evidence has been presented that recognition between glucagon and its receptor depends on the ordered lipid structures surrounding the receptor site in the membrane (Rodbell *et al.*, 1971; Rubalcava & Rodbell, 1973). A possible avenue to further insights into the mode of action would be *via* knowledge of the conformations adopted by glucagon in the different environments encountered on the way from the sites of its synthesis in the islets of Langerhans to the complex formation with the receptor site. Previously, an α-helical conformation of glucagon was determined in single crystals (Sasaki *et al.*, 1975). However, ^{1}H n.m.r.† studies showed that the α-helical form was not preserved in aqueous solutions of monomeric glucagon (Bösch *et al.*, 1978), which coincides with earlier conclusions from circular dichroism experiments that monomeric glucagon in solution adopts predominantly a flexible "random

† Abbreviations used: n.m.r., nuclear magnetic resonance; p.p.m., parts per million; NOE, nuclear Overhauser enhancement; NOESY, 2-dimensional NOE spectroscopy; COSY, 2-dimensional correlated spectroscopy.

0022–2836/82/070367–22 $02.00/0 © 1982 Academic Press Inc. (London) Ltd.

coil" conformation (see e.g. Panijpan & Gratzer, 1974, and references therein). In view of the functional properties of glucagon, it appears of considerable interest to complement these structural data with a determination of the molecular conformation in a lipid/water interface, for example along the surface of lipid micelles. In this paper we describe individual ^{1}H n.m.r. assignments for the entire amino acid sequence of glucagon bound to perdeuterated dodecylphosphocholine micelles.

In previous studies it was shown by a variety of physicochemical methods that mixed micelles of glucagon and dodecylphosphocholine in a sufficiently concentrated solution for n.m.r. studies contained one molecule of glucagon and ~40 detergent molecules, which corresponds to a molecular weight of ~17,000 (Bösch *et al.*, 1980). High resolution ^{1}H n.m.r. studies with one-dimensional techniques indicated that the micelle-bound glucagon adopted a predominantly extended conformation (Bösch *et al.*, 1980). The polypeptide backbone was found to be roughly parallel to the micelle surface, with the depth of immersion corresponding approximately to the average length of an amino acid side-chain (Brown *et al.*, 1981). From combined use of distance constraints obtained from nuclear Overhauser enhancement experiments and a distance geometry algorithm, a spatial structure for the glucagon segment residues 19 to 27 was determined (Braun *et al.*, 1981). Since at that stage the ^{1}H n.m.r. lines of only a small number of amino acid side-chain hydrogen atoms had been assigned individually, only low resolution could be attained for this structure. The resonance assignments in this paper are the first step in the determination of the spatial structure of micelle-bound glucagon at higher resolution (Wüthrich *et al.*, 1982).

2. Materials and Methods

Sequential resonance assignments were obtained with combined use of 2-dimensional correlated spectroscopy (Aue *et al.*, 1976) and 2-dimensional nuclear Overhauser enhancement spectroscopy (Jeener *et al.*, 1979; Anil Kumar *et al.*, 1980*a*,*b*). For a brief description of these 2 experiments, the reader is referred to the preceding paper (Wagner & Wüthrich, 1982). In contrast to the basic pancreatic trypsin inhibitor, where a sizeable proportion of the sequential assignments was obtained from studies in ^2H_2O solutions (Dubs *et al.*, 1979; Wagner *et al.*, 1981), all the amide protons in micelle-bound glucagon exchange quite rapidly, and hence the sequential assignments had to rely entirely on studies in H_2O solution.

The spin systems of the individual amino acid side-chains were identified in a COSY spectrum recorded in a ^2H_2O solution of micelle-bound glucagon. In several cases, ambiguities in the COSY connectivities were resolved by comparison with a spin echo correlated spectrum (Nagayama *et al.*, 1979,1980; Nagayama & Wüthrich, 1981) recorded under identical conditions.

Two-dimensional ^{1}H n.m.r. spectra at 360 MHz were recorded on a Bruker HX 360 spectrometer. The spectra were obtained from 256 measurements with t_1 values from 0 to 40 ms. Special care had to be taken for the suppression of the H_2O solvent resonance, which was much broader than, for example, in a dilute solution of a protein. Satisfactory results were obtained by selective, continuous irradiation of the H_2O line at all times except during data acquisition: 1024 data points were used to store the data for each value of t_1. In the different experiments, between 192 and 256 transients were accumulated for each t_1 value. To end up with a 512×512 point data matrix in the frequency domain, which corresponded to a digital resolution of 6·3 Hz/point for the spectra recorded in H_2O and 5·8 Hz/point for ^2H_2O spectra, the time domain matrix was expanded to 1024 points in t_1 and 2048 points in t_2 by zero filling. Otherwise, data acquisition and data handling were done as described in the preceding paper (Wagner & Wüthrich, 1982).

Bovine glucagon was purchased from SERVA, Heidelberg. [$^2H_{38}$]dodecylphosphocholine was synthesized as described (Brown, 1979). Two samples were prepared with a mixed solvent of 90% H_2O and 10% ^2H_2O, and with ^2H_2O. The samples contained 0·015 M-glucagon, 0·7 M-[$^2H_{38}$]dodecylphosphocholine and 0·05 M-phosphate buffer (pH 6·0). In the ^2H_2O solution, the pH meter reading was used without correction for isotope effects (Kalinichenko, 1976; Bundi & Wüthrich, 1979*a*), and the concentration of residual solvent protons was minimized by repeated lyophilization from ^2H_2O. Both samples were carefully degassed and sealed under a nitrogen atmosphere. Chemical shifts are quoted relative to external 3-trimethylsilyl-[2,2,3,3-$^2H_4$]propionate, where the ϵ-methyl resonance of Met27 was taken to be at 2·04 p.p.m. at 37°C and was used as an internal reference.

Mixed micelles of glucagon and dodecylphosphocholine were previously characterized by a variety of physicochemical methods (Bösch *et al.*, 1980; Brown *et al.*, 1981). For glucagon concentrations between 0·001 and 0·006 M and a 50-fold excess of detergent molecules, it was found that an homogeneous population of micelles consisting of 1 glucagon molecule and ~40 detergent molecules prevailed, with a molecular weight of ~17,000. For the 2-dimensional n.m.r. experiments, the glucagon concentration was increased to 0·015 M. Identical ^{1}H n.m.r. spectra were obtained to those for solutions with 0·004 M-glucagon (Bösch *et al.*, 1980; Braun *et al.*, 1981). Since other methods, for example, light-scattering and the analytical ultracentrifuge, could not be used reliably at these high concentrations, we relied on the n.m.r. evidence that similar mixed micelles were formed over the entire concentration range.

3. Results and Discussions

(a) *Identification of the amino acid side-chain spin systems*

Previously, the complete spin systems of the non-labile protons of Ala19, Val23 and Trp25 in glucagon bound to perdeuterated dodecylphosphocholine micelles were identified with one-dimensional n.m.r. experiments (Bösch *et al.*, 1980). In addition, the imidazole resonances of His1, the δ-methyl resonances of the two leucine residues in positions 14 and 26, the ϵ-methyl line of Met27 and a Thr $C^{\beta}H$–$C^{\gamma}H_3$ fragment were described. In what follows, we describe how the complete spin systems of all but two of the 29 amino acid residues were identified in a COSY spectrum recorded in ^2H_2O solution (Fig. 1).

In the COSY spectrum of Figure 1, peaks corresponding to the resonance positions in the one-dimensional spectrum are located on the diagonal from the upper right to the lower left corner. Non-diagonal "cross peaks" manifest through-bond J-connectivities between distinct diagonal peaks. In Figure 1, two well-separated spectral regions from 0·5 to 5·0 p.p.m., and from 6·5 to 7·5 p.p.m. can be distinguished, which are not connected by J-couplings. The low-field region contains the resonances of the aromatic rings, the high-field region includes those of all the other protons. In the lower right triangle of Figure 1, the connectivities for seven aliphatic spin systems and the aromatic rings of tryptophan and the two tyrosine residues are indicated.

The COSY connectivity patterns for the amino acid residues in glucagon are shown in Table 1. They represent a different presentation of the previously

370 G. WIDER, K. H. LEE AND K. WÜTHRICH

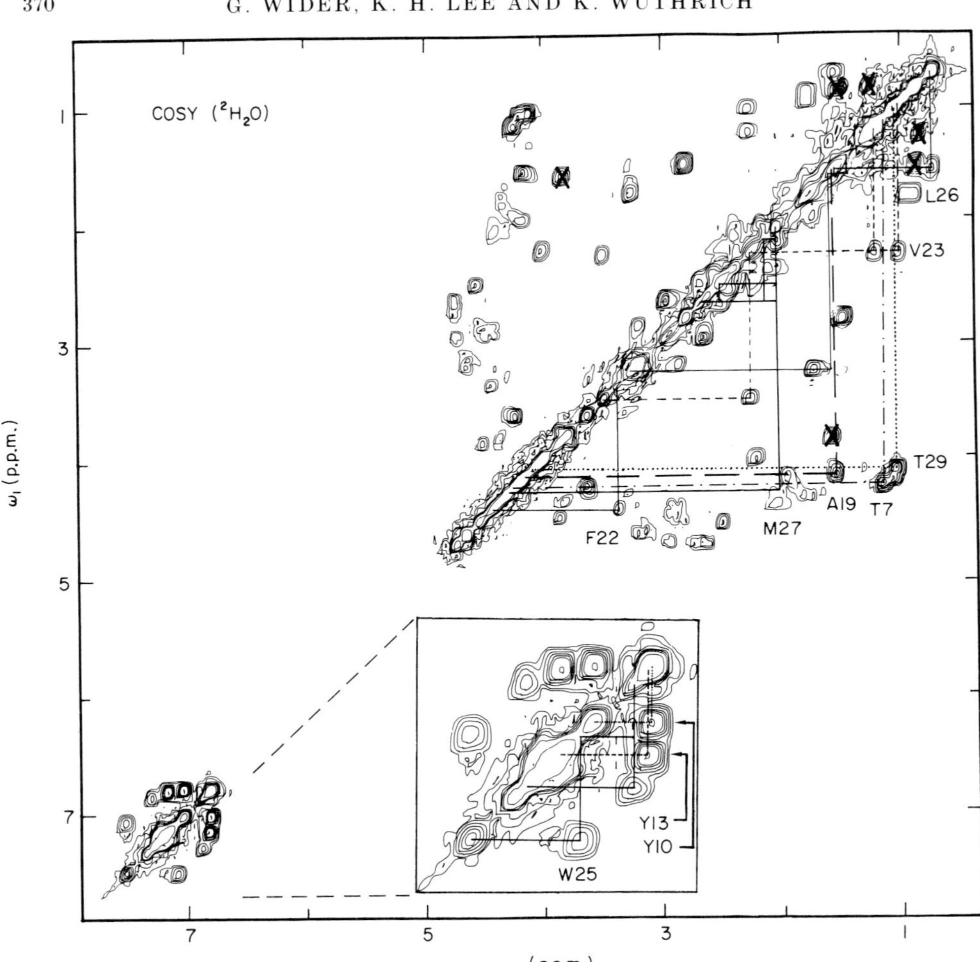

FIG. 1. Contour plot of a 360 MHz ^{1}H COSY spectrum of glucagon bound to perdeuterated dodecylphosphocholine micelles in ^2H_2O. The sample contained 0·015 M-glucagon, 0·7 M-[$^2H_{38}$]dodecylphosphocholine, 0·05 M-phosphate buffer, p^{2}H 6·0, t = 37°C. Under these conditions the predominant species in the solution are mixed micelles of 1 glucagon molecule and ~40 detergent molecules, with a molecular weight of about 17,000 (Bösch *et al.*, 1980). The spectrum was recorded in 24 h, the digital resolution is 5·88 Hz/point. The symmetrized (Baumann *et al.*, 1981) absolute value spectrum is shown. The aromatic region is also presented on an expanded scale. Proton–proton J-connectivities are indicated for the following residues: Thr7 (—·—·), Ala19 (— —), Phe22 (——), Val23 (– – – –), Leu26 (——), Met27 (——), Thr29 (····) and the aromatic rings of Tyr10 (····), Tyr13 (– – – –) and Trp25 (——). In order not to overcrowd the Figure, only the C$^\alpha$H connectivity with the lower field C$^\beta$H line is shown even for amino acid residues where 2 non-degenerate β-methylene resonances were observed (Table 2). Cross peaks originating from residual protons in the perdeuterated dodecylphosphocholine are marked (X).

described connectivity patterns in spin echo correlated spectroscopy (Nagayama & Wüthrich, 1981). Unique connectivities, which can provide the information needed for unambiguous identification, prevail for valine, isoleucine (not shown in Table 1), leucine, proline (not shown in Table 1), threonine, lysine and arginine.

TABLE 1

COSY connectivity diagrams for the weakly coupled spin systems of the non-labile, aliphatic protons in the common amino acid residues†

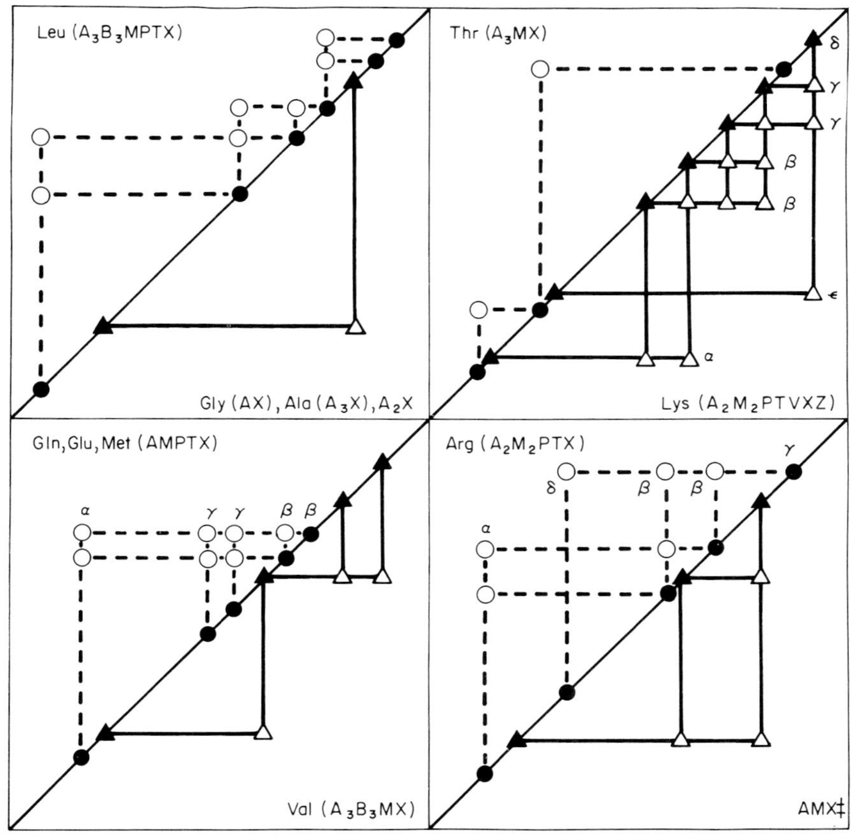

† In each of the 4 squares, 2 connectivity diagrams are shown. For the diagram in the upper left area of each square, diagonal peaks are indicated with filled circles, cross peaks with open circles and connectivities with broken lines. In the lower right area, filled and open triangles and continuous lines are used. The amino acid types and spin systems represented by the diagrams are also indicated. For better readability, the diagonal peaks of the AMPTX system, Arg and Lys, are identified. Ile and Pro, which are not present in glucagon, are not shown.

‡ A_2X or AMX spin systems may arise from Asn, Asp, Cys, His, Phe, Ser, Trp and Tyr.

For the other common amino acids, the COSY patterns allow only a classification into groups of several residues, and additional information is required to further distinguish the individual amino acid types. For instance, Gly, A_2X spin systems and Ala can be distinguished from the resonance intensities and chemical shifts, and the $C^{\alpha}H$–$C^{\beta}H_2$ fragments of the aromatic side-chains can be distinguished from the other AMX or A_2X spin systems on the basis of the NOE connectivities with the aromatic rings (see below). Alternatively, combination of the COSY connectivities with sequential assignment of the polypeptide backbone resonances can also be used to identify the individual residue types (see below).

To illustrate the analysis, we consider the Val spin system in Figure 1. From the diagonal peak at 3·49 p.p.m., a broken, horizontal line leads to the C$^{\alpha}$H–C$^{\beta}$H cross peak at (ω_1 = 3·49, ω_2 = 2·26 p.p.m.). From there a vertical line leads to the diagonal peak of C$^{\beta}$H at 2·26 p.p.m. A horizontal line makes the connections to the two cross peaks with the C$^{\gamma}$H$_3$ resonances at (ω_1 = 2·26, ω_2 = 1·21 p.p.m.) and (ω_1 = 2·26, ω_2 = 1·02 p.p.m.). Two vertical lines then lead to the diagonal peaks of

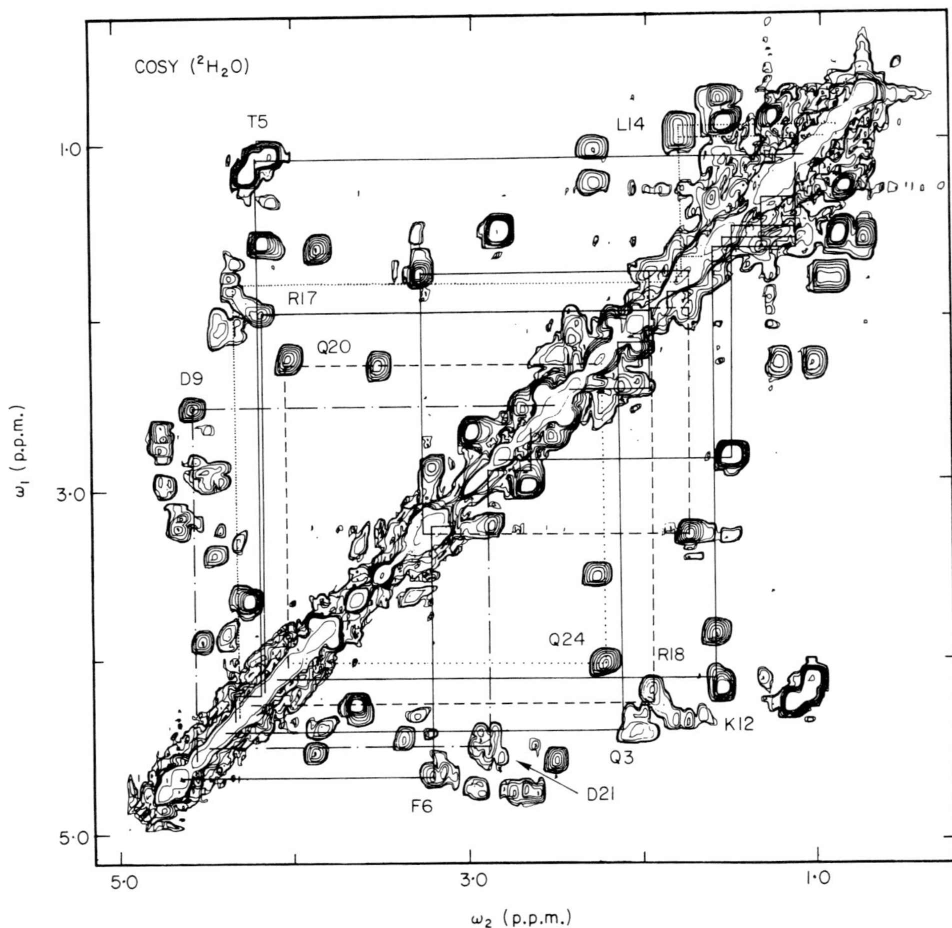

FIG. 2. Spectral region from 0·2 to 5·1 p.p.m. of the 360 MHz ^{1}H COSY spectrum of micelle-bound glucagon in Fig. 1. J-connectivities for the following residues are indicated in the upper left triangle of the spectrum: Thr5 (——), Asp9 (—·—·), Leu14 (····), Arg17 (——), Gln20 (— — —). The lower right triangle contains the J-connectivities for Gln3 (——), Phe6 (——), Lys12 (——), Arg18 (— — —), Asp21 (—·—), Gln24 (····). In order not to overcrowd the Figure, only the C$^{\alpha}$H connectivity with the lower field C$^{\beta}$H line is shown even for amino acid residues where 2 non-degenerate β-methylene resonances were observed (Table 2).

the γ-methyl groups. Similarly, the connectivities for the other 28 amino acid residues are documented in Figures 1 to 3. When comparing the connectivities in these Figures with those in Table 1, one should consider that for practical reasons only one of the connectivities between $C^{\alpha}H$ and $C^{\beta}H_2$ is outlined, even when the β-methylene protons are not degenerate.

For the aromatic side-chains, COSY provided two separate spin systems of $C^{\alpha}H$–$C^{\beta}H_2$ (Figs 1 to 3) and the ring protons (Fig. 1), respectively. Except for His1, the connectivity with the ring protons could be established *via* NOEs, as predicted by theoretical considerations (Billeter *et al.*, 1982). These results are documented in

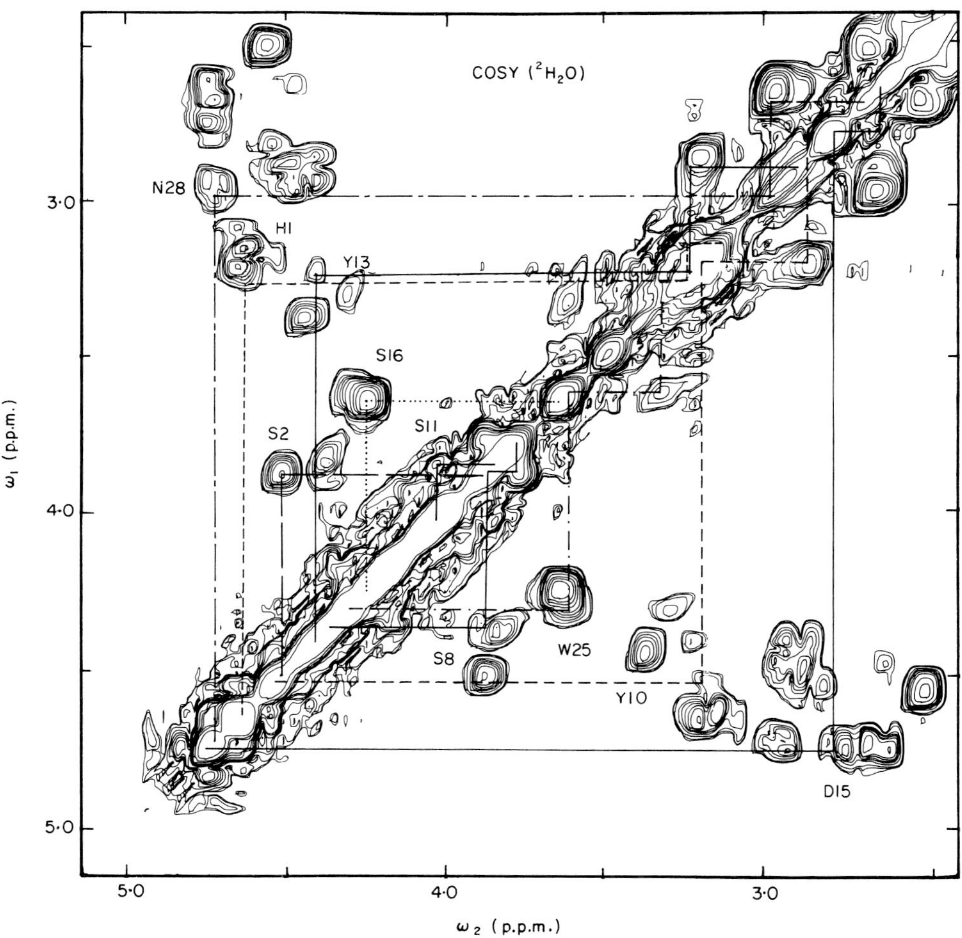

FIG. 3. Spectral region from 2·4 to 5·1 p.p.m. of the 360 MHz ^{1}H COSY spectrum of micelle-bound glucagon in Fig. 1. Connectivities for the following residues are indicated in the upper left triangle: His1 (– – –), Ser2 (—–—), Ser11 (——), Tyr13 (——), Ser16 (····), Asn28 (–·–). The lower right triangle contains the J-connectivities for Ser8 (——), Tyr10 (– – –), Asp15 (——), Trp25 (–·–·). In order not to overcrowd the Figure, only the $C^{\alpha}H$ connectivity with the lower field $C^{\beta}H$ line is shown even for amino acid residues where 2 non-degenerate β-methylene resonances were observed (Table 2).

Figure 4, which shows a NOSEY spectrum recorded in the same ^2H_2O solution of micelle-bound glucagon as the COSY spectrum in Figure 1. A NOESY spectrum (Anil Kumar *et al.*, 1980*a*) has a similar appearance to that of a COSY spectrum (Fig. 1), but the non-diagonal peaks manifest through-space NOE connectivities between distinct resonances on the diagonal. In Figure 4, only a small spectral region is shown, which contains all the cross peaks between $C^{\beta}H_2$ and ring protons within the individual aromatic side-chains.

With regard to the sequential resonance assignments, it was important that one could at the outset rely on 23 completely identified side-chain spin systems. The spin systems of Gly4, one Thr, one Leu and three of the four five-spin systems of Gln and Met could be identified only after the sequential assignments were completed. It should be added that, with the exception of the residue types that occur only once in the amino acid sequence of glucagon and give rise to unique

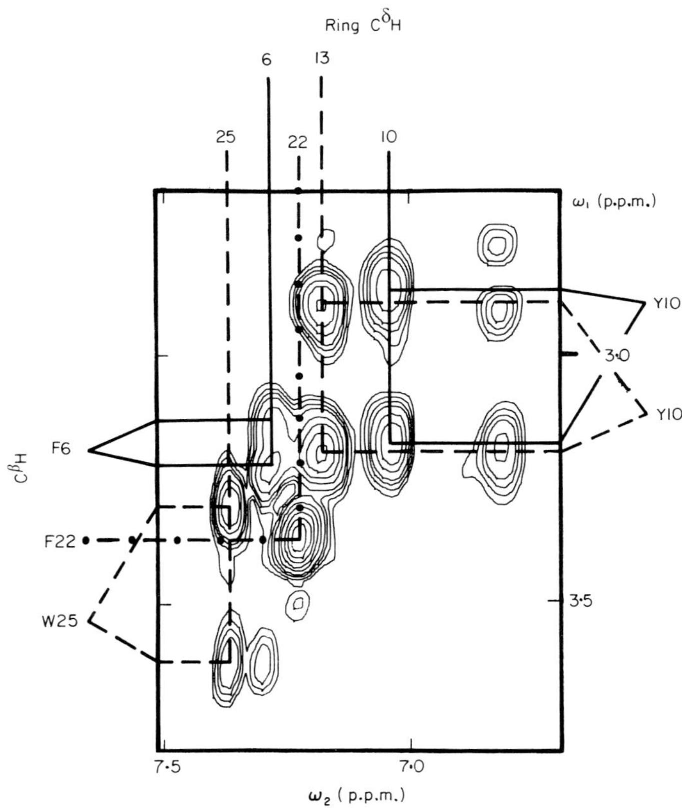

FIG. 4. Contour plot of the spectral region (2·6 to 3·8 p.p.m.) × (6·7 to 7·5 p.p.m.) of a 360 MHz ^1H NOESY spectrum of glucagon bound to perdeuterated dodecylphosphocholine micelles recorded in the sample described in the legend to Fig. 1. The temperature was 37°C. The spectrum was recorded in 26 h, using a mixing time of 200 ms. The digital resolution is 5·8 Hz/point. The symmetrized absolute value spectrum is shown. The connectivities between $C^{\beta}H_2$ and the C^{δ} ring protons of the aromatic residues are indicated as follows: Phe6 (——), Phe22 (·–·), Tyr10 (——), Tyr13 (– – –), Trp25 (– – –).

COSY patterns (Table 1; Fig. 6), i.e. Lys12, Ala19, Val23 and Trp25, the assignments to specific sequence locations indicated in Figures 1 to 5 were obtained only after completion of the sequential resonance assignments (see below).

(b) *Strategy used to assign the ^1H n.m.r. spectrum of micelle-bound glucagon*

Compared to the resonance assignments for basic pancreatic trypsin inhibitor described by Wagner *et al.* (1981) and in the preceding paper (Wagner & Wüthrich, 1982), a somewhat different strategy had to be selected for glucagon. For example, since the labile protons of micelle-bound glucagon exchange too rapidly with the solvent to be observed in the ^{1}H n.m.r. spectra recorded in ^2H_2O solution (Bösch *et al.*, 1980), all the sequential assignments had to be obtained with spectra recorded in H_2O. Furthermore, sequential assignments *via* NOEs between NH_{i+1} and $C^{\beta}H_i$ (d_3) played a much more important role than in basic pancreatic trypsin inhibitor. Otherwise the fundamental aspects of the procedures used were closely similar. Neighbouring amino acid residues were identified *via* one or two of the NOE connectivities d_1 (from NH_{i+1} to $C^{\alpha}H_i$), d_2 (from NH_{i+1} to either NH_{i+2} or NH_i) or d_3 (from NH_{i+1} to $C^{\beta}H_i$) (Billeter *et al.*, 1982), the J-connectivities with the labile protons were obtained from a COSY spectrum recorded in H_2O and, as far as possible, the sequential assignments were checked step by step against the amino acid sequence.

Figure 5 shows the region of the COSY spectrum in H_2O that contains all the NH_i–$C^{\alpha}H_i$ cross peaks in micelle-bound glucagon. As discussed in more detail in the preceding paper (Wagner & Wüthrich, 1982), each amino acid residue in glucagon (Fig. 6) should give rise to one such peak, except for Gly4, which might give two peaks, and the N-terminal residues His1 and Ser2, which might not be seen because of rapid exchange of the labile backbone protons (Bundi & Wüthrich, 1979*b*). From the resonance identifications in Figures 1 to 3, it was further known that the cross peaks with the C^{α} protons of two AMX spin systems at 4·66 and 4·64 p.p.m. would not be seen for experimental reasons, since at 37°C they coincided with the water line and were therefore bleached out by the water irradiation (Anil Kumar *et al.*, 1980*b*). Figure 5 contained all the expected cross peaks, and thus provided a complete "fingerprint" of the polypeptide chain. From the side-chain identifications, 23 peaks could be assigned to specific spin systems. Among these, Lys12, Ala19, Val23 and Trp25 could be assigned immediately to a specific location in the amino acid sequence, since glucagon contains only one residue of each of these four types (Fig. 6). These four residues provided ideal starting points for the sequential assignments, which could then be checked against the amino acid sequence from the first step.

Figure 5 affords a convincing illustration of the improved resolution achieved when going from unidimensional n.m.r. to two-dimensional n.m.r. However, because of the inherently broad lines and the small dispersion of the $C^{\alpha}H$ and NH chemical shifts, some overlap of different peaks persisted in the COSY and NOESY spectra of micelle-bound glucagon, which limited the analysis of the cross peaks among C^{α} and amide protons in some instances. Therefore, and because of the superior spectral resolution in the $C^{\beta}H$ region, d_3 connectivities between NH_{i+1} and

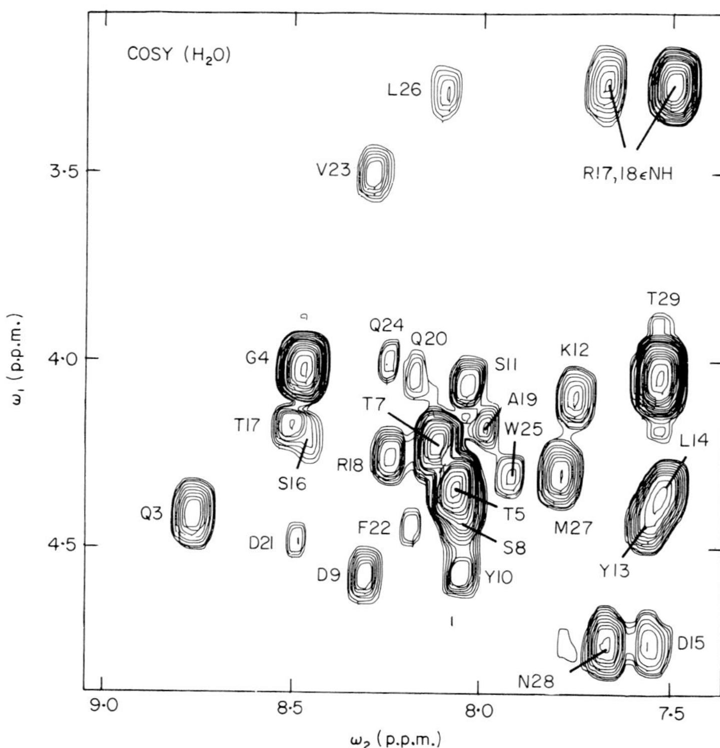

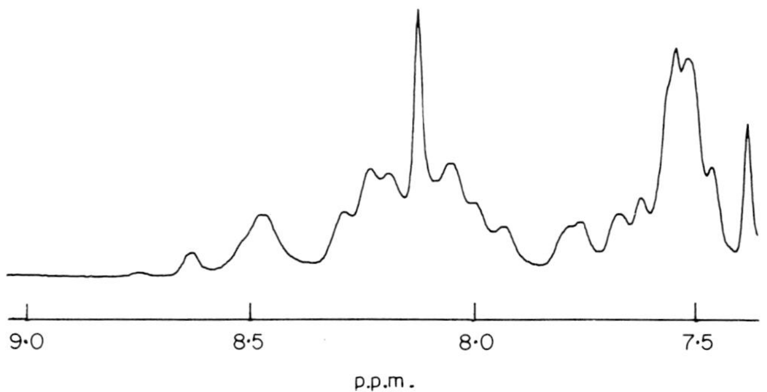

FIG. 5. Spectral region (3·1 to 4·9 p.p.m.) × (7·4 to 9·1 p.p.m.) of a 360 MHz ^1H COSY spectrum of glucagon bound to perdeuterated dodecylphosphocholine micelles. The composition of the sample was identical to that in Fig. 1, except that a mixed solvent of 90% H_2O and 10% ^2H_2O was used. The temperature was 37°C. The letters and numbers indicate the resonance assignments for the $C^{\alpha}H_i$–NH_i cross peaks for residues 3 to 29. In addition, the cross peaks connecting $C^{\delta}H_2$ and the guanidinium protons of the 2 arginine side-chains are seen in the upper right corner. For comparison, a conventional, unidimensional spectrum of the same sample is shown at the bottom of the Figure. It contains the well-separated resonance of the amide proton of Phe6 at 8·63 p.p.m. In the COSY spectrum, the cross peak for Ph6 was bleached out by the irradiation of the solvent line (Anil Kumar *et al.*, 1980*b*), since the chemical shift of $C^{\alpha}H$ coincides with that of the H_2O line (Table 2).

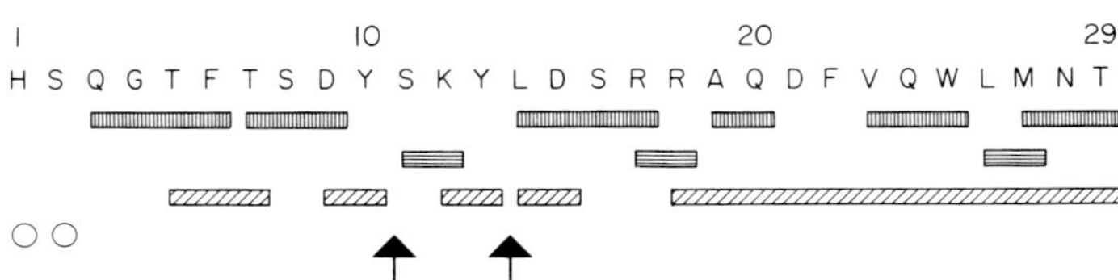

FIG. 6. Amino acid sequence of bovine glucagon and survey of the experimental data by which individual resonance assignments were obtained for the micelle-bound polypeptide. (▥) Sequential assignments *via* d_1 (NOE from NH_{i+1} to $C^{\alpha}H_i$); (▤) sequential assignments *via* d_2 (NOE from NH_i to NH_{i+1}); (▨) sequential assignments *via* d_3 (NOE from NH_{i+1} to $C^{\beta}H_i$); (○) assignment in the sequence relied on the identification of the spin system in the COSY spectrum, whereby the amide proton resonances of these residues were not observed. The arrows indicate locations where all the resonances were assigned but the connectivity between 2 neighbouring residues was not established.

$C^{\beta}H_i$ played an important role for the sequential assignments (see Fig. 6 for a survey of the experiments used), even though their inherent reliability is considerably lower than that for the connectivities among amide and C^{α} protons (Billeter *et al.*, 1982). To illustrate this practically important aspect, all three connectivities *via* d_1, d_2 and d_3 are documented in what follows for the C-terminal segment residues 18 to 29 of glucagon (Figs 8 to 11). Only one connectivity for each assignment is documented for the other regions (Figs 11 to 13).

(c) *Sequential connectivities* via d_1, d_2 *and* d_3 *in glucagon residues 18 to 29*

The sequential assignments were unravelled in a H_2O NOESY spectrum of micelle-bound glucagon recorded at 37°C (Fig. 7) and in a corresponding COSY spectrum. In the NOESY spectrum, only the region ($\omega_1 = 1{\cdot}4$ to $9{\cdot}0$ p.p.m., $\omega_2 = 7{\cdot}5$ to $9{\cdot}0$ p.p.m.) was needed for the assignments, so that the analysis was not impeded by the strong vertical bands of spurious noise that arose between 1 and 5 p.p.m. from the water irradiation and from the sharp diagonal peaks of residual protons in the deuterated lipid. In Figures 8 to 13, small regions of the NOESY spectrum in Figure 7 and in some cases of the corresponding COSY spectrum are displayed. The chemical shifts of the assigned resonances are listed in Table 2.

As indicated in Figure 6, a continuous line of sequential assignments *via* d_3 was obtained for glucagon residues 18 to 29. These data are documented in Figure 8. The connectivity diagram starts at the chemical shift of the Thr29 amide proton in the lower right corner. A vertical line at this position goes through the centre of only one peak, which was therefore assigned to $C^{\beta}H$ of Asn28 (Billeter *et al.*, 1982). The amide proton of Asn28 was then identified in the COSY spectrum *via* the J-connectivities in the fragment $-C^{\beta}H_2-C^{\alpha}H-NH-$. The COSY connectivity is not shown; instead, the ($NH_{29}-C^{\beta}H_{28}$) cross peak in the NOESY spectrum of Figure 8 is connected with the ($C^{\beta}H_{28}-NH_{28}$) cross peak. A vertical line at the chemical shift of Asn28 NH goes through the centre of only one peak, which was therefore assigned to $C^{\beta}H$ of Met27. The horizontal connection from ($NH_{28}-C^{\beta}H_{27}$) to ($C^{\beta}H_{27}-NH_{27}$) was again based on the identification of the amide proton chemical shift of Met27 in the COSY spectrum. The next step was more ambiguous, since

378 G. WIDER, K. H. LEE AND K. WÜTHRICH

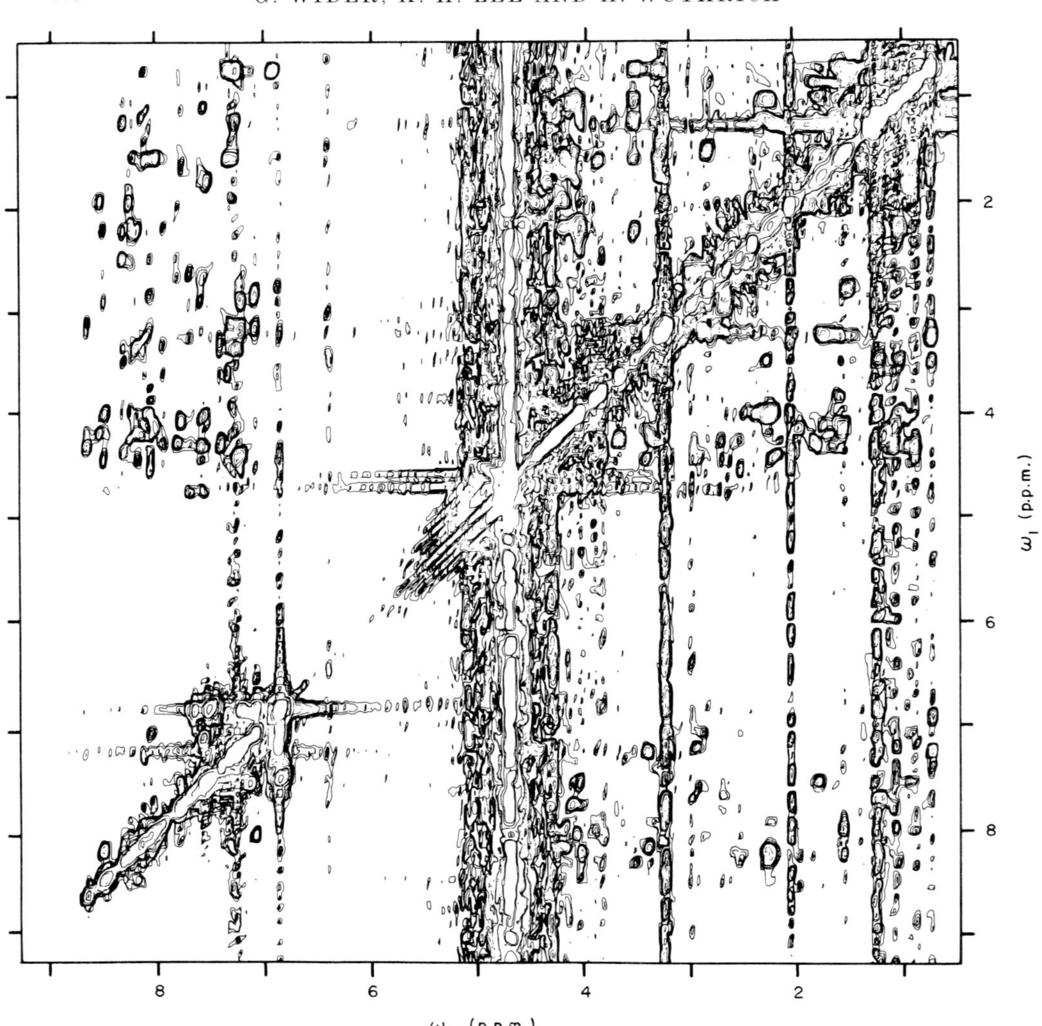

FIG. 7. Contour plot of a 360 MHz ^{1}H NOESY spectrum observed in the same sample of micelle-bound glucagon as the COSY spectrum of Fig. 5. The spectrum was recorded in 26 h, using a mixing time of 100 ms with random modulation of amplitude 6 ms (Macura *et al.*, 1981). The digital resolution is 6·3 Hz/point. An absolute value spectrum is shown.

there are three cross peaks on the vertical line through the chemical shift position of Met27 NH. The correct assignment was obtained on the basis of the amino acid sequence and the identified side-chain spin systems. Continuing in this way, assignments were obtained from Thr29 all the way to C$^\beta$H of Arg18.

Statistical studies of protein crystal structures have shown that the inherent reliability of sequential assignments *via* d_3 is only ~75%, as compared to ~88% for d_1 or d_2, and between 92 and 99% when two of the three connectivities can be established simultaneously. Therefore, we looked for further connectivities in

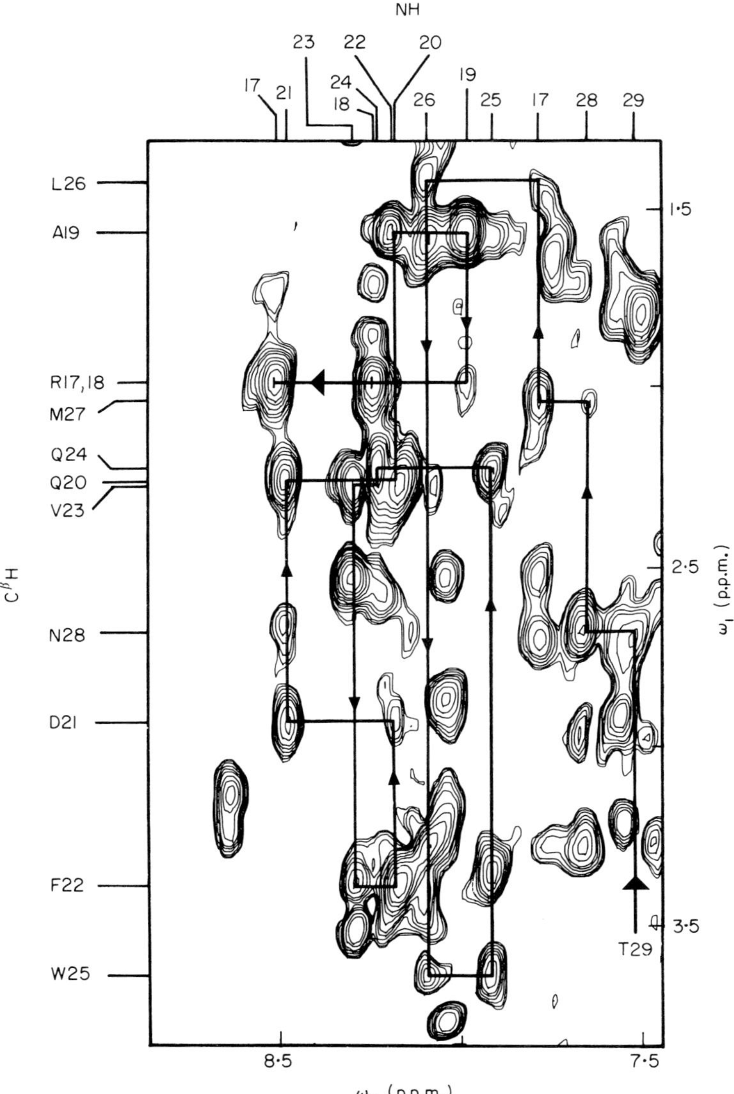

FIG. 8. Spectral region (1·3 to 3·8 p.p.m.) × (7·4 to 8·8 p.p.m.) of the 360 MHz ^{1}H NOESY spectrum of micelle-bound glucagon shown in Fig. 7. The continuous lines with arrows indicate the sequential assignments for the polypeptide segment residues 17 to 29, which were obtained from NOEs between amide protons and the C^β protons of the preceding residues (d_3). The numbers at the top of the Figure indicate the amide proton chemical shifts of the corresponding residues, those on the left margin indicate the chemical shifts of 1 C^β proton for each residue. The amide proton chemical shifts for Arg17 and Arg18 were identified *via* d_1 and d_2 connectivities (Figs 11 and 12) and have from there been added to the present Figure.

TABLE 2

Chemical shifts, δ†, of the assigned ¹H n.m.r. lines of glucagon bound to perdeuterated dodecylphosphocholine micelles, pH 6·0, t = 37°C

Amino acid residue	δ (±0·01 p.p.m.)†			
	NH	$C^{\alpha}H$	$C^{\beta}H$	Others
His1‡	n.o.	4·66‡	3·07, 3·21‡	$C^{\delta}H$ 7·22, $C^{\epsilon}H$ 8·11
Ser2‡	n.o.	4·51‡	3·86, 3·86‡	
Gln3§	8·75	4·41	2·01, 2·17	$C^{\gamma}H_2$, 2·38, 2·38
Gly4	8·47	4·02 4·02		
Thr5	8·06	4·34	4·19	$C^{\gamma}H_3$ 1·07
Phe6	8·63	4·64	3·13, 3·22	Ring 7·27
Thr7	8·11	4·23	4·24	$C^{\gamma}H_3$ 1·17
Ser8	8·04	4·37	3·78, 3·86	
Asp9	8·30	4·57	2·52, 2·52	
Tyr10	8·05	4·56	2·85, 3·18	$C^{\delta}H$ 7·06, 7·06 $C^{\epsilon}H$ 6·82, 6·82
Ser11	8·03	4·06	3·96, 3·96	
Lys12	7·75	4·09	1·31, 1·59	$C^{\gamma}H_2$ 1·10, 1·52 $C^{\delta}H_2$ 1·50, 1·50 $C^{\epsilon}H_2$ 2·80, 2·80
Tyr13	7·55	4·42	2·89, 3·23	$C^{\delta}H$ 7·19, 7·19 $C^{\epsilon}H$ 6·84, 6·84
Leu14	7·52	4·33	1·61, 1·81	$C^{\gamma}H$ 1·77 $C^{\delta}H_3$ 0·88, 0·96
Asp15	7·55	4·75	2·63, 2·76	
Ser16	8·45	4·23	3·63, 3·63	
Arg17‖	8·50	4·17	1·96, 1·96	$C^{\gamma}H_2$ 1·73, 1·73 $C^{\delta}H_2$ 3·25, 3·25
Arg18‖	8·24	4·25	1·86, 1·94	$C^{\gamma}H_2$ 1·71, 1·71 $C^{\delta}H_2$ 3·27, 3·27
Ala19	7·99	4·17	1·56	
Gln20§¶	8·17	4·02	2·25	
Asp21	8·48	4·47	2·62, 2·92	
Phe22	8·18	4·44	3·38, 3·38	Ring 7·23
Val23	8·28	3·49	2·26	$C^{\gamma}H_3$ 1·02, 1·21
Gln24§¶	8·24	4·00	2·22	
Trp25	7·92	4·30	3·33, 3·62	$C^{\delta}H$ 7·37 $N^{\epsilon}H$ 10·53 $C^{\epsilon}H$ 7·33 $C^{\zeta2}H$ 7·54 $C^{\zeta3}H$ 6·89 $C^{\eta}H$ 7·11
Leu26	8·09	3·27	1·41, 1·57	$C^{\gamma}H$ 1·56 $C^{\delta}H_3$ 0·72, 0·72
Met27	7·79	4·30	2·03, 2·13	$C^{\gamma}H_2$ 2·53, 2·70 $C^{\epsilon}H_3$ 2·04
Asn28§	7·66	4·74	2·68, 2·97	
Thr29	7·53	4·04	4·11	$C^{\gamma}H_3$ 1·07

† The chemical shifts, δ, are relative to external sodium 3-trimethylsilyl-[2,2,3,3-$^2H_4$]propionate. The ε-methyl resonance of Met27 was taken to be at 2·04 p.p.m. and was used as internal reference.

‡ After sequential assignments for the amino acid residues 3 to 29, the remaining resonances were assigned to His1 and Ser2 based on comparison with the corresponding random coil values (Bundi & Wüthrich, 1979*a*). The amide protons for these residues were not observed (n.o.).

glucagon residues 18 to 29. From Figure 8, the chemical shifts of the amide protons of residues 19 to 29 and at least one C^β proton of residues 18 to 29 are known, and the C^αH chemical shifts were established from the COSY spectrum. Accurate positions for the cross peaks needed to establish d_1 and d_2 connectivities could thus be predicted. In Figures 9 to 11, the predicted sequential d_1 and d_2 connectivities have been drawn into the experimental spectra.

The d_1 connectivities between NH_{i+1} and $C^\alpha H_i$ are presented in a combined COSY–NOESY connectivity diagram (Wagner *et al.*, 1981; Wagner & Wüthrich, 1982), which manifests the through-space NOE connectivities between NH_{i+1} and $C^\alpha H_i$, as well as the through-bond J-connectivities between $C^\alpha H_i$ and NH_i. A record of sequential assignments *via* d_1 then consists of a spiral-like connectivity pattern. In Figure 9, the spectral regions containing the C^αH–NH cross peaks have been combined in such a way that the connectivities are *via* a virtual diagonal, exactly analogous to the connectivities through the real diagonal peaks when the entire halves of the NOESY and COSY spectra are joined along the diagonal (Wagner *et al.*, 1981). The first step starts with the COSY peak of Thr29, from where a horizontal line leads to the virtual diagonal position of the Thr29 amide proton. A vertical line then leads to the NOESY cross peak position with C^αH of Asn28, which was predicted from the NH and C^αH chemical shifts known on the basis of the previous assignments (Fig. 8). Next, a horizontal line leads to the virtual, diagonal position of C^αH of Asn28 and a vertical line to the COSY peak of Asn28. Continuing on, we follow the d_1 connectivities for residues 29 to 23 in Figure 9 and for residues 23 to 17 in Figure 10. When comparing the predicted positions for the cross peaks with the actual spectra, one finds that the NOESY spectrum provides clear evidence only for the d_1 connectivities between residues 29 to 27, 25 to 23 and 20 to 19. The predicted cross peak positions connecting residues 26 to 25, 23 to 22, 19 to 18 and 18 to 17 do not coincide with cross peaks in the spectrum. For the connectivities 27 to 26, 22 to 21, and 21 to 20, it could not be determined unambiguously whether they are manifested by a cross peak, since they fall into crowded locations.

The d_2 connectivities are indicated in Figure 11. Since d_2 is symmetrical with respect to the direction of the polypeptide chain (Billeter *et al.*, 1982), a non-terminal residue may be connected *via* two NH–NH cross peaks with the residues preceding it and following it in the sequence. The positions for all these cross peaks predicted from the assignments in Figure 8 are connected. The first three steps

§ In the NOESY spectrum recorded in H_2O, 3 cross peaks were observed at (ω_1 = 6·83 p.p.m., ω_2 = 7·63 p.p.m.), (6·82 p.p.m., 7·50 p.p.m.) and (6·77 p.p.m., 7·45 p.p.m.), which must come from the side-chain amide groups of Gln3, Gln20, Gln24 and Asn28. They have so far not been assigned individually.

|| Two cross peaks between Arg $C^\delta H_2$ and N^ϵH were observed at (ω_1 = 3·26 p.p.m., ω_2 = 7·67 p.p.m.) and (3·26 p.p.m., 7·49 p.p.m.) in the COSY spectrum (Fig. 5). Since both $C^\gamma H_2$ and $C^\delta H_2$ have the same chemical shifts for the 2 residues, the resonances of the guanidinium groups could not be assigned individually.

¶ Two cross peaks at (ω_1 = 2·44 p.p.m., ω_2 = 2·59 p.p.m.) and (2·46 p.p.m., 2·52 p.p.m.) in the COSY spectrum recorded in ^2H_2O must correspond to the C^γ protons of Gln20 and Gln24. For these 2 residues, the connectivities between the 2 C^β protons and between C^β and C^γ protons could not be established.

382 G. WIDER, K. H. LEE AND K. WÜTHRICH

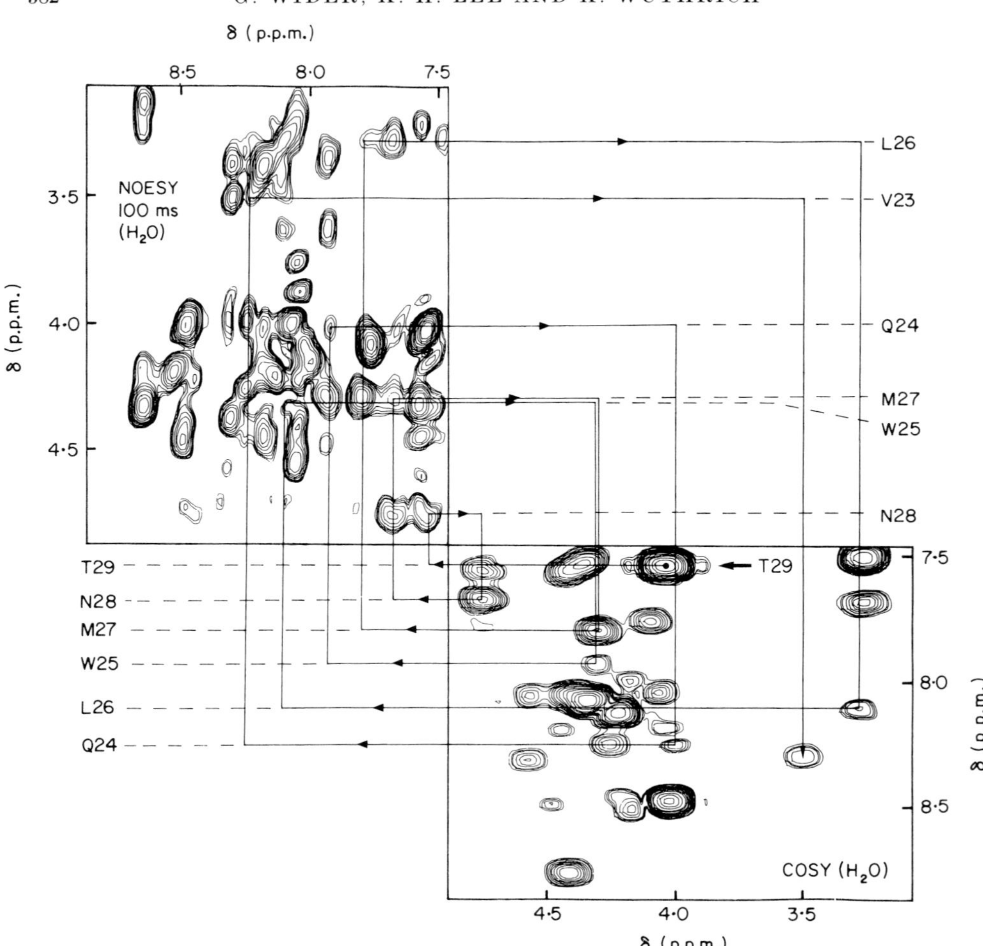

FIG. 9. Combined COSY–NOESY connectivity diagram for sequential resonance assignments *via* NOEs between amide protons and the C$^\alpha$ protons of the preceding residue (d_1) (Wagner *et al.*, 1981). In the upper left, the region (3·1 to 4·9 p.p.m.) × (7·4 to 8·9 p.p.m.) from the 360 MHz ^{1}H NOESY spectrum of micelle-bound glucagon in Fig. 7 is presented. In the lower right, the corresponding region from a COSY spectrum recorded from the same sample under identical conditions is shown. The straight lines and arrows indicate the connectivities between neighbouring residues in the segment residues 23 to 29 of micelle-bound glucagon. "←T29" identifies the starting point of the d_1 connectivity pattern. The amide proton chemical shifts are indicated by the assignments in the lower left corner, those for the C$^\alpha$ protons by the assignments in the upper right corner of the Figure.

from Thr29 to Leu26 are easy to follow, since they lead in each case to a lower field position. Leu26 connects to Trp25, which is at higher field. In these somewhat more complex cases, the connectivities to the diagonal peaks are indicated with broken lines. Continuing on, the connectivity pattern ends at Arg18. Comparison of the predicted peak positions with the experiment shows that the two are fully compatible, since none of the predicted cross peaks falls into an empty region of the spectrum. On the other hand, because of the small chemical shift differences

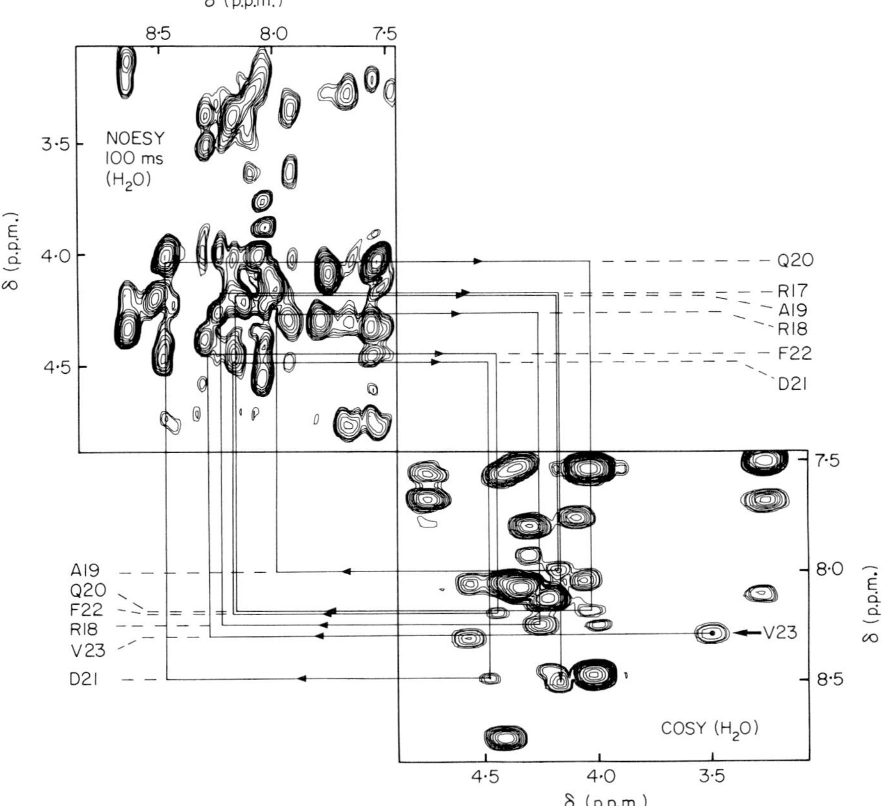

FIG. 10. Combined COSY–NOESY connectivity diagram for sequential resonance assignments *via* NOEs between amide protons and the C^α protons of the preceding residues (d_1) (same spectra as in Fig. 9). The straight lines and arrows indicate the connectivities between neighbouring residues in the segment residues 17 to 23 of micelle-bound glucagon. "←V23" identifies the starting point of the d_1 connectivity pattern. The amide proton chemical shifts are indicated by the assignments in the lower left corner, those for the C^α protons by the assignments in the upper right corner of the Figure.

between the amide protons of the different residues, most of the predicted NOE cross peak locations are so near the diagonal that it can hardly be seen whether there is indeed a cross peak. Overall, it appears that in the segment residues 18 to 29 only the d_2 connectivity between Met27 and Leu26 could have been established unambiguously from this spectrum.

The comparison of the d_1, d_2 and d_3 connectivities in glucagon residues 18 to 29 provides a valuable complementation of the theoretical and statistical evaluations of the reliability of sequential assignments *via* the different connectivities (Billeter *et al.*, 1982). It is seen that an inherently highly reliable parameter, such as d_2 in the present example, may be virtually uninformative, depending on the spectral characteristics of the protein. The intrinsically least reliable d_3 connectivity may thus become a very valuable parameter in many actual systems.

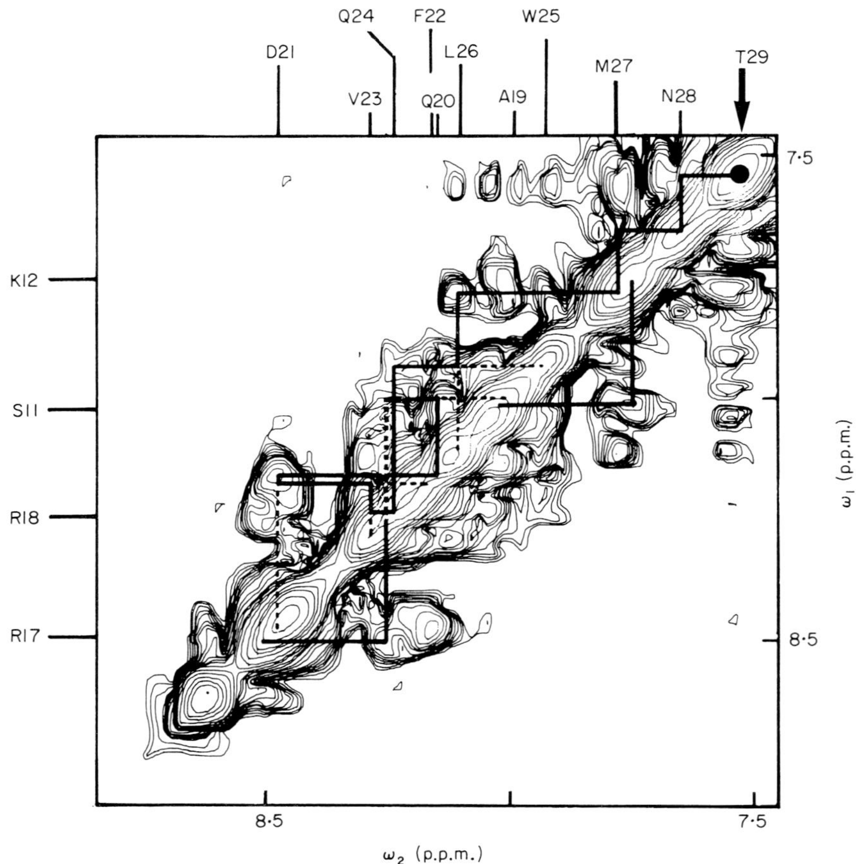

FIG. 11. Region from 7·4 to 8·8 p.p.m. of the 360 MHz ^{1}H NOESY spectrum of micelle-bound glucagon shown in Fig. 7, after symmetrization (Baumann *et al.*, 1981). In this spectral region, the diagonal peaks include all the backbone amide proton resonances, and the cross peaks manifest NOEs between different amide protons. The straight continuous and broken lines in the upper left triangle indicate the NOEs between the amide protons of neighbouring residues (d_2) in the polypeptide residues 18 to 29. In the lower right triangle, the sequential assignments obtained for the segment residues 11 to 12 and 17 to 18 are indicated. The chemical shifts for the amide protons in the segment residues 19 to 29 are indicated at the top of the Figure, those for the 2 short segments on the left margin.

(d) *Individual assignments for glucagon residues 1 to 18*

As outlined in Figure 6, the major part of the sequential assignments in this region were obtained *via* d_1 connectivities, with some missing links filled by d_2 or d_3 connections. The results are documented in Figures 11 to 13 in diagrams similar to those used for glucagon residues 29 to 18. Adding to this previously assigned segment, we first have the d_2 connectivity Arg18–Arg17 in Figure 11. Next we have d_1 connectivities from Arg17 to Leu14, which also resolved the ambiguity in the assignments of Arg18 and Arg17 (see Fig. 8). Connectivities within the tripeptide

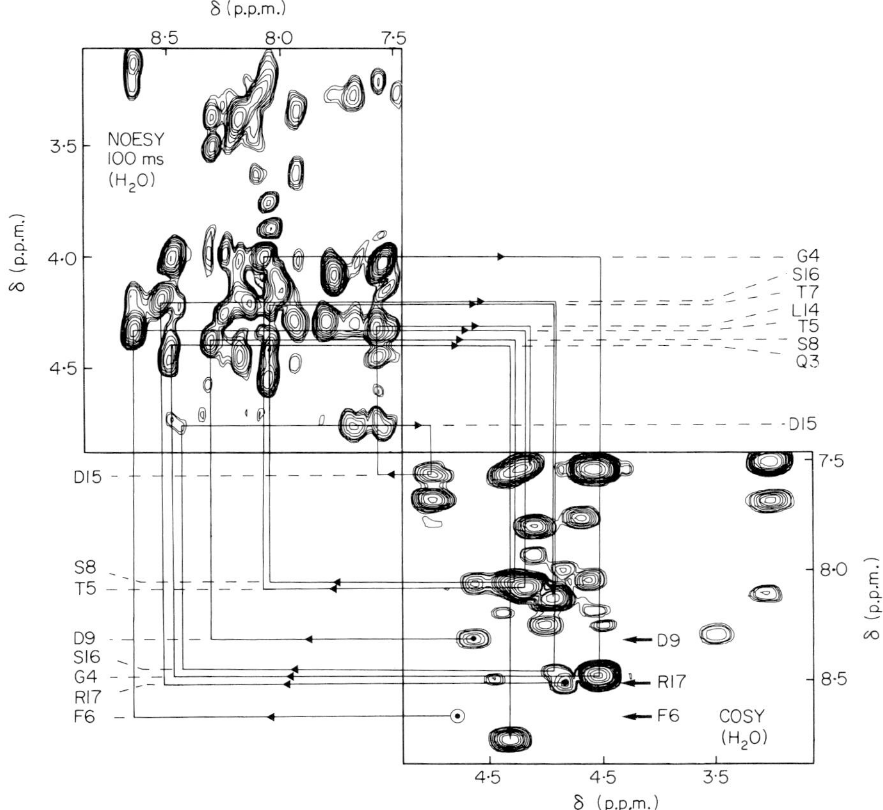

FIG. 12. Combined COSY-NOESY connectivity diagram for sequential resonance assignments *via* NOEs between amide protons and the C^{α} protons of the preceding residues (d_1) (same spectra as in Fig. 9). The straight lines and arrows indicate the sequential resonance assignments obtained for the segment residues 3 to 6, 7 to 9 and 14 to 17. At 37°C, the C^{α}H–NH COSY peak of Phe6 coincides with the water resonance and was therefore bleached out by the solvent irradiation. Its location (⊙) was taken from the data in Figs 2 and 13.

Tyr13-Lys12-Ser11 were obtained *via* d_3 (Fig. 13) and *via* d_2 (Fig. 11). A d_3 connection from Tyr10 to Asp9 (Fig. 13) is followed by d_1 connectivities from Asp9 to Tyr7 and from Phe6 to Gln3 (Fig. 12), with a d_3 link between residues 7 and 6 (Fig. 13). Finally, the two N-terminal residues were assigned on the basis of the spin system identifications (Fig. 3) and comparison with the random coil chemical shifts for these residues (Bundi & Wüthrich, 1979*a*). The imidazole ring proton lines of His1 were identified independently from their chemical shifts (Bösch *et al.*, 1980; Table 2).

4. Conclusions

The assignment of the ^{1}H n.m.r. spectrum of micelle-bound glucagon presented in this paper is almost complete (Table 2). All the non-labile protons were observed

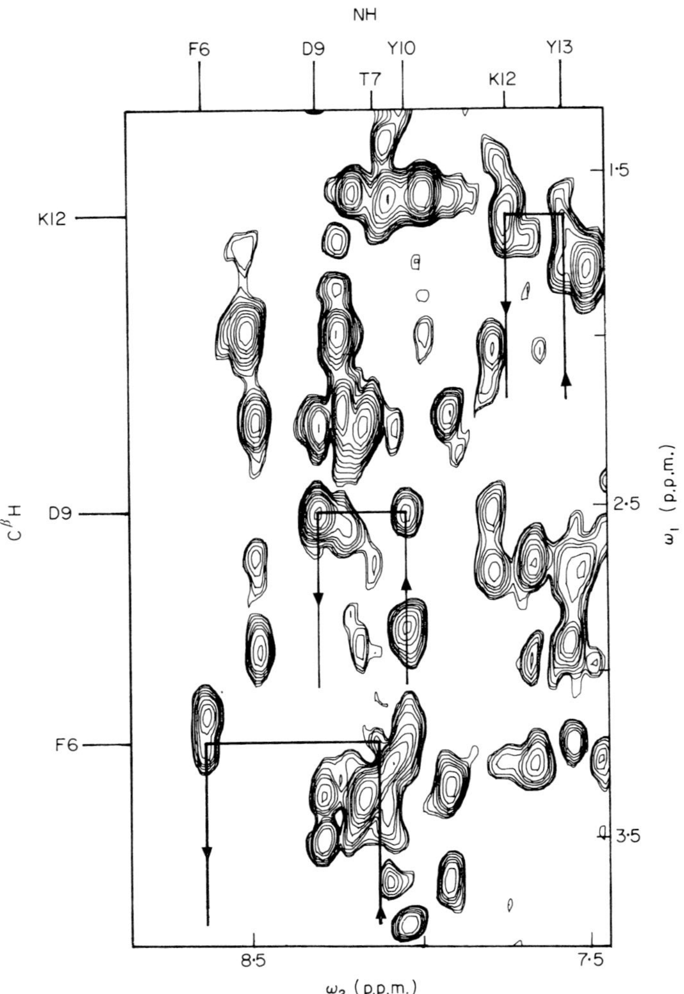

FIG. 13. Spectral region (1·3 to 3·8 p.p.m.) × (7·4 to 8·8 p.p.m.) of the 360 MHz ^1H NOESY spectrum of micelle-bound glucagon shown in Fig. 7. The continuous lines with arrows indicate the sequential assignments for the polypeptide segment residues 6 to 7, 9 to 10 and 12 to 13, which were obtained from NOEs between amide protons and the C^β protons of the preceding residues (d_3). The numbers at the top of the Figure indicate the amide proton chemical shifts of the corresponding residues, those on the left margin indicate the chemical shifts of one C^β proton for each residue.

and assigned, with the exception only of the C^γ protons of Gln20 and Gln24. These two residues have nearly identical $C^\alpha H$ and $C^\beta H$ chemical shifts, which was probably the main reason for failure to establish the connectivities to $C^\gamma H_2$. Among the labile protons, those of the two arginine side-chains were observed (Fig. 5), but could not be assigned individually because of the degeneracy of the $C^\delta H_2$ resonances. Three NH–NH NOE cross peaks were identified as originating from side-chain amide groups of the single asparagine and the three glutamine residues

(Table 2). So far, no unambiguous individual assignment on the basis of NOE connectivities with $C^{\beta}H_2$ in Asn or $C^{\gamma}H_2$ in Gln (Billeter *et al.*, 1982) could be obtained. The labile backbone protons of His1 and Ser2 and the labile side-chain protons of His1, Lys12 and all the serine, threonine and tyrosine residues were not observed, presumably because they exchange too rapidly with the solvent to be seen as separate lines (Wüthrich, 1976; Bundi & Wüthrich, 1979*a*,*b*).

Previously, a three-dimensional structure for residues 19 to 27 of micelle-bound glucagon was determined on the basis of intramolecular distance constraints between a small number of assigned resonances, which were measured by NOE experiments and analysed with the use of a distance geometry algorithm (Braun *et al.*, 1981). The new assignments described here provide the basis for a much more accurate characterization of the polypeptide conformation. In the earlier work, only very few of the observed NOEs in ^2H_2O solution (Bösch *et al.*, 1981) could be used for the structure determination, since the others had not been assigned (Braun *et al.*, 1981). For the new structure determination, essentially all the previously reported NOEs (Bösch *et al.*, 1981) and, in addition, all those with the backbone amide protons 3 to 29 will be assigned to specific pairs of closely spaced protons. The work on the determination of the three-dimensional structure is in progress.

We thank Drs L. R. Brown and A. Chrzeszczyk for help with the synthesis of perdeuterated dodecylphosphocholine, Dr W. Braun, M. Billeter and Dr G. Wagner for stimulating discussions on sequential resonance assignments, Mrs E. Huber for the careful preparation of the manuscript, Mrs E. H. Hunziker for some of the illustrations, and the Schweizerischer Nationalfonds for financial support (project 3.528.79).

REFERENCES

Anil Kumar, Ernst, R. R. & Wüthrich, K. (1980*a*). *Biochem. Biophys. Res. Commun.* **95**, 1–6.

Anil Kumar, Wagner, G., Ernst, R. R. & Wüthrich, K. (1980*b*). *Biochem. Biophys. Res. Commun.* **96**, 1156–1163.

Aue, W. P., Bartholdi, E. & Ernst, R. R. (1976). *J. Chem. Phys.* **64**, 2229–2246.

Baumann, R., Wider, G., Ernst, R. R. & Wüthrich, K. (1981). *J. Magn. Reson.* **44**, 402–406.

Billeter, M., Braun, W. & Wüthrich, K. (1982). *J. Mol. Biol.* **155**, 321–346.

Bösch, C., Bundi, A., Oppliger, M. & Wüthrich, K. (1978). *Eur. J. Biochem.* **91**, 209–214.

Bösch, C., Brown, L. R. & Wüthrich, K. (1980). *Biochim. Biophys. Acta*, **603**, 298–312.

Bösch, C., Anil Kumar, Baumann, R., Ernst, R. R. & Wüthrich, K. (1981). *J. Magn. Reson.* **42**, 159–163.

Braun, W., Bösch, C., Brown, L. R., Go, N. & Wüthrich, K. (1981). *Biochim. Biophys. Acta*, **667**, 377–396.

Brown, L. R. (1979). *Biochim. Biophys. Acta*, **557**, 135–148.

Brown, L. R., Bösch, C. & Wüthrich, K. (1981). *Biochim. Biophys. Acta*, **642**, 296–312.

Bundi, A. & Wüthrich, K. (1979*a*) *Biopolymers*, **18**, 285–298.

Bundi, A. & Wüthrich, K. (1979*b*). *Biopolymers*, **18**, 299–312.

Dubs, A., Wagner, G. & Wüthrich, K. (1979). *Biochim. Biophys. Acta*, **577**, 177–194.

Jeener, J., Meier, B. H., Bachmann, P. & Ernst, R. R. (1979). *J. Chem. Phys.* **71**, 4546–4553.

Kalinichenko, P. (1976). *Stud. Biophys.* **58**, 235–240.

Macura, S., Huang, Y., Suter, D. & Ernst, R. R. (1981). *J. Magn. Reson.* **43**, 259–281.

Nagayama, K. & Wüthrich, K. (1981). *Eur. J. Biochem.* **114**, 365–374.

Nagayama, K., Wüthrich, K. & Ernst, R. R. (1979). *Biochem. Biophys. Res. Commun.* **90**, 305–311.

Nagayama, K., Anil Kumar, Wüthrich, K. & Ernst, R. R. (1980). *J. Magn. Res.* **40**, 321–334.
Panijpan, B. & Gratzer, W. B. (1974). *Eur. J. Biochem.* **45**, 547–553.
Pohl, S. L., Birnbaumer, L. & Rodbell, M. (1969). *Science*, **164**, 566–569.
Rodbell, M., Birnbaumer, L., Pohl, S. L. & Sundby, F. (1971). *Proc. Nat. Acad. Sci., U.S.A.* **68**, 909–913.
Rubalcava, B. & Rodbell, M. (1973). *J. Biol. Chem.* **248**, 3831–3837.
Sasaki, K., Dockerill, S., Adamiak, D. A., Tickle, I. J. & Blundell, T. (1975). *Nature (London)*, **257**, 751–757.
Wagner, G. & Wüthrich, K. (1982). *J. Mol. Biol.* **155**, 347–366.
Wagner, G., Anil Kumar & Wüthrich, K. (1981). *Eur. J. Biochem.* **114**, 375–384.
Wüthrich, K. (1976). *NMR in Biological Research: Peptides and Proteins*, North-Holland Publishing Company, Amsterdam.
Wüthrich, K., Wider, G., Wagner, G. & Braun, W. (1982). *J. Mol. Biol.* **155**, 311–319.

Edited by S. Brenner

Eur. J. Biochem. *143*, 659 – 667 (1984)
© FEBS 1984

^{113}Cd-^{1}H spin-spin couplings in homonuclear ^{1}H correlated spectroscopy of metallothionein

Identification of the cysteine ^{1}H spin systems

David NEUHAUS, Gerhard WAGNER, Milan VAŠÁK, Jeremias H. R. KÄGI, and Kurt WÜTHRICH

Institut für Molekularbiologie und Biophysik, Eidgenössische Technische Hochschule, Zürich; and Biochemisches Institut der Universität Zürich, Zürich

(Received April 19, 1984) – EJB 84 0419

Heteronuclear spin-spin couplings between ^{113}Cd and C^{β} protons of the metal-bound cysteines were observed in phase-sensitive, double-quantum filtered, homonuclear two-dimensional correlated (COSY) ^{1}H NMR spectra of ^{113}Cd-metallothionein-2 from rabbit liver. Comparison of ^{113}Cd- and ^{112}Cd-metallothionein-2 spectra revealed that 19 ^{1}H spin systems show heteronuclear couplings to at least one ^{113}Cd and were thus identified as 19 of the 20 cysteines in this protein. From a detailed analysis of the manifestations of heteronuclear couplings in the homonuclear ^{1}H COSY spectra, two cysteines could be identified as 'bridging cysteines', with spin-spin couplings to two different ^{113}Cd nuclei. The observed ^{113}Cd-^{1}H coupling constants vary between ≤5 Hz and 80 Hz.

Mammalian metallothioneins consist of a single polypeptide chain with 61 amino acid residues. 20 residues are cysteines, which are involved in the complexation of 7 diamagnetic metal ions, such as Zn^{2+}, Cd^{2+}, Cu^{+} or Hg^{2+} [1]. Metallothionein has been much studied using a variety of physicochemical techniques [2], but so far the investigations of structure-function correlations have had to be carried out in ignorance of the three-dimensional structure. In the absence of crystallographic data nuclear magnetic resonance (NMR) has already provided important clues to various structural features. In particular, observation of ^{113}Cd-^{113}Cd spin-spin couplings indicated the presence of two distinct 'metal thiolate clusters' [3,5]. Since only cysteine residues are involved in metal binding, a partition into terminal (60%) and bridging (40%) cysteines was suggested [2 – 5]. The present paper describes some new NMR results, obtained during a study intended to determine the conformation of the polypeptide chain of metallothionein and the location of the metal ions in the three-dimensional structure.

Primarily the structure determination follows the previously outlined strategy in which sequence-specific proton resonance assignments for the polypeptide chain are first established and then a network of short ^{1}H-^{1}H distances is measured, from which the spatial structure may be derived [6 – 9]. However, in contrast to other proteins the possibility of incorporating $^{113}Cd^{2+}$ ions into metallothionein makes accessible additional sources of information from NMR. Here, we use the observation of ^{113}Cd-^{1}H spin-spin couplings for identification of the spin systems of the metal-bound cysteines. It is to be expected that these data will also be crucial for the elucidation of the three-dimensional protein structure in solution.

Abbreviations. NMR, nuclear magnetic resonance; *J*, spin-spin coupling constants in Hz; COSY, two-dimensional correlated spectroscopy; MT-2, metallothionein isoprotein 2.

MATERIALS AND METHODS

Rabbit liver metallothionein (MT) was isolated from rabbits injected subcutaneously 15 times in 2 – 3-day intervals with 1 mg Cd^{2+}/kg body weight as $CdCl_2$ [10]. The protein was purified by a procedure adapted from Kägi et al. [11] and Kimura et al. [10], resulting in the separation of the two electrophoretically different forms MT-1 and MT-2. The latter was further purified by additional ion-exchange chromatography on CM Bio-Gel A (Bio-Rad) in 20 mM Tris/HCl (pH 8.6) using a linear gradient with 40 mM Tris/HCl (pH 8.6) as limiting buffer. The purity of each preparation was documented by amino acid analysis (Durrum 500) and by metal analysis using atomic absorption spectrometry (Instrumentation Laboratory, Model IL 157). The protein concentration was determined spectrophotometrically by measuring the absorbance of the apoprotein at 220 nm in 0.01 M HCl (ε_{220} = 47300 M^{-1} cm^{-1}) [12].

Cd^{2+}-MT-2 was prepared either by exchange of Zn^{2+} in $(Cd,Zn)_7$-MT-2 in the presence of excess Cd^{2+} or by reconstitution from apoprotein with appropriate amounts of Cd^{2+} [13]. Both procedures yield ^{1}H COSY patterns in $^{2}H_2O$ and H_2O, which are closely similar to each other and to that obtained with native Zn_7-MT-2 (Vašák, M., Wagner, G., Kägi, J. H. R., and Wüthrich, K., unpublished work). ^{113}Cd-enriched and ^{112}Cd-enriched MT were prepared by the reconstitution procedure [13]. $^{113}Cd^{2+}$ and $^{112}Cd^{2+}$ (95% enriched) salts were products of Harwell. Apo-MT was obtained by dialyzing native MT against three changes of 0.1 M HCl using Spectrapor Co. dialyzing tube with a molecular mass cut-off of 3500 Da.

Prior to the NMR measurements the Tris/HCl buffer in the MT-2 samples was replaced by 20 mM [$^{2}H_{11}$]Tris/HCl (Merck und Sharp) and 20 mM KCl in H_2O (pH 7.0), using an Aminco ultrafiltration apparatus (YM-2 membrane). The protein concentration was adjusted at 8 – 12 mM. For ex-

660

Fig. 1. *Structure of cysteine bound to* ^{112}*Cd (A) or* ^{113}*Cd (B)*. The cysteine sulfur is deprotonated. The atoms are identified following the IUPAC-IUB convention [17]. In the NMR spectra, where no stereospecific assignments to the methylene protons $H^{\beta 2}$ and $H^{\beta 3}$ have been obtained, the $C^{\beta}H_2$ resonances are identified as $H^{\beta a}$ and $H^{\beta b}$, whereby $H^{\beta a}$ is always the proton with the larger homonuclear spin-spin coupling $^3J_{\alpha\beta}$

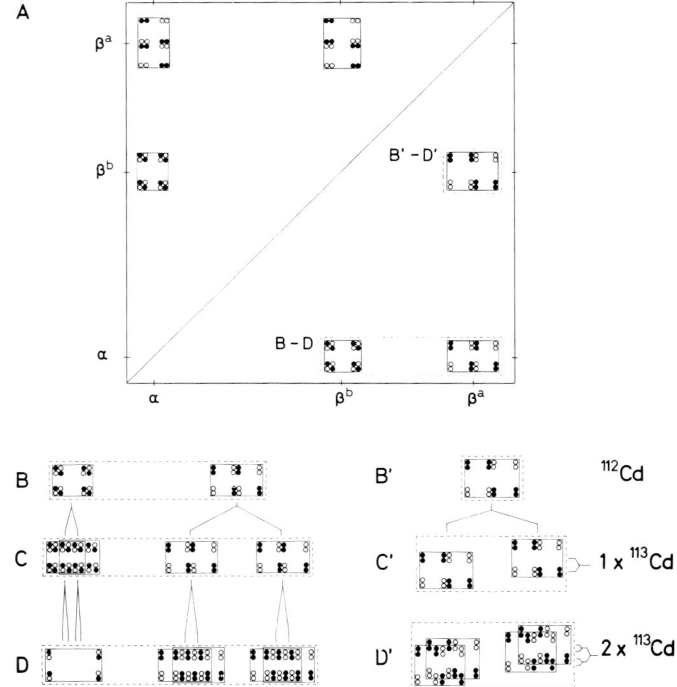

Fig. 2. *Schematic representation of the influence of heteronuclear* ^1H-^{113}Cd *J-couplings on the cross-peaks of a cysteine spin system in a* ^1H *COSY spectrum*. (A) Location and appearance of cross-peaks in ^{112}Cd-MT-2. No diagonal peaks are drawn. All components belonging to one multiplet are contained within a solid frame. Positive (●) and negative (○) multiplet components are indicated. The chemical shifts of $C^{\alpha}H$, $H^{\beta a}$ and $H^{\beta b}$ are indicated with α, β^a and β^b respectively. The 'active coupling', which gives rise to the cross-peak, is manifested as an anti-phase splitting, whereas all 'passive couplings' are manifested as in-phase splittings. Depending on the size of the different coupling constants, cancellation of some multiplet components may occur when positive and negative components fall on top of each other. The two $C^{\alpha}H$-$C^{\beta}H$ cross-peaks and the $H^{\beta a}$ $H^{\beta b}$ cross-peak in the broken frames are shown again in B – D and B' – D' respectively. B, B': no heteronuclear coupling (^{112}Cd-MT-2). C, C': heteronuclear coupling to one ^{113}Cd spin, with a large coupling constant $^3J_{^{113}CdH^{\beta a}}$ and a small $^3J_{^{113}CdH^{\beta b}}$ D, D': heteronuclear coupling to two ^{113}Cd spins. In the situation shown the couplings lead to an almost complete cancellation of the H^{α}-$H^{\beta b}$ cross-peak. All four heteronuclear coupling constants can readily be obtained, however, from the $H^{\beta a}$-$H^{\beta b}$ cross-peak

periments in ^2H_2O the labile protons in MT were exchanged by twice freeze drying and redissolving the material in ^2H_2O. All samples were sealed under argon to avoid oxidation of cysteine sulfur. The pH values given refer to that of the buffered sample in H_2O.

Double-quantum-filtered COSY spectra were recorded on a Bruker WM 500 spectrometer as described previously [14 – 16]. Quadrature detection was used in both dimensions, and the carrier was placed in the center of the spectrum. 512 experiments with t_1 values from 1 μs to 99 ms were recorded, each free induction decay consisting of 2048 data points with an acquisition time of 0.4 s. Prior to Fourier transformation the time-domain data were multiplied with phase-shifted sine bell window functions, with phase shifts of $\pi/4$ and $\pi/8$ in the ω_1 and ω_2 directions respectively; then the time-domain data were extended to twice the original size by zero-filling in both

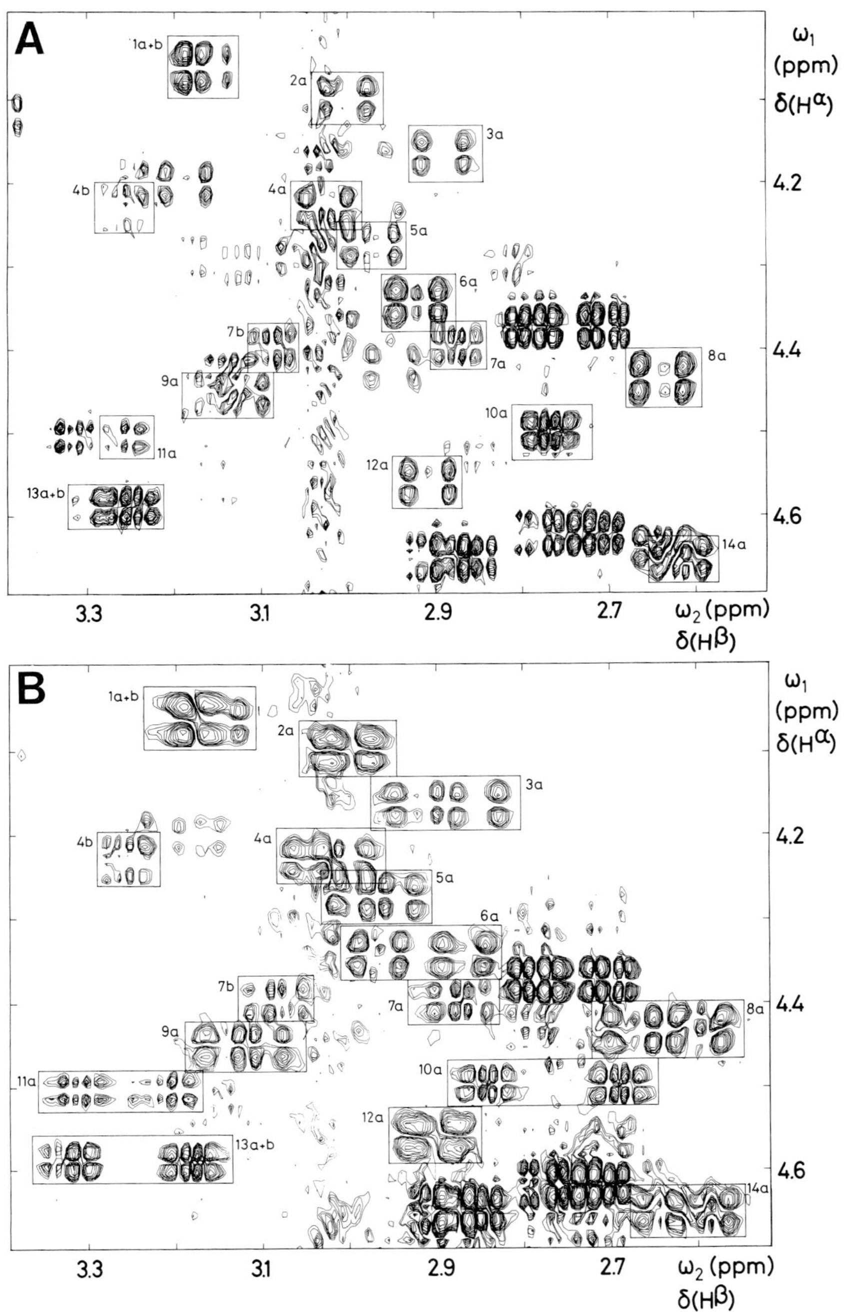

Fig. 3. *Illustration of the effects of* J-*couplings between cysteine* C^{β} *protons and* ^{113}Cd *in* ^{1}H *COSY spectra of MT-2.* Phase-sensitive, double-quantum-filtered COSY spectra are shown, and both positive and negative levels are plotted. The ^{1}H-^{113}Cd couplings can be detected from a comparison of the $C^{\alpha}H$-$C^{\beta}H$ cross-peak patterns in COSY spectra of MT-2 reconstituted with ^{112}Cd (A) and ^{113}Cd (B) respectively. The spectral region ($\omega_1 = 4.0 - 4.7$ ppm, $\omega_2 = 2.4 - 3.3$ ppm) is shown, which contains $C^{\alpha}H$-$C^{\beta}H$ cross-peaks of 14 cysteinyl residues. The chemical shifts of the C^{α} protons can be read from the ω_1 axis, those of the C^{β} protons from the ω_2 axis. In spectrum B coupling of cysteine C^{β} protons to ^{113}Cd results in a splitting or broadening along ω_2 of the multiplet patterns, which originate from ^{1}H-^{1}H coupling (spectrum A). $C^{\alpha}H$-$C^{\beta}H$ cross-peaks from 14 cysteines are framed and labelled 1 – 14 in the order of increasing C^{α} proton chemical shift. For a specified spin system, a and b indicate the cross-peaks of the C^{β} proton with the larger and smaller homonuclear coupling $^{3}J_{\alpha\beta}$ respectively. In some cases only the cross-peak a was observed, apparently because $^{3}J_{\alpha\beta b}$ is very small. It should be remembered when examining these spectra (and also those of Fig. 5 below) that the distinction between positive and negative levels (see Fig. 2), vital in identifying cross-peaks in crowded areas and originally made clear by plotting in different colours, has had to be sacrificed when preparing the black and white figures for publication

the t_1 and t_2 directions, and a phase-sensitive Fourier transformation was carried out. The digital resolution in the ω_1 and ω_2 directions was 5 Hz/point and 1.3 Hz/point respectively. All chemical shifts are quoted relative to the internal standard sodium 3-trimethylsilyl-(2,2,3,3-$^2H_4$)propionate.

FUNDAMENTAL CONSIDERATIONS AND RESULTS

The identification of ^{113}Cd-^{1}H spin-spin couplings in the ^{1}H COSY spectra of metallothionein relies on a comparison of spectra from different samples, prepared with isotopically enriched ^{112}Cd or ^{113}Cd respectively (Fig. 1). ^{112}Cd has a zero nuclear spin and, therefore, does not affect the fine structure of the proton resonances. In contrast, ^{113}Cd has a nuclear spin of 1/2, so that J-coupling between the β protons in metal-bound cysteines and a single ^{113}Cd nucleus can cause a doublet splitting of the H^{β} resonances if a suitable pathway exists to transmit such ^{113}Cd-^{1}H spin-spin coupling. As with many other types of scalar spin-spin coupling, it seems that a pathway of not more than three chemical bonds is required for appreciable ^{113}Cd-^{1}H spin-spin coupling in MT-2, so that detectable splittings (≥ 5 Hz, see below) are only observed for the β protons of the metal-bound cysteines (Fig. 1). The splitting patterns depend on the number of cadmium ions involved; a single ^{113}Cd coupling partner produces a simple doublet pattern in the β proton resonances, whereas a β proton coupled to two ^{113}Cd ions should show a doublet of doublets or a triplet pattern, depending on the sizes of the two ^{113}Cd-^{1}H coupling constants [18]. In the following the manifestation of ^{113}Cd-^{1}H spin-spin couplings in ^{1}H COSY spectra is discussed in more detail and is illustrated with the data obtained for metallothionein.

In order to analyse the ^{113}Cd-^{1}H couplings in detail it is necessary first to consider the features expected in a ^{1}H COSY spectrum for a cysteine in the absence of ^{113}Cd-^{1}H coupling, i.e. from a sample of ^{112}Cd-MT-2. Provided that the H^{β} chemical shifts are not degenerate, each cysteine gives rise, on each side of the diagonal, to a pattern of two (H^{α}, H^{β}) cross-peaks and one ($H^{\beta a}$, $H^{\beta b}$) cross peak (Fig. 2A); degeneracy of the H^{β} chemical shifts would cause the ($H^{\beta a}$, $H^{\beta b}$) cross-peaks to merge with the diagonal, while the (H^{α}, H^{β}) cross-peaks would merge with each other. The multiplet structures of the cross-peaks are due entirely to the homonuclear couplings between H^{α}, $H^{\beta a}$ and $H^{\beta b}$. These may be analysed, at least in first-order cases, by a two-dimensional analogue of the familiar method of 'successive splittings' [18,19]. For a given COSY cross-peak, the 'active' coupling, that is the coupling between those two nuclei connected by the cross-peak, gives rise to an anti-phase doublet, both in the ω_1 and ω_2 directions, appearing in the two-dimensional contour plots as a quadratic array of two positive and two negative peaks (Fig. 2B). Further, 'passive', couplings with the third proton then split each of these four peaks into in-phase components. Examples of cross peaks of the general type of Fig. 2B are displayed in Fig. 3A, which shows a region from below the diagonal in a phase-sensitive, double-quantum-filtered COSY spectrum of ^{112}Cd-MT-2.

The effects of further spin-spin coupling with one or two ^{113}Cd nuclei on the (H^{α}, H^{β}) COSY cross-peaks are schematically drawn in Fig. 2, C and D respectively. It is seen that these additional 'passive' couplings cause further in-phase splitting along the β proton shift dimension, i.e. ω_2 for the cross-peaks below the diagonal. Application of the method of 'successive splittings' readily yields the magnitude of $^3J_{^{113}CdH^{\beta}}$ for each β proton (Fig. 2, C and D). Examples of cross-peaks of the general type of Fig. 2C are shown in Fig. 3B. The spectrum of ^{113}Cd-MT-2 contains no (H^{α}, H^{β}) cross-peak, which would present direct evidence for a bridging cysteine with couplings to two ^{113}Cd spins (Fig. 2D). Such evidence for bridging cysteines has been obtained only from ($H^{\beta a}$, $H^{\beta b}$) cross-peaks (see below).

All the cysteine (H^{α}, H^{β}) cross-peaks identified in the spectral region shown in Fig. 3 are marked using frames and are numbered from 1 to 14 according to increasing H^{α} chemical shift. The (H^{α}, H^{β}) cross-peaks of the remaining five identified cysteines fall outside the spectral region of Fig. 3 and are shown separately in Fig. 4. The 20th cysteine could not be identified on the basis of J couplings to ^{113}Cd. For each cysteine in Fig. 3 and 4, the (H^{α}, H^{β}) cross-peak having the larger active coupling constant is labelled 'a', while the other cross-peak is labelled 'b'. Some noncysteine residues (e.g. Asp and Asn) also give rise to cross-peaks within the region of Fig. 3, but as these will not be discussed here they are not identified. The appearance of the individual cross-peaks in ^{113}Cd-MT-2, at least in first-order cases, varies predictably with the magnitude of $^3J_{^{113}CdH^{\beta}}$. With the presently used digital resolution in ω_2, the method detects couplings down to a lower limit of about 5 Hz. Small couplings appear to broaden the original ^{112}Cd-MT-2 cross-peak patterns (e.g. peaks 2a and 12a), and in such cases the number of ^{113}Cd coupling partners cannot be determined. Intermediate couplings, from about 10 to about 30 Hz, produce more complex, wider patterns in which the total multiplet structure can be analysed to yield $^3J_{\alpha\beta}$, $^2J_{\beta\beta}$ and $^3J_{^{113}CdH^{\beta}}$ (e.g. peaks 5a and 16a). Couplings larger than approximately 30 Hz produce ^{113}Cd satellites, which do not overlap, so that the original ^{112}Cd multiplet structure is preserved in the two, separately resolved, cross-peaks from ^{113}Cd-MT-2 (e.g. peaks 10a and 19a).

For those cysteines in which the two β proton shifts are nearly degenerate, the first-order analysis (Fig. 2) breaks down, and the effects of the spin-spin couplings between ^{113}Cd and both C^{β} protons are then exerted on the combined multiplet due to both $H^{\beta a}$ and $H^{\beta b}$. Although an exact analysis of these strongly coupled systems would require spectral simulations, just as in one-dimensional spectra [18, 19], some key features at least allow such situations to be distinguished from first-order cases. The most obvious is asymmetry within the multiplet structure of the cross-peak (e.g. peaks 1a + b and 13a + b); such second-order systems would only be expected to give a symmetrical multiplet structure if $^3J_{\alpha\beta^a} = {}^3J_{\alpha\beta^b}$ and $^3J_{^{113}CdH^{\beta a}} = {}^3J_{^{113}CdH^{\beta b}}$, which is a very unlikely circumstance. Peak 18a + b illustrates a second-order pattern which, in ^{112}Cd-MT-2, is nearly symmetrical since $^3J_{\alpha\beta^a} \approx {}^3J_{\alpha\beta^b}$, but which becomes unsymmetrical in ^{113}Cd-MT-2 because $^3J_{^{113}CdH^{\beta a}} \neq {}^3J_{^{113}CdH^{\beta b}}$. [It may be added that, in principle, a spin-system might be first-order in ^{112}Cd-MT-2 but second-order in ^{113}Cd-MT-2, if one of the (H^{α}, H^{β}) cross-peaks experiences a sufficiently large splitting from its coupling to ^{113}Cd, this may cause one of the resulting satellite peaks to be close enough to a satellite of the order (H^{α}, H^{β}) cross-peak for second-order effects to become apparent. There appear to be no examples of this behaviour in the MT-2 spectra, however.]

In the spectra of Fig. 3 and 4 several cysteine (H^{α}, H^{β}) cross-peaks are very weak or absent, since the cross-peak intensity diminishes rapidly when the value of J for the active coupling becomes comparable to or smaller than the line-width. As one can readily anticipate from Fig. 2B, loss of intensity results because the anti-phase components mutually cancel in this situation. In MT-2 the β^b proton chemical shift

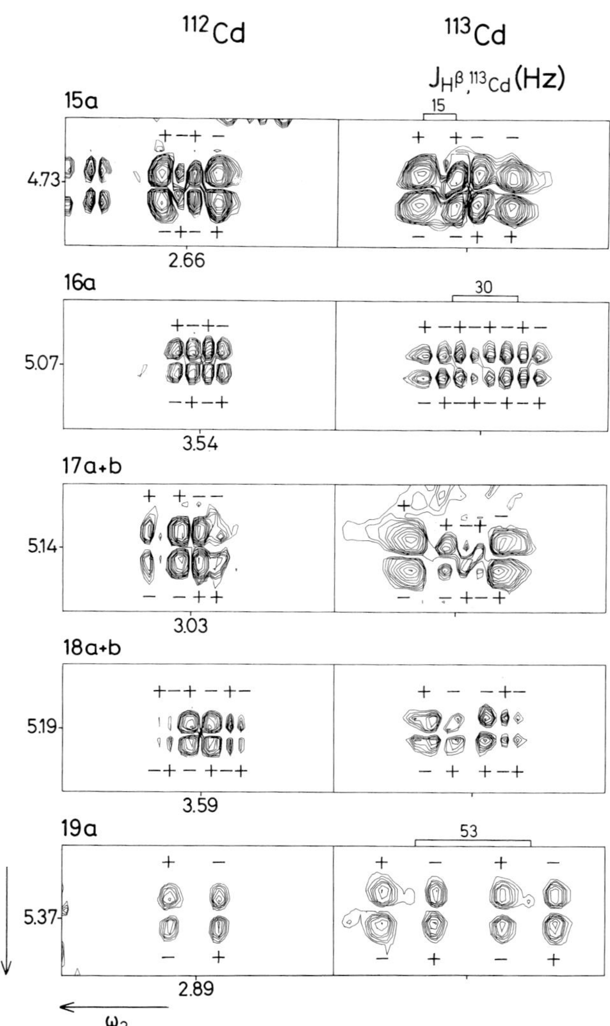

Fig. 4. *Enlarged plots of the $C^{\alpha}H$-$C^{\beta}H$ cross-peaks of five cysteine residues, labelled 15 – 19, which are outside the spectral range of Fig. 3.* Left: ^{112}Cd-MT-2, right ^{113}Cd-MT-2. + and − indicate positive and negative intensities, respectively, of the multiplet components. The ^{113}Cd doublet splittings along ω_2 are indicated for the cross peaks 15a, 16a and 19a. For the strong coupling patterns 17a + b and 18a + b, coupling constants cannot readily be indicated in the figure. Estimates of the values are given in Table 1

could in all such cases be determined using the strong ($H^{\beta a}$, $H^{\beta b}$) cross-peaks (Fig. 2A). These fall in a more crowded spectral region and are more complex than the (H^{α}, H^{β}) cross-peaks (Fig. 2, B′ – D′). On the other hand, in practice no second-order effects need to be considered, since only first-order ($H^{\beta a}$, $H^{\beta b}$) cross-peaks are sufficiently removed from the diagonal to be observed. The relevant regions of phase-sensitive double-quantum-filtered COSY spectra of ^{112}Cd-MT-2 and ^{113}Cd-MT-2 are shown in Fig. 5A and B respectively, and the enumeration of individual cross-peaks corresponds to that used in Fig. 3 and 4. The general principles of the method of 'successive splittings', given above, apply equally well to an ($H^{\beta a}$, $H^{\beta b}$) cross-peak, but one further point needs consideration when both $H^{\beta a}$ and $H^{\beta b}$ are coupled to ^{113}Cd; such situations are depicted schematically in Fig. 2C′ and D′. The initially surprising feature in Fig. 2C′ is that the ($H^{\beta a}$, $H^{\beta b}$) cross-peak is split into only two components, rather than the four, which would have resulted had these 'passive' couplings been to another proton. The origin of this difference is that, unlike the protons, the ^{113}Cd nuclei experience no mixing pulse during the COSY pulse sequence. The two components detected thus correspond to the two possible spin-orientations of the ^{113}Cd nucleus, while the two 'missing' components correspond to impossible situations in which $H^{\beta a}$ would see the ^{113}Cd spin in one orientation at the same time as $H^{\beta b}$ sees it in the opposite one. Had a 90° mixing pulse been applied to the ^{113}Cd spins, this would have equalised the intensities of all four components, as in the homonuclear coupling case. The extension of this reasoning to the case of a bridging cysteine, where either or both of $H^{\beta a}$ and $H^{\beta b}$ may

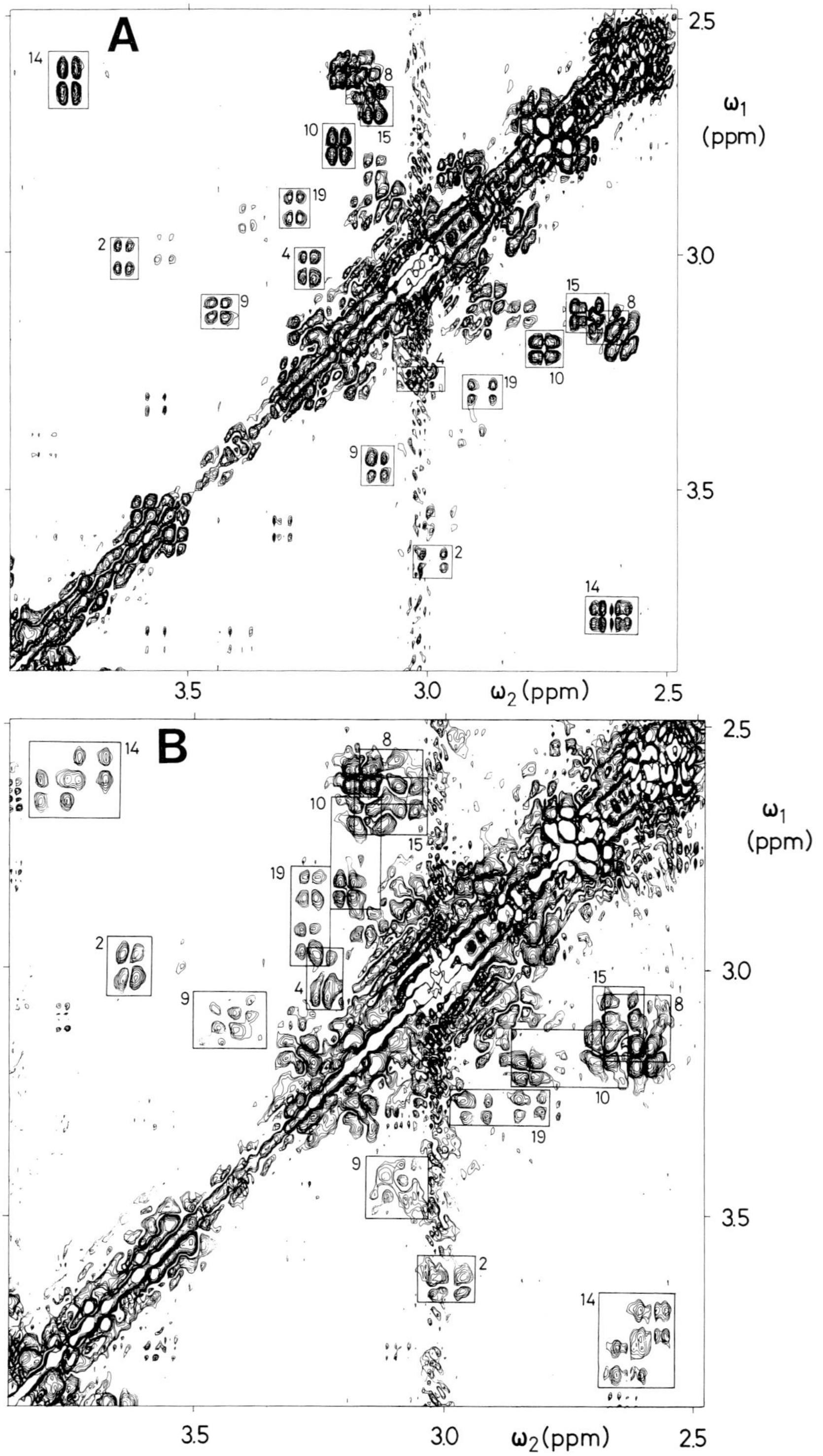

Fig. 5. *Illustration of the effects of* J-*coupling between cysteine* C^{β} *protons and* ^{113}Cd *on* $C^{\beta}H$-$C^{\beta}H$ *cross-peaks in* ^{1}H *COSY spectra of MT-2.* The spectral regions shown are from the same experiments as those in Fig. 3. (A) ^{112}Cd-MT-2, (B) ^{113}Cd-MT-2. Cross-peaks of 8 cysteines, which are sufficiently well resolved, are framed and labelled with the same numbers as in Fig. 3. Spectral patterns as schematically discussed in Fig. 2 can readily be recognized. The multiplet structure of the spin systems 9 and 14 indicates spin coupling to two ^{113}Cd nuclei. The $C^{\beta}H$-$C^{\beta}H$ cross peak 14 located below the diagonal is shown enlarged in Fig. 6

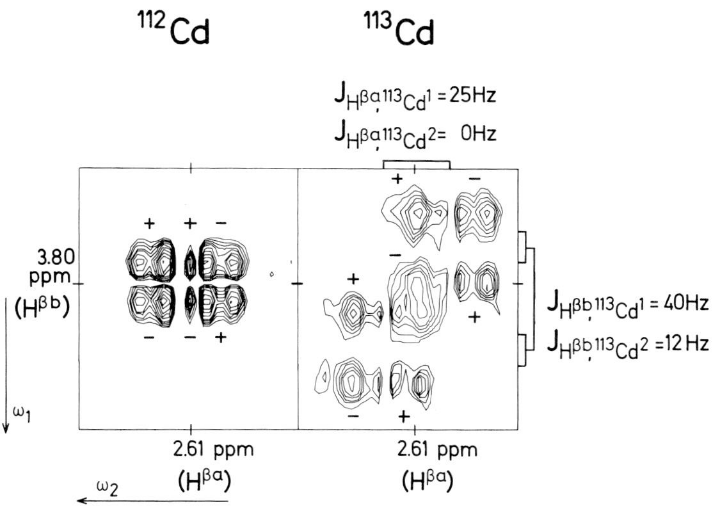

Fig. 6. *Enlarged plot of the ($H^{\beta a}$, $H^{\beta b}$) cross peak of the cysteine spin system 14 in ^{112}Cd-MT-2 (left) and ^{113}Cd-MT-2 (right).* The $H^{\beta a}$ chemical shift is along ω_2, the $H^{\beta b}$ chemical shift along ω_1. + and − indicate positive and negative contour levels

have two couplings to ^{113}Cd, is straightforward (Fig. 2D′), and all these qualitative arguments are fully supported by a more rigorous treatment based on the operator-product formalism of Sørensen et al. [20]. Inspection of Fig. 5B reveals cross-peak fine structures corresponding to the general type of Fig. 2C′, and there are also two cases, cysteines 9 and 14, where coupling to two ^{113}Cd spins is clearly manifested.

The $H^{\beta a}H^{\beta b}$ cross-peak of cysteine-14 is shown on an expanded scale in Fig. 6. The multiplet pattern in ^{112}Cd-MT-2 shows no splitting due to ^{113}Cd. Going to ^{113}Cd-MT-2, we first consider couplings to only one metal ion (^{113}Cd^1), where the multiplet pattern arising from ^{1}H-^{1}H coupling is split into two components corresponding to the two orientations of the ^{113}Cd spin. One component is at $\omega_1 = (3.80\text{ ppm} - J_{H^{\beta b}\,^{113}Cd^1})$ and $\omega_2 = (2.61\text{ ppm} - J_{H^{\beta a}\,^{113}Cd^1})$; the other component is at $\omega_1 = (3.80\text{ ppm} + J_{H^{\beta b}\,^{113}Cd^1})$ and $\omega_2 = (2.61\text{ ppm} + J_{H^{\beta a}\,^{113}Cd^1})$. These two components are readily visible in the lower left and the upper right corners, respectively, in the ^{113}Cd-MT-2 spectrum. The coupling to the second ^{113}Cd is detected when analyzing the vertical splitting within the two multiplet components discussed above. This vertical splitting is twice as large as the one in the ^{112}Cd-MT-2 pattern. This comes about as a result of the in-phase splitting by $J_{H^{\beta b}\,^{113}Cd^2} \approx 12$ Hz along ω_1 of the anti-phase ^{1}H-^{1}H doublet split by $^3J_{H^{\beta a}H^{\beta b}} \approx 15$ Hz. Thus a triplet multiplet pattern arises where the two central components have opposite sign and cancel each other. The coupling constant $J_{H^{\beta a}\,^{113}Cd^2}$ appears to be ≈ 0 Hz. Otherwise the two square-shaped major components in the right-hand figure would appear rhombicly distorted, as is observed for some other ($H^{\beta a}$, $H^{\beta b}$) cross-peaks in ^{113}Cd-MT-2. Comparison with the schematics in Fig. 2B′ and 2D′ shows that an additional complication in the experimental spectra arises because two cross-peaks with slightly different $H^{\beta a}$ chemical shifts, 2.61 ppm and 2.63 ppm, are overlapped. This is best seen in the ^{112}Cd-MT-2 spectrum, where it causes the appearance of the two weak components in the centre of the cross peak along ω_2 and a spreading of the major components along ω_2. This 'doubling' of the cross-peak is due to artifacts of the protein preparation, which could be eliminated in subsequent experiments. It can also be observed in the H^{α}-$H^{\beta a}$ cross-peak for cysteine-14 (Fig. 3A).

DISCUSSION

The experiments described in this paper were recorded in the course of the procedures used for obtaining sequence-specific resonance assignments for rabbit liver MT-2. The first step in the assignment is the identification of the J-coupling networks of the non-labile protons of individual amino acid residues [6–9]. At this stage ^{1}H NMR alone can only distinguish between spin systems with different symmetries [8,9,21]. All amino acid residues with $C^{\alpha}H$-$C^{\beta}H_2$ fragments are, therefore, classified as AMX systems [8,9,22], and additional criteria are needed to distinguish between different AMX residues (in metallothionein: Cys, Ser, Asp, Asn). According to the available sequence information MT-2 contains 33 AMX spin systems [10]. With regard to the completion of the ^{1}H NMR assignments it will be a great help that 19 cysteine spin systems could be identified from the ^{113}Cd-^{1}H couplings.

All published preparations of MT-2 are microheterogeneous in their sequence [10], and further heterogeneities may result from preparation artifacts. Given the additional complication that multiple conformations may also be possible, it is therefore not surprising that all ^{1}H NMR spectra of MT-2 include more spin systems than would be expected for a single conformation of a unique 61-residue polypeptide chain. We are quite confident, however, that the 19 cysteine spin systems, identified in Fig. 3–6 and Table 1, all belong to the same major species present in MT-2 solutions, since comparative studies show that all other components are not conserved between different sample preparations (Vašák, M., Wagner, G., Kägi, J. H. R. and Wüthrich, K., unpublished results).

Table 1 lists the chemical shifts and the spin-spin coupling data for the 19 identified cysteines. Whenever 'doubling' of the resonances was observed (see above and following section) only the parameters for the species seen in all the different

666

Table 1. *Chemical shifts, ^1H-^1H and ^1H-^{113}Cd coupling constants for 19 cysteines of ^{113}Cd-MT-2 at 24 °C, pH 7.0*

Number of spin system	Chemical shift			^1H-^1H coupling constants			^1H-^{113}Cd coupling constants			
	H^α	$H^{\beta a}$	$H^{\beta b}$	$^3J_{\alpha\beta^a}$	$^3J_{\alpha\beta^b}$	$^3J_{\beta\beta}$	$J_{\beta^a\,^{113}Cd^1}$	$J_{\beta^b\,^{113}Cd^1}$	$J_{\beta^a\,^{113}Cd^2}$	$J_{\beta^b\,^{113}Cd^2}$
	ppm			Hz						
1	4.05	3.11 —	3.23	—[a]	—[a]	—[a]	10	—[a]		
2	4.10	2.99	3.64	10	<2	15	10	5		
3	4.16	2.88	3.10	10	<2	15	37			
4	4.23	3.01	3.25	10	3	15	17	5		
5	4.27	2.95	3.05	10	<2	15	16			
6	4.34	2.91	3.01	11	4	14	48			
7	4.39	2.86	3.08	7	7	15	12	13		
8	4.43	2.63	3.14	10	4	15	37			
9	4.44	3.11	3.44	12	<2	15	15	15	15	15
10	4.49	2.75	3.18	8	<2	14	76	15		
11	4.49	3.24	3.14	8	<2	14	48	10		
12	4.55	2.89	3.14	10	<2	15	5	25		
13	4.58	3.21 —	3.28	—[a]	—[a]	—[a]	70	—[a]		
14	4.63	2.61	3.80	12	4	14	25	40	0	12
15	4.73	2.66	3.11	12	3	15	15	40		
16	5.06	3.53	3.59	6	<2	14	30	20		
17	5.14	2.98 —	3.08	—[a]	—[a]	—[a]	≈20	—[a]		
18	5.18	3.53 —	3.63	—[a]	—[a]	—[a]	≈20	—[a]		
19	5.36	2.88	3.27	10	<2	15	53	5		

[a] Second order $H^{\beta a}$-$H^{\beta b}$ multiplet pattern; information on coupling constants is difficult to obtain.

preparations are listed. Values for $^3J_{^{113}CdH^\beta}$, varying over the range 0 – 80 Hz, were found, which is the range of values observed also in low-molecular-mass compounds [21].

One perhaps surprising point from these data is the paucity of clear evidence concerning the existence of bridging cysteines, of which eight should be present according to the proposed metal cluster structures [2 – 5]. For only two cysteine spin systems (9 and 14) does an interpretation based on the method of 'successive splittings' definitely indicate that they have two ^{113}Cd coupling partners, and in both cases it is only in the ($H^{\beta a}$, $H^{\beta b}$) cross-peak multiplet structure that this evidence is clear. There are a number of residues, however, for which the available evidence is ambiguous, usually because the coupling in ^{113}Cd-MT-2 is observed only as a broadening rather than a resolved splitting. Nonetheless, even if all such cases should actually be bridging cysteines, this total would still fall short of the predicted eight, and as yet we have no explanation for this apparent discrepancy.

The existence of large (^{113}Cd, ^1H) couplings in ^{113}Cd-MT-2 clearly offers a novel source of structural information concerning both the overall conformation and details of the metal-binding in this protein. Although an interpretation of the very considerable variations in size of $^3J_{^{113}CdH_\beta}$ must await determination of the local conformation and stereospecific β-proton assignments for each cysteine, the couplings can nonetheless already be exploited to yield valuable conformational constraints. Each cadmium in the proposed cluster structures [2,4] would be tetrahedrally co-ordinated by four cysteinyl residues, so that an assignment of the individual ^{113}Cd coupling partners for each cysteine should provide connectivities of the type Cys-Cd-Cys. We are at present tackling this problem using a variety of heteronuclear two-dimensional NMR techniques, and have already assigned ^{113}Cd coupling partners for most of the cysteines. Obviously, combining such connectivities with sequence-specific resonance assignments will dramatically reduce the choice of possible topologies for the three-dimensional structure of MT-2.

The authors thank Miss M. Sutter for the preparation of the biological material and Mr R. Marani for the careful preparation of the manuscript. This research was supported by a fellowship from the Science and Engineering Research Council (UK) (Overseas post-doctoral fellowship to D. N.) and by the *Schweizerischer Nationalfonds* (projects 3.284-82 and 3.207-82).

REFERENCES

1. Nordberg, M. & Kojima, Y. (1979) in *Metallothionein* (Kägi, J. H. R. & Nordberg, M., eds) pp. 41 – 124, Birkhäuser, Basel.
2. Vašák, M. & Kägi, J. H. R. (1983) in *Metal Ions in Biological Systems*, vol. 15 (Sigel, H., ed.) pp. 213 – 273, Marcel Dekker, New York, Basel.
3. Otvos, J. D. & Armitage, I. M. (1980) *Proc. Natl Acad. Sci. USA 77*, 7094 – 7098.
4. Vašák, M. & Kägi, J. H. R. (1981) *Proc. Natl Acad. Sci. USA 78*, 6709 – 6713.
5. Armitage, I. M. & Otvos, J. D. (1982) in *Biological Magnetic Resonance*, vol. 4 (Berliner, L. J. & Reuben, J., eds) pp. 79 – 144, Plenum, New York.
6. Wüthrich, K., Wider, G., Wagner, G. & Braun, W. (1982) *J. Mol. Biol. 155*, 311 – 319.
7. Billeter, M., Braun, W. & Wüthrich, K. (1982) *J. Mol. Biol. 155*, 321 – 346.
8. Wagner, G. & Wüthrich, K. (1982) *J. Mol. Biol. 155*, 347 – 366.
9. Wider, G., Lee, K. H. & Wüthrich, K. (1982) *J. Mol. Biol. 155*, 367 – 388.
10. Kimura, M., Otaki, N. & Imano, M. (1979) in *Metallothionein* (Kägi, J. H. R. & Nordberg, M., eds) pp. 163 – 168, Birkhäuser, Basel.
11. Kägi, J. H. R., Himmelhoch, S. R., Whangner, P. D., Bethune, J. L. & Vallee, B. L. (1974) *J. Biol. Chem. 249*, 3537 – 3542.

12. Bühler, R. H. O. & Kägi, J. H. R. (1979) in *Metallothionein* (Kägi, J. H. R. & Nordberg, M., eds) pp. 211–220, Birkhäuser, Basel
13. Vašák, M., Nicholson, J. K., Hawkes, G. E., Sadler, P. J. (1984) *Biochemistry*, in press.
14. Piantini, O. W., Sørensen, O. W. & Ernst, R. R. (1982) *J. Am. Chem. Soc. 104*, 6800–6801.
15. Shaka, A. J. & Freeman, R. (1983) *J. Magnet. Resonance 51*, 169–173.
16. Rance, M., Sørensen, O. W., Bodenhausen, G., Wagner, G., Ernst, R. R. & Wüthrich, K. (1984) *Biochem. Biophys. Res. Commun. 117*, 479–485.
17. IUPAC-IUB Commission on Biological Nomenclature (1970) *Eur. J. Biochem. 17*, 193–201.
18. Wüthrich, K. (1976) *NMR in Biological Research: Peptides and Proteins*, North Holland, Amsterdam.
19. Lynden-Bell, R. M. & Harris, R. K. (1969) *Nuclear Magnetic Resonance Spectroscopy*, Nelson, London.
20. Sørensen, O. W., Eich, G. W., Levitt, M. H., Bodenhausen, G. & Ernst, R. R. (1983) *Prog. Nucl. Magnet Resonance Spectros. 16*, 163–192.
21. Turner, C. J. & White, F. M. (1977) *J. Magnet. Resonance 26*, 1–5.
22. Nagayama, K. & Wüthrich, K. (1981) *Eur. J. Biochem. 114*, 365–374.

D. Neuhaus, G. Wagner, and K. Wüthrich,
Institut für Molekularbiologie und Biophysik der Eidgenössischen Technischen Hochschule Zürich,
Zürich-Hönggerberg, CH-8093 Zürich, Switzerland

M. Vašák and J. H. R. Kägi, Biochemisches Institut der Universität Zürich,
Winterthurerstrasse 190, CH-8057 Zürich, Switzerland

J. Mol. Biol. (1984) **180**, 715–740

Polypeptide Secondary Structure Determination by Nuclear Magnetic Resonance Observation of Short Proton–Proton Distances

KURT WÜTHRICH, MARTIN BILLETER AND WERNER BRAUN

Institut für Molekularbiologie und Biophysik
Eidgenössische Technische Hochschule-Hönggerberg
CH-8093 Zürich, Switzerland

(Received 3 January 1984, and in revised form 20 July 1984)

The use of proton–proton nuclear Overhauser enhancement (NOE) distance information for identification of polypeptide secondary structures in non-crystalline proteins was investigated by stereochemical studies of standard secondary structures and by statistical analyses of the secondary structures in the crystal conformations of a group of globular proteins. Both regular helix and β-sheet secondary structures were found to contain a dense network of short ^1H–^1H distances. The results obtained imply that the combined information on all these distances obtained from visual inspection of the two-dimensional NOE (NOESY) spectra is sufficient for determination of the helical and β-sheet secondary structures in small globular proteins. Furthermore, *cis* peptide bonds can be identified from unique, short sequential proton–proton distances. Limitations of this empirical approach are that the exact start or end of a helix may be difficult to define when the adjoining residues form a tight turn, and that unambiguous identification of tight turns can usually be obtained only in the hairpins of antiparallel β-structures. The short distances between protons in pentapeptide segments of the different secondary structures have been tabulated to provide a generally applicable guide for the analysis of NOESY spectra of proteins.

1. Introduction

Recent experience (Wagner *et al.*, 1981; Braun *et al.*, 1983; Zuiderweg *et al.*, 1983; Wemmer & Kallenbach, 1983; Arseniev *et al.*, 1983; Williamson *et al.*, 1984) has shown that determination of polypeptide secondary structures may be based on recognition of characteristic cross peak patterns by visual inspection of the NOESY† spectra. This paper describes fundamental considerations on polypeptide structure and conformation, which provide the basis for n.m.r. determinations of the secondary structure.

NOESY spectra are capable of providing information on short proton–proton

† Abbreviations used: n.m.r., nuclear magnetic resonance; NOE, nuclear Overhauser enhancement; NOESY, 2-dimensional nuclear Overhauser enhancement spectroscopy.

0022–2836/84/350715–26 $03.00/0 © 1984 Academic Press Inc. (London) Ltd.

distances in the range from the van der Waals' distance of 2·0 Å to approximately 4·5 Å (Anil Kumar *et al.*, 1981; Braun *et al.*, 1981,1983). Here we use stereochemical studies of standard secondary structures and statistical studies of experimentally determined secondary structures in globular proteins to investigate which ^1H–^1H contacts in polypeptide chains could be used for identification of specified secondary structures in the amino acid sequence and for distinguishing between different secondary structures. These procedures are in some aspects related to earlier work on the identification of secondary structure elements in crystal structures of proteins (e.g. see Levitt & Greer, 1977; Chou & Fasman, 1977; Kuntz, 1972; Crawford *et al.*, 1973; Rose & Seltzer, 1977; Isogai *et al.*, 1980; Kolaskar *et al.*, 1980). However, since these earlier methods rely either entirely on the heavy-atom co-ordinates or on combined use of heavy-atom co-ordinates and known location of hydrogen bonds, they could not provide the information needed in conjunction with structural studies by ^1H n.m.r.

Determination of the global spatial structure of non-crystalline polypeptide chains by n.m.r. requires extensive use of numerical and graphical computer facilities (Braun *et al.*, 1981,1983; Havel & Wüthrich, 1984 and unpublished results; Zuiderweg *et al.*, 1984). In contrast, the presently discussed procedures for secondary structure determination rely entirely on the results of visual inspections of the NOESY spectra. Such direct and straightforward access to the secondary structure is of considerable interest as a first step in the determination of the protein conformation and also with regard to future practical applications of n.m.r. for characterization of protein products obtained, for example, with gene technological procedures.

2. Methods

(a) *Notation for proton–proton distances*

Based on initial results obtained in this study and on experience gained in practical applications (Braun *et al.*, 1983; Williamson *et al.*, 1984; Zuiderweg *et al.*, 1983), this paper concentrates on investigations of distances between different backbone hydrogen atoms and between backbone hydrogens and $C^\beta H_n$ groups. For these, the following notation is used. The distance between the hydrogen atoms A and B located in the amino acid residues in positions i and j, respectively, is denoted by $d_{AB}(i,j)$†. A and B are N for amide protons, α for $C^\alpha H$ and β for $C^\beta H$. With the exception of certain contacts involving glycine or proline (see Appendix), this notation provides for an unambiguous description of all intramolecular distances between backbone and C^β protons:

$$d_{\alpha N}(i,j) \equiv d(C^\alpha H_i, NH_j), \tag{1}$$

$$d_{NN}(i,j) \equiv d(NH_i, NH_j) \equiv d_{NN}(j,i), \tag{2}$$

$$d_{\beta N}(i,j) \equiv \min\{d(C^\beta H_i, NH_j)\}, \tag{3}$$

$$d_{\alpha\alpha}(i,j) \equiv d(C^\alpha H_i, C^\alpha H_j) \equiv d_{\alpha\alpha}(j,i), \tag{4}$$

$$d_{\alpha\beta}(i,j) \equiv \min\{d(C^\alpha H_i, C^\beta H_j)\}. \tag{5}$$

† This corresponds to the notation introduced by Billeter *et al.* (1982), except that for improved clarity the numerical subscripts in the previously used notation are replaced by identification of the atom types. Thus the previously used $d_1(i,j)$ is replaced by $d_{\alpha N}(i,j)$, $d_2(i,j)$ is now $d_{NN}(i,j)$, $d_3(i,j)$ is $d_{\beta N}(i,j)$, $d_4(i,j)$ is $d_{\alpha\alpha}(i,j)$ and $d_5(i,j)$ is $d_{\alpha\beta}(i,j)$.

In the definitions (3) and (5), min indicates that in the case of β-methylene or methyl groups, the shortest distance to any of the C^β protons is taken.

Distances between backbone and C^β protons located in residues that are nearest-neighbours in the amino acid sequence are referred to as sequential distances. For simplicity, the indices i and j are omitted for the sequential distances; for example, $d_{\alpha N}(i, i+1) \equiv d_{\alpha N}$ and $d_{NN}(i, i+1) \equiv d_{NN}$. Medium range distances are all non-sequential and non-intraresidue distances between backbone and C^β protons within a fragment of 5 consecutive residues. For clarity, we assume for sequential and medium range distances that $j \geq i$, so that for the distances that are asymmetric with respect to the chain propagation we also use $d_{N\alpha}(i,j)$, $d_{N\beta}(i,j)$ and $d_{\beta\alpha}(i,j)$. Long range backbone distances are between backbone protons in residues that are at least 6 positions apart in the sequence, i.e. $|j-i| \geq 5$.

(b) *Computation of proton–proton distances in standard polypeptide secondary structures*

Standard secondary structures were generated on an Evans and Sutherland MPS computer graphics system, using a polyalanine chain with ECEPP standard geometry for the individual residues (Momany *et al.*, 1975)† and with *trans* peptide bonds. A standard α-helix was obtained by adjusting the dihedral angles ϕ and ψ to $-57°$ and $-47°$, respectively. Similarly, a regular 3_{10} helix was obtained with $\phi = -60°$ and $\psi = -30°$.

For tight turns, we adopted the definitions of 2 major classes, I and II, given by Richardson (1981). The molecular geometries are defined by the dihedral angles ϕ and ψ of 2 sequentially neighbouring residues (see Fig. 1), but for the present considerations the protons in the 2 residues preceding and following this dipeptide segment are also of interest.

Double-stranded β-sheets were constructed using a procedure described by Salemme & Weatherford (1981). Short segments of single strands were obtained with the dihedral angles $\phi = -139°$ and $\psi = -135°$ for antiparallel β, $\phi = -119°$ and $\psi = -113°$ for parallel β. Two identical strands were then combined to obtain, respectively, a double-stranded antiparallel or parallel β-sheet (see Fig. 2). A best fit procedure (McLachlan, 1979) was used to define the relative positions of the 2 strands when they form linear hydrogen bonds C=O | | | | H–N of length 2·8 Å. From the secondary structures thus generated, all proton–proton distances were extracted with a computer program.

(c) *Statistics of short ^1H–^1H distances in the secondary structures of globular proteins*

From among the structures in the files of the Protein Data Bank in Cambridge (Bernstein *et al.*, 1977) that were determined at a resolution $\leq$2·0 Å, a group of 19 proteins was selected for the present study. These are actinidine, parvalbumin B, carbonic anhydrase C, carboxypeptidase A, concanavalin A, ferricytochrome b_5, ferrocytochrome c, haemoglobin, flavodoxin, high potential iron protein, immunoglobulin F_{ab}, insulin, lactate dehydrogenase, lysozyme, plastocyanine, prealbumin, ribonuclease S, trypsin and basic pancreatic trypsin inhibitor. The same proteins were used previously for investigations on sequential ^1H–^1H distances (Billeter *et al.*, 1982), and additional criteria used for the selection of the proteins are described there. Hydrogen atoms were added to the crystal structures as described (Billeter *et al.*, 1982).

The 19 proteins contain 3227 amino acid residues. For the present work, it was of particular interest to obtain separate counts of short ^1H–^1H distances for the residues in different secondary structure types. However, in the original publications on the structure determinations and in the data sets submitted to the protein data bank (Bernstein *et al.*, 1977), different authors used quite different criteria for identification of secondary

† After these calculations were completed, we became aware of a recent updating of the ECEPP parameters for amino acid residues (Némethy *et al.*, 1983). However, the proposed modifications do not affect the numbers given in Tables 1 to 4 by more than $\pm$0·1 Å, nor do they influence the conclusions drawn in this paper.

structure, so that no consistent overall secondary structure determination is obtained from this source. An objective and consistent compilation of secondary structure elements in a large number of globular proteins has been presented by Kabsch & Sander (1983), and these secondary structure identifications were used in our studies. Accordingly, 796 residues were attributed to helical structure, 767 to parallel and antiparallel β-sheets, 818 to turns and bends, and 846 to random coil segments.

A computer program extracted all short sequential and medium range ^{1}H–^{1}H distances from the different secondary structure types in the 19 proteins. Thereby, the counts for α-helix and 3_{10} helix were added, type I and type II turns were added, and parallel and antiparallel β-sheets were considered as one structure type. No separate counts were made for distances with glycine or proline. Instead, these contacts were counted with the corresponding contacts for L-amino acids, whereby the δ-methylene protons of Pro were used in the place of the amide proton (see Appendix). For example, the counts for $d_{\alpha N}(i,j)$ include $d_{\alpha N}^{GX}(i,j)$ and $d_{\alpha\delta}^{XP}(i,j)$, those for $d_{NN}(i,j)$ include $d_{N\delta}^{XP}(i,j)$ and $d_{\delta N}^{PX}(i,j)$, etc. For all contacts involving methylene groups, only the shortest distance was counted. Since χ^1 for alanyl residues is not determined in the protein crystal structures, the methyl protons of alanine were replaced by a reference point for distance measurements, M, which is located on the extension of the C$^\alpha$–C$^\beta$ bond and 0·36 Å away from C$^\beta$. This primarily affects the counts for the distances $d_{\alpha\beta}(i, i+3)$. In a regular α-helix, the distance $d_{\alpha\beta}(i, i+3)$ to the reference point M is 3·5 Å, as compared to a range of from 2·5 to 3·1 Å for the distance to the nearest methyl proton (Table 1). For the regular 3_{10} helix, the corresponding numbers are 4·1 Å and 3·1 to 3·7 Å.

It may be helpful to add some comments on the evaluation of the "total number" of times a specified ^{1}H–^{1}H distance is defined within peptide segments assigned to a particular secondary structure. First, distances to the amide proton of the first residue following a secondary structure segment are counted with this secondary structure, since they are entirely determined by the backbone dihedral angles within this segment. The counts for the 4 secondary structure types (helix, β sheets, turns and bends, random coil) then account for all sequential connectivities in a protein. Second, for the medium range distances we have further a "mixed" category, since these may be defined also between residues attributed to segments of different secondary structure. Third, the total number is, in general, different for the different distances. For example, there are obviously a larger number of sequential distances than medium range distances; the total number of distances d_{NN} may be smaller than that for $d_{\alpha N}$, because d_{NN} is not defined for the N-terminal residue of a polypeptide chain; the total number of distances $d_{\alpha\beta}(i, i+3)$ is smaller than that for $d_{\alpha N}(i, i+3)$, since the contacts to the first residue following the secondary structure segment are not counted for $d_{\alpha\beta}(i, i+3)$ and since $d_{\alpha\beta}(i, i+3)$ is not defined for glycyl residues.

3. Results and Discussion

(a) *^{1}H–^{1}H distances in standard regular polypeptide secondary structures*

Sequential and medium range distances between backbone and C$^\beta$ protons in the standard secondary structures generated as described in Methods, section (b), are listed in Tables 1 to 4. In a regular α-helix (Table 1), there is a dense network of short ^{1}H–^{1}H distances within segments of four successive residues and, in addition, there is a single distance to the fifth residue, $d_{\alpha N}(i, i+4)$, which is sufficiently short to be within reach of observation by NOE experiments. All ^{1}H–^{1}H distances between α-helical residues that are six or more sequence positions apart are too long to be observed by n.m.r. Table 2 presents the corresponding data for a tetrapeptide segment of 3_{10} helix, where all ^{1}H–^{1}H distances between residues that are five or more sequence positions apart are too long to be seen by n.m.r.

TABLE 1

Proton–proton distances in regular α-helical polypeptide chains. A polyalanine α-helix was constructed with ECEPP standard geometry for the amino acid residues and with the dihedral angles $\phi = -57°$, $\psi = -47°$

	j‡			
Distance†	1	2	3	4
$d_{\alpha N}(i, i+j)$	3·5	4·4	3·4	4·2
$d_{N\alpha}(i, i+j)$	5·2	6·8	7·2	8·6
$d_{NN}(i, i+j)$	2·8	4·2	4·8	6·1
$d_{\alpha\alpha}(i, i+j)$	4·7	6·4	5·3	6·4
$d_{\alpha\beta}(i, i+j)C^\beta H$	5·4–6·5	5·1–7·0	2·5–4·4	5·0–7·0
$C^\beta H_2$	5·4–6·3	5·1–6·5	2·5–4·0	5·0–6·5
$C^\beta H_3$	5·4–5·7	5·1–5·6	2·5–3·1	5·0–5·5
$d_{\beta\alpha}(i, i+j)C^\beta H$	3·8–5·6	7·4–8·3	7·3–8·0	6·7–8·4
$C^\beta H_2$	3·8–5·2	7·4–8·1	7·3–7·8	6·7–8·0
$C^\beta H_3$	3·8–4·3	7·4–7·6	7·3–7·5	6·7–7·2
$d_{N\beta}(i, i+j)C^\beta H$	4·5–6·3	4·9–7·0	4·8–6·7	6·7–8·8
$C^\beta H_2$	4·5–5·9	4·9–6·5	4·8–6·3	6·7–8·3
$C^\beta H_3$	4·5–5·0	4·9–5·5	4·8–5·3	6·7–7·3
$d_{\beta N}(i, i+j)C^\beta H$	2·5–4·1	5·1–6·1	5·3–5·7	5·0–6·2
$C^\beta H_2$	2·5–3·8	5·1–5·8	5·3–5·6	5·0–6·0
$C^\beta H_3$	2·5–3·0	5·1–5·3	5·3–5·4	5·0–5·3

† For C^β protons, the lower and upper limits to the range of distances covered as a function of the torsion angle χ^1 are indicated. For $C^\beta H_2$ and $C^\beta H_3$, the upper limit is shorter than for $C^\beta H$ because the nearest hydrogen atom is considered in all cases.

‡ There are also 2 intra-residue ^1H–^1H contacts: $d_{N\alpha}(i, i)$ varies between 2·2 and 2·8 Å when ϕ_i goes from −180° to +180° and is equal to 2·7 Å for an α-helix. $d_{N\beta}(i, i)$ varies between 2·0 and 4·1 Å for a single proton when ϕ_i and χ_i^1 go from −180° to +180° and covers the range from 2·0 to 3·4 Å for an α-helix.

Table 3 presents the proton–proton distances in type I and type II tight turns, which are determined uniquely by the torsion angles ϕ_2, ψ_2, ϕ_3 and ψ_3 (Fig. 1) and can adopt values ≤4·5 Å. It is seen that, similar to the helical structures, there is a dense network of short ^1H–^1H distances within the tripeptide segment 2–4 of both types of turns. In addition, the distance $d_{\alpha N}(i, i+3)$ between residues 1 and 4 may be observable by NOEs. In β-structures (Table 4), all intra-strand ^1H–^1H distances other than those between sequentially neighbouring residues are too long for observation by NOEs. However, there are additional long range ^1H–^1H contacts between neighbouring strands in β-sheets (Fig. 2) that are sufficiently short for observation by NOE experiments.

In the footnotes to Tables 1 to 4 the intra-residue distances $d_{N\alpha}(i, i)$ and $d_{N\beta}(i, i)$ are also given. These adopt values between 2·0 and 4·0 Å in all the different secondary structures.

The data in Tables 1 to 4 are strictly valid only for polypeptide chains with L-amino acids linked by *trans* peptide bonds. The results of corresponding studies with polypeptide chains that include glycine, proline or *cis* peptide bonds are described in the Appendix. For proline, similar results are obtained as for L-amino acids, if $C^\delta H_2$ is used in the place of the amide proton. With glycine or with *cis*

TABLE 2

Proton–proton distances in a regular 3_{10} helical polypeptide chain. A polyalanine 3_{10} helix was constructed with ECEPP standard geometry for the amino acid residues and with the dihedral angles $\phi = -60°$, $\psi = -30°$

Distance†	j‡ 1	2	3
$d_{\alpha N}(i, i+j)$	3·4	3·8	3·3
$d_{N\alpha}(i, i+j)$	5·1	6·9	7·7
$d_{NN}(i, i+j)$	2·6	4·1	5·2
$d_{\alpha\alpha}(i, i+j)$	4·7	6·1	5·4
$d_{\alpha\beta}(i, i+j)C^{\beta}H$	5·4–6·5	4·4–6·3	3·1–5·1
$C^{\beta}H_2$	5·4–6·2	4·4–5·9	3·1–4·6
$C^{\beta}H_3$	5·4–5·7	4·4–4·9	3·1–3·7
$d_{\beta\alpha}(i, i+j)C^{\beta}H$	4·0–5·7	7·4–8·2	6·8–8·0
$C^{\beta}H_2$	4·0–5·3	7·4–8·0	6·8–7·7
$C^{\beta}H_3$	4·0–4·4	7·4–7·6	6·8–7·1
$d_{N\beta}(i, i+j)C^{\beta}H$	4·3–6·1	5·0–7·1	5·6–7·6
$C^{\beta}H_2$	4·3–5·7	5·0–6·6	5·6–7·2
$C^{\beta}H_3$	4·3–4·9	5·0–5·6	5·6–6·2
$d_{\beta N}(i, i+j)C^{\beta}H$	2·9–4·4	4·8–5·7	5·0–5·7
$C^{\beta}H_2$	2·9–4·0	4·8–5·5	5·0–5·5
$C^{\beta}H_3$	2·9–3·3	4·8–5·0	5·0–5·2

† For C^{β} protons the lower and upper limits to the range of distances covered as a function of the torsion angle χ^1 are indicated. For $C^{\beta}H_2$ and $C^{\beta}H_3$ the upper limit is shorter than for $C^{\beta}H$ because the nearest hydrogen atom is considered in all cases.

‡ The intra-residue contacts for the 3_{10} helix are $d_{N\alpha}(i, i) = 2·7$ Å, and $d_{N\beta}(i, i)$ varies from 2·0 to 3·4 Å as a function of χ_i^1.

peptide bonds, the following three sequential distances adopt considerably shorter values than in poly-L-alanine: $d_{\alpha N}^{GX}$ is <3·0 Å in all secondary structure types. $d_{N\alpha}^{cis}$ can be as short as the van der Waals' distance of 2·0 Å and $d_{\alpha\alpha}^{cis}$ is between 2·2 and 4·0 Å for most of the sterically allowed, intervening torsion angles. These two distances can therefore be used for identification of *cis* peptide bonds by n.m.r. (Arseniev *et al.*, 1983).

Some of the proton–proton distances in tight turns and β-sheet structures (Tables 3 and 4) were previously investigated in connection with studies of oligopeptides, in particular cyclopeptides, which form antiparallel β-sheet conformations (Schwyzer, 1958). The numbers in Tables 3 and 4 coincide closely with the results obtained by these earlier studies (Leach *et al.*, 1977; Kuo & Gibbons, 1979,1980).

(b) *Implications for the analysis of NOESY spectra*

While this is not the main purpose of this paper, it should be pointed out that Tables 1 to 4 and the Appendix present a collection of data that has been most helpful in our interpretations of protein NOESY spectra (e.g. see Braun *et al.*, 1983; Williamson *et al.*, 1984; Zuiderweg *et al.*, 1983). At the outset of an n.m.r.

TABLE 3

Proton–proton distances in standard tight turns of type I and II constructed for polyalanine with ECEPP parameters for the amino acid residues and the dihedral angles indicated in Fig. 1

Distance†	Type I turn	Type II turn
$d_{\alpha N}(2, 3)$	3·4	2·2
$d_{\alpha N}(3, 4)$	3·2	3·2
$d_{\alpha N}(2, 4)$	3·6	3·3
$d_{NN}(2, 3)$	2·6	4·5
$d_{NN}(3, 4)$	2·4	2·4
$d_{NN}(2, 4)$	3·8	4·3
$d_{\alpha\alpha}(2, 3)$	4·5	4·3
$d_{\alpha\beta}(2, 3)$	5·4–6·5	4·1–5·7
	5·4–6·2	4·1–5·3
	5·4–5·7	4·1–4·6
$d_{\beta\alpha}(2, 3)$	4·1–5·7	4·8–6·3
	4·1–5·3	4·8–5·9
	4·1–4·5	4·8–5·2
$d_{N\beta}(2, 3)$	4·4–6·2	5·0–6·9
	4·4–5·8	5·0–6·4
	4·4–4·9	5·0–5·5
$d_{\beta N}(2, 3)$	2·9–4·4	3·6–4·6
	2·9–4·1	3·6–4·4
	2·9–3·4	3·6–3·8
$d_{\beta N}(3, 4)$	3·6–4·6	3·6–4·6
	3·6–4·4	3·6–4·4
	3·6–3·9	3·6–3·9
$d_{\alpha N}(1, 4)$‡	3·1–4·2	3·8–4·7

† For C^{β} protons the lower and upper limits to the range of distances covered as a function of the torsion angle χ^1 are indicated. For $C^{\beta}H_2$ and $C^{\beta}H_3$ the upper limit is shorter than for $C^{\beta}H$ because the nearest hydrogen atom is considered in all cases.

The intra-residue contacts are: for type I turns, $d_{N\alpha}(2, 2) = 2{\cdot}7$ Å, $d_{N\alpha}(3, 3) = 2{\cdot}8$ Å, $d_{N\beta}(2, 2) = 2{\cdot}0$ to 3·4 Å, $d_{N\beta}(3, 3) = 2{\cdot}1$ to 3·5 Å. For type II turns, $d_{N\alpha}(2, 2) = 2{\cdot}7$ Å, $d_{N\alpha}(3, 3) = 2{\cdot}2$ Å, $d_{N\beta}(2, 2) = 2{\cdot}0$ to 3·4 Å, $d_{N\beta}(3, 3) = 3{\cdot}2$ to 4·0 Å.

‡ The range of distances corresponds to that covered as a function of the torsion angle ψ_1.

investigation, one is faced with an abundance of cross peaks in the NOESY spectra of a protein. Initially, a large proportion of the cross peaks can be attributed to short intra-residue and sequential ^{1}H–^{1}H distances. These in turn provide first indications of the locations of particular secondary structure elements in the amino acid sequence (Wagner *et al.*, 1981; Billeter *et al.*, 1982). Using the information on short, medium and long range ^{1}H–^{1}H contacts in specified secondary structures (Tables 1 to 4 and Appendix), a systematic search for identification of additional NOESY cross peaks can then be initiated. Since regular secondary structures contain a particularly dense network of short ^{1}H–^{1}H distances, one is, after this step of the spectral analysis, usually left with a relatively small number of unassigned backbone–backbone and backbone–C^{β}H NOEs in the NOESY spectra, most of which must arise from long range ^{1}H–^{1}H contacts other than those in regular β-structures (Fig. 2).

TABLE 4

Short intra- and inter-strand proton–proton distances in regular, non-twisted β-sheet structures. The double-stranded sheets were constructed as described in the text

Distance†	β Antiparallel ($\phi = -139°$, $\psi = 135°$)	β Parallel ($\phi = -119°$, $\psi = 113°$)
$d_{\alpha N}$	2·2	2·2
$d_{N\alpha}$	4·7	4·8
d_{NN}	4·3	4·2
$d_{\alpha\alpha}$	4·3	4·3
$d_{\alpha\beta}C^{\beta}H$	4·0–5·6	4·2–5·7
$C^{\beta}H_2$	4·0–5·2	4·2–5·3
$C^{\beta}H_3$	4·0–4·4	4·2–4·6
$d_{\beta\alpha}C^{\beta}H$	4·2–5·7	4·4–5·8
$C^{\beta}H_2$	4·2–5·4	4·4–5·5
$C^{\beta}H_3$	4·2–4·6	4·4–4·8
$d_{N\beta}C^{\beta}H$	4·6–6·4	4·8–6·6
$C^{\beta}H_2$	4·6–6·0	4·8–6·2
$C^{\beta}H_3$	4·6–5·1	4·8–5·3
$d_{\beta N}C^{\beta}H$	3·2–4·5	3·7–4·7
$C^{\beta}H_2$	3·2–4·2	3·7–4·4
$C^{\beta}H_3$	3·2–3·6	3·7–4·0
$d_{\alpha N}(i,j)$‡	3·2	3·0
$d_{NN}(i,j)$‡	3·3	4·0
$d_{\alpha\alpha}(i,j)$‡	2·3	4·8

† For C^{β} protons the lower and upper limits to the range of distances covered as a function of the torsion angle χ^1 are indicated. For $C^{\beta}H_2$ and $C^{\beta}H_3$ the upper limit is shorter than for $C^{\beta}H$ because the nearest hydrogen atom is considered in all cases.

The intra-residue contacts are: For parallel β, $d_{N\alpha}(i,i) = 2·8$ Å and $d_{N\beta}(i,i) = 2·4$ to 3·7 Å. For antiparallel β, $d_{N\alpha}(i,i) = 2·8$ Å and $d_{N\beta}(i,i) = 2·6$ to 3·8 Å.

‡ These are inter-strand distances. The indices (i,j) indicate the residues that give the shortest distance of a given type.

(c) *Selection of suitable short ^1H–^1H distances for use as experimental parameters in secondary structure determinations by n.m.r.*

With regard to applications for identification of secondary polypeptide structures by n.m.r., one has two principal criteria for the selection of "useful" ^{1}H–^{1}H distances. First, they should be sufficiently short that they can be observed in NOESY experiments, i.e. shorter than about 4·5 Å (Anil Kumar *et al.*, 1981; Braun *et al.*, 1981,1983; Dubs *et al.*, 1979; Poulsen *et al.*, 1980; Williamson *et al.*, 1984). Second, they should be different in different secondary structures, so that they present a valid criterion for distinguishing between different structures. In Tables 1 to 4 one thus finds that the seven sequential and medium range distances $d_{\alpha N}$, $d_{\alpha N}(i, i+2)$, $d_{\alpha N}(i, i+3)$, $d_{\alpha N}(i, i+4)$, d_{NN}, $d_{NN}(i, i+2)$ and $d_{\alpha\beta}(i, i+3)$ are of potential interest for n.m.r. studies. Figure 3 and Table 5 present a survey of the locations of these distances in the polypeptide chain and of their values in the different secondary structures. Additional useful distances in β-structures are the three interstrand distances $d_{\alpha N}(i,j)$, $d_{NN}(i,j)$ and $d_{\alpha\alpha}(i,j)$ (Fig. 2), which are all suitable for the delineation of double-stranded or multiple-stranded sheets (Table 4).

Type I

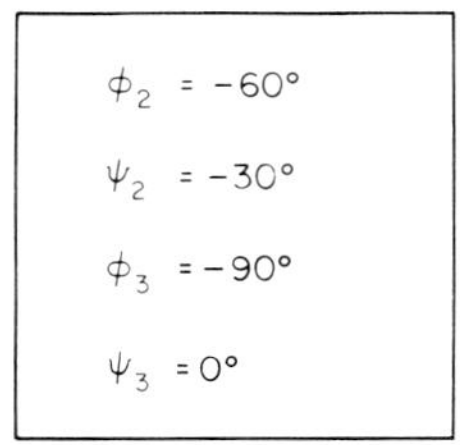

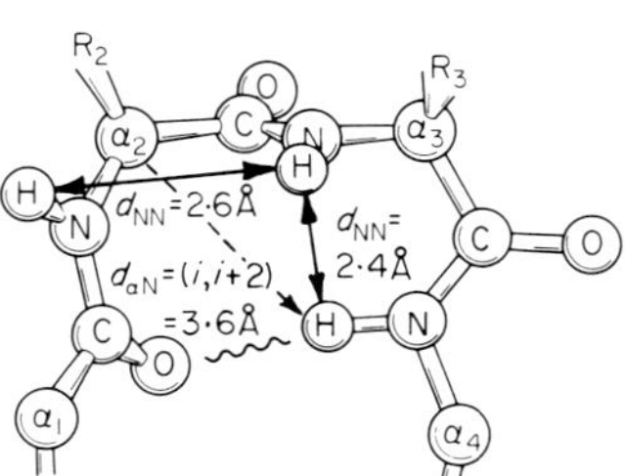

Type II
(R3 usually Gly)

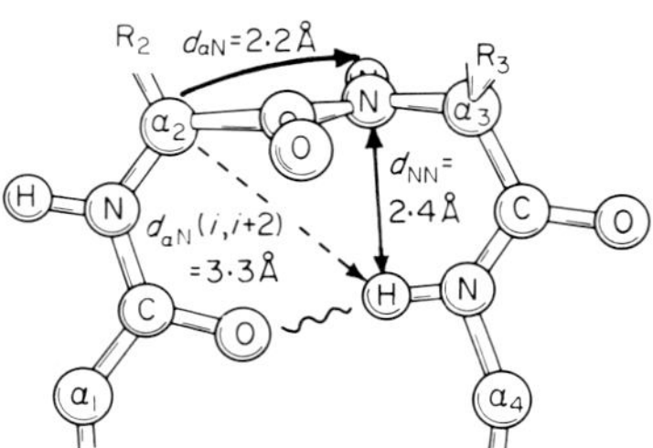

FIG. 1. Standard tight turns of type I and type II (Richardson, 1981). The wavy lines indicate the location of hydrogen bonds. Continuous and broken arrows identify sequential and medium range ^{1}H–^{1}H contacts that are of particular interest for the identification of tight turns by n.m.r. (see Tables 1 to 4 and the text).

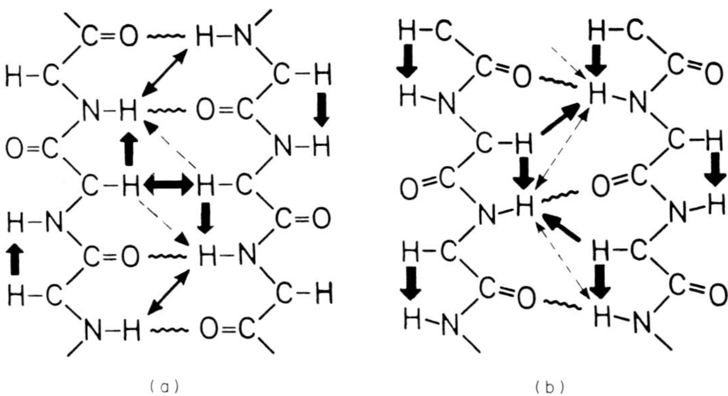

FIG. 2. Standard double stranded β-sheets. The sequential distances $d_{\alpha N}$ are indicated by thick vertical arrows. (a) Antiparallel β. The thick horizontal arrow indicates a long range backbone distance $d_{\alpha\alpha}(i,j)$, thin continuous arrows indicate $d_{NN}(i,j)$ and broken arrows identify $d_{\alpha N}(i,j)$. (b) Parallel β. Continuous arrows indicate $d_{\alpha N}(i,j)$, broken arrows $d_{NN}(i,j)$. All contacts indicated by thick arrows are very short, i.e. $\leq 2{\cdot}3$ Å.

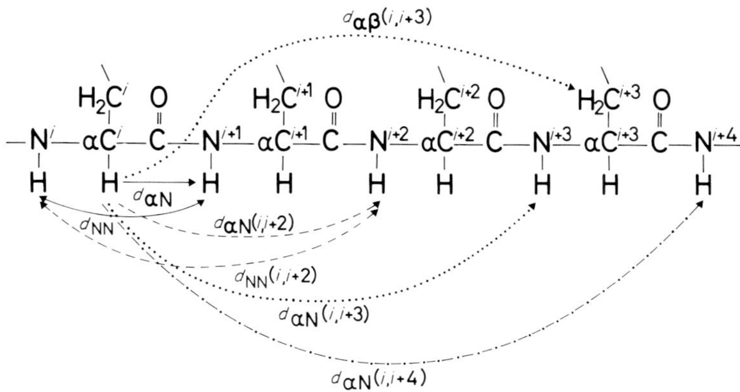

FIG. 3. Short ^{1}H–^{1}H connectivities in secondary structures. Pentapeptide segment with indication of the sequential and medium range ^{1}H–^{1}H contacts that are useful for characterization of the secondary structure by ^{1}H n.m.r. (see the text and Tables 1 to 4).

Further inspection of Tables 1 to 4 shows that Figure 3 contains all inter-residue ^{1}H–^{1}H distances within a pentapeptide fragment that can be shorter than 4·5 Å, with the following exceptions. The sequential connectivities with C^{β} protons were not retained, since they adopt similar values for different secondary structures. $d_{\alpha\beta}(i, i+2)$ was discarded, since it can be expected to be longer than 4·5 Å in all but a very few instances (Table 2). $d_{\alpha\alpha}$ is not further considered, since in *trans* peptides it is slightly shorter than 4·5 Å only in β-structures and type II tight turns, which are much more prominently manifested by the outstandingly short distances $d_{\alpha N}$ (Tables 3 and 4). In addition, the intra-residue ^{1}H–^{1}H distances $d_{N\alpha}(i, i)$ and $d_{N\beta}(i, i)$ were not retained, even though they are always shorter than 4·0 Å (see footnotes to Tables 1 to 4); since they adopt closely similar values for extended and helical secondary structures, practical uses for distinguishing between different structures would be rather limited. A notable

TABLE 5

Useful sequential and medium range ^{1}H–^{1}H distances (in Å) for identification of secondary structure in polypeptide chains

	α-Helix	3_{10} Helix	β Antiparallel	β Parallel	Type I turn	Type II turn
$d_{\alpha N}$	3·5	3·4	2·2	2·2	3·4/3·2	2·2/3·2
$d_{\alpha N}(i, i+2)$	4·4	3·8			3·6	3·3
$d_{\alpha N}(i, i+3)$	3·4	3·3			3·1–4·2	3·8–4·7
$d_{\alpha N}(i, i+4)$	4·2					
d_{NN}	2·8	2·6	3·3	4·0	2·6/2·4	4·5/2·4
$d_{NN}(i, i+2)$	4·2	4·1			3·8	4·3
$d_{\alpha\beta}(i, i+3)$	2·5–4·4	3·1–5·1				

For the helices and β sheets, no number is given for distances >4·5 Å.

For the turns, no number is given when the specified distance is >4·5 Å or when it is not uniquely determined by the torsion angles ϕ_2, ψ_2, ϕ_3 and ψ_3 (Fig. 1). Where 2 numbers are given, the first applies to the distance between residues 2 and 3, the second to that between residues 3 and 4.

exception is the very short distance $d_{N\alpha}(3, 3) = 2{\cdot}2$ Å in type II tight turns (Table 3), which could be useful for identification of this structure element.

(d) *Extent and uniqueness of secondary structure identification by specified short sequential and medium range ${}^1H$–${}^1H$ distances*

In globular proteins, the secondary structures are usually somewhat distorted (Schulz & Schirmer, 1979; Richardson, 1981). Therefore, it was of interest to complement the stereochemical studies of regular model secondary structures by investigating the distribution of ^{1}H–^{1}H distances in experimental protein structures. For this, the distribution of short ^{1}H–^{1}H distances was evaluated separately for the helical, β-sheet, tight turn and random coil segments identified by Kabsch & Sander (1983) in a selected group of proteins.

The procedures used for these statistical studies are illustrated in Figure 4 and described in detail in Methods, section (c). Figure 4 indicates the secondary

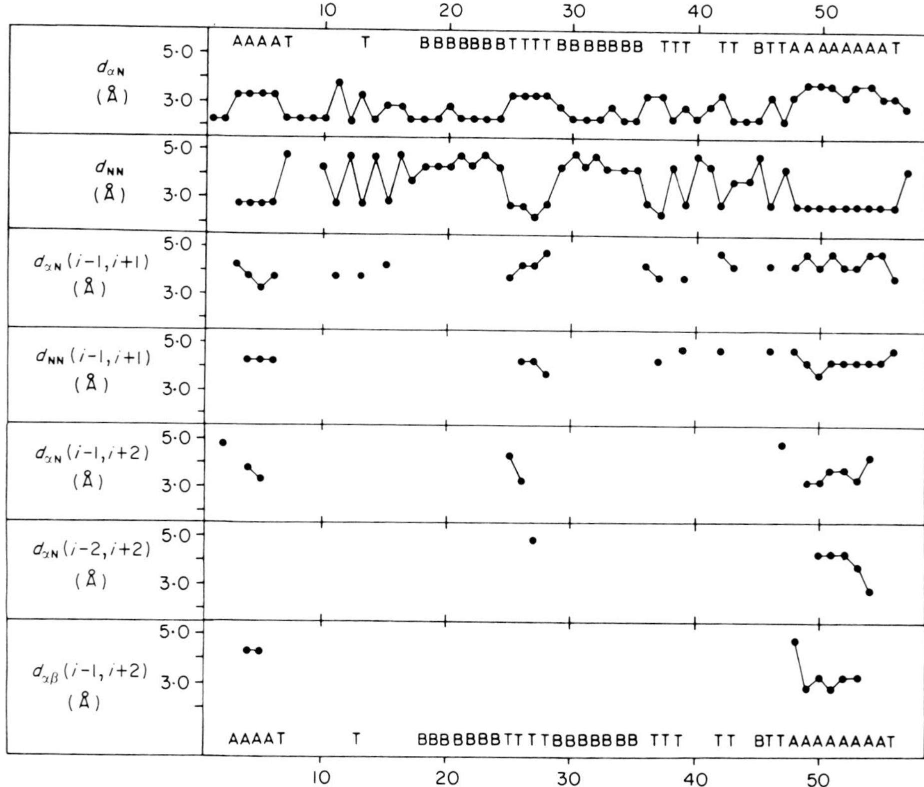

FIG. 4. Plot of short sequential and medium range ^{1}H–^{1}H contacts in the crystal structure of the basic pancreatic trypsin inhibitor (Deisenhofer & Steigemann, 1975) *versus* the amino acid sequence. Distances within the range 2·0 to 5·0 Å are indicated with increments of 0·5 Å. For the sequential connectivities, the values given for residue i indicate the distances $d_{\alpha N}(i, i+1)$ and $d_{NN}(i, i+1)$, respectively. For the medium range contacts, the distances given for residue i correspond, respectively, to $d_{\alpha N}(i-1, i+1)$, $d_{NN}(i-1, i+1)$, $d_{\alpha N}(i-1, i+2)$, $d_{\alpha N}(i-2, i+2)$ and $d_{\alpha\beta}(i-1, i+2)$. At the top and the bottom, the secondary structure identifications by Kabsch & Sander (1983) are indicated, where A stands for helix, B for β-sheets and T for tight turns or bends.

structure identifications of Kabsch & Sander (1983) for the basic pancreatic trypsin inhibitor and shows the distribution of relevant, short ^1H–^1H distances along the sequence of this small globular protein. It is readily apparent that two quantities are of interest, i.e. the extent and the uniqueness of secondary structure identification by a specified distance. For example, the β structures would be identified extensively by very short distances $d_{\alpha N}$, but $d_{\alpha N} \leq 2{\cdot}5$ Å is not unique for β-strands, since it occurs also in random coil segments and tight turns. Similarly, the helical residues are connected throughout by short distances d_{NN}, but again $d_{NN} \leq 3{\cdot}0$ Å is also found in random coil segments and tight turns. Compared to isolated short $d_{\alpha N}$ or d_{NN} values, patterns of several consecutive short values for these distances provide a more reliable, but not yet unique identification of β-strands or helical structures, respectively. On the other hand, distances $d_{\alpha\beta}(i, i+3)$ and $d_{\alpha N}(i, i+4)$ shorter than 4·5 Å provide a unique identification of helical structure in this protein, but the extent to which the helical residues are recognised is only about 50%.

Similar analyses to that in Figure 4 were conducted for the 19 proteins listed in Methods, section (c), and the results were added up. Plots of the distances

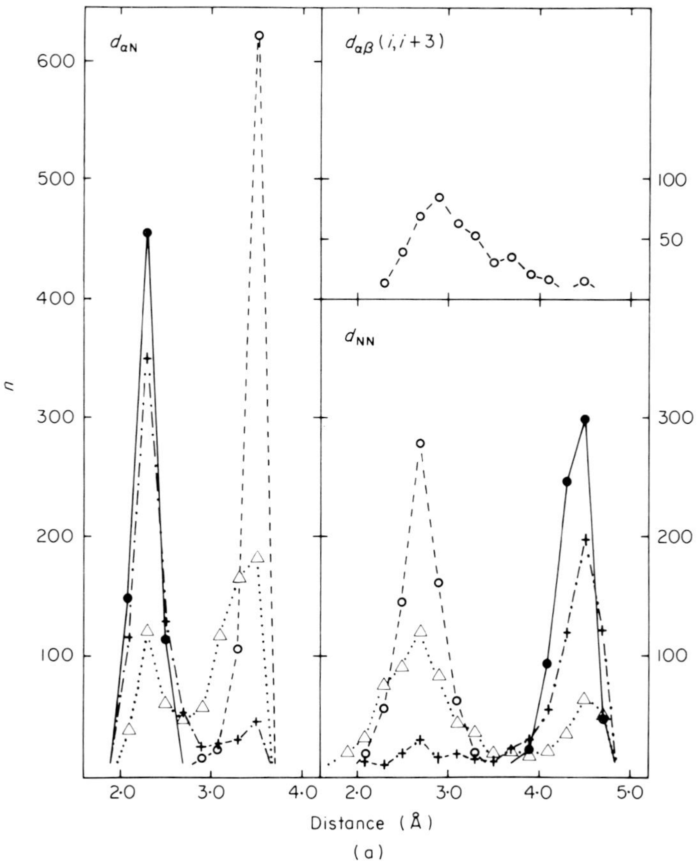

(a)

FIG. 5.

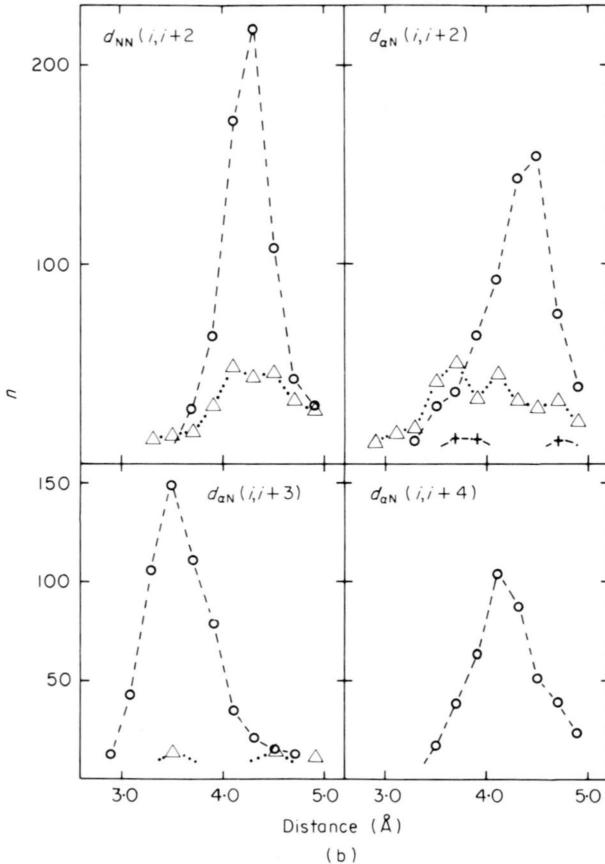

FIG. 5. Distribution of short distances adopted by sequential and medium range ^{1}H–^{1}H contacts in the secondary structures of the group of 19 protein crystal structures described in the text. The identification of secondary structure elements was taken from Kabsch & Sander (1983). The total number of times a specified ^{1}H–^{1}H contact occurs in each secondary structure type is contained in Table 5. On the vertical axis, n indicates how often the distance given on the horizontal axis was observed. Each data point represents the distance values over a range of 0·2 Å; for example, the point at 3·1 Å corresponds to the sum of ^{1}H–^{1}H distances larger or equal to 3·0 Å and smaller than 3·2 Å (n values smaller than 10 are not indicated). For clarity of presentation, the data points were connected with lines. The different secondary structures are identified by: (●) β-sheets (parallel and antiparallel sheets added up); (○) helix (α-helix and 3_{10} helix added up); (△) tight turns (all different types added up); (+) random coil.

adopted by the proton–proton contacts of Figure 3 in the different secondary structures are shown in Figure 5. This Figure shows that the distribution of short ^{1}H–^{1}H distances in the experimental protein structures coincides well with the expectations from the stereochemistry of standard secondary structures. The sequential distances $d_{\alpha N}$ and d_{NN} in helix and β structure are represented by narrow peaks centred about the theoretical values (Table 5). Short medium range ^{1}H–^{1}H distances are observed only in helical structures and tight turns, except for the appearance of some short distances $d_{\alpha N}(i, i+2)$ in random coil segments.

The data on $d_{\alpha\beta}(i, i+3)$ deserve special mention. Tables 1 and 2 predict a spread over the distance range 2·5 to 5·1 Å in helical structures. Figure 5(a) shows that this spread is indeed observed. However, there is a pronounced bias towards the lower end of the stereochemically allowed distance range, which makes $d_{\alpha\beta}(i, i+3)$ an interesting parameter for practical use in n.m.r. studies.

A quantitative evaluation of the extent to which the secondary structures determined by Kabsch & Sander (1983) would be identified with the use of specified $^1H-^1H$ distances is provided in Table 6. This Table lists the total number of times a given type of $^1H-^1H$ distance occurs in a particular secondary structure in the 19 proteins (Methods, section (c)) and the percentages of these distances that are shorter than three specified limits. These percentages were calculated by taking the sums of the counts in Figure 5 up to the limits indicated and dividing them by the total number. The distance limits used were chosen for the following reasons: 4·4 Å represents approximately the longest distance that one can presently expect to observe by NOESY; 3·6 Å corresponds to the longest, stereochemically allowed value for the sequential distance $d_{\alpha N}$ (Billeter *et al.*, 1982). In an experiment where $d_{\alpha N}$ is observed between all pairs of sequentially neighbouring residues in the protein, one can thus expect to detect all other distances up to a limit of $\leq 3{\cdot}6$ Å. Similarly, a calibration is available for the limit of $\leq 3{\cdot}0$ Å, since the maximum value for the intra-residue distance $d_{N\alpha}(i, i)$ is 2·8 Å (Leach *et al.*, 1977).

To compute the uniqueness of identification of a secondary structure type by one of the $^1H-^1H$ distances in Figure 3, the total number of times this distance is shorter than a predetermined limit in the 19 proteins was counted. The uniqueness was then obtained as the percentage relative to this total number of counts of short distances observed in the secondary structure of interest, as it was identified by Kabsch & Sander (1983) in these 19 proteins. In practice, the secondary structure identification must rely on a compromise for acceptably high extent (Table 6) combined with acceptably high uniqueness. Table 7 lists these two quantities for a selection of particularly useful distance parameters. (Corresponding data for all 7 distances of Fig. 3 in all different secondary structures can be computed from the data presented in Table 6.) As is further discussed in the following, the potentialities of n.m.r. for secondary structure determination do not rely on an optimal coincidence of high extent and high uniqueness for a single distance parameter, but rather on the combined data for a specific secondary structure obtained from a variety of different parameters, such as all the distances listed in Table 7 for helical structures.

(e) *Identification of helical and β-sheet secondary structure by segments of several subsequent short sequential distance constraints*

The sequential distances $d_{\alpha N}$ and d_{NN} are usually extensively investigated in the early phases of a n.m.r. study of a protein, since they play a pivotal role in the procedures used for obtaining sequence-specific resonance assignments (Billeter *et al.*, 1982). Furthermore, with the use of $d_{\alpha N}$ and d_{NN}, β-sheets and helical structures, respectively, can be identified to an extent of nearly 100% (Table 7).

TABLE 6

Extent of identification by sequential and medium range ^1H–^1H distances for the secondary structures attributed by Kabsch & Sander (1983) in a group of 19 proteins with 3227 residues

^1H–^1H contact	α-Helix + 3_{10} helix				β-Sheets				Turns + bends				Random coil				Mixed			
	Total no.	Short distances (%)			Total no.	Short distances (%)			Total no.	Short distances (%)			Total no.	Short distances (%)			Total no.	Short distances (%)		
		≤3·0 Å	≤3·6 Å	≤4·4 Å		≤3·0 Å	≤3·6 Å	≤4·4 Å		≤3·0 Å	≤3·6 Å	≤4·4 Å		≤3·0 Å	≤3·6 Å	≤4·4 Å		≤3·0 Å	≤3·6 Å	≤4·4 Å
$d_{\alpha N}$	796	4	99	100	767	98	100	100	818	41	99	100	824	87	100	100	—	—	—	—
d_{NN}	796	86	98	99	767	2	3	52	818	57	70	82	802	14	20	50	—	—	—	—
$d_{\alpha N}(i, i+2)$	698	0	7	55	614	0	0	2	478	4	19	53	420	1	4	13	973	1	8	32
$d_{\alpha N}(i, i+3)$	600	3	52	94	461	0	0	0	239	3	13	26	216	0	0	0	1645	0	2	8
$d_{\alpha N}(i, i+4)$	502	0	6	65	341	0	0	0	123	2	4	11	112	0	0	0	2083	0	1	2
$d_{NN}(i, i+2)$	698	0	2	72	614	0	0	0	478	3	9	38	401	0	0	2	970	0	3	11
$d_{\alpha\beta}(i, i+3)$	486	44	74	90	319	0	0	0	96	9	20	32	106	0	0	0	1880	2	4	8

For each secondary structure type, the total number of ^1H–^1H contacts indicated in the first column (see the text) and the percentage of the actual counts for 3 distance limits with respect to the total number are given.

The mixed category includes the distances between residues attributed to segments of different secondary structure (see Methods).

TABLE 7

^1H–^1H distances representing an acceptable compromise for high extent and high uniqueness of regular secondary polypeptide structure identification

Secondary structure	^1H–^1H distance constraint†	Extent of Identification‡ (%)	Uniqueness of identification§ (%)
α-Helix + 3_{10} helix	$d_{NN} \leq 3{\cdot}6$ Å	98	51
	$d_{\alpha N}(i, i+3) \leq 3{\cdot}6$ Å	52	81
	$d_{\alpha N}(i, i+3) \leq 4{\cdot}4$ Å	94	74
	$d_{\alpha N}(i, i+4) \leq 4{\cdot}4$ Å	65	84
	$d_{NN}(i, i+2) \leq 4{\cdot}4$ Å	72	62
	$d_{\alpha\beta}(i, i+3) \leq 3{\cdot}6$ Å	74	79
	$d_{\alpha\beta}(i, i+3) \leq 4{\cdot}4$ Å	87	71
$\beta\uparrow\downarrow + \beta\uparrow\uparrow$	$d_{\alpha N} \leq 3{\cdot}0$ Å	98	41

† In practice, secondary structure determinations by n.m.r. do not rely on a single one of these distance parameters, but either on a combination of different distances or on segments of several subsequent short distances of the same type (see the text).

‡ Indicates what percentage of the total number of residues in the specified secondary structure are recognized by this distance constraint.

§ Indicates what percentage of the residues recognized by this distance constraint are located in the specified secondary structure.

Since the uniqueness of identification of these secondary structures is, however, only of the order of 50% or less, we have further investigated how regular secondary structures could be identified by segments of a specified number of successive short distances $d_{\alpha N}$ or d_{NN} (see also Fig. 4). The results of this study (Tables 8 and 9) show that, as one would expect, the extent of secondary structure identification decreases and the uniqueness increases with increasing length of the segment. For identification of helical structures, segments of three to five successive distance constraints $d_{NN} \leq 3{\cdot}6$ Å give excellent results. For β-structures, an optimal compromise for satisfactory extent and uniqueness is obtained with segments of three to five distances $d_{\alpha N} \leq 2{\cdot}6$ Å.

TABLE 8

Extent and uniqueness of the identification of helical secondary structure by several subsequent short sequential distance constraints d_{NN}

Length of segment†	$d_{NN} \leq 3{\cdot}0$ Å		$d_{NN} \leq 3{\cdot}6$ Å	
	Extent‡	Uniqueness§	Extent‡	Uniqueness§
1	86	54	98	51
3	80	74	97	68
5	68	83	91	78
7	55	85	85	80

† Indicates the number of subsequent short distances d_{NN}.

‡ Indicates what percentage of the total number of helical residues are recognized by the specified sequence of distance constraints d_{NN}.

§ Indicates what percentage of the residues recognized by the specified sequence of distance constraints d_{NN} are located in helices.

TABLE 9

Extent and uniqueness of the identification of β-sheet secondary structure by several subsequent short sequential distance constraints $d_{\alpha N}$

Length of segment†	$d_{\alpha N} \leq 2{\cdot}6$ Å		$d_{\alpha N} \leq 3{\cdot}0$ Å	
	Extent‡	Uniqueness§	Extent‡	Uniqueness§
1	95	46	98	41
3	90	55	95	48
5	79	63	88	53
7	59	65	74	57

† Indicates the number of subsequent short distances $d_{\alpha N}$.

‡ Indicates what percentage of the total number of β-sheet residues are recognized by the specified sequence of distance constraints $d_{\alpha N}$.

§ Indicates what percentage of the residues recognized by the specified sequence of distance constraints $d_{\alpha N}$ are located in β-sheets.

(f) *Alignment of the individual strands in β-sheets using short interstrand distances between backbone protons*

Tables 4 to 7 and 9 show that an extended polypeptide segment, which is the typical conformation of the individual strands in β structures (Fig. 2), can be recognized from the prevalence of several subsequent, very short sequential distances $d_{\alpha N}$, and that additional, indirect evidence comes from the absence of short medium-range ^{1}H–^{1}H distances. However, the relatively low uniqueness of β-sheet identification by segments of three to seven subsequent short $d_{\alpha N}$ values (Table 9) further implies that all extended polypeptide segments are not integrated into double-stranded or multiple-stranded β-structures. Therefore, additional, direct evidence for the formation of β-sheets from observation of NOEs between protons in neighbouring strands (Fig. 2) is needed (Wagner *et al.*, 1981; Williamson *et al.*, 1984).

In regular antiparallel β-sheets, the long range distances $d_{\alpha N}(i,j)$, $d_{NN}(i,j)$ and $d_{\alpha\alpha}(i,j)$ are all sufficiently short for observation by NOESY, and in parallel β-sheets $d_{\alpha N}(i,j)$ and $d_{NN}(i,j)$ can be used (Table 4). Inspection of Figure 2 shows that the total number of contacts $d_{\alpha N}(i,j)$ between opposite residues in neighbouring strands is equal to the number of interstrand hydrogen bonds, and the total number of $d_{NN}(i,j)$ and $d_{\alpha\alpha}(i,j)$ contacts corresponds to one half of the number of hydrogen bonds. For a regular, untwisted antiparallel double-stranded β-sheet consisting of two tetrapeptide segments one thus expects two short contacts $d_{\alpha\alpha}(i,j)$, two short $d_{NN}(i,j)$ values and four short $d_{\alpha N}(i,j)$ values, and in a corresponding parallel β-sheet one expects two short $d_{NN}(i,j)$ values and four short $d_{\alpha N}(i,j)$ values. The total number of, respectively, eight or six short interstrand distances between individual assigned backbone protons would obviously be sufficient evidence for correct alignment of the two strands. Furthermore, observation of these interstrand ^{1}H–^{1}H distances provides direct evidence for the formation of the typical β-sheet hydrogen bonds.

The short interstrand ^{1}H–^{1}H distances in the 19 protein structures used in Table 6 were counted separately for parallel and antiparallel β-sheets, as is described in the legend to Table 10. The results obtained imply that the number of short inter-strand ^1H–^1H distances in the β-structures of globular proteins is quite generally sufficiently high for correct alignments of neighbouring strands to be achieved from observation of the corresponding NOEs. For antiparallel β-sheets, these alignments have to rely primarily on observation of short distances $d_{\alpha\alpha}(i,j)$ and $d_{NN}(i,j)$, whereas for parallel β-sheets short distances $d_{\alpha N}(i,j)$ are most useful.

4. Conclusions for Determination of Polypeptide Secondary Structure using NOE Distance Constraints

The following conclusions are valid when sequence-specific resonance assignments are available for the backbone and C^β protons of a protein (Wüthrich *et al.*, 1982). Furthermore, it is assumed that from measurements of NOEs with different mixing times, spin diffusion (Kalk & Berendsen, 1976) was excluded as the cause for the observed NOEs (Anil Kumar *et al.*, 1981) and very short distances between backbone protons, say $\lesssim 2{\cdot}8$ Å, were distinguished from distances in the range approximately 2·8 to $\lesssim 4{\cdot}5$ Å (Braun *et al.*, 1981,1983). It has been demonstrated that such data can be obtained for small proteins (e.g. see Wagner *et al.*, 1981; Braun *et al.*, 1983; Zuiderweg *et al.*, 1983; Williamson *et al.*, 1984).

(a) *General conclusions*

For all secondary structure types, the data given in Results and Discussion show that identification from NOE data relies almost entirely on distance constraints with amide protons. Therefore, the n.m.r. data must be collected under conditions where the backbone amide protons can be observed, i.e. in non-deuterated aqueous solution with pH values $\lesssim 6$, or possibly in organic solvents (Wüthrich *et al.*, 1982).

TABLE 10

Short long-range, inter-strand distances between backbone hydrogen atoms in β-sheets

^1H–^1H contact	Antiparallel β			Parallel β		
	≤3·0 Å	≤3·6 Å	≤4·0 Å	≤3·0 Å	≤3·6 Å	≤4·0 Å
$d_{\alpha N}(i,j)$	0	2	5	3	5	5
$d_{NN}(i,j)$	2	3	3	0	1	2
$d_{\alpha\alpha}(i,j)$	2	3	3	0	0	1

For this Table, all β-strands in the 19 proteins of Table 5 that contain 4 or more residues were considered. There are 53 pairs of such antiparallel neighbouring strands, with an average length of 5·7 residues, and 17 pairs of parallel neighbouring strands, with an average length of 4·6 residues. For the specified ^1H–^1H distances, total counts over all these pairs of strands were made and then divided by the number of pairs. The Table thus indicates the average number of times a specified short distance occurs between a pair of strands in a β-sheet (rounded to integers).

Observation of short sequential distances $d_{\alpha\alpha}$ or $d_{N\alpha}$ provides direct evidence for *cis* peptide bonds (Arseniev *et al.*, 1983).

Both helical and sheet regular secondary structures contain a dense network of short ^1H–^1H distances. The identification of these structures will, in practice, rely on the combination of observations on all available distance parameters (Fig. 4), possibly complemented further by data on spin–spin couplings and location of hydrogen bonds (Williamson *et al.*, 1984). While there is no single observable parameter that would lead to extensive and unique identification of a particular structure type, it is highly unlikely that a regular secondary structure would escape detection when the search extends over all available parameters.

(b) *Helical structures*

Tables 7 and 9 show that there are several different distance criteria that can provide helix identifications with high extent and high uniqueness. Only a small proportion of these short distances needs to be observed to characterize the main body of a helix. The reliability of determination of the residues where the helix starts and ends will, however, depend on the availability of more complete sets of distance constraints for these residues (see also section (c), below), and can be improved greatly by additional data on the amide proton–C^α proton spin–spin coupling constants (Pardi *et al.*, 1984) and the location of hydrogen bonds (Zuiderweg *et al.*, 1983; Williamson *et al.*, 1984).

The α-helix is by far the most abundant helical structure in globular proteins, but short segments of 3_{10} helix have been observed also in protein crystal structures (Richardson, 1981). The possibilities for distinction of the two helix types from NOE distance constraints are rather limited (Table 5). Primarily, the different periodicities of the two structures are manifested by short $d_{\alpha N}(i, i+2)$ values in the 3_{10} helix and short $d_{\alpha N}(i, i+4)$ values in the α-helix. Furthermore, observation of short $d_{\alpha\beta}(i, i+3)$ values would be indicative of the prevalence of α-helical structure.

(c) *β-Sheets*

The outstandingly short sequential distances $d_{\alpha N}$ in extended peptide segments (Table 5) give rise to particularly prominent NOEs (Kuo & Gibbons, 1980; Dubs *et al.*, 1979), so that these structure elements are usually readily identified by n.m.r. The distinction of the extended strands that are located in β-sheets must come from observation of interstrand ^1H–^1H NOEs (Tables 4 and 10). In principle, observation of two short inter-strand distances is sufficient to define the relative polarity of two adjoining strands, and observation of short $d_{\alpha\alpha}(i, j)$ values provides an independent criterion for identification of antiparallel β structure (Tables 4 and 10). If a more complete set of interstrand short contacts is observed, the extent of the β-sheet and local distortions, such as β-bulges (Richardson, 1981) will also be accessible for determination by n.m.r. (Wagner *et al.*, 1981; Williamson *et al.*, 1984). Independent checks on the results obtained can be obtained from measurements of the amide proton–C^αproton spin–spin couplings (Pardi *et al.*, 1984) and from identification of the inter-strand hydrogen bonds (Fig. 2) by amide proton exchange studies (Wagner & Wüthrich, 1982).

(d) *Tight turns*

The most favourable situation for identification of a tight turn with the use of local NOE distance constraints is in the hairpins formed in antiparallel β-sheets, since both the sequential and medium range ^1H–^1H distances in the turn are characteristically different from those in the adjoining extended β-strands (Table 5). It should then also be possible to distinguish between different types of turns (Fig. 1). In contrast, it may be difficult to identify a tight turn located at either end of a helix, because the local distance constraints are overall closely similar in helices and turns (Table 5). In general, with the exception of the hairpin turns in antiparallel β-structures, it appears advisable to defer locating tight turns to the tertiary structure determination with a complete set of local and long range NOE distance constraints, for example with the use of distance geometry calculations (Braun *et al.*, 1981,1983; Havel & Wüthrich, 1984 and unpublished results).

APPENDIX

Short Intramolecular ^1H–^1H Distances in Polypeptide Chains Containing Glycine, Proline or *cis* Peptide Bonds

(a) *Proton–proton distances with C^α protons of glycyl residues*

The C^α proton resonances of glycine cannot usually be stereospecifically assigned (Wüthrich *et al.*, 1983). The definitions of the ^1H–^1H contacts with C^α protons ((1), (4) and (5) in the main text) are therefore modified so as to select always the shortest distance:

$$d_{\alpha N}^{GX}(i,j) \equiv \min\{d(C^\alpha H_i, NH_j)\}, \tag{A1}$$

$$d_{\alpha\alpha}^{GX}(i,j) \equiv \min\{d(C^\alpha H_i, C^\alpha H_j)\}, \tag{A2}$$

$$d_{\alpha\beta}^{GX}(i,j) \equiv \min\{d(C^\alpha H_i, C^\beta H_j)\}. \tag{A3}$$

Corresponding definitions hold for $d_{N\alpha}^{XG}(i,j)$, $d_{\alpha\alpha}^{XG}(i,j)$ (or $d_{\alpha\alpha}^{GG}(i,j)$) and $d_{\beta\alpha}^{XG}(i,j)$, where X indicates one of the common L-amino acids.

Data sets corresponding to Tables 1 to 4 were computed for the distances (A1) to (A3) in the different secondary structures. The following description covers all the distances involving C^α protons of glycine that are shorter than the corresponding distances in Tables 1 to 4 and that are also shorter than 4·5 Å (and therefore of practical interest for n.m.r. studies). Considering the distances in the order they are listed in Tables 1 to 4, we first observe that the sequential connectivity $d_{\alpha N}^{GX}$ is not a reliable parameter for characterization of the conformation. $d_{\alpha N}^{GX}$ adopts short distances over a wide range of values of the torsion angle ψ, which correspond to extended forms of the polypeptide chain as well as to helical structures (Billeter *et al.*, 1982). In the α-helix, we have that $d_{\alpha\alpha}^{GX} = 4{\cdot}3$ Å, as compared to $d_{\alpha\alpha} = 4{\cdot}7$ Å, and $d_{\alpha\alpha}^{XG}(i, i+3) = 3{\cdot}9$ Å, as compared to $d_{\alpha\alpha}(i, i+3) = 5{\cdot}3$ Å (Table 1). In the 3_{10} helix, we find $d_{\alpha\alpha}^{GX} = 4{\cdot}3$ Å, as compared to $d_{\alpha\alpha} = 4{\cdot}7$ Å, and $d^{XG}_{\alpha\alpha}(i, i+3) = 4{\cdot}4$ Å, as compared to $d_{\alpha\alpha}(i, i+3) = 5{\cdot}4$ Å (Table 2). In type I turns, $d^{GX}_{\alpha\alpha}(2, 3) = 4{\cdot}3$ Å, as compared to $d_{\alpha\alpha}(2, 3) = 4{\cdot}5$ Å; in type II turns, $d^{GX}_{\alpha\beta}(2, 3) \geq 3{\cdot}7$ Å, as compared to $d_{\alpha\beta}(2, 3) \geq 4{\cdot}1$ Å, and $d^{GX}_{\beta\alpha}(2, 3) \geq 4{\cdot}3$ Å, as compared to $d_{\beta\alpha}(2, 3) \geq 4{\cdot}8$ Å (Table 3). The long range backbone distance $d^{GX}_{\alpha\alpha}(i, j)$ between neighbouring strands in parallel β-sheets can be as short as 3·5 Å, as compared to $d_{\alpha\alpha}(i, j) = 4{\cdot}8$ Å (Table 4). In all other instances, the distances (A1) to (A3) deviate by less than 0·1 Å from the corresponding distances in Tables 1 to 4 (or they are longer than 4·5 Å). Essentially identical values prevail for distances between pairs of glycyl residues as between Gly and an L-amino acid.

(b) *Sequential proton–proton distances with prolyl residues*

Compared to L-amino acid residues, the most obvious new aspects in ^1H–^1H contacts with prolyl residues arise from the absence of the amide proton and the steric constraints on rotations about the torsion angles ϕ and χ^1. A set of ^1H–^1H distances corresponding to those of equations (1) to (5) can be defined, when the distances are referred to $C^{\delta}H_2$ of proline instead of an amide proton and the shorter of the two distances to the C^{δ} methylene protons is taken. The ranges of values adopted by all the different sequential proton–proton distances with proline are listed in Table A1. In each case, the dipeptide fragment considered is specified by superscripts, i.e. d^{XX}, d^{PX}, d^{XP} and d^{PP}, where P stands for proline and X for one of the common L-amino acid residues.

In as far as it is manifested in the data of Table A1, the angular dependence of the sequential distances in proline peptides is qualitatively similar to that for the corresponding proton–proton distances with L-amino acid residues (Billeter *et al.*, 1982). In the following, this is discussed in more detail for the distances corresponding to d_{NN} and $d_{\alpha N}$. Figure A1 shows the dependence of $d^{XP}_{N\delta}$ on the two torsion angles ϕ_i and ψ_i. These data were computed with the same procedure as the corresponding sequential connectivity d_{NN} (Appendix of Billeter *et al.* (1982)),

TABLE A1

Sequential proton–proton distances in dipeptides with trans-*proline*

Dipeptide fragment†	Range of sequential ^1H–^1H distances (in Å)‡			
X-X	d_{NN}: –4·7	$d_{N\alpha}$: 3·5–6·5	$d_{\alpha N}$: 2·2–3·6	$d_{\alpha\alpha}$: 4·2–5·2
Pro-X	$d^{PX}_{\delta N}$: 2·8–5·7	$d^{PX}_{\delta\alpha}$: 4·8–7·1	$d^{PX}_{\alpha N}$: 2·2–3·6	$d^{PX}_{\alpha\alpha}$: 4·2–5·2
X-Pro	$d^{XP}_{N\delta}$: –4·8	$d^{XP}_{N\alpha}$: 3·8–6·1	$d^{XP}_{\alpha\delta}$: 2·1–3·8	$d^{XP}_{\alpha\alpha}$: 4·3–4·8
Pro-Pro	$d^{PP}_{\delta\delta}$: –5·7	$d^{PP}_{\delta\alpha}$: 5·3–6·5	$d^{PP}_{\alpha\delta}$: 2·1–3·8	$d^{PP}_{\alpha\alpha}$: 4·3–4·8

† X stands for one of the common L-amino acids.

‡ When $C^{\delta}H_2$ of proline is involved, the shorter of the distances to the 2 C^{δ} methylene protons is taken. Only an upper limit is indicated when the lower limit would be shorter than the van der Waals' limit of 2·0 Å.

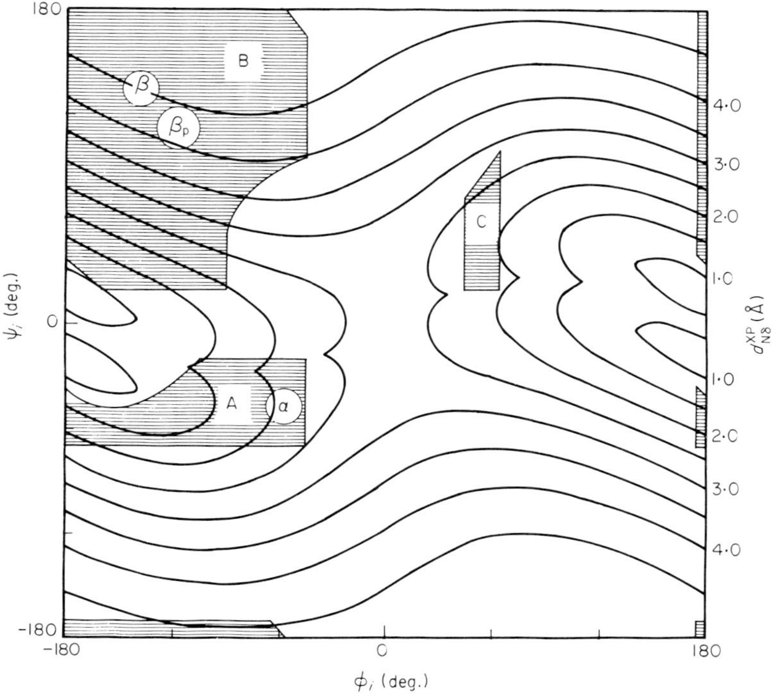

FIG. A1. $\phi_i - \psi_i$ plane with contour lines for fixed values of the distance $d^{XP}_{N\delta}$ between NH_i and proline $C^{\delta}H_{i+1}$ in *trans* X-Pro. The contour lines were computed with equation (A16) of Billeter *et al.* (1982), using the ECEPP parameters for Pro in position $i+1$ and the average of the ECEPP parameters for the other common amino acid residues for position i. The $d^{XP}_{N\delta}$ values for the individual contour lines are given on the right. The areas in the $\phi_i - \psi_i$ plane that are sterically allowed for an alanylalanine dipeptide (Ramachandran & Sasisekharan, 1968) are hatched and labelled A, B and C. The $\phi_i - \psi_i$ combinations for the regular α-helix and parallel and antiparallel β-structures are indicated by α, β_p and β, respectively.

whereby a standard planar structure of the proline ring was used (Momany *et al.*, 1975) and all peptide bonds were assumed to be in the *trans* form (see section (c), below). The computations used molecular structures with dimensionless atoms, so that distances were also obtained for sterically forbidden combinations of the dihedral angles. In Figure A1, the ϕ_i–ψ_i plane contains, in addition to the contour lines for predetermined values of $d^{XP}_{N\delta}$, also an outline of the sterically allowed regions, A, B and C, for an alanylalanine dipeptide (Ramachandran & Sasisekharan, 1968). It is seen that for most of the region A and a small part of region B the distance $d^{XP}_{N\delta}$ is shorter than 2·0 Å and therefore sterically not allowed. Otherwise, the dependence on ϕ_i and ψ_i is closely similar for $d^{XP}_{N\delta}$ and for d_{NN} (Billeter *et al.*, 1982). For β-sheets, both distances are near 4·2 Å, and for the ϕ_i–ψ_i combination corresponding to a regular α-helix, $d_{NN} = 2{\cdot}8$ Å and $d^{XP}_{N\delta} = 2{\cdot}1$ Å.

Figure A2 contains a plot of $d^{PP}_{\alpha\delta} (\equiv d^{XP}_{\alpha\delta})$ *versus* the dihedral angle ψ. The curve is closely similar to a corresponding presentation of $d_{\alpha N}$ *versus* ψ (Billeter *et al.*,

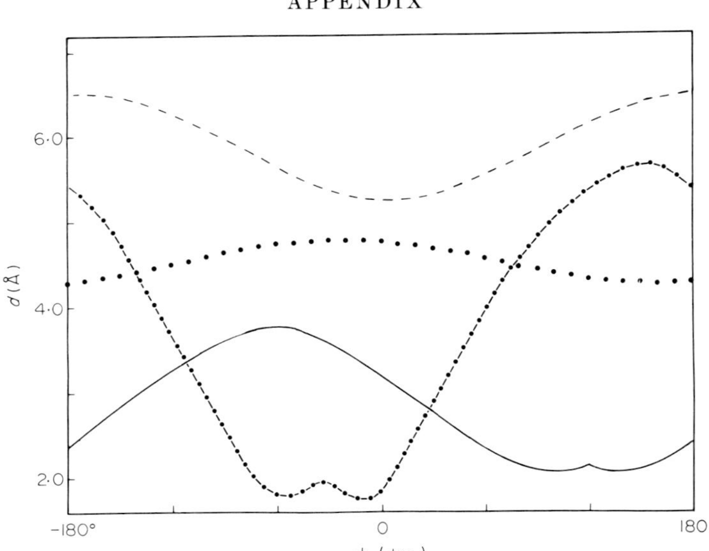

FIG. A2. Plot of the distances $d^{PP}_{\alpha\delta}$ (————), $d^{PP}_{\delta\alpha}$ (— — — —), $d^{PP}_{\alpha\alpha}$ (····) and $d^{PP}_{\delta\delta}$ (—·—·—) *versus* the dihedral angle ψ_i for *trans* Pro-Pro. The curves were computed with equation (A7) of Billeter *et al.* (1982), using a rigid, planar conformation for the proline ring.

1982). $d^{XP}_{\alpha\delta}$ would be approximately 2·1 Å for a regular β structure and approximately 3·7 Å for a regular α-helix, as compared to 2·2 Å and 3·5 Å, respectively, for $d_{\alpha N}$ (Tables 1 and 4). Overall, the sequential connectivities $d^{XP}_{\alpha\delta}$ and $d^{XP}_{N\delta}$ thus provide a clear discrimination between helical and extended polypeptide conformations.

When a rigid planar proline ring is assumed, the sequential distance $d^{PP}_{\delta\delta}$ depends on a single torsion angle, ψ. As can be seen from Figure A2, $d^{PP}_{\delta\delta}(\psi)$ varies between the van der Waals' distance of 2·0 Å and 5·7 Å, and would be approximately 2·0 Å for a regular α-helix and 5·5 Å for an extended chain. A similar dependence on ψ prevails for $d^{PX}_{\delta N}$, which varies between 2·8 and 5·7 Å, and is approximately 2·9 Å for an α-helix and 5·7 Å for an extended chain. Thus $d^{PX}_{\delta N}$, or $d^{PP}_{\delta\delta}$, respectively, are also suitable distances for discrimination between extended and helical secondary structures.

(c) *Sequential proton–proton distances across* cis *and* trans *peptide bonds*

In the dipeptide fragments X-Pro (where X stands for any of the common amino acids), the occurrence of *cis* peptide bonds is well-documented (Huber & Steigemann, 1974; Grathwohl & Wüthrich, 1976; Schulz & Schirmer, 1979; Richardson, 1981) and in other dipeptide fragments the *cis* form should not be excluded *a priori* when the polypeptide adopts a globular conformation (Grathwohl & Wüthrich, 1976; Hetzel & Wüthrich, 1979). It is therefore of interest to investigate whether *cis* and *trans* peptide bonds can be distinguished also on the basis of different, short ^{1}H–^{1}H contacts.

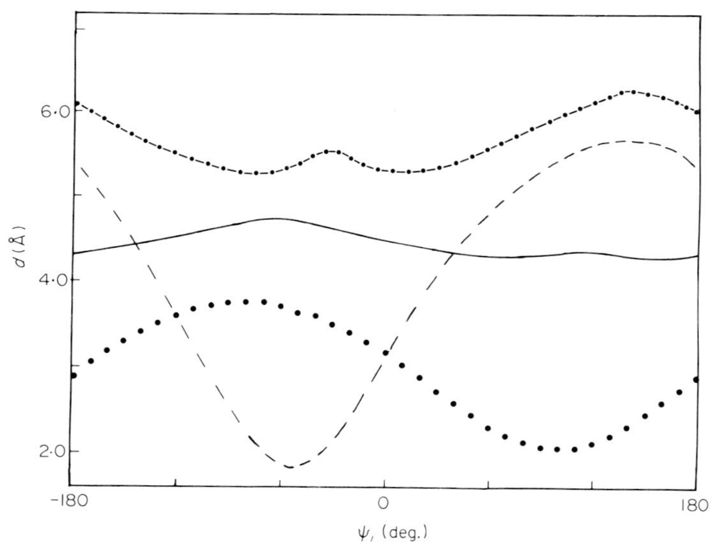

FIG. A3. Plot of the distances $d_{\alpha\delta}^{\mathrm{PPcis}}$ (———), $d_{\delta\alpha}^{\mathrm{PPcis}}$ (— — — —), $d_{\alpha\alpha}^{\mathrm{PPcis}}$ (····) and $d_{\delta\delta}^{\mathrm{PPcis}}$ (—·—·—) *versus* the dihedral angle ψ for *cis* Pro-Pro. The curves were computed with the same procedures as Fig. A2, and using a planar conformation for the proline ring.

Since "regular secondary structures" do not include *cis* peptide bonds, the following considerations are limited to sequential distances. $d_{\alpha\alpha}$, $d_{\alpha\delta}^{\mathrm{XP}}$ (or $d_{\alpha\delta}^{\mathrm{PP}}$, or $d_{\alpha\mathrm{N}}$), $d_{\delta\alpha}^{\mathrm{PX}}$ (or $d_{\delta\alpha}^{\mathrm{PP}}$, or $d_{\mathrm{N}\alpha}$) and $d_{\mathrm{N}\delta}^{\mathrm{XP}}$ (or $d_{\delta\delta}^{\mathrm{PP}}$, or d_{NN}) were found to be most useful for discriminating between *cis* and *trans* peptide bonds. For simplicity of presentation, plots of these four distances *versus* the dihedral angle ψ for the *trans* and *cis* forms of the peptide bond are shown for Pro-Pro (Figs A2 and A3). Comparison of the two Figures shows that the curve describing the dependence of $d_{\alpha\alpha}^{\mathrm{PP}}$ on ψ for the *cis* peptide bond is nearly identical to the curve for $d_{\alpha\delta}^{\mathrm{PP}}$ of the *trans* peptide bond, and *vice versa*. For these two distances, a similar presentation for *cis* and *trans* X-Pro was presented by Arseniev *et al.* (1983). Since $d_{\alpha\alpha}$ across a *trans* peptide bond cannot be shorter than 4·2 Å (Fig. A2), not even in glycine peptides, observation of a short $d_{\alpha\alpha}$ distance provides direct evidence for a *cis* peptide bond. (If residue $i+1$ is not proline, $d_{\alpha\alpha}$ will depend on two dihedral angles, ψ_i and ϕ_{i+1}, and for *cis* peptide bonds it will adopt values shorter than 3·0 Å for a sizeable area in the ψ_i–ϕ_{i+1} plane.) Similarly, qualitatively nearly identical ψ dependences are observed for $d_{\delta\alpha}^{\mathrm{PP}}$ in the *cis* peptide (Fig. A3) and $d_{\delta\delta}^{\mathrm{PP}}$ in the *trans* peptide (Fig. A2), and *vice versa*. Again, since for *trans* peptide bonds $d_{\delta\alpha}^{\mathrm{PP}}$ is always longer than 5·2 Å (Fig. A2) and the corresponding distance $d_{\mathrm{N}\alpha}$ in non-proline peptides is always longer than 3·5 Å (and longer than 4·7 Å in any of the regular secondary structures (Tables 1 to 4)), observation of a short, sequential contact $d_{\delta\alpha}^{\mathrm{PP}}$ (or $d_{\delta\alpha}^{\mathrm{PX}}$, or $d_{\mathrm{N}\alpha}$) constitutes direct evidence for the presence of a *cis* peptide bond.

We thank Dr G. Wagner, Dr M. P. Williamson, Dr E. R. P. Zuiderweg and Mrs J. Richardson for helpful discussions on various aspects of this investigation and Mrs E.

Huber, Mrs E. H. Hunziker and Mr R. Marani for the careful preparation of the manuscript and the illustrations. Financial support was obtained from the Schweizerischer Nationalfonds (project 3.284.82) and through a special grant of the Eidgenössische Technische Hochschule (ETH) Zürich. Use of the facilities at the Zentrum für Interaktives Rechnen (ZIR) of the ETH is gratefully acknowledged.

REFERENCES

Anil Kumar, Wagner, G., Ernst, R. R. & Wüthrich, K. (1981). *J. Amer. Chem. Soc.* **103**, 3654–3658.

Arseniev, A. S., Kondakov, V. I., Maiorov, V. N., Volkova, T. M., Grishin, E. V., Bystrov, V. F. & Ovchinnikov, Yu. A. (1983). *Bioorgan. Khim.* **9**, 768–793.

Bernstein, F. C., Koetzle, T. F., Williams, G. J. B., Meyer, E. F. Jr., Brice, M. D., Rodgers, J. R., Kennard, D., Shimanouchi, T. & Tasumi, M. (1977). *J. Mol. Biol.* **112**, 535–542.

Billeter, M., Braun, W. & Wüthrich, K. (1982). *J. Mol. Biol.* **155**, 321–346.

Braun, W., Bösch, C., Brown, L. R., Gō, N. & Wüthrich, K. (1981). *Biochim. Biophys. Acta,* **667**, 377–396.

Braun, W., Wider, G., Lee, K. H. & Wüthrich, K. (1983). *J. Mol. Biol.* **169**, 921–948.

Chou, P. Y. & Fasman, G. D. (1977). *J. Mol. Biol.* **115**, 135–175.

Crawford, J. L., Lipscomb, W. N. & Schellman, C. G. (1973). *Proc. Nat. Acad. Sci., U.S.A.* **70**, 538–542.

Deisenhofer, J. & Steigemann, W. (1975). *Acta Crystallogr.* **B31**, 238–250.

Dubs, A., Wagner, G. & Wüthrich, K. (1979). *Biochim. Biophys. Acta,* **577**, 177–194.

Grathwohl, C. & Wüthrich, K. (1976). *Biopolymers,* **15**, 2042–2057.

Havel, T. F. & Wüthrich, K. (1984). *Bull. Math. Biol.* In the press.

Hetzel, R. & Wüthrich, K. (1979). *Biopolymers,* **18**, 2589–2606.

Huber, R. & Steigemann, W. (1974). *FEBS Letters,* **48**, 235–237.

Isogai, Y., Nemethy, G., Rackovsky, S., Leach, S. J. & Scheraga, H. A. (1980). *Biopolymers,* **19**, 1183–1210.

Kabsch, W. & Sander, C. (1983). *Biopolymers,* **22**, 2577–2637.

Kalk, A. & Berendsen, H. J. C. (1976). *J. Magn. Reson.* **24**, 343–366.

Kolaskar, A. S., Ramabrahmam, V. & Soman, K. V. (1980). *Int. J. Pept. Protein Res.* **16**, 1–11.

Kuntz, I. D. (1972). *J. Amer. Chem. Soc.* **94**, 4009–4012.

Kuo, M. & Gibbons, W. A. (1979). In *Peptides: Structure and Biological Function, Proc. 6th Amer. Peptide Symp.* (Gross, E. & Meienhofer, J., eds), pp. 229–232, Pierce Chem. Co., Rockford.

Kuo, M. & Gibbons, W. A. (1980). *Biophys. J.* **32**, 807–836.

Leach, S. J., Némethy, G. & Scheraga, H. A. (1977). *Biochem. Biophys. Res. Commun.* **75**, 207–215.

Levitt, M. & Greer, J. (1977). *J. Mol. Biol.* **114**, 181–293.

McLachlan, A. D. (1979). *J. Mol. Biol.* **128**, 49–79.

Momany, F. A., McGuire, R. F., Burgess, A. W. & Scheraga, H. A. (1975). *J. Phys. Chem.* **79**, 2361–2381.

Némethy, G., Pottle, M. S. & Scheraga, H. A. (1983). *J. Phys. Chem.* **87**, 1883–1887.

Pardi, A., Billeter, M. & Wüthrich, K. (1984). *J. Mol. Biol.* **180**, 741–751.

Poulsen, F. M., Hoch, J. C. & Dobson, G. M. (1980). *Biochemistry,* **19**, 2597–2607.

Ramachandran, G. N. & Sasisekharan, V. (1968). *Advan. Protein Chem.* **23**, 283–437.

Richardson, J. S. (1981). *Advan. Protein Chem.* **34**, 167–339.

Rose, G. D. & Seltzer, J. P. (1977). *J. Mol. Biol.* **113**, 153–164.

Salemme, F. R. & Weatherford, D. W. (1981). *J. Mol. Biol.* **146**, 101–117.

Schulz, G. E. & Schirmer, R. H. (1979). *Principles of Protein Structure,* Springer-Verlag, New York.

Schwyzer, R. (1958). In *Amino Acids and Peptides with Antimetabolic Activity* (Wolstenholme, G. E. W., ed.), p. 171, J. and A. Churchill, Ltd., London.

Wagner, G. & Wüthrich, K. (1982). *J. Mol. Biol.* **160**, 334–361.

Wagner, G., Anil Kumar & Wüthrich, K. (1981). *Eur. J. Biochem.* **114**, 375–384.

Wemmer, D. & Kallenbach, N. R. (1983). *Biochemistry* **22**, 1901–1906.

Williamson, M. P., Marion, D. & Wüthrich, K. (1984). *J. Mol. Biol.* **173**, 341–359.

Wüthrich, K. Billeter, M. & Braun, W. (1983). *J. Mol. Biol.* **169**, 949–961.

Wüthrich, K., Wider, G., Wagner, G. & Braun, W. (1982). *J. Mol. Biol.* **115**, 311–319.

Zuiderweg, E. R. P., Kaptein, R. & Wüthrich, K. (1983). *Proc. Nat. Acad. Sci., U.S.A.* **80**, 5837–5841.

Zuiderweg, E. R. P., Billeter, M., Boelens, R., Scheek, R. M., Wüthrich, K. & Kaptein, R. (1984). *FEBS Letters*, **174**, 243–247.

Edited by C. R. Cantor

Reprinted from *Biochimica et Biophysica Acta*, Vol. 667, pp. 377–396 (1981)
Copyright © 1981, with kind permission from
Elsevier Science B. V. Amsterdam, The Netherlands.

BBA 38613

COMBINED USE OF PROTON-PROTON OVERHAUSER ENHANCEMENTS AND A DISTANCE GEOMETRY ALGORITHM FOR DETERMINATION OF POLYPEPTIDE CONFORMATIONS

APPLICATION TO MICELLE-BOUND GLUCAGON

WERNER BRAUN, CHRIS BÖSCH, LARRY R. BROWN, NOBUHIRO GŌ * and KURT WÜTHRICH

Institut für Molekularbiologie und Biophysik, Eidgenössische Technische Hochschule, 8093 Zürich (Switzerland)

(Received August 7th, 1980)

Key words: ^{1}H-NMR; Polypeptide conformation; Distance geometry algorithm; Glucagon; Nuclear Overhauser enhancement

Summary

In a new approach for the determination of polypeptide conformation, experimental data on intramolecular distances between pairs of hydrogen atoms obtained from nuclear Overhauser enhancement studies are used as input for a distance geometry algorithm. The algorithm determines the limits of the conformation space occupied by the polypeptide chain. The experimental data are used in such a way that the real conformation should in all cases be within these limits. Two important features of the method are that the results do not depend critically on the accuracy of the distance measurements by nuclear Overhauser enhancement studies and that internal mobility of the polypeptide conformation is explicitly taken into consideration. The use of this new procedure is illustrated with a structural study of the region 19—27 of glucagon bound to perdeuterated dodecylphosphocholine micelles.

Introduction

The potential of high resolution proton nuclear magnetic resonance (NMR) for the determination of spatial structures of polypeptide chains in proteins has long been recognized. However, practical uses have been curtailed by the limited resolution of protein ^{1}H-NMR spectra and by the lack of quantitative

* Permanent address: Department of Physics, Kyushu University 33, Fukuoka 812, Japan.

correlations between NMR parameters and polypeptide conformation [1,2]. Recent rapid progress in NMR techniques has dramatically improved spectral resolution to the point that a large number of spectral parameters can be obtained from ^{1}H-NMR spectra of small and medium sized proteins [2—6]. Further investigations of correlations between NMR parameters and spatial arrangement of polypeptide chains are therefore particularly interesting. This paper describes a new concept in which systematic measurements of intramolecular proton-proton Overhauser effects [5,7—9] are used in conjunction with a distance geometry algorithm [10,11] to determine limits to the conformation space occupied by a polypeptide chain. Special consideration was given to the effects of internal mobility of protein structure [1,2,12].

The combined use of nuclear Overhauser effects and a distance geometry algorithm is illustrated with a determination of the conformation space occupied by a fragment of micelle-bound glucagon [13]. Since different molecular conformations were observed for glucagon in single crystals [14] and in dilute aqueous solution [15] it would appear difficult to predict from these data the glucagon structure to be found in a lipid/water interface. Investigations of the spatial structure of lipid-bound glucagon are therefore of special interest with regard to possible modes of action of this polypeptide hormone at the receptor site.

Methods

When adverse effects of spin diffusion are eliminated by suitable selection of the experimental parameters, nuclear Overhauser effects can in principle be used to measure the distances between different groups of protons in a protein [5,8,9,16,18]. In practice, however, the inherent internal flexibility of protein molecules [1,2,12] tends to make accurate distance measurements difficult. The proton-proton distances may vary with time and different effective rotational correlation times may prevail for the modulation of dipole-dipole coupling between different groups of protons in the molecule [1]. Rather than attempting accurate measurements of the distances between individual protons, we used the nuclear Overhauser data to determine upper limits for selected proton-proton distances. By use of a distance geometry algorithm these data were then combined with the steric limitations imposed by the covalent structure [17] to determine the bounds of the conformation space which may be occupied by the polypeptide chain. In the following, details of this approach are described.

Aquisition of proton-proton Overhauser enhancement data for use with the distance geometry algorithm

The nuclear Overhauser effect is the fractional change in intensity by cross-relaxation of one NMR line, b, when another resonance, a, is perturbed. In a system which contains, besides a and b, additional weakly coupled spins denoted by i, the time course of the magnetization may be written as [7,9]

$$\frac{\mathrm{d}M_b(t)}{\mathrm{d}t} = -\rho_b M_b(t) - \sum_i \sigma_{bi} M_i(t) - \sigma_{ba} M_a(t) \qquad (1)$$

M_a, M_b and M_i are the differences between the actual magnetization, M_z, and the equilibrium magnetization, $\tilde{M}_0$, of the respective spins, ρ_b is the local spin-lattice relaxation rate of spin b, and σ_{bi} and σ_{ba} are the cross-relaxation rates given by

$$\sigma_{ij} = f(\tau_{ij})/r_{ij}^6 \tag{2}$$

where r_{ij} is the distance between spins i and j, and $f(\tau_{ij})$ is a function of the correlation time τ_{ij} for the dipole-dipole interactions between spins i and j. When macromolecular systems are studied at high fields, spin diffusion by cross-relaxation tends to mask the distance information contained in nuclear Overhauser effects [7—9]. In truncated driven Overhauser experiments [9] the free induction decay is accumulated immediately after pre-irradiation of a selected resonance during a time period, t_1, which is short compared to the irradiation time necessary to obtain a steady-state Overhauser effect [18]. In a typical experiment of this type, the irradiated resonance can be saturated essentially instantaneously [9]. Since $M_b(0) = M_i(0) = 0$, Eqn. 1 reduced to a simpler form (Eqn. 3) for short pre-irradiation times t_1

$$M_b(t) = -\sigma_{ba} M_a^0 t \tag{3}$$

where M_a^0 is the steady-state magnetization of the pre-irradiated spin a. The cross relaxation rate σ_{ba} between the irradiated and the observed spin can thus be measured under conditions where the effects of cross-relaxation with other spins and local spin-lattice relaxation are not important. In truncated driven Overhauser experiments with small and medium sized proteins, Eqn. 3 presents a good approximation for the time course of $M_b(t)$ for pre-irradiation times, t_1, shorter than approx. 0.5 s [16]. Eqn. 3 provides also an adequate basis for the interpretation of transient [8] and two-dimensional [5] nuclear Overhauser experiments.

In a rigid protein structure the effective correlation time for the modulation of dipole coupling between all the different pairs of protons would be identical and equal to the correlation time for the overall rotational tumbling of the molecule, τ_R. With the experimental values for σ_{ba} from Eqn. 3, the relative distances between different pairs of protons could thus be determined with Eqn. 2. A useful standard is the distance between two methylene hydrogen atoms, i.e. $r_0 = 1.75$ Å. Since this distance is less than the van der Waals contact distance between two hydrogen atoms, the cross-relaxation between geminal methylene protons [8] provides an estimate of the maximum possible cross-relaxation rate, $\sigma(r_0)$, which could occur in a rigid protein structure. Fig. 1 shows a plot of the ratio of the cross-relaxation rates for two protons at distances r_{ij} and $\sigma(r_0)$, i.e. $Q(r_{ij}) = \sigma(r_{ij})/\sigma(r_0)$ vs. r_{ij} ($\equiv R$ in Fig. 1). The smallest cross-relaxation rate $\sigma(r_{ij})$ which may be observed is determined in practice by the sensitivity with which truncated driven Overhauser effects can be measured [9,16]. It is unlikely that in the experiments discussed in this paper, cross-relaxation rates of less than approx. 5% of $\sigma(r_0)$ can be seen. Fig. 1 thus indicates that in a rigid macromolecule selective nuclear Overhauser effects can be detected for pairs of protons which are at a distance of approx. 3.0 Å or less.

In a more realistic approach the inherent flexibility of protein structures (see, for example, Refs. 2, 12, 18) has to be taken into account in the interpre-

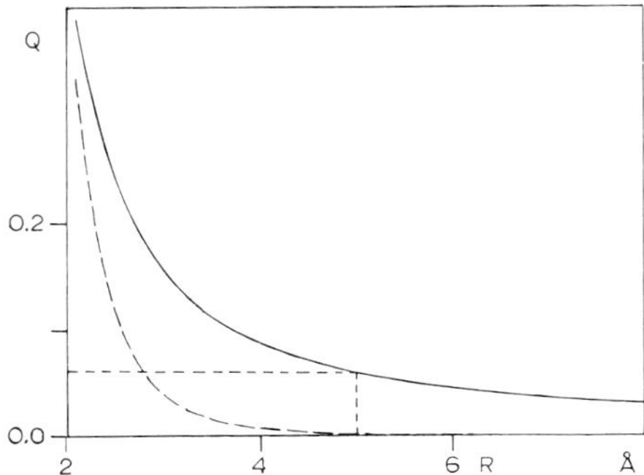

Fig. 1. The solid curve indicates the dependence of the magnitude of cross-relaxation rates on interatomic ^{1}H-^{1}H distances to be expected in flexible protein structures. $Q(R_m) = \sigma^e_{ij}/\sigma(r_0)$, i.e. the ratio of the cross-relaxation rates between protons i and j and the two protons of a methylene group, is plotted as a function of the maximum distance between i and j, assuming uniform averaging of the interatomic distance between the van der Waals contact of 2.0 Å for two hydrogen atoms and R_m (see text). For comparison, the usual $1/r_{ij}{}^6$ dependence which would apply for rigid structures is indicated by the broken curve, where $Q(r_{ij}) = \sigma(r_{ij})/\sigma(r_0)$. The straight dotted lines show the value of Q which corresponds to $R_m = 5$ Å in a flexible protein (see text).

tation of the nuclear Overhauser effects. The intramolecular motions may lead to variations in the orientation of interatomic vectors relative to the external magnetic field as well as to variations in the proton-proton distances r_{ij}. If stochastic variations of the vector orientation occur at frequencies comparable to or faster than the overall rotational correlation time, τ_R, which is of the order $1 \cdot 10^{-8}$ s to $1 \cdot 10^{-9}$ s for a medium sized protein, this will affect the correlation function $f(\tau_{ij})$ in Eqn. 2. From the form of $f(\tau_{ij})$ [19] it is apparent that for macromolecular systems, where the Overhauser enhancement is negative [18], the intramolecular angular changes can only reduce the magnitude of the Overhauser effect, i.e., $f(\tau_{ij})/f(\tau_R) \leqslant 1$. Because of the time variations of r_{ij} the proton-proton Overhauser effects in a flexible protein structure do not correspond to a fixed interatomic distance. To account for the flexibility of the molecular structure we used a uniform averaging model in which the distance between hydrogen atoms i and j was allowed to vary uniformly between a minimum distance r_m and a maximum distance R_m. This leads to an effective cross-relaxation rate, σ^e_{ij}, given by

$$\sigma^e_{ij} = \left\langle \frac{f(\tau_{ij})}{r^6} \right\rangle = \frac{f(\tau_{ij})}{5(R_m - r_m)} \left(\frac{1}{r_m^5} - \frac{1}{R_m^5} \right) \tag{4}$$

In Eqn. 4 the influence of internal motions with frequencies comparable to or higher than $1/\tau_R$ is accounted for in $f(\tau_{ij})$, where $f(\tau_{ij})/f(\tau_R) \leqslant 1$, while slower fluctuations are included in the explicit average taken over the time variations of r_{ij} [18]. To obtain a quantity which can readily be compared to $Q(r_{ij})$ in Fig. 1 we define

$$\begin{aligned} Q(R_m) &\equiv \frac{r_0^6}{5(R_m - r_m)} \left(\frac{1}{r_m^5} - \frac{1}{R_m^5} \right) \\ &\geqslant \frac{f(\tau_{ij})}{f(\tau_R)} \frac{r_0^6}{5(R_m - r_m)} \left(\frac{1}{r_m^5} - \frac{1}{R_m^5} \right) = \frac{\sigma^e_{ij}}{\sigma(r_0)} \end{aligned} \tag{5}$$

In Fig. 1 a plot of $Q(R_m)$ vs. R_m ($\equiv R$ in the figure), was computed with r_m equal to the van der Waals contact distance of 2.0 Å between two hydrogen atoms. It should be emphasized that $r_m = 2.0$ Å is the distance of closest non-bonding approach for protons in a polypeptide chain. We have not at present attempted to determine r_m empirically. Assumption of this lowest possible limit for r_m in the present treatment leads to very conservative values for the experimentally determined upper limit for the distances, R_m. Fig. 1 shows that in flexible proteins Q is larger than 0.05 when R_m is smaller or equal to 5.0 Å. Hence the nuclear Overhauser effects seen in our experiments have been attributed to pairs of protons with $R_m \leqslant 5.0$ Å. No further distance discrimination was attempted with the Overhauser experiments. Comparison with the plot of $Q(r_{ij})$ vs. r_{ij} obtained for the rigid body model (Fig. 1) indicates that neglect of internal motion leads to too restrictive estimates of interatomic distances from Overhauser enhancement experiments.

The distance-geometry algorithm

The problem to be solved is the following. Given upper limits, U_{ij}, and lower limits, L_{ij}, on the distances between N points, what are the possible conformations $\{r_j = (x_i, y_i, z_i), i = 1, ..., N\}$ which are consistent with the constraints

$$L_{ij} \leqslant |r_i - r_j| \leqslant U_{ij} \tag{6}$$

where $|r_i - r_j|$ denotes the ordinary Euclidean distance between the points r_i and r_j? One way of attacking this problem is to minimize the error function

$$F_{err}(r_1, ..., r_N) = \sum_{i<j}{}' (|r_i - r_j| - U_{ij})^2 + \sum_{k<l}{}' (L_{kl} - |r_k - r_l|)^2 \tag{7}$$

where Σ' indicates that only the terms violating the bounds in Eqn. 6 are summed. $F_{err}(r_l, ..., r_N) = 0$ and Eqn. 6 are then simply different formulations of the same problem.

When searching for solutions to this problem, one encounters the same difficulties as in empirical energy calculations [20], i.e., how to find a reasonable starting conformation and how to avoid the problem of local minima. Crippen et al. [10,11,21] introduced the metric matrix $G_{ij} = r_i \cdot r_j$, which we also apply, in order to tackle this problem. The following considerations prompted us, however, to extend the previously described procedures. Firstly, by definition the metric matrix, G, should have at most three non-zero eigenvalues, none of which is negative. However, for a matrix, G, which is defined via randomly chosen distances satisfying the constraints in Eqn. 6, these properties of the metric matrix G are, in general, not satisfied. A second difficulty is that for a N-dimensional space, where N is of the order of 100, truncation of the G matrix to the subspace with the three largest eigenvalues is a dramatic step. The previous formulation [10,11,21] was therefore extended to allow the triangle inequalities for the distances to be checked and corrected within the given distance constraints. In all our calculations this had the effect that for the resulting matrix G the three largest eigenvalues were all positive, whereas without these corrections one of the eigenvalues was sometimes negative. In addition, a convex mixing factor has been introduced in order to allow gradual contraction of the G matrix to a metrix matrix, G^*, with the properties noted above. Since the resulting algorithm involves considerable extensions of the algorithm described in the literature [10,11,21], the individual steps of the modification are described in some detail in the following.

(i) Selection of the bounds

The initial choice of upper and lower bounds is determined by bond angles, bond lengths and van der Waals contact distances which we took from the amino acid library of ECEPP [17,22]. For upper limits where no direct distance information is available from the covalent structure or from the Overhauser effect data, a distance large enough to cover all eventualities, e.g. 40 Å, has been assumed. It is possible to improve the limits by lowering the upper bounds and raising the lower limits without restricting the allowed conformation space. It is easy to check that the limits have to satisfy the following inequalities:

$$U_{ij} \leqslant U_{ik} + U_{jk}$$

$$L_{ij} \geqslant L_{ik} - U_{jk} \text{ for all } (i, j, k) \tag{8}$$

This is a consequence of the triangle inequaltities for the distances. A corresponding inequality for the lower bounds,

$$L_{ij} \leqslant L_{ik} + L_{jk} \tag{9}$$

is not valid. If the inequalities (Eqn. 8) are not satisfied by the given bounds we set

$$\tilde{U}_{ij} = U_{ik} + U_{jk} \tag{10}$$

and

$$\tilde{L}_{ij} = L_{ik} - U_{jk} \tag{11}$$

respectively. Improvement of the bounds with Eqns. 10 and 11 was achieved by an exhaustive search over all triangles (i, j, k) with $1 \leqslant j < k \leqslant N$; within each triangle a check of all inequalities was made. The pointers to the corresponding matrix elements of the lower and upper bounds, which were stored in one-dimensional arrays, could thereby be calculated without multiplication. This procedure was repeated iteratively until a self-consistent set of bounds was obtained. In practice the bounds were improved in two steps. First the bond angles, bond lengths and van der Waals contact distances were used to obtain improved bounds which were consistent with the covalent structure. Further improvement of the bounds was then obtained by inclusion of the Overhauser effect data. For the 109 atoms treated in the glucagon calculation, this second step resulted in about 4000 changes in the upper bounds.

(ii) Random choice of distances

To sample randomly the allowed conformation space, the distances D_{ij} are chosen randomly within the allowed bounds,

$$\tilde{L}_{ij} \leqslant D_{ij} \leqslant \tilde{U}_{ij} \tag{12}$$

This means that at this stage the $N(N-1)/2$ distances are treated as independent variables even though the system has only $3N-6$ degrees of freedom. Some of the additional correlations between the distances are included in the next step.

(iii) Consistency of the distances with the triangle inequality

Even if the improved bounds $\tilde{U}$ and $\tilde{L}$ satisfy the triangle inequalities (Eqn. 8), this does not mean that the randomly chosen distances will be consistent with the triangle inequalities. For the glucagon calculation, about 10 000—20 000 violations of the triangle inequality were typically found. It is desirable to change the chosen distances in such a way that they become consistent with the inequalities and still lie within the given bounds. For the case

$$D_{ij} > D_{ik} + D_{jk} \tag{13}$$

we set

$$\tilde{D}_{ik} = D_{ik} + (\tilde{U}_{ik} - D_{ik})\Delta/\Gamma \tag{14}$$

$$\tilde{D}_{jk} = D_{jk} + (\tilde{U}_{jk} - D_{jk})\Delta/\Gamma \tag{15}$$

and

$$\tilde{D}_{ij} = D_{ij} \tag{16}$$

$$\text{with } \Delta = D_{ij} - D_{ik} - D_{jk} \tag{17}$$

$$\text{and } \Gamma = (\tilde{U}_{ik} - D_{ik}) + (\tilde{U}_{jk} - D_{jk}) \tag{18}$$

The crucial inequality

$$0 \leqslant \Delta/\Gamma \leqslant 1 \tag{19}$$

then implies that

$$\tilde{D}_{ij} = \tilde{D}_{ik} + \tilde{D}_{jk} \tag{20}$$

$$D_{ik} \leqslant \tilde{D}_{ik} \leqslant \tilde{U}_{ik} \tag{21}$$

and

$$D_{jk} \leqslant \tilde{D}_{jk} \leqslant \tilde{U}_{jk} \tag{22}$$

In this procedure all changes are to larger distances, since the gradual contraction in step (v) generally decreases the distances.

It is empirically found that these triangle correlations ensure that the three largest eigenvalues of the G matrix are positive.

(iv) Calculation of the metric matrix, G

For an actual conformation, the metric matrix, G, is defined by

$$G_{ij} = r_i \cdot r_j \tag{23}$$

where the origin of the coordinate system is located at the geometric center of mass. The unique relation between the matrices G and D,

$$G_{ii} = \frac{1}{N}\sum_{j=1}^{N} D_{ij}^2 - \frac{1}{2N^2}\sum_{j,k} D_{jk}^2 \tag{24}$$

384

$$G_{ij} = \frac{1}{2}(G_{ii} + G_{jj} - D_{ij}^2) \tag{25}$$

is now generalized to the $\tilde{D}$ matrix calculated in (iii) to allow calculation of a matrix $\tilde{G}$. By definition the calculated $\tilde{G}$-matrix is symmetric and hence possesses a spectral representation of the form

$$\tilde{G}_{ij} = \sum_{\alpha=1}^{N} \lambda_\alpha E_{i,\alpha} E_{j,\alpha} \ . \tag{26}$$

Truncation to the subspace with the three largest eigenvalues

$$G_{ij}^{\star} = \sum_{\alpha=1}^{3} \lambda_\alpha E_{i,\alpha} E_{j,\alpha} \tag{27}$$

defines the coordinates

$$r_{i,\alpha}^{\star} = \sqrt{\lambda_\alpha}\, E_{i,\alpha} \qquad (i = 1, ..., N \text{ and } \alpha = 1, 2, 3) \tag{28}$$

with

$$G_{ij}^{\star} = r_i^{\star} \cdot r_j^{\star} \tag{29}$$

Calculation of the three largest eigenvalues and the corresponding eigenvectors was done with Tschebychew polynomials, instead of the more common use of powers, in the exhaustion method [23]. The Tschebychew polynomials of the metric matrix $\tilde{G}$, $T_n(\tilde{G})$, applied to an arbitrary starting vector u are calculated by the well known recurrence relation

$$T_0(\tilde{G})u = u \tag{30}$$

$$T_1(\tilde{G})u = \tilde{G}u \tag{31}$$

$$T_{n+1}(\tilde{G})u = 2\tilde{G}T_n(\tilde{G})u - T_{n-1}(\tilde{G})u \tag{32}$$

After scaling with

$$s = 1/N \text{ trace } \tilde{G} \tag{33}$$

$$\tilde{G}_s = 1/s\tilde{G} \tag{34}$$

one obtaines:

$$q_n = T_n(\tilde{G}_s)\, u = \sum_{a=1}^{N} T_n(\lambda_\alpha^{\star})(e_\alpha^{\star}, u)\, e_\alpha^{\star} \tag{35}$$

with

$$\frac{q_n}{|q_n|} \to e_{\max}^{\star} \tag{36}$$

and

$$\frac{(q_n, \tilde{G}_s q_n)}{(q_n, q_n)} \to \lambda_{\max}^{\star} \tag{37}$$

$\lambda_\alpha^{\star}$ and $e_\alpha^{\star}$ are eigenvalues and eigenvectors of the matrix $\tilde{G}_s$, $\lambda_{\max}^{\star}$ and $e_{\max}^{\star}$ are the largest eigenvalue and the corresponding eigenvector. Scaling with s has the effect that

$$\lambda^\star_{\max} = \frac{1}{s}\lambda_{\max} \geqslant 1 \,. \tag{38}$$

Outside [−1, 1] the Tschebychew polynomials [24,15] grow more efficiently than powers. Therefore convergence in Eqn. 36 and 37 is faster than it would be in a corresponding formalism based on the use of powers. After $\lambda^\star_{\max}$ and $e^\star_{\max}$ have been obtained, the matrix $(\tilde{\tilde{G}}_{ij})$ given by

$$(\tilde{\tilde{G}})_{ij} = (\tilde{G}_s)_{ij} - \lambda^\star_{\max} e^\star_{\max,i} \cdot e^\star_{\max,j} \tag{39}$$

is calculated and the procedure of Eqns. 30—32 and 35—37 is repeated to get the second largest eigenvalue λ_2. Further repetition then gives λ_3. In practice scaling with s was sufficient to make all three largest eigenvalues greater than 1. Therefore scaling was done only once for all three eigenvalues.

(v) Gradual contraction

The coordinates $r^\star_{i,\alpha}$ defined by Eqn. 28 may not be a solution to the problem because truncation of the N-dimensional G matrix can lead to coordinates which violate the distance constraints. To circumvent this problem, we have used a convex combination of the old distances and the calculated coordinates to give a new set of distances:

$$(D^{\text{new}}_{ij})^2 = f(D^{\text{old}}_{ij})^2 + (1-f)\,|r^\star_i - r^\star_j|^2 \tag{40}$$

The new distances D^{new}_{ij} are then checked for violation of the bounds and, if necessary, corrected according to the procedure in (iii). Steps (iv) and (v) are then repeated. For each cycle through steps (iii) to (v), the convex mixing factor f is linearly decreased stepwise from 1 to 0. In our experience, this iterative procedure is convergent and reduced the number and magnitude of violations of the distance constraints contained in the calculated coordinates. Although as many cycles as desired may be included, five to ten cycles have been found to give a good compromise with calculation time.

(vi) Chirality and refinement

Since the chirality of the asymmetric C_α-atoms cannot be defined by distances alone, this stereochemical feature was included in the process of refinemet of the coordinates *. The refinement was carried out with the use of the conjugate gradient method of Fletcher and Reeves [26]. Instead of using the error function F_{err} as defined in Eqn. 7, technical reasons favoured a slightly different function $\tilde{F}_{\text{err}}$:

$$E(r_1, \ldots, r_N) = \tilde{F}_{\text{err}}(r_1, \ldots, r_N) + C(r_1, \ldots, r_N) \tag{41}$$

$$\tilde{F}_{\text{err}}(r_1, \ldots, r_N) = \tfrac{1}{2}\sum_{i<j}{}'\,(|r_i - r_j|^2 - U^2_{ij})^2 + \tfrac{1}{2}\sum_{k<l}{}'\,(L^2_{kl} - |r_k - r_l|^2)^2 \tag{42}$$

* Consideration of chirality was independently also included in the latest version of the distance geometry algorithm by Crippen, G.M., et al. (personal communication).

In Eqn. 42 Σ' indicates, in analogy to Eqn. 7, that only the terms which violate the bounds $|r_i - r_j| > U_{ij}$ or $|r_k - r_l| < L_{kl}$, respectively, are summed.

$$C(r_1, ..., r_N) = \sum_{\nu} (r_{HN}^{(\nu)} \cdot (r_{HC}^{(\nu)} \times r_{HC\beta}^{(\nu)}) - c^{(\nu)})^2 \tag{43}$$

The sum in Eqn. 43 runs over all asymmetric C_α-atoms. The ideal value $c^{(\nu)}$ for the pseudoscalar $ps = r_{HN} \cdot (r_{HC} \times r_{HC\beta})$ (in case of L-amino acids ps is positive, in case of D-amino acids negative; $r_{HN} = r_N - r_H$, $r_{HC} = r_C - r_H$, $r_{HC\beta} = r_{C\beta} - r_H$) has been calculated for each amino acid by taking the standard coordinates of ECEPP [17,22].

Experimental procedure

Sample preparation

Glucagon was purchased from Calbiochem and used without further purification. The synthesis of [$^2H_{38}$]dodecylphosphocholine has been described previously [27].

For the ^{1}H-NMR studies a ^2H_2O solution containing $4 \cdot 10^{-3}$ M glucagon, 0.2 M [$^2H_{38}$]dodecylphosphocholine and 0.05 M phosphate buffer, p^{2}H 7.0, was prepared. Prior to the NMR experiments all the labile protons were exchanged with deuterium of ^2H_2O. The sample was then lyophilized repeatedly from ^2H_2O, dissolved in 99.979% ^2H_2O and sealed under an N_2 atmosphere. It was shown previously [13] that such a solution contained well defined mixed micelles consisting of one glucagon molecule and approx. 40 detergent molecules. The spectral properties of this sample were stable over a period of several months.

NMR measurements

^{1}H-NMR spectra were recorded on a Bruker HX 360 spectrometer equipped with an Aspect 2000 data system. Truncated driven nuclear Overhauser enhancement difference spectra [9,16] were recorded with the pulse sequence

$$[-t_1(\omega_A)-\overset{\text{Observation}}{\text{pulse}}-t_2-t_1(\omega_{\text{off-res}})-\overset{\text{Observation}}{\text{pulse}}-t_2-]_n \; .$$

In the two experiments used to obtain the difference spectrum, selective pre-irradiation for a time period t_1 was applied at the frequency A and sufficiently far off-resonance to record a spectrum without nuclear Overhauser effect. t_2 is a waiting time and n indicates the number of transients accumulated.

To obtain an extensive set of input data for the distance geometry algorithm, a series of truncated driven nuclear Overhauser enhancement difference spectra were recorded in which the pre-irradiation frequency was varied in steps of 20 Hz through the regions from 0.5 to 5.0 ppm and from 6.5 to 8.0 ppm of the ^{1}H-NMR spectrum of micelle-bound glucagon. Each of these experiments used a pre-irradiation time, t_1, of 0.4 s, so that selective Overhauser effects were obtained [9,16]. While the resolution of the resonances which showed Overhauser effects corresponded to the usual spectral resolution, i.e., approx. 0.01 ppm, the pre-irradiation in any single experiment was less selective and corresponded usually to a resolution of approx. 0.07 ppm. However, with the handling of the data set described in the following, which was based on the sym-

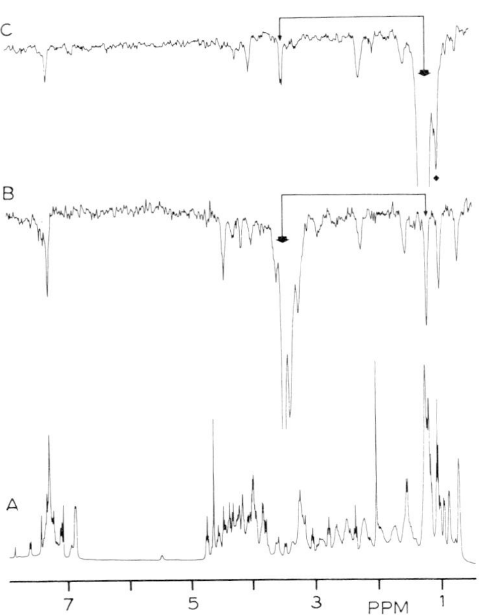

Fig. 2. Truncated driven nuclear Overhauser enhancement difference spectra obtained for glucagon bound to fully deuterated dodecylphosphocholine micelles. (A) Normal 360 MHz ^{1}H-NMR spectrum recorded with 4 mM glucagon and 0.2 M [$^2H_{38}$]dodecylphosphocholine in 0.05 M phosphate buffer at p_2H 7.0 and 37° C. (B) Difference spectrum obtained by pre-irradiation of the Val-23 αCH resonance at 3.50 ppm for 0.4 s prior to data acquisition. (C) Difference spectrum obtained by pre-irradiation of the Val-23 βCH_3 resonance at 1.21 ppm for 0.4 s prior to data acquisition.

metry of the cross-relaxation terms for spins 1/2 [18], an effective resolution of 0.01 ppm was obtained throughout.

Each individual truncated driven Overhauser difference spectrum was screened for peaks indicating the occurrence of an Overhauser effect, whereby the only discrimination, made by visual inspection, was 'Overhauser effect' or 'no Overhauser effects'. As an illustration, Fig. 2B shows the peaks obtained with pre-irradiation of the Val-23 αCH resonance at 3.50 ppm. These included the Val-23 γCH_3 resonance at 1.21 ppm, which appeared with a resolution of approx. 0.01 ppm, whereas the selectivity of pre-irradiation at 3.50 ppm was not better than 0.07 ppm. However, irradiation of the Val-23 γ-CH_3 resonance at 1.21 ppm led to a strong Overhauser enhancement at the Val-23 αCH resonance at 3.50 ppm, and there was no other strong Overhauser effect within the chemical shift range 3.50 ± 0.07 ppm (Fig. 2C). From the combination of the two experiments in Fig. 2B and C, the effective resolution is thus 0.01 ppm for both resonances and it may be concluded that there is cross-relaxation between the αCH and γCH_3 protons of Val-23 at 3.50 and 1.21 ppm. Obviously, by variation of the pre-irradiation frequency across the entire spectrum similar data are obtained for all the peaks in Fig. 2B. For the practical evaluation of the entire data set recorded for micelle-bound glucagon, the results from visual inspection were collected from all the Overhauser difference spectra and arranged in a two-dimensional plot. The horizontal axis corresponds to the chemical shift at which the pre-irradiation field ω_A was applied and the vertical axis to the chemical shifts of the observed Overhauser effects. A highly resolved spectrum was thus obtained along the vertical axis, whilst the resolution in the horizontal direction was approx. 0.07 ppm, i.e., at this stage the two-dimensional array consisted of short lines parallel to the horizontal axis and located along the vertical axis at the chemical shift positions of the observed Overhauser effects. Folding of this two-dimenstional data matrix about the diagonal produced crossings of lines which manifest effects between identical pairs of protons. A symmetrical array of crosses with respect to the diagonal was thus obtained (Fig. 3A), where each cross manifests an Overhauser effect with a resolution of 0.01 ppm in both directions. For example, the cross-relaxation between the αCH and γCH_3 protons of Val-23 manifested in Fig. 2 corresponds to crosses at (3.50 ppm, 1.21 ppm) and (1.21 ppm, 3.50 ppm) in Fig. 3A.

Results

^1H-^1H Overhauser enhancements in micelle-bound glucagon

In an earlier study, numerous ^{1}H-NMR lines of the region 19—27 of the glucagon amino acid sequence were individually assigned [13]. Here, truncated driven nuclear Overhauser experiments are used to investigate the distances between the assigned resonances. In all, 130 spectra were recorded, whereby the pre-irradiation frequency was systematically varried in steps of 20 Hz across the spectral regions which contain resonance lines [13]. Following the procedures outlined in the preceding section the data thus obtained were arranged in a two-dimensional plot of the cross-relaxation effects on micelle-bound glucagon (Fig. 3A). In Fig. 3A each cross denotes a pair of chemical shifts for which the

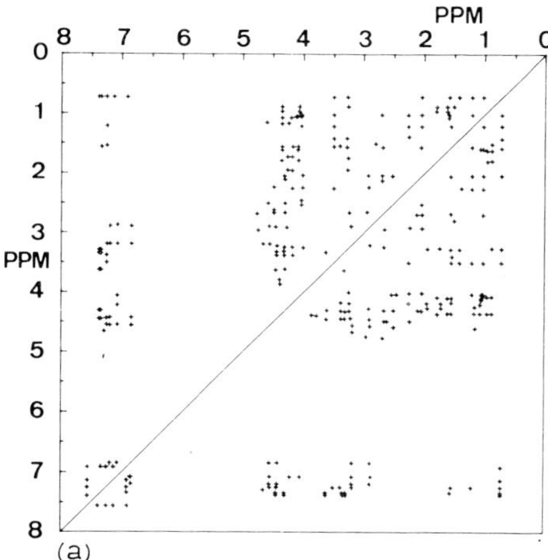

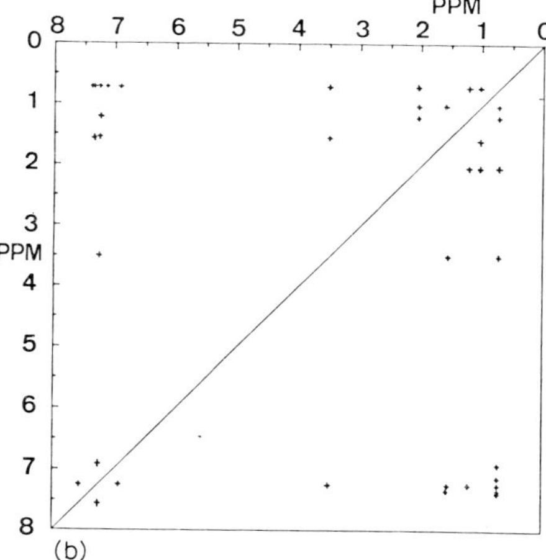

Fig. 3. Two-dimensional presentation of truncated driven ^1H-^1H Overhauser effects observed for micelle-bound glucagon with a pre-irradiation time of 0.4 s. Each cross indicates that a nuclear Overhauser effect between the two corresponding chemical shift positions had been observed (see text). (A) Overhauser effects observed in a series of experiments where the pre-irradiation frequency was varied in steps of approx. 20 Hz across the entire glucagon spectrum. (B) Only those crosses from A are retained which indicate cross-relaxation between protons of different amino acid residues in the region 19—27 of the glucagon sequence and which had been individually assigned [13].

corresponding protons show cross-relaxation.

Fig. 3A shows that a large number of individual cross-relaxation effects can be measured for micelle-bound glucagon. Since these were obtained under conditions where spin diffusion can be neglected [7,9] they contain information on the spatial structure of micelle-bound glucagon. This information can be fully exploited only when the pairs of protons connected by cross-relaxation have been individually assigned. In Fig. 3B only those data were retained from Fig. 3A which show cross-relaxation between protons of different amino acid residues in the glucagon segment 19—27 and which had been individually assigned [13]. A list of these resonances is given in Table I. In those instances where the assigned resonances and other resonances had similar chemical shifts, the identity of the peaks seen in the Overhauser enhancement difference spec-

TABLE I

UPPER BOUNDS FOR THE DISTANCES USED IN THE DISTANCE-GEOMETRY ALGORITHM AS OBTAINED FROM TRUNCATED DRIVEN OVERHAUSER EFFECT MEASUREMENTS

All observed Overhauser effects were attributed to proton-proton distances of 5.0 Å or less (see Fig. 1). Where individual protons could not resolved in the ^{1}H-NMR spectrum, for example, the ring protons of Phe-22 or δ_1CH_3 and δ_2CH_3 of Leu-26, a proton-proton distance sufficiently large to include all indistinguishable protons was used (see text).

Observed Overhauser effects		Maximum distance (R_m) (Å)
between	and	
Ala-19 βCH_3	Phe-22 ring	10.0
Phe-22 ring	Val-23 αCH	10.0
	Val-23 $\gamma_1 CH_3$	10.0
	Trp-25 C5H	10.0
	Trp-25 C7H	10.0
	Leu-26 δ_1, δ_2 CH_3	12.5
Val-23 αCH	Leu-26 γCH	5.0
	Leu-26 δ_1, δ_2 CH_3	7.5
Val-23 $\gamma_1 CH_3$	Leu-26 δ_1, δ_2 CH_3	7.5
	Met-27 ϵCH_3	5.0
Val-23 $\gamma_2 CH_3$	Leu-26 γCH	5.0
	Leu-26 δ_1, δ_2 CH_3	7.5
	Met-27 ϵCH_3	5.0
Val-23 αCH *	Val-23 βCH	2.87 *
	Val-23 γ_1, γ_2 CH_3	2.71 *
Val-23 βCH *	Val-23 N	2.65 *
	Val-23 C$'$	2.74 *
Trp-25 C2H	Leu-26 δ_1, δ_2 CH_3	7.5
Trp-25 C4H	Leu-26 γCH	5.0
	Leu-26 δ_1, δ_2 CH_3	7.5
Trp-25 C5H	Leu-26 δ_1 δ_2 CH_3	7.5
Trp-25 C6H	Leu-26 δ_1, δ_2 CH_3	7.5
Leu-26 δ_1, δ_2 CH_3	Met-27 ϵCH_3	7.5

* For Val-23, the $^3J_{\alpha\beta}$ coupling constant was 10.5 Hz, which indicates that αCH and βCH are *trans*. This information has been included in the input data as fixed values of the distances from the α-proton to the β-proton and the γ-carbons, and from the β-proton to the backbone amide nitrogen and carbonyl carbon of Val-23.

tra were verified by spin-decoupling in the Overhauser difference spectra [28].

As was discussed in the preceding section, each cross in Fig. 3B was taken to indicate a proton-proton distance of less than 5.0 Å (see Fig. 1) and no further distance discrimination was made from the Overhauser data. However, when the list of upper bounds for interatomic distances to be used as input for the distance geometry algorithm was prepared (Table I), the following additional aspects were also taken into consideration.

(i) In some cases, protons which are distinct in the covalent structure could not be distinguished in the ^{1}H-NMR spectra; for example, the resonances of the δ_1CH_3 and δ_2CH_3 groups of Leu-26 are overlapped in the NMR spectra. Here, we have explicitly taken into acount the spatial separation of 2.5 Å between the δ_1CH_3 and δ_2CH_3 groups of leucine, and have taken an upper limit to the interatomic distances which includes all indistinguishable protons. Thus, while the Overhauser effect experiments indicated that at least one of the δCH_3 groups of Leu-26 was within 5.0 Å of the C2-H proton of the indole ring of Trp-25, we have included into Table I the less stringent result that both δCH_3 groups of Leu-26 are within 7.5 Å of the C2-H proton of Trp-25. Similar corrections have been made for the aromatic ring of Phe-22 (Table I).

(ii) For Val-23, the $^3J_{\alpha\beta}$ coupling constant was observed to be 10.5 Hz, which is consistent with a *trans* orientation of the αCH and βCH protons. This was included in the distance information as fixed distances between the α-proton, the β-proton and other atoms of Val-23 (Table I).

Computation of molecular geometrics for micelle-bound glucagon

Figs. 4—6 show the molecular geometries obtained for residues 19—27 of micelle-bound glucagon in three different computer runs with ten cycles for contraction of the G matrix and 1000 cycles for the refinement. In all three conformations, all violations of the distance constraints were less than 0.1 Å. To economize on computer time, the residues Gln-20, Asp-21 and Gln-24, where no Overhauser data were obtained which could be used to characterize the spatial arrangement of the side-chains (Table I), were substituted with glycines for these computations. The following common features of the three structures are readily apparent. The molecule has a roughly cylindrical shape. The backbone is in the form of a distorted S, which consists of two loops from residues 19—22 and 23—27 connected by a nick at residue 22 (Fig. 6). In all three structures the same amino acid side-chains are in close proximity to one another (Figs. 4 and 5). The aromatic rings of Phe-22 and Trp-25 are on the same side of the backbone, forming a hydrophobic 'patch' (Fig. 5). Leu-26 is not part of this patch, and Val-23 is not in the immediate neighbourhood of the side-chain of Trp-25.

In additional computer runs, general aspects of the performance of the distance-geometry algorithm were tested. When the upper and lower limits for C_α—C_α distances were taken equal to the distances in the crystal structures of lysozyme or basic pancreatic trypsin inhibitor, the algorithm was able to reconstruct the three-dimensional arrangement of the α-carbon atoms in these proteins. This verifies that when sifficient distance information is available, the algorithm leads to convergence to the appropriate conformations. When only the stereochemical constraints, i.e. bond angles, bond lengths and van der Waals

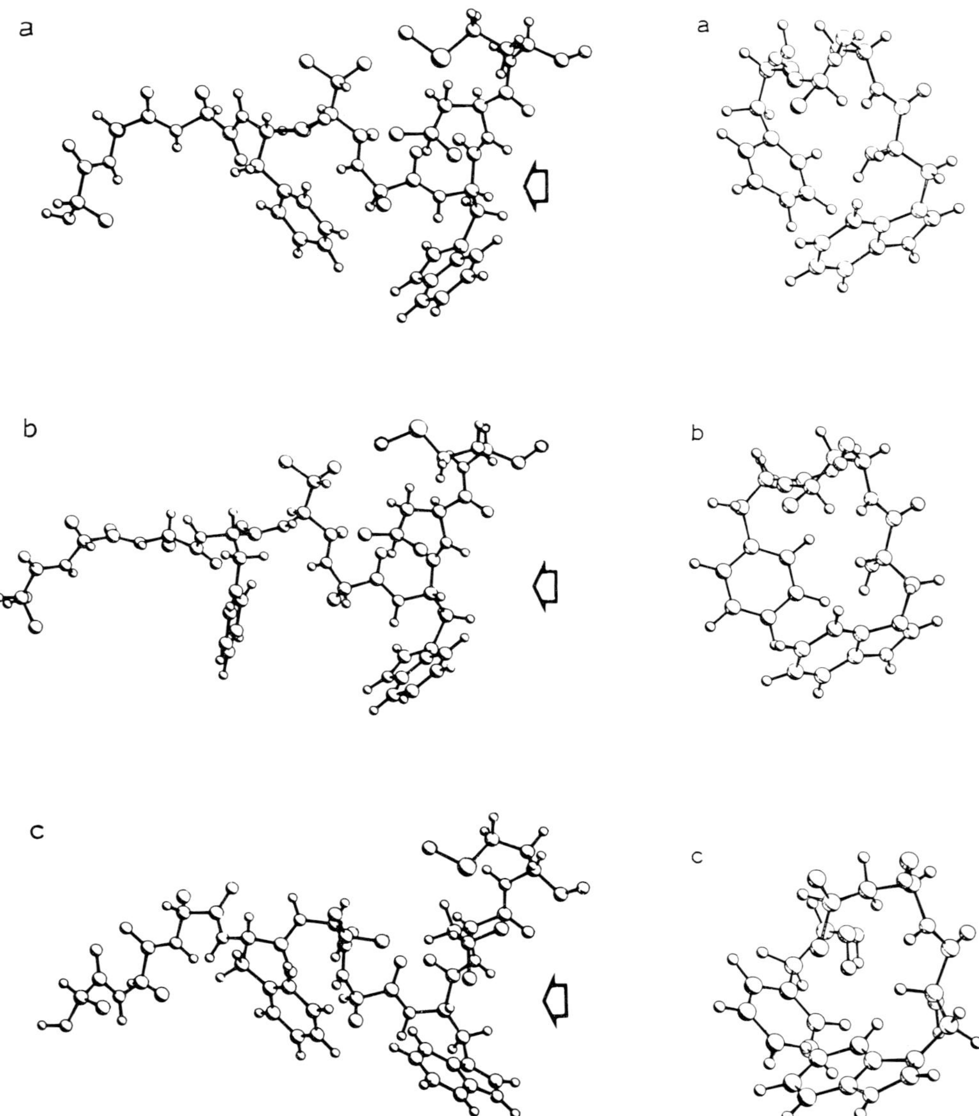

Fig. 4. Computer drawings of three molecular geometries of the glucagon region 19—27 which satisfy the distance constraints imposed by the Overhauser data of Table I. The drawings show the sequence —Ala-19—Gln-20—Asp-21—Phe-22—Val-23—Gln-24—Trp-25—Leu-26—Met-27—, with Ala-19 in the lower left and Met-27 in the upper right in each figure. The complete molecular structure including the hydrogen atoms is shown, with the following exceptions. The residues Gln-20, Asp-21 and Gln-24, for which no Overhauser enhancement with side-chain hydrogen atoms was obtained (Table I) were replaced by glycine; the methyl groups of Ala-19, Val-23, Leu-26 and Met-27 are represented by the carbon atom only. The arrows indicate the direction along which the same structures are viewed in Fig. 5.

Fig. 5. Computer drawings of the residue Phe-22 and the dipeptide fragment Trp-25—Leu-26 in the three structures of glucagon 19—27 in Fig. 4 viewed along the direction indicated by the arrows in Fig. 4. In this presentation the locations of the hydrophobic rings of Phe-22 and Trp-25 and the orientation of the side-chain of Leu-26 relative to the aromatic ring are nicely illustrated.

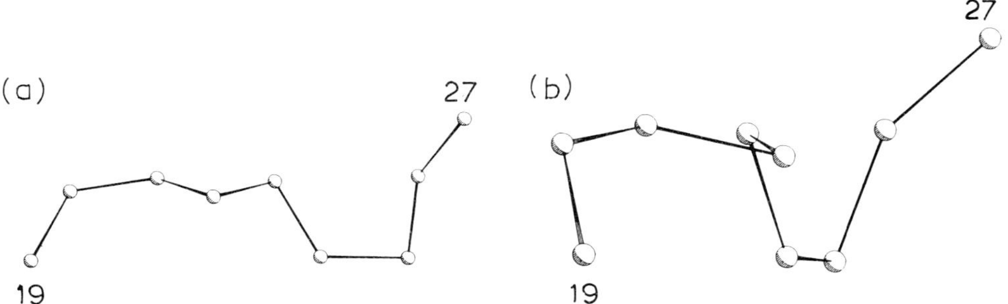

Fig. 6. Positions of the α-carbon atoms in the glucagon 19—27 molecular geometry of Fig. 4a. In presentation a, the structure is viewed along the same direction as in Fig. 4. In presentation b it was rotated by +60° about the vertical axis through the center of gravity of the peptide fragment shown, ie., residues 19 and 27 were moved towards and away from the reader, respectively. The spatial folding of the backbone is characterized by the formation of two loops, 19—22 and 23—27, which are connected by a 'nick'. This is best seen in presentation b. The nick at residue 22 is found in all three structures of Fig. 4.

radii, were used as input to the algorithm, very different 'random coil' conformations for glucagon 19—27 were obtained in different computer runs. This indicates that random choice of initial distances in different computer runs correponds to a random sampling of the allowed conformational space. Finally, that similar conformations for glucagon 19—27 were obtained in the three experiments of Fig. 4 and additional computer runs, which all used both the stereochemical constraints and the Overhauser effect data, inplies that sufficient distance information was available in Table I to restrict effectively the allowed conformation space of these residues.

Discussion

Physical meaning of the molecular geometries derived from nuclear Overhauser enhancement data by the distance geometry algorithm

In the method described in this paper the search for molecular structures which would be compatible with the experimental NMR data starts without any assumptions on the structure to be found. The entire input to the distance geometry algorithm consists of distance constraints imposed by the polypeptide covalent structure and the Overhauser enhancement data. Furthermore, the NMR data are used only to provide upper limits for distance constraints on the flexible molecular conformation [12,29]. Finally, there is no assumption that a single, unique, spatial structure exists. As a consequence, the molecular geometries obtained (Figs. 4—6) are to be regarded as typical members of the conformation space characterized by the present experiments rather than as average spatial structures. Since the upper limit distances for the Overhauser effects were chosen very conservatively, the conformation space characterized by the molecular geometries of Figs. 4—6 should be less restricted than the space actually occupied by the molecule. This ensures that common features found in the different calculated structures represent structural features which are significantly populated in the real physical conformation space. The extent to which the calculated molecular geometries adquately characterize the real physical conformation space obviously depends on the quality of the experimental data available, as will be further considered below in the discussion of the results on glucagon.

The present approach is fundamentally different from previously suggested methods for the structural interpretation of nuclear Overhauser data on polypeptide chains. Overhauser effects have been used to distinguish between the two possible solutions implicit in Karplus type curves for vicinal coupling constants as a function of dihedral angle [30,31]. In a different approach, steady-state Overhauser effects for an assumed structure were calculated and statistically compared with the experimental data [32,33]. While these approaches may yield useful information, they both depend rather critically on quantitative measurements of Overhauser effects, which are difficult to obtain in macromolecular species [7—9].

Application to micelle-bound glucagon

A previous investigation showed that from the point of view of the spatial folding, three regions could be distinguished in micelle-bound glucagon [13]. These are the N-terminus, which is flexible, the central residues which adopt an extended structure with restricted flexibility, and the C-terminal residues 19—27 where there is evidence for a tightly folded conformation with numerous contacts between different amino acid side-chains [13]. In the present paper, combination of Overhauser enhancement data and the distance geometry algorithm is used to investigate further the spatial folding of the C-terminal part 19—27 of glucagon.

The data on glucagon 19—27 in Figs. 4—6 and Table II allow more detailed

TABLE II

COMPARISON OF THE MOLECULAR GEOMETRIES FOR RESIDUES 19—27 OF GLUCAGON IN CRYSTALS AND BOUND TO MICELLES

For those proton-proton distances in the X-ray crystal structures of glucagon [14] which exceed the upper bound distances obtained for micelle-bound glucagon from nuclear Overhauser enhancements (Table I), the distances in the crystal structure and in the three molecular geometries of Fig. 4 are listed.

Proton-proton distances		Proton-proton distances			
		in the structure (a)—(c) of Fig. 4 which were computed from the Overhauser data			in the X-ray structure [14]
between	and	a	b	c	
Val-23 γ_1	Leu-26-δ_1	7.4	7.2	6.4	8.5
Val-23 γ_1	Met-27-ϵ	5.0	5.0	5.0	7.1
Val-23 γ_2	Leu-26-γ	4.6	4.8	3.0	6.9
Val-23 γ_2	Leu-26-δ_1	6.0	6.1	4.0	8.6
Val-23 γ_2	Met-27-ϵ	4.2	4.6	4.5	9.6
Trp-25 ring C2H	Leu-26-δ_1	7.2	7.2	6.0	9.0
Trp-25 ring C2H	Leu-26-δ_2	7.5	7.5	7.5	9.8
Trp-25 ring C4H	Leu-25-γ	4.7	4.9	5.0	5.8
Trp-25 ring C6H	Leu-26-δ_1	6.5	6.7	5.2	7.6
Trp-25 ring C6H	Leu-26-δ_2	7.5	7.5	7.5	8.7
Leu-26 δ_1	Met-27-ϵ	5.5	5.0	5.8	10.4
Leu-26 δ_2	Met-27-ϵ	4.3	4.2	4.6	10.8

assessment of the potential and the limitations of the method. There are several independent checks for the validity of the calculated structures for micelle-bound glucagon. Thus, although the intensities of the observed Overhauser effects were not explicitly used, the relative intensities were generally consistent with the calculated molecular geometries. For example, the maximum interatomic distances between the two Leu-26 δCH_3 groups and the C4, C5 and C6 indole ring protons of Trp-25 were taken as 7.5 Å (Table I). However, the Overhauser effect intensities indicated that the distances from the Leu-26 δCH_3 groups increase in the order C4H, C5H and C6H of Trp-25, which coincides with all the calculated molecular geometries. Further, for all the pairs of assigned resonances for which no Overhauser effect could be observed, the interatomic distances are greater than 4.4 Å (Phe-22 ring to Trp-25C6H), 5.4 Å (Leu-26 δCH_3 to Met-27 ϵCH_3) and 3.9 Å (Phe-22 ring to Trp-25 C6H) in all of the three structures a, b and c in Fig. 4. Overall it thus appears that the calculated conformations do not of necessity imply the existence of any Overhauser effect which was not observable. Finally, the calculated conformations were quite similar when the upper bound distances for the observation of Overhauser effects were chosen to be either 4.0, 5.0 or 6.0 Å, indicating that the exact value of this parameter was not critical. Although the different calculated structures in Fig. 4 are qualitatively similar, there are still some distances which show appreciable changes between the different structures (some examples are listed in Table II). The most pronounced difference amongst the calculated molecular geometries is reflected in the Trp-25 C5H to Phe-22 C4H interatomic distance, which is 5.9 Å, 9.9 Å and 2.7 Å for a, b and c (Fig; 4), respectively. This variability arises because the resonances of the aromatic ring of Phe-22 could not be individually assigned, thereby necessitating the use of a large upper bound distance in the input to the algorithm (Table I). In the calculated molecular geometries this appears to be largely expressed as rotation of the two aromatic rings relative to one another, since the distances between the centres of mass of the two rings are 6.6 Å, 8.4 Å and 7.8 Å for conformations a, b and c (Fig. 4), respectively.

Comparison of glucagon in aqueous solution, in single crystals and when bound to micelles

The conformation of glucagon has previously been studied for monomeric glucagon in aqueous solution by high resolution NMR [15] and for trimeric glucagon in crystals by X-ray diffraction [14]. For monomeric glucagon in aqueous solution, Val-23 was found to be close to the indole ring of Trp-25 [15] which indicated the conformation was different from the crystal structure. Since all calculated conformations for micelle-bound glucagon indicated that this proximity was not present (Fig. 4), the possibility that aqueous, monomeric glucagon and micelle-bound glucagon have similar conformations near residues 23 and 25 can be excluded. In the crystal structure, residues 19—27 of glucagon form a distorted α-helix [14]. Comparison of the calculated molecular geometries for micelle-bound glucagon with the crystal structure of trimeric glucagon shows that there is a considerable number of interatomic distances in the X-ray structure which are not consistent with the present Overhauser effect measurements (Tables I and II). From Table II, the most pronounced inconsistency between the Overhauser effect data and the crystal structure would appear to be the position of the side-chain of Leu-26. However, the inconsistencies between the crystal structure and the Overhauser data appear to involve more than alterations in side-chain positions. When the backbone geometry from the X-ray structure and the Overhauser data in Table I were used as input to the distance-geometry algorithm, the resulting molecular geometries included a large number of violations of the covalent structure, thus suggesting that there are appreciable conformational differences between crystalline and micelle-bound glucagon.

Conclusions for future work

A major advantage of the present method is that it provides a realistic estimate of the extent to which a set of experimental data is sufficient to define a molecular conformation. In the case of micelle-bound glucagon, the calculated structures still show considerable variability (Figs. 4 and 5). Although it is not possible to exclude that this variability reflects molecular flexibility, a consequence of the present procedures is that the real physical conformation space is probably more restricted than the conformational limits defined by the present experiments. We therefore regard the glucagon results as preliminary structures subject to further refinement. Such refinement will depend on the availability of more precise Overhauser data and possibly additional distance constraints obtained, for example, with the use of paramagnetic shift reagents, measurement of spin-spin coupling constants or analysis of ring current shifts in the ^{1}H-NMR spectrum [13]. That the new approach has worked sensibly with the use of the relatively sparse data on glucagon provides additional impetus for further improvements of the collection of NMR data. Already the recent development of a two-dimensional experiment for measurements of selective proton-proton Overhauser effects [5] promises to add greatly to the experimental data sets which will be available in future studies.

Acknowledgements

Use of the facilities at the Zentrum für Interaktives Rechnen (ZIR) of the ETH is gratefully acknowledged. Financial support was obtained from the Schweizerischer Nationalfonds (projects 3.0046.76 and 3.528.79) and through special grants of the ETH. We would like to thank Dr. G. Crippen for providing us with the distance geometry program [21] which was of great help at the outset of this project.

References

1 Wüthrich, K. (1976) NMR in Biological Research: Peptides and Proteins, North-Holland, Amsterdam
2 Jardetsky, O. (1980) Biochim. Biophys. Acta 621, 227—232
3 Nagayama, K., Bachmann, P., Wüthrich, K. and Ernst, R.R. (1978) J. Magn. Reson. 31, 133—148
4 Nagayama, K., Wüthrich, K. and Ernst, R.R. (1979) Biochem. Biophys. Res. Commun. 90, 305—311
5 Anil Kumar, Ernst, R.R. and Wüthrich, K. (1980) Biochem. Biophys. Res. Commun. 95, 1—6
6 Wüthrich, K. and Wagner, G. (1979) J. Mol. Biol. 130, 1—18
7 Kalk, A. and Berendsen, H.J.C. (1976) J. Magn. Reson. 24, 343—366
8 Gordon, S.L. and Wüthrich, K. (1978) J. Am. Chem. Soc. 100, 7094—7096
9 Wagner, G. and Wüthrich, K. (1979) J. Magn. Reson. 33, 675—680

10 Crippen, G.M. (1979) Int. J. Peptide Protein Res. 13, 320—326
11 Havel, T.F., Crippen, G.M. and Kuntz, I.D. (1979) Biopolymers 18, 73—81
12 Wüthrich, K. and Wagner, G. (1979) Trends Biol. Sci. 3, 227—230
13 Bösch, C., Brown, L.R. and Wüthrich, K. (1980) Biochim. Biophys. Acta 603, 298—312
14 Sasaki, K., Dockerill, S., Adamiak, D.A., Tickle, I.J. and Blundell, T. (1975) Nature 257, 751—757
15 Bösch, C., Bundi, A., Oppliger, M. and Wüthrich, K. (1978) Eur. J. Biochem. 91, 209—214
16 Dubs, A., Wagner, G. and Wüthrich, K. (1979) Biochim. Biophys. Acta 577, 177—194
17 Momany, F.A., McGuire, R.F., Burgess, A.W. and Scheraga, H.A. (1975) J. Phys. Chem. 79, 2361—2381
18 Noggle, J.H. and Schirmer, R.E. (1971) The Nuclear Overhauser Effect, Academic Press, New York
19 Woessner, D.E., Snowden, B.S., Jr. and Meyer, G.H. (1969) J. Chem. Phys. 50, 719—721
20 $G\overline{O}$, N. and Scheraga, H.A. (1978) Macromolecules 11, 552—559
21 Crippen, G.M. and Havel, T.F. (1978) Acta Cryst. A34, 282—284
22 Scheraga, H.A. (1975) Quantum Chemistry Program Exchange, Indiana University, Vol. 10, Program No. 286
23 Faddejwe, D.K. and Faddejewa, W.N. (1973) Numerische Methoden der Linearen Algebra, Olfenburg, München
24 Engeli, M., Ginsburg, T., Rutishauser, H. and Stiefel, E. (1959) Refined Iterative Methods for Computation of the Solution and the Eigenvalues of Self-Adjoint Boundary Values Problems, Birkhäuser, Basel
25 Wilkinson, J.H. (1969) The Algebraic Eigenvalue Problem, Clarendon Press, Oxford
26 Fletcher, R. and Reeves, C.M. (1964) Computer J. 7, 149—154
27 Brown, L.R. (1979) Biochim. Biophys. Acta 557, 135—148
28 Richarz, R. and Wüthrich, K. (1978) J. Magn. Reson. 30, 147—150
29 McCammon, J.A. and Karplus, M. (1979) Proc. Natl. Acad. Sci. U.S.A. 76, 3585—3589
30 Leach, S.J., Némethy, G. and Scheraga, H. (1977) Biochem. Biophys. Res. Commun. 75, 207—215
31 Jones, C.R., Sikakana, C.T. Hehir, S., Kuo, M.C. and Gibbons, W.A. (1978) Biophys. J. 24, 815—832
32 Bothner-By, A.A. and Johner, P.E. (1978) Biophys. J. 24, 779—790
33 Krishna, N.R., Agresti, D.G., Glickson, J.D. and Walter, R. (1978) Biophys. J. 24, 791—814

J. Mol. Biol. (1983) **169**, 921–948

Conformation of Glucagon in a Lipid–Water Interphase by ¹H Nuclear Magnetic Resonance

W. Braun†, G. Wider‡, K. H. Lee and K. Wüthrich

Institut für Molekularbiologie und Biophysik
Eidgenössische Technische Hochschule
CH-8093 Zürich-Hönggerberg, Switzerland

(Received 30 March 1983)

A determination of the spatial structure of the polypeptide hormone glucagon bound to perdeuterated dodecylphosphocholine micelles is described. A map of distance constraints between individually assigned hydrogen atoms of the polypeptide chain was obtained from two-dimensional nuclear Overhauser enhancement spectroscopy. These data were used as the input for a distance geometry algorithm for computing conformations that would be compatible with the experiments. In the region from residues 5 to 29 the mobility of the polypeptide backbone and most of the amino acid side-chains was found to be essentially restricted to the overall rotational tumbling of the micelles. The secondary structure in this region includes three turns of irregular α-helix in the segment of residues 17 to 29 near the C terminus, a stretch of extended polypeptide chain from residues 14 to 17, an α-helix-like turn formed by the residues 10 to 14 and another extended region from residues 5 to 10. In the N-terminal tetrapeptide H-His-Ser-Gln-Gly- the two terminal residues are highly mobile, indicating that they extend into the aqueous phase, and the mobility of the residues Gln3 and Gly4 appears to be only partially restricted by the binding to the micelle. The absence of long range nuclear Overhauser effects between the peptide segments 5–9 and 11–29, and between 5–16 and 19–29 shows that the polypeptide chain does not fold back on itself and hence that micelle-bound glucagon does not adopt a globular tertiary structure. Previously it was shown that the polypeptide backbone of glucagon is located close to and runs roughly parallel to the micelle surface. Combination of these observations suggests that the overall spatial arrangement of the glucagon polypeptide chain in a lipid–water interphase is largely determined by the topology of the lipid support, in the present case the curvature of the dodecylphosphocholine micelles. The tertiary structure is further characterized by the formation of two hydrophobic patches by the side-chains of Phe6, Tyr10 and Leu14, and the side-chains of Ala19, Phe22, Val23, Trp25 and Leu26, respectively.

1. Introduction

Glucagon is a hormone which consists of a linear polypeptide chain of 29 amino acid residues and has a molecular weight of 3500. The primary target organ for

† Present address: Dept of Physics, Kyushu University 33, Fukuoka 812, Japan.
‡ Present address: Spectrospin AG, Industriestrasse 26, CH-8117 Fällanden, Switzerland.

0022–2836/83/280921–28 $03.00/0 © 1983 Academic Press Inc. (London) Ltd

glucagon is the plasma membrane of liver and other cells, where binding to a specific receptor site mediates activation of glycogenolysis (Pohl *et al.*, 1969). Evidence has been presented that recognition between glucagon and its receptor depends on the ordered lipid structures surrounding the receptor site in the membrane (Rodbell *et al.*, 1971; Rubalcava & Rodbell, 1973). Otherwise, structural data on the receptor system are scarce and therefore much effort has been invested to characterize the glucagon–receptor interactions through studies of the conformational properties of the hormone (Sasaki *et al.*, 1975; Blundell & Wood, 1982).

Early studies of glucagon conformation by circular dichroism and other physical-chemical techniques indicated a tendency of this polypeptide to adopt different spatial structures in different environments. For example, for monomeric glucagon in aqueous solution a flexible "random coil" structure was indicated (Panijpan & Gratzer, 1974). For self-aggregated glucagon in aqueous solution evidence was presented that it could adopt either an α-helical (Gratzer *et al.*, 1967; Srere & Brooks, 1969) or β-sheet (Epand, 1971; Moran *et al.*, 1977) secondary structure and that the species formed depended critically on the peptide concentration (Wagman *et al.*, 1980). Furthermore, interactions with lipids and detergents were found to induce changes of the glucagon conformation (Schneider & Edelhoch, 1972; Epand *et al.*, 1977; Bösch *et al.*, 1980). These observations indicated that other techniques, which would be capable of providing more detailed structural information based on measurements of a large number of parameters, should be applied to glucagon in different milieus. X-ray studies of glucagon trimers in single crystals showed that the individual peptide molecules had an α-helical conformation (Sasaki *et al.*, 1975). High resolution ^{1}H n.m.r.† studies confirmed the earlier observations that monomeric glucagon in aqueous solution adopts a predominantly flexible random coil form and further revealed a structured region involving the residues 22 to 25, with a conformation different from α-helix-type secondary structure (Bösch *et al.*, 1978). The present paper reports on the conformation of monomeric glucagon bound to dodecylphosphocholine micelles.

Previous studies using spectroscopy and other physical-chemical methods showed that with the experimental conditions used for the present experiments, glucagon-containing DPC micelles contain one molecule of glucagon and ~40 detergent molecules and have a molecular weight of ~17,000 (Bösch *et al.*, 1980; Wider *et al.*, 1982). When perdeuterated DPC is used (Brown, 1979) the ^{1}H n.m.r. lines of MB-glucagon are sufficiently well resolved for a detailed spectral analysis. Studies with conventional, one-dimensional n.m.r. techniques showed that the peptide has a non-globular, predominantly extended form and a low resolution conformation based on individual assignments for a limited number of amino acid side-chain protons was obtained for the segment of residues 19 to 27 (Braun *et al.*, 1981). In the meantime almost all resonances in the ^{1}H n.m.r. spectrum of MB-

† Abbreviations used: MB-glucagon, glucagon bound to dodecylphosphocholine micelles; DPC, [$^2H_{38}$]dodecylphosphocholine; n.m.r., nuclear magnetic resonance; p.p.m., parts per million; NOE, nuclear Overhauser enhancement; NOESY, two-dimensional nuclear Overhauser enhancement spectroscopy; r.m.s.d., root-mean-square distance.

glucagon were assigned to specific residues in the amino acid sequence (Wider *et al.*, 1982). In this paper these resonance assignments provide the basis for structural interpretations of NOESY spectra.

2. Nuclear Magnetic Resonance Measurements

The structure determination for MB-glucagon relies on the evaluation of cross peak intensities in two-dimensional nuclear Overhauser enhancement spectra. For the NOESY experiments we used two samples, one in H_2O and one in ^2H_2O, which contained 0·015 M-glucagon, 0·7 M-[$^2H_{38}$]-dodecylphosphocholine and 0·05 M-phosphate buffer, with a pH meter reading of 6·0. The sample temperature during the n.m.r. measurements was 37°C. Under these conditions the solutions contain mixed micelles of molecular weight ~17,000, which consist of one molecule of glucagon and ~40 molecules of DPC (Bösch *et al.*, 1980; Wider *et al.*, 1982).

NOESY spectra were recorded with the pulse sequence (Jeener *et al.*, 1979; Anil Kumar *et al.*, 1980*a*):

$$(90^\circ - t_1 - 90^\circ - \tau_m - 90^\circ - t_2)_n,$$

where t_1 is the evolution period, τ_m the mixing time and t_2 the observation period. To obtain a two-dimensional spectrum the measurement is repeated for a set of equidistant t_1 values. The signal-to-noise ratio is improved by accumulation of n transients for each value of t_1. After each observation the system is allowed to reach equilibrium during a fixed delay time. For all the experiments needed to record a complete NOESY spectrum, the same value for τ_m is used.

The NOESY spectra were recorded at 500 MHz on a Bruker WM 500 spectrometer. Quadrature detection was used, with the carrier frequency at the low field end of the spectrum. To eliminate experimental artifacts, groups of 16 recordings with different phases were added for each value of t_1 (Nagayama *et al.*, 1979,1980). To suppress contributions from coherent magnetization transfer to the cross peak intensities, mixing times shorter than 100 ms were stochastically modulated with a modulation amplitude of ± 5 ms (Macura *et al.*, 1981). For measurements in H_2O the solvent resonance was suppressed by selective, continuous irradiation at all times except during data acquisition (Anil Kumar *et al.*, 1980*b*; Wider *et al.*, 1983). To end up with a 1024×1024 point frequency domain data matrix, which corresponds to the digital resolution given in the legends to Figures 1 and 2, the time domain matrix was expanded to 1024 points in t_1 and to 2048 points in t_2 by "zero-filling". Prior to Fourier transformation the time domain data matrix was multiplied in the t_1 direction with a phase-shifted sine bell, sin $(\pi(t+t_o)/t_s)$, and in the t_2 direction with a phase-shifted sine squared bell, $\sin^2$ $(\pi(t+t_o)/t_s)$. The length of the window functions, t_s, was adjusted for the bells to reach zero at the last experimental data point in the t_1 or t_2 direction, respectively. The phase shifts, t_o/t_s, were 1/64 and 1/128 in the t_1 and t_2 directions, respectively. The spectra were obtained in the absolute value mode.

Exploratory investigations of the dependence on τ_m of the intensities of several well separated NOESY cross peaks showed that after a rapid initial growth

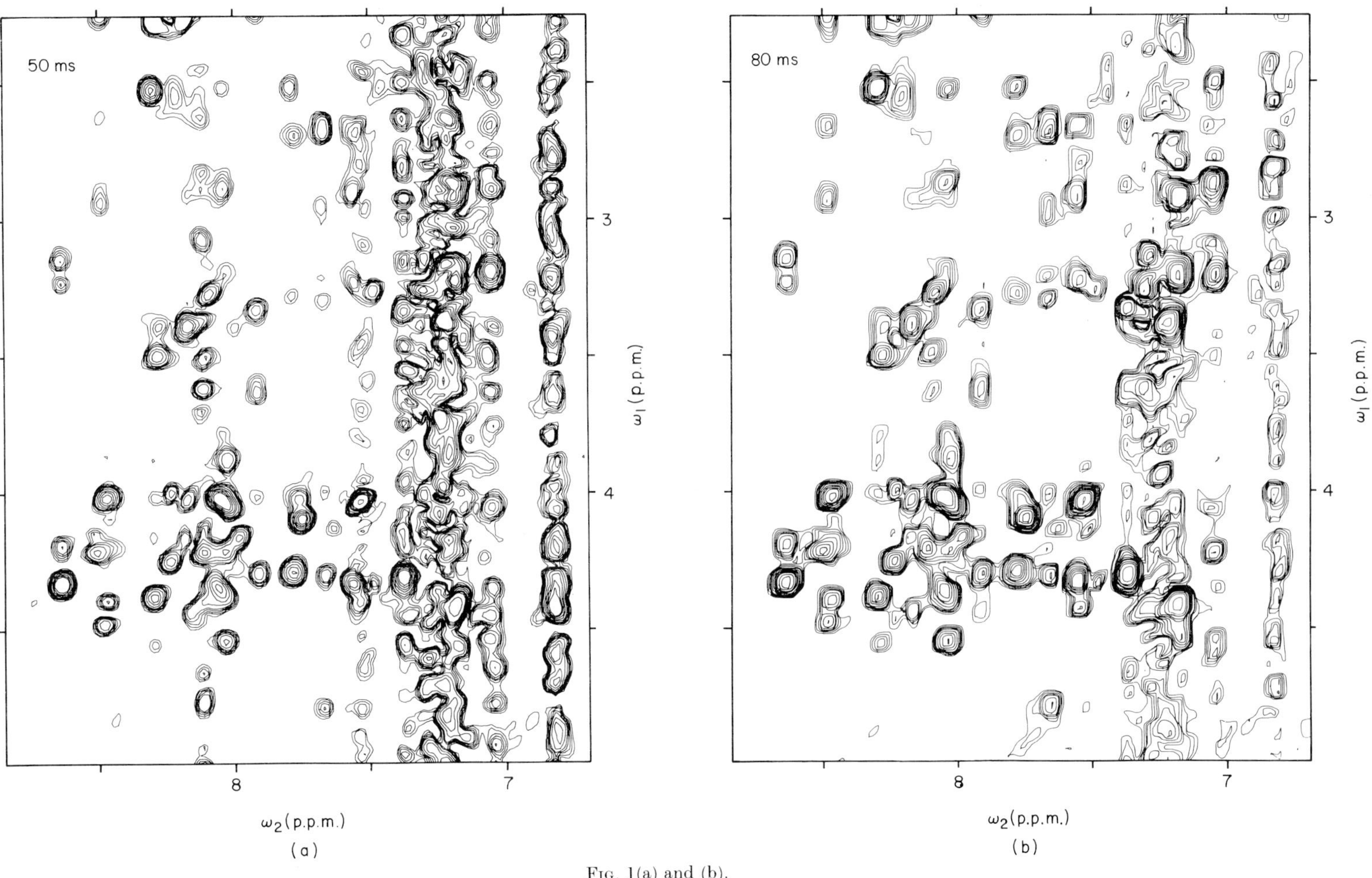

50 ms
3
4
ω_1(p.p.m.)
8
7
ω_2(p.p.m.)
(a)
80 ms
3
4
ω_1(p.p.m.)
8
7
ω_2(p.p.m.)
(b)

FIG. 1(a) and (b).

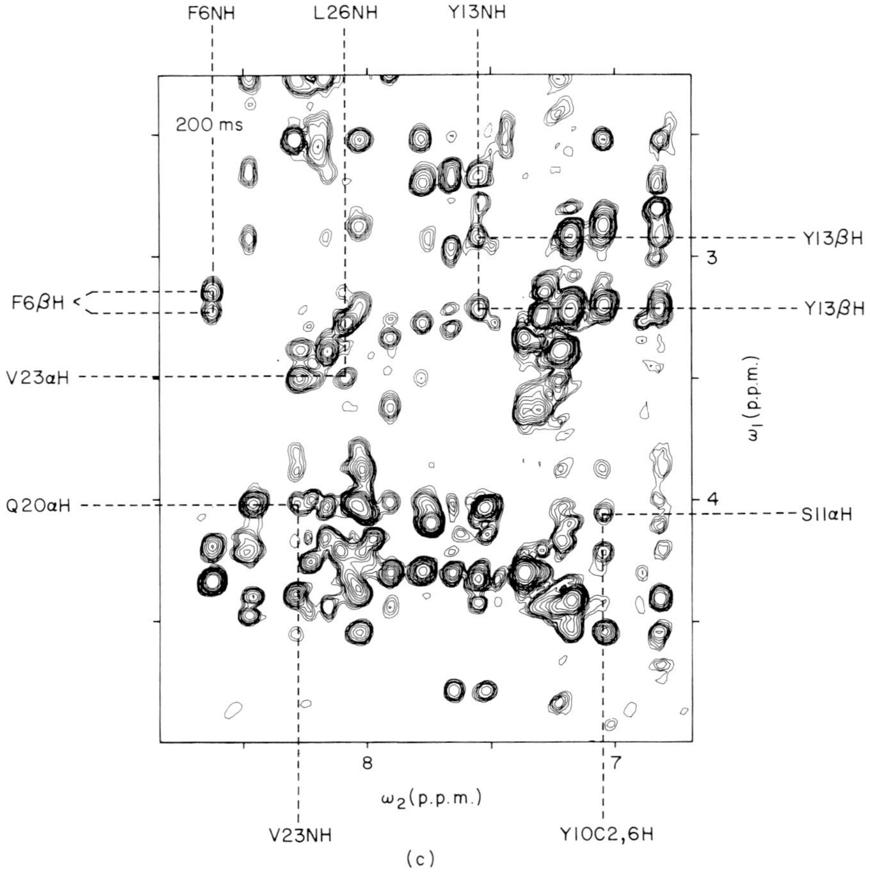

(c)

FIG. 1. (a) to (c) Contour plots of the spectral region ($\omega_1 = 2{\cdot}3$ to $4{\cdot}8$, $\omega_2 = 6{\cdot}7$ to $8{\cdot}8$) from 3 absolute value 500 MHz ^{1}H NOESY spectra of glucagon bound to perdeuterated dodecylphosphocholine micelles in H_2O solution. The 5 spectra were recorded under identical conditions (see the text), except for the different mixing times indicated in the upper left-hand corner. The digital resolution is 7·5 Hz/point. The contour levels are 2000, 2500, 3000, 3500, 4000, 5000, 6250, 7500, 8750, 10,000, 12,500, 15,000, 20,000, 25,000, 35,000, 50,000, 75,000, 100,000. As an illustration selected cross peaks are identified in spectrum (c), where the broken lines connect the cross peaks with the assignments of the 2 interacting resonances.

during the period $\tau_m = 0$ to ~ 150 ms, the peak heights varied only little when τ_m was further increased. The intensity at $\tau_m = 1000$ ms was between 20 and 60% of the maximum intensity attained near $\tau_m = 200$ ms, depending on the cross peak considered. On the basis of these observations, mixing times of 30, 50, 80, 130 and 200 ms were selected for the NOESY spectra used for the present study (Figs 1 and 2; complete NOESY spectra of MB-glucagon in ^2H_2O (Bösch *et al.*, 1981) and in H_2O (Wider *et al.*, 1982) were previously presented). To ensure that the five spectra for each sample could be directly compared, they were recorded on five consecutive days without removing the sample from the spectrometer between measurements. Subsequently, identical data handling was used for each of the five data sets and contour plots with identical contour levels were obtained (legends to Figs 1 and 2). The spectra in H_2O were recorded from 400 measurements, with t_1 values from 0·3 to 38 ms. 128 transients were accumulated for each value of t_1. The spectra in ^2H_2O were obtained from 420 measurements, with t_1 values from 0·3 to 50 ms. 144 transients were accumulated for each value of t_1.

Figure 1 illustrates the variation of the NOESY spectra when different mixing times were employed. The spectrum with $\tau_m = 50$ ms contains intense, vertical bands of noise at 6·8 p.p.m., from 7·0 to 7·6 p.p.m. and at 8·1 p.p.m. These artifacts originate from the presence of intense, sharp diagonal peaks of aromatic protons (Wider *et al.*, 1982) and represent "t_1 noise" as well as peak intensity smeared out along ω_1 by the random modulation of the mixing time τ_m (Macura *et al.*, 1981). In the spectra with longer mixing times these perturbations become less important, primarily because of the decrease of the diagonal peak intensities relative to the cross peak intensities (Anil Kumar *et al.*, 1981).

3. Regular Secondary Structures from Typical Patterns of NOESY Cross Peaks Between Backbone Hydrogen Atoms

Regular secondary structures in proteins contain characteristic patterns of short distances between amide, C^α and C^β protons (Billeter *et al.*, 1982; unpublished results). For example the distance d_1† is 2·2 Å in the extended polypeptide chain of a β-structure and 3·5 Å in the α-helix. In β-structures there are no other intrachain distances between backbone protons which would be shorter than 4·0 Å (Billeter *et al.*, 1982). In a regular α-helix one has further the short distances $d_2 = 2{\cdot}8$ Å and $d_1(i,i+3) = 3{\cdot}4$ Å, and $d_5(i,i+3)$ varies between 2·5 Å and 4·4 Å, depending on χ^1_{i+3}. For practical use we have, overall, that a succession of very short ($\lesssim 2{\cdot}5$ Å) distances d_1 in a peptide segment is indicative of an extended chain and a succession of short distances d_2 and/or $d_1(i,i+3)$ and/or $d_5(i,i+3)$ indicates that the polypeptide adopts a helical secondary structure.

The NOESY spectra of MB-glucagon in H_2O and in ^2H_2O were screened for cross peaks that correspond to d_1, d_2, $d_1(i,i+3)$ and $d_5(i,i+3)$. For this we used the previously published chemical shifts (Wider *et al.*, 1982) to determine the locations (ω_1,ω_2) where all the constraints d_1, d_2, $d_1(i,i+3)$ and $d_5(i,i+3)$ would be manifested in the NOESY spectra. From inspection of the experimental spectra (Figs 1 and 2) three situations were distinguished. (1) Connectivities

† The symbols used for characteristic ^1H–^1H distances in polypeptides were introduced elsewhere (Billeter *et al.*, 1982). i and j identify two residues in the same polypeptide chain. $d_1(i,j)$ is the distance between the C^α proton of residue i and the amide proton of residue j; for the sequential connectivity $d_1(i,i+1)$ the abbreviated symbol d_1 is used. $d_2(i,j)$ is the distance between the amide protons of residues i and j; for the sequential connectivity $d_2(i,i+1) \equiv d_2(i+1,i)$ the abbreviated symbol d_2 is used. $d_3(i,j)$ is the distance between the amide proton of residue j and the nearest C^β proton of residue i; for the sequential connectivity $d_3(i,i+1)$ the abbreviated symbol d_3 is used. $d_5(i,j)$ is the distance between the C^α proton of residue i and the nearest C^β proton of residue j.

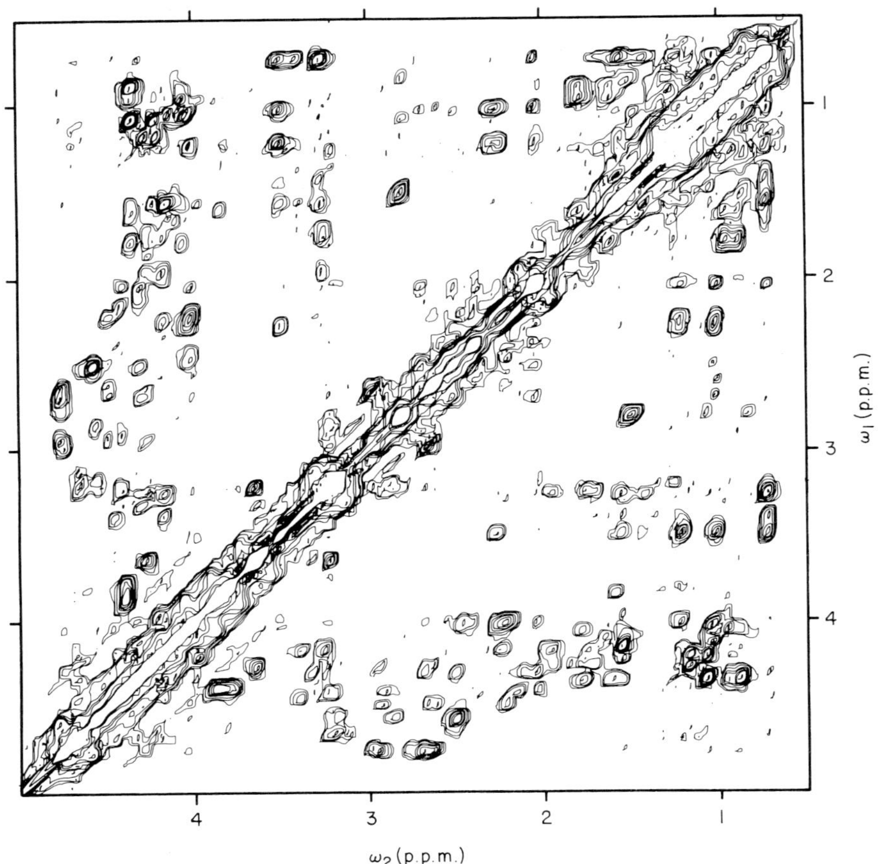

FIG. 2. Contour plot of the spectral region ($\omega_1 = 0{\cdot}5$ to $5{\cdot}0$ p.p.m., $\omega_2 = 0{\cdot}5$ to $5{\cdot}0$ p.p.m.) from a symmetrized, absolute value 500 MHz ^{1}H NOESY spectrum of glucagon bound to perdeuterated DPC micelles in ^2H_2O solution. The mixing time was 200 ms. Identical spectra were recorded with mixing times of 30, 50, 80 and 130 ms (see the text). The digital resolution is 6·4 Hz/point. The contour levels are 2000, 3000, 4000, 5000, 6500, 9000, 10,500, 12,000, 15,000 18,000, 23,000, 30,000, 40,000, 60,000, 80,000, 100,000.

manifested by resolved and unambiguously assigned cross peaks in the NOESY spectra (indicated in Fig. 3 by + or |—|, respectively). (2) Connectivities not manifested by cross peak intensity in the NOESY spectra. (3) Connectivities that would be overlapped with other peaks in a crowded region of the spectrum (indicated in Fig. 3 by ○ or | · · · · |, respectively). The connectivity patterns thus obtained (Fig. 3) indicate that the peptide segments from residues 3 to 9 and 14 to 17 adopt a predominantly extended secondary structure. For residues 12 to 14 formation of a loop or a turn is indicated and for the segment from residues 18 to 27 there is evidence for a helical structure.

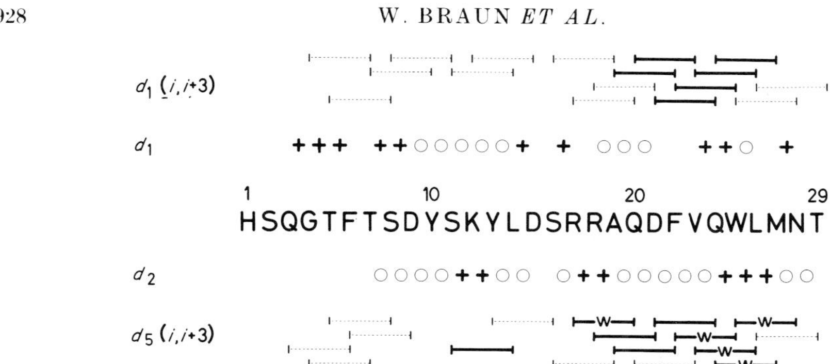

FIG. 3. Amino acid sequence of glucagon and NOESY connectivities d_1, d_2, $d_1(i, i+3)$ and $d_5(i, i+3)$ used for an initial, qualitative delineation of the secondary structure of MB-glucagon. The following symbols are used for d_1 and d_2, and for $d_1(i, i+3)$ and $d_5(i, i+3)$, respectively: (+) and (⊢—⊣) indicate that the connectivity was positively identified and the intensity of the cross peak was at least 2 contours in the NOESY spectrum recorded with a mixing time of 80 ms. Since the distance d_5 in an α-helix can vary appreciably because of the dependence on χ^1 (see the text) weak connectivities manifested by cross peaks with less than 2 contours after a mixing time of 80 ms but higher intensities at $\tau_m = 130$ and 200 ms are also indicated (⊢w⊣); these connectivities are not included in Table 2. (○) and (⊢ · · · ⊣) indicate that the presence or absence of the connectivity could not be determined unambiguously because of overlap in crowded spectral regions; absence of a symbol indicates that there is no connectivity manifested in the NOESY spectra.

4. Structural Interpretation of the NOESY Spectra With the Use of Distance Geometry Calculations

(a) *Correlation between intensity of the NOESY cross peaks and ^1H–^1H distance constraints*

When macromolecular species are studied at high magnetic fields, spin diffusion can become quite efficient (Hull & Sykes, 1975; Kalk & Berendsen, 1976) so that the information on proton–proton distances in a NOESY spectrum recorded with a fixed mixing time may be masked. On the other hand the initial build-up rates of NOEs in macromolecules are simply related to the inverse sixth power of the distance between the different groups of hydrogen atoms under consideration and can therefore provide the information needed for studies of the spatial molecular structure (Gordon & Wüthrich, 1978; Wagner & Wüthrich, 1979; Anil Kumar *et al.*, 1981; Bothner-By & Noggle, 1979; Braun *et al.*, 1981; Wüthrich *et al.*, 1982; Dobson *et al.*, 1982). When working with complex systems such as MB-glucagon, certain practical aspects must also be considered. For example, since the signal-to-noise ratio of NOESY spectra recorded with short mixing times is inherently poor (Fig. 1), it was not practical to measure initial NOE build-up rates by a series of experiments with different short τ_ms. Instead a NOESY spectrum with a long mixing time and correspondingly favourable signal-to-noise ratio was employed to properly define the location of the cross peaks in the two-dimensional spectrum (Figs 1(c) and 2). The peak intensities at these locations in spectra recorded with shorter mixing times but otherwise identical conditions (Fig. 1(a) and (b)) were then used as an approximate manifestation of the initial build-up rates.

When one correlates ^1H–^1H distances in proteins with NOEs the internal motions of the macromolecular structure (Wüthrich & Wagner, 1978; Gurd & Rothgeb, 1979; Karplus & McCammon, 1981) must also be considered. These may affect both the effective rotational correlation time for dipole–dipole coupling between different protons and the distance between the protons (Braun *et al.*, 1981). For MB-glucagon negative NOEs could be observed between neighbouring protons within any of the residues 3 to 29 (Bösch *et al.*, 1980) and it was demonstrated that the negative NOEs seen by one-dimensional n.m.r. experiments correspond to the cross peaks in NOESY (Bösch *et al.*, 1981). This shows that with the exception of the N-terminal residues His1 and Ser2, the glucagon polypeptide chain is immobilized by the binding to the micelle and indicates that the rotational motions of the vectors joining different hydrogen atoms are effectively restricted to the overall tumbling motions of the micelles. In the semi-quantitative interpretation of the NOESY spectra used to prepare the input for the distance geometry calculations we therefore assumed a common correlation time for all the observed dipole–dipole interactions. It can be shown that in those locations of the molecular structure where the effective rotational correlation time is further affected by intramolecular motions, the proton–proton distances derived from the NOE data with the use of this assumption are upper limits to the actual distances (Braun *et al.*, 1981).

In a completely rigid structure the relative NOE build-up rates between different pairs of hydrogen atoms would correspond to $1/d^6$, where d is the distance between the interacting protons (Noggle & Schirmer, 1971). We have used this "rigid model" treatment for NOEs between hydrogen atoms that are separated by a sufficiently small number of bonds in the covalent polypeptide structure so that the relative spatial locations are determined by at most three torsion angles. For hydrogen atoms that are further separated in the covalent

TABLE 1

Correlations between intensity of the cross peaks in the NOESY spectra of MB-glucagon recorded in H_2O and constraints on ${}^1H$–${}^1H$ distances, which were used as the input for the distance geometry calculations

NOESY		d (Å)†	d_{max} (Å)†
τ_m (ms)	Intensity‡	(rigid model)	(uniform averaging model)
50	≥10	≤2·4	≤3·0
50	6–9	≤2·7	≤3·0
50	2–5	≤3·1	≤4·0
80	≥2	≤4·0	≤5·0

† The rigid model was applied for NOEs between hydrogen atoms that are separated by a sufficiently small number of bonds in the covalent structure so that the ${}^1H$–${}^1H$ distance can be determined by 3 or less torsion angles about single bonds. In all other cases the uniform averaging model was applied (see the text). In the distance geometry calculations d or d_{max} were used as upper limits for the distance between the 2 groups of protons that are connected by the cross peak.

‡ The intensity of the NOESY cross peaks is given as the number of contour lines in the contour plots of the spectra. The contour levels are listed in the legend to Fig. 1.

structure the NOEs were interpreted with a previously described "uniform averaging model", which accounts for time variations of the proton–proton distances d in a flexible structure (Braun *et al.*, 1981). In this treatment the distance between two hydrogen atoms is allowed to vary uniformly between a minimum distance d_{min}, which is taken to be the sum of the van der Waals' radii, and a maximum distance d_{max}, which is determined by the NOE data. For a given NOE build-up rate d_{max} obtained with the uniform averaging model is always longer than the corresponding d obtained with the rigid model. Examples of corresponding distance constraints computed from the same NOE data for the rigid and the flexible model are 2·3 and 2·7 Å, 2·6 and 3·7 Å, and 2·9 and 4·5 Å (Braun *et al.*, 1981).

Based on the fundamental considerations outlined above and using NOESY cross peaks between neighbouring hydrogen atoms in the covalent structure for calibration purposes, the rules in Table 1 were used for the interpretation of the NOESY data recorded in H_2O solution of MB-glucagon. A similar set of rules (with different numbers to account for the different contour levels in the ^2H_2O and H_2O spectra, see legends to Figs 1 and 2) was used for the ^2H_2O spectra. The resulting distance constraints are listed in Table 2, where the numbers accompanied by one or two lower case letters include systematic corrections that account for the use of "pseudo-structures" for the amino acids (see the text below, footnotes to Table 2 and Wüthrich *et al.* (1983)).

(b) *Spatial structure determination with distance geometry calculations*

The problem to be solved by distance geometry calculations is the following (Crippen & Havel, 1978; Havel *et al.*, 1979; Braun *et al.*, 1981; Wako & Scheraga, 1982). Given upper limits and lower limits for the distances between the N atoms in a molecular structure, what are the conformations that are compatible with these distance constraints? In the present application to a polypeptide chain, the upper and lower limits for the distances between covalently linked atoms correspond to a set of standard geometries for the common amino acid residues (Momany *et al.*, 1975). For non-bonding interactions the lower limits correspond to the sum of the van der Waals' radii of the atoms considered and the upper limits on the interatomic distances are obtained from the analysis of the NOESY spectra (Table 2). Because of the limited resolution of n.m.r. spectra and the limitation of NOESY distance measurements to distances smaller than ~5·0 Å, n.m.r. provides an incomplete set of relatively inaccurate distance constraints. Therefore, distance geometry calculations will usually not provide a unique conformation, but repeated computations using the same set of n.m.r. distance constraints will provide a group of somewhat different molecular geometries that are all compatible with the experimental data (Fig. 4).

The distance geometry algorithm used was previously described in detail (Braun *et al.*, 1981). However, since a much more extensive set of distance constraints was available than for the earlier applications, new pseudo-structures for the amino acid residues were used for the input (Wüthrich *et al.*, 1983). In the pseudo-structures groups of hydrogen atoms for which no stereospecific

TABLE 2

Distance constraints between groups of hydrogen atoms in the three-dimensional structure of glucagon bound to perdeuterated dodecylphosphocholine micelles obtained from NOESY spectra in H_2O and in ^2H_2O

Distance constraints in Å†			
Sequential	Intra-residue	Long range backbone	Long range
His1			
Ser2			
Gln3 HA G4 HN 2·7			
Gly4 HN Q3 HA 2·7			
Thr5 HN T5 HA 2·7 HA F6 HN 2·4 HB F6 HN 2·7	HN T5 MG 4·1 m (8·06, 1·07) HA T5 MG 3·0 i (4·34, 1·07)		HB F6 QR 7·4 q (4·19, 7·27) HB T7 MG 4·0 m (4·19, 7·17) MG F6 HN 6·0 m (1·07, 8·63)
Phe6 HN T5 HA 2·4 HN T5 HB 2·7 HN F6 PB 3·3 i	HN F6 PB 3·3 i (8·63, 3·13), (8·63, 3·22) HA F6 QR 4·7 r (4·64, 7·27)		HN T5 MG 6·0 m (8·63, 1·07) HA Y10 C1 7·0 r (4·64, 7·06) QR T5 HB 7·4 q (7·27, 4·19) QR T7 HA 7·4 q (7·27, 4·23) QR T7 MG 8·4 qm(7·27, 1·17)
Thr7 HN T7 HA 2·7 HN T7 HB 3·1 HA S8 HN 2·7	HN T7 HB 3·1 (8·11, 4·24) HN T7 MG 3·4 m (8·11, 1·17) HA T7 MG 3·7 i (4·23, 1·17)		HA F6 QR 7·4 q (4·23, 7·27) HA Y10 C1 6·0 r (4·23, 7·06) MG T5 HB 4·0 m (1·17, 4·19) MG F6 QR 8·4 qm(1·17, 7·27) MG S8 HN 5·0 m (1·17, 8·04) MG Y10 C1 8·0 mr (1·17, 7·06)
Ser8 HN T7 HA 2·7 HN S8 PB 3·3 i HA D9 HN 2·7	HN S8 PB 3·3 i (8·04, 3·86)		HN T7 MG 5·0 m (8·04, 1·17)
Asp9 HN S8 HA 2·7 HN D9 PB 3·3 i PB Y10 HN 4·1 m	HN D9 PB 3·3 i (8·30, 2·52)		PB Y10 C1 7·0 mr (2·52, 7·06)

Tyr10 HN D9 PB 4·1 m HN Y10 PB 3·3 i	HN Y10 PB 3·3 i (8·05, 2·85), (8·05, 3·18)		HA Y13 C1 6·0 r (4·56, 7·19) C1 F6 HA 7·0 r (7·06, 4·64) C1 T7 HA 6·0 r (7·06, 4·23) C1 T7 MG 8·0 mr (7·06, 1·17) C1 D9 PB 7·0 mr (7·06, 2·52) C1 S11 HA 7·0 r (7·06, 4·06) C1 L14 QD 7·4 rq (7·06,0·88), (7·06, 0·96) C4 L14 QD 7·4 rq (6·82, 0·96)
Ser11 HN S11 HA 2·4 HN K12 HN 3·1		HA L14 PB 6·0 m (4·06, 1·61)	HA Y10 C1 7·0 r (4·06, 7·06) HA L14 HG 5·0 (4·06, 1·77) HA L14 QD 5·4 q (4·06, 0·88), (4·06, 0·96)
Lys12 HN S11 HN 3·1 HN K12 HA 2·4 HN Y13 HN 4·0 PB Y13 HN 4·1 m	HN K12 PG 4·1 m (7·75, 1·10) HA K12 PD 5·0 m (4·09, 1·50)		HA Y13 C1 7·0 r (4·09, 7·19) PB Y13 C1 8·0 mr (1·59, 7·19)
Tyr13 HN K12 HN 4·0 HN K12 PB 4·1 m HN Y13 PB 3·7 i	HN Y13 PB 3·7 i (7·55, 2·89), (7·55, 3·23)		C1 Y10 HA 6·0 r (7·19, 4·56) C1 K12 HA 7·0 r (7·19, 4·09) C1 K12 PB 8·0 mr (7·19, 1·59) C1 L14 QD 9·4 rq (7·19, 0·88)
Leu14 HA D15 HN 2·7	HN L14 MD1 5·0 m (7·52, 0·96) HN L14 MD2 5·0 m (7·52, 0·88) HA L14 QD 4·1 s (4·33, 0·88), (4·33, 0·96)	PB S11 HA 6·0 m (1·61, 4·06)	HG S11 HA 5·0 (1·77, 4·06) QD Y10 C4 7·4 rq (0·96, 6·82) QD Y10 C1 7·4 rq (0·88, 7·06), (0·96, 7·06) QD S11 HA 5·4 q (0·88, 4·06), (0·96, 4·06) QD Y13 C1 9·4 rq (0·88, 7·19) QD D15 HN 7·4 q (0·88, 7·55)
Asp15 HN L14 HA 2·7 HN D15 PB 3·3 i	HN D15 PB 3·3 i (7·55, 2·76)		HN L14 QD 7·4 q (7·55, 0·88)
Ser16 HA R17 HN 2·7			
Arg17 HN S16 HA 2·7 HN R17 PB 3·3 i HN R18 HN 3·1	HN R17 PB 3·3 i (8·50, 1·96)	HA Q20 PB 5·0 m (4·17, 2·25)	

TABLE 2 *(continued)*

Distance constraints in Å†			
Sequential	Intra-residue	Long range backbone	Long range
Arg18			
HN R17 HN 3·1	HN R18 PB 3·3 i (8·24, 1·86)		
HN R18 PB 3·3 i	HN R18 PG 4·1 m (8·24, 1·71)		
HN R18 HA 2·7			
HN A19 HN 3·1			
Ala19			
HN R18 HN 3·1	HN A19 MB 3·0 i (7·99, 1·56)	HA F22 HN 4·0 (4·17, 8·18)	HA F22 QR 7·4 q (4·17, 7·23)
HN A19 MB 3·0 i		HA F22 PB 6·0 m (4·17, 3·38)	MB F22 QR 8·4 mq(1·56, 7·23)
HN A19 HA 2·7		MB F22 PB 6·0 m (1·56, 3·38)	
HA Q20 HN 3·1		MB V23 HA 5·0 m (1·56, 3·49)	
MB Q20 HN 3·7 m			
Gln20			
HN A19 MB 3·7 m	HN Q20 PB 3·3 i (8·17, 2·25)	HA V23 HN 4·0 (4·02, 8·28)	HA V23 QG 6·4 q (4·02, 1·21)
HN A19 HA 3·1		PB R17 HA 5·0 m (2·25, 4·17)	
HN Q20 HA 2·7			
HN Q20 PB 3·3 i			
HA D21 HN 3·1			
PB D21 HN 4·1 m			
Asp21			
HN Q20 PB 4·1 m		HA Q24 HN 4·0 (4·47, 8·24)	
HN D21 HA 2·7		HA Q24 PB 5·0 m (4·47, 2·22)	
PB F22 HN 4·1 m			
Phe22			
HN D21 PB 4·1 m	HN F22 PB 3·3 i (8·18, 3·38)	HN A19 HA 4·0 (8·18, 4·17)	HA W25 H4 5·0 (4·44, 7·33)
HN F22 PB 3·3 i		HA W25 HN 5·0 (4·44, 7·92)	QR A19 HA 7·4 q (7·23, 4·17)
PB V23 HN 4·1 m		PB A19 HA 6·0 m (3·38, 4·17)	QR A19 MB 8·4 mq(7·23, 1·56)
		PB A19 MB 6·0 m (3·38, 1·56)	QR V23 QG 8·8 qq (7·23, 1·21)
			QR V23 HN 7·4 q (7·23, 8·28)
			QR V23 HA 7·4 q (7·23, 3·49)
			QR L26 QD 7·8 qq (7·23, 0·72)
			PB L26 QD 6·4 mq(3·38, 0·72)
Val23			
HN F22 PB 4·1 m	HN V23 HB 2·7 (8·28, 2·26)	HN Q20 HA 4·0 (8·28, 4·02)	HN F22 QR 7·4 q (8·28, 7·23)
HN V23 HB 2·7	HA V23 MG1 3·7 m (3·49, 1·21)	HA A19 MB 5·0 m (3·49, 1·56)	HA F22 QR 7·4 q (3·49, 7·23)
HN V23 HA 2·7	HA V23 MG2 3·7 m (3·49, 1·02)	HA L26 HN 3·0 (3·49, 8·09)	HA L26 QD 5·4 q (3·49, 0·72)
HA Q24 HN 3·1			QG Q20 HA 6·4 q (1·21, 4·02)

HB Q24 HN 3·1			QG F22 QR 8·8 qq (1·02, 7·23), (1·21, 7·23)
			QG Q24 HN 7·4 q (1·01, 8·24), (1·21, 8·24)
Gln24			
HN V23 HA 3·1	HN Q24 PB 3·3 m (8·24, 2·22)	HN D21 HA 4·0 (8·24, 4·47)	HN V23 QG 7·4 q (8·24, 1·02), (8·24, 1·21)
HN V23 HB 3·1		HA M27 HN 4·0 (4·00, 7·99)	
HN Q24 HA 2·7		HA N28 HN 4·0 (4·00, 7·66)	
HN Q24 PB 3·3 m		PB D21 HA 5·0 m (2·22, 4·47)	
HN W25 HN 3·1			
HA W25 HN 3·1			
PB W25 HN 4·1 m			
Trp25			
HN Q24 HN 3·1	HN W25 PB 3·3 i (7·92, 3·33), (7·92, 3·62)	HN F22 HA 5·0 (7·92, 4·44)	H2 L26 QD 6·4 q (7·37, 0·72)
HN Q24 HA 3·1	HN W25 H4 5·0 (7·92, 7·33)		H4 F22 HA 5·0 (7·33, 4·44)
HN Q24 PB 4·1 m	HA W25 H2 2·4 (4·30, 7·37)		H4 L26 HA 5·0 (7·33, 3·27)
HN W25 PB 3·3 i	PB W25 H2 3·7 m (3·33, 7·37), (3·67, 7·37)		H4 L26 PB 6·0 m (7·33, 1·57)
HN W25 HA 2·7	PB W25 H4 4·1 m (3·33, 7·33), (3·67, 7·33)		H4 L26 QD 5·4 q (7·37, 0·72)
HN L26 HN 4·0			H4 L26 HN 4·0 (7·33, 8·09)
PB L26 HN 4·1 m			H5 L26 PB 6·0 m (6·89, 1·57)
			H5 L26 QD 6·4 q (6·89, 0·72)
			H5 L26 HA 5·0 (6·89, 3·27)
			H6 L26 QD 7·4 q (7·11, 0·72)
Leu26			
HN W25 HN 4·0	HN L26 PB 3·7 m (8·09, 1·41)	HN V23 HA 3·0 (8·09, 3·47)	HN W25 H4 4·0 (8·09, 7·33)
HN W25 PB 4·1 m	HA L26 QD 4·1 s (3·27, 0·72)		HA W25 H4 5·0 (3·27, 7·33)
HN L26 HA 2·7			HA W25 H5 5·0 (3·27, 6·89)
HN L26 PB 3·7 m			QD F22 QR 7·8 qq (0·72, 7·23)
HN M27 HN 3·1			QD F22 PB 6·4 mq (0·72, 3·38)
PB M27 HN 4·1 m			QD V23 HA 5·4 q (0·72, 3·49)
			QD W25 H4 5·4 q (0·72, 7·33)
			QD W25 H2 6·4 q (0·72, 7·37)
			QD W25 H5 6·4 q (0·72, 6·89)
			QD W25 H6 7·4 q (0·72, 7·11)
			QD M27 HN 7·4 q (0·72, 7·79)
			QD M27 ME 8·4 qm (0·72, 2·04)
			PB W25 H4 6·0 m (1·57, 7·33)
			PB W25 H5 6·0 m (1·57, 6·89)
Met27			
HN L26 HN 3·1	HN M27 PB 3·7 i (7·79, 2·13)	HN Q24 HA 4·0 (7·79, 4·00)	HN L26 QD 7·4 q (7·79, 0·72)
HN L26 PB 4·1 m	HN M27 PG 4·1 m (7·79, 2·53), (7·79, 2·70)		ME L26 QD 8·4 qm (2·04, 0·72)
HN M27 HA 2·7			PG N28 HN 6·0 m (2·03, 7·66)
HN M27 PB 3·7 i			
HN N28 HN 3·1			
HA N28 HN 3·1			

TABLE 2 *(continued)*

Distance constraints in Å†			
Sequential	Intra-residue	Long range backbone	Long range
Asn28			
HN M27 HA 3·1	HN N28 PB 3·3 i (7·66, 2·68), (7·66, 2·97)	HN Q24 HA 4·0 (7·66, 4·00)	HN M27 PG 6·0 m (7·66, 2·03)
HN N28 PB 3·3 i			
PB T29 HN 3·7 m			
Thr29			
HN N28 PB 3·7 m	HN T29 HB 4·0 (7·53, 4·11)		
HN T29 HB 4·0	HN T29 MG 4·1 m (7·53, 1·09)		
HN T29 HA 2·4	HA T29 MG 3·0 i (7·53, 1·07)		

† The amino acid residues are identified either by the 3-letter symbol or the 1-letter symbol followed by a number indicating the position in the amino acid sequence. HN stands for amide proton, HA for C^{α} proton, HB for C^{β} proton, HG for C^{γ} proton. C1 and C4 are the ring carbon atoms in positions 1 and 4 of the aromatic rings of Phe and Tyr, H2 and H4 to H7 are the protons attached to the indole ring carbons 2 and 4 to 7 of Trp. The following upper case letters indicate pseudo-atoms which substitute for a group of 2 or more protons and the lower-case letters indicate that the distance constraint indicated has been modified to account for the use of the pseudo-atom (see the text and Wüthrich *et al.*, 1983): M replaces the 3 protons of a methyl group and is located in the centre of the 3 proton positions, with the corrections factors $m = 1{\cdot}0$ Å (for long range constraints) and $i = 0{\cdot}6$ Å (for certain short range constraints). P replaces the 2 protons of a methylene group and is located in the centre between the 2 proton positions, m and i apply as for the methyl group. Q replaces the 2 methyl groups of Val or Leu and is located centrally with respect to the 2 methyl groups. The correction q is 2·4 Å, except that $s = 1{\cdot}7$ Å applies for the intraresidue constraint from C^{α}H to QD in Leu. QR is at the centre of the ring protons C2H, C3H, C5H and C6H in Phe or Tyr, the correction q us 2·4 Å. The correction $r = 2{\cdot}0$ Å applies for Tyr or Phe when one replaces either the 2,6 ring protons by C1 or the 3,5 ring protons by C4, and for the intraresidue constraint from C^{α}H to QR.

The rows following the 3-letter symbol for the amino acid list all the distance constraints which involve hydrogen atoms of this residue. For each residue the distance constraints are grouped into 4 classes, which are presented in the 4 columns. *Sequential* constraints are those between amide, C^{α} and C^{β} protons within each residue and in sequentially neighbouring residues. *Intraresidue* constraints are between side-chain hydrogen atoms and any other proton of the same residue. *Long range backbone* constraints are between amide, C^{α} and C^{β} protons in residues that are not nearest neighbours in the sequence. *Long range* constraints are all those that have not been listed in the first 3 columns. Note that the distance constraints between the backbone hydrogens and C^{β}H of the same residue are listed in both the first and second column and that with the exception of the intra-residue constraints all entries appear twice, for 2 different residues. In each column the first entry identifies a hydrogen atom in the amino acid indicated by the 3-letter symbol. The second entry indicates a different residue and the position in this residue to which a short distance constraint exists. The third entry is the distance constraint in Å, possibly with symbols indicating that one or several correction factors were added to the NOESY distance constraint to allow for the use of pseudo-atoms. The fourth entry indicates the location of the NOESY cross peak in the ω_1–ω_2 plane (Figs 1 and 2), where (ω_1, ω_2) are given in p.p.m. This information has been omitted for the sequential constraints, since these were previously described in detail (Wider *et al.*, 1982).

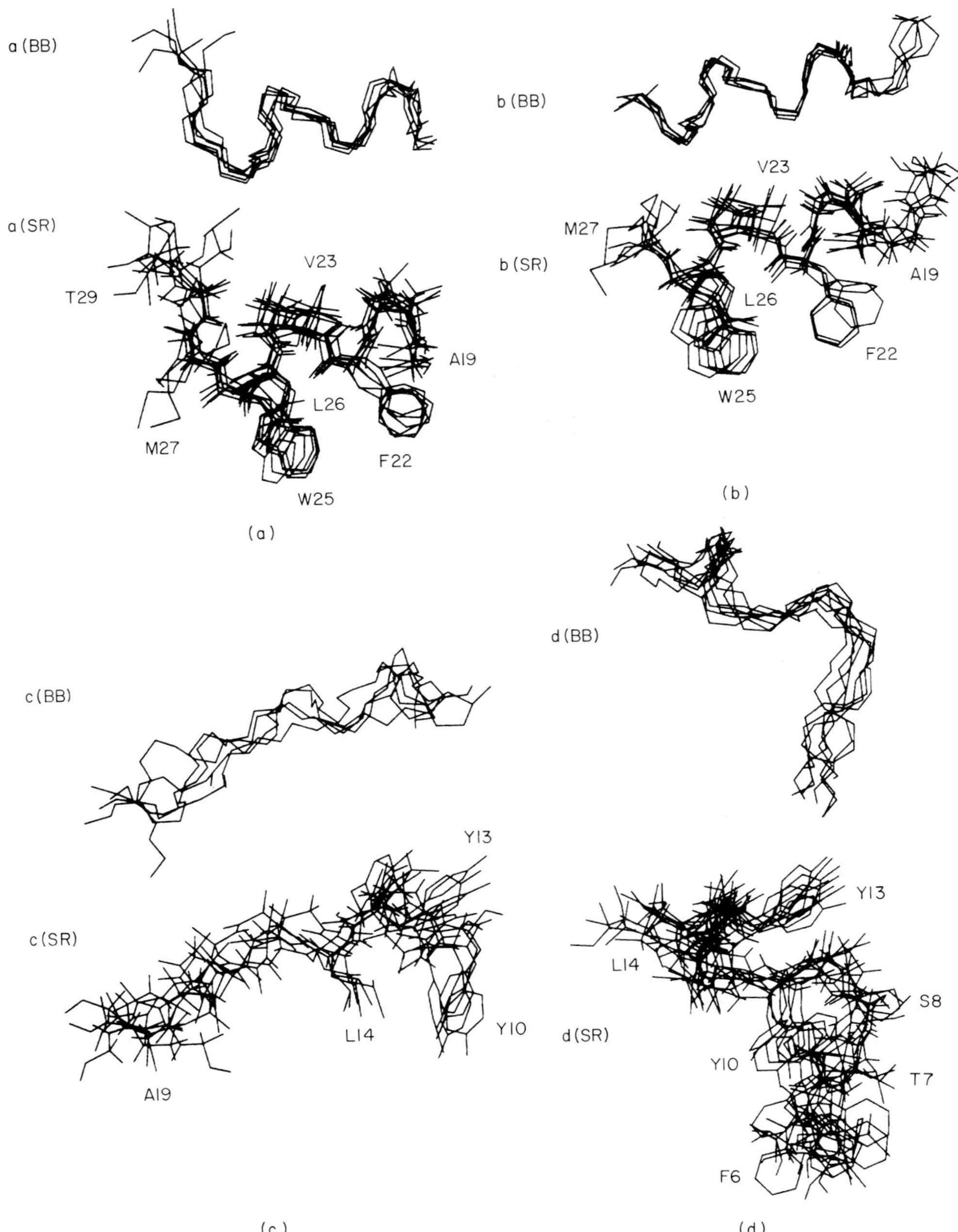

FIG. 4. Computer drawings of the spatial structure for 4 segments of MB-glucagon. From the final group of 10 computations for each segment all those structures that have satisfactory stereochemistry (see the text) are shown superimposed on each other. For each segment drawings of the backbone (BB) and of the "restrained side-chain" presentation (SR) are shown. SR includes the complete side-chains for the residues that are identified in the drawings and for which NOE constraints were observed for the peripheral protons (Table 2). For the other residues only the backbone atoms, including the C^{α} and amide protons and C^{β} are shown. CH_3 and CH_2 groups are represented by the spherical pseudo-atoms M and L, respectively, and the CH groups in the aromatic rings of Phe and Tyr by the spherical pseudo-atom K (see the text). (a) Segment 19–29, 6 structures are superimposed; (b) segment 17–27, 5 structures; (c) segment 10–20, 5 structures; (d) segment 5–15, 8 structures.

assignments of the n.m.r. lines were obtained (Wider *et al.*, 1982) are replaced by a pseudo-atom located in a central position relative to the protons for which it substitutes, and the pseudo-atom position is then used in the distance geometry calculations. The pseudo-atoms P, M, Q and QR are described in the footnotes to Table 2. Since the distance constraints manifested in the NOESY spectra are of course always between real hydrogen atoms, the use of pseudo-atoms must be accounted for by the introduction of corrections to the experimentally determined distance constraints. A list of the corrections used is given in the footnotes to Table 2 and a complete description of the dimensions of the pseudo-structures for amino acids to be used in n.m.r. structure determinations is contained in the following paper (Wüthrich *et al.*, 1983).

To reduce the computing time a different type of pseudo-atoms, M, L and K, was introduced to represent, respectively, the van der Waals' dimensions of the CH_3 and CH_2 groups and the CH fragments in positions 2, 3, 5 and 6 of the aromatic rings of Phe and Tyr. M is a sphere of radius 1·8 Å located in the centre of the three methyl protons. (Its centre coincides with the pseudo-atom position M used as the point of reference for NOESY distance constraints involving methyl groups.) L is a sphere of radius 1·6 Å located at the position of the methylene carbon. K is a sphere of radius 1·5 Å located at the position of the ring carbon atom. K, L and M are also used in the representations of the MB-glucagon structures in Figures 4 to 8.

Inspection of the distance constraints in Table 2 shows that no NOEs were observed between residues that would be further apart than five positions in the amino acid sequence. In the peptide segment 14–17 there are actually exclusively intra-residue and sequential NOEs. To save computer time the distance geometry calculations were therefore performed separately for four segments of the polypeptide chain, i.e. 5–15, 10–20, 17–27 and 19–29. This seemed justified, since in the absence of real "long range" distance constraints the overall shape of the molecule could anyway not be determined in a meaningful way. Approximate structures for MB-glucagon 5–29 were then obtained by a program that would minimize the root-mean-square-distance between the overlapping parts of the four segments.

For each peptide segment ten independent computer runs were made, with five G-cycles and 2000 refinement-cycles (Braun *et al.*, 1981). The weight of the NOE distance constraints relative to the distance constraints imposed by the covalent structure and the van der Waals' radii was taken to be 0·01. The criterion for proper convergence of each individual computer run was that none of the distance constraints by covalent bonds or by the van der Waals' radii was violated by more than 0·1 Å. Structures with larger violations were discarded. For those structures that had reasonable stereochemical properties (typically 5 to 8 out of a group of 10 computations) a record was made of all short distances between hydrogen atoms. Using the previously-determined (Wider *et al.*, 1982) chemical shifts the corresponding cross peak positions in a hypothetical NOESY spectrum of the computed conformation were calculated and compared with the input data (Table 2) and with the experimental spectra. This procedure was repeated three times to search for and subsequently eliminate inconsistencies between the molecular geometry, the input data and the experimental spectra. Quite generally, most of the short distance constraints that were not contained in the input data were found to correspond to peaks in poorly resolved regions of the NOESY spectra, so that they could not be reliably assigned and evaluated in the initial spectral analysis. The checks on internal consistency between experiment and structures obtained included also a search for large violations of NOE distance constraints that would occur in the majority of the results. For the structures with satisfactory stereochemistry at least 90% of the input NOE constraints were typically fulfilled with violations of the order of 0·1 to 0·4 Å, which seems reasonable in view of the semi-quantitative data analysis (Tables 1 and 2). Consistently larger violations were found for the d_1-connectivities from Ala19 to Gln20, Val23 to Gln24 and Gln24 to Trp25, which were all estimated to be 3·1 Å (Table 2). These constraints were subsequently omitted in the last group of ten computations for the segments 17–27 and 19–29, which produced no significant changes in the molecular geometries. This showed quite convincingly that the structure determination in the region of residues 19 to 25 is dominated by the longer range constraints, such as $d_1(i,i+3)$ and $d_5(i,i+3)$ (Fig. 3).

5. The Spatial Structure of MB-Glucagon from n.m.r. Distance Constraints and Distance Geometry Calculations

The results of the above described structural interpretation of the NOESY distance constraints in MB-glucagon (Table 2) with the use of a distance geometry algorithm are presented in Figures 4 to 6 and Table 3. From the final ten computer runs for each of the segments 5–15, 10–20, 17–27 and 19–29, eight conformers for the segment 5–15, five conformers for each of the segments 10–20 and 17–29, and six conformers for 19–29 had stereochemically acceptable spatial structures (Fig. 4 and Table 3). Four different presentations of the molecular structures showing different amounts of detail are employed. In the presentation BB only the backbone atoms N, C^α and C′ are included. SB contains the backbone and all those atoms that are directly bonded to one of the backbone atoms, i.e. it includes the fragments N—H, C^αH—C^β and C′=O. SR stands for "restricted side-chain representation", i.e. in addition to the atoms shown in SB the complete side-chains are shown for all those residues where the conformation is determined by NOEs involving the peripheral hydrogen atoms (Table 2). These residues are Thr5, Phe6, Thr7, Ser8, Tyr10, Tyr13, Leu14, Ala19, Phe22, Val23, Trp25, Leu26, Met27 and Thr29. In this presentation CH_3, CH_2 and the CH fragments of the rings of Tyr and Phe are replaced by the spherical pseudo-atoms M, L and K, respectively. In the HA (heavy atom) presentation all heavy atoms of all side-chains are included. This presentation is mainly added in Table 3 as a check for the sampling property of the algorithm. The structures in Figure 4 correspond to the presentations BB and SR. For each of the four segments in Figure 4 the conformer with the smallest deviations of the covalent structure from the standard ECEPP geometry (Momany *et al.*, 1975) was selected and mono- and stereo-drawings of these species are shown in Figure 5. In Figure 6 the four partial structures of MB-glucagon in Figure 5 were combined so as to minimize

TABLE 3

Comparison of all stereochemically acceptable conformers obtained from the final group of 10 computer runs for each of the 4 fragments of MB-glucagon

MB-glucagon segment	Number of conformers	Average r.m.s.d. (Å)†			
		BB‡	SB‡	SR‡	HA‡
5–15	8	1·52	2·03	2·63	2·79
10–20	5	2·08	2·67	3·06	4·09
17–27	5	0·85	1·18	1·59	1·94
19–29	6	1·15	1·47	1·77	2·07

The average of the root-mean-square distances between any 2 of the conformers is indicated.

† r.m.s.d. $= \left\{\frac{1}{N}\sum_{j=1}^{N}|Rx_j^u - x_j^v|^2\right\}^{\frac{1}{2}}$, where N is the number of atoms in the structure, x_j are the atomic co-ordinates, u and v indicate the 2 conformers which are compared, R is the rotation matrix which affords the best match in space between the 2 conformers. R was obtained using an algorithm proposed by McLachlan (1979).

‡ The following presentations of the molecular structures are used (see the text): BB, only the backbone atoms N, C^{α} and C′ are considered. SB, The backbone atoms with the directly bonded atoms are included in the calculation, i.e. N—H, C^{α}H—C^{β} and C′=O. SR, Restricted side-chain representation. Besides atoms of SB these structures contain all side-chain atoms (without hydrogens) for those residues, where NOE distance constraints were available up to the peripheral protons. These residues are T5, F6, T7, S8, Y10, Y13, L14, A19, F22, V23, W25, L26, M27, T29. HA, All heavy atoms are included.

the r.m.s.d. between the overlapping residues of any two segments. This Figure affords a survey of the secondary structure from residues 5 to 29 of MB-glucagon.

For the reasons discussed in section 4, above, the presently used procedures cannot be expected to provide a unique spatial polypeptide structure, but the result consists of a group of structures that are all compatible with the experimental data of Table 2. Each individual conformer is to be regarded as a typical member of the group of structures occupying the conformation space within the confines imposed by the NOE distance constraints, and not as an "average spatial structure" (Braun *et al.*, 1981). Therefore significant information on the quality of the results obtained comes primarily from comparison of the different structures in Figure 4 and Table 3.

Inspection of Table 2 shows that the number of NOE distance constraints used for the structure determination of the four segments 5–15, 10–20, 17–27 and 19–29 are, respectively, 51, 39, 77 and 79. There is a good correlation between the number of distance constraints used and the r.m.s.d. values in Table 3, i.e. smaller r.m.s.d. values prevail for the segments with a larger number of distance constraints per residue. The best constrained parts of the polypeptide chain are the backbone structures (BB) 17–27 and 19–29, which is also clearly seen in Figure 4. The r.m.s.d. values of 0·85 Å and 1·15 Å for BB 17–27 and BB 19–29, respectively, indicate that the atom positions in these fragments are determined nearly within the limiting uncertainty expected from the thermal fluctuations of

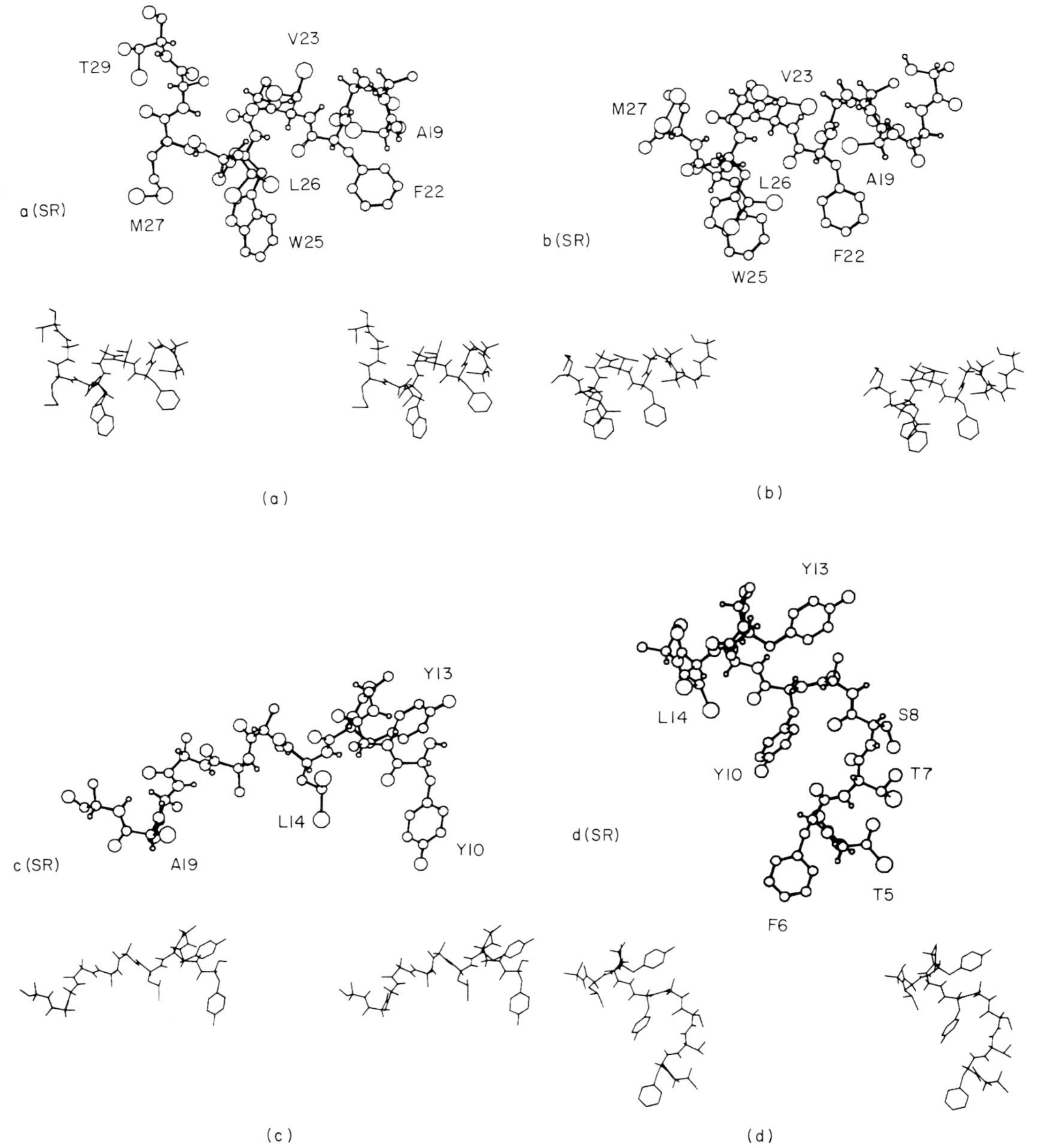

FIG. 5. Computer drawing of the stereochemically best structure (see the text) for each segment from the group of molecular geometries shown in Fig. 4. The SR presentation is shown and the same pseudo-atoms are used as in Fig. 4. For each segment a mono and a stereo drawing is shown. (a) Segment 19–29; (b) 17–27; (c) 10–20; (d) 5–15.

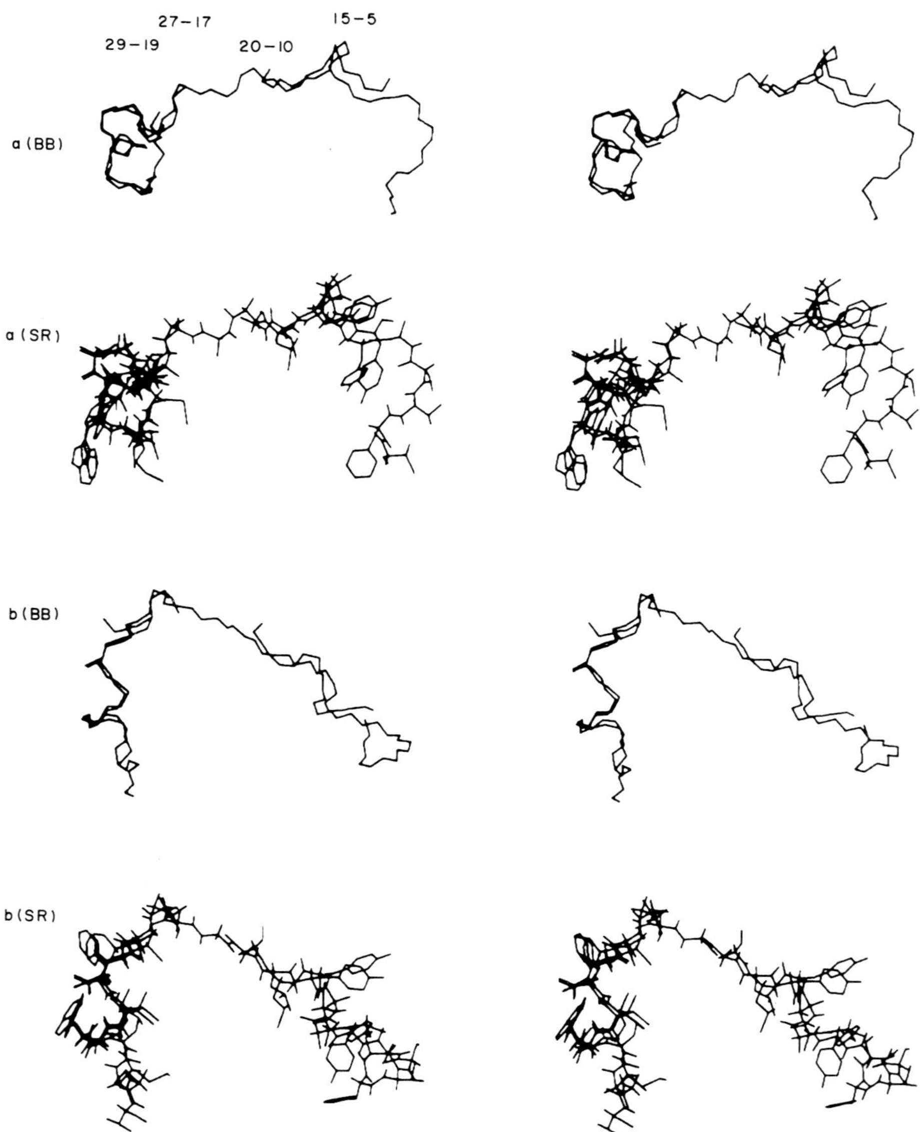

FIG. 6. Stereo drawings of the structure of MB-glucagon 5–29 obtained by combination of the structures for the 4 individually computed segments 5–15, 10–20, 17–27 and 19–29 in Fig. 5. The segments were fitted together so as to minimize the root-mean-square distance between the overlapping residues for each pair of segments. The r.m.s.d. values for the overlapping segments 11–14, 17–20 and 20–26 in the SR presentation are 1·7 Å, 1·7 Å and 1·1 Å, respectively. The BB and SR presentations of the structure are shown (see Fig. 4). The drawings (b) show the same structure as (a) after a 90° rotation about a horizontal axis aligned parallel to the projection plane.

the polypeptide chain. (For example, average root-mean-square fluctuations of 0·6 Å were observed for the C^α atom positions in simulations of the thermal motions in small globular proteins by molecular dynamics calculations (McCammon & Karplus, 1980).) The least restrained segment is 10–20. Here, the jump of the r.m.s.d. value from 3·06 Å to 4·09 Å between SR and HA indicates good sampling of the allowed conformation space, since half of the side-chains in this segment are only poorly constrained. Overall the numbers in Table 3 indicate and Figure 4 shows that there are many common features in all stereochemically acceptable conformers obtained from the same n.m.r. data in different computer runs, indicating that these are characteristic structural traits of MB-glucagon. These are discussed in the following in more detail.

The backbone conformation includes a predominantly extended polypeptide segment from residues 5 to 9, one helix-like turn formed by residues 10 to 14, another stretch of extended chain between residues 14 and 17 and three turns of a distorted α-helix from residues 17 to 29 (Fig. 6). In the segment 5–15 (Figs 4(d) and 5(d)) there is a clear spatial separation of the hydrophobic residues Phe6, Tyr10 and Leu14 from the charged or polar residues. If one connects the C^β atoms (which are indicated for all residues in the SR representations of Figs 4 and 5) separately for polar and non-polar residues, one finds two lines that are always on opposite sides of the polypeptide backbone and never cross each other. Tyr13 seems not to be included in the hydrophobic patch formed by residues 6, 10 and 14. In the C-terminal dodecapeptide a hydrophobic patch is formed by the side-chains of Ala19, Phe22, Val23, Trp25 and Leu26 (Figs 4(a) and (b) and 5(a) and (b)).

The overall shape of the molecule in Figure 6 is not reliably characterized by the NOE data, since no long range distance constraints extending over more than five residues were observed. However, the backbone outlines approximately the curvature of the dodecylphosphocholine micelles (Bösch *et al.*, 1980), and since MB-glucagon was found to be located near the micelle surface (Brown *et al.*, 1981), the overall shape of the molecule in Figure 6 might nonetheless coincide rather closely with the micelle-bound polypeptide.

In the structures of Figure 4, which were computed with relative weights of 0·01 for distance constraints by NOEs and 1 for distance constraints by the covalent structure, the three peptide bonds between residues 20 and 21, 23 and 24, and 26 and 27 deviate quite markedly from planarity, with ω-angles between 160 and 165°. To improve the planarity of the peptide bonds we have applied further refinement cycles to the stereochemically most satisfactory conformers in Figure 5, with a relative weight of 0·001 of the NOE distance constraints to the stereochemical distance constraints. After this regularization no ω-angle deviates more than 15° from planarity, which corresponds to an energy of about 2RT. This regularization involved only small changes in the structures (Fig. 7).

A plot of the ϕ–ψ torsion angles in the regularized structures of Figure 7 is presented in Figure 8. The sterically allowed regions in the ϕ–ψ plane (Ramachandran & Sasisekharan, 1968) are also indicated. The data points for MB-glucagon fall within or near to the allowed areas, as is also generally observed in X-ray structures of proteins (Richardson, 1981). Compared to typical

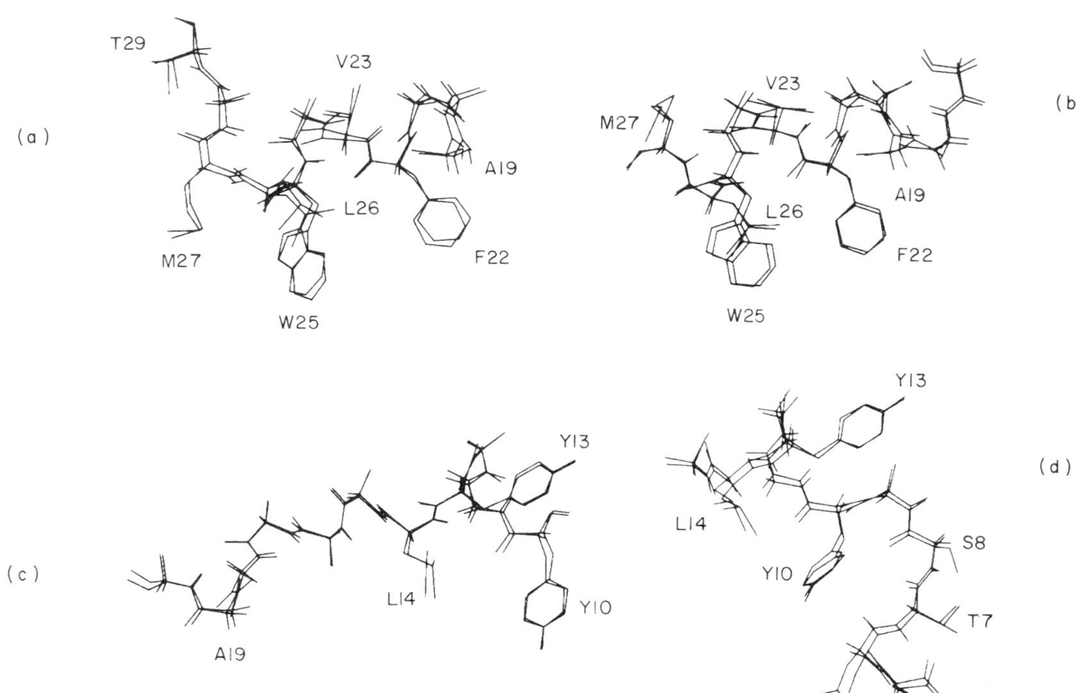

FIG. 7. Structural changes obtained when the stereochemically best structures of the 4 segments of MB-glucagon shown in Fig. 5 (see the text) were subjected to the process of "regularization" to get energetically acceptable peptide bonds (see the text). After the regularization the dihedral angles ω for all the peptide bonds deviate at most by 15° from planarity. The root-mean-square distances between the 2 structures before and after regularization are (d) 0·5 Å for segment 5–15, (c) 0·3 Å for segment 10–20, (b) 0·5 Å for segment 17–27 and (a) 0·6 Å for segment 19–29. The SR presentation of the molecular structures with pseudo-atoms K, L and M described in Fig. 4 is used.

distributions of ϕ–ψ values in globular proteins, MB-glucagon contains relatively many data points near $\phi = +60°$. The ϕ–ψ values for the residues 18 to 29 emphasize that the helical structure formed by this region of the polypeptide chain is pronouncedly irregular.

Using the criteria that the carbonyl oxygen-to-amide proton distance in a hydrogen bond should be between 1·8 and 2·5 Å, and the distance from the carbonyl oxygen to the peptide nitrogen between 2·8 and 3·5 Å, evidence was obtained that C=O . . . H—N hydrogen bonds could be formed between residues 11 and 13, 11 and 14, 17 and 19, 17 and 20, 22 and 25, 23 and 26, 23 and 27, 25 and 27. The irregular helix from residues 17 to 29 thus contains relatively few intramolecular hydrogen bonds. Since the polypeptide is located near the dodecylphosphocholine head groups in the micelles, it is tempting to speculate that the lipid might compete with the polar groups of the polypeptide for the proton acceptor and donor sites and that many of the backbone groups of MB-glucagon are actually hydrogen bonded with lipid head groups.

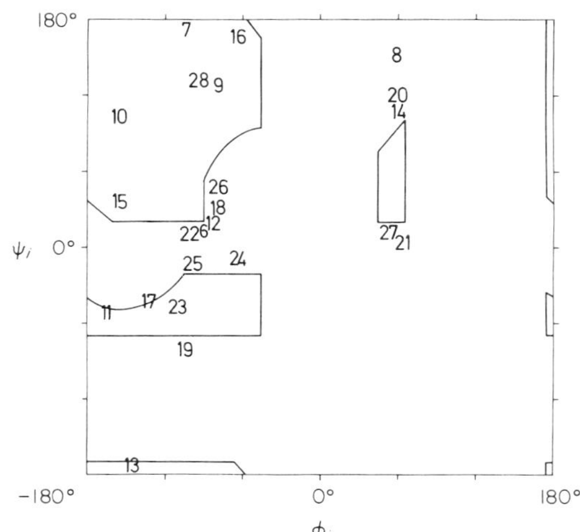

FIG. 8. Plot of the backbone torsion angles ϕ and ψ (Ramachandran & Sasisekharan, 1968) for the regularized (see the text) structure of MB-glucagon in Fig. 7. The data for residues 5 to 12, 13 to 18, 19 to 24 and 25 to 29 are, respectively, from the calculations for the segments 5–15, 10–20, 17–27 and 19–29.

6. Discussion

(a) *Limiting factors in the n.m.r. structure determination of MB-glucagon*

When the present study of MB-glucagon is compared with conceptually similar studies reported previously for MB-glucagon (Braun *et al.*, 1981) and micelle-bound melittin (Brown *et al.*, 1982) it is clear that a much more thorough and extensive structure determination was possible because of the availability of nearly complete resonance assignments to specific residues in the primary structure. However, the text of sections 2 to 5 should also have shown that there is room for further improvements in the future. Thus, while the present structure determination relied on the availability of a large number of relatively inaccurate ^1H–^1H distance constraints, improved accuracy of the measurements of the individual distances will be of interest, e.g. by improvement of the signal-to-noise ratio in NOESY spectra recorded with short mixing times (Fig. 1(a)) and replacement of the uniform averaging model (Braun *et al.*, 1981) with a more realistic treatment for NOE distance constraints in flexible structures. A more quantitative treatment of the NOE distance measurements would probably also have to include a more thorough investigation of the possibility that different parts of the molecular structure might have different effective motional correlation times.

When compared with similar studies of small globular proteins in aqueous solution (unpublished results) certain limitations of the n.m.r. structure determination resulted from the inherent complexity of the MB-glucagon system. For example, the structure determination had to rely entirely on the NOE distance constraints in Table 2 because it was for practical reasons not possible to obtain complementary information either from spin–spin coupling constants or from studies of the amide proton exchange with the solvent (Wüthrich, 1976) and exploitation of the chemical shifts was limited because of the scarcity of reference data on polypeptide ^{1}H n.m.r. shifts in a lipid–water interphase. The polypeptide backbone is more precisely constrained than most of the side-chains (Table 3), since in most regions there is a dense network of distance constraints between backbone hydrogen atoms (Table 2) and none of the backbone–backbone distance constraints had to be referred to pseudo-atoms (Wüthrich *et al.*, 1983). It is also worth noting that the determination of regular secondary structure elements based on recognition of characteristic NOESY cross peak patterns (Fig. 3) coincides well with the structure obtained from the distance geometry calculations. On the other hand, because of the absence of a globular structure NOE distance constraints between side-chain hydrogens are relatively scarce.

(b) *Comparison of glucagon conformations in different environments*

Three structures are available for comparative studies, i.e. the glucagon trimers studied in single crystals (Sasaki *et al.*, 1975), a conformation of monomeric glucagon in aqueous solution (Bösch *et al.*, 1978) and the present structure of MB-glucagon, where the molecule is in a lipid–water interphase (Brown *et al.*, 1981). Figure 9 shows a comparison of the polypeptide backbone structures of the three segments 5–15, 10–20 and 19–29 of micelle-bound glucagon with the glucagon single crystal X-ray structure (Sasaki *et al.*, 1975). The backbone of each

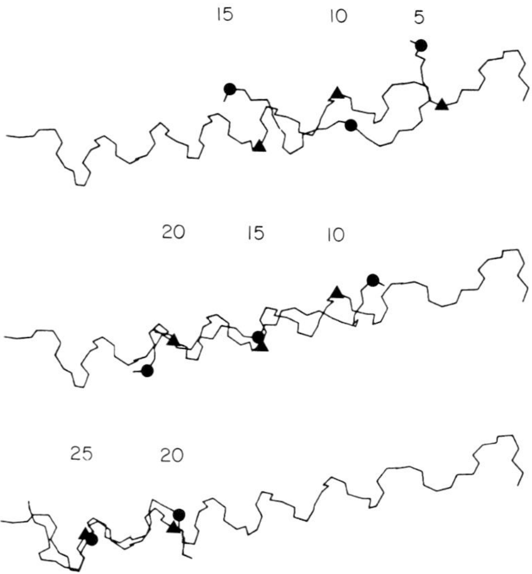

FIG. 9. Comparison of the polypeptide backbone of the structures for the 4 segments of MB-glucagon shown in Fig. 5 with the X-ray structure of glucagon in single crystals (Sasaki *et al.*, 1975). The symbols (●) for MB-glucagon and (▲) for the crystal structure of glucagon mark the C^α positions indicated by the numbers. Root-mean-square distances are 3·5 Å for segment 5–15, 3·0 Å for segment 10–20, 2·1 Å for segment 17–27 and 1·6 Å for segment 19–29.

segment was individually translated and rotated to give a best fit with the X-ray structure. Corresponding C^{α} positions are marked with the symbol (●) for MB-glucagon and with the symbol (▲) for the X-ray structure. There is obviously a trend to better coincidence towards the C-terminal region. The length of the segments 5–10 and 10–20 and the overall length of the sum of the different segments of MB-glucagon exceed the length of the helical crystal structure.

In monomeric glucagon in aqueous solution the residues 22 to 25 form a stable, non-random conformation, which differs from a helical structure by major rearrangements in the backbone (Bösch *et al.*, 1978). For the other regions of the glucagon polypeptide chain the data would be compatible with an extended, flexible conformation.

Earlier studies (Bösch *et al.*, 1980) showed that the conformation of glucagon bound to DPC micelles is representative for glucagon bound to various different micellar lipids and appears to be similar to the structure of glucagon bound to lipid bilayers. It may also be added that a previously published "low resolution" n.m.r. structure for the segment 19–27 of MB-glucagon (Braun *et al.*, 1981) indicated the formation of a hydrophobic cluster with the side-chains of Ala19, Phe22, Val23, Trp25 and Leu26 and that this structural feature is confirmed by the present study.

(c) *Glucagon conformation in a lipid–water interphase and biological function*

The following considerations rely on the hypothesis that glucagon adopts similar conformations in the lipid–water interphase near the surface of dodecylphosphocholine micelles and on the surface of the target organ. Fundamental considerations make it appear rather unlikely that the initial contact of glucagon with its target cell would be the formation of the specific complex with the receptor site. Rather, one could expect non-specific binding to the cell surface, followed by diffusion to the specific receptor site in the two-dimensional space provided by the lipid–water interphase on the surface of the cell membrane. With regard to this hypothetical scheme for the initial phase of the glucagon action on the surface of the target organ, two features in the above comparison of the glucagon conformations in different environments are particularly intriguing. First, a major structural rearrangement in the polypeptide region 22–24, which is primarily responsible for the binding of the hormone to the receptor site (Wright & Rodbell, 1979), is implicated for the transfer from dilute aqueous solution to the lipid–water interphase of the micellar surface. If specific binding of glucagon is determined by the conformation of the binding region, the structural features responsible for this specificity would thus be formed only after the hormone has been incorporated into the lipid–water interphase of the cell surface. Second, with regard to the observation that the activity of glucagon depends on the presence of the N-terminal histidine residue while the region near residues 20 to 26 is responsible for specific binding of the hormone to the receptor site (Rodbell *et al.*, 1971; Wright *et al.*, 1978; Wright & Rodbell, 1979), it is interesting that the polypeptide chain between these two locations is more extended in the MB-glucagon conformation than in the crystal structure (Fig. 9).

Use of the facilities at the Zentrum für Interaktives Rechnen (ZIR) of the ETH Zürich is gratefully acknowledged. Financial support was obtained from the Schweizerischer Nationalfonds (project 3.528.79) and through a special grant of the ETH Zürich. We thank Dr T. Blundell and Mrs J. Richardson for helpful discussions and correspondence on this work, and Mrs E. Huber and Mrs E. H. Hunziker for the careful preparation of the manuscript and the illustrations.

REFERENCES

Anil Kumar, Ernst, R. R. & Wüthrich, K. (1980*a*). *Biochem. Biophys. Res. Commun.* **95**, 1–6.

Anil Kumar, Wagner, G., Ernst, R. R. & Wüthrich, K. (1980*b*). *Biochem. Biophys. Res. Commun.* **96**, 1156–1163.

Anil Kumar, Wagner, G., Ernst, R. R. & Wüthrich, K. (1981). *J. Amer. Chem. Soc.* **103**, 3654–3658.

Billeter, M., Braun, W. & Wüthrich, K. (1982). *J. Mol. Biol.* **155**, 321–346.

Blundell, T. L. & Wood, S. (1982). *Annu. Rev. Biochem.* **51**, 123–154.

Bösch, C., Bundi, A., Oppliger, M. & Wüthrich, K. (1978). *Eur. J. Biochem.* **91**, 204–214.

Bösch, C., Brown, L. R. & Wüthrich, K. (1980). *Biochim. Biophys. Acta*, **603**, 298–312.

Bösch, C., Anil Kumar, Baumann, R., Ernst, R. R. & Wüthrich, K. (1981). *J. Magn. Reson.* **42**, 159–163.

Bothner-By, A. A. & Noggle, J. H. (1979). *J. Amer. Chem. Soc.* **101**, 5152–5153.

Braun, W., Bösch, C., Brown, L. R., Gō, N. & Wüthrich, K. (1981). *Biochim. Biophys. Acta*, **667**, 377–396.

Brown, L. R. (1979). *Biochim. Biophys. Acta*, **557**, 135–148.

Brown, L. R., Bösch, C. & Wüthrich, K. (1981). *Biochim. Biophys. Acta*, **642**, 296–312.

Brown, L. R., Braun, W., Anil Kumar & Wüthrich, K. (1982). *Biophys. J.* **37**, 319–328.

Crippen, G. M. & Havel, T. F. (1978). *Acta Crystallogr.* **34**, 282–284.

Dobson, C. M., Olejniczak, E. T., Poulsen, F. M. & Ratcliffe, R. G. (1982). *J. Magn. Reson.* **48**, 97–110.

Epand, R. M. (1971). *Can. J. Biochem.* **49**, 166–169.

Epand, R. M., Jones, A. J. S. & Schreier, S. (1977). *Biochim. Biophys. Acta*. **491**, 296–304.

Gordon, S. L. & Wüthrich, K. (1978). *J. Amer. Chem., Soc.* **100**, 7094–7096.

Gratzer, W. B., Bailey, E. & Beaver, G. H. (1967). *Biochem. Biophys. Res. Commun.* **28**, 914–919.

Gurd, F. R. N. & Rothgeb, T. M. (1979). *Advan. Protein Chem.* **33**, 74–165.

Havel, T. F., Crippen, G. M. & Kuntz, I. D. (1979). *Biopolymers*, **18**, 73–81.

Hull, W. A. & Sykes, B. D. (1975). *J. Chem. Phys.* **65**, 867–880.

Jeener, J., Meier, B. H., Bachman, P. & Ernst, R. R. (1979). *J. Chem. Phys.* **71**, 4546–4553.

Kalk, A. & Berendsen, H. J. C. (1976). *J. Magn. Reson.* **24**, 343–366.

Karplus, M. & McCammon, J. A. (1981). *Crit. Rev. Biochem.* **9**, 293–349.

Macura, S., Huang, Y., Suter, D. & Ernst, R. R. (1981). *J. Magn. Reson.* **43**, 259–281.

McCammon, J. A. & Karplus, M. (1980). *Annu. Rev. Phys. Chem.* **31**, 29–45.

McLachlan, A. D. (1979). *J. Mol. Biol.* **128**, 49–79.

Momany, F. A., McGuire, R. F., Burgess, A. W. & Scheraga, H. A. (1975). *J. Phys. Chem.* **79**, 2361–2381.

Moran, E. C., Chou, P. Y. & Fassman, G. D. (1977). *Biochem. Biophys. Res. Commun.* **77**, 1300–1306.

Nagayama, K., Wüthrich, K. & Ernst, R. R. (1979). *Biochem. Biophys. Res. Commun.* **90**, 305–311.

Nagayama, K., Anil Kumar, Wüthrich, K. & Ernst, R. R. (1980). *J. Magn. Reson.* **40**, 321–334.

Noggle, J. H. & Schirmer, R. E. (1971). *The Nuclear Overhauser Effect*, Academic Press, New York.

Panijpan, B. & Gratzer, W. B. (1974). *Eur. J. Biochem.* **45**, 547–553.

Pohl, S. L., Birnbaumer, L. & Rodbell, M. (1969). *Science*, **164**, 566–569.

Ramachandran, G. N. & Sasisekharan, V. (1968). *Advan. Protein Chem.* **25**, 283–437.
Richardson, J. (1981). *Advan. Protein Chem.* **34**, 167–339.
Rodbell, M., Birnbaumer, L., Pohl, S. L. & Sundby, F. (1971). *Proc. Nat. Acad. Sci., U.S.A.* **68**, 909–913.
Rubalcava, B. & Rodbell, M. (1973). *J. Biol. Chem.* **248**, 3831–3837.
Sasaki, K., Dockevill, S., Ackmiak, D. A., Tickle, I. J. & Blundell, T. L. (1975). *Nature (London)*, **257**, 751–757.
Schneider, A. B. & Edelhoch, H. (1972). *J. Biol. Chem.* **247**, 4986–4991.
Srere, P. A. & Brooks, G. C. (1969). *Arch. Biochem. Biophys.* **129**, 708–710.
Wagman, M. E., Dobson, C. M. & Karplus, M. (1980). *FEBS Letters*, **119**, 256–270.
Wagner, G. & Wüthrich, K. (1979). *J. Magn. Reson.* **33**, 675–680.
Wako, H. & Scheraga, H. A. (1982). *J. Protein Chem.* **1**, 5–45.
Wider, G., Lee, K. H. & Wüthrich, K. (1982). *J. Mol. Biol.* **155**, 367–388.
Wider, G., Hosur, R. V. & Wüthrich, K. (1983). *J. Magn. Reson.* **52**, 130–135.
Wright, D. E. & Rodbell, M. (1979). *J. Biol. Chem.* **254**, 268–269.
Wright, D. E., Hruby, V. J. & Rodbell, M. (1978). *J. Biol. Chem.* **253**, 6338–6340.
Wüthrich, K. (1976). *NMR in Biological Research: Peptides and Proteins*, North-Holland Publishing Company, Amsterdam.
Wüthrich, K. & Wagner, G. (1978). *TIBS*, **3**, 227–230.
Wüthrich, K., Bösch, C. & Brown, L. R. (1980). *Biochem. Biophys. Res. Commun.* **95**, 1504–1509.
Wüthrich, K., Wider, G., Wagner, G. & Braun, W. (1982). *J. Mol. Biol.* **155**, 311–319.
Wüthrich, K., Billeter, M. & Braun, W. (1983). *J. Mol. Biol.* **169**, 949–961.

Edited by J. C. Kendrew

J. Mol. Biol. (1983) **169**, 949–961

Pseudo-structures for the 20 Common Amino Acids for Use in Studies of Protein Conformations by Measurements of Intramolecular Proton–Proton Distance Constraints with Nuclear Magnetic Resonance

K. Wüthrich, M. Billeter and W. Braun

Institut für Molekularbiologie und Biophysik
Eidgenössische Technische Hochschule
CH-8093 Zürich-Hönggerberg, Switzerland

(Received 30 March 1983)

"Pseudo-structures" of the 20 common amino acid residues are introduced for use in protein spatial structure determinations, which rely on the use of intramolecular proton–proton distance constraints determined by nuclear Overhauser effects as input for distance geometry calculations. The proposed structures satisfy requirements for the initial structural interpretation of the nuclear magnetic resonance data that arise from the absence of stereospecific assignments and/or limited spectral resolution for certain resonance lines. The pseudo-atoms used as reference points for the experimental distance constraints can be used in conjunction with the real amino acid structures representing the van der Waals' constraints on the spatial molecular structure, or with simplified models in order to reduce the computing time for the distance geometry calculations.

1. Introduction

We have previously outlined a strategy for determination of the spatial structures of non-crystalline polypeptide chains by n.m.r.† (Wüthrich *et al.*, 1982) and in the preceding paper (Braun *et al.*, 1983) this procedure was applied for a study of the conformation of the polypeptide hormone glucagon. The method relies on the assignments of the ^{1}H n.m.r. lines to specific amino acid residues in the primary structure and on measurements of intramolecular distance constraints between distinct hydrogen atoms or groups of hydrogen atoms. However, the "sequential resonance assignments" (Wagner & Wüthrich, 1982; Wider *et al.*, 1982) do not include stereospecific assignments for all the protons within the spin systems of the individual amino acid residues (Wüthrich, 1976). Therefore, it is necessary for the initial analysis of the n.m.r. data to replace certain groups of atoms by "pseudo-atoms". The present note describes a suitable set of "pseudo-structures" for the common amino acid residues.

† Abbreviations used: n.m.r., nuclear magnetic resonance; NOE nuclear Overhauser enhancement; NOESY, two-dimensional nuclear Overhauser enhancement spectroscopy.

0022–2836/83/280949–13 $03.00/0 © 1983 Academic Press Inc. (London) Ltd.

Simplified structures of polypeptide chains were previously used to reduce the computing time in theoretical studies of protein folding (e.g. Levitt & Warshel, 1975; Levitt, 1976; Dunfield *et al.*, 1978). These earlier schemes would not be suitable for the present purpose. However, for reducing the computing time for the structural interpretation of the n.m.r. data, the n.m.r. pseudo-structures may be complemented with similar schemes aimed at a reduction of the number of atoms in the molecular structures.

2. Fundamental Considerations

(a) *Pseudo-atoms as points of reference for proton–proton distance constraints in the absence of stereospecific resonance assignments*

Nuclear Overhauser enhancement spectroscopy provides distance constraints between distinct hydrogen atoms in a polypeptide chain (e.g. Anil Kumar *et al.*, 1981). When such NOE distance constraints are used for determination of spatial protein structures, an inherent limitation arises since stereospecific resonance assignments cannot be obtained for the individual protons in methylene groups or the individual methyl groups in the isopropyl fragments of Val and Leu†, and since the resonances of the individual protons in methyl groups and in the symmetry-related positions of the aromatic rings of Phe and Tyr are not usually resolved. Pseudo-atoms are introduced in the 20 common amino acids to replace all those groups of protons for which inspection of the covalent structure shows that stereoselective resonance assignments cannot be obtained with ^{1}H n.m.r. alone.

(b) *Introduction of pseudo-atoms for reducing the computing time for the distance geometry calculations*

The dimensionless pseudo-atoms introduced as reference points for NOE distance constraints can be used in conjunction with the real amino acid structures representing the constraints on the molecular conformations that arise from van der Waals' interactions. For practical reasons, however, it may be advisable to represent the van der Waals' dimensions of selected molecular fragments by a simplified model. In the pseudo-structures presented in this paper CH_3, CH_2 and certain CH fragments are replaced by single spheres M, L and K, respectively, with radii corresponding to the volumes for the CH_3, CH_2 and CH fragments given by Richards (1974), i.e. $r_M = 1{\cdot}8$ Å, $r_L = 1{\cdot}6$ Å and $r_K = 1{\cdot}5$ Å. Within the scheme of n.m.r. pseudo-structures for amino acids the above values for r_M, r_L and r_K can readily be substituted by different definitions of the van der Waals' radii, e.g. following the criteria employed in UNICEPP (United Atom Conformational Energy Program for Peptides; Dunfield *et al.*, 1978).

† Stereospecific assignments in small peptides have been obtained by combination of ^{1}H n.m.r. either with heteronuclear double resonance experiments (Feeney *et al.*, 1974) or with stereospecific isotope labelling (Fischman *et al.*, 1978). At present these procedures are not yet practical for work with proteins.

3. Pseudo-structures of Amino Acids for Use in n.m.r. Studies

A suitable set of pseudo-structures for the 20 common amino acids is presented in the second column of Table 1. The pseudo-atoms used are explained in the footnotes to the Table. The following are some additional comments on particular features of the pseudo-structures. (The entries in the third and fourth columns of Table 1 are discussed in section 4, below.)

Methylene groups are represented by two pseudo-atoms, i.e. the sphere L which represents the location and the volume of the CH_2 fragment and the dimensionless reference point P for experimental distance constraints. Methyl groups are represented by a single pseudo-atom M, located centrally with respect to the three methyl protons. In valine and leucine a dimensionless reference point for distance constraints to the methyl protons, Q, is located in the centre of the six methyl protons. The van der Waals' volume is represented by the two spheres M.

In the side-chains of Ser, Thr, Asp, Asn, Glu, Gln, Lys, Arg, Cys and Tyr the labile protons are omitted to simplify the polypeptide structure for the distance geometry calculations. Most of these protons are usually not observed in the ^{1}H n.m.r. spectra (Bundi & Wüthrich, 1979). Otherwise, the distance constraints would be referred to the heavy-atom positions, e.g. the nitrogen atom for the amide NH_2 groups of Asn and Gln.

In the aromatic rings of Phe and Tyr two different situations are considered.

TABLE 1

Pseudo-structures for amino acids used for initial structural interpretation of NOE distance constraints in proteins

Amino acid	Pseudo-structure[a]	Intra-residue correction[b,c]	Long range correction[b,d]
G	 NH—LA—CO ⋮ PA		*m*
A	CH | MB	*i*(N → β)	*m*
V	CH | CH / ⋮ \\ MG1 QG MG2	*m*(N → Mγ) *s*(N → Qγ) *i*(α → Mγ) *i*(α → Qγ)	*q*, *m*
I	CH | CH / \\ PG⋯LG MG | MD	*m*(N → Mγ, Pγ) *i*(α → Mγ, Pγ) *m*(α → Mδ)	*m* *m*

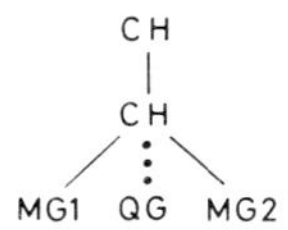

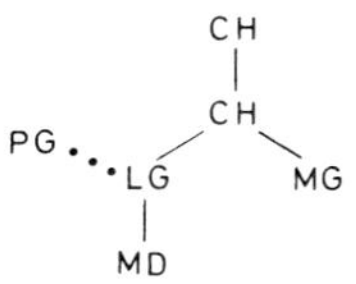

TABLE 1 *(continued)*

Amino acid	Pseudo-structure[a]	Intra-residue correction[b,c]	Long range correction[b,d]
L	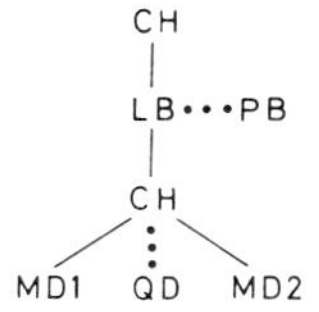 CH – LB···PB – CH (MD1, QD, MD2)	i(N → Pβ) m(α → Mδ) s(α → Qδ)	m q, m
P	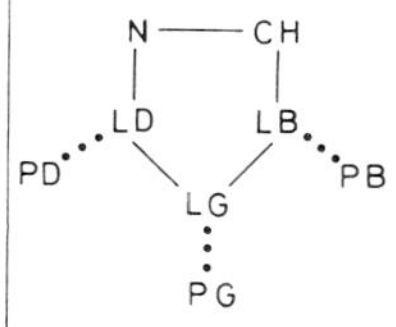 N — CH; LD···PD; LB···PB; LG···PG		m m
S	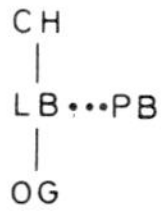 CH – LB···PB – OG	i(N → Pβ)	m
T	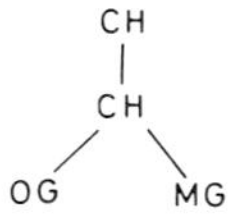 CH – CH (OG, MG)	m(N → Mγ) i(α → Mγ)	m
D, N	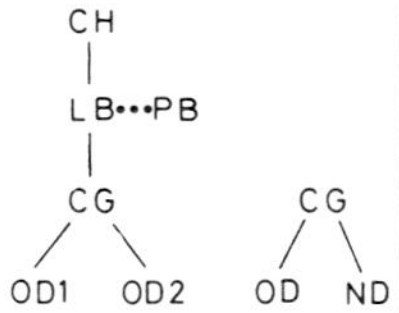 CH – LB···PB – CG (OD1, OD2) / CG (OD, ND)	i(α → Pβ)	m
E, Q	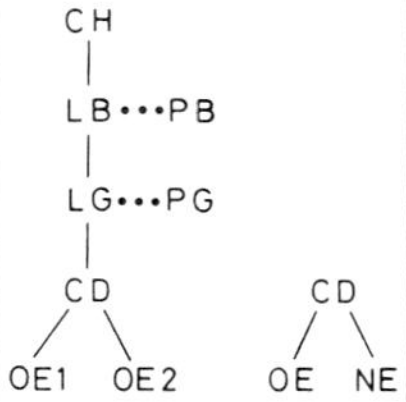 CH – LB···PB – LG···PG – CD (OE1, OE2) / CD (OE, NE)	i(N → Pβ) m(N → Pγ) i(α → Pγ)	m m
K	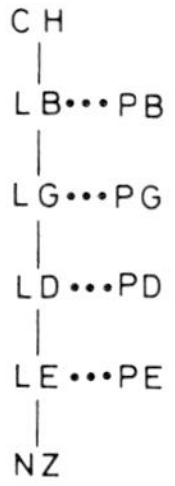 CH – LB···PB – LG···PG – LD···PD – LE···PE – NZ	i(N → Pβ) m(N → Pγ) i(α → Pγ) m(α → Pδ)	m m m m

TABLE 1 *(continued)*

Amino acid	Pseudo-structure[a]	Intra-residue correction[b,c]	Long range correction[b,d]
R	CH		
	LB···PB	*i*(N → Pβ)	*m*
	LG···PG	*m*(N → Pγ) *i*(α → Pγ)	*m*
	LD···PD	*m*(α → Pδ)	*m*
	NE		
	CZ		
	NI1 NI2		
M	CH		
	LB···PB	*i*(N → Pβ)	*m*
	LG···PG	*m*(N → Pγ) *i*(α → Pγ)	*m*
	SD		
	ME		*m*
C	CH		
	LB···PB	*i*(N → Pβ)	*m*
	SG		
F, Y	CH		
	LB···PB	*i*(N → Pβ)	*m*
	C1 (F) C1 (Y)	*r*(N → C1) *m*(α → C1)	*r*
	K6 K2 QR K5 K3 (F); K6 K2 QR K5 K3 (Y)	*q*(N → QR) *r*(α → QR)	*q*
	C4 (F) C4 (Y)	*m*(N → C4) *i*(α → C4)	*r*
	H4 (F) O4 (Y)		
H	CH		
	LB···PB	*i*(N → Pβ)	*m*
	C5		
	H4—C4 N1—H1		
	N3—C2—H2		

TABLE 1 (*continued*)

Amino acid	Pseudo-structure[a]	Intra-residue correction[b,c]	Long range correction[b,d]
W	CH – PB•••LB; LB–C3; H4–C4; C3–C2, C3–C9, C4–C9, C4–C5; H2–C2; C5–H5; C2–N1; C9–C8; C5–C6; H1–N1–C8; C6–H6; C8–C7, C6–C7; C7–H7	i(N→Pβ)	m

[a] The following pseudo-atoms are used (see the text): P, (≡QP) dimensionless pseudo-atoms in the centre between the 2 methylene protons. All distances to methylene groups are referred to P. (In our computer input all the dimensionless pseudo-atoms are identified by the first letter Q.) M, central location with respect to the 3 hydrogen atoms of a methyl group, 0·36 Å away from the methyl carbon atom. All distance constraints to methyl groups are referred to this position. As a sphere of radius 1·8 Å M further represents the van der Waals' volume of the CH_3 fragment. Q, dimensionless pseudo-atom in the centre between the 2 methyl groups of Val and Leu. In the initial interpretation distances to the methyl protons of Val and Leu are referred to Q. QR, dimensionless pseudo-atom in the centre of the aromatic rings of Phe and Tyr. Distances to unresolved aromatic protons are referred to QR. C1, distances to unresolved C2,6H of Phe and Tyr are referred to the position of the ring carbon atom C1. C4, distances to unresolved C3,5H of Phe and Tyr are referred to the position of the ring carbon atom C4. K, a sphere of radius 1·5 Å, which represents the van der Waals' volume of certain CH fragments. Its centre coincides with the location of the carbon atom. L, a sphere of radius 1·6 Å, which represents the van der Waals' volume of CH_2. Its centre coincides with the location of the carbon atom.

[b] The following symbols are used to describe the corrections to the ^1H–^1H distances determined by NOESY, which account for the replacement of hydrogen atoms by pseudo-atoms (see also Table 2): $i = 0{\cdot}6$ Å, $m = 1{\cdot}0$ Å, $s = 1{\cdot}7$ Å, $r = 2{\cdot}0$ Å, $q = 2{\cdot}4$ Å.

[c] Intra-residue corrections are given for distance constraints between NH or Hα and the side-chain atoms that are separated from these by 2 or 3 torsion angles (see the text). In each case the connectivity for which the intra-residue correction applies is indicated in parentheses.

[d] Long range corrections are always the actual distances from the hydrogen atoms to the pseudo-atoms by which they are replaced. These are approximated by:

methylene protons → P; methyl protons → M:	$m = 1{\cdot}0$ Å;
C2,6H of F or Y → C1; C3,5H of F or Y → C4:	$r = 2{\cdot}0$ Å;
methyl protons of V or L → Q; C2,3,5,6H of F or Y → QR:	$q = 2{\cdot}4$ Å.

First, when the 2,6 proton resonances are separated from the 3,5 proton lines in the n.m.r. spectrum, distance constraints are referred to the position of the carbon atom C-1, or C-4, respectively, which is approximately in the centre between the two protons. Second, when the 2,6H and 3,5H n.m.r. lines cannot be individually resolved, distance constraints are referred to a dimensionless pseudo-atom QR at the centre of the proton positions 2,3,5 and 6. To simplify the molecular structures the ring CH fragments 2,3,5 and 6 are replaced by spherical pseudo-atoms K.

4. Dimensions of the Pseudo-structures for Amino Acids and Corrections to the Experimental Distance Constraints

The practical use of the pseudo-structures in Table 1 for structural interpretations of ^1H–^1H distance constraints from NOESY experiments requires that the dimensions of the native amino acid residues and the pseudo-structures are known. These are then used to calculate the corrections that must be applied to the experimental data when the protons in the molecular models are replaced by pseudo-atoms. Different approaches for obtaining these correction factors are in the following described for "long range" distance constraints which depend on more than three intervening torsion angles about covalent single bonds and for "short range" distance constraints between selected hydrogen atoms which depend on three or less torsion angles.

For all the distance calculations ECEPP standard geometries with planar *trans* peptide bonds (Momany *et al.*, 1975) were used for the amino acid residues, and the pseudo-atoms were located in these standard covalent structures as indicated in Table 1. For long range distance constraints the actual correction would in each instance depend on the orientation relative to the distance vector of the vector joining the hydrogen atom with the pseudo-atom by which it was replaced. The maximum correction corresponds to addition of the full distance from the hydrogen atom to the pseudo-atom. For the following reasons this maximum correction is used throughout (Table 2). First, in an unknown conformation there is no direct way to find out what the actual correction should be. Second, as long as NOE distance constraints are used as upper bounds to the allowed distances (Braun *et al.*, 1981,1983), use of the largest possible correction appears to be the most conservative approach.

For the amino acid residue in position i all the short range distances which involve any of the backbone hydrogen atoms NH_i, HN_{i+1} and HC_i^α were computed. In the standard covalent structures the through-space interatomic distances are unambiguously determined by the choice of the intervening torsion angles about the single bonds (Ramachandran & Sasisekharan, 1968). For example the distance from the amide proton N_iH to $C_i^\gamma H$ depends on the three torsion angles, ϕ_i, χ_i^1 and χ_i^2 (IUPAC-IUB Commission on Biochemical Nomenclature, 1970). Short range interatomic distances were computed for all combinations of the intervening torsion angles. The mathematical procedures used are described in the Appendix to the paper by Billeter *et al.* (1982). For distances depending on one or two torsion angles equations (A7) and (A18) (Billeter *et al.*, 1982), respectively, were used. For calculations with three torsion angles an extension of these equations was employed. In all these computations the atoms were treated as dimensionless points in space and no consideration was given to steric hindrance by overlap of different atoms.

The results are listed in Table 2. The first and second columns identify the atoms between which the distances were evaluated. The third column describes the range of values that a particular distance can adopt. A lower limit is explicitly indicated only when the computations gave a larger value than 2·0 Å, which is taken to represent the van der Waals' distance between two hydrogen atoms. It is

TABLE 2

Distances between hydrogen atoms and pseudo-atoms in the common 20 amino acid residues and the pseudo-structures used for structural interpretations of NOE distance constraints

Type of hydrogen atoms[a]	Short range connectivities[b]			Long range distance correction[f]
	Atoms connected[c]	Range of distances[d]	Distance correction[e]	
Backbone NH, $C^{\alpha}H$	HN–HA	2·16–2·85		
	HN_i–HN_{i+1}	4·68		
	HA_i–HN_{i+1}	2·22–3·56		
Val, Ile, Thr $C^{\beta}H$	HB–HN	4·12		
	HB–HA	2·20–2·90		
	HB_i–HN_{i+1}	4·68		
Leu $C^{\gamma}H$	HG–HN	5·35		
	HG–HA	2·02–4·06		
	HG_i–HN_{i+1}	5·97		
Gly $C^{\alpha}H_2$	HA–HN	2·21–2·74		
	CA–	2·08	0·65	1·0
	PA–	2·25–2·62	0·17	0·8
	HA_i–HN_{i+1}	2·22–3·28		
	CA_i–	2·59	0·69	1·0
	PA_i–	2·47–3·19	0·52	0·8
$C^{\beta}H_2$				
	HB–HN	4·01		
	CB–	2·46–3·36	0·75	1·09
	PB–	2·31–3·94	0·62	0·88
	HB–HA	2·23–2·81		
	CB–	2·09	0·71	1·09
	PB–	2·29–2·68	0·21	0·88
	HB_i–HN_{i+1}	4·64		
	PB_i–	4·58	0·86	0·88
$C^{\gamma}H_2$	HG–HN	5·25		
	CG–	2·00–4·59	0·93	1·09
	PG–	5·22	0·87	0·88
	HG–HA	4·04		
	CG–	2·46–3·38	0·76	1·09
	PG–	2·39–3·99	0·65	0·88
	HG_i–HN_{i+1}	5·79		
	PG_i	5·75	0·88	0·88
$C^{\delta}H_2$	HD–HA	5·25		
	CD–	4·58	0·92	1·09
	PD–	5·18	0·85	0·88
Ala CH_3	HB–HN	3·62		
	MB–	2·65–3·68	0·69	1·03
	HB–HA	2·13–2·36		
	MB–	2·34	0·21	1·03
	HB_i–HN_{i+1}	4·49		
	MB_i–	2·32–4·44	0·97	1·03

TABLE 2 *(continued)*

Type of hydrogen atoms[a]	Short range connectivities[b]			Long range distance correction[f]
	Atoms connected[c]	Range of distances[d]	Distance correction[e]	
Thr CH_3	HG–HN	4·80		
	MG–	2.00–4·92	0·97	1·03
	HG–HA	2·03–3·69		
	MG–	2·71–3·73	0·68	1·03
	HG_i–HN_{i+1}	5·30		
	MG_i–	5·44	1·03	1·03
Ile γ-CH_3	HG–HN	4·77		
	MG–	2·04–4·90	0·98	1·03
	HG–HA	3·66		
	MG–	2·67–3·71	0·69	1·03
	HG_i–HN_{i+1}	5·28		
	MG_i–	5·42	1·03	1·03
Ile δ-CH_3	MD–HN	6·18	1·03	1·03
	HD–HA	4·78		
	MD–	2·02–4·91	0·98	1·03
	MD_i–HN_{i+1}	6·81	1·03	1·03
Val CH_3	HG–HN	4·42		
	MG–	2·02–4·90	0·98	1·03
	QG–	2·36–4·35	1·73	2·31
	HG–HA	3·35		
	MG–	2·65–3·70	0·69	1·03
	QG–	2·52–3·08	0·76	2·31
	HG_i–HN_{i+1}	4·93		
	MG_i–	5·42	1·03	1·03
	QG_i–	4·95	2·23	2·31
Leu CH_3	MD–HN	6·19	1·03	1·03
	QD–HN	5·64	2·31	2·31
	HD–HA	4·43		
	MD–	2·17–4·89	0·98	1·03
	QD–	2·48–4·37	1·68	2·31
	MD_i–HN_{i+1}	6·85	1·03	1·03
	QD_i–HN_{i+1}	6·21	2·31	2·31
Phe, Tyr C2,6H	H2,6–HN	5·20		
	C1–	4·61	1·89	2·13
	H2,6–HA	4·16		
	C1–	2·60–3·44	0·98	2·13
	$H2,6_i$–HN_{i+1}	5·68		
	$C1_i$	5·16	2·13	2·13
Phe, Tyr C3,5H	H3,5–HN	2·81–7·41		
	C4–	3·50–7·21	1·11	2·13
	H3,5–HA	4·14–6·35		
	C4–	4·76–6·09	0·62	2·13
	$H3,5_i$–HN_{i+1}	2·28–7·89		
	$C4_i$–	2·57–7·72	1·32	2·13

TABLE 2 *(continued)*

Type of hydrogen atoms[a]	Short range connectivities[b] Atoms connected[c]	Range of distances[d]	Distance correction[e]	Long range distance correction[f]
Phe, Tyr C2,3,5,6H	H2,3,5,6–HN	5·20		
	QR–	2·46–5·89	2·43	2·48
	H2,3,5,6–HA	4·16		
	QR–	3·57–4·57	1·96	2·48
	H2,3,5,6_i–HN_{i+1}	5·68		
	QR_i–	6·40	2·47	2·48
Phe C4H	H4–HN	4·45–8·27		
	H4–HA	5·76–7·16		
	$H4_i$–HN_{i+1}	3·60–8·77		
Trp C2H	H2–HN	6·28		
	H2–HA	4·97		
	$H2_i$–HN_{i+1}	6·97		
Trp C4H	H4–HN	6·44		
	H4–HA	5·14		
	$H4_i$–HN_{i+1}	7·17		
Trp C5H	H5–HN	2·65–8·73		
	H5–HA	3·73–7·42		
	$H5_i$–HN_{i+1}	2·14–9·46		
Trp C6H	H6–HN	4·28–9·82		
	H6–HA	5·43–8·52		
	$H6_i$–HN_{i+1}	3·77–10·41		
Trp C7H	H7–HN	4·56–9·04		
	H7–HA	5·75–7·85		
	$H7_i$–HN_{i+1}	3·85–9·55		
His C2H	H2–HN	3·09–7·87		
	H2–HA	4·29–6·63		
	$H2_i$–HN_{i+1}	2·42–8·39		
His C4H	H4–HN	6·36		
	H4–HA	2·02–5·06		
	$H4_i$–HN_{i+1}	7·05		

[a] For the 20 common amino acid residues all the different types of hydrogen atoms and the groups of hydrogen atoms that may be lumped together by the limitations on spectral resolution and stereospecific assignment in the ^{1}H n.m.r. spectra are listed.

[b] Short range connectivities are between atoms separated by a sufficiently small number of covalent bonds so that the intervening distance depends on 3 or less torsion angles about single bonds.

[c] Lists the combinations of atoms or groups of atoms in the 20 common amino acid residues and the pseudo-structures (Table 1) between which the distances have been evaluated. Data are given for the hydrogen atoms, selected carbon atoms and the following pseudo-atoms: P, centre between the 2 protons in a methylene group; M, centre of the 3 protons in a methyl group; Q, centre between the 2 pseudo-atoms M, which represent the hydrogen atoms in the 2 methyl groups of Val and Leu; QR, centre of the hydrogen atom positions 2,3,5 and 6 in the aromatic rings of the Phe and Tyr. Atom combinations within a residue i and to the amide proton of the subsequent residue $i+1$ are included. Where no subscripts are given, the combinations involve atoms of the same residue. The first entry always refers to atoms in the molecular fragment listed in the first column. When several successive distances are to the same atom the second entry is not repeated.

to be expected that in many instances the lower limits indicated in the Table cannot be attained because of steric hindrance. In contrast, the upper limits should in all cases be meaningful numbers. The fourth column lists the largest difference between the distances to corresponding protons and pseudo-atoms, which would arise for any combination of the intervening one, two or three torsion angles. For steric reasons this correction is in many cases appreciably shorter than the actual distance from the proton to the pseudo-atom (fifth column). No data for proline are included, since n.m.r. data are presently of little practical interest for proline side-chain conformations in proteins. Combined use of the first two columns in Table 1 and of Table 2 should in principle allow one to handle all situations which may come up in protein structure determinations by n.m.r. Depending on the accuracy of the n.m.r. data obtained for a particular protein, simplified schemes for the corrections to the experimental distance constraints may be derived from Table 2. For example, considering the presently available limited accuracy of the experimental NOESY distance constraints (Braun *et al.*, 1981,1983) it seems adequate to represent the results of the computations in Table 2 by five correction terms $i = 0{\cdot}6$ Å, $m = 1{\cdot}0$ Å, $s = 1{\cdot}7$ Å, $r = 2{\cdot}0$ Å and $q = 2{\cdot}4$ Å. The applicable corrections for the individual pseudo-atoms are listed in Table 1, third and fourth column. The selection of these five correction terms resulted from the following considerations. Inspection of Table 2 showed that distances between vicinal protons, i.e. those determined by a single torsion angle, cover such a small range that experimental determination by NOESY with the use of pseudo-atoms is presently not warranted. Furthermore, distance constraints to methylene carbons replacing the methylene protons are not considered because this substitution is not satisfactory for our purposes. The remaining long range distance corrections and corrections for intra-residue constraints which depend on two or three torsion angles can then be grouped into five classes with a relatively small spread of the values in each class (Table 2). Class i includes corrections in the range from 0·62 to 0·69 Å, class m from 0·85 to 1·11 Å, class s from 1·68 to 1·73 Å, class r from 1·89 to 2·13 Å and class q from 2·31 to 2·48 Å. There is a single distance correction which does not fall into any of these ranges, i.e. 0·76 Å for substitution of the distance Hα to Hγ in valine by Hα to Q (Table 2). This correction was included into class i.

---

[d] Indicates the range of values for the distance defined in the second column. For distance constraints to methylene, methyl or isopropyl groups the distances are always measured to the nearest hydrogen atom. The lower limit is explicitly listed only when the computed shortest distance is longer than 2·0 Å, which is taken to represent the van der Waals' distance between 2 hydrogen atoms. For steric reasons neither this value nor the lower limits listed may in some cases actually be attained (see the text).

[e] Correction which must be added to the experimental ^{1}H–^{1}H NOE constraints when these are interpreted with the use of pseudo-structures for the amino acids. It corresponds to the largest difference between the distance to the hydrogen atoms and to the pseudo-atom by which these protons are replaced, respectively, for any combination of the intervening torsion angles about single bonds. For steric reasons this maximum difference is often smaller than the actual distance between corresponding hydrogens and pseudo-atoms.

[f] For long range distance constraints the correction for the use of pseudo-structures is made by adding the distance between the hydrogen atoms and the pseudo-atom by which they are replaced to the experimental NOE constraints.

5. Conclusions

For long range distance constraints the correction terms in Table 1 are certainly adequate for the initial structure determination. For short range constraints the situation is more delicate, since the NOEs are used to discriminate between different distances, which can be within a relatively narrow range (Table 2). Furthermore, the theoretical upper and lower bounds on certain short range ^1H–^1H distances may be useful for empirical calibrations of the correlations between ^1H–^1H distances and NOESY parameters. In this context it may be warranted in the future to recalculate lower bounds to selected distances with more elaborate computations, which take account of the steric hindrance due to the van der Waals' volumes of the individual atoms.

In globular proteins one can expect that certain stereospecific assignments will be obtained in the course of the structure determination, similar to the determination of some sequential resonance assignments in micelle-bound melittin which were not accessible in the initial n.m.r. studies (Brown *et al.*, 1982). For example individual assignments should result for the methyl groups of Val and Leu when a sufficient number of NOE distance constraints to the individual methyl groups and the methine proton can be obtained, and similar considerations apply for the 2,6 and 3,5 ring protons of Phe and Tyr in the situation where these rings are immobilized in the spatial protein structure (Wüthrich, 1976). A certain flexibility of the library of amino acid structures used for the distance geometry calculations is therefore advisable, so that substitutions of the pseudo-structures in Table 1 by structures that are more closely related to or identical with the real amino acids is readily feasible in the course of the structure determination.

Financial support by the Schweizerischer Nationalfonds (Project 3.528.79) and the use of the facilities of the Zentrum für Interaktives Rechnen (ZIR) of the ETH Zürich is gratefully acknowledged. We thank Mrs E. Huber and Mrs E. H. Hunziker for the careful preparation of the manuscript.

REFERENCES

Anil Kumar, Wagner, G., Ernst, R. R. & Wüthrich, K. (1981). *J. Amer. Chem. Soc.* **103**, 3654–3658.

Billeter, M., Braun, W. & Wüthrich, K. (1982). *J. Mol. Biol.* **155**, 321–346.

Braun, W., Bösch, C., Brown, L. R., Gō, N. & Wüthrich, K. (1981). *Biochim. Biophys. Acta*, **667**, 377–396.

Braun, W., Wider, G., Lee, K. H. & Wüthrich, K. (1983). *J. Mol. Biol.* **169**, 921–948.

Brown, L. R., Braun, W., Anil Kumar & Wüthrich, K. (1982). *Biophys. J.* **37**, 319–328.

Bundi, A. & Wüthrich, K. (1979). *Biopolymers*, **18**, 285–298.

Dunfield, L. G., Burgess, A. W. & Scheraga, H. A. (1978). *J. Phys. Chem.* **82**, 2609–2616.

Feeney, J., Hansen, P. E. & Roberts, G. C. K. (1974). *Chem. Commun.* 465–466.

Fischman, A. J., Wyssbrod., H. R., Agosta, W. C. & Cowburn, D. (1978). *J. Amer. Chem. Soc.* **100**, 54–58.

IUPAC-IUB Commission on Biochemical Nomenclature (1970). *J. Mol. Biol.* **52**, 1–17.

Levitt, M. (1976). *J. Mol. Biol.* **104**, 59–197.

Levitt, M. & Warshel, A. (1975). *Nature (London)*, **253**, 694–698.

Momany, F. A., McGuire, R. F., Burgess, A. W. & Scheraga, H. A. (1975). *J. Phys. Chem.* **79**, 2361–2381.

Ramachandran, G. N. & Sasisekharan, V. (1968). *Advan. Protein Chem.* **250**, 283–437.
Richards, F. M. (1974). *J. Mol. Biol.* **82**, 1–14.
Wagner, G. & Wüthrich, K. (1982). *J. Mol. Biol.* **155**, 347–366.
Wider, G., Lee, K. H. & Wüthrich, K. (1982). *J. Mol. Biol.* **155**, 367–388.
Wüthrich, K. (1976). *NMR in Biological Research: Peptides and Proteins,* North-Holland Publishing Company, Amsterdam.
Wüthrich, K., Wider, G., Wagner, G. & Braun, W. (1982). *J. Mol. Biol.* **155**, 311–319.

Edited by J. C. Kendrew

J. Mol. Biol. (1985) **182**, 281–294

An Evaluation of the Combined Use of Nuclear Magnetic Resonance and Distance Geometry for the Determination of Protein Conformations in Solution

Timothy F. Havel and Kurt Wüthrich

*Institut für Molekularbiologie und Biophysik
Eidgenössische Technische Hochschule-Hönggerberg
CH-8093 Zürich, Switzerland*

(Received 14 August 1984)

An evaluation of the potential of nuclear magnetic resonance (n.m.r.) as a means of determining polypeptide conformation in solution is performed with the aid of a new distance geometry program which is capable of computing complete spatial structures for small proteins from n.m.r. data. Ten sets of geometric constraints which simulate the results available from n.m.r. experiments of varying precision and completeness were extracted from the crystal structure of the basic pancreatic trypsin inhibitor, and conformers consistent with these constraints were computed. Comparison of these computed structures with each other and with the original crystal structure shows that it is possible to determine the global conformation of a polypeptide chain from the distance constraints which are available from n.m.r. experiments. The results obtained with the different data sets also provide a standard by which the quality of protein structures computed from n.m.r. data can be evaluated when no crystal structure is available, and indicate directions in which n.m.r. experiments for protein structure determination could be further improved.

1. Introduction

Once sequence-specific resonance assignments could be obtained for the ^{1}H nuclear magnetic resonance spectra of small proteins, it was clear that the information on intramolecular distances contained in two-dimensional nuclear Overhauser enhancement spectra would to some extent characterize the spatial structure. While the possibility that this might eventually include the complete determination of the conformation of small proteins was discussed (Wüthrich *et al.*, 1982), these expectations have so far been realized only to a limited extent. For example, regular secondary structures were identified in the n.m.r.† spectra of small proteins (Zuiderweg *et al.*, 1983; Williamson *et al.*, 1984; Arseniev *et al.*, 1983; Wemmer & Kallenbach, 1983), and complete structures were computed for decapeptide segments of glucagon using a distance geometry algorithm (Braun *et al.*, 1983). For larger polypeptide segments, calculations which would take account of all the constraints imposed by the covalent structure and the n.m.r. measurements had to be postponed, because no program capable of handling polypeptides with more than 150 atoms at a time was available.

DISGEO is a new distance geometry program which is capable of computing complete spatial structures for polypeptide chains of up to about 100 amino acid residues, which are consistent with a wide variety of experimental data (Havel & Wüthrich, 1984). In the present paper DISGEO is used to evaluate the efficacy of n.m.r. experiments of varying precision and completeness in determining the conformations of small proteins in solution. Previous studies of this sort (Havel *et al.*, 1979; Wako & Scheraga, 1981) have not explicitly considered the types of geometric constraints which are available from n.m.r. experiments, and have used highly simplified, one point per residue

† Abbreviations used: n.m.r., nuclear magnetic resonance; NOESY, two-dimensional nuclear Overhauser enhancement spectroscopy; COSY, two-dimensional correlated spectroscopy; DISGEO, name for the presently used distance geometry program; BPTI, basic pancreatic trypsin inhibitor; RMS, root-mean-square; RMSD, root-mean-square difference; DHAD, dihedral angle difference; NOE, nuclear Overhauser enhancement; ECEPP, empirical conformational energy program for peptides (Momany *et al.*, 1975).

0022–2836/85/060281–14 $03.00/0

© 1985 Academic Press Inc. (London) Ltd.

molecular models. In contrast, we use here the complete molecular structure of the polypeptide chain, augmented by the addition of "pseudo atoms" to which geometric constraints involving non-stereospecifically assigned protons can be referred (Wüthrich *et al.*, 1983).

For our evaluation, we extracted sets of intramolecular, interproton distance constraints from the crystal structure of the protein BPTI (Deisenhofer & Steigemann, 1985). These data sets were designed to stimulate the information content of NOESY spectra of varying quality, and one set included supplementary information available from other n.m.r. experiments as well. Conformations consistent with each data set were computed, and compared with each other and with the crystal structure from which the data were obtained to establish how faithfully a spatial protein structure could in principle be determined from the type of experiment being simulated. From these model studies, we hope first to assess the precision with which polypeptide conformations in solution can be determined using the information available from n.m.r. experiments, and second to formulate guidelines by which the collection of n.m.r. data for protein spatial structure determination can be most efficiently performed.

2. Methods

All computations reported here were done on a DEC 10 computer running the TOPS 10 operating system. All programs have been written in the PASCAL programming language.

(a) *Distance geometry calculations with n.m.r. data*

The input for a distance geometry calculation consists of geometric constraints imposed on a spatial molecular structure in order that it be consistent with known experimental data†. There are 2 distinct types of constraints involved in our calculations. "Distance constraints" are lower and upper bounds on the distances between selected pairs of atoms in the molecule. 'Chirality constraints" consist of lower and upper bounds on the oriented volumes spanned by selected sets of 4 atoms in the molecule (Fig. 1). Areal constraints could in principle also be used, but have not been required in the problems we have considered.

In distance geometry calculations with the n.m.r. data on proteins in solution, the distance and chirality constraints are obtained both from the primary structure of the protein and from the actual n.m.r. measurements. From the primary structure, one obtains essentially exact values for the distances between all covalently bonded and geminal pairs of atoms in the molecule (Némethy *et al.*, 1983), as well as lower bounds on the distances between all pairs of atoms more than 3 bonds apart in the covalent chain, since these distances must be greater than the sum of the atomic hard sphere radii. In addition, the magnitude and sign of the oriented volume spanned by the tetrahedron of atoms around each sp^3 carbon follows from the structures of the naturally occurring amino acids. Two n.m.r. experiments are particularly useful in providing geometric constraints for use in a distance geometry calculation. The NOESY experiment (Anil Kumar *et al.*, 1980, 1981) yields information on the spatial proximities of pairs of protons in the molecule, which can be formulated as distance constraints. The COSY experiment yields information on the dihedral angles in the molecule (Marion & Wüthrich, 1983), which may under some circumstances be formulated as chirality constraints (see section (b), below).

One important feature of the chirality constraints used here, as well as all distance constraints obtained from the primary structure, is that they are short-range in nature, i.e. the atoms involved in these constraints are always covalent neighbors in the molecule. In contrast, the distance constraints obtained from the NOESY experiment are often long-range constraints between atoms which are separated by many covalent bonds. Therefore, the NOE distance constraints play a dominant role in the determination of the global conformation of the polypeptide chain (Havel *et al.*, 1979; Wako & Scheraga, 1981).

(b) *A summary of the distance geometry program*

DISGEO is a distance geometry program based on the EMBED alogrithm of Havel *et al.* (1983). Its input consists exclusively of bounds on the distances and chiralities in the molecule, as described above. It produces as output the Cartesian co-ordinates of a random sample from the ensemble of spatial molecular structures which are consistent with these constraints. The calculations of the EMBED algorithm can be roughly divided into 2 parts. First, a calculation is done using the distances themselves as the primary variables, which yields Cartesian co-ordinates that are an approximate fit to the input distance constraints. This process is sometimes referred to as "embedding" the structure. Second, an optimization is performed *versus* an "error" function of these co-ordinates which increases rapidly with deviations of the co-ordinates from the input distance and chirality constraints, and which is zero if and only if all the constraints are satisfied exactly. Because the initial distance calculation produces co-ordinates which are already an approximate fit to the input data, this optimization converges rapidly and reliably when compared to most other macromolecular optimization procedures.

Randomness is obtained in the starting structures for the optimization by making a random guess at the unknown interatomic distances in a conformation of the protein which is consistent with the data. The embedding procedure by which a starting structure is obtained from each such set of random distances has not been proven to produce a truly random sample of the conformation space as a whole. However, it is clear from the results obtained here that the EMBED algorithm is capable of producing a disperse collection of structures consistent with a given set of data. The experience accrued so far implies that the chances are low that substantially different conformations exist which would also be consistent with the same input data.

The chirality constraints are not used during the initial embedding, so that the starting structure for the

† By constraints, we mean logical conditions that the computed structures should satisfy, and not an objective function which measures deviations of the observables from their estimated values. See section 2 of Havel & Wüthrich (1984) for details.

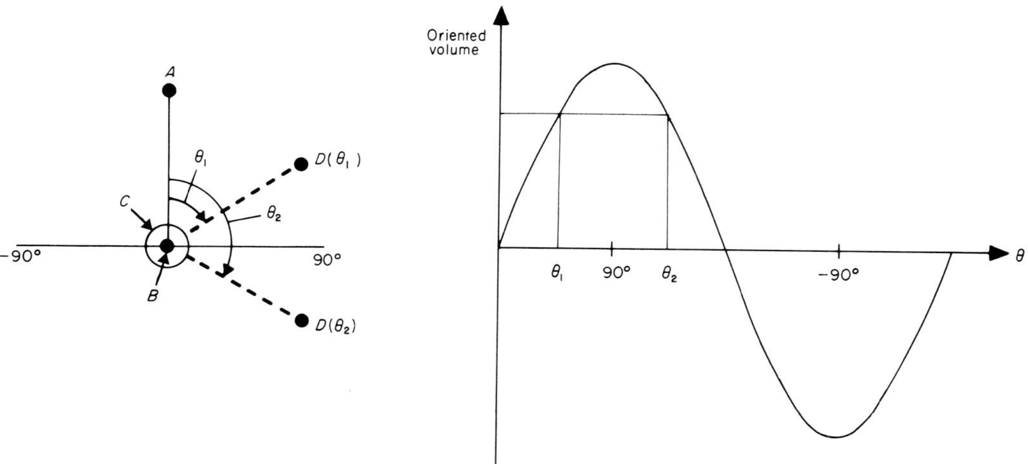

Figure 1. Diagram illustrating the oriented volume spanned by the 4 atoms which define a dihedral angle (*A–D* in the Newman projection of a molecular fragment on the right), and its relation to the dihedral angle. The oriented volume is defined as the triple product of the vectors from any 1 of these 4 atoms to the other 3. Two possible values of the dihedral angle, denoted as θ_1 and θ_2, which correspond to a given value of the oriented volume, are defined in the molecular fragment on the left. On the right, the oriented volume is plotted as a function of the dihedral angle θ.

optimization does not take account of them except to the extent that they are redundant with the distance information. The use of chirality constraints during the optimization ensures that one obtains structures in which all sp^2 hybridized groups are properly planar, and in which all asymmetric carbons have the correct stereochemistry (Braun *et al.*, 1981; Crippen *et al.*, 1981; Havel *et al.*, 1983). Chirality constraints are also a suitable technique for imposing bounds on the values of dihedral angles about single bonds. The oriented volume spanned by the 4 atoms defining a dihedral angle, however, is symmetric about plus and minus 90° (see Fig. 1). If, for example, we attempt to fix a dihedral angle at 0° by constraining the oriented volume to be zero, we find that an angle of 180° also satisfies this constraint. Fortunately, the dihedral angles of interest are about single carbon–carbon or carbon–nitrogen bonds, so that the rotational state about a given bond can always be defined by fixing the values of multiple dihedral angles about the bond, where each dihedral angle relates to a different pair of vicinal substituents. A unique definition of the rotational state of the bond can thus in general be obtained†.

A heuristic procedure employed by the DISGEO program consists of initially embedding a "substructure" containing only a relatively small subset of the atoms in the complete structure. This subset consists of the atoms adjacent to the α-carbon of each amino acid residue, together with selected side-chain atoms. The distances obtained from the co-ordinates of this substructure may then be used to guide the selection of the remaining distances in the complete structure, and thus to obtain an improved starting structure for the final optimization. A detailed description of this procedure, and of the DISGEO program itself, was given by Havel & Wüthrich (1984).

In the calculations that we have performed here, we usually embedded a total of 10 substructures for each data set, and selected the 3 which converged best for the subsequent calculations. With the less constraining data sets, it was sometimes found that the global chain fold in the substructure was the mirror image of that in the crystal structure, although the chiralities at the α-carbons were correct. These inviable starting structures were explicitly checked for, and likewise discarded.

(c) *Techniques used for the comparison of structures*

The most widely used method of comparing the global conformations of 2 polypeptides is a difference measure known as the root-mean-square co-ordinate deviation, or RMSD (Nyburg, 1974). In this method, the centroids of the 2 conformations are superimposed, and the rotation $\mathbb{R}$ which minimizes the r.m.s. difference in the vectors $\vec{r}_i$ and $\vec{r}'_i$ from the common centroid to each of the n atoms is found. Then:

$$\mathrm{RMSD} = \left[\frac{1}{n}\min_{\mathbb{R}} \sum_{i=1}^{n} \|\vec{r}_i - \mathbb{R}\,\vec{r}'_i\|^2\right]^{1/2}. \quad (1)$$

The superposition of the 2 structures for which this minimum is obtained also allows one to visualize the differences in the conformations. The RMSD between the α-carbons alone will be referred to as the RMSD^a, while that between all heavy-atoms (i.e. non-hydrogen atoms) in the backbone and side-chains will be referred to as the RMSD^h. To further quantitate the size and shape of the molecule, we have also computed the radii of gyration R_G^x, R_G^y and R_G^z of the α-carbon positions (assuming unit mass) about the 3 principle axes of the molecule (i.e. the axes of the inertial ellipsoid), as well as the RMS value thereof, R_G (i.e. the radius of an equivalent sphere). R_G^x, R_G^y and R_G^z are equal to the square-roots of the eigen-

† The conformation could also be unambiguously constrained by a combination of distance and chirality constraints, but for technical reasons we have not done so here.

values of the normalized inertial tensor, which is given in terms of the atomic co-ordinates x, y and z by (Goldstein, 1950):

$$\hat{\mathrm{I}} = \frac{3}{2n}\begin{bmatrix} \sum(y_i^2+z_i^2) & -\sum x_i y_i & -\sum x_i z_i \\ -\sum x_i y_i & \sum(x_i^2+z_i^2) & -\sum y_i z_i \\ -\sum x_i z_i & -\sum y_i z_i & \sum(x_i^2+y_i^2) \end{bmatrix}, \quad (2)$$

where the summations are over all α-carbons.

The medium range conformation of a polypeptide chain (i.e. the detailed structure of segments consisting of about 2 to 5 residues in the sequence) can be very different in conformations with the same overall chain fold. For example, a protein conformation consisting entirely of random coil can be arranged so as to have a low RMSD when compared to another conformation consisting largely of regular secondary structure. To measure differences in the medium range structure of 2 polypeptide conformations, we have plotted the RMSD between segments of 4 consecutive α-carbons, which we call the RMSD_i^a, as a function of i, the sequence position of the second residue of the tetrapeptide segment. The RMSD_i^a correlates well with other measures based on the curvature and torsion of the polypeptide chain (Rackovsky & Scheraga, 1980), and using the analytic method of McLaughlan (1979), it may be computed just as rapidly. The curvature/torsion measure, however, is insensitive to the variations in virtual bond angles which can occur between structures, and hence we feel that the RMSD_i^a is better suited to our purposes.

The conformations of the backbone and side-chains of the individual amino acid residues are best compared directly in terms of their dihedral angle differences. To examine the backbone conformation, we have computed the RMS of the differences between the ϕ and ψ angles of corresponding residues in 2 different conformers (i.e. these differences are the smallest rotations necessary to align the atoms about corresponding bonds, irrespective of the direction of the rotation), which we call the backbone dihedral angle difference, i.e.:

$$\mathrm{DHAD}^{\mathrm{b}} = \left[\frac{1}{m}\sum_{i=1}^{m}\frac{1}{2}\left((\phi_i-\phi_i')^2+(\psi_i-\psi_i')^2\right)\right]^{1/2}, \quad (3)$$

where m is the number of amino acid residues, and the primed and unprimed torsion angles are in different conformers. The RMS backbone dihedral angle difference of a single residue, i, in 2 separate structures will be denoted as $\mathrm{DHAD}_i^{\mathrm{b}}$. In addition, we have computed the RMS of the dihedral angle differences between corresponding side-chain χ^1 angles in a pair of conformations, which we denote as the DHAD^1. This is given by:

$$\mathrm{DHAD}^1 = \left[\frac{1}{m}\sum_{i=1}^{m}(\chi_i^1-\chi_i^{1\prime})^2\right]^{1/2}. \quad (4)$$

(d) *Simulation of n.m.r. data from the crystal structure of BPTI*

The most important source of geometric constraints from n.m.r. measurements are short ^{1}H–^{1}H distances manifested as nuclear Overhauser effects (Wüthrich *et al.*, 1982). For proteins these data are presently most efficiently collected by NOESY experiments (Anil Kumar *et al.*, 1980). For various fundamental and technical reasons the precision of NOESY distance measurements is still limited, so that in practical applications with proteins we have so far interpreted the NOE data in terms of distance constraints rather than as exact distance values (Braun *et al.*, 1981, 1983). When a "strong" NOE is observed, the distance between the 2 hydrogen atoms involved is fixed in the range between the van der Waals' limit of 2·0 Å and an upper bound of 2·5 Å. A "medium" NOE fixes the corresponding distance between 2·0 and 3·0 Å, and a "weak" NOE imposes an allowed distance range of 2·0 to 4·0 Å. We have not so far accepted the absence of a NOESY cross peak as conclusive evidence for a certain minimum distance between the protons.

To simulate geometric constraints which could in principle be obtained from n.m.r. experiments on BPTI, hydrogen atoms were added to its crystal structure (Deisenhover & Steigemann, 1975) using a procedure similar to that described by Billeter *et al.* (1982), which assures a close fit of the resulting structure to the ECEPP standard geometry of the individual amino acid residues (Némethy *et al.*, 1983). In the structure thus obtained, 508 of the ^{1}H–^{1}H distances between NH, CH, CH_2 and CH_3 groups located in different residues were found to be shorter than 4·0 Å†. When more than one ^{1}H–^{1}H distance between a pair of these groups was found to be less than 4·0 Å, only the shortest such distance was used. After rounding these distances from machine precision to $\pm 0{\cdot}1$ Å, the following simulated n.m.r. data sets were prepared from them.

For the first data set (I in the Tables and Figures), all of the 508 distances above which were shorter than 2·5 Å were constrained to a range between 2·0 and 2·5 Å, while those from 2·5 to 3·0 Å were constrained between 2·0 and 3·0 Å, and the remaining distances were constrained between 2·0 and 4·0 Å. These distance constraints correspond to the interpretation of experimental NOE values as strong, medium and weak, as described above. The upper bounds on the distances between groups of protons for which stereospecific assignments could not generally be obtained were modified with the necessary corrections to yield constraints between suitably chosen pseudo atoms, in exactly the same way as would be done with experimental NOEs (Wüthrich *et al.*, 1983)‡. With these pseudo structures for the amino acid residues of BPTI, we had a total of 666 geometric points in each spatial structure.

Since all the NOEs corresponding to the complete list of 508 short interresidue ^{1}H–^{1}H distances from the crystal structure of BPTI could not in practice be resolved in a NOESY spectrum, incomplete data sets were used for the remaining calculations. In data set II (Table 1) we retained only those NOEs from data set I which involved at least one amide or aromatic proton, and which would therefore generally give rise to well-resolved cross peaks in NOESY spectra (Wagner & Wüthrich, 1982). In addition, the "strong" NOEs between pairs of aliphatic protons were retained, since they could in general be singled out by proper experimentation (Anil Kumar *et al.*, 1981).

Further data sets (III through VIII in Table 1) were prepared in order to investigate the influence of the precision with which the different NOE distance constraints were specified on the resulting computed structures. In the data sets III and IV, only the

† We did not consider short intraresidue ^{1}H–^{1}H distances, and thus our calculations never included any local constraints within the amino acid side-chains.

‡ The only exceptions were the methyl groups of the single valine and the 2 leucines in BPTI, which were treated as if they had been stereospecifically assigned.

Table 1

Survey of the interresidue distance constraints used to simulate n.m.r. data sets for the structure calculations

Data set	Total constraints†	$d_{\alpha N}$, d_{NN}, $d_{\beta N}$‡ Number	Limits¶	Medium-range§ Number	Limits¶	Long-range‖ Number	Limit¶
I	508 (119)	38	2·5	9	2·5	26	2·5
		44	3·0	35	3·0	57	3·0
		40	4·0	116	4·0	143	4·0
II	356 (112)	38	2·5	9	2·5	26	2·5
		44	3·0	20	3·0	33	3·0
		40	4·0	67	4·0	79	4·0
III	356 (112)	38	2·5	96	5·0	138	5·0
		44	3·0				
		40	4·0				
IV	356 (112)	38	2·5	96	4·0	138	4·0
		44	3·0				
		40	4·0				
V	234 (38)			96	4·0	138	4·0
VI	234 (38)			96	±0·1	138	±0·1
VII	356 (112)	122	±0·1	96	±0·1	138	±0·1
VIII	356 (112)	122	±0·5	96	±0·5	138	±0·5
IX	170 (59)	38	2·5	29	4·0	59	4·0
		44	3·0				
X	170 (59)	38	2·5	29	4·0	59	4·0
		44	3·0				
		Plus the supplementary constraints of Fig. 2.					

† The left number for each data set indicates the total number of NOE distance constraints used. The number in parentheses indicates how many of these constraints are between backbone hydrogen atoms, i.e. NH and $C^{\alpha}H$.

‡ $d_{\alpha N}$, d_{NN} and $d_{\beta N}$ are, respectively, the distances between the C^{α} proton of residue i and the amide proton of the sequentially adjoining residue $i+1$, between NH_i and NH_{i+1}, and between $C^{\beta}H_i$ and NH_{i+1}. These sequential proton–proton distances are used for obtaining sequence-specific resonance assignments (Billeter *et al.*, 1982).

§ Medium-range distances are all interresidue distances between hydrogen atoms located within a pentapeptide segment, with the exception of $d_{\alpha N}$, d_{NN} and $d_{\beta N}$.

‖ Long-range distances are between protons located in residues which are separated by at least 4 intervening residues in the primary structure.

¶ Unsigned numbers define upper bounds on the distances in ångström units (in these cases the lower bounds are 2·0 Å). Where numbers are preceded by ±, the exact distance in the crystal structure was taken to be in the center of the indicated range.

sequential distances $d_{\alpha N}$, d_{NN} and $d_{\beta N}$ were quantitated and classified as strong, medium or weak NOEs. All of the medium-range and long-range distances used in data set II were attributed an upper bound of 5·0 Å (in data set III) or 4·0 Å (in data set IV). Such a set of NOE data would be considerably easier to prepare than one in which all of the NOEs had been quantitated.

In order to assess the importance of the sequential NOEs $d_{\alpha N}$, d_{NN} and $d_{\beta N}$ in protein structure determinations, the 122 NOEs of this type were excluded from data set IV to obtain data set V. The same NOEs as in data set V were used again for data set VI, but now a narrow range of ±0·1 Å about the actual distance in the crystal structure was used for the medium and long-range constraints†. To obtain data set VII, the sequential constraints $d_{\alpha N}$, d_{NN} and $d_{\beta N}$ were added to the distance constraints of data set VI, with the same precision of ±0·1 Å about the actual distances in the crystal structure. Data set VIII is identical to data set VII, except that now the range about the actual distance in the crystal structure was chosen to be ±0·5 Å.

† For those distance constraints which involved a pseudo atom, the position of the pseudo atom in the crystal structure was calculated from the positions of the hydrogen atoms, as described by Wüthrich *et al.* (1983).

In data sets IX and X (Table 1) we prepared a relatively "poor" set of NOESY input data, and investigated how the structures obtained with this NOE data set could be improved by the inclusion of supplementary information obtained from other n.m.r. experiments. Data set IX was obtained from data set IV by arbitrarily eliminating all constraints corresponding to weak NOEs. The supplementary information (Fig. 2) included in data set X consisted of the 21 backbone hydrogen bonds of BPTI implied by the secondary structure assignments of Kabsch & Sander (1983), together with constraints on the values of the 27 ϕ angles for which the corresponding coupling constants $^3J_{HN\alpha}$ could be unambiguously interpreted in terms of the dihedral angle in an n.m.r. experiment (Pardi *et al.*, 1984). The hydrogen bonds were defined by setting the amide proton-to-carbonyl oxygen distance between 1·8 and 2·0 Å, and the nitrogen-to-oxygen distance between 2·7 and 3·0 Å. All ϕ angles with values in the crystal structure between −150° and −100° were constrained to lie between −160° and −80°, and those between −70° and −50° were constrained to lie between −90° and −40°. The dihedral angle ranges were imposed as chirality constraints, as described in section (b), above.

Implicitly included in all of the above "data sets" are all distance and chirality constraints necessary to fix the covalent structure, including the 3 disulfide bonds which are present in BPTI. There were a total of 3290 such

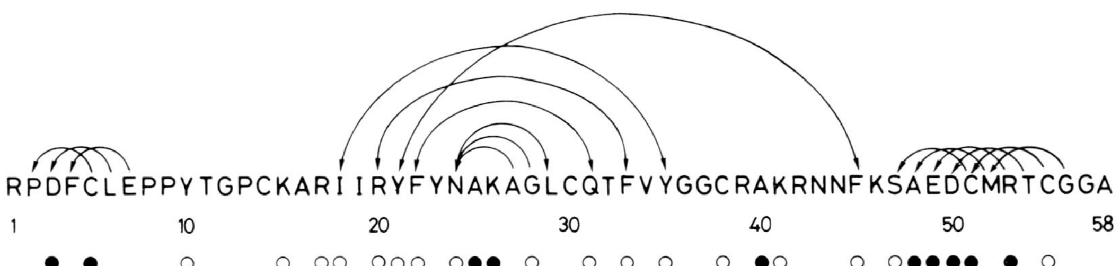

Figure 2. Diagram showing the sequence locations of the hydrogen bonds (arrows) and ϕ angle chirality constraints (circles) used in data set X (Table 1). The sequence of BPTI is given in the standard one-letter amino acid code. The arrows point from the amide proton to the carbonyl oxygen of each hydrogen bond (2-headed arrows identify an antiparallel β-bridge). A filled circle indicates that the ϕ angle was constrained between $-90°$ and $-40°$, and an open circle indicates that the ϕ angle was constrained between $-160°$ and $-80°$ (see the text).

"covalent distance constraints", and 450 "covalent chirality constraints", which consist of essentially exact values for the distances and oriented volumes involved. In addition, many thousands of lower bounds were implicit in the hard sphere radii used for the individual atoms.

It is immediately apparent that in all data sets I to X the total number of constraints used exceeded the $3N-6 = 1995$ minimum number of constraints which would be necessary to define the conformation uniquely. Yet in every case there remained multiple conformations consistent with the data. The following are 3 main reasons for this. First, $3N-6$ gives only the *minimum* number of constraints which could be found to determine the structure. For example, if $3N-6$ exact distance constraints are unevenly distributed so that some subsets of M atoms have more than $3M-6$ constraints amongst them, then these parts of the structure will be overdetermined, while the structure as a whole remains underdetermined. Second, even if the given constraints are sufficient to hold the structure rigid, multiple rigid realizations may exist. For example, if the $N-1$ bond lengths, $N-2$ geminal distances, and $N-3$ vicinal distances of a straight chain of N atoms are given, the signs of the dihedral angles about the bonds remain undefined. Third, the $3N-6$ argument would be strictly applicable only to exact distance constraints, while many of the constraints used here consist only of distinct lower and upper bounds on the distances. In this case, the need for more than the minimum number of constraints is apparent.

3. Results and Discussion

When using n.m.r. data as input for a distance geometry calculation of the spatial structure of a protein for which no crystal structure is available (see the accompanying paper, Williamson *et al.*, 1985), or in cases in which the solution conformation differs markedly from the crystal structure (Braun *et al.*, 1983), the following three questions are of particular interest. First, how does the DISGEO program (Havel & Wüthrich, 1984) function when different n.m.r. data sets are used as input? Second, can the geometric constraints which are available in practice from n.m.r. experiments in solution be sufficient to enable a protein conformation to be determined by means of distance geometry calculations? Third, how could the collection of n.m.r. data for use in protein structure determination by distance geometry be most efficiently performed?

To investigate these questions, the ten sets of geometric constraints listed in Table 1 were used as input for the DISGEO program. As described in Methods section (d), these data sets were derived from the crystal structure of BPTI, and were designed to simulate results of n.m.r. experiments of varying completeness and precision. The advantage of using constraints derived from a known crystal structure is that we can be sure that they are consistent with at least one three-dimensional structure. Thus, the convergence properties of the DISGEO program can be analyzed without fear that the constraints themselves are inconsistent, and comparison with the crystal structure provides a standard by which results obtained with real experimental n.m.r. data can be judged.

For each of the ten data sets evaluated (Table 1), we had in the end a total of four spatial structures consistent with the input constraints: the three structures computed with DISGEO (denoted as A, B and C), and the original crystal structure (denoted as X). In Figure 3, the α-carbons of A, B and C have been superimposed on those of X so as to minimize the RMS co-ordinate difference (see Methods section (c)) for each of the ten data sets. Although substantial variations in the quality of the fit with the crystal structure can be seen among the different data sets, the overall size and shape of the computed structures are nonetheless all very similar to that of the crystal structure. In the following, the convergence obtained with the DISGEO program will be described, and the similarities and differences among the structures A, B, C and X from each of the individual data sets will be quantitatively analyzed.

(a) *The convergence obtained in the computations with DISGEO*

The quality of the convergence obtained with the DISGEO program in the individual computations

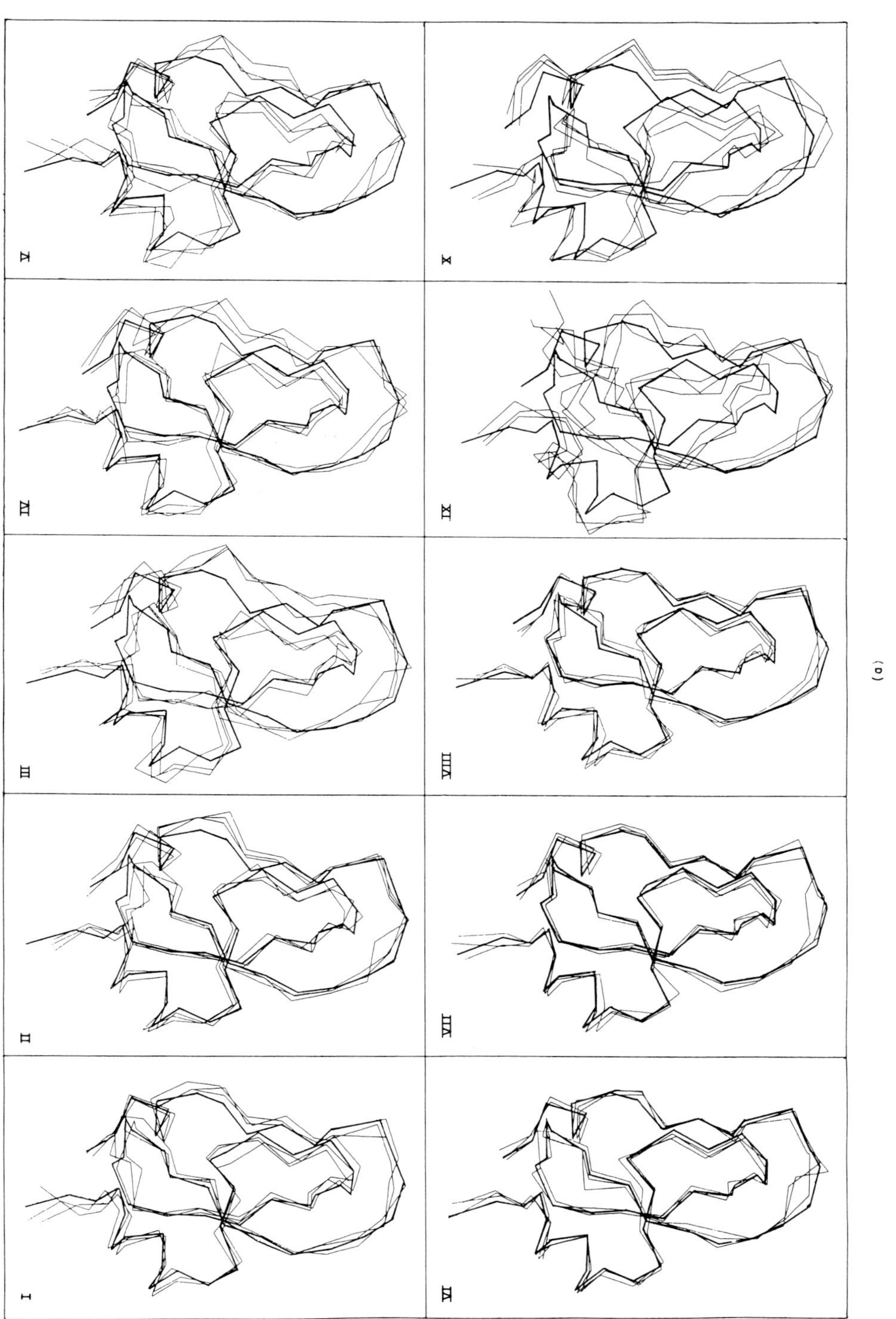

(a)

Fig. 3

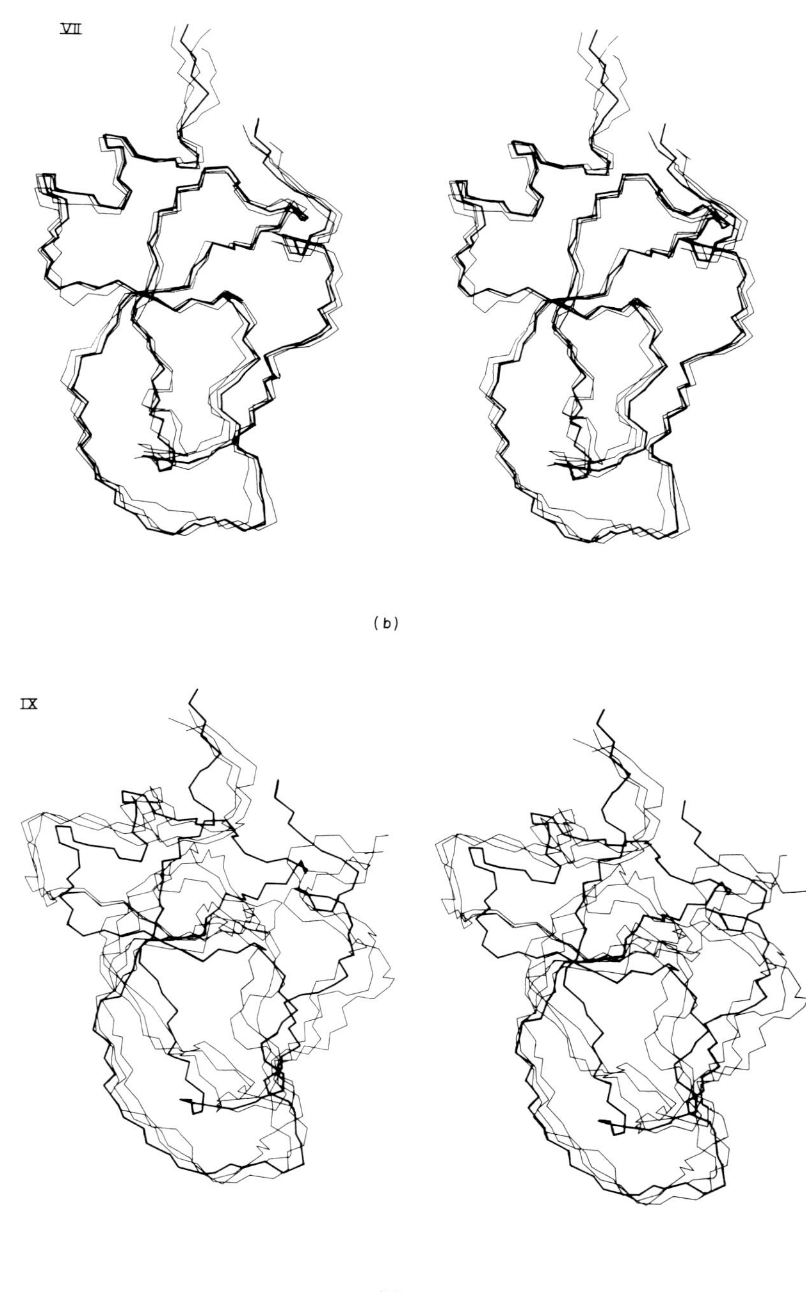

Figure 3. Pictures of the 3 structures computed for each of the data sets I to X (see Table 1). Each of the 3 computed structures has been individually superimposed on the crystal structure of BPTI (drawn with a heavier line) so as to minimize the RMA difference in the α-carbon co-ordinates. In (a), only the α-carbons and the virtual bonds connecting them are shown. (b) Stereo views of complete backbones (N, C^{α} and C′) of the structures computed from the most constraining data set (VII) and the least constraining data set (IX).

with the data sets I to X (Table 1) may be evaluated from Figure 4, which shows histograms in 0·1 ångström increments of all violations of the n.m.r. and covalent distance constraints in each of the 30 structures generated. The program reliably produced structures having no violations of the input distance constraints greater than 1·0 Å. No structure contained more than eight violations in the range of 0·5 to 1·0 Å, and in nine out of the 30 computations all violations were below 0·5 Å. More than 90% of all violations were always below 0·1 Å, and the residual violations of the covalent bond lengths and hard sphere radii were almost always well below this limit. The peptide bonds and other sp^2 hybridized groups were usually within 1° of planarity. The larger residual violations usually occurred in clusters confined to short segments of two to four residues in the chain. With the exception of such occasional local "snags", this convergence is good enough to ensure that, in general, the differences observed between the different computed structures or between the computed structures and the crystal structure are those allowed by the constraints and are not due to differences in the nature or the magnitude of the residual constraint violations.

(b) *The global conformation of the polypeptide backbone*

One structural feature of particular interest in this study was the overall size and shape of the computed structures, which is visualized in Figure 3, and described quantitatively by the radii of gyration which are listed in Table 2. Given that all the distance information considered here involves only short distances and is often quite imprecise as well, one might expect that a sum of many small variations in these distances could lead to considerably larger variations in the sizes and shapes of the computed structures than those encountered in Table 2. In the less constraining data sets, there is a slight tendency for the computed structures to become more expanded than the original crystal structure. For example, in data set III (Table 1) the average radius of gyration of the α-carbons about the third principal axis was about 10% larger than it is in the crystal structure (the third axis corresponds roughly to the vertical in Fig. 3). In data sets IX and X the average isotropic radius of gyration was about 5% larger than the crystal structure, with most of the increase again occurring about the third principal axis. Figure 3 shows that this tendency towards expansion occurs primarily at strands and loops on the surface of the molecule, which are no longer held tightly against the central core, while the core itself remains relatively constant.

The observed tendency towards expansion is an indication that the DISGEO program exhibits good sampling properties. For example, when we deliberately used an upper bound of 5·0 Å for all medium and long-range distances in the crystal structure

I-A	1869	60	11	6	2	4				
I-B	1860	49	12	5	1	4	2	0	2	
I-C	1808	16	4	1	1	1				
II-A	1826	56	8	6	1	0	0	0	1	
II-B	1813	25	5	2	1					
II-C	1827	25	3	6	1					
III-A	1734	12	1	2						
III-B	1777	21	11	2	0	0	0	1		
III-C	1859	52	9	5	0	1	1	1		
IV-A	1856	54	12	7	0	1	0	0	0	1
IV-B	1860	53	24	3	2	2	1	1	0	1
IV-C	1859	23	8	2						
V-A	1906	33	9	1	1					
V-B	1857	24	6							
V-C	1885	40	7	3						
VI-A	2029	69	11	6	4	1	1	1		
VI-B	2043	120	25	12	4	0	0	1		
VI-C	1988	97	20	10	6	1	2	1		
VII-A	2046	108	30	17	7	2	1			
VII-B	1996	58	12	4	2	1				
VII-C	2089	77	18	7	4	2	2			
VIII-A	1897	41	8	3	0	0	0	1		
VIII-B	1813	37	14	4	2					
VIII-C	1927	76	22	5	3	2				
IX-A	1802	38	4	1	0	1				
IX-B	1862	63	15	6	3	0	1			
IX-C	1820	25	8	3	1					
X-A	1539	21	2	1	0	1				
X-B	1709	38	6	2	0	0	1			
X-C	1716	62	29	12	7	0	3	2		

0·1 0·3 0·5 0·7 0·9
Distance (Å)

Figure 4 Histograms of the residual violations of the distance bounds imposed by the covalent structure and by the simulated n.m.r. data from each of the 3 structures, A to C, computed with data sets I to X. The number of violations which fall within each 0·1 Å interval are listed. For the individual computed structures, the total number of constraints with residual violations is about 40 to 60% of the total number of input distance constraints.

that were less than 4·0 Å (data set III, Table 1), we essentially introduced a bias in the data towards longer distances. Since the isotropic radius of gyration of the molecule is proportional to the sum of the squares of all interatomic distances (Havel *et al.*, 1983), we see that structures consistent with these data should in fact tend to be more expanded than the crystal structure. Since the lack of an upper bound on a distance is equivalent to an infinite upper bound, we see that this argument

Table 2

Average over the three structures A, B and C computed for each of the ten data sets of Table 1 of the radii of gyration of the C^α positions in ångström units

Structures	R_G^x†	R_G^y†	R_G^z†	R_G‡
Crystal	12·13	11·65	7·51	10·64
I	11·90	11·56	7·73	10·57
II	11·82	11·29	8·03	10·51
III	12·02	11·39	8·30	10·69
IV	12·17	11·58	8·04	10·75
V	12·04	11·50	8·06	10·68
VI	12·01	11·58	7·54	10·57
VII	12·06	11·58	7·64	10·62
VIII	11·96	11·54	7·75	10·59
IX	12·54	11·68	8·61	11·07
X	12·45	11·86	8·20	11·00

† R_G^x, R_G^y and R_G^z are the radii of gyration about the 3 principal axes, which are calculated from the α-carbon positions. The z-axis is roughly parallel to the vertical in the presentations of the structures shown in Fig. 3.

‡ R_G is the isotropic radius of gyration of an equivalent sphere, and is equal to the RMS value of R_G^x, R_G^y and R_G^z.

applies to any data set in which many of the short distances are unconstrained. In particular, this argument explains why the differences between the structures computed with the less constraining data sets and the crystal structure tend to be greater than the differences among the computed structures themselves (Fig. 3 and Table 3).

All 30 of the conformations that we have computed contain the same elements of secondary structure as the crystal structure (see Fig. 3). This was not unexpected, since the secondary structure of polypeptides can be determined rather directly from n.m.r. measurements (Williamson *et al.*, 1984; Wüthrich *et al.*, 1984). It is particularly pleasing to find that the "open-face sandwich" topology characteristic of the secondary structure packing in BPTI (Richardson, 1981) is also apparent in all of the computed structures. Even when the constraints used were imprecise (as in III) or sparse (as in IX), the right-handed twist of the β sheet was always present. We conclude that the global chain fold and secondary structure topology are structural features of proteins which can be determined particularly reliably by the n.m.r. experiments considered here.

It is interesting that the overall sizes, shapes and chain folds of the computed structures are preserved over a wide range of precision in the input distance constraints (Fig. 3 and Tables 2 and 3). For example, on going from data set VII to VIII, the precision with which the distances were specified was decreased from $\pm0{\cdot}1$ to $\pm0{\cdot}5$ Å, while the average RMSD^a with the crystal structure increased from 0·70 to only 0·85 Å (Table 3). When we explicitly loosened the upper bounds on the medium and long-range NOEs by between 1·0 and 2·5 Å to convert data set II to III (Table 1), the average RMSD^a with the crystal structure increased from 1·2 to 1·5 Å. On the other hand, the average RMSD^a with the crystal structure increased from 1·5 to 2·5 Å when approximately half the distance constraints were left out entirely to obtain data set IX from data set IV. It thus appears that, given the present limits on the quantitation of NOESY distance measurements, efforts should be directed towards maximizing the number of medium and long-range NOEs which can be resolved and assigned in the NOESY spectra before attempting to quantitate those already found. It is also worth noting that short-range distance constraints, such as the sequential distances $d_{\alpha N}$, d_{NN} and $d_{\beta N}$, contribute relatively little to determination of the global conformation even when they are available with high precision (compare, for example, the RMSD values shown in Table 3 for data sets IV and V, or VI and VII). These observations are in accord with earlier work (Havel *et al.*, 1979).

Table 3

Average minimum r.m.s. differences between atomic co-ordinates for each of the 10 sets of computed structures

Data set	$RMSD^a_{A-C,X}$	$RMSD^a_{A,B,C}$	$RMSD^h_{A-C,X}$	$RMSD^h_{A,B,C}$
I	1·17	1·25	2·08	1·91
II	1·19	1·06	2·21	1·89
III	1·91	1·63	3·05	2·53
IV	1·51	1·65	2·71	2·55
V	1·65	1·52	2·88	2·27
VI	0·79	0·94	2·11	1·84
VII	0·70	0·89	1·82	1·78
VIII	0·87	0·92	2·18	1·75
IX	2·48	1·84	3·32	2·72
X	2·22	1·61	3·11	2·51

The RMSD values are in ångström units. The superscript a refers to a comparison between the α-carbons alone, the superscript h to a comparison of all heavy (non-hydrogen) atoms. The subscript A–C,X refers to the average of the differences between each of the computed structures A to C and the crystal structure X, while the subscript A,B,C refers to the average of the differences between the 3 pairs of computed structures.

(c) *The local conformation of the polypeptide backbone*

To see how well the backbone conformation was determined locally by each of the ten data sets, we have made additional comparisons of the computed structures, both at the level of short segments of the polypeptide chain, and in terms of the dihedral angles ϕ and ψ about individual single bonds. In Figure 5, we have plotted the average for each data set of the three RMSD^a values between corresponding tetrapeptide segments in the three pairs of structures (A, X), (B, X) and (C, X) *versus* the sequence position of the second residue of the four. The mean value of the RMSD^a for each plot is also shown. Figure 6 shows plots of the average RMS dihedral angle difference DHAD^b (see Methods, section (c)) between the ϕ and ψ angles of corresponding residues in the pairs of structures

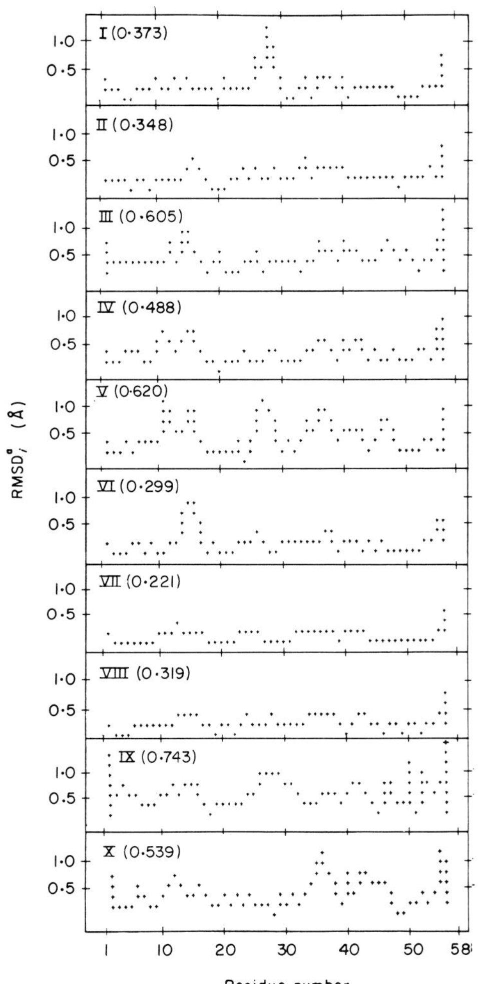

Figure 5. Plots of the average of the 3 RMSD^a_i values between the individual computed structures and the crystal structure as a function of the sequence position i for data sets I to X. The RMSD^a_i value is the RMSD between 4 successive α-carbons along the polypeptide chain, where i is the sequence number in the second residue of the 4 (see Methods section (c)). The numbers in parentheses are the average heights of each of the plots.

(A, X), (B, X) and (C, X) as a function of sequence number. The corresponding plots of RMSD^a and DHAD^b that were obtained when the average was taken over the pairs of structures (A, B), (B, C) and (A, C) were very similar, and are not shown. Both averages of the DHAD values have been included in Table 4 for each data set.

The results presented in Table 4 and Figures 5 and 6 show that the local backbone conformation is not directly affected by distance constraints which connect pairs of atoms both of which lie within a segment of the length being considered. Unlike the global structure, the local conformation depends significantly on the precision with which such local constraints are specified. For example, when the upper bounds of all medium-range distances were increased from between 2·5 and 4·0 Å to 5·0 Å on going from data set II to III (Table 1), the average heights of the plots of tetrapeptide RMSD^a values in Figure 5 rose sharply from 0·35 to 0·61 Å. On eliminating the 122 $d_{\alpha N}$, d_{NN} and $d_{\beta N}$ sequential constraints from data set IV to obtain data set V, the average height increased from 0·49 to 0·62 Å. Even when all of the medium and long-range distances were specified to within ±0·1 Å, a large peak remained in the tetrapeptide RMSD^a plot at residues 14 to 17 (Fig. 5, data set VI) unless the sequential constraints were also given (data set VII).

Overall, the DHAD^b values in Table 4 and Figure 6 indicate a high degree of variability in the individual amino acid residues even with those data sets for which the global RMSD^a value was quite low (Table 3: the reference for comparisons involving the DHAD^b is $180/\sqrt{3} = 104°$, which is the value that would be obtained if the residues were randomly distributed over the ϕ, ψ map). The most striking case of this occurs in data set VI, where the sequential $d_{\alpha N}$, d_{NN} and $d_{\beta N}$ connectivities were left out completely, while the remaining medium and long-range distances were given to ±0·1 Å. The global RMSD^a value was always well under 0·1 Å, while the DHAD^b value was as much as 75°. On fixing the sequential distances with a precision of ±0·1 Å, the global RMSD^a value remained virtually unchanged, whereas the DHAD^b value fell below 40° (Tables 3 and 4). In fact, the two remaining peaks in the plots of the DHAD^b_i between individual residues (VIII in Fig. 6) are probably due to local convergence problems (a large peak can result when even only one of the three structures gets caught in a local minimum in which a pair of ϕ, ψ angles across a peptide bond are turned by 180°).

Part of the reason for the large variability in the local backbone conformation is that the vicinal van der Waals' repulsions which are largely responsible for the characteristic patterns of Ramachandran plots (Ramachandran & Sasisekharan, 1968) were not included in these calculations (Havel & Wüthrich, 1984). Another reason has to do with the short-range distances on which the local backbone conformation depends most directly. Because the range of values accessible to the short-range distances is small compared to the long-range distances (Billeter *et al.*, 1982), a meaningful geometric constraint can be obtained only from measurements with high enough precision to confine the distance to a small fraction of this range. For example, the quantitation of the sequential distances at 2·5, 3·0 and 4·0 Å, which was used in many of the data sets of Table 1, does not in most instances impose effective constraints on $d_{\alpha N}$, since it can vary only between 2·2 and 3·6 Å.

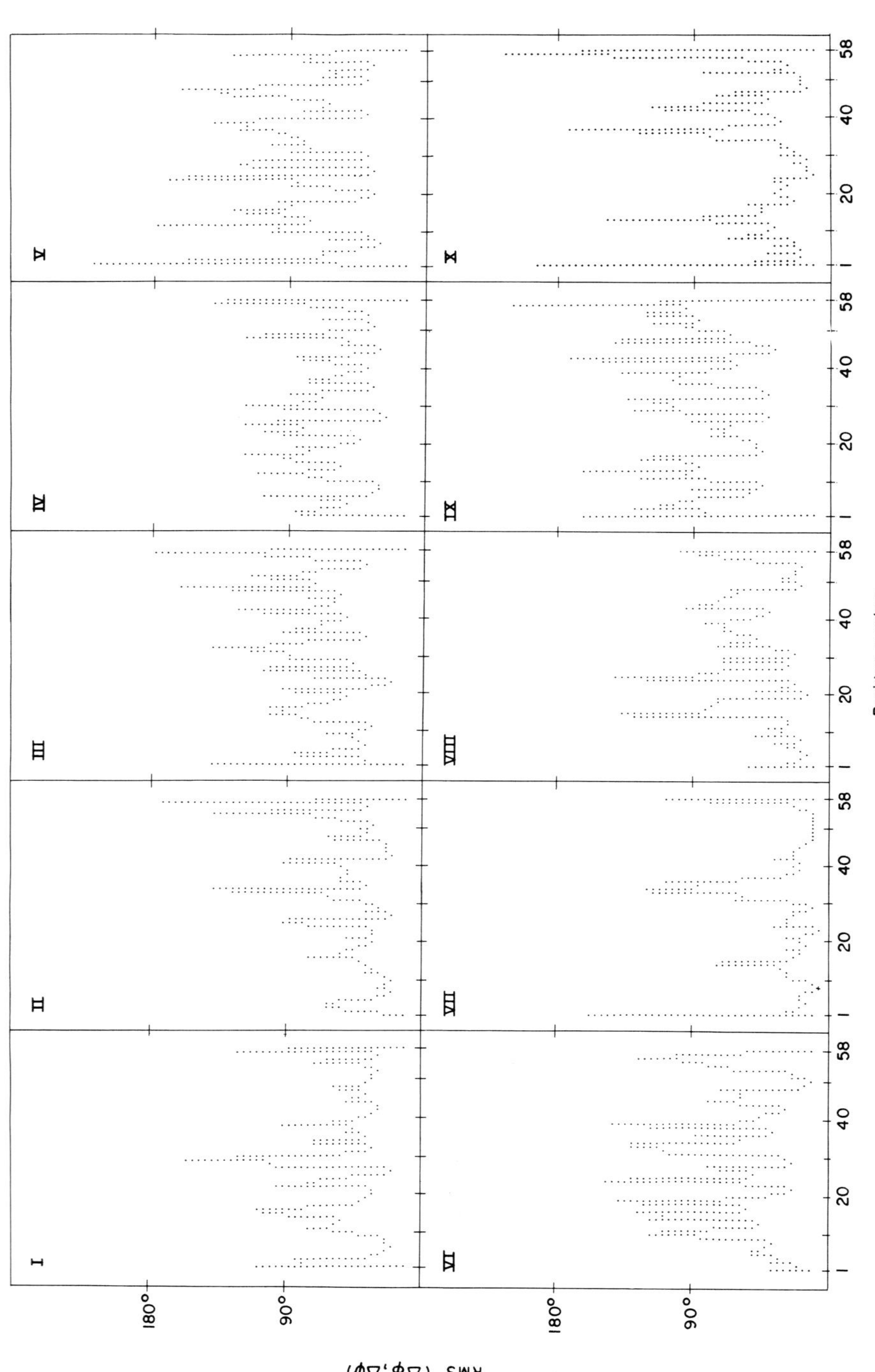

Figure 6. Plots of the average of the 3 DHAD$_i^b$ values between the individual computed structures and the crystal structure as a function of the sequence position i for data sets I to X. The DHAD$_i^b$ value is the RMS difference between the ϕ and ψ angles of the ith residue in 2 polypeptide conformations (see Methods, section (c)).

Table 4

Average RMS differences between selected dihedral angles for each of the ten sets of computed structures

Data set	$DHAD^b_{A-C,X}$	$DHAD^b_{A,B,C}$	$DHAD^1_{A-C,X}$	$DHAD^1_{A,B,C}$
I	49·7	63·1	59·8	56·1
II	47·7	51·4	70·7	72·8
III	63·0	74·2	84·9	85·8
IV	55·8	68·3	73·9	87·9
V	71·0	72·3	81·8	83·0
VI	62·2	73·0	63·7	61·8
VII	34·9	38·5	50·8	57·9
VIII	50·3	61·2	65·0	62·5
IX	74·6	84·5	85·9	86·0
X	52·9	54·1	79·0	75·1

The DHAD values are in degrees. The superscript b refers to a comparison of the backbone ϕ,ψ dihedral angles; the superscript 1 refers to a comparison of the side-chain χ^1 dihedral angles. The subscript A–C, X refers to the average of the DHAD values with the crystal structure, while A, B, C refers to the average of the DHAD values among computed structures, as defined in the footnote to Table 3.

As a consequence of the poorly determined local backbone conformation, the secondary structure elements in the conformers computed from the less-constrained data sets (e.g. III and IX in Fig. 3) are distorted so that they lack some of their characteristic backbone hydrogen bonds, as judged by an O-to-NH distance greater than 2·5 Å. In data set X, we have investigated the extent to which the regular secondary structures in the conformers computed for data set IX could be improved by the inclusion of the standard hydrogen bond distances which are available in those instances in which the secondary structure elements can be independently determined (Williamson *et al.*, 1984; Pardi *et al.*, 1984; Wüthrich *et al.*, 1984). Also included in data set X were ϕ angle constraints like those inferred from the coupling constants $^3J_{HN\alpha}$ measured by COSY experiments (Marion & Wüthrich, 1983). From all the available criteria it is readily apparent that the helical and β-sheet structures are considerably better confined in the structures of data set X than in those of data set IX. In addition, the $DHAD^b$ values between the ϕ and ψ angles *versus* sequence number decreased to just over 50° on going from IX to X (Table 4), and the remaining peaks in the plots of the $DHAD^b_i$ (Fig. 6) were found to correlate well with the absence of the ϕ angle constraints (see Fig. 2). This result is somewhat surprising, because the ϕ angle constraints used here are quite imprecise, and the ψ angles have not been constrained at all. Apparently, a combination of distance and angle information can have a synergistic effect in restraining the conformation of the individual amino acids, even when the individual constraints are rather imprecise.

(d) *The conformations of the individual side-chains*

Given that no intraresidue distance constraints were employed in this study, and that relatively few interresidue constraints between non-aromatic side-chains were retained in data sets II to X, one could not expect the side-chain conformations to be as well-defined in general as the backbone conformation. The main constraints on the side-chain conformations arise indirectly from the combined effects of the packing restrictions and the well-defined C^α to C^β bond orientations which follow from having a relatively precise backbone conformation. The dihedral angles of the side-chains were not well defined as a whole, as indicated by the averages of the $DHAD^1$ between χ^1 angles, which were always greater than 50° (Table 4). The averages of the $RMSD^h$ (Table 2) in some cases exceeded the value of 2·40 Å which was obtained between structures that were generated from the crystal structure by randomizing its side-chain χ angles (i.e. by setting its χ angles to random values between $\pm 180°$). However, variations in the backbone conformations also contribute significantly to the $RMSD^h$, and in fact the average $RMSD^h$ with the crystal structure always increased further on randomizing the side-chain χ angles of the computed conformations. From this we conclude that the interior side-chains are generally in the approximately correct spatial locations, even though their dihedral angles are not well defined by any of the data sets I to X (see also Appendix to Williamson *et al.* (1985).

4. Conclusions

The results given here show that the DISGEO program (Havel & Wüthrich, 1984) is capable of producing complete spatial structures for small proteins which are consistent with extensive sets of complex geometric constraints of the types which are available from n.m.r. experiments in solution. The strength of the program lies in its ability to produce globular structures with the correct overall dimensions, chain threading and secondary structure. In order to do this, the program ignores the details of local conformation which are determined by energetic considerations such as torsional potentials and van der Waals' interactions across rotatable bonds. The ability to reliably obtain the correct global structure is an important advance, and methods based on energy minimization are available by which the local structure may be further improved once the global chain fold is known (Fitzwater & Scheraga, 1982). The energies of the refined structures thus obtained from the different conformers computed with DISGEO may then be used as a criterion for the selection of a single "best" conformation from the ensemble consistent with the experimental data.

Another important conclusion from these model calculations, which is also implicit in earlier work of Gō & Scheraga (1970), is that in the calculation of protein conformation generally, the determination of the conformations of the individual residues and the determination of the global conformation must be considered as two separate, largely independent problems. Thus, the global conformation can be reconstructed with a suitable number of rather imprecise medium and long-range distance constraints, whereas determination of the local conformation requires the use of either relatively precise measurements of the short-range distances, or the direct determination of the dihedral angles about single bonds. The general indication is that the determination of the solution structures of macromolecules by n.m.r. and distance geometry will generally consist of two distinct phases requiring different experimental measurements and different computational procedures.

For the collection of n.m.r. data for protein structure determination, some helpful guidelines may be derived from this study. To begin with, the results obtained from the data sets I to X (Table 1) provide a standard by which the precision of a structure computed from an experimental n.m.r. data set can be judged. It is particularly reassuring that acceptable global structures were obtained from incomplete sets of quite imprecise medium and long-range distance constraints (Fig. 3), since such data sets can in practice be obtained from NOESY experiments. The spatial secondary structures are also relatively well determined by the NOESY data, and can be further improved by the incorporation of hydrogen bonds inferred from these and other n.m.r. data (Williamson *et al.*, 1984; Pardi *et al.*, 1984; Wüthrich *et al.*, 1985). Unless the short-range distances can be measured with much higher precision than has heretofore been possible in NOESY experiments with proteins (Anil Kumar *et al.*,1981; Keepers & James, 1984), complete characterization of the local structure of the backbone and side-chains will have to rely on theoretical procedures based on conformational energy calculations. Additional information may also be derived from other n.m.r. measurements which have traditionally been used for solution studies with proteins of known crystal structure (e.g. see Wüthrich, 1976).

Use of the facilities at the Zentrum für Interaktives Rechnen (ZIR) of the ETH Zürich is gratefully acknowledged. Financial support was obtained from the Schweizerischer Nationalfonds (project 3.284.82). We thank Mr M. Billeter for providing us with an interactive molecular computer graphics program specialized for the structural analysis of NOE data, and Mrs E. H. Hunziker-Kwik for the careful preparation of the Figures.

References

Anil Kumar, Ernst, R. R. & Wüthrich, K. (1980). *Biochem. Biophys. Res. Commun.* **95**, 1–6.

Anil Kumar, Wagner, G., Ernst, R. R. & Wüthrich, K. (1981). *J. Amer. Chem. Soc.* **103**, 3654–3658.

Arseniev, A. S., Kondakov, V. I., Maiorov, V. N., Volkova, T. M., Grishin, E. V., Bystrov, V. F. & Ovchinnikov, Yu. A. (1983). *Bioorgan, Khim.* **9**, 768–793.

Billeter, M., Braun, W. & Wüthrich, K. (1982). *J. Mol. Biol.* **155**, 321–346.

Braun, W., Bösch, C., Brown, L., Gō, N. & Wüthrich, K. (1981). *Biochim. Biophys. Acta,* **667**, 377–396.

Braun, W., Wider, G., Lee, K. H. & Wüthrich, K. (1983). *J. Mol. Biol.* **169**, 921–948.

Crippen, G. M., Oppenheimer, N. J. & Connolly, M. L. (1981). *Int. J. Pept. Protein Res.* **17**, 156–169.

Deisenhofer, J. & Steigemann, W. (1975). *Acta Crystallogr. sect. B,* **31**, 238–250.

Fitzwater, S. & Scheraga, H. A. (1982). *Proc. Nat. Acad. Sci., U.S.A.* **79**, 2133–2137.

Gō, N. & Scheraga, H. A. (1970). *Macromolecules,* **3**, 178–187.

Goldstein, H. (1950). *Classical Mechanics,* Addison-Wesley, Reading, Mass.

Havel, T. F. & Wüthrich, K. (1984). *Bull. Math. Biol.* **46**, 673–698.

Havel, T. F., Crippen, G. M. & Kuntz, I. D. (1979). *Biopolymers,* **18**, 73–81.

Havel, T. F., Kuntz, I. D. & Crippen, G. M. (1983). *Bull. Math. Biol.* **45**, 665–720.

Kabsch, W. & Sander, C. (1983). *Biopolymers,* **22**, 2577–2637.

Keepers, J. W. & James, T. L. (1984). *J. Magn. Reson.* **57**, 404–426.

Marion, D. & Wüthrich, K. (1983). *Biochem. Biophys. Res. Commun.* **113**, 967–974.

McLaughlan, A. D. (1979). *J. Mol. Biol.* **128**, 49–79.

Momany, F. A., McGuire, R. F., Burgess, A. W. & Scheraga, H. A. (1975). *J. Phys. Chem.* **79**, 2361–2381.

Némethy, G., Pottle, M. & Scheraga, H. A. (1983). *J. Phys. Chem.* **87**, 1883–1887.

Nyburg, S. C. (1974). *Acta Crystallogr. sect. B,* **30**, 251–253.

Pardi, A., Billeter, M. & Wüthrich, K. (1984). *J. Mol. Biol.* **180**, 741–752.

Rackovsky, S. & Scheraga, H. A. (1980). *Macromolecules,* **13**, 1440–1453.

Ramachandran, G. N. & Sasisekharan, V. (1968). *Advan. Protein Chem.* **23**, 284–438.

Richardson, J. S. (1981). *Advan. Protein Chem.* **34**, 167–339.

Wagner, G. & Wüthrich, K. (1982). *J. Mol. Biol.* **155**, 347–366.

Wako, H. & Scheraga, H. A. (1981). *Macromolecules,* **9**, 961–969.

Wemmer, D. & Kallenbach, N. R. (1983). *Biochemistry,* **22**, 1901–1906.

Williamson, M. P., Marion, D. & Wüthrich, K. (1984). *J. Mol. Biol.* **173**, 341–359.

Williamson, M. P., Havel, T. F. & Wüthrich, K. (1985). *J. Mol. Biol.* **182**, 295–315.

Wüthrich, K. (1976). *NMR in Biological Research: Peptides and Proteins,* North-Holland Publishing Company, Amsterdam.

Wüthrich, K., Wider, G., Wagner, G. & Braun, W. (1982). *J. Mol. Biol.* **155**, 311–319.

Wüthrich, K., Billeter, M. & Braun, W. (1983). *J. Mol. Biol.* **169**, 949–961.

Wüthrich, K., Billeter, M. & Braun, W. (1985). *J. Mol. Biol.* In the press.

Zuiderweg, E. R. P., Kaptein, R. & Wüthrich, K. (1983). *Proc. Nat. Acad. Sci., U.S.A.* **80**, 5837–5841.

Edited by M. F. Moody

J. Mol. Biol. (1985) **182**, 295–315

Solution Conformation of Proteinase Inhibitor IIA from Bull Seminal Plasma by ^{1}H Nuclear Magnetic Resonance and Distance Geometry

Michael P. Williamson, Timothy F. Havel and Kurt Wüthrich

Institut für Molekularbiologie und Biophysik
Eidgenössische Technische Hochschule-Hönggerberg
CH-8093 Zürich, Switzerland

(Received 14 August 1984)

A determination of the solution conformation of the proteinase inhibitor IIA from bull seminal plasma (BUSI IIA) is described. Two-dimensional nuclear Overhauser enhancement spectroscopy (NOESY) was used to obtain a list of 202 distance constraints between individually assigned hydrogen atoms of the polypeptide chain, to identify the positions of the three disulfide bridges, and to locate the single *cis* peptide bond. Supplementary geometric constraints were derived from the vicinal spin–spin couplings and the locations of certain hydrogen bonds, as determined by nuclear magnetic resonance (n.m.r.). Using a new distance geometry program (DISGEO) which is capable of computing all-atom structures for proteins the size of BUSI IIA, five conformers were computed from the NOE distance constraints alone, and another five were computed with the supplementary constraints included. Comparison of the different structures computed from the n.m.r. data among themselves and with the crystal structures of two homologous proteins shows that the global features of the conformation of BUSI IIA (i.e. the overall dimensions of the molecule and the threading of the polypeptide chain) were well-defined by the available n.m.r. data. In the Appendix, we describe a preliminary energy refinement of the structure, which showed that the constraints derived from the n.m.r. data are compatible with a low energy spatial structure.

1. Introduction

This paper describes the determination of the solution conformation of proteinase inhibitor IIA from bull seminal plasma in aqueous solution. BUSI IIA† is a protein of 57 amino acid residues and with a molecular weight of 6000, and was first described by Čechová *et al.* (1979). It inhibits proteinases such as trypsin, and may play a part in the protection of tissues against inflammatory processes (Fritz *et al.*, 1978), and possibly also in the control of fertilization (Schiessler *et al.*, 1976).

The structure determination described here relies on the use of nuclear magnetic resonance to derive a list of proton–proton distance constraints, followed by application of a distance geometry program, DISGEO (Havel & Wüthrich, 1984), to compute structures consistent with those constraints. In the preceding paper (Havel & Wüthrich, 1985), the DISGEO program was evaluated using n.m.r. data which had been simulated from the known crystal structure of BPTI. Comparison of the results of

† Abbreviations used: BUSI IIA, proteinase inhibitor IIA from bull seminal plasma; BPTI, basic pancreatic trypsin inhibitor; PSTI, porcine secretory trypsin inhibitor; OMJPQ3, third domain of the Japanese quail ovomucoid; n.m.r., nuclear magnetic resonance; NOE, nuclear Overhauser enhancement; NOESY, 2-dimensional nuclear Overhauser enhancement spectroscopy; COSY, 2-dimensional correlated spectroscopy; $^3J_{HN\alpha}$, vicinal spin–spin coupling constant between the amide proton and the C^α proton of the same amino acid residue, in units of Hertz (Hz); p.p.m., parts per million; DISGEO, name of distance geometry program; RMS, root-mean-square; RMSD, root-mean-square co-ordinate difference; DHAD, dihedral angle difference.

0022–2836/85/060295–21 $03.00/0

© 1985 Academic Press Inc. (London) Ltd.

these computations with the crystal structure provided an estimate of how well a protein spatial structure can be determined from the geometric constraints which are available from n.m.r. These results are used here to assess the quality of the solution conformations which we have computed for BUSI IIA, for which no crystal structure is available. Further evidence for the correctness of the structure determined by n.m.r. is obtained by comparison with the crystal structures of two homologous proteins, porcine secretory inhibitor (Bolognesi *et al.*, 1982), and the third domain of the Japanese quail ovomucoid (Papamokos *et al.*, 1982).

The structural features of BUSI IIA elucidated by this study are the direct result of distance geometry calculations using geometric constraints derived exclusively from its primary structure and from the n.m.r. experiments. In the Appendix, we describe preliminary results from an energy minimization by which the side-chain packing and local structure were further improved, and which demonstrates the existence of low energy conformations that are consistent with the experimental data.

2. Materials and Methods

(a) *n.m.r. measurements*

BUSI IIA, purified by the method of Čechová *et al.* (1979), was obtained as a gift from Drs D. Čechová and P. Štrop. For the n.m.r. experiments, approx. 0·016 M solutions of BUSI IIA were prepared, both in ^2H_2O and in a mixture of 90% H_2O and 10% ^2H_2O, at pH 5·3. No buffer was used, and the spectra were run at 45°C. For the ^2H_2O spectra, the labile protons of the protein were replaced by ^2H by heating the solution to 40°C for 20 min at neutral p^2H, and the residual water protons were removed by lyophilization from ^2H_2O.

NOESY spectra (Anil Kumar *et al.*, 1980) were recorded at 500 MHz on a Bruker WM 500 spectrometer equipped with an Aspect 2000 computer and a Control Data Corp. disk drive. To reduce contributions from coherent magnetization transfer to the cross peak intensities, mixing times shorter than 200 ms were randomly modulated with a modulation amplitude of ± 5 to 10 ms, depending on the mixing time (Macura *et al.*, 1981). For measurements in H_2O the solvent resonance was suppressed by selective, continuous irradiation at all times except during data acquisition (Wider *et al.*, 1983). Absorption mode spectra were obtained as described (Williamson *et al.*, 1984), using the time proportional phase incrementation scheme (Redfield & Kunz, 1975; Bodenhausen *et al.*, 1980; Marion & Wüthrich, 1983).

A high resolution NOESY spectrum in H_2O was acquired with 4096 points in t_2 and 800 points in t_1, with a mixing time τ_m of 200 ms. It was processed as described by Williamson *et al.* (1984), to give a digital resolution of 2·4 Hz in ω_1 and 0·6 Hz in ω_2. Lower resolution NOESY spectra with τ_m values of 50 ms, 80 ms, 100 ms and 120 ms had 1024 points in t_2 and 512 points in t_1. The time domain data matrix was expanded to 1024 points in t_1 by zero filling, and the window function used in both directions was a Lorentz-to-Gaussian transformation with a line-broadening factor of -1 and a maximum at 0·3 t_2. This window function was used as it was felt to be the least discriminatory for different linewidths, consistent with eliminating most of the t_1 noise and the dispersion parts of slightly incorrectly phased peaks (Ferrige & Lindon, 1978). Similarly, a high resolution NOESY spectrum with $\tau_m = 150$ ms and several lower resolution spectra with shorter mixing times were acquired in ^2H_2O.

(b) *^1H–^1H distance constraints derived from NOESY spectra*

In the high resolution NOESY spectra of BUSI IIA recorded in H_2O and in ^2H_2O, 556 cross peaks were assigned with the help of the previously reported chemical shifts (Štrop *et al.*, 1983). Distance constraints were then derived from 202 of these cross peaks, using NOESY spectra recorded with lower digital resolution and a shorter mixing time (τ_m) in order to avoid spin diffusion (Kalk & Berendsen, 1976; Gordon & Wüthrich, 1978; Anil Kumar *et al.*, 1981). Among the assigned cross peaks that were not used for the structure determination are the majority of the intraresidue NOEs. Many of these manifest ^1H–^1H distances which are fixed by the covalent structure, and others are not informative at the present level of quantification. Most of the remaining unused NOEs are sequential NOEs between $C^{\beta}H_i$ and NH_{i+1}. They were not used because the absence of stereospecific assignments (Wüthrich *et al.*, 1983) made them uninformative. Roughly 20 additional NOEs could be resolved but not assigned; by comparing these to expected close proton–proton contacts in the computed structures, we conclude that most of them probably arise from unassigned C^{β} and C^{γ} protons.

For the calibration of correlations between NOESY cross peak heights and ^1H–^1H distance constraints we proceeded in a manner similar to that used by Braun *et al.* (1983). Provided that a specified cross peak could be seen at high digital resolution to be sufficiently well-resolved from all other peaks, the height at the centre of this peak was measured in the spectra recorded with $\tau_m = 50$ ms, 80 ms, 100 ms and 120 ms, and the peak heights plotted *versus* τ_m. To check that the observed peak intensities were due to direct NOEs (Anil Kumar *et al.*, 1981), a smooth curve was drawn through these points. Subsequently, the NOE build-up was assumed to be linear up to 80 ms, and the peak heights at this value of τ_m were used to determine relative initial NOE build-up rates for the individual pairs of dipolar-coupled protons.

For connectivities between protons separated by more than 3 single bonds in the covalent structure, differences between the apparent build-up rates for different pairs of protons were not further interpreted. All such "long-range" connectivities between backbone amide and/or C^{α} protons which could be shown to arise from direct NOEs and which had a sufficiently high intensity to be unambiguously identified in the spectrum with $\tau_m = 80$ ms were attributed a distance range from the van der Waals' contact of 2·0 Å to an upper bound of 4·0 Å. Similarly, long-range connectivities which also involved side-chain protons were attributed a distance range from 2·0 to 5·0 Å. The different upper bounds for these 2 classes of constraints were chosen on the basis of the implications from a uniform averaging model (Braun *et al.*, 1981) for the influence of internal mobility on the observed NOESY cross peaks, assuming a somewhat higher overall flexibility for the amino acid side-chains than for the backbone segments.

For the intraresidue and sequential connectivities between amide, C^{α} and C^{β} protons with at most 3 intervening torsion angles about C–C or C–N single bonds, a semiquantitative relation between NOESY cross peak intensities and upper bounds on the corresponding distances was established on the basis of empirical calibrations using the following known sequential distances (Billeter *et al.*, 1982): $d_{\alpha N} = 2{\cdot}3$ Å in β-sheets; $d_{NN} = 2{\cdot}8$ Å in α or 3_{10} helices; $(d_{\alpha N})_{max} = 3{\cdot}6$ Å; $d_{NN} = 4{\cdot}3$ Å in β-sheets. For example, all cross peaks corresponding to $d_{\alpha N}$ distances in β-sheets (2·3 Å) had a relative intensity of greater than 150. An upper bound of 2·5 Å was therefore set for all cross peaks of intensity greater than 150. The resulting calibration of the relative cross peak intensities *versus* upper bound constraints is given in Table 1.

NOEs to hydrogen atoms which could not be stereospecifically assigned were referred to pseudo atom positions, using the previously reported corrections for the experimental NOE constraints (Wüthrich *et al.*, 1983).

(c) *Additional geometric constraints derived from NOESY and other n.m.r. experiments*

As described in Results, the location of the 3 disulfide bonds in BUSI IIA could be unambiguously identified from the observed NOEs. The presence of these disulfide bonds was then fixed directly by imposing a range of 2·0 to 2·1 Å on the S–S distance, and of 3·0 to 3·1 Å on the S–C^β distances across each bridge.

The locations of backbone amide proton to carbonyl oxygen hydrogen bonds in certain positions of the regular secondary structures could be inferred from the observed NOEs and the slow amide proton exchange rates (Wagner *et al.*, 1981; Williamson *et al.*, 1984). These hydrogen bonds were defined by specifying a range of 1·8 to 2·0 Å for the H–O distance, and 2·7 to 3·0 Å for the N–O distance, which forces the hydrogen bonds to be approximately linear. Other hydrogen bonds between the backbone amide proton and the γ-carbonyl oxygen of certain glutamic acid residues were identified from pH titrations of the amide proton chemical shifts (Bundi & Wüthrich, 1979; Ebina & Wüthrich, 1984). These were fixed by imposing the same constraints as used for the backbone hydrogen bonds on the distances between the amide proton, and the nitrogen and carbonyl oxygen atoms.

Table 1

Calculated correlations between intensity of the NOESY cross peaks manifesting sequential or intra-residue connectivities between amide, C^α and C^β protons in BUSI IIA and upper bounds on the ^1H–^1H distance constraints

Protons connected†	I_{rel}‡	Upper distance bound§ (Å)
NH, C^αH	>150	2·5
NH, C^βH	60–150	3·0
C^αH, C^βH	20–60	4·0
NH, NH	>100	3·0
	20–100	4·0

† Different calibrations were used for cross peaks involving only amide protons and those involving C^αH or C^βH, since these have characteristically different fine structures.

‡ Relative cross peak heights in the NOESY spectra recorded with a mixing time of 80 ms, in arbitrary units.

§ The lower bound on the distance range is in all cases the van der Waals' distance of 2·0 Å.

Spin–spin coupling constants were measured in phase-sensitive, high resolution COSY spectra (Marion & Wüthrich, 1983; Williamson *et al.*, 1984). An unambiguous interpretation in terms of the dihedral angle can be made only for extremely small or extremely large couplings (Pardi *et al.*, 1985). Accordingly, the ϕ angles corresponding to $^3J_{HN\alpha} \geq 8{\cdot}0$ Hz were constrained in the range of $-160°$ to $-80°$, and the ϕ angles corresponding to $^3J_{HN\alpha} \leq 5{\cdot}5$ Hz were constrained in the range of $-90°$ to $-40°$. In addition, the χ^1 angles of Val, Ile and Thr corresponding to $^3J_{\alpha\beta} \geq 8{\cdot}0$ Hz were confined to a range of $\pm 30°$ about the *trans* position of C^αH and C^βH. These dihedral angle ranges were imposed using chirality constraints, as described by Havel & Wüthrich (1985).

(d) *Geometric constraints derived from the primary structure*

Besides the n.m.r. constraints described above, the constraints imposed by the primary structure of the polypeptide play an important role in the distance geometry calculations. In the present work, the ECEPP amino acid geometries (Némethy *et al.*, 1983) were used to obtain the bond lengths and geminal distances which determine bond angles. Chirality constraints (Havel *et al.*, 1983; Havel & Wüthrich, 1985) were used to fix the chiralities of the amino acids and the planarity of all sp^2 centers. The atomic hard sphere radii described by Wüthrich *et al.* (1983) were used for all atoms, except between pairs of vicinal atoms, where the van der Waals' repulsion was ignored to facilitate the convergence of the distance geometry program (Havel *et al.*, 1983; Havel & Wüthrich, 1985). The total number of constraints thus obtained exceeds by far the minimum number of $3N-6$ constraints needed to define a structure. However, as discussed by Havel & Wüthrich (1985), the exact number of constraints available cannot be related in any simple way to the precision with which the structure is determined.

(e) *Analysis of the computed structures*

For the calculation of the spatial molecular structure, the DISGEO distance geometry program (Havel & Wüthrich, 1984) was used with 2 different sets of experimental input data. The calculations were repeated 5 times for each of these data sets, using different initial values for the random number seed, to obtain a total of 10 different conformers each consistent with its respective input data. These conformers were then compared with one another in a variety of ways in order to decide to what extent specific structural features were uniquely determined by the experimental data (Havel *et al.*, 1979; Braun *et al.*, 1983; Havel & Wüthrich, 1985).

First, the global arrangements of the polypeptide chain in the 5 conformers were compared using the root-mean-square co-ordinate deviation (RMSD) between all pairs of spatial structures. The RMSD is given by the minimum over all rotations about the superimposed centroids, $\mathbb{R}$, of the RMS difference in the n atomic co-ordinates between the 2 conformers (Nyburg, 1974):

$$\text{RMSD} = \left[\frac{1}{n}\min_{\mathbb{R}}\sum_{i=1}^{n}\|\vec{r}_i - \mathbb{R}\vec{r}_i'\|^2\right]^{1/2}, \qquad (1)$$

where $\vec{r}_i$ and $\vec{r}_i'$ are vectors from the centroids to the ith atom of each of the 2 conformers. When computed with respect to the α-carbon positions alone, this quantity will be given the superscript a, and when computed *versus* the positions of all heavy atoms (i.e. non-hydrogen and non-pseudo atoms), it will be given the superscript h. Drawings of the superpositions of the structures obtained by minimizing the RMS co-ordinate difference are also used to visualize the differences between the computed structures.

Further comparisons are necessary to detect differences between conformers which are local with respect to the polypeptide chain. Thus, RMSD values were also computed between corresponding tetrapeptide segments along the chain (Havel & Wüthrich, 1985). To compare the conformations of corresponding pairs of individual amino acid residues in 2 different computed structures, RMS differences (modulo 180°) in their dihedral angles (DHAD) were evaluated. In the case of the backbone angles ϕ and ψ, the DHAD value between 2 conformers is given by:

$$\mathrm{DHAD} = \left[\frac{1}{m}\sum_{i=1}^{m}\frac{1}{2}((\phi_i-\phi_i')^2+(\psi_i-\psi_i')^2)\right]^{1/2}, \quad (2)$$

where the primed and unprimed quantities refer to the 2 structures being compared.

3. Results

(a) *The experimental data used as input for the distance geometry calculations*

An illustration of the NOESY experiments which provided the distance constraints used to determine the spatial structure of BUSI IIA is presented in Figures 1 to 3. Figure 1 shows an expanded plot of a small region of a phase-sensitive NOESY spectrum of BUSI IIA, which was recorded with high digital resolution. Using NOESY spectra of this quality, where the resolved fine structure enables unambiguous identification of cross peaks even when they have nearly identical chemical shifts, 556 NOEs were identified. Figures 2 and 3 show data on the build-up rates of NOESY cross peaks, as they were collected from spectra recorded at lower resolution. Analysis of the complete set of NOESY spectra recorded with variable τ_m in H_2O and ^2H_2O using the procedures described in Materials and Methods, section (b), resulted in the distance constraints presented in Table 2. This Table was laid out so that the density of distance constraints on specified parts of the polypeptide chain becomes readily apparent. Local constraints on the backbone conformation appear in column 1, local constraints on the conformation of individual amino acid side-chains appear in column 2, medium and long-range constraints on the backbone in column 3, and long-range constraints affecting both backbone and side-chains in column 4. The previously described system of pseudo structures for amino acid residues (Wüthrich *et al.*, 1983) was used to obtain proper reference points for distance constraints to groups of protons which could not be stereospecifically assigned. An overview of the locations in the amino acid sequence of BUSI IIA which are connected by the different types of NOE distance constraints listed in Table 2 is presented by the diagonal plot of Figure 4.

The DISGEO distance geometry program used here has been described and tested in previous publications (Havel & Wüthrich, 1984, 1985). It accepts input in the form of lower and upper bounds on the interatomic distances (distance constraints) and lower and upper bounds on the oriented volumes spanned by sets of four atoms (chirality constraints: Havel *et al.*, 1983). These constraints were obtained both from the primary structure of the protein and from the n.m.r. experiments. The n.m.r. data consisted primarily of those listed in Table 2. However, some qualitative features of the polypeptide conformation can be readily derived by inspection of Table 2 and Figure 4, and use of this information can facilitate the computations considerably.

First, a qualitative analysis of the NOESY spectra provided the information needed to establish the presence of the disulfide bridges Cys7–Cys39, Cys17–Cys36 and Cys25–Cys57. For example, there are NOEs between Ala8 NH and Cys39 $C^{\beta}H_2$, Cys17 $C^{\beta}H_2$ and Cys36 NH, and between Thr31 $C^{\gamma}H_3$ and Cys57 NH (Thr31 forms a β-bridge with Cys25: Table 2). The only NOE which could be taken to indicate the existence of an alternative disulfide bridge is between $C^{\alpha}H$ of Cys36 and NH of Cys39. However, these two residues are in a regular α-helix (Williamson *et al.*, 1984), and could not be linked by a disulfide bridge without severe distortion of the helix. The disulfide linkages 7–39, 17–36 and 25–57 are also consistent with those in numerous homologous proteins (Kato *et al.*, 1976). These disulfide bonds were fixed by imposing the distance constraints described in Materials and Methods, section (c).

Secondly, *cis* and *trans* peptide bonds can be distinguished from the sequential NOEs (Arseniev *et al.*, 1983; Wüthrich *et al.*, 1985). Using the previously established criteria, we concluded from the short sequential distance $d_{\alpha\alpha}$ between residues Asp12 and Pro13 that they are linked by a *cis* peptide bond. Similarly, the short contact from $C^{\alpha}H$ of Asn22 to $C^{\delta}H_2$ of Pro23 shows that these two residues are linked by a *trans* peptide group. Direct evidence for a *trans* conformation in all other peptide bonds comes from the observed sequential NOEs, except for the segments from residues 1 to 2, 3 to 6 and 31 to 32, where no data on the sequential connectivities was obtained (Table 2). For the preparation of the input we assumed that with the exception of Asp12–Pro13, all peptide bonds were *trans*.

The first set of structure calculations used only the NOE constraints listed in Table 2, the locations of the disulfide bonds given above, and these *cis–trans* isomers of the peptide bonds as input (i.e. the NOE data set, N). A second set of structure calculations used all the constraints of the data set N together with certain supplementary experimental constraints as input (i.e. the data set complemented with supplementary constraints, C). These supplementary constraints were derived from the locations of hydrogen bonds in the secondary structures, from the vicinal spin–spin couplings

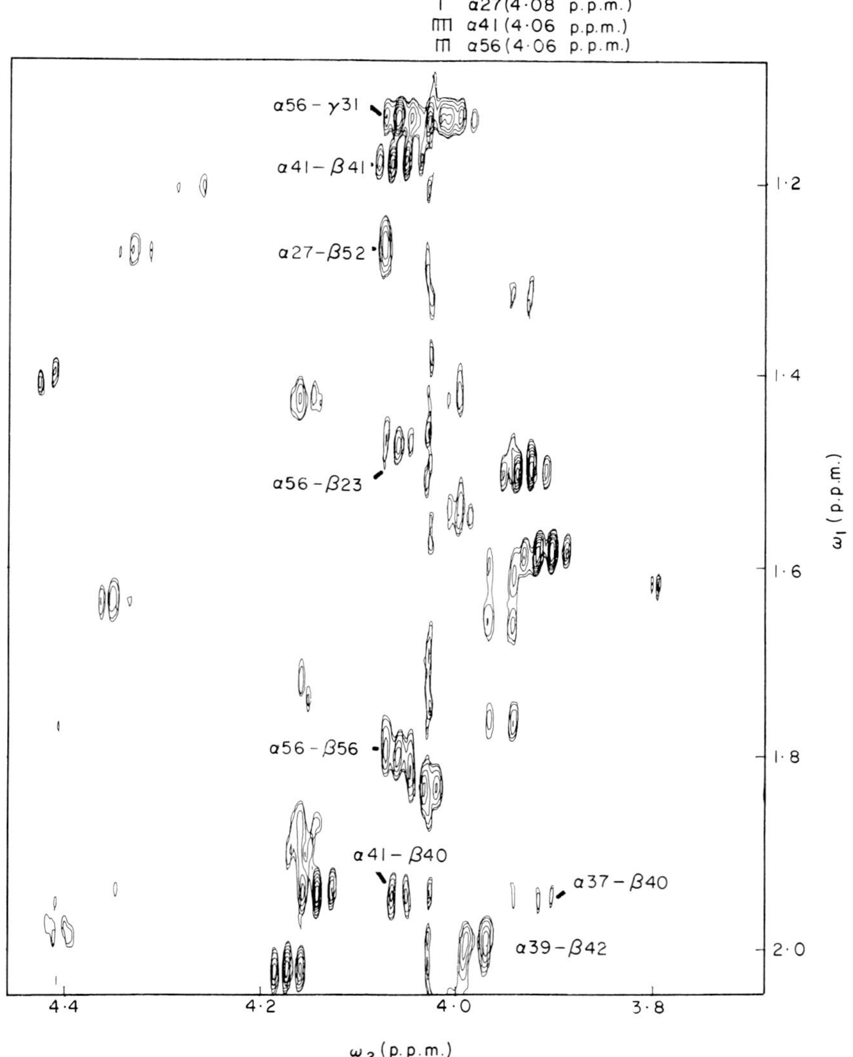

Figure 1. High-resolution absorption mode NOESY spectrum recorded at 500 MHz in a 0·016 M-solution of BUSI IIA in ^2H_2O, p^{2}H = 5·3, t = 45°C, τ_m = 150 ms. The region (ω_1 = 1·1 to 2·0 p.p.m., ω_2 = 3·7 to 4·4 p.p.m.) is shown. The digital resolution after zero filling was 3·7 Hz in ω_1 and 0·45 Hz in ω_2. Some of the cross peaks have been assigned to the pair of protons which are connected by the corresponding NOE, and the first proton listed is the one of which the chemical shift is given by ω_2. For the C$^\alpha$ protons of S27, A41 and K56 the ω_2 positions and the multiplet fine structures are indicated at the top of the Figure. Although the 3 resonances fall within 0·03 p.p.m. of each other all NOEs involving these protons could be readily assigned by the spin–spin splitting observable in ω_2. All of the peaks in this region for which the assignments have not been shown here arise from intraresidue NOEs.

$^3J_{HN\alpha}$ (Williamson *et al.*, 1984), and from $^3J_{\alpha\beta}$, as given in Figure 5. Only those spin–spin couplings which could be unambiguously interpreted in terms of the dihedral angle, and those hydrogen bonds simultaneously implied by the presence of at least two NOEs and by slow amide proton exchange rates (Wüthrich *et al.*, 1984), were used for the supplementary constraints. Figure 5 provides a survey of the sequence locations affected by these constraints, which also include hydrogen bonds between the backbone amide protons and the side-chain carboxyl groups of Glu9 and Glu20 (Ebina & Wüthrich, 1984). The geometric conditions by which these supplementary constraints were

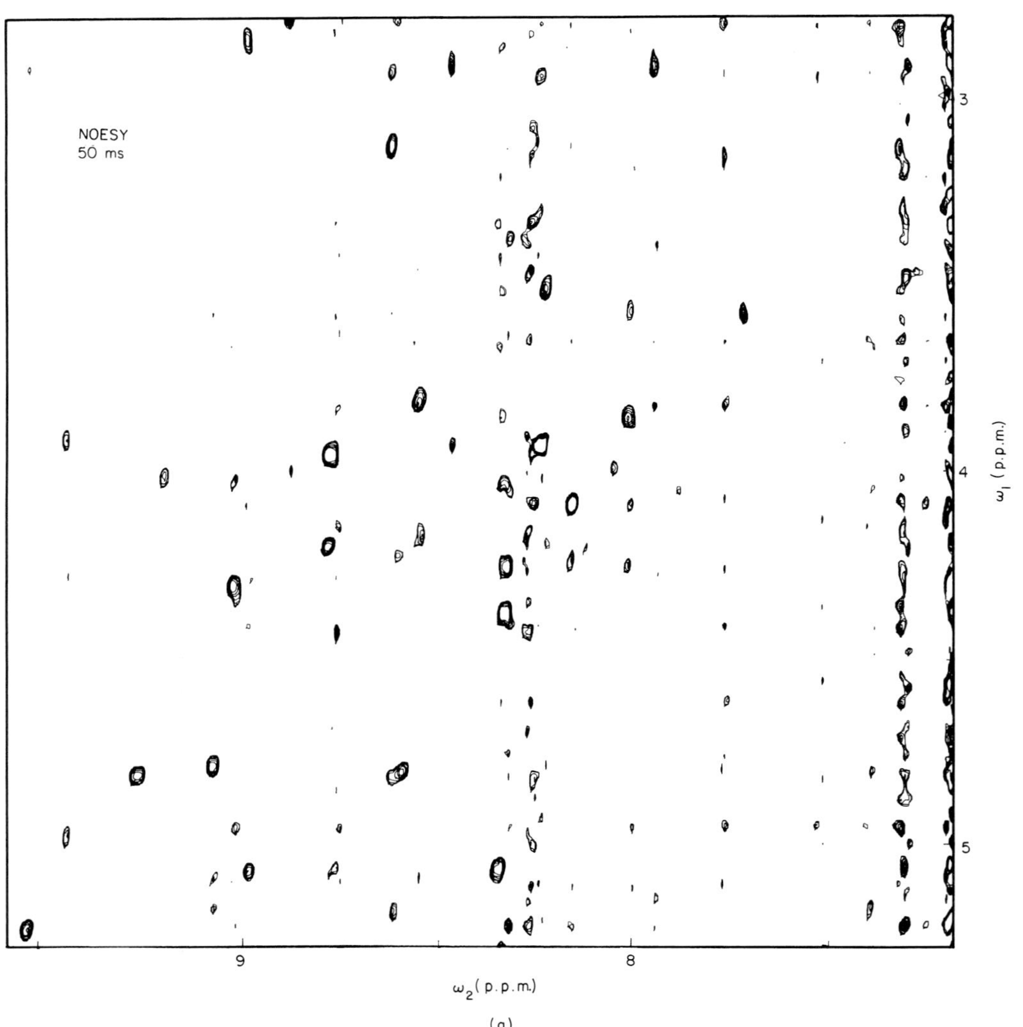

Figure 2. Low-resolution (see the text) absorption mode NOESY spectra recorded at 500 MHz in a 0·016 M-solution of BUSI IIA in H_2O, pH 5·3, $t = 45°C$, with the mixing times indicated in Figures. The spectral region $\omega_1 = 2{\cdot}7$ to 5·3 p.p.m., $\omega_2 = 7{\cdot}2$ to 9·6 p.p.m. is shown. The digital resolution after zero filling is 4·8 Hz in both directions. Approximately 120 NOE cross peaks were resolved and assigned in this region of the spectrum. Of these, roughly 80 were intense enough and sufficiently well-resolved to be used for measurements of the τ_m dependence of the NOE intensities (see Fig. 3).

actually imposed on the computed structures have been described in Materials and Methods, section (c).

(b) *Spatial structures of BUSI IIA computed from the n.m.r. data*

In all, ten random structures† were computed using the two sets of geometric constraints as input, as described in the preceding section. The five conformers N1 to N5 computed from data set N (above) are shown in Figure 6(a), and the five conformers C1 to C5 computed from the data set C are shown in Figure 6(b). The schematic drawing of the main features of the BUSI IIA structure (Fig. 7) may be helpful as a guide to Figure 6.

The convergence of the DISGEO program with

† These random structures were obtained by embedding random sets of distances, as described by Havel & Wüthrich (1984, 1985).

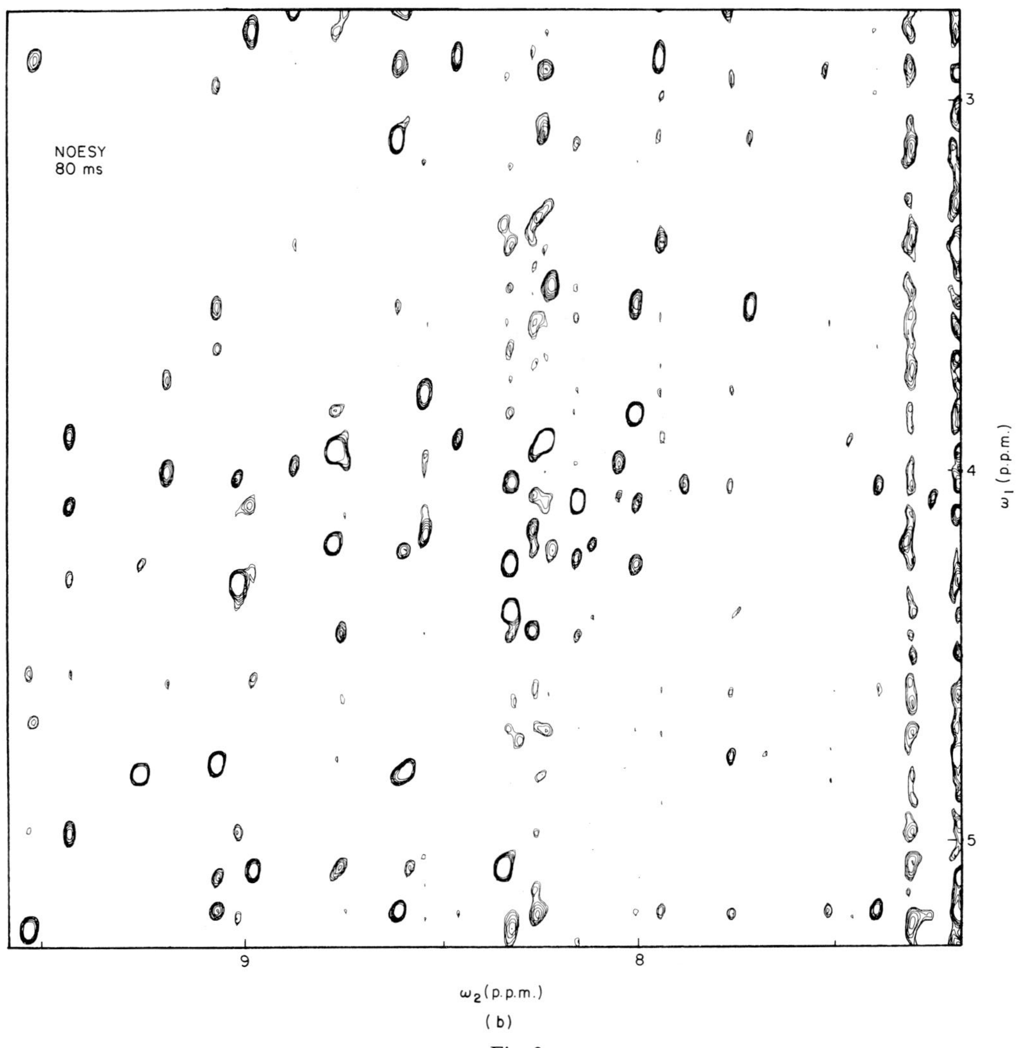

(b)

Fig. 2

both input data sets N and C was comparable to the experience gained with similar data sets which had been simulated from the crystal structure of BPTI (Havel & Wüthrich, 1985). This implies that there are no serious inconsistencies in the geometric constraints used as input. In particular, there were no significant violations of the chirality constraints in any of the ten structures. Violations of covalent bond lengths by more than 0·05 Å were rare, as were van der Waals' violations of more than 0·1 Å. Peptide bonds were always within 3° of planarity. Figure 8 shows histograms in 0·1 Å increments of the violations of all of the input distance constraints found in each of the ten structures. In the best case, C5, there were only three violations exceeding 0·2 Å, and in the worst case there were 24 such violations. No distance constraints exceeding 1·0 Å were found in any of the computed structures, and the size of the largest violation was about 0·5 Å on average.

A list of all pairs of backbone protons that were less than 5·0 Å apart in all five of the structures computed for each data set was prepared, and the NOESY spectra checked for the presence of the corresponding NOEs. Only two of these predicted NOEs were not observed, one of which involves the $C^{\alpha}H$ of Thr31, which resonates at the same frequency as the water signal and was consequently bleached out by the water irradiation. The other is an NOE across the β-sheet between the $C^{\alpha}H$ of Cys25 and the NH of Tyr32, which was probably too weak to be observed.

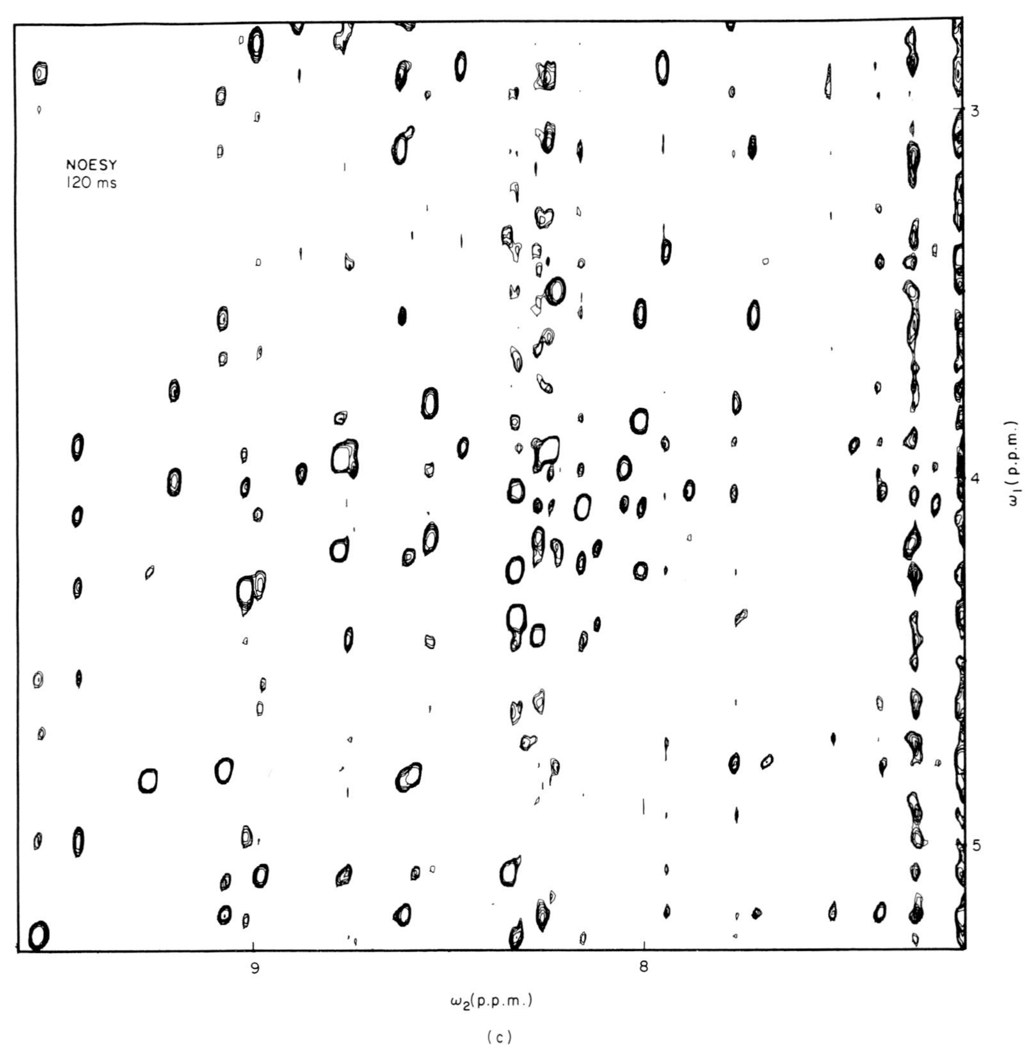

(c)

FIG. 2.

4. Discussion

The input for the present distance geometry calculations of the solution conformation of BUSI IIA consisted exclusively of geometric constraints determined by n.m.r. experiments or imposed by the primary structure of the protein. The search for spatial molecular structures compatible with these constraints started without any additional assumptions on the structures to be found, nor was any assumption made as to the existence of a single unique solution (Braun *et al.*, 1981; Havel & Wüthrich, 1984). The BUSI IIA conformers obtained from different computations with the same geometric constraints are therefore to be regarded as a random selection from the ensemble of all conformations of BUSI IIA which are consistent with the experimental data. It may be added that this method is suitable for studies of "uniquely structured" polymers, in the sense that the atoms are confined to discrete spatial locations about which they undergo limited thermal motions (see Braun *et al.*, 1981). In systems where multiple discrete conformers would be substantially populated in a rapid exchange situation, the structures obtained with the present analysis of the NOE intensities would not necessarily correspond to conformations contributing to the true equilibrium.

Since the geometric constraints imposed by the experimental data were generally rather imprecise,

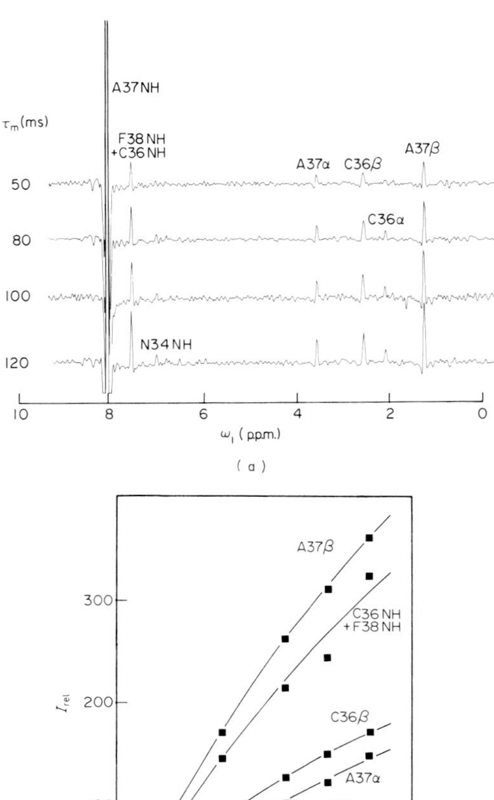

Figure 3. Build-up of NOEs with τ_m, using spectra from the same series as Fig. 2. (a) Cross-sections parallel to ω_1 at the chemical shift of the amide proton of A37 (ω_2 = 8·45 p.p.m.). The peaks have been identified by the IUPAC-IUB one-letter symbol for amino acids, the position in the amino acid sequence, and the proton type involved. (b) Plot of the height of the peaks in (a) against τ_m. The vertical scale is arbitrary, since the cross peak heights were only measured relative to each other.

a certain amount of variation between the computed structures was to be expected. Nevertheless, inspection of the superpositions of the backbones of the individual conformations (Fig. 6) shows that the structures computed from each data set have many features in common. In the following, we first describe the global features of the BUSI IIA solution conformation. We then examine the high resolution crystal structures of two homologous proteins, and quantitatively compare them with the conformers of BUSI IIA computed from the n.m.r. data. Finally, the conformers of BUSI IIA are compared with one another in order to decide what features of the conformation are uniquely determined by them.

(a) *The solution conformation of BUSI IIA*

The elements of regular secondary structure, shown schematically in Figure 7, include a segment of distorted 3_{10}-helix from residues 8 to 11, a quite regular α-helix from residues 34 to 45, and a triple-stranded, antiparallel β-sheet with strands extending from residues 23 to 27, from 29 to 33 and from 50 to 55, with a β-bulge formed by residues 52 and 55. The locations of these regular secondary structure elements along the amino acid sequence coincide with the results of earlier work, where the sequence locations of the helical and β-sheet structures were determined by an empirical approach relying on visual inspection of the n.m.r. data (Williamson *et al.*, 1984). It is particularly pleasing to observe this good agreement with the earlier predictions even in the structures N1 to N5, which were calculated from the NOE distance constraints alone (Table 2). Among the five structures C1 to C5 computed with the inclusion of hydrogen bond and ϕ angle constraints (Fig. 5), the α-helix and the strand of the β-sheet from residues 23 to 27 show a lesser spatial dispersion than among N1 to N5 (Fig. 6).

For both sets of conformers, N1 to N5 and C1 to C5, the molecule is roughly an oblate ellipsoid in shape, with maximum dimensions of about 25 Å × 25 Å × 15 Å, and average radii of gyration of the C^α positions about the three principal axes equal to 11·7 Å, 11·5 Å and 9·7 Å. The variations in these radii of gyration between the different computed structures were less than 5%. The topology of the regular secondary structure elements and the other regions of the polypeptide chain is also very similar in all ten of the computed conformers. In the orientation of the protein chosen in Figures 6 and 7, the polypeptide segment 5–22 describes a large loop on the right. Except for the distorted helical segment near residue 10, this loop consists of irregular random coil secondary structure, and also includes a *cis* peptide bond between residues 12 and 13. The loop is held against the α-helix from residues 34 to 45 by the disulfide bonds 7–39 and 17–36, as well as by several long-range NOEs (Table 2). In contrast, there are only three NOE contacts between this loop and the β-sheet, all of which involve the side-chain of Val5 (Table 2). Figures 6 and 7 indicate that there could be additional contacts between the N-terminal residues 1 to 4 and the β-sheet. Since only a single sequential NOE was observed in this region, a likely explanation for the absence of further NOEs is that this segment is rather mobile in the solution conformation.

In the triple-stranded β-sheet, the antiparallel alignment of the three strands is unambiguously

Table 2

Distance constraints (Å) between groups of hydrogen atoms in the three-dimensional structure of BUSI IIA obtained from NOESY spectra in H_2O and ^2H_2O

Sequential	Intraresidue	Long-range backbone	Long-range
Pyr1			
Gly2			
HN A3 HN 4·0			
Ala3			
HN G2 HN 4·0			
Gln4			
Val5			QG H24 H4 7·4 *q* (1·01, 6·12)
			QG F38 QR 9·8 *qq* (1·01, 6·87)
			QG L51 HN 7·4 *q* (1·01, 8·95)
			QG L51 QD 9·8 *qq* (1·01, 0·67)
Asp6			
HA C7 HN 3·0			
Cys7			
HN D6 HA 3·0		HA F10 HN 4·0 (5·16, 7·64)	HA F10 C1 7·0 *r* (5·16, 7·16)
HA A8 HN 3·0			
HN A8 HN 4·0			
Ala8			
HN C7 HA 3·0	HN A8 MB 3·6 *i* (8·23, 1·51)	HA K11 HN 4·0 (3·93, 7·36)	
HN C7 HN 4·0		HN C39 PB 6·0 *m* (8·23, 2·71)	
HN E9 HN 4·0			
Glu9			
HN A8 HN 4·0	HN E9 PB 3·6 *i* (8·95, 1·57)		PB F10 C4 8·0 *mr* (1·41, 7·27)
HN F10 HN 4·0	(8·95, 1·41)		
Phe10			
HN E9 HN 4·0		HN C7 HA 4·0 (7·64, 5·16)	C1 C7 HA 7·0 *r* (7·16, 5·16)
HN K11 HN 3·0			C4 E9 PB 8·0 *rm* (7·27, 1·41)
			HA V15 QG 7·4 *q* (4·73, 1·04)
			C1 V15 QG 9·4 *rq* (7·16, 1·04)
			C4 V15 QG 9·4 *rq* (7·27, 1·04)
			C1 C36 HA 7·0 *r* (7·16, 2·44)
			C4 C36 HA 7·0 *r* (7·27, 2·44)
			C4 C36 HN 7·0 *r* (7·27, 7·92)
			C4 C36 PB 8·0 *rm* (7·27, 2·93)
Lys11			
HN F10 HN 3·0	HN K11 PB 3·6 *i* (7·36, 1·83)	HN A8 HA 4·0 (7·36, 3·93)	
HN D12 HN 3·0	(7·36, 1·51)		
Asp12			
HN K11 HN 3·0	HN D12 PB 3·6 *i* (7·71, 2·78)	PB V15 HN 6·0 *m* (2·57, 7·43)	HN V15 QG 7·4 *q* (7·71, 1·04)
HA P13 HA 4·0	(7·71, 2·57)		PB V15 QG 8·4 *mq* (2·78, 1·04)
Pro13			
HA D12 HA 4·0			
Lys14			
HN V15 HN 3·0	HN K14 PB 3·6 *i* (8·51, 1·79)		HN V15 QG 7·4 *q* (8·51, 1·04)
	(8·51, 1·43)		
Val15			
HN K14 HN 3·0	HN V15 HB 3·0 (7·43, 2·12)	HN D12 PB 6·0 *m* (7·43, 2·57)	QG F10 HA 7·4 *q* (1·04, 4·73)
HN Y16 HN 4·0			QG F10 C4 9·4 *qr* (1·04, 7·27)
HA Y16 HN 3·0			QG F10 C1 9·4 *qr* (1·04, 7·16)
			QG D12 HN 7·4 *q* (1·04, 7·71)
			QG D12 PB 8·4 *qm* (1·04, 2·78)
			QG K14 HN 7·4 *q* (1·04, 8·51)
			QG Y16 HN 7·4 *q* (1·04, 8·19)
Tyr16			
HN V15 HN 4·0	HN Y16 PB 3·6 *i* (8·19, 3·06)		HN V15 QG 7·4 *q* (8·19, 1·04)
HN V15 HA 3·0	(8·19, 2·93)		C1 C17 PB 8·0 *rm* (7·15, 2·80)
Cys17			
HA T18 HN 2·5	HN C17 PB 3·6 *i* (8·58, 3·35)	PB C36 HN 6·0 *m* (3·35, 7·92)	PB Y16 C1 8·0 *mr* (2·80, 7·15)
	(8·58, 2·80)	PB A37 HN 6·0 *m* (3·35, 8·45)	
Thr18			
HN C17 HA 2·5			
Arg19			
HN E20 HN 4·0			
Glu20			
HN R19 HN 4·0			HN N34 ND 6·0 *m* (8·07, 7·78)
HA S21 HN 2·5			PB N34 ND 7·0 *mm* (2·02, 7·78)
Ser21			
HN E20 HA 2·5			PB H24 H4 6·0 *m* (3·93, 6·12)
HA N22 HN 2·5			
Asn22			
HN S21 HA 2·5		HN N34 HA 4·0 (9·05, 5·13)	ND H24 H2 6·0 *m* (7·76/7·01, 7·53)
HA P23 PD 4·0 *m*		HN N34 PB 6·0 *m* (9·05, 3·55/3·16)	ND K35 HN 6·0 *m* (7·76, 8·60)
HN P23 PD 5·0 *m*		PB N34 HA 6·0 *m* (2·96, 5·13)	
		PB K35 HN 6·0 *m* (3·02, 8·60)	
Pro23			
PD N22 HA 4·0 *m*		HA G33 PA 5·0 *m* (4·78, 3·75)	HA T31 MG 6·0 *m* (4·78, 1·13)
PD N22 HN 5·0 *m*		PB K56 HA 6·0 *m* (1·48, 4·06)	PB T31 MG 7·0 *mm* (2·14, 1·13)
HA H24 HN 2·5			
His24			
HN P23 HA 2·5		HN Y32 HN 4·0 (9·25, 9·50)	H4 V5 QG 7·4 *q* (6·12, 1·01)
HA C25 HN 2·5		HA R54 HA 4·0 (4·22, 3·83)	H4 S21 PB 6·0 *m* (6·12, 3·93)
		HA G55 HN 4·0 (4·22, 8·00)	H2 N22 ND 6·0 *m* (7·53, 7·76/7·01)
			HN T31 MG 6·0 *m* (9·25, 1·13)
			H2 N34 HA 5·0 (7·53, 5·13)
			H2 K35 HA 5·0 (7·53, 2·93)
			H2 K35 HN 5·0 (7·53, 8·60)
			H2 K35 PB 6·0 *m* (7·53, 1·72)
			H4 F38 QR 7·4 *q* (6·12, 6·87)
			HA L51 QD 7·4 *q* (4·22, 0·19)

Table 2 *(continued)*

Sequential	Intraresidue	Long-range backbone	Long-range
			H4 L51 QD 7·4 *q* (6·12, 0·67/0·19)
			PB L51 QD 8·4 *mq* (2·31, 0·19)
Cys25			
HN H24 HA 2·5	HN C25 PB 3·6 *i* (8·29, 2·57)	HA T31 HA 4·0 (5·19, 4·65)	HN L51 QD 7·4 *q* (8·29, 0·19)
HA G26 HN 2·5	(8·29, 1·85)	HN H53 HN 4·0 (8·29, 7·06)	
		HN R54 HA 4·0 (8·29, 3·83)	
Gly26			
HN C25 HA 2·5		HN E30 HN 4·0 (9·53, 7·72)	PA F38 C4 8·0 *mr* (4·94, 7·03)
		PA L51 HA 5·0 *m* (4·94, 4·26)	PA I49 MG 7·0 *mm* (4·94, 0·88)
Ser27			
HN N28 HN 4·0	HN S27 PB 3·6 *i* (9·43, 4·27)	HA G29 HN 4·0 (4·08, 8·52)	PB F38 QR 8·4 *mq* (4·27, 6·87/7·03)
	(9·43, 3·89)	HN I49 HB 5·0 (9·43, 1·32)	HN I49 MG 6·0 *m* (9·43, 0·88)
		HN N50 HN 4·0 (9·43, 8·74)	
		HA K52 PB 6·0 *m* (4·08, 1·28)	
Asn28			
HN S27 HN 4·0		HN E30 HN 4·0 (8·95, 7·72)	HN I49 MG 6·0 *m* (8·95, 0·88)
HN G29 HN 3·0			
Gly29			
HN N28 HN 3·0		HN S27 HA 4·0 (8·52, 4·08)	
HN E30 HN 3·0			
Glu30			
HN G29 HN 3·0	HN E30 PB 3·6 *i* (7·72, 1·94)	HN G26 HN 4·0 (7·72, 9·53)	
HA T31 HN 3·0	(7·72, 1·65)	HN N28 HN 4·0 (7·72, 8·95)	
Thr31			
HN E30 HA 3·0	HA T31 MG 3·1 *i* (4·65, 1·13)	HA C25 HA 4·0 (4·65, 5·19)	MG P23 HA 6·0 *m* (1·13, 4·78)
			MG P23 PB 7·0 *mm* (1·13, 2·14)
			MG H24 HN 6·0 *m* (1·13, 9·25)
			MG Y32 HN 6·0 *m* (1·13, 9·50)
			MG G33 PA 7·0 *mm* (1·13, 3·75)
			MG K56 HA 6·0 *m* (1·13, 4·06)
			MG K56 HN 6·0 *m* (1·13, 7·98)
			MG K56 PB 7·0 *mm* (1·13, 1·51)
			MG C57 HN 6·0 *m* (1·13, 8·13)
Tyr32			
	HN Y32 PB 3·6 *i* (9·50, 3·02)	HN H24 HN 4·0 (9·50, 9·25)	HN T31 MG 6·0 *m* (9·50, 1·13)
	(9·50, 2·91)		QR A37 MB 8·4 *qm* (7·10, 1·58)
			PB A37 MB 7·0 *mm* (3·02, 1·58)
			PB F38 QR 8·4 *qr* (3·02, 6·87)
			QR F38 HA 7·4 *q* (7·10, 3·40)
			QR A41 MB 8·4 *qm* (7·10, 1·16)
			QR I49 MG 8·4 *qm* (7·10, 0·88)
Gly33			
HN N34 HN 4·0		PA P23 HA 5·0 *m* (3·75, 4·78)	PA T31 MG 7·0 *mm* (3·75, 1·13)
		HN A37 MB 6·0 *m* (9·18, 1·58)	
Asn34			
HN G33 HN 4·0		HA N22 HN 4·0 (5·13, 9·05)	ND E20 HN 6·0 *m* (7·78, 8·07)
HA K35 HN 2·5		HA N22 PB 6·0 *m* (5·13, 2·96)	ND E20 PB 7·0 *mm* (7·78, 2·02)
HN K35 HN 4·0		PB N22 HN 6·0 *m* (3·55/3·16, 9·05)	HA H24 H2 5·0 (5·13, 7·53)
PB K35 HN 3·5 *m*		PB C36 HN 6·0 *m* (3·16, 7·92)	
		HN A37 HN 4·0 (7·37, 8·45)	
		HN A37 MB 6·0 *m* (7·37, 1·58)	
		HN F38 HN 4·0 (7·37, 7·92)	
Lys35			
HN N34 HA 2·5		HN N22 PB 6·0 *m* (8·60, 3·02)	HN N22 ND 6·0 *m* (8·60, 7·76)
HN N34 HN 4·0		HA F38 HN 4·0 (2·93, 7·92)	HA H24 H2 5·0 (2·93, 7·53)
HN N34 PB 3·5 *m*		HA F38 PB 6·0 *m* (2·93, 2·78/2·72)	HN H24 H2 5·0 (8·60, 7·53)
HN C36 HN 3·0			PB H24 H2 6·0 *m* (1·72, 7·53)
PB C36 HN 4·0 *m*			
Cys36			
HN K35 HN 3·0		HN C17 PB 6·0 *m* (7·92, 3·35)	HA F10 C4 7·0 *r* (2·44, 7·27)
HN K35 PB 4·0 *m*		HN N34 PB 6·0 *m* (7·92, 3·16)	HA F10 C1 7·0 *r* (2·44, 7·16)
HN A37 HN 3·0		HA C39 HN 4·0 (2·44, 8·83)	HN F10 C4 7·0 *r* (7·92, 7·27)
PB A37 HN 4·0 *m*			PB F10 C4 8·0 *mr* (2·93, 7·27)
Ala37			
HN C36 HN 3·0		HN C17 PB 6·0 *m* (8·45, 3·35)	MB Y32 PB 7·0 *mm* (1·58, 3·02)
HN C36 PB 4·0 *m*		MB G33 HN 6·0 *m* (1·58, 9·18)	MB Y32 QR 8·4 *mq* (1·58, 7·10)
HN F38 HN 3·0		HN N34 HN 4·0 (8·45, 7·37)	
MB F38 HN 4·0 *m*		MB N34 HN 6·0 *m* (1·58, 7·37)	
		HA K40 HN 4·0 (3·91, 8·72)	
		HA K40 PB 6·0 *m* (3·91, 1·97)	
		MB K40 HN 6·0 *m* (1·58, 8·72)	
Phe38			
HN A37 HN 3·0	HN F38 PB 3·1 *i* (7·92, 2·78)	HN N34 HN 4·0 (7·92, 7·37)	QR V5 QG 9·8 *qq* (6·87, 1·01)
HN A37 MB 4·0 *m*	(7·92, 2·72)	HN K35 HA 4·0 (7·92, 2·93)	QR H24 H4 7·4 *q* (6·87, 6·12)
HN C39 HN 3·0		PB K35 HA 6·0 *m* (2·78/2·72, 2·93)	C4 G26 PA 8·0 *rm* (7·03, 4·94)
		HA A41 HN 4·0 (3·40, 7·23)	QR S27 PB 8·4 *qm* (6·87/7·03, 4·27)
		HA A41 MB 6·0 *m* (3·40, 1·16)	HA Y32 QR 7·4 *q* (3·40, 7·10)
			QR Y32 PB 8·4 *qm* (6·87, 3·02)
			QR C39 HN 7·4 *q* (6·87, 8·83)
			QR V42 QG 9·8 *qq* (6·87/7·03, 0·10)
			QR I49 MG 8·4 *qm* (6·87/7·03, 0·88)
			QR N50 HA 7·4 *q* (6·87, 5·03)
			QR L51 HN 7·4 *q* (6·87/7·03, 8·95)
			QR L51 QD 9·8 *qq* (6·87/7·03, 0·19)
Cys39			
HN F38 HN 3·0		PB A8 HN 6·0 *m* (2·71, 8·23)	HN F38 QR 7·4 *q* (8·83, 6·87)
HN K40 HN 3·0		HN C36 HA 4·0 (8·83, 2·44)	HA V42 QG 7·4 *q* (3·97, 0·10)
		HA V42 HB 5·0 (3·97, 2·01)	
		HA V42 HN 4·0 (3·97, 8·21)	
Lys40			
HN C39 HN 3·0	HN K40 PB 3·6 *i* (8·72, 1·97)	HN A37 HA 4·0 (8·72, 3·91)	

Table 2 *(continued)*

Sequential	Intraresidue		Long-range backbone		Long-range	
HN A41 HN 3·0		(8·72, 1·77)	HN A37 MB 6·0 *m*	(8·72, 1·58)		
PB A41 HA 6·0			PB A37 HA 6·0 *m*	(1·97, 3·91)		
			HA M43 HN 4·0	(3·95, 8·13)		
Ala41						
HA K40 PB 6·0 *m*			HN F38 HA 4·0	(7·23, 3·40)	MB Y32 QR 8·4 *mq*	(1·16, 7·10)
HN K40 HN 3·0			MB F38 HA 6·0 *m*	(1·16, 3·40)	MB I49 MD 7·0 *mm*	(1·16, 0·79)
HN V42 HN 3·0			HN M43 HN 4·0	(7·23, 8·13)		
MB V42 HN 4·0 *m*			HA K44 HN 4·0	(4·06, 8·26)		
Val42						
HN A41 HN 3·0	HN V42 HB 2·5	(8·21, 2·01)	HB C39 HA 5·0	(2·01, 3·97)	QG F38 QR 9·8 *qq*	(0·10, 6·87/7·03)
HN A41 MB 4·0 *m*			HN C39 HA 4·0	(8·21, 3·97)	QG C39 HA 7·4 *q*	(0·10, 3·97)
HB M43 HN 3·0			HA S45 HN 4·0	(3·12, 7·86)	QG M43 HA 7·4 *q*	(0·78, 4·22)
HN M43 HN 3·0			HA I49 HB 5·0	(3·12, 1·32)	QG G47 HN 7·4 *q*	(0·78, 8·19)
					QG G47 PA 8·4 *qm*	(0·78, 4·18)
					HN I49 MG 6·0 *m*	(8·21, 0·88)
					QG I49 HB 7·4 *q*	(0·78, 1·32)
					QG N50 HA 7·4 *q*	(0·10, 5·03)
Met43						
HN V42 HB 3·0	HN M43 PB 3·6 *i*	(8·13, 2·13)	HN K40 HA 4·0	(8·13, 3·95)	HA V42 QG 7·4 *q*	(4·22, 0·78)
HN V42 HN 3·0		(8·13, 2·62)	HN A41 HN 4·0	(8·13, 7·23)		
HN K44 HN 4·0						
Lys44						
HN M43 HN 4·0	HN K44 PB 3·1 *i*	(8·26, 1·95)	HN A41 HA 4·0	(8·26, 4·06)		
HN S45 HN 4·0						
Ser45						
HN K44 HN 4·0	HN S45 PB 3·6 *i*	(7·86, 4·23)	HN V42 HA 4·0	(7·86, 3·12)	HA I49 MG 6·0 *m*	(4·53, 0·88)
HN G46 HN 4·0		(7·86, 4·01)			PB I49 MD 7·0 *mm*	(4·01, 0·79)
Gly46						
HN S45 HN 4·0						
HN G47 HN 4·0						
Gly47						
HN G46 HN 4·0					HN V42 QG 7·4 *q*	(8·19, 0·78)
HN K48 HN 3·0					PA V42 QG 8·4 *mq*	(4·18, 0·78)
Lys48						
HN G47 HN 3·0	HN K48 PB 3·6 *i*	(7·29, 1·89)				
HN K48 PB 3·6 *i*		(7·29, 1·73)				
HN I49 HN 4·0						
Ile49						
HN K48 HN 4·0	HA I49 MD 4·0 *m*	(3·94, 0·79)	HB S27 HN 5·0	(1·32, 9·43)	MG G26 PA 7·0 *mm*	(0·88, 4·94)
HA N50 HN 3·0	HA I49 MG 3·5 *m*	(3·94, 0·98)	HB V42 HA 5·0	(1·32, 3·12)	MG S27 HN 6·0 *m*	(0·88, 9·43)
	HB I49 MD 3·5 *m*	(1·32, 0·79)			MG N28 HN 6·0 *m*	(0·88, 8·95)
					MG Y32 QR 8·4 *mq*	(0·88, 7·10)
					MG F38 QR 8·4 *mq*	(0·88, 6·87/7·03)
					MD A41 MB 7·0 *mm*	(0·79, 1·16)
					HB V42 QG 7·4 *q*	(1·32, 0·78)
					MG V42 HN 6·0 *m*	(0·88, 8·21)
					MD S45 PB 7·0 *mm*	(0·79, 4·01)
					MG S45 HA 6·0 *m*	(0·88, 4·53)
					MG S45 PB 7·0 *mm*	(0·88, 4·23/4·01)
					MG N50 HN 5·0 *m*	(0·88, 8·74)
Asn50						
HN I49 HA 3·0			HN S27 HN 4·0	(8·74, 9·43)	HA F38 QR 7·4 *q*	(5·03, 6·87)
HA L51 HN 3·0					HA V42 QG 7·4 *q*	(5·03, 0·10)
PB L51 HN 4·0 *m*					HN I49 MG 5·0 *m*	(8·74, 0·88)
Leu51						
HN N50 HA 3·0	HA L51 QD 4·2 *s*	(4·26, 0·19)	HA G26 PA 5·0 *m*	(4·26, 4·94)	HN V5 QG 7·4 *q*	(8·95, 1·01)
HN N50 PB 4·0 *m*	HN L51 PB 3·6 *i*	(8·95, 1·98)			QD V5 QG 9·8 *qq*	(0·67, 1·01)
HA K52 HN 2·5		(8·95, 1·21)			QD H24 HA 7·4 *q*	(0·19, 4·22)
					QD H24 H4 7·4 *q*	(0·19/0·67, 6·12)
					QD H24 PB 8·4 *qm*	(0·19, 2·31)
					QD C25 HN 7·4 *q*	(0·19, 8·29)
					HN F38 QR 7·4 *q*	(8·95, 7·03/6·87)
					QD F38 QR 9·8 *qq*	(0·19, 7·03/6·87)
					QD K52 HN 7·4 *q*	(0·19, 8·98)
					QD R54 HA 7·4 *q*	(0·19, 3·83)
Lys52						
HN L51 HA 2·5	HN K52 PB 3·6 *i*	(8·98, 1·55)	PB S27 HA 6·0 *m*	(1·28, 4·08)	HN L51 QD 7·4 *q*	(8·98, 0·19)
HN H53 HN 3·0		(8·98, 1·28)				
His53						
HN K52 HN 3·0			HN C25 HN 4·0	(7·06, 8·29)	H2 C57 HA 5·0	(8·83, 4·41)
HA R54 HN 3·0						
Arg54						
HN H53 HA 3·0			HA H24 HA 4·0	(3·83, 4·22)	HA L51 QD 7·4 *q*	(3·83, 0·19)
HA G55 HN 2·5			HA C25 HN 4·0	(3·83, 8·29)		
Gly55						
HN R54 HA 2·5			HN H24 HA 4·0	(8·00, 4·22)		
Lys56						
HA C57 HN 2·5			HA P23 PB 6·0 *m*	(4·06, 1·48)	HA T31 MG 6·0 *m*	(4·06, 1·13)
PB C57 HN 4·0 *m*					HN T31 MG 6·0 *m*	(7·98, 1·13)
					PB T31 MG 7·0 *mm*	(1·51, 1·13)
Cys57						
HN K56 HA 2·5					HN T31 MG 6·0 *m*	(8·13, 1·13)
HN K56 PB 4·0 *m*					HA H53 H2 5·0	(4·41, 8·83)

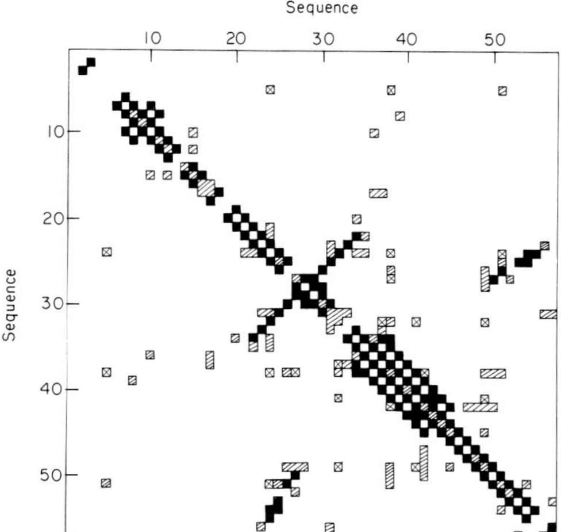

Figure 4. Diagonal plot of the NOEs used for the calculation of the tertiary structure of BUSI IIA. Both axes are calibrated with the sequence of the protein. A filled square at the position (x, y) in the plane indicates that a NOE between backbone protons (NH or $C^{\alpha}H$) of the 2 residues in the sequence locations x and y was observed. A hatched square indicates a NOE between the backbone amide or protons of one residue and any of the side-chain protons in another residue, while a square containing a cross indicates a NOE between protons in the side-chains of 2 different residues. Where 2 residues are connected by more than 1 NOE, only the one which involves the greater number of backbone protons is shown. Further details on the NOEs may be found in Table 2.

determined by numerous interstrand NOEs (Table 2; Williamson *et al.*, 1984). The sheet has a pronounced, right-handed twist in all ten of the computed structures. In Figures 6 and 7, the sheet lies to the left of the α-helix 34–45, and numerous NOEs between residues located in the α-helix and β-sheet show that these two principal secondary structure elements are tightly packed together in the tertiary structure. The C-terminal residue Cys57 is tied to the β-sheet by the disulfide bond with Cys25. The observation of numerous NOEs (Table 2) provided direct evidence that the C-terminal region of the polypeptide chain has a unique spatial arrangement in the solution conformation. The architecture of the β-sheet region further includes two tight turns, one from residues 27 to 29, where the first two strands form a "hairpin", and the other from residues 47 to 49.

Overall, the topology of the solution conformation of BUSI IIA is thus characterized by the α-helix 34 to 45 being sandwiched between the extensive loop from residues 5 to about 20 and the β-sheet. In this spatial arrangement, the α-helix forms the core of the molecule, which provides an explanation for the outstandingly slow exchange of the hydrogen-bonded amide protons in the helix (Wüthrich *et al.*, 1984).

(b) *Comparison with the crystal structures of homologous proteins*

Although no crystal structure exists for BUSI IIA, high resolution crystal structures do exist for three homologous proteins, e.g. the porcine pancreatic secretory trypsin inhibitor (PSTI) (Bolognesi *et al.*, 1982), the third domain of Japanese quail ovomucoid (OMJPQ3) (Papamokos *et al.*, 1982), and the third domain of turkey ovomucoid (Fujinaga *et al.*, 1982). Since the two ovomucoids are very similar in sequence and tertiary structure, we have chosen to compare BUSI IIA with PSTI and OMJPQ3. OMJPQ3 has

The amino acid residues are identified either by the 3-letter symbol or the 1-letter symbol followed by a number indicating the position in the amino acid sequence. HN stands for amide proton, HA for C^{α} proton, HB for C^{β} proton, HG for C^{γ} proton and HD for C^{δ} proton. H2 and H4 are the protons attached to the ring carbons 2 and 4 of His. The following upper case letters indicate pseudo atoms which substitute for a group of 2 or more protons, and lower case letters indicate that the distance constraint indicated has been modified to account for the use of the pseudo atom (Wüthrich *et al.*, 1983). M replaces the 3 protons of a methyl group, and is located in the center of the 3 proton positions, with the correction factors $m = 1{\cdot}0$ Å for long-range constraints and $i = 0{\cdot}6$ Å for certain intraresidue constraints. P replaces the protons of a methylene group and is located in the center between the 2 proton positions, with the same corrections m and i as for methyl groups. Q replaces the 2 methyl groups of Val and Leu and is located centrally with respect to the 2 methyl groups, with corrections of $q = 2{\cdot}4$ Å, or $s = 1{\cdot}7$ Å, respectively, for long-range constraints or for certain intraresidue constraints. QR is at the center of the ring protons C2H, C3H, C5H and C6H in Phe or Tyr, also with a correction of $q = 2{\cdot}4$ Å. Where the C2H and C6H resonances could be resolved from those of C3H and C5H, distances to the ring protons were referred to the ring carbon atoms C-1 and C-4, respectively, with a correction of $r = 2{\cdot}0$ Å. Distance constraints to the amide protons in the side-chains of Asn22 and Asn34 are referred to the amide nitrogen atom ND, with a correction of $m = 1{\cdot}0$ Å.

The rows following the 3-letter symbol for the amino acid list all the distance constraints that involve hydrogen atoms of this residue. For each residue the distance constraints are grouped into 4 classes (see the text), which are presented in the 4 columns. Sequential constraints are those between backbone protons or between a backbone and a C^{β} proton in sequentially neighboring residues. Intraresidue constraints are between side-chain hydrogen atoms and any other proton of the same residue. Long-range backbone constraints are between 2 backbone C^{α} or amide protons, or between a backbone proton and the C^{β} protons in residues that are not nearest-neighbors in the sequence. Long-range constraints are all those that have not been listed elsewhere. With the exception of the intraresidue constraints, all entries appear twice, once for each residue concerned. In each column the first entry identifies a hydrogen atom in the amino acid that is identified by the 3-letter code. The second and third entries indicate a different residue and the hydrogen atom (or pseudo atom) in this residue to which a distance constraint has been observed. The fourth entry is the distance constraint in Å, possibly with symbols indicating that 1 or 2 correction factors were added to the NOESY distance constraints to allow for the use of pseudo atoms. The fifth entry indicates the location of the NOESY cross peak in the ω_1–ω_2 plane in p.p.m. (Figs 1 and 2). (This fifth entry has been omitted for the sequential constraints, since these were previously described in detail (Štrop *et al.*, 1983).) Where two p.p.m. values are given separated by an oblique stroke, 2 resonance lines were resolved for the group of protons indicated.

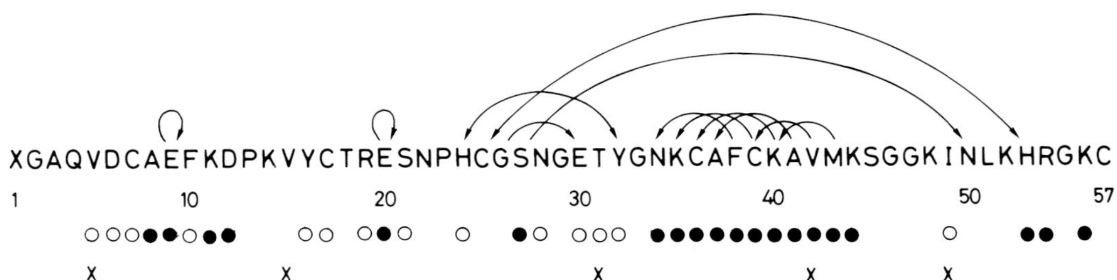

Figure 5. Diagram showing the sequence locations of the hydrogen bonds (arrows), ϕ angle constraints (circles) and χ^1 angle constraints (crosses) used as supplementary constraints in the computations of the structures C1 to C5 (see the text). The sequence of BUSI IIA is given in the standard one-letter amino acid code. The arrows point from the amide proton to the carbonyl oxygen of each hydrogen bond (2-headed arrows identify an antiparallel β-bridge). A filled circle (●) indicates a $^3J_{HN\alpha} \leq 5{\cdot}5$ Hz, so that the ϕ angle was constrained between $-90°$ and $-40°$, and an open circle (○) indicates a $^3J_{HN\alpha} \geq 8{\cdot}0$ Hz, so that the ϕ angle was constrained between $-160°$ and $-80°$. (The present Figure differs from Fig. 7 of Williamson *et al.* (1984) in that residues K14, V15, T18, C25, L51 and K52, which have $^3J_{HN\alpha} = 7{\cdot}5$ Hz, do not carry open circles because of the different cutoff applied; the ^3J value of 5·5 Hz for C57 was not exploited because the ϕ angle could not be constrained properly in the amino acid residue set used.) The crosses indicate that the $^3J_{\alpha\beta}$ values were $\geq 8{\cdot}0$ Hz (i.e. V5, $^3J_{\alpha\beta} = 8{\cdot}0$ Hz; V15, $^3J_{\alpha\beta} = 9{\cdot}0$ Hz; T31, $^3J_{\alpha\beta} = 10{\cdot}0$ Hz; V42, $^3J_{\alpha\beta} = 10{\cdot}0$ Hz; I49, $^3J_{\alpha\beta} = 10{\cdot}0$ Hz), and that the χ^1 angle was therefore constrained so as to keep the H^α and H^β protons within $\pm 30°$ of the *trans* orientation.

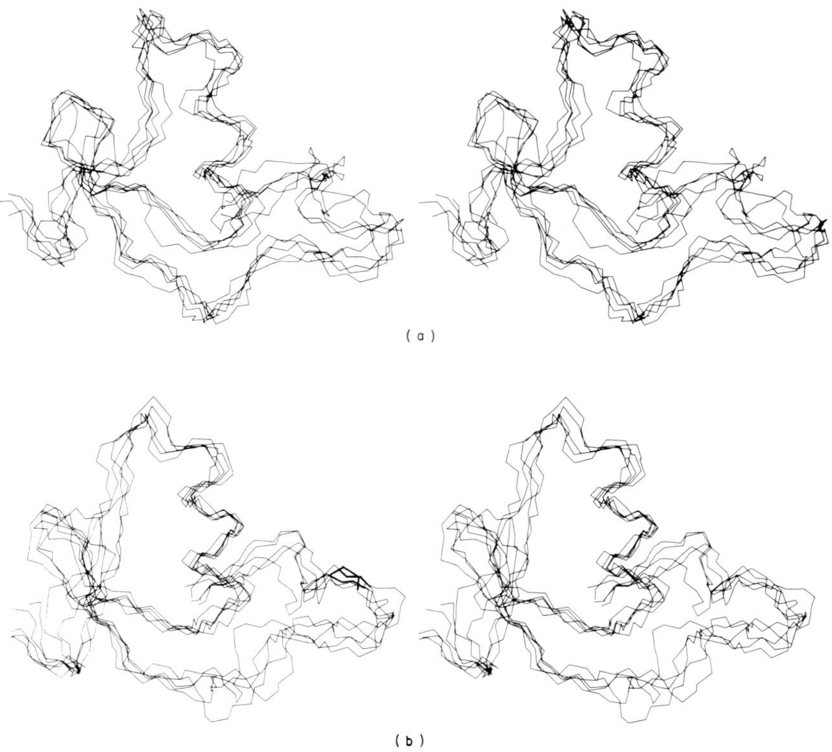

Figure 6. (a) Stereo view of the 5 structures N1 to N5 calculated using the NOE distance constraints alone (Table 2). The structures were superimposed so as to minimize the RMSD with respect to the structure N1. All main-chain backbone atoms (C^α, C', N) are shown. For clarity of presentation the residues 1 to 4, which are essentially unconstrained by the n.m.r. data (Table 2), have been omitted, and the disulfide bonds are not shown. (b) Stereo view of the 5 structures calculated from the NOE constraints used in (a) plus the supplementary constraints shown in Fig. 5. As in (a), all main-chain backbone atoms are shown, the first 4 residues and the disulfide bonds have been omitted, and the 5 structures have been superimposed for best fit with the structure C1.

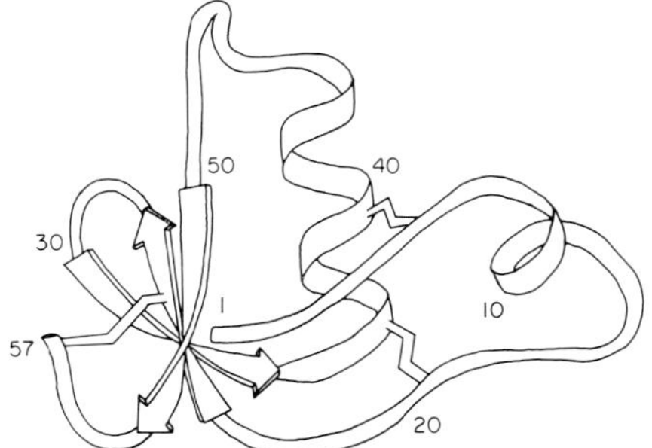

Figure 7. Schematic drawing of the polypeptide backbone "topology" which was obtained in all 10 of the BUSI IIA structures computed from the n.m.r. data presented.

45% homology and a deletion of two residues with respect to BUSI IIA. PSTI has 26% homology and a deletion of three residues. The deletions are both in the N-terminal part of the molecules, which is generally of variable length in this class of secretory proteinase inhibitors (Kato *et al.*, 1976). We have therefore used only the C-terminal part of the molecules (from residues 23 to 57 in BUSI IIA and 22 to 56 in the two homologues) for the purpose of comparison.

The average RMSD values between the α-carbons of these segments in PSTI and the BUSI IIA structures are 2·20 Å and 2·24 Å for the N and C data sets, respectively. For OMJPQ3 the corresponding averages are 1·89 Å and 1·75 Å. The RMSD between the α-carbons in the segments 22 to 56 of PSTI and OMJPQ3 is 1·34 Å. These RMSD values indicate a high degree of similarity between the computed BUSI IIA structures and the crystal structures of these homologous proteins. This similarity is also evident in the superpositions of this region of the backbone of the BUSI IIA structure C5 on the corresponding residues of the homologues (Fig. 9), which shows that the α-helix and the β-sheet have both identical sequence locations and the same relative spatial orientations in the three proteins. Figure 9 shows further that the conformations of the N-terminal segments (residues 1 to 22) are markedly different in the three proteins. Since there are deletions of two and three residues, respectively, between positions 7 and 20 in OMJPQ3 and PSTI, and since furthermore the N-terminal loop is tied to the rest of the molecule by the two disulfide bonds 7–39 and 17-36, these differences between the computed BUSI conformations and the conformations of the homologous proteins are not unexpected and are undoubtedly significant. The segment of residues 1 to 5 is not constrained by the available n.m.r. data, and in the crystal structure of OMJPQ3 it is known to be involved in close intermolecular contacts (Papamokos *et al.*, 1982).

N1	1517	47	5	5	2	1			
N2	1564	43	9	3	4	1	0	1	
N3	1502	66	16	2	0	2	1	1	2
N4	1510	57	12	2	6	0	0	1	
N5	1499	48	4	6	3	1			
C1	1772	33	8	5	3				
C2	1658	27	6	2	0	1			
C3	1660	38	10	5	0	3	1	0	2
C4	1706	26	6	2	5				
C5	1588	14	2	0	1				

0·1 0·3 0·5 0·7

Å

Figure 8. Histograms presenting the residual violations of the distance constraints found in each of the 10 structures of BUSI IIA computed from the n.m.r. data presented. The 5 structures which were obtained using the NOE constraints alone (Table 2) are labeled as N1 to N5, while the 5 structures computed using the NOE distance constraints together with the supplementary constraints of Fig. 5 are labeled as C1 to C5. In each histogram, the number of distance constraint violations which fall within each 0·1 Å interval is given. The total number of distance constraint violations is about 40 to 60% of the total number of input distance constraints.

(c) *Quantiative comparison of the computed structures*

All ten of the structures computed for BUSI IIA have nearly identical overall dimensions and contain the same global chain fold as described in Results, section (a). No single structure is markedly different from the others computed from the same data set. The average values of the RMSD between the C^α positions of pairs of structures (Table 3) are 1·93 Å and 2·13 Å for the data sets N and C, respectively. These are slightly larger than those obtained with comparable simulated data sets (e.g. those labeled as III, IV, IX and X) in the previous model studies with BPTI (Havel & Wüthrich, 1985), primarily because of the uneven distribution of NOE distance constraints obtained with BUSI IIA, which are concentrated in the C-terminal half of the molecule (see Table 2 and Fig. 5). When the RMSD values were calculated between the α-carbons of residues 23 to 57 alone, they averaged 1·59 Å and 1·84 Å for the data sets N and C, respectively. The average RMSD values between all heavy-atoms in the BUSI IIA structures (Table 3) were 2·98 Å and 3·34 Å for the data sets N and C (2·55 Å and 2·84 Å for the residues 23 to 57 alone), which are also of the same order of magnitude as those obtained in the above-mentioned studies with BPTI. They indicate that the location of a large proportion of the side-chains is similar in the different computed structures. This is probably largely a consequence of the stringent packing requirements in the interior of the protein.

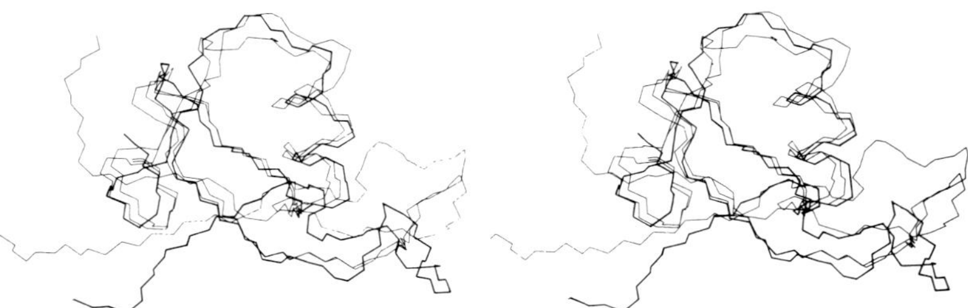

Figure 9. A comparison of the structure of BUSI IIA (heavy line) obtained from the n.m.r. data shown in Table 2 and Fig. 5 with the crystal structures of 2 homologous proteins: the porcine pancreatic secretory inhibitor and the third domain of the Japanese quail ovomucoid. The BUSI IIA structure C5 (Fig. 8) was used for this comparison. Only the backbone atoms and bonds connecting them are shown. The structures have been aligned for a minimum RMS co-ordinate difference between the α-carbons of residues 23 to 57 in BUSI IIA and the homologous residues 22 to 56 of the other 2 proteins.

It may be seen (Table 3) that the RMSD values between structures computed from the different data sets N and C are usually somewhat larger than between those computed from the same constraint set. A rather surprising result is that even though the C data set included all of the constraints of the N data set plus additional supplementary constraints (see Results, section (a)), the RMSD values between pairs of structures computed from the C data set are slighly larger in general than those between pairs computed from the N data set. Most of this increase is due to increased variability in the first strand of the β-sheet and the adjoining residues (numbers 15 through 30: Fig. 6). It can be explained by large-scale changes in the position of this segment relative to the rest of the molecule which are concomitant with changes in the twist of the β-sheet. Such "concerted motions" in the β-sheet were less pronounced in the structures computed from the N data set, presumably because the strands of the sheet were not tied together as tightly without the addition of hydrogen bonds. There is ample space for such "motions" in the β-sheet of BUSI IIA, since it is located near the surface of the protein and, when compared for example to the β-sheet of BPTI, tied down by a lesser number of long-range NOEs to other parts of the molecule.

Like the experimental constraints available for BUSI IIA, the aforementioned data sets III, IV, IX and X used with BPTI (Havel & Wüthrich, 1985) consisted of constraints which were both incomplete and imprecise, but still allowed the global features of the crystal structure from which they had been extracted to be faithfully reconstructed. The experience gained in these model studies further showed that such data are not sufficient to define precisely the local structure of the polypeptide chain, which would require a dense network of very precise local constraints. The data used here do not have the necessary precision because of the current limitations on our ability to quantitate the distance measurements obtained by NOESY, and a lack of stereospecific assignments for the methylene protons. In accord with these expectations, the measures of the differences between the local conformations of the computed structures were relatively large. The RMSD values between the α-carbons of corresponding pairs of tetrapeptide segments in the different structures averaged 0·73 Å and 0·69 Å for the data sets N and C, respectively. The DHAD values between the ϕ and ψ angles (see Materials and Methods, section (e)) averaged 77° and 51°. An exception to this general observation is the α-helix (residues 34 to 45), for which the local conformation was

Table 3

RMS deviations between all pairs formed from the ten structures computed for BUSI IIA

N1	2·04	1·80	1·76	1·57	2·45	2·29	2·57	2·33	2·10
3·08	N2	2·17	2·15	2·12	2·38	2·37	2·38	2·20	1·94
2·71	2·94	N3	1·94	1·78	2·50	2·53	2·60	2·43	2·22
2·79	2·93	2·71	N4	2·03	2·33	2·45	2·92	2·31	2·18
2·39	3·13	2·73	2·83	N5	2·52	2·29	2·50	2·30	2·12
3·90	3·95	4·08	3·67	3·89	C1	2·25	2·44	1·99	2·00
3·45	3·38	3·57	3·20	3·37	3·35	C2	2·20	2·28	2·04
3·71	3·74	3·84	3·96	3·57	3·22	3·12	C3	2·05	2·32
3·44	3·43	3·72	3·61	3·37	3·04	3·38	2·73	C4	1·71
3·09	2·96	3·25	3·19	3·20	3·26	3·05	3·27	2·96	C5

The above diagonal entries are between α-carbons alone, the below diagonal entries are between all heavy atoms. The symbols on the diagonal are those used in Fig. 8.

relatively well-defined by the n.m.r. data, as shown by average RMSD values between sets of four consecutive C^{α} atoms under 0·3 Å even in data set N.

Very few NOEs were observed to side-chains on the surface of the protein (Table 2). A likely explanation is that most of these side-chains NOEs were quenched by pronounced segmental mobility (Nagayama & Wüthrich, 1981). Those residues for which the χ^1 angle had been fixed by use of chirality constraints in the C data set (see Fig. 5) had relatively well-defined side-chain conformations in the computed structures, as did His24 and Phe38, both of which are buried in the interior and participate in numerous long-range NOEs (see Table 2). For His24, this fits well with the earlier observation (Štrop & Wüthrich, 1983) that this residue is unchanged over the pH range from 4·0 to about 13, and is spatially close to a titrating group which must be either a lysine or tyrosine side-chain. In the computed structures, His24 is invariably close to both Tyr32 and Lys35. Since it was previously found that Tyr32 does not titrate between pH 4 and pH 12 (Štrop & Wüthrich, 1983), this group can now be identified as Lys35. Many other side-chains in the interior were significantly restrained in their position and orientation (see the Appendix).

The investigations in this paper show that the information available from n.m.r. spectroscopy is sufficient to define the overall tertiary structure of small proteins in solution, and that algorithms based on distance geometry are a suitable means of determining the global conformation from these data. The close coincidence between the different structures computed for BUSI IIA, and between these structures and the crystal structures of two homologous proteins, is particularly gratifying when one considers that the experimental observations were used with a very conservative structural interpretation, so that the corresponding geometric constraints were quite imprecise. Since a large proportion of the available measurements concern the relative positions of protons that are near to one another in the amino acid sequence, it is perhaps surprising at a first glance to find that the local structure of both the polypeptide backbone and the side-chains is less well-defined than the global conformation. The relatively high precision with which we can determine the global conformation from imprecise distance information is a consequence of the redundancy of the long-range distance information. The local structure depends more directly on a relatively small number of short-range distances, and it can be changed substantially with small changes in these distances. As a consequence, the short-range distances would have to be measured with relatively high precision in order to determine the local conformation reasonably exactly. Alternatively, precise geometric constraints on the local structure could also be obtained from other n.m.r. experiments (Wüthrich, 1976). Because distance geometry calculations do not take account of vicinal van der Waals' repulsions (Materials and Methods, section (d)), substantial improvement in the local conformation could also be expected to follow from subsequent energy minimization. This is further explored in the Appendix.

Appendix

Here we describe a pilot study of the use of computer graphics (Langridge *et al.*, 1982) and restrained energy minimization (Fitzwater & Scheraga, 1982; Levitt, 1983) as a means of improving the local structure (ϕ, ψ angles and side-chain conformations) of the conformations computed with DISGEO from the experimental n.m.r. data. Since the DISGEO program explicitly ignores energetic considerations while attempting to obtain a good fit to the geometric constraints, it was also of interest to demonstrate that the structures are nevertheless close to good energy minima in which all of the experimental constraints are satisfied. We selected the best convergent conformation, C5 (Table 3, main text), for the present trial refinement.

For the energy function, we used the ECEPP/2 parameters (Némethy *et al.*, 1983) with a dielectric constant of four, and left all acidic and basic groups on the side-chains unionized. Considering the preliminary nature of this work, we simplified programming by not including the internal energies of the proline residues (all prolines have the down conformation), neglecting the torsional potential about the beta-to-gamma bond to cystine, and leaving out the N-terminal amide hydrogen and the C-terminal hydroxyl group. These omissions will have a minor effect on the energies we report, but will not significantly alter the location of the energy minima. The minimizations themselves were performed with a standard conjugate gradient algorithm (Luenberger, 1973), using the dihedral angles about all rotatable bonds (including the omega angles) as the variables.

The refinement procedure consisted of six steps. First, the missing hydrogen atoms were added to the C5 structure by a method similar to that described by Billeter *et al.* (1982). The resultant structure had a high energy due to the presence of isolated van der Waals' overlaps involving some of the added protons, unfavorable dihedral interactions, and a general lack of accommodation of internal polar and charged groups in hydrogen bonds and salt bridges. In the second step of the refinement procedure, we manually adjusted the dihedral angles of long side-chains on the exterior of the molecule so as to eliminate energetically unfavorable eclipsed conformations and isolated contacts with adjacent sections of the backbone. This was done with a special computer graphics program, CONFOR, which allowed us to simultaneously monitor the NOE constraints and so avoid making changes which would violate them (M. Billeter, M. Engeli & K. Wüthrich, unpublished

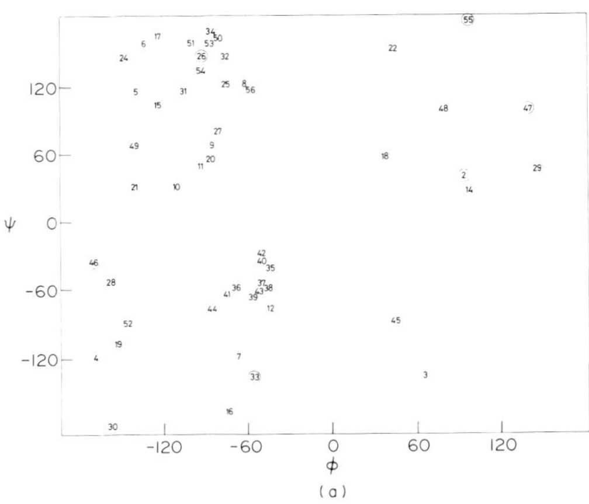

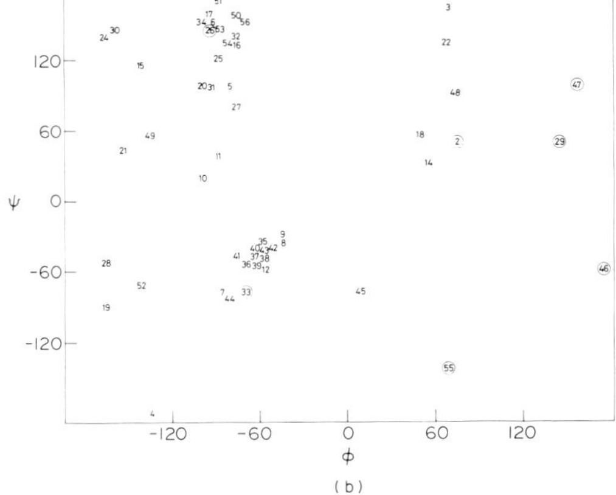

Figure A1. The Ramachandran map of the C5 pseudo structure of BUSI IIA (a) before and (b) after the refinement procedure. The sequence number of each residue in the conformation is printed at the position of its ϕ (horizontal) and ψ (vertical) angles on the axes. The sequence numbers of the glycyl residues have been circled.

work). Further optimization of the van der Waals' packing and perfection of the covalent geometry was achieved by the third step, in which a numerical optimization of the resulting structure was performed with respect to the same error function as used by the DISGEO program (Havel *et al.*, 1983), with the Cartesian co-ordinates of the atoms as the variables. The only constraints now used in calculating the error function, however, were the ECEPP covalent geometries (Némethy *et al.*, 1983) and the full van der Waals' radii of all of the atoms. The n.m.r. constraints were not included in the calculation, except for the hydrogen bonds indicated in Figure 5 of the main text. In our experience, structures subjected to the pre-processing procedure of steps two and three attain substantially lower energies on subsequent minimization.

During the optimization of covalent geometry and van der Waals' packing, approximately 10% of the NOE constraints became violated by between 1 and 3 Å. The only large dihedral angle violations were in the Glu20 and Asn34 ϕ angles, which were off by about 50°. In order to restore consistency with the geometric constraints and at the same time adequately restrain the structure during the initial steps of the energy minimization, the error function (Havel *et al.*, 1983) was calculated from the

Table A1

Minimum RMS co-ordinate differences between the C^{α} atoms of homologous residues in BUSI IIA, OMJPQ3 and PSTI before and after energy minimization of the BUSI IIA structure

BUSI IIA structure C5	OMJPQ3		PSTI	
	1–56	22–56	1–56	22–56
From DISGEO	4·22 Å	1·67 Å	5·16 Å	1·44 Å
Energy minimized	3·99 Å	1·52 Å	4·62 Å	1·23 Å

The RMS difference between the C^{α} atoms of OMJPQ3 and PSTI is 3·88 Å for residues 1 to 56, and 1·34 Å for residues 22 to 56.

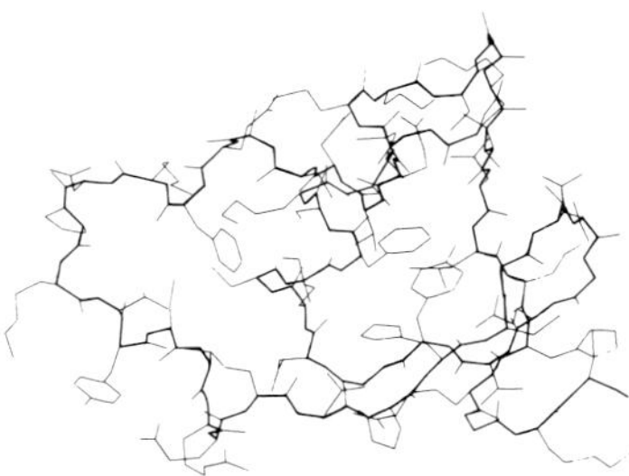

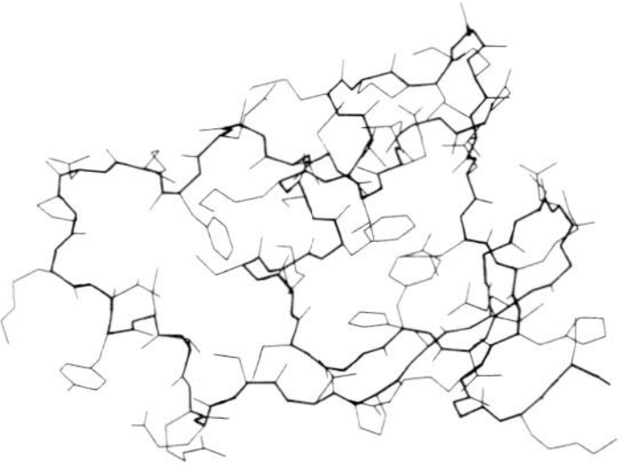

Figure A2. Stereo pair of all heavy (i.e. non-hydrogen) atoms in the energy-minimized structure of BUSI IIA obtained by refinement of the C5 pseudo structure. The bonds connecting the backbone atoms have been drawn with a heavier line. The orientation of the structure has been chosen to allow a good view of the side-chains, and is not the same as that used in Figs 6 and 7 of the main text. The disulfide bonds are not shown.

experimental distance constraints given in Table 2 and Figure 5 of the main text, and added to the energy as a pseudo potential. To do this, the interproton constraints between methylene and methyl groups were referred with the necessary corrections to the carbons of these groups. The ϕ angles indicated in Figure 5 (main text) were also restrained by means of a single-fold pseudo torsional potential which was zero between the corresponding angle bounds, and which was given a barrier height of 400 kcal (1 kcal = 4·184 kJ). The sum of the two pseudo potentials was given an initial weight of 200, and this weight decreased linearly to zero as the minimization progressed for 200 iterations, so as to attain a physically meaningful energy minimum at the end.

After this fourth step, there was only one violation of the NOE constraints exceeding 1·0 Å, and relatively few violations within the α-helix and β-sheet. In the fifth step, the remaining 1·1 Å violation between the amide proton of Ala8 and the β-carbon of Cys39 was alleviated by interactively rotating the Cys7 to Cys39 disulfide bond to the other side of its 10 kcal torsional barrier. Other, smaller NOE violations were also relieved by interactive manipulation of the Thr31, Ser45 and His53 side-chains. In the sixth and final step, the resultant structure was subjected to another 100 cycles of restrained energy minimization with an initial pseudo potential weight of 100. The Ala8 to Cys39 violation was completely eliminated during this second minimization, but a new violation of similar magnitude appeared between the β-carbon of Val5 and the δ-ring proton of His24. Attempts at further improvement were abandoned, and the structure was accepted with 11 NOE violations exceeding 0·5 Å, and an energy of −166 kcal.

The RMS between the α-carbons of the C5 structure from DISGEO and the final energy-minimized structure was 1·69 Å. This shows that the global conformation was not changed greatly by the refinement procedure. The RMS change in the ϕ, ψ angles on refinement was 19°. Although these dihedral angle changes were not large enough to completely correct the local structure of the polypeptide, an examination of the ϕ, ψ maps before and after refinement (Fig. A1) shows that a significant improvement was obtained. The low energy of the refined structures was primarily due to good interior side-chain packing, which resulted in a van der Waals' contribution to the total energy of −207 kcal. This can be clearly seen in the stereo view of the final energy-minimized structure (Fig. A2). Five new backbone hydrogen bonds

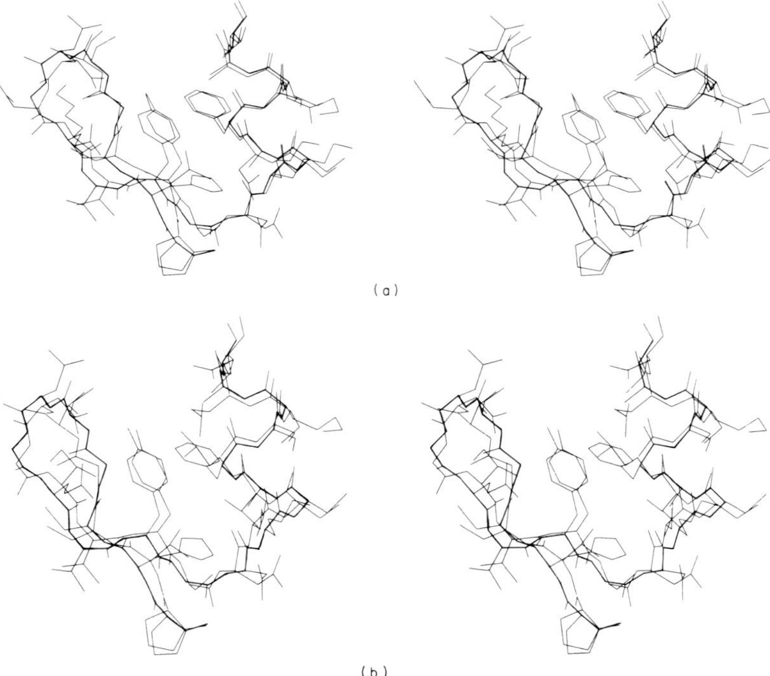

Figure A3. Stereo pairs showing the superpositions of the segment of residues from 23 to 42 in the energy-minimized structure C5 and the homologous segments of (a) the Japanese quail ovomucoid, and (b) the porcine pancreatic secretory inhibitor. The bonds connecting the backbone atoms of the BUSI IIA structure have been drawn with a heavier line. The segments have been aligned so as to minimize the RMSD in their α-carbons.

formed as a result of the energy minimization, which are given here as (carbonyl oxygen, amide proton) pairs: (Cys7, Phe10), (Ala8, Lys11), (Ala41, Ser45), (Cys25, Lys52), (Pro23, Gly55). Although the hydroxyl of Tyr32 does not titrate to a pH of at least 12 (Štrop & Wüthrich, 1983), it did not form a hydrogen bond on energy minimization. However, the side-chain oxygens of Asn28, Glu30 and Ser45 were all less than 6·0 Å away from the Tyr32 hydroxyl oxygen atom.

The RMSD between corresponding α-carbons in the BUSI structure and the homologues, PSTI and OMJPQ3, decreased during the refinement of the C5 structure (Table A1). A remarkable degree of similarity was found in the spatial locations of corresponding side-chains in the fragment of the energy-minimized BUSI structure from residues 23 to 42, and residues 22 to 41 of the homologous proteins OMJPQ3 and PSTI (Fig. A3). The RMSD between all heavy atoms in residues 23 to 42 before and after the refinement was only 0·98 Å. This shows that in the interior of the C5 structure computed by the DISGEO program directly from the n.m.r. data, the side-chains were already in approximately the correct spatial positions†, even though there were substantial uncertainties in the χ dihedral angles.

We thank Dr D. Marion for the measurements of the $^3J_{\alpha\beta}$ values given in Fig. 5. Financial support by the Schweizerischer Nationalfonds (project 3.284.82) and by the S.E.R.C. (U.K.: fellowship to M.P.W.) is gratefully acknowledged. We thank Mrs E. H. Hunziker-Kwik for the careful preparation of the Figures, and Drs D. Čechová and P. Štrop for providing the sample of BUSI IIA.

References

Anil Kumar, Ernst, R. R. & Wüthrich, K. (1980). *Biochem. Biophys. Res. Commun.* **95**, 1–6.

Anil Kumar, Wagner, G., Ernst, R. R. & Wüthrich, K. (1981). *J. Amer. Chem. Soc.* **103**, 3654–3658.

Arseniev, A. S., Kondakov, V. I., Maiorov, V. N., Volkova, T. M., Grishin, E. V., Bystrov, V. F. & Ovchinnikov, Yu. A. (1983). *Bioorgan. Khim.* **9**, 768–793.

Billeter, M., Braun, W. & Wüthrich, K. (1982). *J. Mol. Biol.* **155**, 321–346.

Bodenhausen, G., Vold, R. L. & Vold, R. R. (1980). *J. Magn. Reson.* **37**, 93–106.

Bolognesi, M., Gatti, G., Menegatti, E., Guarneri, M., Marquart, M., Papamokos, E. & Huber, R. (1982). *J. Mol. Biol.* **162**, 839–868.

Braun, W., Bösch, C., Brown, L. R., Gō, N. & Wüthrich, K. (1981). *Biochim. Biophys. Acta,* **667**, 377–396.

Braun, W., Wider, G., Lee, K. H. & Wüthrich, K. (1983). *J. Mol. Biol.* **169**, 921–948.

Bundi, A. & Wüthrich, K. (1979). *Biopolymers,* **18**, 285–298.

Čechová, D., Jonáková, V., Sedláková, E. & Mach, O. (1979). *Hoppe-Seyler's Z. Physiol. Chem.* **360**, 1753–1758.

Ebina, S. & Wüthrich, K. (1984). *J. Mol. Biol.* **179**, 283–288.

Ferrige, A. G. & Lindon, J. C. (1978). *J. Magn. Reson.* **31**, 337–340.

Fitzwater, S. & Scheraga, H. A. (1982). *Proc. Nat. Acad. Sci., U.S.A.* **79**, 2133–2137.

Fritz, H., Schiessler, H., Geiger, R., Ohlsson, K. & Hochstrasser, K. (1978). *Agents Act.* **8**, 57–64.

Fujinaga, M., Read, R. J., Sielecki, A., Ardelt, W., Laskowski, M. Jr & James, M. N. G. (1982). *Proc. Nat. Acad. Sci., U.S.A.* **79**, 4868–4872.

Gordon, S. L. & Wüthrich, K. (1978). *J. Amer. Chem. Soc.* **100**, 7094–7096.

Havel, T. F. & Wüthrich, K. (1984). *Bull Math. Biol.* **46**, 673–698.

Havel, T. F. & Wüthrich, K. (1985). *J. Mol. Biol.* **182**, 281–294.

Havel, T. F., Crippen, G. M. & Kuntz, I. D. (1979). *Biopolymers,* **18**, 73–81.

Havel, T. F., Kuntz, I. D. & Crippen, G. M. (1983). *Bull. Math. Biol.* **45**, 665–720.

Kalk, A. & Berendsen, H. J. C. (1976). *J. Magn. Reson.* **24**, 343–366.

Kato, I., Schrode, J., Wilson, K. A. & Laskowski, M. Jr (1976). *Prot. Biol. Fluids,* **23**, 235–243.

Langridge, R., Ferrin, T., Kuntz, I. & Connolly, M. (1982). *Science,* **211**, 661–666.

Levitt, M. (1983). *J. Mol. Biol.* **170**, 723–764.

Luenberger, D. G. (1973). *Introduction to Linear and Nonlinear Programming,* Addison-Wesley Publishing Co., Reading, Mass.

Macura, S., Huang, Y., Suter, D. & Ernst, R. R. (1981). *J. Magn. Reson.* **43**, 259–281.

Marion, D. & Wüthrich, K. (1983). *Biochem. Biophys. Res. Commun.* **113**, 967–974.

Nagayama, K. & Wüthrich, K. (1981). *Eur. J. Biochem.* **115**, 653–657.

Némethy, G., Pottle, M. & Scheraga, H. (1983). *J. Phys. Chem.* **87**, 1883–1887.

Nyburg, S. C. (1974). *Acta Crystallogr. sect. B,* **30**, 251–253.

Papamokos, E., Weber, E., Bode, W., Huber, R., Empie, M. W., Kato, I. & Laskowski, M. Jr (1982). *J. Mol. Biol.* **158**, 515–537.

Pardi, A., Billeter, M. & Wüthrich, K. (1985). *J. Mol. Biol.* In the press.

Redfield, A. G. & Kunz, S. D. (1975). *J. Magn. Reson.* **19**, 250–254.

Schiessler, H., Arnold, M., Ohlsson, K. & Fritz, H. (1976). *Hoppe-Seyler's Z. Physiol. Chem.* **357**, 1251–1260.

Štrop, P. & Wüthrich, K. (1983). *J. Mol. Biol.* **166**, 631–640.

Štrop, P., Wider, G. & Wüthrich, K. (1983). *J. Mol. Biol.* **166**, 641–667.

Wagner, G., Anil Kumar & Wüthrich, K. (1981). *Eur. J. Biochem.* **114**, 375–384.

Wider, G., Hosur, R. V. & Wüthrich, K. (1983). *J. Magn. Reson.* **52**, 130–135.

Williamson, M. P., Marion, D. & Wüthrich, K. (1984). *J. Mol. Biol.* **173**, 341–360.

Wüthrich, K. (1976). *NMR in Biological Research: Peptides and Proteins,* North-Holland Publishing Company, Amsterdam.

† During the interactive computer graphics manipulations of step 2 above, only the following dihedral angles in the segment from residues 23 to 42 were changed: χ^2 of His24, 55° to 35°; χ^1 and χ^2 of Asn34, −49° to 0° and 126° to 70°; χ^4 of Lys35, 14° to 180°. The dihedral angles changed during the step 5 were: χ^1 of Thr31, −61° to −75°; χ^1 of Cys39, −163° to −85°.

Wüthrich, K., Billeter, M. & Braun, W. (1983). *J. Mol. Biol.* **169**, 949–961.

Wüthrich, K., Štrop, P., Ebina, S. & Williamson, M. P. (1984). *Biochem. Biophys. Res. Comm.* **122**, 1174–1178.

Wüthrich, K., Billeter, M. & Braun, W. (1985). *J. Mol. Biol.* In the press.

Edited by M. F. Moody

III

ESTABLISHING CREDIBILITY FOR NMR STRUCTURES OF PROTEINS

Introduction to papers 26 through 29

The four papers presented in this part relate to NMR structure determinations of the α-amylase inhibitor Tendamistat and mammalian metallothioneins that were performed in 1984–85. In hindsight their special importance for us is mainly anecdotal, although they make reference to a rather emotional time period.

When I presented the NMR structure of BUSI (paper **25**) in some lectures in the spring of 1984, the reaction was one of disbelief and suggestions that our structure must have been modeled after the crystal structure of a homologous protein. Apparently the structural biology community had thoroughly adjusted to the role of NMR as a method that could provide some exotic supplementary data, but which would not be suitable for *de novo* structure determination at atomic resolution. Unexpectedly, this opinion was also voiced by individual eminent colleagues in the NMR field. In the discussion following a seminar in Munich on May 14, 1984, Robert Huber proposed to settle the matter by independently solving a new protein structure in his laboratory by X-ray crystallography and in my laboratory by NMR. For this purpose, each one of us received 100 g (!) of the pure α-amylase inhibitor Tendamistat from Hoechst AG. Complete NMR assignments and the secondary structure determination in solution were obtained by the end of the summer 1984 (paper **26**), and eventually virtually identical complete three-dimensional structures were obtained in solution (paper **27**, Ref. 1) and in crystals.[2] As a complement to the detailed comparison of the two structures in

[1] Kline, A.D., Braun, W. and Wüthrich, K. (1988) *J. Mol. Biol. 204,* 675–724. Determination of the complete three-dimensional structure of the α-amylase inhibitor Tendamistat in aqueous solution by nuclear magnetic resonance and distance geometry.

[2] Pflugrath, J., Wiegand, E., Huber, R. and Vértesy, L. (1986) *J. Mol. Biol. 189,* 383–386. Crystal structure determination, refinement and the molecular model of the α-amylase inhibitor Hoe-467A.

[3] Billeter, M., Kline, A.D., Braun, W., Huber, R. and Wüthrich, K. (1989) *J. Mol. Biol. 206,* 677–687. Comparison of the high-resolution structures of the α-amylase inhibitor Tendamistat determined by nuclear magnetic resonance in solution and by X-ray diffraction in single crystals.

ref. 3, the paper **28** describes a particularly direct demonstration of the near-identity of the molecular architectures in crystals and in solution, since the crystal structure was solved again by molecular replacement using the NMR structure. Some editorials then caused many to believe that Tendamistat rather than the earlier work by Timothy Havel and Michael Williamson with BUSI (paper **25**) was the first complete *de novo* protein structure determination by NMR. The fact that a novel type of polypeptide fold had been determined in the NMR structure of Tendamistat apparently convinced many of the critics that NMR could actually do the job.

The Tendamistat experience was comforting when the structure determination of rabbit metallothionein was completed in the spring of 1985.[4] I presented this structure at Yale University on June 10 and at the University of Pittsburgh on June 12, where I was confronted with a crystal structure of rat metallothionein that was very different from our NMR structure.[5] We subsequently found that the NMR structures of three homologous metallothioneins from rabbit,[4,6] rat (paper **29**) and man[7] are virtually identical. Several years later the crystal structure of rat metallothionein was redetermined and found to coincide very closely with our NMR structure.[8,9]

---

[4] Braun, W., Wagner, G., Wörgötter, E., Vasak, M., Kägi, J.H.R. and Wüthrich, K. (1986) *J. Mol. Biol. 187,* 125–129. Polypeptide fold in the two metal clusters of metallothionein-2 by nuclear magnetic resonance in solution.

[5] Furey, W.F., Robbins, A.H., Clancy, L.L., Winge, D.R., Wang, B.C. and Stout, C.D. (1986) *Science 231,* 704–710. Crystal structure of Cd,Zn metallothionein.

[6] Arseniev, A., Schultze, P., Wörgötter, E., Braun, W., Wagner, G., Vasak, M., Kägi, J.H.R. and Wüthrich, K. (1988) *J. Mol. Biol. 201,* 637–657. Three-dimensional structure of rabbit liver [Cd_7]-metallothionein-2a in aqueous solution determined by nuclear magnetic resonance.

[7] Messerle, B.A., Schäffer, A., Vasak, M., Kägi, J.H.R. and Wüthrich, K. (1990) *J. Mol. Biol. 214,* 765–779. Three-dimensional structure of human [$^{113}Cd_7$]-metallothionein-2 in solution determined by nuclear magnetic resonance spectroscopy.

[8] Robbins, A.H., McRee, D.E., Williamson, M., Collett, S.A., Xuong, N.H., Furey, W.F., Wang, B.C. and Stout, C.D. (1991) *J. Mol. Biol. 221,* 1269–1293. Refined crystal structure of Cd,Zn metallothionein at 2.0 Å resolution.

[9] Braun, W., Vasak, M., Robbins, A.H., Stout, C.D., Wagner, G., Kägi, J.H.R. and Wüthrich, K. (1992) *Proc. Natl. Acad. Sci. USA 89,* 10124–10128. Comparison of the NMR solution structure and the X-ray crystal structure of rat metallothionein-2.

J. Mol. Biol. (1985) **183**, 503–507

Secondary Structure of the α-Amylase Polypeptide Inhibitor Tendamistat from *Streptomyces tendae* Determined in Solution by ^{1}H Nuclear Magnetic Resonance

Complete sequence-specific ^{1}H nuclear magnetic resonance assignments were obtained for the backbone hydrogen atoms in Tendamistat, a protein with 74 residues. From NOESY observation of ^{1}H–^{1}H short distance constraints, measurements of the spin–spin couplings $^3J_{HN\alpha}$ and a qualitative identification of slowly exchanging amide protons, two antiparallel β-sheets containing three and four strands, respectively, were identified. The peptide segments outside the β-sheets do not form regular secondary structure. Preliminary data were obtained on the relative spatial arrangements of the two β-sheets.

Tendamistat (HOE 467) is an α-amylase polypeptide inhibitor, the inhibitory action of which is characterized by tight binding to mammalian α-amylases (Vértesy *et al.*, 1984). The sequence of 74 amino acid residues (Aschauer *et al.*, 1981) shows no apparent homology to other known inhibitor sequences, thus precluding the use of comparisons with presently available protein spatial structures that might indicate the conformation adopted by Tendamistat. Studies of the biochemical properties reveal that the most important structural feature responsible for the inhibitory action of Tendamistat is not a specific set of functional groups in the sequence, but rather a disulfide-linked conformation (Vértesy *et al.*, 1984). It is apparent that further insights into the mechanism of action must await experimental determination of the three-dimensional structure. We have started a structure determination in solution by n.m.r.†, and a crystal structure determination by X-ray methods is also underway (R. Huber, personal communication).

Initial studies on the stability of Tendamistat with respect to temperature and pH as well as its solubility in aqueous media led us to select the following conditions for the 2D n.m.r. experiments: protein concentration of approximately 0·007 M, solvent H_2O or ^2H_2O and pH 3·2. The following experiments were recorded. (1) In ^2H_2O solution after complete exchange of all labile protons: DQF-COSY (Rance *et al.*, 1983), RELAYED-COSY (Wagner, 1983) and NOESY (Anil Kumar *et al.*, 1980). (2) In freshly prepared ^2H_2O solutions where the slowly exchanging amide protons were not exchanged: DQF-COSY and NOESY. (3) In a mixed solvent of 90% H_2O and 10% ^2H_2O, where all amide protons are preserved: DQF-COSY, RELAYED-COSY, NOESY and a double quantum spectrum (Wagner & Zuiderweg, 1983). All experiments were performed at 500 MHz with phase-sensitive data presentation (Marion & Wüthrich, 1983). To illustrate the quality of the spectra obtained, Figures 1 and 2 show regions of a NOESY spectrum and a DQF-COSY spectrum, respectively, recorded in ^2H_2O after complete exchange of the labile protons. Additional experimental details are given in the Figure legends. The spectrum of Figure 1 provided pivotal information towards the secondary structure determination described below. From the analysis of the 2D n.m.r. spectra, complete sequence-specific assignments of the backbone protons were obtained. The assignment work is also nearly complete for the side-chain protons and will be published in a forthcoming paper.

Several recent publications have shown that once the sequence-specific resonance assignments are available, several independent lines of evidence for locating regular secondary structure can be gathered from inspection of the 2D n.m.r. spectra. This includes the manifestation of close approach between specified backbone hydrogen atoms in the NOESY spectra (Wüthrich *et al.*, 1984), measurement of the spin–spin couplings $^3J_{NH\alpha}$ in phase-sensitive COSY spectra (Marion & Wüthrich, 1983; Pardi *et al.*, 1984), and delineation of hydrogen-bonding networks in helical and β-sheet structures (Wagner & Wüthrich, 1982; Wüthrich *et al.*, 1984). These data for Tendamistat are presented in Figures 3 and 4. Figure 4 also shows schematic drawings of two antiparallel β-sheets that could be characterized by the use of the n.m.r. data.

In Figure 3 the prevalence of peptide segments with several subsequent short $d_{\alpha N}$ distances indicated that a large portion of the polypeptide is in an extended form (Wüthrich *et al.*, 1984). Independent support for this conclusion came from the observation of coupling constants $^3J_{HN\alpha} > 8$ Hz for most of these residues (Pardi *et al.*, 1984). Seven peptide segments could then be assembled into two antiparallel β-sheets on the basis of the short interstrand ^{1}H–^{1}H contacts between residues that

† Abbreviations used: n.m.r., nuclear magnetic resonance; 2D, two-dimensional; COSY, 2D correlated spectroscopy; NOESY, 2D nuclear Overhauser enhancement spectroscopy; DQF-COSY, double quantum filtered COSY; RELAYED-COSY, relayed coherence transfer spectroscopy; p.p.m., parts per million.

0022-2836/85/110503-05 $03.00/0

© 1985 Academic Press Inc. (London) Ltd.

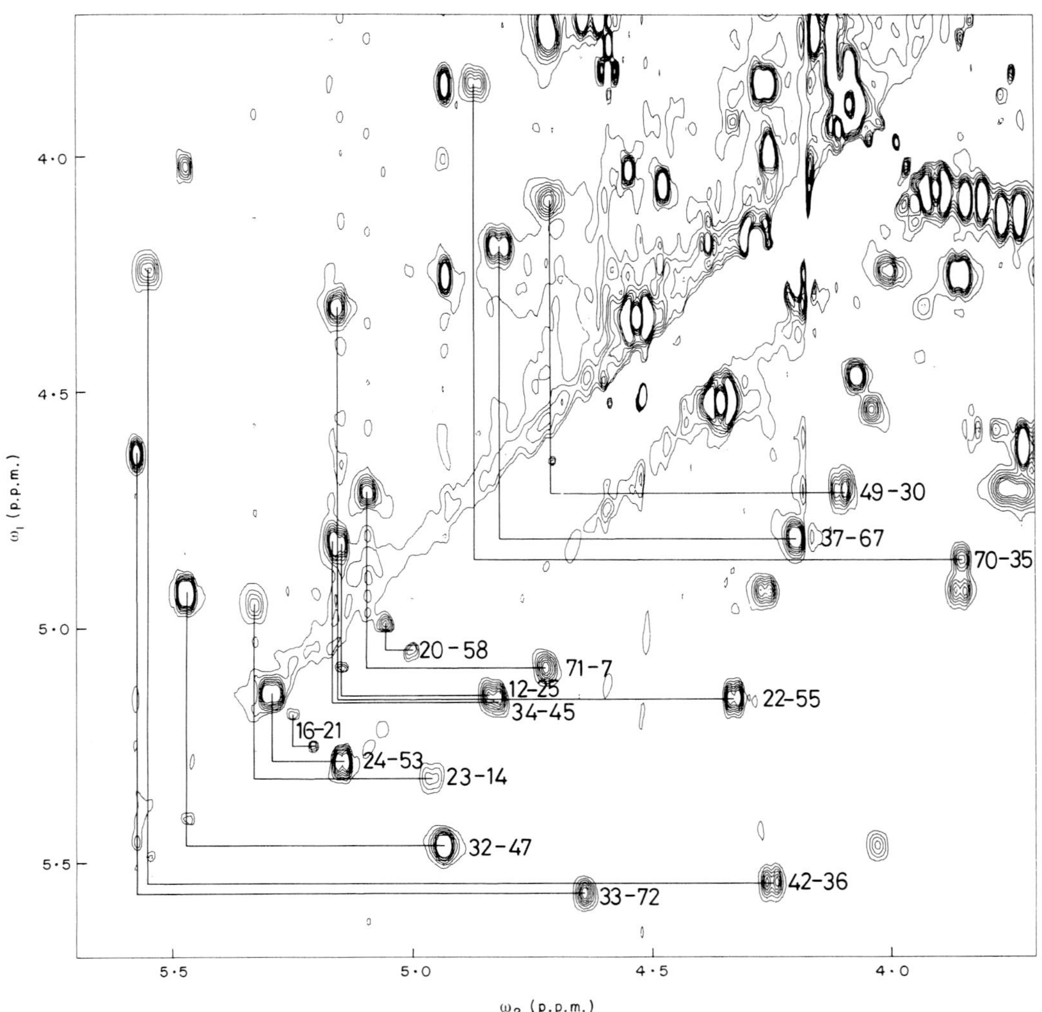

Figure 1. Contour plot of the region (ω_1 = 3·7 to 5·7 p.p.m., ω_2 = 3·7 to 5·7 p.p.m.) from an absorption mode 500 MHz ^{1}H NOESY spectrum of Tendamistat recorded with suppression of the diagonal (Denk *et al.*, 1985). This region contains mainly crosspeaks connecting different C^α protons. The protein was dissolved in ^2H_2O and all the labile protons were replaced by ^{2}H, by heating the sample to 70°C for 10 min; residual water protons were subsequently removed by lyophilization; protein concentration, 0·007 M; p^{2}H = 3·2; t = 50°C. The mixing time was 200 ms; 341 values were collected in t_1 and the data matrix was zero-filled to 2K in t_1 and 4K in t_2 before Fourier transformation. The resulting digital resolution was 4·48 Hz/point in ω_1 and 2·24 Hz/point in ω_2. The p.p.m. scale was calibrated using the highest field methyl resonance at 0·12 p.p.m. as an internal reference. Only positive contour levels have been plotted. Horizontal and vertical lines connect the symmetry-related pairs of crosspeaks that manifest close approach between the C^α protons of the residues indicated by the sequence positions. Occurrence of these crosspeaks implies that the distance between the specified C^α protons is shorter than approximately 4·0 Å.

are far apart in the sequence (Fig. 4). Furthermore, evidence was found that three β-bulges (Richardson, 1981) prevailed near residues 43 to 44, 56 to 57 and 68 to 69. Finally, the patterns of slowly exchanging amide protons are fully compatible with the β-sheet supersecondary structure from the distance constraints. Slow exchange for several subsequent residues was observed exclusively for the non-peripheral strands, since all amide protons are involved in hydrogen bonds (Fig. 4). Only isolated residues were found to exchange slowly in the peripheral strands, where the amide protons of nearest next residues are hydrogen-bonded (Wagner & Wüthrich, 1982).

In conclusion, we have obtained experimental evidence from four different independent avenues

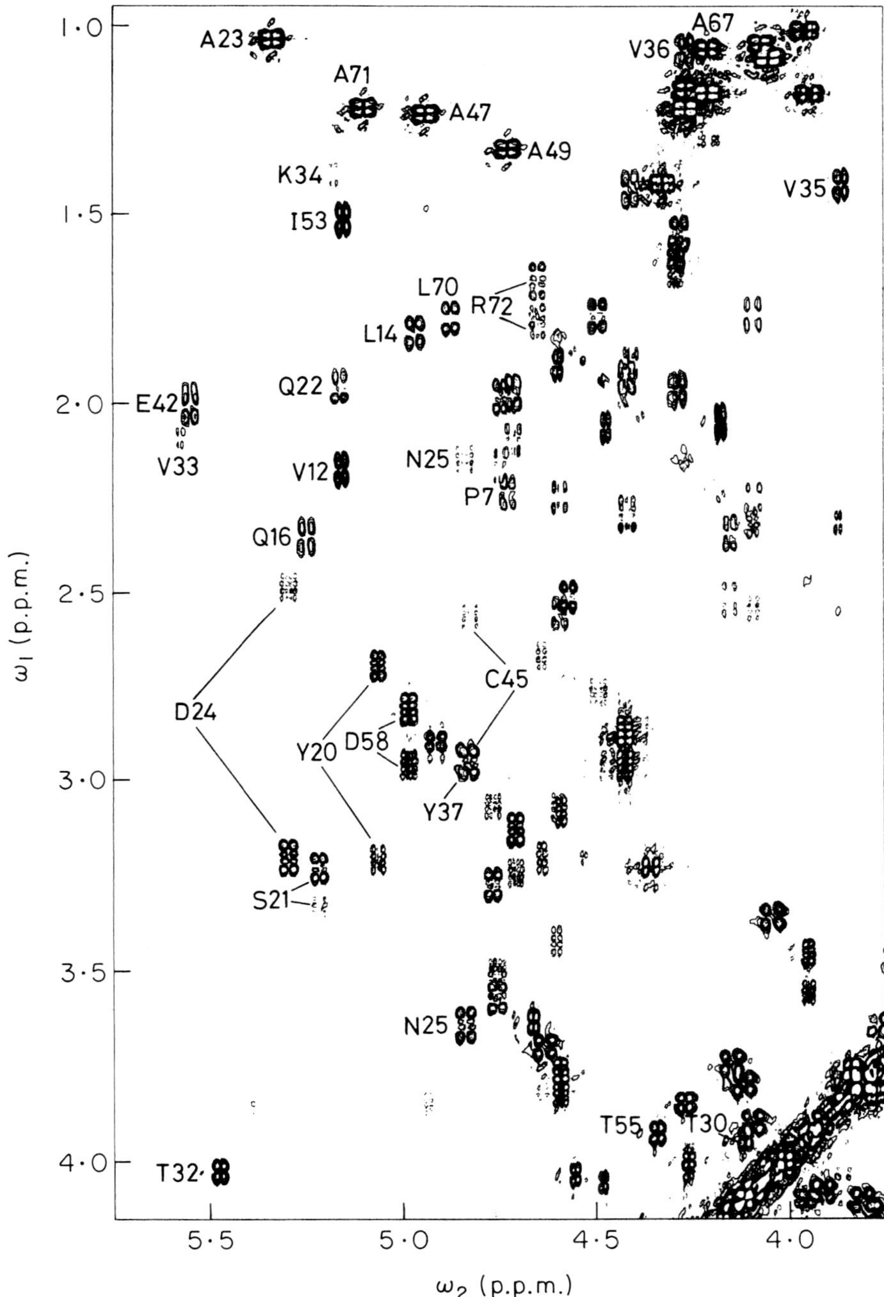

Figure 2. Contour plot of the region ($\omega_1 = 0{\cdot}95$ to $4{\cdot}15$ p.p.m., $\omega_2 = 3{\cdot}75$ to $5{\cdot}75$ p.p.m.) from an absorption mode 500 MHz ^{1}H DQF-COSY. Sample preparation and experimental conditions were identical to those for Fig. 1; 956 values were collected in t_1 and the data matrix was zero-filled to 2K in t_1 and 4K in t_2 before Fourier transformation. The resulting digital resolution was 3·49 Hz/point in ω_1 and 1·74 Hz/point in ω_2. The p.p.m. scale was calibrated as in Fig. 1. Both positive and negative levels have been plotted. This region contains the connectivities between the C^α protons identified in Fig. 1 and the C^β protons of the same residues (these crosspeaks have been labeled). This Figure illustrates the quality of the spectra used for obtaining the resonance assignments in Tendamistat.

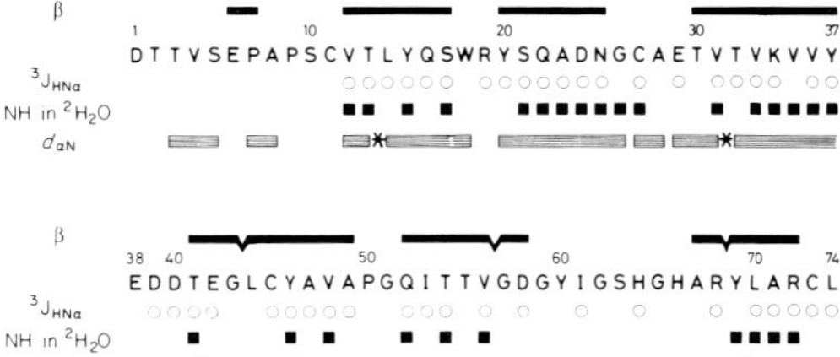

Figure 3. Sequence of Tendamistat and survey of n.m.r. data used for the characterization of the secondary structure. Below the sequences, hatched bars are drawn linking those residues between which a short sequential distance ($\lesssim 2{\cdot}5$ Å) between the C^α proton of residue i and the amide proton of the following residue, $i+1$, was observed ($d_{\alpha N}$). * Indicates locations for which $d_{\alpha N}$ could not be studied due to overlap of the $C^\alpha H$ position with the decoupler frequency. Open circles identify those residues for which $^3J_{HN\alpha} \gtrsim 8$ Hz. No spin–spin coupling data are given for the glycines or, of course, for the prolines. Filled squares identify the residues for which the amide protons exchange sufficiently slowly so that the $C^\alpha H$–NH crosspeaks could be observed in a DQF-COSY spectrum recorded during 12 h in a freshly prepared ^2H_2O solution of the protein; $p^2H = 3{\cdot}2$, $t = 50°C$. Above the sequence, filled bars indicate the sequence positions of the 7 β-strands. Nicks identify the locations of 3 β-bulges.

for characterizing the secondary structure in Tendamistat. The protein contains the two β-sheets shown in Figure 4. In addition there are five chain reversals, two of which are between adjacent hydrogen-bonded strands. The arrangement of the polypeptide chain in the two sheets corresponds to a greek key β-barrel (Richardson, 1981). One loop of eight residues from positions 59 to 66 and both chain termini appear to have no regular secondary structure. As a consequence of the β-sheet formation the side-chains are oriented above and below the plane of the sheet in an alternating fashion. Upon examining the nature of the residues, both sheets are found to have an amphipathic distribution of the side-chains. In the presentation of Figure 4 the predominantly lipophilic side of the β-sheet (a) faces the reader, whereas for sheet (b) it is directed away from the reader. Determination of the relative locations of the two β-sheets requires the collection of a complete set of NOESY distance constraints, including side-chain protons, and a structural analysis using distance geometry calculations (Havel & Wüthrich, 1984; Williamson *et al.*, 1985). However, intuitively, one would expect that the two hydrophobic sides face each other and form the interior of the protein, while the hydrophilic sides interact with the solvent, thus forming a compact, "globular" molecular structure. Initial support for this general orientation of the two β-sheets has been obtained from observation of several NOEs between the lipophilic sides of the two β-sheets.

Since we submitted this manuscript Dr R. Huber has informed us that the crystal structure determination of Tendamistat at the MPI für

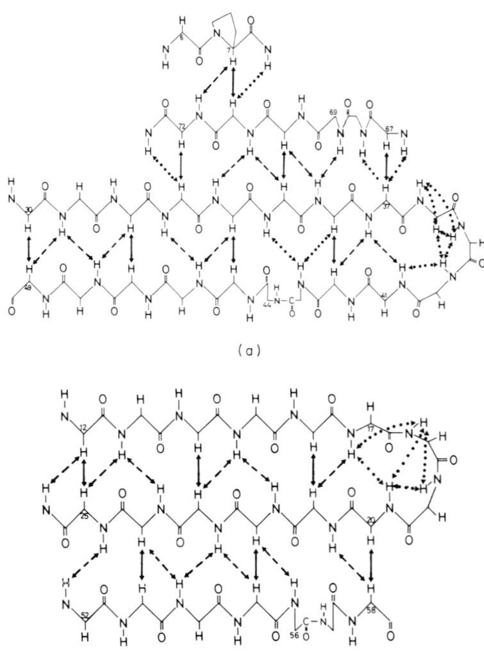

Figure 4. Schematic drawings of 2 antiparallel β-structures in Tendamistat, which were characterized by ^1H n.m.r. in solution. The complete backbone is drawn for the peptide segments involved in the β-strands and the hairpin tight turns. Filled arrows indicate the locations of the short $C^\alpha H$–$C^\alpha H$ distances manifested in Fig. 1. Dotted arrows show the locations of short interstrand ^1H–^1H distances observed only in H_2O solution, i.e. they involve at least one relatively rapidly exchanging amide proton. Broken arrows indicate short ^1H–^1H contacts involving exclusively slowly exchanging amide protons and C^α protons, so that they could be observed in H_2O and in a freshly prepared ^2H_2O solution (see Fig. 3).

Biochemie, Munich, has in the meantime resulted in the construction of a molecular model. According to Dr Huber's comparison of this model with Fig. 4 of this work, identical β-sheet secondary structures prevail in the crystal and in solution.

An ample supply of Tendamistat (HOE 467) was obtained from Hoechst Aktiengesellschaft, Frankfurt am Main. Financial support by the Schweizerischer Nationalfonds is gratefully acknowledged (project 3.284.82 and International Postdoctoral Fellowship to A.K.).

Allen D. Kline
Kurt Wüthrich

Institut für Molekularbiologie und Biophysik
Eidgenössische Technische Hochschule
Zürich-Hönggerberg, CH-8093 Zürich, Switzerland

Received 2 January 1985, and in revised form
11 February 1985

References

Anil Kumar, Ernst, R. R. & Wüthrich, K. (1980). *Biochem. Biophys. Res. Commun.* **95**, 1–6.

Aschauer, H., Vértesy, L. & Braunitzer, G. (1981). *Hoppe-Seyler's Z. Physiol. Chem.* **362**, 465–467.

Denk, W., Wagner, G., Rance, M. & Wüthrich, K. (1985). *J. Magn. Reson.* **62**, 336–340.

Havel, T. & Wüthrich, K. (1984). *Bull. Math. Biol.* **46**, 673–698.

Marion, D. & Wüthrich, K. (1983). *Biochem. Biophys. Res. Commun.* **113**, 967–974.

Pardi, A., Billeter, M. & Wüthrich, K. (1984). *J. Mol. Biol.* **180**, 741–751.

Rance, M., Sørensen, O., Bodenhausen, G., Wagner, G., Ernst, R. R. & Wüthrich, K. (1983). *Biochem. Biophys. Res. Commun.* **117**, 479–485.

Richardson, J. S. (1981). *Advan. Protein Chem.* **34**, 167–339.

Vértesy, L., Oeding, V., Bender, R., Zepf, K. & Nesemann, G. (1984). *Eur. J. Biochem.* **141**, 505–512.

Wagner, G. (1983). *J. Magn. Reson.* **55**, 151–156.

Wagner, G. & Wüthrich, K. (1982). *J. Mol. Biol.* **160**, 343–361.

Wagner, G. & Zuiderweg, E. R. P. (1983). *Biochem. Biophys. Res. Commun.* **113**, 854–860.

Williamson, M., Havel, T. & Wüthrich, K. (1985). *J. Mol. Biol.* **182**, 295–316.

Wüthrich, K., Štrop, P., Ebina, S. & Williamson, M. P. (1984). *Biochem. Biophys. Res. Commun.* **122**, 1174–1178.

Wüthrich, K., Billeter, M. & Braun, W. (1984). *J. Mol. Biol.* **180**, 715–740.

Edited by M. F. Moody

J. Mol. Biol. (1986) **189**, 377–382

Studies by ^{1}H Nuclear Magnetic Resonance and Distance Geometry of the Solution Conformation of the α-Amylase Inhibitor Tendamistat

This is a preliminary report on the determination of the solution conformation of the α-amylase inhibitor Tendamistat by nuclear magnetic resonance and distance geometry calculations. A characterization is given of the complete polypeptide backbone fold and the side-chains of the presumed active site in this protein. These results are based on complete sequence-specific resonance assignments, a list of 401 distance constraints from nuclear Overhauser effects, 168 distance constraints from hydrogen bonds and disulphide bridges, and 50 torsion angle constraints from measurements of spin–spin coupling constants.

Tendamistat (Hoe-467) is a polypeptide inhibitor that binds tightly to mammalian α-amylases (Aschauer *et al.*, 1983; Vértesy *et al.*, 1984). The only information available on the conformation of this class of proteins (Murai *et al.*, 1985; Vértesy & Tripier, 1985) comes from earlier n.m.r.† studies (Kline & Wüthrich, 1985), which showed that nearly the entire polypeptide chain in Tendamistat is in two antiparallel β-pleated sheets. This letter presents the polypeptide backbone fold of Tendamistat in aqueous solution and the side-chain conformations of the proposed active-site residues.

The spatial structure determination relies on sequence-specific ^{1}H n.m.r. assignments (Wüthrich *et al.*, 1982), a list of NOESY cross peaks between pairs of protons that are close in space (Anil Kumar *et al.*, 1981), and a computational method to calculate tertiary structures compatible with these distance constraints (Braun *et al.*, 1981, 1983; Havel & Wüthrich, 1984, 1985; Braun & Gō, 1985). This letter gives a survey of the n.m.r. input constraints and the distance geometry calculations, and a characterization of the chain fold and the conformation of the active site.

Tendamistat was studied in 5 mM-solution in H_2O or ^2H_2O, with the pH meter reading at room temperature adjusted to 3·2 with perchloric acid. All n.m.r. experiments were run on a Bruker WM-500 spectrometer, with the probe temperature regulated to 50°C. After complete exchange of the labile protons in ^2H_2O, diagonal-suppressed NOESY spectra (Denk *et al.*, 1985) were collected, while "regular" NOESY spectra were recorded in a mixed solvent of 90% H_2O/10% ^2H_2O. Three different mixing times were used, i.e. 200, 100 and 50 milliseconds. The quality of the spectra obtained with these experimental conditions was shown by Kline & Wüthrich (1985).

The approach adopted for determining the upper distance limits corresponding to the NOESY cross peaks was similar to that used previously (Braun *et al.*, 1983; Williamson *et al.*, 1985). The 200 millisecond spectra were used for the NOESY peak assignments, those at 50 and 100 milliseconds were helpful in eliminating spin diffusion peaks from the analysis (Anil Kumar *et al.*, 1981). Using the 100 millisecond data, the NOESY cross peak intensities were grouped into three separate categories, and an upper distance limit was defined for each category. In the H_2O spectra the cross peaks with amide protons had similar line shapes throughout, so that the peak height was used as a measure of the NOE intensity. Among the cross peaks with non-labile protons measured in the ^2H_2O spectra, the variation of the line shapes was far greater. Therefore, the area of the ω_2 cross sections was used as the intensity measurement. The choice of upper distance limits for the three categories was made for the cross peaks in the ^2H_2O spectra by comparing with the intensities of a number of peaks corresponding to distances fixed by the covalent structure, such as geminal methylene protons, or αH–βCH_3 in Ala. For the amide protons no covalently fixed distances exist, so the maximum and minimum intensities observed for $d_{\alpha N}$ and $d_{N\alpha}(i, i)$ cross peaks were used to calibrate the range of allowed distances, 2·2 to 3·6 Å and 2·2 to 2·8 Å, respectively. Further slight adjustments of these calibrations resulted from observations made during the distance geometry calculations. Also, at this stage two NOEs with Trp18 were eliminated from the input (see below). Because no stereospecific resonance assignments had been made, the pseudo-atom representation was used as required (Wüthrich *et al.*, 1983). The NOE data used in the calculations have been summarized in Figure 1 as a diagonal plot. The lower left triangle shows all backbone–backbone NOE constraints, while the

† Abbreviations used: n.m.r., nuclear magnetic resonance; NOE, nuclear Overhauser enhancement; NOESY, 2-dimensional nuclear Overhauser enhancement spectroscopy; r.m.s.d., root-mean-square distance.

0022–2836/86/100377–06 $03.00/0

© 1986 Academic Press Inc. (London) Ltd.

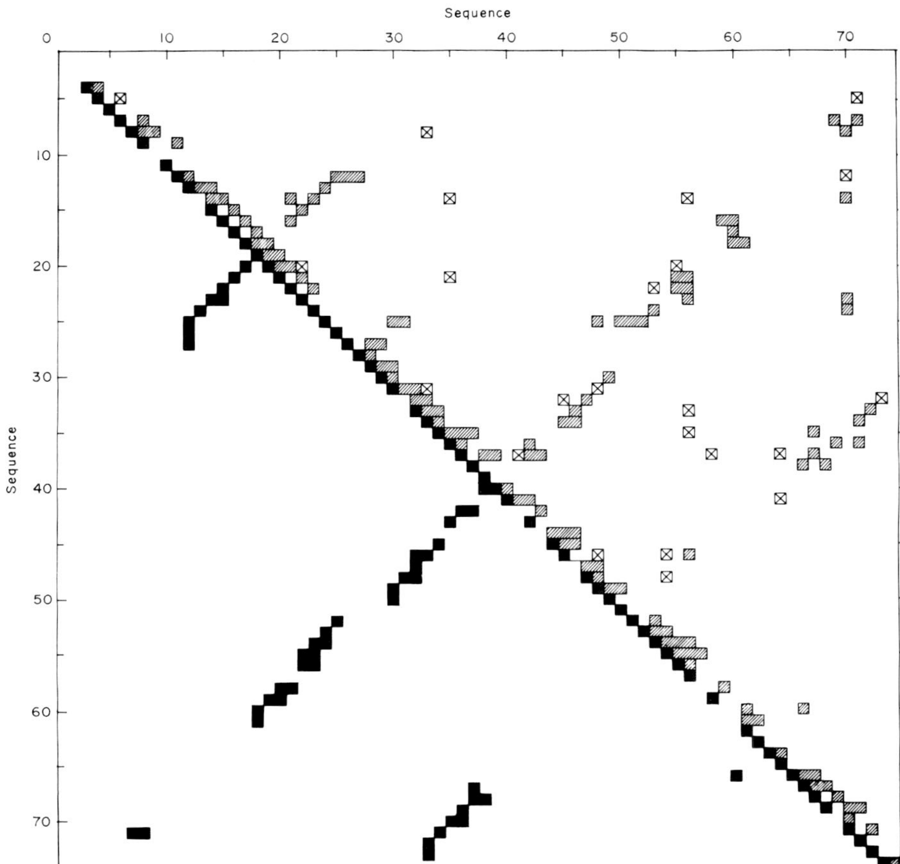

Figure 1. Diagonal plot of the NOE distance constraints for Tendamistat. Both axes are calibrated with the sequence of the protein. The squares connect pairs of residues linked by one or more NOE constraints. The filled squares in the lower left part indicate constraints between backbone protons. The hatched squares and crossed squares in the upper right half indicate backbone–side-chain and side-chain–side-chain NOE constraints, respectively. Whenever a hatched and a crossed square occur at the same place, only the hatched square is shown.

upper right half contains the backbone–side-chain and the side-chain–side-chain constraints. Between sequentially neighbouring residues, 65 backbone–backbone, 48 backbone–side-chain and six side-chain–side-chain NOE constraints were identified. There are a further 51 backbone–backbone, 83 backbone–side-chain and 73 side-chain–side-chain medium-range and long-range NOE constraints, as well as 75 intra-residue constraints.

The NOE data were supplemented with additional experimental observations using the same corresponding constraints as used previously (Williamson *et al.*, 1985). This includes the two disulphide bonds 11–27 and 45–73 (Aschauer *et al.*, 1983), 26 hydrogen bonds identified in the two antiparallel β-sheets, 43 backbone dihedral ϕ angles corresponding to spin–spin coupling constants $^3J_{HN\alpha} \geq 8{\cdot}0$ Hz (Fig. 3 of Kline & Wüthrich, 1985), and the torsion angles χ^1 of Val12, Thr30, Thr32, Val36, Val48, Ile53 and Thr55, which all correspond to spin–spin coupling constants $^3J_{\alpha\beta} \geq 8{\cdot}0$ Hz. As was discussed by Williamson *et al.* (1985), some of these additional constraints are redundant with those imposed by ^{1}H–^{1}H NOEs (for example the locations of the disulphide bonds and the hydrogen bonds could be identified independently on the basis of the NOE data), but the additional experimental observations impose tighter constraints than the NOEs alone.

In the calculations using the DISMAN program (Braun & Gō, 1985), only the dihedral angles were varied, with the bond angles and bond lengths fixed at the standard ECEPP values (Momany *et al.*, 1975; Némethy *et al.*, 1983). It is implicit in these computations that additional constraints arise because of the intrinsic rigidity of the polypeptide covalent structure and the van der Waals' volumes of the individual atoms. "Core radii" as defined by

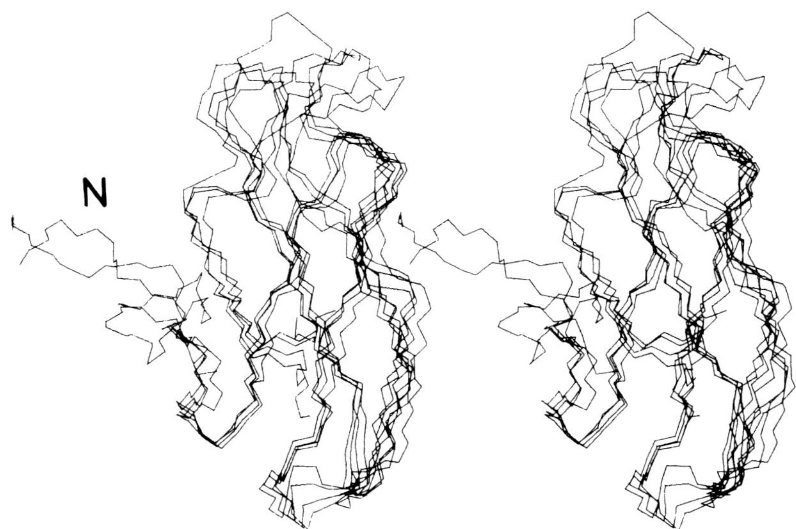

Figure 2. Stereo view of the 4 solutions I to IV calculated by DISMAN, plotting only the backbone atoms N, C^{α}, C′. The structures were superimposed so as to minimize the r.m.s.d. of the heavy atoms of the backbone with respect to structure I. Region 62 to 66 was eliminated from the best fit to better align the constrained areas. To simplify the presentation, the disulphide bridges were not drawn. In this particular orientation the 3 C^{α} positions of Ser17, Lys34, and Asn25 form a vertical plane perpendicular to the projection plane. Lys34 is in the back, and its projection is centred between the projections of the other 2 α-carbons.

Braun & Gō (1985) were used to account for the repulsive forces between different atoms or groups of atoms. To avoid local minima the program uses a variable target function, which considers the shorter range constraints before proceeding to longer range constraints. The performance of this program was tested with calculations on the basic pancreatic trypsin inhibitor, using input data obtained from the crystal structure or from experimental n.m.r. distance constraints (Braun & Gō, 1985; G. Wagner, W. Braun, T. F. Havel, T. Schaumann, N. Gō & K. Wüthrich, unpublished results). These tests showed that, independent of the starting conformation, the algorithm converges to a solution if a complete set of short-range and long-range distance information is used. If less complete constraint sets were used, the computations with some of the initial conformations were occasionally trapped in local minima; these could then usually be excluded on the basis of the large residual distance and core violations.

For Tendamistat the experimental n.m.r. constraints were used as input for ten separate structure computations, which started from different initial conformations and were pursued in parallel. In the initial phase, with the target function covering up to eight residues, local minima problems arising in the individual calculations were generally located in different regions of the polypeptide chain. On this level of local conformations, these problems could be solved for all ten computations by re-randomization of the torsion angles weighted using the ranges in the untrapped structures. This approach was particularly helpful in the regions of the tight turns. Subsequently, four calculations so far converged to a solution of the complete constraint list. These are the basis for the following discussions on the solution conformation of Tendamistat.

Figures 2 and 3 show, respectively, the backbone conformations of the four solutions and a schematic drawing of structure "I". This orientation illustrates the two-sheet architecture of the protein, with one sheet consisting of the strands 12 to 17, 20 to 25 and 52 to 58 in the foreground, and the other with the strands 30 to 37, 41 to 49 and 67 to 73 behind. In the previous study of the secondary structure, Tendamistat was given the domain classification of a Greek-key β-barrel (Kline & Wüthrich, 1985). The +1, +3, −1, −1, +3 topology pattern recognized in Figures 2 and 3 supports the classification as a Greek-key barrel (Richardson, 1981). Though similar barrel topologies have been identified, it appears that none of the other proteins has an identical chain fold to that of Tendamistat. The β-barrel domain structure is also readily apparent from the diagonal presentation of the backbone–backbone NOEs in Figure 1. It is also reflected in the NOEs involving the side-chain protons (upper right in Fig. 1).

The quality of the structures obtained was judged from the residual violations of the n.m.r. and core radii constraints, and from the r.m.s.d. between the different solutions in Figure 2. Table 1 lists the average violation for n.m.r. distance constraints in both the initial and final structures, and thus shows that on average the

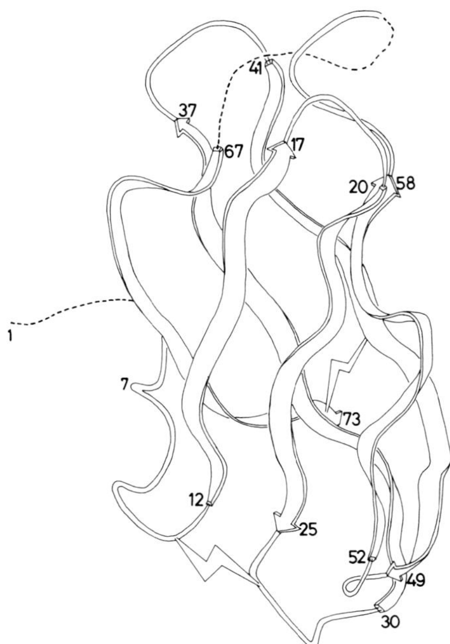

Figure 3. Schematic drawing of the backbone topology obtained using the co-ordinates of structure I. The ribbon drawing was generated by the CONFOR program (Billeter *et al.*, 1985), using the same orientation as in Fig. 2. Modifications to this computer drawing were added by hand to give the picture a 3-dimensional effect. The arrowed ribbons indicate the position and direction of the β-sheet strands. The rope structure was used for the loop areas, with the broken lines indicating those places where the data did not give a well-confined solution. The disulphide bridges are indicated by lightning bolts.

calculations had reduced the violations by a factor of about 500. The four solutions all had similar average violations, indicating that each structure was an equally credible solution. Table 1 also shows that in the initial structures a large number of constraints were violated by 10·0 Å or more, indicating that the starting conformations were not at all solutions to the constraints. In the final structures, all but one of the residual violations were smaller than 0·5 Å. This indicates that after the elimination of two NOEs with the indole ring of Trp18 (see below) the experimental distance constraints contained no serious inconsistencies. Relative to the constraints by the core radii, all four solutions were adequate as there were no violations of the contact limits over 0·5 Å, and only two violations exceeded 0·3 Å in all four of the solutions. This is a good result considering that there are at least 2500 potential core contacts in the folded molecule. Further refinement of the van der Waals' interactions is envisaged with the use of a proper energy function (Momany *et al.*, 1975; Némethy *et al.*, 1983).

In Figure 2 the four conformers are superimposed for minimum r.m.s.d. between the backbone heavy atoms of structure I and the other structures. Clearly, a larger spread is observed for the terminal residues 1 to 5 and 74, which are not confined by any long-range NOEs (Fig. 1). Similarly, the increased variability of the loop region 62 to 66 is consistent with this region's low density of interconnecting NOE data (Fig. 1). For the residues 6 to 73 in the four structures, the average r.m.s.d. for the backbone heavy atoms of the six pair-wise comparisons is 1·6 Å. This compares with 15·2 Å for the same comparison of the corresponding starting structures. The average r.m.s.d. values thus show that the starting structures were not related to each other, whereas the final structures have a similar global conformation. Apart from the terminal regions and the loop 62 to 66, the distance geometry calculations indicate that the conformation of the polypeptide backbone is well-defined by the n.m.r. constraints.

The following are some additional features of the Tendamistat backbone fold. The previously identified antiparallel β-sheets were found to be consistent with the additional, newly acquired n.m.r. constraints and the covalent structure of the molecule. A right-handed twist is seen for both sheets, which is consistent with the handedness in

Table 1

Computations of the Tendamistat conformation using DISMAN: comparison of the violations of distance constraints imposed by the n.m.r. data for the initial and final structures

Computation	Initial structure: Average violation† (Å)	Initial structure: Number of violations >10·0 Å	Initial structure: Number of violations >0·5 Å	Final structure: Average violation† (Å)	Final structure: Number of violations >0·5 Å
I	12·5	236	314	0·026	0
II	12·5	228	301	0·024	0
III	10·4	226	304	0·026	0
IV	15·0	234	303	0·029	1‡

† Sum of all distance violations divided by the total number of n.m.r. distance constraints, i.e. 569.
‡ The single violation was 0·53 Å.

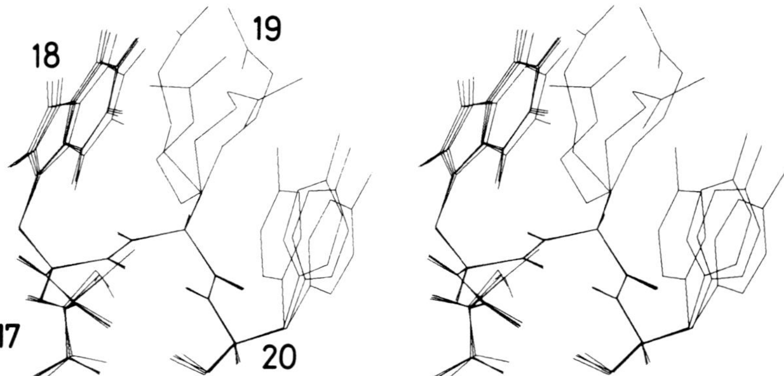

Figure 4. Stereo view taken from the 4 structures in Fig. 2, showing the complete residues Ser17, Trp18, Arg19, and Tyr20. The structures were superimposed so as to minimize the r.m.s.d. of the backbone atoms with respect to structure I. For the backbone and the indole ring all atoms are shown, for the rest of the structure the hydrogen atoms have been omitted. (This orientation of the indole ring is not unique, see the text.)

other proteins (Schulz & Schirmer, 1979; Richardson, 1981). In the structures in Figures 2 and 3 the predominantly hydrophobic sides of the two β-sheets face each other, forming a hydrophobic interior. A tentative identification of the sequence locations and types of tight turns was obtained from the dihedral angles of the corresponding residues. Two non-hairpin β-turns occur at positions 10–11 and 50–51 (lower left and lower right, respectively, in Fig. 3). The 10–11 turn was classified as a type I turn, the 50–51 turn as type II (Richardson, 1981). Two hairpin turns at 18–19 and 37–41 are at the top of the molecule in Figure 3. The two-residue turn appears to be of type I rather than the more commonly observed type I′ for hairpin turns (Sibanda & Thornton, 1985). A classification of the turn at 37–41 would be premature, even though the NOE pattern is consistent with a class of recently identified five-residue turns (Sibanda & Thornton, 1985). Another interesting observation is the formation of the 45–73 disulphide link to the outside of the β-sheet rather than on the inside.

The understanding of α-amylase inhibition by Tendamistat and related proteins centres on a highly conserved sequence of Trp-Arg-Tyr (Aschauer *et al.*, 1983; Murai *et al.*, 1985; Vértesy & Tripier, 1985). These amino acid types have also been implicated by inactivation studies on one member of the inhibitor family (Arai *et al.*, 1985). In Tendamistat these residues are in positions 18 to 20 and form part of the protein surface. Figure 4 shows this peptide segment for the four structures of Figure 2. This specific conformation is consistent with the majority of the NOEs. The location of the Arg between the two rings is also manifested in the high field shifts of the Arg resonances (NH, 6·73; $C^{\alpha}H$, 3·74; $C^{\beta}H_2$, 0·41, −0·17; $C^{\gamma}H_2$, 0·37, 0·37). Despite the surface location, the Trp18 side-chain is confined by nine NOEs, so that its conformation appears particularly well-defined in Figure 4. However, there were two additional NOEs that were found not to be consistent with the ensemble of all other NOEs by exploratory initial distance geometry calculations. Therefore, it cannot be excluded that there are multiple conformers that differ locally by the orientation of the indole ring of Trp18.

This study on the solution conformation of Tendamistat has been carried out without any information from crystallographic studies on this or homologous proteins. To facilitate the comparison of the solution and crystal structures of this protein, the orientation of the molecule described in Figure 2 is also used in the Figures of the accompanying paper on the X-ray data on Tendamistat (Pflugrath *et al.*, 1986).

This letter is a progress report on the n.m.r. determination of the solution conformation of Tendamistat. Additional work will include a re-examination of the NOESY spectra using the present molecular model in conjunction with the sequence-specific resonance assignments to complete the assignments of the NOESY cross peaks, more quantitative evaluation of the NOE intensities, stereospecific assignments of the interior leucine and valine methyl groups and collection of more complete data on side-chain orientations by further measurements of J-couplings. Combined with energy minimization these additional data can be expected to yield refinements of the backbone conformation, characterization of the interior amino acid side-chains, and provide a basis for further studies of the protein surface.

An ample supply of Tendamistat (Hoe-467) was obtained from Höchst Aktiengesellschaft, Frankfurt am Main. All calculations were done at the Zentrum für Interaktives Rechnen, ETH-Hönggerberg. Financial support by the Schweizerischer Nationalfonds is gratefully acknowledged (project 3.284.82 and international postdoctoral fellowship to A.K.). We thank Mrs E. Huber and Mrs E.-H. Hunziker for the careful preparation of the manuscript and the illustrations.

Allen D. Kline
Werner Braun
Kurt Wüthrich

Institut für Molekularbiologie und Biophysik
Eidgenössische Technische Hochschule-Hönggerberg
CH-8093 Zürich, Switzerland

Received 19 December 1985

References

Anil Kumar, Wagner, G., Ernst, R. R. & Wüthrich, K. (1981). *J. Amer. Chem. Soc.* **103**, 3654–3658.

Arai, M., Oouchi, N., Goto, A., Ogura, S. & Murao, S. (1985). *Agric. Biol. Chem.* **49**, 1523–1524.

Aschauer, H., Vértesy, L., Nesemann, G. & Braunitzer, G. (1983). *Hoppe-Seyler's Z. Physiol. Chem.* **364**, 1347–1356.

Billeter, M., Engeli, M. & Wüthrich, K. (1985). *J. Mol. Graph.* **3**, 79–83.

Braun, W. & Gō, N. (1985). *J. Mol. Biol.* **186**, 611–626.

Braun, W., Bösch, C., Brown, L. R., Gō, N. & Wüthrich, K. (1981). *Biochim. Biophys. Acta,* **667**, 377–396.

Braun, W., Wider, G., Lee, K. H. & Wüthrich, K. (1983). *J. Mol. Biol.* **169**, 921–948.

Denk, W., Wagner, G., Rance, M. & Wüthrich, K. (1985). *J. Magn. Res.* **62**, 336–340.

Havel, T. & Wüthrich, K. (1984). *Bull. Math. Biol.* **46**, 673–698.

Havel, T. & Wüthrich, K. (1985). *J. Mol. Biol.* **182**, 281–294.

Kline, A. D. & Wüthrich, K. (1985). *J. Mol. Biol.* **183**, 503–507.

Momany, F. A., McGuire, F. F., Burgess, A. W. & Scheraga, H. A. (1975). *J. Phys. Chem.* **79**, 2361–2381.

Murai, H., Hara, S., Ikenaka, T., Goto, A., Arai, M. & Murao, S. (1985). *J. Biochem.* **97**, 1129–1133.

Némethy, G., Pottle, M. S. & Scheraga, H. A. (1983). *J. Phys. Chem.* **87**, 1883–1887.

Pflugrath, J., Wiegand, E., Huber, R. & Vértesy, L. (1986). *J. Mol. Biol.* **189**, 383–386.

Richardson, J. S. (1981). *Advan. Protein Chem.* **34**, 167–339.

Schulz, G. E. & Schirmer, R. H. (1979). *Principles of Protein Structure,* Springer-Verlag, New York.

Sibanda, B. L. & Thornton, J. M. (1985). *Nature (London),* **316**, 170–174.

Vértesy, L. & Tripier, D. (1985). *FEBS Letters,* **185**, 187–190.

Vértesy, L., Oeding, V., Bender, R., Zepf, K. & Nesemann, G. (1984). *Eur. J. Biochem.* **141**, 505–512.

Williamson, M. P., Havel, T. F. & Wüthrich, K. (1985). *J. Mol. Biol.* **182**, 295–315.

Wüthrich, K., Wider, G., Wagner, G. & Braun, W. (1982). *J. Mol. Biol.* **155**, 311–319.

Wüthrich, K., Billeter, M. & Braun, W. (1983). *J. Mol. Biol.* **169**, 949–961.

Edited by M. F. Moody

J. Mol. Biol. (1989) **206**, 669–676

Solution of the Phase Problem in the X-ray Diffraction Method for Proteins with the Nuclear Magnetic Resonance Solution Structure as Initial Model

Patterson Search and Refinement for the α-Amylase Inhibitor Tendamistat

Werner Braun[1], Otto Epp[2], Kurt Wüthrich[1] and Robert Huber[2]

[1] *Institut für Molekularbiologie und Biophysik Eidgenössische Technische Hochschule-Hönggerberg CH-8093 Zürich, Switzerland*

[2] *Max-Planck-Institut für Biochemie 8033 Martinsried, F.R.G.*

(Received 18 July 1988, and in revised form 6 December 1988)

Patterson search calculations using the three-dimensional structure of the α-amylase inhibitor from *Streptomyces tendae* obtained from experimental nuclear magnetic resonance (n.m.r.) data were performed to study the possibility of solving the phase problem in the X-ray diffraction method with protein structures determined by n.m.r. Using all heavy atoms (C, N, O, S) of the residues 5 to 73 in the best n.m.r. structure of the α-amylase inhibitor (520 out of the 558 heavy atoms in the complete polypeptide chain), the maximum of the rotation function corresponded to the correct solution obtained by the previous independent determination of the crystal structure. However, additional local maxima, which are not significantly lower than the global maximum, also showed up. Performing the Patterson search with a model containing the backbone atoms and the heavy atoms of only the interior side-chains (399 atoms), which are much better defined by the n.m.r. data, the correct maximum was significantly higher than all other maxima. A translation search for the best orientation of the latter model yielded the correct solution. The energy-restrained crystallographic refinement was performed with this model to an R-factor of 26%. This corresponds approximately to the R-factor calculated for the X-ray crystal structure previously determined using the isomorphous replacement technique, if the residues 1 to 4 and 74 and all localized solvent molecules were removed from this structure. During the refinement the root-mean-square deviation between the two structures decreased from 1·03 Å to 0·26 Å for the polypeptide backbone and from 1·64 Å to 0·73 Å for all heavy atoms. There are no major local conformational differences between the two structures, with the single exception of the side-chain of Gln52.

1. Introduction

Recently, the three-dimensional structure of the α-amylase inhibitor Tendamistat† has been solved independently by X-ray crystallography in single crystals (Pflugrath *et al.*, 1986) and by n.m.r. spectroscopy in solution (Kline *et al.*, 1986, 1988). The first comparison showed a close coincidence of the global folds of the polypeptide chain. A detailed comparison of the structures, which includes the side-chains, is described in an accompanying paper (Billeter *et al.*, 1989). The present study examines the possibility of combining the n.m.r. structure in solution and the observed X-ray diffraction intensities in single crystals to solve the phase problem in the X-ray method for protein structure determination. An obvious approach is by Patterson search techniques (Hoppe, 1957; Huber, 1965; Lattman, 1985) with the n.m.r. structure as a search model. This procedure was previously tested

† Abbreviations used: Tendamistat, α-amylase inhibitor from *Streptomyces tendae*; n.m.r., nuclear magnetic resonance; r.m.s., root-mean-square; R-factor,

$$\sum \frac{|F_o - F_c|}{F_o},$$

F_o, observed, and F_c, calculated, structure factor amplitudes, respectively.

0022–2836/89/080669–08 $03.00/0

© 1989 Academic Press Limited

with model calculations on crambin, where the search models were calculated from interproton distance constraints derived from the crystal structure of crambin (Brünger *et al.*, 1987).

The experimental n.m.r. data of Tendamistat consisted of 842 distance constraints from nuclear Overhauser effects and 100 supplementary constraints from spin–spin coupling constraints, identified hydrogen bonds and disulfide bonds. From these, nine protein structures with similar residual error functions were calculated with the distance geometry program DISMAN (Braun & Gō, 1985). The average of the pairwise r.m.s. deviations in this ensemble of nine structures is an indication of the accuracy of the protein structure determination by the n.m.r. data (Wüthrich, 1986; Braun, 1987). For the Patterson search, the n.m.r. structure with the least residual error function was chosen (structure I in Table 7 of Kline *et al.*, 1988). We also tried to determine whether the initial model for the crystal structure thus obtained could be refined to the same R-factor of the crystal structure obtained by multiple isomorphous replacement methods and whether there are differences in the crystal structure determined by conventional X-ray methods.

The study should answer the question of whether a protein structure obtained by the present state of the n.m.r. technique (Wüthrich, 1986) is useful in practice as a model for the determination of the initial phases in the X-ray structure determination. This would allow us to replace multiple isomorphous replacement methods in those cases where a n.m.r. structure determination is possible and the preparation of heavy-atom derivatives is difficult.

2. Methods

(a) *Rotation search*

In the Patterson search for the correct orientation of the molecular model in the unit cell, the Patterson functions $P_c(\mathbf{r})$ and $P_o(\mathbf{r})$:

$$P(\mathbf{r}) = \frac{1}{V}\sum |F(\mathbf{h})|^2 \exp[-2\pi i \mathbf{hr}] \qquad (1)$$

are calculated from the structure factor amplitudes $|F_c(hkl)|$ computed from 2 n.m.r. models A and B, and the observed X-ray intensities $|F_o(hkl)|$, respectively. V is the volume of the unit cell. As the n.m.r. models, 2 subsets of the best n.m.r. structure (structure I of Table 7 of Kline *et al.*, 1988) were chosen. In model A we used all heavy atoms of the residues 5 to 73. These are 520 of the 558 heavy atoms of the residues 1 to 74. In model B the atoms, N, C^α, C′ C^β and O′ of the residues 5 to 73 and the heavy atoms of the "interior" side-chains as defined in Table 10 of Kline *et al.* (1988) were used, which corresponds to 399 atoms. In model B the atom positions are on average significantly better determined by the n.m.r. method than in model A. The average r.m.s. deviation for a pairwise comparison among the 9 calculated DISMAN structures is 1·5 Å for the atoms in model A as compared to 1·0 Å for the atoms in model B (Kline *et al.*, 1988).

The n.m.r. structure I was first best fit to the X-ray structure of Tendamistat in orientation and translation, and then translated so that the center of gravity was at the origin. This ensures that the correct orientation is expected near the rotation angles (0,0,0) and the translation corresponds to the position in the crystal. This procedure defines the orientation and translation of the molecule for zero values of the parameters for rotation and translation, as described below. It obviously does not bias the correlation calculations in any way. To calculate the Fourier transforms of the n.m.r. models, the molecules were placed in a triclinic cell with the cell dimensions set to 70 Å and the cell angles set to 90°. Structure factor calculations were done with the crystallographic program system PROTEIN (Steigemann, 1974). The electron density with a grid spacing of 1·0 Å was calculated using an overall B-value of 25 Å^2 for all atoms of the forementioned models. The structure factors were calculated in the resolution limit of 3·0 Å to 7·0 Å, which resulted in 25,110 structure factors. The calculated Patterson function was screened for peak maxima. Among them only the most prominent peaks were selected, with a cutoff value of 700 and chosen radii limits around the origin of 3·0 Å to 15·0 Å. The number of peaks for the models A and B were 1690 and 1832, respectively.

The rotational correlation function R in real space (Hoppe, 1957; Huber, 1965; see also the reciprocal space formulation by Rossmann & Blow, 1962):

$$R(\Omega) = \int_U P_o(\mathbf{r}) P_c(\Omega\mathbf{r})\, dV \qquad (2)$$

was used to find the rotation matrix Ω giving the best agreement between P_c and P_o. The rotations were parameterized with the angles, ψ, θ and ϕ, defined as follows: at first, one rotates about the z-axis with the angle ψ, then follows a rotation around the new x'-axis with the angle θ, and finally there is the rotation around the new y''-axis using the angle ϕ. With this parametrization the singularity of these rotations at the expected solution near (0,0,0) is avoided. The correlation function was calculated in steps of 5° for each of the 3 angles over the interval from 0° to 180°, which represents the asymmetric unit of the space group $P2_12_12_1$. After this grid search, a fine search in 1° steps was done.

(b) *Translation search*

The translation function of Crowther & Blow (1967) was used in the translation search with programs by Lattman (1985) as modified by Deisenhofer & Huber:

$$T(\mathbf{x}) = \sum_{\mathbf{h}} I(\mathbf{h}) F_{M1}(\mathbf{h}) F^*_{M2}(\mathbf{h}) \exp[-2\pi i \mathbf{hx}]. \qquad (3)$$

The function $T(\mathbf{x})$ has the maximum value when $\mathbf{x}$ represents a vector between a pair of symmetry-related molecules. $T(\mathbf{x})$ is calculated as the Fourier summation of the triple product between the observed intensities $I(\mathbf{h})$, the Fourier transform of the molecule corresponding to symmetry operation 1, $F_{M1}(\mathbf{h})$, and the complex conjugate of the Fourier transform corresponding to the symmetry operation 2, $F^*_{M2}(\mathbf{h})$. The symmetry in the space group $P2_12_12_1$ with equivalent positions: (1),x,y,z; (2),$-x+1/2,-y,1/2+z$; (3),$1/2+x,1/2-y,-z$; (4),$-x,1/2+y,1/2-z$; has 3 independent translation functions of the 3 pairs of molecules (1,2), (1,3) and (1,4). These were calculated and the Harker sections were plotted for both the refined X-ray model and the optimally oriented n.m.r. model.

Table 1

The five highest peaks in the rotational correlation calculation for n.m.r. models A and B

Model A†				Model B‡			
ψ	θ	ϕ	Maxima	ψ	θ	ϕ	Maxima
0·0	5·0	0·0	10·2§	0·0	5·0	0·0	13·3§
155·0	145·0	45·0	9·9	135·0	165·0	75·0	11·7
35·0	130·0	120·0	9·8	100·0	25·0	30·0	11·5
125·0	80·0	170·0	9·5	50·0	130·0	135·0	11·5
65·0	140·0	75·0	9·4	180·0	160·0	60·0	11·5

† All heavy atoms of residues 5 to 73.

‡ All atoms N, C^{α}, C, C^{β}, O of the residues 5 to 73 and all heavy atoms of the interior side-chains, i.e. those of residues 9, 11, 12, 14, 16, 21, 25, 27, 31, 33, 35, 37, 41, 44, 46, 48, 52, 54, 56, 70, 72.

§ Both models were first optimally fit to the X-ray model such that the orientation of the X-ray structure is defined by the angles 0,0,0. The values for the rotation function at these angles are 10·0 and 13·2, respectively, for the models A and B.

(c) *Refinement*

The n.m.r. model was oriented and positioned in the unit cell according to the result of the rotation and translation search, and refined as a rigid body using the Fourier transform fitting program TRAREF (Huber & Schneider, 1985). Refinements of the co-ordinates were then done alternatingly with the program EREF (Jack & Levitt, 1978), an energy-restrained crystallographic program implemented by Remington *et al.* (1982) and the interactive graphics program FRODO (Jones, 1978) for manual fitting of the polypeptide chain to the electron density map.

Details of the refinement procedure are given in the legend to Table 2. In the model-building phases with FRODO, the strategy consisted of rather conservative manual movements of atoms. Only those side-chains that were clearly outside the $2F_o-F_c$ map, were set into density. We also tried to avoid any main-chain movements until stage 4, when it became clear that for the carbonyl group of G43 and for the backbone segment I61–S63, modeling of side-chains only and refinement with the EREF program was insufficient to achieve a good fit between the electron density calculated from the n.m.r. model and the observed electron density.

3. Results and Discussion

Table 1 shows the five highest peaks found in a grid search with step size 5° in the rotation angles ψ, θ and ϕ of the rotation function for the n.m.r. models A and B. First, the n.m.r. models were best fitted to the X-ray model in the range of residues 5 to 73, so that the correct orientation corresponded to the angles 0,0,0. At these angles the rotation function was scaled to 10·0 for model A, and the same scaling factor was applied for model B. The mean values of the rotation function over all angles were 5·9 and 7·1 for the models A and B, respectively, with standard deviations of 1·2 and 1·4, respectively. In both models the correct orientation was indeed positioned near the highest peak of the rotation function. However, while in model A the values of the rotation function at the next four highest peaks were within a standard deviation of 1·2 from the maximum peak, in model B the second highest peak was decreased by more

Table 2

Protocol of the model building and refinement with the n.m.r. model B after the rigid body orientation

Stage	*R*-factor† (%)	RMSD(BB)‡ (Å)	RMSD(SC)§ (Å)
1	48	1·03	1·64
2	37	0·83	1·52
3	30	0·70	1·31
4	26	0·26	0·73

The 4 stages of the refinement procedure are as follows:

(1) n.m.r. structure with the residue range of 5 to 73. For all atoms the *B* factor was set uniformly to 25 Å^2. The atoms N, C^{α}, C′, O, C^{β} of the residues 5 to 73 and the heavy atoms of the interior side-chains as defined in Table 10 of Kline *et al.* (1988) were set as real atoms.

(2) Structure factor and $2F_o-F_c$ map calculations were done using this model, and inspected with the computer graphics program FRODO (Jones, 1978). Those dummy side-chain atoms for which density appeared at the position of the model were set as real atoms. Side-chain atoms of residues S10, T13, S17, E29, K34, E38, I53, D58, I61 and H64 were interactively moved into density with FRODO. Side-chain atoms of R72 were set as dummy atoms. Thirty cycles of automatic refinement were performed using the program EREF (Jack & Levitt, 1978). The scaling factor weighting the contribution of the X-ray data relative to the energy terms was 0·0005. *B*-factor refinement was done after the EREF cycles keeping the atom positions fixed.

(3) Check of the $2F_o-F_c$ map of the previous model and manual moves of the side-chains E6, S10, R19, S21, Q22, E29, V31, K34, V35, E38, T41, E42, I53, T54, T55, I61 and R72. Several combined cycles of manual model building with FRODO and automatic refinement with EREF were done. The scale factors used were in the range of 0·0005 to 0·0060. The standard deviation of bond lengths from ideal geometry at this stage was between 0·02 Å and 0·09 Å, and the standard deviation from ideal bond angles was in the range of 4·0 to 9·8°. At the end of stage 3, these standard deviations were 0·02 Å for bond lengths and 4·0° for bond angles. No movements of main-chain atoms were done up to this point.

(4) Interactive model building with the side-chains of S5, S10, Y15, R19, Y20, K34, V35, I53, R68 and R72, and the following main-chain atoms: the carbonyl group of G43 and all backbone atoms of residues I61, G62 and S63. The scale factor used was 0·005. The standard deviations of the bond lengths from ideal geometry was 0·02 Å, and of the bond angles it was 3·2°.

† The *R*-factor is calculated with a rejection factor of 1·4.

‡ Only the backbone atoms N, C^{α} and C′ of residues 5 to 73 are included.

§ All heavy atoms of residues 5 to 73, except for the atom pairs that are related by a 2-fold symmetry operation, such as two carboxylate oxygens of Asp and Glu, or the δ and ε carbons of the aromatic ring of Tyr.

than 1σ relative to the peak height for the correct solution.

In both models the maximum of the rotation function is slightly shifted from the expected values (0,0,0) by 5° in the angle θ. But the peak is rather broad at the maximum, and for both models A and B the values of the rotation function at (0,0,0) are only slightly decreased compared to the maximum. That the maximum at (0,5,0) is real was also shown by TRAREF, which converged to (0,5,0) when starting from (0,0,0). This slight misorientation might be a consequence of the fact that the residues 1 to 4, residue 74, and the exterior amino acid side-chains were omitted in model B.

Model B was used in the translation search after rotation to the optimal position. In the three translation functions for the pairs of molecules (1,2), (1,3) and (1,4), the Harker sections $z = 1/2$, $x = 1/2$ and $y = 1/2$ were screened for peak maxima. For comparison, these functions were also calculated for the previously determined X-ray crystal structure. In Figure 1 these functions are

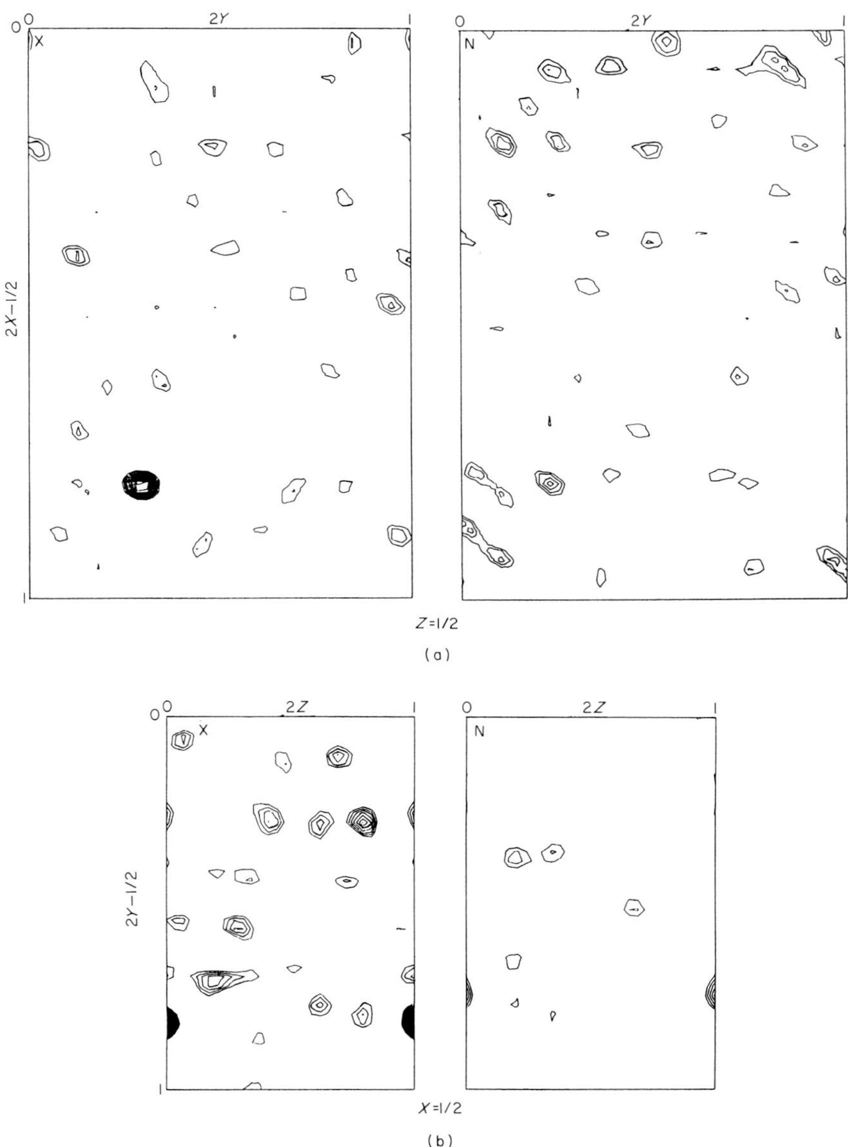

Fig. 1.

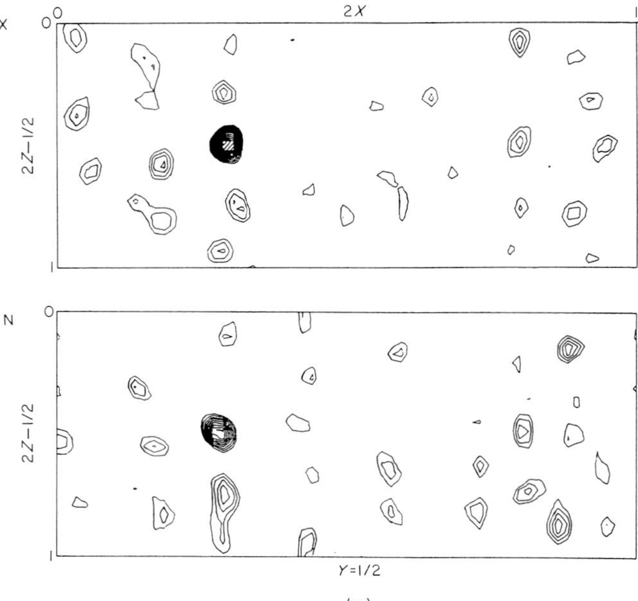

Figure 1. (a) Contour plot of the translation function of the pair of structures (1,2) located on the Harker section $z = 1/2$ for the X-ray structure of Tendamistat (X) and the oriented n.m.r. model B of structure I of Kline *et al.* (1988) (N). Contour levels are plotted in both Figures at intervals of 200 (relative values), starting at the level 600. (b) Contour plot of the Harker section $x = 1/2$ for the X-ray (X) and the n.m.r. structure (N) for the pair (1,3). Plot levels as in (a). (c) Contour plot of the Harker section $y = 1/2$ for the X-ray (X) and n.m.r. structure (N) for the pair (1,4). Plot levels as in (a).

plotted for the X-ray structure (X) and the n.m.r. model (N). In all three sections the correct solutions showed up as the highest peak in the translation functions of the n.m.r. model. This was confirmed by a comparison with the X-ray structure, and by the fact that the translation search gave a unique result, since each parameter was independently determined twice. These pairs of values were in good agreement and, expressed in fractional co-ordinates, gave the translation vector $x = 0{\cdot}14$, $y = 0{\cdot}11$ and $z = 0{\cdot}48$.

After the optimal positioning of the n.m.r. model in the unit cell, refinement of the structure against the 2·1 Å resolution X-ray data was started. At the outset of the refinement procedure, the R-factor was 48%, and it dropped to 26% at the end (Table 2). This value is similar to the R-factor of 25% for the X-ray structure previously determined by diffraction methods, which is obtained when the residues 1 to 4 and 74 and all solvent molecules are excluded from consideration. The small difference appears not to be significant and may be due to different scaling factors. At the end of the refinement, the standard deviation from the ideal geometry was 0·02 Å for the bond lengths and 3·2° for the bond angles.

The r.m.s. deviation for backbone atoms show that during the refinement process the polypeptide backbone converged toward the X-ray crystal structure, with a final deviation of 0·26 Å. All individual deviations of the backbone atoms were smaller than 1·0 Å, except for the residues Ile61 and Gly62, where a different carbonyl orientation of Ile61 caused displacements of the N and C^{α} atoms of Gly62 of 1·4 Å and 1·0 Å, respectively. Inspection of the $2F_o - F_c$ map in this region shows that both carbonyl orientations are consistent with the electron density map.

Figure 2 illustrates the convergence of the polypeptide backbone in the n.m.r. solution structure toward the X-ray structure during the refinement against the X-ray crystal data. In Figure 2(a) a thick line represents the n.m.r. solution structure before the refinement against the crystal data and in Figure 2(b) after the refinement. For comparison, in both Figure 2(a) and (b) the X-ray crystal structure obtained with conventional methods is plotted with a thin line. The small deviation at the residues Ile61 and Gly62 is seen at the top of Figure 2(b).

For some of the side-chains, the final deviations are larger than for the polypeptide backbone. Deviations of around 2 Å were observed for the methyl and hydroxyl groups of the branched side-chains of Leu14, Val35, Thr41 and Thr55. In each case the calculated electron density map with the

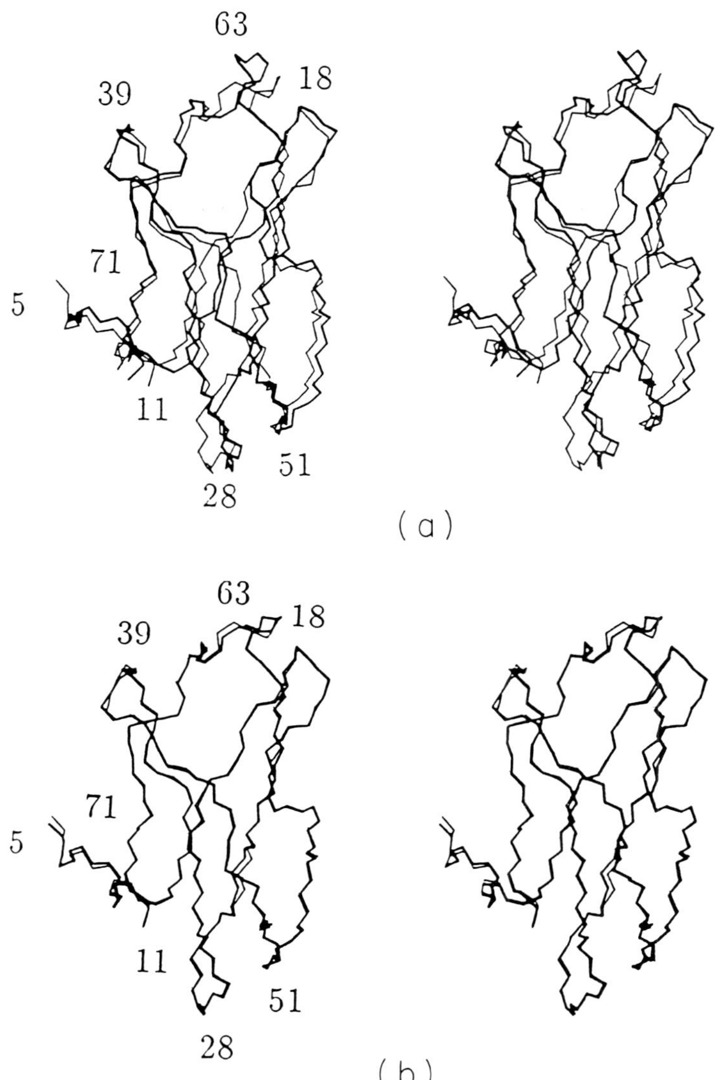

Figure 2. (a) Stereo view of the backbone atoms of the residues 5 to 73 of the X-ray structure determined by initial phasing with heavy-metal derivatives and crystallographic refinement (Pflugrath *et al.*, 1987) (thin line) and the best n.m.r. structure (Kline *et al.*, 1988) (thick line). The orientation of the n.m.r. structure is as found by the Patterson search. (b) Backbone atoms of the same range after the refinement procedure of the n.m.r. structure 1 with the diffraction data up to resolution 2·1 Å (thick line). The thin line is the same as in (a).

X-ray data at 2·1 Å resolution was not sufficiently detailed to position the two branches of the side-chains. The residues for which these differences were observed are somewhat more mobile, and the ambiguities cannot be resolved with the available X-ray data at 2·1 Å resolution.

The largest deviations between the crystal structures determined with the multiple isomorphous replacement method, and with the n.m.r. solution structures, respectively, amount to about 5 Å and were observed at the external side-chains of Glu29 and Gln52. For Glu29 no electron density was observed for the peripheral parts of the side-chain, so that this difference is fortuitous. For Gln52, electron density at the position of the n.m.r. side-chain conformation (Fig. 3(a)) appeared already early in the refinement procedure, and persisted in the further refinement (Fig. 3(b)). As a

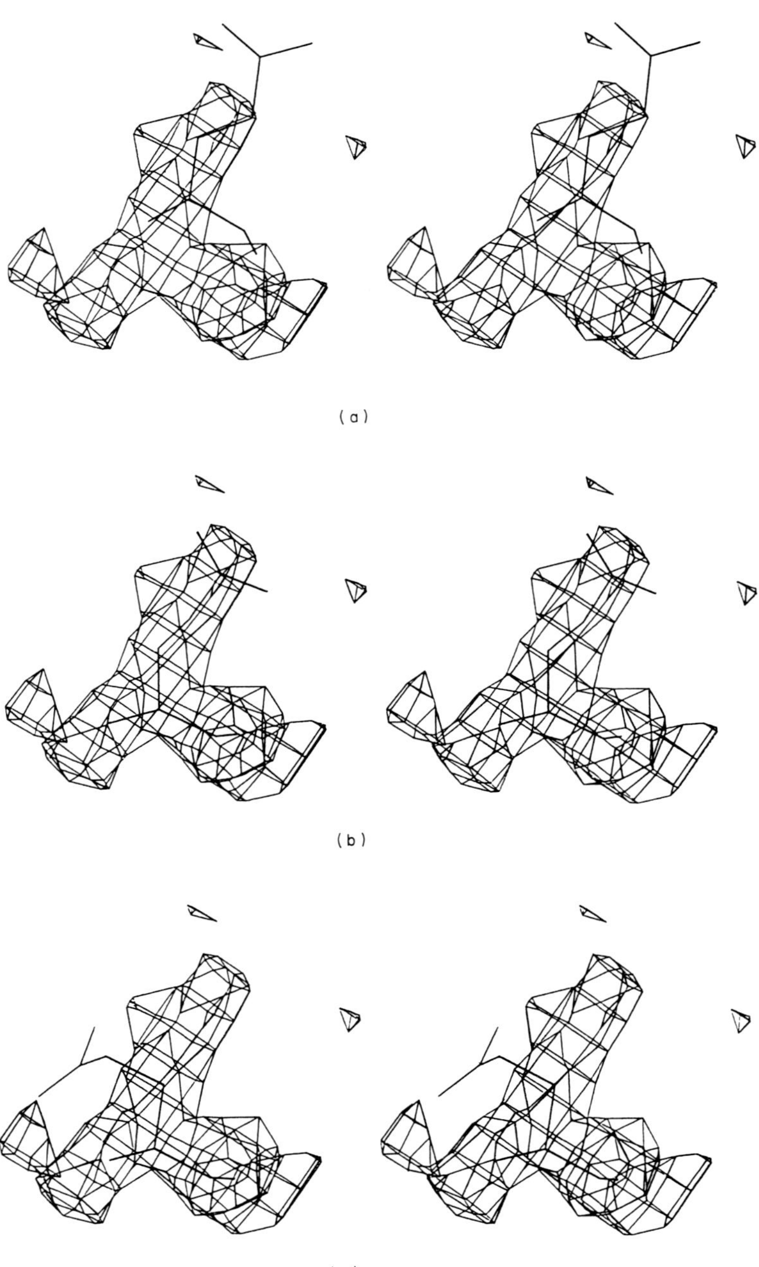

Figure 3. Electron density for the residue Q52 at the final stage of refinement of the n.m.r. model B against the X-ray crystal data. Superimposed is the side-chain conformation of the n.m.r. structure (a) before refinement and (b) after refinement. The side-chain conformation in the X-ray crystal structure model is shown in (c). It was clearly incorrectly positioned.

comparison, Figure 3(c) shows the conformation of this side-chain in the X-ray crystal structure. It is certainly outside the electron density. The $2F_o - F_c$ electron density maps calculated for both structures are very similar at this side-chain and again corroborates the n.m.r. position. This is clearly a case of an error in the X-ray model, which could have been avoided by more careful inspection of the electron density map.

Some of the remaining differences seen in Figure 2(b) might be consequences of using an incomplete model of the crystal structure, which lacked the N and C termini and the ordered solvent. Furthermore, the temperature parameters were not completely refined, which may have some effects on distant parts of the molecule and influence their convergence behavior.

The result of this study clearly shows that a protein structure in solution determined at high resolution by the n.m.r. method can be used as a model for the determination of the initial phases in the determination of the crystal structure of the same protein by X-ray diffraction, provided that the overall molecular architecture is the same in the two states. In contrast to the test calculations obtained with simulated n.m.r. interproton distance constraints for the protein crambin (Brünger *et al.*, 1987), an individual distance geometry solution was sufficiently accurate to solve uniquely the rotation and translation search, so that one does not have to refer to an average of the distance geometry solutions. A careful selection of the atom set to be used in the Patterson search seems to be important in obtaining a successful result in practice. It also shows that instead of the complete molecule, certain well-determined parts of the n.m.r. solution structure might already be sufficient to determine the initial phases. These observations are promising for a practical combination of n.m.r. and X-ray methods to determine the three-dimensional structure of a protein in an efficient way, in particular in situations where isomorphous heavy-atom derivatives are difficult to obtain.

Financial support by the Schweizerischer Nationalfonds (project 3.198.85) is gratefully acknowledged. We thank Dr M. Billeter for helpful discussions.

References

Billeter, M., Kline, A. D., Braun, W., Huber, R. & Wüthrich, K. (1989). *J. Mol. Biol.* **207**, 677–687.

Braun, W. (1987). *Quart. Rev. Biophys.* **19**, 115–157.

Braun, W. & Gō, N. (1985). *J. Mol. Biol.* **186**, 611–626.

Brünger, A. T., Campbell, R. L., Clore, G. M., Gronenborn, A. M., Karplus, M., Petsko, G. A. & Teeter, M. M. (1987). *Science,* **235**, 1049–1053.

Crowther, R. A. & Blow, D. M. (1967). *Acta Crystallogr.* **23**, 544–548.

Hoppe, W. (1957). *Acta Crystallogr.* **10**, 750–751.

Huber, W. (1965). *Acta Crystallogr.* **19**, 353–356.

Huber, R. & Schneider, M. (1985). *J. Appl. Crystallogr.* **18**, 165–169.

Jack, A. & Levitt, M. (1978). *Acta Crystallogr. sect. A,* **29**, 291–295.

Jones, T. A. (1978). *J. Appl. Crystallogr.* **11**, 268–272.

Kline, A. D., Braun, W. & Wüthrich, K. (1986). *J. Mol. Biol.* **189**, 377–382.

Kline, A. D., Braun, W. & Wüthrich, K. (1988). *J. Mol. Biol.* **204**, 675–724.

Lattman, E. E. (1985). *Methods Enzymol.* **115**, 55–77.

Pflugrath, J., Wiegand, E., Huber, R. & Vértesy, L. (1986). *J. Mol. Biol.* **189**, 383–386.

Remington, S. J., Wiegand, G. & Huber, R. (1982). *J. Mol. Biol.* **158**, 111–152.

Rossmann, M. G. & Blow, D. M. (1962). *Acta Crystallogr.* **15**, 24–31.

Steigemann, W. (1974). Doctoral thesis, Technische Universität, München.

Wüthrich, K. (1986). *NMR of Proteins and Nucleic Acids*, Wiley, New York.

Edited by W. Hendrickson

J. Mol. Biol. (1988) **203**, 251–268

Conformation of [Cd_7]–metallothionein-2 from Rat Liver in Aqueous Solution Determined by Nuclear Magnetic Resonance Spectroscopy

Peter Schultze[1], Erich Wörgötter[1]†, Werner Braun[1], Gerhard Wagner[1]‡ Milan Vašák[2], Jeremias H. R. Kägi[2] and Kurt Wüthrich[1]

[1]*Institut für Molekularbiologie und Biophysik Eidgenössische Technische Hochschule-Hönggerberg CH-8093 Zürich, Switzerland*

and

[2]*Biochemisches Institut der Universität Zürich Winterthurerstrasse 190, CH-8057 Zürich, Switzerland*

(Received 26 January 1988, and in revised form 19 April 1988)

The three-dimensional structure of [Cd_7]–metallothionein-2 from rat liver was determined in aqueous solution, using nuclear magnetic resonance spectrometry and distance geometry calculations. The experimental data provided proton–proton distance constraints from measurements of nuclear Overhauser effects, constraints on the geometry of the metal–cysteine clusters determined by heteronuclear correlation spectroscopy, and dihedral angle constraints derived from both coupling constants and nuclear Overhauser effects. The structure calculations were performed with the program DISMAN. As in previous studies with rabbit liver metallothionein-2a, the structure calculations were performed separately for the α and β-domains containing the 4 and 3-metal clusters, respectively, since no interdomain constraints were found. For both domains, the global polypeptide fold, the location of polypeptide secondary structure elements, the architecture of the metal–sulfur cluster and the local chirality of the metal co-ordination are very similar to the solution structure of rabbit metallothionein-2a, but show considerable difference relative to the crystal structure of rat metallothionein-2.

1. Introduction

Mammalian metallothioneins consist of a single polypeptide chain of approximately 60 amino acid residues. They contain 20 cysteinyl residues, which can bind seven divalent metal ions, such as Zn^{2+}, Cd^{2+} or Hg^{2+} (Vašák & Kägi, 1983). Recently, MT-2§ from rat liver was investigated in single crystals by X-ray diffraction studies (Furey *et al.*, 1986, 1987), and rabbit liver MT-2a was studied in solution by n.m.r. spectroscopy (Neuhaus *et al.*, 1984, 1985; Frey *et al.*, 1985; Wagner *et al.*, 1986*a*, 1987). The two amino acid sequences differ in ten substitutions and one insertion. The three-dimensional structures obtained from the two investigations are markedly different (Braun *et al.*, 1986; Arseniev *et al.*, 1988). To obtain a more direct comparison with the available crystal structure that could not be affected by the primary structure differences, we started a determination of the solution conformation of rat MT-2. In a previous paper on this project, we reported the sequence-specific ^1H n.m.r. assignments for rat MT-2 (Wörgötter *et al.*, 1987). We further demonstrated that the metal co-ordination for rat [Cd_7]–MT-2, and for the most abundant form of the 4-metal cluster in aqueous solution of biosynthetic rat [Cd_5, Zn_2]–MT-2 coincides with the metal–cysteine bonds in rabbit [Cd_7]–MT-2a in solution (Vašák *et al.*, 1987). This paper describes the three-dimensional

† Present address: Biochemie Ges.m.b.H., A-6250 Kundl, Austria.

‡ Present address: Institute of Science and Technology, Biophysics Research Division, University of Michigan, 2200 Bonisteel Boulevard, Ann Arbor, MI 48109, U.S.A.

§ Abbreviations used: MT, metallothionein; n.m.r., nuclear magnetic resonance; NOE, nuclear Overhauser enhancement; 3D, 3-dimensional; NOESY, 2-dimensional NOE spectroscopy; RMSD, root-mean-square difference; ECEPP, empirical conformational energy program for peptides (Némethy, *et al.*, 1983).

0022-2836/88/170251-18 $03.00/0

© 1988 Academic Press Limited

structure of the polypeptide chain and the metal–sulfur clusters in rat [Cd_7]–MT-2.

In addition to enabling a direct comparison with the only known crystal structure of a metallothionein (Furey *et al.*, 1986, 1987), the present solution structure analysis of rat MT-2 and its comparison with rabbit MT-2a (Arseniev *et al.*, 1988) is also of interest, since the two proteins differ in nearly 20% of the amino acid residues. Structure comparisons of this kind may shed light on the chemical and biological significance of the multiple differences between the known primary structures of metallothioneins (Kägi & Kojima, 1987).

2. Materials and Methods

The preparation of rat [Cd_7]–MT-2 was as described (Vašák *et al.*, 1987). For the n.m.r. measurements, the following conditions were used: protein concentration 10 mM, 20 mM-[$^2H_{11}$]Tris · HCl (pH 7·0), 50 mM-KCl; sample temperature, 10°C.

NOESY spectra were recorded at 500 MHz with a Bruker WM 500 spectrometer. Two series of spectra in H_2O and ^2H_2O, respectively, were recorded with mixing times of 60, 120 and 250 ms. A similar number of NOESY cross-peaks as in the earlier work with rabbit MT-2a (Arseniev *et al.*, 1988) could be identified and analyzed using the following cross-peak intensity *versus* distance-bound relations. Following the arguments presented by Arseniev *et al.* (1988; see also Wüthrich *et al.*, 1984; Wüthrich, 1986), intraresidual cross-peaks corresponding to $d_{N\alpha}$ (i,i) with an intensity of 3 contour levels at 60 ms were related with an upper distance bound of 3·0 Å (1 Å = 0·1 nm). The other cross-peaks in the spectrum recorded in H_2O with a 60 ms mixing time were then referenced relative to this relation, with the intensities corresponding to 1 or 2, 4 and 5 exponentially spaced contour levels equated with upper distance bounds of 4·0, 2·6 and 2·3 Å, respectively. In all, 77% of the distance constraints were obtained from this spectrum. The upper distance limits for cross-peaks observed only in the 120 ms NOESY spectrum in H_2O were set to 4·0 Å, regardless of the number of contour levels. This spectrum yielded 8% of all distance constraints. The remaining 15% of the distance constraints came from the 2 NOESY spectra recorded in ^2H_2O solution with mixing times of 60 ms and 120 ms. Irrespective of the number of contour levels, all cross-peaks identified at 60 ms were equated with a distance limit of 3·0 Å, and those seen only at 120 ms with a limit of 4·0 Å.

Except for the aforementioned somewhat different NOESY cross-peak intensity *versus* upper distance-bound calibration, the procedures used for the collection of the n.m.r. data and the structural analysis with the program DISMAN (Braun & Gō, 1985) were virtually identical with those described in detail for rabbit MT-2a (Arseniev *et al.*, 1988).

3. Results

(a) *Input for the distance geometry calculations*

The input for the distance geometry calculations included constraints ensuring the experimentally determined Cd–Cys co-ordinative bonds with tetrahedral co-ordination geometry at each metal ion. The connectivities between the seven cadmium ions and the 20 cysteinyl residues were determined from heteronuclear couplings between ^{113}Cd and Cys H^β (Vašák *et al.*, 1987). The results were the same for both rabbit MT-2a and rat MT-2, except for one ambiguity. Since the chemical shifts of the β-protons of Cys24 and Cys29 in rat MT-2 were degenerate, and the heteronuclear relayed connectivities to the α-protons of these two residues could not be observed, it was not possible to determine experimentally which of these two cysteinyl residues formed the bridging ligand between CdII and CdIV. To further investigate the sequence position of this bridging cysteine, we repeated the structure calculations for both possible co-ordinations in the 3-metal domain. In the first input, the metal–cysteine bonds were included exactly as in Table 1 of Arseniev *et al.* (1988), with Cys24 linked to both CdII and CdIV, and Cys29 linked to CdII only, and in the second input with the roles of Cys24 and Cys29 exchanged but otherwise identical parameters.

Table 1 lists the upper distance constraints obtained from NOESY at 10°C. This set of distance constraints was derived from the experiments as described in Materials and Methods. Thereby, some uncertainty about the intensity–distance calibration for individual cross-peaks arose because of sizeable variations between the intensities of different amide proton resonances. Most probably, these intensity variations resulted from saturation transfer from the preirradiated water line, since this transfer can be quite efficient for some protons at the neutral pH value used for the structure determination of MT-2 (Wüthrich, 1986). To eliminate inconsistencies in the initial set of distance constraints, we started a first round of distance geometry calculations with the program DISMAN (Braun & Gō, 1985) for 200 starting structures of each domain with randomized dihedral angles. After these structures had been calculated up to level 2 of the variable target function, 40 of them with the lowest target values were selected and minimized using all further target function levels. The ten best solutions thus obtained were then analyzed for distance constraints that were consistently violated by more than 0·3 Å. After rechecking the experimental spectra, these constraints were either changed or eliminated from the input, yielding the dataset of Table 1, which presents the distance constraints separately for the β and α-domains, since no interdomain NOEs could be identified with the present experiments. For the same reason, all calculations were performed independently for both domains.

Table 2 lists the dihedral angle constraints obtained from suitable combinations of spin-spin coupling constants, sequential NOEs and intraresidual NOEs, as described for rabbit MT-2a (Arseniev *et al.*, 1988). These had also been checked for residual violations during the first round of DISMAN calculations and, as a result, the bounds on three dihedral angle constraints were modified. The numbers of constraints in Tables 1 and 2 are

Table 1

Distance constraints (in Å) for the domains of rat MT-2 obtained from NOESY spectra in H_2O and ^2H_2O at 10°C and pH 7·0

A. *The β-domain*

	Sequential backbone constraints (11 constraints)	Medium-range backbone and long-range backbone constraints (7 constraints)	Interresidual constraints with wide-chain protons (28 constraints)
Met1	HA D2 HN 2·6 (4·43, 8·66)		
Asp2	HN M1 HA 2·6 (8·66, 4·43)		PB C5 HN 5·0m (3·18/2·60, 7·51)
Asn4	HN C5 HN 4·0 (8·78, 7·51)	HA K22 HN 4·0 (4·74, 8·85)	PB C5 HN 5·0m (2·95/2·78, 7·51)
		HA Q23 HN 4·0 (4·74, 8·99)	PB Q23 HN 4·0m (2·95/2·78, 8·99)
			PB Q23 PB 5·0mm (2·95/2·78, 1·95/1·77)
			ND Q23 PB 6·0mm (8·00/7·07, 1·95/1·77)
Cys5	HN N4 HN 4·0 (7·51, 8·78)	HA C7 HN 4·0 (5·36, 8·49)	HN D2 PB 5·0 m (7·51, 3·18/2·60)
	HA S6 HN 2·6 (5·36, 8·68)	HA C21 HA 4·0 (5·36, 4·05)	HN N4 PB 5·0 m (7·51, 2·95/2·78)
		HA Q23 HN 4·0 (5·36, 8·99)	HN Q23 PB 5·0 m (7·51, 1·95/1·77)
		HA C24 HA 4·0 (5·36, 4·24)	HA C21 PB 4·0 m (5·36, 3·65/3·00)
			PB S6 HN 5·0m (2·87/3·36, 8·68)
			PB C7 HN 5·0m (2·87/3·36, 8·49)
			PB Q23 HN 5·0m (2·87/3·36, 8·99)
			PB C24 HN 4·0m (2·87/3·36, 4·24)
			PB K25 HN 4·0m (2·87/3·36, 9·47)
Ser6	HN C5 HA 2·6 (8·68, 5·36)		HN C5 PB 5·0 m (8·68, 2·87/3·36)
			HN C21 PB 5·0 m (8·68, 3·65/3·00)
Cys7	HA A8 HN 2·3 (4·27, 8·91)	HN C5 HA 4·0 (8·49, 5·36)	HN C5 PB 5·0 m (8·49, 2·87/3·36)
			PB A8 HN 5·0m (3·04/2·90, 8·91)
Ala8	HN C7 HA 2·3 (8·91, 4·27)		HN C7 PB 5·0 m (8·91, 3·04/2·90)
			MB T9 HN 5·0m (1·49, 8·61)
			MB D10 HN 4·0m (1·49, 7·91)
			MB C13 HN 5·0m (1·49, 8·13)
Thr9			HN A8 MB 5·0 m (8·61, 1·49)
			MG D10 HN 5·0m (1·24, 7·91)
Asp10			HN A8 MB 4·0 m (7·91, 1·49)
			HN T9 MG 5·0 m (7·91, 1·24)
Cys13			HN A8 MB 5·0 m (8·13, 1·49)
Cys15			PB C19 HA 5·0m (3·32/3·05, 4·34)
Ala16			MB G17 HN 5·0m (1·40, 8·64)
Gly17			HN A16 MB 5·0 m (8·64, 1·40)
Ser18	HA C19 HN 3·0 (4·55, 8·26)		
Cys19	HN S18 HA 3·0 (8·26, 4·55)	HA C21 HN 4·0 (4·34, 8·61)	HA C15 PB 5·0 m (4·34, 3·32/3·05)
	HA K20 HN 3·0 (4·34, 9·14)		
Lys20	HN C19 HA 3·0 (9·14, 4·34)		PB C21 HN 5·0m (2·12/1·78, 8·61)
	HN C21 HN 4·0 (9·14, 8·61)		
Cys21	HN K20 HN 4·0 (8·61, 9·14)	HN C19 HA 4·0 (8·61, 4·34)	HN K20 PB 5·0 m (8·61, 2·12/1·78)
	HA K22 HN 2·3 (4·05, 8·85)	HA C5 HA 4·0 (4·05, 5·36)	PB C5 HA 4·0m (3·65/3·00, 5·36)
			PB S6 HN 5·0m (3·65/3·00, 8·68)
			PB K22 HN 5·0m (3·65/3·00, 8·85)
Lys22	HN C21 HA 2·3 (8·85, 4·05)	HN N4 HA 4·0 (8·85, 4·74)	HN C21 PB 5·0 m (8·85, 3·65/3·00)
			PB Q23 HN 5·0m (1·98/1·84, 8·99)
Gln23		HN N4 HA 4·0 (8·99, 4·74)	HN N4 PB 4·0 m (8·99, 2·95/2·78)
		HN C5 HA 4·0 (8·99, 5·36)	HN C5 PB 5·0 m (8·99, 2·87/3·36)
			HN K22 PB 5·0 m (8·99, 1·98/1·84)
			PB N4 PB 5·0mm (1·95/1·77, 2·95/2·78)
			PB N4 ND 6·0mm (1·95/1·77, 8·00/7·07)
			PB C5 HN 5·0m (1·95/1·77, 7·51)

Table 1 *cont.*

A. *The β-domain*

	Sequential backbone constraints (11 constraints)	Medium-range backbone and long-range backbone constraints (7 constraints)	Interresidual constraints with side-chain protons (28 constraints)
Cys24	HA K25 HN 2·3 (4·24, 9·47)	HA C5 HA 4·0 (4·24, 5·36)	HA C5 PB 4·0 m (4·24, 2·87/3·36)
Lys25	HN C24 HA 2·3 (9·47, 4·24) HN C26 HN 3·0 (9·47, 8·63)		HN C5 PB 4·0 m (9·47, 2·87/3·36)
Cys26	HN K25 HN 3·0 (8·63, 9·47)		PB T27 HN 5·0m (3·17/3·02, 8·99) PB C29 HN 5·0m (3·17/3·02, 7·42)
Thr27			HN C26 PB 5·0 m (8·99, 3·17/3·02) HA K30 PB 5·0 m (4·04, 1·85/1·73)
Ser28			PB K30 HN 5·0m (4·04/3·94, 7·57)
Cys29	HA K30 HN 3·0 (4·39, 7·57)		HN C26 PB 5·0 m (7·42, 3·17/3·02) PB K30 HN 5·0m (3·11/2·86, 7·57)
Lys30	HN C29 HA 3·0 (7·57, 4·39)		HN S28 PB 5·0 m (7·57, 4·04/3·94) HN C29 PB 5·0 m (7·57, 3·11/2·86) PB T27 HA 5·0m (1·85/1·73, 4·04)

B. *The α-domain*

	Sequential backbone constraints (27 constraints)	Medium-range backbone and long-range backbone constraints (4 constraints)	Interresidual constraints with side-chain protons (49 constraints)
Lys31	HA S32 HN 3·0 (4·51, 8·89)		HN V39 QG 6·4 q (8·46, 0·98/0·92) HA V39 QG 6·4 q (4·51, 0·98/0·92)
Ser32	HN K31 HA 3·0 (8·89, 4·51)		HN V39 QG 6·4 q (8·89, 0·98/0·92) HA V39 QG 6·4 q (4·45, 0·98/0·92) PB C34 HN 5·0m (4·05/3·93, 8·39)
Cys33			HN V39 QG 6·4 q (8·17, 0·98/0·92) HA C48 PB 5·0 m (4·47, 2·97/2·90) PB C48 HN 5·0m (3·29/3·14, 8·89) PB C48 PB 6·0mm (3·29/3·14, 2·97/2·90)
Cys34			HN S32 PB 5·0 m (8·39, 4·05/3·93) PB S35 HN 5·0m (3·59/3·51, 8·94) PB C36 HN 5·0m (3·59/3·51, 8·55) PB C37 HN 5·0m (3·59/3·51, 7·26)
Ser35	HN C36 HN 4·0 (8·94, 8·55)	HA C37 HN 4·0 (4·42, 7·26)	HN C34 PB 5·0 m (8·94, 3·59/3·51)
Cys36	HN S35 HN 4·0 (8·55, 8·94) HN C37 HN 2·6 (8·55, 7·26)		HN C34 PB 5·0 m (8·55, 3·59/3·51) PB C37 HN 5·0m (2·83/3·23, 7·26) PB K56 HN 5·0m (2·83/3·23, 7·88) PB K56 HA 4·0m (2·83/3·23, 4·68) PB C57 HA 5·0m (2·83/3·23, 5·18)
Cys37	HN C36 HN 2·6 (7·26, 8·55)	HN S35 HA 4·0 (7·26, 4·42)	HN C34 PB 5·0 m (7·26, 3·59/3·51) HN C36 PB 5·0 m (7·26, 2·83/3·23) HN P38 PD 5·0 m (7·26, 3·83/3·78) HA C41 PB 5·0 m (5·16, 3·19/3·13)
Pro38	HA V39 HN 2·3 (4·71, 8·62)		PB V39 HN 4·0m (2·33/2·07, 8·62) PD C37 HN 5·0m (3·83/3·78, 7·26)
Val39	HN P38 HA 2·3 (8·62, 4·71) HA G40 HN 2·6 (3·82, 8·98)	HA C41 HN 4·0 (3·82, 7·05)	HN P38 PB 4·0 m (8·62, 2·33/2·07) QG K31 HN 6·4q (0·98/0·92, 8·46) QG K31 HA 6·4q (0·98/0·92, 4·51) QG S32 HN 6·4q (0·98/0·92, 8·89) QG S32 HA 6·4q (0·98/0·92, 4·45) QG C33 HN 6·4q (0·98/0·92, 8·17) QG G40 HN 5·4q (0·98/0·92, 8·98) QG C41 HN 6·4q (0·98/0·92, 7·05)
Gly40	HN V39 HA 2·6 (8·98, 3·82) HN C41 HN 3·0 (8·98, 7·05)		HN V39 QG 5·4 q (8·98, 0·98/0·92)

Table 1 *cont.*

B. *The α-domain*

Residue	Sequential backbone constraints (27 constraints)	Medium-range backbone and long-range backbone constraints (4 constraints)	Interresidual constraints with side-chain protons (49 constraints)
Cys41	HN G40 HN 3·0 (7·05, 8·98) HA A42 HN 3·0 (4·08, 9·46)	HN V39 HA 4·0 (7·05, 3·82)	HN V39 QG 6·4 q (7·05, 0·98/0·92) PB C37 HA 5·0m (3·19/3·13, 5·16) PB C44 HN 5·0m (3·19/3·13, 7·69)
Ala42	HN C41 HA 3·0 (9·46, 4·08) HN K43 HN 4·0 (9·46, 8·39) HA K43 HN 3·3 (4·17, 8·39)		MB C44 HN 5·0m (1·58, 7·69) MB S45 HN 5·0m (1·58, 7·41)
Lys43	HN A42 HN 4·0 (8·39, 9·46) HN A42 HA 3·3 (8·39, 4·17) HN C44 HN 2·6 (8·39, 7·69)		PB C44 HN 4·0m (2·16/2·04, 7·69)
Cys44	HN K43 HN 2·6 (7·69, 8·39) HN S45 HN 2·3 (7·69, 7·41) HA S45 HN 3·0 (4·70, 7·41)	HA C48 HA 4·0 (4·70, 4·36)	HN C41 PB 5·0 m (7·69, 3·19/3·13) HN A42 MB 5·0 m (7·69, 1·58) HN K43 PB 4·0 m (7·69, 2·16/2·04) HN I49 PG 5·0 m (7·69, 1·44/1·02) HA I49 PG 4·0 m (4·70, 1·44/1·02) PB S45 HN 5·0m (3·77/2·65, 7·41) PB I49 HN 5·0m (3·77/2·65, 7·23)
Ser45	HN C44 HN 2·3 (7·41, 7·69) HN C44 HA 3·0 (7·41, 4·70)		HN A42 MB 5·0 m (7·41, 1·58) HN C44 PB 5·0 m (7·41, 3·77/2·65) PB Q46 HN 5·0m (4·07/4·00, 8·30)
Gln 46	HA G47 HN 3·0 (4·63, 7·40)		HN S45 PB 5·0 m (8·30, 4·07/4·00) HN I49 PG 6·4 m (8·30, 1·44/1·02) PB G47 HN 5·0m (2·43/1·94, 7·40)
Gly47	HN Q46 HA 3·0 (7·40, 4·63) HN C48 HN 4·0 (7·40, 8·89)		HN Q46 PB 5·0 m (7·40, 2·43/1·94) HN I49 MG 5·0 m (7·40, 1·05)
Cys48	HN G47 HN 4·0 (8·89, 7·40) HA I49 HN 2·3 (4·36, 7·23)	HA C44 HA 4·0 (4·36, 4·70)	HN C33 PB 5·0 m (8·89, 3·29/3·14) PB C33 HA 5·0m (2·97/2·90, 4·47) PB C33 PB 6·0mm (2·97/2·90, 3·29/3·14) PB C50 HN 5·0m (2·97/2·90, 9·17)
Ile49	HN C48 HA 2·3 (7·23, 4·36) HN C50 HN 2·6 (7·23, 9·17) HA C50 HN 3·0 (4·69, 9·17)		HN C44 PB 5·0 m (7·23, 3·77/2·65) PG C44 HN 5·0m (1·44/1·02, 7·69) PG C44 HA 4·0m (1·44/1·02, 4·70) PG Q46 HN 6·4m (1·44/1·02, 8·30) MG G47 HN 5·0m (1·05, 7·40) MD C59 PB 5·0mm (0·96, 3·29/3·24)
Cys50	HN I49 HN 2·6 (9·17, 7·23) HN I49 HA 3·0 (9·17, 4·69) HA K51 HN 2·3 (4·42, 8·56)		HN C48 PB 5·0m (9·17, 2·97/2·90) PB K51 HN 5·0m (3·10/2·73, 8·56)
Lys51	HN C50 HA 2·3 (8·56, 4·42) HA E52 HN 3·0 (4·24, 8·52)		HN C50 PB 5·0m (8·56, 3·10/2·73)
Glu52	HN K51 HA 3·0 (8·52, 4·24) HA A53 HN 2·6 (4·32, 8·41)		PG A53 HN 5·0m (2·27, 8·41) PG S54 HN 5·0m (2·27, 8·19) PG C57 HA 5·0m (2·27, 5·18)
Ala53	HN E52 HA 2·6 (8·41, 4·32) HN S54 HN 3·0 (8·41, 8·19)		HN E52 PG 5·0m (8·41, 2·27) MB C57 HA 4·0m (1·44, 5·18)
Ser54	HN A53 HN 3·0 (8·19, 8·41) HA D55 HN 3·0 (4·64, 8·64)		HN E52 PG 5·0 m (8·19, 2·27) PB K56 HN 5·0m (3·93/3·85, 7·88)
Asp55	HN S54 HA 3·0 (8·64, 4·64) HN K56 HN 4·0 (8·64, 7·88) HA K56 HN 3·0 (4·42, 7·88)		PB K56 HN 4·0m (2·72, 7·88)
Lys56	HN D55 HN 4·0 (7·88, 8·64) HN D55 HA 3·0 (7·88, 4·42) HA C57 HN 2·3 (4·68, 8·56)		HN C36 PB 5·0 m (7·88, 2·83/3·23) HN S54 PB 5·0 m (7·88, 3·93/3·85) HN D55 PB 4·0 m (7·88, 2·72) HA C36 PB 4·0 m (4·68, 2·83/3·23) PB C57 HN 4·0m (1·82/1·73, 8·56) PG C57 HN 4·0m (1·39, 8·56)

Table 1 *cont.*

B. *The α-domain*	Sequential backbone constraints (27 constraints)	Medium-range backbone and long-range backbone constraints (4 constraints)	Interresidual constraints with wide-chain protons (49 constraints)
Cys57			
	HN K56 HA 2·3 (8·56, 4·68)	HA C59 HN 4·0 (5·18, 8·44)	HN K56 PB 4·0 m (8·56, 1·82/1·73) HN K56 PG 4·0 m (8·56, 1·39) HA C36 PB 5·0 m (5·18, 2·83/3·23) HA E52 PG 5·0 m (5·18, 2·27) HA A53 MB 4·0 m (5·18, 1·44) PB C59 HN 4·0m (3·69/3·59, 8·44) PB C60 HN 5·0m (3·69/3·59, 7·78)
Ser58			
			PB C59 HN 5·0m (3·98/3·91, 8·44)
Cys59			
	HN C60 HN 3·0 (8·44, 7·78)	HN C57 HA 4·0 (8·44, 5·18)	HN C57 PB 4·0 m (8·44, 3·69/3·59) HN S58 PB 5·0 m (8·44, 3·98/3·91) PB I49 MD 5·0mm (3·29/3·24, 0·96) PB C60 HN 5·0m (3·29/3·24, 7·78)
Cys60			
	HN C59 HN 3·0 (7·78, 8·44) HN A61 HN 3·0 (7·78, 7·23)		HN C57 PB 5·0 m (7·78, 3·69/3·59) HN C59 PB 5·0 m (7·78, 3·29/3·24) PB A61 HN 5·0m (3·14/2·66, 7·23)
Ala61			
	HN C60 HN 3·0 (7·23, 7·78)		HN C60 PB 5·0 m (7·23, 3·14/2·66)

The amino acid residues are identified either by the 3-letter symbol or the 1-letter symbol followed by a number indicating the position in the amino acid sequence. HN stands for amide proton, HA for C^{α} proton, HB for C^{β} proton, HG for C^{γ} proton and HD for C^{δ} proton. The following upper case letters indicate pseudoatoms that substitute for a group of 2 or more protons, and lower case letters indicate that the distance constraint has been modified to account for the use of the pseudoatom (Wüthrich *et al.*, 1983). M replaces the 3 protons of a methyl group and is located in the center of the 3 proton positions, with the correction M = 1·0 Å. P replaces the protons of a methylene group and is located in the center between the 2 proton positions, with the same correction, $m = 1{\cdot}0$ Å, as for methyl groups. Q replaces the 2 methyl groups of Val and is located at the center of gravity of all 6 methyl protons, with the correction $q = 2{\cdot}4$ Å. Distance constraints to the amide protons in the side-chain of Asn are referred to the amide nitrogen atom ND, with the correction m. The rows following the 3-letter symbol for the amino acids list all the distance constraints that involve hydrogen atoms of this residue. For each residue, the distance constraints are grouped into 3 classes, which are presented in 3 columns. Sequential backbone constraints are those between backbone protons (for Pro also C^{δ} protons) in sequentially neighboring residues. Medium-range backbone and long-range backbone constraints are between the same proton types in residues that are not nearest-neighbors in the sequence. Constraints with side-chains are all those that have not been listed elsewhere. All entries appear twice, once for each residue concerned. In each column, the 1st entry identifies a hydrogen atom (or pseudoatom) in the amino acid identified by the 3-letter symbol. The 2nd and 3rd entries indicate a hydrogen atom in a different residue, to which a distance constraint has been observed. The 4th entry is the distance constraint (in Å), possibly with symbols indicating that 1 or 2 corrections were added to the NOESY distance constraint to allow for the use of pseudoatoms. The 5th entry, in parentheses, lists the chemical shifts of the interacting protons.

nearly identical with those obtained for rabbit MT-2a (Arseniev *et al.*, 1988).

(b) *Structure calculations with the program DISMAN*

After the aforementioned preliminary calculations, the input of Tables 1 and 2 and Table 1 from Arseniev *et al.* (1988) was used for a computation of 200 structures up to level 6 of the variables DISMAN target function. For each of the two domains, the 40 best structures were selected and computed to completion. The ten best structures for each domain are used for the following discussions. For the β-domain, an additional, identical sequence of calculations and structure selections was performed using input data that were identical except for an exchange of the metal-binding sites of Cys24 and Cys29 relative to Table 1 of Arseniev *et al.* (1988).

(c) *Cadmium–cysteine co-ordination in the 3-metal cluster*

The ambiguity in the metal binding by Cys24 and Cys29 that remained after the experimental determination of the metal–sulfur bonds was resolved by a comparison of DISMAN calculations for both possible co-ordination schemes. For the ten best DISMAN structures of the β-domain with Cys24 forming a bridge CdII and CdIV, and Cys29 singly bound to CdII, the sum of the residual distance constraint violations was between 2·8 Å and 5·4 Å (Table 3). With Cys29 in the bridging position, and Cys24 singly bound to CdII, the sum of the residual violations ranged from 13·3 Å to 19·7 Å, indicating clearly inferior solutions. A metal co-ordination identical with that in rabbit MT-2a in solution (Arseniev *et al.*, 1988) was confirmed by its superior compatibility with the ensemble of all other constraints for the β-domain (Tables 1 and 2) when compared with the alternative co-ordination scheme that would have been allowed by the cadmium–proton correlation experiments (Vašák *et al.*, 1987).

(d) *The MT-2 structures obtained from the DISMAN calculations*

Table 3 shows that, for identical metal co-ordination as in rabbit MT-2a, the β-domain

Table 2

Dihedral angle constraints used as supplementary input for the DISMAN calculations with rat liver MT-2

Residue	Dihedral angle constraints (deg.)† ϕ‡	ψ§	χ^1‖
Met1	[−180, −60]		
Asp2	[−160, −80]	[60, 180]	[150, −150]
Pro3			
Asn4			
Cys5	[−170, −60]		[150, −150]
Ser6	[−160, −80]		[30, 90]
Cys7	[−180, −60]		[150, −30]¶
Ala8	[−170, −70]		
Thr9			
Asp10	[−160, −80]		[30, 90]
Gly11			
Ser12			
Cys13			
Ser14			
Cys15	[−180, −60]		
Ala16			
Gly17			
Ser18			
Cys19			[150, −150]
Lys20			
Cys21	[−180, −60]		[−90, −30]
Lys22	[−180, −60]		
Gln23	[−170, −70]		
Cys24			[150, −150]
Lys25	[−170, −70]		
Cys26	[−180, −40]		
Thr27			
Ser28			
Cys29	[−170, −70]		[150, −150]
Lys30	[−180, −60]		
Lys31	[−170, −70]		
Ser32			
Cys33	[−180, −60]		
Cys34	[−180, −60]		
Ser35			
Cys36	[−170, −70]		
Cys37	[−160, −80]	[60, 180]	
Pro38			
Val39	[−180, −40]		
Gly40			
Cys41	[−180, −40]		
Ala42			
Lys43	[−180, −40]		
Cys44	[−170, −70]		[150, −30]¶
Ser45	[−180, −40]		
Gln46	[−150, −90]		[150, −150]
Gly47			
Cys48			[150, −150]
Ile49	[−150, −90]		[30, 90]
Cys50	[−180, −60]		[150, −30]¶
Lys51	[−180, −60]		
Glu52	[−180, −60]		
Ala53			
Ser54	[−170, −70]		
Asp55			
Lys56	[−170, −70]		
Cys57	[−170, −70]		
Ser58			
Cys59	[−180, −60]		
Cys60	[−160, −80]		[150, −30]¶
Ala61	[−180, −40]		

† The dihedral angle constraints are given in the form of allowed ranges bounded by the 2 values listed; [150, −150] stands for an allowed range comprising 150° to 180° *and* −180° to −150°

‡ The ϕ angle constraints were derived from $^3J_{HN\alpha}$ measured in a COSY spectrum in H_2O at 10°C, using the following relations:

$$^3J_{HN\alpha} < 7\text{ Hz } \phi = [-180°, -40°]$$
$$= 7 \text{ to } 8\text{ Hz} = [-180°, -60°]$$
$$= 8 \text{ to } 9\text{ Hz} = [-170°, -70°]$$
$$= 9 \text{ to } 10\text{ Hz} = [-160°, -80°].$$

§ ψ angle constraints in X-Pro dipeptide segments resulting from sequential NOEs (Arseniev *et al.*, 1988).

‖ χ^1 angle constraints were obtained from combining the data on $^3J_{\alpha\beta}$ (only those residues were considered where at least 1 coupling constant $^3J_{\alpha\beta}$ was either <5 Hz or >9 Hz), intra-residual and sequential NOEs (see Arseniev *et al.* (1988) for a detailed description of the procedures used).

¶ For these 4 cysteinyl residues, the conformations g^+t and tg^- could not be distinguished from the experimental data, but g^+g^- could be excluded. Therefore, the allowed range from 150° to 180° *and* −180° to −30° covers both g^+t and tg^-.

converged almost ideally in all of the ten best calculations, with a single NOE constraint violation between 0·3 and 0·5 Å, and very small numbers of violations for the other classes of constraints. Somewhat more important residual violations occur in the 4-metal domain, but these compare favorably with the results obtained with rabbit MT-2a (Arseniev *et al.*, 1988).

Table 4 includes average RMSD values for the ten best structures of each domain. These present a second, essential criterion for assessing the quality of the structures obtained, since they can be taken as a measure of how well a structure is defined by the given set of constraints. As was to be expected from the higher number of constraints, the α-domain has a lower RMSD average than the β-domain. However, when the least constrained part of the β-domain, residues 1 to 12, is excluded from the best-fit calculation, the RMSD values become comparable for the two domains (Table 4B). An impression of the variance among the ten best DISMAN structures for the two domains is afforded by Figure 1. In the β-domain, the amino-terminal end and the region between residues 7 and 12 are visibly less well-defined than the rest of the chain. Almost all of the α-domain agrees quite well in the superposition, except for the amino-terminal residues near the junction between the two domains.

The Cd and S atoms of the two clusters are superimposed in Figure 2. All ten structures of the 3-metal cluster show a nearly ideal boat conformation of the six-membered ring formed by the three cadmium ions and the three bridging sulfur atoms. The two fused, six-membered rings in the α-domain both show somewhat distorted boat conformations. The chiralities defined by the four sulfur atoms bound to the individual cadmium ions are identical in all ten structures (Fig. 2). At some of the bridging sulfur atoms, which are co-ordinated tetrahedrally to two different cadmium ions, one C^β atom, and a lone electron pair, both possible chiral configurations occur; for example, at Cys37 and Cys50 (Fig. 2).

4. Discussion

To provide an overall view of the molecular structure of rat MT-2, Figure 3 shows the DISMAN

Table 3
Residual constraint violations in the ten best DISMAN structures of the two domains of rat MT-2

Sum (Å)	Number of violations†					
	NOE constraints		Cd–S co-ordination‡		Non-bonded contacts§	Dihedral angles‖
	0·3–0·5 Å	>0·5 Å	0·3–0·5 Å	>0·5 Å	>0·1 Å	>5°
A. *The β-domain* ¶						
2·8	0	0	0	0	0	0
2·8	0	0	0	0	0	0
4·0	0	0	1	0	1(0·1)	0
4·3	0	0	1	0	0	0
4·5	1	0	1	0	1(0·1)	0
4·7	0	0	1	0	0	0
4·8	0	0	0	0	1(0·2)	0
4·9	0	0	2	1(0·6)	0	0
5·2	0	0	2	0	0	0
5·4	0	0	2	0	0	0
B. *The α-domain*						
14·3	4	1(0·6)	10	1(0·7)	2(0·2)	1(9)
15·0	6	0	2	5(1·0)	6(0·3)	0
15·3	4	1(0·7)	7	4(0·6)	4(0·3)	0
15·6	6	0	9	3(1·0)	1(0·2)	2(6)
15·6	4	1(0·8)	4	6(1·0)	5(0·1)	1(8)
16·0	2	3(0·6)	7	5(0·8)	5(0·1)	0
16·8	3	0	9	7(0·6)	5(0·2)	1(6)
16·8	5	3(0·7)	9	3(0·9)	2(0·2)	1(6)
17·0	4	1(0·7)	11	3(1·0)	4(0·2)	0
17·2	3	1(0·5)	13	5(1·0)	0	0

† In each column, the number of constraint violations in the ranges indicated at the top of the columns are listed. The numbers in parentheses are the maximal violations.
‡ Number of upper and lower distance constraint violations arising from the assumption of tetrahedral Cd–S co-ordination with a Cd–S distance of 2·6 Å.
§ Violations of the van der Waals radii used in the DISMAN target function.
‖ Violations of the angular constraints listed in Table 2.
¶ Metal co-ordination identical with that in the β-domain of rabbit MT-2a (Arseniev *et al.*, 1988).

structures with smallest residual constraint violations (Table 3) for the β-domain with residues 1 to 30 and the 3-metal cluster, and the α-domain with the polypeptide segment 31–61 and the 4-metal cluster. The polypeptide backbone and all Cys side-chains are drawn, and the cadmium ions are shown as dotted spheres with a radius of 0·9 Å. The polypeptide chain forms a right-handed loop around the 3-metal cluster in the β-domain, and a left-handed loop around the 4-metal cluster in the α-domain. Details of the metal co-ordination are shown in Figure 2.

Table 4
Average RMSD values for pairwise comparisons of the ten DISMAN structures in Table 3

Domain	Atoms considered	Average RMSD (Å)	Standard deviation
β	C^{α}, C′, N	2·0	0·5
(residues 1–30)	C^{α}, C′, N, C^{β}, S^{γ}, Cd†	2·0	0·5
	S^{γ}, Cd	0·4	0·2
β	C^{α}, C′, N	1·6	0·5
(residues 13–30)	C^{α}, C′, N, C^{β}, S^{γ}, Cd†	1·5	0·4
α	C^{α}, C′, N	1·7	0·3
	C^{α}, C′, N, C^{β}, S^{γ}, Cd†	1·8	0·4
	S^{γ}, Cd	0·8	0·2

† This corresponds to all structure elements that are constrained by the n.m.r. data (Tables 1 to 3), except for the side-chains of Asn4, Thr9, Pro38, Val39, Ile49, Glu52 and Lys56 beyond C^{β}.

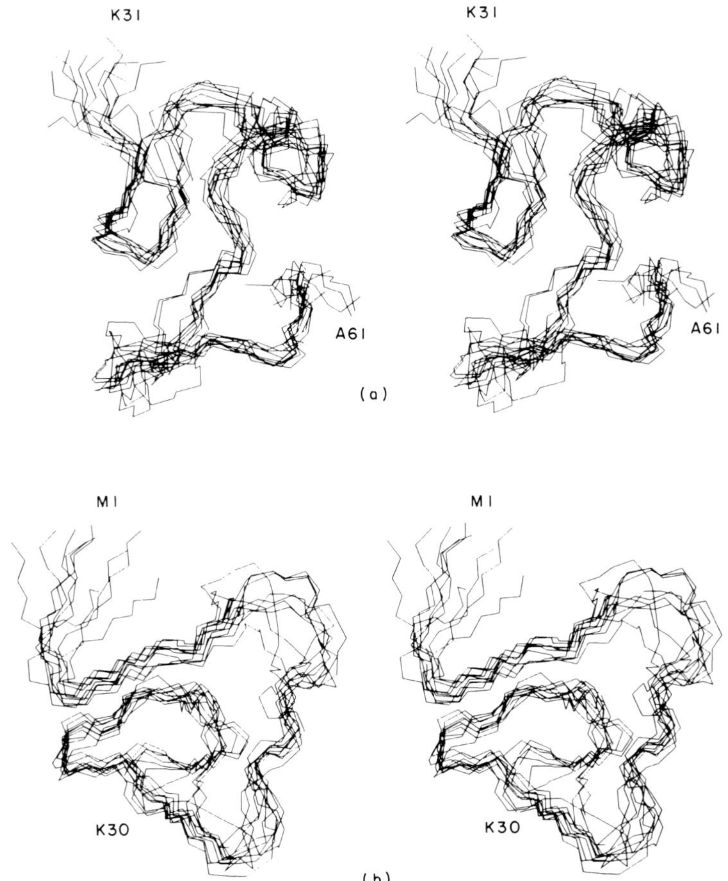

Figure 1. (b) Stereo view of a superposition for the β-domain (residues 1 to 30) in rat MT-2 of the 10 DISMAN structures with the lowest residual error functions (Table 3). The superposition is for minimum pairwise RMSD of the best structure with each of the others, whereby only residues 13 to 30 were considered. Note that residues 5 to 9, although excluded from the best-fit calculation, nevertheless coincide quite well (see the text). Only the backbone is shown. The residues Met1 and Lys30 at the 2 ends of the domain are labeled. (a) Same for the α-domain from Lys31 to Ala61. The orientation of the 2 domains relative to each other could not be determined from the n.m.r. data.

(a) *Comparison of the solution structures of rat MT-2 and rabbit MT-2a*

From a comparison of Figures 2 and 3 with the corresponding presentations for rabbit MT-2a (Figs 7 and 8 of Arseniev *et al.*, 1988), it is readily apparent that both the global polypeptide fold and the local environment of the cadmium ions are closely similar in the two proteins. A more detailed comparisons of the two proteins is based on the information presented in Figures 4 to 6 and Table 5.

Figure 4 was obtained by pairwise superposition for minimum RMSD of the best structure of rat MT-2 with the next best four structures (Table 3) and the best DISMAN structure of rabbit MT-2a (Arseniev *et al.*, 1988). Except for the N termini of both domains, where there are also large deviations among the five rat MT-2 structures, no global differences in the chainfolds between the two species can be recognized. With the exception of a few limited segments discussed below, the rabbit MT-2a structure lies well within the spread of the ensemble of the five rat MT-2 structures.

Figure 5 presents a statistical analysis of the ϕ and ψ angles as a function of the amino acid sequence, aimed at the recognition of secondary structural elements in the ensembles of the 20 best DISMAN structures of rabbit MT-2a (Arseniev *et al.*, 1988) and the ten best DISMAN structures of rat MT-2 (Table 3). The procedure used here checks in an objective way the occurrence of the secondary structure elements 3_{10}-helix and half-turn (Arseniev *et al.*, 1988). Figure 5(a) is a plot of the deviations of ϕ and ψ from the ideal values for half-turns

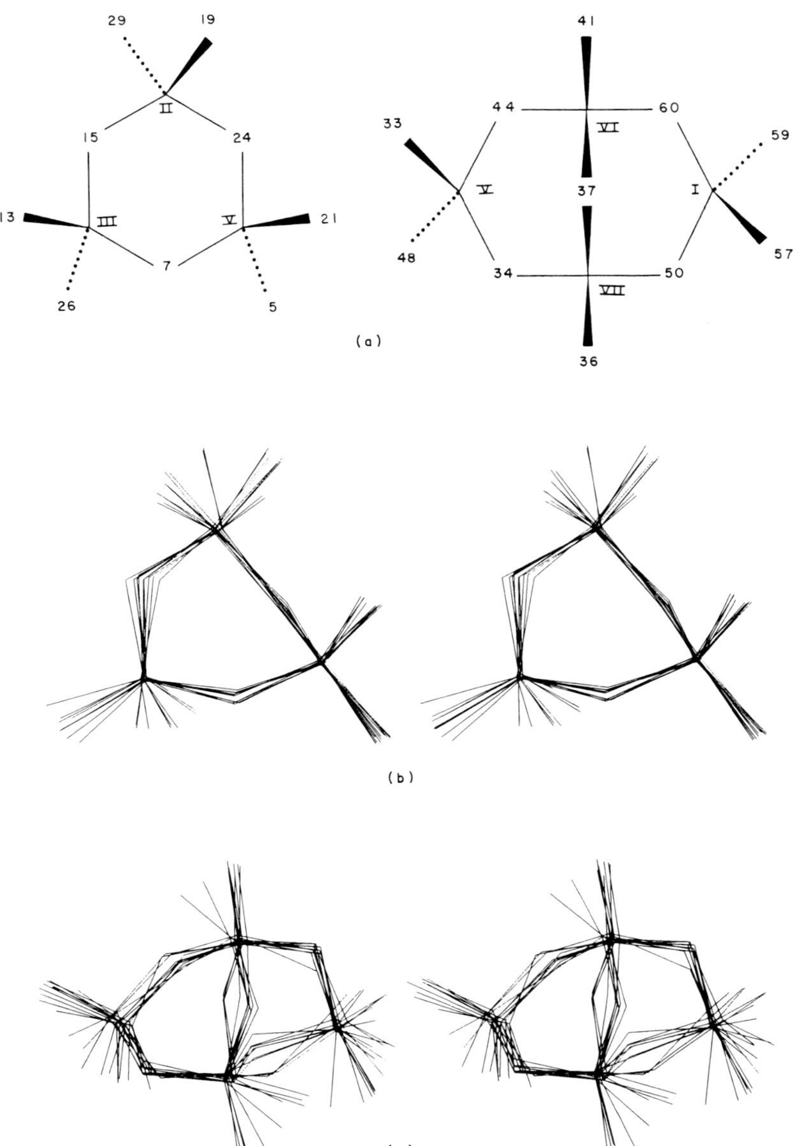

Figure 2. Cadmium–sulfur clusters in rat MT-2 in solution. (a) A presentation of the cadmium–sulfur connectivities and the chirality at the cadmium ions in the 2 domains. The cadmium ions are identified with arbitrary roman numbers, corresponding to increasing ^{113}Cd chemical shifts (Frey *et al.*, 1985). The sulfur atoms of the Cys ligands are identified by the sequence positions. (b) and (c) Superposition of the 10 best DISMAN structures of the 3-metal and 4-metal clusters (Table 3), which are located in the β-domain and the α-domain, respectively. Only the cadmium ions and the sulfur positions of the bridging cysteinyl residues were considered in the optimization of the superpositions.

(Wagner *et al.*, 1986*a*). A single low value of Δ_i is indicative of a half-turn involving the residues i-1 to $i+2$. Clearly, there is evidence that rat MT-2 and rabbit MT-2a contain half-turns in the same locations; namely, for $i=19$, 21, 24, 41 and 48. In Figure 5(b), several adjoining low values of Δ_i are indicative of a stretch of 3_{10}-helix. For both proteins, 3_{10}-helices are revealed at residues 42 to 45 and 58 to 60.

Figure 5(c) shows plots of the variability of the backbone dihedral angles ϕ and ψ *versus* the amino acid sequence in the 20 best DISMAN structures of

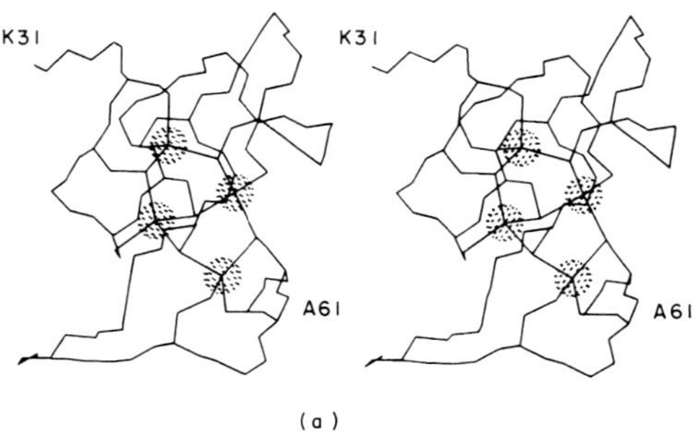

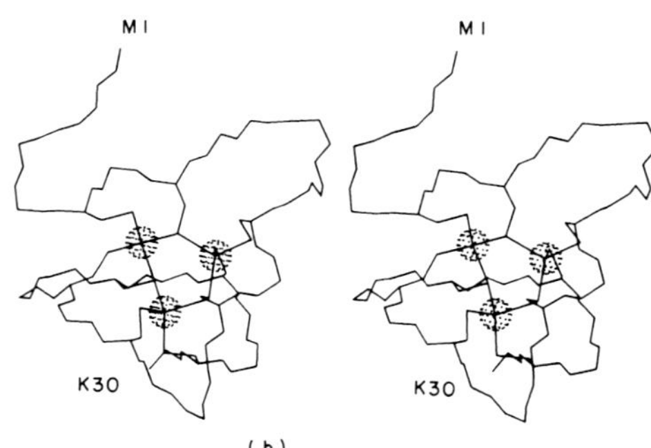

Figure 3. (b) Stereo view of the polypeptide backbone and the heavy-atom side-chain positions of the 9 cysteinyl residues at positions 5, 7, 13, 15, 19, 21, 24, 26 and 29 in the β-domain structure with the lowest residual error function (Table 3). The cadmium ions are represented as dotted spheres of radius 0·9 Å. (a) Polypeptide backbone and heavy-atom side-chain positions of the 11 cysteinyl residues at positions 33, 34, 36, 37, 41, 44, 48, 50, 57, 59 and 60 in the best structure obtained for the α-domain (Table 3). The orientation of the 2 domains is the same as that in Fig. 1.

rabbit MT-2a (Arseniev *et al.*, 1988) and the ten best structures of rat MT-2 (Table 3). See the Figure legend for details. These data, which describe the precision with which the local conformation is defined, coincide particularly closely between the two proteins for residues 1 to 6, 10 to 14, 19 to 30, 33 to 36, 40 to 43 and 46 to 59. Throughout, the standard deviation, σ, is low at the cysteine positions, which demonstrates that the chain fold is considerably constrained by the metal cluster formation. Between the cysteine positions, the largest deviations between the two proteins occur at residues 7 to 10 and 37 to 39. Residues 7 to 10 are part of a large disordered loop (Fig. 1). In this loop, the rabbit MT-2a structure diverges somewhat from rat MT-2 (Fig. 4), which is expected from the fact that the rabbit sequence contains the insertion Ala8′ (Wagner *et al.*, 1986*b*). Interestingly, the mutation frequency in the known mammalian MT sequences is high for all of the residues 8 to 12 (Kägi & Kojima, 1987), and in this region both rat MT-2 (Fig. 1) and rabbit MT-2a (Fig. 6(a) of Arseniev *et al.*, 1988) showed outstandingly large disorder in the solution structure. At positions 37 to 39, rat MT-2 has a higher variability of the local backbone conformation. This might be related to the fact that rat MT-2 contains valine in position 39, whereas this position in rabbit MT-2a contains proline, which has intrinsically lower conformational flexibility, since its ϕ angle is fixed in the DISMAN calculations. Overall, the conclusions resulting for both structures coincide with the earlier observation in rabbit MT-2a, that the variance among the different DISMAN structures increases with increasing size of the loops between neighbouring metal-bound cysteines (Arseniev *et al.*, 1988). Additional evidence for the importance of the metal co-ordination for the rigidity of the polypeptide structure comes from the following. In Figures 1 and 4, only residues 13 to 30 were included in the computation of the minimal RMSD for the superposition. Nevertheless, there is very good agreement also for residues 4 to 7, indicating that their conformation is determined mainly by the participation of Cys5 and Cys7 in the 3-metal cluster.

Table 5 lists the RMSD values for all possible pairs among the five best DISMAN structures each of rat MT-2 and rabbit MT-2a. For the β-domain, the average RMSDs within the sets of rabbit MT-2a and rat MT-2 structures are similar, but the average of the RMSD values between the two sets is considerably higher (Table 5A). However, when the predominantly disordered segment 1 to 12 (Figs 1 and 4), which includes the insertion in the rabbit MT-2a sequence, is omitted from the computation, then the RMSDs over the residues 13 to 30 between the two β-domains become comparable to those between the different DISMAN structures of the same species (Table 5B). This again demonstrates that the only sizeable differences between the β-domains of rat MT-2 and rabbit MT-2a are in the disordered loop region 7 to 11 and the poorly constrained N-terminal segment 1 to 4 (see also Fig. 4). For the α-domain, nearly identical RMSDs were obtained between rat MT-2 and rabbit MT-2a, or between different, individual DISMAN structures of either species (Table 5C). The RMSDs are smaller than for the β-domain, which reflects the fact that a considerably larger number of NOE distance constraints could be obtained for the α-domain than for the β-domain; namely, 80 *versus* 46 in rat MT-2 (Table 1), and 80 *versus* 49 in rabbit MT-2a (Table 1, of Arseniev *et al.*, 1988). The observation of a smaller number of NOE constraints in the β-domain might be a consequence of its higher conformational flexibility, which is indicated also by the experimental observations that the metal binding is weaker and the metal exchange faster in the 3-metal cluster than in the 4-metal cluster (Kägi & Kojima, 1987).

Table 5

Pairwise RMSD values between the five best structures of rabbit MT-2a (Arseniev et al., *1988) and rat MT-2 (Table 3)*

A. *The β-domain (residues 1 to 30)*†

		Rat MT-2					Rabbit MT-2a				
		1	2	3	4	5	1	2	3	4	5
	1		1·35	1·50	1·91	1·46	2·41	2·84	2·86	2·62	2·08
	2			1·87	2·22	1·99	1·95	2·34	2·32	2·26	1·60
Rat	3				1·38	1·23	2·82	2·72	3·43	2·54	2·38
MT-2	4		[1·68]			1·87	2·99	2·79	3·66	2·61	2·39
	5						2·74	2·45	3·39	2·25	2·31
	1							1·88	2·03	1·80	1·42
	2								2·78	1·03	1·73
Rabbit	3			[2·59]						2·61	2·03
MT-2a	4							[1·91]			1·74
	5										

B. *The β-domain (residues 13 to 30)*

		Rat MT-2					Rabbit MT-2a				
		1	2	3	4	5	1	2	3	4	5
	1		0·96	0·97	1·17	0·92	1·26	1·44	1·52	1·31	1·34
	2			0·89	1·31	1·06	1·08	1·00	1·05	1·06	0·98
Rat	3				0·97	0·86	0·80	1·16	1·28	1·04	1·08
MT-2	4		[1·03]			1·20	1·19	1·25	1·57	1·23	1·34
	5						1·26	1·20	1·39	1·11	1·12
	1							1·08	1·18	0·89	1·13
	2								0·97	0·97	0·76
Rabbit	3			[1·20]						0·99	0·96
MT-2a	4							[1·00]			1·05
	5										

C. *The α-domain (residues 31 to 61)*

		Rat MT-2					Rabbit MT-2a				
		1	2	3	4	5	1	2	3	4	5
	1		1·32	1·36	1·54	1·55	1·54	1·81	1·92	1·74	1·59
	2			1·56	1·45	1·69	1·79	1·54	1·92	1·89	1·55
Rat	3				1·15	0·98	1·42	1·51	1·82	1·64	1·52
MT-2	4		[1·42]			1·59	1·51	1·45	2·02	1·81	1·55
	5						1·48	1·56	1·78	1·52	1·62
	1							1·62	1·18	0·83	1·22
	2								1·84	1·74	1·24
Rabbit	3			[1·66]						1·03	1·27
MT-2a	4							[1·33]			1·36
	5										

The following atoms were included in the computation: C^{α}, C, N of all residues and C^{β}, S^{γ}, Cd^{δ} of all cysteine residues. The averages of the values in each submatrix are given in square brackets in the symmetric positions relative to the diagonal.

† Ala8′ in the rabbit MT-2a sequence was not included in the comparison.

Overall, within the precision of the present structure determination for rat MT-2 and the earlier work with rabbit MT-2a (Arseniev *et al.*, 1988), the solution conformations of these two proteins show no significant differences, despite their different primary structures. The following picture emerges. The two metallothioneins bind metal ions with their perfectly conserved cysteinyl residues, forming two clusters with identical binding topologies and highly similar spatial structures. Each cluster is tightly wrapped into one half of the polypeptide chain, so that two flexibly linked domains are formed. Conformational disorder, which may be static or dynamic, or some of both, is observed mainly in loops formed by several consecutive non-cysteinyl residues, and it is interesting that the same regions also show the highest sequence variability between different species (Kägi & Kojima, 1987).

(b) *Comparison of the structures observed for rat MT-2 by n.m.r. in solution and by X-ray diffraction in single crystals*†

The global molecular architectures of rat MT-2 in solution (Fig. 3) and in single crystals (Furey *et al.*, 1986) display similar triats. In both environments, one observes two domains with identical components. The β-domain includes the amino acid residues 1 to 30 and a 3-metal cluster, and the α-domain is composed of the polypeptide segment 31 to 61 and a 4-metal cluster. Only the individual domains can be compared, since the relative spatial locations of the two domains have not been characterized in solution. For the β-domain, the

† This comparison of the rat MT-2 structures in solution and in single crystals was added at the request of one of the referees.

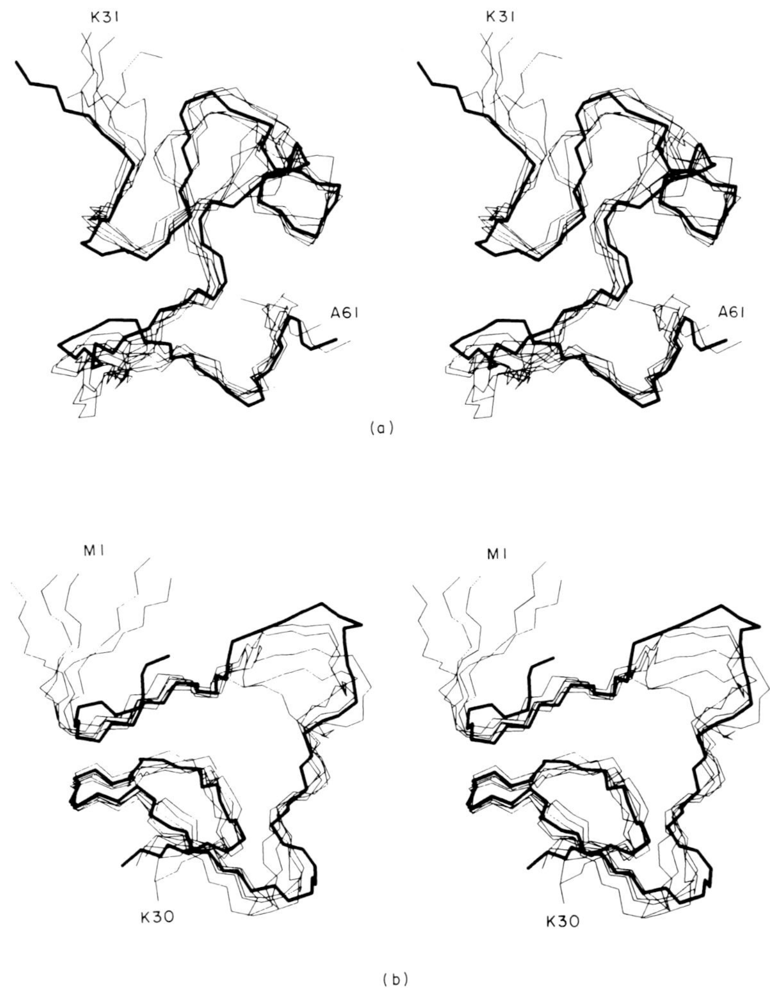

Figure 4. Stereo views of the best DISMAN structure of the α and β-domain in rabbit MT-2a from Arseniev *et al.* (1988) (thick line) superimposed on the 5 best structures of rat MT-2 (see the text for a description of the fitting procedure). For the β-domain, only residues 13 to 30 were used for the best-fit computation.

overall handedness of the polypeptide fold is right-handed in both structures, and for the α-domain it is left-handed. Furthermore, at a first glance it appears that the polypeptide backbone runs a largely parallel course in the β-domains of both structures, and that there are also sizeable regions in the α-domains where backbone segments are in closely similar locations and orientations. A closer look at the detailed features of the two domains, however, reveals that in spite of these apparent similarities there are fundamental differences between the rat MT-2 structures in solution and in single crystals.

As was pointed out previously (Vašák *et al.*, 1987), rat MT-2 in solution contains different metal–cysteine co-ordinative bonds than the presently known crystal structure (Furey *et al.*, 1986). The closest fit is obtained by matching the metal ions in the two structures as shown in Figure 6 (see also Fig. 2(a)). The differences then include that, in the β-domain, Cys13 and Cys15 are exchanged between a bridging and a singly bound ligand site, and furthermore Cys19 and Cys26 are in different single-bond sites. In the α-domain there are four exchanges of singly bound with bridging cysteines, involving the residue pairs 33–34, 41–44, 48–50 and 59–60. In the following, we document and visualize the changes of the polypeptide conformation that occur in parallel with the different metal-binding arrangements in the two different ways.

First, an overview of the similarities and differences in the polypeptide backbone folds of the

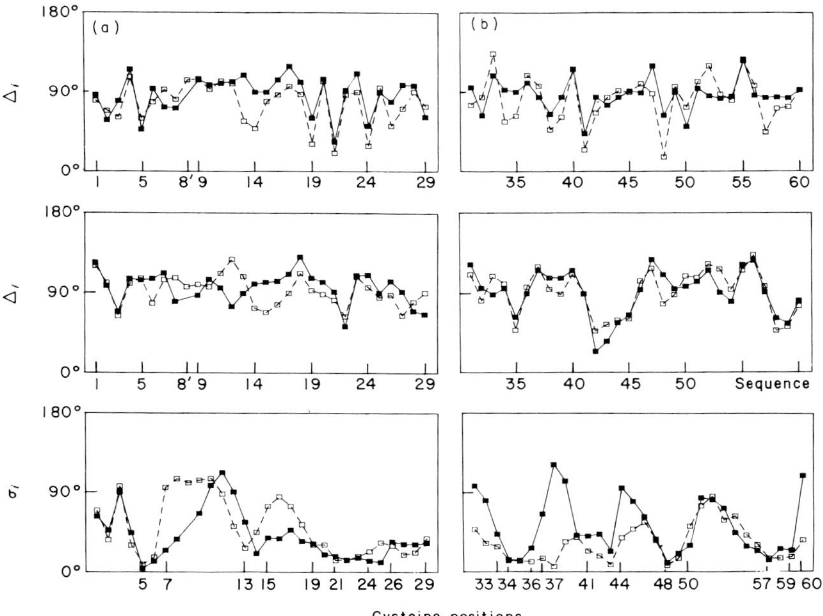

Figure 5. (a) and (b) Comparison of the identification of half-turns and 3_{10}-helices in rat MT-2 and rabbit MT-2a based on a statistical analysis of the ϕ and ψ backbone dihedral angles. Standard deviations Δ_i of the actual dihedral angles in the 10 best DISMAN structures of rat MT-2 (Table 3) to the dihedral angles for standard half-turns and 3_{10}-helices are plotted with filled squares and continuous lines. The corresponding data for the best 20 rabbit MT-2a structures taken from Arseniev *et al.* (1988) are drawn with open squares and broken lines. The standard deviations Δ_i were computed as:

$$\Delta_i = \sqrt{\frac{\sum_{j=1}^{n} \frac{\delta_i^j}{4}}{n-1}},$$

where n is the total number of DISMAN structures and δ_i^j is given by the following expressions. (a) Half-turns: $\delta_i^j = |\phi_i^j - (-60°)|^2 + |\psi_i^j - 120°|^2 + |\phi_{i+1}^j - (-90°)|^2 + |\psi_{i+1}^j - 0°|^2$, where $-60°$, $120°$, $-90°$ and $0°$ are the standard backbone dihedral angles ϕ and ψ of the 2nd and 3rd residues in a half-turn (Wagner *et al.*, 1986*a*). Single low values of Δ_i indicate the presence of a half-turn involving the residues $i-1$ to $i+2$. Note the insertion of residue 8′ in rabbit MT-2a (Wagner *et al.*, 1986*b*). (b) 3_{10}-helix: $\delta_i^j = |\phi_i^j - (-60°)|^2 + |\psi_i^j - (-30°)|^2 + |\phi_{i+1}^j - (-60°)|^2 + |\psi_{i+1}^j - (-30°)|^2$, where $\phi = -60°$ and $\psi = -30°$ are the standard values for the 3_{10}-helix. Several subsquent low values of Δ_i indicate the presence of 3_{10}-helical secondary structure. (c) Comparison of the precision with which the local conformations of rat MT-2 and rabbit MT-2a were determined by the n.m.r. data. Plots of the standard deviation $\sigma(\phi_{i+1} + \psi_i)$ are shown for the 10 best structures of rat MT-2 (filled boxes and continuous lines) and the 20 best structures of rabbit MT-2a (open boxes and broken lines). The numbers at the bottom identify the cysteine positions in the sequence.

two structures is afforded by Figure 7, which shows a least RMSD superposition of the backbones of the best DISMAN solution structure and the crystal structure of rat MT-2. The RMSD values for all C^α atoms are 3·65 Å in the β-domain, and 5·63 Å in the α-domain. In the β-domain, the two chains run nearly parallel from residue 1 to 16, except for a local divergence at Ser12 and Cys13. The peptide segment from Gly17 to Cys21 shows the major differences in the β-domain. As shown in Figure 7, it forms a loop that extends to the left in the solution structure, and to the right in the crystal structure. For the rest of this domain, the two structures are rather similar, except that the segment from 25 to 27 is displaced in a parallel fashion by approximately 3 Å. For the α-domain, a comparison of the two structures is less straightforward. The apparent quite close fit between certain segments of the polypeptide chain indicated by Figure 7 is deceptive, since the nearby residues are not in corresponding sequence positions in the two structures. For example, in Figure 7 the pentapeptide segment 33 to 37 in the solution structure matches quite closely with residues 36 to 40 in the crystal structure, or the segment 47 to 57 in solution matches with 46 to 56 in the crystal. For the other parts of the polypeptide chain, extensive divergence is clearly apparent in the superposition

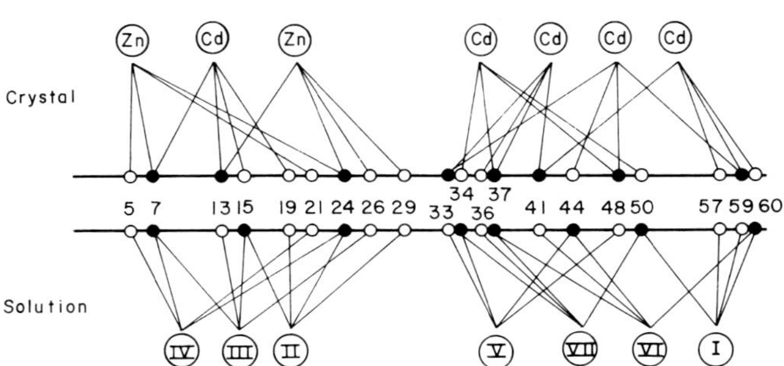

Figure 6. Comparison of the metal co-ordination in the solution structure and the crystal structure of rat MT-2. The positions of bridging cysteinyl residues are identified by filled circles, those of singly bound cysteines by open circles. The 7 Cd ions in the solution structure are identified with roman numerals as in Fig. 2(a). The metal co-ordination was obtained as the direct result of a $^{1}H-^{113}Cd$ correlation experiment (Vašák *et al.*, 1987) and the sequence-specific ^{1}H n.m.r. assignments (Wörgötter *et al.*, 1987). The closest coincidence between the 2 structures is attained when the metal ions are matched 1 by 1 in the order shown here from left to right.

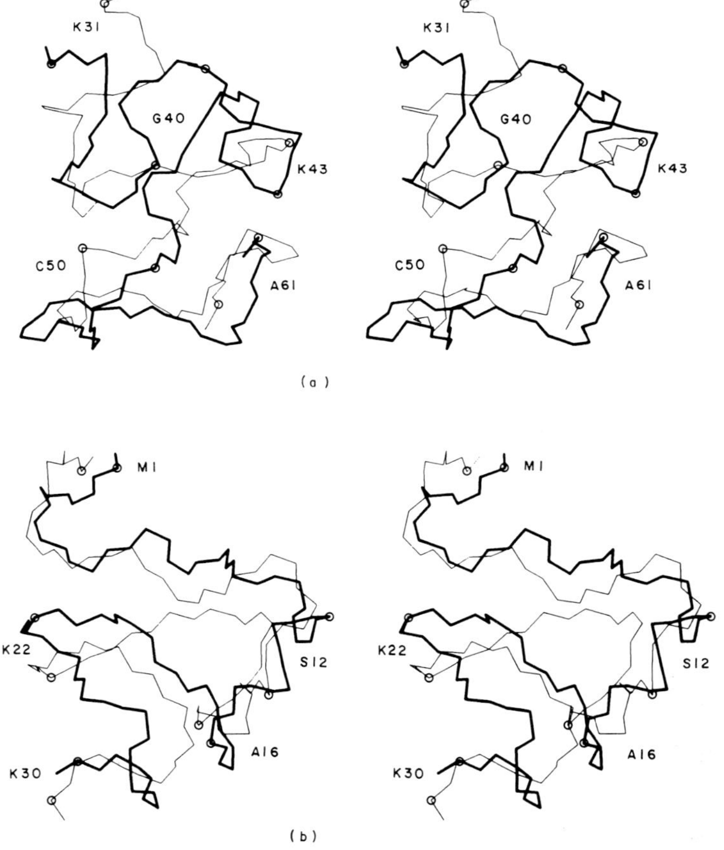

Figure 7. Stereo views of the best DISMAN structure (same as in Fig. 3; thick lines) superimposed for minimal RMSD with the crystal structure (entry 2MT2 from the Brookhaven Protein Data Bank; thin lines) for the (a) α and (b) β-domains of rat MT-2. Selected C^{α} positions are marked with small circles and labeled with the amino acid 1-letter code and the sequence position.

Table 6

Proton–proton distances in the crystal structure of rat MT-2 that exceed the corresponding upper distance limit measured in solution by n.m.r. by more than 3·0 Å

Proton 1†	Proton 2‡	NOE upper distance limit (Å)	Crystal structure distance (Å)
A. *The α-domain*			
42 Ala MB	45 Ser HN	5·0	8·9
46 Gln HN	49 Ile PG	6·4	10·5
47 Gly HN	49 Ile MG	5·0	10·3
31 Lys HA	39 Val QG	6·4	16·8
31 Lys HN	39 Val QG	6·4	17·9
32 Ser HA	39 Val QG	6·4	14·6
32 Ser HN	39 Val QG	6·4	15·5
33 Cys HN	39 Val QG	6·4	14·7
33 Cys PB	48 Cys HN	5·0	8·8
36 Cys PB	56 Lys HA	4·0	16·4
36 Cys PB	56 Lys HN	5·0	14·2
36 Cys PB	57 Cys HA	5·0	12·6
37 Cys HA	41 Cys PB	5·0	8·3
44 Cys HA	49 Ile PG	4·0	12·7
44 Cys PB	49 Ile HN	5·0	8·1
44 Cys HN	49 Ile PG	5·0	10·9
49 Ile MD	59 Cys PB	5·0	9·5
52 Glu PG	57 Cys HA	5·0	11·2
53 Ala MB	57 Cys HA	4·0	9·5
B. *The β-domain*			
27 Thr HA	30 Lys PB	5·0	10·0
4 Asn HA	22 Lys HN	4·0	7·9
4 Asn HA	23 Gln HN	4·0	9·0
4 Asn PB	23 Gln HN	4·0	8·2
5 Cys PB	25 Lys HN	4·0	7·5
15 Cys PB	19 Cys HA	5·0	8·8

Protons or pseudoatoms were attached to the crystal structure using the standard ECEPP geometry.

† PB, PG, MG, MD and QG are pseudoatoms representing a group of protons (Wüthrich *et al.*, 1983).

of Figure 7. It is worth noting that the carboxy-terminal residues form loops of opposite orientation in the two structures, which seems to reflect directly the exchanged roles of Cys59 and Cys60 in the co-ordination as singly bound or bridging ligands to the same metal ions.

Second, as an attempt to obtain a more quantitative assessment of the structural differences seen in Figure 7, the upper limits on distances between distinct pairs of protons measured by n.m.r. spectroscopy in solution were compared with the corresponding proton–proton distances in the crystal structure. Table 6 lists all those distances in the crystal structure that exceed the n.m.r. upper limit by 3·0 Å or more. In the β-domain, these are six distances, with the differences ranging from 3·5 to 5·0 Å. In the α-domain, there are 19 such distances, with differences between 3·1 and 12·4 Å. These numbers clearly support the qualitative impression gained from Figure 7, that more pronounced differences occur in the α-domain than in the β-domain. In Figure 8 the contents of Table 6 are visualized as straight lines drawn into the crystal structure of rat MT-2, which represent the direction and the size of the differences between intramolecular interatomic distances and corresponding distance constraints in solution. In the β-domain, there is one bundle of such lines between the residues near Asn4 on one hand, and those near Gln23 on the other. In the α-domain, there are three bundles of lines in the approximate directions from Lys31 to Val39, Cys36 to Lys56, and Cys44 to Ile49, respectively. In addition to the identification and localization of the major structural differences, Table 6 and Figure 8 clearly demonstrate that the crystal structure of rat MT-2 would not be an acceptable solution to a distance geometry calculation based on the n.m.r. data measured in solution (Tables 1 and 2).

In conclusion, the studies described in this paper resulted in two contrasting comparisons. On one hand, we have the near-identity of the solution structures of rat MT-2 and rabbit MT-2a (Figs 4 and 5), in spite of numerous amino acid substitutions. On the other hand, there are significant differences between the rat MT-2 structures in solution and in single crystals, both on the level of the metal co-ordination (Fig. 6) and in the polypeptide fold (Figs 7 and 8). There appears to be a correlation between the extent of the differences on these two levels of the structural hierarchy, as evidenced by the fact that a larger number of differences in the metal co-ordination of the α-domain is accompanied by more pronounced

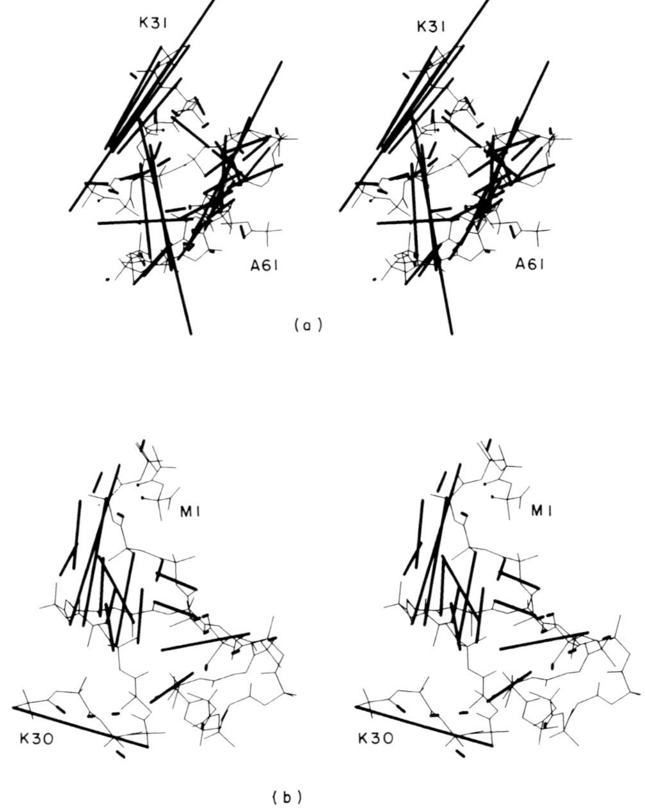

Figure 8. Stereo view of the crystal structure of rat MT-2 (entry 2MT2 from the Brookhaven Protein Data Bank) affording a visual display of the violations of NOE distance constraints observed by n.m.r. in solution (Table 6) in (a) the α-domain and (b) the β-domain. The thick lines indicate the direction and the length of the movements for individual pairs of hydrogen atoms, which would be required to satisfy the corresponding upper distance limits in solution. Some lines are not attached to a visible atom since, for clarity, only the C^{β} atoms of the side-chains are shown. Protons or, where applicable, pseudoatoms (Wüthrich *et al.*, 1983) were attached to the crystal structure using the standard geometry of the library from the empirical conformational energy program for peptides (ECEPP; Némethy *et al.*, 1983).

differences between the polypeptide conformations in this domain. In view of the fact that more extensive structural differences relative to the crystal structure are observed for the α-domain, it should be emphasized again that the α-domain is more precisely defined by the n.m.r. data than the β-domain in solutions of both rat MT-2 and rabbit MT-2a.

We thank Ms M. Sutter for the preparation of the biological material, and Ms E.-H. Hunziker-Kwik for the illustrations. Financial support by the Schweizerischer Nationalfonds (projects 3.198.0.85 and 3.164.0.85) and EMBO (fellowship to E.W.) is gratefully acknowledged.

References

Arseniev, A., Schultze, P., Wörgötter, E., Braun, W., Wagner, G., Vašák, M., Kägi, J. H. R. & Wüthrich, K. (1988). *J. Mol. Biol.*, **201**, 637–657.

Braun, W. & Gō, N. (1985). *J. Mol. Biol.* **186**, 611–626.

Braun, W., Wagner, G., Wörgötter, E., Vašák, M., Kägi, J. H. R. & Wüthrich, K. (1986). *J. Mol. Biol.* **187**, 125–129.

Frey, M. H., Wagner, G., Vašák, M., Sørensen, O. W., Neuhaus, D., Wörgötter, E., Kägi, J. H. R. & Wüthrich, K. (1985). *J. Amer. Chem. Soc.* **107**, 6847–6851.

Furey, W. F., Robbins, A. H., Clancy, L. L., Winge, D. R., Wang, B. C. & Stout, C. D. (1986). *Science*, **231**, 704–710.

Furey, W. F., Robbins, A. H,. Clancy, L. L., Winge, D. R., Wang, B. C. & Stout, C. D. (1987). In *Metallothionein II* (Kägi, J. H. R. & Kojima, Y., eds), pp. 139–148, Birkhäuser Verlag, Basel.

Kägi, J. H. R. & Kojima, Y. (1987). In *Metallothionein II* (Kägi, J. H. R. & Kojima, Y., eds), pp. 25–61, Birkhäuser Verlag, Basel.

Némethy, G., Pottle, M. S. & Scheraga, H. A. (1983). *J. Phys. Chem.* **87**, 1883–1887.

Neuhaus, D., Wagner, G., Vašák, M., Kägi, J. H. R. & Wüthrich, K. (1984). *Eur. J. Biochem.* **143**, 659–667.
Neuhaus, D., Wagner, G., Vašák, M., Kägi, J. H. R. & Wüthrich, K. (1985). *Eur. J. Biochem.* **151**, 257–273.
Vašák, M. & Kägi, J. H. R. (1983). In *Metal Ions in Biological Systems* (Sigel, H., ed.), vol. 15, pp. 213–273, Marcel Dekker, New York.
Vašák, M., Wörgötter, E., Wagner, G., Kägi, J. H. R., & Wüthrich, K. (1987). *J. Mol. Biol.* **196**, 711–719.
Wagner, G., Neuhaus, D., Wörgötter, E., Vašák, M., Kägi, J. H. R., & Wüthrich, K. (1986*a*). *J. Mol. Biol.* **187**, 131–135.
Wagner, G., Neuhaus, D., Wörgötter, E., Vašák, M., Kägi, J. H. R. & Wüthrich, K. (1986*b*). *Eur. J. Biochem.* **157**, 275–289.
Wagner, G., Frey, M. H., Neuhaus, D. Wörgötter, E., Braun, W., Vašák, M., Kägi, J. H. R. & Wüthrich, K. (1987). In *Metallothionein II* (Kägi, J. H. R. & Kojima, Y., eds), pp. 149–157, Birkhäuser Verlag, Basel.
Wörgötter, E., Wagner, G., Vašák, M., Kägi, J. H. R. & Wüthrich, K. (1987). *Eur. J. Biochem.* **167**, 457–466.
Wüthrich, K. (1986). *NMR of Proteins and Nucleic Acids*, Wiley, New York.
Wüthrich, K., Billeter, M. & Braun, W. (1983). *J. Mol. Biol.* **169**, 921–948.
Wüthrich, K., Billeter, M. & Braun, W. (1984). *J. Mol. Biol.* **180**, 715–740.

Edited by M. F. Moody

IV

EVOLUTION OF NMR STRUCTURE DETERMINATION WITH BIOLOGICAL MACROMOLECULES 1985–94

Introduction to papers 30 through 38

This part contains nine reprints of papers on methods development, which by necessity represent a limited and therefore somewhat arbitrary selection of our work on aspects of techniques during the decade 1985–94. These papers have their place in two different frames of reference. Firstly, our research during this period was primarily focussed on structure determinations with systems of biological and biomedical interest (see Part V), and on the physical-chemical characterization of the state of protein molecules in physiological fluids as mimicked by the aqueous solutions used for NMR experiments (see Parts VI, VII and VIII). Methods development by my group was therefore mainly motivated by the requirements for successful pursuance of these projects. Secondly, starting about 1986, a rapidly increasing number of research groups in the area of NMR studies with biological macromolecules was established worldwide. Many of these groups have made important contributions to methods development, and over the years we saw a steadily increasing flow of publications on NMR techniques for macromolecular studies. From 1991 onwards, much of this work has been published in the newly founded *Journal of Biomolecular NMR*. The general scope of the present volume allows only scant reference to the developments outside of my laboratory.

During the time span covered, important advances have been made in instrument design, including the introduction of superconducting magnets producing polarizing magnetic fields that correspond to ^{1}H resonance frequencies of 600 and 750 MHz, the introduction of novel experimental schemes for multi-dimensional NMR, new strategies for computer-supported resonance assignment and data collection, and increased efficiency of structure calculation from NMR data. With this general background of improved technology, major advances resulted from the use of uniform, partial or residue-specific isotope labeling of biopolymers with ^{13}C or/and ^{15}N, and the development of NMR experiments tailored specifically for studies with labeled molecules. These technical advances can readily be accommodated in the general scheme of Fig. 1 in paper **1**, which is in the following used as a guideline for the presentation of the papers **30** to **38.**

In the sample preparation for NMR studies my group focussed primarily on techniques for residue-specific (paper **31**), site-specific (paper **19**) and biosynthetically-directed fractional (paper **36**) isotope labeling. In addition, partially labeled binary or multimolecular complexes were prepared by combining components with and without uniform isotope labeling (see Part V).

In paper **30** we discovered an oversight in a commonly used routine for Fourier transformation of digitized time domain recordings. This paper has been included here because it had far-reaching consequences on our laboratory installations. Once the comparatively large base line instabilities arising from incorrect digitization of the free induction decay were eliminated, it became apparent that other factors needed to be closely monitored. For example, even minor fluctuations of the room temperature during an experiment turned out to be deleterious. In the papers **31** and **32** we introduced heteronuclear half-filters and double-half-filters, respectively. These experiments represent a major extension of the principle of heteronuclear editing,[1,2] and they presently find widespread application in structure determinations of multimolecular complexes (see Part V). The paper **33** is representative of a series of publica-

[1] Griffey, R.H. and Redfield, A.G. (1987) *Q. Rev. Biophys. 19,* 51–82. Proton-detected heteronuclear edited and correlated NMR and nuclear Overhauser effect in solution.

[2] Otting, G. and Wüthrich, K. (1990) *Q. Rev. Biophys. 23,* 39–96. Heteronuclear filters in two-dimensional [^{1}H,^{1}H]-NMR spectroscopy: combined use with isotope labelling for studies of macromolecular conformation and intermolecular interactions.

tions on the use of spin-lock pulses to purge NMR spectra from unwanted strong signals.[3] This technique was used in all the early NMR observations of hydration water (see Part VII). For improved efficiency of triple-resonance experiments,[4,5] which have a crucial role in resonance assignments of ^{13}C,^{15}N-doubly-labeled compounds,[6] the paper **34** describes a projection technique enabling reduction of the dimensionality of experiments with a predetermined, fixed number of correlation steps.

In paper **23** a general approach was introduced for the treatment of nuclear Overhauser effects with multiple hydrogen atoms that had not been individually assigned. This "pseudoatom" approach brings about a loss of information in the input for the structure calculations, which can be recovered once individual assignments for the individual hydrogen atoms are available. The papers **35** and **36** describe two different approaches for obtaining individual, stereospecific assignments for pairs of diastereotopic substituents. The determination of stereospecific assignments was one of the key factors which enabled significant improvement of the quality of NMR structures of proteins during the decade 1985–94.

Our contribution to more efficient structure calculation from NMR data was to implement the variable target function approach introduced by Braun and Gō[7] in a new, fully vectorized program DIANA (paper **37**) and to achieve a decisive improvement

---

[3] Otting, G. and Wüthrich, K. (1988) *J. Magn. Reson. 76,* 569–574. Efficient purging scheme for proton-detected heteronuclear two-dimensional NMR.

[4] Montelione, G.T. and Wagner, G. (1990) *J. Magn. Reson. 87,* 183–188. Conformation-independent sequential NMR connections in isotope-enriched polypeptides by ^{1}H–^{13}C–^{15}N triple-resonance experiments.

[5] Bax, A. and Grzesiek, S. (1993) *Acc. Chem. Res. 26,* 131–138. Methodological advances in protein NMR.

[6] Clore, G.M. and Gronenborn, A.M. (1991) *Science 252,* 1390–1399. Structures of larger proteins in solution: three- and four-dimensional heteronuclear NMR spectroscopy.

[7] Braun, W. and Gō, N. (1985) *J. Mol. Biol. 186,* 611–626. Calculation of protein conformations by proton–proton distance constraints. A new efficient algorithm.

of the convergence of such calculations with the REDAC strategy (paper **38**). DIANA has then become the core of a line of mutually fully compatible programs, including PROSA[8] for rapid Fourier transformation of higher-dimensional NMR experiments, and EASY [9] (subsequently XEASY) for computer-supported resonance assignments and preparation of the input for the structure calculation.

[8] Güntert, P., Dötsch, V., Wider, G. and Wüthrich, K. (1992) *J. Biomol. NMR 2,* 619–629. Processing of multi-dimensional NMR data with the new software PROSA.

[9] Eccles, C., Güntert, P., Billeter, M. and Wüthrich, K. (1991) *J. Biomol. NMR 1,* 111–130. Efficient analysis of protein 2D NMR spectra using the software package *EASY.*

JOURNAL OF MAGNETIC RESONANCE **66,** 187–193 (1986)

Origin of t_1 and t_2 Ridges in 2D NMR Spectra and Procedures for Suppression

GOTTFRIED OTTING, HANS WIDMER, GERHARD WAGNER, AND KURT WÜTHRICH

Institut für Molekularbiologie und Biophysik, Eidgenössische Technische Hochschule–Hönggerberg, CH-8093 Zurich, Switzerland

Received September 18, 1985

The appearance of "t_1 noise" has long been a limiting factor in practical applications of 2D NMR techniques (*1*). It is particularly detrimental in 2D spectroscopy with biological macromolecules in aqueous solution (*2, 3*). Techniques for suppression of t_1 noise therefore attracted considerable interest (*4, 5*) and improvement of the spectra was, for example, achieved with symmetrization (*6*). These efforts relied on the assumption that general spectrometer instabilities were the primary cause of t_1 noise (*1, 3, 5*), until it was recently recognized that "t_1 noise" also contains spurious bands of constant offset from the base plane (*7*). These are particularly severe in 2D nuclear Overhauser enhancement (NOESY) spectra, where the diagonal peaks are usually much more intense than the cross peaks, and it was suggested that systematic digitization errors or base line distortions in the data recorded with the first few t_1 values were probably the cause of such "t_1 ridges" (*7*). Procedures proposed up to now for their elimination in NOESY are a difference method for combined suppression of the diagonal peaks and the t_1 ridges (*7*), and subtraction of rows without either cross or diagonal peaks from all other rows of the spectrum (*8*). The present communication gives a more precise description of the origins of t_1 ridges, and simple and efficient techniques for suppression of these artifacts are proposed.

The origin of t_1 ridges is in the ways that Fourier transformations of 2D NMR data sets are commonly performed. Usually the time domain signal, $S(t)$, is recorded as a cosine-modulated free induction decay (FID), $S^c(t)$. For a single resonance A the time dependence of $S^c(t)$ is

$$S^c(t) = \cos(\omega_A t)e^{-t/T_2}. \quad [1]$$

An analytical cosine Fourier transformation (cos FT) of $S^c(t)$ yields a Lorentzian absorption lineshape, $L^{cc}(\omega)$. The same absorption lineshape is obtained as $L^{ss}(\omega)$ by sine Fourier transformation (sin FT) of a time-domain signal recorded as a sine-modulated free induction decay by changing the phase of the receiver by 90°. With a discrete Fourier transformation of the usual type, spectrum simulations using the program SPHINX (H. Widmer and K. Wüthrich, unpublished) showed that a cos FT of cosine-modulated exponentials leads to Lorentzians which are displaced from the base line by a constant offset proportional to the integral of the spectrum.

0022-2364/86 $3.00
Copyright © 1986 by Academic Press, Inc.
All rights of reproduction in any form reserved.

In Fig. 1A, ridges resulting from this offset are clearly visible along both the ω_1 and ω_2 frequency axis. In contrast, sin FT of sine-modulated exponentials resulted in a much better approximation of the analytical Lorentz function (Fig. 1B).

We conclude from Figs. 1A and B, that the ridges along ω_1 and ω_2 are due to an improper weighting of the initial time-domain data point in the commonly used discrete FT. In the spectrometer, $S(t)$ is sampled at discrete times (Fig. 2):

$$S_i^c = S^c(t_0 + i\Delta t) \qquad i = 0, 1, \ldots, (n-1). \tag{2}$$

The discrete cos FT of this experimental record is commonly performed as (*9*)

$$L_r^{cc} = \sum_{i=0}^{n-1} S_i^c \cos(r\Delta\omega i\Delta t)\Delta t \tag{3}$$

where $\Delta\omega$ is the frequency increment, and L_r^{cc} is the frequency-domain data point at $\omega = r\Delta\omega$ in a cos FT of a cosine-modulated time signal. Inspection of Fig. 2 shows that S_0 would correctly represent a time interval Δt if t_0 were exactly $0.5\Delta t$, and in this case use of Eq. [3] would be appropriate. In the practice of 2D NMR, however, the initial delay along t_1 is chosen as short as possible to avoid the necessity of phasing the spectra along ω_1, and the discrete FT [4] must be used in the place of [3].

$$L_r^{cc} = 0.5S_0^c(1 + 2t_0/\Delta t)\cos(r\Delta\omega t_0)\Delta t + \sum_{i=1}^{n-1} S_i^c \cos(r\Delta\omega(t_0 + i\Delta t))\Delta t. \tag{4}$$

The important feature of Eq. [4] when compared to Eq. [3] is the scaling of the first time-domain data point. For $t_0 = 0$, this scaling corresponds to a multiplication of the first time-domain data point with 0.5. The use of Eq. [4] is illustrated in Fig. 1C, which was obtained by the following procedure: The first digital data point of each FID was divided by two before the cos FT along t_2, and the spectrum corresponding to the first value of t_1 was divided by two before cos FT along t_1. Clearly the ridges along ω_1 and ω_2 were thus efficiently suppressed. This confirms that if the discrete FT of Eq. [3] is used and t_0 is smaller than $\Delta t/2$, the first time-domain data point will be

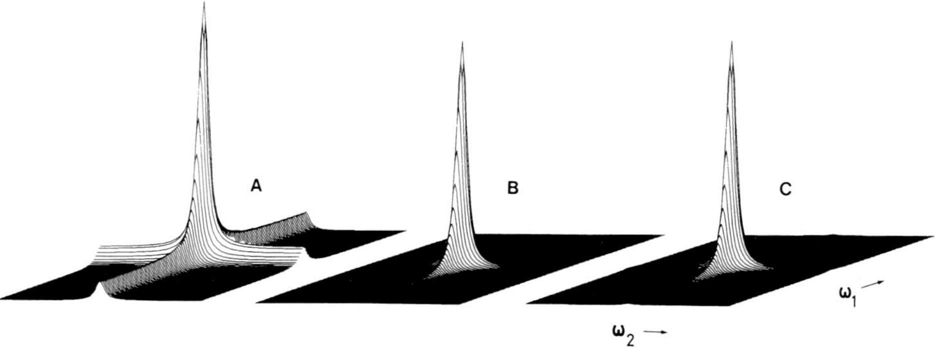

FIG. 1. Simulated 2D Lorentzians. (A) From cos FT along ω_1 and ω_2 of cosine-modulated data. (B) From sin FT along ω_1 and ω_2 of sine-modulated data. (C) Same data set as (A), but corrected for offsets before the cos FT (see text).

overestimated. Inspection of Fig. 2 shows that the resulting base line distortion, d_r^c, is related to the first time-domain point, S_0^c, and the initial delay, t_0, by

$$d_r^c = 0.5S_0^c(1 - 2t_0/\Delta t)\cos(r\Delta\omega t_0). \quad [5]$$

For $t_0 = 0$, d_r^c is a constant offset with the value $S_0^c/2$.

If the time-domain signal is recorded as a sine-modulated exponential, the signal for a single resonance A has the time dependence

$$S^s(t) = \sin(\omega_A t)e^{-t/T_2} \quad [6]$$

which is digitized as

$$S_i^s = S^s(t_0 + i\Delta t) \qquad i = 0, 1, \ldots, (n - 1). \quad [7]$$

The commonly used discrete sin FT is performed as (9)

$$L_r^{ss} = \sum_{i=0}^{n-1} S_i^s \sin(r\Delta\omega i\Delta t)\Delta t. \quad [8]$$

In analogy to Eq. [4], the correct transformation would be

$$L_r^{ss} = 0.5S_0^s(1 + 2t_0/\Delta t)\sin(r\Delta\omega t_0)\Delta t + \sum_{i=1}^{n-1} S_i^s \sin(r\Delta\omega(t_0 + i\Delta t))\Delta t. \quad [9]$$

Using Eq. [8] instead of Eq. [9] leads to a base line distortion

$$d_r^s = 0.5S_0^s(1 - 2t_0/\Delta t)\sin(r\Delta\omega t_0). \quad [10]$$

This base line distortion is zero for $t_0 = 0$, and it is always much smaller than d_r^c, since S_0^s is close to zero.

For experimental verification of the predictions in Fig. 1 and Fig. 2 and Eq. [1] to [10], two phase-sensitive NOESY spectra were recorded with identical conditions except for the following: To record the absorption-mode spectrum of Fig. 3A, the receiver phases were adjusted so that the time-domain signal was cosine-modulated in t_1 and t_2, and a cos FT according to Eq. [3] was applied in both dimensions. For the spectrum

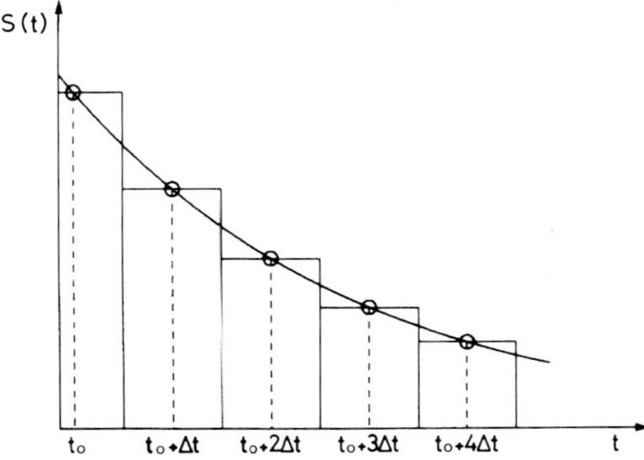

FIG. 2. Scheme illustrating a source of deviations between discrete FT and continuous analytical FT (see text).

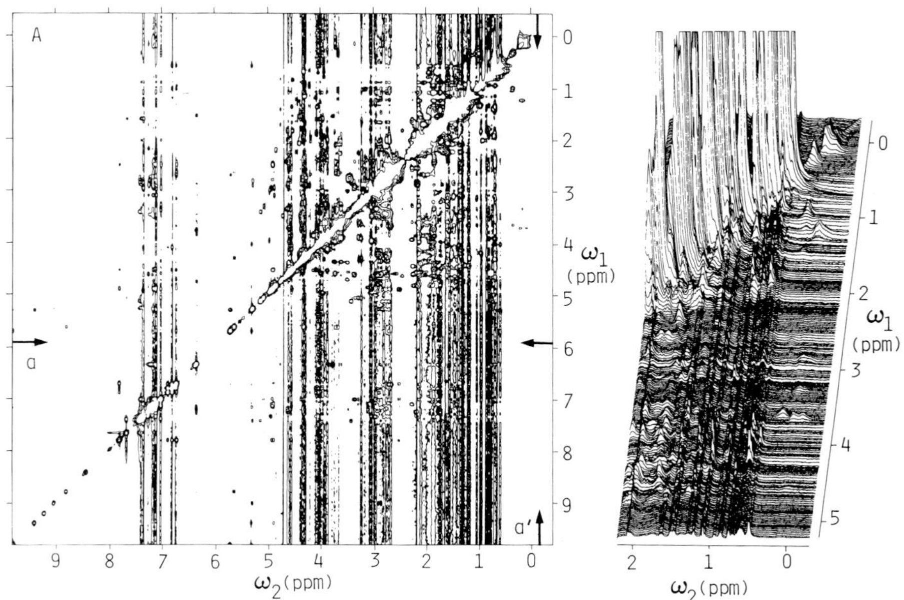

FIG. 3. Contour plots and stacked plots of three phase-sensitive NOESY spectra recorded with a 20 m*M* solution of the protein BPTI in D_2O at 36°C and pH 4.6 on a Bruker AM-360 spectrometer. (A) From cos FT along ω_1 and ω_2 of a cosine-modulated data set. (B) From sin FT along ω_1 and ω_2 of a sine-modulated data set. (C) Same data set as (A), but with offset corrections applied prior to cos FT (see text). The two data sets (A) and (B) were recorded in immediate succession, with an accumulation time of 6.5 h each. 512 t_1 values were sampled with 32 scans each, acquiring 4K points per FID. The mixing time was 150 ms. Before FT, both data sets were multiplied in both dimensions with a sine-bell window shifted by $\pi/2$, and were zero filled from 512 to 2048 points in t_1. In all three spectra identical levels are drawn, with an exponentially increasing scale. The contour plots show the entire spectrum, the stacked plots the region (ω_1 = −0.3–5.2 ppm, ω_2 = −0.3–2.2 ppm). Arrows in the contour plots identify the locations where the cross sections of Fig. 4 were taken.

of Fig. 3B the time-domain signal was sine-modulated in t_1 and t_2, and a sin FT with Eq. [8] was applied in both dimensions. The phase shift in ω_2 was achieved by a 90° shift of the receiver reference phase. Along ω_1 the phase shift was obtained by adding 90° to the phase of the first pulse of the usual NOESY pulse sequence.

Figures 3 and 4 show that there are strong t_1 ridges as well as t_2 ridges in spectrum A, whereas only "real" t_1 noise and weaker t_2 ridges are recognizable in the spectrum B, which was obtained with sin FT. The appearance of the t_1 ridges in spectrum A corresponds to an attenuated projection of the 1D spectrum along ω_1 (Fig. 4a). In contrast to the prediction of positive t_2 ridges (see Fig. 1) the spectrum of Fig. 3A contains negative t_2 ridges (Fig. 4a′), which correspond to an attenuated projection of the 1D spectrum along ω_2. The negative sign of the t_2 ridges can be attributed to the effect of the analog filters applied to the incoming signal before digitization, which reduce the value of the first sampled point. This reduction is actually more pronounced than the desired factor 0.5 (Eq. [4]), so that in the 1D spectra along ω_2 at each value of t_1 a negative base line offset is observed. To check on this effect we recorded 1D

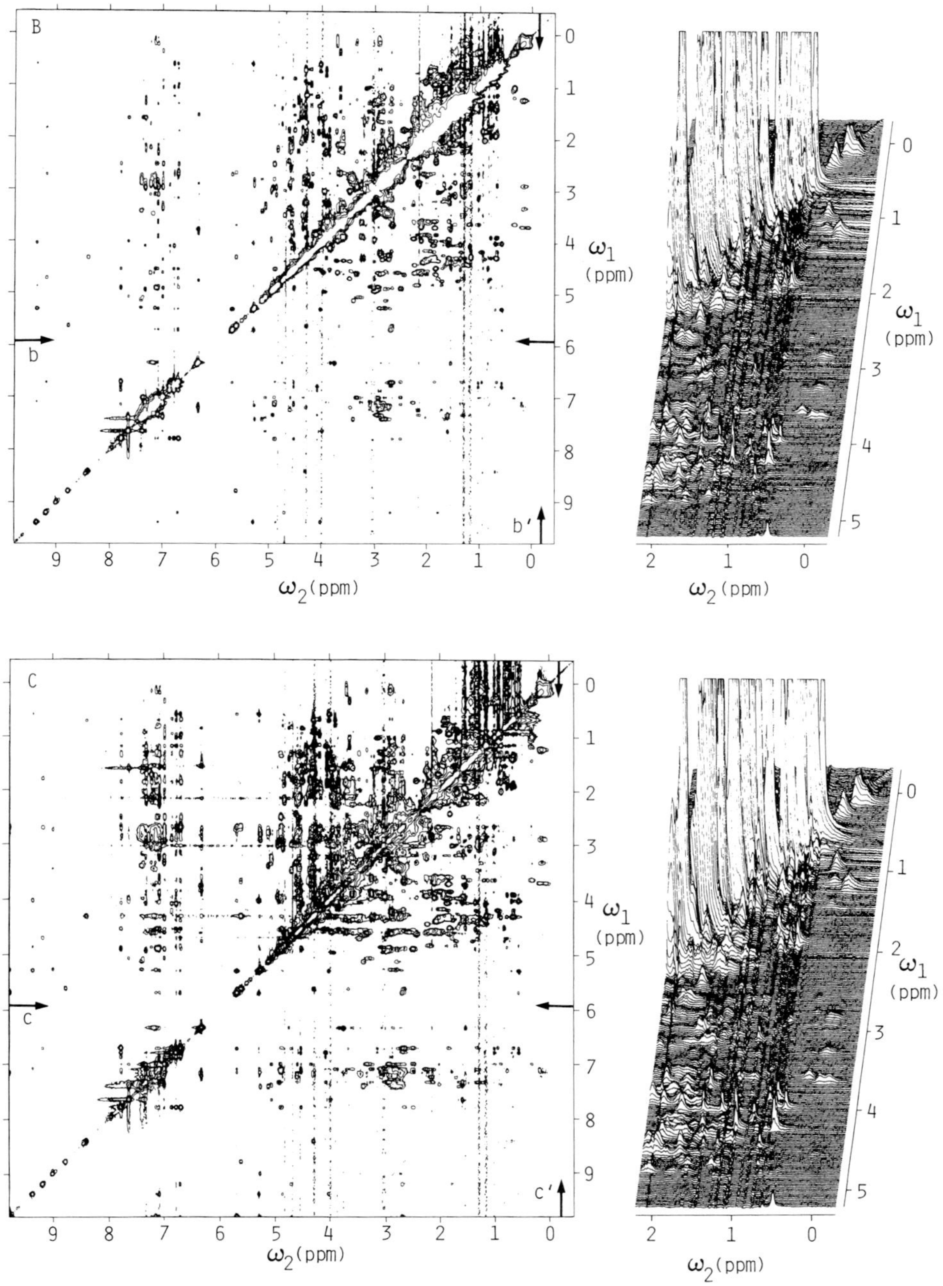

FIG. 3—*Continued.*

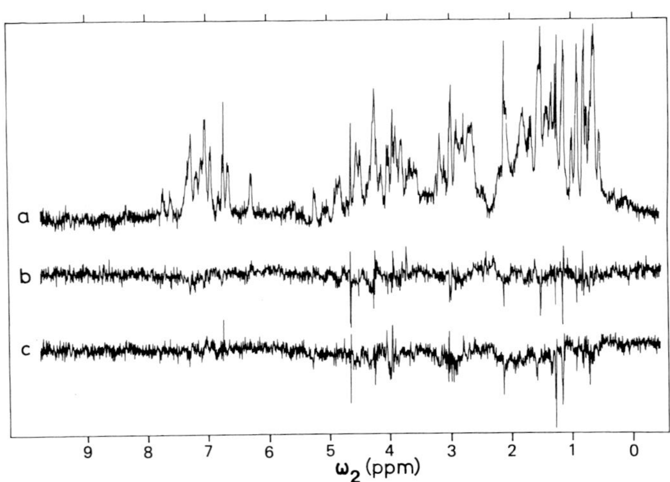

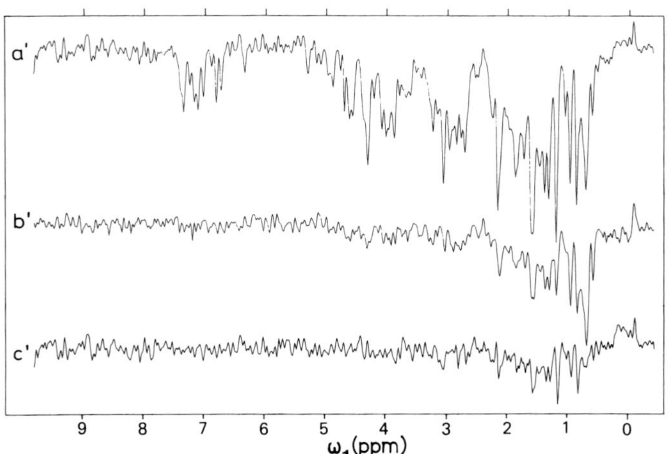

FIG. 4. Cross sections through the spectra of Fig. 3. (a)–(c) Cross sections along ω_2 at $\omega_1 = 5.87$ ppm. (a′)–(c′) Cross sections along ω_1 at $\omega_2 = -0.17$ ppm. The same scale was used for plotting the cross sections along ω_1 and ω_2.

experiments with the width of the analog filter set to 20 times the sweep width, whereupon the predicted positive constant offset was observed according to Eq. [5].

The negative t_2 ridges lower the base plane to such a degree that many cross peaks appear to be broader in the sin FT spectrum of Fig. 3B than in the cos FT spectrum of Fig. 3A merely because the contour levels plotted correspond to a lower distance above the effective base plane in the sin FT spectrum. This is best seen by comparing the stacked plots of Figs. 3A and B. The spectrum of Fig. 3C was obtained from the same data matrix as Fig. 3A after a twofold correction of the accumulated data set. First, for the elimination of the negative t_2 ridges, the first point of each FID was multiplied by a factor of 6.6 before cos FT along t_2, which was empirically found to lift the base line of the transformed FIDs to a value close to zero. Second, to eliminate the positive t_1 ridges the spectrum belonging to the first t_1 value was divided by two before cos FT along t_1. Comparison of the stacked plots of Figs. 3A and C shows that the base plane of the spectrum was thus greatly improved. Comparison of the cross sections of Figs. 4a and c further shows that t_1 ridges are nearly completely eliminated. The suppression of the t_2 ridges is not as successful over the entire spectrum either by the sin FT or by the corrected cos FT (Figs. 4b′ and c′). The reasons for this are being investigated.

Thus the use of Eq. [3] for discrete FT of cosine modulated FIDs has the following consequences for frequency-domain 2D NMR spectra: The first cos FT along t_2 produces an offset in each 1D spectrum along ω_2, which is proportional to the integral of all resonances in this spectrum. In the 2D data matrix, this corresponds to a modulation of the time-domain signal along t_1 which is identical in each column, and which contains all resonance frequencies. After the second cos FT along t_1 this modulation shows up as t_2 ridges. Independently, t_1 ridges appear as an artifact of the second cos FT along t_1 due to the improper weighting of the first time-domain point along t_1. The preferred technique for new NOESY experiments appears to be acquiring the data with sine modulation in t_1 and use of sin FT with Eq. [9]. Alternatively, data recorded with cosine modulation can be improved by corrections based on Eq. [4] prior to cos FT. This can be applied to already existing data sets. Similar techniques for ridge elimination can be applied for all other 2D NMR experiments which yield in-phase resonance peaks, as, for example, MQ spectroscopy.

ACKNOWLEDGMENTS

Financial support was provided by the Schweizerischer Nationalfonds (Project 3.284.82). We thank Mr. R. Baumann for software modifications.

REFERENCES

1. K. Nagayama, P. Bachmann, K. Wüthrich, and R. R. Ernst, *J. Magn. Reson.* **31,** 133 (1978).
2. W. Braun, G. Wider, K. H. Lee, and K. Wüthrich, *J. Mol. Biol.* **169,** 921 (1983).
3. G. Wider, S. Macura, Anil Kumar, R. R. Ernst, and K. Wüthrich, *J. Magn. Reson.* **56,** 207 (1984).
4. H. Santos, D. L. Turner, and A. V. Xavier, *J. Magn. Reson.* **55,** 463 (1983).
5. A. F. Mehlkopf, D. Korbee, T. A. Tiggelman, and R. Freeman, *J. Magn. Reson.* **58,** 315 (1984).
6. R. Baumann, G. Wider, R. R. Ernst, and K. Wüthrich, *J. Magn. Reson.* **44,** 402 (1981).
7. W. Denk, G. Wagner, M. Rance, and K. Wüthrich, *J. Mag. Reson.* **62,** 350 (1985).
8. R. E. Klevitt, *J. Magn. Reson.* **62,** 551 (1985).
9. J. W. Cooper, "The Minicomputer in the Laboratory," Wiley, New York, 1977.

JOURNAL OF MAGNETIC RESONANCE **70**, 500–505 (1986)

Editing of 2D ^{1}H NMR Spectra Using X Half-Filters. Combined Use with Residue-Selective ^{15}N Labeling of Proteins

GOTTFRIED OTTING, HANS SENN,* GERHARD WAGNER, AND KURT WÜTHRICH

Institut für Molekularbiologie und Biophysik, Eidgenössische Technische Hochschule-Hönggerberg, CH-8093 Zurich, Switzerland

Received July 28, 1986

The introduction of efficient sequential assignment techniques for obtaining sequence-specific ^{1}H NMR assignments in proteins (*1, 2*) has provided a basis for determination of the three-dimensional molecular structure in solution (*3, 4*), as well as for detailed studies of the molecular dynamics (e.g., (*5*)) and of intermolecular interactions (*4*). Limitations on the use of this approach arise if two or several cross peaks in the 2D ^{1}H NMR spectra overlap. The degree of spectral overlap increases quite naturally with increasing molecular size, but if the ^{1}H resonances are not well dispersed complete sequence-specific assignment of the spectrum may be difficult even for small proteins. As was also pointed out by others (*6–9*), the situation can be improved by residue-selective isotope labeling with ^{15}N or ^{13}C. In such preparations the protons directly bound to ^{15}N or ^{13}C can be singled out on the basis of the large heteronuclear spin–spin coupling constants $^1J_{XH}$. Simplified spectra have thus been obtained which correspond either to the difference between recordings with and without broadband decoupling of the X nucleus (*6, 7*), or with and without $\pi(^{15}N)$ refocusing pulse (*8*), or were recorded with a heteronuclear zero-quantum filter (*9*). In this communication we describe the use of $X(\omega_1)$ and $X(\omega_2)$ *half-filters* for editing the 2D ^{1}H NMR spectra of X-labeled molecules. Compared to difference spectroscopy techniques the half-filters have the advantage of providing *two* subspectra corresponding, respectively, to the X-labeled ^{1}H peaks and to all other peaks. Compared to the heteronuclear zero-quantum filter (*9*) the sensitivity of the corresponding half-filter is improved by a factor 2. X half-filters can be employed with a wide variety of 2D NMR experiments, including 2D correlated spectroscopy (COSY), 2D total correlation spectroscopy (TOCSY), and 2D nuclear Overhauser enhancement spectroscopy (NOESY) (Fig. 1). Clearly, their use is not limited to proteins and there is a wide spectrum of potential applications with other macromolecules and with small molecules, in particular for studies of intermolecular interactions.

In the experimental schemes for obtaining the desired simplified spectra with ^{15}N-labeled proteins, we inserted ^{15}N half-filters consisting of a $[-\tau/2-\pi(^1H, ^{15}N)-\tau/2-\pi(^{15}N)-]$ pulse sequence, either before the evolution period (ω_1 half-filter) or immediately before detection (ω_2 half-filter) of the conventional 2D ^{1}H NMR experiments

* Present address: Sandoz Ltd., Pharmaceutical Division, Preclinical Research, CH-4002 Basel, Switzerland.

0022-2364/86 $3.00

Copyright © 1986 by Academic Press, Inc.
All rights of reproduction in any form reserved.

A

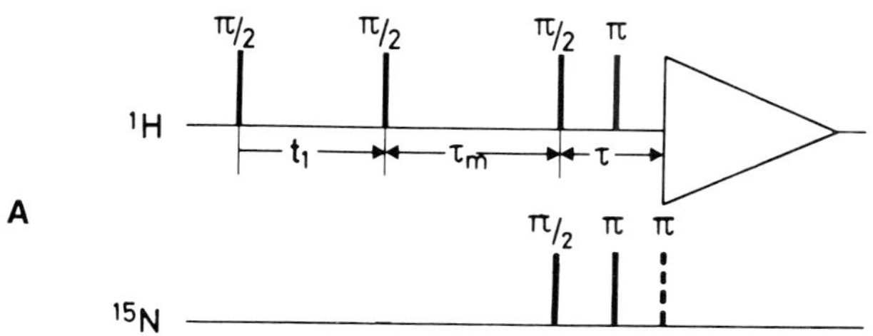

B

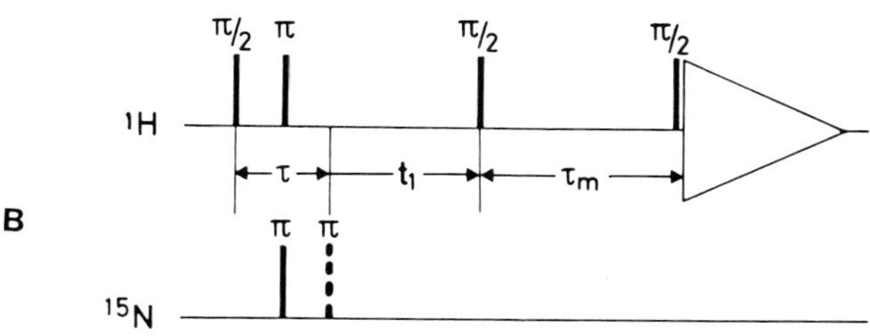

C

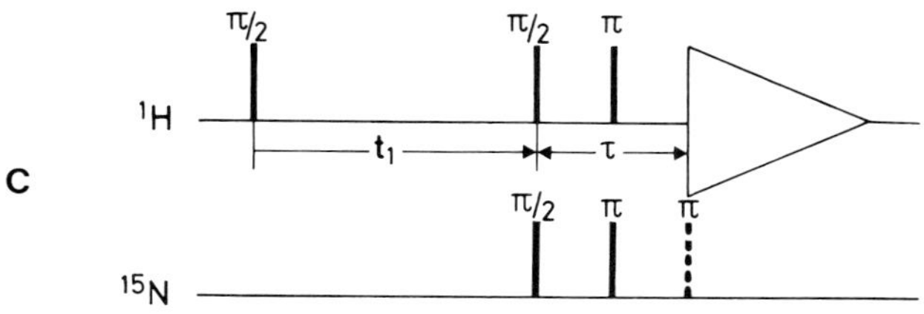

D

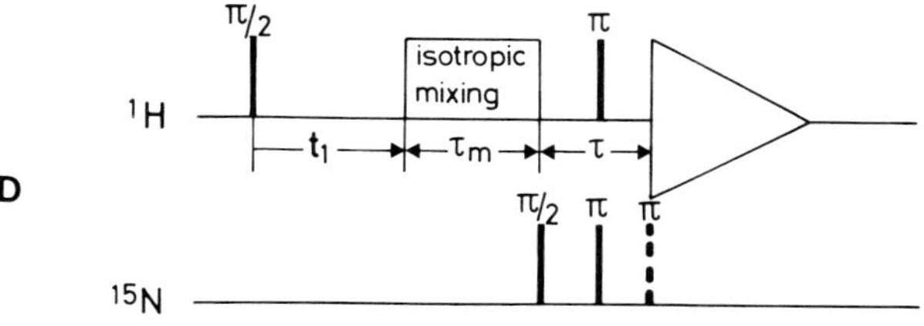

FIG. 1. Experimental schemes for (A) ^{1}H NOESY with ^{15}N(ω_2) half-filter, (B) ^{1}H NOESY with ^{15}N(ω_1) half-filter, (C) ^{1}H COSY with ^{15}N(ω_2) half-filter, (D) ^{1}H TOCSY with ^{15}N(ω_2) half-filter. The same phase cycling of the ^{1}H pulses is used as in the conventional ^{1}H 2D NMR experiments (e.g., (*10, 11*)). The π(^{1}H) pulse of the half-filter is cycled with the immediately preceding ^{1}H pulse and may further be phase alternated. The ^{15}N pulses require no phase cycling. The editing π(^{15}N) pulse, which is drawn as a broken line, is applied in every second scan, and the sum as well as the difference of the spectra with and without editing pulse are used (see text).

(Fig. 1). Efficient editing relies on the facts that the direct coupling constants $^1J(^{15}$N, ^{1}H) are much larger than any other relevant coupling constants and vary only by a few percent between the different residues in a polypeptide chain. Typically, $^1J(^{15}$N,

^{1}H) = 90 ± 5 Hz, while ^{3}J(NH, αH) or ^{2}J(^{15}N, αH) are smaller than 12 Hz. (Similar situations would be encountered for the X-bound protons in ^{13}C-labeled compounds.)

Figures 1A, C, and D show explicit examples for the expansion of the regular homonuclear NOESY, TOCSY, and COSY pulse sequences with a ^{15}N(ω_2) half-filter. Figure 1B contains the ^{1}H NOESY pulse sequence with a ^{15}N(ω_1) half-filter. The half-filter includes a (π/2)(^{15}N) purging pulse either at the end of t_1 or at the beginning of the filter delay τ, which is discussed below, a refocusing π(^{1}H, ^{15}N) pulse to eliminate effects from chemical shift dispersion, and the π(^{15}N) editing pulse. To explain the functioning of the X half-filters with a product operator description (*12*) we denote ^{1}H spins directly attached to X as H, X spins as X, and protons not coupled to X as I. We start with H_x, which denotes in-phase coherence with respect to the heteronuclear coupling, and with I_x. Then, if the delay τ is tuned to $1/2J_{HX}$ and all other coupling constants are much smaller than the direct coupling J_{HX}, we have

$$H_x \xrightarrow{\tau} H_yX_z, \quad [1]$$

and

$$I_x \xrightarrow{\tau} I_x. \quad [2]$$

In Eqs. [1] and [2], signs and coefficients were omitted for simplicity. Repeating the same pulse sequence with an additional π(X) pulse (*editing* pulse) gives the same product operator terms, but with sign inversion for the terms containing the X operator. The difference between the data sets recorded with and without the editing π(X) pulse contains only the H_yX_z terms arising from X-bound protons, whereas addition of the two data sets selects the I_x terms, which come from all other hydrogens. This selection is only along one frequency axis, either ω_1 or ω_2 (Fig. 1), and therefore the name *half-filter* was chosen. The *sum* spectrum and the *difference* spectrum are 90° out of phase relative to each other.

In Figs. 2 and 3 we illustrate the use of a ^{15}N(ω_2) filter with two regions from a ^{1}H TOCSY spectrum (Fig. 1D) of the protein c2 repressor 1–76 from *Salmonella* phage P22, where all 10 leucyl residues were labeled with ^{15}N in the extent of >85% (*13*). As is generally advisable for applications of these experiments, two data sets with and without the π(^{15}N) pulse were recorded in an interleaved manner, with alternating storage of the scans with and without the editing pulse in two different memory locations. Furthermore, the π(^{15}N) editing pulse was applied by using alternatively the sequence (π/2)(^{15}N)$_x$–(π/2)(^{15}N)$_x$, which corresponds to an effective π(^{15}N) pulse, and (π/2)(^{15}N)$_x$–(π/2)(^{15}N)$_{-x}$, which corresponds to an effective zero rotation (*14*). Additional experimental details are given in the caption to Fig. 2.

The following rules on the appearance of the difference spectrum derive from the product operator analysis. On the diagonal there are exclusively the peaks corresponding to the ^{15}N-bound protons (Fig. 2A). These contain four fine-structure components separated by ^{1}J(^{15}N, ^{1}H) along both ω_1 and ω_2. Cross peaks with protons bound to ^{15}N are selected along ω_2, with antiphase fine structure components separated by ^{1}J(^{15}N, ^{1}H), whereas there is no filter effect along ω_1. Figure 3A thus contains only the direct peaks between the ^{15}N-labeled amide protons and the nonlabeled α-protons of the 10 Leu residues. Obviously, the spectra recorded with a half-filter are not symmetric with respect to the diagonal, so that the ^{15}NH–αH fingerprint of Fig. 3A has

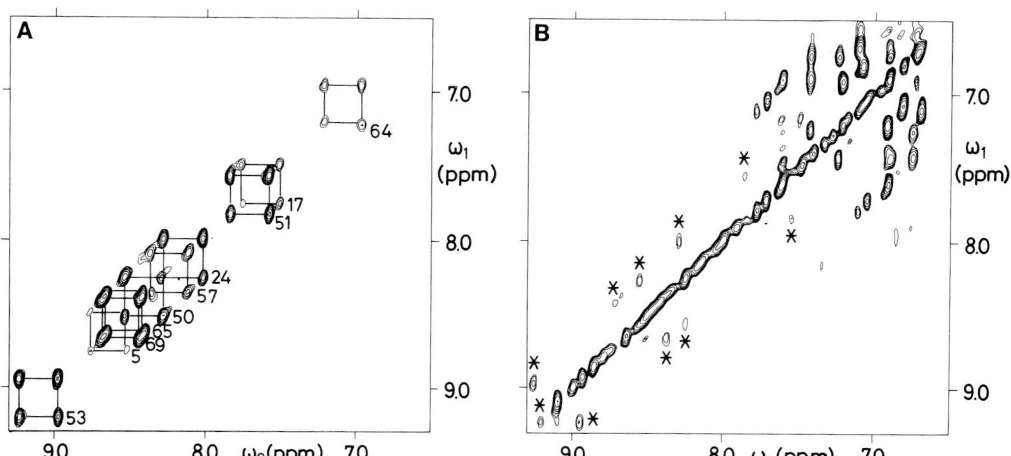

FIG. 2. [1]H TOCSY with $^{15}N(\omega_2)$ half-filter of the DNA-binding domain 1–76 of P22 c2 repressor with all 10 leucyl residues labeled with ^{15}N. The spectral region (ω_1 = 6.5–9.3 ppm, ω_2 = 6.5–9.3 ppm), which contains the resonances of the labile NH's and the aromatic protons, is shown. Protein concentration 8 m*M*, solvent H_2O, pH 4.8, $T = 20°C$. The spectrum was recorded on a Bruker AM 360 spectrometer with the pulse sequence of Fig. 1D, but without the $(\pi/2)(^{15}N)$ purge pulse before τ. The total measuring time was about 30 h. A tuned delay of 5.55 ms was used. t_{1max} was 27.7 ms, t_{2max} was 57.4 ms, τ_m was 60 ms. The MLEV-17 pulse sequence was used during the mixing time (*15*). (A) Difference spectrum. A trapezoidal window was applied along t_2, with the plateau region from 5.55 to 24.6 ms. Along t_1 a sine-bell window shifted by $\pi/3$ was used. The diagonal peaks of the 10 leucyl amide protons are identified with the sequence positions. The peaks are in-phase along ω_1 and antiphase along ω_2 with respect to $^1J(^{15}N, {}^1H)$. (B) Sum spectrum. Prior to Fourier transformation the data matrix was multiplied with a sine bell along t_2 and with a sine bell shifted by $\pi/3$ along t_1. The weak peaks identified with asterisks originate from the leucyl residues (see text).

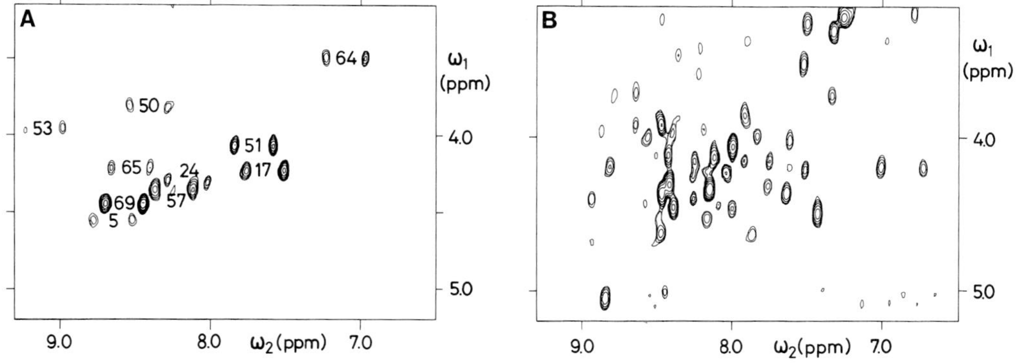

FIG. 3. Same spectrum as Fig. 2, region (ω_1 = 3.1–5.1 ppm, ω_2 = 6.5–9.3 ppm), which contains the direct amide proton to α-proton cross peaks, and possibly some remote peaks of Ser, Thr, Arg, and Lys. (A) Difference spectrum. The peaks of the Leu residues are identified with the sequence locations. (B) Sum spectrum.

no counterpart in the region (ω_1 = 6.5–9.3 ppm, ω_2 = 3.1–5.1 ppm). Cross peaks between two ^{15}N-labeled protons have a fine structure similar to that of the diagonal peaks (Fig. 2A), except that the splittings along ω_1 and ω_2 may be slightly different. Such cross peaks arise in the low-field region of the ^{15}N(ω_2) half-filter ^{1}H NOESY spectrum of the c2 repressor. It is also clear that these peaks must occur twice, in symmetrical positions with respect to the diagonal; see Ref. (*16*).

The sum spectrum contains the diagonal peaks of all hydrogen atoms which are not bound to ^{15}N, and all cross peaks between protons which are not bound to ^{15}N. In Fig. 2B we thus have a nearly continuous line of diagonal peaks from NH and aromatic protons, and numerous cross peaks between different aromatic protons and between amide protons of the side chains of Asn and Gln. Figure 3B contains the NH–αH cross peaks of the nonlabeled amino acid residues, and possibly some remote connectivities. In addition the diagonal peaks of some of the ^{15}N-labeled leucines are weakly seen in Fig. 2A. They can be completely suppressed by the $(\pi/2)(^{15}\text{N})$ purging pulse (Fig. 1D, see also below).

Analogous editing, but with selection of cross peaks to ^{15}N-bound protons along ω_1, can be achieved with an X(ω_1) half-filter, as in Fig. 1B. For a two-spin system ^{1}H–^{15}N with the one-bond heteronuclear coupling constant J one has the following relevant product operator terms at the end of evolution, when starting with H_y:

$$[-\cos(\pi J\tau)\cos(\pi Jt_1) \pm \sin(\pi J\tau)\sin(\pi Jt_1)]\cos(\omega_H t_1)H_y + [\cos(\pi J\tau)\sin(\pi Jt_1) \pm \sin(\pi J\tau)\cos(\pi Jt_1)]\sin(\omega_H t_1)2H_yN_z, \quad [3]$$

where $\pm$ refers to the pulse sequence with and without the editing $\pi(^{15}\text{N})$ pulse. The difference spectrum obtained with an X(ω_1) half-filter corresponds to that obtained with the heteronuclear zero-quantum filter of Ref. (*9*), but since the heteronuclear zero-quantum filter produces unobservable multiple-quantum coherence in every second scan the half-filter experiment has a twofold improved sensitivity. Similar to the heteronuclear zero-quantum filter (Fig. 2 of Ref. (*9*)) the diagonal peaks in a difference spectrum recorded with an X(ω_1) half-filter contain only two of the four multiplet components seen in Fig. 2A, which are in opposite corners of the quadrangle. This results from the second term in Eq. [3], which describes coherence in antiphase with respect to the heteronuclear coupling at the end of the evolution period. A $(\pi/2)(^{15}\text{N})$ purge pulse at the end of t_1 would restore the full square pattern by turning this second term into unobservable multiple-quantum coherence.

Compared with difference spectroscopy using broadband decoupling of X (*6, 7*), the half-filters have the advantage of providing simpler spectra and, if the transverse relaxation during the filter delay τ can be neglected, improved sensitivity. For work with systems with very short T_2's, the sensitivity may become a limiting factor because of the loss of signal during τ.

Compared to the recently described X-filter technique (*17*) the half-filter method is fully complementary. The X-filter selects for peaks connecting protons which are *both* bound to the *same* X spin, and it is not tuned for a particular range of spin–spin coupling constants. Clearly, the main advance made with the introduction of the half-filters is that they enable observation of correlations between X-bound protons and protons which are not involved in heteronuclear couplings.

Finally, two points relating to the appearance of spectra recorded with half-filters must be taken up. First, there is the presence of the "forbidden" diagonal peaks of [2-^{15}N]Leu in Fig. 2B. For this let us consider a two-spin system ^{1}H–^{15}N with the coupling constant J in a ^{1}H TOCSY experiment. The density matrix obtained with a phase cycle selecting the cosine modulated signal during t_1 contains the following observable product operators before the $\pi(^{15}\text{N})$ editing pulse, when starting with H_y:

$$-\cos(\pi J t_1)\cos(\omega_H t_1)[\cos(\pi J\tau)H_y - \sin(\pi J\tau)2H_xN_z] + \sin(\pi J t_1)\cos(\omega_H t_1)\sin(\pi J\tau)H_x. \quad [4]$$

The last term is not affected by the editing pulse and is responsible for the appearance of residual leucyl ^{15}N–H diagonal peaks in the sum spectra. The $(\pi/2)(^{15}\text{N})$ purge pulse either at the end of t_1 or immediately before the filter delay τ (Fig. 1) would change this undesirable term into unobservable heteronuclear multiple quantum coherences. Cross talk in the sum spectrum is also expected to arise if τ deviates from $1/2J$. This cross talk would not be suppressed by the $(\pi/2)(^{15}\text{N})$ purge pulse. Second, minor multiplet distortions could result from the presence of proton–proton couplings during the filter delay τ. These effects will hardly become evident in the spectra of macromolecules, but they should be kept in mind when working with small molecules giving rise to very narrow NMR lines.

ACKNOWLEDGMENTS

Financial support was provided by the Schweizerischer Nationalfonds (Project 3.198-9.85). We thank Mr. A. Eugster for hardware modifications on the spectrometer.

REFERENCES

1. M. Billeter, W. Braun, and K. Wüthrich, *J. Mol. Biol.* **155,** 321 (1982).
2. G. Wagner and K. Wüthrich, *J. Mol. Biol.* **155,** 347 (1982).
3. K. Wüthrich, G. Wider, G. Wagner, and W. Braun, *J. Mol. Biol.* **155,** 311 (1982).
4. K. Wüthrich, "NMR of Proteins and Nucleic Acids," Wiley, New York, 1986.
5. G. Wagner and K. Wüthrich, *J. Mol. Biol.* **160,** 343 (1982).
6. D. M. LeMaster and F. M. Richards, *Biochemistry* **24,** 7263 (1985).
7. M. A. Weiss, A. G. Redfield, and R. H. Griffey, *Proc. Natl. Acad. Sci. USA* **83,** 1325 (1986).
8. R. H. Griffey, A. G. Redfield, R. E. Loomis, and F. W. Dahlquist, *Biochemistry* **24,** 817 (1985).
9. P. H. Bolton, *J. Magn. Reson.* **62,** 143 (1985).
10. G. Bodenhausen, H. Kogler, and R. R. Ernst, *J. Magn. Reson.* **58,** 370 (1984).
11. G. Wider, S. Macura, Anil Kumar, R. R. Ernst, and K. Wüthrich, *J. Magn. Reson.* **56,** 207 (1984).
12. O. W. Sørensen, G. W. Eich, M. H. Levitt, G. Bodenhausen, and R. R. Ernst, *Prog. Nucl. Magn. Reson. Spectrosc.* **16,** 163 (1983).
13. H. Senn, A. Eugster, G. Otting, F. Suter, and K. Wüthrich, *Eur. Biophysics J.*, submitted.
14. R. Freeman, T. H. Mareci, and G. A. Morris, *J. Magn. Reson.* **42,** 341 (1981).
15. A. Bax and D. G. Davis, *J. Magn. Reson.* **65,** 355 (1985).
16. H. Senn, G. Otting, and K. Wüthrich, *J. Am. Chem. Soc.*, submitted.
17. E. Wörgötter, G. Wagner, and K. Wüthrich, *J. Am. Chem. Soc.*, in press.

JOURNAL OF MAGNETIC RESONANCE **85**, 586–594 (1989)

COMMUNICATIONS

Extended Heteronuclear Editing of 2D ^{1}H NMR Spectra of Isotope-Labeled Proteins, Using the X(ω_1, ω_2) Double Half Filter

G. OTTING AND K. WÜTHRICH

Institut für Molekularbiologie und Biophysik, Eidgenössische Technische Hochschule-Hönggerberg, CH-8093 Zurich, Switzerland

Received July 27, 1989

The standard procedure for protein structure determination by NMR (*1, 2*) relies exclusively on results obtained by homonuclear ^{1}H NMR experiments. In particular, the quality of a structure determination depends critically on the number of ^{1}H–^{1}H distance constraints that can be derived from ^{1}H NOE measurements (*3*). However, with increasing size of the protein beyond a molecular weight of 10,000 to 15,000, spectral overlap in the ^{1}H NMR spectra tends to become a limiting factor. It has been recognized for some time that this size limitation can be shifted to significantly larger proteins by suitable isotope labeling combined with heteronuclear editing of the ^{1}H NMR spectra (*4–6*). Today, two heteronuclear editing procedures appear most promising, i.e., heteronuclear 3D NMR with nonselectively labeled compounds (*7, 8*), and the use of heteronuclear-filtered 2D ^{1}H NMR experiments combined with residue-selective isotope labeling (*9–15*). Heteronuclear filters are particularly attractive also for studies of intermolecular complexes with biological macromolecules, where only one of the interacting components is labeled with isotopes (*16, 17*).

Among a variety of heteronuclear filters (*9–15*), X half filters seem to have the best potentialities for a wide range of practical applications. X half filters have so far been designed for the purpose of separating the contents of a 2D ^{1}H NMR spectrum into two subspectra, where the subspectrum of prime interest contains in one dimension only resonances from X-bound protons, whereas all protons appear in the second dimension (*11–15*). The sum of the two subspectra contains all the peaks which would be present in the corresponding unfiltered 2D ^{1}H NMR experiment. Since the editing effect is limited to one of the two frequency dimensions, a conventional X half filter does not discriminate between X-bound protons and unlabeled protons in the unfiltered dimension of the spectrum. In the present Communication we introduce the X(ω_1, ω_2) double half filter, where a X(ω_1) half filter is used in conjunction with a X(ω_2) half filter in the same 2D ^{1}H NMR experiment. The resonances of X-bound protons are thus distinguished in both dimensions from those of the other protons, so that the cross peaks between two X-bound protons can be separated from those between X-labeled and unlabeled proton resonances. The spectral analysis is further facilitated by efficient suppression of the diagonal peaks in some of the subspectra. X(ω_1, ω_2) double half filters are universally applicable with 2D ^{1}H NMR

0022-2364/89 $3.00

Copyright © 1989 by Academic Press, Inc.
All rights of reproduction in any form reserved.

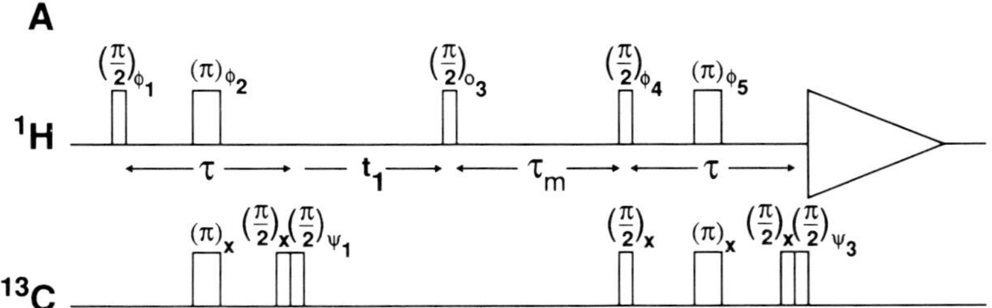

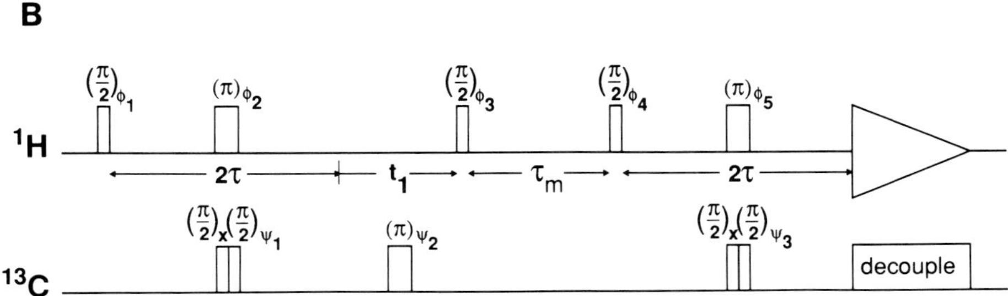

FIG. 1. Experimental schemes for ^{1}H NOESY spectroscopy with a ^{13}C(ω_1, ω_2) double half filter. (A) Without heteronuclear decoupling. (B) With heteronuclear decoupling during the evolution and the detection periods. The phases ϕ_1 to ϕ_5 and ψ_2 are independently alternated between x and $-x$, which results in phase cycles of 32 and 64 steps, respectively, for the schemes (A) and (B). The receiver phase is inverted whenever the phase of a ($\pi/2$)(^{1}H) pulse is alternated. This basic phase cycle is repeated four times with the different phases ψ_1 and ψ_3 listed in Table 1. In the experiment (B) the π(^{13}C) pulse in the middle of t_1 could be replaced by broadband decoupling, which would shorten the phase cycle.

experiments, and could also be adapted for use with heteronuclear or homonuclear 3D NMR experiments. As with the standard X-half-filter techniques there is no loss of sensitivity compared to the corresponding experiment without filters, other than that arising from relaxation during the filter delays.

For a NOESY experiment with ^{13}C(ω_1, ω_2) double half filter without heteronuclear decoupling (Fig. 1A) the filter element is [$-\tau/2-\pi(^1\text{H}, ^{13}\text{C})-(\tau/2)-(\pi/2)_x(^{13}\text{C})-(\pi/2)_{\pm x}(^{13}\text{C})-$]. Depending on the phase of the last pulse of the filter element the combined effect of the two $\pi/2$(^{13}C) pulses is either a zero degree or a 180° rotation, corresponding to the absence or presence of an effective π(^{13}C) editing pulse (*11*). The delay τ is chosen as $\tau = 1/[2\,^1J(^{13}\text{C}, ^1\text{H})]$. Since the direct coupling constants $^1J(^{13}\text{C}, ^1\text{H})$ are nearly the same for all ^{13}C-bound protons and are much larger than any other relevant coupling constants, the π(^{13}C) editing pulse of the filters acts only on heteronuclear antiphase coherence H_xC_z, where H represents a ^{13}C-bound proton, and leaves the magnetization of all other protons I unaffected:

$$
\begin{array}{ccccc}
-\text{H}_y & \xrightarrow{\tau/2-\pi_x(^1\text{H},\, ^{13}\text{C})-\tau/2} & 2\text{H}_x\text{C}_z & \xrightarrow{(\pi/2)_x(^{13}\text{C})-(\pi/2)_{\pm x}(^{13}\text{C})} & \mp 2\text{H}_x\text{C}_z \\
-\text{I}_y & \longrightarrow & \text{I}_y & \longrightarrow & \text{I}_y.
\end{array}
\quad [1]
$$

TABLE 1

Phase Cycling of ψ_1 and ψ_3 Used for the Four Recordings in the X(ω_1, ω_2)-Double-Half-Filter Experiment of Fig. 1

Recording	ψ_1	ψ_3
I	x	x
II	$-x$	x
III	x	$-x$
IV	$-x$	$-x$

The upper and lower sign represent the presence or absence of the effective $\pi(^{13}\mathrm{C})$ editing pulse, respectively. In addition, a $\pi/2(^{13}\mathrm{C})$ pulse at the end of the mixing period purges heteronuclear two-spin order (*11*). In the experiment with heteronuclear broadband decoupling (Fig. 1B), an additional refocusing delay τ is required after the $\pi(^{13}\mathrm{C})$ editing pulse, and the filter element is $[-\tau-\pi(^1\mathrm{H})-\pi/2(^{13}\mathrm{C})-\pi/2(^{13}\mathrm{C})-\tau-]$. The following equation describes only the evolution under $^1J(^{13}\mathrm{C}, ^1\mathrm{H})$, since the additional evolution under the ^{1}H chemical-shift Hamiltonian is refocused by the $\pi_x(^1\mathrm{H})$ pulse:

$$\begin{array}{l} -\mathrm{H}_y \xrightarrow{\tau} 2\mathrm{H}_x\mathrm{C}_z \xrightarrow{\pi_x(^1\mathrm{H})-(\pi/2)_x(^{13}\mathrm{C})-(\pi/2)_{\pm x}(^{13}\mathrm{C})} \mp 2\mathrm{H}_x\mathrm{C}_z \xrightarrow{\tau} \mp \mathrm{H}_y \\ -\mathrm{I}_y \longrightarrow -\mathrm{I}_y \xrightarrow{\qquad\qquad\qquad} \mathrm{I}_y \longrightarrow \mathrm{I}_y. \end{array} \quad [2]$$

In a X(ω_1, ω_2)-double-half-filter experiment four data files I–IV are recorded with the phases ψ_1 and ψ_3 (Fig. 1). The data are listed in Table 1. The informative spectra are then obtained as the four linear combinations of I–IV given in Table 2.

The sum of all four data files, (A), contains all those signals that are not affected by the ^{13}C editing pulses, i.e., all diagonal peaks and cross peaks from protons not bound to ^{13}C. In an ideal experiment the peaks involving resonances from ^{13}C-bound protons would be completely absent from this $^{13}\mathrm{C}(\omega_1)$–$^{13}\mathrm{C}(\omega_2)$ doubly filtered spectrum. In the combination (B) (Table 2) the selected coherences are insensitive to the

TABLE 2

Desired Subspectra Obtained as Suitable Linear Combinations of the Four Recordings with an X(ω_1, ω_2) Double Half Filter Defined in Table 1

Subspectrum	Linear combination of recordings	Filter pass characteristics
(A) $^{13}\mathrm{C}(\omega_1)$–$^{13}\mathrm{C}(\omega_2)$ doubly filtered	I + II + III + IV	Unlabeled resonances in ω_1 and ω_2
(B) $^{13}\mathrm{C}(\omega_1)$-filtered/$^{13}\mathrm{C}(\omega_2)$-selected	(I + II) − (III + IV)	^{13}C labeled in ω_2, unlabeled in ω_1
(C) $^{13}\mathrm{C}(\omega_2)$-filtered/$^{13}\mathrm{C}(\omega_1)$-selected	(I − II) + (III − IV)	^{13}C labeled in ω_1, unlabeled in ω_2
(D) $^{13}\mathrm{C}(\omega_1)$–$^{13}\mathrm{C}(\omega_2)$ doubly selected	(I − II) − (III − IV)	^{13}C labeled in ω_1 and ω_2

phase ψ_1 but change sign depending on the phase ψ_3 (Fig. 1). The resulting spectrum corresponds to that obtained with a conventional $^{13}C(\omega_2)$ half filter, except that it contains in ω_1 only signals from protons not bound to ^{13}C. The combination of $^{13}C(\omega_1)$ filtration with $^{13}C(\omega_2)$ selection efficiently suppresses the diagonal peaks. The combination (C) (Table 2) gives the same result as (B), except that the roles of ω_1 and ω_2 are interchanged. The spectra (B) and (C) are nonetheless complementary, because the digital resolution is usually different in ω_1 and ω_2, and artifacts such as t_1 noise are not symmetric with respect to the diagonal. The combination (D) (Table 2) finally leads to a subspectrum that contains exclusively signals between resonances of different ^{13}C-bound protons, which may be coupled to either the same or two different ^{13}C spins. This contrasts with the X-filter techniques, which select for the cross peaks between protons coupled to the same heterospins (*9, 10*).

For experimental verification we recorded a ^1H NOESY experiment of the protein c2 repressor(1–76) from Salmonella phage P22 using a $^{13}C(\omega_1, \omega_2)$ double half filter. In the protein preparation used each methyl group of the four valyl residues was ^{13}C-labeled in the extent of close to 50% in such a way that there was no double labeling of any individual valine side chain. This preparation was obtained by expressing the P22 c2 repressor in *Escherichia coli* grown on a minimal medium (*18*) supplemented with 99% L-[4-^{13}C]valine containing the ^{13}C label randomly distributed between the two diastereotopic methyl positions. To ensure optimum subtraction for the spectral combinations of Table 2, the four data sets I to IV were recorded in an interleaved mode; i.e., the free induction decays of all four data sets were recorded and stored separately for each t_1 value before incrementation of the t_1 delay. The four data sets thus obtained were combined according to Table 2, and the four subspectra (A)–(D) were then Fourier transformed, baseline corrected, and plotted with identical parameters.

Figure 2A shows the $^{13}C(\omega_1)$–$^{13}C(\omega_2)$ doubly filtered spectrum (A) (Table 2), which contains the resonances from ^{12}C-bound protons in both dimensions. Since the individual valyl methyl groups are enriched with ^{13}C in the extent of only approximately 50%, this subspectrum contains all the cross peaks that would be observed in a conventional ^1H NOESY spectrum of the unlabeled protein. Pronounced overlap is observed for many cross peaks, and diagonal peaks with narrow lineshapes are accompanied by t_1 noise. The extensive overlap of cross peaks results in further artifacts introduced by the automated baseline correction routine, for example, the edge running from $\omega_2 = 1.4$ to 2.2 ppm at $\omega_1 = 1.05$ ppm.

Figure 2B shows the $^{13}C(\omega_1)$-filtered/$^{13}C(\omega_2)$-selected subspectrum (B) (Table 2). All the observed cross peaks are between resonances of ^{12}C-bound protons in ω_1 and of ^{13}C-bound protons in ω_2. The diagonal peaks are suppressed to the extent that for all eight valyl methyl groups they are smaller than the intraresidual NOE cross peaks between the ^{13}C-labeled and the unlabeled methyl group. The residual diagonal peak intensity comes from incomplete separation from the $^{13}C(\omega_1)$–$^{13}C(\omega_2)$ doubly selected subspectrum (Fig. 2D). Contributions to the residual diagonal peak intensity may come from imperfect matching of the delay τ because of small variations in $^1J(^{13}C, ^1H)$, from small amounts of heteronuclear antiphase magnetization present at the end of the filter delays due to small heteronuclear couplings, and from imperfect editing pulses. Compared to Fig. 2A the extent of overlap between different cross

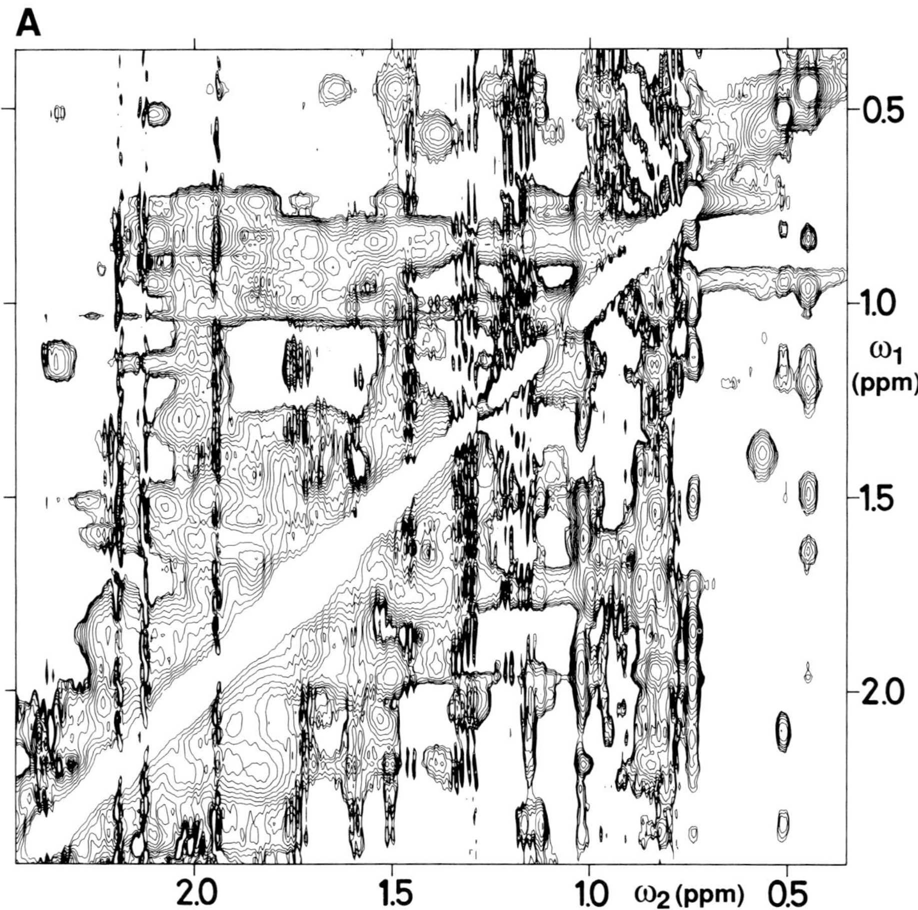

FIG. 2. ^{1}H NOESY spectra recorded with the $^{13}C(\omega_1, \omega_2)$-double-half-filter experiment of Fig. 1B of the DNA binding domain 1–76 of P22 c2 repressor. The repressor was selectively enriched with ^{13}C so that each methyl group of the four valines 28, 30, 33, and 74 contained approximately 50% ^{13}C, and there was no double labeling of the individual valyl side chains (see text). The protein concentration was 5 mM, solvent H_2O, pH 4.8, T = 28°C. The spectrum was recorded on a Bruker AM500 spectrometer; the total measuring time was 80 h, τ = 3.94 ms, t_{1max} = 53 ms, t_{2max} = 254 ms, τ_m = 100 ms. Prior to Fourier transformation the data were multiplied with a sine-squared bell in both dimensions, which was shifted by $\pi/3$ and $\pi/5$ along ω_1 and ω_2, respectively. The rows and columns were individually baseline-corrected by fitting a polynomial of third order to the baseline, using a computer routine supplied by the spectrometer manufacturer. The same contour lines have been drawn in all spectra, using an exponential scale with a factor $\sqrt{2}$ separating adjoining contours. The spectral region (ω_1 = 0.4–2.4 ppm, ω_2 = 0.4–2.4 ppm), which contains the diagonal peaks of all eight methyl groups of valine, is shown. (A) $^{13}C(\omega_1)$–$^{13}C(\omega_2)$ doubly filtered, (B) $^{13}C(\omega_1')$-filtered/$^{13}C(\omega_2)$-selected, (C) $^{13}C(\omega_2)$-filtered/$^{13}C(\omega_1)$-selected, (D) $^{13}C(\omega_1)$–$^{13}C(\omega_2)$ doubly selected. In the subspectra (B) and (C) the eight valyl methyl chemical shifts are identified at the top and on the left, respectively, with the one-letter amino acid symbol and the sequence number. Intraresidual cross peaks between the two methyl groups of the same valine side chain are identified by arrows, and the interresidual cross peaks with unlabeled protons are identified with the assignments of the latter. In (D) the cross peaks between different ^{13}C-bound protons are identified with the sequence positions of the two interacting valines.

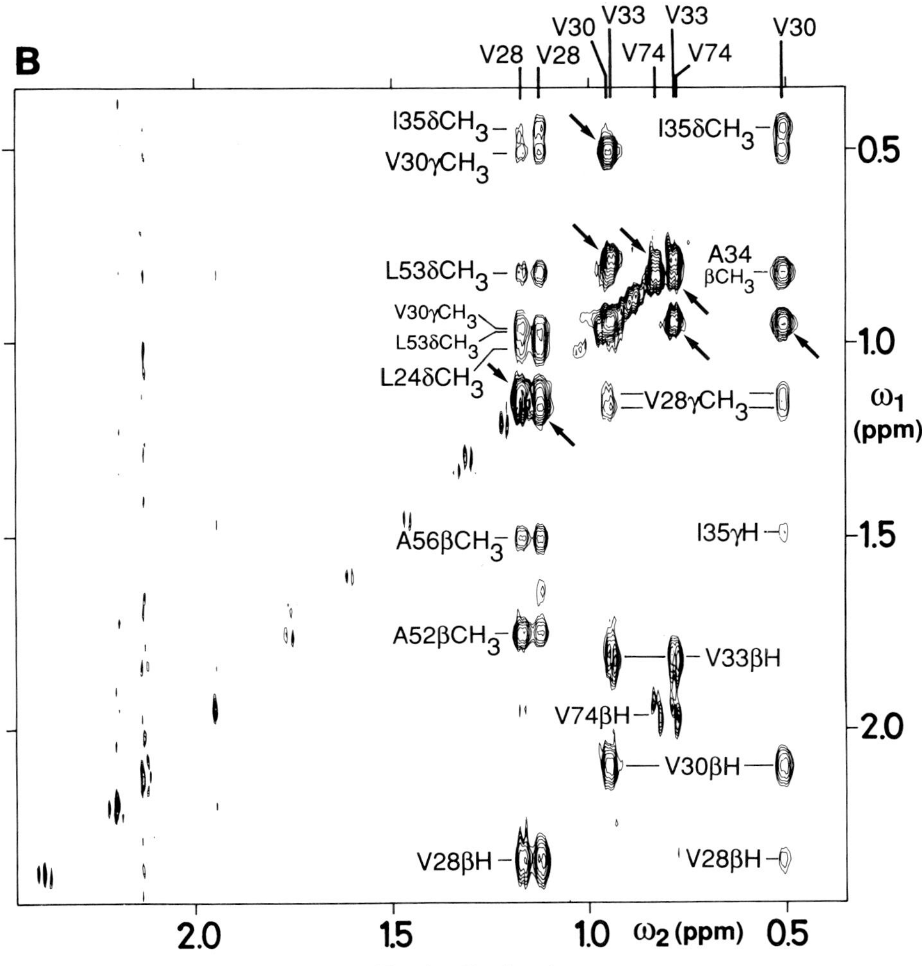

FIG. 2—*Continued*

peaks is dramatically reduced and the t_1 noise is nearly completely suppressed. Along ω_2 there are indications of resolved cross-peak fine structure.

The $^{13}C(\omega_2)$-filtered/$^{13}C(\omega_1)$-selected subspectrum Fig. 2C contains the same cross peaks as Fig. 2B, but the frequency axes are interchanged. Again there is some partial resolution of the cross-peak fine structure along ω_2.

Signals in the $^{13}C(\omega_1)$–$^{13}C(\omega_2)$ doubly selected subspectrum (D) (Table 2) are observed only if the proton that precesses during the evolution period as well as the proton observed during the detection period is bound to ^{13}C. With the protein preparation used here each cross peak in this subspectrum (Fig. 2D) therefore manifests an interresidual NOE between $^{13}CH_3$ groups of different valyl residues. Figure 2D shows only two well-resolved cross peaks, which represent NOEs between the two methyl groups of Val 28 and one of the methyl groups of Val 30. (These cross peaks are also present in Figs. 2A–2C, since the level of ^{13}C enrichment is only approxi-

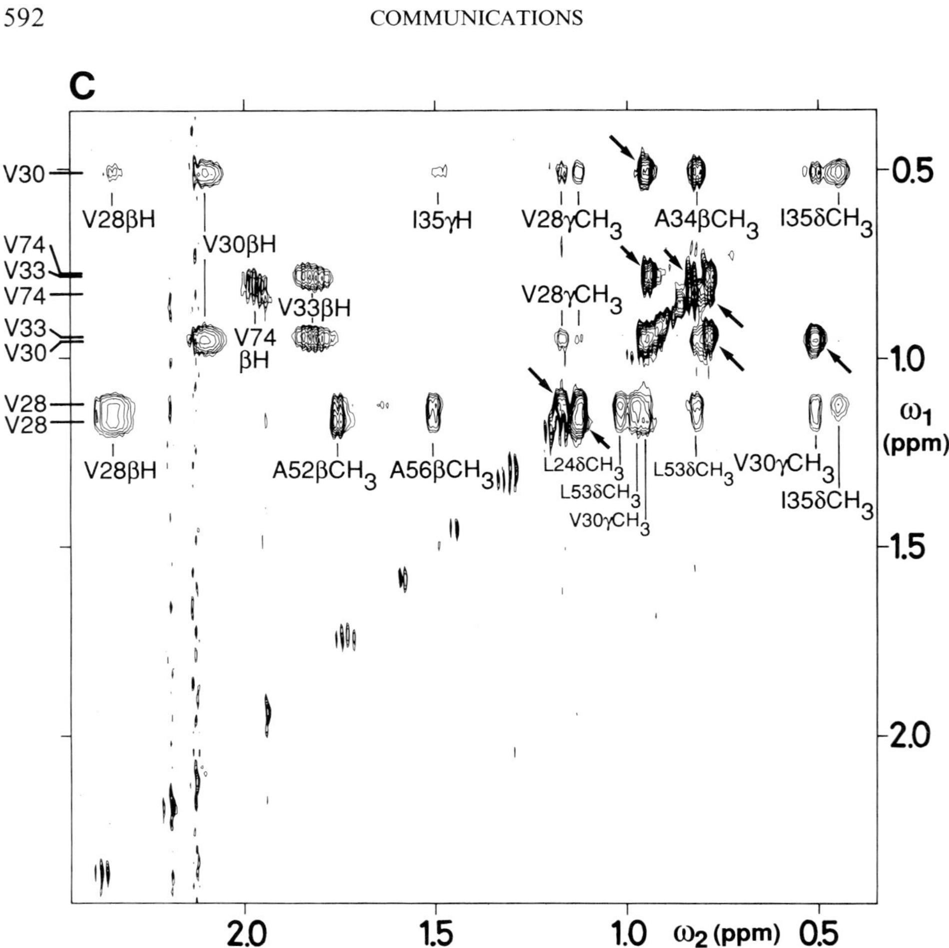

FIG. 2—*Continued*

mately 50%.) In addition, Fig. 2D contains the diagonal peaks of the eight valyl $^{13}CH_3$ groups, and t_1 noise associated with these intense diagonal peaks is also transferred into this subspectrum. The continuous line of diagonal peaks between 1.3 and 2.4 ppm comes primarily from ^{13}C at natural abundance, as is indicated by observations in corresponding experiments with unlabeled proteins.

In conclusion the extensive subspectral editing potential inherent in the $X(\omega_1, \omega_2)$-double-half-filter technique makes it an attractive tool for the investigation of biopolymers with residue-specific or site-specific isotope labeling, or complexes of biopolymers with isotope-enriched ligands. All cross peaks seen in the corresponding conventional, unfiltered 2D ^1H NMR experiment are contained in at least one of the four subspectra (A)–(D) (Table 2, Fig. 2) so that no information is lost by the editing procedure. (There may be some loss of signal intensity due to relaxation during the additional delays of the filter elements (Fig. 1).) The $X(\omega_1)$–$X(\omega_2)$ doubly selected subspectrum (D in Table 2 and Fig. 2) is unique in its potential for resolution and identification of cross peaks between protons bound to different heterospins. The

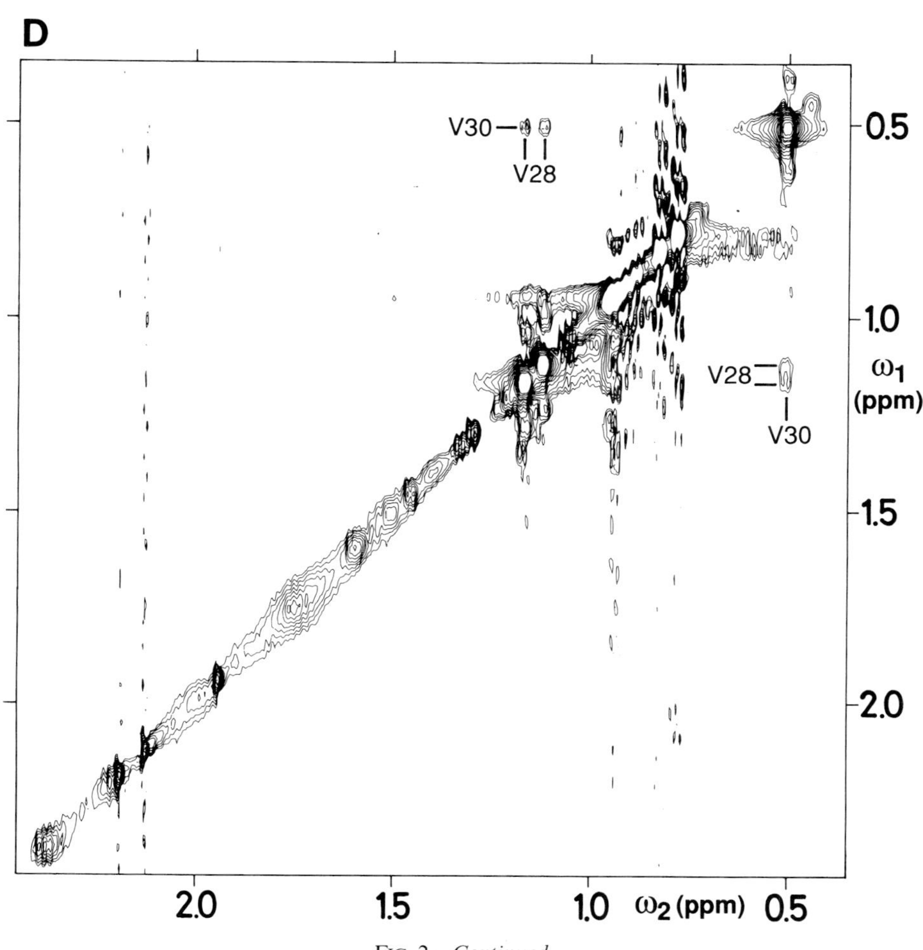

FIG. 2—*Continued*

ensemble of these features and the additional diagonal suppression in two of the four subspectra should make $X(\omega_1, \omega_2)$ double half filters a useful tool for work with isotope-labeled macromolecular systems.

ACKNOWLEDGMENTS

We thank A. Eugster and D. Neri for the preparation of ^{13}C-labeled P22 c2 repressor(1–76), Drs. M. Ptashne and O. Koudelka for the plasmid used for the protein preparation, and R. Marani for the careful processing of the manuscript. Financial support by the Schweizerischer Nationalfonds (Project 31.25174.88) is gratefully acknowledged.

REFERENCES

1. K. WÜTHRICH, "NMR of Proteins and Nucleic Acids," Wiley, New York, 1986.
2. K. WÜTHRICH, *Science* **243,** 45 (1989).
3. T. F. HAVEL AND K. WÜTHRICH, *J. Mol. Biol.* **182,** 281 (1985).

4. R. H. Griffey and A. G. Redfield, *Q. Rev. Biophys.* **19,** 51 (1987).
5. D. M. LeMaster and F. M. Richards, *Biochemistry* **24,** 7263 (1985).
6. J. A. Wilde, P. H. Bolton, N. J. Stolowich, and J. A. Gerlt, *J. Magn. Reson.* **68,** 168 (1986).
7. S. W. Fesik and E. R. P. Zuiderweg, *J. Magn. Reson.* **78,** 588 (1988).
8. D. Marion, L. E. Kay, S. W. Sparks, D. A. Torchia, and A. Bax, *J. Am. Chem. Soc.* **111,** 1515 (1989).
9. E. Wörgötter, G. Wagner, and K. Wüthrich, *J. Am. Chem. Soc.* **108,** 6162 (1986).
10. E. Wörgötter, G. Wagner, H. Vašák, J. H. R. Kägi, and K. Wüthrich, *J. Am. Chem. Soc.* **110,** 2388 (1988).
11. G. Otting, H. Senn, G. Wagner, and K. Wüthrich, *J. Magn. Reson.* **70,** 500 (1986).
12. S. W. Fesik, R. T. Gampe, Jr., and T. W. Rockway, *J. Magn. Reson.* **74,** 366 (1987).
13. L. P. McIntosh, F. W. Dahlquist, and A. G. Redfield, *J. Biomol. Struct. Dyn.* **5,** 21 (1987).
14. P. H. Bolton, *J. Magn. Reson.* **62,** 143 (1985).
15. A. Bax and M. Weiss, *J. Magn. Reson.* **71,** 571 (1987).
16. S. W. Fesik, J. R. Luly, J. W. Erickson, and C. Abad-Zapatero, *Biochemistry* **27,** 8297 (1988).
17. H. Senn, G. Otting, and K. Wüthrich, *J. Am. Chem. Soc.* **109,** 1090 (1987).
18. H. Senn, A. Eugster, G. Otting, F. Suter, and K. Wüthrich, *Eur. Biophys. J.* **14,** 301 (1987).

JOURNAL OF MAGNETIC RESONANCE **85**, 608–613 (1989)

Solvent Suppression Using a Spin Lock in 2D and 3D NMR Spectroscopy with H_2O Solutions

BARBARA A. MESSERLE, GERHARD WIDER, GOTTFRIED OTTING, CHRISTOPH WEBER, AND KURT WÜTHRICH

Institut für Molekularbiologie und Biophysik, Eidgenössische Technische Hochschule-Hönggerberg, CH-8093 Zurich, Switzerland

Received July 27, 1989

The ability to observe all polypeptide backbone amide protons in ^{1}H NMR spectra is a preliminary condition for the use of NMR to determine the structure of proteins in solution (*1*). The acquisition of homonuclear 2D ^{1}H NMR spectra of proteins in H_2O solution (*2*) was therefore a turning point in the methodological development which made it possible to obtain complete ^{1}H NMR assignments of proteins (*3, 4*), and on this basis to determine three-dimensional protein structures (*5*). For proteins at slightly acidic pH, where the amide proton exchange rate is sufficiently slow, the most widely used solvent suppression method is by selective presaturation of the H_2O line (*1, 2, 6*). However, under certain conditions of pH and temperature, amide protons may exchange sufficiently fast with the bulk water for their signal intensity to be significantly reduced on presaturation of H_2O. In addition, bleaching of spectral regions near the water line and concomitant loss of information in the two-dimensional spectra may also result from presaturation of the water resonance (*2*). As an alternative to solvent presaturation, one of many selective excitation methods (*7–10*) may be used to suppress the water signal. The main disadvantages of selective excitation techniques lie in the facts that they bring about phase distortions in 2D NMR spectra and that they cannot be combined with uniform spectral excitation across the entire spectrum. Today heteronuclear NMR experiments are increasingly used in protein structure determinations to provide supporting information for the resonance assignments and to facilitate the collection of the ^{1}H NMR data needed as input for the structure determination (*11–13*). In this paper we present a technique whereby the water signal can be suppressed in heteronuclear NMR experiments with proton detection, without the need for either water presaturation or selective excitation methods.

It has been previously shown that suppression of undesired proton magnetization in heteronuclear spectra can be greatly enhanced by the use of spin-lock pulses (*14*). Here, spin-lock purge pulses are used during the application of heteronuclear pulse sequences to suppress the magnetization due to water protons. Figures 1A and 1B show, respectively, the experimental schemes for 2D [^{15}N, ^{1}H]-COSY (*15*) and 3D NOESY-[^{15}N, ^{1}H]-COSY (*12*) supplemented by spin-lock purge pulses for solvent suppression.

0022-2364/89 $3.00
Copyright © 1989 by Academic Press, Inc.
All rights of reproduction in any form reserved.

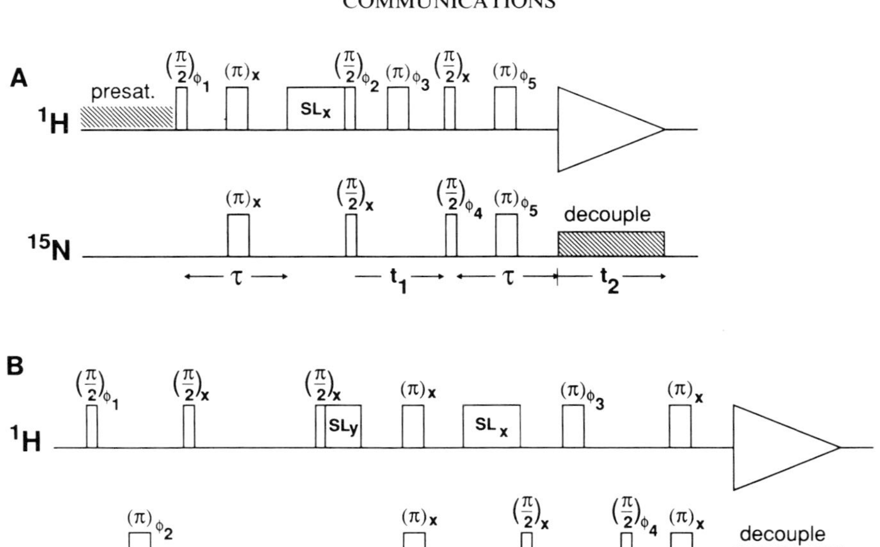

FIG. 1. (A) Experimental scheme for [^{15}N, ^{1}H]-COSY (*15*) extended by a spin-lock pulse (*14*) for solvent suppression. This scheme was used with and without presaturation of H_2O (indicated by dashed lines during the preparation period). The phases ϕ_1, ϕ_3, ϕ_4, and ϕ_5 are independently alternated between x and $-x$, and the phase ϕ_2 between y and $-y$, resulting in a 32-step phase cycle. The receiver phase is alternated together with ϕ_1, ϕ_2, and ϕ_4. Quadrature detection in ω_1 is achieved by applying TPPI (*18*) to the first two ^{15}N pulses. τ equals $1/[2\,^1J(^1\mathrm{H}, ^{15}\mathrm{N})]$. SL denotes the spin-lock pulse. (B) Three-dimensional NOESY-[^{15}N, ^{1}H]-COSY (*12*) with water suppression using spin-lock pulses. The phases ϕ_1 to ϕ_4 are independently alternated between x and $-x$, resulting in a 16-step phase cycle. The receiver phase is alternated together with ϕ_1 and ϕ_4. Quadrature detection in ω_1 is achieved by subjecting ϕ_1 to TPPI, and in ω_2 by applying TPPI to the first three ^{15}N pulses.

The purging effect of the spin-lock pulse is best explained using the product operator notation of Sørensen *et al.* (*16*), where I and S denote proton spins and ^{15}N spins, respectively. Considering the scheme of Fig. 1A we start with equilibrium z magnetization I_z. After the sequence $[(\pi/2)_x(^1\mathrm{H})-\tau/2-(\pi)_x(^1\mathrm{H}, ^{15}\mathrm{N})-\tau/2]$ with $\tau = 1/[2\,^1J(^{15}\mathrm{N}, ^1\mathrm{H})]$, we have heteronuclear antiphase magnetization, $2I_xS_z$, aligned along the x axis for all ^{15}N-bound protons. The protons of the water resonance do not couple to other protons and are represented by in-phase magnetization I_y. The following spin-lock pulse along the x axis retains the magnetization of the ^{15}N-bound protons and randomizes the water magnetization by its radiofrequency field inhomogeneity. Efficient water suppression is achieved with spin-lock pulses of 2 ms duration and the same RF amplitude as the other proton pulses. This suppression scheme is quite insensitive to pulse imperfections of the refocusing $(\pi)_x(^1\mathrm{H})$ pulse if the carrier frequency is set at the frequency of the water resonance. The same principles can be used for water suppression in NOESY and TOCSY with ^{15}N- or ^{13}C-labeled compounds. Figure 1B shows an example of the use of this suppression scheme with 3D NOESY-[^{15}N, ^{1}H]-COSY. The pulse sequence selects for z magnetization I_z during the mixing time τ_m. The following $(\pi/2)_x(^1\mathrm{H})$ pulse converts I_z into transverse magnetization I_y. Then it is spin-locked by the spin-lock pulse SL_y, which randomizes all magnetization that was not aligned along the z axis during the mixing period. After this "trim pulse" the sequence $[\tau/2-\pi(^1H, ^{15}N)-\tau/2]$ converts the magnetization I_y from ^{15}N-bound protons into antiphase magnetization $2I_xS_z$, while the water magnetization remains aligned along the y axis. The following spin-lock purge pulse SL_x randomizes the water magnetization and retains the desired proton magnetization $2I_xS_z$. In order to avoid refocusing effects from the use of more than one spin lock, the two spin-lock pulses should be of different duration (*17*). Compared to the 3D NOESY-[^{15}N, ^1H]-COSY pulse sequence devised by Fesik and Zuiderweg (*12*), the use of the spin-lock pulses for water suppression requires additional refocusing $(\pi)_x(^1H, ^{15}N)$ pulses in the middle of each delay τ to prevent the evolution of proton magnetization under the chemical-shift Hamiltonian.

The water suppression schemes of Fig. 1 were applied with cyclophilin. This protein cannot be studied at pH values below 6.0, so that numerous rapidly exchanging amide protons could not be observed when the water line was presaturated. Cyclophilin contains 165 amino acids and has a molecular weight of about 17,900. The cDNA of the human protein was expressed in *Escherichia coli*, and 99% labeling with ^{15}N was achieved by growing the bacterial cells on a minimal medium containing ^{15}N ammonium sulfate as the sole nitrogen source. For the NMR experiments in Figs. 2 and 3 a 2 m*M* solution of cyclophilin in a mixed solvent of 90% H_2O/10% D_2O at pH 6.0 was used. All measurements were made at a temperature of 26°C on a Bruker AM 600 instrument.

As an initial test of the proposed water suppression with a spin lock, [^{15}N, ^1H]-COSY spectra of cyclophilin were recorded with and without presaturation of the water resonance, and otherwise identical conditions (Fig. 2). In Fig. 2A those resonances are identified for which the signal intensity is clearly reduced by the water presaturation. Although the majority of the resonances were not or only slightly affected, certain resonances lost a significant amount of intensity in spectrum B. In a different presentation, the comparison of corresponding cross sections from spectra A and B clearly shows that the water presaturation caused a reduction of the peak intensities II and III of more than 50% relative to peak I (Figs. 2C and 2D). These observations thus demonstrate that the experiment of Fig. 1A can yield clean and complete [^{15}N, ^1H]-COSY fingerprints of a protein in H_2O solution under conditions where water preirradiation causes significant loss of signal intensity. The latter is probably due in part to saturation transfer by chemical exchange of amide protons, and in part to spin diffusion starting from water protons or α protons located at the water proton chemical shift (*1*).

The spin-lock suppression technique was also applied in a 3D NOESY-[^{15}N, ^1H]-COSY experiment (Fig. 1B) with ^{15}N-labeled cyclophilin. For comparison the corresponding region of a conventional, homonuclear 2D ^1H NOESY spectrum is also shown (Fig. 3A). The extensive overlap of resonances in Fig. 3A makes clear that ^{15}N-resolved 3D NMR was indicated as a means to facilitate the spectral analysis. The 2D ^1H-NOESY experiment was acquired with selective water presaturation, and the band of bleached resonances along ω_2 at $\omega_1 \approx 4.75$–4.85 ppm is readily apparent in Fig. 3A. In contrast, the two cross sections taken from the corresponding 3D spectrum acquired without water presaturation show several resonances at and very near

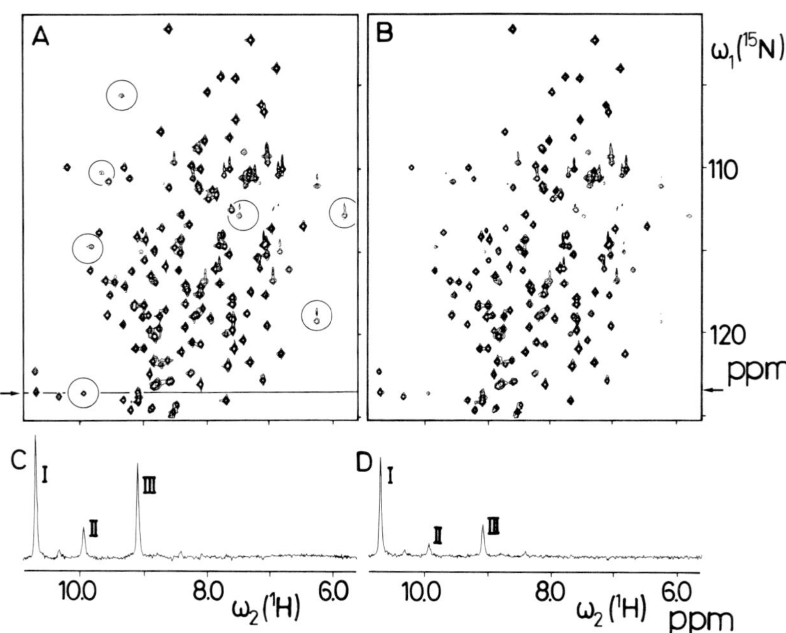

FIG. 2. Part of two phase-sensitive [^{15}N, ^1H]-COSY spectra of ^{15}N-labeled cyclophilin (2 mM protein in a mixed solvent of 90% H_2O/10% D_2O, with 10 mM [$^2H_3$]sodium acetate, 10 mM disodium phosphate, and 2 mM sodium dithionite, pH 6.0, 26°C, recorded at 600 MHz proton frequency on a Bruker AM600 spectrometer). Spectra (A) and (B) were acquired using the pulse sequence in Fig. 1A, without and with presaturation of the water resonance for 1 s during the preparation period, respectively. τ = 5.6 ms, t_{1max} = 134 ms, and t_{2max} = 127 ms; ^{15}N decoupling during acquisition using WALTZ with suppression of cycling sidebands (*19*). (C) and (D) Cross sections taken from (A) and (B), respectively, at the position indicated by the arrows. In spectrum (A) some of the resonances with clearly decreased intensity in (B) are circled. ^{15}N chemical shifts were referenced to liquid NH_3 (*20*).

to the chemical shift of water in ω_1 (Figs. 3B and 3C). These are mainly due to NOE interactions of amide protons with α protons that have chemical shifts near that of the water. Additional resonances at the water chemical shift may arise either from chemical exchange of NH protons with the bulk water, or from NOEs between H_2O and protons in the protein (*21*). Comparison of Fig. 3A with Figs. 3B and 3C also presents a nice illustration of the improved resolution achieved with the heteronuclear 3D NMR experiment (*12*).

In conclusion, the experimental data and their interpretation presented in this Communication clearly demonstrate the advantages of obtaining spectra of proteins in water using nonselective ^1H pulses without the need for presaturation of the water signal. Using the experimental schemes of Fig. 1, uniform excitation of the entire spectrum is achieved and a complete spectrum of the amide protons can be observed under conditions of pH and temperature, where water presaturation would cause the disappearance of the resonance lines of solvent-exposed labile protons and α protons with chemical shifts near that of the water (*1*). The 3D NMR experiment of Fig. 1B could further facilitate investigations of protein hydration in aqueous solu-

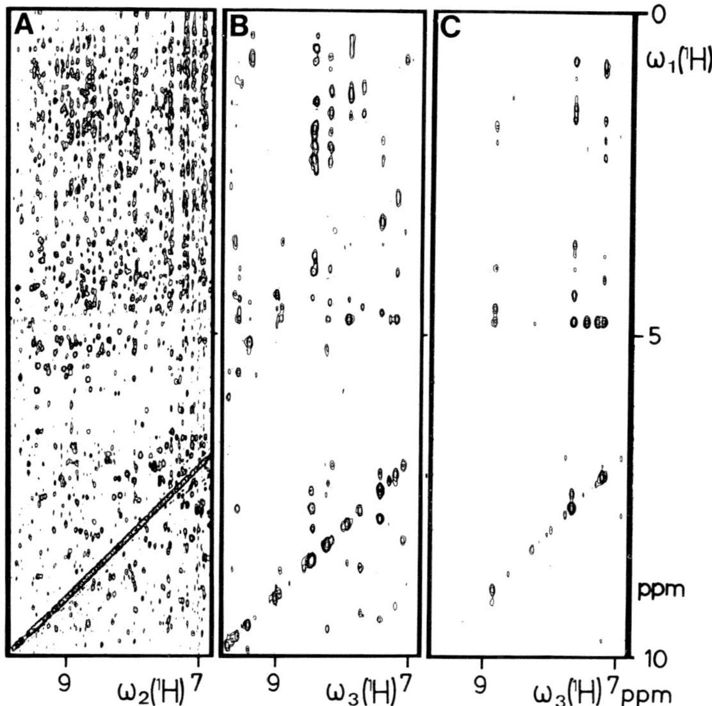

FIG. 3. (A) Two-dimensional ^1H NOESY spectrum of unlabeled cyclophilin (same sample conditions as those in Fig. 2, mixing time = 120 ms, total experimental time about 44 hours). Presaturation of the water resonance for 1 s was used during the preparation period, and also for the whole duration of the mixing time. (B) and (C) Cross sections from a three-dimensional NOESY-[^{15}N, ^1H]-COSY spectrum of ^{15}N-labeled cyclophilin (same sample conditions as those in Fig. 2). The pulse sequence of Fig. 1B was used, with spin-lock pulses of 0.5 ms (SL_y) and 2 ms (SL_x) duration. The cross sections were taken perpendicular to the $\omega_2(^{15}N)$ axis; (B) plane 28 and (C) plane 42 from a total of 64 planes. The total experimental time was 5 days, τ_m = 100 ms, τ = 5.6 ms, t_{1max} = 13.4 ms, t_{2max} = 16.9 ms, and t_{3max} = 63.5 ms, and ^{15}N decoupling during acquisition was achieved using WALTZ (*19*). The sweep width used in ω_1 and ω_3 was 8064 Hz, and in ω_2 it was 2840 Hz; 214 points were acquired in t_1, 76 points were acquired in t_2, and 1024 points were acquired in t_3. Prior to Fourier transformation, zero filling to 1024 points in ω_1, to 128 points in ω_2, and to 2048 points in ω_3 was used. Sine-bell weighting functions were applied, with phase shifts of $\pi/3$ in ω_3 and $\pi/2$ in both ω_1 and ω_2. Polynomial baseline correction was applied along ω_3 following the initial 2D Fourier transformation of the data from (t_1, t_2, t_3) to $(t_1, \omega_2, \omega_3)$.

tion, following up on the recently described procedures using homonuclear 2D ^1H NMR (*21*).

ACKNOWLEDGMENTS

We thank Sandoz AG, Basel, for help with the production of cyclophilin and financial support, Drs. C. Griesinger and R. R. Ernst for a software routine which enables processing 3D NMR data on a Bruker Aspect 1000 computer, and the Schweizerischer Nationalfonds for financial support (Project 31.25174.88).

REFERENCES

1. K. Wüthrich, "NMR of Proteins and Nucleic Acids," Wiley, New York, 1986.
2. Anil-Kumar, G. Wagner, R. R. Ernst, and K. Wüthrich, *Biochem. Biophys. Res. Commun.* **96,** 1156 (1980).
3. G. Wagner and K. Wüthrich, *J. Mol. Biol.* **155,** 347 (1982).
4. G. Wider, K. H. Lee, and K. Wüthrich, *J. Mol. Biol.* **155,** 367 (1982).
5. K. Wüthrich, G. Wider, G. Wagner, and W. Braun, *J. Mol. Biol.* **155,** 311 (1982).
6. G. Wider, R. V. Hosur, and K. Wüthrich, *J. Magn. Reson.* **52,** 130 (1983).
7. V. Sklenář, R. Tschudin, and A. Bax, *J. Magn. Reson.* **75,** 352 (1987).
8. A. Redfield and R. K. Gupta, *J. Chem. Phys.* **54,** 1418 (1971).
9. P. J. Hore, *J. Magn. Reson.* **55,** 283 (1983).
10. P. Plateau and M. Guéron, *J. Am. Chem. Soc.* **104,** 7310 (1982).
11. R. H. Griffey and A. G. Redfield, *Q. Rev. Biophys.* **19,** 51 (1987).
12. S. W. Fesik and E. R. P. Zuiderweg, *J. Magn. Reson.* **78,** 588 (1988).
13. H. Senn, G. Otting, and K. Wüthrich, *J. Am. Chem. Soc.* **109,** 1090 (1987).
14. G. Otting and K. Wüthrich, *J. Magn. Reson.* **76,** 569 (1988).
15. G. Bodenhausen and D. Ruben, *Chem. Phys. Lett.* **69,** 185 (1980).
16. O. W. Sørensen, G. W. Eich, M. H. Levitt, G. Bodenhausen, and R. R. Ernst, *Prog. NMR Spectrosc.* **16,** 163 (1983).
17. C. J. R. Counsell, M. H. Levitt, and R. R. Ernst, *J. Magn. Reson.* **64,** 470 (1985).
18. D. Marion and K. Wüthrich, *Biochem. Biophys. Res. Commun.* **113,** 967 (1983).
19. A. J. Shaka, P. B. Barker, C. J. Bauer, and R. Freeman, *J. Magn. Reson.* **67,** 396 (1986).
20. D. H. Live, D. G. Davis, W. C. Agosta, and D. Cowburn, *J. Am. Chem. Soc.* **106,** 1939 (1984).
21. G. Otting and K. Wüthrich, *J. Am. Chem. Soc.* **111,** 1871 (1989).

Reprinted from the Journal of the American Chemical Society, 1993, *115*.
Copyright © 1993 by the American Chemical Society and reprinted by permission of the copyright owner.

Reduced Dimensionality in Triple-Resonance NMR Experiments

T. Szyperski, G. Wider, J. H. Bushweller, and K. Wüthrich*

Institut für Molekularbiologie und Biophysik
Eidgenössische Technische Hochschule-Hönggerberg
CH-8093 Zürich, Switzerland

Received April 30, 1993

As an alternative to the conventional assignment of protein NMR[1] spectra by observation of sequential NOEs[2] in homonuclear 2D [^{1}H,^{1}H]-NOESY or 3D and 4D heteronuclear-resolved [^{1}H,^{1}H]-NOESY spectra, 3D and 4D triple-resonance experiments have been proposed for establishing intra- and interresidual connectivities via heteronuclear scalar couplings.[3-12] 4D experiments of this type are conceptually particularly attractive, since only a single experiment is needed for intraresidual correlation of the four backbone spins ^{1}H^N, ^{15}N, ^{13}C$^\alpha$, and ^{1}H$^\alpha$, and suitable combinations of two 4D experiments can provide sequential assignments.[11,13] However, because of the short T_2 relaxation times of ^{13}C$^\alpha$ in bigger molecules, the use of 4D triple resonance experiments[6,9,11] is in practice limited to proteins with molecular weights below approximately 15 000,[6] where such spectra are only sparsely populated with cross peaks and hence dispersion in four dimensions is not really needed. Therefore, the development of variant triple-resonance experiments that provide the same connectivity information in spectra with reduced dimensionality is attractive, since larger values of t_{max} can then be chosen for the indirect dimensions, more extensive phase cycling is feasible within the same accumulation time, and the smaller data sets facilitate data handling and processing. Recently, we presented an experimental scheme that used ^{13}C$^\alpha$–^{15}N heteronuclear two-spin coherence to obtain spectra with reduced dimensionality.[14] The present communication introduces a more general projection technique which does not require the generation of two-spin coherence and can readily be used with all presently available triple-resonance NMR schemes.

To demonstrate the utility of the proposed projection technique, we recorded a 3D HA <u>CA</u> <u>N</u> HN experiment (the underlined letters indicate that the ^{15}N and ^{13}C$^\alpha$ chemical shifts evolve simultaneously as single-quantum coherences) (Figure 1), which was derived from the 4D pulse sequence developed by Boucher *et al.*,[11] except that in view of the long T_2 relaxation time for the carbonyl carbon, ^{15}N and ^{13}C$^\alpha$ are decoupled from ^{13}C=O with selective 180° pulses on ^{13}C=O instead of a WALTZ-16 sequence (Figure 2). Following Boucher *et al.*,[11] the transfer amplitude that produces the peak patterns observed in 3D HA <u>CA</u> <u>N</u> HN spectra was evaluated using the product-operator formalism.[15] Thereby, only terms resulting in observable magnetization during the detection period were retained, relaxation terms and constant multiplicative factors were omitted, and interresidue transfer of magnetization via the two-bond ^{13}C$^\alpha_i$ – ^{15}N$_{i+1}$ coupling was neglected. Provided that $2\tau_1 = \tau_2 = {}^1/_2[^1J(^{13}\text{C}^\alpha, {}^1\text{H}^\alpha)]$ and $\tau_5 = 2\tau_6 = {}^1/_2[^1J(^{15}\text{N}, {}^1\text{H}^N)]$ (Figure 2), the observable magnetization at the beginning of the acquistion is given by (1),

$$\sigma(t_3 = 0) = I_x^N \cos[\Omega(^1\text{H}^\alpha)t_1] \cos[\Omega(^{13}\text{C}^\alpha)t_2] \cos[\Omega(^{15}\text{N})t_2] \quad (1)$$

where I^N is the spin operator for the amide proton and Ω(X) denotes the chemical shift of spin X. As described previously in the context of two-spin coherence spectroscopy,[14] $\sigma(t_3 = 0)$ contains the sum and the difference of the chemical shifts of ^{13}C$^\alpha$ and ^{15}N, which can be detected in a phase-sensitive manner by

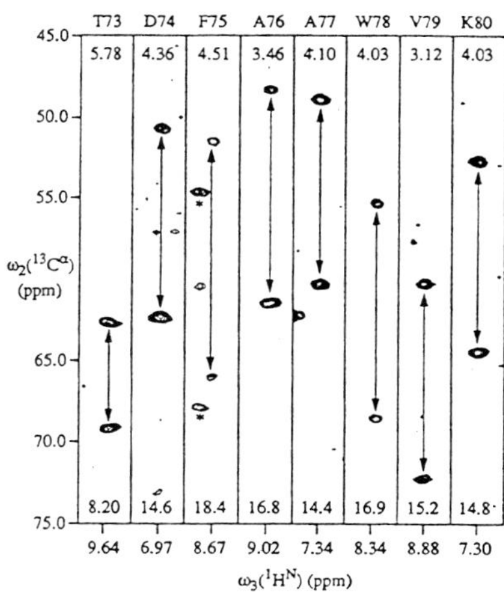

Figure 1. Contour plots of (ω_1(^{13}C$^\alpha$) – ω_3(^{1}H^N))-strips from a 3D HA <u>CA</u> <u>N</u> HN spectrum obtained with a 2.5 mM sample of the uniformly ^{13}C- and ^{15}N-labeled mixed disulfide of *E. coli* glutaredoxin-(C14S) with glutathione[17] in 90% H$_2$O/10% D$_2$O, 100 mM potassium phosphate, pH 6.5, at T = 20 °C. A Bruker AMX 600 spectrometer equipped with four channels was used. 24 (t_1) * 98 (t_2) * 512 (t_3) complex points were accumulated, with t_{1max}(^{1}H$^\alpha$) = 16.8 ms, t_{2max}(^{15}N,^{13}C$^\alpha$) = 11.2 ms, and t_{3max}(^{1}H^N) = 65.5 ms. 32 scans per increment were acquired, resulting in a total measuring time of 3.5 days. The carrier frequencies of the ^{15}N and ^{13}C$^\alpha$ pulses were set to 105.1 and 56 ppm, respectively. Phase-sensitive detection was achieved using States-TPPI[16] in t_1 and t_2, so that the peak positions along ω_2 are at Ω(^{13}C$^\alpha$) ± Ω(^{15}N). In addition to the purge pulse (Figure 2), the water signal was further reduced with the convolution method of Marion *et al.*[21] The digital resolution after zero-filling was 22 Hz along ω_1, 34 Hz along ω_2, and 7.6 Hz along ω_3. Prior to Fourier transformation, the data were multipled with a sine bell window shifted by 45° in t_1, and a cosine window in t_2 and t_3.[22] No linear prediction or maximum entropy processing was applied. The spectrum was processed using the program PROSA.[23] The strips were taken at the ^{1}H$^\alpha$ chemical shifts of the residues 73–80. The sequence-specific assignments are indicated at the top of each strip by the one-letter amino acid symbol and the sequence position, the ^{1}H$^\alpha$ chemical shift along ω_1 is given in ppm below the resonance assignment, the ^{1}H^N chemical shifts around which the strips are centered along ω_3 are indicated in ppm below each strip, and the axes along ω_2 gives the chemical shift of the ^{13}C$^\alpha$ nuclei. The chemical shifts of the ^{15}N nuclei relative to the ^{15}N carrier frequency (105.1 ppm), which were extracted from the in-phase splittings indicated by the arrows, are given at the bottom of each strip. Additional peaks in the strip of F75 (marked with an asterisk) belong to V64.

(1) Abbreviations used: NMR, nuclear magnetic resonance; 2D, two-dimensional; 3D, three-dimensional; 4D, four-dimensional; NOE, nuclear Overhauser effect; NOESY, 2D NOE spectroscopy; TPPI, time-proportional phase incrementation; glutaredoxin(C14S), mutant glutaredoxin with Cys 14 replaced by Ser.

(2) Wüthrich, K. *NMR of Proteins and Nucleic Acids*; Wiley: New York, 1986.
(3) Ikura, M.; Kay, L. E.; Bax, A. *Biochemistry* **1990**, *29*, 4659–4667.
(4) Kay, L. E.; Ikura, M.; Tschudin, R.; Bax, A. *J. Magn. Reson.* **1990**, *89*, 496–514.
(5) Kay, L. E.; Ikura, M.; Bax, A. *J. Magn. Reson.* **1991**, *91*, 84–92.
(6) Kay, L. E.; Wittekind, M.; McCoy, M. A.; Friedrichs, M. S.; Mueller, L. *J. Magn. Reson.* **1992**, *98*, 443–450.
(7) Bax, A.; Ikura, M. *J. Biomol. NMR* **1991**, *1*, 99–104.
(8) Powers, R.; Gronenborn, A. M.; Clore, G. M.; Bax, A. *J. Magn. Reson.* **1991**, *94*, 209–213.
(9) Boucher, W.; Laue, E. D. *J. Am. Chem. Soc.* **1992**, *114*, 2262–2264.
(10) Clubb, R. T.; Thanabal, V.; Wagner, G. *J. Biomol. NMR* **1992**, *2*, 203–210.
(11) Boucher, W.; Laue, E. D.; Campbell-Burk, S. L.; Domaille, P. J. *J. Biomol. NMR* **1992**, *2*, 631–637.
(12) Olejniczak, E. T.; Xu, R. X.; Petros, A. M.; Fesik, S. W. *J. Magn. Reson.* **1992**, *100*, 444–450.
(13) Bax, A.; Grzesiek, S. *Acc. Chem. Res.* **1993**, *26*, 131–138.
(14) Szyperski, T.; Wider, G.; Bushweller, J. H.; Wüthrich, K. *J. Biomol. NMR* **1993**, *3*, 127–132.
(15) Sørensen, O. W.; Eich, G. W.; Levitt, M. H.; Bodenhausen, G.; Ernst, R. R. *Prog. Nucl. Magn. Reson. Spectrosc.* **1983**, *16*, 163–192.

0002-7863/93/1515-9307$04.00/0 © 1993 American Chemical Society

9308 J. Am. Chem. Soc., Vol. 115, No. 20, 1993

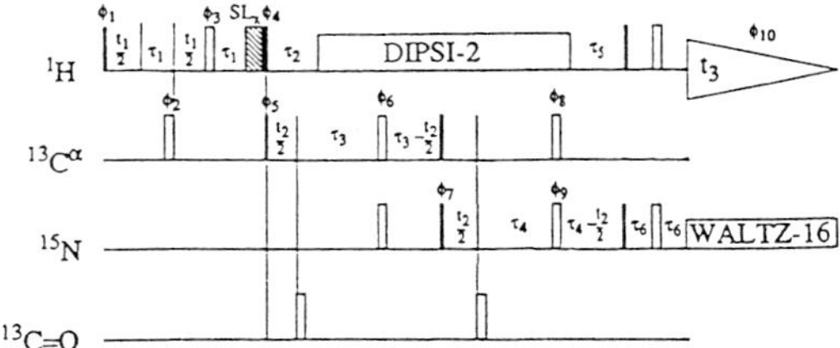

Figure 2. Experimental scheme for the 3D HA CA N HN experiment derived from the 4D HA CA N HN experiment of Boucher *et al.*[11] 90° and 180° pulses are indicated by thin and thick vertical bars, respectively, and the phases are indicated above the pulses. Where no radio frequency phase is marked, the pulse is applied along x. A spin-lock pulse, SL_x, of 2-ms duration is used to suppress the water signal.[24] The delays were set to the following values: $\tau_1 = 1.5$ ms, $\tau_2 = {}^1/_2[{}^1J({}^1H^{\alpha},{}^{13}C^{\alpha})] = 3.4$ ms, $\tau_3 = 12.564$ ms, $\tau_4 = 11.664$ ms, $\tau_5 = {}^1/_2[{}^1J({}^1H^N,{}^{15}N)] = 5.4$ ms, $\tau_6 = 2.5$ ms. A DIPSI-2[25] sequence is used to decouple ^{1}H during the heteronuclear magnetization transfer from $^{13}C^{\alpha}$ to ^{15}N, and a WALTZ-16 decoupling sequence is used during proton detection.[26] Phase cycling: $\phi_1 = 16(x)$; $\phi_2 = 8\{2x, 2(-x)\}$; $\phi_3 = 16(x, -x)$; $\phi_4 = 8\{2y, 2(-y)\}$; $\phi_5 = 2\{8(x), 8(-x)\}$; $\phi_6 = 4\{4(x), 4(y), 4(-x), 4(-y)\}$; $\phi_7 = \phi_8 = 16(x, -x)$; $\phi_9 = 2\{16(x), 16(-x)\}$; ϕ_{10} (receiver) $= \{x,-x,-x,x, 2(-x,x,x,-x), x,-x,-x,x\}$. Quadrature detection in t_1 and t_2 is accomplished by altering the phases ϕ_1 and ϕ_5, respectively, according to States-TPPI.[16]

applying the States-TPPI method[16] either to $^{13}C^{\alpha}$ or to ^{15}N. In the experiment of Figure 1, States-TPPI[16] was applied to $^{13}C^{\alpha}$, yielding resonances at $\Omega(^{13}C^{\alpha}) \pm \Omega(^{15}N)$ along the frequency axis ω_2 in the 3D HA CA N HN spectrum. Since the ^{15}N chemical shift is extracted from the difference between $\Omega(^{13}C^{\alpha}) - \Omega(^{15}N)$ and $\Omega(^{13}C^{\alpha}) + \Omega(^{15}N)$, the ^{15}N carrier must be at the edge of the ^{15}N spectral range to obtain unambiguous ^{15}N assignments. As the sweep width for ^{15}N (<2000 Hz at 14.1 T) is significantly smaller than that for $^{13}C^{\alpha}$ (~4500 Hz at 14.1 T), the thus required sweep width along ω_2 is only about one-third larger than in a corresponding 4D experiment. (This increase in sweep width could be circumvented if the in-phase splitting due to $\Omega(^{15}N)$ were scaled down using a smaller increment for ^{15}N than for $^{13}C^{\alpha}$, which would, however, also reduce $t_{max}(^{15}N)$).

Figure 1 shows contour plots of $(\omega_2(^{13}C^{\alpha}),\omega_3(^1H^N))$-strips at given $^1H^{\alpha}$ chemical shifts from a 3D HA CA N HN spectrum of a 2.5 mM solution of the ^{13}C,^{15}N-doubly-labeled mixed disulfide between glutaredoxin(C14S) and glutathione,[17] which has a molecular weight of 11 kDa. Each pair of peaks encodes the four backbone resonance frequencies of $^1H^{\alpha}$, $^1H^N$, $^{13}C^{\alpha}$, and ^{15}N for a particular residue: the $^1H^{\alpha}$ and $^1H^N$ chemical shifts were directly obtained from the positions of the peak pairs in the 3D spectrum, the center of the peak pair yields the $^{13}C^{\alpha}$ chemical shift, and the separation of the two peaks is equal to twice the offset of the amide nitrogen resonance from the ^{15}N carrier frequency. In the presently studied molecule, all backbone resonances could thus be assigned, with the sole exceptions of the six Gly and the three Pro, for obvious reasons, and Gln 66, which is not observable due to exchange broadening.[18] This result is equivalent to what one could expect to obtain from a 4D HA CA N HN experiment.[11]

In spite of the reduced dimensionality and the ensuing ease of both data processing and optimizing the experimental parameters, the 3D HA CA N HN experiment has thus been shown to retain the full potentialities of a 4D HA CA N HN data set for identification of all intraresidual backbone connectivities. In principle, if the interresidual two-bond scalar couplings $^2J(^{13}C_i^{\alpha},{}^{15}N_{i+1})$ are also included in the derivation of (1), then two pairs of peaks will be expected for each residue i, representing respectively the intraresidual connectivities and the sequential $^{13}C_i^{\alpha} - {}^{15}N_{i+1}$ connectivities. In a 3D HA CA N HN spectrum, both pairs of peaks would be centered about $\Omega(^{13}C_i^{\alpha})$ in the plane belonging to $\Omega(^1H_i^{\alpha})$, with the intraresidual connectivity located at $\Omega(^1H_i^N)$ and split by $\Omega(^{15}N_i)$ and the sequential one at $\Omega(^1H_{i+1}^N)$ and split by $\Omega(^{15}N_{i+1})$. However, due to the smaller magnitude of $^2J(^{13}C_i^{\alpha},{}^{15}N_{i+1})$ when compared to $^1J(^{13}C_i^{\alpha},{}^{15}N_i)$, the sequential connectivities have usually much smaller intensities; in the mixed disulfide of glutaredoxin(C14S) and glutathione, they were observed only for a few residues located in flexible parts of the molecular structure.[19]

Considering that the desired information can also be obtained either by a 4D HA CA N HN experiment or a combination of two 3D triple-resonance experiments, the sensitivity of the 3D HA CA N HN measurement is lower by a factor of $\sqrt{2}$. This results because $\Omega(^{15}N)$ is encoded in the in-phase splitting. However, the fact that peak pairs (rather than single peaks) must be identified in 3D HA CA N HN spectra (Figure 1) greatly facilitates the identification of weak signals, which partly compensates for the intrinsic loss in sensitivity.

When working with smaller molecules, *e.g.*, in protein folding studies with labeled polypeptides or investigations of receptor-bound ligands, similar advantages may result from reducing 3D triple-resonance experiments to two dimensions. It is then recommended to implement the experiments in such a way that $\Omega(^1H^N) \pm \Omega(^{13}C^{\alpha})$ is observed along the heteronuclear frequency axis. Since one has for most non-glycyl residues that $\Omega(^{13}C^{\alpha}) \gg \Omega(^{15}N)$, this results in the appearance of well-separated high-field and low-field regions, which facilitates the spectral analysis. As an illustration, a 2D HN N CA experiment derived from the *ct*-HNNCA scheme of Grzesiek and Bax[20] is presented as supplementary material.

Acknowledgment. Financial support was obtained from the Schweizerischer Nationalfonds (project 31.32033.91) and the Kommission zur Förderung der wissenschaftlichen Forschung (project 2223.1).

Supplementary Material Available: Figure S1, displaying a 2D HN N CA spectrum of the mixed disulphide of *E. coli* gutaredoxin(C14S) with glutathione (2 pages). Ordering information is given on any current masthead page.

(16) Marion, D.; Ikura, K.; Tschudin, R.; Bax, A. *J. Magn. Reson.* **1989**, *85*, 393–399.
(17) Bushweller, J. H.; Aslund, F.; Wüthrich, K.; Holmgren, A. *Biochemistry* **1992**, *31*, 9288–9293.
(18) Bushweller, J. H.; Holmgren, A.; Wüthrich, K. *Eur. J. Biochem.*, submitted for publication.
(19) Bushweller, J. H.; Billeter, M.; Holmgren, A.; Wüthrich, K. *J. Mol. Biol.*, submitted for publication.
(20) Grzesiek, S.; Bax, A. *J. Magn. Reson.* **1992**, *96*, 432–440.
(21) Marion, D.; Ikura, M.; Bax, A. *J. Magn. Reson.* **1989**, *84*, 425–430.
(22) DeMarco, A.; Wüthrich, K. *J. Magn. Reson.* **1976**, *24*, 201–204.
(23) Güntert, P.; Dötsch, V.; Wider, G.; Wüthrich, K. *J. Biomol. NMR* **1992**, *2*, 619–629.
(24) Otting, G.; Wüthrich, K. *J. Magn. Reson.* **1988**, *76*, 569–574.
(25) Shaka, A. J.; Lee, C. J.; Pines, A. *J. Magn. Reson.* **1988**, *77*, 274–293.
(26) Shaka, A. J.; Keeler, J.; Frenkiel, T.; Freeman, R. *J. Magn. Reson.* **1983**, *52*, 335–336.

Reprinted from the Journal of the American Chemical Society, 1989, *111*, 3997.
Copyright © 1989 by the American Chemical Society and reprinted by permission of the copyright owner.

Automated Stereospecific ^{1}H NMR Assignments and Their Impact on the Precision of Protein Structure Determinations in Solution

Peter Güntert, Werner Braun, Martin Billeter, and Kurt Wüthrich*

Contribution from the Institut für Molekularbiologie und Biophysik, Eidgenössische Technische Hochschule—Hönggerberg, CH-8093 Zürich, Switzerland. *Received June 30, 1988*

Abstract: Two sets of constraints on proton–proton distances and dihedral angles, which mimic data that can be obtained from nuclear magnetic resonance experiments in solution, were derived from the crystal structure of the protein basic pancreatic trypsin inhibitor (BPTI). In one of these data sets, all prochiral groups of protons were replaced by pseudoatoms. In the second set, stereospecific assignments were used for all β-methylene groups, all protons of glycine and proline, the methyl groups of valine and leucine, and the ring protons of phenylalanine and tyrosine. Comparison of the BPTI structures calculated from these data with the distance geometry program DISMAN showed that, with otherwise identical distance constraints, the use of stereospecific assignments results in significantly improved precision of the structure determination for the polypeptide backbone as well as the amino acid side chains. The paper further describes the program HABAS, which determines stereospecific assignments by a systematic analysis of tne proton–proton scalar couplings and the intraresidual and sequential proton–proton nuclear Overhauser effects. To investigate to what extent stereospecific assignments could be obtained for a predetermined completeness and precision of the input data set, HABAS was used for test calculations with a standard dipeptide unit and a database derived from a group of high-resolution protein crystal structures. From these data we estimate that with the precision presently achieved for NMR measurements with proteins, stereospecific assignments can be obtained for approximately half of the β-methylene protons. Quite generally, this ratio can be expected to be higher for β-proteins than for those that contain predominantly α-helical secondary structure.

I. Introduction

It has by now been quite widely accepted that as a second method besides X-ray diffraction in single crystals, nuclear magnetic resonance (NMR)[1] in solution can be used for the determination of the complete three-dimensional structure of proteins.[2–6] Presently, considerable effort is directed at improvements of the efficiency as well as the precision of such structure determinations, for example, with the use of ever more sophisticated NMR techniques[6,7] and mathematical methods for the structural interpretation of the NMR data.[6,8–13] It is a fundamental advantage of the NMR method for protein structure determination that it can depend on qualitative experimental constraints on the conformation,[2,10,12] which makes it both robust and efficient in practical applications. Besides experimental limitations, quantitative distance measurements would be intrinsically difficult because the observed NOEs depend not only on the proton–proton distances but also on the effective rotational correlation times,[14–16] which may be variable for different locations in a protein molecule.[6,16] Stereospecific assignments for prochiral groups of protons can yield more precise structures without requirements for more quantitative distance measurements. It had already been demonstrated that the precision of protein solution conformations determined from qualitative NMR constraints can be comparable to that of a refined high-resolution crystal structure,[17] provided that a sufficiently large number of constraints is available, and that as far as possible stereospecific assignments were determined for the prochiral groups of protons.[18] The present paper describes investigations of the improvements in the precision of protein structure determinations that can be anticipated from stereospecific assignments, using input data sets derived from the crystal structure of the protein BPTI. It further introduces the program HABAS, which performs an automated analysis of the experimentally accessible, local NMR parameters to obtain stereospecific ^{1}H NMR assignments before the start of the distance geometry calculations.

The generally used sequential resonance assignment procedure for proteins[2,6] does not yield stereospecific assignments for the individual protons in prochiral groups. To deal with this situation a set of pseudoatoms replacing the prochiral groups was introduced.[19] This is inevitably a compromise, since the use of these pseudoatoms reduces the precision of the experimental conformational constraints.[6] In special situations some stereospecific assignments resulted in the course of the three-dimensional structure determination,[18,20,21] and recently, a procedure for obtaining stereospecific assignments during metric matrix distance geometry calculations was proposed.[22] Conversely, empirical procedures for obtaining stereospecific assignments before the structure determination have also been described, which use

(1) Abbreviations used: NMR, nuclear magnetic resonance; NOE, nuclear Overhauser enhancement; BPTI, basic pancreatic trypsin inhibitor; RMSD, root-mean-square distance; DISMAN, distance geometry program for proteins; HABAS, program for obtaining stereospecific resonance assignments for α- and β-protons in proteins.
(2) Wüthrich, K.; Wider, G.; Wagner, G.; Braun, W. *J. Mol. Biol.* **1982**, *155*, 311–319.
(3) Braun, W.; Wider, G.; Lee, K. H.; Wüthrich, K. *J. Mol. Biol.* **1983**, *169*, 921–948.
(4) Williamson, M. P.; Havel, T. F.; Wüthrich, K. *J. Mol. Biol.* **1985**, *182*, 295–315.
(5) Kline, A. D.; Braun, W.; Wüthrich, K. *J. Mol. Biol.* **1986**, *189*, 377–382.
(6) Wüthrich, K. *NMR of Proteins and Nucleic Acids*; Wiley: New York, 1986.
(7) Ernst, R. R.; Bodenhausen, G.; Wokaun, A. *Principles of Nuclear Magnetic Resonance in One and Two Dimensions*; Clarendon Press: Oxford, U.K., 1987.
(8) Havel, T. F.; Wüthrich, K. *Bull Math. Biol.* **1984**, *46*, 673–698.
(9) Kaptein, R.; Zuiderweg, E. R. P.; Scheek, R. M.; Boelens, R.; van Gunsteren, W. F. *J. Mol.* Biol. **1985**, *182*, 179–182.
(10) Braun, W.; and Gō, N. *J. Mol. Biol.* **1985**, *186*, 611–626.
(11) Clore, G. M.; Gronenborn, A. M.; Brünger, A. T.; Karplus, M. *J. Mol. Biol.* **1985**, *186*, 435–455.
(12) Havel, T. F.; Wüthrich, K. *J. Mol. Biol.* **1985**, *182*, 281–294.
(13) Braun, W. *Q. Rev. Biophys.* **1987**, *19*, 115–157.
(14) (a) Solomon, I. *Phys. Rev.* **1955**, *99*, 559–565. (b) Noggle, J. H.; Schirmer, R. E. *The Nuclear Overhauser Effect*; Academic Press: New York, 1971.
(15) Wagner, G.; Wüthrich, K. *J. Magn. Reson.* **1979**, *33*, 675–680.
(16) Olejniczak, E. T.; Dobson, C. M.; Karplus, M.; Levy, R. M. *J. Am. Chem. Soc.* **1984**, *106*, 1923–1930.
(17) Billeter, M.; Kline, A. D.; Braun, W.; Huber, R.; Wüthrich, K. *J. Mol. Biol.*, in press.
(18) (a) Kline, A. D.; Braun, W.; Wüthrich, K. *J. Mol. Biol.* **1988**, *204*, 675–724. (b) Billeter, M.; Schaumann, Th.; Braun, W.; Wüthrich, K. *Biopolymers*, in press.
(19) Wüthrich, K.; Billeter, M.; Braun, W. *J. Mol. Biol.* **1983**, *169*, 949–961.
(20) Senn, H.; Billeter, M.; Wüthrich, K. *Eur. Biophys. J.* **1984**, *11*, 3–5.
(21) Zuiderweg, E. R. P.; Boelens, R.; Kaptein, R. *Biopolymers* **1985**, *24*, 601–611.
(22) Weber, P. L.; Morrison, R.; Hare, D. *J. Mol. Biol.* **1988**, *204*, 483–487.

0002-7863/89/1511-3997$01.50/0 © 1989 American Chemical Society

manual screening of the spin–spin couplings and the intraresidual and sequential NOEs.[18,23,24] The program HABAS replaces these empirical approaches by an unbiased screening of the local constraints, whereby for the torsion angle χ^1 either a continuous population of all values from 0 to +180°, or of limited ranges near the staggered rotamers, for example, 60 ± 20, 180 ± 20, and −60 ± 20°, can be assumed.[25] In addition to the determination of stereospecific assignments, HABAS analyses the local NMR constraints in terms of allowed regions in local conformation space, rather than individual, discrete points thereof. The starting structures for distance geometry calculations, e.g., using the program DISMAN,[10,13] can then be chosen randomly from this locally constrained conformation space.

In the first part of this paper the study of the influence of stereospecific assignments on the precision of protein structure determination by NMR uses the assumption that stereospecific assignments are available for distinct classes of prochiral groups of protons. While this provides a useful, general guideline, it is further of interest to assess the extent to which stereospecific assignments can be derived, depending on the completeness and precision of the available NMR data. To this end the program HABAS is applied for test calculations using simulations of NMR input data derived from a group of high-resolution crystal structures of small proteins.

II. Test Calculations on the Impact of Stereospecific Assignments on the Precision of a Protein Structure Determination

To investigate the influence of stereospecific resonance assignments on the precision of protein structure determinations by ^{1}H NMR, test calculations with the small globular protein BPTI (58 residues) were carried out. Two sets of conformational constraints were derived from the regularized crystal structure of BPTI, which differed only in stereospecific assignments. In the data set NOST no stereospecific assignments were used and all prochiral groups of protons were represented by pseudoatoms.[19] When preparing the data set WIST it was assumed that stereospecific assignments were available for the following groups: β-methylene protons, α-methylene protons of glycine, γ- and δ-methylene protons of proline, methyl groups of valine and leucine, and ring protons of tyrosine and phenylalanine. The remaining prochiral groups of protons were again represented by pseudoatoms. The data set WIST thus contained stereospecific assignments for 73 of the total of 101 prochiral groups of protons in BPTI. With each of the two constraint sets, four structures were calculated by using the distance geometry program DISMAN.[10] These two groups of structures were then compared with each other and with the regularized crystal structure from which the input data had been obtained.

Simulated Input Data Sets. Because the program DISMAN works with fixed bond lengths and bond angles, the simulated constraint sets for the test calculations with DISMAN were extracted from a structure with standard geometry of the amino acid residues. For this the BPTI crystal structure[26] (code of the Protein Data Bank: 4PTI)[27] was regularized with the program DISMAN, by using 2990 exact distances from the unregularized crystal structure as constraints. The resulting regularized crystal structure, XRAY, with the desired standard geometry and all hydrogen atoms attached, coincided closely with the unregularized structure, with RMSD values[28] of 0.27 Å for the backbone atoms and 0.35 Å for all heavy atoms. (This difference is small compared to the difference between the two crystal forms I and II of BPTI, where the RMSD for the backbone atoms is 0.4 Å.[29]) The regularized structure contained seven violations of steric constraints greater than 0.2 Å; the maximal violation was 0.32 Å. The advantage of using a regularized structure as the source of the distance constraints is that this ensures a clearcut distinction between possible effects arising either from the limited accuracy of the simulated data or from distortions of the standard geometry.[10,13]

Distance constraints were derived from the regularized crystal structure following the strategy previously used for test calculations without stereospecific assignments,[8,12] whereby for the steric constraints the standard parameters employed with DISMAN were used.[10] All interresidual proton–proton distances shorter that 4.0 Å were considered, as well as the intraresidual distances shorter than 4.0 Å from backbone amide or α-protons to the side-chain protons attached to C^γ or beyond. These precise distances were substituted by corresponding upper limits on the distances in order to mimic a typical NMR input for a structure calculation.[6] For the long-range constraints, the upper limit was 4.0 Å throughout. For the intraresidual constraints and for constraints between protons in sequentially adjacent residues, upper limits of 2.5, 3.0, 3.5, and 4.0 Å were used, where the limit <2.5 Å applies to all distances shorter than 2.5 Å, the limit <3.0 Å to all distances in the range from 2.5 to 3.0 Å, etc. Whenever a prochiral group of protons was represented by a centrally located pseudoatom, the appropriate correction was added to these upper bounds.[6,19] A survey of the distance constraints used is afforded by Table I. [In addition, the constraints on the intraresidual distances $d_{N\beta2}(i, i)$

Table I. Survey of the Distance Constraints Used in the Input for the Distance Geometry Calculations with BPTI and Analysis of the Residual Constraint Violations

	NOST			WIST	
		violations[b]			
type of constraint[a]	no. of constraints	>0.2 Å	>0.5 Å	no. of constraints	violations[b] >0.2 Å
intraresidual	78	0.5	0.3 (0.76)	88	0.3 (0.22)
neighbor residue	172	1.3	0.5 (0.93)	211	1.5 (0.31)
long range	379	4.3	1.0 (0.80)	528	2.0 (0.37)
steric		6.3	0.3 (0.58)		2.5 (0.43)

[a] *Neighbor residue* distance constraints are between atoms in sequentially neighboring residues. All other interresidual constraints are *long range*. The steric lower limit distance constraints are those imposed by the van der Waals volumes of the atoms as described in ref 10. [b] Four structures were calculated for each of the two input data sets, NOST and WIST. Among the four structures the average number of residual violations exceeding the indicated limit is given, and the values in parentheses are the largest individual residual violations.

Table II. Survey of the Dihedral Angle Constraints Used in the Input for the Distance Geometry Calculations with BPTI[a] and Analysis of the Residual Violations

	NOST		WIST	
type of constraint[b]	no. of constraints	violations[c] >5°	no. of constraints	violations[c] >5°
$0° < \Delta\phi \leq 90°$	32	0.5 (10.8)	32	
$90° < \Delta\phi \leq 300°$	9		9	
$0° < \Delta\psi \leq 90°$	35	0.3 (16.4)	35	
$90° < \Delta\psi \leq 300°$	6		6	
$0° < \Delta\chi^1 \leq 90°$	21	0.5 (6.8)	38	
$90° < \Delta\chi^1 \leq 300°$	20	0.3 (5.5)	3	

[a] These dihedral angle constraints resulted from a combined analysis of the spin–spin coupling constants $^3J_{HN\alpha}$, $^3J_{\alpha\beta2}$, and $^3J_{\alpha\beta3}$, the intraresidual distance constraints $d_{N\beta2}(i, i)$ and $d_{N\beta3}(i, i)$, and the sequential distance constraints $d_{\alpha N}$, d_{NN}, $d_{\beta2N}$, and $d_{\beta3N}$ using the program HABAS. [b] $\Delta\phi$, $\Delta\psi$, and $\Delta\chi^1$ indicate the full size of the allowed dihedral angel intervals. [c] Four structures were calculated for each of the two input data sets, NOST and WIST. Among the four structures the average number of residual violations exceeding 5° is given, and in parentheses, the largest of these violations is indicated.

(23) Hyberts, S.; Märki, W.; Wagner G., *Eur. J. Biochem.* **1987**, *164*, 625–635.
(24) (a) Arseniev, A.; Schultze, P.; Wörgötter, E.; Braun, W.; Wagner, G.; Vašák, M.; Kägi, J. H. R.; Wüthrich, K. *J. Mol. Biol.* **1988**, *201*, 637–657. (b) Wagner, G.; Braun, W.; Havel, T. F.; Schaumann, T.; Gō, N.; Wüthrich, K. *J. Mol. Biol.* **1987**, *196*, 611–639.
(25) (a) Ponder, J. W.; Richards, F. M. *J. Mol. Biol.* **1987**, *193*, 755–791. (b) McGregor, M. J.; Islam, S. A.; Sternberg, M. J. E. *J. Mol. Biol.* **1987**, *198*, 295–310.
(26) Marquart, M.; Walter, J.; Deisenhofer, J.; Bode, W.; Huber, R. *Acta Crystallogr. B* **1983**, *39*, 480–490.
(27) Bernstein, F. C.; Koetzle, T. F.; Williams, G. J. B.; Meyer, E. F., Jr.; Brice, M. D.; Rodgers, J. R.; Kennard, O.; Shimanouchi, T.; Tasumi, M. *J. Mol. Biol.* **1977**, *112*, 535–542.
(28) (a) Nyburg, S. C. *Acta Crystallogr. B* **1974**, *30*, 251–253. (b) McLachlan, A. D. *J. Mol. Biol.* **1979**, *128*, 49–79.
(29) Wlodawer, A.; Walter, J.; Huber, R.; Sjolin, L. *J. Mol. Biol.* **1984**, *180*, 301–329.

Table III. Average and Standard Deivations of the Pairwise RMSD Values among the BPTI Structures Used in This Study

structures compared[a]	RMSD[b] backbone (N, C^{α}, C′), Å			RMSD[b] all heavy atoms, Å		
	3–55	β-sheet	α-helix	3–55	β-sheet	α-helix
XRAY/NOST	0.9 ± 0.2	0.2 ± 0.1	0.7 ± 0.3	1.6 ± 0.3	1.4 ± 0.3	1.3 ± 0.3
XRAY/WIST	0.4 ± 0.1	0.2 ± 0.1	0.2 ± 0.1	1.1 ± 0.1	1.1 ± 0.1	0.7 ± 0.1
NOST/WIST	1.0 ± 0.3	0.3 ± 0.1	0.7 ± 0.3	1.7 ± 0.2	1.3 ± 0.3	1.4 ± 0.2
NOST/WIST	1.1 ± 0.2	0.3 ± 0.1	0.9 ± 0.3	1.8 ± 0.3	1.4 ± 0.2	1.4 ± 0.3
WIST/WIST	0.5 ± 0.1	0.2 ± 0.1	0.2 ± 0.1	1.2 ± 0.1	0.9 ± 0.2	0.8 ± 0.1

[a] XRAY is the structure 4PTI from the Protein Data Bank[26,27] after regularization. NOST and WIST are the two groups of four structures calculated from the corresponding input data sets (see text). For example, the comparisons XRAY/NOST and NOST/WIST yield 4 and 16 pairwise RMSDs, respectively. [b] RMSDs were calculated for the residues 3–55, rather than for the complete structure (residues 1–58), to exclude chain termination effects. The β-sheet and the α-helix comprise the residues 18–35 and 48–55, respectively.

and $d_{N\beta3}(i, i)$ are included implicitly in the constraints on the dihedral angles ϕ, ψ, and χ^1 listed in Table II].

Allowed intervals for the dihedral angles ϕ, ψ, and χ^1 were determined by the program HABAS (See section III below. Note that in this application HABAS has not been used to obtain stereospecific assignments. The determinations of allowed dihedral angle intervals is a part of the program that is independent from the stereospecific assignment part. Restrictions on the allowed dihedral angle ranges may result even if no unambiguous stereospecific assignments can be derived from the available data). For this, spin–spin coupling constants were calculated from the regularized crystal structure by using Karplus-type relations calibrated for use with proteins.[30,31] To mimic the precision of a typical NMR experiment, these J values were taken to define the center of an interval of half-width 2.0 Hz. These intervals were then combined with the sequential distance constraints $d_{\alpha N}$, d_{NN}, $d_{\beta 2N}$, and $d_{\beta 3N}$ and the intraresidual constraints $d_{N\beta 2}(i, i)$ and $d_{N\beta 3}(i, i)$ (see ref 6 for the notation used) to define allowed ranges for the dihedral angles. For the residues for which stereospecific assignments had been assumed in WIST, these allowed ranges were further confined so as to include only the values that were compatible with the correct assignments. A survey of all dihedral angle constraints thus obtained is afforded by Table II. In addition to the data in Tables I and II, the input for the DISMAN calculations contained three constraints for each of the disulfide bonds, using the parameters described by Williamson et al.[4]

Results. The DISMAN program has several options for generating starting structures. The option used here for the calculations with both NOST and WIST was to choose the variable dihedral angles within those limits that are allowed by the $d_{\alpha N}$, d_{NN}, and ϕ distance and dihedral angle constraints. Different starting structures were generated for the calculations with NOST and WIST, respectively. Convergent structures for the constraint sets NOST and WIST were selected according to their residual constraint violations. In Tables I and II it is shown that the four final structures of each group satisfy nearly all distance and dihedral angle constraints perfectly. The number of violations of distance constraints by more than 0.2 Å or angle constraints by more than 5° is always small relative to the total number of constraints, whereby the structures obtained from the data set WIST converged slightly better than the structures obtained without stereospecific assignments.

In the following, two criteria are used to evaluate the calculated structures. One is the average RMSD relative to the regularized crystal structure from which the input data were taken, which indicates how faithfully this structure was reproduced. The second criterion is the average RMSD among the four structures in each group, which indicates how precisely the atom positions are determined and further provides information on the sampling by the DISMAN program.

RMSD values were calculated separately for the backbone, and for the complete structure including the amino acid side chains (Table III). Table III shows that the improvement of the structures with stereospecific assignments is particularly pronounced for the polypeptide backbone. The backbone atoms of the WIST structures are significantly nearer to the corresponding atoms in the regularized crystal structure than those of the NOST structures, with an RMSD value of 0.4 Å as compared to 0.9 Å. The RMSDs among the different structures of the group NOST are bigger than among the different structures WIST. The RMSD values for all heavy atoms show a similar reduction, 0.6 Å, between the structures NOST and WIST. As the absolute RMSD values of the backbone atoms are smaller than those for all heavy atoms, the relative improvement with the use of stereospecific assignments is particularly striking for the backbone. The differences between the average RMSD values for the WIST and NOST structures are in all comparisons larger than the standard deviations for the pairwise RMSDs among the individual structures within each group, which emphasizes that the improvements achieved with the stereospecific assignments are indeed significant. To study the influence of stereospecific assignments on the regular secondary structures, separate RMSD values were calculated for the β-sheet region 18–35 and for the α-helical region 48–55. The improvement in the α-helical region is especially pronounced. The RMSD values dropped from 0.9 Å among all NOST structures to 0.2 Å among all WIST structures (Table III). The polypeptide backbone fold of the NOST structures in the β-sheet region is already well-defined, and here the improvement with stereospecific assignments is only marginal. When the RMSDs for all heavy atoms are considered, one finds similar improvements with stereospecific assignments for the β-sheet, the α-helix, and the complete molecule (Table III).

Visual impressions of the results in Table III are afforded by Figures 1–3, which show molecular models produced with the molecular graphics program CONFOR.[32] Figure 1A shows the regularized crystal structure (thick line) superimposed with the four NOST structures. Figure 1B shows the spread of the four WIST structures (thin lines) around the regularized crystal structure. The distribution of the structures calculated with the NOST and WIST input data around the regularized crystal structure, from which the input data were taken, is consistent with unbiased sampling (Figure 1 and Table III). The regions that contribute most to the improvements of the backbone conformations in the WIST structures with respect to the NOST structures are the segments 5–10, the β-turn region near 25, the loop at 36–40, and the α-helical region 48–55. In Figure 2 the side chains are also shown. While the improvement in the backbone of the β-sheet is only marginal, some side chains that were not well-defined by the NOST data set were significantly more tightly constrained, e.g., Ile-18, Ile-19, Leu-29, Phe-33, and Tyr-35. The aromatic side chains of Tyr-21, Phe-22, and Tyr-23 are already quite well confined in the NOST structures, which is probably largely due to the internal packing restrictions.[6] For these side chains the improvement by the stereospecific assignments is less pronounced.

Figure 3 presents two direct comparisons of the four structures NOST with the four structures WIST. Figure 3A illustrates the improved precision of the backbone structure determination by

(30) Pardi, A.; Billeter, M.; Wüthrich, K. *J. Mol. Biol.* **1984**, *180*, 741–751.

(31) DeMarco, A.; Llinás, M.; Wüthrich, K. *Biopolymers* **1978**, *17*, 617–636.

(32) Billeter, M.; Engeli, M.; Wüthrich, K. *Mol. Graphics* **1985**, *3*, 79–83, 97–98.

4000 J. Am. Chem. Soc., Vol. 111, No. 11, 1989

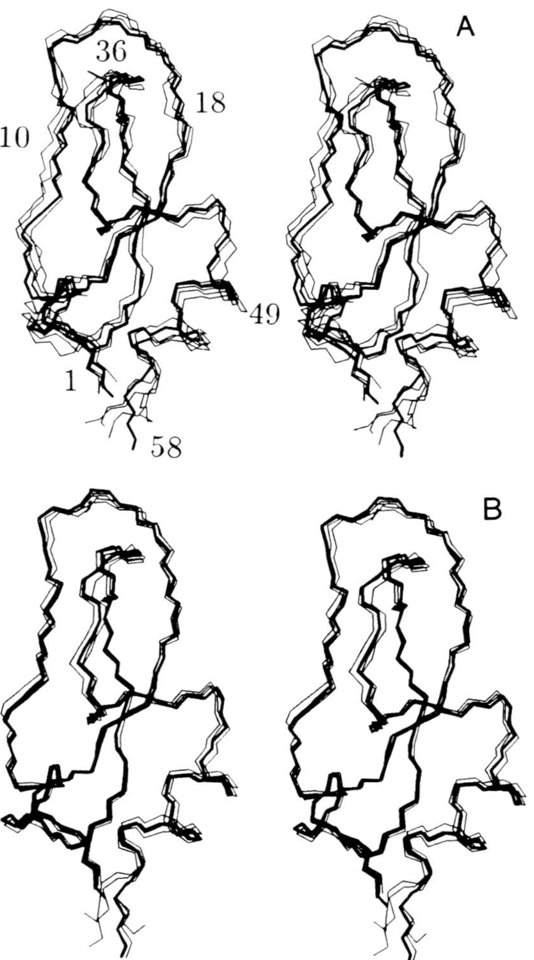

Figure 1. Stereo views affording comparisons of the regularized crystal structure of BPTI (thick line) with (A) four structures calculated from the data set NOST (thin lines) and (B) four structures calculated from the data set WIST (thin lines). The bonds connecting the backbone atoms N, C^{α} and C′ are shown, and the structures were superimposed for minimal pairwise RMSD of these atoms between the individual calculated structures and the regularized crystal structure. In (A) the locations of selected residues are indicated.

the four structures WIST. It further reveals a tendency of the BPTI structures calculated from the data with stereospecific assignments to be somewhat contracted relative to NOST, i.e., to have slightly reduced global dimensions. Figure 3B identifies in a clear fashion the sequence regions with least well determined spatial backbone structure. Thereby it is quite striking that the regions near residues 27 and 37 have not only the largest dispersion among the four NOST structures, but show also the most pronounced improvement when stereospecific assignments are used.

III. The Program HABAS for Automated Determination of Stereospecific ^{1}H NMR Assignments

The program HABAS systematically scans experimentally determined sets of structural constraints in proteins in order to obtain stereospecific assignments for β-methylene protons in amino acid side chains, and for the γ-methyl groups of valine. It is applied before the start of the structure calculations and uses experimental data corresponding to local conformational constraints that are available after determination of the sequence-specific ^{1}H NMR

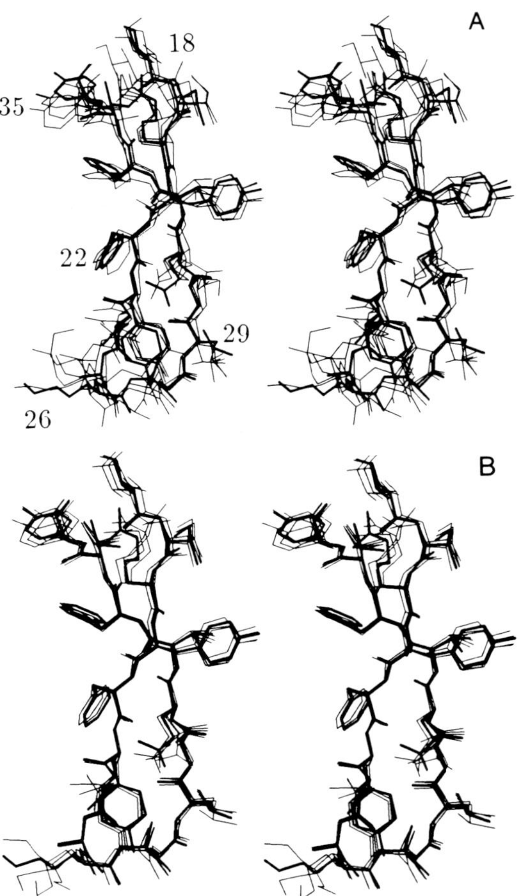

Figure 2. Same as Figure 1, except that all heavy atoms of the β-sheet formed by the residues 18–35 in BPTI are shown. In (A) the regularized crystal structure (thick line) is shown superimposed with four structures calculated from NOST and in (B) with four structures calculated from WIST. In (A) the locations of some residues are identified.

assignments.[6] No medium-range or long-range NOEs are considered. For a particular residue i, HABAS analyzes the constraints that depend only on the three dihedral angles ϕ_i, ψ_i, and χ_i^1 (Figure 4). This includes steric constraints, allowed ranges for proton–proton distances, relations between pairs of such distances, and spin–spin coupling constants. The following describes how these input data are handled in the preparation of the input. (HABAS is available upon request addressed to the authors.)

To describe steric constraints, a repulsive core radius is assigned to each atom in the polypeptide chain. A pair of atoms violates a steric constraint if the distance between the two atoms is smaller than the sum of their repulsive core radii. The same core radii were used as in the program DISMAN.[10] Upper bounds on ^{1}H–^{1}H distances are obtained from the corresponding ^{1}H–^{1}H NOEs, where HABAS makes use of the constraints on the intraresidual distances $d_{N\beta2}(i, i)$ and $d_{N\beta3}(i, i)$ and the sequential distances $d_{\alpha N}$, d_{NN}, $d_{\beta2N}$, and $d_{\beta3N}$ (see ref 6 for the notation used). In addition, for valine the intraresidual and sequential distances between amide protons and the γ-methyl groups are also considered.

Besides the constraints on individual distances, HABAS also accepts relations between the two distances from a proton A to the two protons B and B′ of a methylene group. The relational constraint is fulfilled if $d(A, B) > d(A, B') + \Delta d$, where Δd is an arbitrary parameter (usually $\Delta d = 0$). This option of HABAS takes into account that relative values for two NOEs are often

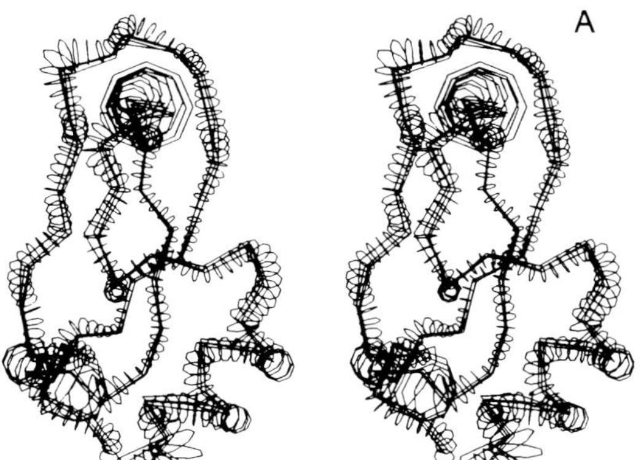

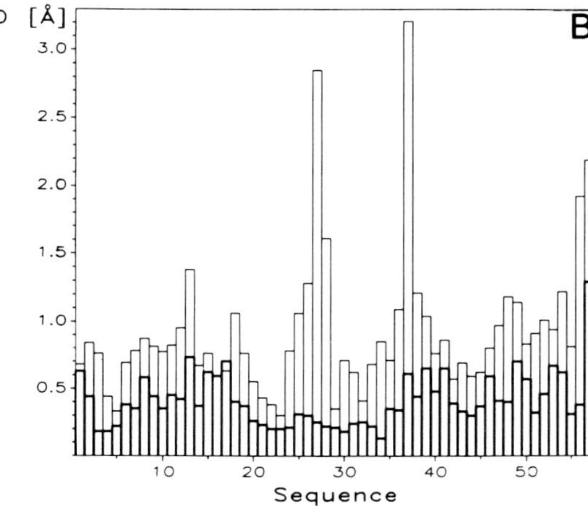

Figure 3. (A) Stereo view affording a comparison of the two groups of four BPTI structures calculated from the data sets NOST and WIST, respectively. All structures were superimposed so as to minimize the RMSD of the C^α atoms with respect to those in the regularized crystal structure. The four structures WIST are represented by straight lines connecting the C^α positions. The structures NOST are represented by a single set of circles, with five circles per residue. The centers of these circles were obtained by calculating the average of the C^α positions in the four structures NOST and fitting a spline function through these average positions. The planes of the circles are perpendicular to this spline function. At the position of a given C^α atom the radius of the circle is equal to the largest of the four distances between this C^α atom in the individual structures and in the average, andd the radii of the four circles between two neighboring C^α positions are smoothly interpolated. (B) Plot versus the amino acid sequence of the largest of the four displacements, D, between a given C^α atom in the individual structures and the corresponding average for the structures calculated from the data set NOST (thin line) and those from WIST (thick line). The values for the displacements plotted here for NOST correspond to the radii of the circles around the C^α positions in (A). For the C-terminal residue Ala-58 the displacements among the four structures NOST are very large, and they were therefore not included in the drawings.

more easily accessible than estimates of absolute distance values.

To determine allowed intervals for individual dihedral angles, HABAS makes use of the following Karplus-type relations, which were calibrated with experimental data in peptides and proteins.[30,31]

$$^3J_{HN\alpha}(\phi) = 6.4 \cos^2 (\phi - 60°) - 1.4 \cos (\phi - 60°) + 1.9 \quad (1)$$

$$^3J_{\alpha\beta_2}(\chi^1) = 9.5 \cos^2 (\chi^1 - 120°) - 1.6 \cos (\chi^1 - 120°) + 1.8 \quad (2)$$

$$^3J_{\alpha\beta_3}(\chi^1) = 9.5 \cos^2 \chi^1 - 1.6 \cos \chi^1 + 1.8 \quad (3)$$

H H$^\alpha$ O
| ϕ | ψ ||
$\cdots C'_{i-1}$ — N_i — C^α_i — C'_i — $N_{i+1} \cdots$
|| χ^1 |
O H^{β_2}— C^β — H^{β_3} H

Figure 4. Dipeptide segment examined by the program HABAS in each step of the calculation. A residue with a β-methylene group is shown.

The experimental data, $^3J^{exp}$, are supplemented with an arbitrary parameter, Δ^3J, defining the precision of the experiment (Δ^3J is usually chosen in the range 1.0–2.0 Hz). The corresponding dihedral angle is then constrained within the range that corresponds to the interval of spin–spin coupling constants from ($^3J^{exp} - \Delta^3J$) to ($^3J^{exp} + \Delta^3J$). Note that eq 3 applies for the β-methine proton in Val and eq 2 for the β-methine proton in Ile or Thr.

To determine stereospecific assignments for a pair of β-methylene protons, HABAS goes through a process that corresponds to two subsequent, independent grid searches of the three-dimensional space defined by ϕ, ψ, and χ^1 for conformations that fulfill all experimental constraints. The two grid searches are for the two possible stereospecific assignments, and for each assignment, the number of conformations that are consistent with all constraints is computed. If all those conformations fulfil the constraints for only one of the two possible stereospecific assignments, then this stereospecific assignment is considered to be unambiguously identified by the input data used. In the present form, the program applies a grid search with steps of $\Delta\phi = \Delta\psi = \Delta\chi^1 = 10°$. The values for proton–proton distances and spin–spin couplings, which are needed for the grid search, are obtained from the peptide segments Ala-Ser-Ala, representing all non-proline residues, and Ala-Pro-Ala in the ECEPP standard geometry.[33,34] Of the 36^3 = 46 656 conformations generated in the course of this grid search, many are not allowed due to steric hindrance (see ref 10 for the core radii used). Therefore, only 13 050 conformations need to be checked against the experimental data.

Valine is treated as a special case of the non-proline residues. In the place of C^γ and $H^{\beta 2}$ two pseudo atoms $Q^{\gamma 1}$ and $Q^{\gamma 2}$ representing the γ-methyl groups[19] are attached to C^β. The scalar coupling between the α-proton and the β-methine proton is analyzed with eq 3, and the resulting information on χ^1 is combined with the distance constraints for the intraresidual and sequential distances $d_{NQ^{\gamma 1}}(i, i)$, $d_{NQ^{\gamma 2}}(i, i)$, $d_{N\beta}(i, i)$, $d_{Q^{\gamma 1}N}(i, i + 1)$, $d_{Q^{\gamma 2}N}(i, i + 1)$, $d_{\alpha N}$, d_{NN}, and $d_{\beta N}$ to determine the stereospecific assignments of the γ-methyl groups.

(33) Momany, F. A.; McGuire, R. F.; Burgess, A. W.; Scheraga, H. A. *J. Phys. Chem.* **1975**, *79*, 2361–2381.

(34) Nêmethy, G.; Pottle, M. S.; Scheraga, H. A. *J. Phys. Chem.* **1983**, *87*, 1883–1887.

(35) Teeter, M. M. *Proc. Natl. Acad. Sci. U.S.A.* **1984**, *81*, 6014–6018.

(36) Leijonmarck, M.; Liljas, A. *J. Mol. Biol.* **1987**, *195*, 555–579.

(37) Carter, C. W., Jr.; Kraut, J.; Freer, S. T.; Xuong, N.-H.; Alden, R. A.; Bartsch, R. G. *J. Biol. Chem.* **1974**, *249*, 4212–4225.

(38) Blundell, T. L.; Pitts, J. E.; Tickle, I. J.; Wood, S. P.; Wu, C.-W. *Proc. Natl. Acad. Sci. U.S.A.* **1981**, *78*, 4175–4179.

(39) Almassy, R. J.; Fontecilla-Camps, J. C.; Suddath, F. L.; Bugg, C. E. *J. Mol. Biol.* **1983**, *170*, 497–527.

(40) Mathews, F. S.; Argos, P.; Levine, M. *Cold Spring Harbor Symp. Quant. Biol.* **1972**, *36*, 387–395.

(41) Bourne, P. E.; Sato, A.; Corfield, P. W. R.; Rosen, L. S.; Birken, S.; Low, B. W. *Eur. J. Biochem.* **1985**, *153*, 521–527.

(42) Bode, W.; Epp, O.; Huber, R.; Laskowski, M., Jr.; Ardelt, W. *Eur. J. Biochem.* **1985**, *147*, 387–395.

(43) Matsuura, Y.; Takano, T.; Dickerson, R. E. *J. Mol. Biol.* **1982**, *156*, 389–395.

(44) Watenpaugh, K. D.; Sieker, L. C.; Jensen, L. H. *J. Mol. Biol.* **1980**, *138*, 615–633.

(45) Guss, J. M.; Harrowell, P. R.; Murata, M.; Norris, V. A.; Freeman, H. C. *J. Mol. Biol.* **1986**, *192*, 361–387.

4002 *J. Am. Chem. Soc., Vol. 111, No. 11, 1989*

With its grid search approach, HABAS in principle never misses possible stereospecific assignments, nor will it make erroneous assignments. In practice, however, if one uses "hard limits" and discards all conformations that cause any violation of at least one constraint, experimental errors and the internal mobility of proteins may distort the results. Initial experience showed that with these hard limits one may draw incorrect conclusions if the input data contain small inconsistencies, or if the covalent structure deviates somewhat from the standard ECEPP geometry. Therefore, the program contains a "soft limit" option, with which small violations of the constraints can be tolerated. The size of these allowed violations can be specified by the user and adjusted to the quality of the experimental data.

HABAS goes through a complete search with all values for the dihedral angles χ^1. In some of the empirical approaches to stereospecific assignments using local constraints,[18,24] it was assumed that χ^1 would be within narrow ranges about the staggered rotamers with $\chi^1 = -60$, 60, and 180°. To enable investigations on the impact of this assumption, HABAS includes the option to produce a separate output corresponding to a search of the χ^1 values over a range of ±20° about the three staggered rotamers. Comparison of the results obtained with and without this restriction showed that it does not significantly improve the outcome of the analysis, but that such restricted screening of the χ^1 values may even lead to inconsistencies and erroneous stereospecific assignments (see section IV and Table VI).

As a side product of the grid search for obtaining stereospecific resonance assignments, HABAS yields constraints on the torsion angles ϕ, ψ, and χ^1 which are compatible with all the local NOE distance constraints and the measured spin–spin coupling constants. If the experimental data are not sufficient for obtaining stereospecific assignments, the program still identifies the smallest intervals for the three dihedral angles, which include all conformations that are consistent with the structural constraints for either of the two possible stereospecific assignments. Obviously, these constraints are redundant with the experimental constraints from which they are computed. However, there are indications that the use of these supplementary constraints on the dihedral angles improves convergence for the first stages of structure calculations with the program DISMAN. For example, such dihedral angle constraints were used in section II for the test calculations with BPTI.

The program was applied with a data set derived from the regularized crystal structure of BPTI (see footnotes to Table IV). Table IV illustrates the output format of HABAS and presents the different types of results that one may obtain in applications of the program. For both possible stereospecific assignments of each prochiral group of protons the number of allowed conformations, the allowed values of the dihedral angles ϕ, ψ, and χ^1, and for each dihedral angle the smallest interval containing all these allowed values are given. For the first example listed in the table, Tyr-10, an unambiguous stereospecific assignment was obtained, since $N_r = 0$ (see following section). Pro-13 was stereospecifically assigned, as were all other prolines. Quite generally, since the dihedral angles ϕ and χ^1 are fixed in proline, stereospecific assignments for βCH_2 can be obtained from a minimum of experimental constraints, e.g., when the spin–spin coupling constants $^3J_{\alpha\beta2}$ and $^3J_{\alpha\beta3}$ are available, or from the relative intensities of the NOEs from H^α to the two β-protons, using that in Pro the distance H^α–$H^{\beta2}$ is always longer than the distance H^α–$H^{\beta3}$. Phe-22 and Phe-33 are examples of residues for which no stereospecific assignments were obtained, since $N_r \neq 0$. For Phe-22 the allowed values for the χ^1 angle constitute two nonoverlapping, well-separated intervals, each of which corresponds to one of the two stereospecific assignments for βCH_2. In this situation, if one can determine by independent, additional procedures that χ^1 falls into only one of the two separated intervals (in Phe-22 either near −120° or near +90°), the stereospecific assignments can be determined from these additional measurements (The additional data would usually be longer range NOE distance constraints). Val-34 demonstrates that stereospecific assignments can also be obtained for the γCH_3 groups of valine. Thr-54 was included to illustrate that for residues with a β-methine proton, the values of the dihedral angles ϕ, ψ, and χ^1 that are consistent with all constraints can be determined with the program HABAS, even though in this case no problem of stereospecific assignments must be solved.

Table IV. HABAS Results for Selected Residues of BPTI from a Test Calculation Using Input Data Derived from the Regularized Crystal Structure 4PTI[a,26,27]

residue[b]	N_i[c]	N_r[c]		allowed dihedral angle values[d] (-180 … -90 … 0 … 90 …)	HABAS constraints on ϕ, ψ, and χ^1[e]
<u>Tyr 10</u>	131	0	ϕ :	/////	-125°...-75°
			ψ :	////////////	55°...175°
			χ^1 :	// … /	165°..-165°
<u>Pro 13</u>	9	0	ψ :	/////// … //	135°...-35°
Phe 22	167	65	ϕ :	\XXXXXXXX … X	55°...-75°
			ψ :	//XXXX	115°...175°
			χ^1 :	\\ … ////	65°..-125°
Phe 33	262	88	ϕ :	XXXXX XXX// … XXX//	55°...-75°
			ψ :	//XXXX	115°...175°
			χ^1 :	\\ … \\\ ////	15°..-125°
<u>Val 34</u>	68	0	ϕ :	////	-95°...-55°
			ψ :	///////	55°...125°
			χ^1 :	// … //	155°..-165°
Thr 54	50		ϕ :	/////	-135°...-85°
			ψ :	///	-75°...-45°
			χ^1 :	/////	-95°...-45°

[a] For the distances $d_{N\beta2}(i, i)$, $d_{N\beta3}(i, i)$, $d_{\alpha N}$, d_{NN}, $d_{\beta2N}$, and $d_{\beta3N}$, as well as for the pairs of distances $d_{N\alpha}(i, i)$ for glycine, and $d_{NCH_3}(i, i)$ and $d_{CH_3N}(i, i+1)$ for valine (see ref 6 for the notation used), upper limits were extracted from the regularized crystal structure whenever the distances were shorter than 4.0 Å. The values 2.5, 3.0, 3.5, or 4.0 Å were used as upper limits, where the limit <2.5 Å replaced all distances shorter than this value, <3.0 Å all distances in the range 2.50–2.99 Å, etc. Spin–spin coupling constants $^3J_{HN\alpha}$, $^3J_{\alpha\beta2}$, and $^3J_{\alpha\beta3}$ were calculated from the crystal structure by using eq 1–3, and Δ^3J was always set to 2.0 Hz. Violations of distance constraints and steric constraints up to 0.1 Å, and of coupling constant constraints up to 0.5 Hz, were tolerated (soft limit option; see text). [b] For underlined residues stereospecific assignments i for βCH_2, or for $\beta CH(CH_3)_2$ in valine, were obtained. [c] The two possible stereospecific assignments are i and r. N_i and N_r are the numbers of conformations found in the grid search that fulfil the constraints for i and r, respectively. [d] /// indicates values that are allowed for the stereospecific assignment i, \\\ is the same for the reversed stereospecific assignment r, and X indicates values that are allowed for both stereospecific assignments. [e] For each dihedral angle the two numbers are the bounds enclosing the smallest interval that contains all allowed values. When a stereospecific assignment was obtained by HABAS, this interval includes only the values that are compatible with this assignment. Otherwise, for example, for Phe-22 and Phe-33, this interval extends over all values that would be compatible with either of the two possible stereospecific assignments.

IV. Extent to Which Stereospecific Assignments in Proteins Can Be Determined

Any approach to the determination of stereospecific assignments for prochiral groups of protons will be limited on several different levels. The most obvious limitation in practice arises if the chemical shifts of the different prochiral protons are accidentally degenerate, so that they cannot be individually observed in the NMR experiments, or if there is high rotational mobility about the single bonds. These situations are not explicitely considered here, but it should be kept in mind that their occurrence will always lower the percentage of accessible stereospecific assignments relative to the results of test calculations, which assume that the prochiral protons can be individually observed and that one deals with immobilized prochiral groups.

With methods such as HABAS, which use exclusively local constraints, the stereospecific assignment can in principle always be obtained, provided that a sufficient amount of exact data is available and that the assumption of standard geometry is strictly valid. In the present practice of NMR studies with proteins, however, it is not easy, for example, to obtain reliable, nontrivial lower bounds on ^1H–^1H distances from NOESY experiments, to

Table V. Results of a Systematic Test of the Program HABAS with the Dipeptide Segment of Figure 4

constraints[a]	Δ^3J,[b] Hz	fraction (%) of the conformations with[c] $N_w = 0$	$N_c \geq 10N_w$	$N_c \geq 2N_w$	$N_w > N_c$
A	1.0	73	74	82	12
	2.0	52	52	70	12
	3.0	32	37	60	16
B	1.0	81	82	87	8
	2.0	57	62	80	8
	3.0	42	50	72	12

[a] In the data set A it was assumed that all ^1H–^1H distances used by HABAS (see section III) are constrained in the intervals between the sum of the two core radii and an upper limit of 2.5, 3.0, 3.5, and 4.0 Å, where for each distance the shortest constraint that includes its actual value is used. The spin–spin coupling constraints $^3J_{HN\alpha}$, $^3J_{\alpha\beta2}$, and $^3J_{\alpha\beta3}$ were derived from the molecular conformation by using eq 1–3. In set B the two constraints with the allowed ranges extending from the sum of the two core radii to 3.5 and 4.0 Å, respectively, were replaced by 2.5 Å < d < 3.5 Å and 3.0 Å < d < 4.0 Å, and a new lower bound, d > 4.0 Å, was used for all distances longer than this limit. All other constraints were the same as in A. [b] Δ^3J defines the assumed accuracy of the ^3J measurements, with the allowed range extending from ($^3J - \Delta^3J$) to ($^3J + \Delta^3J$). [c] The result of the grid search, which was conducted in steps of 10° over the range from −180 to 180° for all three dihedral angles ϕ, ψ, and χ^1: N_c is the number of sterically allowed conformations for which the *correct* stereospecific assignment was obtained, and N_w is the corresponding number for the reverse, *wrong* stereospecific assignment.

measure upper distance bounds more accurately than within approximately 0.2 Å, or to measure ^3J coupling constants more accurately than to ±1.0 Hz. As a consequence, when using conformational constraints with the precision that is accessible in NMR measurements with proteins, the occurrence of certain local conformations may preclude unambiguous determination of stereospecific assignments. To get an estimate of the percentage of assignments that can be expected with real experimental data, we used two kinds of test calculations. In the first test, the polypeptide segment of Figure 4 was subjected to grid searches using different assumptions about the precision of the experimental data, and the percentage of the conformations enabling unambiguous stereospecific assignments was evaluated. In the second approach, HABAS was used with a database derived from a selection of high-resolution protein structures in single crystals and in solution.

Table V lists the results obtained by systematic screening of all sterically allowed conformations of the peptide segment in Figure 4 with two different sets of constraints. The constraint set A corresponds to data that can presently routinely be obtained from NMR experiments.[6] In the set B, more stringent constraints were introduced (for details see footnotes to Table V). Constraint sets of the types A and B were derived from each of the 13 050 sterically allowed conformations of the peptide in a grid search with 10° intervals for the three dihedral angles ϕ, ψ, and χ^1, and the number of allowed conformations for the two possible stereospecific assignments for βCH_2, N_c and N_w, respectively, were calculated by HABAS. Table V shows that the constraints chosen were not sufficient to establish stereospecific assignments for all of the 13 050 conformations. For example, using the constraint set A with Δ^3J = 2.0 Hz, one gets unambiguous stereospecific assignments for the βCH_2 group (with $N_w = 0$) for just over 50% of the conformations. Somewhat higher percentages of stereospecific assignments were obtained if Δ^3J in A was reduced to 1.0 Hz, or when more stringent NOE distance constraints were assumed with the data set B (Table V). For approximately 10% of the conformations (those with $N_w > N_c$) the program had a tendency to indicate the wrong stereospecific assignment, but there was not a single case where unambiguous evidence for an erroneous assignment (with $N_c = 0$) was obtained.

Table V provides an additional result of interest with respect to the criteria to be used in defining unambiguous stereospecific assignments by HABAS. It shows that similar percentages of conformations with $N_w = 0$ or $N_c \geq 10N_w$ are obtained. Therefore, the determination of stereospecific assignments can be based on the more stringent criterion that $N_w = 0$ without running the risk of substantial reduction in the extent of the assignments achieved.

The results of Table V can be applied to proteins only in an indirect fashion, since they do not account for the preferential population of certain local conformations in globular protein structures. Therefore a second test was performed using as input the constraints derived from a group of 13 protein crystal structures taken from the Protein Data Bank[27] and the solution structure of Tendamistat.[18] The first four columns in Table VI describe the origin and the resolution of these protein structures, and the fifth column lists the number of β-methylene groups contained in them. Hydrogen atoms were attached to the crystal structures, and two simulated NMR input data sets of similar precision to those used in Table V were generated. L is a low-precision input data set as it can be obtained routinely from present NMR experiments. H is a higher precision input data set with tighter constraints for ^1H–^1H distances and more accurate ^3J coupling constant values (L corresponds to data set A of Table V with Δ^3J = 2.0 Hz and H to data set B of Table V with Δ^3J = 1.0 Hz. See also footnotes to Table VI). In Table VI the columns LF and HF list the results obtained with a grid search over the entire χ^1 range and LS and HS those from a limited search near the three staggered rotamers, i.e., from 40 to 80, 160 to −160 and −80 to −40°.

The third row from the bottom in Table VI summarizes the result of these test calculations: Among all 496 non-proline, nonterminal β-methylene groups of the database, HABAS yielded 42% stereospecific assignments with the input LF and 77% with HF. With the assumption that all χ^1 values are near the staggered rotamers, the corresponding results are 49% for LS and 67% for HS. As the first and most important result we thus see that the extent of stereospecific assignments is limited and depends critically on the precision of the input data. The results obtained further imply that it is preferable to use HABAS with an unrestrained grid search of the χ^1 dihedral angle space, since the limitation to values near the staggered rotamers produced only slightly better results for the input L, and for input H, a smaller number of stereospecific assignments were actually obtained with the restricted χ^1 angle range. This reduction of the level of unambiguous stereospecific assignments stems from those residues in the proteins for which the χ^1 values are outside of the ranges of ±20° about the staggered rotamers.[25] Furthermore, for 10% and 15% of the β-methylene groups, respectively, either no sterically allowed conformations or only conformations consistent with the wrong stereospecific assignment were found with the restricted grid search. In contrast, the number of inconsistencies encountered with the complete grid search amounted to less than 0.5%.

The lowest two rows in Table VI show that with the input L there is a significantly higher percentage of stereospecific assignments for residues located in β-strands than for those in helices. With the higher precision input data H this difference disappears. The dependence on the secondary structure presents an explanation for at least part of the sizable variations in the extent of stereospecific assignments obtained for the individual proteins in Table VI. It also adds to earlier observations that β-proteins are generally more readily amenable to structural studies by NMR than α-proteins.[6]

V. Conclusions

There have been indications previously from practical experience with protein structures calculated from experimental NMR data that the use of stereospecific assignments for prochiral groups of protons contributes to improved precision of the structure determination.[18,24] However, since in these projects the input for the structure calculations was at the same time changed in other ways, the influence of the stereospecific assignments could not be properly assessed. The test calculations in section II of this paper now show that the inclusion of stereospecific assignments yields substantial improvements when combined with the usual qualitative conformational constraints corresponding to those that can be obtained from NMR experiments.[6] Thereby the improved

Table VI. Extent of Stereospecific Assignments for β-Methylene Groups Achieved with the Program HABAS by Using Simulated Input Data Derived from 13 Known Protein Structures

protein[a]	code[b]	resolution, Å	R factor, %	βCH_2 groups[c]	stereospecific assignments for βCH_2,[d] % LF	LS	HF	HS
crambin[35]	1CRN	1.5	11	18	28	44 (–)	89	89 (–)
L7/L12 50S ribosomal protein[36]	1CTF	1.7	17	33	45	45 (6)	88	82 (6)
HIPIP[37]	1HIP	2.0	24	43	33	35 (21)	70	44 (33)
avian pancreatic polypeptide[38]	1PPT	1.4		22	41	45 (5)	73	73 (5)
scorpion neurotoxin[39]	1SN3	1.8	16	43	44	56 (5)	74	72 (9)
cytochrome b_5 (oxidized)[40]	2B5C	2.0		58	38	40 (16)	81	55 (28)
erabutoxin b[41]	2EBX	1.4	22	40	50	65 (5)	83	75 (13)
ovomucoid third domain[42]	2OVO	1.5	20	35	43	46 (3)	80	74 (11)
cytochrome c_{551} (red.)[43]	451C	1.6	19	42	38	43 (7)	76	71 (10)
BPTI[26]	4PTI	1.5	16	35	37	54 (–)	69	71 (–)
rubredoxin (oxidized)[44]	5RXN	1.2	12	31	52	52 (10)	74	74 (10)
plastocyanin[45]	6PCY	1.9	15	59	41	58 (15)	70	58 (24)
tendamistat[18]				37	43	51 (19)	86	73 (22)
residues in these proteins	870			496	42	49 (10)	77	67 (15)
residues in helices[e]	225			133	38	41 (10)	89	74 (17)
residues in β-sheets[e]	230			132	57	69 (8)	83	77 (11)

[a] For this analysis all but one of the crystal structures from the Protein Data Bank[27] were used that contain between 30 and 99 amino acid residues and for which data were collected to 2.0 Å or higher resolution (ferredoxin was not included because the refinement method used makes this structure unsuitable for the present study). In addition the solution structure of tendamistat (structure I of ref 18) was included. The root-mean-square deviation with respect to the ECEPP/2 standard geometry[33,34] of the lengths of all covalent bonds between N, C^α, C^β, C′, and O is less than 0.038 Å, and the root-mean-square deviation of all bond angles involving these atoms is less than 4° in all these structures. [b] File indentification code of the Protein Data Bank. [c] All β-methylene groups are considered, except those of all prolines and the chain-terminal residues. In HIPIP Ser-26 and Gln-50 were not used because the atoms O^γ and C^γ, respectively, are not listed in the coordinate file of the Protein Data Bank. [d] Stereospecific assignments were obtained by using two sets of simulated input constraints corresponding to different precision of the NMR measurements, L and H, as described in detail below. Each of the two data sets was used with the assumption that either all values for χ^1 are accessible (indicated by F), or that χ^1 could adopt only values within a limited range about the three staggered rotamers, i.e., 60 ± 20, 180 ± 20, and −60 ± 20° (S). For the low-precision input data (L) it was assumed that all ^1H–^1H distances used by HABAS (see section III) are constrained in the intervals between the sum of the two core radii and an upper limit of 2.5, 3.0, 3.5, and 4.0 Å, where for each distance the shortest constraint that includes its actual value is used. The spin–spin coupling constants $^3J_{HN\alpha}$, $^3J_{\alpha\beta 2}$, and $^3J_{\alpha\beta 3}$ were derived from the molecular conformations by using eq 1–3 with allowed deviations Δ^3J = 2.0 Hz. The higher precision data (H) were the same with the following exceptions: The two constraints with the allowed ranges extending from the sum of the two core radii to 3.5 and 4.0 Å, respectively, were replaced by 2.5 Å $< d <$ 3.5 Å and 3.0 Å $< d <$ 4.0 Å, and a new lower bound, $d >$ 4.0 Å, was used for all distances longer than this limit. The allowed deviation of spin–spin coupling constants Δ^3J was 1.0 Hz. In parentheses the percentages of inconsistencies and erroneous assignments occurring when the investigation is done with the χ^1 angles restricted near the three rotamers (S) is given. [e] The secondary structure identification in the Protein Data Bank were used.[27]

precision of the polypeptide backbone conformation is an outstanding, and not necessarily expected result (Table III). These test calculations thus provided new motivation to work on improved techniques for obtaining stereospecific assignments and to investigate the extent to which stereospecific assignments could a priori be obtained.

The program HABAS screens all available local constraints in a grid search of the dipeptide conformation space and thus performs an unbiased search of stereospecific assignments. It is an improved alternative to the previously applied manual screening of local constraints prior to the three-dimensional structure calculations.[23,24] As such it might in the future be further improved by substituting part or all of the grid search by an analytical analysis. HABAS does not take account of long-range constraints in the three-dimensional structure, which can also lead to stereospecific assignments.[18,20–22] For practical purposes we recommend that HABAS, or in the future perhaps some improved version of this program, is applied to obtain the maximum possible number of stereospecific assignments prior to the start of the structure calculations. These results can then be supplemented by a suitable search procedure for additional assignments based on long-range constraints, which will foreseeably consist of several rounds of screening of the three-dimensional structure at different stages of refinement.[18,20,22]

The results obtained in section IV indicate that the comparison in Figures 1–3 and Table III of corresponding structures calculated with or without stereospecific assignments represents an upper limit to the improvements of protein structure determinations that can in practice be expected from stereospecific resonance assignments. This is a consequence of the fact that these test calculations assumed that 100% of the β-methylene groups were assigned, and that stereospecific assignments were also available for selected additional groups of protons. In contrast, the Tables V and VI predict that with the use of experimental NMR data stereospecific assignments for βCH_2 groups can be expected in the extent of approximately 30–80%, depending primarily on the precision of the NMR measurements. An important message from the present study then is that significant further improvement of the precision of protein structure determination by NMR in solution may be achieved through further progress in quantitative measurements of the local conformational constraints.

Acknowledgment. Financial support by the Schweizerischer Nationalfonds (project no. 3.198.85) and by special funds of the ETH Zürich, as well as the use of the facilities of the Zentrum für Interaktives Rechnen (ZIR) of the ETH Zürich, is gratefully acknowledged. We thank Mrs. E. Huber for the careful processing of the typescript.

Reprinted from Biochemistry, 1989, 28, 7510.
Copyright © 1989 by the American Chemical Society and reprinted by permission of the copyright owner.

Stereospecific Nuclear Magnetic Resonance Assignments of the Methyl Groups of Valine and Leucine in the DNA-Binding Domain of the 434 Repressor by Biosynthetically Directed Fractional ^{13}C Labeling[†]

Dario Neri,[‡] Thomas Szyperski,[‡] Gottfried Otting,[‡] Hans Senn,[§] and Kurt Wüthrich*[,‡]

Institut für Molekularbiologie und Biophysik, ETH-Hönggerberg, CH-8093 Zürich, Switzerland, and Präklinische Forschung, Sandoz Ltd., CH-4002, Basel, Switzerland

Received June 14, 1989; Revised Manuscript Received July 11, 1989

ABSTRACT: Stereospecific ^1H and ^{13}C NMR assignments were made for the two diastereotopic methyl groups of the 14 valyl and leucyl residues in the DNA-binding domain 1–69 of the 434 repressor. These results were obtained with a novel method, biosynthetically directed fractional ^{13}C labeling, which should be quite widely applicable for peptides and proteins. The method is based on the use of a mixture of fully ^{13}C-labeled and unlabeled glucose as the sole carbon source for the biosynthetic production of the protein studied, knowledge of the independently established stereoselectivity of the pathways for valine and leucine biosynthesis, and analysis of the distribution of ^{13}C labels in the valyl and leucyl residues of the product by two-dimensional heteronuclear NMR correlation experiments. Experience gained with the present project and a previous application of the same principles with the cyclic polypeptide cyclosporin A provides a basis for the selection of the optimal NMR experiments to be used in conjunction with biosynthetic fractional ^{13}C labeling of proteins and peptides.

Nuclear magnetic resonance (NMR)[1] spectroscopy in solution is by now quite well established as a method for the determination of the three-dimensional structure of proteins [for recent reviews see, for example, Wemmer and Reid (1985), Clore and Gronenborn (1987), Kaptein et al. (1988), and Wüthrich (1989a,b)], and there is keen interest in additional refinements of the method to further improve the precision of the structure determinations. One avenue toward this goal is the use of stereospecific assignments for diastereotopic groups of protons (Wüthrich, 1986; Kline et al., 1988; Driscoll et al., 1989; Güntert et al., 1989), which are not obtained by the generally used sequential resonance assignment procedure for proteins (Wüthrich et al., 1982; Billeter et al., 1982; Wagner & Wüthrich, 1982; Wider et al., 1982). For structure determinations without stereospecific assignments, a set of pseudoatoms replacing the diastereotopic hydrogen atoms was introduced (Wüthrich et al., 1983). This is inevitably a compromise, since the use of these pseudoatoms reduces the precision of the experimental conformational constraints (Wüthrich, 1986). More recently, systematic manual and automated procedures were introduced for obtaining stereospecific assignments for β-methylene groups (Arseniev et al., 1988; Güntert et al., 1989; Hyberts et al., 1987; Wagner et al., 1987; Weber et al., 1988). For more peripheral side-

† Financial support was obtained from the Schweizerischer Nationalfonds (Project Nr. 31.25174.88).

‡ ETH-Hönggerberg.

§ Sandoz Ltd.

[1] Abbreviations: NMR, nuclear magnetic resonance; COSY, two-dimensional correlated spectroscopy; TOCSY, two-dimensional total correlation spectroscopy; Tris, tris(hydroxymethyl)aminomethane; EDTA, ethylenediaminetetraacetic acid; HEPES, 4-(2-hydroxyethyl)-piperazine-1-ethanesulfonic acid; biosynthetic fractional ^{13}C labeling, biosynthetically directed fractional ^{13}C labeling.

0006-2960/89/0428-7510$01.50/0 © 1989 American Chemical Society

 Biochemistry, Vol. 28, No. 19, 1989 7511

chain protons, however, only few stereospecific assignments were so far obtained, primarily during the final stages of the three-dimensional structure determination by reference to spatially proximate protons (Kline et al., 1988; Senn et al., 1984; Zuiderweg et al., 1985). Among these peripheral groups of protons the isopropyl moieties of Val and Leu have outstandingly large pseudoatom corrections (Wüthrich et al., 1983; Wüthrich, 1986). Furthermore, their methyl resonance lines are prominent features in the ^{1}H NMR spectra of proteins, so that a large number of NOE's with the methyl protons of Val and Leu can usually be observed and identified. Overall, stereospecific assignments of the isopropyl groups in a protein can therefore have an important influence on the precision of the entire structure determination. The present paper describes a novel approach for obtaining such stereospecific NMR assignments in proteins, which uses biosynthetic fractional ^{13}C labeling of the polypeptide chain and heteronuclear NMR experiments, and does not depend on any prior knowledge of the three-dimensional structure. As a practical application, the stereospecific ^{1}H and ^{13}C assignments of the 28 methyl groups of Val and Leu in the DNA-binding domain 1–69 of the 434 repressor are described.

Methods

Stereospecific Assignment of the Methyl Groups of Valine and Leucine by Nonrandom ^{13}C Labeling. This approach for obtaining stereospecific NMR assignments derives from the fact that the biosynthesis of the amino acids valine and leucine from glucose is known to be stereoselective (Crout et al., 1980; Gough & Murray, 1983; Hill & Yan, 1971; Hill et al., 1973, 1979). Thereby, as is shown in Figure 1, the isopropyl group is made up of a two-carbon fragment from one pyruvate unit, while the second methyl groups is transferred from another pyruvate unit. This methyl migration has been shown to be stereoselective, and the migrating methyl group becomes *pro-S* in both valine and leucine; i.e., it is γ^2CH_3 or δ^2CH_3, respectively. Direct proof for this stereoselective biosynthetic pathway was obtained for *Escherichia coli* (Sylvester & Stevens, 1979; Hill et al., 1973), which was also used to express the protein studied in this paper.

Biosynthetic fractionally ^{13}C-labeled proteins can be obtained from microorganisms grown on minimal media containing a mixture of roughly 10% [$^{13}C_6$]glucose and 90% unlabeled glucose as the sole carbon source. The carbon positions in such preparations are uniformly ^{13}C labeled to an extent of about 10%. Disregarding the natural ^{13}C abundance of 1.1% in the unlabeled glucose, the probability that two adjacent carbon positions are labeled in the same molecule is then 1%, unless the two carbon atoms originate from the same carbon source molecule so that this probability becomes 10%. These two distinct situations prevail for the isopropyl group in valine and leucine (Figure 1): The *pro-R* methyl group (γ^1 and δ^1, respectively) and the adjacent >CH– group originate from the same pyruvate molecule and are, in the absence of isotope scrambling, labeled with ^{13}C in the same molecules. On the other hand, the *pro-S* methyl group and the adjacent carbon atom originate from two different pyruvate molecules (Figure 1). Therefore, if the *pro-S* methyl group is enriched with ^{13}C, there is a probability of only 1% that the adjoining >CH– group in the same molecule is also labeled.

The stereospecific distinction between the pairs of isopropyl methyl groups in a fractionally ^{13}C-labeled biosynthetic protein is most clearly evidenced in ^{1}H-decoupled ^{13}C NMR spectra, where the ^{13}C resonance of the *pro-R* methyl group is a doublet with a splitting of about 33 Hz due to the one-bond ^{13}C–^{13}C coupling with the neighboring ^{13}C spin, while the ^{13}C NMR signal of the *pro-S* methyl group is a singlet. For work with proteins, 2D NMR experiments (Ernst et al., 1987) are usually employed to ensure a workable spectral resolution (Wüthrich, 1986). For the present project a two-dimensional ^{1}H-detected heteronuclear correlation experiment, [^{13}C,^{1}H]-COSY, is a good choice. The pulse sequence originally devised by Bodenhausen and Ruben (1980) ensures both ^{1}H–^{1}H and ^{1}H–^{13}C decoupling in the ^{13}C dimension, so that the forementioned differences between the *pro-R* and *pro-S* methyl groups can readily be observed along the ω_1 frequency axis (Figure 2).

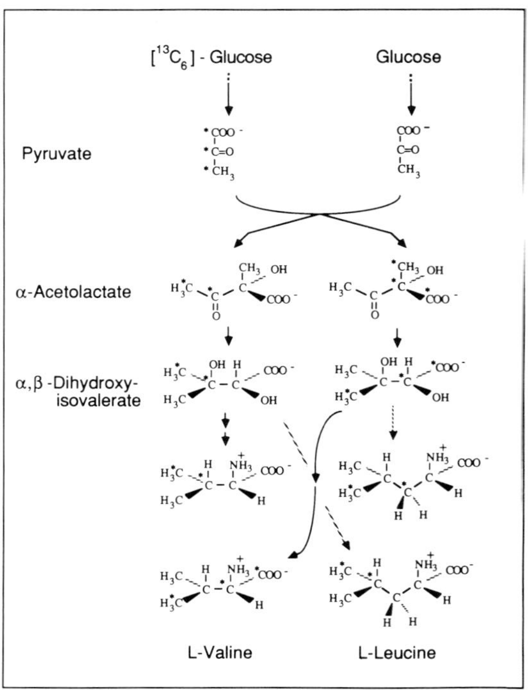

FIGURE 1: Reaction pathways for the biosynthesis of valine and leucine from a mixture of fully ^{13}C-labeled and unlabeled glucose showing the stereochemistry and the principal labeling patterns (an asterisk indicates a ^{13}C label; the absence of the asterisk indicates the natural ^{13}C abundance of 1.1%).

Identification of the Pairs of Methyl NMR Lines That Originate from the Same Isopropyl Group. In principle, sequence-specific ^{1}H NMR assignments can be extended to the ^{13}C lines by [^{13}C,^{1}H]-COSY. However, in crowded spectral regions such assignments based entirely on alignment of individual ^{1}H chemical shifts may not all be unambiguous. As an additional assignment criterion TOCSY-relayed [^{13}C,^{1}H]-COSY at natural abundance of ^{13}C provides particularly clear evidence for identifying the pairs of methyl resonances in the isopropyl groups of Val and Leu (Otting & Wüthrich, 1988). In this experiment the proton magnetization is first transferred to directly bound ^{13}C atoms, from there back to the protons, and finally by a TOCSY pulse sequence to all other protons of the spin system. The resulting unique cross-peak pattern for the isopropyl groups of Val and Leu is shown in Figure 3.

Experimental Procedures

Preparation of Nonrandomly ^{13}C-Labeled 434 Repressor 1–69. The 434 repressor was produced from an overexpression system consisting of NM522 *E. coli* cells (Gough & Murray, 1983) bearing a pRW190 plasmid kindly provided by M.

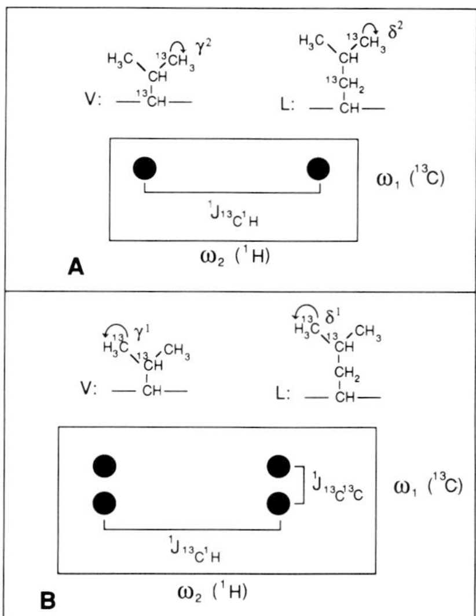

FIGURE 2: Schematic representation of the dominant multiplet fine structures expected for the methyl ^{13}C–^1H cross peaks of Val and Leu in [^{13}C,^1H]-COSY spectra recorded with a protein preparation fractionally labeled with ^{13}C according to Figure 1. The arrows indicate the coherence transfer that is relevant for the present study. (A) For γ^2CH_3 of Val and δ^2CH_3 of Leu the multiplet consists of two components along ω_2 separated by the $^1J_{^{13}C,^1H}$ coupling constant. No splitting is observed along ω_1. This pattern is the same as for methyl groups in a protein with natural abundance of ^{13}C (Figure 4A). (B) In addition to the $^1J_{^{13}C,^1H}$ splitting along ω_2, γ^1CH_3 of Val and δ^1CH_3 of Leu have a splitting of $^1J_{^{13}C,^{13}C}$ along ω_1.

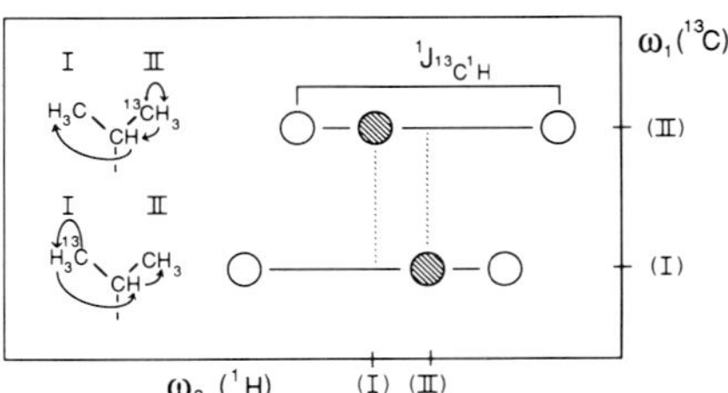

FIGURE 3: Schematic representation of the cross-peak fine structure patterns in TOCSY-relayed [^{13}C,^1H]-COSY at natural ^{13}C abundance that result in the identification of the pairs of diastereotopic methyl resonances belonging to the same isopropyl group of Val or Leu. The two relevant coherence transfer pathways starting from ^{13}C(I) and ^{13}C(II), respectively, are indicated by arrows. In the spectral region displayed in Figure 5, these give rise to two groups of three cross-peak components with identical ^{13}C shifts. At the ^{13}C shift (II) the transfer starting from ^{13}C(II) yields a doublet from the direct [^{13}C,^1H]-COSY cross peak and a singlet from the double relay to the methyl protons I, and vice versa. (Note that the relayed cross peaks from the methyl ^{13}C resonances to the methine proton resonance are outside of the spectral region shown here and in Figure 5.) The two groups of three peaks originating from the same isopropyl group are related by the fact that along ω_2 the singlet with the ^{13}C shift (II) is centrally located relative to the two doublet components at the ^{13}C chemical shift I, and vice versa.

Ptashne and G. Koudelka, Harvard University, Cambridge, MA. The cells were grown at 34 °C with aeration in 8.5 L of a minimal medium containing 10 g of $(NH_4)_2SO_4$, 90 g of K_2HPO_4, 42 g of KH_2PO_4, 50 g of glucose monohydrate, 20 mg of vitamin B1, 10 mg of biotin, 1 g of $MgSO_4{\cdot}7H_2O$, 30 mg of $FeSO_4{\cdot}7H_2O$, and 68 mg of tetracycline hydrochloride. After 150 min from the start of culture growth, 4 g of [$^{13}C_6$]glucose monohydrate (CIL, Cambridge) was added. After another 60 min, when the A_{600} of the culture was approximately 0.6, isopropyl 1-thio-β-D-galactopyranoside was added to a final concentration of 0.1 mM, and growth was continued for 4 h. The cells were harvested by centrifugation and resuspended in 100 mL of lysis buffer containing 100 mM Tris-HCl, 0.5 M NaCl, 1 mM EDTA, 2 mM $CaCl_2$, 10 mM $MgCl_2$, 1.4 mM 2-mercaptoethanol, and 5% glycerol at pH 7.9. To this suspension we added 5 mL of 70–100 mesh glass beads and 5 mg of phenylmethanesulfonyl fluoride in 0.2 mL of dioxane. Cells were then lysed by sonication at 4 °C; the lysate was diluted with 200 mL of lysis buffer and centrifuged for 45 min at 8000 rpm in a GSA rotor of a Sorval centrifuge. The pellet was discarded and 18 mL of 10% poly(ethylenimine) in H_2O was added to the supernatant, which was stirred for 10 min at 4 °C. After centrifugation for 15 min at 7000 rpm with a GSA rotor at 4 °C, the pellet was again discarded and 160 g of ammonium sulfate was added to the supernatant. After being stirred for 45 min at 4 °C, the precipitate was collected by centrifugation at 8000 rpm for 45 min in a GSA rotor. The pellet was then resuspended in 150 mL of buffer A (50 mM Tris-HCl, 0.1 mM EDTA, 1.4 mM 2-mercaptoethanol, 10% glycerol at pH 7.9) plus 50 mM KCl and dialyzed against several changes of the same buffer. Following dialysis this material was loaded onto an S-Sepharose (Pharmacia) column equilibrated with buffer A plus 50 mM KCl. The 434 repressor was bound to the column and was then eluted with a linear gradient from buffer A plus 50 mM KCl to buffer A plus 600 mM KCl. The fractions containing the repressor, which was at this point approximately 70% pure, were passed through an Amicon 8200 ultrafiltration apparatus equipped with a Diaflo YM-10 membrane to obtain a final concentration of 0.4 mg/mL in buffer B (20 mM HEPES, pH 6.9, 0.1 mM dithiothreitol, 0.1 mM EDTA, 5% glycerol) plus 60 mM NaCl. The N-terminal domain of the 434 repressor (residues 1–69) was obtained through limited proteolytic digestion of this material (Anderson et al., 1984). Papain (Fluka, 2.2 units/mg activity) was added to the repressor solution under continuous stirring at room temperature to a final protease/protein ratio of 1:600, and the reaction was allowed to proceed for 2 h; antipain (Fluka) was then added in 100-fold excess with respect to papain to stop digestion. The resulting solution was applied to an S-Sepharose (Pharmacia) column equilibrated with buffer B plus 60 mM NaCl. The C-terminal fragment of the 434 repressor flowed through this column, whereas the N-terminal domain was retained. The latter was eluted with a linear gradient from buffer B plus 60 mM NaCl to buffer B plus 400 mM NaCl. The fractions containing the N-terminal domain 1–69 of the 434 repressor were shown to be greater than 90% pure. By use of a YM-2 Amicon Diaflo membrane the buffer was replaced with distilled water, and the protein was lyophilized.

The NMR sample was prepared by dissolving 7 mg of the ^{13}C-labeled N-terminal domain 1–69 of the 434 repressor in 0.45 mL of a ^2H_2O solution containing 25 mM KH_2PO_4 and 100 mM KCl at p^{2}H 6.0. The protein was again lyophilized from this solution and then redissolved in 0.45 mL of ^2H_2O.

NMR Measurements. All NMR measurements were performed at 28 °C. [^{13}C,^1H]-COSY spectra were recorded on a Bruker AM-600 spectrometer with the pulse sequence of Bodenhausen and Ruben (1980) extended by a 2-ms spin lock purge pulse (Otting & Wüthrich, 1988). To improve the suppression of t_1 noise originating from magnetization of ^{12}C-bound protons, which are incompletely relaxed during the preparation time between successive scans, the phase cycle given by Otting and Wüthrich (1988) was extended 2-fold by

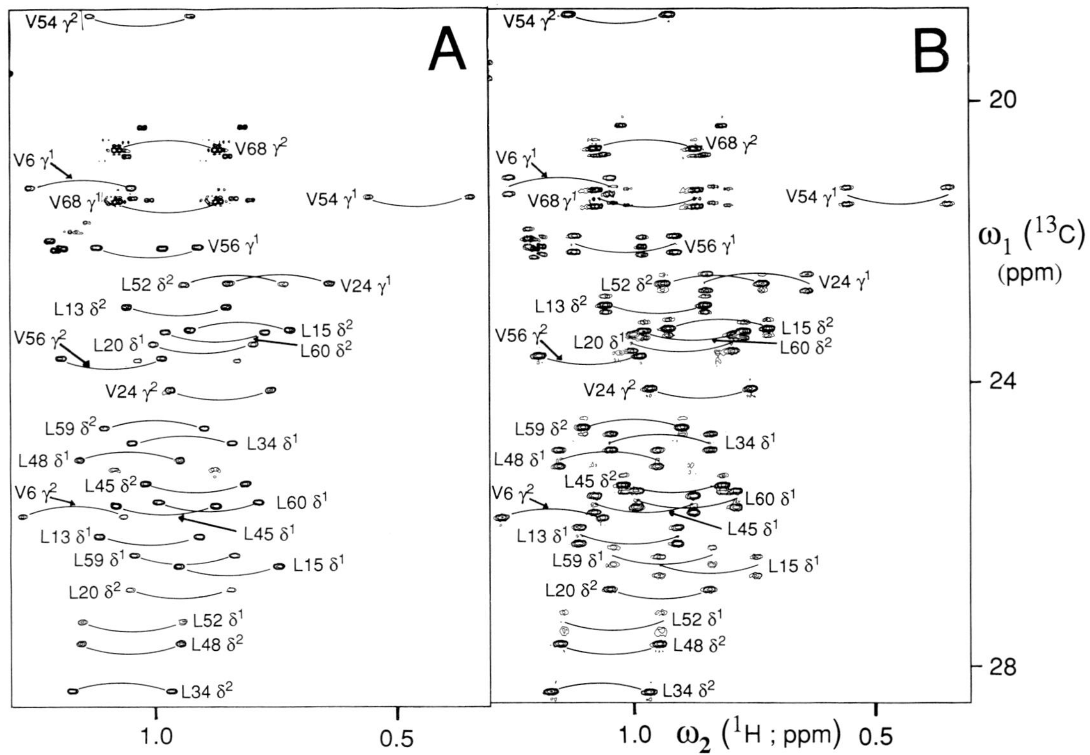

FIGURE 4: Spectral region of two $[^{13}C,^1H]$-COSY spectra of the N-terminal domain 1–69 of the 434 repressor at a proton frequency of 600 MHz containing the cross peaks between the directly bonded carbon and hydrogen atoms of the methyl groups of valine and leucine. (A) Natural abundance of ^{13}C, protein concentration 5 mM. (B) Biosynthetic fractionally ^{13}C labeled in the extent of 15%, protein concentration 2 mM. For each methyl resonance of Val and Leu the fine structure components separated by $^1J_{^{13}C,^1H}$ along ω_2 are connected by a curved line. The resonance assignments are given by the one-letter code for the amino acid, the sequence position, and the stereospecific identification of the methyl group according to the standard IUB–IUPAC nomenclature; γ^1 and δ^1 are *pro-R* and γ^2 and δ^2 are *pro-S*.

simultaneous permutation of the second 90° (^{13}C) pulse and the receiver phase (Wörgötter et al., 1988). Both spectra shown in Figure 4 were recorded and processed with identical parameters: $t_{1,max}$ = 102 ms; $t_{2,max}$ = 340 ms; total recording time about 60 h. Before Fourier transformation the time domain data were multiplied with sine-bell windows (De Marco & Wüthrich, 1976) along t_1 and t_2, with phase shifts of $\pi/5$ and $\pi/9$, respectively. The digital resolution after zero filling was 2.9 Hz along ω_1 and 1.5 Hz along ω_2.

A TOCSY-relayed $[^{13}C,^1H]$-COSY spectrum (Otting & Wüthrich, 1988) was recorded on a Bruker AM-500 spectrometer. The mixing sequence of the clean-TOCSY experiment (Griesinger et al., 1988) was used: $t_{1,max}$ = 49 ms; $t_{2,max}$ = 246 ms; total recording time about 58 h. Before Fourier transformation the time domain data were multiplied with sine-bell windows along t_1 and t_2, with phase shifts of $\pi/5$ and $\pi/9$, respectively.

Results and Discussion

By comparison of the ^{13}C satellites with the main peaks in a ^1H NMR spectrum, the extent of ^{13}C labeling in the 434 repressor 1–69 prepared as described in the preceding section was found to be about 15%. The methyl resonances of the five Val and nine Leu in this protein are all in the spectral region shown in Figure 4. At natural abundance of ^{13}C (Figure 4A), all methyl cross peaks have the fine structure of Figure 2A, as expected. The resonance assignments given in the figure are based on conventional sequential ^1H NMR assignments (D. Neri, G. Otting, and K. Wüthrich, unpublished results) and the experiment of Figure 5 below. In the labeled protein (Figure 4B) some of the methyl cross peaks have the four-component fine structure of Figure 2B, which identifies these methyls as γ^1 of Val or δ^1 of Leu. Figure 4 shows that all 28 ^{13}C–^1H methyl cross peaks of Val and Leu in 434 repressor 1–69 were resolved in $[^{13}C,^1H]$-COSY, and complete stereospecific assignments were obtained (Table I).

In Figure 4B the two-component multiplet patterns observed for the *pro-S* methyl groups are superimposed by a weak four-component multiplet indicative of the presence of about 2% of $^{13}C^{\delta 2}$–$^{13}C^{\gamma}$. This coincides with the expectations. Similarly, weak two-component signals are superimposed on the strong four-component peaks of the *pro-R* methyl groups. It can also be seen that the methyl ^{13}C chemical shift in $>^{12}CH$–$^{13}CH_3$ is always about 0.01 ppm downfield from the center of the $[^{13}C]$methyl doublet in $>^{13}CH$–$^{13}CH_3$, which is due to the isotope effect on the chemical shift (Hansen, 1983).

Figure 5 shows the methyl region of the TOCSY-relayed $[^{13}C,^1H]$-COSY spectrum of 434 repressor 1–69. The typical isopropyl patterns of Figure 3 are identified for all 14 Val and Leu residues. The resulting identification of the 14 methyl pairs is fully compatible with Figure 4, since in each case a *pro-R* methyl is correlated with a *pro-S* methyl group.

Conclusions

The presently described procedure for obtaining stereospecific assignments of the diastereotopic methyl groups of Val

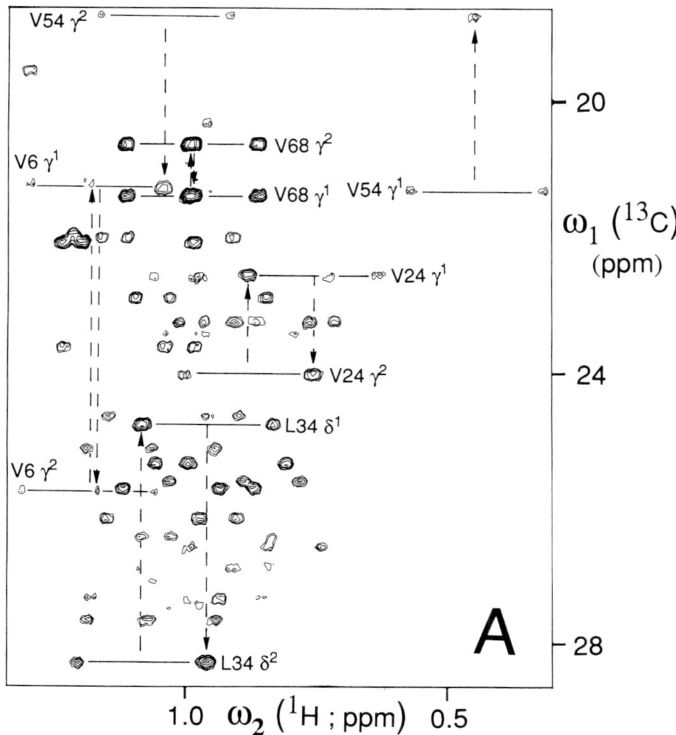

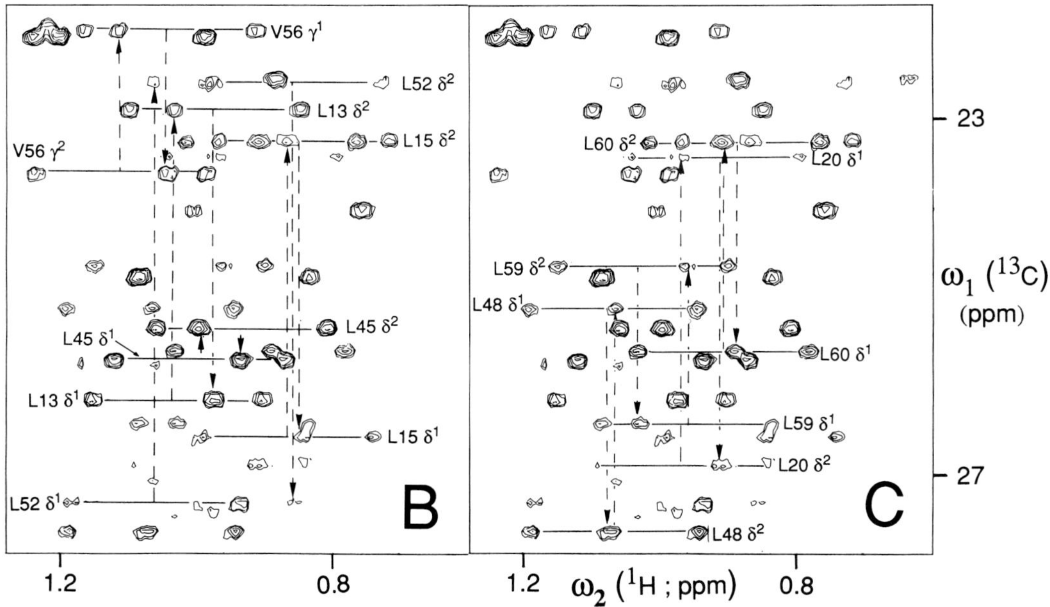

FIGURE 5: TOCSY-relayed [^{13}C,^{1}H]-COSY spectrum of the N-terminal domain 1–69 of the 434 repressor at 500 MHz. The same sample with natural ^{13}C abundance was used as in Figure 4A. (A) Spectral region containing the cross peaks relating the carbon-13 resonance of one methyl group to the proton signal of the other methyl group in the same Val or Leu residue. The solid horizontal and broken vertical lines outline the typical peak patterns for five isopropyl groups, as explained in Figure 3. For improved clarity, arrows pointing at the double-relayed cross peaks have been added to the broken lines. (B and C) The region (ω_1 = 22–28 ppm, ω_2 = 0.6–1.3 ppm) from spectrum A replotted on an expanded scale. The peak patterns leading to the identification of the remaining nine isopropyl groups of Val and Leu in 434 repressor 1–69 are indicated as in (A).

Table I: Chemical Shifts and Stereospecific Assignments Obtained for the ^{13}C and ^{1}H Resonances of the Methyl Groups of Valine and Leucine in the N-Terminal Domain 1–69 of the 434 Repressor

amino acid		chemical shift[a] ^{13}C	chemical shift[a] ^{1}H	amino acid		chemical shift[a] ^{13}C	chemical shift[a] ^{1}H
Val-6	γ^1	21.3	1.15	Leu-20	δ^1	23.5	0.89
	γ^2	25.9	1.16		δ^2	26.9	0.95
Val-24	γ^1	22.6	0.75	Leu-34	δ^1	24.8	0.94
	γ^2	24.1	0.88		δ^2	28.3	1.07
Val-54	γ^1	21.3	0.46	Leu-45	δ^1	25.7	0.98
	γ^2	21.2	1.04		δ^2	25.4	0.92
Val 56	γ^1	22.1	1.02	Leu-48	δ^1	25.1	1.05
	γ^2	23.6	1.09		δ^2	27.7	1.05
Val-68	γ^1	21.4	0.97	Leu-52	δ^1	27.3	1.05
	γ^2	21.7	0.97		δ^2	22.6	0.85
Leu-13	δ^1	26.2	1.02	Leu-59	δ^1	26.4	0.94
	δ^2	22.9	0.96		δ^2	24.6	1.00
Leu-15	δ^1	26.6	0.85	Leu-60	δ^1	25.6	0.89
	δ^2	23.3	0.83		δ^2	23.3	0.87

[a] The stereospecific assignments were obtained by using the nonrandom labeling experiments described in this paper. The *pro-R* methyl groups of Val and Leu are denominated γ^1 and δ^1, according to the standard IUB–IUPAC nomenclature; the *pro-S* methyl groups are denominated γ^2 and δ^2.

and Leu based on biosynthetic fractional ^{13}C labeling (Figure 1) is generally applicable for biosynthetic peptides and proteins. It is a special advantage of this approach that the assignments can be completed before the start of the structure calculations, and hence a reduction of the computation time can be expected to result as well. Since [$^{13}C_6$]glucose is quite expensive, we tried to minimize its use by adding it only 1 h before induction of the overexpression system (see Experimental Procedures). The result obtained indicates that, by careful selection of the time at which the labeled material is added to the cell culture, the technique can probably be further optimized with regard to reducing the amount of labeled glucose used.

In a previous paper the biosynthetically directed fractional labeling approach was used with the cyclic peptide cyclosporin A obtained from *Tolypocladium inflatum* (Senn et al., 1989). The experience gained there and in the present work shows that different NMR experiments will give optimal results with fractionally labeled biosynthetic peptides or proteins, respectively. For example, although in 434 repressor 1–69 the chemical shift dispersion was sufficiently large to enable a complete analysis of the methyl region in [^{13}C,^{1}H]-COSY (Figure 4), lack of conformation-dependent chemical shift dispersion (Wüthrich, 1986) in cyclosporin A prevented a detailed analysis of the same spectral region for this compound. Conversely, TOCSY-relayed [^{13}C,^{1}H]-COSY gave complete sets of cross peaks for the spin systems of all Val and Leu residues in cyclosporin A, so that data corresponding to those obtained here from Figure 4 could be extracted from the cross peaks $\gamma\underline{C}H_3$–$\alpha\underline{H}$ and $\delta\underline{C}H_3$–$\alpha\underline{H}$, respectively (Senn et al., 1989). With the 434 repressor, the sensitivity of this same experiment was sufficient to observe the methyl–methyl correlations within the isopropyl groups (Figure 5), but, presumably because of the shorter spin relaxation time T_2 when compared to cyclosporin A, the cross peaks δCH_3–αH were not observed for all Leu residues.

A comparative analysis of the crystal and solution structures of the protein Tendamistat (Billeter et al., 1989) has shown that, even at 2.0-Å resolution, the orientation of the isopropyl groups of Val and Leu about the C^α–C^β or C^β–C^γ bonds, respetively, is difficult to determine by X-ray crystallography. In contrast, once stereospecific assignments of the two methyls are available, a precise determination of the side-chain conformations of Val and Leu usually results from NMR measurements in solution. This analysis emphasizes further the importance of stereospecific assignments of the isopropyl groups for structural analysis of proteins by NMR spectroscopy.

References

Anderson, J., Ptashne, M., & Harrison, S. C. (1984) *Proc. Natl. Acad. Sci. U.S.A. 181*, 1307–1311.

Arseniev, A., Schultze, P., Wörgötter, E., Braun, W., Wagner, G., Vašák, M., Kägi, J. H. R., & Wüthrich, K. (1988) *J. Mol. Biol. 201*, 637–657.

Billeter, M., Braun, W., & Wüthrich, K. (1982) *J. Mol. Biol. 155*, 321–346.

Billeter, M., Kline, A. D., Braun, W., Huber, R., & Wüthrich, K. (1989) *J. Mol. Biol. 206*, 677–687.

Bodenhausen, G., & Ruben, D. (1980) *Chem. Phys. Lett. 69*, 185–188.

Clore, G. M., & Gronenborn, A. M. (1987) *Protein Eng. 1*, 275–288.

Crout, D. H. G., Hedgecock, J. R., Lipscomb, E. L., & Armstrong, F. B. (1980) *J. Chem. Soc., Chem. Commun.*, 304–305.

De Marco, A., & Wüthrich, K. (1976) *J. Magn. Reson. 24*, 201–204.

Driscoll, P. C., Gronenborn, A. M., & Clore, G. M. (1989) *FEBS Lett. 243*, 223–233.

Ernst, R. R., Bodenhausen, G., & Wokaun, A. (1987) *Principles of Nuclear Magnetic Resonance in One and Two Dimensions*, Clarendon Press, Oxford.

Gough, J., & Murray, N. (1983) *J. Mol. Biol. 166*, 1–19.

Griesinger, C., Otting, G., Wüthrich, K., & Ernst, R. R. (1988) *J. Am. Chem. Soc. 110*, 7870–7872.

Güntert, P., Braun, W., Billeter, M., & Wüthrich, K. (1989) *J. Am. Chem. Soc. 111*, 3997–4004.

Hansen, P. E. (1983) *Annu. Rep. NMR Spectrosc. 15*, 105–238.

Havel, T. F., & Wüthrich, K. (1985) *J. Mol. Biol. 182*, 281–294.

Hill, R. K., & Yan, S. (1971) *Bioorg. Chem. 1*, 446–456.

Hill, R. K., Yan, S., & Arfin, S. M. (1973) *J. Am. Chem. Soc. 95*, 7857–7859.

Hill, R. K., Sawada, S., & Arfin, S. M. (1979) *Bioorg. Chem. 8*, 175–189.

Hyberts, S., Märki, W., & Wagner, G. (1987) *Eur. J. Biochem. 164*, 625–635.

Kaptein, R., Boelens, R., Scheek, R. M., & van Gunsteren, W. F. (1988) *Biochemistry 27*, 5389–5395.

Kline, A. D., Braun, W., & Wüthrich, K. (1988) *J. Mol. Biol. 204*, 675–724.

Otting, G., & Wüthrich, K. (1988) *J. Magn. Reson. 76*, 569–574.

Senn, H., Billeter, M., & Wüthrich, K. (1984) *Eur. Biophys. J. 11*, 3–5.

Senn, H., Werner, B., Messerle, B., Weber, C., Traber, R., & Wüthrich, K. (1989) *FEBS Lett. 249*, 113–118.

Sylvester, S. R., & Stevens, C. M. (1979) *Biochemistry 18*, 4529–4531.

Wagner, G., & Wüthrich, K. (1982) *J. Mol. Biol. 155*, 347–366.

Wagner, G., Braun, W., Havel, T. F., Schaumann, T., Gō, N., & Wüthrich, K. (1987) *J. Mol. Biol. 196*, 611–639.

Weber, P. L., Morrison, R., & Hare, D. (1988) *J. Mol. Biol. 204*, 483–487.

Wemmer, D. E., & Reid, B. R. (1985) *Annu. Rev. Phys. Chem. 36*, 105–137.

7516

Wider, G., Lee, K. H., & Wüthrich, K. (1982) *J. Mol. Biol. 155*, 367–388.
Wörgötter, E., Wagner, G., Vašák, M., Kägi, J. H. R., & Wüthrich, K. (1988) *J. Am. Chem. Soc. 110*, 2388–2393.
Wüthrich, K. (1986) *NMR of Proteins and Nucleic Acids*, Wiley, New York.
Wüthrich, K. (1989a) *Science 243*, 45–50.
Wüthrich, K. (1989b) *Acc. Chem. Res. 22*, 36–44.
Wüthrich, K., Wider, G., Wagner, G., & Braun, W. (1982) *J. Mol. Biol. 155*, 311–319.
Wüthrich, K., Billeter, M., & Braun, W. (1983) *J. Mol. Biol. 169*, 949–961.
Zuiderweg, E. R. P., Boelens, R., & Kaptein, R. (1985) *Biopolymers 24*, 601–611.

J. Mol. Biol. (1991) **217**, 517–530

Efficient Computation of Three-dimensional Protein Structures in Solution from Nuclear Magnetic Resonance Data Using the Program DIANA and the Supporting Programs CALIBA, HABAS and GLOMSA

Peter Güntert, Werner Braun and Kurt Wüthrich

Institut für Molekularbiologie und Biophysik
Eidgenössische Technische Hochschule-Hönggerberg
CH-8093 Zürich, Switzerland

(Received 22 June 1990; accepted 2 October 1990)

A novel procedure for efficient computation of three-dimensional protein structures from nuclear magnetic resonance (n.m.r.) data in solution is described, which is based on using the program DIANA in combination with the supporting programs CALIBA, HABAS and GLOMSA. The first part of this paper describes the new programs DIANA, CALIBA and GLOMSA. DIANA is a new, fully vectorized implementation of the variable target function algorithm for the computation of protein structures from n.m.r. data. Its main advantages, when compared to previously available programs using the variable target function algorithm, are a significant reduction of the computation time, and a novel treatment of experimental distance constraints involving diastereotopic groups of hydrogen atoms that were not individually assigned. CALIBA converts the measured nuclear Overhauser effects into upper distance limits and thus prepares the input for the previously described program HABAS and for DIANA. GLOMSA is used for obtaining individual assignments for pairs of diastereotopic substituents by comparison of the experimental constraints with preliminary results of the structure calculations. With its general outlay, the presently used combination of the four programs is particularly user-friendly. In the second part of the paper, initial results are presented on the influence of the novel DIANA treatment of diastereotopic protons on the quality of the structures obtained, and a systematic study of the central processing unit times needed for the same protein structure calculation on a range of different, commonly available computers is described.

1. Introduction

The early stages in the development of the presently widely used n.m.r.† method for the determination of three-dimensional biomacromolecular structures in solution (for a review, see Wüthrich, 1989) made it clear that the key data measured by n.m.r. would consist of a network of distance constraints between spatially proximate hydrogen atoms (Gordon & Wüthrich, 1978; Dubs *et al.*, 1979; Keller & Wüthrich, 1980; Wüthrich *et al.*, 1982). It immediately followed that the techniques for structure determination from other experimental data, in particular X-ray diffractions, could not be adapted for the structural analysis of the n.m.r. data, and hence new ways had to be developed. Initially, algorithms were used that combined metric matrix distance geometry (Blumenthal, 1970), which had been applied by the groups of Crippen and Kuntz for systematic studies on protein structures (Crippen, 1977; Kuntz *et al.*, 1976; Havel *et al.*, 1983), with a detailed description of the interplay of constraints imposed by the covalent polypeptide structure and those from the n.m.r. measurements (Braun *et al.*, 1981, 1983; Havel & Wüthrich, 1984, 1985). Subsequent work included a variable target function algorithm (Braun & Gō, 1985), interactive molecular modeling using computer graphics (Billeter *et al.*, 1985), restrained molecular dynamics calculations either applied directly with the n.m.r. data (Brünger *et al.*, 1986) or in conjunction with

† Abbreviations used: n.m.r., nuclear magnetic resonance; BPTI, basic pancreatic trypsin inhibitor; NOE, nuclear Overhauser enhancement; NOESY, 2-dimensional nuclear Overhauser enhancement spectroscopy; c.p.u., central processing unit; MFLOPS, million floating point operations per second; r.m.s.d., root-mean-square deviation.

0022–2836/91/030517–14 $03.00/0

© 1991 Academic Press Limited

model building (Kaptein *et al.*, 1985) or distance geometry calculations (Clore *et al.*, 1985), and an ellipsoid algorithm (Billeter *et al.*, 1987). Inspection of the recent literature shows that the following procedures are currently mostly employed. (1) Embedding using a metric matrix distance geometry program, e.g. DISGEO (Havel & Wüthrich, 1984) or DSPACE (Hare Research, Woodinville, WA 98072, U.S.A.), followed by simulated annealing using molecular dynamics (Driscoll *et al.*, 1989; Lee *et al.*, 1989). (2) Structure determination using a variable target function algorithm, e.g. DISMAN (Braun & Gō, 1985), which directly generates acceptable structures (Kline *et al.*, 1988; Schultze *et al.*, 1988; Zuiderweg *et al.*, 1989) or can be supplemented by a molecular mechanics energy minimization (e.g. Billeter *et al.*, 1990; Qian *et al.*, 1989; Widmer *et al.*, 1989).

Once it had been established that n.m.r. measurements could provide sufficient data for the determination of globular protein structures at atomic resolution (Havel & Wüthrich, 1984, 1985; Williamson *et al.*, 1985), the main interest shifted to the development of procedures ensuring both high efficiency of structure calculations and minimal bias of the results by the algorithms used. This paper presents a further step in this development by describing a new implementation of the variable target function algorithm of Braun & Gō (1985) in the program DIANA. This program was primarily designed to be efficient with respect to the c.p.u. time used and, in combination with the supporting programs CALIBA, HABAS and GLOMSA, to be user-friendly in routine structure determinations (Güntert *et al.*, 1990, accompanying paper). Furthermore, thanks to the high efficiency of the fully vectorized program, systematic large-scale investigations on the course of variable target function calculations could be started with DIANA (P. Güntert, W. Braun & K. Wüthrich, unpublished results), which should provide a basis for further optimization of structure calculations.

With regard to improving the quality of structure determinations by n.m.r., the treatment of distance constraints with diastereotopic groups of protons (Wüthrich *et al.*, 1983) can be of crucial importance (Güntert *et al.*, 1989). Recent work in this regard focused mainly on establishing individual assignments for diastereotopic pairs of protons (Weber *et al.*, 1988; Neri *et al.*, 1989; Nilges *et al.*, 1990). The program DIANA includes a novel treatment of constraints with prochiral centers for which the diastereotopic ligands were not individually assigned. In calculations with BPTI, the results of this new treatment are compared with corresponding results obtained with the original pseudoatom concept (Wüthrich *et al.*, 1983).

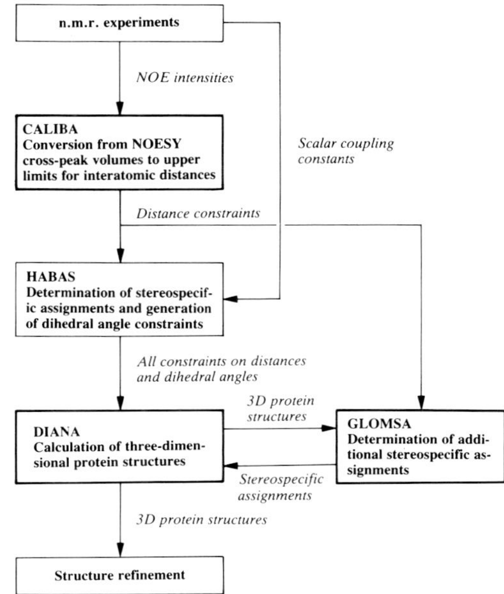

Figure 1. Schematic representation of the functions of the programs CALIBA, DIANA and GLOMSA and the input and output of these programs. See the text for details.

2. Methods

This section describes the 3 programs CALIBA ("*cali*bration of NOE intensity *versus* distance constraints"), DIANA ("*di*stance geometry *a*lgorithm for *n*.m.r. *a*pplications") and GLOMSA ("*glo*bal *m*ethod for obtaining *s*tereospecific *a*ssignments"), which were all written in Fortran-77. Fig. 1 affords a survey of the functions of these programs, which are used in the order CALIBA, DIANA and then GLOMSA. The program CALIBA accepts the experimental NOE intensities as input and performs the calibration of NOESY cross-peaks, i.e. the conversion from peak volumes to upper distance limits. It thus prepares the principal input for the program HABAS (Güntert *et al.*, 1989), which in turn adds stereospecific assignments to the input for DIANA. DIANA is used for efficient calculation of protein conformations based on distance and dihedral angle constraints that can be obtained by n.m.r. measurements (Wüthrich, 1986). It is a new, improved and vectorized implementation of the variable target function algorithm that has first been used in the program DISMAN (Braun & Gō, 1985) and subsequently in other programs (Vásquez & Scheraga, 1988; Kohda *et al.*, 1988). In addition to the structure calculations, DIANA screens the experimental distance constraints and eliminates irrelevant constraints from the input, and it applies a novel adjustment routine to distance constraints with pairs of diastereotopic protons for which no individual assignments are available. The program GLOMSA accepts as input the structures calculated with DIANA and the conformational constraints list produced by CALIBA. It is used to obtain additional stereospecific assignments based on the comparison of upper distance limit pairs or relational constraints (Güntert *et al.*, 1989) involving the diastereotopic substituents of prochiral centers with a set of preliminary conformers (Kline *et al.*, 1988). The additional stereospecific assignments thus obtained are included in the input for a new DIANA calculation, which in turn produces the structures used for further refinements, e.g. by energy minimization (Brooks *et al.*, 1983; Brünger *et al.*, 1986; Schaumann *et al.*, 1990; Weiner & Kollman, 1981).

(a) *The program CALIBA*

In the present version of the program CALIBA, the calibration of NOE intensity *versus* the corresponding upper distance bound is based on either of 2 model assumptions. The 1st assumes that the NOESY cross-peak volume, V, is inversely proportional to the power n of the corresponding upper distance bound b:

$$V = \frac{C}{b^n}, \tag{1}$$

where C is a constant, and the values of n are typically in the range from 4 to 6. Clearly the value $n = 6$ is an upper limit for n obtained theoretically by assuming a rigid structure. Exponents $n < 6$ were found empirically to afford improved representations of the relation between cross-peak volumes and distances for peaks that involve peripheral side-chain protons. The 2nd possibility uses the uniform averaging model (Braun *et al.*, 1981):

$$V = \frac{C'}{b-d_0}\left(\frac{1}{d_0^5} - \frac{1}{b^5}\right). \tag{2}$$

Here, C' is a constant, and $d_0 = 1{\cdot}9$ Å is the shortest sterically allowed distance between 2 protons (1 Å = 0·1 nm).

The input for CALIBA consists of cross-peak volumes measured in 1 or several NOESY spectra, for example, with the program EASY (Eccles *et al.*, 1989). The peak volumes from different spectra, which may have been recorded with different experimental conditions, can be multiplied with user-specified weighting factors. The volumes of peaks that correspond to 2 or more protons with degenerate chemical shifts are divided by the number of protons they contain. If more than 1 peak intensity corresponding to the same distance is retained from the analysis of the spectra, only the strongest (after the aforementioned weighting) is considered for the conversion. For an optimal empirical calibration, the curves (1) or (2) and the constants C or C' can be chosen independently for different classes of interatomic distances, e.g. intraresidual, sequential, medium-range and long-range backbone, and long-range constraints (Wüthrich, 1986), and for cross-peaks that do or do not involve methyl groups. To obtain reasonable upper bounds also for very strong or very weak cross-peaks, the values for the upper bounds b are restricted to a limited range $b^{\min} \leq b \leq b^{\max}$, with typical values for $b^{\min}$ of 2·4 Å, and $b^{\max}$ of 5·0 Å. Constraints corresponding to cross-peaks relating resonance lines from multiple protons with degenerate chemical shifts are referred to pseudoatoms, and the appropriate pseudoatom corrections (Wüthrich *et al.*, 1983) are automatically applied. The program produces upper distance limit files that can be read directly by the programs HABAS (Güntert *et al.*, 1989), DIANA and GLOMSA.

To optimize the choice of the calibration curves at different points during a structure calculation where one already has a set of preliminary structures for the protein under investigation, CALIBA has the option to produce a doubly logarithmic plot of peak volumes *versus* the average of the corresponding distances in this set of structures. Using this plot, the calibration curves can then be adjusted such that most of the resulting upper distance limits are fulfilled in the given preliminary structures without being unnecessarily loosened.

(b) *The program DIANA*

The algorithm used by the program DIANA is based on the minimization of a variable target function $T(\phi_1, \ldots, \phi_n)$, where the n degrees of freedom are the dihedral angles $\phi_1, \ldots, \phi_n$ about single (rotatable) bonds of the polypeptide chain. During the calculation the bond lengths, bond angles and chiralities of the covalent structure are kept fixed at the ECEPP standard values (Momany *et al.*, 1975). The target function T, with $T \geq 0$, (for an explicit definition see eqn (6) below) is defined such that $T = 0$ if all experimental distance and dihedral angle constraints are fulfilled and all non-bonded atom pairs satisfy a check for the absence of steric overlap. $T(\phi_1, \ldots, \phi_n) \leq T(\theta_1, \ldots, \theta_n)$ if the conformation $(\phi_1, \ldots, \phi_n)$ satisfies the constraints better than the conformation $(\theta_1, \ldots, \theta_n)$. The problem to be solved is to find the values $(\phi_1, \ldots, \phi_n)$ that yield low values of the target function. To reduce the danger of becoming trapped in a local minimum with a function value much higher than the global minimum, the target function is varied during a structure calculation. At the outset, only local constraints with respect to the polypeptide sequence are considered, and in subsequent rounds of calculations, constraints between atoms further apart with respect to the primary structure are included in a stepwise fashion. Consequently, in the 1st stages of a structure calculation, the local features of the conformation will be established, and the global fold of the protein will be obtained only toward the end of the calculation. Similar strategies of avoiding local minima by variation of the pseudoenergy function during a structure calculation have been used with restrained molecular dynamics techniques (Holak *et al.*, 1987; Nilges *et al.*, 1988).

(i) *The variable target function*

Two different kinds of constraints are considered by the target function, i.e. upper and lower bounds on interatomic distances, and restraints on individual dihedral angles in the form of an allowed interval (Wüthrich, 1986; Braun, 1987). An upper or lower limit, b, on the distance between the 2 atoms α and β is denoted by the triple (α,β,b) or, if there is no danger of ambiguity, simply by b. A direct constraint on the dihedral angle a that restricts its value ϕ_a to an allowed interval $[\phi^{\min},\phi^{\max}]$, with $\phi^{\min} < \phi^{\max} < \phi^{\min} + 2\pi$, is denoted by $(a, \phi^{\min}, \phi^{\max})$. In the definition of the variable target function we use further the half-width, Γ, of the forbidden interval of dihedral angle values:

$$\Gamma = \pi - \frac{\phi^{\max} - \phi^{\min}}{2}, \tag{3}$$

and the signed dihedral angle constraint violation

$$\Delta = \begin{cases} 0, & \text{if } \phi_a \in [\phi^{\min},\phi^{\max}]; \\ -\Delta^{\min}, & \text{if } \phi_a \notin [\phi^{\min},\phi^{\max}] \text{ and } \Delta^{\min} \leq \Delta^{\max}; \\ \Delta^{\max}, & \text{if } \phi_a \notin [\phi^{\min},\phi^{\max}] \text{ and } \Delta^{\min} > \Delta^{\max}; \end{cases} \tag{4}$$

with

$$\Delta^{\min} = \min\{|\hat\phi^{\min} - \hat\phi_a|, 2\pi - |\hat\phi^{\min} - \hat\phi_a|\},$$

and

$$\Delta^{\max} = \min\{|\hat\phi^{\max} - \hat\phi_a|, 2\pi - |\hat\phi^{\max} - \hat\phi_a|\}.$$

$\hat\phi$ denotes the equivalent value of ϕ in the interval $[0, 2\pi[$, which can be obtained in all instances by the addition of an integer multiple of 2π to ϕ. The sign of Δ will be important only in the calculation of the gradient of the target function; it is positive if a small increase of ϕ_a also increases the violation of the angle constraints, and negative otherwise.

To formulate the target function, we assume that there are n_u experimental upper limits, n_l experimental lower limits, and n_v van der Waals' repulsion lower limits on interatomic distances, and n_a direct dihedral angle constraints:

$$\begin{array}{ll} (\alpha_i^u, \beta_i^u, b_i^u), & i = 1, \ldots, n_u; \\ (\alpha_i^l, \beta_i^l, b_i^l), & i = 1, \ldots, n_l; \\ (\alpha_i^v, \beta_i^v, b_i^v), & i = 1, \ldots, n_v; \\ (a_i, \phi_i^{\min}, \phi_i^{\max}), & i = 1, \ldots, n_a. \end{array} \tag{5}$$

The target function, T, then is:

$$T = \sum_{c=u,l,v} w_c \sum_{i \in I_c} \left(\Theta_c\left(\frac{d_i^{c2} - b_i^{c2}}{2b_i^c} \right) \right)^2 + w_a \sum_{i=1}^{n_a} \left(1 - \frac{1}{2}\left(\frac{\Delta_i}{\Gamma_i} \right)^2 \right) \Delta_i^2, \tag{6}$$

with:

$$\Theta_c(t) = \begin{cases} \max(0,t), & \text{if } c = u; \\ \min(0,t), & \text{if } c = l,v; \end{cases}$$

Here, d_i^c denotes the distance between the 2 atoms α_i^c and β_i^c, $w_c \geq 0$ are weighting factors for the 4 types of constraints ($c = u,l,v,a$), and $I_c \subseteq \{1, \ldots, n_c\}$ with $c = u,l,v$ are the subsets of distance constraints included in the target function. In the present version of the program DIANA the subsets I_c cannot be chosen arbitrarily but consist of all distance constraints between the atoms α and β in those residues between which the sequence numbers, R_α and R_β, respectively, differ by not more than a given minimization level L_c:

$$I_c = \left\{ i \in \{1, \ldots, n_c\} \,\middle|\, |R_{\alpha_i^c} - R_{\beta_i^c}| \leq L_c \right\}, \quad c = u,l,v. \tag{7}$$

It is usual to choose the same minimization level for all 3 kinds of distance constraints; in this case, we denote the common minimization level simply by L.

In general, a complete structure calculation with the program DIANA includes several minimization steps (not to be confused with individual iterations of the conjugate gradient minimizer), i.e. the minimization of several forms of the variable target function that differ in the minimization levels L_c, and in the weighting factors w_c (see the Appendix). An optimal strategy for selecting the minimization steps is not known, but we found it essential (1) to increase the minimization level gradually in a stepwise fashion, starting with $L_c = 0$ or 1, and (2) to use a weighting factor w_v for steric constraints that is small with respect to the weighting factors w_u and w_l for experimental distance constraints, e.g. $w_v = 0{\cdot}2 w_u$, except toward the end of the calculation, where one usually increases w_v to 2 to 3 times w_u in order to minimize steric overlaps.

The target function of eqn (6) is continuously differentiable over the entire conformation space, and is chosen such that the contribution of a single small violation δ_c is given by $w_c\delta_c^2$ for all types of constraints ($c = u,l,v,a$). Because only squared interatomic distances and no square-roots have to be computed, the target function can be calculated rapidly.

(ii) *Comparison of the variable target functions used in DIANA and DISMAN*

In the notation used here, the target function in the program DISMAN, T', which corresponds to the target function of eqn (6) used in DIANA, is defined by (Braun & Gō, 1985; Braun, 1987);

$$T' = \sum_{c=u,l} w_c \sum_{i \in I_c} \left(\Theta_c\left(\frac{d_i^{c2} - b_i^{c2}}{2b_i^c} \right) \right)^2 + \frac{w_v}{4} \sum_{i \in I_v} (\Theta_v(d_i^{v2} - b_i^{v2}))^2 + 4w_a \sum_{i=1}^{n_a} \left(1 - \frac{1}{2}\frac{|\Delta_i|}{\Gamma_i} \right)^2 \left(\frac{\Delta_i}{\Gamma_i} \right)^2. \tag{8}$$

The treatment of experimental upper and lower distance constraints is the same in the 2 programs. Steric constraints and dihedral angle constraints, however, are treated somewhat differently; because other normalization factors are used in DISMAN, the contribution of a small violation, δ, of a steric or dihedral angle constraint is not simply equal to the weighting factor multiplied with the squared violation. Rather, this contribution is $w_v b_i^{v2} \delta^2$ for a steric constraint and $4w_a(\delta/\Gamma_i)^2$ for a dihedral angle constraint. Furthermore, for a dihedral angle constraint violation, the maximal contribution to the DISMAN target function equals w_a and is independent of the width of the forbidden dihedral angle range, whereas in DIANA it equals $(w_a/2)\Gamma_i^2$ and is proportional to the squared width of the forbidden region.

(iii) *The input and output formats*

There are several different input files and some interactively entered parameters. The nomenclature in the standard residue library follows the IUPAC rules (IUPAC-IUB Commission on Biochemical Nomenclature, 1970), and the covalent structure is that of the ECEPP force field (Momany *et al.*, 1975; Némethy *et al.*, 1983) for the 20 proteinogenic amino acid residues. The primary structure is entered in the amino acid sequence file, which also identifies *cis*-peptide bonds. Pairs of diastereotopic substituents for which individual assignments are available are identified in the stereospecific assignments input file. If this file is missing, one has to provide the information that stereospecific assignments are available either for all or for none of the prochiral centers. Upper distance limits, and lower distance limits and dihedral angle constraints are read from input files. The minimization parameters input file contains details about the minimization procedure. The start conformations for the structure calculations can be generated by the program, or read from input files.

The results output file records the interactive input, includes information on the course of the minimization, and lists the constraint violations exceeding given threshold values. The overview output file includes a complete list of the numbers, the sums, and the maximal values of the residual constraint violations for each calculated structure. Furthermore, a table of the important violations in all structures with final target function values less than a user-defined cutoff is written whenever a structure calculation is finished. For the DIANA user, it is important that the overview file can be inspected during the operation of the program, whereas the results file can usually be examined only after completion of the current job. The dihedral angles and Cartesian co-ordinates files of the calculated structures can be written either at intermediate stages or at the end of the minimization. The r.m.s.d. values file includes pairwise global or local r.m.s.d. values between calculated structures, and the modified upper distance limits and modified lower distance limits files list the experimental constraints after processing by DIANA.

(iv) *Identification of irrelevant constraints and too restrictive constraints*

A distance limit is irrelevant if (1) the corresponding interatomic distance is independent of the conformation, (2) there exists no conformation ($\phi_1, \ldots, \phi_n$) that violates the given limit or (3) a lower distance limit is smaller than the steric limit automatically imposed by DIANA. Conditions (1) and (3) are easy to check, whereas a complete check of condition (2) is difficult. Therefore, this condition is checked only for all constraints on distances that depend on a single dihedral angle, and for some constraints relating to 2 dihedral angles. If the distance between 2 atoms α and β, $|\mathbf{r}_\alpha - \mathbf{r}_\beta|$, depends on 1 dihedral angle a, the range of its values is given by:

$$A - B \leq |\mathbf{r}_\alpha - \mathbf{r}_\beta|^2 \leq A + B, \qquad (9)$$

where:

$$A = |\mathbf{d}_\alpha|^2 + |\mathbf{d}_\beta|^2 - 2(\mathbf{e}_a \cdot \mathbf{d}_\alpha)(\mathbf{e}_a \cdot \mathbf{d}_\beta)$$

$$B = 2\sqrt{[\mathbf{d}_\alpha^2 - (\mathbf{e}_a \cdot \mathbf{d}_\alpha)^2][\mathbf{d}_\beta^2 - (\mathbf{e}_a \cdot \mathbf{d}_\beta)^2]}$$

with

$$\mathbf{d}_\alpha = \mathbf{r}_\alpha - \mathbf{r}_a \quad \text{and} \quad \mathbf{d}_\beta = \mathbf{r}_\beta - \mathbf{r}_a.$$

$\mathbf{r}_\alpha$ and $\mathbf{r}_\beta$ denote the position vectors of the atoms α and β for an arbitrary conformation, $\mathbf{r}_a$ is the position vector of the start point of the rotatable bond a, and $\mathbf{e}_a$ is a unit vector along the rotatable bond a. In the notation of eqn (9), an upper distance limit (α, β, b) is irrelevant if $b \geq A + B$, and too restrictive if $b < A - B$. Irrelevant constraints are removed from the input used for the calculation. Too restrictive distance constraints that cannot be fulfilled by any conformation will thus be identified in the results file, but they will not be removed from the input used for the calculation. For distances depending on 2 dihedral angles, the relation corresponding to eqn (9) is somewhat more complicated.

(v) *Processing of distance constraints involving pairs of diastereotopic substituents without stereospecific resonance assignments*

Because the standard sequential assignment procedure for proteins (Wüthrich, 1986) does not assign individually the diastereotopic substituents of prochiral groups, and additional techniques used for this purpose (e.g. Güntert *et al.*, 1989; Neri *et al.*, 1989) can provide stereospecific assignments for only part of the prochiral centers, programs used for structure calculations from n.m.r. data must contain routines to process distance constraints with pairs of diastereotopic substituents β_1 and β_2, (α, β_1, b_1) and (α, β_2, b_2), to a pseudoatom β_Q located centrally with respect to the 2 diastereotopic substituents β_1 and β_2, and to add a correction to the distance limit that equals the distance between the diastereotopic substituents and the pseudoatom (Wüthrich *et al.*, 1983):

$$b_Q = \min(b_1, b_2) + |\mathbf{r}_{\beta_1} - \mathbf{r}_{\beta_Q}|. \qquad (10)$$

In the program DIANA, we replaced these pseudoatom corrections with a combination of 2 approaches, which has the advantage that a lesser part of the information contained in the experimental data is lost by the data processing.

The 1st approach by DIANA uses the conventional pseudoatom concept, but with variable corrections, depending on the available experimental data. The upper distance limit for the pseudoatom constraint, (α, β_Q, b_Q), is then calculated as:

$$b_Q = \sqrt{\frac{b_1^2 + b_2^2}{2} - |\mathbf{r}_{\beta_1} - \mathbf{r}_{\beta_Q}|^2}. \qquad (11)$$

If only 1 of the 2 constraints can be measured, say (α, β_1, b_1), no improvement of the original pseudoatom correction can be attained, and $b_Q = b_1 + |\mathbf{r}_{\beta_1} - \mathbf{r}_{\beta_Q}|$. If there are 4 constraints between the diastereotopic substituents of 2 prochiral centers, $(\alpha_1, \beta_1, b_{11})$, $(\alpha_1, \beta_2, b_{12})$, $(\alpha_2, \beta_1, b_{21})$, $(\alpha_2, \beta_2, b_{22})$, the corresponding pseudoatom upper distance limit (α_Q, β_Q, b_Q) is given by:

$$b_Q = \sqrt{\frac{b_{11}^2 + b_{12}^2 + b_{21}^2 + b_{22}^2}{4} - |\mathbf{r}_{\alpha_1} - \mathbf{r}_{\alpha_Q}|^2 - |\mathbf{r}_{\beta_1} - \mathbf{r}_{\beta_Q}|^2}. \qquad (12)$$

In the frequently encountered situation where 1 or several of the 4 constraints are missing, redundant upper distance limits are generated by application of the triangle inequality, so that eqn (12) can still be applied.

In the 2nd approach used by DIANA, no pseudoatom is introduced. The same distance limit:

$$b = \min[\max(b_1, b_2), \min(b_1, b_2) + |\mathbf{r}_{\beta_1} - \mathbf{r}_{\beta_2}|]$$

is applied for both diastereotopic substituents. Obviously, the application of 2 identical limits is, in general, not equivalent to the use of a pseudoatom, and the 2nd approach is applied only if it yields additional information. For example, if only 1 of the 2 diastereotopic substituents has an experimental constraint to an outside proton, only the pseudoatom constraint will be used. On the other hand, if the n.m.r. experiments show that $b_1 \approx b_2$, there is no advantage to the introduction of a pseudoatom distance limit.

Table 1 lists some results obtained by processing distance constraints with prochiral centers with DIANA, or with the conventional pseudoatom correction method (Wüthrich *et al.*, 1983). The 1st example resulted from an input of 2 nearly equal upper distance limits; DIANA imposed the higher limit on both distances, but left the pseudoatom distance unconstrained because the upper bound resulting from eqn (11) would be meaningless besides b_1 and b_2. The 2nd example involves 2 significantly different upper bounds; DIANA imposes constraints on both individual distances and the pseudoatom distance constraint calculated by eqn (11). In both cases, the resulting constraints are significantly tighter than with the conventional pseudoatom corrections. In contrast, in the 3rd example in Table 1, where there is only 1 experimental upper distance limit, the 2 methods yield nearly equivalent results. In the 4th example, DIANA detected that the pseudoatom constraint would be irrelevant and hence dropped it from the input. The final example shows the result of a treatment of constraints between 2 pairs of diastereotopic protons. Using the triangle inequality, DIANA first generated the smallest possible redundant constraints of the 2 distances that were not constrained by the experimental upper bounds. Next, it imposed the biggest of the 4 upper bounds thus obtained on all four individual distances, and an upper bound on the distance between the 2 pseudoatoms was obtained from eqn (12). Overall, Table 1 confirms that the loss of constraining information is often smaller when the experimental input is processed by DIANA than by the conventional pseudoatom approach. Analogous modifications to those shown here for upper distance constraints result for lower limit distance constraints.

(vi) *Checks for steric overlap*

In molecular mechanics programs, non-bonded interactions between atoms are usually treated by a Lennard-Jones potential (Momany *et al.*, 1975; Weiner & Kollman, 1981; Brooks *et al.*, 1983; van Gunsteren *et al.*,

Table 1

Examples of results obtained by processing experimental distance constraints involving pairs of diastereotopic substituents without individual assignments either by DIANA or with the original pseudoatom concept

Constrained distances†	Input upper bounds (Å)‡	Modified upper bounds (Å)	
		DIANA§	Conventional pseudoatom concept¶
$H^{\beta 2,3}_{Arg1}$–H^{ε}_{Arg1}	$b_1 = 4{\cdot}9$, $b_2 = 5{\cdot}0$	$b_1 = b_2 = 5{\cdot}0$	$b_Q = 5{\cdot}9$
$H^{\gamma 2,3}_{Glu7}$–H^{N}_{Glu7}	$b_1 = 3{\cdot}7$, $b_2 = 3{\cdot}1$	$b_1 = b_2 = 3{\cdot}7$, and $b_Q = 3{\cdot}3$	$b_Q = 4{\cdot}1$
$H^{\beta 2,3}_{Tyr10}$–$H^{\beta 2}_{Lys41}$	$b_1 = 3{\cdot}5$	$b_Q = 4{\cdot}4$	$b_Q = 4{\cdot}5$
$H^{\beta 2,3}_{Tyr10}$–H^{N}_{Thr11}	$b_1 = 4{\cdot}5$		$b_Q = 5{\cdot}5$
$H^{\beta 2,3}_{Tyr10}$–$H^{\delta 2,3}_{Lys42}$	$b_{11} = 4{\cdot}7$, $b_{12} = 4{\cdot}6$	$b_{11} = b_{12} = b_{21} = b_{22} = 6{\cdot}5$, and $b_Q = 5{\cdot}5$	$b_Q = 6{\cdot}6$

† Taken from an experimental n.m.r. data set collected with BPTI.

‡ b_1 and b_2 denote upper distance bounds from the 2 substituents of a diastereotopic pair to the same proton outside the prochiral center. b_{11} denotes an upper distance bound from the 1st atom of one diastereotopic pair to the first atom of another diastereotopic pair, etc.

§ b_Q denotes the upper limit imposed on the pseudoatom distance.

¶ The numbers given result from adding the distance from the pseudoatom to the protons that it replaces to the smaller of the 2 experimental constraints with the prochiral center. For methylene groups, this correction is $m = 1{\cdot}0$ Å (Wüthrich *et al.*, 1983; Wüthrich, 1986).

1983; Wako & Gō, 1987; Schaumann *et al.*, 1990). In a distance geometry approach for structure determination of proteins from n.m.r. data, only the most dominant part of the energy function is kept, i.e. the steric repulsion (Havel & Wüthrich, 1984; Braun & Gō, 1985). In the program DIANA, the steric repulsion between 2 atoms is treated as a lower distance limit for the corresponding interatomic distance, the distance bound being set equal to the sum of the repulsive core radii of the 2 atoms. In the present version of DIANA, the same values for the repulsive core radii are used as by Braun & Gō (1985), i.e. 0·95 Å for amine or amide hydrogen atoms, 1·0 Å for all other hydrogen atoms, 1·35 Å for aromatic carbon atoms, 1·40 Å for all other carbon atoms, 1·30 Å for nitrogen atoms, 1·20 Å for oxygen atoms and 1·60 Å for sulfur atoms. If the distance between 2 atoms exceeds the sum of their repulsive core radii, no contribution to the target function results from this atom pair.

A straightforward implementation of a check for the aforementioned steric overlaps would require the calculation of almost all interatomic distances, and would therefore be very inefficient. Therefore, DIANA stores all atom pairs with reasonably small interatomic distances in a list of potential non-bonded interactions (Verlet, 1967; Allen & Tildesley, 1987). This list is updated only after a notable conformation change or after several iterations of the conjugate gradient minimization, and a fast algorithm for this update ensures that most interatomic distances need not be computed (Hockney & Eastwood, 1981; Braun & Gō, 1985; Grest *et al.*, 1989). In a protein molecule with $m \approx 1000$ atoms, the list of potential non-bonded interactions will usually contain less than 30,000 atom pairs, whereas the total number of atom pairs is of the order of $m(m-1)/2 \approx 500{,}000$. The number of atom pairs that actually give non-vanishing contributions to the target function at the end of the minimization is again much smaller, typically of the order of 100 for a "good" conformation.

In the program DIANA, the list of potential non-bonded contacts is divided into 2 parts. The 1st part is invariant during a structure calculation, i.e. it is set up only once at the start of a calculation and will not be affected by subsequent updates; it comprises all intraresidual and sequential distances (Wüthrich, 1986). Most steric lower limits that are irrelevant will be excluded from the list, which obviously includes all distances that are independent of the conformation. The 2nd part of the list is subject to an updating procedure and includes interatomic distances that are not already included in the invariant part. To create this list, the present conformation of the protein molecule is placed into a cubic lattice with a lattice constraint g equal to twice the biggest repulsive core radius, i.e. $g = 3{\cdot}2$ Å in the present version of the program DIANA, and only distances between atoms located within the same or in neighboring cells of the lattice are added to the list (Hockney & Eastwood, 1981; Braun & Gō, 1985; Grest *et al.*, 1989). Thus, it is ascertained that the list contains all non-bonded contacts that yield a non-vanishing contribution to the target function for the conformation present at the time the list is computed. Slight changes to this conformation will presumably change the list only slightly. Therefore, the list is updated only if, since the last update, at least 1 dihedral angle was changed by more than a preset limit, e.g. 10°, or if this limit is not reached, after a preset number of iterations, e.g. 50.

Special treatments are required for hydrogen bonds, disulfide bridges, and possibly other non-standard covalent links, because these bonds are not represented by the tree structure of rotatable bonds (Abe *et al.*, 1984), since the latter does not allow for flexible, closed rings. As a consequence, the steric lower limits for acceptor–donor distances in potential hydrogen bonds, and the distances between C^{β}_i and S^{γ}_j in disulfide Cys_i–Cys_j are reduced by 1·0 Å, and the steric lower limits between the cysteine sulfur atoms are decreased by 2·0 Å relative to the sum of the corresponding repulsive core radii. The bond lengths and angles of hydrogen bonds and disulfide bridges are fixed by explicit upper and lower distance limits (Williamson *et al.*, 1985). The proline rings are rigid structures in the ECEPP force field (Momany *et al.*, 1975; Némethy *et al.*, 1983) in the sense that there are no internal degrees of freedom within them. In contrast, in the program DIANA, one can allow for flexibility in such rings by "cutting" one of the covalent bonds. In the case of proline, this creates 4 new rotatable bonds. The ring is then closed only by explicit distance constraints, and the necesary elimination or decrease of some steric lower distance limits is done automatically. In the input for DIANA, a flexible proline residue is entered *via* an additional entry in the residue library, where the new rotatable bonds have to be defined and the closure of the ring is inherent only in the connectivity list, but not in the tree structure of the rotatable bonds.

(vii) *The minimization procedure*

The gradient of the target function defined by eqn (6) can be calculated with a fast algorithm because the target function can be written as a sum of functions of individual interatomic distances and individual dihedral angles (Noguti & Gō, 1983; Abe *et al.*, 1984). The partial derivative of the function T of eqn (4) with respect to a dihedral angle a' is given by:

$$\frac{\partial T}{\partial \phi_{a'}} = -(\mathbf{e}_{a'}, \mathbf{e}_{a'} \wedge \mathbf{r}_{a'}) \cdot \sum_{c=u,l,v} w_c \sum_{\substack{i \in I_c \\ \alpha_i^c \in M_{a'}}} \Theta_c \left(\frac{d_i^{c2} - b_i^{c2}}{b_i^{c2}} \right) \begin{pmatrix} \mathbf{r}_{\alpha_i^c} \wedge \mathbf{r}_{\beta_i^c} \\ \mathbf{r}_{\alpha_i^c} - \mathbf{r}_{\beta_i^c} \end{pmatrix} + 2w_a \sum_{i=1}^{n_a} \left(1 - \left(\frac{\Delta_i}{\Gamma_i} \right)^2 \right) \Delta_i \delta_{a_i a'}. \quad (13)$$

$r_{\alpha_i^c}$ and $\mathbf{r}_{\beta_i^c}$ are the position vectors of the atoms α_i^c and β_i^c, respectively, $\mathbf{e}_{a'}$ denotes the unit vector along the rotatable bond a',$\mathbf{r}_{a'}$ the start point of it, and $M_{a'}$ the set of all atoms for which the positions are affected by a change of the dihedral angle a' if the N-terminal part of the protein molecule is kept fixed.

The minimization algorithm used in the program DIANA is the well-known method of conjugate gradients (Powell, 1977). At each minimization step, conjugate gradient iterations are done until either the norm of the gradient vector is smaller than some preset value, or the maximal number of iterations at this step (as given in the minimization parameters input file) is exceeded. Because the minimization routine assumes a continuously differential target function, problems may arise if an update of the list of potential non-bonded contacts (see above) results in a discontinuous change of the target function. Therefore, the conjugate gradient minimization is automatically restarted after premature termination due to a jump in the target function.

(viii) *Optimization of structure calculations with DIANA*

In order to achieve a high level of efficiency of the target function and gradient evaluation, the calculation is divided up into several parts, and each part is executed only if it is necessary. At the outset of a structure calculation, the static list of potential non-bonded contacts is set up, irrelevant constraints are eliminated, and the distance limits with prochiral centers are processed. Then, at the start of each minimization step, a list of all currently used distance constraints according to the subsets I_u, I_l, I_v and I_a is prepared, where the list of non-static potential non-bonded contacts is subject to the aforementioned updating procedure. The parts of the calculation that have to be excluded once for each combined computation of the target function and its gradient are the generation of the Cartesian atomic co-ordinates from given dihedral angles, the identification of violated constraints, and the evaluation of some terms that are present in both eqns (6) and (13), and will therefore be needed for the evaluation of the target function and the gradient. Here, the time-limiting step is the computation of the interatomic distances corresponding to all distance constraints in the list, of which the great majority are steric constraints.

The program DIANA has been optimized for the vectorization capabilities of the CRAY X-MP, and it has been implemented on other UNIX machines and on VAX computers (see Table 2). Since standard Fortran-77 has been used as far as possible, the additional implementation of DIANA on other computers will be straightforward.

(c) *The program GLOMSA*

The input for the program GLOMSA includes a group of m 3-dimensional protein structures, and the list of conformational constraints from which these structures were calculated by DIANA. In a typical situation analyzed by GLOMSA, an atom α outside the prochiral center considered has 2 upper distance limits (α, β_1, b_1) and (α, β_2, b_2) to the diastereotopic substituents β_1 and β_2. We denote with $d_{k,l}$ $(k = 1,2;\ l = 1, \ldots, m)$ the distance between the atoms α and β_k in the lth conformation. Then the program GLOMSA computes the sum of the residual violation of the 2 upper distance limits in the m conformations, V, for either of the 2 possible stereospecific assignments I and R (I, is the assignment used arbitrarily in the input, R is the reversed one).

$$V^I = \sum_{l=1}^{m} [\Theta(d_{1,l}-b_1)(d_{1,l}-b_1) + \Theta(d_{2,l}-b_2)(d_{2,l}-b_2)]$$
$$V^R = \sum_{l=1}^{m} [\Theta(d_{1,l}-b_2)(d_{1,l}-b_2) + \Theta(d_{2,l}-b_1)(d_{2,l}-b_1)], \quad (14)$$

and the minimum value, v, that the larger of the violations of the 2 constraints b_1 and b_2 has in any of the conformations:

$$v^I = \min_{l=1,\ldots,m} \max [\Theta(d_{1,l}-b_1)(d_{1,l}-b_1),\ \Theta(d_{2,l}-b_2)(d_{2,l}-b_2)]$$
$$v^R = \min_{l=1,\ldots,m} \max [\Theta(d_{1,l}-b_2)(d_{1,l}-b_2),\ \Theta(d_{2,l}-b_1)(d_{2,l}-b_1)]. \quad (15)$$

In eqns (14) and (15) as well as in eqn (17) below, Θ denotes the Heaviside function that equals 1 if the argument is positive, and vanishes otherwise. The program GLOMSA further calculates the average of the differences $\Delta d_l = d_{1,l} - d_{2,l}$ in the m structures:

$$\overline{\Delta d} = \frac{1}{m} \sum_{l=1}^{m} \Delta d_l \quad (16)$$

and the signed maximal number of conformations where Δd_l has the same sign:

$$n_{\Delta d} = \begin{cases} s & s > m/2; \\ -(m-s) & s \le m/2; \end{cases} \qquad s = \sum_{l=1}^{m} \Theta(\Delta d_l). \quad (17)$$

$n_{\Delta d}$ is constructed such that $|n_{\Delta d}| \ge m/2$, and that its sign is positive if $d_{1,l}$ is bigger than $d_{2,l}$ in the majority of the conformations, and negative otherwise. The sign of $n_{\Delta d}$ is not necessarily the same as the sign of $\overline{\Delta d}$.

To identify stereospecific assignments, GLOMSA correlates the signs $\overline{\Delta d}$ and $n_{\Delta d}$ with the sign of $\Delta b = b_1 - b_2$. Matching signs confirm that the stereospecific assignment I assumed in the input is correct, and opposite signs indicate the stereospecific assignments R. In order to exclude cases where the available data do not clearly distinguish between the 2 possibilities, $|\Delta b|$, $|\overline{\Delta d}|$ and $|n_{\Delta d}|$ are further required to exceed user-defined threshold values for an unambiguous stereospecific assignment by the program GLOMSA. If a relation of the type $d_1 > d_2$ or $d_1 < d_2$ was unambiguously established by the experiments, this relation is accepted without imposing the threshold condition on $|\Delta b|$.

If there are several pairs of upper distance constraints from different hydrogen atoms to the same prochiral center, the above procedure is repeated independently for each distance constraint pair. The user then manually combines the individual stereospecific assignments obtained, since potential inconsistencies could be hidden in the output resulting from an automated combination of the individual stereospecific assignments. On the basis of the quantities defined in eqns (14) and (15), another method for obtaining stereospecific assignments is conceivable, where an unambiguous assignment I would be assumed if $v^I = 0$ and $v^R > 0$. However, it turns out that such a criterion would be very restrictive and would usually not yield a significant number of stereospecific assignments. Therefore, GLOMSA calculates V and v only as informative output.

Examples of the results obtained with the program GLOMSA in an application with experimental n.m.r. data are given elsewhere (Güntert *et al.*, 1990).

3. Results and Discussion

(a) *Computing time for a protein structure calculation with DIANA*

The computation speed of the program DIANA on ten different computers (Table 2) was measured for the calculation of one structure of BPTI using the data set WIST, which was derived from the regularized crystal structure of BPTI (Marquardt *et al.*, 1983) so as to mimic an experimental n.m.r. input (Güntert *et al.*, 1989). The same random start conformation was used on all computers. A total number of 5900 target function evaluations was allowed, and updates of the list of potential non-bonded contacts were made when a dihedral angle was changed by more than 10° since the previous update, or after 50 iterations without update.

Table 2

Central processing unit times required by the program DIANA for the calculation of one BPTI structure

Computer type	c.p.u. time (min)†	Factor‡
Cray X-MP/28§	0·82 (3·2)	1·0 (3·9)
VAX 8650	20	25
VAX 6000-420	18	22
SUN 386i	111	136
SUN 3/260¶	44 (188)	54 (229)
SUN 4/390	14	17
SUN 4/60	18	22
Silicon Graphics Personal IRIS 4D/25	13	16
Convex C1	18	22
CDC Cyber 855	12	15

One completely folded conformation of the protein BPTI was calculated starting from random dihedral angles. The same starting structure and the same constraints were used on all computers. The same code, which was optimized for the Cray X-MP, and single precision floating point arithmetics were used on all machines (see the text for further details).

† c.p.u. times were measured using the library routines SECOND on the Cray X-MP and the CDC Cyber, ETIME on the other UNIX computers SUN, Silicon Graphics Personal IRIS, and Convex, and LIB$STAT_TIMER on the VAX machines.

‡ Ratio between the c.p.u. times required on the given machine and on the Cray X-MP using vectorization.

§ The numbers in parentheses were obtained when vectorization was deliberately inhibited. In either case only 1 c.p.u. was used.

¶ The numbers in parentheses were obtained when the floating point coprocessor MC 68881 was used instead of the floating point accelerator.

A clear-cut result from the measurements of the total c.p.u. time used is that the program DIANA runs by more than one order of magnitude faster on the Cray X-MP than on any of the other computers included in the test (Table 2). Even when the vectorization was completely inhibited, which increased the c.p.u. time by a factor of 3·9, the Cray X-MP remained the fastest machine. Following the Cray X-MP, there is a group of computers with similar performances, i.e. the VAX 8650, VAX 6000-420, SUN 4, Silicon Graphics Personal IRIS, Convex C1 and CDC Cyber 855, which used 12 to 20 minutes of c.p.u. time. Finally, the smaller machines SUN 3 and SUN 386i required again significantly more computer time to solve the test problem. It should be pointed out that the same code has been used on all machines, which is optimized with regard to the vectorization capabilities of the Cray X-MP.

(b) *Comparison of the structures calculated with different computers*

Even though exactly the same input of conformational constraints and identical starting conformations have been supplied to the program on each computer, the final conformation obtained at the end of the minimization was in general different on the different computers (only the SUN 4/60 and the SUN 4/390 produced exactly identical structures). Repeating the calculation on the same computer yielded identical results. On all computers, the target function value of the starting conformation was the same, and the target function values gradually started to diverge with increasing minimization level, with final target function values ranging from 1·28 to 3·81 Å^2. These are small target function values when compared to "bad" structures obtained from different starting conformations, which often end up in high local minima with target function values above 100 Å^2. The average of the pairwise r.m.s.d. values (McLachlan, 1979) between 11 structures obtained from the same starting conformation on different computers, or on the same computer with different compiler settings (the result of the CDC Cyber was for technical reasons not used in this comparison) are 0·59 and 0·84 Å for all backbone atoms and all heavy atoms, respectively, the maximal values being 0·98 and 1·41 Å. If only the residues 3 to 55 were taken into account (i.e. the loose chain ends are discarded), the average of the pairwise r.m.s.d. values were 0·47 and 0·78 Å for the backbone and heavy atoms, respectively, with maximal values of 0·72 and 1·30 Å. These r.m.s.d. values are comparable to those between the four conformations obtained with the program DISMAN (Braun & Gō, 1985) starting from widely different initial structures, as described previously (Güntert *et al.*, 1989). Overall, the implication is that the structures obtained with the same starting conformation when using the different computers are distributed within the bounds of the conformation space defined by the experimental constraints.

This result can be rationalized from the following. Because the experimental data restrict distances and dihedral angles to allowed ranges rather than to unique values (Wüthrich, 1986), one cannot expect the target function to have a well-defined global minimum. Instead there is an allowed region in conformation space where the target function has low values throughout, so that conformation changes within this region alter the target function only slightly. Therefore, although the aforementioned allowed region of the conformation space is defined by the experimental constraints (note, however, that the boundaries of this allowed region are not sharp, since they are in practice also influenced by the somewhat arbitrary selection of a set of "good" structures from among those calculated with distance geometry from the same input with different starting conformations), the exact conformation attained within this allowed region is heavily influenced by, for example, round-off errors. Overall, it is thus not really a surprise that the r.m.s.d. values between the structures obtained from different computers and an identical starting conformation are similar to those between structures obtained from widely different starting conformations when using the same computer.

(c) *Influence of different treatments of distance constraints with pairs of diastereotopic substituents*

In order to assess the influence of different treatments of distance constraints with pairs of diastereotopic substituents on the convergence of the variable target function algorithm and on the quality of the structures obtained, DIANA calculations were carried out using experimental n.m.r. data for BPTI. Four input data sets that differed only in the treatment of the distance constraints with pairs of diastereotopic substituents were compared (Table 3). Datasets I and II include all stereospecific assignments present in the experimental n.m.r. data set (L. Orbons, P. Güntert & K. Wüthrich, unpublished results), whereas datasets III and IV include no stereospecific assignments. Upper distance limits involving pairs of diastereotopic substituents without individual n.m.r. assignments were treated by the method implemented in the program DIANA for datasets I and III, or by the conventional pseudoatom method (Wüthrich *et al.*, 1983) for datasets II and IV. Note that the different numbers of upper distance limits in datasets I to IV (Table 3) are a direct consequence of the different treatments of the NOE distance constraints with prochiral centers. The lower limit distance constraints and dihedral angle constraints were identical in datasets I to IV.

For each of the four datasets, 150 structure calculations were started with random start conformations. The final minimization level was $L = 58$, and the maximal number of target function evaluations per conformation was 9900. The calculations were done on a Cray X-MP computer, and the total c.p.u. time required was 18·6 hours. (Note that the time used per structure calculation is longer than in Table 2, because more target function evaluations were performed.)

A statistical analysis of the residual constraint violations in the 20 conformations with smallest final target function values in each of the groups I to IV is afforded by Table 4. Overall, the structure groups I to IV have similar quality in terms of the target function values and the sums of distance

Table 3

Characterization of the four BPTI data sets used to investigate the influence of different treatments of distance constraints with pairs of diastereotopic substituents

	Datasets			
Quantity	I	II	III	IV
Stereospecifically assigned prochiral centers†	42	42	0	0
Individually assigned NH_2 groups of Asn†	3	3	0	0
Treatment of diastereotopic pairs without individual assignments‡	DIANA	Conventional pseudoatom concept	DIANA	Conventional pseudoatom concept
Upper distance limits	866	781	924	637
Lower distance limits§	31	31	31	31
Dihedral angle restraints¶	140	140	140	140

† The total number of prochiral centers is 87. 14 β-methylene groups were stereospecifically assigned using the program HABAS, and all other stereospecific assignments (25 for methylene groups, and 3 for isopropyl groups) were obtained with GLOMSA. In addition, individual assignments for 3 NH_2 groups of asparagine were established from the intraresidual NOEs to $C^{\gamma}H_2$.

‡ DIANA refers to the novel treatment of diastereotopic pairs without individual assignments as explained in the text. Conventional pseudoatom concept refers to the method of Wüthrich *et al.* (1983).

§ Exactly the same experimental lower distance limits for disulfide bridges and experimentally established hydrogen bonds (Williamson *et al.*, 1985) were included in the datasets I to IV.

¶ Exactly the same dihedral angle restraints were included in the datasets I to IV. They were derived from a combined analysis of ^3J scalar coupling constants and short-range distance constraints using the program HABAS (Güntert *et al.*, 1989).

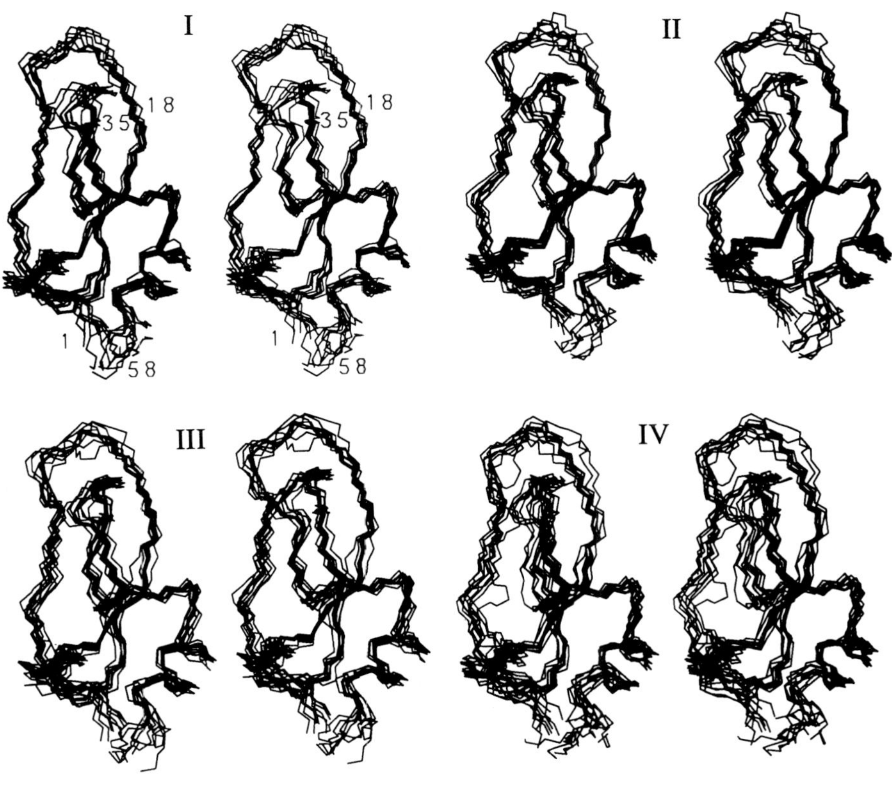

(a)

Figure 2. Stereo views of the 10 BPTI conformations with smallest target function values in each of the groups I to IV (Table 4). (a) Backbone of the whole protein (residues 1 to 58). (b) Heavy atoms of the β-sheet (residues 18 to 35).

constraint violations, but nonetheless clearly improved convergence of the calculations was obtained for the datasets with stereospecific assignments (I and II). Thus, using a less constraining dataset does not generally yield lower final target function values, even though the global minimum of the target function cannot be higher than for a "better" data set. This observation probably results because the folding pathway is also less determined in the absence of stereospecific assignments.

The different precision of the structure determinations with datasets I to IV is visualized with the aid of stereoviews of superpositions of the ten conformations with smallest final target function values in each group (Fig. 2). These images were produced with the program CONFOR (Billeter *et al.*, 1985). For the backbone of the whole protein (Fig. 2(a)) as well as for the all heavy-atom presentation of the β-sheet (Fig. 2(b)) more precisely defined structures were obtained in groups I and II, which include stereospecific assignments, than in III and IV. The fact that the structures from group IV are clearly less well determined than the structures from group III demonstrates the advantage of the novel DIANA processing of distance constraints with pairs of diastereotopic substituents that were not individually assigned.

For a more quantitative assessment of the observations in Figure 2, we calculated r.m.s.d. values (McLachlan, 1979) among the conformations of each group (Table 5). In a first comparison we included the 20 conformations with smallest final target function values in each of the four structure groups I to IV (Table 3). Because conformations with higher target function values tend to exhibit higher r.m.s.d. values, we made a second comparison using only the conformations with final target function values less than 5 Å^2. In this second comparison, possible effects from the poorer convergence found for groups III and IV, which use no stereospecific assignments, are eliminated. There are 17, 19, 10 and 6 such conformations in structure groups I, II, III and IV, respectively. In both comparisons, the average pairwise r.m.s.d. values given in Table 5 show a clear tendency to increase from the structure groups with stereospecific assign-

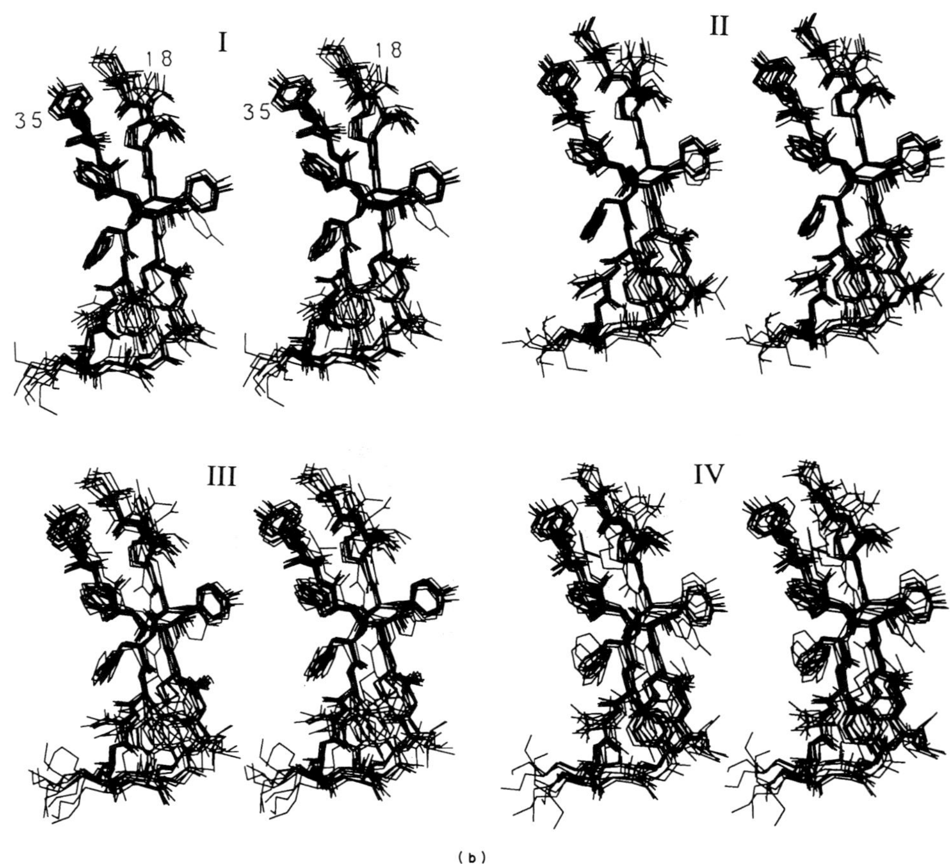

(b)

Fig. 2.

ments (I and II) to those without stereospecific assignments. A smaller increase of the r.m.s.d. is obtained also when going from a structure group calculated using the novel DIANA processing of distance constraints with pairs of diastereotopic substituents to the corresponding group calculated with conventional pseudoatom concept (Wüthrich *et al.*, 1983). Very similar results were obtained with the best 20 conformations from each group, or only those with final target function values less than 5 Å^2 (Table 5).

As was shown previously (Güntert *et al.*, 1989; Nilges *et al.*, 1990) the precision of the distance geometry structures can be significantly improved when stereospecific assignments are available. This observation is confirmed by the data in Table 5, which were obtained using experimental input data rather than test situations derived from known structures. The results obtained encourage continuation of studies on experimental methods, e.g. biosynthetically directed fractional ^{13}C labeling (Senn *et al.*, 1989; Neri *et al.*, 1989), for obtaining stereospecific assignments prior to the structure calculation. When the novel processing of upper distance constraints with pairs of diastereotopic substituents implemented in DIANA is compared to the more conservative pseudoatom concept (Wüthrich *et al.*, 1983), only a slight improvement of the calculated structures can be registered (Table 5 and Fig. 2, compare I and II, or III and IV). From a practice-oriented viewpoint, it is perhaps a more important advantage of the DIANA processing that it is fully automated, and integrated into the structure calculations. It is thus less laborious than obtaining a maximum number of stereospecific assignments. Therefore, the combined use of the automated HABAS routine (Güntert *et al.*, 1989), which can provide a limited number of stereospecific assignments (see the footnote to Table 3), and the DIANA processing of the distance constraints might become a viable alternative. This idea receives support from the observation that the advantages of the DIANA processing relative to the conventional pseudoatom treatment (Wüthrich *et*

Table 4

Statistics of the final target function values and residual constraint violations of the 20 BPTI structures with smallest final target function values in each of the four structure groups calculated from datasets I to IV

Quantity	Average value ± standard deviation†			
	I	II	III	IV
Target function value (Å^2)‡	2·9±1·3	2·6±1·3	4·8±2·5	6·0±3·1
Upper distance limits:				
Violations > 0·2 Å	8·0±4·2	7·0±3·4	11·6±6·9	11·2±6·0
Sum of violations (Å)	8·1±2·4	7·1±2·0	9·4±3·2	8·3±3·5
Maximal violation (Å)	0·51±0·16	0·48±0·17	0·71±0·26	0·78±0·32
Lower distance limits:				
Violations > 0·2 Å	0·9±1·0	0·4±0·6	0·6±0·7	0·8±0·9
Sum of violations (Å)	0·7±0·3	0·5±0·2	0·7±0·2	0·6±0·3
Maximal violation (Å)	0·24±0·12	0·17±0·06	0·22±0·08	0·19±0·10
Steric constraints:				
Violations > 0·2 Å	1·0±1·3	1·5±1·3	3·7±3·1	5·2±3·7
Sum of violations (Å)	3·6±1·0	3·8±1·2	5·2±1·8	6·3±2·5
Maximal violation (Å)	0·28±0·19	0·27±0·11	0·39±0·20	0·43±0·16
Dihedral angle restraints:				
Violations > 5°	1·2±1·2	1·1±1·5	1·6±1·1	2·7±1·8
Sum of violations (°)	27±13	29±15	40±13	50±24
Maximal violation (°)	6·9±2·6	7·3±4·2	9·2±4·4	14·3±10·7

The data sets I to IV are defined in Table 3. A total of 150 structures were calculated with each of the datasets I to IV, and the 20 structures with smallest final target function value were included in this analysis.

† Of the individual values for the 20 conformations of each structure group.

‡ The weighting factors for experimental upper and lower limit distance constraints were $w_u = w_l = 1$, the weighting factor for steric lower distance limits was $w_v = 2$, and the weighting factor for dihedral angle restraints was $w_a = 5$ Å^2 for the final minimization step.

Table 5

Average values and empirical standard deviations of the pairwise r.m.s.d. within the BPTI structure groups calculated from datasets I to IV to monitor the influence of stereospecific assignments

Atom set	r.m.s.d. within structure group (Å)			
	I	II	III	IV
A. *The 20 conformations with smallest final target function*				
Backbone 1–58	1·12±0·22	1·15±0·19	1·29±0·22	1·52±0·24
Backbone 3–55†	0·89±0·16	0·93±0·15	1·09±0·18	1·32±0·20
Backbone 18–35 (β-sheet)	0·54±0·16	0·57±0·16	0·79±0·28	0·75±0·24
Backbone 48–55 (α-helix)	0·27±0·10	0·26±0·09	0·32±0·10	0·36±0·11
Heavy atoms 1–58	1·80±0·18	1·87±0·19	2·10±0·23	2·24±0·22
Heavy atoms 3–55	1·72±0·17	1·80±0·19	2·04±0·24	2·16±0·22
Heavy atoms 18–35 (β-sheet)	1·10±0·16	1·12±0·14	1·46±0·29	1·41±0·28
Heavy atoms 48–55 (α-helix)	1·28±0·32	1·19±0·29	1·44±0·33	1·39±0·32
B. *Conformations with final target function value* $<5{\cdot}0$ *Å*^{2}‡				
Backbone 1–58	1·10±0·23	1·16±0·19	1·14±0·15	1·43±0·27
Backbone 3–55	0·86±0·16	0·94±0·16	0·97±0·15	1·27±0·26
Backbone 18–35 (β-sheet)	0·55±0·16	0·56±0·16	0·64±0·23	0·76±0·28
Backbone 48–55 (α-helix)	0·27±0·10	0·26±0·09	0·29±0·08	0·38±0·09
Heavy atoms 1–58	1·79±0·18	1·88±0·19	1·96±0·18	2·18±0·24
Heavy atoms 3–55	1·72±0·17	1·81±0·20	1·93±0·20	2·13±0·23
Heavy atoms 18–35 (β-sheet)	1·10±0·16	1·11±0·14	1·33±0·25	1·39±0·28
Heavy atoms 48–55 (α-helix)	1·30±0·31	1·17±0·26	1·36±0·27	1·59±0·40

The data sets I to IV that were used to calculate the structure groups I to IV are described in Table 3 (see the text for further details).

† Residues 3 to 55 are chosen in order to exclude the less well determined terminal parts of the polypeptide chain.

‡ We obtained 17, 19, 10 and 6 conformations with final target function values less than 5 Å^2 from datasets I, II, III and IV, respectively.

al., 1983) are most clear-cut when no stereospecific assignments are available (III and IV in Table 5 and Fig. 2).

Appendix

Choice of Parameters for Protein Structure Determinations with DIANA

Table A1 affords a complete list of the minimization levels, the weighting factors for steric constraints and the iteration limits used for the protein structure calculations described in Results and Discussion, sections (a) and (b). The choice of these parameters is typical for structure calculations with proteins of the size of BPTI, and agrees with the general recommendations given in the main text following equation (7).

Table A1

Minimization steps used for a structure calculation of BPTI

Step	Minimization level, L†	Weight of steric constraints, w_v‡	Maximal number of iterations§
1	0	0·2	300
2	1	0·2	500
3	2	0·2	300
4	3	0·2	200
5	4	0·2	100
6	5	0·2	100
7	6	0·2	100
8	7	0·2	100
9	9	0·2	200
10	10	0·2	100
11	11	0·2	100
12	12	0·2	100
13	13	0·2	200
14	15	0·2	100
15	17	0·2	100
16	18	0·2	100
17	19	0·2	100
18	21	0·2	300
19	22	0·2	100
20	23	0·2	100
21	24	0·2	300
22	25	0·2	100
23	26	0·2	200
24	27	0·2	100
25	30	0·2	100
26	31	0·2	100
27	32	0·2	100
28	36	0·2	100
29	38	0·2	100
30	39	0·2	100
31	45	0·2	100
32	50	0·2	100
33	54	0·2	100
34	56	0·2	100
35	58	0·2	100
36	58	0·6	300
37	58	2·0	500

BPTI consists of a polypeptide chain with 58 residues.

† For each minimization step, the same minimization levels were used for the 3 kinds of distance constraints, $L = L_u = L_l = L_v$.

‡ Throughout the structure calculation, the weighting factors for experimental distance constraints were $w_u = w_l = 1$, and the weighting factor for dihedral angle restraints was $w_a = 5$ Å^2.

§ The numbers of iterations are usually chosen based on the number of upper distance limits on the given level that were not already included at the preceding level.

We thank Dr M. Billeter for helpful discussions, Mr F. Suter for the use of a SUN 386i work station, and Mr R. Marani for the careful processing of the typescript. We acknowledge financial support by the Schweizerischer Nationalfonds (project 31.25174.88), the use of the Cray X-MP/28 of the ETH Zürich, and the use of a Silicon Graphics Personal IRIS of the Institut für Zellbiologie of the ETH Zürich.

References

Abe, H., Braun, W., Noguti, T. & Gō, N. (1984). *Comput. Chem.* **8**, 239–247.

Allen, M. P. & Tildesley, D. J. (1987). *Computer Simulation of Liquids*, pp. 147–152, Clarendon Press, Oxford.

Billeter, M., Engeli, M. & Wüthrich, K. (1985). *J. Mol. Graphics*, **3**, 79–83 and 97–98.

Billeter, M., Havel, T. F. & Wüthrich, K. (1987). *J. Comp. Chem.* **8**, 132–141.

Billeter, M., Schaumann, T., Braun, W. & Wüthrich, K. (1990). *Biopolymers*, **29**, 695–706.

Blumenthal, L. M. (1970). *Theory and Applications of Distance Geometry*, Chelsea, New York.

Braun, W. (1987). *Quart. Rev. Biophys.* **19**, 115–157.

Braun, W. & Gō, N. (1985). *J. Mol. Biol.* **186**, 611–626.

Braun, W., Bösch, C., Brown, L. R., Gō, N. & Wüthrich, K. (1981). *Biochim. Biophys. Acta*, **667**, 377–396.

Braun, W., Wider, G., Lee, K. H. & Wüthrich, K. (1983). *J. Mol. Biol.* **169**, 921–948.

Brooks, B. R., Bruccoleri, R. E., Olafson, B. D., States, D. J., Swaminathan, S. & Karplus, M. (1983). *J. Comp. Chem.* **4**, 187–217.

Brünger, A. T., Clore, G. M., Gronenborn, A. M. & Karplus, M. (1986). *Proc. Nat. Acad. Sci., U.S.A.* **83**, 3801–3805.

Clore, G. M., Gronenborn, A. M., Brünger, A. T. & Karplus, M. (1985). *J. Mol. Biol.* **186**, 435–455.

Crippen, G. M. (1977). *J. Comp. Phys.* **24**, 96–107.

Driscoll, P. C., Gronenborn, A. M., Beress, L. & Clore, G. M. (1989). *Biochemistry*, **28**, 2188–2198.

Dubs, A., Wagner, G. & Wüthrich, K. (1979). *Biochim. Biophys. Acta*, **577**, 177–194.

Eccles, C., Billeter, M., Güntert, P. & Wüthrich, K. (1989). *Abstracts Xth Meeting of the International Society of Magnetic Resonance*, Morzine, France, July 16–21, 1989, p. S50.

Gordon, S. L. & Wüthrich, K. (1978). *J. Amer. Chem. Soc.* **100**, 7094–7096.

Grest, G. S., Dünweg, B. & Kremer, K. (1989). *Comput. Phys. Comm.* **55**, 269–285.

Güntert, P., Braun, W., Billeter, M. & Wüthrich, K. (1989). *J. Amer. Chem. Soc.* **111**, 3997–4004.

Güntert, P., Qian, Y. Q., Otting, G., Müller, M., Gehring, W. & Wüthrich, K. (1991). *J. Mol. Biol.* **217**, 531–540.

Havel, T. F. & Wüthrich, K. (1984). *Bull. Math. Biol.* **46**, 673–698.

Havel, T. F. & Wüthrich, K. (1985). *J. Mol. Biol.* **182**, 281–294.

Havel, T. F., Kuntz, I. D. & Crippen, G. M. (1983). *Bull. Math. Biol.* **45**, 665–720.

Hockney, R. W. & Eastwood, J. W. (1981). *Computer Simulations Using Particles*, McGraw-Hill, New York.

Holak, T. A., Prestegard, J. H. & Forman, J. D. (1987). *Biochemistry*, **26**, 4652–4660.
IUPAC-IUB Commission on Biochemical Nomenclature (1970). *J. Mol. Biol.* **52**, 1–17.
Kaptein, R., Zuiderweg, E. R. P., Scheek, R. M., Boelens, R. & van Gunsteren, W. F. (1985). *J. Mol. Biol.* **182**, 179–182.
Keller, R. M. & Wüthrich, K. (1980). *Biochim. Biophys. Acta*, **621**, 204–217.
Kline, A. D., Braun, W. & Wüthrich, K. (1988). *J. Mol. Biol.* **204**, 675–724.
Kohda, D., Gō, N., Hayashi, K. & Inagaki, F. (1988). *J. Biochem.* **103**, 741–743.
Kuntz, I. D., Crippen, G. M., Kollman, P. A. & Kimmelman, D. (1976). *J. Mol. Biol.* **106**, 983–994.
Lee, M. S., Gippert, G. P., Soman, K. V., Case, D. A. & Wright, P. E. (1989). *Science*, **245**, 635–637.
Marquardt, M., Walter, J., Deisenhofer, J., Bode, W. & Huber, R. (1983). *Acta Crystallogr. sect. B*, **39**, 480–490.
McLachlan, A. D. (1979). *J. Mol. Biol.* **128**, 49–79.
Momany, F. A., McGuire, R. F., Burgess, A. W. & Scheraga, H. A. (1975). *J. Phys. Chem.* **79**, 2361–2381.
Némethy, G., Pottle, M. S. & Scheraga, H. A. (1983). *J. Phys. Chem.* **87**, 1883–1887.
Neri, D., Szyperski, T., Otting, G., Senn, H. & Wüthrich, K. (1989). *Biochemistry*, **28**, 7510–7516.
Nilges, M., Clore, G. M. & Gronenborn, A. M. (1988). *FEBS Letters*, **239**, 129–136.
Nilges, M., Clore, G. M. & Gronenborn, A. M. (1990). *Biopolymers*, **29**, 813–822.
Noguti, T. & Gō, N. (1983). *J. Phys. Soc. Jpn*, **52**, 3685–3690.
Powell, M. J. D. (1977). *Math. Program.* **12**, 241–254.
Qian, Y. Q., Billeter, M., Otting, O., Müller, M., Gehring, W. J. & Wüthrich, K. (1989). *Cell*, **59**, 573–580.
Schaumann, T., Braun, W. & Wüthrich, K. (1990). *Biopolymers*, **29**, 679–694.
Schultze, P., Wörgötter, E., Braun, W., Wagner, G., Vašák, M., Kägi, J. H. R. & Wüthrich, K. (1988). *J. Mol. Biol.* **203**, 251–268.
Senn, H., Werner, B., Messerle, B. A., Weber, C., Traber, R. & Wüthrich, K. (1989). *FEBS Letters*, **249**, 113–118.
van Gunsteren, W. F., Berendsen, H. J. C., Hermans, J., Hol, W. G. J. & Postma, J. P. M. (1983). *Proc. Nat. Acad. Sci., U.S.A.* **80**, 4315–4319.
Vásquez, M. & Scheraga, H. A. (1988). *J. Biomol. Struct. Dynam.* **5**, 757–784.
Verlet, L. (1967). *Phys. Rev.* **159**, 98–103.
Wagner, G., Braun, W., Havel, T. F., Shaumann, T., Gō, N. & Wüthrich, K. (1987). *J. Mol. Biol.* **196**, 611–639.
Wako, H. & Gō, N. (1987). *J. Comp. Chem.* **8**, 625–635.
Weber, P. L. Morrison, R. & Hare, D. (1988). *J. Mol. Biol.* **204**, 483–487.
Weiner, P. K. & Kollman, P. A. (1981). *J. Comp. Chem.* **2**, 287–303.
Widmer, H., Billeter, M. & Wüthrich, K. (1989). *Proteins*, **6**, 357–371.
Williamson, M. P., Havel, T. F. & Wüthrich, K. (1985). *J. Mol. Biol.* **182**, 295–315.
Wüthrich, K. (1986). *NMR of Proteins and Nucleic Acids*, Wiley, New York.
Wüthrich, K. (1989). *Acc. Chem. Res.* **22**, 36–44.
Wüthrich, K., Wider, G., Wagner, G. & Braun, W. (1982). *J. Mol. Biol.* **155**, 311–319.
Wüthrich, K., Billeter, M. & Braun, W. (1983). *J. Mol. Biol.* **169**, 949–961.
Zuiderweg, E. R. P., Nettesheim, D. G., Mollison, K. W. & Carter, G. W. (1989). *Biochemistry*, **28**, 175–185.

Edited by P. E. Wright

Journal of Biomolecular NMR, 1 (1991) 447–456
ESCOM

J-Bio NMR 035

Improved efficiency of protein structure calculations from NMR data using the program DIANA with redundant dihedral angle constraints

Peter Güntert and Kurt Wüthrich

Institut für Molekularbiologie und Biophysik, Eidgenössische Technische Hochschule-Hönggerberg, CH-8093 Zürich, Switzerland

Dedicated to the memory of Professor V.F. Bystrov

Received 29 July 1991
Accepted 5 August 1991

Keywords: NMR structures of proteins; Structure calculation from NMR data; Variable target function method; Program DIANA; Local minimum problem; Dihedral angle constraints

SUMMARY

A new strategy for NMR structure calculations of proteins with the variable target function method (Braun, W. and Gō, N. (1985) *J. Mol. Biol.*, **186**, 611) is described, which makes use of redundant dihedral angle constraints (REDAC) derived from preliminary calculations of the complete structure. The REDAC approach reduces the computation time for obtaining a group of acceptable conformers with the program DIANA 5–100-fold, depending on the complexity of the protein structure, and retains good sampling of conformation space.

INTRODUCTION

NMR structures of proteins in solution are commonly presented as a group of conformers, each of which has been calculated individually from the same experimental input data. In a high-quality structure determination each individual conformer has small residual violations of the experimental conformational constraints, and the root-mean-square deviations (RMSDs) among all conformers in the group are small (Braun, 1987; Wüthrich, 1986,1989). In this paper we describe

Abbreviations: RMSD, root-mean-square deviation; REDAC, use of redundant dihedral angle constraints; HD, mutant *Antennapedia* homeodomain with Cys^{39} replaced by Ser; BPTI, basic pancreatic trypsin inhibitor; ADB, activation domain from porcine procarboxypeptidase B.

0925-2738/$ 5.00 © 1991 ESCOM Science Publishers B.V.

a new strategy for the use of the variable target function program DIANA (Güntert et al., 1991a) that enables efficient calculation of such groups of conformers.

The basic idea of the variable target function algorithm (Braun and Gō, 1985) is to *gradually* fit an initially randomized starting structure to the conformational constraints collected with the use of NMR experiments, starting with intraresidual constraints only, and increasing the 'target size' stepwise up to the length of the complete polypeptide chain. Since 1986 different implementations of the variable target function algorithm in the programs DISMAN (Braun and Gō, 1985), DADAS (Kohda et al., 1988) and DIANA (Güntert et al., 1991a) have been used for numerous structure determinations (e.g., Wagner et al., 1987; Arseniev et al., 1988; Kline et al., 1988; Qian et al., 1989; Widmer et al., 1989; Güntert et al., 1991b; Ikura et al., 1991), so that its performance in practice can be quite reliably evaluated. Advantages of the method are its conceptual simplicity and the fact that it works in dihedral angle space, so that the covalent geometry is preserved during the entire calculation. A drawback is that for all but the most simple molecular topologies (see below) only a small percentage of the calculations converge with small residual constraint violations, which is a typical local minimum problem (Li and Scheraga, 1987). Because of the low yield of acceptable conformers, calculations have typically been started with a large number of randomized starting conformers in order to obtain a group of good solutions, and sometimes a compromise had to be made between the requirements of small residual violations, the availability of approximately 10–20 'good' conformers to represent the solution conformation, and the available computing time (Kline et al., 1988; Widmer et al., 1989). With the introduction of the highly optimized program DIANA, which significantly reduced the computation time needed for the calculation of a single conformer, a workable situation was achieved for α-proteins (Güntert et al., 1991b), but for β-proteins with more complex topology the situation remained unsatisfactory. With the use of redundant dihedral angle constraints (REDAC) described in this paper, a greatly improved yield of converged conformers is now obtained also for β-proteins.

METHOD

In Fig. 1 the new strategy for the use of DIANA with REDAC is outlined and placed in perspective with the 'direct' variable target function method as proposed originally by Braun and Gō (1985) and used here as a reference for evaluating the merits of the new approach. In the direct approach, n start conformers with randomized dihedral angles are selected, and the program HABAS (Güntert et al., 1989) is applied for an initial analysis of the intraresidual and sequential NMR constraints (A in Fig. 1). The n conformers are then subjected to DIANA minimization against the experimental NMR constraints ($B^{(0)}$). Experience has shown that for well-converged solutions, the target function can be further reduced by repeating the DIANA refinement at L_{max} with variable weights for the van der Waals constraints. A limited number of k conformers ($m \leqslant k \leqslant n$) is subjected to this refinement in step D. Among the resulting solutions, m conformers with the smallest final target function values are selected to represent the solution structure. In practice, n is adjusted so as to obtain $m = 10$–20 acceptable conformers.

To use REDAC, one or several cycles $C^{(i)}$–$B^{(i)}$ are added to the calculation, providing a partial feedback of structural information from all conformers that were calculated up to the maximal level L_{max} (for a definition of L, see Güntert et al., 1991a) in the step $B^{(i-1)}$. In the step $C^{(i)}$, a particular amino acid residue is considered to have an acceptably well-defined conformation if the tar-

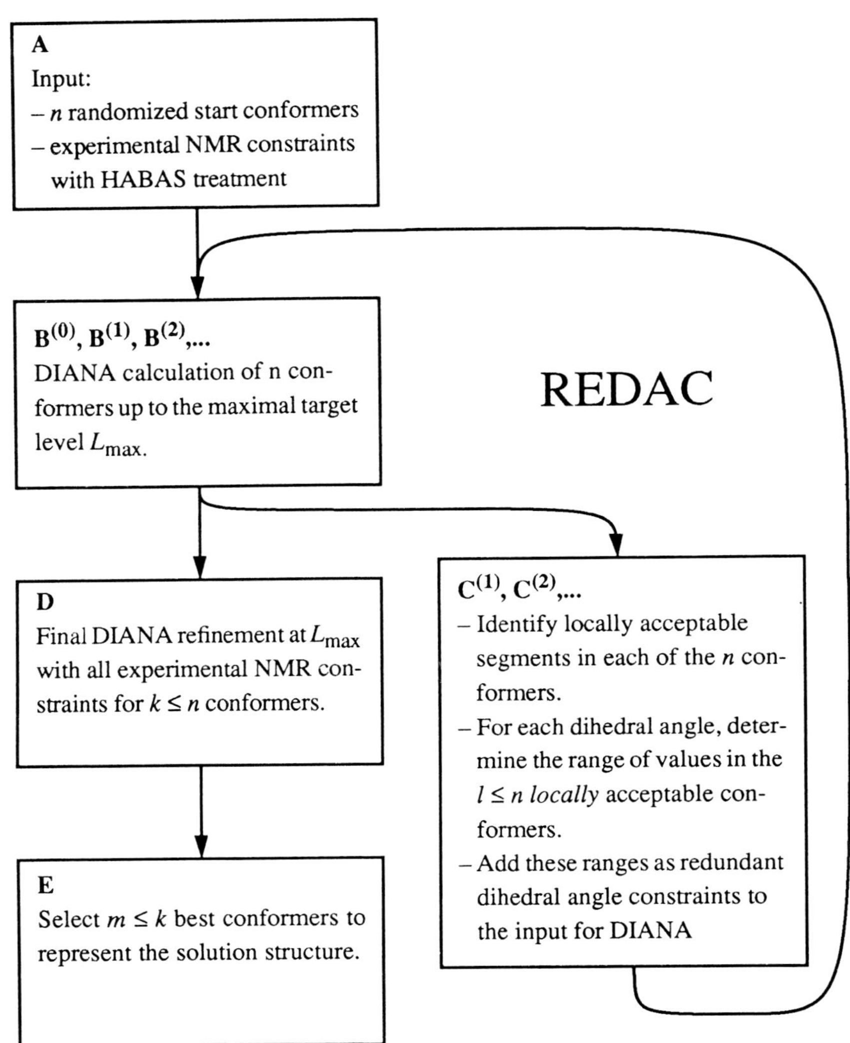

Fig. 1. Flowchart outlining the course of a protein structure calculation with the program DIANA using either the 'direct' way (A-$B^{(0)}$-D-E) or REDAC (A-$B^{(0)}$-[$C^{(1)}$-$B^{(1)}$-...]-D-E). Typically, the number of REDAC-cycles is 1 or 2.

get function value due to constraint violations that involve atoms or dihedral angles of this residue is less than a predefined value, typically 0.4 Å^2, and if the same condition holds for the two sequentially neighboring residues. Redundant dihedral angle constraints are generated for all those residues that were found to be acceptable in at least a predefined minimal number of conformers, typically 10 if the calculation is started with $n = 50$ randomized conformations (see Fig. 1), by taking the two extreme dihedral angle values in the group of acceptable conformers as upper and lower bounds. If the dihedral angle interval defined by these bounds is larger than a predefined maximal width, typically 270°, the redundant dihedral angle constraint is discarded, otherwise it is added to the input for the DIANA structure calculation in step $B^{(i)}$. The automated use of REDAC is implemented in version 1.14 of the program DIANA.

RESULTS

To compare the efficiency of DIANA calculations with and without use of REDAC, we calculated structures from high-quality experimental NMR data sets for the *Antp*(C39→S) homeodomain (HD; Güntert et al., 1991b), the basic pancreatic trypsin inhibitor (BPTI; K.D. Berndt, P. Güntert, L. Orbons and K. Wüthrich, to be published), and the activation domain from porcine procarboxypeptidase B (ADB; Vendrell et al., 1991). The HD is a typical α-protein, BPTI contains α and β secondary structure, and ADB has a more complex topology including a four-stranded β-sheet, two α-helices, and three loops that are only poorly determined by the NMR data (Vendrell et al., 1990). For each protein we performed a structure calculation starting with $n=50$ randomized conformations and using REDAC, and selected the $m=20$ conformers with the smallest final target function values for further analysis. In step $B^{(i)}$ we used the standard selection of minimization levels and parameters of DIANA, i.e., a maximal number of 150 conjugate gradient iterations and a weighting factor w_v of 0.2 for the van der Waals constraints (for a definition, see Güntert et al., 1991b) at all but the final level L_{max}, and three times 400 iterations with van der Waals weights of 0.2, 0.6, and 2.0 at L_{max}. The other weights had the same values throughout the entire calculation, with $w_u=w_l=1$, and $w_a=5$ Å^2. In step D we used $k=50$, and we allowed for a maximal number of three times 1000 iterations at L_{max}, using the van der Waals weights $w_v=0.2$, 0.6, and 2.0, respectively. For the HD and BPTI more than 40 conformers with final target function values at L_{max} below 2.1 Å^2 and 1.3 Å^2, respectively, were obtained after one REDAC cycle (Fig. 2). For ADB two REDAC cycles were needed to yield a group of 20 conformers with target function values below 2.9 Å^2 (Fig. 2).

For the HD and BPTI the DIANA calculations were repeated with the direct approach (Fig. 1), with the aim of producing a group of 20 conformers of equal quality, i.e., with final target function values in the same range as the 20 best conformers obtained from 50 starting conformers with the use of REDAC. For the HD this was achieved with $n=400$ starting conformers, for BPTI with $n=2000$ (Table 1). To obtain a fair comparison, the maximally allowed number of iterations for each target level in step $B^{(0)}$ was doubled when compared with the aforementioned parameters for the calculations with REDAC, and only the $k=50$ conformers with lowest target function values at the end of step $B^{(0)}$ were further refined in step D. For the ADB it was found that calcula-

TABLE 1
EFFICIENCY OF DIANA CALCULATIONS WITH AND WITHOUT USE OF REDAC

	HD		BPTI		ADB	
	Direct	REDAC	Direct	REDAC	Direct[a]	REDAC
n[b]	400[b]	50	2000[b]	50	≈8000[b]	50
CPU time (h)[c]	3.8	0.66	17.7	0.61	≈140	1.48

[a] Estimated (see text).
[b] n is the number of randomized starting conformers (see Fig. 1). For the direct approach, n values were chosen so as to obtain the same number of acceptable conformers as with $n=50$ and use of REDAC.
[c] Measured on a Cray Y/MP using one processor.

451

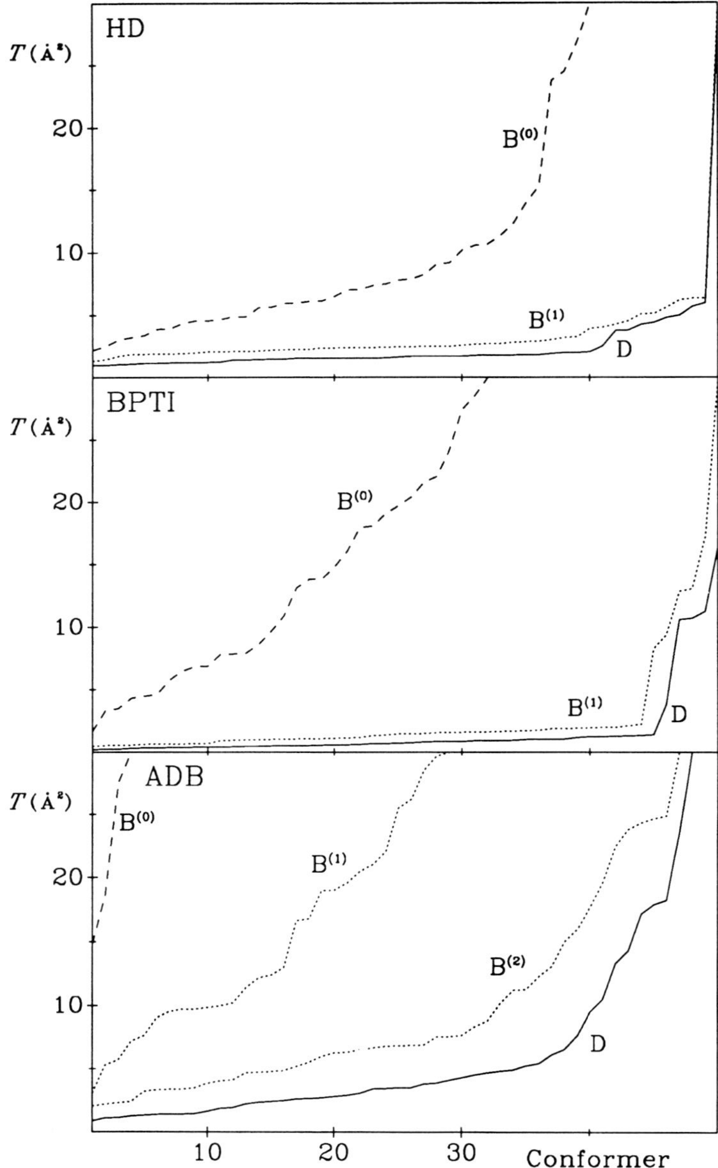

Fig. 2. Distribution of the DIANA target function values, T, among the 50 conformers calculated using REDAC for each of the three proteins *Antp*($C39 \rightarrow S$) homeodomain (HD), BPTI, and activation domain B (ADB). Along the horizontal axis the 50 conformers are ordered according to increasing target function value. For the HD and for BPTI the calculation consisted of the sequence of steps A-$B^{(0)}$-$C^{(1)}$-$B^{(1)}$-D; for ADB the sequence of steps was A-$B^{(0)}$-$C^{(1)}$-$B^{(1)}$-$C^{(2)}$-$B^{(2)}$-D. The letters identify the results at the end of the respective step in Fig. 1.

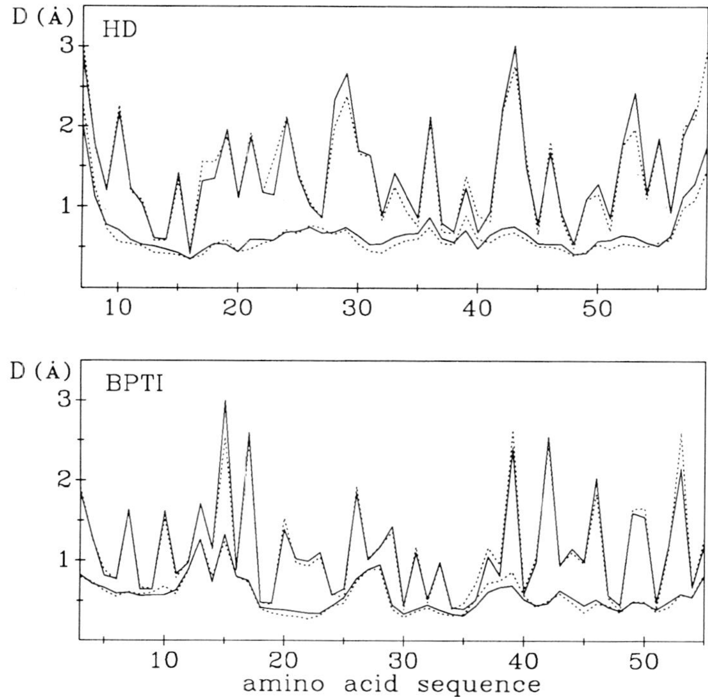

Fig. 3. Mean displacements (Billeter et al., 1989), i.e., averages of the pairwise RMSD values for the backbone atoms N, C^{α}, and C′ (lower curves) and for all heavy atoms (upper curves) of the individual amino acid residues after global superposition of the 20 best DIANA conformers obtained with (solid lines) or without (dashed lines) use of REDAC. The global superposition was made for the regions of the polypeptide chain that are well defined in the solution structure, i.e. residues 7-59 for the HD and residues 3-55 for BPTI, and displacements are presented only for the residues located within these regions.

tions without use of REDAC produced only one acceptable converged conformer from 400 starting conformers, so that we would have had to compute of the order of 8000 conformers in order to obtain a comparable result to that shown in Fig. 2 after the steps $B^{(2)}$ and D. Table 1 shows that for the three proteins the effective overall CPU time was reduced through the use of REDAC by factors of 5.7, 29, and about 100, respectively. The improved efficiency when using REDAC is particularly pronounced for the β-proteins, i.e., those proteins where the direct approach gives the lowest yields of acceptable structures.

To further evaluate the relevancy of the data in Table 1, we compared the quality of the DIANA structure calculations with and without REDAC on the basis of the parameters in Table 2 and Fig. 3. For both the HD and BPTI both types of calculations gave nearly identical values for the final target function, the different types of residual constraint violations, and the global pairwise RMSDs (McLachlan, 1979) among the 20 best conformers. Nearly identical values were also obtained for the displacements (Billeter et al., 1989) of the individual amino acid residues (Fig. 3). In addition, for both proteins the global pairwise RMSDs between conformers calculated with and without use of REDAC are for all practical purposes identical to the RMSDs between different conformers calculated using the same protocol (Tables 2 and 3).

TABLE 2
ANALYSIS OF THE 20 BEST CONFORMERS OBTAINED WITH THE DIANA CALCULATIONS OF TABLE 1

Quantity[a]	HD		BPTI	
	Direct[b]	REDAC[b]	Direct[b]	REDAC[b]
Final target function values (Å^2)	1.31 ± 0.19	1.29 ± 0.19	0.40 ± 0.12	0.39 ± 0.11
Distance constraint violations[c]:				
Number > 0.2 Å	3 ± 2	3 ± 2	0	0
Maximum (Å)	0.28 ± 0.06	0.28 ± 0.06	0.18 ± 0.02	0.19 ± 0.03
Sum (Å)	9.8 ± 0.9	9.8 ± 0.8	3.1 ± 0.6	3.1 ± 0.6
Dihedral angle constraint violations:				
Number > 5°	0	1 ± 1	0	0
Maximum (°)	4.5 ± 1.5	5.0 ± 1.4	1.9 ± 1.0	1.8 ± 1.0
Sum (°)	25.8 ± 4.8	24.7 ± 4.7	5.9 ± 2.3	5.7 ± 2.0
Average pairwise RMSDs (Å)[d]:				
Backbone atoms N, C^α, C′	0.76 ± 0.14	0.80 ± 0.16	0.67 ± 0.12	0.67 ± 0.13
All heavy atoms	1.70 ± 0.13	1.76 ± 0.16	1.49 ± 0.12	1.48 ± 0.14

[a] The average value and the standard deviation are given.
[b] Without (direct) or with use of REDAC.
[c] These include both violations of the distance constraints in the NMR input to DIANA and violations of the van der Waals lower distance limits imposed by DIANA.
[d] Only the well-defined parts of the protein structures were used for the superposition and the RMSD calculation, i.e., the residues 7-59 in the HD, and 3-55 in BPTI.

DISCUSSION

The empirically found higher yield of good conformers with the use of REDAC can be rationalized as follows: In many regions of a protein structure, in particular in β-strands, the local conformation is determined not only by the local conformational constraints derived from intraresidual, sequential and medium-range NOEs (Wüthrich, 1986), but also by longer-range constraints, e.g., interstrand distance constraints in β-sheets. Therefore, the local constraints alone may allow for

TABLE 3
COMPARISON OF THE 20 BEST CONFORMERS OBTAINED USING THE DIANA CALCULATIONS OF FIG. 1 WITH AND WITHOUT REDAC

	Average pairwise RMSD ± standard deviation (Å)[a]	
	HD	BPTI
Backbone atoms N, C^α, C′	0.77 ± 0.15	0.67 ± 0.14
All heavy atoms	1.74 ± 0.15	1.47 ± 0.16

[a] Each of the 20 conformers calculated with REDAC was compared with each of the 20 conformers obtained with the direct approach (see text). The numbers given are the average and the standard deviation for the resulting 400 pairwise RMSDs, calculated for residues 7-59 in the HD and 3-55 in BPTI.

multiple different local conformations at low target levels in a DIANA calculation, of which some may be incompatible with the longer-range constraints taken into account at higher minimization levels. Obviously, incorrect local conformations that satisfy the experimentally available local constraints are potential local minima, which could only be ruled out from the beginning if the information contained in the long-range constraints were already available at low levels of the mini-

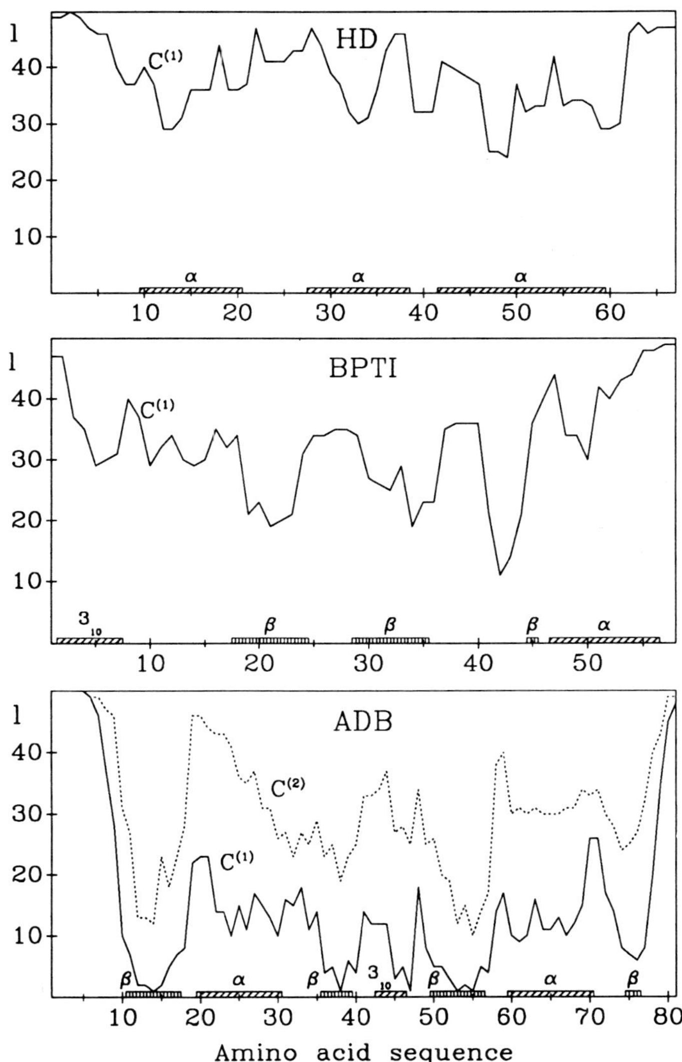

Fig. 4. Plot versus the amino acid sequence of the number of locally acceptable conformers, l, found in step $C^{(i)}$ of the DIANA structure calculations using REDAC. An amino acid residue is acceptable if the target function value due to constraint violations that involve atoms or dihedral angles of this residue is less than 0.4 Å^2, and if the same condition holds for the immediately preceding and following residues in the sequence. Helices (α,3_{10}) and β-strands (β) are indicated by hatched bars.

mization. The use of REDAC achieves this: information contained in the complete data set is translated into (by definition intraresidual) dihedral angle constraints. The same argument also explains why earlier attempts to use redundant dihedral angle constraints taken from calculations with L up to about 5 (Kline et al., 1988; Widmer et al., 1989; Billeter et al., 1990) had only limited success. It further makes clear why the yield of good solutions with the direct strategy was in general higher for α-proteins than for β-proteins, since the conformation of an α-helix is particularly well-determined by sequential and medium-range constraints (Wüthrich et al., 1984; Wüthrich, 1986). The plots of the number of locally acceptable conformers versus the amino acid sequence in Fig. 4 show that on the average the highest number of locally acceptable conformers for the generation of REDAC was obtained for the HD, even though the *final* target function values for BPTI were lower. Both in BPTI and in ADB particularly low numbers of locally acceptable conformers are observed in the β-strands, whereas loops, helices, and (often poorly determined) segments with non-regular secondary structure gave, in general, a higher number of locally acceptable conformers. This finding strongly supports the aforementioned explanation of the higher yield of good solutions with the use of REDAC.

In conclusion, the success of DIANA structure calculations using REDAC is primarily due to the feedback of useful structural information derived from conformers calculated up to the maximal level L_{max} into a subsequent round of structure calculations, which starts with local constraints only. In this way information gathered during the entire duration of the structure calculation is used in obtaining the final result, whereas most of this information (up to 95%) is discarded in the direct approach. Figure 3 and Tables 2 and 3 show that groups of conformers of equal quality are obtained with and without use of REDAC, and that the only significant effect of the use of REDAC is a large reduction of the overall computation time (Table 1). The use of REDAC should therefore become the standard strategy for protein structure calculations with the program DIANA and, more generally, with all implementations of the variable target function algorithm.

ACKNOWLEDGEMENTS

We thank Drs. Kurt Berndt and Leonard Orbons for the use of unpublished data of BPTI, and Drs. Martin Billeter and Werner Braun for helpful discussions. Financial support by the Schweizerischer Nationalfonds (project 31.25174.88) and the use of the Cray Y/MP of the ETH Zürich are gratefully acknowledged.

REFERENCES

Arseniev, A., Schultze, P., Wörgötter, E., Braun, W., Wagner, G., Vašák, M., Kägi, J.H.R. and Wüthrich, K. (1988) *J. Mol. Biol.*, **201**, 637–657.

Billeter, M., Kline, A.D., Braun, W., Huber, R. and Wüthrich, K. (1989) *J. Mol. Biol.*, **206**, 677–687.

Billeter, M., Qian, Y.Q., Otting, G., Müller, M., Gehring, W.J. and Wüthrich, K. (1990) *J. Mol. Biol.*, **214**, 183–197.

Braun, W. (1987) *Q. Rev. Biophys.*, **19**, 115–157.

Braun, W. and Gō, N. (1985) *J. Mol. Biol.*, **186**, 611–626.

Güntert, P., Braun, W., Billeter, M. and Wüthrich, K. (1989) *J. Am. Chem. Soc.*, **111**, 3997–4004.

Güntert, P., Braun, W. and Wüthrich, K. (1991a) *J. Mol. Biol.*, **217**, 517–530.

Güntert, P., Qian, Y.Q., Otting, G., Müller, M., Gehring, W. and Wüthrich, K. (1991b) *J. Mol. Biol.*, **217**, 531–540.

Ikura, T., Gō, N. and Inagaki, F. (1991) *Proteins*, **9**, 81–89.

Kline, A.D., Braun, W. and Wüthrich, K. (1988) *J. Mol. Biol.*, **204**, 675–724.

Kohda, D., Gō, N., Hayashi, K. and Inagaki, F. (1988) *J. Biochem.*, **103**, 741–743.

456

Li, Z. and Scheraga, H.A. (1987) *Proc. Natl. Acad. Sci. U.S.A.*, **84**, 6611–6615.
McLachlan, A.D. (1979) *J. Mol. Biol.*, **128**, 49–79.
Qian, Y.Q., Billeter, M., Otting, G., Müller, M., Gehring, W.J. and Wüthrich, K. (1989) *Cell*, **59**, 573–580.
Vendrell, J., Wider, G., Avilés, F.X. and Wüthrich, K. (1990) *Biochemistry*, **29**, 7515–7522.
Vendrell, J., Billeter, M., Wider, G., Avilés, F.X. and Wüthrich, K. (1991) *EMBO J.*, **10**, 11–15.
Wagner, G., Braun, W., Havel, T.F., Schaumann, T., Gō, N. and Wüthrich, K. (1987) *J. Mol. Biol.*, **196**, 611–639.
Widmer, H., Billeter, M. and Wüthrich, K. (1989) *Proteins*, **6**, 357–371.
Wüthrich, K. (1986) *NMR of Proteins and Nucleic Acids*, Wiley, New York.
Wüthrich, K. (1989) *Science*, **243**, 45–50.
Wüthrich, K., Billeter, M. and Braun, W. (1984) *J. Mol. Biol.*, **180**, 715–740.

V

SELECTED NMR STRUCTURE DETERMINATIONS

Introduction to papers 39 through 44

This part contains six reprints which relate to two research themes that have been of central interest to us during the past decade, *i.e.,* transcriptional regulation of gene expression and molecular aspects of immune suppression.

Work on proteins involved in transcriptional regulation started in my laboratory in 1983 as a collaboration with Prof. R. Kaptein, who provided us with a sample of the DNA-binding domain of the *lac* repressor. Dr. E. Zuiderweg joined my laboratory in Zürich, and the two of us obtained complete sequence-specific resonance assignments[1] and a determination of the secondary structure[2] of the "*lac* headpiece" quite efficiently. The result of the secondary structure determination was then combined with a limited number of long-range NOE distance constraints to determine the molecular architecture with the use of an interactive computer graphics program, CONFOR,[3] which was written by Martin Billeter as part of his Ph.D. thesis. Although the extent of the data collection at the time was not sufficient for obtaining an atomic-resolution structure, the paper **39** represents the molecular architecture with the functionally essential helix–turn–helix motif. NMR work on the *lac* project was subsequently continued for many years in Prof. Kaptein's laboratory, both to refine the protein structure to atomic resolution and to investigate the interaction of the *lac* repressor with its operator DNA.

---

[1] Zuiderweg, E.R.P., Kaptein, R. and Wüthrich, K. (1983) *Eur. J. Biochem. 137,* 279–292. Sequence-specific resonance assignments in the ^{1}H nuclear magnetic resonance spectrum of the *lac* repressor DNA-Binding domain 1–51 from *Escherichia coli* by two-dimensional spectroscopy.

[2] Zuiderweg, E.R.P., Kaptein, R. and Wüthrich, K. (1983) *Proc. Natl. Acad. Sci. USA 80,* 5837–5841. Secondary structure of the *lac* repressor DNA-binding domain by two-dimensional ^{1}H nuclear magnetic resonance in solution.

[3] Billeter, M., Engeli, M. and Wüthrich, K. (1985) *J. Mol. Graphics 3,* 79–83, 97–98. Interactive program for investigation of protein structures based on ^{1}H NMR experiments.

The papers **40** to **42** resulted from a most fruitful collaboration with Prof. W. Gehring on the structural foundations of transcriptional regulation by homeodomains. The NMR part of the project was the subject of the Ph.D. thesis of Mrs. Yan Qiu Qian, with important support by Martin Billeter and Gottfried Otting. Papers **40** and **41** describe, respectively, the initial structure determinations of the *Antennapedia* homeodomain and of a complex of this homeodomain with its operator DNA. The paper **42** is a review which summarizes the results obtained from structure determinations of several mutant forms of the *Antennapedia* homeodomain (see Table 2) and from the structure refinement of the *Antennapedia* homeodomain–DNA complex,[4,5,6] and discusses the structural data in the context of biochemical and genetic experiments.

The work with the immunosuppressant cyclosporin A and one of its immunophilins, human cyclophilin A, was pursued as a collaboration with a group of scientists at Sandoz AG. From the Sandoz side it was initiated by my former student and associate Dr. Hans Senn, who then handed the project over to Dr. Hans Widmer, also one of my Ph.D. students. In my laboratory the project was the subject of the Ph.D. theses of Christoph Weber and Claus Spitzfaden, who were both involved in the biochemical work as well as in the NMR structure determination, where they had crucial support from Gerhard Wider and Werner Braun. Paper **43** describes the structure determination of the receptor-bound cyclosporin A, which had immediate practical interest since the structure turned out to be very different from the previously determined X-

---

[4] Qian, Y.Q., Otting, G., Billeter, M., Müller, M., Gehring, W.J. and Wüthrich, K. (1993) *J. Mol. Biol. 234,* 1070–1083. Nuclear magnetic resonance spectroscopy of a DNA complex with the uniformly ^{13}C-labeled *Antennapedia* homeodomain and structure determination of the DNA-bound homeodomain.

[5] Billeter, M., Qian, Y.Q., Otting, G., Müller, M., Gehring, W. J. and Wüthrich, K. (1993) *J. Mol. Biol. 234,* 1084–1093. Determination of the nuclear magnetic resonance solution structure of an *Antennapedia* homeodomain–DNA complex.

[6] Qian, Y.Q., Otting, G. and Wüthrich, K. (1993) *J. Am. Chem. Soc. 115,* 1189–1190. NMR detection of hydration water in the intermolecular interface of a protein–DNA complex.

ray crystal and NMR solution structures of free cyclosporin A. It was, for all involved, a completely unexpected and for many reasons surprising result! Paper **44** was selected for inclusion in this volume since it presents a particularly nice illustration of combined use of data obtained by NMR in solution and by X-ray diffraction in single crystals: Based on complete sequence-specific NMR assignments of the polypeptide backbone, the secondary structure with all connections between regular secondary structure elements was determined for cyclophilin A as well as the cyclophilin A–cyclosporin A complex.[7] Using X-ray methodology, an electron density map at 2.8 Å resolution was obtained by Dr. M. Walkinshaw and his colleagues, which could then be traced very efficiently with the use of the NMR information on the secondary structure. Using the resulting crystal structure and additional NMR data collected with the cyclophilin–cyclosporin A complex, a detailed model of the latter was obtained[8] long before the structure determinations of this complex were completed either with X-ray crystallography or with NMR.[9,10]

The papers **39** through **44** had to be selected somewhat arbitrarily for presentation in this volume. A more comprehensive survey of our activities in the practical application of NMR structure determination is afforded by Table 2, which lists all the structures that we have so far deposited in the Brookhaven Protein Data Bank. In addition to transcriptional regulation and immune suppression, these relate to the areas of

---

[7] Wüthrich, K., Spitzfaden, C., Memmert, K., Widmer, H. and Wider, G. (1991) *FEBS Lett. 285,* 237–247. Protein secondary structure determination by NMR: application with recombinant human cyclophilin.

[8] Spitzfaden, C., Weber, H.P., Braun, W., Kallen, J., Wider, G., Widmer, H., Walkinshaw, M.D. and Wüthrich, K. (1992) *FEBS Lett. 300,* 291–300. Cyclosporin A–cyclophilin complex formation. A model based on X-ray and NMR data.

[9] Thériault, Y., Logan, T.M., Meadows, R., Yu, L., Olejniczak, E.T., Holzman, T.F., Simmer, R.L. and Fesik, S.W. (1993) *Nature 361,* 88–91. Solution structure of the cyclosporin A / cyclophilin complex by NMR.

[10] Spitzfaden, C., Braun, W., Wider, G., Widmer, H. and Wüthrich K. (1994) *J. Biomol. NMR 4,* 463–482. Determination of the NMR solution structure of the cyclophilin A–cyclosporin A complex.

enzymology, toxicology and intercellular signal transduction. In most instances the results leading to the complete structure determination were reported in several papers covering the sequence-specific resonance assignments, the secondary structure determination and possibly an initial, low quality three-dimensional structure determination (see Part II). For the earlier projects, completion of these different steps sometimes extended over several years, and different students and research associates would be engaged in successive phases of the work. To properly reflect on this situation, multiple references have been included with the entries in Table 2, except in those cases where the preliminary papers are reprinted in this volume. The table is self-explanatory, but a comment should be added on the fact that the bovine pancreatic trypsin inhibitor (BPTI) is not among the proteins for which a complete NMR structure determination was reported early-on, although this protein was used in most of the methods development work in my group, and a high-resolution crystal structure has been available since 1975. In the early attempts to solve this structure, we realized from the commonly used criteria to evaluate a NMR structure determination that we were not obtaining a good quality input of conformational NMR constraints. Eventually it turned out that the generally available BPTI preparations contain about 10% of closely related BPTI analogs. Although this did not interfere either with the use of BPTI for the development of NMR techniques or with the resonance assignments, it made the collection of an extensive set of NOE distance constraints as input for a structure determination difficult, since the assignments of many NOESY cross-peaks remained ambiguous. Furthermore, work with a recombinant, isotope-labeled mutant BPTI also confirmed earlier observations by Gerhard Wagner that some resonance lines were not observed because of conformational exchange (see Part VI). With the use of a homogeneous protein sample, a high quality structure determination was finally completed in 1992.

The advent of NMR structure determination opened an avenue for direct comparison of atomic resolution protein structures in crystals and in solution, and such comparative studies have been an important part of our work. The comparisons focused primarily on the molecular architecture of globular proteins for which both crystal and solution structures are available, and were either included in the papers describing the NMR structure determination,[11–16] or published separately as joint ventures with the colleagues who determined the crystal structure.[17–21] Much additional effort went into investigations of structural features in solution that are complementary to the crystallographic data, as described in more detail in the following Parts VI and VII.

---

[11] Williamson, M.P., Havel, T.F. and Wüthrich, K. (1985) *J. Mol. Biol. 182,* 295–315. Solution conformation of proteinase inhibitor IIA from bull seminal plasma by ^{1}H nuclear magnetic resonance and distance geometry. (Paper **25**)

[12] Schultze, P., Wörgötter, E., Braun, W., Wagner, G., Vasak, M., Kägi, J.H.R. and Wüthrich, K. (1988) *J. Mol. Biol. 203,* 251–268. Conformation of [Cd_7]-metallothionein-2 from rat liver in aqueous solution determined by nuclear magnetic resonance spectroscopy. (Paper **29**)

[13] Neri, D., Billeter, M. and Wüthrich, K. (1992) *J. Mol. Biol. 223,* 743–767. Determination of the NMR solution structure of the DNA-binding domain 1–69 of the 434 repressor and comparison with the X-ray crystal structure.

[14] Berndt, K.D., Güntert, P., Orbons, L.P.M. and Wüthrich, K. (1992) *J. Mol. Biol. 227,* 757–775. Determination of a high-quality nuclear magnetic resonance solution structure of the bovine pancreatic trypsin inhibitor and comparison with three crystal structures.

[15] O'Connell, J.F., Bougis, P.E. and Wüthrich, K. (1993) *Eur. J. Biochem. 213,* 891–900. Determination of the nuclear-magnetic-resonance solution structure of cardiotoxin CTX IIb from *Naja mossambica mossambica.*

[16] Sevilla-Sierra, P., Otting, G. and Wüthrich, K. (1994) *J. Mol. Biol. 235,* 1003–1020. Determination of the nuclear magnetic resonance structure of the DNA-binding domain of the *P22 c*2 repressor (1 to 76) in solution and comparison with the DNA-binding domain of the *434* repressor.

[17] Billeter, M., Kline, A.D., Braun, W., Huber, R and Wüthrich, K. (1989) *J. Mol. Biol. 206,* 677–687. Comparison of the high-resolution structures of the α-amylase inhibitor Tendamistat determined by nuclear magnetic resonance in solution and by X-ray diffraction in single crystals.

[18] Braun, W., Epp, O., Wüthrich, K. and Huber, R. (1989) *J. Mol. Biol. 206,* 669–676. Solution of the phase problem in the X-ray diffraction method for proteins with the nuclear magnetic resonance solution structure as initial model.

[19] Billeter, M., Vendrell, J., Wider, G., Avilés, F.X., Coll, M., Guasch, A., Huber, R. and Wüthrich, K. (1992) *J. Biomol NMR 2,* 1–10. Comparison of the NMR solution structure with the X-ray crystal structure of the activation domain from procarboxypeptidase B.

Table 2. *Chronological list of the NMR solution structures of biological macromolecules deposited in the Brookhaven Protein Data Bank (PDB)*[22] *by the research group of Kurt Wüthrich up to December 1994.*

Compound (source / organism used for recombinant protein)	PDB Codes
Proteinase inhibitor IIA from bull seminal plasma (bovine)[11,23–25]	1,2BUS
α-Amylase inhibitor Tendamistat (*Streptomyces tendae*)[26,27]	2,3,4AIT
[Cd_7^{2+}]-Metallothionein (rabbit)[28–30]	1,2MRB
[Cd_7^{2+}]-Metallothionein (rat)[12,31]	1,2MRT
Toxin ATX Ia (*Anemonia sulcata*)[32,33]	1ATX
Antennapedia homeodomain (*Drosophila* / rec. *E. coli*)[34,35]	1HOM
[Cd_7^{2+}]-Metallothionein (human)[36]	1,2MHU
Antennapedia homeodomain(C39S) (*Drosophila* / rec. *E. coli*)[37]	2HOA
Activation domain of procarboxypeptidase B (porcine)[38,39]	1PBA
Glutaredoxin, reduced (*E. coli* / rec. *E. coli*)[40]	1EGR
Cyclosporin A bound to cyclophilin A (*Tolypocladium inflatum;* human T-cells / rec. *E. coli*)[41]	2CYS
Glutaredoxin, oxidized (*E. coli* / rec. *E. coli*)[42,43]	1EGO
434 repressor DNA-binding domain (phage 434 / rec. *E. coli*)[13]	1PRA
Hirudin(1–51) (*Hirudo medicinalis* / rec. *E. coli*)[44,45]	1HIC
BPTI (bovine pancreatic trypsin inhibitor)[14,46]	1PIT
α-Neurotoxin (*Dendroaspis polylepis polylepis*)[47,48]	1NTX
Epidermal growth factor (murine)[49–51]	1,3EGF
Antennapedia homeodomain–DNA complex (*Drosophila* / rec. *E. coli;* chemical synthesis)[4,5]	1AHD
Cardiotoxin CTX IIb (*Naja mossambica mossambica*)[15,52,53]	2CCX

Table 2. *(continued)*

Compound (source / organism used for recombinant protein)	PDB Codes
Dendrotoxin K (*Dendroaspis polylepis polylepis*)[54,55]	1DTK
Trypsin inhibitor (*Stychodactyla helianthus*)[56]	1SHP
Pheromone E*r*-10 (*Euplotes raikovi*)[57]	1ERP
Glutaredoxin–glutathione complex (*E. coli* / rec. *E. coli*)[58]	1GRX
P22 c2 repressor DNA-binding domain (phage P22 / rec. *E. coli*)[16]	1ADR
Fushi tarazu homeodomain (*Drosophila* / rec. *E. coli*)[59]	1FTZ
des(1–6)Antennapedia homeodomain (*Drosophila* / rec. *E. coli*)[60]	1SAN
Cyclosporin A–cyclophilin A complex (*Tolypocladium inflatum; human T-cells* / rec. *E. coli*)[10,61]	3CYS
Pulmonary surfactant-associated polypeptide C (pig)[62]	1SPF
Tick anticoagulant protein (*Ornithodoros moubata* / rec. yeast)[63]	1TAP
Pheromone E*r*-1 (*Euplotes raikovi*)[64]	1ERC
Pheromone E*r*-2 (*Euplotes raikovi*)[65]	1ERD

[20] Braun, W., Vasak, M., Robbins, A.H., Stout, C.D., Wagner, G., Kägi, J.H.R. and Wüthrich, K. (1992) *Proc. Natl. Acad. Sci. USA 89,* 10124–10128. Comparison of the NMR solution structure and the X-ray crystal structure of rat metallothionein-2.

[21] Szyperski, T., Güntert, P., Stone, S.R., Tulinsky, A., Bode, W., Huber, R. and Wüthrich, K. (1992) *J. Mol. Biol. 228,* 1206–1211. Impact of protein–protein contacts on the conformation of thrombin-bound hirudin studied by comparison with the nuclear magnetic resonance solution structure of hirudin(1–51).

[22] Bernstein, F.C., Koetzle, T.F., Williams, G.J.B., Meyer, E.F. Jr., Brice, M.D., Rodgers, J.R., Kennard, O., Shimamouchi, T. and Tasumi, M. (1977) *J. Mol. Biol. 112,* 535–542. The protein data bank: a computer-based archival file for macromolecular structures.

[23] Strop, P. and Wüthrich, K. (1983) *J. Mol. Biol. 166,* 631–640. Characterization of the proteinase inhibitor IIA from bull seminal plasma by ^{1}H nuclear magnetic resonance: stability, amide proton exchange and mobility of aromatic residues.

[24] Strop, P., Wider, G. and Wüthrich, K. (1983) *J. Mol. Biol. 166,* 641–667. Assignment of the ^{1}H nuclear magnetic resonance spectrum of the proteinase inhibitor IIA from bull seminal plasma by two-dimensional nuclear magnetic resonance at 500 MHz.

[25]Williamson, M.P., Marion, D. and Wüthrich, K. (1984) *J. Mol. Biol. 173,* 341–359. Secondary structure in the solution conformation of the proteinase inhibitor IIA from bull seminal plasma by nuclear magnetic resonance.

[26] Kline, A.D. and Wüthrich, K. (1986) *J. Mol. Biol. 192,* 869–890. Complete sequence-specific ^{1}H nuclear magnetic resonance assignments for the α-amylase polypeptide inhibitor Tendamistat from *Streptomyces tendae.*

[27] Kline, A.D., Braun, W. and Wüthrich, K. (1988) *J. Mol. Biol. 204,* 675–724. Determination of the complete three-dimensional structure of the α-amylase inhibitor Tendamistat in aqueous solution by nuclear magnetic resonance and distance geometry.

[28] Braun, W., Wagner, G., Wörgötter, E., Vasak, M., Kägi, J.H.R. and Wüthrich, K. (1986) *J. Mol. Biol. 187,* 125–129. Polypeptide fold in the two metal clusters of metallothionein-2 by nuclear magnetic resonance in solution.

[29] Wagner, G., Neuhaus, D., Wörgötter, E., Vasak, M., Kägi , J.H.R. and Wüthrich, K. (1986) *Eur. J. Biochem. 157,* 275–289. Sequence-specific ^{1}H NMR assignments in rabbit liver metallothionein-2.

[30] Arseniev, A., Schultze, P., Wörgötter, E., Braun, W., Wagner, G., Vasak, M., Kägi, J.H.R. and Wüthrich, K. (1988) *J. Mol. Biol. 201,* 637–657. Three-dimensional structure of rabbit liver [Cd_7]-metallothionein-2a in aqueous solution determined by nuclear magnetic resonance.

[31] Wörgötter, E., Wagner, G., Vasak, M., Kägi, J.H.R. and Wüthrich, K. (1987) *Eur. J. Biochem. 167,* 457–466. Sequence-specific ^{1}H NMR assignments in rat liver metallothionein-2.

[32] Widmer, H., Wagner, G., Schweitz, H., Lazdunski, M. and Wüthrich, K. (1988) *Eur. J. Biochem. 171,* 177–192. The secondary structure of the toxin ATX Ia from *Anemonia sulcata* in aqueous solution determined on the basis of complete sequence-specific ^{1}H NMR assignments.

[33] Widmer, H., Billeter, M. and Wüthrich, K. (1989) *Proteins 6,* 357–371. Three-dimensional structure of the neurotoxin ATX Ia from *Anemonia sulcata* in aqueous solution determined by nuclear magnetic resonance spectroscopy.

[34] Otting, G., Qian, Y.Q., Müller, M., Affolter, M., Gehring, W.J. and Wüthrich, K. (1988) *EMBO J. 7,* 4305–4309. Secondary structure determination for the *Antennapedia* homeodomain by nuclear magnetic resonance and evidence for a helix–turn–helix motif.

[35] Billeter, M., Qian, Y.Q., Otting, G., Müller, M., Gehring, W.J. and Wüthrich, K. (1990) *J. Mol. Biol. 214,* 183–197. Determination of the three-dimensional structure of the *Antennapedia* homeodomain from *Drosophila* in solution by ^{1}H nuclear magnetic resonance spectroscopy.

[36] Messerle, B.A., Schäffer, A., Vasak, M., Kägi, J.H.R. and Wüthrich, K. (1990) *J. Mol. Biol. 214,* 765–779. Three-dimensional structure of human [$^{113}Cd_7$]-metallothionein-2 in solution determined by nuclear magnetic resonance spectroscopy.

[37] Güntert, P., Qian, Y.Q., Otting, G., Müller, M., Gehring, W.J. and Wüthrich, K. (1991) *J. Mol. Biol. 217,* 531–540. Structure determination of the *Antp(C39→S)* homeodomain from nuclear magnetic resonance data in solution using a novel strategy for the structure calculation with the programs DIANA, CALIBA, HABAS and GLOMSA.

[38] Vendrell, J., Wider, G., Avilés, F.X. and Wüthrich, K. (1990) *Biochemistry* **29**, 7515–7522. Sequence-specific ^{1}H NMR assignments and determination of the secondary structure for the activation domain isolated from pancreatic procarboxypeptidase B.

[39] Vendrell, J., Billeter, M., Wider, G., Avilés, F.X. and Wüthrich, K. (1991) *EMBO J. 10,* 11–15. The NMR structure of the activation domain isolated from porcine procarboxypeptidase B.

[40] Sodano, P., Xia, T., Bushweller, J.H., Björnberg, O., Holmgren, A., Billeter, M. and Wüthrich, K. (1991) *J. Mol. Biol. 221,* 1311–1324. Sequence-specific ^{1}H NMR assignments and determination of the three-dimensional structure of reduced *Escherichia coli* glutaredoxin.

[41] Weber, C., Wider, G., von Freyberg, B., Traber, R., Braun, W., Widmer, H. and Wüthrich, K. (1991) *Biochemistry 30,* 6563–6574. The NMR structure of cyclosporin A bound to cyclophilin in aqueous solution. (Paper **43**)

[42] Sodano, P., Chary, K.V.R., Björnberg, O., Holmgren, A., Kren, B., Fuchs, J.A. and Wüthrich, K. (1991) *Eur. J. Biochem. 200,* 369–377. Nuclear magnetic resonance studies of recombinant *Escherichia coli* glutaredoxin: sequence-specific assignments and secondary structure determination of the oxidized form.

[43] Xia, T., Bushweller, J.H., Sodano, P., Billeter, M., Björnberg, O., Holmgren, A. and Wüthrich, K. (1992) *Protein Science 1,* 310–321. NMR structure of oxidized *Escherichia coli* glutaredoxin: comparison with reduced *E. coli* glutaredoxin and functionally related proteins.

[44] Haruyama, H. and Wüthrich, K. (1989) *Biochemistry 28,* 4301–4312. Conformation of recombinant desulfatohirudin in aqueous solution determined by nuclear magnetic resonance.

[45] Szyperski, T. Güntert, P., Stone, S.R. and Wüthrich, K. (1992) *J. Mol. Biol. 228,* 1193–1205. Nuclear magnetic resonance solution structure of hirudin(1–51) and comparison with corresponding three-dimensional structures determined using the complete 65-residue hirudin polypeptide chain.

[46] Wagner, G., Braun, W., Havel, T.F., Schaumann, T., Gō, N. and Wüthrich, K. (1987) *J. Mol. Biol. 196,* 611–639. Protein structures in solution by nuclear magnetic resonance and distance geometry: the polypeptide fold of the basic pancreatic trypsin inhibitor determined using two different algorithms, DISGEO and DISMAN.

[47] Labhardt, A.M., Hunziker-Kwik, E.-H. and Wüthrich, K. (1988) *Eur. J. Biochem. 177,* 295–305. Secondary structure determination for α-neurotoxin from *Dendroaspis polylepis polylepis* based on sequence-specific ^{1}H-nuclear-magnetic-resonance assignments.

[48] Brown, L.R. and Wüthrich, K. (1992) *J. Mol. Biol. 227,* 1118–1135. Nuclear magnetic resonance solution structure of the α-neurotoxin from the black mamba *(Dendroaspis polylepis polylepis).*

[49] Montelione, G.T., Wüthrich, K., Nice, E.C., Burgess, A.W. and Scheraga, H.A. (1987) *Proc. Natl. Acad. Sci. USA 84,* 5226–5230. Solution structure of murine epidermal growth factor: determination of the polypeptide backbone chain-fold by nuclear magnetic resonance and distance geometry.

[50] Montelione, G.T., Wüthrich, K. and Scheraga, H.A. (1988) *Biochemistry 27,* 2235–2243. Sequence-specific ^{1}H NMR assignments and identification of slowly exchanging amide protons in murine epidermal growth factor.

[51] Montelione, G.T., Wüthrich, K., Burgess, A.W., Nice, E.C., Wagner, G., Gibson, D. and Scheraga, H.A. (1992) *Biochemistry 31,* 236–249. Solution structure of murine epidermal growth factor determined by NMR spectroscopy and refined by energy minimization with restraints.

[52] Otting, G., Steinmetz, W.E., Bougis, P.E., Rochat, H. and Wüthrich, K. (1987), *Eur. J. Biochem. 168,* 609–620. Sequence-specific ^{1}H-NMR assignments and determination of the secondary structure in aqueous solution of the cardiotoxins CTXIIa and CTXIIb from *Naja mossambica mossambica.*

[53] Steinmetz, W.E., Bougis, P.E., Rochat, H., Redwine, O.D., Braun, W. and Wüthrich, K. (1988) *Eur. J. Biochem. 172,* 101–116. ^{1}H nuclear magnetic resonance studies of the three-dimensional structure of the cardiotoxin CTXIIb from *Naja mossambica mossambica* in aqueous solution and comparison with the crystal structures of homologous toxins.

[54] Keller, R.M., Baumann, R., Hunziker-Kwik, E.H., Joubert , F.J. and Wüthrich, K. (1983) *J. Mol. Biol. 163,* 623–646. Assignment of the ^{1}H nuclear magnetic resonance spectrum of the trypsin inhibitor homologue K from *Dendroaspis polylepis polylepis*: two-dimensional nuclear magnetic resonance at 360 and 500 MHz.

[55] Berndt, K.D., Güntert, P. and Wüthrich, K. (1993) *J. Mol. Biol. 234,* 735–750. Nuclear magnetic resonance solution structure of dendrotoxin K from the venom of *Dendroaspis polylepis polylepis.*

[56] Antuch, W., Berndt, K.D., Chavez, M.A., Delfin, J. and Wüthrich, K. (1993) *Eur. J. Biochem. 212,* 675–684. The NMR solution structure of a Kunitz-type proteinase inhibitor from the sea anemone *Stichodactyla helianthus.*

[57] Brown, L.R., Mronga, S., Bradshaw, R.A., Ortenzi, C., Luporini, P. and Wüthrich, K. (1993) *J. Mol. Biol. 231,* 800–816. Nuclear magnetic resonance solution structure of the pheromone E*r*-10 from the ciliated protozoan *Euplotes raikovi.*

[58] Bushweller, J.H., Billeter, M., Holmgren, A. and Wüthrich, K. (1994) *J. Mol. Biol. 235,* 1585–1597. The nuclear magnetic resonance solution structure of the mixed disulfide between *Escherichia coli* glutaredoxin(C14S) and glutathione.

[59] Qian, Y.Q. , Furukubo-Tokunaga, K., Resendez-Perez, D., Müller, M., Gehring, W.J. and Wüthrich K. (1994) *J. Mol. Biol. 238,* 333–345. Nuclear magnetic resonance solution structure of the *fushi tarazu* homeodomain from *Drosophila* and comparison with the *Antennapedia* homoedomain.

[60] Qian, Y.Q., Resendez-Perez, D., Gehring, W.J. and Wüthrich, K. (1994) *Proc. Natl. Acad. Sci. USA 91,* 4091–4095. The *des(1–6)Antennapedia* homeodomain: comparison of the NMR solution structure and the DNA-binding affinity with the intact *Antennapedia* homeodomain.

[61] Wüthrich, K., Spitzfaden, C., Memmert, K., Widmer, H. and Wider, G. (1991) *FEBS Lett. 285,* 237–247. Protein secondary structure determination by NMR: application with recombinant human cyclophilin.

[62] Johansson, J., Szyperski, T., Curstedt, T. and Wüthrich, K. (1994) *Biochemistry 33,* 6015–6023. The NMR Structure of the pulmonary surfactant-associated polypeptide SP-C in an apolar solvent contains a valyl-rich α-helix.

[63] Antuch, W., Güntert, P., Billeter, M., Hawthorne, T., Grossenbacher, H. and Wüthrich, K. (1994) *FEBS Lett. 352,* 251–257. NMR solution structure of the recombinant tick anticoagulant protein (rTAP), a factor XA inhibitor from the tick *Ornithodoros moubata.*

[64] Mronga, S., Luginbühl, P., Brown, L.R., Ortenzi, C., Luporini, P., Bradshaw, R.A. and Wüthrich, K. (1994) *Protein Science 3,* 1527–1536. The NMR solution structure of the pheromone E*r*-1 from the ciliated protozoan *Euplotes raikovi.*

[65] Ottiger, M., Szyperski, T., Luginbühl, L., Ortenzi, C., Luporini, P., Bradshaw, R.A. and Wüthrich, K. (1994) *Protein Science 3,* 1515–1526. The NMR solution structure of the pheromone E*r*-2 from the ciliated protozoan *Euplotes raikovi.*

Reprinted from *Progress in Bioorganic Chemistry and Molecular Biology* (Yu. A. Ovchinnikov, ed.), pp. 65–70.
Copyright © 1984, with kind permission from Elsevier Science B. V. Amsterdam, The Netherlands.

SOLUTION CONFORMATION OF E. COLI LAC REPRESSOR DNA BINDING DOMAIN BY 2D NMR: SEQUENCE LOCATION AND SPATIAL ARRANGEMENT OF THREE α-HELICES

ERIK R.P. ZUIDERWEG*, MARTIN BILLETER†, ROBERT KAPTEIN*, ROLF BOELENS*, RUUD M. SCHEEK* AND KURT WÜTHRICH†

† Institut für Molekularbiologie und Biophysik, Eidgenössische Technische Hochschule, ETH-Hönggerberg, CH-8093 Zürich, Switzerland

* Department of Physical Chemistry, University of Groningen, Nijenborgh 16, NL-9747 Groningen, The Netherlands

INTRODUCTION

The lac repressor of E. coli is a tetrameric protein of molecular weight 150,000. It regulates the expression of genes coding for enzymes envolved in the lactose metabolism by specific binding to the lac operator (1,2). Each of the four identical monomeric subunits consists of two domains. The DNA-binding domain (3,4,5) or "headpiece" obtained by limited proteolytic digestion of the intact repressor (6) retains its native three-dimensional structure and sequence specific DNA binding capacity (4,7-9). Single crystal structures are not available either for intact repressor or for the headpiece. However, due to its small size the isolated headpiece (51 amino acid residues) is accessible for high resolution ^{1}H NMR (9-13).

PROTEIN ISOLATION AND NMR EXPERIMENTS

Headpiece was obtained as an enzymatic digest of lac repressor from E. coli strain BMH 74-12 as described previously (9). Ca. 5 mM solutions of headpiece in a buffer containing 0.4 M KCl, 0.05 M potassium phosphate and 0.02% NaN_3 were prepared, with the pH between 4.0 and 6.0, depending on the experiment. Two-dimensional (2D) NMR spectra were recorded on Bruker HX 360 and WM 500 spectrometers at the ETH Zürich and at the Dutch National NMR Facilities at Groningen and Nijmegen. The standard pulse sequences for 2D correlated spectroscopy (COSY), 2D spin echo correlated spectroscopy (SECSY), relayed COSY and 2D nuclear Overhauser enhancement spectroscopy (NOESY) were used, with phase cycling and data processing similar to previously described procedures (12,14,15). For studies of ^{1}H-^{1}H distance constraints, NOESY spectra were recorded with mixing times of 50, 80, 120 and 200 ms both in ^2H_2O and in H_2O (16).

66

SEQUENCE SPECIFIC RESONANCE ASSIGNMENTS

Initially sequence specific assignments for the ring protons of the single histidine and the four tyrosines in lac repressor headpiece were obtained by combined use of genetic experiments, chemical modification and NMR (10, 11). Using the recently introduced, generally applicable method of sequential resonance assignments (17-20), the NMR lines of over 200 additional hydrogen atoms in lac repressor headpiece were assigned, which included nearly all hydrogens of the polypeptide backbone (12).

SECONDARY STRUCTURE DETERMINATION

On the basis of the sequence specific resonance assignments for the polypeptide backbone in lac repressor headpiece (12), the locations of regular secondary structure elements in the primary structure could be identified (13). From observation of short sequential and medium range ^1H-^1H distances and identification of the slowly exchanging, hydrogen bonded backbone amide protons, three helices extending approximately from the residues 6-15, 17-25 and 34-45 (referred to as helix I, II and III respectively) were characterized (Fig. 1) (This "pattern recognition" approach for the determination of regular secondary structure cannot usually distinguish between residues at the ends of a helix and those in tight turns immediately following the helix (21). As a consequence of further work on the tertiary structure, described below and in ref. (22), the length of helix I was subsequently modified to extend from residues 6-13).

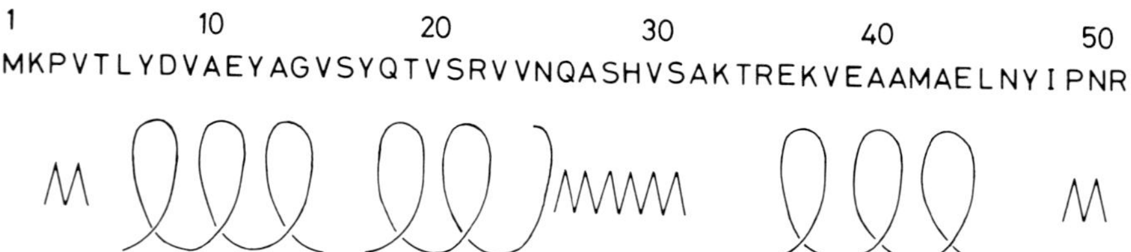

Fig. 1 Amino acid sequence of E. coli lac repressor headpiece 1-51 (23) and schematic indication of the locations of helical and extended segments of the polypeptide chain.

SPATIAL ARRANGEMENT OF THE THREE α-HELICES

The relative spatial orientations of the three helical segments shown in Fig. 1 were determined using distance constraints obtained from "long range" NOE's (NOE's between residues which are five or more positions apart in the sequence)between protons which are not both located in the same helix. The following qualitative interpretation of the NOESY cross peaks was used. First, the cross peaks in spectra recorded with a mixing time of 200 ms were identified on the basis of the previously obtained resonance assignments (12). Secondly, only those NOE's were retained which could also be observed in spectra recorded with a shorter mixing time of 50 or 80 ms (16). These were taken to manifest that the corresponding proton-proton distance is $\leqslant$ 4.0 Å (16,24). Fig. 2 illustrates between which residues such distance constraints were observed.

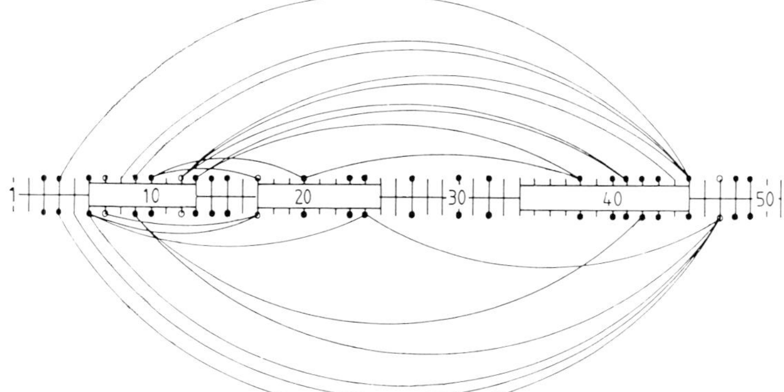

Fig. 2 Survey of the NOE connectivities used to determine the spatial arrangement of the three α-helices in the lac repressor DNA binding domain 1 — 51. The three helices are indicated with boxes. Non-polar side chains are marked with filled circles and tyrosyl residues with open circles. The lines above and below the sequence indicate for which pairs of residues close spatial proximity was indicated by NOESY cross peaks.

In the structural interpretation of the NOE distance constraints of Fig.2 the helices I-III were preserved as regular α-helices. Furthermore, since the available data were not sufficient to determine the conformation of the amino acid side chains, all experimental distance constraints were referred to backbone hydrogens or C^β atoms in the following way: For NOE's to backbone amide and C^α protons the distance constraint of 4.0 Å was used without correction. For NOE's to amino acid side chains the maximum sterically allowed distance from the observed hydrogen atom to C^β was added to the distance of 4.0 Å and this corrected constraint was referred to the C^β

position (22). The resulting set of distance constraints (Table 1 in ref.22)) was used in two independent model building procedures. In one, Nicholson molecular models were used to position the three helices in accordance with a number of key NOE's. Then the intervening loops were attached, which provided further constraints on the relative helix orientation. The molecular model obtained was further adjusted to be compatible with the other NOE constraints listed in Table 1. The other procedure makes use of an interactive computer graphics program written for an Evans & Sutherland picture system connected to a PDP 11-34 computer. The program allows to generate different conformations of molecular structures with up to 1000 atoms by real time variation of the torsion angles about single bonds. Since NOESY distance constraints may be between widely separated positions in the sequence, it is important that the complete structure was stored in memory. As a result, up to 8 independent torsion angles could be varied simultaneously to generate different relative spatial locations of the helices I-III, and the complete resulting conformations could be inspected on the screen (a detailed description of this program by M. Billeter, M. Engeli and K. Wüthrich is in preparation). The topology of the lac repressor headpiece obtained from the NMR measurements with the use of these procedures is shown in Fig. 3.

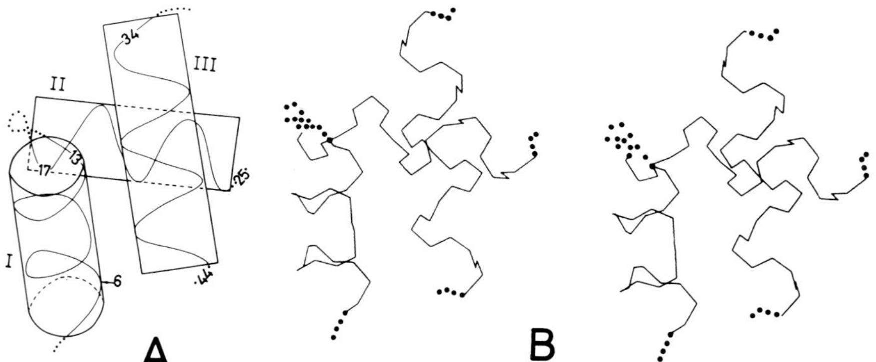

Fig. 3 Spatial arrangement of the three helices in lac repressor headpiece. The helices II and III are parallel to the projection plane and helix I runs from back to front at an angle of ca.50^0 relative to that plane. A. A smooth line is drawn through the C^{α} positions within the helices; the C^{α}'s 6,13,17, 25,34 and 44 are identified. Outside of the helices the backbone is indicated by dotted lines; the loop from residues 26-33 and the chain terminal segments are omitted. B. Stereo view of the same orientation as in A. For each helical residue the backbone atoms N, C^{α} and C' have been drawn.

The majority of the identified NOE's occur between protons of hydrophobic amino acid residues (Fig. 2). Many of these residues make multiple contacts, which indicates that they are clustered in the interior of the protein. Tyr 47 is in contact with at least 5 non-polar residues and therefore is part of the hydrophobic cluster. This concurs with genetic and biochemical data indicating that Tyr 47 can neither be mutated nor removed without disrupting the three dimensional structure of the protein (3,10).

SPATIAL STRUCTURE HOMOLOGY WITH OTHER DNA BINDING PROTEINS

Comparison of the location of helical segments in the amino acid sequence of _E. coli_ _lac_ repressor headpiece determined by ^{1}H NMR with the presently available three crystal structures of DNA binding proteins (CAP(25), cro(26), cI(27)) revealed a close homology on the level of the secondary structure (13). Following Matthews and Steitz (25,26) the primary structures of the four proteins were aligned on the basis of sequence homologies in the structural genes of the DNA-binding domains (13). Most interesting is that all four molecules contain two helices connected by a short peptide of two or three residues in the region that corresponds to sequence positions 5-29 in the _lac_ repressor. The third helix in the _lac_ repressor headpiece, which extends between residues 34 and 45, has a homologous counterpart in the cI repressor, whereas no helical structures were reported for the corresponding segments in CAP and cro. This homology between _lac_ repressor and cI repressor extends further to the molecular topology. The relative orientation of the helices of _lac_ repressor headpiece shown in Fig. 3 is very similar to that of the central three helices of the DNA-binding domain of the cI repressor, which is a protein that, as _lac_ repressor, recognizes a specific DNA sequence (27).

ACKNOWLEDGEMENTS

We acknowledge financial support by the Netherlands Foundation for Chemical Research (SON), the Netherlands Organization for the Advancement of Pure Research (ZWO), the Roche Research Foundation (Basel,Switzerland), the Schweizerischer Nationalfonds (project 3.284.82) and a special grant of the ETH Zürich, as well as the use of the facilities at the Zentrum für Interaktives Rechnen (ZIR) of the ETH, Zürich.

70

REFERENCES

1. Jacob F, Monod J (1961) J Mol Biol 3:318-356
2. Caruthers MH (1980) Acc Chem Res 13:155-160
3. Miller JH, Coulondre C, Hofer M, Schmeissner U, Sommer H, Smitz A, Lu P (1982) Proc Nat Acad Sci USA 79:218-222
4. Ogata RT, Gilbert W (1979) J Mol Biol 132:709-728
5. Schlottmann M, Beyreuther K (1979) Eur J Biochem 95:39-49
6. Geisler N, Weber K (1977) Biochemistry 16:938-943
7. Wade-Jardetzky NG, Bray RP, Conover WE, Jardetzky O, Geisler N, Weber K (1979) J Mol Biol 128:259-264
8. Nick H, Arndt K, Boschelli F, Jarema MA, Lillis M, Sadler J, Caruthers M, Lu P (1982) Proc Nat Acad Sci USA 79:218-222
9. Scheek RM, Zuiderweg ERP, Klappe KJM, van Boom JH, Kaptein R, Rüterjans H, Beyreuther K (1983) Biochemistry 22:228-235
10. Ribeiro AA, Wemmer D, Bray RP, Wade-Jardetzky NG, Jardetzky O (1981) Biochemistry 20:818-823
11. Arndt KT, Boschelli F, Lu P, Miller JH (1981) Biochemistry 20:6109-6118
12. Zuiderweg ERP, Kaptein R, Wüthrich K (1983) Eur J Biochem 137:279-292
13. Zuiderweg ERP, Kaptein R, Wüthrich K (1983) Proc Nat Acad Sci USA 80:5837-5841
14. Wider G, Macura S, Anil Kumar, Ernst RR, Wüthrich K (1984) J Magn Reson 56:207-234
15. States DJ, Haberkorn RA, Reuben DJ (1982) J Magn Reson 48:286-292
16. Anil Kumar, Wagner G, Ernst RR, Wüthrich K (1981) J Am Chem Soc 103: 3654-3658
17. Wüthrich K, Wider G, Wagner G, Braun W (1982) J Mol Biol 155:311-319
18. Billeter M, Braun W, Wüthrich K (1982) J Mol Biol 155:321-346
19. Wagner G, Wüthrich K (1982) J Mol Biol 155:347-366
20. Wüthrich K (1983) Biopolymers 22:131-138
21. Billeter M, Braun W, Wüthrich K (1984) submitted
22. Zuiderweg ERP, Billeter M, Boelens R, Scheek RM, Wüthrich K, Kaptein R (1984) FEBS Lett in press
23. Beyreuther K (1978) In: The Operon (Miller JH, Reznikoff W, eds), Cold Spring Harbor Laboratory, Cold Spring Harbor, N.Y., pp 123-154
24. Braun W, Wider G, Lee KH, Wüthrich K (1983) J Mol Biol 169:921-948
25. Weber IT, McKay DB, Steitz TA (1982) Nucleic Acids Res 10:5085-5102
26. Matthews BW, Ohlendorf DH, Anderson WF, Takeda Y (1982) Proc Nat Acad Sci USA 79:1428-1432
27. Sauer RT, Yocum RR, Doolittle RF, Lewis M, Pabo CO (1982) Nature (London) 298:447-451

Cell, Vol. 59, 573–580, November 3, 1989, Copyright © 1989 by Cell Press

The Structure of the *Antennapedia* Homeodomain Determined by NMR Spectroscopy in Solution: Comparison with Prokaryotic Repressors

Y. Q. Qian,* M. Billeter,* G. Otting,* M. Müller,† W. J. Gehring,† and K. Wüthrich*
* Institut für Molekularbiologie und Biophysik
Eidgenössische Technische Hochschule–Hönggerberg
CH-8093 Zürich
Switzerland
† Biozentrum der Universität Basel
Abteilung Zellbiologie
Klingelbergstrasse 70
CH-4056 Basel
Switzerland

Summary

The structure of the *Antennapedia* homeodomain from Drosophila melanogaster was determined by nuclear magnetic resonance spectroscopy in solution. It includes three well-defined helices (residues 10–21, 28–38, and 42–52) and a more flexible fourth helix (residues 53–59). Residues 30–50 form a helix-turn-helix motif virtually identical to those observed in various prokaryotic repressors. Further comparisons of the homeodomain with prokaryotic repressors showed that there are also significant differences in the molecular architectures. Overall, these studies support the view that the third helix of the homeodomain may function as the DNA recognition site. The elongation of the third helix by the fourth helix is a structural element that so far appears to be unique to the *Antennapedia* homeodomain.

Introduction

The homeobox is a highly conserved DNA segment that was first found in homeotic genes of Drosophila (McGinnis et al., 1984a; Scott and Weiner, 1984) and subsequently in many other eukaryotic organisms, including vertebrates and man (Carrasco et al., 1984; McGinnis et al., 1984b; Levine et al., 1984). The homeobox encodes the homeodomain, which represents a conserved protein domain that is shared by the various homeotic proteins. The homeodomain encoded by the *Antennapedia* (*Antp*) gene of Drosophila consists of about 60 residues located close to the C-terminus of the *Antp* protein. It was shown that the homeodomain polypeptide binds specifically to certain DNA sequences and that it protects the same DNA segment from DNAase I attack as the complete homeoprotein encoded by the *fushi tarazu* (*ftz*) gene, or a truncated *Antp* protein (Müller et al., 1988). Structure predictions based on sequence homologies have led to the assumption that DNA binding by the homeodomain involves a helix-turn-helix motif similar to that found in prokaryotic DNA binding proteins (Shepherd et al., 1984; Laughon and Scott, 1984; Gehring, 1987; Tsonis et al., 1988).

For a direct experimental investigation of the DNA binding modes of homeodomains, a project to determine the three-dimensional structure of the *Antp* homeodomain in aqueous solution was initiated. In a first step a recombinant plasmid (pAop2) containing the *Antp* homeobox was constructed, coding for amino acids 297 to 363 of the *Antp* protein (Schneuwly et al., 1986). The encoded polypeptide consists of the *Antp* homeodomain with one additional methionine at the N-terminus and an extra heptapeptide at the C-terminal end. Using the T7 expression system (Studier and Moffat, 1986; Rosenberg et al., 1987), the homeodomain was overexpressed in Escherichia coli and purified to homogeneity (Müller et al., 1988). Next, nuclear magnetic resonance (NMR) spectra were recorded for this homeodomain polypeptide, and sequence-specific resonance assignments were obtained; at the same time the regular secondary structures were determined, which provided the first direct experimental evidence for the presence of a helix-turn-helix motif (Otting et al., 1988).

In this paper we present the three-dimensional structure of the *Antp* homeodomain in solution, and compare this structure with the DNA binding domain of the phage 434 repressor (Mondragón et al., 1989) and with the *trp* repressor from E. coli (Lawson et al., 1988).

Results and Discussion

Presentation of Protein Structures Determined with NMR Spectroscopy

Typically, a protein structure determination by NMR yields a group of conformers, which characterize the solution conformation. Each individual conformer is the result of a separate structure calculation, which all use the same experimental input but start from different, randomly chosen starting conditions. The extent of the deviations between different conformers calculated from the same experimental input is determined by the number of conformational constraints and their precision, which are also related to internal mobility of the protein. It is customarily expressed either as the average of the pairwise root mean square deviations (RMSDs; McLachlan, 1979) among the conformers in the group or as the average of the RMSDs between the individual conformers and the mean structure, where a successful structure determination of good quality is characterized by small values of these deviations (Wüthrich, 1986). The presentation of a protein solution conformation by such a group of conformers can be qualitatively related to the characterization of a crystal structure by experimental, apparent temperature factors (Billeter et al., 1989), which also combine effects from protein mobility, static disorder, and experimental errors (Glusker and Trueblood, 1985).

For the *Antp* homeodomain a calculation procedure similar to that employed with the sea anemone toxin ATX-I (Widmer et al., 1989) was used to obtain a group of 19 conformers, all of which are in good agreement with the experimental NMR data and have favorable conformation energies. The input for the structure calculations con-

Cell
574

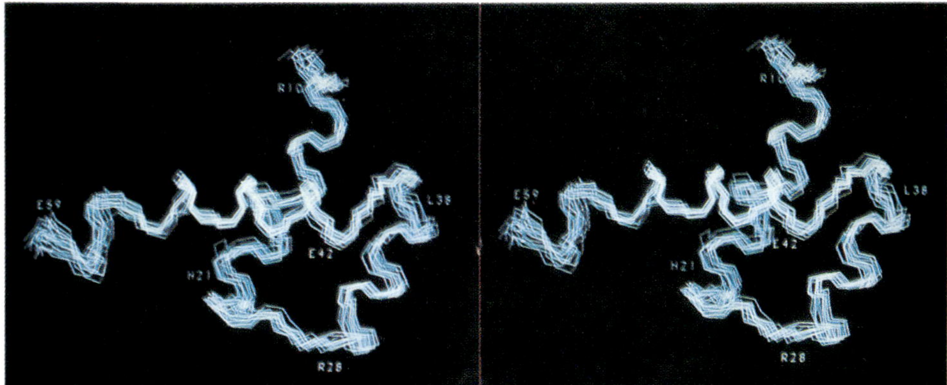

Figure 1. Partial Result of the Structure Calculations from NMR Data for the *Antp* Homeodomain

Nineteen conformers obtained from distance geometry calculations and subsequent restrained energy refinements have been superimposed for minimal pairwise RMSDs of the polypeptide backbone atoms N, C^{α}, and C′ of residues 7–59 with respect to structure 1. A stereo view of the bonds connecting these backbone atoms is shown.

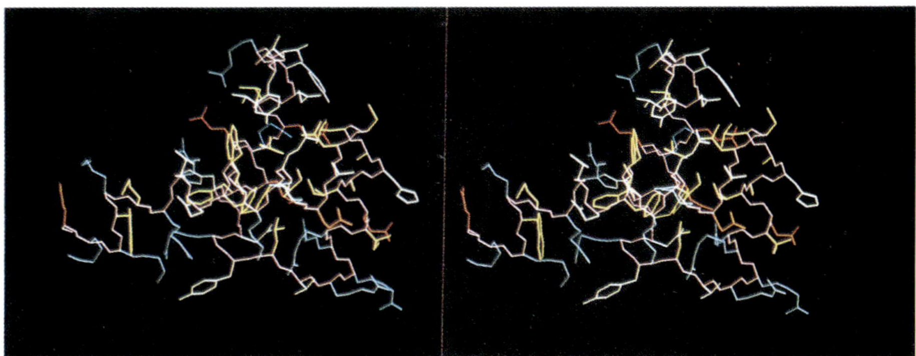

Figure 2. Stereo View Including All Heavy Atoms of Residues 7–59 in One of the 19 Structures of Figure 1 (Same Orientation)

The protein backbone is drawn in purple. For the side chains the following colors are used: blue for Arg and Lys; red for Glu; yellow for Ala, Cys, Ile, Leu, Met, Phe, Trp; white for Asn, Gln, Ser, Thr, Tyr, His. (Asp, Val, Gly, and Pro are absent.)

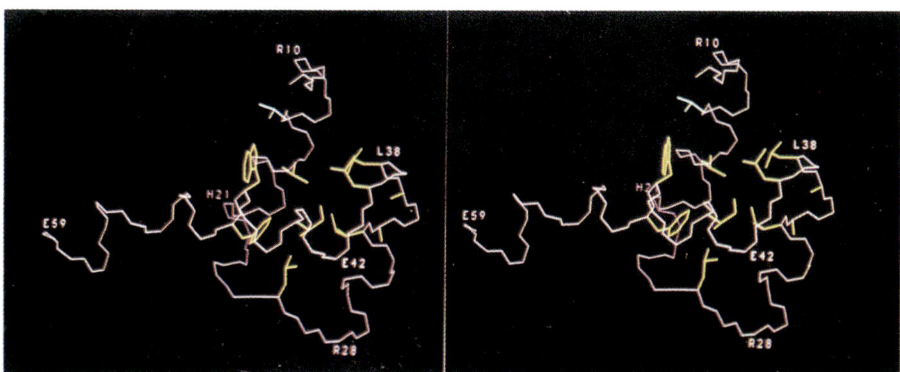

Figure 3. Same Stereo View as in Figure 2, Except Fewer Side Chains Displayed

Only the following side chains are displayed, which form a well-defined core of the protein (see text and Table 1): Thr-13, Leu-16, Leu-26, Ile-34, Ala-35, Ala-37, Leu-38, Leu-40, Ile-45, Trp-48, and Phe-49.

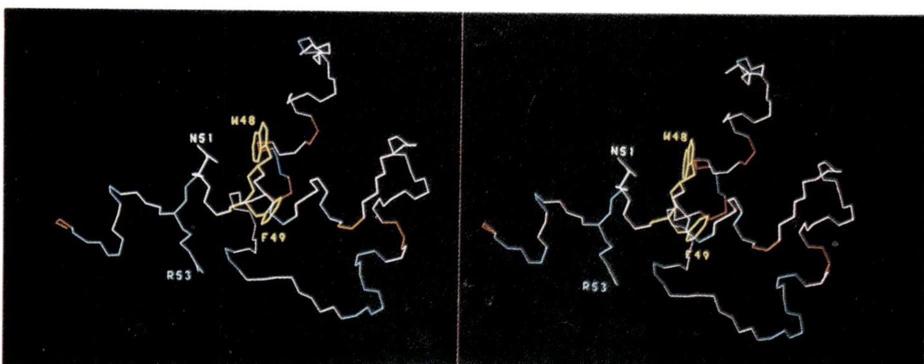

Figure 4. Same Stereo View as in Figure 2, Except Only Strictly Conserved Side Chains Displayed
Trp-48, Phe-49, Asn-51, and Arg-53 are displayed, and the backbone is colored blue for Arg and Lys, and red for Glu.

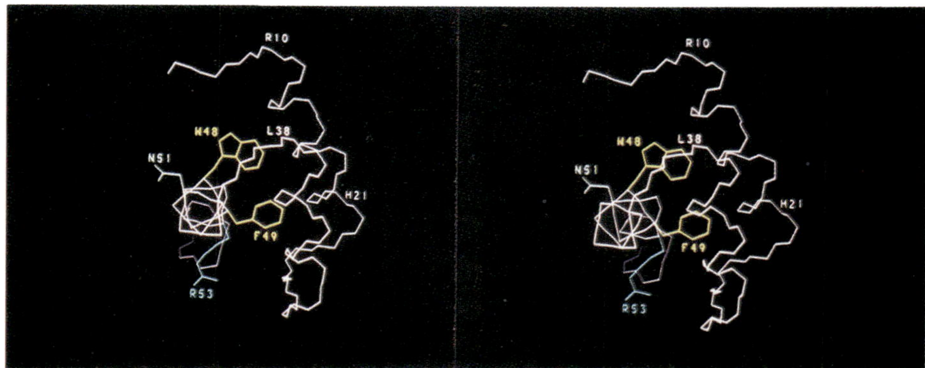

Figure 5. Stereo View of the Same Structure as in Figure 4 after Rotation by 90° about a Vertical Axis
The resulting view is along the axis of helix III. The entire backbone is shown in purple.

tained about 700 distance constraints determined with two-dimensional nuclear Overhauser enhancement spectroscopy (NOESY), and 118 scalar coupling constants that determine allowed ranges for the dihedral angles φ and χ^1. From such an extensive set of structural constraints one may expect to obtain a structure comparable in quality to that resulting from X-ray diffraction in single crystals at a resolution of about 2.0 Å, as was recently demonstrated by a comparative study of the structure of the α-amylase inhibitor Tendamistat in crystals and in solution (Billeter et al., 1989).

The *Antp* Homeodomain Structure

A partial result of the structure calculation for the 68 residue form of the *Antp* homeodomain (Müller et al., 1988) in solution is presented in Figure 1, which shows the backbone folds of 19 distance geometry solutions for residues 7-59. The conformations of the remaining N- and C-terminal segments, 0-6 and 60-67, are not well defined by the NMR data. This may be due in part to the fact that the intact *Antp* protein has 15 additional amino acid residues at the C-terminus (Schneuwly et al., 1986). For the N-terminus, flexibility may be expected for the peptide link that connects the homeodomain to the rest of the *Antp* protein.

The well-defined part of the homeodomain conformation includes three helical regions. Helix I contains residues 10–21, helix II the residues 28–38, and the two directly adjoining helices III and IV the residues 42–52 and 53–59, respectively. The three helical regions are connected by the hexapeptide 22–27 and the tripeptide 39–41 (Otting et al., 1988). The three helices I, II, and III are particularly well defined and were previously identified using an empirical approach (Wüthrich et al., 1984) of recognizing patterns of sequential and medium-range NOE constraints and slow amide proton exchange (Otting et al., 1988). The stability of helix IV is clearly smaller than that of the other three helices, as is indicated by the larger values of the local RMSDs, rapid exchange of the amide

Table 1. Average RMSDs between the 19 Energy-Refined Distance Geometry Solutions for the *Antp* Homeodomain and, Respectively, the Mean of these 19 Homeodomain Structures, the Crystal Structure of the 434 Repressor, and the Crystal Structure of the *trp* Repressor

Reference Structure	Atoms Considered[a]	Average RMSD (Å)
Mean *Atnp* homeodomain solution structure[b]	backbone 7–59	0.6 ± 0.1[c]
	backbone 7–59 plus "core" side chains[d]	0.6 ± 0.1[c]
	all heavy atoms 7–59	1.4 ± 0.1[c]
Crystal structure of 434 repressor	backbone h-t-h[e]	0.9 ± 0.1
	backbone helix I + h-t-h[e]	3.9 ± 0.1
Crystal structure of *trp* repressor	backbone h-t-h[f]	1.0 ± 0.1
	backbone helix I + h-t-h[f]	1.5 ± 0.1

[a] Backbone atoms are N, C^{α}, and C′. The numbers indicate the amino acid residues considered. (The polypeptide studied by NMR consists of residues 0–67. The conformation of the terminal segments 0–6 and 60–67 is not well defined by the NMR data [see text].)

[b] The mean structure was obtained by taking the average of the atom coordinates in the 19 refined distance geometry solutions.

[c] In other structure characterizations (e.g., Wüthrich, 1986; Kline et al., 1988) we used the average of the pairwise RMSD values among all distance geometry solutions instead. For easy, direct comparison, the corresponding values for this quantity are 0.9 ± 0.2, 0.9 ± 0.2, and 2.0 ± 0.2 Å.

[d] The following 11 residues are denoted as "core" residues (see text): 13, 16, 26, 34, 35, 37, 38, 40, 45, 48, 49.

[e] The h-t-h (helix-turn-helix motif) includes residues 30–50 of the *Antp* homeodomain, and 16–36 of the 434 repressor. Helix I includes residues 10–21 of the homeodomain, and 1–12 of the 434 repressor.

[f] The h-t-h includes residues 30–50 of the *Antp* homeodomain, and residues 67–87 of the *trp* repressor. Helix I includes residues 10–21 of the homeodomain, and residues 51–62 of the *trp* repressor. The coordinates of the *trp* repressor used are from the file 2WRP of the protein data bank.

protons involved in the presumed helix hydrogen bonding network, and the narrower linewidths of the helix IV proton resonances in the NMR spectra (Otting et al., 1988).

In the three-dimensional molecular architecture helices I and II are aligned in almost exactly antiparallel fashion, and helices III and IV are arranged approximately perpendicular to the other two helices. The axis of helix IV forms an angle of about 30° with respect to the axis of helix III, and it points away from the globular core of the protein. The global precision of the determination of the backbone fold is characterized in a more quantitative way by the RMSD values in Table 1, which show that for the backbone heavy atoms N, C^{α}, and C′ in the 19 distance geometry structures the average deviation from their mean positions is 0.6 Å. This value is a measure of the radius of the circular band that is spanned by the group of 19 conformers in Figure 1; it can be related to the B-factors in crystal structures (Glusker and Trueblood, 1985; Billeter et al., 1989).

Based on sequence comparisons with prokaryotic DNA binding proteins it was previously hypothesized that residues 28 to 52 in the *Antp* homeodomain form a helix-turn-helix motif (Shepherd et al., 1984; Laughon and Scott, 1984; Gehring, 1987). This prediction is substantiated by the present structure of the homeodomain. In all structurally known wild-type prokaryotic repressor proteins the first residue of the tripeptide forming the "turn" between the two helices is glycine, and it assumes a conformation defined by the dihedral angle values $\varphi \approx 60°$ and $\psi \approx 60°$. This conformation corresponds to a left-handed α helix and is energetically favorable only for glycine (Richardson, 1981). One mutant of λ repressor with this glycine replaced by glutamic acid was shown to be fully functional (Hochschild et al., 1983). Only a few homeodomains, for example, those of *engrailed* and *invected* (Poole et al., 1985; Fjose et al., 1985; Coleman et al., 1987), have a glycine residue at this position. In the *Antp* homeodomain the first residue of the turn is cysteine, but nonetheless the local conformation corresponds again to $\varphi \approx 60°$, $\psi \approx 60°$. Various residues other than glycine are found in this position in different homeodomains, including the sterically strongly restricted amino acids valine and threonine (Scott et al., 1989; their numbering differs from the one used here by a shift of one unit).

Figure 2 is an all-heavy-atom representation of one of the structures from the group shown in Figure 1. The molecule is in the same orientation as in Figure 1. The colors indicate the electrostatic charges and the water affinity of the side chains: blue is used for Arg and Lys, red for Glu, yellow for the hydrophobic side chains of Ala, Cys, Ile, Leu, Met, Phe, and Trp, and white for Asn, Gln, Tyr, Thr, and His. The backbone is drawn in purple. This all-atom display illustrates the globular character of the structure formed by residues 7–52 of the protein, and shows the protruding helix IV in the lower left of the picture. Characteristic features of the homeodomain are a hydrophobic core that holds helices I to III together and a pronounced anisotropy of the charge distribution on the surface. These two features can be seen more clearly in Figures 3 and 4.

Figure 3 shows the backbone in the same orientation as in Figures 1 and 2 together with those 11 amino acid side chains for which the smallest variations among the 19 distance geometry structures were computed. Similar to other protein solution structures determined by NMR (e.g., Wüthrich, 1986; Kline et al., 1988), these well-defined side chains are located in the interior part of the homeodomain. The ten residues Leu-16, Leu-26, Ile-34, Ala-35, Ala-37, Leu-38, Leu-40, Ile-45, Trp-48, and Phe-49 form a hydrophobic core, and the Thr-13 side chain further stabilizes this protein core by a hydrogen bond to the indole nitrogen of Trp-48. Table 1 shows that the conformation of these core side chains is as well defined by the NMR data as the polypeptide backbone, with an RMSD value of 0.6 Å, whereas the RMSD value for all heavy atoms of residues 7–59 is 1.4 Å.

In Figure 4 the distribution of charged residues is indicated by the coloring of the backbone, with Arg and Lys colored blue, and Glu red. Two features of the charge distribution stand out, both of which could contribute to favorable interactions of helix III (and possibly helix IV) with the DNA. One is that in the orientation of Figure 4, the negative charges are grouped together in the back of the molecule to the right, and the positive charges are grouped together in the front. As a result of this asymmetric charge distribution, the molecule should have a sizable global electric dipole moment. Furthermore, a total of seven positive charges are distributed along helices III and IV, which could contribute to energetically favorable interactions of these helices with the major groove of the target DNA (Aggarwal et al., 1988; Otwinowski et al., 1988; Jordan and Pabo, 1988; Wolberger et al., 1988).

In addition to the polypeptide backbone, Figure 4 displays those four side chains that were found to be strictly conserved in the 83 homeodomains of metazoa known to date (Scott et al., 1989). These are Trp-48 and Phe-49, which are located in helix III and are part of the hydrophobic core, and Asn-51 and Arg-53, which are located on the surface of helices III and IV and are thus likely candidates for interactions with the DNA. (In addition, Gln-50 is invariant in all known homeodomains of the *Antp* gene family [Scott et al., 1989]). These four residues are also shown in a view of the molecule obtained after a 90° rotation about a vertical axis (Figure 5). In Figure 5 one looks along the axis of helix III, with helices I and II on the right and parallel to the projection plane. Assuming a binding mode of the type observed in prokaryotic repressors (Aggarwal et al., 1988; Otwinowski et al., 1988; Jordan and Pabo, 1988; Wolberger et al., 1988; Lamerichs et al., 1989), the DNA would be bound to the lower left of the protein, with the axis of the double helix at an angle of about 45° relative to the vertical in Figure 5. The view of Figure 5 clearly illustrates that Trp-48 and Phe-49 are part of the protein core and that the other two conserved residues, Asn-51 and Arg-53, are favorably located for potential interactions with a DNA molecule. It is well documented in the case of the 434 repressor–operator complex that hydrogen bonds from glutamine side chains to bases of the DNA and electrostatic interactions of arginine side chains with phosphate groups are important details of the intermolecular interactions (Aggarwal et al., 1988).

The kink of about 30° between helices III and IV introduces a curvature into the helical segment from residues 42 to 59. It is tempting to speculate that this kink allows tight contacts between the entire length of helices III and IV and the target DNA, which could result in increased specificity and stability of the homeodomain–DNA complexes formed. The possibility that helix III might be extended has previously been postulated by Porter and Smith (1986) on the basis of their mutational analysis of the yeast repressor protein MATα2.

Comparison of the *Antp* Homeodomain with the DNA Binding Domains of the 434 Repressor and the *trp* Repressor

In a preliminary structural comparison based primarily on the identification of regular secondary structure elements, we reported that a helix-turn-helix motif of the type responsible for DNA recognition in numerous prokaryotic repressors (Aggarwal et al., 1988; Otwinowski et al., 1988; Jordan and Pabo, 1988; Wolberger et al., 1988; Lamerichs et al., 1989; Brennan and Matthews, 1989) is also present in the *Antp* homeodomain (Otting et al., 1988). Based on homeodomain sequence comparisons with prokaryotic

Table 2. Structural Correspondence between Helix-Turn-Helix Motifs found in the λ Cro Protein, Several Prokaryotic Repressors, and the *Antp* Homeodomain

Protein	Residues	RMSD (Å)
λ Cro	16–36	–
Phage 434 repressor	17–37	0.4
Phage 434 Cro	19–39	0.6
λ repressor	33–53	0.6
Antp homeodomain	31–51	0.8
trp repressor	68–88	0.9
CAP	169–189	0.9

Adapted from Brennan and Matthews (1989).

repressor proteins, it has independently been hypothesized that both the complete structure and the DNA binding mode of the homeodomain encoded by the *ftz* gene might be the same as in the 434 repressor (Tsonis et al., 1988; the modeled sequence corresponds to residues 14–50 in the *Antp* homeodomain, where 32 of the 37 residues are identical to those in the *ftz* homeodomain and the remaining 5 residues are replaced by functionally related residues).

On the basis of the complete three-dimensional structure of the *Antp* homeodomain (Figures 1–5), we have now continued these comparative studies with various prokaryotic DNA binding proteins for which three-dimensional structures have also been determined. For this we adopted procedures for structural comparisons that were previously employed with different prokaryotic repressor proteins. The uniqueness of the helix-turn-helix motif among these known protein structures was initially demonstrated by Steitz et al. (1982). More recently, Brennan and Matthews (1989) compared the helix-turn-helix motif of the λ Cro protein with five other known helix-turn-helix motifs by superimposing the C^{α} coordinates of 21 residue segments containing the motif for optimal fit. These comparisons resulted in all five cases in values smaller than 1.0 Å, while the smallest RMSDs to other protein fragments of the same length taken from the protein data bank were, with three exceptions in the range 1.4–1.6 Å, larger than 2.0 Å. Table 2 presents a recalculation of these data with the *Antp* homeodomain added to the list of helix-turn-helix containing proteins. In this comparison of six proteins to λ Cro, the homeodomain ranks fourth, with an RMSD value comparable to those of *trp* repressor and CAP. For the individual 19 homeodomain conformers shown in Figure 1, the RMSD values relative to the λ Cro protein range from 0.6–1.1 Å. The presence of a helix-turn-helix motif in the *Antp* homeodomain is thus clearly substantiated.

As the 434 repressor was previously selected for comparisons based on molecular modeling (Tsonis et al., 1988), we have also initially focused on this molecule. In Figure 6A the *Antp* homeodomain is superimposed with the N-terminal DNA binding domain of the 434 repressor for optimal fit between the helix-turn-helix segments. An initial inspection of Figure 6A shows that although the polypeptide backbone arrangement in the region of the helix-turn-helix segments is nearly identical in the two proteins, there are no additional similarities between the architec-

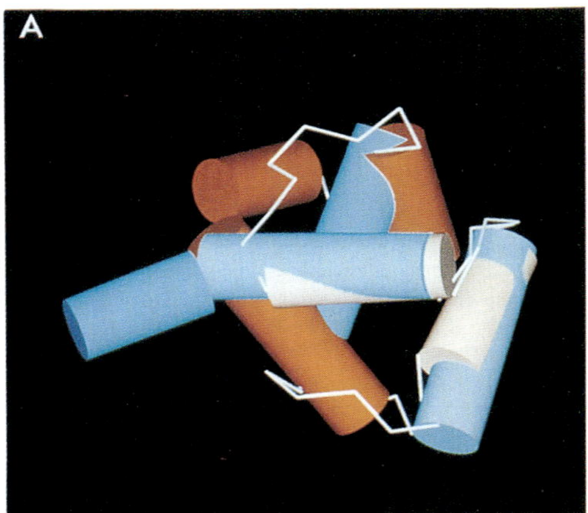

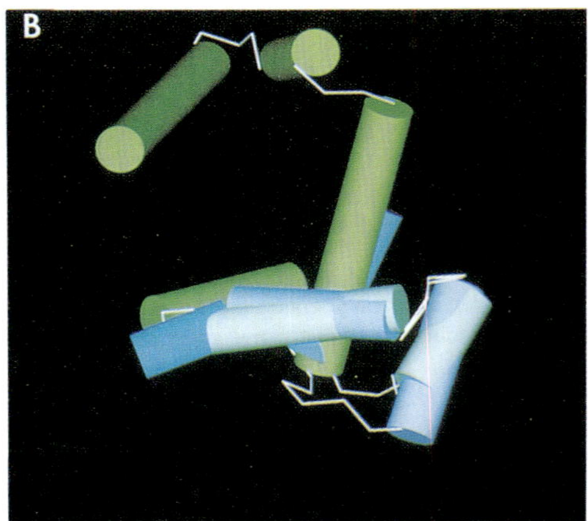

Figure 6. Comparison of the *Antp* Homeodomain with Prokaryotic Repressors

Comparison of the solution structure of the *Antp* homeodomain (blue) with (A) the crystal structure of the 434 repressor (red) and (B) the crystal structure of the *trp* repressor (green). In each case the two structures were superimposed for minimum RMSD between the 21 residue segments containing the helix-turn-helix motif, i.e., residues 30–50 of the *Antp* homeodomain, 16–36 of the 434 repressor, and 67–87 of the *trp* repressor (see Table 1). The helices are represented as cylinders, and the C^{α} atoms of the connecting backbone segments are linked by straight white lines. The helix-turn-helix motif is shown in lighter color than the other helices. The *Antp* homeodomain is represented in the same orientation as in Figures 1–4.

tures of the two molecules. This visual impression is confirmed by the numbers in Table 1, with an RMSD value of 0.9 for the backbone atoms of the helix-turn-helix segment and an RMSD value of 3.9 for the helix-turn-helix segment plus the N-terminal helix. The principal differences between the two structures in Figure 6A are the different number of helices and the spatial arrangement of the additional helices relative to the helix-turn-helix segment. In the homeodomain, helix III in the helix-turn-helix segment is effectively elongated by the addition of helix IV, and there is only one additional helix (residues 10–21) toward the N-terminus. In contrast, the 434 repressor N-terminal domain (residues 1–69) contains three helices outside the helix-turn-helix segment, one on the N-terminal and two on the C-terminal side of the latter (Mondragón et al., 1989). As we have seen in Figures 1–5, helices I, II, and III enclose a well-structured hydrophobic core region in the globular homeodomain structure. In the 434 repressor the N-terminal helix is oriented nearly perpendicular to the corresponding helix I of the homeodomain (Figure 6A). Combined with the helix-turn-helix segment of the 434 structure it forms a flat, open structure. A globular structure with an interior core is obtained only with the participation of the two C-terminal helices (Figure 6A). Overall, we find that the only common structural feature of these two proteins is the 21 residue helix-turn-helix segment.

Good fits for the helix-turn-helix segment where also achieved between the *Antp* homeodomain and the prokaryotic DNA binding proteins λ repressor (Pabo and Lewis, 1982), *lac* repressor (Zuiderweg et al., 1983, 1984), CAP (McKay et al., 1981), and *trp* repressor (Lawson et al., 1988) (Tables 1 and 2). Extension of the comparisons to include helix I of the homeodomain and the corresponding polypeptide segments in the prokaryotic proteins showed that the packing of this helix against the helix-turn-helix motif coincides quite closely in the *Antp* homeodomain, CAP, and *trp* repressor. Figure 6B and Table 1 present in more detail the superposition of the *Antp* homeodomain with the *trp* repressor obtained upon optimal fitting of the respective helix-turn-helix segments. The N-terminus of the *trp* repressor is at the top left of Figure 6B, and the structure starts with two helical segments that have no counterpart in the homeodomain structure. (The native *trp* repressor is a homodimer in which the first three helices are in intimate contact through the dimer interface. Figure 6B shows only a single subunit.) In the superposition of Figure 6B, the third helix of the *trp* repressor has the same orientation as helix I of the *Antp* homeodomain. Although the third helix in the *trp* repressor is nearly twice as long as helix I in the *Antp* homeodomain, the C-terminal ends of the two helices coincide closely in the superposition shown. In the helix-turn-helix region the orientations of helices II and III in the homeodomain are nearly identical to those of the corresponding helices in *trp* repressor, but the lengths of these helices are somewhat different in the two proteins. The spatial arrangements of the C-terminal polypeptide segments following the helix-turn-helix motif are again clearly different in the two proteins. While helix IV of the *Antp* homeodomain effectively elongates helix III (Figures 1–5), the *trp* repressor structure ends with a long helix that is folded back against the core of the protein.

Conclusions

The solution structure of the *Antp* homeodomain obtained from a complete analysis of the NMR data (Figures 1–5) confirms and expands the previously reported results of a preliminary NMR study (Otting et al., 1988). The homeodomain structure includes a helix-turn-helix segment nearly identical to that found to be responsible for specific DNA binding in the phage 434 repressor (Aggarwal et al., 1988), but otherwise these two proteins have different molecular architectures (Figure 6A). The structural similarity with the *trp* repressor from E. coli (Lawson et al., 1988) also includes the packing arrangement of helix I in the *Antp* homeodomain and the corresponding helix in the repressor against the helix-turn-helix motif (Figure 6B).

Compared with the well-known prokaryotic repressor proteins (Zuiderweg et al., 1983, 1984; Mondragón et al., 1989; Lawson et al., 1988; McKay et al., 1981; Pabo and Lewis, 1982; Ohlendorf et al., 1983), only little is known about the DNA binding modes of homeodomains. It is therefore of interest that the locations of two of the four conserved residues, Asn-51 and Arg-53, and the distribution of charged residues (Figure 4) provide support for the hypothesis that the homeodomain may bind to DNA by close contact of helix III with the target DNA, which would be similar to the binding mode of the prokaryotic repressors (Aggarwal et al., 1988; Otwinowski et al., 1988; Jordan and Pabo, 1988; Wolberger et al., 1988; Lamerichs et al., 1989). Possible roles of helix IV in these interactions are particularly intriguing. This helix differs from the other three helices in the fact that although typical NOE distance constraints for α helices were observed (Otting et al., 1988) and there are 20 NOEs between protons of helix IV and the rest of the protein, the amide protons involved in the hydrogen bonds of the helix were found to exchange too rapidly with the solvent to be seen in D_2O solution. One explanation for these observations is that the polypeptide segment 53–59 is predominantly in a helical conformation that is in rapid equilibrium with small populations of unfolded structures. Quite possibly the stability of this helix would be increased in the complex with DNA. To obtain more definite information on the DNA binding properties of homeodomains, we are currently working to determine the structure of a complex formed between the *Antp* homeodomain and a synthetic DNA duplex containing one of the homeodomain consensus binding sites (Müller et al., 1988). These experiments can be expected to shed more light also on the significance of the structural similarities and differences observed between the *Antp* homeodomain and the known structures of prokaryotic repressors (Figure 6, Tables 1 and 2).

Finally, a general remark concerning the prediction of three-dimensional protein structures can be made. In the absence of direct experimental data on the structure of a protein, modeling procedures using the known structures of sequentially homologous molecules are often applied to predict the unknown structure. Using such an approach, Tsonis et al. (1988) modeled the homeodomain encoded by the *ftz* gene. Although the helix-turn-helix segment was correctly predicted by this modeling study, all the global structural differences observed in Figure 6A between the homeodomain and the N-terminal domain of the 434 repressor are also present when the experimental and predicted homeodomain structures are compared. For example, the two conserved interior residues Trp-48 and Phe-49 (Figures 4 and 5) were predicted to be located on the protein surface (Tsonis et al., 1988; note that these workers use a different numbering of residues). This is yet another example indicating that much care should be exercised in protein tertiary structure predictions in cases where one cannot independently ascertain that the global molecular architecture is homologous to that of the proteins used as a basis for the structure prediction.

Experimental Procedures

The *Antp* homeodomain was purified to homogeneity as described in Müller et al. (1988). Basically, the purification involves Polymin P precipitation, Biorex A-70 chromotography, and FPLC on a Mono S 10/10 column (Pharmacia). The purified protein was desalted by ultrafiltration and lyophilized.

For the preparation of the NMR samples, the protein was dissolved either in a mixed solvent of 90% H_2O and 10% D_2O or in 100% D_2O after exchange of all the amide protons. All NMR measurements were performed at pH 4.3, T = 20°C on a Bruker AM 600 spectrometer. Under these conditions virtually identical spectra were obtained with 4 mM and 11 mM protein concentrations. For the collection of the input for the structure calculations, we used primarily NOESY (Anil-Kumar et al., 1980) and soft-NOESY (Brüschweiler et al., 1988) spectra recorded with mixing times of 40 ms. NOESY spectra of the *Antp* homeodomain were previously presented (Figure 2 in Otting et al., 1988); a detailed description of the spectral analysis including the integration of the peak intensities and the calibration of NOE intensities versus ^1H–^1H distances is in preparation. A complete listing of the input for the structure calculations will be submitted to the Brookhaven protein data bank, together with the atomic coordinates of the *Antp* homedomain structure.

The main programs used for the structure calculation were: HABAS for the stereospecific assignment of prochiral groups (Güntert et al., 1989), DISMAN for the generation of geometrically correct structures (Braun and Gō, 1985), and a modified version of AMBER (Singh et al., 1986) for the restrained energy minimization. The strategy for the structure calculations followed closely the procedures described in Widmer et al. (1989).

For quantitative comparisons of different structures, RMSD values were used (McLachlan, 1979). The RMSD is a measure of the differences in atomic positions of corresponding atoms in two structures after these structures have been superimposed to minimize the sum of the squares of these differences:

$$\mathrm{RMSD} = \min_{\mathbf{R}} \, [1/n \sum_i (\mathbf{x}_i - \mathbf{y}_i)^2]^{1/2}.$$

$\mathbf{x}_i$ and $\mathbf{y}_i$ are the position vectors of n corresponding atoms in the two structures, $\mathbf{R}$ is a rotation of the second structure relative to the first one, and i runs over all n atoms selected for the calculation for the best fit.

Acknowledgments

We thank Dr. S. Harrison for the coordinates of 434 repressor residues 1–69, Mr. T. H. Xia for help with the picture display, and Mrs. E. Huber and Mr. R. Marani for the careful processing of the typescript. Financial support by the Schweizerischer Nationalfonds (projects 31.25174.88 and K O3.613.87) and the use of the Cray X-MP/28 of the ETH Zürich are gratefully acknowledged.

The costs of publication of this article were defrayed in part by the payment of page charges. This article must therefore be hereby marked "*advertisement*" in accordance with 18 U.S.C. Section 1734 solely to indicate this fact.

Received July 31, 1989; revised September 20, 1989.

References

Aggarwal, A. K., Rodgers, D. W., Drottar, M., Ptashne, M., and Harrison, S. C. (1988). Recognition of a DNA operator by the repressor of phage 434: a view at high resolution. Science *242*, 899–907.

Anil-Kumar, Wagner, G., Ernst, R. R., and Wüthrich, K. (1980). Studies of J-connectivities and selective ^1H–^1H Overhauser effects in H_2O solutions of biological macromolecules by two-dimensional NMR experiments. Biochem. Biophys. Res. Commun. *96*, 1156–1163.

Billeter, M., Kline, A. D., Braun, W., Huber, R., and Wüthrich, K. (1989).

Comparison of the high-resolution structures of the α-amylase inhibitor Tendamistat determined by nuclear magnetic resonance in solution and by X-ray diffraction in single crystals. J. Mol. Biol. *206*, 677–687.

Braun, W., and Go, N. (1985). Calculation of protein conformations by proton–proton distance constraints. A new efficient algorithm. J. Mol. Biol. *186*, 611–626.

Brennan, R. G., and Matthews, B. W. (1989). The helix-turn-helix DNA binding motif. J. Biol. Chem. *264*, 1903–1906.

Brüschweiler, R., Griesinger, C., Sorensen, O. W., and Ernst, R. R. (1988). Combined use of hard and soft pulses for ω_1 decoupling in two-dimensional nmr spectroscopy. J. Magnet. Reson. *78*, 178–185.

Carrasco, A. E., McGinnis, W., Gehring, W. J., and De Robertis, E. M. (1984). Cloning of an X. laevis gene expressed during early embryogenesis coding for a peptide region homologous to Drosophila homeotic genes. Cell *37*, 409–414.

Coleman, K. G., Poole, S. J., Weir, M. P., Soeller, W. C., and Kornberg, T. (1987). The *invected* gene of *Drosophila*: sequence analysis and expression studies reveal a close kinship to the *engrailed* gene. Genes Dev. *1*, 19–28.

Fjose, A., McGinnis, W. J., and Gehring, W. J. (1985). Isolation of a homoeo box-containing gene from the *engrailed* region of *Drosophila* and the spatial distribution of its transcripts. Nature *313*, 284–289.

Gehring, W. J. (1987). Homeo boxes in the study of development. Science *236*, 1245–1252.

Glusker, J. P., and Trueblood, K. N. (1985). Crystal Structure Analysis (New York: Oxford University Press).

Güntert, P., Braun, W., Billeter, M., and Wüthrich, K. (1989). Automated stereospecific ^{1}H nmr assignments and their impact on the precision of protein structure determinations in solution. J. Am. Chem. Soc. *111*, 3997–4004.

Hochschild, A., Irwin, N., and Ptashne, M. (1983). Repressor structure and the mechanism of positive control. Cell *32*, 319–325.

Jordan, S. R., and Pabo, C. O., (1988) Structure of the lambda complex at 2.5 Å resolution: details of the repressor–operator interactions. Science *242*, 893–899.

Kline, A. D., Braun, W., and Wüthrich, K. (1988). Determination of the complete three-dimensional structure of the α-amylase inhibitor Tendamistat in aqueous solution by nuclear magnetic resonance and distance geometry. J. Mol. Biol. *204*, 675–724.

Lamerichs, R. M. J. N., Boelens, R., van der Marel, G. A., van Boom, J. H., Kaptein, R., Buck, F., Fera, B., and Rüterjans, H. (1989). ^{1}H NMR study of a complex between the *lac* repressor headpiece and a 22 base pair symmetric *lac* operator. Biochemistry *28*, 2985–2991.

Laughon, A., and Scott, M. P. (1984). Sequence of a *Drosophila* segmentation gene: protein structure homology with DNA-binding proteins. Nature *310*, 25–31.

Lawson, C. L., Zhang, R.-G., Schevitz, R. W., Otwinowski, Z., Joachimiak, A., and Sigler, P. B. (1988). Flexibility of the DNA-binding domains of *trp* repressor. Proteins *3*, 18–31.

Levine, M., Rubin, G. M., and Tjian, R. (1984). Human DNA sequences homologous to a protein coding region conserved between homeotic genes of Drosophila. Cell *38*, 667–673.

McGinnis, W., Levine, M. S., Hafen, E., Kuroiwa, A., and Gehring, W. J. (1984a). A conserved DNA sequence in homoeotic genes of the *Drosophila* Antennapedia and bithorax complexes. Nature *308*, 428–433.

McGinnis, W., Hart, C. P., Gehring, W. J., and Ruddle, F. H. (1984b). Molecular cloning and chromosome mapping of a mouse DNA sequence homologous to homeotic genes of Drosophila. Cell *38*, 675–680.

McKay, D. B., Weber, I. T., and Steitz, T. A. (1981). Structure of catabolite gene activator protein at 2.9 Å resolution. J. Biol. Chem. *257*, 9518–9524.

McLachlan, A. D. (1979). Gene duplications in the structural evolution of chymotrypsin. J. Mol. Biol. *128*, 49–79.

Mondragón, A., Subbiah, S., Almo, S. C., Drottar, M., and Harrison, S. C. (1989). Structure of the amino-terminal domain of phage 434 repressor at 2.0 Å resolution. J. Mol. Biol. *205*, 189–200.

Müller, M., Affolter, M., Leupin, W., Otting, G., Wüthrich, K., and Gehring, W. J. (1988). Isolation and sequence-specific DNA binding of the *Antennapedia* homeodomain. EMBO J. *7*, 4299–4304.

Ohlendorf, D. H., Anderson, W. F., Lewis, M., Pabo, C. O., and Matthews, B. W. (1983). Comparison of the structures of *cro* and λ repressor proteins from bacteriophage λ. J. Mol. Biol. *169*, 757–769.

Otting, G., Qian, Y. Q., Müller, M., Affolter, M., Gehring W. J., and Wüthrich, K. (1988). Secondary structure determination for the *Antennapedia* homeodomain by nuclear magnetic resonance and evidence for a helix-turn-helix motif. EMBO J. *7*, 4305–4309.

Otwinowski, Z., Schevitz, R. W., Zhang, R.-G., Lawson, C. L., Joachimiak, A., Marmorstein, R. Q., Luisi, B. F., and Sigler, P. B. (1988). Crystal structure of *trp* repressor/operator complex at atomic resolution. Nature *335*, 321–329.

Pabo, C. O., and Lewis, M. (1982). The operator-binding domain of λ repressor: structure and DNA recognition. Nature *298*, 443–447.

Poole, S. J., Kauvar L. M., Drees, B., and Kornberg, T. (1985). The *engrailed* locus of Drosophila: structural analysis of an embryonic transcript. Cell *40*, 37–43.

Porter, S. D., and Smith, M. (1986). Homoeo-domain homology in yeast MATα2 is essential for repressor activity. Nature *320*, 766–768.

Richardson, J. (1981). The anatomy and taxonomoy of protein structure. Adv. Protein Chem. *34*, 167–339.

Rosenberg, A. H., Lade, B. N., Chui, D.-S., Lin, S.-W., Dunn, J. J., and Studier, F. W. (1987). Vectors for selective expression of cloned DNAs by RNA polymerase. Gene *56*, 125–135.

Schneuwly, S., Kuroiwa, A., Baumgartner, P., and Gehring, W. J. (1986). Structural organization and sequence of the homeotic gene *Antennapedia* of *Drosophila melanogaster*. EMBO J. *5*, 733–739.

Scott, M. P., and Weiner, A. J. (1984). Structural relationship among genes that control development: sequence homology between the *Antennapedia*, *Ultrabithorax*, and *fushi tarazu* loci of *Drosophila*. Proc. Natl. Acad. Sci. USA *81*, 4115–4119.

Scott, M. P., Tamkun, J. W., and Hartzell, G. W., III (1989). The structure and function of the homeodomain. BBA Rev. Cancer *989*, 25–49.

Shepherd, J. C. W., McGinnis, W., Carrasco, A. E., De Robertis, E. M., and Gehring, W. J. (1984). Fly and frog homoeo domains show homologies with yeast mating type regulatory proteins. Nature *310*, 70–71.

Singh, U. C., Weiner, P. K., Caldwell, J. W., and Kollman, P. A. (1986). AMBER 3.0. University of California, San Francisco.

Steitz, T. A., Ohlendorf, D. H., McKay, D. B., Anderson, W. F., and Matthews, B. W. (1982). Structural similarity in the DNA-binding domains of catabolite gene activator and *cro* repressor proteins. Proc. Natl. Acad. Sci. USA *79*, 3097–3100.

Studier, F. W., and Moffatt, B. A. (1986). Use of bacteriophage *T7* RNA polymerase to direct selective high-level expression of cloned genes. J. Mol. Biol. *189*, 113–130.

Tsonis, P. A., Carperos, V., and Shiahaan, T. (1988). Modeling of the homeo domain suggests similar structure to repressors. Biochem. Biophys. Res. Commun. *157*, 100–105.

Widmer, H., Billeter, M., and Wüthrich, K. (1989). Determination and refinement of the solution structure of the polypeptide neurotoxin ATX Ia from *Anemonia sulcata*. Proteins, in press.

Wolberger, C., Dong, Y., Ptashne, M., and Harrison, S. C. (1988). Structure of phage 434 Cro/DNA complex. Nature *335*, 789–795.

Wüthrich, K. ('986). NMR of Protein and Nucleic Acids (New York: John Wiley & Sons).

Wüthrich, K., Billeter, M., and Braun, W. (1984). Polypeptide secondary structure determination by nuclear magnetic resonance observation of short proton–proton distances. J. Mol. Biol. *180*, 715–740.

Zuiderweg, E. R. P., Kaptein, R., and Wüthrich, K. (1983). Secondary structure of the *lac* repressor DNA-binding domain by two-dimensional ^{1}H nuclear magnetic resonance in solution. Proc. Natl. Acad. Sci. USA *80*, 5837–5841.

Zuiderweg, E. R. P., Billeter, M., Boelens, R., Scheek, R. M., Wüthrich, K., and Kaptein, R. (1984). Spatial arrangement of the three α helices in the solution conformation of *E. coli lac* repressor DNA-binding domain. FEBS Lett. *174*, 243–247.

The EMBO Journal vol.9 no.10 pp.3085–3092, 1990

Reprinted by permission of Oxford University Press

Protein–DNA contacts in the structure of a homeodomain–DNA complex determined by nuclear magnetic resonance spectroscopy in solution

Gottfried Otting, Yan Qiu Qian, Martin Billeter, Martin Müller[1], Markus Affolter[1], Walter J.Gehring[1] and Kurt Wüthrich

Institut für Molekularbiologie und Biophysik, ETH-Hönggerberg, CH-8093 Zürich, Switzerland and [1]Biozentrum der Universität Basel, Abt. Zellbiologie, Klingelbergstr. 70, CH-4056 Basel, Switzerland

Communicated by K.Wüthrich

The 1:1 complex of the mutant *Antp(C39→S)* homeodomain with a 14 bp DNA fragment corresponding to the BS2 binding site was studied by nuclear magnetic resonance (NMR) spectroscopy in aqueous solution. The complex has a molecular weight of 17 800 and its lifetime is long compared with the NMR chemical shift time scale. Investigations of the three-dimensional structure were based on the use of the fully ^{15}N-labelled protein, two-dimensional homonuclear proton NOESY with $^{15}N(\omega_2)$ half-filter, and heteronuclear three-dimensional NMR experiments. Based on nearly complete sequence-specific resonance assignments, both the protein and the DNA were found to have similar conformations in the free form and in the complex. A sufficient number of intermolecular $^1H-^1H$ Overhauser effects (NOE) could be identified to enable a unique docking of the protein on the DNA, which was achieved with the use of an ellipsoid algorithm. In the complex there are intermolecular NOEs between the elongated second helix in the helix-turn-helix motif of the homeodomain and the major groove of the DNA. Additional NOE contacts with the DNA involve the polypeptide loop immediately preceding the helix-turn-helix segment, and Arg5. This latter contact is of special interest, both because Arg5 reaches into the minor groove and because in the free *Antp(C39→S)* homeodomain no defined spatial structure could be found for the apparently flexible N-terminal segment comprising residues 0–6.

Key words: Antennapedia homeodomain/DNA binding/gene expression/homeodomain–DNA complex/three-dimensional structure/two-dimensional NMR

Introduction

The homeobox is a 180 bp DNA sequence that encodes a 60 amino acid protein domain, the homeodomain. The homeobox was first found in the *Antennapedia* (*Antp*) and the *fushi tarazu* (*ftz*) genes, which are involved in the determination of segmental identity and segment number, respectively, in *Drosophila* (McGinnis *et al.*, 1984a,b; Scott and Weiner, 1984). Subsequently, the homeobox was found in many other *Drosophila* development regulatory genes (Gehring, 1987; Scott *et al.*, 1989). The homeobox has also been isolated from vertebrate genomes, which indicates extensive evolutionary conservation, and possible functional conservation (McGinnis *et al.*, 1984b). Evidence is now accumulating that homeodomain proteins act as transcription factors in which the homeodomain is involved in sequence-specific recognition of DNA (for reviews, see Levine and Hoey, 1988; Scott *et al.*, 1989; Affolter *et al.*, 1990b).

Amino acid sequence comparison of the homeodomain with prokaryotic DNA binding proteins indicated that the homeodomain might contain a helix-turn-helix motif (Laughon and Scott, 1984; Shepherd *et al.*, 1984). The determination of the three-dimensional structure of the *Antp* homeodomain by nuclear magnetic resonance (NMR) spectroscopy indeed demonstrated the existence of the postulated helix-turn-helix motif (Otting *et al.*, 1988; Qian *et al.*, 1989; Billeter *et al.*, 1990). However, recent results obtained by site-directed mutagenesis imply a direct involvement in DNA binding for different amino acid residues of the helix-turn-helix motif than those that one might have anticipated by analogy to prokaryotic repressor proteins, for example the 434 repressor. In particular, a single amino acid substitution of the ninth residue of helix III seems to be sufficient to switch the DNA binding specificities of different homeodomains (Hanes and Brendt, 1989; Treisman *et al.*, 1989).

In a project aimed at the elucidation of the molecular basis of DNA recognition by homeodomains we previously determined the structure of the *Antp* homeodomain in aqueous solutions using NMR (Qian *et al.*, 1989; Billeter *et al.*, 1990). The protein fragment used contained 68 amino acid residues, which correspond to residues 297–363 of the *Antp* protein and an additional N-terminal methionine introduced by the overexpression system used (Müller *et al.*, 1988). This study was now extended to the complex of the *Antp(C39→S)* homeodomain with a 14 bp DNA fragment. The DNA fragment used corresponds to the BS2 site, which had been identified as a specific binding site by DNA footprinting and gel retardation assays (Müller *et al.*, 1988). Residue 39 of the *Antp* homeodomain was changed from cysteine to serine to prevent oxidative dimerization of the protein (Müller *et al.*, 1988). DNA binding studies showed that the mutant protein has the same DNA binding affinity as the wild type polypeptide. It binds to the BS2 site as a monomer with a binding affinity of $\sim 10^9$/M and half-life of the resulting complex of ~90 min (Affolter *et al.*, 1990a). Furthermore, a full structure determination of the *Antp(C39→S)* homeodomain by NMR confirmed that the mutation does not significantly affect the protein conformation (Güntert *et al.*, 1990).

Results and discussion

Selection of the DNA fragment used to prepare the complex

A systematic search was carried out to determine the shortest possible DNA fragment which shows a binding constant

© Oxford University Press

comparable to the 26-mer used in earlier binding studies, i.e. d(AGCTGAGAAAAAGCCATTAGAGAAGC) (Müller *et al.*, 1988). G≡C base pairs were placed at both ends of the DNA fragment for improved stability of the two ends of the duplex. The sequences d(GAAAAAGCCATTAGAG) (16-mer), d(GAGAAAGCCATTAGAG) (16-mer$_2$), d(GGAAAGCCATTAGAG) (15-mer), d(GAAAAAGC-CATTAGG) (15-mer$_2$), d(GAAAGCCATTAGAG) (14-mer) and d(GAAGCCATTAGAG) (13-mer) were synthesized, and after combination with their complementary strands the binding constants with the *Antp(C39→S)* homeodomain were determined. The substitution of A3 to G3 between 16-mer and 16-mer$_2$, and the subsequent deletions of A2 between 16-mer$_2$ and 15-mer and G1 between 15-mer and 14-mer did not significantly alter the binding constant with respect to that of the 26-mer. In contrast, any further deletion at the 5′ end, for example, between 14-mer and 13-mer, or at the 3′ end, for example, between 16-mer and 15-mer$_2$ lead to reduced binding affinities. Therefore we decided to use the 14-mer for the present study.

Preparation of a 1:1 complex between Antp(C39→S) and the DNA 14-mer

A 0.5 mM aqueous solution (90% H_2O/10% D_2O, buffer: 25 mM sodium phosphate, 100 mM KCl, 2 mM NaN_3, pH 6.0) of the DNA 14-mer was titrated at 20°C with uniformly ^{15}N-labelled *Antp(C39→S)* homeodomain. Figure 1 shows the imino proton resonances of the DNA at different ratios of protein:DNA. For titration ratios between 0.0 and 1.0 the imino proton spectrum is a superposition of those of the free DNA (titration ratio 0.0) and the 1:1 complex (titration ratio 1.0), indicating that the exchange between free and bound DNA is slow on the chemical shift time scale. This is in accordance with the half-life of the complex as measured by gel mobility shift (Affolter *et al.*, 1990a). At titration ratios >1.0 a precipitate formed, which could be redissolved by further addition of 14-mer. The imino proton spectrum observed for the 1:1 complex did not change upon the addition of an excess of the protein, except that the amplitude decreased with increasing amounts of protein added. A plausible explanation for these observations would be that the precipitate formed was an insoluble complex with two molecules of *Antp(C39→S)* bound to one 14-mer.

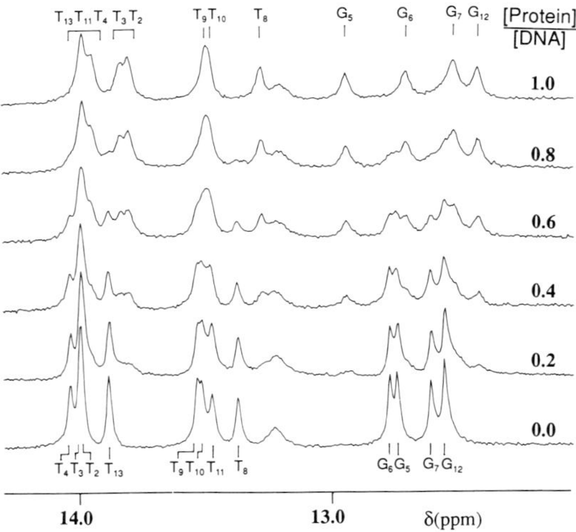

Fig. 1. Imino proton region from 12.0−14.3 p.p.m. of the one-dimensional ^{1}H NMR spectra obtained upon stepwise addition of *Antp(C39→S)* homeodomain to a solution of the DNA 14-mer in 90% H_2O/10% D_2O, pH 6.0, 20°C. The molar ratios of protein:DNA are indicated on the right. The assignments of the chemical shifts for the free and complexed DNA are given below and above the corresponding imino proton spectra, respectively. Bars indicate groups of resonances which have not been assigned individually at 20°C.

A solution of the 1:1 complex in the aforementioned solvent medium showed broad line-widths at concentrations >1 mM. In contrast, only little line-broadening relative to the 0.5 mM solution used to record the spectra of Figure 1 was observed for a 3.5 mM solution of the 1:1 complex prepared without the addition of KCl and phosphate buffer. Comparison of [^{15}N,^{1}H]-COSY spectra recorded, respectively, with the 3.5 mM salt-free protein solution and with the 0.5 mM buffered protein solution under otherwise identical conditions did not reveal any significant changes in chemical shifts, indicating that the conformation of the complex is insensitive to the presence of the salt and the buffer used in the more dilute protein solution.

Circular dichroism (CD) measurements with a 0.1 mM solution of the *Antp(C39→S)*−14-mer complex at pH 6.0 showed no evidence for denaturation of the complex in the temperature range 18−50°C.

^{1}H NMR assignments for Antp(C39→S) and the DNA 14-mer in the 1:1 complex

Sequence specific ^{1}H NMR assignments were previously obtained for both the free *Antp(C39→S)* homeodomain (Güntert *et al.*, 1990) and the free DNA 14-mer. However, the spectral changes observed upon complex formation were so extensive that new assignments had to be worked out for both the protein and the DNA in the complex. This was achieved using [^{1}H,^{1}H]-NOESY with $^{15}N(\omega_2)$-half-filter (Otting and Wüthrich, 1990) and three-dimensional ^{15}N-correlated [^{1}H,^{1}H]-NOESY (Messerle *et al.*, 1989). To avoid transfer of saturation from the water to the solute molecules, the experimental schemes for both experiments were designed for measurements in H_2O solution without water saturation by preirradiation (Otting and Wüthrich, 1989; Messerle *et al.*, 1989).

Figure 2A and B shows the sum spectrum and the difference spectrum, respectively, of a [^{1}H,^{1}H]-NOESY experiment with $^{15}N(\omega_2)$-half-filter (see Materials and methods for experimental details). The $^{15}N(\omega_2)$-half-filter discriminates between protons bound directly to ^{15}N and all other protons, where this discrimination is applied only along the ω_2 frequency axis. Thus, the sum spectrum (Figure 2A) contains only peaks with ω_2 frequencies of unlabelled protons, while the difference spectrum (Figure 2B) contains only peaks which correlate with the resonances of ^{15}N-bound protons along the ω_2 frequency axis. Both spectra contain the peaks correlating with all proton resonances along ω_1. Since the protein is uniformly enriched with ^{15}N to the extent of >95%, the difference spectrum (Figure 2B) contains the diagonal peaks and cross peaks with the amide protons of the protein, while the sum spectrum contains the diagonal peaks and cross peaks with all DNA resonances and with those protons of the protein which are not bound to ^{15}N. In the spectral region shown in Figure 2 these are the protons of the aromatic side chains and some hydroxyl protons of amino acid side chains. Thus the $^{15}N(\omega_2)$-half-filter technique effectively separates the bulk of the protein spectrum from the DNA spectrum. A conventional [^{1}H,^{1}H]-NOESY spectrum would contain the peaks from both subspectra shown in Figure 2A and B.

The improvement in resolution gained with the use of the $^{15}N(\omega_2)$-half-filter was sufficient to obtain sequence-specific assignments for the DNA 14-mer from the sum spectrum (Figure 2A), and to identify a large number of sequential connectivities in the protein (Wüthrich, 1986) in the difference spectrum (Figure 2B). However, most of the protein assignments could only be ascertained with the use of a three-dimensional (3D) ^{15}N-correlated [^{1}H,^{1}H]-NOESY experiment. The frequency axes of this 3D NMR spectrum are ^{1}H in ω_1, ^{15}N in ω_2 and ^{1}H in ω_3. The spectrum was analysed in plots of the $\omega_1 - \omega_3$ cross sections. Figure 2C shows the 22nd out of a total of 64 planes obtained, which are separated by the ^{15}N chemical shifts corresponding to the ^{15}N-bound protons observed in the ω_3 dimension. It illustrates the dramatic improvement in resolution gained from the development of the spectrum in an additional dimension. (The difference spectrum of Figure 2B corresponds to the projection of the peaks in all 64 planes of the 3D NMR spectrum onto a single $\omega_1(^1H) - \omega_3(^1H)$ frequency plane (Otting and Wüthrich, 1990).)

For the presently studied protein – DNA complex the usual strategy of sequential resonance assignments of proteins, which involves spin-system identifications by scalar [^{1}H,^{1}H] couplings prior to the use of NOESY for delineation of sequential connectivities (Wüthrich, 1986) could not be applied because the COSY and TOCSY spectra were of poor quality. Therefore a strategy relying exclusively on NOE data was adopted, which started from the assumption that the conformations of both the protein and the DNA are similar in the 1:1 complex and in the free compounds. The patterns of intramolecular NOE connectivities would then be expected to be conserved, even though the chemical shifts may be sizeably different. The assignments achieved on this basis turned out to be self-consistent, and eventually confirmed the starting assumptions made. A few otherwise unassigned intramolecular NOEs could finally be identified as signals from the slowly exchanging hydroxyl protons of Tyr8, Thr9 and Thr41, which were not observed in the free protein due to rapid exchange with H_2O.

In summary the sequence-specific resonance assignments obtained for the *Antp(C39→S)* homeodomain and the DNA 14-mer in the complex are almost complete. All proton resonances of the polypeptide backbone were assigned, with the exception of Met0 and Arg1, where the amide protons exchange too rapidly with the solvent to be observable. β-Proton and γ-proton resonances were assigned for 60 and 40 residues, respectively, and for about half of the 68 amino acid residues the resonance assignments include all non-exchangeable side chain protons. For the 14-mer all non-exchangeable base protons and all 1′ sugar protons were assigned. With two exceptions all 2′H, 2″H and 3′H resonances were also assigned. Furthermore, intramolecular

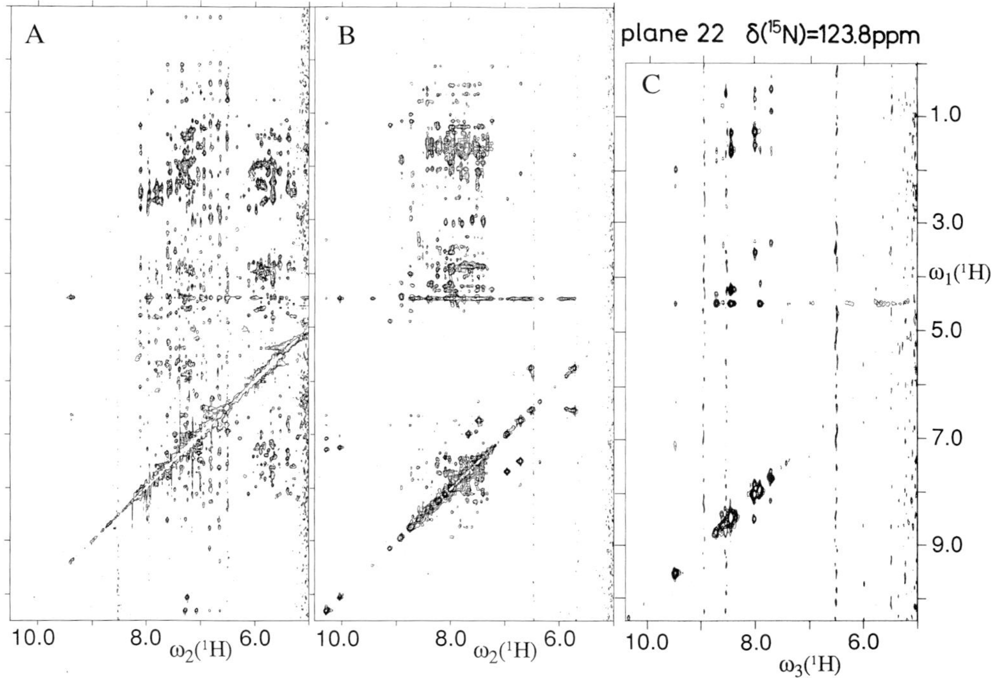

Fig. 2. Spectral region (ω_1 = −1.0 – 10.5 p.p.m., ω_2 = 5.0 – 10.5 p.p.m.) of ^{15}N-edited [^{1}H,^{1}H]-NOESY spectra of a 3.5 mM solution of the 1:1 complex formed by uniformly ^{15}N-labelled *Antp(C39→S)* homeodomain and DNA 14-mer at 36°C. (**A**) Sum spectrum and (**B**) difference spectrum of [^{1}H,^{1}H]-NOESY with $^{15}N(\omega_2)$-half-filter, pH 6.8, mixing time 110 ms. (**C**) $\omega_1 - \omega_3$ cross plane taken at ω_2 = 123.8 p.p.m. through the three-dimensional ^{15}N-correlated [^{1}H,^{1}H]-NOESY spectrum, pH 6.0, mixing time 40 ms. The ω_1 dimension is shown only up to 0.0 p.p.m.

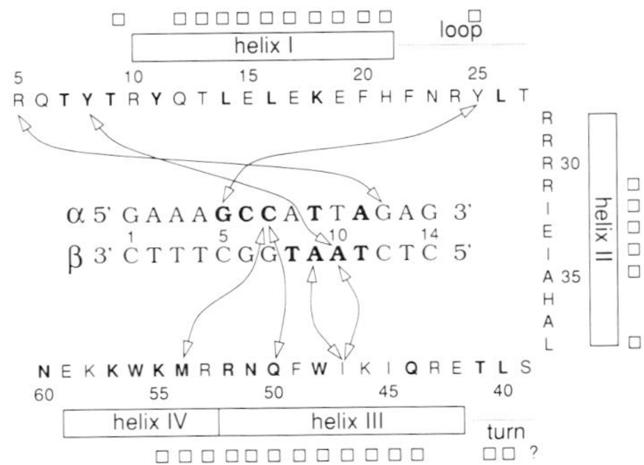

Fig. 3. Survey of the experimental data on the 1:1 complex between the *Antp(C39→S)* homeodomain and the DNA 14-mer. The sequence of the 14-mer is given in the centre with the numeration used for the base pairs. The two strands are arbitrarily denoted α and β. The *Antp(C39→S)* amino acid sequence is arranged clockwise around the 14-mer. The terminal residues 0–4 and 61–67 have been omitted, since no reliable experimental data were obtained for these polypeptide segments in the free protein, and thus no comparisons with the complex were possible. The secondary structure of the protein observed in the free form and in the complex is indicated alongside the amino acid sequence. Bold letters identify those amino acid residues or nucleotides for which changes in chemical shifts larger than or equal to |0.2|p.p.m. were observed upon complex formation. Squares alongside the polypeptide sequence identify the residues with slow amide proton exchange in the complex, where ? indicates that a spectral artifact prevented the measurement of the NH exchange for Ser39. Finally, the arrows identify intermolecular contacts evidenced by NOEs between the protein and the DNA (Table I).

NOEs with the group of resonances involving the 4′H, 5′H and 5″H protons were assigned for most of the nucleotides, without attempting to assign these protons individually. Similar ambiguities also arose for some of the side chain resonances of the protein, where the lack of scalar coupling connectivities prevented the unambiguous distinction between resonances with similar chemical shifts, e.g. $C^{\gamma}H_2$ and $C^{\delta}H_2$ of lysyl side chains.

Conformations of the Antp(C39→S) homeodomain and the DNA 14-mer in the 1:1 complex

Based on the essentially complete sequence-specific resonance assignments for the *Antp(C39→S)* homeodomain and the DNA 14-mer in the complex, several independent lines of experimental evidence show that in spite of the extensive chemical shift changes observed upon complex formation, the conformations of both components are very similar in the free state and in the complex. The most direct implications come from the fact that the intramolecular NOEs in the protein-bound 14-mer are typical of those for B-DNA (Wüthrich, 1986), and that the patterns of NOEs in the DNA-bound protein correspond closely to those of the free *Antp(C39→S)* homeodomain (Güntert *et al.*, 1990). Furthermore, for the majority of the resonance lines the chemical shifts are virtually unchanged after complexation, and as it eventually turned out (see below) the lines with shift differences exceeding |0.2|p.p.m. correspond almost exclusively to groups of protons located at the protein–DNA interface.

Independent additional evidence for similar local conformation in the free protein and the complex was obtained from different experiments for the helical regions and the polypeptide segments with non-regular secondary structure, respectively. For the latter this information came from measurements of the vicinal coupling constants $^3J_{Hn\alpha}$ obtained by recording a series of J-modulated $^{15}N,^1H$-COSY experiments (Neri *et al.*, 1990). In the free protein these data were collected at pH 4.3 and 20°C, and in the complex at pH 6.0 and 36°C. In the complex the $^3J_{HN\alpha}$ coupling constants could be measured with a precision of ±1 Hz for the residues 3, 4, 6–9, 12, 23, 24, 26, 28, 39, 41, 42, 60–64 and 66. In spite of the somewhat different solvent conditions used for the two measurements the $^3J_{HN\alpha}$ values did not indicate any significant conformational change. Even for the three residues where a difference between the free and complexed states of the protein could be detected within the accuracy of the measurements, i.e. Thr7, Tyr8 and Asn23, the variations were small, i.e. the corresponding coupling constants were ~1 Hz smaller in the complex. For most residues in the helices I, II and III the $^3J_{HN\alpha}$ values were found to be <5.0 Hz in the free protein. The measurements obtained for the corresponding residues in the complex are indicative of equally small $^3J_{HN\alpha}$ values. Except for Arg10 the quality of the data was sufficient to determine upper limits of 4.6–6.7 Hz for these couplings. More direct information on the helical segments came from amide protein exchange measurements (Wüthrich, 1986). Overall these indicated decreased flexibility and/or solvent accessibility for both components in the complex, i.e. the amide protons of the protein and the imino protons of the DNA in the complex exchanged significantly slower with the bulk water than in the free components. Nevertheless, the *relative* exchange rates of the individual amide protons were found to be very similar in the free and the complexed *Antp(C39→S)* homeodomain. In particular the slowed exchange associated with the hydrogen bonds in the helices of the free protein (Otting *et al.*, 1988) was also found in the DNA-bound *Antp(C39→S)* homeodomain (Figure 3). Exceptions are the residues 30, 36 and 37, for which the amide proton exchange in the complex was too fast for the resonance lines to be seen in a freshly prepared ^2H_2O solution (Wüthrich, 1986). This appears to indicate that helix II may be somewhat distorted by the binding of the homeodomain to the DNA.

The structure of the Antp(C39→S) DNA–14-mer complex

Contacts between the two molecules in the 1:1 complex are directly manifested by the intermolecular NOEs indicated by arrows in Figure 3 and listed in Table I. Figure 3 further identifies the residues with variations of the ^1H chemical shifts ≥|0.2|p.p.m. between the free and complexed forms of both components, and indicates the locations of slowly exchanging backbone amide protons in the complex. Some qualitative features of the 1:1 complex are readily apparent from inspection of Figure 3, Table I, and a molecular model of B-DNA. With one exception all the observed contact sites are in the major groove of the DNA, the exception being the side chain of Arg5, which interacts with a sugar proton in the minor groove. Specific contacts of the homeodomain with the DNA are indicated for the amino acid residues Ile47, Gln50 and Met54, which contact base protons. All other intermolecular NOEs are with sugar protons of the DNA. The intermolecular contacts are distributed over a large part of the molecular surfaces, and their numbers should be

Table I. Intermolecular NOEs observed between protons of the *Antp(C39→S)* homeodomain and the DNA 14-mer in the 1:1 complex[a]

Antp(C39→S)		14-mer[b]	
Arg 5	δCH_2	G 12	H1′
Tyr 8	C3,5H	A 10	H3′
Tyr 8	C4OH	A 10	H3′
Tyr 25	C2,6H	G 5	H5′,H5″
Tyr 25	C3,5H	G 5	H3′
		G 5	H5′,H5″
Ile 47	γCH_3	A 10	H8
Ile 47	δCH_3	A 9	H2′,H2″
		A 9	H3′
		A 9	H8
Gln 50	γCH_2	C 7	H5
Met 54	γCH_2	C 7	H5

[a]All NOEs in this table were observed in a NOESY spectrum recorded in D_2O with a mixing time of 40 ms, with the single exception of the NOE with the OH of Tyr8 which was recorded in a H_2O solution with a mixing time of 110 ms.
[b]This numbering refers to the base pairs (see Figure 3).

sufficient to position and orient the homeodomain with respect to the DNA. Helix III has the largest number of contacts with the major groove, and they are all on the same side of the helix.

To obtain a more precise characterization of the complex we used the data of Figure 3 and Table I as input for an ellipsoid algorithm calculation (Billeter *et al.*, 1987a,b) to dock the protein against the DNA 14-mer. During the entire calculation a B-DNA conformation was maintained for the 14-mer, and the solution structure of the free *Antp(C39→S)* homeodomain polypeptide (Güntert *et al.*, 1990) used for the docking, i.e. residues 7–54, was also maintained. The use of these structures for the docking is based on the experimental evidence presented in the preceding section, which supports that a B-type DNA and the conformation of the free protein are also present in the complex. In a first step, atomic coordinates for the two strands of the DNA were obtained by using standard covalent geometry and torsion angle values corresponding to B-DNA (Arnott and Chandrasekaran, private communication). These two strands were then docked with the ellipsoid algorithm using distance constraints for the formation of the interstrand hydrogen bonds and for steric repulsion of any two atoms. Next, coordinates for the residues 7–54 of the protein were taken from the best structure of the *Antp(C39→S)* homeodomain obtained by distance geometry and energy refinement (Güntert *et al.*, 1990). All amino acid side chains were replaced by alanine to exclude steric effects by the peripheral parts of the surface side chains, for which the conformation is not yet known in the complex. Docking of this rigid protein molecule to the 14-mer was again achieved with the ellipsoid algorithm, using as input all but one of the NOE distance constraints listed in Table I and all possible steric repulsions between the two molecules. The NOE with Arg5 was not used because the structure of residues 0–6 is not defined in the free protein (Qian *et al.*, 1989; Billeter *et al.*, 1990; Güntert *et al.*, 1990). (There was no problem later on to accommodate this NOE by an appropriate selection of the conformation of residues 5 and 6.) All NOEs were translated into distance constraints of 5.0 Å; these were further corrected to account for the fact that NOEs to side chain protons were referred to the β carbon atoms. An initial crude structure of the complex was obtained by manual placement of the protein near the contact site of the DNA, using interactive molecular graphics. This structure was then modified by random translation of up to 5.0 Å along the three coordinate axes, and random rotations of up to 45° about these coordinate axes. From 100 calculations each started with a different randomly modified structure, five structures of the complex were obtained that contained no residual violations of any NOE distance constraint, and no steric contact violations exceeding 0.1 Å. The average time used per structure calculation was <2 min on a VAX 8650.

Figure 4 shows the five *Antp(C39→S)* homeodomain structures docked to the DNA without any residual NOE violations. In this presentation the DNA molecules in the five structures were exactly superimposed, and the polypeptide backbone is drawn for residues 7–54. The average atomic displacement for the backbone of the polypeptide chain 7–54 was 3.1 Å. For the residues 42–54 from helices III and IV (Figure 3) the average displacement was 2.5 Å.

The NOE distance constraints used in the docking calculations are indicated in Figure 5. All NOEs involve side chain protons of the polypeptide (Table I). The NOEs from Tyr8 to the βDNA strand and from Tyr25 to the α DNA strand contact the backbone of the DNA on opposite sides of the major groove. Other NOE contacts are between Ile47 and sugar protons of the DNA as well as a base proton inside the major groove, and from Gln50 and Met54 to a base proton inside the major groove (the individual protons are identified in Table I). In addition, there is a NOE from Arg5 to the minor groove (H1′ of G12, not shown in Figure 5). As one may conclude from Figure 5 and further test by model building, a conformation of residues 5 and 6 can readily be found that satisfies this constraint. The spread of the NOE contacts over both the protein and the DNA molecular surfaces is sufficiently large to fix the relative orientation of the two molecules; in particular, the position of the elongated helix III in the major groove is determined by NOEs from three subsequent windings of this α-helix. It is worth noting that Gln50 in position 9 of helix III shows a NOE with C7 of the DNA α strand (see Introduction; Hanes and Brendt, 1989; Treisman *et al.*, 1989). In addition, residues 5–8 are in contact with the DNA, and a particularly noteworthy feature is the antiparallel arrangement of the polypeptide segment comprising residues Tyr25, Leu26 and Thr27 with respect to the backbone of the DNA α strand near A4 and G5.

The distribution of the NOE contacts coincides well with the spatial distribution of hydrogen atoms with large differences between the chemical shifts in the 1:1 complex and in the free molecules. All nucleotides involved in the major groove contacts (Figure 5) show such shift differences $>|0.2|$p.p.m. (bold letters in Figure 3) and the minor groove contact causes a sizable shift difference for the nucleotide A11. For the protein, differences are again observed mainly for the NOE contact sites, i.e. residues 7–9, various residues on helix I, residue 26, and the helices III and IV. No significant shift differences were observed for helix II, which is furthest away from the DNA. Figure 5 shows only the polypeptide backbone up to the last residue involved in a NOE with the 14-mer, i.e. Met54, which is part of helix IV. It is not clear at this stage if there are further contacts between helix IV and the DNA. Although there are chemical shift differences for residues between positions 55 and 60 which would be compatible with such contacts, most amide

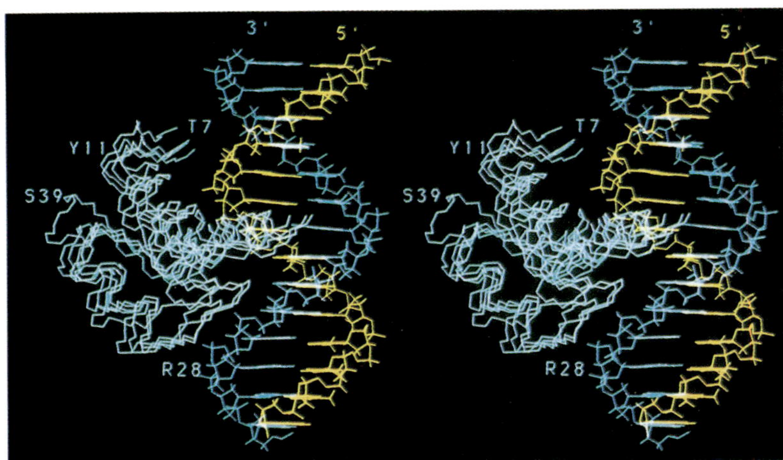

Fig. 4.

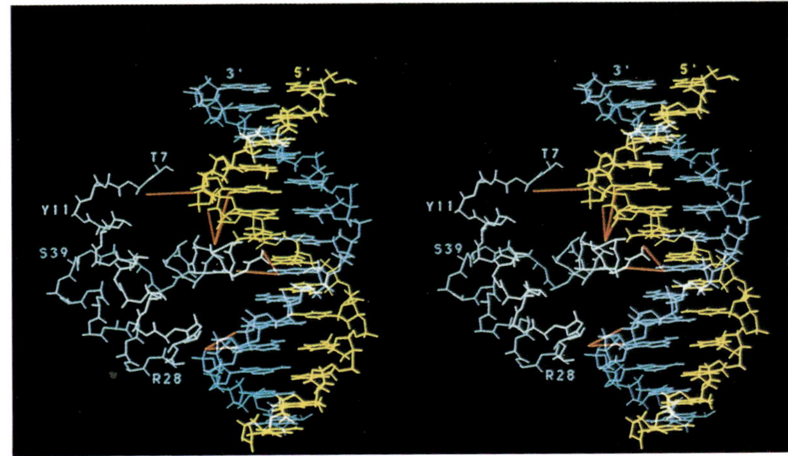

Fig. 5.

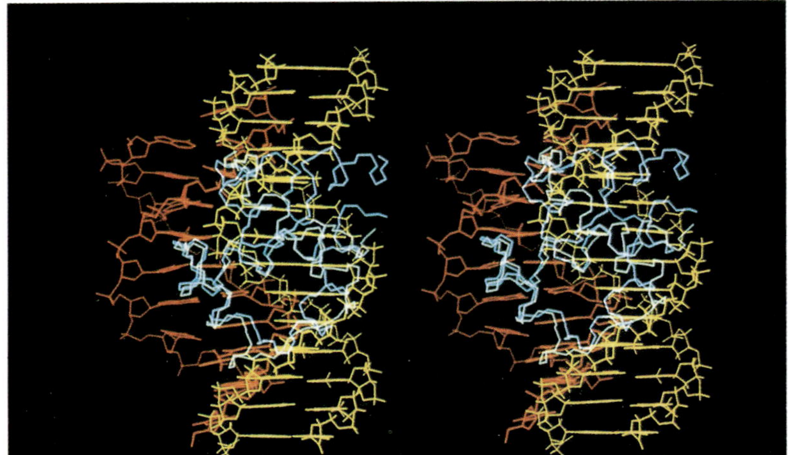

Fig. 6.

protons of helix IV exchange rapidly with the solvent, indicating that unlike helix III, helix IV is not more stable in the complex than in the free *Antp(C39→S)* homeodomain.

One of the most interesting features of the homeodomain–DNA complex is the NOE between Arg5 and the sugar of G12 (Table I, Figure 3). This indicates that Arg5, which is a strongly conserved amino acid residue in the different homeodomains characterized so far (Scott *et al.*, 1989), reaches into the minor groove. This coincides with the results of methylation and ethylation interference studies, which suggested the possibility of minor groove interactions in this region (Affolter *et al.*, 1990a). In particular, the phosphate of A13 interfered with binding when ethylated; this phosphate is in close proximity to the contact of Arg5 with the G12 sugar manifested by a NOE. The NMR and interference data are thus both consistent with the N-terminus of the *Antp* homeodomain interacting with this region of the binding site in the minor groove.

Comparison of the Antp(C39→S)–DNA complex with the 434 repressor–DNA complex

We selected the 434 repressor–DNA crystal structure (Aggarwal *et al.*, 1988) for an initial comparison of the *Antp(C39→S)*–DNA complex with a prokaryotic system. A single N-terminal domain 1–63 of the 434 repressor and the 10 bp forming the binding half-site for this monomeric subunit were selected for the comparison. To this complex the homeodomain–DNA complex was superimposed by optimal fit of the corresponding helix-turn-helix motifs, i.e. residues 17–36 of the repressor and residues 31–50 of the homeodomain (Figure 6). The structures of these motifs are very similar, with a r.m.s.d. of the backbone atoms of 0.9 Å (Qian *et al.*, 1989). When compared with the 434 repressor–DNA complex, the DNA in the homeodomain complex is shifted by about two turns of the helix III toward the helix IV. This same translation relative to the 434 repressor–DNA complex was observed for all five structures in Figure 4. In contrast, the apparent tilt between the DNA duplexes bound, respectively, to the *Antp(C39→S)* homeodomain or the 434 repressor, is not present in all five structures. Otherwise it appears that with the exception of the minor groove NOE with Arg5 (Table I), the principal contacts of the homeodomain with the 14-mer have counterparts in the 434 repressor–DNA complex. In particular, the NOEs from Tyr8 of the homeodomain to the backbone of the DNA β strand are paralleled by interactions of the residues preceding the fourth helix of the 434 repressor with phosphate groups of the DNA, and the NOEs from Tyr25 to the backbone of the DNA α strand have a counterpart in the hydrogen bonds from the first residues of the second helix of the 434 repressor to DNA phosphates (see Agarwal *et al.*, 1988; Pabo *et al.*, 1990).

Materials and methods

Preparation of the Antp(C39→S) homeodomain

The cysteinyl residue in position 39 of the *Antp* homeodomain peptide as expressed from the expression vector pAop2 (Müller *et al.*, 1988) was changed to Ser using site-directed mutagenesis (Amersham oligonucleotide-directed *in vitro* mutagenesis system version 2). An appropriate fragment of pAop2 was subcloned into Bluescript M13+ (STRATAGENE). ssDNA was prepared according to the manufacturer's protocol. The mutagenesis of oligonucleotide was purified over an oligonucleotide purification cartridge (Applied Biosystems). Its sequence was (mismatch is underlined): 5′-CACGCCCTGT<u>C</u>CCTCACGGAG-3′. The correct sequence of the mutagenized vector was confirmed by sequencing. A 170 bp *Kpn*I–*Hpa*I fragment was cloned back into pAop2 to obtain the new expression vector pAop2CS.

The homeodomain peptide expressed from pAop2CS was purified exactly as described previously (Müller *et al.*, 1988). In addition, a ^{15}N-labelled *Antp(C39→S)* homeodomain was isolated from bacteria grown in 5 mM $(^{15}NH_4)_2SO_4$/22 mM KH_2PO_4/39 mM Na_2HPO_4/8.5 mM NaCl/0.5% glucose/0.1 mM $CaCl_2$/1 mM $MgSO_4$/0.5 μM $FeCl_3$/5 μg/ml thiamin. The *Antp(C39→S)* homeodomain no longer dimerizes via an S–S bridge, as has been observed for the native homeodomain (Müller and Gehring, unpublished). The *in vitro* DNA binding properties of the *Antp* homeodomain are not altered by the (C39→S) exchange (Affolter *et al.*, 1990a). The stability of the folded conformation of the two proteins with respect to thermal denaturation was found to be identical at pH 6.9, with a denaturation temperature of 60°C.

Preparation of the 14-mer

The single strands d(GAAAGCCATTAGAG) and d(CTCTAATGGCTTTC) were synthesized on a DNA synthesizer (Applied Biosystems 380B) on a 10 μM scale. After cleavage of the protection groups with 25% aqueous NH_3 at 55°C overnight, the single strands were purified by preparative FPLC (Pharmacia mono-Q HR5/5, gradient 0.3–0.8 M NaCl, pH 12.0) and desalted by ultrafiltration (YM-2 diaflow filter). The final yields were 14 mg for the purine rich strand and 36 mg for the second strand, respectively. For the formation of the duplex 8 mg of one strand were dissolved in 0.4 ml D_2O (no salt added) and titrated with a D_2O solution of the second strand. The titration was monitored by NMR at room temperature.

NMR sample preparation and NMR measurements

Initial exploratory 1D NMR spectra (Figure 1) were recorded with a 0.5 mM sample in a buffered solution of 90% H_2O/10% D_2O (25 mM sodium phosphate, 100 mM KCl, 2 mM NaN_3, pH 6.0). The spectra were recorded without water presaturation using a selective Gaussian-shaped 90°_x pulse at the water frequency followed by a non-selective 90°_{-x} pulse (Sklenář *et al.*, 1987). The same buffer was used to prepare the NMR sample for the assignments of the free 14-mer. This DNA sample was ~2 mM

Fig. 4. Stereo view of a superposition for a perfect fit of the B-DNA duplex of the five homeodomain–DNA complexes obtained by docking with an ellipsoid algorithm (see text). The protein structures (green) contain the backbone atoms of residues 7–54. The DNA α strand (see Figure 3) is shown in blue and the β strand in yellow. Selected residue positions of the protein are labelled and the orientation of the DNA strands is indicated.

Fig. 5. One of the five structures of the homeodomain–DNA complex from Figure 4 with the NOE distance constraints used for the docking shown in red. Note that these constraints connect in all instances to the β carbons on the protein, even when the NOEs have been observed with more peripheral side chain protons (Table I).

Fig. 6. Comparison of the DNA complex of the *Antp(C39→S)* homeodomain obtained in solution by NMR and of the 434 repressor DNA complex measured by X-ray diffraction in single crystals (Aggarwal *et al.*, 1988). In this stereo view the relative orientation of the two complexes is determined by an optimal fit of the helix-turn-helix motifs, i.e. residues 17–36 of the repressor and 31–50 of the homeodomain. The colours are green for the homeodomain, yellow for the homeodomain-bound DNA, blue for the 434 repressor, and red for the repressor-bound DNA.

in duplex. The assignments for the free DNA were obtained from NOESY (Anil-Kumar et al., 1980), clean TOCSY (Griesinger et al., 1988) and two-quantum experiments (Braunschweiler et al., 1983) recorded in D_2O and H_2O solution. The resonances of the free *Antp(C39→S)* homeodomain were assigned in a 5.8 mM salt-free solution (Güntert et al., 1990).

All the resonance assignments of the 1:1 complex were obtained with a single 3.5 mM salt-free solution in 90% H_2O/10% D_2O at 36°C. NMR spectra of the complex were recorded at 600 MHz proton frequency on a Bruker AM600 spectrometer. Measurements in H_2O were performed without water saturation by pre-irradiation. In [^{14}N,^{1}H]-COSY a spin lock purge pulse of 2.0 ms duration was sufficient for water suppression (Messerle et al., 1989). In [^{1}H,^{1}H]-NOESY with $^{15}N(\omega_2)$-half-filter the water was suppressed by the combination of selective Gaussian pulses at the water frequency and a spin lock purge pulse of 2.5 ms duration (Otting and Wüthrich, 1989). The Gaussian shape of each of the selective pulses was approximated by a series of 16 pulses which were separated by 200 μs delays. The durations of these pulses were varied between 0.6 μs and 36.3 μs according to a Gaussian function. Two data sets were recorded with and without effective $\pi(^{15}N)$ editing pulse, respectively, in an FID-interleaved manner, where the acquisition of each FID for one data set was followed by the corresponding acquisition for the other data set prior to incrementation of the t_1 delay (Otting and Wüthrich, 1990). Summation and subtraction of both data sets yielded the sum spectrum (Figure 2A) and the difference spectrum (Figure 2B), repectively. Spectral parameters: t_{1max} = 44 ms, t_{2max} = 164 ms, τ = 5.6 ms, mixing time τ_m = 110 ms, total recording time ~51 h. Before Fourier transformation the data were multiplied with shifted sine bell functions, using shifts of $\pi/3$ in ω_1 and $\pi/5$ in ω_2. A three-dimensional ^{15}N-correlated [^{1}H,^{1}H]-NOESY spectrum was recorded with the experimental scheme described by Messerle et al. (1989). The water was sufficiently suppressed by two spin lock purge pulses, SL_y and SL_x, of 0.5 ms and 2 ms length, respectively. Spectral parameters: t_{1max} = 11.9 ms, t_{2max} = 16 ms, t_{3max} = 68 ms, τ = 5.6 ms, mixing time τ_m = 40 ms, total recording time ~80 h. Prior to Fourier transformation the data were multiplied in all three dimensions with sine bell functions shifted by $\pi/2$. Due to the short evolution periods, the signal-to-noise ratio observed in each individual [^{1}H−^{1}H] plane (Figure 2C) of the 3D NMR spectrum was better than in the corresponding 2D [^{1}H,^{1}H]-NOESY spectrum with $^{15}N(\omega 2)$-half-filter, even though a shorter mixing time was used. Two [^{1}H,^{1}H]-NOESY spectra were recorded in D_2O solution using mixing times of 40 ms and 100 ms. The total recording times were ~62 h and 36 h, respectively.

Amide proton exchange rates were measured by dissolving the lyophilized complex in D_2O and monitoring of the intensities of the amide proton signals in a series of [^{15}N,^{1}H]-COSY spectra, where each spectrum took 25 min to measure. For measurements in D_2O, complete exchange of the amide protons was achieved when the sample was kept at 36°C for 4 h, lyophilized and redissolved in D_2O.

Coupling constants $^3J_{NH\alpha}$ were measured with a series of J-modulated [^{15}N,^{1}H]-COSY experiments using the 0.5 mM sample. The experimental scheme used was a variant of the original experiment (Neri et al., 1990) in which heteronuclear multiple-quantum coherence evolves during the J-modulation period (Neri et al., in preparation). Seven experiments with J-modulation periods of 10.9, 15, 30, 45, 60, 75 and 90 ms were recorded. Each individual recording took ~9 h. The coupling constants $^3J_{HN\alpha}$ were evaluated from the measured cross peak intensities by a best fit procedure using a theoretical time evolution function, which accounts for the $^3J_{HN\alpha}$ modulation and for relaxation effects.

Acknowledgements

We thank Mr R.Marani for the careful processing of the text, and Dr A.Pervical-Smith for critical comments on the manuscript. Financial support by the Schweizerischer Nationalfonds (projects 31.25174.88 and K03.613.87) is gratefully acknowledged.

References

Affolter,M., Percival-Smith,A., Müller,M., Leupin,W. and Gehring,W.J. (1990a) *Proc. Natl. Acad. Sci. USA*, **87**, 4093−4097.
Affolter,M., Schier,A. and Gehring,W.J. (1990b) *Curr. Opinion Cell Biol.*, **2**, 485−495.
Aggarwal,A.K., Rodgers,D.W., Drottar,M., Ptashne,M. and Harrison,S.C. (1988) *Science*, **242**, 899−907.
Anil-Kumar, Ernst,R.R. and Wüthrich,K. (1980) *Biochem. Biophys. Res. Commun.*, **95**, 1−6.
Billeter,M., Havel,T.F. and Wüthrich,K. (1987a) *J. Comp. Chem.*, **8**, 132−141.
Billeter,M., Havel,T.F. and Kuntz,I.D. (1987b) *Biopolymers*, **26**, 777−793.
Billeter,M., Qian,Y.Q., Otting,G., Müller,M., Gehring,W.J. and Wüthrich,K. (1990) *J. Mol. Biol.*, **214**, 183−197.
Braunschweiler,L., Bodenhausen,G. and Ernst,R.R. (1983) *Mol. Phys.*, **48**, 535−560.
Gehring,W.J. (1987) *Science*, **236**, 1245−1252.
Griesinger,C., Otting,G. Wüthrich,K. and Ernst,R.R. (1988) *J. Am. Chem. Soc.*, **110**, 7870−7872.
Güntert,P., Qian,Y.Q., Otting,G., Müller,M., Gehring,W.J. and Wüthrich,K. (1990) submitted.
Hanes,S.D. and Brendt,R. (1989) *Cell*, **57**, 1275−1283.
Laughon,A. and Scott,M.P. (1984) *Nature*, **310**, 25−31.
Levine,M. and Hoey,T. (1988) *Cell*, **55**, 537−540.
McGinnis,W., Levine,M., Hafen,E., Kuroiwa,A. and Gehring,W.J. (1984a) *Nature*, **308**, 428−433.
McGinnis,W., Garber,R.L., Wirz,J., Kuroiwa,A. and Gehring,W.J. (1984b) *Cell*, **37**, 403−408.
Messerle,B.A., Wider,G., Otting,G., Weber,C. and Wüthrich,K. (1989) *J. Magn. Reson.*, **85**, 608−613.
Müller,M., Affolter,M., Leupin,W., Otting,G., Wüthrich,K. and Gehring,W.J. (1988) *EMBO J.*, **7**, 4299−4304.
Neri,D., Otting,G. and Wüthrich,K. (1990) *J. Am. Chem. Soc.*, **112**, 3663−3665.
Otting,G. and Wüthrich,K. (1989) *J. Am. Chem. Soc.*, **111**, 1871−1875.
Otting,G. and Wüthrich,K. (1990) *Q. Rev. Biophys.*, **23**, 39−96.
Otting,G., Qian,Y.Q., Müller,M., Affolter,M., Gehring,W.J. and Wüthrich,K. (1988) *EMBO J.*, **7**, 4305−4309.
Pabo,C.O., Aggarwal,A.K., Jordan,S.R., Beamer,L.J., Obeysekare,U.R. and Harrison,S.C. (1990) *Science*, **247**, 1210−1213.
Qian,Y.Q., Billeter,M., Otting,G., Müller,M., Gehring,W.J. and Wüthrich,K. (1989) *Cell*, **59**, 573−580.
Scott,M.P. and Weiner,A.J. (1984) *Proc. Natl. Acad. Sci. USA*, **81**, 4115−4119.
Scott,M.P., Tamkun,J.W. and Hartzell,G.W. (1989) *BBA Rev. Cancer*, **989**, 25−48.
Shepherd,J.W.C., McGinnis,W., Carrasco,A.E., DeRobertis,E.M. and Gehring,W.J. (1984) *Nature*, **310**, 70−71.
Sklenář,V., Tschudin,R. and Bax,A. (1987) *J. Magn. Reson.*, **75**, 352−357.
Treisman,J., Gönczy,P., Vahishtha,M., Harris,E. and Desplan,C. (1989) *Cell*, **59**, 553−562.
Wüthrich,K. (1986) *NMR of Proteins and Nucleic Acids*. Wiley, New York.

Received on June 19, 1990

Cell, Vol. 78, 211–223, July 29, 1994, Copyright © 1994 by Cell Press

Homeodomain–DNA Recognition

Review

Walter J. Gehring,* Yan Qiu Qian,†‡ Martin Billeter,† Katsuo Furukubo-Tokunaga,*§ Alexander F. Schier,*‖ Diana Resendez-Perez,* Markus Affolter,* Gottfried Otting,†# and Kurt Wüthrich†
*Biozentrum der Universität Basel
Klingelbergstrasse 70
CH-4056 Basel
Switzerland
†Institut für Molekularbiologie und Biophysik
Eidgenössische Technische Hochschule-Hönggerberg
CH-8093 Zürich
Switzerland

Introduction

The regulation of gene transcription is based upon specific interactions between transcription factors and their target genes. Generally these transcription factors are sequence-specific DNA-binding proteins, activators, and repressors. In eukaryotic gene regulatory proteins, several different types of DNA-binding domains with characteristic structural motifs have been identified, which are encoded by families of regulatory genes. One of these, the homeobox gene family, has been extensively studied and plays a fundamental role in development and evolution. Homeobox genes are involved in the genetic control of development, in particular in the specification of the body plan, pattern formation, the determination of cell fate, and several other basic developmental processes (for reviews see Gehring, 1987; Scott et al., 1989; McGinnis and Krumlauf, 1992; Kornberg, 1993). The homeobox encodes a 60 amino acid residue polypeptide, called the homeodomain, that represents the DNA-binding domain of the respective proteins. Since homeodomain proteins have been found not only in metazoa but also in fungi and plants, we may assume that they arose early during the evolution of eukaryotes. Some structural features of the homeodomain and some sequence similarities are even found in prokaryotic DNA–binding proteins. In the course of evolution, the amino acid sequence of the homeodomain has been conserved to a high degree (see Wüthrich and Gehring, 1992). For example, the human Hox-A7 homeodomain differs in only 1 out of 60 positions from that of the Antennapedia (Antp) homeodomain, which is its putative homolog (ortholog) in Drosophila, even though vertebrates and insects separated more than 500 million years ago. This indicates that there is strong evolutionary pressure to preserve the amino acid sequence of the homeodomain.

‡Present address: Memorial Sloan–Kettering Cancer Center, 1275 York Avenue, New York, New York 10021.

§Present address: Zoologisches Institut der Universität Basel, Rheinsprung 9, CH-4051 Basel, Switzerland.

‖Present address: Cardiovascular Research Center, Harvard Medical School, East 4 13th Street, Charlestown, Massachusetts 02129.

#Present address: Department of Medical Biochemistry and Biophysics, Karolinska Institute, S-17177 Stockholm.

The Three-Dimensional Structure of the Homeodomain

To elucidate the three-dimensional structure of a homeodomain, the Antp homeodomain was expressed as a 68 amino acid residue polypeptide in Escherichia coli (Müller et al., 1988), and its solution structure was determined by nuclear magnetic resonance (NMR) spectroscopy (Qian et al., 1989; Billeter et al., 1990). The Antp homeodomain consists of three helical regions folded into a tight globular structure: helix I is preceeded by a flexible N-terminal arm and separated by a loop from helix II, which forms with helix III a helix-turn-helix motif (Figure 1). This is a structure motif that was previously described for several prokaryotic gene-regulatory proteins (see Steitz, 1990; Pabo and Sauer, 1992). A comparison of the backbone structure of the helix-turn-helix motif of five prokaryotic gene-regulatory proteins with that of the Antp homeodomain reveals that these motifs are readily superimposable (Qian et al., 1989; Gehring et al., 1990). Therefore, the three-dimensional structure of the helix-turn-helix motif is highly conserved among otherwise very different species. However, in comparison with prokaryotic repressor proteins such as λ or 434, the third helix of the homeodomain is extended by two turns of a more flexible helix, termed helix IV, which contains several basic side chains. Furthermore, the global fold of the homeodomain is clearly different from that of the DNA-binding domains of these prokaryotic gene-regulatory proteins (Qian et al., 1989).

There are no X-ray crystallographic studies of free homeodomain polypeptides available, but structures of homeodomains complexed to operator DNA sequences were obtained both by NMR for the Antp homeodomain (Otting et al., 1990; Billeter et al., 1993) and by X-ray crystallography for the engrailed (en) (Kissinger et al., 1990) and MATα2 (Wolberger et al., 1991) homeodomains. The complex of the Antp system showed that the structure of the free homeodomain was largely preserved also in the DNA-bound polypeptide. The structures of the en and Antp homeodomains are very similar, and rather unexpectedly, the MATα2 homeodomain, which shares only 28% sequence identity and carries an insertion of three amino acid residues when compared to Antp, also has a very similar three-dimensional structure. The additional tripeptide segment is accomodated at the C-terminal end of helix I in the loop between helices I and II without affecting the global structure significantly. An insertion of three amino acid residues at the same position of the homeodomain has also been found in the human *prl* (=*pbx*) genes (Kamps et al., 1990; Nourse et al., 1990; Monica et al., 1991) and in the Drosophila gene *extradenticle* (*exd*) (Rauskolb et al., 1993). The rat liver transcription factor LFB1 (=HNF1) includes a more extreme variation in that it contains 81 rather than 60 amino acids in its atypical homeodomain (De Simeone and Cortese, 1991; Mendel and Crabtree, 1991). In LFB1 the three amino acid residues in the turn of the helix-turn-helix motif are replaced by a 24 residue linker region between the two helices, accomodating the extra amino acids without any major

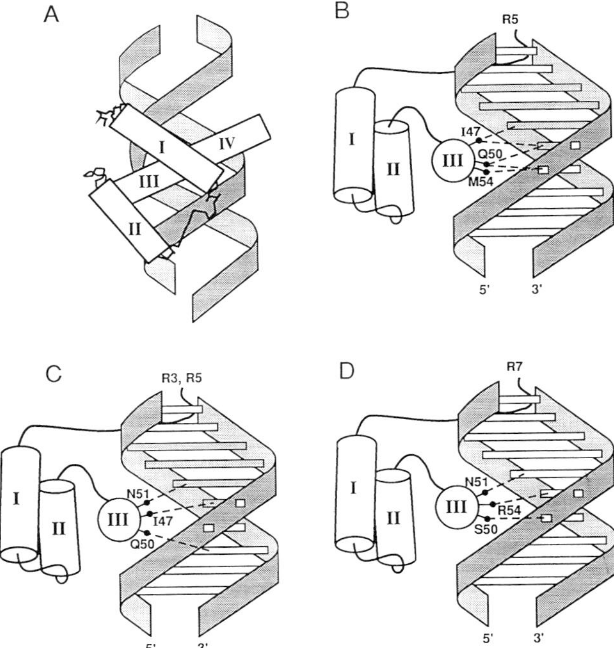

Figure 1. Schematic Comparison of the Homeodomain–DNA Complexes of Antp, en, and MATα2

(A) Antp. View perpendicular to the axis of the recognition helix (III, IV) located in the major groove. The α helices I, II and III, IV are indicated by cylinders. For the N-terminal arm (upper left), the loop between helices I and II, and the turn of the helix-turn-helix motif (II-III) only the polypeptide backbone is drawn as a solid line.

(B), Antp; (C), en; (D), MATα2. View along the axis of the recognition helix (III). Amino acid residues that establish contacts to specific bases are indicated. For residues in the recognition helix, these contacts, which are indicated by dotted lines, are in the major groove; for those in the N-terminal arm, they are in the minor groove of the DNA. In (B) and (C), the TAAT core motif (see text) is shaded.

effects on the overall conformation of the molecule (Leiting et al., 1993; Ceska et al., 1993). These findings indicate that there is strong selective pressure to maintain the overall three-dimensional structure of the homeodomain.

In the cocrystals of the en and MATa2 homeodomain with DNA duplexes, the entire recognition helix appears as a regular long α helix (Kissinger et al., 1990; Wolberger et al., 1991), and the distinction of the aforementioned helices III and IV in the solution structure of the Antp homeodomains is not apparent. Although in an early cartoon of the Antp homeodomain structure the transition from helix III to helix IV has been visualized as a distinct kink (Qian et al., 1989), the important difference between the two helical segments is that helix III is a precisely defined standard α helix, whereas helix IV is flexibly disordered with rapid amide proton exchange. This distinction of helices III and IV has been clearly evidenced in all subsequent studies of the solution structures of the Antp homeodomain (Billeter et al., 1990) and mutant forms thereof (Güntert et al., 1991; Qian et al., 1992, 1994b). Its significance is emphasized by the recent observation that although the global folds of the Antp and ftz homeodomains are very similar, the free ftz homeodomain in solution does not contain helix IV, and the corresponding polypeptide segment is flexibly disordered (Qian et al., 1994a). On the basis of comparative DNA binding studies, it cannot be excluded that formation of helix IV is induced by DNA contacts in the ftz–DNA complex (see below).

Sequence-Specific DNA Binding

The DNA binding properties of homeodomain-containing polypeptides were analyzed by footprinting, mobility shift assays, and transactivation assays (see Laughon, 1991; Gehring et al., 1994; Kalionis and O'Farrell, 1993). The majority of the homeodomains thus characterized recognize DNA sequences containing a 5'-TAAT-3' core motif (or ATTA on the other strand). These include members of the Drosophila homeotic gene complexes as well as a number of more diverged homeodomains. In the following, we have focused our discussion on those homeodomains and binding sites for which structural studies of protein–DNA complexes are available.

The binding site BS2, from which the sequence of the DNA 14mer used in the NMR structural studies of the Antp homeodomain–DNA complex is derived, is recognized by the purified Antp homeodomain, a partial Antp protein, and the full-length ftz protein (Müller et al., 1988). Genetic experiments discussed below indicate that this binding site is functional in vivo (Schier and Gehring, 1993a). To avoid artificial dimerization of the homeodomain polypeptide, subsequent studies were carried out with a mutant Antp(C39S) homeodomain, in which the single cysteinyl residue at position 39 is replaced by serine. The three-dimensional structure of the Antp(C39S) homeodomain is essentially identical to that of wild–type Antp (Güntert et al., 1991). Binding studies indicated that the Antp(C39S) polypeptide interacts as a monomer with its BS2 target DNA (Affolter et al., 1990), which is in agreement with the

NMR studies (Otting et al., 1990). Equilibrium DNA binding studies using gel mobility shift assays indicated that stable DNA complexes are formed with an equilibrium dissociation constant (K_D) of 1.6×10^{-9} M, and kinetic binding studies gave an even lower value of 1.8×10^{-10} M (Affolter et al., 1990). Direct comparison between the Antp(C39S) and ftz homeodomains showed that ftz has an approximately 3-fold higher affinity for BS2 (Qian et al., 1994a).

For en, a similar K_D value of approximately 1×10^{-9} to 2×10^{-9} M was obtained for the binding site used in the X-ray crystallographic studies (Kissinger et al., 1990), but it is not yet known whether this binding site is functional in vivo. The MATα2 homeodomain binds to its recognition site with only modest affinity (Johnson, 1992). Its target specificity is greatly increased by combinatorial interactions with other transcription factors (see below).

Mode of DNA Binding

The binding mode of the Antp homeodomain monomer to a 14 bp DNA duplex was determined by NMR spectroscopy (Otting et al., 1990). The recognition helix (III/IV) lies in the major groove of the DNA, and there are additional specific contacts to bases in the minor groove by the N-terminal arm. The loop between helices I and II interacts with the DNA backbone. A similar mode of DNA binding has subsequently been found for the homeodomains of en and MATα2 by X-ray diffraction at 2.8 Å and 2.7 Å resolution, respectively (Kissinger et al., 1990; Wolberger et al., 1991). The solution structure of the Antp(C39S) homeodomain–DNA complex has then been refined by NMR spectroscopy using uniformly ^{13}C-labeled homeodomain (Billeter et al., 1993; Qian et al., 1993a).

The overall arrangement of the three homeodomain–DNA complexes is quite similar (Figure 1). Helices I and II are aligned in an antiparallel arrangement above the DNA, each spanning the major groove nearly perpendicular to the local direction of the DNA backbones. The recognition helix (III/IV) is positioned in the major groove of the DNA, roughly parallel to the groove, where most of the specific intermolecular contacts occur. Additional base-specific contacts are established by the flexible N-terminal arm in the minor groove, whereas the loop between helices I and II, as well as residues at the start of helix II, contacts the DNA backbone. The DNA is in its B form with only minor distortions in the double helix. The recent X-ray structure determination of an Oct-1 POU domain bound to an octamer DNA site indicates that the docking of the POU homeodomain is also quite similar to that of Antp, en, and MATα2 (Klemm et al., 1994). This indicates that the mode of docking of the polypeptide to the DNA has been tightly conserved among the four homeodomains analyzed, even though they belong to only distantly related homeobox families.

The DNA-binding domains of such prokaryotic gene-regulatory proteins as λ repressor, cro, CAP, and others differ in several respects from homeodomains. Both their amino acid sequences and their three-dimensional global folds are quite variable, and only the three-dimensional structure of the prototypical helix-turn-helix motif is conserved. Furthermore, the recognition helix tends to be shorter than in the homeodomains and lacks the basic residues at the elongated C-terminus (Suzuki, 1993). As a consequence, the mode of docking on the DNA is not conserved, since the recognition helix of different prokaryotic gene-regulatory proteins is oriented at different angles relative to the DNA (Steitz, 1990). Another major difference concerns the fact that prokaryotic gene-regulatory proteins generally bind as homodimers (or homotetramers) to palindromic DNA sequences, and their monomers have only low affinity to their half-binding sites. However, there is a prokaryotic DNA-binding protein that shows some similarity in its binding mode to the homeodomains, even though it has a short recognition helix of only eight residues. This is the Hin recombinase (Hin) of Salmonella, a site-specific recombinase that controls the alternate expression of two flagellin genes by reversibly switching the orientation of a promoter. On the basis of indirect genetic and biochemical evidence, it was proposed that an N-terminal arm of the Hin DNA-binding domain interacts with the minor groove of the DNA (Sluka et al., 1987, 1990) in a manner similar to the mode of DNA binding of the homeodomain (Affolter et al., 1991). Indeed, the amino acid sequence of the Hin DNA-binding domain can be aligned with that of the en homeodomain and clearly shows some sequence similarity (Affolter et al., 1991). Recent structural studies of the Hin recombinase–DNA complex have confirmed the similarity of the mode of binding and demonstrated minor groove interactions for both the N-terminal arm and the C-terminal tail (Feng et al., 1994). Therefore, the Hin recombinase represents in many respects an intermediate between the prototypical bacterial helix-turn-helix proteins and the eukaryotic homeodomain proteins. It is interesting to note that the Hin recombinase in Salmonella, MATα2 in yeast, and the homeodomain proteins of the fungus Ustilago (Schulz et al., 1990) are all used to control primitive forms of cellular differentiation, which is similar to the function of homeodomain proteins in higher organisms.

Protein–DNA Contacts: Comparison between Structural and Biochemical Studies

The determination of the three-dimensional structure of the Antp(C39S) homeodomain–DNA complex (Billeter et al., 1993; Qian et al., 1993a) was based on the use of the ^{13}C- and ^{15}N-labeled protein in the complex combined with heteronuclear half-filter experiments and heteronuclear three-dimensional NMR. These experiments yielded a set of 1045 nuclear Overhauser effect (NOE) upper distance constraints connecting pairs of protons within the protein, within the DNA duplex, and between the protein and the DNA. Starting from 20 coordinate sets with random relative positions of the protein and the DNA, and using the aforementioned input of conformational constraints, simulated annealing and molecular dynamics calculations in a water bath yielded a set of 16 well-converged conformers, which are used to represent the solution structure of the Antp(C39S) homeodomain–DNA complex. The complete

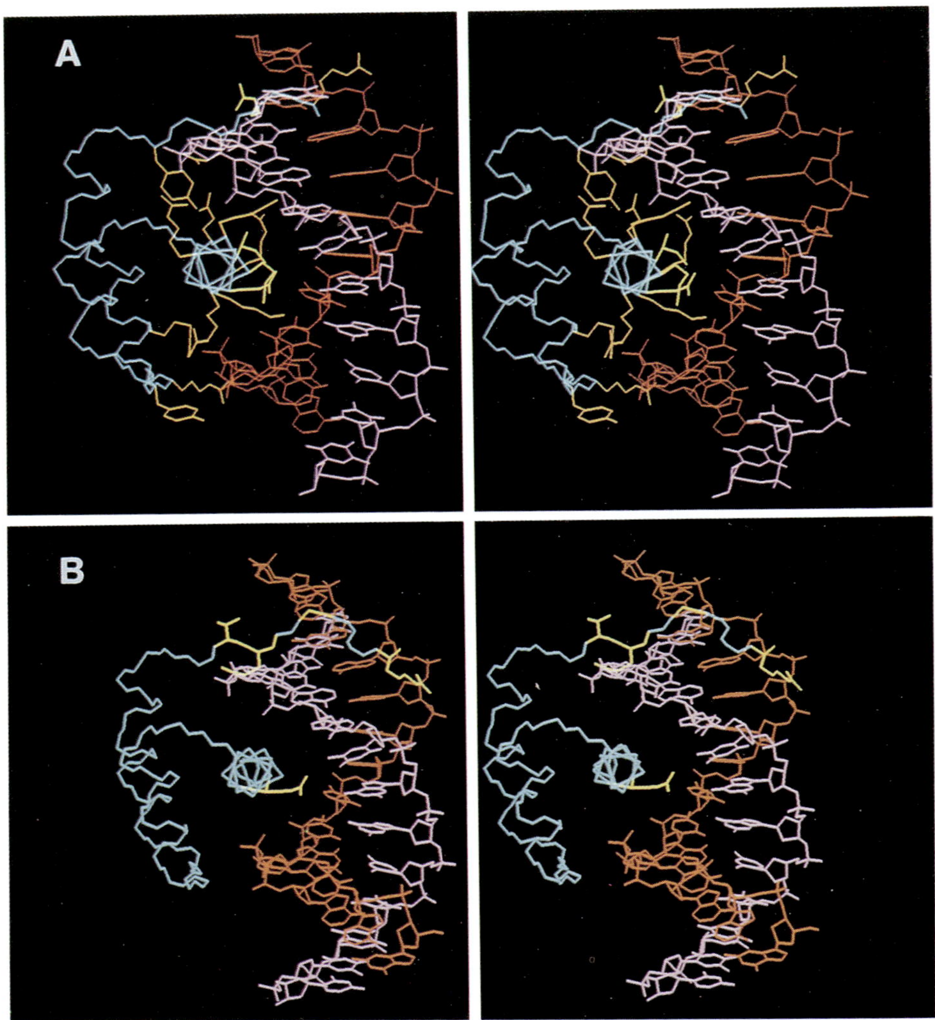

Figure 2. Stereo Views of One of the 16 Conformers Used to Describe the NMR Solution Structure of the Antp(C39S) Homeodomain–DNA Complex

(A) The drawing shows the backbone of the homeodomain residues 3–55 (cyan), all side chains of the homeodomain that contact the DNA (yellow; for the selection criteria used for inclusion of individual side chains, see Tables 4–6 of Billeter et al. [1993]), and the base pairs 3–13 of the DNA (red for the α strand with the sequence d-GAAAGCCATTAGAG, magenta for the complementary β strand).

(B) Stereo view of one conformer, emphasizing those amino acid residues (shown in yellow) that are implicated in the DNA binding specificity (Gln-50) and in the functional specificity (Arg-1, Gly-4, Gln-6, and Thr-7) by genetic experiments.

structure of the complex has been discussed elsewhere (Billeter et al., 1993). Here we emphasize the close interatomic contacts in the protein–DNA interface of the NMR structure and their relations with biochemical and genetic data on homeodomain–DNA interactions. As an introduction to this theme, Figure 2A shows a picture of one of the 16 conformers that emphasizes the homeodomain side chains that contact the DNA. A more detailed survey of the numerous intermolecular contacts is provided by the schematic representations of Figure 3, where salt bridges (Figure 3A) and other short contacts (Figure 3B) between the homeodomain and the DNA are indicated by arrows. Three regions of the Antp(C39S) sequence are involved in DNA interactions in the NMR structure of the homeodomain–DNA complex; i.e., the N-terminal arm of the protein with residues 1–6 contacts the DNA base pairs 11–13 in the minor groove, the loop between helices I and II as well as the residues Arg-28 and Arg-31 of helix II are

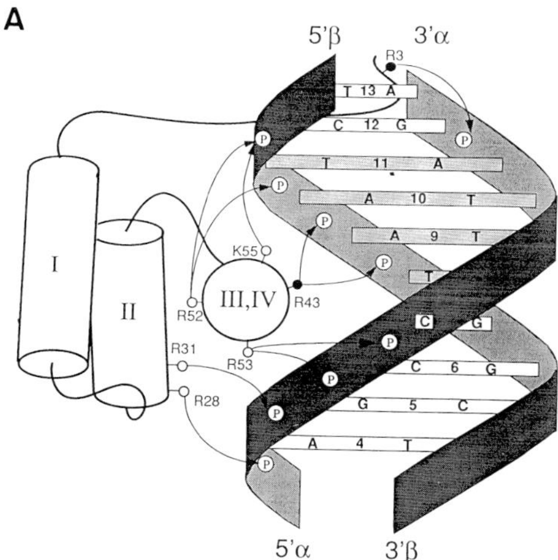

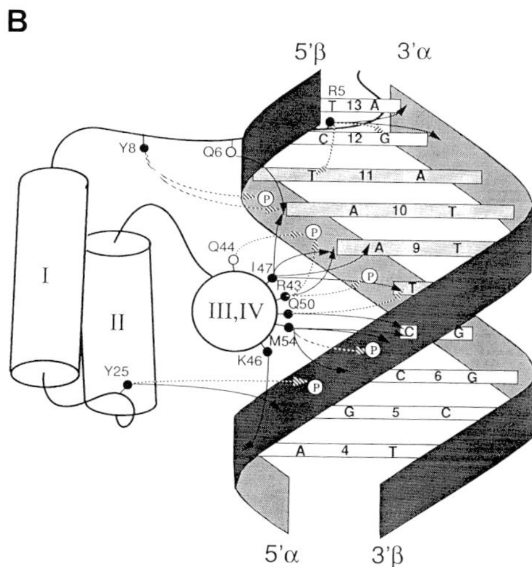

Figure 3. Schematic Drawings of the Antp(C39S) Homeodomain–DNA Complex

The view is along the axis of helix III from the C- to the N-terminus. The helices of the homeodomain are represented by cylinders, and the amino acid side chains that contact the DNA are identified with circles, where filled circles identify residues for which direct intermolecular NOEs with the DNA have been identified (Billeter et al., 1993). The DNA backbone is represented by ribbons and the DNA base pairs by horizontal bars with the single-letter symbols of the two bases. The four bars for the consensus base pairs 5′-TAAT-3′·5′-ATTA-3′ are shaded. Selected phosphate groups are identified by open circles with the letter "P." Arrows indicate short contacts between individual amino acid residues of the homeodomain and the DNA 14-mer that were identified in at least 9 of the 16 conformers representing the NMR structure of the Antp(C39S) homeodomain–DNA complex (Billeter et al., 1993). Arrows ending within the bars indicate specific contacts with the DNA bases, those pointing to the open circles with the letter P describe contacts to the phosphate groups, and all other arrows indicate contacts with the deoxyribose moieties of the DNA.

The interaction of Lys-57 with the DNA is not consistently observed in all NMR conformers representing the structure of the complex (Billeter et al., 1993) and is therefore not included in Figure 3, but this contact is independently implied by the nondegeneracy of the chemical shifts of the εCH_2 group of Lys-57 (Qian et al., 1993a).

(A) Intermolecular electrostatic interactions with interatomic distances shorter than 5.0 Å (see Table 4 of Billeter et al., 1993).

(B) Hydrophobic contacts with interatomic distances shorter than 3.5 Å (see Table 5 of Billeter et al., 1993) are shown with solid arrows, and other, miscellaneous contacts with interatomic distances shorter than 3.5 Å, including intermolecular hydrogen bonds (see Table 6 of Billeter et al., 1993), are shown with dashed arrows.

located close to the backbone of the α strand of the DNA (d-GAAAGCCATTAGAG has arbitrarily been labeled the α strand, and the complementary sequence the β strand), and the recognition helix is located in the major groove of the DNA near the core consensus sequence TAAT on the β strand. In the crystal structures of the en and MATα2 homeodomain–DNA complexes (Kissinger et al., 1990; Wolberger et al., 1991), the corresponding segments of the homeodomain sequences are also involved in DNA contacts. These three contacting regions are separately discussed below.

Ethylation and methylation interference experiments (Affolter et al., 1990; Percival-Smith et al., 1990) provided independent evidence of how the homeodomain docks onto the DNA. The results of these studies using both the Antp(C39S) and the ftz homeodomains are summarized in Figure 4. The phosphate groups shown are those for which ethylation interference is observed. The distribution of these sites indicates that the homeodomain contacts the DNA mainly from one side in the major groove. This interpretation is supported by the methylation interference data, which show strong interference upon methylation of the bases A9 and A10 (circled in Figure 4), which are located in the core consensus sequence TAAT on the β strand. A putative contact to the phosphate group of A13 on the α strand was detected by ethylation interference, and methylation interference was observed at A11 and G12 in the minor groove. Upon deletion of the N-terminal arm (amino acid residues 1–6) of the ftz homeodomain, the ethylation interference at A13 and the methylation interference at G12 and A11 were abolished (Percival-Smith et al., 1990). This indicates that the N-terminal arm is involved in these minor groove contacts, which is in agreement with the NMR data. Ethylation interference is also observed at the phosphate group of A8 on the α strand, and weak interference was detected upon methylation at G7 on the β strand, where no contacts were found by NMR. This apparent discrepancy may be due to indirect interference effects.

DNA Contacts of the N-Terminal Arm of the Homeodomain

The N-terminal arm formed by residues 1–6 of the Antp homeodomain is flexibly disordered in solution (Qian et

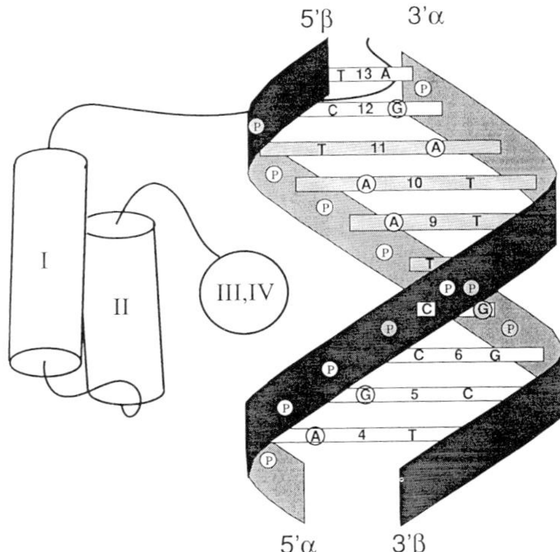

Figure 4. Ethylation and Methylation Sites on BS2 DNA That Interfere with Homeodomain Binding

Backbone phosphate groups which upon ethylation interfere with the binding of the Antp(C39S) and ftz homeodomains are indicated by open circles with the letter P. Weakly interfering sites are shaded. There are only two differences between Antp(C39S) and ftz: ethylation of A13 on the α strand interferes strongly in ftz and weakly in Antp, and ethylation of C12 on the β strand interferes weakly in ftz, but has no effect on Antp homeodomain binding. A and G residues interfering with binding upon methylation are encircled. Weakly interfering sites are shaded (data from Affolter et al., 1990; Percival-Smith et al., 1990).

al., 1989), and upon binding to DNA it establishes contacts to the minor groove. In the NMR structure, all contacts of the homeodomain to the DNA minor groove are formed by the two arginyl residues in positions 3 and 5. While Arg-3 forms a salt bridge to the phosphate group of G12, Arg-5 makes hydrophobic contacts to the sugar moieties of G12 and A13 and has further short contacts to the bases of T11 and G12. The contact to T11 might take the form of a hydrogen bond, but the local precision of the structure in this region is not sufficient to confirm this clearly. Other DNA contacts of the polypeptide segment preceeding helix I involve Gln-6 and Tyr-8 and backbone atoms of A10 in the DNA β strand (see Figure 3).

The analysis of the NMR structure of a truncated (des 1–6)Antp(C39S) homeodomain, lacking the N-terminal arm, showed that the overall structure of the homeodomain remains unchanged (Qian et al., 1994b). In particular, the removal of residues 1–6 does not noticeably affect helix I from residue 10 to residue 21. However, the DNA binding constant of the truncated homeodomain is reduced about 10-fold relative to the intact homeodomain. This reduced binding affinity can be attributed to the absence of the DNA contacts formed by the N-terminal arm of the intact homeodomain in the minor groove.

Genetic experiments indicate that the N-terminal arm of the homeodomain contributes significantly to the functional specificity of the homeodomain proteins. The homeodomains encoded by the *Antp* and *Sex combs reduced* (*Scr*) genes are very similar and differ in five amino acids only, in the C-terminal position 60 and at positions 1, 4, 6, and 7 in the N-terminal arm. Experiments in which parts of the coding region of these two genes were exchanged clearly showed that in transgenic flies the functional specificity of the proteins resides in or very close to the homeodomain (Gibson et al., 1990). In a heat shock assay in which chimeric Antp–Scr proteins are expressed ectopically in transgenic flies, it was demonstrated that the difference in the functional specificity between *Antp* and *Scr* is determined by the four amino acid residues located in the N-terminal arm of the homeodomain, which differ between Antp and Scr (Furukubo-Tokunaga et al., 1993; Zeng et al., 1993). Of these four amino acid residues, only Gln-6 contacts the DNA in the NMR structure of the Antp complex (Billeter et al., 1993), and it is not known whether the functional specificity is determined by differences in DNA binding, different protein–protein interactions with other transcription factors, or the combined influence of both. The functional specificity of the homeotic Drosophila genes *Deformed* (*Dfd*) and *Ultrabithorax* (*Ubx*) was also shown to reside in the N-terminal arm of the homeodomain (Lin and McGinnis, 1992; Mann and Hogness, 1990; Chan and Mann, 1993). For the mammalian POU homeodomain proteins Pit-1 and Oct-1 it has been demonstrated that phosphorylation of Ser-7 in the N-terminal arm prevents DNA binding (Kapiloff et al., 1991; Segil et al., 1991). Furthermore, the I-POU homeodomain protein, which lacks the two highly conserved Lys residues at positions 3 and 4, fails to bind to DNA, whereas the twin of I-POU, which is a product of alternative splicing that contains these two residues, is capable of specific DNA binding (Treacy et al., 1992). All these results emphasize the functional importance of the flexible N-terminal arm of the homeodomains.

In the en homeodomain–DNA complex, Arg-3 and Arg-5 also establish DNA contacts in the minor groove, but the flexible N-terminal arm bends more backward toward the TAAT core motif: Arg-3 forms hydrogen bonds to T at base pair 12 (or with A at base pair 13, or both), whereas Arg-5, which is the most highly conserved residue in the N-terminal arm, hydrogen bonds with T at base pair 11. In the MATα2 homeodomain–DNA complex, Arg-7 contacts the minor groove at two T residues of base pairs 8 and 9 (see Figure 1) outside of the core motif. Thus, minor groove contacts by the flexible N-terminal arm represent a general feature of homeodomain–DNA recognition and contribute to the high DNA binding affinity.

DNA Contacts of Residues near the Start of Helix II of the Homeodomain

The structure of the Antp(C39S) homeodomain–DNA complex reveals a second area of contacts, which involves the DNA backbone and homeodomain residues in the loop between helices I and II and in the N-terminal part of helix II. The loop between helices I and II runs approximately parallel to the α strand of the DNA. The side chain of Tyr-25 contacts the sugar moiety of G5 and the phosphate group of C6, and the side chains of Arg-28 and Arg-31 form salt bridges to the phosphate groups of A4 and G5, respectively (see Figures 2 and 3).

The functional importance of the DNA contacts near the start of helix II was demonstrated genetically by helix-turn-helix swapping experiments (Furukubo-Tokunaga et al., 1992) using the *ftz* gene. *ftz* is more suitable for genetic analysis, since it allows complementation studies in transgenic flies that are not possible in *Antp* because of the large size (~100 kb) of the *Antp* gene. By swapping the helix-turn-helix motif, chimeric proteins were generated between ftz and Scr, which have similar amino acid sequences of their helix-turn-helix motifs, and between ftz and the muscle segment homeobox (msh) protein, which has a very different amino acid sequence in its helix-turn-helix motif (Furukubo-Tokunaga et al., 1992). By complementation tests in transgenic flies, cotransfection assays in cultured Drosophila cells, and in vitro DNA binding studies, it was found that ftz activity is retained in a ftz–Scr chimera, carrying the helix-turn-helix motif of Scr, whereas it is lost in the ftz–msh chimera, which differs in 19 amino acid positions from that of ftz. By systematic back mutation of the ftz–msh chimera, a set of residues was identified in the helix-turn-helix motif, including Arg-28 and Arg-43, which are required for efficient target site recognition and hence for full ftz activity both in vitro and in vivo. The NMR data (see Figure 3) show that both of these residues, Arg-28 (located at the N-terminus of helix II) and Arg-43 (in helix III), are involved in DNA backbone contacts in the Antp(C39S) homeodomain–DNA complex.

In the en homeodomain, Arg-28 is replaced by Glu, but the conserved Tyr-25 and Arg-31 also establish backbone contacts that are very similar to Antp. In MATα2, only the Tyr-25–DNA backbone contact is observed, since all of the four Arg residues 28 to 31 are substituted by other amino acids. This might explain in part or entirely the reduced DNA binding affinity of MATα2.

DNA Contacts of the Recognition Helix of the Homeodomain

Numerous contacts are observed between the recognition helix and both strands of the DNA. It is interesting to note that the residues of the recognition helix of Antp that have positively charged side chains interact with the two DNA strands in a strictly alternating way, i.e., Arg-43 with the β strand, Lys-46 with the α strand, Arg-52 with the β strand, Arg-53 with the α strand, Lys-55 with the β strand, and Lys-57 with the α strand. Because of the perpendicular relative orientation of the recognition helix and the DNA double helix (see Figure 1), alternating interactions from the protein helix to the DNA offer the most efficient way of stabilizing the protein–DNA complex. The first two interactions connect the first turn of the helix with the DNA. This is followed by two helix turns with DNA base contacts in the major groove, and finally by four salt bridges formed by the C-terminal turn of the recognition helix to the DNA backbone. Additional contacts of the recognition helix with the DNA backbone include hydrophobic interactions from Arg-43 and Ile-47 to the β strand, and from Lys-46 and Met-54 to the α strand. These are supplemented by contacts of Gln-44 and Met-54 to phosphate groups of both DNA strands (see Figure 3). Finally, specific DNA recognition is provided by amino acid side chain contacts to DNA bases, i.e., from Ile-47 to T8 and A9, from Gln-50 to C7 and T8, and from Met-54 to C7 (see Figures 1 and 2).

As indicated in Figure 1, the en homeodomain makes a very similar set of base-specific contacts through Ile-47 and Gln-50. In both en and MATα2, Asn-51 makes a hydrogen bond to A of base pair 9 in the TAAT core, whereas the DNA contacts of Asn-51 in Antp are water mediated (see below). Residue 54, Met in Antp, is substituted by Ala in en and Arg in MATα2. Arg-54 in MATα2 also projects into the major groove and makes a base-specific contact to G in base pair 4, whereas Ala-54 in en has too short a side chain to contact the DNA directly. Since Asn-51 is invariant, the sequence specificity of the DNA binding is likely to reside primarily in residues 47, 50, and 54, which are more variable among different homeodomains.

Changing the DNA Binding Specificity

On the basis of their amino acid sequences, homeodomains can be subdivided into different classes that differ in some characteristic residues. For example, homeodomains of the Antp class, such as Antp and ftz, have a Gln residue at position 50 in the recognition helix, whereas bicoid (bcd), which belongs to a different class, has a Lys residue at this position. Residue 50 was suspected to be crucial for DNA sequence recognition, since in a heterologous expression system in yeast the substitution of Lys by Gln (K50Q) in the recognition helix of a bcd fusion protein switched its DNA target specificity from bcd to an Antp class protein (Hanes and Brent, 1989), and similar results were obtained by Treisman et al. (1989) using in vitro DNA binding assays. However, it was not known which nucleotide(s) were contacted by residue 50. A sequence comparison between the consensus-binding site for bcd, which is GGGATTAGA (Driever and Nüsslein-Volhard, 1989), and the Antp class–binding site BS2, GCCATTAGA (Müller et al., 1988) (sequences for the α strand in the notation used here), indicated that there are only two differences, i.e., GG versus CC at positions 6 and 7. This observation suggested that Lys-50 might require two G nucleotides for high affinity binding, whereas Gln-50 might preferentially bind to two C residues at positions 6 and 7 of the binding site. This hypothesis was tested in the context of the ftz homeodomain in order to enable subsequent complementation analyses in vivo. For this purpose a ftz mutant homeodomain, ftz(Q50K), was constructed, and its binding affinities were compared to that of the wild-type ftz homeodomain with Gln at position 50. Measurement of the binding constants for the four possible binding sites with the sequences CC, CG, GC, and GG at positions 6 and 7 clearly showed that wild-type ftz has the highest affinity for CC at positions 6 and 7, whereas ftz(Q50K) preferentially binds to GG. The differences in K_D are 39- and 62-fold, respectively (Percival-Smith et al., 1990). The NMR analysis showed that there are NOEs between Gln-50 and base protons of both C6 and C7 in the Antp homeodomain–DNA complex (Billeter et al., 1993). Structural data on a Lys-50-containing homeodomain–DNA complex are still elusive, but the results obtained from a genetic dissection of the DNA binding specificity of the *bcd* gene product in yeast support the view that Lys hydrogen bonds with the guanine base immediately 5′ to the ATTA core motif (Hanes and Brent, 1991).

On the basis of this structural and biochemical information, we designed experiments to investigate the questions of whether the homeodomain interacts directly with its binding sites in vivo and whether the mode of DNA binding is similar in vitro and in vivo (Schier and Gehring, 1992). For this purpose the *ftz* gene was used, since it contains an upstream autoregulatory control element with enhancerlike properties (Hiromi and Gehring, 1987) that lends itself to an in vivo analysis in transgenic embryos. This upstream region contains a 430 bp minimal autoregulatory element, in which one high affinity and five medium affinity binding sites for the ftz protein were detected in vitro (Pick et al., 1990). In transgenic embryos, this autoregulatory element, when fused to a β-galactosidase reporter gene, directs the expression of β-galactosidase in a characteristic *ftz*-like pattern of seven stripes. This expression is dependent on the ftz protein itself, indicating that *ftz* is autoregulated, but the possibility of an indirect effect of the ftz protein had to be considered. Deletion analysis of the in vitro ftz-binding sites first indicated that these multiple binding sites are functionally redundant. However, when three or more of these binding sites were deleted or point mutated, reporter gene expression was essentially abolished.

To test whether the ftz protein binds directly to the autoregulatory element in vivo, we asked whether point mutations in the binding sites can be suppressed by compensating mutations in the homeodomain of a *ftz* transgene. Three of the in vitro binding sites in the autoregulatory element were point mutated to GGATTA, and one was deleted entirely. Transgenic embryos with a wild-type ftz protein carrying this mutated autoregulatory element show hardly any reporter gene expression. However, this mutant defect can be suppressed by introducing a *ftz(Q50K)* transgene. This homeodomain mutation affects the DNA binding specificity and restores the expression pattern in seven stripes. This suppression experiment demonstrates a direct in vivo interaction of the ftz protein with its binding sites in the autoregulatory element and shows that the modes of DNA binding in vivo and in vitro must be very similar.

Effects of the Elongation of the Recognition Helix by the Flexible Helix IV

As mentioned above, the recent structure determination of the ftz homeodomain (Qian et al., 1994a) revealed a very similar global fold as for Antp, but no elongation of the recognition helix by helix IV is present, i.e., the corresponding residues are flexibly disordered. This appears to be related to the substitution of the Trp residue at position 56, which anchors helix IV in the hydrophobic core of the Antp homeodomain, by Ser in the ftz homeodomain. In spite of the absence of helix IV, which makes important DNA contacts in the Antp homeodomain–DNA complex (see above), direct comparison of the K_D of the two homeodomains indicates that the ftz homeodomain has an even somewhat higher affinity for BS2 than the Antp(C39S) homeodomain. This indicates that the polypeptide segment corresponding to helix IV, which is flexibly disordered in ftz, contributes significantly to the binding affinity. To measure the contribution of this polypeptide segment to the binding affinity, a truncated ftz homeodomain lacking the C-terminal amino acid residues from Met-54 at the start of helix IV to His-68 was expressed in E. coli, purified, and the K_D directly compared to that of the full-length ftz homeodomain. This C-terminally truncated ftz homeodomain has an 18-fold lower DNA binding affinity relative to the full-length ftz homeodomain (D. R.-P. and W. J. G., unpublished data). Considering that the pattern of DNA contacts by residues 54–57 in the Antp homeodomain–DNA complex is largely dictated by the helical structure of this polypeptide segment (see Figure 3), one is tempted to speculate on the basis of the DNA binding affinities that the corresponding flexibly disordered segment of the free ftz homeodomain might also form a helix when bound to the DNA.

Hydration and Internal Mobility of the Homeodomain–DNA Complex

Notwithstanding the large number of direct protein–DNA contacts, the amino acid side chains do not completely fill the space between the recognition helix and the DNA duplex in the Antp(C39S) homeodomain–DNA complex. Three-dimensional heteronuclear-resolved [^{1}H,^{1}H]-NOE spectroscopy experiments showed that at least some of the space in the protein–DNA interface is filled by one or several molecules of hydration water (Qian et al., 1993b), which have lifetimes with respect to exchange with the bulk solvent in the range between about 2 ns and 20 ms (Otting et al., 1991). This time scale overlaps with that for the independently evidenced slow-motional averaging for the side chain of Asn-51 of the Antp(C39S) homeodomain in the DNA complex (Billeter et al., 1993; Qian et al., 1993a), which suggests that this side chain moves between two or several different contact sites in the protein–DNA interface on a time scale of milliseconds. These experimental findings on mobility in the protein–DNA interface on a time scale much shorter than the lifetime of the complex are also reflected by the structural variations observed among the 16 conformers that represent the NMR structure of the complex (Billeter et al., 1993). Up to five water molecules are present in the protein–DNA interface in the different individual conformers that were subjected to the final step of the structure refinement. In 15 of the 16 conformers, these waters are well separated from the bulk solvent, without any hydrogen bonds with bulk water molecules. This group of 15 conformers was analyzed for the location of cavity space in the protein–DNA interface. Figures 5A and 5B show the water positions found in two of the conformers that represent the solution structure. Closer examination of all conformers revealed that in some conformers, the functionally important side chains of Gln-50 and Asn-51 occupy the same space that is occupied by interior water molecules in other conformers. We adopt the hypothesis that these structural variations represent actual conformational fluctuations

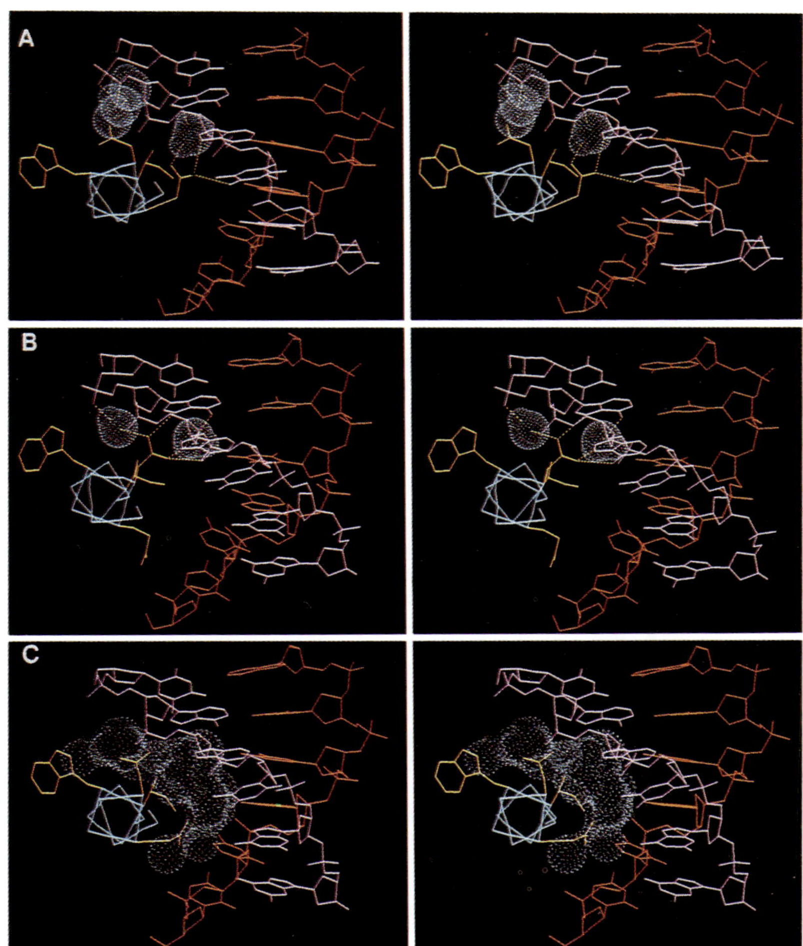

Figure 5. Hydration Water Molecules at the Homeodomain–DNA Interface

(A) and (B) show two of the conformers used to represent the NMR structure of the Antp(C39S) homeodomain–DNA complex obtained after molecular dynamics refinement in a water bath (Billeter et al., 1993). The coloring scheme for the protein and the DNA is the same as for Figure 1, and water molecules are represented in blue dots indicating the van der Waals surface. The polypeptide backbone of residues 43–55 in the recognition helix and the side chains of Ile-47, Trp-48, Gln-50, and Asn-51 of the Antp(C39S) homeodomain are shown, as well as the base pairs 6–11 of the DNA. Hydration waters enclosed in the cavity at the protein–DNA interface (see text) are displayed (four waters in the conformer of [A], two waters in the conformer of [B]). Intermolecular hydrogen bonds involving amino acid side chains of the protein, hydration water molecules, and the DNA are indicated by broken yellow lines.

(C) One of the conformers used to represent the NMR structure of the Antp(C39S) homeodomain–DNA complex, in which the dotted surface surrounds all the water positions found in the 15 conformers, each of which contained between one and five water molecules in the interfacial cavity (see text). The same fragments of the protein and the DNA are displayed as in (A) and (B), and the same coloring scheme was used.

with time. This is illustrated in Figure 5C with an envelope surrounding the locations of all internal waters found in the 15 conformers (for an exact definition of this envelope, see Billeter et al., 1993). The implication is that the individual hydration water molecules not only exchange with the bulk water, but that their locations within the cavity also vary with time. Overall, on the basis of combining the results of NMR spectroscopy and molecular dynamics refinement, the essential interactions with the side chains of the functionally important residue Gln-50 and the invariant residue Asn-51 may be rationalized as a fluctuating network of short-lived, weak-bonding contacts involving DNA bases as well as interfacial hydration water molecules. The water molecules might mediate specific hydrogen bonds between Gln-50 and Asn-51 and polar groups on the DNA (Figures 5A and 5B) whenever the interatomic distances across the interfacial cavity become too long to enable direct hydrogen bonds.

A closer look at the two presently known crystal structures of homeodomain–DNA complexes (Kissinger et al., 1990; Wolberger et al., 1991) indicates for both the presence of interfacial cavities in locations corresponding to the cavity space in the Antp(C39S)–DNA complex. There is the implication from model studies that these cavities are filled with water molecules in much the same way as in the DNA complex with Antp(C39S) (Billeter and Wüthrich, 1993). No hydration water was reported in the aforementioned crystal structures, which is either owing to the limited resolution of 2.8 Å and 2.7 Å achieved in these studies, or to the fact that these waters are disordered and therefore not observable by X-ray diffraction experiments. Water-mediated interactions have earlier been reported for trp repressor/operator cocrystals (Otwinowski et al., 1988). Recent X-ray crystallographic studies indicate that the hydration sites are already fully occupied in the free DNA (Shakked et al., 1994).

DNA Binding Specificity and Combinatorial Interactions among Different Transcription Factors

The DNA binding specificity of the ftz homeodomain can be switched to that of bcd by substituting a single amino acid in the recognition helix, as discussed above. However, this change in DNA binding specificity does not convert the ftz protein into a functional bcd protein. In fact, the ftz(Q50K) mutant protein retains partial ftz activity and, when tested in transgenic flies in a *ftz*⁻ background, is capable of partial complementation of the *ftz*⁻ phenotype (Schier and Gehring, 1993a). Also, ftz(Q50K) does not activate bcd target genes in the realms of *ftz* expression. These observations indicate that DNA binding specificity alone does not account for the functional specificity. In addition to the multiple ftz-binding sites in the autoregulatory element of *ftz*, there are conserved binding sites for other transcription factors that are required for the proper expression of the striped pattern (Schier and Gehring, 1993b). This strongly suggests that combinatorial interaction with other transcription factors is required for functional specificity.

Such cooperative action has been well established for the yeast homeodomain protein MATα2 (Keleher et al., 1988; Ammerer, 1990), where it has been found that the binding characteristics are dependent not only on its own binding specificity, which is rather weak, but depend also on its association with other transcription factors (see Johnson, 1992). In combination with the MCM1 protein, MATα2 forms a heterotetramer and recognizes a set of operator sites associated with a-specific genes with improved specificity and affinity (Passmore et al., 1989). In combination with the MATa1 protein, MATα2 forms a heterodimer which binds to a set of different operator sequences associated with haploid-specific genes (Dranginis, 1990; Goutte and Johnson, 1993). Recent studies of the MATα2 protein indicate that the interaction with MCM1 is mediated by an apparently unstructured region of the protein located close to the N-terminus of the homeodomain (Vershon and Johnson, 1993). This region corresponds to a flexible linker at the N-terminus of the Antp homeodomain (Qian et al., 1992). These observations raise the possibility that the flexible N-terminal linker in Antp, which includes the YPWM motif that is conserved from Drosophila to man, may be involved in similar protein–protein interactions that increase both the affinity and specificity of DNA binding.

Investigation of POU homeodomain proteins in the developing Drosophila nervous system has revealed a functionally important interaction between I-POU and Cf1-a, two POU proteins containing both a POU domain and a homeodomain. Cf1-a binds to a specific regulatory element in the *dopa decarboxylase* (*Ddc*) gene and activates its transcription (Johnson and Hirsh, 1990). I-POU forms a heterodimer with Cf1-a in solution, thereby inhibiting Cf1-a binding to the *Ddc* regulatory element and preventing transactivation (Treacy et al., 1991). The protein–protein interaction between I-POU and Cf1-a is highly specific and depends upon the cluster of basic amino acids at the N-terminus and in the first two helices of the homeodomain. I-POU lacks residues 3 and 4 of the homeodomain and therefore does not bind to DNA. A splicing variant transcript, referred to as twin of I-POU, restores these two amino acid residues. Twin of I-POU is no longer capable of interacting with Cf1-a, but its capacity of specific DNA binding is restored. This shows that a two amino acid difference in the N-terminal arm of a homeodomain distinguishes an inhibitor from an activator of transcription.

Protein–protein interactions have also been postulated to occur between the homeodomain proteins of *exd* and *Ubx* or *Antp* in Drosophila (Wieschaus and Noell, 1986; Peifer and Wieschaus, 1990; Rauskolb et al., 1993). At early developmental stages, *exd* is expressed ubiquitously, and loss-of-function mutants cause homeotic transformations by altering the morphological consequences of homeotic selector gene activity (e.g., Ubx) without affecting their expression patterns. This raises the possibility that exd could interact directly with Ubx proteins. It remains to be shown whether such interactions are direct, leading to the formation of heterodimers with a DNA binding specificity different from that of the respective monomers or homodimers.

Conclusions and Perspectives

The structural studies summarized in this review show that the three-dimensional structure and the mode of DNA binding of various homeodomains have been highly conserved during evolution. Thus, disparate homeodomains such as Antp, en, MATα2, and Oct-1 show similar global folds and dock onto the DNA in essentially identical ways. The majority of the specific contacts with the operator site in the homeodomain–DNA complexes involve amino acid residues located in the recognition helix. The biological relevance of the structural analysis and the in vitro DNA binding studies was established by genetic experiments: a mutation of two base pairs in the binding site of the ftz homeodomain can actually be suppressed in transgenic embryos by a single amino acid substitution in the recognition helix, as was predicted on the basis of the structural and biochemical data. This crucial experiment indicates that the mode of DNA binding in vivo and in vitro must be very similar.

The refined NMR analysis gives much more emphasis to dynamic features of the protein–DNA interaction than the crystal structures, with amino acid side chains moving between two or several contact sites on the DNA on a time scale of milliseconds. This internal mobility is reminiscent of a continuously ongoing scanning process. Furthermore, water molecules present in a cavity at the interface between protein and DNA may mediate transient formation of specific hydrogen bonds between the functionally important amino acid side chains and polar groups on the DNA.

The DNA binding and the functional specificity of various homeodomains does not reside exclusively in the recognition helix but to a large extent also in the N-terminal arm of the homeodomain. In the free protein, this arm does not acquire a precisely defined structure, but is rather flexible. This flexibility is also observed in an N-terminally elongated Antp homeodomain. In the crystal structure of the Oct-1–DNA complex, in which the N-terminal arm of the homeodomain forms part of a linker to a second DNA-binding domain, the POU-specific domain, this linker is not visible, presumably owing to the fact that it is flexibly disordered (Klemm et al., 1994). The flexible N-terminal arm establishes contacts to the bases in the minor groove of the DNA. Deletion or phosphorylation of specific residues in the N-terminal arm strongly reduces the DNA binding affinity. Genetic experiments show that reciprocal substitution of the four amino acids in the N-terminal arm, which differ between the Antp and Scr homeodomains, converts the functional specificity of the hybrid proteins from Antp to Scr, and vice versa. These experiments clearly demonstrate the functional importance of the flexible N-terminal arm of the polypeptide.

The DNA binding specificity and the functional activity of homeodomain proteins also depend upon combinatorial interactions with other transcription factors. These interactions may occur between the free proteins in solution or alternatively when the proteins are bound to the DNA. The operator in this case may not be simply a DNA segment, but rather a DNA–protein complex (see Johnson, 1992). The DNA binding specificity can also be increased in proteins containing two or more separate DNA-binding domains. So far, the structural analysis has been focused on the homeodomain as a module. Future studies will have to consider this module in the context of the entire protein and its interactions with other regulatory proteins.

Acknowledgments

We would like to thank Martin Müller for his advice on the purification of the homeodomain polypeptides and Erika Marquardt-Wenger for the processing of the manuscript. Financial support by the Schweizerischer Nationalfonds (Projects 31.32033.91, to K. W., and 31.28707.90, to W. J. G.) and by the Kantons of Basel (W. J. G.) is gratefully acknowledged.

References

Affolter, M., Percival-Smith, A., Müller, M., Leupin, W., and Gehring, W. J. (1990). DNA binding properties of the purified *Antennapedia* homeodomain. Proc. Natl. Acad. Sci. USA *87*, 4093–4097.

Affolter, M., Percival-Smith, A., Müller, M., Billeter, M., Qian, Y. Q., Otting, G., Wüthrich, K., and Gehring, W. J. (1991). Similarities between the homeodomain and the *Hin* recombinase DNA-binding domain. Cell *64*, 879–880.

Ammerer, G. (1990). Identification, purification, and cloning of a polypeptide (PRTF/GRM) that binds to mating-specific promoter elements in yeast. Genes Dev. *4*, 299–312.

Billeter, M., Qian, Y. Q., Otting, G., Müller, M., Gehring, W. J., and Wüthrich, K. (1990). Determination of the three-dimensional structure of the *Antennapedia* homeodomain from *Drosophila* in solution by ^{1}H nuclear magnetic resonance spectroscopy. J. Mol. Biol. *214*, 183–197.

Billeter, M., Qian, Y. Q., Otting, G., Müller, M., Gehring, W., and Wüthrich, K. (1993). Determination of the NMR solution structure of an Antennapedia homeodomain–DNA complex. J. Mol. Biol. *234*, 1084–1093.

Billeter, M., and Wüthrich, K. (1993). Appendix: model studies relating nuclear magnetic resonance data with the three-dimensional structure of protein–DNA complexes. J. Mol. Biol. *234*, 1094–1097.

Ceska, T. A., Lamers, M., Monaci, P., Nicosia, A., Cortese, R., and Suck, D. (1993). The X-ray structure of an atypical homeodomain present in the rat liver transcription factor LFB1/HNF1 and implications for DNA binding. EMBO J. *12*, 1805–1810.

Chan, S. K., and Mann, R. S. (1993). The segment identity functions of *Ultrabithorax* are contained within its homeo domain and carboxy-terminal sequences. Genes Dev. *7*, 796–811.

De Simeone, V., and Cortese, R. (1991). Transcriptional regulation of liver-specific gene expression. Curr. Opin. Cell Biol. *3*, 960–965.

Dranginis, A. M. (1990). Binding of yeast a1 and α2 as a heterodimer to the operator DNA of a haploid-specific gene. Nature *347*, 682–685.

Driever, W., and Nüsslein-Volhard, C. (1989). The *bicoid* protein is a positive regulator of *hunchback* transcription in the early *Drosophila* embryo. Nature *337*, 138–143.

Feng, J. A., Johnson, R. C., and Dickerson, R. E. (1994). *Hin* recombinase bound to DNA: the origin of specificity in major and minor groove interactions. Science *263*, 348–355.

Furukubo-Tokunaga, K., Müller, M., Affolter, M., Pick, L., Kloter, U., and Gehring, W. J. (1992). *In vivo* analysis of the helix-turn-helix motif of the *fushi tarazu* homeo domain of *Drosophila melanogaster*. Genes Dev. *6*, 1082–1096.

Furukubo-Tokunaga, K., Flister, S., and Gehring, W. J. (1993). Functional specificity of the Antennapedia homeodomain. Proc. Natl. Acad. Sci. USA *90*, 6360–6364.

Gehring, W. J. (1987). Homeo boxes in the study of development. Science *236*, 1245–1252.

Gehring, W. J., Müller, M., Affolter, M., Percival-Smith, A., Billeter, M., Qian, Y. Q., Otting, G., and Wüthrich, K. (1990). The structure of the homeodomain and its functional implications. Trends Genet. *6*, 323–329.

Gehring, W. J., Affolter, M., and Bürglin, K. (1994). Homeodomain proteins. Annu. Rev. Biochem. *63*, 437–526.

Gibson, G., Schier, A., LeMotte, P., and Gehring, W. J. (1990). The specificites of *Sex combs reduced* and *Antennapedia* are defined by a distinct portion of each protein that includes the homeodomain. Cell *62*, 1087–1103.

Goutte, C., and Johnson, A. D. (1993). Yeast a1 and α2 homeodomain proteins form a DNA-binding activity with properties distinct from those of either protein. J. Mol. Biol. *233*, 359–371.

Güntert, P., Qian, Y. Q., Otting, G., Müller, M., Gehring, W., and Wüthrich, K. (1991). Structure determination of the *Antp*(C39→S) homeodomain from nuclear magnetic resonance data in solution using a novel strategy for the programs DIANA, CALIBA, HABAS, and GLOMSA. J. Mol. Biol. *217*, 531–540.

Hanes, S. D., and Brent, R. (1989). DNA specificity of the bicoid activator protein is determined by homeodomain recognition helix residue 9. Cell *57*, 1275–1283.

Cell
222

Hanes, S. D., and Brent, R. (1991). A genetic model for interaction of the homeodomain recognition helix with DNA. Science *251*, 426–430.

Hiromi, Y., and Gehring, W. J. (1987). Regulation and function of the Drosophila segmentation gene *fushi tarazu*. Cell *50*, 963–974.

Johnson, W. W., and Hirsh, J. (1990). Binding of a *Drosophila POU* domain protein to a sequence element regulating gene expression in specific dopminergic neurons. Nature *343*, 467–470.

Johnson, A. (1992). A combinatorial regulatory circuit in budding yeast. In Transcriptional Regulation, S. L. McKnight and K. R. Yamamoto, eds. (Cold Spring Harbor, New York: Cold Spring Harbor Laboratory Press), pp. 975–1006.

Kalionis, B., and O'Farrell, P. H. (1993). A universal sequence is bound in vitro by diverse homeodomains. Mech. Dev. *43*, 57–70.

Kamps, M. P., Murre, C., Sun, X. H., and Baltimore, D. (1990). A new homeobox gene contributes the DNA binding domain of the t(1;19) translocation protein in pre-B ALL. Cell *60*, 547–555.

Kapiloff, M. S., Farkash, Y., Wegner, M., and Rosenfeld, M. G. (1991). Variable effects of phosphorylation of Pit-1 dictated by the DNA response elements. Science *253*, 786–789.

Keleher, C. A., Goutte, C., and Johnson, A. D. (1988). The yeast cell-type-specific repressor α2 acts cooperatively with a non-cell-type-specific protein. Cell *53*, 927–935.

Kissinger, C. R., Liu, B., Martin-Blanco, E., Kornberg, T. B., and Pabo, C. O. (1990). Crystal structure of an engrailed homeodomain–DNA complex at 2.8 Å resolution: a framework for understanding homeodomain–DNA interactions. Cell *63*, 579–590.

Klemm, J. D., Rould, M. A., Aurora, R., Herr, W., and Pabo, C. O. (1994). Crystal structure of the Oct-1 POU domain bound to an octamer site: DNA recognition with tethered DNA-binding modules. Cell *77*, 21–32.

Kornberg, T. B. (1993). Understanding the homeodomain. J. Biol. Chem. *268*, 26813–26816.

Laughon, A. (1991). DNA binding specificity of homeodomains. Biochemistry *30*, 11357–11367.

Leiting, B., De Francesco, R., Tomei, L., Cortese, R., Otting, G., and Wüthrich, K. (1993). The three-dimensional NMR-solution structure of the polypeptide fragment 195–286 of the LFB1/HNF1 transcription factor from rat liver comprises a non-classical homeodomain. EMBO J. *12*, 1797–1803.

Lin, L., and McGinnis, W. (1992). Mapping functional specificity in the *Dfd* and *Ubx* homeo domains. Genes Dev. *6*, 1071–1081.

Mann, R., and Hogness, D. S. (1990). Functional dissection of Ultrabithorax proteins in Drosophila melanogaster. Cell *60*, 597–610.

McGinnis, W., and Krumlauf, R. (1992). Homeobox genes and axial patterning. Cell *68*, 283–302.

Mendel, D. B., and Crabtree, G. R. (1991). HNF-1, a member of a novel class of dimerizing homeodomain proteins. J. Biol. Chem. *266*, 677–680.

Monica, K., Galili, N., Nourse, J., Saltman, D., and Cleary, M. L. (1991). *PBX2* and *PBX3*, new homeobox genes with extensive homology to the human proto-oncogene *PBX1*. Mol. Cell. Biol. *11*, 6149–6157.

Müller, M., Affolter, M., Leupin, W., Otting, G., Wüthrich, K., and Gehring, W. J. (1988). Isolation and sequence-specific DNA binding of the *Antennapedia* homeodomain. EMBO J. *7*, 4299–4304.

Nourse, J., Melletin, J. D., Galili, N., Wilkinson, J., Stanbridge, E., Smith, S. D., and Cleary, M. L. (1990). Chromosomal translocation t(1;19) results in the synthesis of a homeobox fusion mRNA that codes for the potential chimeric transcription factor. Cell *60*, 535–545.

Otting, G., Qian, Y. Q., Billeter, M., Müller, M., Affolter, M., Gehring, W. J., and Wüthrich, K. (1990). Protein–DNA contacts in the structure of a homeodomain–DNA complex determined by nuclear magnetic resonance spectroscopy in solution. EMBO J. *9*, 3085–39092.

Otting, G., Liepinsh, E., and Wüthrich, K. (1991). Proton exchange with internal water molecules in the protein BPTI in aqueous solution. J. Am. Chem. Soc. *113*, 4363–4364.

Otwinowski, Z., Schevitz, R. W., Zhang, R.-G., Lawson, C. L., Joachimiak, A., Marmorstein, R. Q., Luisi, B. F., and Sigler, P. B. (1988). Crystal structure of *trp* repressor/operator complex at atomic resolution. Nature *335*, 321–329.

Pabo, C. O., and Sauer, R. T. (1992). Transcription factors: structural families and principles of DNA recognition. Annu. Rev. Biochem. *61*, 1053–1095.

Passmore, S., Elbie, R., and Tye, B.-R. (1989). A protein involved in minichromosome maintenance in yeast binds a transcriptional enhancer conserved in eukaryotes. Genes Dev. *3*, 921–935.

Peifer, M., and Wieschaus, E. (1990). Mutations in the *Drosophila* gene *extradenticle* affect the way specific homeo domain proteins regulate segmental identity. Genes Dev. *4*, 1209–1223.

Percival-Smith, A., Müller, M., Affolter, M., and Gehring, W. J. (1990). The interaction with DNA of wild-type and mutant *fushi tarazu* homeodomains. EMBO J. *9*, 3967–3974. Corrigendum, EMBO J. *11*, 382.

Pick, L., Schier, A., Affolter, M., Schmidt-Glenewinkel, T., and Gehring, W. J. (1990). Analysis of the *ftz* upstream element: germ layer-specific enhancers are independently autoregulated. Genes Dev. *4*, 1224–1239.

Qian, Y. Q., Billeter, M., Otting, G., Müller, M., Gehring, W. J., and Wüthrich, K. (1989). The structure of the *Antennapedia* homeodomain determined by NMR spectroscopy in solution: comparison with prokaryotic repressors. Cell *59*, 573–580.

Qian, Y. Q., Otting, G., Furukubo-Tokunaga, K., Affolter, M., Gehring, W. J., and Wüthrich, K. (1992). NMR structure determination reveals that the homeodomain is connected through a flexible linker to the main body in the Drosophila Antennapedia protein. Proc. Natl. Acad. Sci. USA *89*, 10738–10742.

Qian, Y. Q., Otting, G., Billeter, M., Müller, M., Gehring, W., and Wüthrich, K. (1993a). NMR spectroscopy of a DNA complex with the uniformly ^{13}C-labeled Antennapedia homeodomain and structure determination of the DNA-bound homeodomain. J. Mol. Biol. *234*, 1070–1083.

Qian, Y. Q., Otting, G., and Wüthrich, K. (1993b). NMR detection of hydration water in the intermolecular interface of a protein–DNA complex. J. Am. Chem. Soc. *115*, 1189–1190.

Qian, Y. Q., Furukubo-Tokunaga, K., Müller, M., Resendez-Perez, D., Gehring, W. J., and Wüthrich, K. (1994a). Nuclear magnetic resonance solution structure of the *fushi tarazu* homeodomain from *Drosophila* and comparison with the *Antennapedia* homeodomain. J. Mol. Biol. *238*, 333–345.

Qian, Y. Q., Resendez-Perez, D., Gehring, W. J., and Wüthrich, K. (1994b). The *des(1–6)* Antennapedia homeodomain: comparison of the NMR solution structure and the DNA binding affinity with the intact Antennapedia homeodomain. Proc. Natl. Acad. Sci. USA *91*, 4091–4095.

Rauskolb, C., Peifer, M., and Wieschaus, E. (1993). *extradenticle*, a regulator of homeotic gene activity, is a homolog of the homeobox-containing human proto-oncogene *pbx1*. Cell *74*, 1101–1112.

Schier, A. F., and Gehring, W. J. (1992). Direct homeodomain–DNA interaction in the autoregulation of the *fushi tarazu* gene. Nature *356*, 804–807.

Schier, A. F., and Gehring, W. J. (1993a). Functional specificity of the homeodomain protein *fushi tarazu*: the role of DNA binding specificity in vivo. Proc. Natl. Acad. Sci. USA *90*, 1450–1454.

Schier, A. F., and Gehring, W. J. (1993b). Analysis of a *fushi tarazu* autoregulatory element: multiple sequence elements contribute to enhancer activity. EMBO J. *12*, 1111–1119.

Schulz, B., Banuett, F., Dahl, M., Schlesinger, R., Schäfer, W., Martin, T., Herskowitz, I., and Kahmann, R. (1990). The b alleles of U. maydis, whose combinations program pathogenic development, code for polypeptides containing a homeodomain-related motif. Cell *60*, 295–306.

Scott, M. P., Tamkun, J. W., and Hartzell, G. W., III (1989). The structure and function of the homeodomain. Biochim. Biophys. Acta *989*, 25–48.

Segil, N., Boseman, R. S., and Heintz, N. (1991). Mitotic phosphorylation of the Oct-1 homeodomain and regulation of Oct-1 DNA binding activity. Science *254*, 1814–1816.

Shakked, Z., Guzikevich-Guerstein, G., Frolow, F., Rabinovich, D., Joachimiak, A., and Sigler, P. B. (1994). Determinants of repressor/operator recognition from the structure of the *trp* operator binding site. Nature *368*, 469–473.

Sluka, J. P., Horvath, S. J., Bruist, M. F., Simon, M. I., and Dervan, P. B. (1987). Synthesis of a sequence-specific DNA-cleaving peptide. Science *238*, 1129–1132.

Sluka, J. P., Horvath, S. J., Galsgow, A. C., Simon, M. I., and Dervan, P. B. (1990). Importance of minor-groove contacts for recognition of DNA by the binding domain of Hin recombinase. Biochemistry *29*, 6551–6561.

Steitz, T. A. (1990). Structural studies of protein–nucleic acid interaction: the sources of sequence-specific binding. Quart. Rev. Biophys. *23*, 205–280.

Suzuki, M. (1993). Common features in DNA recognition helices of eukaryotic transcription factors. EMBO J. *12*, 3221–3226.

Treacy, M. N., He, X., and Rosenfeld, M. G. (1991). I-POU: a POU domain protein that inhibits neuron-specific gene activation. Nature *350*, 577–584.

Treacy, M. N., Neilson, L. I., Turner, E. E., He, X., and Rosenfeld, M. G. (1992). Twin of I-POU: a two amino acid difference in the I-POU homeodomain distinguishes an activator from an inhibitor of transcription. Cell *68*, 491–505.

Treisman, J., Gönczy, P., Vashishtha, M., Harris, E., and Desplan, C. (1989). A single amino acid can determine the DNA binding specificity of homeodomain proteins. Cell *59*, 553–562.

Vershon, A. K., and Johnson, A. D. (1993). A short, disordered protein region mediates interactions between the homeodomain of the yeast α2 protein and the MCM1 protein. Cell *72*, 1–20.

Wieschaus, E., and Noell, E. F. (1986). Specificity of embryonic lethal mutations in *Drosophila* analysed in germ line clones. Roux's Arch. Dev. Biol. *195*, 63–73.

Wolberger, C., Vershon, A. K., Liu, B., Johnson, A. D., and Pabo, C. O. (1991). Crystal structure of a *MATα2* homeodomain–operator complex suggests a general model for homeodomain–DNA interactions. Cell *67*, 517–528.

Wüthrich, K., and Gehring, W. (1992). Transcriptional regulation by homeodomain proteins: structural, functional, and genetic aspects. In Transcriptional Regulation, S. L. McKnight and K. R. Yamamoto, eds. (Cold Spring Harbor, New York: Cold Spring Harbor Laboratory Press), pp. 535–577.

Zeng, W., Andrew, D. J., Mathies, L. D., Horner, M. A., and Scott, M. P. (1993). Ectopic expression and function of the *Antp* and *Scr* homeotic genes: the N terminus of the homeodomain is critical to functional specificity. Development *118*, 339–352.

Reprinted from Biochemistry, 1991, *30*. 6563
Copyright © 1991 by the American Chemical Society and reprinted by permission of the copyright owner.

The NMR Structure of Cyclosporin A Bound to Cyclophilin in Aqueous Solution†

C. Weber,‡ G. Wider,‡ B. von Freyberg,‡ R. Traber,§ W. Braun,‡ H. Widmer,§ and K. Wüthrich*,‡

Institut für Molekularbiologie und Biophysik, Eidgenössische Technische Hochschule—Hönggerberg, CH-8093 Zürich, Switzerland, and Präklinische Forschung, Sandoz Pharma AG, CH-4002 Basel, Switzerland

Received December 31, 1990; Revised Manuscript Received March 28, 1991

ABSTRACT: Cyclosporin A bound to the presumed receptor protein cyclophilin was studied in aqueous solution at pH 6.0 by nuclear magnetic resonance spectroscopy using uniform ^{15}N- or ^{13}C-labeling of cyclosporin A and heteronuclear spectral editing techniques. Sequence-specific assignments were obtained for all but one of the cyclosporin A proton resonances. With an input of 108 intramolecular NOEs and four vicinal $^3J_{HN\alpha}$ coupling constants, the three-dimensional structure of cyclosporin A bound to cyclophilin was calculated with the distance geometry program DISMAN, and the structures resulting from 181 converged calculations were energy refined with the program FANTOM. A group of 120 conformers was selected on the basis of the residual constraint violations and energy criteria to represent the solution structure. The average of the pairwise root-mean-square distances calculated for the backbone atoms of the 120 structures was 0.58 Å. The structure represents a novel conformation of cyclosporin A, for which the backbone conformation is significantly different from the previously reported structures in single crystals and in chloroform solution. The structure has all peptide bonds in the trans form, contains no elements of regular secondary structure and no intramolecular hydrogen bonds, and exposes nearly all polar groups to its environment. The root-mean-square distance between the backbone atoms of the crystal structure of cyclosporin A and the mean of the 120 conformers representing the NMR structure of cyclosporin A bound to cyclophilin is 2.5 Å.

The immunosuppressive cyclic undecapeptide cyclosporin A (CsA)[1] (Figure 1) has found widespread use in clinical organ transplantation (Borel, 1986; Kahan, 1988). Much insight has been gained into its action on lymphoid cells at a subcellular level. Among other things it inhibits lymphokine production by T-helper cells on the level of mRNA transcription in vitro (Krönke et al., 1984; Elliot et al., 1984) by the inhibition of specific transcriptional activators, such as the nuclear factor of activated T-cells, NF-AT (Emmel et al., 1989). Relative to the molecular basis for its in vivo and in vitro properties, Handschumacher et al. (1984) have described the cytosolic binding protein cyclophilin and proposed that it is the cellular receptor. Cyclophilin (CYP) has been found in diverse organisms from *Escherichia coli* to man (Koletsky et al., 1986; Kawamukai et al., 1989), and in all tissues studied it is present at high abundance. Recently it was found that the enzyme peptidyl–prolyl cis–trans isomerase is identical with CYP (Fischer et al., 1989; Takahashi et al., 1989) and that CsA is a potent inhibitor (K_i = 2.6 nM) of the enzymatic action. It was proposed that CsA exerts its effects on cells via the inhibition of the cis–trans isomerization. Harrison and Stein (1990) have presented elegant kinetic evidence for a mechanism based on distortion of the susceptible peptide bond.

Implications for the mode of interaction between CsA and CYP are so far based on indirect evidence. The structure of cyclosporin A in single crystals and in apolar solvents has been determined (Loosli et al., 1985; Kessler et al., 1990). In both environments it consists of a twisted β-sheet involving the residues 11, 1, 2, 3, 4, 5, 6, and 7, a type II′ turn at Sar 3 and MeLeu 4, and the loop of the remaining amino acids involves a cis peptide bond between MeLeu 9 and MeLeu 10. Monoclonal antibodies raised against CsA were used by Quesniaux and co-workers (Quesniaux et al., 1987a, 1988) to identify the recognition sites of CsA. They measured the cross reactivity of cyclosporin analogues modified at various positions and concluded that in aqueous solution the side chain of MeBmt 1 is folded back onto the molecule as in both the crystal structure and the structure in chloroform solution (Kessler et al., 1990), since antibodies recognizing the face of the molecule defined by the residues 6, 8, and 9 also recognize the peripheral atoms of the MeBmt side chain. These workers also investigated the ability of cyclosporin analogues to CYP using ELISA techniques. They found that recognition of cyclosporins by CYP correlates with the immunosuppressive activity and that CYP interacts preferentially with the residues 1, 2, 10, and 11 of cyclosporin A (Quesniaux et al., 1987b). To gain more direct insight into the molecular details of the function of CsA, we investigated its complex with CYP by nuclear magnetic resonance (NMR) spectroscopy in solution, making extensive use of heteronuclear editing techniques (Fesik, 1988; Fesik et al., 1988; Griffey & Redfield, 1987; Otting et al., 1986). In particular, the use of heteronuclear

† Financial support was obtained from the Schweizerischer Nationalfonds (project no. 31.25174.88), a special grant of the ETH Zürich, and Sandoz Pharma AG., Basel.

‡ Institut für Molekularbiologie and Biophysik.
§ Präklinische Forschung.

[1] Abbreviations: NMR, nuclear magnetic resonance; CYP, cyclophilin; CsA, cyclosporin A; MeBmt, (4*R*)-4-[(*E*)-2-butenyl]-4,*N*-dimethyl-L-threonine; Abu, L-α-aminobutyric acid; MeLeu, *N*-methylleucine; MeVal, *N*-methylvaline; FAB-MS, fast atom bombardment mass spectrometry; hplc, high-performance liquid chromatography; Tris, tris(hydroxymethyl)aminomethane; PMSF, phenylmethanesulfonyl fluoride; EDTA, ethylenediaminetetraacetic acid; DTT, dithiothreitol; MES, 4-morpholineethanesulfonic acid; 1D, one dimensional; 2D, two dimensional; COSY, two-dimensional correlation spectroscopy; TOCSY, two-dimensional total correlation spectroscopy; NOE, nuclear Overhauser effect; NOESY, two-dimensional NOE spectroscopy; CD, circular dichroism; $d_{AB}(i,j)$ designates the distance between the proton types A and B located in the amino acid residues *i* and *j* respectively, where N, NQ, α, and β denote the amide proton, the N-methyl protons, C$^\alpha$H, and C$^\beta$H, respectively; RMSD, root-mean-square deviation.

0006-2960/91/0430-6563$02.50/0 © 1991 American Chemical Society

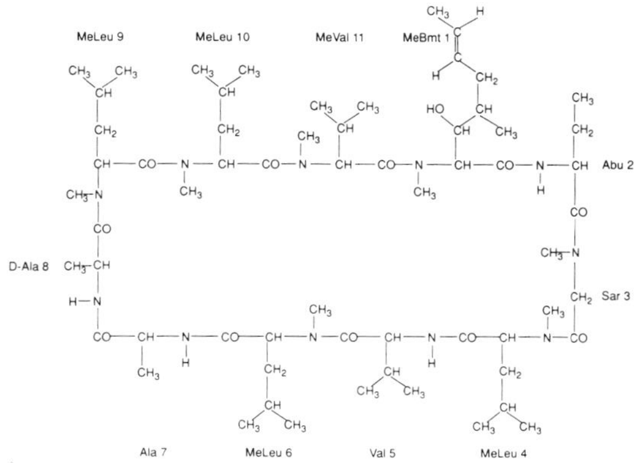

FIGURE 1: Chemical structure of cyclosporin A (CsA) with the standard numeration of the amino acid residues. Symbols: MeBmt, (4*R*)-4-[(*E*)-2-butenyl]-4,*N*-dimethyl-L-threonine; Abu, L-α-aminobutyric acid; MeLeu, *N*-methylleucine; MeVal, *N*-methylvaline.

filters (Otting & Wüthrich, 1990) in conjunction with uniform ^{15}N and ^{13}C enrichment of CsA enabled the spectroscopic distinction of the two molecules in the complex. 2D [^{1}H,^{1}H] NMR spectra recorded with a ^{13}C double half-filter (Otting & Wüthrich, 1989, 1990), were processed into four subspectra, of which one contains only resonances of CsA without interference from CYP, one contains only resonances of CYP without interference from CsA, and two contain exclusively peaks connecting protons of CsA with protons of CYP. For the present work the subspectrum containing the ^{1}H NMR lines of CsA was of most direct importance, as it enabled the use of the well-established protocol for sequential assignment of all CsA proton resonances and collection of a set of conformational constraints to derive a three-dimensional structure of CsA in the bound state (Wüthrich, 1986, 1989). Another study of CsA bound to CYP conducted independently by another group is described in the accompanying paper (Fesik et al., 1991).

MATERIAL AND METHODS

Preparation of Isotope-Labeled CsA. Uniformly ^{15}N-labeled CsA was obtained by growing a high-producing strain of *Tolypocladium inflatum* on a minimal medium consisting of unlabeled glucose (50 g/L), [^{15}N]urea (4 g/L, 99% ^{15}N, Isotech Inc., Miamisburg, OH), and salts and vitamins as described by Kobel and Traber (1982). After 21 days of growth at 27 °C, with shaking at 200 rpm in 500-mL Erlenmeyer flasks, the mycelium was homogenized, extracted with 50% methanol, filtered through celite, and then the liquid phase was evaporated. The crude extract was chromatographed on silicagel, with 9:1 diethylether/methanol as the liquid phase, and the CsA-containing fractions (as judged by thin-layer chromatography) were pooled and analyzed by reverse-phase hplc and FAB-MS. The ^{15}N-enriched CsA eluted as a single peak in reverse-peak hplc and accounted for >97% of the total integrated intensity. From the mass spectrum, we calculated that the ^{15}N enrichment was >98%. Uniformly ^{13}C-labeled CsA was obtained with the same method, by utilizing [$^{13}C_6$]glucose (>98% ^{13}C, Cambridge Isotope Lab., Woburn, MA) and unlabeled urea in an otherwise identical medium. The ^{13}C-enriched CsA was >98% pure as analyzed by reverse-phase hplc. From the position and the width of the molecular ion peak in the mass spectrum, we calculated the ^{13}C enrichment to be >97%. In a control experiment in which the fungus was grown on unlabeled glucose and [^{13}C]urea, there was no evidence for ^{13}C enrichment, indicating that the carbon atom of urea is not used as a carbon source. Biosynthetically directed fractional ^{13}C labeling of CsA was obtained by growing the fungus on a mixture of 10% [$^{13}C_6$]glucose and 90% unlabeled glucose (Senn et al., 1989).

Preparation of Recombinant Human CYP. The protein was overexpressed in *E. coli*. The gene was cloned from Jurkat cells (Haendler et al., 1987) and was under control of a *Taq* promoter. The cells were grown to the stationary phase in a buffer containing casamino acids (25 g/L), yeast extract (10 g/L), and glucose (30 g/L), harvested by centrifugation, and stored at −40 °C. Frozen cells were thawed and suspended in two volumes of a buffer containing 25% sucrose, 50 mM Tris-HCl at pH 7.5, 2 mM EDTA, 1 mM PMSF, 1 mM *o*-phenanthroline, and 2.5 mg/mL lysozyme. After 4 h the solution was supplemented with 20 mg/L deoxyribonuclease I, 4 mg/L $MgCl_2$, 0.2 mg/L $MnCl_2$ and stirred. After 6–12 h at room temperature, the solution was centrifuged at 12000*g* for 40 min and the supernatant decanted. The pellet was resuspended in two volumes of extraction buffer and extracted as above. A total of 40–200 mL of the clear lysate was loaded onto 190 mL of Affi-Gel 10 (Bio-Rad, Richmond, CA) derivatized with 1,1-dihydro-*O*-acetyl-8-D-β-aminoalanine-cyclosporin (K. Stedman, personal communication). The column was eluted with 50 mM Tris-HCl, 2 mM EDTA, 2 mM DTT, pH 7.5, with a longitudinal flow of 0.2 cm/min. CYP eluted as an asymmetric shallow peak at 1100 mL. The UV-positive fractions were pooled and concentrated by ultrafiltration with an Amicon YM-10 membrane. The pooled fractions from the affinity column were diafiltered into 20 mM MES-NaOH, 2 mM EDTA, 2 mM DTT, pH 6.0. Up to 70 mg of CYP were loaded onto a Mono S HR 10-10 column (Pharmacia LKB, Uppsala, Sweden) equilibrated with the same buffer. The column was eluted with a gradient from 50 to 180 mM NaCl in the same buffer in two column volumes at a linear flow rate of 2.5 cm/min. As soon as UV-positive fractions appeared, the gradient was halted, and all CYP fractions eluted isocratically. The main CYP peak was cut at 30% peak height and stored at 4 °C. The main fractions of the previous step containing up to 150 mg of CYP in a volume of 3–10 mL were applied onto 450 mL of Sephacryl S-100 HR (Pharmacia LKB, Uppsala, Sweden) in a 2.6 × 84 cm column. The column was eluted with 20 mM MES-NaOH, 2 mM EDTA, 2 mM DTT, pH 6.0, at a flow rate of 0.2 cm/min. CYP eluted at 293 mL as a single peak, which was cut at 20% peak height. These fractions were sterile filtered and stored at 4 °C.

NMR Samples of the CsA–CYP Complex. The procedure used for the preparation of the complex was largely dictated by the fact that CsA is nearly insoluble in water. In a solution containing 10 mg of CYP, the aforementioned buffer was exchanged against either 99.8% ^2H_2O or a mixture of 90% H_2O/10% D_2O containing 10 mM potassium deuteroacetate (Glaser AG, Basel, Switzerland) and 10 mM potassium phosphate at pH 6.0 (uncorrected pH meter reading). CsA was added from a 60 mg/mL stock solution in deuterated ethanol (99.5% ^{2}H, Glaser AG, Basel, Switzerland) to a final CsA/CYP ratio of 2:1 by slow stepwise infusion through a very fine needle ensuring good dispersion of the resulting solid particles. The suspension was vortexed between subsequent additions of CsA and sonicated in the middle and at the end of the infusion procedure. The complexation was monitored by the disappearance of two methyl resonances at −0.57 and −0.14 ppm in the 1D ^{1}H NMR spectrum of free CYP and the concomitant appearance of two new methyl resonances at −0.90 and −0.69 ppm in the spectrum of the complex. Excess solid CsA was removed by centrifugation and the supernatant twice diluted with 4 mL of the above buffer and concentrated to 500 μL in an Amicon Centriprep 10 ultrafiltration unit to lower the ethanol content. It was then transferred to an NMR tube. Two types of complexes were prepared: ^{15}N-labeled CsA bound to CYP in H_2O solution to study the amide protons of CsA and ^{13}C-labeled CsA bound to CYP in ^2H_2O solution to study the carbon-bound protons of CsA.

NMR Experiments. $^{15}N(\omega_2)$ half-filtered and $^{13}C(\omega_1,\omega_2)$-double-half-filtered 2D [^{1}H,^{1}H] NMR spectra were recorded with the previously described pulse sequences for COSY, TOCSY, and NOESY (Otting & Wüthrich, 1990) on Bruker AM-500 and AM-600 instruments at 25 °C. Data sets of 400 by 2048 points were acquired with sweep widths of 8064 Hz. The filter delays were set to 5.5 ms for ^{15}N-filtered and 3.6 ms for ^{13}C-filtered experiments. Heteronuclear decoupling was accomplished by a 180° proton pulse in the middle of t_1 and WALTZ decoupling in t_2. 90° pulse lengths were 10.5 μs for protons, 21 μs or 190 μs (for WALTZ decoupling) for ^{15}N, and 13.2 μs or 85 μs (for WALTZ decoupling) for ^{13}C. The transmitter was offset to 118 ppm in the ^{15}N dimension and to 40 ppm in the ^{13}C dimension. Heteronuclear correlation spectra were recorded with the pulse sequence of Bodenhausen and Ruben (1980) modified by addition of a spin lock purge pluse to suppress all signals originating from protons not bound to a heterospin (Otting & Wüthrich, 1988). No additional H_2O suppression was needed in these experiments, and there was therefore no loss of NH resonance line intensity (Messerle et al., 1989). For [^{13}C,^{1}H]-COSY of CsA with biosynthetically directed fractional ^{13}C labeling in the complex, the transmitter was offset to 28.5 ppm and the spectral width in the ^{13}C dimension was reduced to 52 ppm. A NOE-relayed [^{13}C,^{1}H]-COSY experiment was recorded with a $^{13}C(\omega_2)$ half-filter. The resulting spectrum contains the same information as [^{1}H,^{1}H]-NOESY recorded with a $^{13}C(\omega_2,\omega_2)$ double half-filter, but it has the advantage that peaks that are overlapped in the latter experiment may be resolved along the ^{13}C frequency axis (Wider et al., 1991). $^3J_{HN\alpha}$ coupling constants were determined with a 1D version of the J-modulated [^{15}N,^{1}H]-COSY experiment (Neri et al., 1990), which used ^{15}N decoupling during acquisition and was acquired with 5632 transients per τ_2 value. A total of 10 τ_2 values with 10-ms intervals ranging from 0 to 90 ms were used, and the peak amplitudes were fitted to the expression

$$V(\tau_2) = A[\cos(\pi J\tau_2)\cos(\pi J\tau_1) - 0.5\sin(\pi J\tau_2)\sin(\pi J\tau_1)]e^{(-\tau_2/T_2)} \quad (1)$$

where $V(\tau_2)$ is the experimental peak amplitude, A the peak amplitude at $\tau_2 = 0$, J the coupling constant $^3J_{HN\alpha}$, and T_2 the transverse relaxation time (D. Neri, M. Billeter, G. Otting, G. Wider, and K. Wüthrich, to be published). The spectra were processed on Bruker Aspect 1000 or X32 computers using standard software, except for homemade software for the addition/subtraction operations.

Determination of the Three-Dimensional CsA Structure. For the collection of the conformational constraints, the NMR spectra were analyzed with the program package EASY (Eccles et al., 1989) on SUN computers. Three-dimensional structures of the bound form of CsA consistent with the NOE distance constraints and dihedral angle constraints were calculated by using the distance geometry program DISMAN (Braun & Gō, 1985). The nonstandard amino acids of CsA, i.e., MeBmt, Abu, Sar, MeLeu, and MeVal, were constructed with the program package GEOM (Sanner et al., 1989) and deposited into the library of DISMAN. Two open chain forms of CsA with Ala 7 at the amino terminal and MeLeu 6 at the carboxy terminal, or with MeBmt 1 at the amino terminal and MeVal 11 at the carboxy terminal, respectively, were assembled from the monomeric building blocks of the library. Conformations of the linear chain with random dihedral angles generated by DISMAN were used as starting structures for the distance geometry calculations. Geometric cyclization of the structures was achieved either with nine distance constraints between N, C$^\alpha$, and HN of Ala 7 and C, O, and C$^\alpha$ of MeLeu 6, or between N, C$^\alpha$, and QCN of MeBmt 1 and C, O, and C$^\alpha$ of MeVal 11. After cyclization, the ω angle of the C-terminal amino acid was thus fixed in the trans conformation. The program DISMAN was modified to selectively exclude repulsive van der Waals checks for cyclic structures (Senn et al., 1990). Method B in the previous structure determination of a cyclic bouvardin analogue (Senn et al., 1990) was used for the distance geometry calculations: the cyclization constraints and the experimental NOE constraints were used simultaneously. Structures with a final target function below a predetermined threshold value were retained and subjected to energy minimization using the program FANTOM (Schaumann et al., 1990), which had been adjusted for use with a cyclic peptide. FANTOM minimizes the ECEPP/2 energy function for proteins (Momany et al., 1975; Nēmethy et al., 1983) using a Newton–Raphson method. The torsion angles are the independent variables, and the ideal covalent geometry of ECEPP/2 is strictly retained during the minimization. Calculations were done on SUN4 and Cray X/MP computers. The resulting structures were graphically analyzed on an E&S PS 390 system using the program CONFOR (Billeter et al., 1985).

Results

Complex Formation between CYP and CsA. CsA is very little soluble in H_2O, and therefore during the aforementioned procedure for the preparation of the complex with CYP the solution contains temporarily up to 5% ethanol (within 24 h after the preparation of the complex, the ethanol was removed by dialysis). Exploratory experiments using CD and 1D NMR spectroscopy revealed no evidence for conformational changes in cyclophilin, and measurements of the fluorescence enhancement of the single Trp in CYP upon binding of CsA showed that the CsA binding properties of CYP were not affected by the ethanol either. In the NMR samples, the concentration of the CYP–CsA complex was 0.8 mM, and because of the low solubility of CsA, there was no detectable concentration of free ligand. In the ^{1}H NMR spectra of aqueous solutions containing an excess of CYP, two sets of resonance lines for free and complexed CYP were observed, showing that chemical exchange of the ligand is slow on the NMR chemical shift time scale.

NMR Assignments for CsA Bound to CYP. For studies of the conformation of a peptide ligand bound to a polypeptide receptor, obtaining ^{1}H NMR assignments (Wüthrich et al., 1982; Wüthrich, 1986) is difficult because many ^{1}H resonances of the ligand are overlapped with the much more numerous NMR lines originating from the receptor. Here, this difficulty was overcome by recording 2D [^{1}H,^{1}H] NMR spectra of complexes with isotope-labeled CsA and unlabeled CYP using heteronuclear half-filters (Fesik, 1988; Otting et al., 1986; Otting & Wüthrich, 1990). Figure 2 affords an illustration of the information contained in the subspectra that were most important for obtaining the resonance assignments. A $^{15}N(\omega_2)$-selected subspectrum from a [^{1}H,^{1}H]-NOESY experiment with an $^{15}N(\omega_2)$-half-filter (Otting et al., 1986) contains only the CsA amide proton lines along ω_2 and all proton lines of CsA and CYP along ω_1 so that all intra- and intermolecular

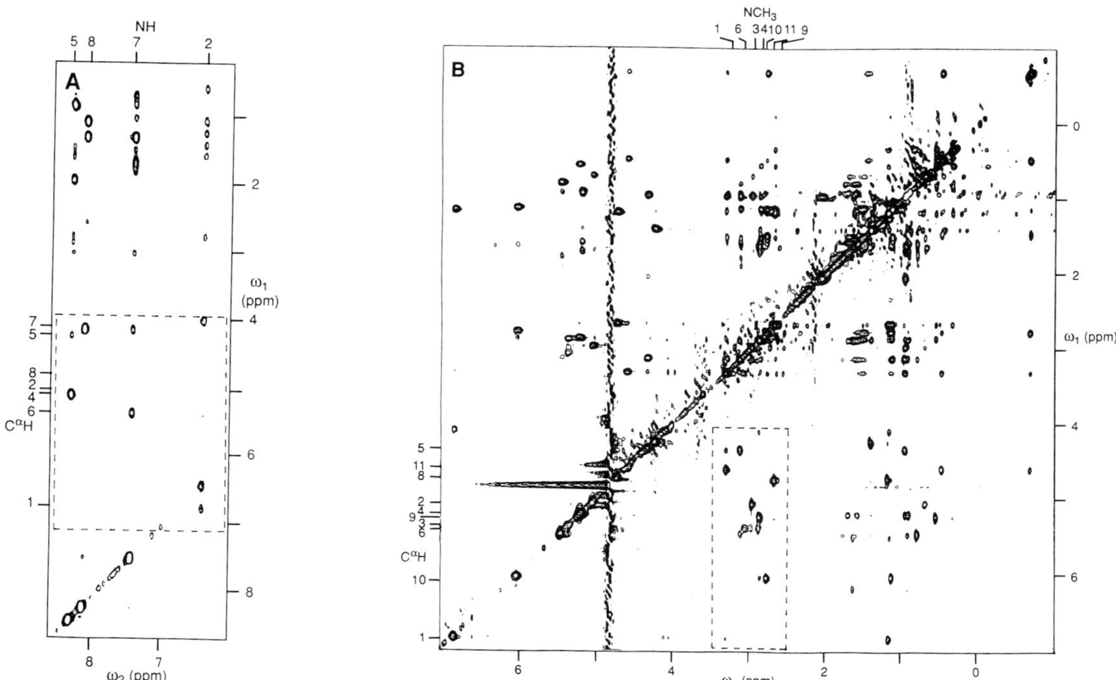

FIGURE 2: [^{1}H,^{1}H]-NOESY spectra of CsA bound to CYP recorded with heteronuclear filters. (A) NOESY spectrum with a $^{15}N(\omega_2)$ half-filter recorded with uniformly ^{15}N-enriched CsA bound to unlabeled CYP. The region (ω_1 = 0.2–8.6 ppm, ω_2 = 6.0–8.6 ppm) that contains all the cross peaks of the $^{15}N(\omega_2)$-selected subspectrum is shown (0.8 mM complex in 500 μL of a mixed solvent of 90% H_2O/10% ^2H_2O, 10 mM potassium deuteroacetate, 10 mM potassium phosphate, 0.04% sodium azide, pH 6.0, 25 °C, 85 h measuring time, ^{1}H frequency 500 MHz, mixing time τ_m = 120 ms). The box drawn with broken lines contains all the $C^\alpha H$–NH cross peaks; this spectral region is reproduced in Figure 3. (B) NOESY spectrum with a $^{13}C(\omega_1,\omega_2)$-double-half-filter recorded with uniformly ^{13}C-enriched CsA bound to unlabeled CYP. The $^{13}C(\omega_1)$-doubly-selected subspectrum is shown (0.8 mM complex in 550 μL of ^2H_2O, 10 mM potassium deuteroacetate, 10 mM potassium phosphate, p^{2}H 6.0, 25 °C, ^{1}H frequency 500 MHz, mixing time τ_m = 80 ms). This spectrum resulted from coaddition of three independent experiments recorded in immediate succession with a measuring time of 82 h each. The region shown contains all the peaks that have been identified. The box drawn with broken lines contains all the $C^\alpha H$–NCH_3 cross peaks and is reproduced in Figure 3. The chemical shift positions of the C^α protons, the amide protons, and the N-methyl groups are indicated on the left and at the top of the spectra.

NOE cross peaks with the ^{15}N-bound CsA amide protons are observed (Figure 2A). For the complex with ^{13}C-labeled CsA, a $^{13}C(\omega_1,\omega_2)$-double-half-filter was used to obtain the doubly selected subspectrum of Figure 2B. This subspectrum contains all the intramolecular NOE's between the ^{13}C-bound protons of CsA. Double half-filters were described in detail elsewhere (Otting & Wüthrich, 1989, 1990), and the complete set of four subspectra obtained with the [^{13}C]CsA–CYP complex was recently presented (Wider et al., 1990).

In homonuclear 2D ^{1}H NMR spectra recorded with a $^{15}N(\omega_2)$ half-filter or a $^{13}C(\omega_1,\omega_2)$-double-half-filter, the number of resonances for amide, C^α, and N-methyl protons expected for CsA could be identified and distinguished from the side-chain resonances (in the NOESY spectra of Figure 2, the spectral regions containing cross peaks between these protons are identified with broken lines). In a first step, the following spin systems were recognized in $^{13}C(\omega_1,\omega_2)$-doubly-selected COSY or TOCSY experiments with complexes containing fully ^{13}C-labeled CsA: The $C^\alpha H$–$C^\beta H_3$ fragments of the two alanines; the cross peak between the geminal C^α protons of Sar; complete side-chain spin systems for Abu, the two valines, and one leucine; an AMX_3 fragment; two isopropyl groups; and a presumed $C^\alpha H$–$C^\beta H$–$C^{\beta'}H$ fragment. The amide protons were connected to their respective side-chain spin systems by TOCSY-relayed [^{15}N,^{1}H]-COSY, which provided a distinction between the Val 5 and MeVal 11 spin systems.

In a second step, we established the cyclic sequential assignment pathway via $d_{\alpha N}$ and $d_{\alpha NQ}$ NOE connectivities. The sequential order of free and methylated amide groups (Figure 1) and the cyclic nature of the resulting assignment path (Figure 3) provided important internal checks on the result obtained, which further had to yield the correct placement for the seven side-chain spin systems that had already been identified; the only identified Leu spin system was found to belong to MeLeu 4. Since the N-methyl groups had to be connected to their corresponding side chains by NOE connectivities, it was crucial to have these multiple checks on the sequential assignment. The sequential assignments obtained with $d_{\alpha N}$ and $d_{\alpha NQ}$ were further confirmed by sequential NOE connectivities $d_{\beta N}$, $d_{\beta NQ}$, d_{NN}, d_{NNQ}, d_{NQN}, and d_{NQNQ} (Figure 4).

In a third step, the side-chain assignments were completed by further analysis of [^{13}C,^{1}H]-COSY spectra and NOESY spectra of a complex containing fully ^{13}C-labeled CsA bound to unlabeled CYP. In [^{13}C,^{1}H]-COSY two protons that were known from the sequential assignments to belong to MeBmt 1 could be assigned as $C^\alpha H$ and $C^\beta H$ from the characteristic ^{13}C chemical shift of 58.8 ppm for C^α and 74.7 ppm for C^β, and the same holds for the two allyl proton–carbon correlation peaks of MeBmt 1. In the [^{13}C,^{1}H]-COSY spectrum of biosynthetically directed fractionally ^{13}C-labeled CsA (Senn et al., 1989; Neri et al., 1989) bound to unlabeled CYP, the

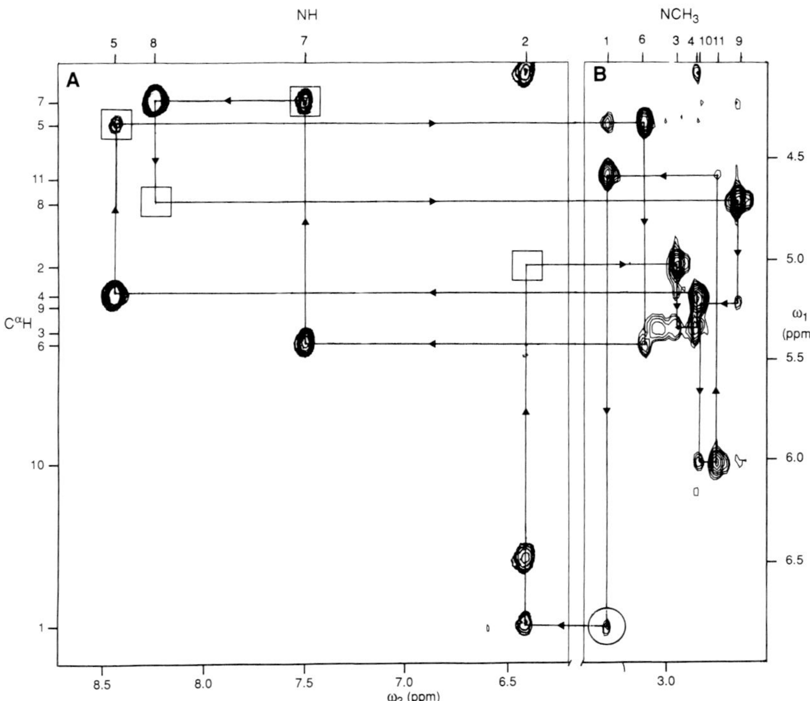

FIGURE 3: Illustration of the sequential assignment procedure via $d_{\alpha N}$ and $d_{\alpha NQ}$ connectivities in CsA bound to CYP. The following parts of the spectra in Figure 2 are used: (A) The region containing cross peaks between amide protons and C^α protons in the $^{15}N(\omega_2)$-selected NOESY subspectrum of Figure 2A. (B) The region containing cross peaks between NCH_3 and C^α protons in the ^{13}C (ω_1,ω_2)-doubly-selected NOESY subspectrum of Figure 2B. The cyclic assignment pathway is indicated with straight lines and arrows, leading from the circled intraresidual NCH_3–$C^\alpha H$ cross peak of MeBmt 1 clockwise around the structure of Figure 1 and back to the point of departure. The chemical shifts of the individual amide protons and N-methyl groups are indicated at the top and identified by the sequence positions (Figure 1). On the left, the same information is given for the C^α protons. The locations of the four intraresidual amide proton–C^α proton COSY cross peaks in the spectrum A are indicated with square frames, which contain in two instances the corresponding intraresidual amide proton–$C^\alpha H$ NOE cross peak.

FIGURE 4: Survey of the sequential assignments of CsA bound to CYP. Below the amino acid sequence, thin, medium, and thick bars indicate sequential NOE connectivities represented by weak, medium, and strong NOESY cross peaks, respectively. Note that four different combinations of sequential NOE's between NH and NCH_3 may be present in CsA (Figure 1).

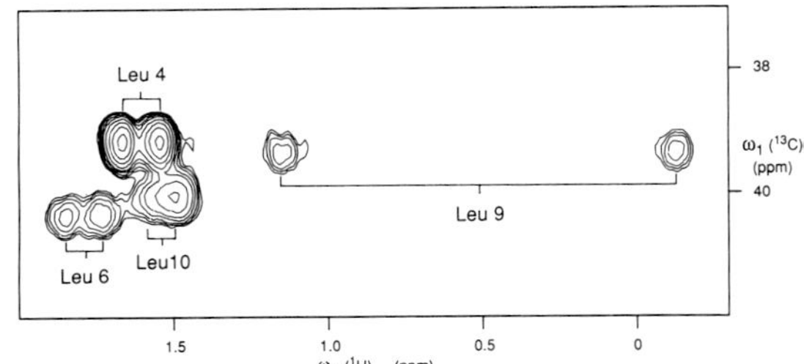

FIGURE 5: Identification of the leucine β-methylene groups in CsA bound to CYP in a [^{13}C,^{1}H]-COSY spectrum of fully ^{13}C-labeled CsA bound to unlabeled CYP. The resonance positions of the individual β-protons and β-carbons in MeLeu 4, MeLeu 6, MeLeu 9, and MeLeu 10 are indicated in the figure (0.8 mM CsA–CYP complex in 550 μL of ^2H_2O, 10 mM potassium deuteroacetate, 10 mM potassium phosphate, p^{2}H 6.0, 25 °C, ^{1}H frequency 500 MHz).

four remaining unassigned methyl groups were distinguished by their carbon–carbon coupling patterns along ω_1 (Figure 6): $C^\eta H_3$ of MeBmt 1 has a distinctly larger one-bond carbon–carbon coupling ($^1J_{CC} \approx 42$ Hz) than all other methyl groups because it is bound to an allylic carbon, and $C^\theta H_3$ of MeBmt 1 is a singlet because it is incorporated into CsA as a one-carbon fragment (Kobel & Traber, 1982). The $C^\beta H$ chemical shift of MeBmt 1 obtained from the sequential assignments was found to coincide with the A spin of the AMX_3 fragment found in COSY, which was thus assigned as $C^\beta H$–$C^\gamma H$–$C^\delta H_3$ in MeBmt. In the $^{13}C(\omega_1,\omega_2)$-doubly-selected NOESY spectrum of a complex containing fully ^{13}C-labeled CsA (Figure 2B), there were two NOE's from a resonance line at 2.31 ppm to both $C^\gamma H$ and $C^\delta H_3$ of MeBmt. This line was therefore assigned to one or both of the δ-protons of MeBmt 1. Of the two allyl proton lines, the one at 6.18 ppm displayed a NOE to the η-methyl resonance of MeBmt 1 at 1.62 ppm and was therefore assigned to the ζ-proton. The C^β-H correlation peaks of the four MeLeu residues were well separated in a distinct spectral region of the [^{13}C,^{1}H]-COSY spectrum and could be assigned from their unique chemical shifts (Figure 5) (Wüthrich, 1976, 1986). MeLeu 6 was found to contain the $C^\alpha H$–$C^\beta H$–$C^{\beta'} H$ fragment and one of the two isopropyl groups that were previously identified with COSY and TOCSY. The two fragments were combined on the basis of NOE's from

Table I: [1]H Chemical Shifts, δ (ppm), for CsA Bound to CYP at pH 6.0 and 25 °C

residue	chemical shift, δ (ppm)[a] NH/NCH_3	αH	βH	others
Meβmt 1	$\underline{3.28}$	$\overline{6.82}$	$\overline{4.05}$	γH 1.49; δCH_2 *2.31*; δCH_3 $\overline{1.14}$; εH 5.54; ζH $\overline{6.18}$; ηCH_3 1.62
Abu 2	$\underline{6.42}$	5.01	1.67, 1.56	γCH_3 0.68
Sar 3	$\underline{2.93}$	$\underline{3.02}$, $\overline{5.32}$[b]		
MeLeu 4	$\underline{2.84}$	5.15	$\underline{1.68}$, 1.56	γH 1.38; $\delta^1 CH_3$ *0.93*; $\delta^2 CH_3$ 0.88
Val 5	$\overline{8.44}$	$\underline{4.30}$	$\underline{2.02}$	$\gamma^1 CH_3$ *0.95*; $\gamma^2 CH_3$ *0.94*
MeLeu 6	3.10	$\overline{5.42}$	$\underline{\mathit{1.85}}$, $\overline{\mathit{1.74}}$	γH 1.58; $\delta^1 CH_3$ *0.90*; $\delta^2 CH_3$ *0.78*
Ala 7	$\underline{7.49}$	$\underline{4.20}$	1.38	
D-Ala 8	$\overline{8.24}$	4.69	1.16	
MeLeu 9	$\underline{2.63}$	$\underline{5.19}$	$\underline{0.88}$, $\underline{-0.12}$	γH $\underline{0.79}$; $\delta^1 CH_3$ $\underline{\mathit{0.32}}$; $\delta^2 CH_3$ $\underline{\mathit{0.54}}$
MeLeu 10	2.83	$\overline{5.99}$	$\underline{1.60}$, $\overline{1.57}$	γH 1.48; $\delta^1 CH_3$ *1.11*; $\delta^2 CH_2$ *1.12*
MeVal 11	2.74	$\underline{4.56}$	$\underline{1.43}$	$\gamma^1 CH_3$ $\underline{\mathit{0.46}}$; $\gamma^2 CH_3$ $\underline{\mathit{-0.69}}$

[a] Measured relative to internal TSP [3-(trimethylsilyl)propanesulfonic acid]. For methylene groups two chemical shifts are given only when two resolved signals were observed. Stereospecific assignments are indicated in italics. For the isopropyl methyl groups of Val and Leu these were obtained by biosynthetically directed fractional ^{13}C labeling, and for $C^\beta H_2$ of MeLeu 6 they were obtained by HABAS ($H^{\beta 2}$ is listed first). Chemical shifts which are more than 0.2 ppm different relative to CsA in chloroform solution (Kessler et al., 1985) are indicated as follows: underlined, shifted to higher field; overlined, shifted to lower field. [b] Probable individual assignments of the C^α protons of Sar 3 were obtained by analysis of structures calculated with either assignment of the two relevant NOE's (Table SI): The sequential NOE from $C^{\alpha 2}H$ of Sar 3 to NCH_3 of MeLeu 4 could only be satisfied with an assignment to $C^{\alpha 2}H$. $C^{\alpha 1}H$ is then coplanar with the carbonyl group of Sar 3, which could explain its high-field shift. Structure calculations using this individual assignment in the place of the pseudoatom did not noticeably improve the final structures, and it was therefore not included in the final input data set.

$C^\alpha H$ to $C^\delta H_3$ and $C^{\delta'}H_3$ and from NCH_3 to the two methyl groups. MeLeu 9 was found on the basis of NOE's from $C^\alpha H$ to $C^\beta H$, $C^{\beta'}H$, $C^\gamma H$, $C^\delta H_3$, and $C^{\delta'}H_3$ and from NCH_3 to $C^{\beta'}H$, $C^\gamma H$, $C^\delta H_3$, and $C^{\delta'}H_3$ to contain the other isopropyl group identified by COSY. The MeLeu 10 side-chain spin system was assigned from NOE's between $C^\alpha H$ and the side-chain protons, by using the two yet unassigned methyl resonances. In summary, all ^{1}H resonances of CsA bound to CYP could be assigned, with the possible exception of $C^\delta H$ of MeBmt 1.

Stereospecific assignments for all six pairs of diastereotopic isopropyl methyls were obtained by the method of biosynthetically directed fractional ^{13}C labeling (Senn et al., 1989), Neri et al., 1989). Although in two of the six isopropyl groups the methyl resonances had nearly degenerate ^{1}H chemical shifts, and additional methyl proton lines were strongly overlapped near 0.93 ppm, all the methyl resonances were well separated in the ^{13}C dimension (Figure 6). The assigment of the *pro-R* methyl groups, γ^1 and δ^1, respectively, was straightforward since they were represented by well-resolved doublets in the ^{13}C dimension (Neri et al., 1989). For each *pro-S* methyl group, we observed a singlet superimposed by a much weaker isotope-shifted doublet, which coalesced into a cross peak with barely resolved fine structure.

The ^{1}H chemical shifts of CsA bound to CYP are compiled in Table I. Also indicated are those protons with shifts of more than 0.2 ppm relative to CsA in chloroform solution (Kessler et al., 1985). Large chemical shift differences were found for the backbone and C^β protons of nearly all residues, but only for the side-chain protons of MeBmt 1, MeLeu 9, and MeVal 11.

Preparation of the Input for Structure Calculations. The general strategy described in Table 10.1 of Wüthrich (1986) was followed to relate NOE intensities with upper distance bounds. Since NOE distance constraints were collected from NOESY spectra recorded with different heteronuclear filters, special care was taken to rely on *empirical calibration* of the relations between NOE intensities and corresponding [1]H–[1]H distance constraints. To obtain additional reference points, the dihedral angle dependence was calculated for the distances $d_{NQ\alpha}(i,i)$ and $d_{\alpha NQ}(i,i+1)$, where Q stands for a pseudoatom located centrally with respect to the three protons of the

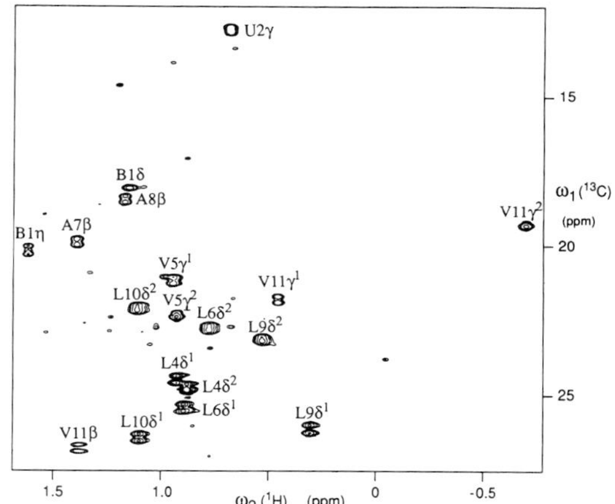

FIGURE 6: Individual assignments of the isopropyl methyl groups of valine and leucine in CsA bound to CYP. The region containing all the CsA methyl resonances is shown for a [^{13}C,[1]H]-COSY spectrum recorded with a complex of biosynthetically directed fractionally ^{13}C-labeled CsA bound to unlabeled CYP. The cross peaks are labeled with the one-letter symbol of the amino acid symbols: B, (4*R*)-4-[(*E*)-2-butenyl]-4,*N*-dimethyl-L-threonine; U, α-aminobutyric acid; V stands for valine and N-methylvaline, A for L- and D-alanine, and L for *N*-methylleucine; a greek letter indicates the carbon atom position and a superscript the branch number of the carbon atom (0.8 mM complex in 550 μL of ^2H_2O, 10 mM potassium deuteroacetate, 10 mM potassium phosphate, p[2]H 6.0, 25 °C, ^{1}H frequency 500 MHz, 20 Hz digital resolution recorded in t_1, zero filled to a 6 Hz final digital resolution in ω_1).

N-methyl group [for the notation used, see p 117 in Wüthrich (1986)]. Standard ECEPP geometry was used for the amino acid residues [for details of the calculation, see Billeter et al. (1982)]. Figure 7 shows that for $d_{NQ\alpha}(i,i)$ both the lower and upper limits are significantly longer than for $d_{N\alpha}$ (Wüthrich, 1986), whereas for $d_{\alpha NQ}(i,i+1)$ the lower limit is nearly the same as for $d_{\alpha N}(i,i+1)$ (Billeter et al., 1982) and only the upper limit is approximately 0.5 Å longer.

Details of the analysis of the [1]H–[1]H NOEs are given in the supplementary materials, which also include Table SI with a complete list of all NOE upper distance constraints. Figure

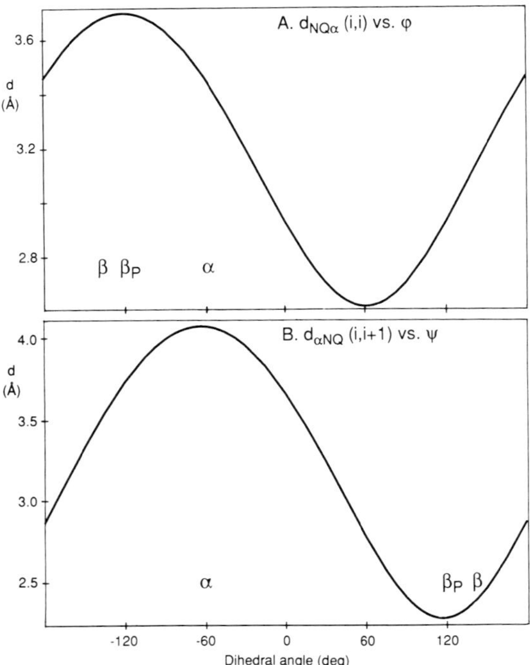

FIGURE 7: Dependence of intraresidual and sequential distances on the intervening dihedral angle in N-methylated polypeptides. (A) Intraresidual distance between the pseudoatom NQ in the center of the three N-methyl protons and the C^α proton, $d_{NQ\alpha}(i,i)$, versus ϕ_i. (B) Sequential distance $d_{\alpha NQ}(i,i+1)$ versus ψ_i. The values of ϕ and ψ that correspond to a regular α-helix or a parallel or antiparallel β-sheet are indicated by the letters α, βp, and β, respectively.

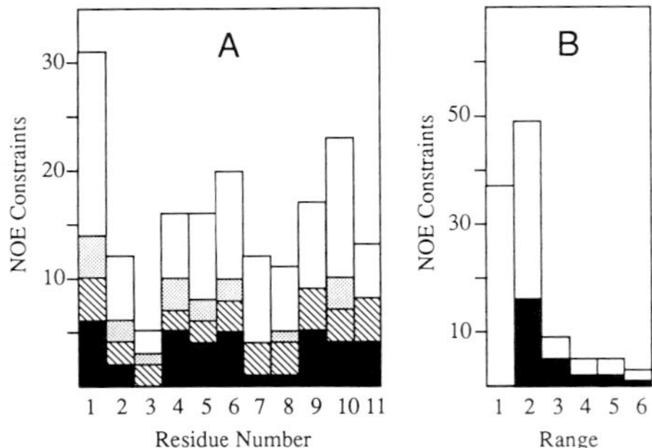

FIGURE 8: (A) Plot of the number of NOE distance constraints per residue versus the amino acid sequence of CsA bound to CYP. All interresidual constraints appear twice, once for each of the two interacting residues. Four types of NOE constraints are specified as follows (Wüthrich, 1986): black, intraresidual; hatched, sequential backbone; dotted, medium- and long-range backbone; white, interresidual with side chain protons. (B) Plot of the number of NOE distance constraints versus their range along the amino acid sequence of CsA bound to CYP. Two types of NOE constraints are specified as follows: black, backbone–backbone constraints; white, backbone–side chain or side chain–side chain constraints. Because of the cyclic structure of CsA, the range cannot exceed 6.

8 shows plots of the number of NOE constraints per residue vs the sequence and of the number of NOE's vs their range along the sequence. Note that because of the cyclic structure, the maximum range is 6.

Supplementary constraints were obtained from the $^3J_{HN\alpha}$ coupling constants measured with a 1D version of J-modulated [^{15}N,^{1}H]-COSY (Neri et al., 1990). The resulting allowed ranges of $^3J_{HN\alpha}$ were Abu 2, >8 Hz; Val 5, 4.1–5.8 Hz; Ala 7, 4.0–5.9 Hz; D-Ala 8, 5.1–6.5 Hz. The combined analysis of NOE's and spin–spin coupling constants by the program HABAS (Güntert et al., 1989) yielded additional dihedral angle constaints for MeLeu 6 of ψ = 170° to −140° and χ^1 = −60° to −10° and a constraint for Val 5 of χ^1 = 40° to −40°.

With regard to the distance geometry calculations, the following qualitative observations made during the resonance assignments and the collection of distance constraints are of importance: All sequential NOE's $d_{\alpha N}$ or $d_{\alpha NQ}$ were strong or medium-strong (Figures 3 and 4), and no sequential NOE's between C^α protons were detected (Figure 2B). Furthermore, to assess whether twisted amide bonds would be compatible with the observed NOE's, the sequential distance $d_{\alpha NQ}$ (Figure 7) was calculated as a function of both intervening angles ψ and ω. It was found that $d_{\alpha NQ}$ distances shorter than 3.0 Å are only compatible with ω angles from 130° to −130°. From this it was concluded that there are no cis amide bonds in CsA bound to CYP, and the structure was calculated assuming that ω = 180° for all peptide bonds.

Calculation of the CsA Structure from the NMR Data. For the distance geometry calculations with the program DISMAN, the input consisted of the NMR constraints discussed in the preceding section and a set of constraints enforcing the cyclic structure as described under Materials and Methods. The influence of the artificial cyclization on the resulting structures was assessed by comparison of calculations starting from randomly chosen starting conformations of the linear chain with MeBmt 1 at the N-terminus and MeVal 11 at the C-terminus or with Ala 7 and MeLeu 6 as the terminal residues. It was found that the location chosen for the ends of the linear chain and the use of the extrinsic cyclization constraints had at most a minimal effect on the calculated structures. We therefore arbitrarily decided to place the cyclization between residues 6 and 7.

For the final structure calculation, 3500 randomly chosen starting conformations of the open chain of CsA with Ala 7 at the N-terminus and MeLeu 6 at the C-terminus were used. The DISMAN calculations were taken through all 11 target levels. At levels one and two, 200 optimization cycles were calculated, at levels three to ten, 100 cycles, and at level 11, 500 cycles to achieve convergence. The relative weight of NOE constraints, van der Waals constraints, and dihedral angle constraints was 1, 0.2, and 10, respectively. At the completion of the calculation at level 11 the van der Waals weighting was increased to 0.5 and another 300 optimization cycles were added. Final structures with a target function value smaller than 1.5 Å^2 were retained. The 181 structures selected by using this criterion were subjected to restrained energy minimization with the program FANTOM (Schaumann et al., 1990), using the same constraint set as was used for the distance geometry. From among the energy-refined structures, 123 species were selected with the following criteria: maximum violation of upper distance limits ≤0.5 Å, of angle constraints ≤5.0°, and of cyclization constraints ≤0.1 Å; ω = 180° ± 20° for the peptide bond enforced by the cyclization; the sum of violations of upper distance limits, the sum of violations of angle constraints, the total physical energy, and the total nonbonding energy all had to be equal to or smaller than the sum of the mean and the standard deviation of the corresponding quantity. Three pairs of structures among these 123 species had identical torsion angles to within 1°. There were therefore only 120 independent solutions to the minimization problem, which are all good solutions with respect to residual violations of the input data (Table II). Since CsA is very

Table II: Analysis of the 120 Energy-Minimized Structures Used To Represent the Solution Conformation of CsA Bound to CYP

quantity	average value ± standard deviation (range)
conformational energy[a]	
total (kcal/mol)	126.0 ± 14.1 (99.6–158.7)
electrostatic (kcal/mol)	65.9 ± 1.1 (62.7–68.3)
H-bond (kcal/mol)	−1.5 ± 1.2 (−3.8 to −0.3)
torsional (kcal/mol)	19.1 ± 3.5 (12.6–27.8)
Lennard-Jones (kcal/mol)	42.6 ± 12.4 (20.9–71.9)
residual NOE distance constraint violations[b]	
number >0.2 Å	5.32 ± 1.48 (2–11)
sum (Å)	2.64 ± 0.49 (1.53–3.69)
maximum (Å)	0.39 ± 0.05 (0.25–0.47)
residual dihedral angle constraint violations[b]	
number >0°	3.22 ± 0.93 (1–6)
sum (deg)	0.83 ± 0.44 (0.10–1.96)
maximum (deg)	0.49 ± 0.28 (0.09–1.39)
residual cyclization constraint violations[b]	
number >0.1 Å	0.0 ± 0.0 (0–0)
sum (Å)	0.14 ± 0.02 (0.10–0.18)
maximum (Å)	0.02 ± 0.00 (0.02–0.03)
peptide angle (deg)	176.4 ± 6.1 (160.2–189.5)
average of the global pairwise RMSDs (Å)	
backbone atoms N, C^α, C′	0.58 ± 0.19 (0.00–1.24)
all heavy atoms	1.19 ± 0.25 (0.01–2.00)
average of the local RMSDs for all tripeptide segments (Å)	
backbone atoms N, C^α, C′	0.21 ± 0.04 (0.13–0.28)

[a] The balance between the conformational energy and the restraint energy is determined by the following selection of the parameters in eq 4 in Schaumann et al. (1990): the exponent was set to $n = 4$, and the distance violation that corresponds to $kT/2$ ($T = 298$ K) was set to 0.2 Å for NOE upper distance limits and 0.03 Å for the cyclization constraints. In the potential accounting for the dihedral angle restraints (Braun, 1987), a violation of 1° corresponds to an energy of $100kT/2$. [b] The input data set consisted of 108 NOE distance constraints, 22 dihedral angle constraints, and 9 cyclization constraints.

hydrophobic but does not form a buried core, there are no sizeable negative energy terms. The total conformational energy is therefore positive (Table II), which is in contrast with typical globular proteins. The 120 conformers represent a well-defined conformation of the polypeptide backbone, as can be seen from the small global and local RMSD values for the backbone atoms (Table II) and in the drawings of Figure 9A. The mean values of the dihedral angles and their standard deviations, which define the solution conformation of the polypeptide backbone of CsA bound to CYP, are listed in Table III.

As a final qualitative check of the compatibility of the structures obtained with the experimental input, we searched the structures for short distances, i.e., <4.0 ± 1.5 Å, that would not have been observed as NOE's. We found no evidence that such short ^{1}H–^{1}H distances would have escaped detection.

Discussion

The Solution Conformation of CsA Bound to CYP. The structure of bound CsA contains no regular secondary structure and no intramolecular hydrogen bonds. The two antiparallel backbone segments from residues 4 to 7 and 9 to 2 have their planar peptide bonds rotated out of the plane defined by the cyclic polypeptide backbone, which contrasts with regular β-strands. In the strand from MeLeu 9 to Abu 2, the residue MeVal 11 bulges out. Five of the seven NCH_3 moieties are buried to some degree inside the cyclosporin ring, while the four amide protons are exposed to the outside. This coincides with the observation that the amide proton exchange is fast for Val 5, Ala 7 and D–Ala 8. For Abu 2 the half-life for NH exchange is approximately 3 days at pH 6.0 and 25 °C. As there is no obvious acceptor group in CsA, we attribute the slow exchange to an intermolecular hydrogen bond of the amide proton of Abu 2 with CYP (see below). Of the 11

Table III: Backbone Dihedral Angles in the Solution Conformation of CsA Bound to CYP

amino acid	dihedral angle ± standard deviation (deg)[a] ϕ	ψ	ω
MeBmt 1	−125 ± 14	−167 ± 15	−176 ± 3
Abu 2	−131 ± 13	99 ± 29	180 ± 4
Sar 3	138 ± 24	−41 ± 14	−179 ± 4
MeLeu 4	−150 ± 22	92 ± 17	180 ± 2
Val 5	−75 ± 3	136 ± 7	177 ± 2
MeLeu 6	−114 ± 6	−159 ± 14	176 ± 6
Ala 7	−80 ± 1	172 ± 9	178 ± 1
D-Ala 8	83 ± 4	−156 ± 6	−177 ± 4
MeLeu 9	−126 ± 7	86 ± 5	180 ± 2
MeLeu 10	−117 ± 17	153 ± 18	180 ± 3
MeVal 11	−131 ± 11	80 ± 15	177 ± 5

[a] The numbers are the mean values and standard deviations of the dihedral angles in the 120 energy-minimized structures used to represent the solution conformation of CsA bound to CYP.

carbonyl oxygen atoms, all but one are to some extent exposed, the exception being MeLeu 4, for which the C=O group is completely buried.

In the view of Figure 9B, the side chains of MeBmt 1, MeLeu 4, MeLeu 6, and MeLeu 10 are in front of the plane occupied by the polypeptide backbone, and the remaining four bulky side chains of Abu 2, Val 5, MeLeu 9, and MeVal 11 are behind the backbone plane and point away from the reader. Except for Val 5 and MeLeu 6, for which the χ^1 angles are well defined at about 177° and −54°, respectively, the other long side chains were mainly constrained by long-range NOE's with peripheral groups of protons. Although the spatial orientation is thus quite well defined (Figure 9B), the individual dihedral angles χ^n vary within rather wide ranges. Most prominent is the side chain of MeBmt 1, which is folded back over the molecule and makes contacts with the isopropyl groups of MeLeu 4 and MeLeu 6 across the ring. It is also apparent that the side chains 1, 4, 6, and 10 form a quite compact cluster over the cyclic backbone. In contrast, the side chains 2, 9, and 11 located behind the backbone plane point away from the core of the molecule.

Comparison of the Structure of CsA Bound to CYP with Free CsA. Overall the NMR structure of CsA bound to CYP represents a new cyclosporin conformation, which to the best of our knowledge has not been observed so far. It contains features that have been predicted from other data, for example that the side chain of MeBmt 1 is folded back onto the molecule (Quesniaux et al., 1988). When compared to the crystal structure of free CsA, there are important differences and only few similarities. The global RMSD between the backbones of the two structures is 2.5 Å, and numerous NOE distance constraints measured for CsA bound to CYP are sizeably violated in the crystal structure of CsA. In the crystal structure of CsA there is a cis amide bond between residues 9 and 10, whereas in the CsA bound to CYP all amide bonds are trans. The crystal structure of CsA contains three transannular hydrogen bonds and one additional hydrogen bond from NH of D-Ala 8 to C′=O of MeLeu 6, but there are no intramolecular hydrogen bonds in the conformation of CsA bound to CYP. Instead, the latter contains 5 N-methyl groups within the ring formed by the polypeptide backbone. In the crystal structure, all but one of the NCH_3 groups are oriented toward the molecular surface, whereas in the bound CsA only two NCH_3 groups are exposed, one of which is the NCH_3 group of MeVal 11, which is buried in the crystal. In both structures the same groups of bulky side chains are clustered together on the two sides of the backbone plane.

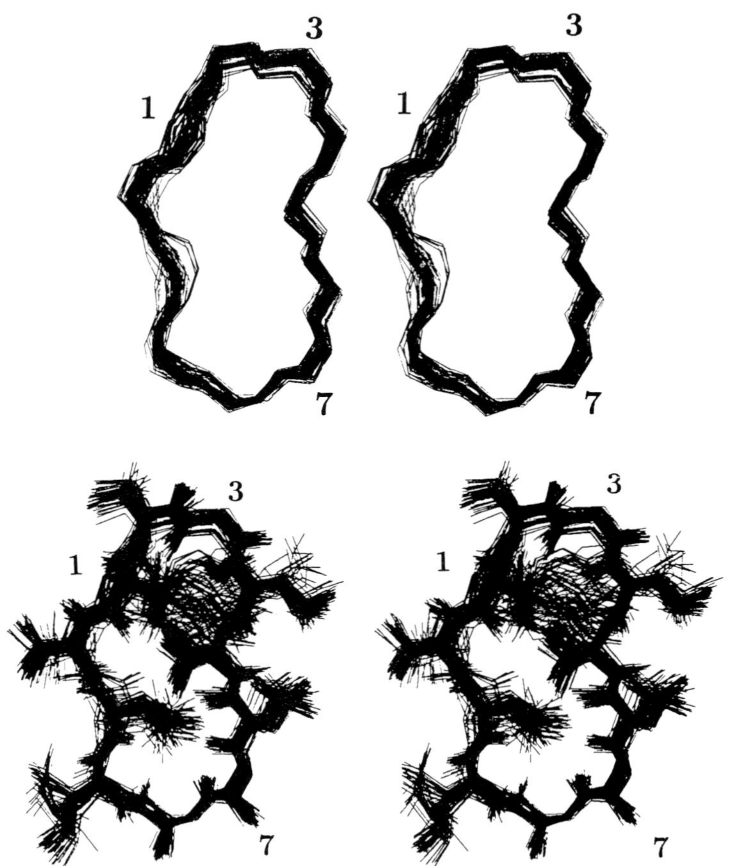

FIGURE 9: NMR structure of CsA bound to CYP in solution. Stereoviews are shown of a superposition of the 120 best energy-minimized structures characterized in Table II. (A) The bonds connecting the backbone atoms N, C^{α}, and C′ are drawn. (B) The bonds connecting all heavy atoms are shown, with the same orientation as in A.

However, in the views of Figure 10 the side chains of MeBmt 1, MeLeu 4, MeLeu 6, and MeLeu 10 are located behind the ring in the crystal structure but above the ring in CsA bound to CYP.

A recently reported NMR structure of CsA in chloroform solution (Kessler et al., 1990) is closely similar to the crystal structure. In particular, the four hydrogen bonds identified in the crystal structure are preserved, and the side chain of MeBmt 1 is also folded back onto the molecule (Figure 10). Overall, a detailed comparison between the NMR structures of CsA in chloroform solution and bound to CYP leads to similar conclusions as the comparison of the bound CsA with the crystal structure, i.e., there are important conformational differences of the type detailed in the preceding paragraph.

Influence of the Intermolecular Contacts between CsA and CYP. In addition to the collection of intramolecular NOE's in CsA, which is the principal subject of this paper, numerous intermolecular NOE's between protons of CsA and of CYP were observed in the $^{13}C(\omega_1)$-selected $^{13}C(\omega_2)$-filtered and $^{13}C(\omega_1)$-filtered $^{13}C(\omega_2)$-selected subspectra from NOESY recorded with a $^{13}C(\omega_1,\omega_2)$-double-half-filter (Wider et al., 1990). Overall, these data show that there are close contacts of CYP with the residues MeBmt 1, Abu 2, Sar 3, MeLeu 9, MeLeu 10, and MeVal 11 of CsA. This is schematically visualized in Figure 11. From Table I it is seen that many protons of these residues (and none of the other side chains) have large chemical shift differences relative to free CsA. Very probably, because of the intermolecular steric constraints that can be expected to create a similar environment to that experienced by interior residues in globular proteins (Wüthrich, 1986), the side chains of residues 9–3, which interact with CYP, are among the most strongly confined parts of the CsA molecule. Nonetheless, these side chains are relatively poorly defined in the present CsA structure (Figure 9B, Table III). This apparent discrepancy must be due largely or entirely to the fact that in the present structure determination neither intermolecular NOE constraints nor intermolecular steric constraints were included in the input for the structure calculations.

The chemical shift differences between free CsA and CsA bound to CYP can be qualitatively rationalized from the present preliminary information on the intermolecular interactions. Many of the intermolecular NOE's are with aromatic rings of CYP. In particular, Trp 121 in CYP interacts with CsA protons. This explains that the shift differences cannot have a clear trend toward high or low field, because both the sign and the magnitude of the ring current shifts depend on the relative orientation of the aromatic rings and the CsA protons (Wüthrich, 1976). We even found CsA protons involved in strong NOE's with aromatic protons of CYP that were hardly shifted relative to CsA in $CDCl_3$ solution, e.g., the NCH_3 group of MeVal 11 (Table I). The intermolecular

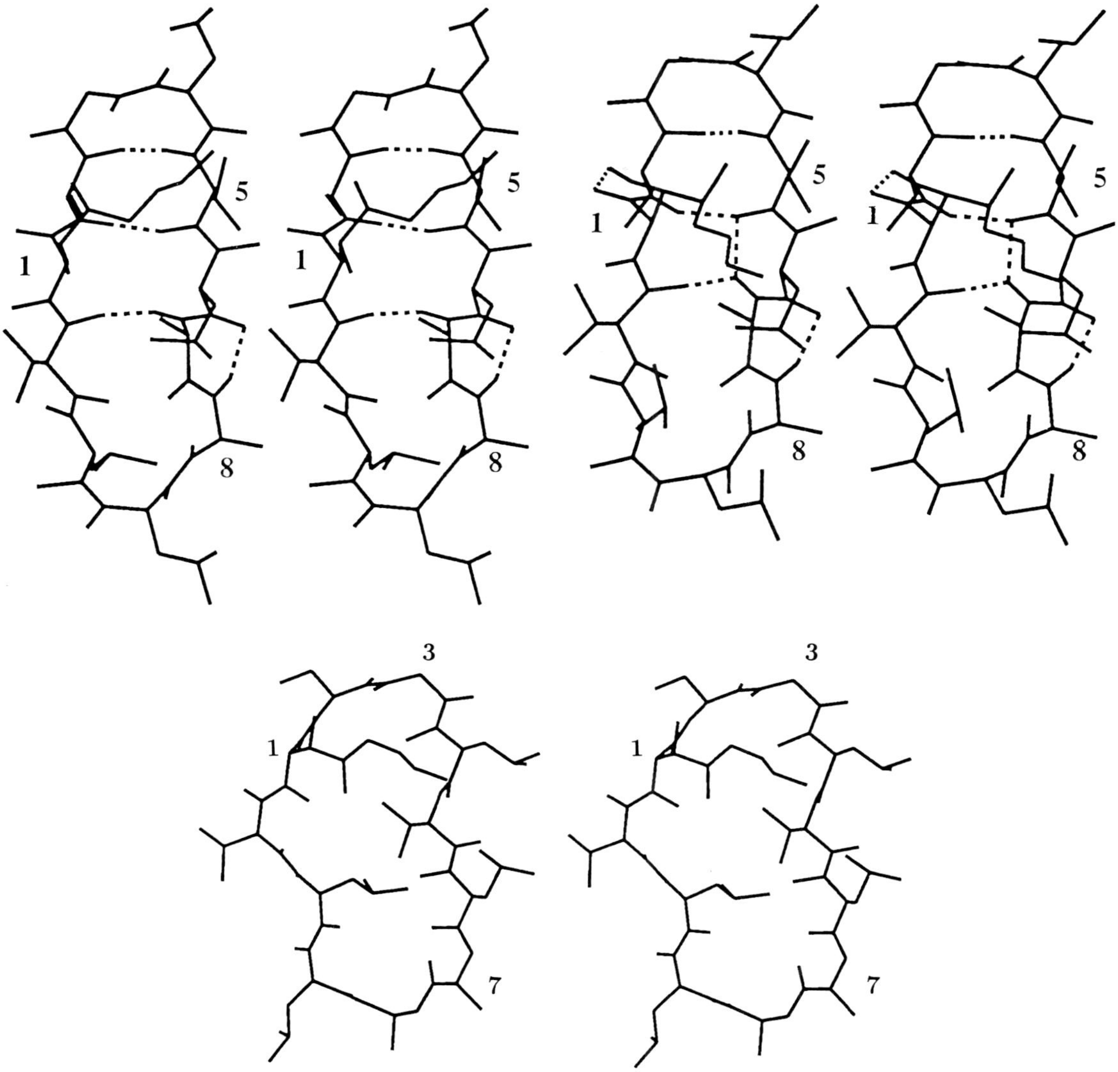

FIGURE 10: Structure of CsA in different environments. Stereoviews are shown of the bonds connecting all heavy atoms. For panels A and B the amide protons of Abu 2, Val 5, Ala 7, and D-Ala 8 are shown, and the hydrogen bonds are indicated by broken lines. (A) X-ray structure of CsA in single crystals (Loosli et al., 1985). (B) NMR structure of CsA in chloroform solution (Kessler et al., 1990). (C) NMR structure of CsA bound to CYP in aqueous solution.

NOE's with aromatic rings of CYP can also explain the observation of an enhancement of the Trp fluorescence upon CsA binding to CYP (Handschumacher et al., 1984). They are further consistent with the hypothesis based on NMR data that the hydrophobic core in CYP contains numerous aromatic rings (Dalgarno et al., 1986), of which many resonance lines shift upon CsA binding (Heald et al., 1990; our unpublished results).

Conclusions

All previous discussions on structure–function correlations in CsA were based on the structure of free CsA determined either by X-ray diffraction in single crystals or by NMR in chloroform solution (Loosli et al., 1985; Kessler et al., 1990). Although the present study accounts only partially for the influence of the receptor protein on the conformation of CsA bound to CYP (see the immediately preceeding section), it clearly demonstrates the importance of direct experimental studies with both the free and the receptor-bound effector molecule as a basis for discussions on its structure–function correlations. Most striking is the observation that in the cyclic structure of CsA, which has a greatly reduced accessible conformation space when compared to a corresponding linear polypeptide, the backbone conformation is largely rearranged in the receptor-bound state. As was explained in detail in the immediately preceding section, a more precise description of the spatial arrangement of the amino acid side chains of CsA that are in direct contact with CYP will have to await a detailed characterization of the entire complex, including CYP and its intermolecular interactions. However, already the results presented in this paper, which were obtained without the use of any direct structural information on the receptor

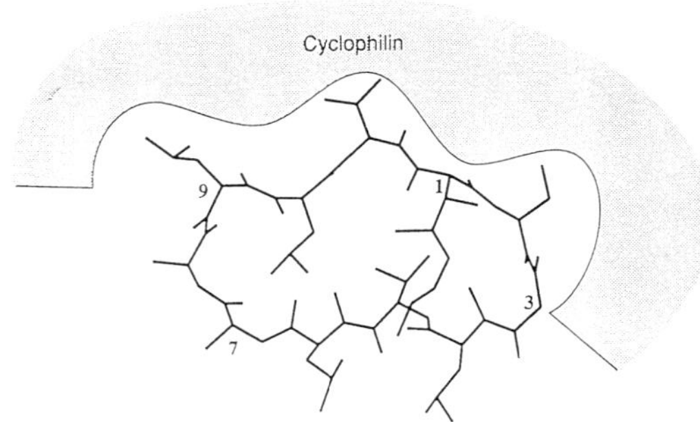

FIGURE 11: Schematic visualization of the intermolecular interactions between CsA and CYP. The best structure of bound CsA from Figure 9 has been docked into a schematic binding site of CYP in such a way that there are contacts with all amino acid residues of CsA for which strong intermolecular NOE's were observed. In the third dimension perpendicular to the planar drawing, the binding site of CYP would have a similar surface and extent as the one depicted here in two dimensions.

protein, should be of interest to those engaged in the design of cyclosporin-related drugs with improved and/or altered activity profiles.

ACKNOWLEDGMENTS

The use of the Cray-XMP/28 of the ETH Zürich is gratefully acknowledged. We thank Prof. H. Kessler for the atom coordinates of the recently published NMR structure of CsA in chloroform solution, and Mr. R. Marani for the careful processing of the manuscript.

SUPPLEMENTARY MATERIAL AVAILABLE

Table SI showing the NOE distance constraints used as input for the calculation of the structure of cyclosporin A bound to cyclophilin (6 pages). Ordering information is given on any current masthead page.

REFERENCES

Billeter, M., Braun, W., & Wüthrich, K. (1982) *J. Mol. Biol. 155*, 321–346.

Billeter, M., Engeli, M., & Wüthrich, K. (1985) *J. Mol. Graphics 3*, 79–83; 97–98.

Bodenhausen, G., & Ruben, D. (1980) *Chem. Phys. Lett. 69*, 185–188.

Borel, J. F., Ed. (1986) *Ciclosporin*, Karger, Basel.

Braun, W. (1987) *Q. Rev. Biophys. 19*, 115–157.

Braun, W., & Go, N. (1985) *J. Mol. Biol. 186*, 611–626.

Dalgarno, D. C., Harding, M. W., Lazarides, A., Handschumacher, R. E., & Armitage, I. M. (1986) *Biochemistry 25*, 6778–6784.

Denk, W., Baumann, R., & Wagner, G. (1986) *J. Magn. Reson. 67*, 386–390.

Eccles, C., Billeter, M., Güntert, P., & Wüthrich, K. (1989) *Abstracts 10th Meeting of the International Society of Magnetic Resonance*, Morzine, France, July 16–21, 1989, p S50.

Elliot, J. F., Lin, Y., Mizel, S. B., Bleakley, R. C., Harnish, D. G., & Paetkau, V. (1984) *Science 226*, 1439–1441.

Emmel, E. A., Verweij, C. L., Durand, D. B., Higgins, K. M., Lacy, E. & Crabtree, G. R. (1989) *Science 246*, 1617–1620.

Fesik, S. W. (1988) *Nature 332*, 865–866.

Fesik, S. W., Luly, J. R., Erickson, J. W., & Abad-Zapatero, C. (1988) *Biochemistry 27*, 8297–8301.

Fesik, S. W., Gampe, R. T., Jr., Eaton, H. L., Gemmacker, G., Olejniczak, E. T., Neri, P., Holzman, T. F., Egan, D. A., Edalji, R., Simmer, R., Helfrich, R., Hochlowski, J., & Jackson, M. (1991) *Biochemistry*, following paper in this issue.

Fischer, G., Wittmann-Liebold, B., Lang, K., Kiefhaber, T., & Schmid, F. X. (1989) *Nature 337*, 476–478.

Griffey, R. H., & Redfield, A. G. (1987) *Q. Rev. Biophys. 19*, 51–82.

Güntert, P., Braun, W., Billeter, M., & Wüthrich, K. (1989) *J. Am. Chem. Soc. 111*, 3997–4004.

Güntert, P., Qian, Y. Q., Otting, G., Müller, M., Gehring, W., & Wüthrich, K. (1991) *J. Mol. Biol. 217*, 531–540.

Haendler, B., Hofer-Warbinek, R., & Hofer, E. (1987) *EMBO J. 6*, 947–950.

Handschumacher, R. E., Harding, M. W., Rice, J., & Drugge, R. J. (1984) *Science 226*, 544–547.

Harrison, R. K., & Stein, R. L. (1990) *Biochemistry 29*, 1684–1689.

Heald, S. L., Harding, M. W., Handschumacher, R. E., & Armitage, I. M. (1990) *Biochemistry 29*, 4466–4478.

Kahan, B. D., Ed. (1988) *Cyclosporine—Nature of the Agent and its Immunologic Actions*, Grune & Stratton, New York.

Kawamukai, M., Matsuda, H., Fuji, W., Utsumi, R., & Komano, T. (1989) *J. Bacteriol. 171*, 4525–4529.

Kessler, H., Loosli, H. R., & Oschkinat, H. (1985) *Helv. Chim. Acta 68*, 661–681.

Kessler, H., Köck, M., Wein, T., & Gehrke, M. (1990) *Helv. Chim. Acta 73*, 1818–1832.

Kobel, H., & Traber, R. (1982) *Eur. J. Appl. Microbiol. Biotechnol. 14*, 237–240.

Koletsky, A. J., Harding, M. W., & Handschumacher, R. E. (1986) *J. Immunol. 137*, 1054–1059.

Krönke, M., Leonard, W. J., Depper, J. M., Arya, S. K., Wong-Staal, F., Gallo, R. C., Waldmann, T. A., & Greene, W. C. (1984) *Proc. Natl. Acad. Sci. U.S.A. 81*, 5214–5218.

Loosli, H. R., Kessler, H., Oschkinat, H., Weber, H. P., Petcher, T. J., & Widmer, A. (1985) *Helv. Chim. Acta 68*, 682–704.

Messerle, B. A., Wider, G., Otting, G., Weber, C., & Wüthrich, K. (1989) *J. Magn. Reson. 85*, 608–613.

Momany, F. A., McGuire, R. F., Burgess, A. W., & Scheraga, H. A. (1975) *J. Phys. Chem. 79*, 2361–2381.

Nemethy, G., Pottle, M. S., & Scheraga, H. A. (1983) *J. Phys. Chem. 87*, 1883–1887.

Neri, D., Szyperski, T., Otting, G., Senn, H., & Wüthrich, K. (1989) *Biochemistry 28*, 7510–7516.

Neri, D., Otting, G., & Wüthrich, K. (1990) *J. Am. Chem. Soc. 112*, 3663–3665.

Otting, G., & Wüthrich, K. (1988) *J. Magn. Reson. 76*, 569–574.

Otting, G., & Wüthrich, K. (1989) *J. Magn. Reson. 85*, 586–594.

Otting, G., & Wüthrich, K. (1990) *Q. Rev. Biophys. 23*, 39–96.

Otting, G., Senn, H., Wagner, G., & Wüthrich, K. (1986) *J. Magn. Reson. 70*, 500–505.

Quesniaux, V. F. J., Wenger, R. M., Schreier, M. H., & Van Regenmortel, M. H. V. (1987a) *Protides Biol. Fluids 35*, 507–510.

Quesniaux, V. F. J., Schreier, M. H., Wenger, R. M. Hiestand, P. C., Harding, M. R., & Van Regenmortel, M. H. V. (1987b) *Eur. J. Immunol. 17*, 1359–1365.

Quesniaux, V. F. J., Wenger, R. M., Schmitter, D., & Van Regenmortel, M. H. V. (1988) *Int. J. Pept. Protein Res. 31*, 173–185.

6574

Sanner, M., Widmer, A., Senn, H., & Braun, W. (1989) *J. Comput. Aided Mol. Des. 3*, 195–210.

Schaumann, T., Braun, W., & Wüthrich, K. (1990) *Biopolymers 29*, 679–694.

Senn, H., Werner, B., Messerle, B. A., Weber, C., Traber, R., & Wüthrich, K. (1989) *FEBS Lett. 249*, 113–118.

Senn, H., Loosli, H. R., Sanner, M., & Braun, W. (1990) *Biopolymer 29*, 1387–1400.

Takahashi, N., Hayano, T., & Suzuki, M. (1989) *Nature 337*, 437–475.

Wider, G., Weber, C., Widmer, H., Traber, H., & Wüthrich, K. (1990) *J. Am. Chem. Soc. 112*, 9015–9017.

Wider, G., Weber, C., & Wüthrich, K. (1991) *J. Am. Chem. Soc.* (in press).

Wüthrich, K. (1976) *NMR in Biological Research: Peptides and Proteins*, North Holland/American Elsevier, New York.

Wüthrich, K. (1986) *NMR of Proteins and Nucleic Acids*, Wiley, New York.

Wüthrich, K. (1989) *Science 243*, 45–50.

Wüthrich, K., Wider, G., Wagner, G., & Braun, W. (1982) *J. Mol. Biol. 155*, 311–319.

Wüthrich, K., Billeter, M., & Braun, W. (1983) *J. Mol. Biol. 169*, 949–961.

LETTERS TO NATURE

Reprinted with permission from *Nature* **353**, 276–279 (1991)
Copyright © 1991 MacMillan Magazines Limited.

Structure of human cyclophilin and its binding site for cyclosporin A determined by X-ray crystallography and NMR spectroscopy

Jörg Kallen, Claus Spitzfaden*, Mauro G. M. Zurini, Gerhard Wider*, Hans Widmer, Kurt Wüthrich* & Malcolm D. Walkinshaw†

Preclinical Research, Sandoz Pharma AG, 4002 Basel, Switzerland
* Institut für Molekularbiologie und Biophysik, ETH-Hönggerberg, 8093 Zürich, Switzerland

THE protein cyclophilin is the major intracellular receptor for the immunosuppressive drug cyclosporin A (ref. 1). Cyclosporin A acts as an inhibitor of T-cell activation and can prevent graft rejection in organ and bone marrow transplantation[2]. Cyclophilin may be responsible for mediating this immunosuppressive response. Cyclophilin also catalyses the interconversion of the *cis* and *trans* isomers of the peptidyl–prolyl amide bonds of peptide and protein substrates[3,4]. Here we report the X-ray crystal structure of human recombinant cyclophilin complexed with a tetrapeptide and the identification, by nuclear magnetic resonance spectroscopy, of the specific binding site for cyclosporin A. Cyclophilin has an eight-stranded antiparallel β-barrel structure. The prolyl isomerase substrate-binding site is coincident with the cyclosporin-binding site. These results may help to provide a structural basis for rationalizing the immunosuppressive function of the cyclosporin–cyclophilin system and will also be important in the design of improved immunosuppressant drugs.

Cyclophilin (relative molecular mass 17,800 (M_r, 17.8K)) consists of a single polypeptide chain with 165 amino-acid residues (Fig. 1*a*). Cyclosporin A (CsA) is a cyclic undecapeptide with the sequence c-(MeBmt-Abu-Sar-MeLeu-Val-MeLeu-Ala-D-Ala-MeLeu-MeLeu-MeVal), where the prefix Me indicates *N*-methylation and the uncommon amino acids in positions 1 and 2 are (4R)-4-((E)-2-butenyl)-4,*N*-dimethyl-L-

TABLE 1 Crystallographic and multiple isomorphous replacement (MIR) data

Data set	Space group $P2_12_12_1$ (a=108.2 Å, b=123.0 Å, c=35.8 Å) Native (1)	Native (2)	EMTS*	PHMPS	K_2PtCl_4 (1)	K_2PtCl_4 (2)
Number of collected reflections	43,082	42,788	43,308	26,385	23,713	23,600
R_{sym}(I) in parentheses†	(6.9%)	(7.6%)	(7.2%)	(8.0%)	(10.2%)	(8.7%)
Number of unique reflections‡	11,254	12,303	9,602	8,354	8,156	7,891
Maximum resolution Å	2.6	2.6	3.0	3.0	3.2	3.2
Mean fraction isomorphous change§			28.4%	27.8%	16.8%	17.5%
Number of heavy atom sites			2	4	6	6
R_c‖			55.8%	50.7%	61.4%	57.9%
Correlation¶			0.47	0.59	0.43	0.49
Number of centric reflections			1,611	1,455	1,415	1,323
Overall phasing power**			1.26	1.47	1.90	1.93

EMTS is ethyl-mercuri-thiosalicylate; PHMPS is *p*-hydroxymercuri-phenyl-sulphonate.
* Data set collected on image-plate area detector at the EMBL outstation, DESY, Hamburg.
† $R_{sym} = \Sigma_i |I_i - \bar{I}| / \Sigma \bar{I}$, where I_i is the intensity of the reflection and $\bar{I}$ is the mean intensity of the i observations.
‡ Number of unique reflections phased: 9,898 (15 Å–2.9 Å); overall figure of merit=0.78.
§ Versus merged data set Native(1)+Native(2).
‖ R_c is Cullis R-factor for centric reflections[19].
¶ Correlation between F_{obs} and F_{calc} for heavy atom structure (program REFINE in CCP4-package[18]).
** Phasing power is $F_H/E_{r.m.s.}$, where F_H is heavy-atom structure factor and $E_{r.m.s.}$ is residual lack of closure[19].

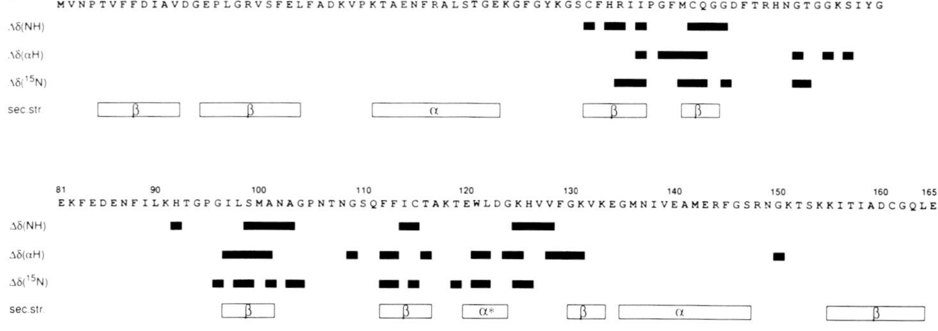

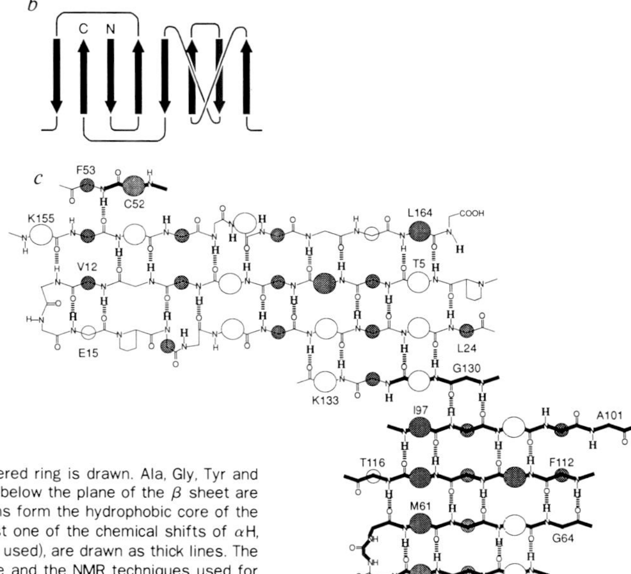

FIG. 1 Summary of the NMR results defining the CsA-binding site and the regular secondary structure and topology of cyclophilin. *a*, Amino-acid sequence of human cyclophilin, sequence locations of the regular α-helical or β-sheet secondary structures (α^* indicates that this structure could either be a short α helix or a turn-like structure) and the chemical shift data used to delineate the binding site for CsA. The residues showing significant backbone chemical shift variations on binding of cyclosporin are identified with black bars. Chemical shift differences are indicated if they exceed the following limits: δ(NH), 0.10 parts per million (p.p.m.); $\delta(\alpha$H), 0.05 p.p.m.; $\delta(^{15}$N), 0.50 p.p.m. *b*, Schematic representation of the antiparallel β sheet in cyclophilin showing a global (+1, −3, −1, −2, +1, −2, −3) topology of the β sheet. *c*, Drawing showing the sequence positions of the individual β strands, the hydrogen bonds, the distribution of the hydrophobic side chains and the location of residues with large chemical shift changes on binding of cyclosporin. The two ends of each β strand are labelled with the one-letter amino-acid code and the sequence location. Slowly exchanging amide protons are printed in bold. The presence of hydrogen bonds is identified with dashed lines. The amino-acid side chains are represented with circles at the Cα carbon positions using the following code: big circles, side chains toward the reader; small circles, side chains pointing away from the reader; shaded circles, hydrophobic residues (C, I, V, L, F, M); empty circles, polar and charged residues (S, T, N, D, Q, E, R, K). For Pro a five-membered ring is drawn. Ala, Gly, Tyr and His have no symbol at the Cα position. Essentially all side chains below the plane of the β sheet are hydrophobic. In the X-ray structure (Figs 2 and 3) these side chains form the hydrophobic core of the β barrel. Residues that experience a significant change of at least one of the chemical shifts of αH, NH or ^{15}N after complexation with cyclosporin (see *a* for the limits used), are drawn as thick lines. The sequence-specific NMR assignments for the polypeptide backbone and the NMR techniques used for the secondary structure determination are described in ref. 7.

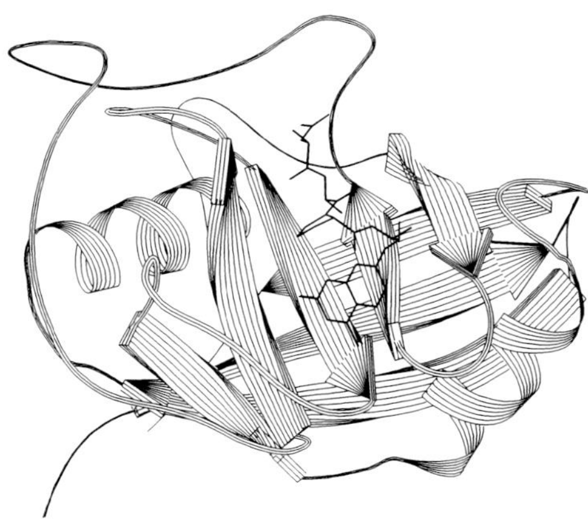

FIG. 2 A ribbon plot[18] illustrating the overall fold of human cyclophilin. The model tetrapeptide prolyl isomerase substrate (*N*-acetyl-Ala-Ala-Pro-Ala-amidomethylcoumarin) is also shown. The crystal structure was solved for a complex of human recombinant cyclophilin with this linear tetrapeptide using phases determined with the multiple isomorphous replacement (MIR) method from two heavy-atom derivatives (Table 1). There are two molecules in the asymmetric unit related to each other by a noncrystallographic twofold rotation axis. The MIR phases provided an electron density map[18,19] good enough for chain tracing[20] for all amino acids of both molecules. No solvent-flattening procedures were used. The secondary structure of cyclophilin as determined by NMR (ref. 7) was of considerable help in the chain tracing and model-building. The two independent molecules differ only slightly in loop regions involved in lattice contacts; the Cα atoms currently have an r.m.s. fit of 0.8 Å. Noncrystallographic symmetry was not imposed during refinement using X-PLOR (ref. 21). The present *R*-factor using data between 2.5 Å and 8 Å is 25.7%. Water molecules have not yet been included and further refinement is underway.

threonine and L-α-aminobutyric acid, respectively. The three-dimensional structure of cyclophilin-bound CsA has recently been determined by nuclear magnetic resonance (NMR) (refs 5, 6). Complete sequence-specific ^{1}H and ^{15}N NMR assignments for the cyclophilin backbone have also been determined and used to define the regular secondary structure[7], which consists of two well defined α helices and an eight-stranded antiparallel β sheet (Fig. 1*b*).

Various crystal forms of human cyclophilin complexed with either a tetrapeptide or CsA have been grown[8]. The crystal complex of cyclophilin with CsA has six molecules in the asymmetric unit and is being studied. The legend for Fig. 2 contains details of the X-ray crystallographic structure determination of a complex of human recombinant cyclophilin with the tetrapeptide *N*-acetyl-Ala-Ala-Pro-Ala-amidomethylcoumarin. This peptide is a model substrate for the prolyl isomerase, which catalyses the *cis–trans* conversion of the alanyl-prolyl amide bond.

Cyclophilin is a roughly spherical molecule with a radius of about 17 Å (Fig. 2). The main structural feature is the eight-stranded antiparallel β barrel that has a +1, −3, −1, −2, +1, −2, −3 topology (Fig. 1*b*). The barrel consists of two roughly perpendicular four-stranded β sheets (Fig. 1*c*) connected by short junctions at residues Leu98–Gly130 and Phe53–Ile156. Inside the barrel, a tightly packed core contains most of the hydrophobic side chains (Fig. 1*c*). Other hydrophobic residues are located in the contact region of the two amphipathic helices with the β barrel and in the cyclosporin-binding site.

There is a structural resemblance between cyclophilin and the superfamily of proteins involved in ligand transport including retinol-binding protein (RBP), bilin-binding protein and β-lactoglobulin[9]. Most of these molecules encapsulate their ligand in the β-barrel core. By contrast, the barrel core in cyclophilin is tightly packed with hydrophobic residues (Fig. 1*c*), and the putative ligand-binding site is on the outside of the barrel. The topology of cyclophilin also differs from the simple $(+1)_n$ up-and-down fold found in the RBP class of proteins, or the (−3, +1, +1) Greek key topology that is most frequently found in antiparallel β-barrel proteins.

LETTERS TO NATURE

In particular the two crossover connections Gly64–Ile97 and Thr116–Gly130 represent an uncommon topological feature. As both loops lie on the outside of the barrel, the [−2, +1, −2] topology requires that the two loops cross each other. The unusual left-handed connection of the 116–130 loop may be rationalized by the length of the loop which may accommodate a variety of local conformations.

Sequence-specific NMR assignments for the ^{1}H and ^{15}N spins of the polypeptide backbone of cyclophilin were also obtained for a complex formed with a water-soluble CsA derivative. The very similar nuclear Overhauser effect (NOE) patterns observed for free and CsA-bound cyclophilin indicate that there is no change in secondary structure on CsA binding. Nonetheless, comparison with free cyclophilin showed that there are residues for which cyclosporin binding caused significant chemical shift changes (Fig. 1*a*). These chemical shift data were used to delineate the CsA-binding site in the three-dimensional cyclophilin structure (Fig. 3).

When the cyclophilin residues thus identified as being involved in cyclosporin binding (Fig. 1*a*) are mapped onto the three-dimensional structure, they are found to cluster on one side of the molecule and incorporate the tetrapeptide-binding site (Figs 1*c*, 2 and 3). In the crystal structure, the tetrapeptide ligand binds in a long deep groove located on the protein surface between one face of the β barrel and the Thr116–Gly130 loop. The best fit in the current model shows the Ala–Pro amide bond in the *trans* conformation. Site-directed mutagenesis studies[10,11] have been used to study enzymatic activity. Both Cys115 and Cys62 are near the peptide-binding groove, however replacing each Cys individually by Ala did not affect prolyl isomerase activity[10]. Site-directed mutagenesis studies have also shown that the sole tryptophan (Trp121), which sits in the middle of the binding loop, is implicated in CsA binding, but has little effect on prolyl isomerase activity[11]. In the crystal structure, this tryptophan is close to the coumarin ring of the tetrapeptide, and intermolecular NOEs observed in the cyclophilin–CsA complex in solution showed that it is in close contact with residues 9 and 11 of CsA (ref. 5). His126 sits on the same binding loop as Trp121, with N^{δ} about 6 Å from the carbonyl carbon of the prolyl amide. A possible mechanism for prolyl isomerase activity could involve His126 acting as proton acceptor for a water molecule involved in nucleophilic attack on the carbonyl carbon of the Ala–Pro amide bond. Further activation of the amide carbonyl group could be provided by the hydrogen bond between the carbonyl oxygen atom and the guanidinium group of Arg55.

The macrolide FK506 is chemically unrelated to CsA but is also a potent immunosuppressant with a very similar biological profile[12]. FK506 binds to the specific cytosolic immunophilin protein receptor FK-binding protein (FKBP) (ref. 12). Human recombinant FKBP has a chain length of 107 amino acids (M_r, 11.8K) and has no significant sequence homology with cyclophilin. The recently determined X-ray and NMR structures of human recombinant FKBP (refs 13–15) show that it exists in the crystal and in solution as a five-stranded antiparallel β-sheet which wraps around a short helix. There is thus no readily apparent similarity with the three-dimensional cyclophilin structure. The FK506 class of drugs do not bind to cyclophilin and CsA does not bind to FKBP. Nonetheless the mode of action of CsA and FK506 in the cell seem to be very similar[12,16]. FKBP also shows a *cis–trans* prolyl isomerase activity. But for both FK506 and CsA, suppression of prolyl isomerase activity alone seems insufficient to explain the biological activity, as drug concentrations necessary to inhibit T-cell activation would not saturate the abundant prolyl isomerases[12,17].

Our results are in line with the earlier NMR reports on CsA bound to cyclophilin[5]: intermolecular NOEs determined between cyclophilin and CsA implicate residues along one face of cyclosporin (residues 9, 10, 11, 1 and 2) as being important for binding. These complementary structural studies using both NMR spectroscopy in solution and X-ray crystallography thus provide new insights into the binding of CsA and, more generally, the molecular basis of immunosuppressive activity.

Note added in proof: The current R-factor is 21.8% for all data from 8 to 2.3 Å. An independent X-ray structure of unliganded cyclophilin has been determined[22]. Despite different unit cells, the molecular architecture seems similar. □

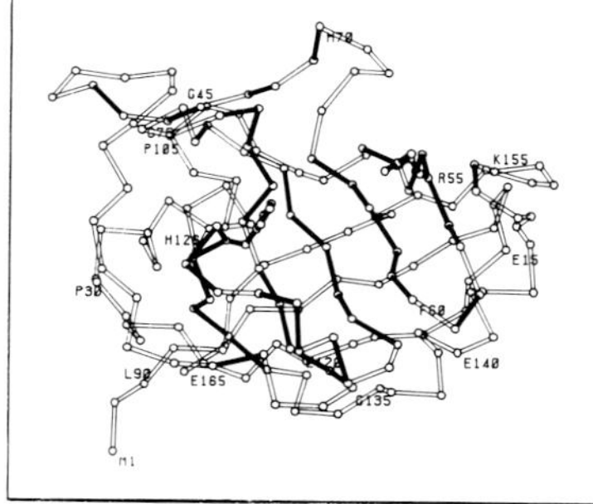

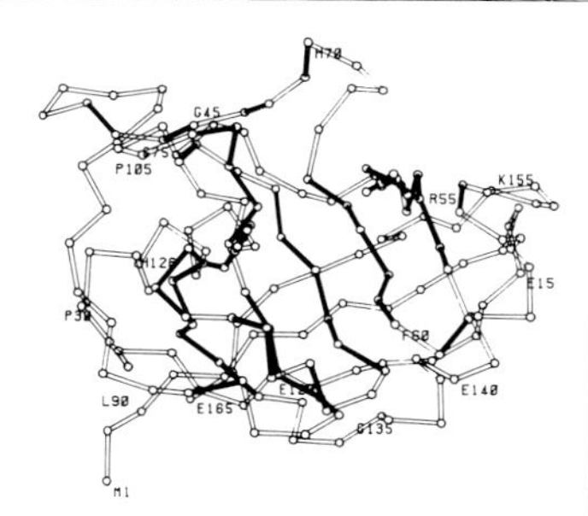

FIG. 3 Stereo PLUTO plot[18] showing the Cα skeleton of cyclophilin. Selected Cα atoms have been labelled with the sequence number and the one-letter amino-acid code. Residues which are implicated in CsA binding based on the chemical shift data in Fig. 1*a*, are shown in this diagram with filled Cα–Cα bonds. The side chains of Arg55 and His126, which may be involved in the prolyl isomerase mechanism, are also shown.

Received 5 July; accepted 1 August 1991.

1. Handschumacher, R. E., Harding, M. W., Rice, J., Drugge, R. J. & Speicher, D. W. *Science* **226**, 544–547 (1984).
2. Borel, J. F. *Pharmac. Rev.* **41**, 259–371 (1989).
3. Takahashi, N., Hayano, T. & Suzuki, M. *Nature* **337**, 473–475 (1989).
4. Fischer, G., Wittmann-Liebold, B., Lang, K., Kiefhaber, T. & Schmid, F. X. *Nature* **337**, 476–478 (1989).
5. Weber, C. *et al. Biochemistry* **30**, 6563–6574 (1991).
6. Fesik, S. W. *et al. Biochemistry* **30**, 6574–6583 (1991).
7. Wüthrich, K., Spitzfaden, C., Memmert, K., Widmer, H. & Wider, G. *FEBS Lett.* (in the press).
8. Zurini, M. *et al. FEBS Lett.* **276**, 63–66 (1990).
9. Cowan, S. W., Newcomer, M. E. & Jones, T. A. *Proteins* **8**, 44–61 (1990).
10. Liu, J., Albers, M. W., Chen, C., Schreiber, S. L. & Walsh, C. T. *Proc. natn. Acad. Sci. U.S.A.* **87**, 2304–2308 (1990).
11. Liu, J., Chen, C. & Walsh, C. T. *Biochemistry* **30**, 2306–2310 (1991).
12. Schreiber, S. L. *Science* **251**, 283–287 (1991).
13. Michnick, S. W., Rosen, M. K., Wandless, T. J., Karplus, M. & Schreiber, S. L. *Science* **252**, 836–839 (1991).
14. Van Duyne, G. D., Standaert, R. F., Karplus, P. A., Schreiber, S. L. & Clardy, J. *Science* **252**, 839–842 (1991).
15. Moore, J. M., Peattie, D. A., Fitzgibbon, M. J. & Thomson, J. A. *Nature* **351**, 248–250 (1991).
16. Tropschug, M., Barthelmess, I. B. & Neupert, W. *Nature* **342**, 953–955 (1989).
17. Bierer, B. E. *et al. Proc. natn. Acad. Sci. U.S.A.* **87**, 9231–9235 (1990).
18. *CCP4; A Suite of Programs for Protein Crystallography* (SERC Daresbury Laboratory, Warrington, UK, 1986).
19. Blundell, T. L. & Johnson, L. N. *Protein Crystallography* 357 (Academic, Oxford, 1976).
20. Jones, T. A., Zou, J. Y., Cowan, S. W. & Kjeldgaard, M. *Acta crystallogr.* **A47**, 110–119 (1991).
21. Brunger, A. T., Kuriyan, J. & Karplus, M. *Science* **235**, 458–460 (1987).
22. Ke, H. *et al. Proc. natn. Acad. Sci. U.S.A.* (in the press).

ACKNOWLEDGEMENTS. We thank R. Wenger for supplying the water-soluble cyclosporin and K. Wilson and C. Betzel of the EMBL outstation, Hamburg, for their help with data collection using synchrotron radiation. This work was supported by the Schweizerischer Nationalfonds (to K.W.).

VI

CONFORMATIONAL EQUILIBRIA AND INTERNAL MOBILITY IN PROTEINS IN SOLUTION

Introduction to papers 45 through 54

The ten papers in this part have been published during a time period spanning more than two decades. Although all these papers address the general theme of NMR investigations of internal mobility in globular proteins, they cover three clearly distinct phases in my research group. Firstly, from 1973 to 1978 there was the observation of rate processes with quite precise temporal resolution but, because of the lack of sequence-specific resonance assignments, absence of spatial resolution (papers **45–47**). Secondly, with the advent of sequence-specific NMR assignments in 1979, spatio-temporal resolution for the description of rate process in proteins was obtained by reference to the X-ray crystal structures (papers **48–51**). Thirdly, although for several years from 1982 onward our research was mainly focused on protein structure determination (see Part II), systematic efforts at complementing NMR structures of proteins with data on the molecular dynamics have again been prominent in my laboratory since the late 1980s (papers **52–54**; see also Part VII).

The paper **45** has been of special importance for us in two ways. It was our first report on work with the protein bovine pancreatic trypsin inhibitor (BPTI), which became the focus of (or, alternatively, the "workhorse" for) a large proportion of our projects during the following two decades. Furthermore, the paper describes the first systematic NMR study in my laboratory of amide proton exchange in a protein (similar experiments with BPTI were published during the same year by Prof. B.D. Sykes and his colleagues[1]). In these experiments the protein was dissolved in D_2O and the replacement of individual amide protons by deuterons was followed by measurement of the decrease of the intensities of the corresponding ^1H NMR lines with time. Amide

[1] Karplus, S., Snyder, G.H. and Sykes, B.D. (1973) *Biochemistry 12*, 1323–1329. A nuclear magnetic resonance study of bovine pancreatic trypsin inhibitor. Tyrosine titrations and backbone NH groups.

proton exchange measurement had of course been a classical approach in protein chemistry long before any such studies were started with BPTI, dating back to the days of the Linderstrøm-Lang laboratory in Copenhagen.[2] The NMR measurements with BPTI added a new element in that the exchange from individual amide groups in the protein could be followed, whereas earlier techniques yielded information only on the percentage of the ensemble of all amide protons that had exchanged after a certain time span. Proton exchange studies were one of few NMR experiments with proteins that could be done well with the instrumentation available in 1973, and we collected exchange data with BPTI under a wide variety of conditions of pH, temperature and ionic strength, and subsequently also with homologs and chemical modifications of BPTI[3–5] (see also paper **47**).

The paper **46** describes NMR observations on ring flips of phenylalanine and tyrosine in BPTI, which in turn provided precise data on the frequencies of concerted, large-amplitude internal motions in the protein core. The observation of these ring flipping motions on the millisecond to microsecond time scale was a genuine surprise for the following reasons: In the refined X-ray crystal structure of BPTI the aromatic rings of phenylalanine and tyrosine are among the side chains with the smallest temperature factors. For each ring the relative values of the B-factors increase toward the periphery, so that the largest positional uncertainty is indicated for the peripheral carbon atom on the symmetry axis through the C^{β}–C^{γ} bond rather than for the four ring

[2] Hvidt, A. and Nielsen, S.O. (1966) *Adv. Protein Chem. 21,* 287–386. Hydrogen exchange in proteins.

[3] Wüthrich, K. and Wagner, G. (1979) *J. Mol. Biol. 130,* 1–18. Nuclear magnetic resonance of labile protons in the basic pancreatic trypsin inhibitor.

[4] Richarz, R., Sehr, P., Wagner, G. and Wüthrich, K. (1979) *J. Mol. Biol. 130,* 19–30. Kinetics of the exchange of individual amide protons in the basic pancreatic trypsin inhibitor.

[5] Wagner, G. and Wüthrich, K. (1979) *J. Mol. Biol. 130,* 31–37. Correlation between the amide proton exchange rates and the denaturation temperatures in globular proteins related to the basic pancreatic trypsin inhibitor.

carbon atoms which undergo extensive movements during the ring flips. Theoretical studies performed in a collaboration of my group with Prof. R. Huber's laboratory[6] then showed that the crystallographic B-factors sample multiple rotation states about the C^{α}–C^{β} bond, whereas the ring flips about the C^{β}–C^{γ} bond seen by NMR are very rapid 180°-rotations connecting two indistinguishable equilibrium orientations of the ring. The B-factors do not manifest these rotational motions because the populations of all non-equilibrium rotational states about the C^{β}–C^{γ} bond are vanishingly small. Indications of aromatic ring mobility were also observed in other proteins.[7,8] Eventually the ring flip phenomenon turned out to be a general feature of globular proteins, manifesting ubiquitous low-frequency internal motions which have activation energies of 60–100 kJ mol^{-1}, amplitudes of $\gtrsim$ 1.0 Å, activation volumes of about 50 Å^3, and involve concerted displacement of numerous groups of atoms.[6,9–11] A survey of the amide proton exchange and ring flip data collected with a series of chemical modifications of BPTI and homologous proteins (paper **47**) showed that the two rate processes are differently related to the thermal stability of the protein, which reflects the

[6] Hetzel, R., Wüthrich, K., Deisenhofer, J. and Huber, R. (1976) *Biophys. Struct. Mech. 2,* 159–180. Dynamics of the aromatic amino acid residues in the globular conformation of the basic pancreatic trypsin inhibitor (BPTI) II: semi-empirical energy calculations.

[7] Campbell, I.D., Dobson, C.M. and Williams, R.J.P. (1975) *Proc. R. Soc. Lond. B. 189,* 503–509. Proton magnetic resonance studies of the tyrosine residues of hen lysozyme–assignment and detection of conformational mobility.

[8] Hull, W.E. and Sykes, B.D. (1975) *J. Mol. Biol. 98,* 121–153. Fluorotyrosine alkaline phosphatase: internal mobility of individual tyrosines and the role of chemical shift anisotropy as a ^{19}F nuclear spin relaxation mechanism in proteins.

[9] Wagner, G., DeMarco, A. and Wüthrich, K. (1976) *Biophys. Struct. Mech. 2,* 139–158. Dynamics of the aromatic amino acid residues in the globular conformation of the basic pancreatic trypsin inhibitor (BPTI) I: ^{1}H NMR studies.

[10] Wagner, G. (1980) *FEBS Lett. 112,* 280–284. Volumes for the rotational motion of interior aromatic rings in globular proteins determined by high resolution ^{1}H NMR at variable pressure.

[11] Gelin, B. R. and Karplus, M. (1975) *Proc. Natl. Acad Sci. USA 72,* 2002–2006. Sidechain torsional potentials and motion of amino acids in proteins: bovine pancreatic trypsin inhibitor.

fact that the proton exchange is sensitive to conformational equilibria while the ring flips manifest directly the frequencies of time fluctuations of the folded conformation.

The papers **48** through **51** are from the period when sequence-specific resonance assignments enabled the generation of spatial maps of equilibrium and rate processes in protein structures, using the assumption that the molecular architectures in the crystals and in solution are closely similar or identical. At this stage it became quite clear that amide proton exchange rates would be a reliable supporting criterion for the identification of regular secondary structures in globular proteins (paper **48**). In the paper **49**, experimental conditions were identified where the amide proton exchange in a chemically modified BPTI derivative proceeds *via* a so-called EX_1 process, where the exchange rates are directly related to the frequencies of time fluctuations of the protein structure. This paper is a direct demonstration that EX_1 processes for amide proton exchange in proteins can be obtained only under extreme conditions of pH and temperature, and that therefore nearly all available exchange data are governed by EX_2 processes. The amide proton exchange rates are thus not directly related to the frequencies of time fluctuations of the protein about an equilibrium structure, but rather to the acid/base-catalyzed exchange reaction at the amide group modulated by conformational equilibria in the protein. The paper **50** describes the completion of the ^{1}H NMR assignments for the aromatic rings in BPTI with the use of heteronuclear NMR techniques, and paper **51** describes the use of ^{13}C spin relaxation measurements for mapping of sequential mobility along the polypeptide backbone in BPTI.

During the past five years, characterization of transient local conformational features has become an integral part of NMR structure determinations in my laboratory. Most of these data can only be collected when a high-quality NMR structure is available. Examples for this currently particularly exciting part of our work are the hydration studies described in Part VII, the identification of slow interconversion between two conformational states of disulfide bonds (papers **52** and **53**), and the characterization of discrete hydrogen bonding networks on the protein surface (paper **54**).

Volume 31, number 1 FEBS LETTERS April 1973

PROTON MAGNETIC RESONANCE INVESTIGATION OF THE CONFORMATIONAL PROPERTIES OF THE BASIC PANCREATIC TRYPSIN INHIBITOR

A. MASSON and K. WÜTHRICH

Institut für Molekularbiologie und Biophysik, Eidg. Technische Hochschule, CH-8049 Zürich, Switzerland

Received 31 January 1973

1. Introduction

The basic pancreatic trypsin inhibitor (BPTI) from bovine pancreas [1] has a molecular weight of 6500 and consists of 58 amino acid residues. The amino acid sequence [2–5] and the molecular conformation in single crystals [6, 7] are known, and it has been found that the solution conformation of BPTI is unusually stable towards denaturing agents and heat [8]. Some time ago we briefly discussed proton nuclear magnetic resonance (NMR) data which revealed yet another rather unexpected structural feature of BPTI, i.e. very slow exchange of some of the amide protons [9]. This paper presents new data on the denaturation of BPTI and the proton exchange in solutions of BPTI in D_2O.

An obvious goal of the investigations of BPTI is to understand in more detail the mode of action of this inhibitor [6–8]. In addition, since BPTI is by its small size amenable to detailed studies by different techniques, work on this molecule could on a more general basis be particularly valuable for the further development of the investigations of molecular conformations in proteins. In view of the temperature stability of BPTI [8], some of the data on this protein might conceivably even be relevant for the investigation of conformational features in proteins from thermophilic organisms [10].

2. Materials and methods

The basic pancreatic trypsin inhibitor (BPTI;

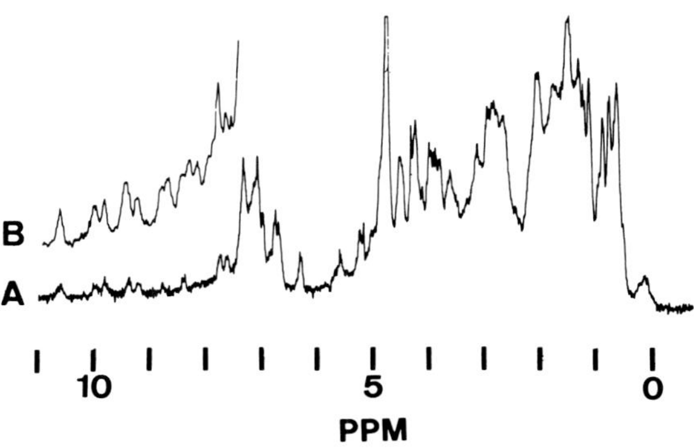

Fig. 1. Proton NMR spectrum at 220 MHz of a 0.01 M solution of the basic pancreatic trypsin inhibitor (BPTI) in D_2O, pD = 7.7, T = 20°. Two spectra were recorded: A) 170 hr and B) 10 min after the preparation of the solution.

Trasylol®,* was obtained from the Farbenfabriken Bayer AG. For most of the experiments ca. 0.01 M solutions of the protein were used. From additional NMR experiments at variable concentrations between 0.0005 and 0.015 M we found no evidence for intermolecular association [11]. The pD of the aqueous solutions was adjusted by the addition of DCl, or NaOD, respectively, and measured in the NMR tube with a combination electrode. The pD-values are given as read from the pH-meter, without correction for the isotope effect [12].

* ® Registered trade mark Bayer Leverkusen, Germany.

Reprinted from *FEBS Letters*, Vol. 31, pp. 114–118 (1973)
Copyright © 1973, with kind permission from
Elsevier Science B. V. Amsterdam, The Netherlands.

Volume 31, number 1 FEBS LETTERS April 1973

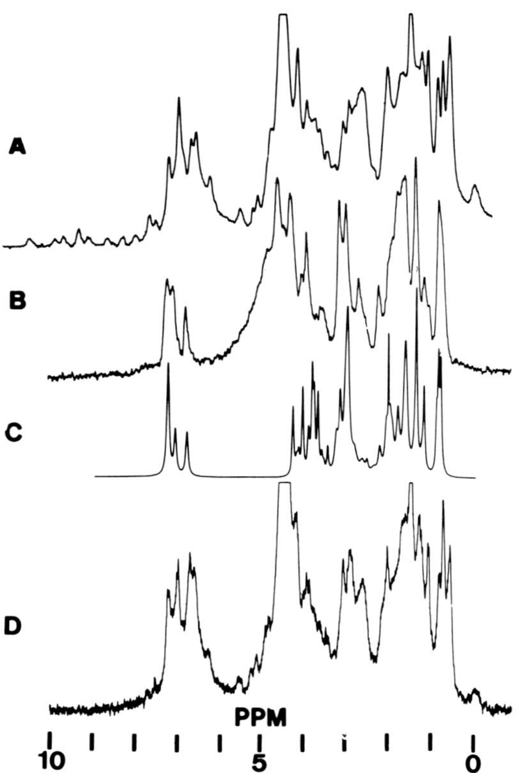

Fig. 2. Proton NMR spectral changes during thermal denaturation of BPTI in D_2O solution containing 0.01 M of the protein and 6 M guanidinium chloride, pD = 7.1. A) T = 20°. B) T = 83°. C) Hypothetical spectrum for BPTI in the random coil form, computed with the amino acid spectra of McDonald and Phillips [14]. D) Spectrum of the solution B after cooling down to 20°.

High resolution proton NMR spectra were recorded on a Varian HR-220 spectrometer. Chemical shifts are in parts per million (ppm) from internal sodium 3-Trimethylsilyl-propionate.

3. Results and discussion

The proton NMR spectrum of BPTI is shown in fig. 1. It contains a series of methyl resonances at around 1 ppm, the remaining resonances of the aliphatic amino acid side chains between 1.5 and 4 ppm, the C_α-proton resonances between 4 and 5 ppm, the solvent resonance at 4.8 ppm, the resonances of the phenylalanyl and tyrosyl residues between 6 and 8 ppm, and 15 readily recognizeable resonances of "exchangeable" protons in the spectral region from 7.5 to 11 ppm. The molecular conformation is manifested in the differences between this spectrum and that of the denatured protein (fig. 2, B and C). Particularly prominent spectral features of native BPTI are the lines at around 0, 5.2, 5.5, and 6.3 ppm, and the appearance of the resonances of a number of "exchangeable" protons (fig. 1B). These qualitative spectral features agree with what we reported previously [9], and seem to coincide essentially with observations made independently by Karplus et al. [13]. In this paper the spectral differences between figs. 1 and 2C are employed for studies of the denaturation of BPTI, and of the exchange of the protons observed between 8 and 11 ppm.

Fig. 2 shows that the thermal denaturation of BPTI in 6 M guanidinium chloride at 83° produces a typical spectrum for a random coil form of the polypeptide chain, as judged from the close similarity in the regions from 0 to 3.5 and 6 to 10 ppm of the spectrum 2B with the computed spectrum 2C [14]. The denaturation is almost completely reversible as judged from the reappearance in the spectral region from 0 to 8 ppm of the prominent spectral features of the native molecule when the solution is cooled down again (fig. 2D). All the exchangeable protons have been replaced by ^2D in the denatured molecule, and hence there are no lines between 8 and 11 ppm in the spectrum D. In a series of measurements of this type the results obtained previously from ORD experiments by Vincent et al. [8] were confirmed, e.g. there were at most very minor changes in the spectral region from 0 to 8 ppm when a neutral solution of BPTI in D_2O was heated to 85°, and in an acidic solution at pD = = 0.7 some spectral changes arose only at temperatures above 70°. In addition we found that a random coil form of BPTI is present at ambient temperature in solutions in trifluoroacetic acid, d_4-methanol, and d_6-DMSO. Upon admixture of D_2O the random coil form of BPTI present in methanol and in DMSO goes over into a molecular conformation with NMR spectral properties which are very similar to those of the native protein. This is illustrated in fig. 3. In DMSO (spectrum A) there are no resonances at 0 ppm and between 5 and 6.5 ppm. The broad resonance at 7 to 9 ppm comes from most of the potentially

Volume 31, number 1 FEBS LETTERS April 1973

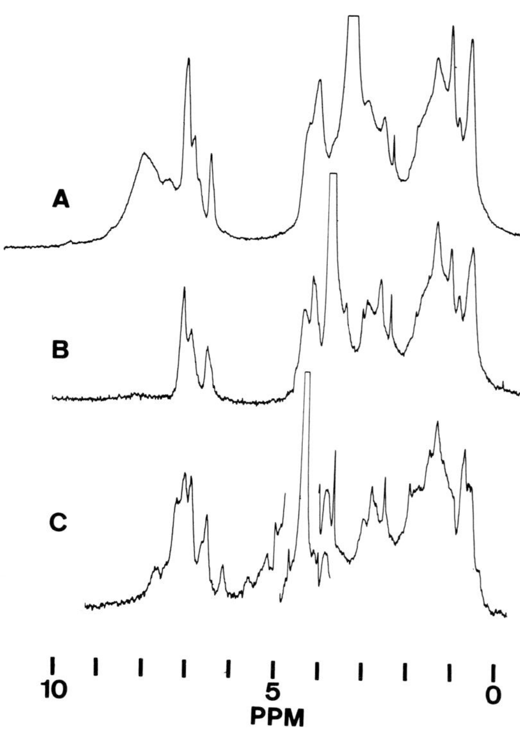

Fig. 3. Proton NMR spectrum at 220 MHz and 20° of 0.01 M solutions of BPTI in mixed solvents of d_6-DMSO and D_2O. A) ≈ 99:1; B) 85:15; C) 65:35, volume percent. The solvent resonances are in the following positions: d_6-DMSO is at 2.5 ppm in all the three spectra, and HDO is observed in A at 3.5 ppm; B, 3.8 ppm; and C, 4.3 ppm.

exchangeable protons. Upon addition of 15% D_2O (fig. 3B) the spectrum is still quite typical for a random coil form of the protein. All the amide protons have been exchanged with ^2D from D_2O. After addition of 35% D_2O (fig. 3C) the NMR spectrum contains the typical features of the native protein except of course for the exchanged protons. A random coil spectrum is again observed if the solution of fig. 3C is heated to ca. 60°.

The appearance of a series of well resolved potentially exchangeable proton resonances between 8 and 11 ppm is a spectral feature which has apparently not been reported for any other protein in D_2O solution. Of the 15 lines of this type which are observed in the freshly prepared solution at neutral pH (fig. 1B and

Table 1
Positions and life-times with respect to chemical exchange of the one-proton resonances in the spectral region 7.5 to 11 ppm of BPTI in solutions in D_2O at 22°.

	$\Delta\nu$ (ppm)	$\tau_{1/2}$ (hr)
A. pD = 7.3	7.55	$<$ 2[a]
	7.79	150
	7.87	$<$ 2[a]
	7.98	20
	8.17	20
	8.28	0.4
	8.43	$>$ 1000
	8.67	1.1
	8.80	270
	9.24	$>$ 1000
	9.41	$>$ 1000
	9.41	9
	9.82	$>$ 1000
	9.98	1000
	10.60	$>$ 1000
B. pD = 0.9	7.95	2
	8.1 ... 9.0	$>$ . . . .[b]
	9.2	$>$ 1000
	9.40	$>$ 1000
	9.45	$>$ 1000
	9.85	$>$ 1000
	10.00	$>$ 1000
	10.60	$>$ 1000
C. pD = 11.0	9.15	15
	9.80	$>$ 100
	10.60	10

[a] This line corresponds possibly to more than one proton.
[b] The resonances between 8.1 and 9 ppm correspond in intensity to 10 to 13 protons. The individual resonances in this spectral region have life-times between 15 hr and $>$ 1000 hr.

table 1), four decrease visibly in intensity within a few hours at ambient temperature (figs. 1B, 4A, and 4B; ▾), for six additional resonances the intensity is clearly reduced after several hundred hours (fig. 4, C; ■) and the remaining five protons exchange "unmeasurably slowly" under these conditions (fig. 4, ○). At 53° exchange of all these protons can be observed within a few hours, but three lines have not completely disappeared after 1 day (fig. 5). In a freshly prepared acidic solution of BPTI the resonances of ca. 17 to 20 protons are in the spectral region from 8 to 11 ppm (table 1). At pD = 0.9 the exchange of at least 10 of these protons is not complete after one day at

Volume 31, number 1 FEBS LETTERS April 1973

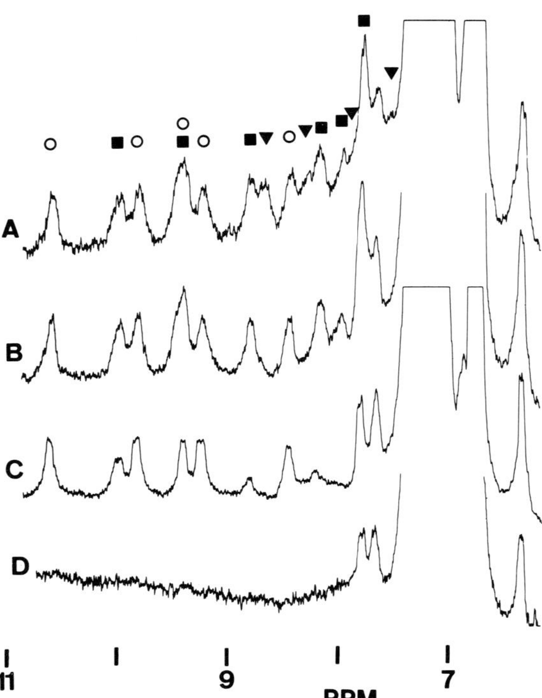

Fig. 4. Spectral region from 7 to 11 ppm of the proton NMR spectrum at 220 MHz of BPTI at different times after preparation of a 0.01 M solution in D_2O, pD = 7.3, T = 22°. A) 25 min; B) 150 min; C) 660 hr; D) after standing at 85° for a few minutes.

53° (fig. 6). In a freshly prepared solution of BPTI at pD = 11, there are only three resonances of slowly exchanging protons (table 1). If a solution of BPTI in neutral H_2O is compared with the spectrum of fig. 1B, no additional resonances can be detected in the spectral region at low field from 9 ppm.

On the basis of what is generally observed in peptides and proteins [15, 16] the above data seem to imply that the slow exchange of some protons in native BPTI comes about because some of the potentially exchangeable protons are shielded from interaction with the solvent. This interpretation would be consistent with the observation that all the labile protons in BPTI are rapidly exchanged when a random coil type NMR spectrum is observed (figs. 2B, 3B). From inspection of the molecular model the resonances between 8 and 11 ppm (fig. 1) have previously been

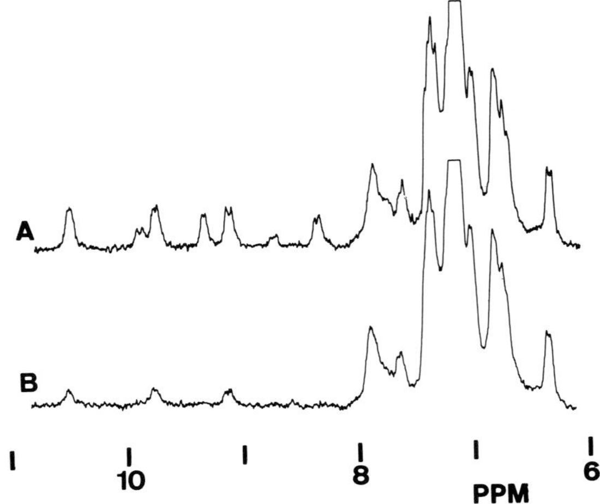

Fig. 5. Spectral region from 7 to 11 ppm of the proton NMR spectrum at 220 MHz of a 0.01 M solution in D_2O of BPTI, pD = 7.3, after being kept at 53° for: A) 2.5 hr; and B) 23 hr.

tentatively assigned to amide protons involved in the hydrogen bonds of the secondary structure elements located in the interior of the BPTI molecule [9]. The doublet structure which can be recognized in several of these resonances (figs. 4–6) provides additional

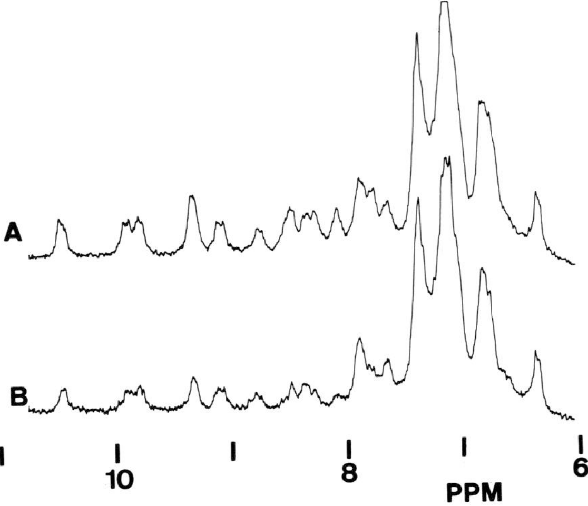

Fig. 6. Spectral region from 7 to 11 ppm of the proton NMR spectrum at 220 MHz of a 0.01 M solution in D_2O of BPTI, pD = 0.9, after being kept at 53° for: A) 2.5 hr; B) 22 hr.

Volume 31, number 1 FEBS LETTERS April 1973

evidence for the assignment to amide protons of the peptide backbone. Overall it is rather surprising that groups of atoms located in the interior of this small protein molecule appear to have an extremely small probability to get in contact with solvent molecules. This seems to indicate that the solution conformation of BPTI is rather rigid, and that there is a small probability for the occurrence of molecular species which differ appreciably from the average form manifested in the NMR spectrum.

It seemed of interest to compare the conditions where the solution conformation of BPTI becomes sufficiently dynamic for all the amide protons to be exchanged rapidly, as judged from the disappearance of the resonances between 8 and 11 ppm, and those where the average conformation is modified, as judged from the spectral changes in the region 0 to 8 ppm. In all these experiments we found that the resonances of all the exchangeable protons disappeared well before there were any modifications in the spectrum of the non-exchangeable protons. The following are a few examples to illustrate this point. In a solution of BPTI in D_2O at neutral pH all the resonances of exchangeable protons disappear within minutes at 65°, and there is no evidence for denaturation of the protein at 85°. In a neutral BPTI solution in 6 M guanidinium chloride, the proton exchange is rapid at 50°, and denaturation begins at ca. 75° until a random coil spectrum is seen at 83° (fig. 2). In a mixed solvent of DMSO and D_2O (1:1) all the resonances of exchangeable protons disappear at 43°, and denaturation to a random coil form is observed at ca. 70°.

In conclusion it was the purpose of this paper to present some unusual features of the solution conformation of BPTI. The present NMR studies have confirmed the previously described stability of this molecule towards certain denaturing agents [8]. On the other hand it was found that BPTI is in a random coil form at ambient temperature in DMSO, TFA, and CH_3OH. NMR has revealed another unusual feature in this molecule, i.e. the extremely slow rate at which certain exchangeable protons are replaced by deuterium in solutions of BPTI in D_2O. Observation of these protons provides a means to investigate the dynamics of the solution conformation of BPTI under variable non-denaturing conditions. Work is in progress to relate these data to particular features in the amino acid sequence [2–5] and the single crystal conformation [6, 7] of the molecule.

Acknowledgements

We would like to thank Dr. R. Schmidt-Kastner, Farbenfabriken Bayer A.G., for a generous gift of BPTI (Trasylol®). Financial support by the Swiss National Science Foundation (project 3.423.70) is gratefully acknowledged.

References

[1] M. Kunitz and J.H. Northrop, J. Gen. Physiol. 19 (1936) 991.

[2] F.A. Anderer and S. Hörnle, J. Biol. Chem. 241 (1966) 1568.

[3] J. Chauvet, G. Nouvel and R. Acher, Biochim. Biophys. Acta 92 (1964) 200.

[4] V. Dlouhá, P. Pospíšilová, B. Meloun and F. Šorm, Coll. Czech. Chem. Commun. 30 (1965) 1311.

[5] B. Kassel and M. Laskowsi, Jr., Biochem. Biophys. Res. Commun. 20 (1965) 463.

[6] R. Huber, D. Kukla, A. Rühmann and W. Steigemann, Proc. Int. Research Conf. on Proteinase Inhibitors, 1970 (Walter de Gruyter, Berlin, 1971) p. 56.

[7] R. Huber, D. Kukla, A. Ruhmann and W. Steigemann, Cold Spring Harbor Symp. Quant. Biol. 36 (1971) 141.

[8] J.P. Vincent, R. Chicheportiche and M. Lazdunski, European J. Biochem. 23 (1971) 401.

[9] K. Wüthrich, Naturwissenschaften 60 (1973) Heft 2.

[10] H. Zuber, Naturwissenschaftl. Rundschau 22 (1969) 16.

[11] W. Scholtan and Sie Ying Lie, Makromol. Chemie 98 (1966) 204.

[12] P.K. Glasoe and F.A. Long, J. Phys. Chem. 64 (1960) 188.

[13] S. Karplus, G. Snyder and B.D. Sykes, private communication.

[14] C.C. McDonald and W.D. Phillips, J. Am. Chem. Soc. 91 (1969) 1513.

[15] A. Hvidt and S.O. Nielsen, Adv. Protein Chem. 21 (1966) 287.

[16] P.S. Molday, S.W. Englander and R.G. Kallen, Biochemistry 11 (1972) 150.

Volume 50, number 2 FEBS LETTERS February 1975

NMR INVESTIGATIONS OF THE DYNAMICS OF THE AROMATIC AMINO ACID RESIDUES IN THE BASIC PANCREATIC TRYPSIN INHIBITOR

K. WÜTHRICH and G. WAGNER

Institut für Molekularbiologie und Biophysik, Eidgenössische Technische Hochschule, 8049 Zürich, Switzerland

Received 5 December 1974

1. Introduction

For quite some time it has been recognized that the combination of single crystal X-ray data with high resolution NMR studies is a promising approach for investigations of the molecular conformations of globular proteins in solution [1]. It is a particular asset of the NMR method that the data on the static average spatial structure obtained from the X-ray measurements can be complemented with information on the dynamic aspects of the protein conformations. Recently, studies of the longitudinal spin relaxation times T_1 of ^{13}C [2], and ^1H NMR investigations of the kinetics of the exchange of labile protons with ^2D of the solvent [3] have been particularly emphasized in this context. Following the most recent advances of the instrumentation for high field ^1H NMR measurements, these studies can now be extended to include investigations of the intramolecular rotational motions of the aromatic amino acid side chains in globular proteins. In the present report this will be illustrated with a proton NMR study at 360 MHz of the basic pancreatic trypsin inhibitor (BPTI).

BPTI from bovine pancreas has a molecular weight of 6500, and consists of one polypeptide chain with 58 amino acid residues. The amino acid sequence includes 4 tyrosines and 4 phenylalanines as the only aromatic residues [4]. The molecular conformation in single crystals is known [5], and it was found that the solution conformation of BPTI is unusually stable towards denaturation by chemicals and by heat [3,6,7]. This outstanding heat stability made it possible in the present experiments to observe the NMR of the aromatic protons in the globular protein conformation over the temperature range from 4°C to 85°C, and from this to characterize the dynamic states of most of the aromatic rings in the protein at variable temperatures.

2. Materials and methods

The basic pancreatic trypsin inhibitor (BPTI, Trasylol ® Registered trade mark Bayer Leverkusen, Germany) was obtained from the Farbenfabriken Bayer AG. A 0.01 M solution of BPTI in D_2O, pD = 1.8, was used for the NMR studies. Sodium 2,2-dimethyl-2-silapentane-5-sulfonate (DSS) was added as an internal reference. Prior to the experiments reported in this paper, the labile protons had been replaced with deuterium of the solvent by heating the solution to 85°C for 5 min [3].

^1H NMR spectra at 360 MHz were obtained on a Bruker HXS-360 spectrometer equipped with a standard temperature unit. The probe temperature was measured with an ethylene glycol standard sample.

3. Results and discussion

Fig.1 shows the ^1H NMR spectrum at 360 MHz and 34°C of the non-labile protons between 5 and 8 ppm in the basic pancreatic trypsin inhibitor. The intensity of the resonance lines between 6.0 and 8.0 ppm was found to correspond to 34±2 protons. As will be shown below, these lines account for all the 36 aromatic protons of the 4 tyrosyl and the 4 phenylalanyl residues in BPTI [8]. The previously described identification of the tyrosine resonances with double resonance techniques [8] indicated that each of the

Reprinted from *FEBS Letters*, Vol. 50, pp. 265–268 (1975)
Copyright © 1975, with kind permission from
Elsevier Science B. V. Amsterdam, The Netherlands.

Volume 50, number 2 FEBS LETTERS February 1975

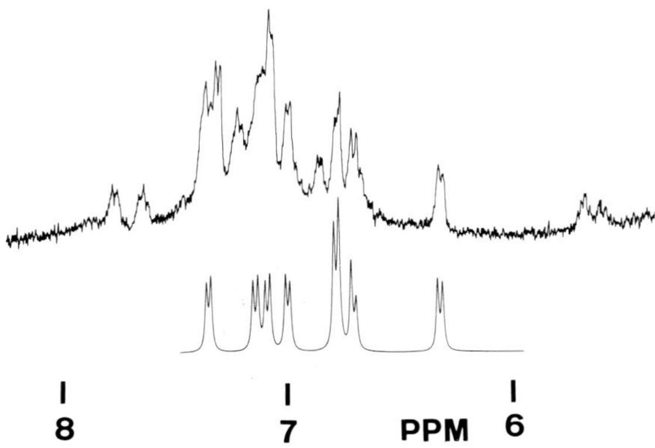

Fig.1. ^{1}H NMR spectrum at 360 MHz between 5 and 8 ppm of the basic pancreatic trypsin inhibitor (BPTI) at 34°C. The spectrum corresponds to a 0.01 M solution of BPTI in D_2O, $pD = 1.8$, where the labile protons had previously been replaced by deuterium by heating the solution to 85°C for 5 min. The lower trace shows the contributions to the spectrum from the 4 tyrosyl residues, which have been computed from the previously reported resonance assignments [8].

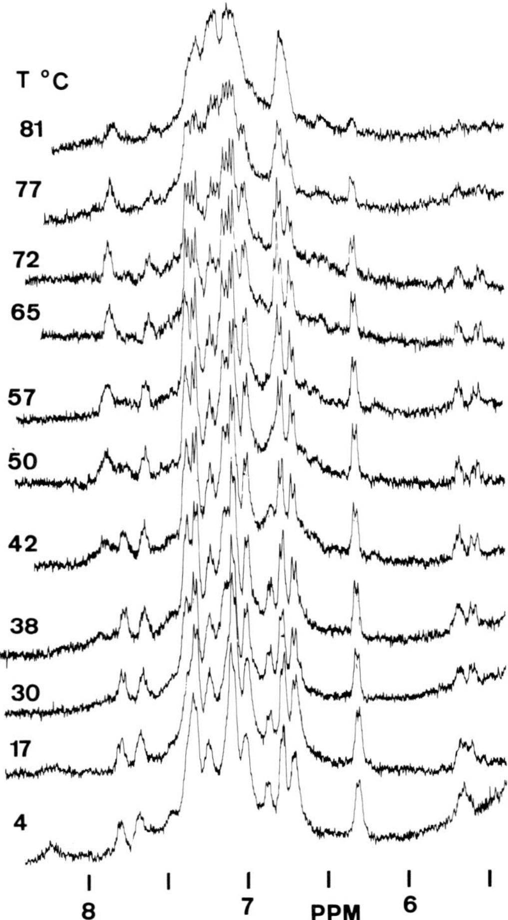

Fig.2. Temperature dependence between 4°C and 81°C of the spectral region from 5 to 9 ppm in the 360 MHz ^{1}H NMR spectrum of a 0.01 M solution of BPTI in D_2O, $pD = 1.8$. Prior to these experiments, the labile protons had been replaced by deuterium.

four tyrosines gives rise to an AA′BB′ type ^{1}H NMR spectrum. The lower trace in fig.1 corresponds to the sum of the resonances of the 4 tyrosyl residues computed from the data in ref. [8]. Subtracting the tyrosine resonances from the experimental spectrum then leaves one with the resonances corresponding to the 4 phenylalanines in BPTI. At 34°C these are thus found to cover the spectral range from 6.6 to 7.8 ppm, and to contain at least 5 lines with the intensity of one proton. This in turn indicates that on the NMR time scale the rotational motions of at least one of the phenylalanine rings in BPTI are essentially restricted to the motions of the entire molecule. This will in the following be confirmed by investigating the temperature dependence of the ^{1}H NMR spectrum, which will also explain why a rather small overall intensity is found for the aromatic resonances observed at ambient temperature (fig.1).

Fig.2 shows the temperature dependence of the resonances of the aromatic protons in BPTI between 4°C and 81°C. It is readily seen that quite extensive changes occur before the overall denaturation sets in at around 80°C [3,6], and that variations with temperature can be observed throughout the spectral region from 6.5 to 8.5 ppm. In the present preliminary discussion of the data of fig.2, we shall, however, concentrate on a few selected temperature dependent features which are particularly well resolved in the spectra, and at the same time pertinent for the description of the solution conformation of BPTI.

The structural information to be derived from the data of fig.2 concerns mainly the mobility of the aromatic rings in BPTI, and is obtained from the following symmetry considerations. Two pairs of

Volume 50, number 2 FEBS LETTERS February 1975

protons in the positions 2 and 6, and 3 and 5, respectively, of the aromatic rings of tyrosine and phenylalanine are related by a C_2 symmetry operation about the axis given by the $C^\beta - C^\gamma$ bond.

$$\text{H}-\underset{|}{\overset{|}{\text{C}^\alpha}}-\text{C}^\beta\text{H}_2 —— \text{C}_6\text{H}_5 \quad (\text{ring positions } 2\ 3 \text{ top},\ 4,\ 6\ 5 \text{ bottom})$$

In view of the non-periodic distribution of structural elements in the interior of globular proteins, chemical shift equivalence of the protons 2 and 6, and 3 and 5, respectively, will in most cases only be compatible with a dynamic situation where the aromatic rings would flip about the $C^\beta - C^\gamma$ axis at a rate which is rapid on the NMR time scale. Observation of an AA′BB′ type ^1H NMR spectrum for tyrosine, or an AA′BB′C type spectrum for phenylalanine is therefore indicative of a mobile aromatic ring. On the other hand the appearance of single proton resonance lines for tyrosine, or more than one single proton line for any given phenylalanine implies that the rings are quite rigidly fixed in the protein molecule.

To the extent that they can be recognized as resolved lines, the tyrosine resonances (fig.1) are almost independent of temperature between 4°C and 72°C (fig.2). The only temperature dependent feature of those resonances between 6.0 and 7.0 ppm which had been assigned to tyrosine protons [8] is a small chemical shift between the two two-proton doublet resonances which are both at 6.76 ppm at temperatures below 50°C, and give rise to a 3 line structure at higher temperature. The observations in fig.2 are thus compatible with the earlier assignments of the tyrosine resonances [8], and support the conclusion that the tyrosine rings are rotating about the $C^\beta - C^\gamma$ axis at a rate which is rapid on the NMR time scale at ambient temperature.

Following fig.1, the resonances between 7.4 and 8.5 ppm correspond to phenylalanine ring protons. At 4°C, this spectral region contains five one-proton lines at 7.48, 7.58, 7.67, 7.78, and 8.22 ppm, where the resonances at 7.58 and 8.22 ppm are markedly broader than the other three lines (fig.2). Whereas the latter are essentially independent of temperature between 4°C and 38°C, the lines at 7.58 and 8.22 ppm first broaden, then disappear, and finally merge into a single resonance of intensity corresponding to two protons at 7.90 ppm. This new resonance at 7.90 ppm is quite broad at 38°C, and sharpens as the temperature is further increased. Spectral variations with temperature which would occur simultaneously with those involving the lines at 7.58, 7.90, and 8.22 ppm, could only be detected between 7.1 and 7.4 ppm, where the bulk of the phenylalanine ring proton resonances are usually located. On the basis of the symmetry considerations presented above we conclude that the two lines which are at 7.58 and 8.22 ppm at 4°C correspond to a pair of 2,6- or 3,5-protons of phenylalanine. The life time with respect to 180° 'flips' about the $C^\beta - C^\gamma$ axis of one phenylalanine ring in BPTI can thus be estimated to be of the order 1×10^{-2} sec at 4°C, and 8×10^{-4} sec at 38°C. Considering the temperature dependence of the lines at 7.58 and 8.22 ppm, the origin of the above mentioned apparently reduced overall intensity of the aromatic resonances at ambient temperature is now also quite apparent.

There are four resonance lines at 7.79, 7.48, 6.87, and 6.67 ppm which consecutively broaden and disappear when the temperature is raised from 38°C to 72°C (fig.2). Homonuclear INDOR experiments indicated that these four lines come probably from the same aromatic ring. However, considering the additional spectral changes between 7.1 and 7.4 ppm when the temperature is raised from 38°C to 72°C, we cannot at this point definitely rule out that there might be two aromatic residues which would accidentally show a very similar temperature dependence. We conclude that there is one, possibly two aromatic rings in BPTI which have at 38°C a life time of the order 1×10^{-1} sec with respect to 180°-flips about the $C^\beta - C^\gamma$ bond.

There is an additional resolved one-proton line at 7.67 ppm which does not vary with temperature between 4°C and 65°C. On the one hand this line could correspond to any proton of an aromatic ring which would be immobilized in the protein over this temperature range. On the other hand it could also be that it corresponds to the C_4 proton of one of the phenylalanines, in which case the dynamic state of the ring would not necessarily be manifested in this NMR line.

With the possible exception of some features at temperatures above 65°C, where denaturation sets

Volume 50, number 2 FEBS LETTERS February 1975

in in acidic solutions of BPTI [3,6], there is no indication in fig.2 of the occurrence of intermediate rotational states of the aromatic rings which would be sufficiently long lived to be manifested in the NMR spectra.

In conclusion the ^1H NMR data described in this paper show that for at least 6 of the 8 aromatic rings in BPTI, the frequency of the rotational motions about the $C^{\beta}-C^{\gamma}$ bond axis exceeds 10^2 sec^{-1} at 60°C. For two, perhaps three, of the phenylalanines the transition from a slow to a fast process on the NMR time scale could be followed between 4°C and 72°C. In a subsequent paper [9], these observations will be analysed in the light of the refined single crystal of BPTI [10].

Acknowledgements

We would like to thank Professor Z. Luz for interesting discussions on the subject of this paper, Dr R. Schmidt-Kastner, Farbenfabriken Bayer AG, for a generous gift of BPTI (Trasylol ®), and the Schweizerischer Nationalfonds (Project 3.1510.73) for financial support.

References

[1] Wüthrich, K. (1974) Experientia 30, 577.
[2] Allerhand, A., Doddrell, D., Glushko, V., Cochran, D. W., Wenkert, E., Lawson, P. J. and Gurd, F. R. N. (1971) J. Amer. Chem. Soc. 93, 544.
[3] Masson, A. and Wüthrich, K. (1973) FEBS Lett. 31, 114.
[4] Kassel, B. and Laskowski, M. (1965) Biochem. Biophys. Res. Commun. 20, 463.
[5] Huber, R., Kukla, D., Rühmann, A. and Steigemann, W. (1971) Cold Spring Harbor Symp. Quant. Biol. 36, 141.
[6] Vincent, J. P., Chicheportiche, R. and Lazdunski, M. (1971) Eur. J. Biochem. 23, 401.
[7] Karplus, S., Snyder, G. H. and Sykes, B. D. (1973) Biochemistry 12, 1323.
[8] Wagner, G. and Wüthrich, K. (1974). Paper presented at the 6th International Conference on Magnetic Resonance in Biological Systems, Kandersteg, Switzerland, Sept. 16–21, 1974. Submitted to J. Magn. Res.
[9] Wüthrich, K., Wagner, G., Huber, R. and Deisenhofer, W. (1974) to be submitted to J. Mol. Biol.
[10] Deisenhofer, H. and Steigemann, W., private communication.

Reprinted with permission from *Nature* **275**, 247–248 (1978)
Copyright © 1978 MacMillan Magazines Limited.

Dynamic model of globular protein conformations based on NMR studies in solution

BASIC PANCREATIC TRYPSIN INHIBITOR (BPTI) is a small globular protein of molecular weight 6,500 (refs 1–17). It consists of one polypeptide chain with 58 amino acid residues, including three disulphide bonds and four phenylalanines and four tyrosines as the only aromatics[18]. The crystal structure at 1.5 Å resolution has been extensively refined[19]. In the ^{1}H- and ^{13}C-NMR (nuclear magnetic resonance) spectra, a large number of resolved resonance lines have been assigned to particular residues in the amino acid sequence[6,8,10–16] and related to particular aspects of the molecular conformation[1–17]. Hence, NMR spectral similarities and differences between BPTI and related proteins can on the basis of these earlier studies be interpreted in terms of structural features in well defined locations of the protein molecule. We report here the use of NMR observations to describe the globular BPTI conformation in terms of a dynamic ensemble of rapidly interconverting molecular structures. BPTI was a suitable compound for this study as its globular form is outstandingly stable in aqueous solution (Table 1)[1,9,20], leaving a large experimentally accessible range for studies of modified BPTI species with reduced stability. The chemically modified and homologous proteins used are characterised in the footnotes to Table 1.

Using the extensively refined X-ray structure[19] and the well resolved NMR spectra it has been demonstrated that the average molecular conformations of BPTI in crystal form and in solution are, with the exception of localised regions on the protein surface[8,17], very similar[1,2,14,16] and the structure type seen in BPTI crystals prevailed also in aqueous solutions of all the proteins in Table 1[13,21–23]. In this study the information on the solution conformations obtained from comparison with the crystal structure of BPTI was complemented by NMR measurements of the denaturation temperature, the amide proton exchange rates with the solvent in D_2O solution and the intramolecular mobility of the aromatic rings. The conclusions on the molecular dynamics are based largly on comparison of these features in the different structurally closely related proteins of Table 1.

The denaturation temperatures (see Table 1) were determined from the transition of the ^{1}H-NMR spectrum of the globular proteins to a random coil spectrum[1,27]. An additional important observation was that at the denaturation temperature, the exchange between globular and denatured protein, was slow on the NMR time scale[14].

The amide proton exchange rates were obtained for numerous individual protons by measurements of the decrease of the resonance intensities with time after dissolving the proteins in D_2O (ref. 27). Approximately 25 interior amide protons could readily be studied using this technique[1,2]. Table 1 shows rate constants for the most slowly exchanging amide proton in the different proteins[13,22,23].

High resolution ^{1}H-NMR spectra provided an unambiguous criterion to determine rotational motions of the aromatic rings of phenylalanine and tyrosine about the C_β–C_γ single bond[27]. In BPTI, three aromatic rings were found to undergo 180° flip motions about the C_β–C_γ bond[28] with frequencies of the order of 50–1,500 s^{-1} at 40° and activation energies ΔG^* of the order of 15 kcal mol^{-1}. One ring was immobilised over the entire accessible temperature range, with $\Delta G^* \approx 20$ kcal mol^{-1} at 80° and four rings were flipping too rapidly to be studied quantitatively[3,5,9]. In Table 1 the flipping motions for two selected rings are characterised for the different proteins.

Three features of the data obtained are worthy of attention. (1) The amide proton exchange rates are outstandingly slow for globular proteins[29,30], with life times of the order of several years for the most slowly exchanging protons in BPTI at neutral *p*D and 25 °C (refs 1, 2). On the other hand, local flexibility of the protein structures is implied by the rotational motions of the aromatic rings[3,9,27,28,31]. (2) The amide proton exchange rates are correlated with the thermal stability of the globular conformation, that is, in proteins with lower denaturation temperature the amide proton exchange is faster. (3) The rotational motions of the aromatic rings are not correlated with the thermal stability or the amide proton exchange, that is, for corresponding rings similar flip frequencies prevailed in the different proteins, unless the immediate ring environment was affected by the protein modification. These observations were used to characterise the globular solution conformations of the proteins studied.

Earlier investigations of amide proton exchange in globular

Table 1 Denaturation temperatures, amide proton exchange rates and intramolecular mobility of aromatic rings in BPTI and six related proteins

			Mobility of aromatic rings§			
			Tyr 35		Phe 45	
Protein*	Denaturation† T_D (°C)	NH exchange‡ k_{EX} (min^{-1})	ν (s^{-1})	ΔG^* (kcal mol^{-1})	ν (s^{-1})	ΔG^* (kcal mol^{-1})
BPTI	>95	2×10^{-7}	6	16.5	30	14.2
CTI	70	6×10^{-5}	~6	~16.5	~300	~13.0
HPI	60	5×10^{-3}	6	16.5	150	13.3
R-BPTI	76	1×10^{-4}	~6	~16.5	30	14.2
BPTI[1]	88	5×10^{-5}	—‖	—‖	60	13.9
Des(A, R)BPTI	>65	7×10^{-5}	—‖	—‖	60	13.9
TRAM-BPTI	>95¶	—¶	6	16.5	30	14.2

* BPTI, basic pancreatic trypsin inhibitor; HPI, isoinhibitor K from *Helix pomatia*[24]; CTI, cow colostrum trypsin inhibitor[25]; R-BPTI, modified BPTI obtained by reduction of the disulphide bond 14–38, with the cysteinyl residues protected by carboxamidomethylation[23]; BPTI[1], modified BPTI obtained by cleavage of the peptide bond Lys 15–Ala 16 (ref. 26); Des(A, R)BPTI, modified BPTI obtained by cleavage of the peptide bond Lys 15–Ala 16 and removal of Ala 16 and Arg 17 (ref. 26); TRAM-BPTI, modified BPTI obtained by transamination of the α-amino group of Arg 1 (ref. 17).

† T_D, temperature at which at *p*D 5.0 equal concentrations of globular and denatured protein were observed.

‡ k_{EX}, Rate constant for the most slowly exchanging amide proton at *p*D 4.5 and 36 °C.

§ ν, Frequency at which the rings undergo 180° flips about the C_β-C_γ bond[5,28], ΔG^*, activation energy for these rotational motions. The data for Tyr 35 were obtained at 27°, those for Phe 45 at 4 °C.

‖ In these proteins, the modifications are immediately adjacent to Tyr 35; as a consequence, the Tyr 35 spectrum corresponds to the situation of rapid exchange over the entire accessible temperature range.

¶ BPTI and TRAM-BPTI were compared in different conditions. In 1M guanidinium chloride solution at *p*H 6.0, T_D was 75° for BPTI and 68° for TRAM-BPTI. At *p*H 6.0 and 55°, k_{EX} was approximately eight times larger for TRAM-BPTI than for native BPTI.

proteins have resulted in the suggestion that the exchange occurs by admixture of 'open' structures, O(H), to the 'closed' globular form of the molecules, C(H) (refs 29, 30). Depending on the relative rates for the closing of the protein, O(H)→C(H), and the acid/base catalysed exchange of exposed amide protons in the open forms O(H), the overall proton exchange rate k_{EX} may be governed either by an EX_1 or an EX_2 mechanism[29,30]. For BPTI, an EX_2 mechanism was indicated by the pH dependence of k_{EX} in the range 0.5–11.0. The correlation between denaturation temperature and amide proton exchange rates for the different proteins in Table 1 further implies that the equilibrium C(H)⇄O(H) involves global variations of the protein structure. This is also supported by the observation that, with few exceptions, the order of the relative exchange rates for the individual amide protons was the same in all the proteins studied[1,13,17,22,23]. On the other hand, the lack of a correlation between the frequencies of the 180° flips of the aromatic rings and the denaturation temperature or the amide proton exchange rates shows that the open forms O(H), which mediate the exchange of interior amide protons, cannot be in the same class as the thermally denatured proteins. This implies that the local environment of the aromatic rings in the closed form C(H) of the proteins is also essentially preserved in the open forms O(H). Independently, admixture among the open species O(H) of molecular structures corresponding to the thermally denatured protein was excluded on the grounds that the transition from denatured to globular BPTI was slow on the NMR time scale[27] even at the denaturation temperature T_D (ref. 14), as a slow closing rate O(H)→C(H) would not be compatible with the observed kinetics of the amide proton exchange[13].

From the above considerations it seems that the 'globular solution conformations' of the proteins in Table 1, and probably proteins in general, are best described as dynamic ensembles of all the structural species C(H) and O(H) involved in the amide proton exchange, where the Boltzmann law favours population of species closely related to C(H) over the population of more extremely opened forms. The data suggest that previously quoted (refs 30, 32) near-equality in several cases of ΔG^0 between the closed and open protein structures involved in amide proton exchange, C(H) and O(H), and ΔG^0 between the globular conformation and the denatured protein is coincidental. In the proteins of Table 1, structural species corresponding to the thermally denatured forms are excluded by a sizeable energy barrier from the rapid conformational equilibria which govern the exchange of interior labile protons. That the hydrophobic pockets which enclose the aromatic rings in the proteins used are apparently preserved in the open structures O(H) is compatible with the assumption that the molecular fluctuations promoting the amide proton exchange consist primarily of intramolecular translational and rotational motions of intact hydrophobic domains relative to each other. Such fluctuations would primarily open the hydrogen bonded secondary structures which link the different hydrophobic pockets[19]. Independently, the mobility of the aromatic rings would be correlated with the internal structural flexibility of the individual hydrophobic domains. Hydrophobic stability domains as evidenced here in one class of globular proteins might quite generally play an essential role in the architecture of protein molecules, a role which may so far have been assigned too one-sidedly to the regular secondary structures adopted by the polypeptide backbone[33]. In particular, the lack of correlation of the preservation of the hydrophobic pockets surrounding the aromatic rings with the overall stability of the globular protein conformation indicates that such hydrophobic domains might be formed at an early stage of polypeptide folding, possibly even as primary nucleation sites for the folding process.

The observed correlation between amide proton exchange rates and denaturation temperature (Table 1) may in this model be rationalised by the assumption that the energy barrier ΔG^* separating the globular conformation from the denatured forms of the protein can be overcome only by cooperative structural fluctuations resulting from overlap of different open structures O(H). With the increase of the population of states O(H) when approaching the denaturation temperature, the probability of such cooperative processes is greatly enhanced. The cooperative fluctuations would also cause destabilisation of the hydrophobic domains, thus causing a sharp transition to the denatured form.

While a thorough comparison of our results with other studies of protein flexibility must be deferred to a subsequent, more detailed paper[36], it seems important to point out that the many-parameter characterisation of internal motions in globular proteins by high resolution NMR provides a considerably more detailed description of interior parts of the molecules than that implied, for example, by the terms 'fluid-like' or 'continuous visco-elastic'[35] suggested in interpretations of theoretical model studies.

Financial support by the Swiss NSF (project 3.0040.76) is gratefully acknowledged.

GERHARD WAGNER

KURT WÜTHRICH

Institut für Molekularbiologie und Biophysik,
Eidgenössische Technische Hochschule,
CH-8093 Zürich-Hönggerberg, Switzerland

Received 29 March; accepted 20 July 1978.

1. Masson, A. & Wüthrich, K. *FEBS Lett.* **31**, 114–118 (1973).
2. Karplus, S., Snyder, G. H. & Sykes, B. D. *Biochemistry* **12**, 1323–1329 (1973).
3. Wüthrich, K. & Wagner, G. *FEBS Lett.* **50**, 265–268 (1975).
4. Wagner, G. & Wüthrich, K. *J. Magn. Resonance* **20**, 435–445 (1975).
5. Wagner, G., De Marco, A. & Wüthrich, K. *J. Magn. Resonance* **20**, 565–569 (1975).
6. Snyder, G. H., Rowan III, R., Karplus, S. & Sykes, B. D. *Biochemistry* **14**, 3765–3777 (1975).
7. Wüthrich, K. & Baumann, R. *Org. Magn. Resonance* **8**, 532–535 (1976).
8. Brown, L. R., De Marco, A., Wagner, G. & Wüthrich, K. *Eur. J. Biochem.* **62**, 103–107 (1976).
9. Wagner, G., De Marco, A. & Wüthrich, K. *Biophys. Struct. Mechanism* **2**, 139–158 (1976).
10. Marinetti, T. D., Snyder, G. H. & Sykes, B. D. *Biochemistry* **15**, 4600–4608 (1976).
11. De Marco, A., Tschesche, H., Wagner, G. & Wüthrich, K. *Biophys. Struct. Mechanism* **3**, 303–315 (1977).
12. Richarz, R. & Wüthrich, K. *FEBS Lett.* **79**, 64–68 (1977).
13. Wagner, G. *Thesis* ETH Zürich (1977).
14. Wüthrich, K., Wagner, G., Richarz, R. & Perkins, S. J. *Biochemistry* **17**, 2253–2263 (1978).
15. Richarz, R. & Wüthrich, K. *Biochemistry* **17**, 2263–2269 (1978).
16. Perkins, S. J. & Wüthrich, K. *Biochim. biophys. Acta* (in the press).
17. Brown, L. R., De Marco, A., Richarz, R., Wagner, G. & Wüthrich, K. *Eur. J. Biochem.* **88**, 87–95 (1978)
18. Kassell, B. & Laskowski Sr., M. *Biochem. biophys. Res. Commun.* **20**, 463–468 (1965).
19. Deisenhofer, J. & Steigemann, W. *Acta crystallogr.* **B31**, 238–250 (1975).
20. Vincent, J. P., Chicheportiche, R. & Lazdunski, M. *Eur. J. Biochem.* **23**, 401–411 (1971).
21. Wüthrich, K., Wagner, G. & Tschesche, H. *Proc. XXIII Coll.-Proteids of the Biological Fluids* 201–204 (Pergamon, Oxford, 1976).
22. Wagner, G., Wüthrich, K. & Tschesche, H. *Eur. J. Biochem* **86**, 67–76 (1978).
23. Wagner, G., Wüthrich, K. & Tschesche, H. *Eur. J. Biochem* (in the press).
24. Tschesche, H. & Dietl, T. *Eur. J. Biochem* **58**, 439–451 (1975).
25. Čechová, D., Jonáková, V. & Šorm, F. in *Proc. Int. Conf. on Proteinase Inhibitors* (eds Fritz, H. & Tschesche, H.) 105–107 (W. de Gruyter, Berlin, 1971).
26. Jering, H. & Tschesche, H. *Eur. J. Biochem.* **61**, 443–452 (1976).
27. Wüthrich, K. *NMR in Biological Research: Peptides and Proteins* (North-Holland, Amsterdam 1976).
28. Hetzel, R., Wüthrich, K., Deisenhofer, J. & Huber, R. *Biophys. Struct. Mechanism* **2**, 159–180 (1976).
29. Hvidt, A. & Nielsen, S. O. *Adv. Protein Chem.* **21**, 287–386 (1966).
30. Englander, S. W., Downer, N. W. & Teitelbaum, H. A. *Rev. Biochem.* **41**, 903–924 (1972).
31. Gelin, B. R. & Karplus, M. *Proc. natn. Acad. Sci. U.S.A.* **72**, 2002–2006 (1975).
32. Tanford, Ch. *Adv. Protein Chem.* **24**, 1–95 (1970).
33. Dickerson, R. E. & Geis, I. *The Structure and Action of Proteins* (Harper and Row, New York 1969).
34. McCammon, J. A., Gelin, B. R. & Karplus, M. *Nature* **267**, 585–590 (1977).
35. Suezaki, Y. & Go, N. *Int. J. Peptide Protein Res.* **7**, 333–334 (1975).
36. Wagner, G. & Wüthrich, K. *J. molec. Biol.* (submitted).

J. Mol. Biol. (1982) **160**, 343–361

Amide Proton Exchange and Surface Conformation of the Basic Pancreatic Trypsin Inhibitor in Solution

Studies with Two-dimensional Nuclear Magnetic Resonance

GERHARD WAGNER AND KURT WÜTHRICH

Institut für Molekularbiologie und Biophysik
Eidgenössische Technische Hochschule
ETH-Hönggerberg, CH-8093 Zürich
Switzerland

(Received 11 February 1982)

A novel approach for studies of amide proton exchange in proteins is presented. It relies on measurements of the amide proton–C^{α} proton cross-peak intensities in the two-dimensional homonuclear correlated ^{1}H nuclear magnetic resonance spectra. The protein is dissolved in $^{2}H_2O$ and the solution is exposed to the conditions of $p^{2}H$ and temperature where the exchange rates are to be measured. After variable intervals, the amide proton exchange in a sample of this protein solution is quenched by lowering the temperature and possibly by $p^{2}H$ variation, and a COSY† spectrum of this sample is then recorded. Comparison of the NH–$C^{\alpha}H$ cross-peak intensities in the spectra recorded after different exchange times yields exchange rates for the individual amide protons. The main advantage compared to previously described techniques is that a much more complete set of individual amide proton exchange rates can be obtained. In the basic pancreatic trypsin inhibitor, where all the amide proton resonances were previously individually assigned, quantitative exchange rates were obtained for 38 of the total of 53 backbone amide protons, and for 14 additional protons lower limits for the exchange rates were established from comparison of the COSY spectra recorded in H_2O and in $^{2}H_2O$. Proton exchange data were thus for the first time obtained for numerous peptide groups that are located near the protein surface in the single crystal structure of BPTI. For some locations on the protein surface, it appears that the amide proton exchange rates cannot be correlated readily with the static accessible surface areas in the crystal structure.

1. Introduction

The stability of biologically active spatial protein structures relies on a complex interplay of a multitude of weak, non-bonding interactions among different atoms of the polypeptide chains and between the polypeptide chain and the surrounding medium. The latter may, for example, be an aqueous solvent, an ordered lipid matrix

† Abbreviations used: n.m.r., nuclear magnetic resonance; 2D n.m.r., 2-dimensional n.m.r.; COSY, 2D correlated spectroscopy, BPTI, basic pancreatic trypsin inhibitor; p.p.m., parts per million.

0022–2836/82/260343–19 $03.00/0 © 1982 Academic Press Inc. (London) Ltd.

in biological membranes, or the ordered aggregation in single crystals used for X-ray studies. Since the contribution of each individual non-bonding interaction to the free energy that stabilizes the protein conformation is typically of the same order of magnitude as the thermal energy at temperatures near 300 K, these "secondary bonds" are constantly broken and reformed. As a result, protein molecules are highly dynamic structures. On the one hand, since they depend on intramolecular interactions and on interactions with their environment, protein conformations adapt readily to local changes of the polypeptide covalent structure and to changes of their surroundings. Examples are the conformational transition from trypsinogen to trypsin (Huber, 1979), the local conformation change in carboxypeptidase A upon substrate binding (Quiocho & Lipscomb, 1971), the quaternary structure transitions in tetrameric hemoglobin (Perutz, 1970) and quite generally the potential adaptability built into multi-domain protein structures with flexible hinge regions (Huber, 1979; Schulz & Schirmer, 1979; Richardson, 1981). On the other hand, protein molecules in thermodynamic equilibrium situations undergo time fluctuations about an average set of atom co-ordinates. These structure fluctuations cover a wide range of frequencies, amplitudes and energies of activation. They have been observed, for example, in nuclear spin relaxation measurements (Allerhand *et al.*, 1971; Ribeiro *et al.*, 1980; Richarz *et al.*, 1980), aromatic ring flips (Wagner *et al.*, 1976) and amide proton exchange studies (Hvidt & Nielsen, 1966; Englander *et al.*, 1972; Richarz *et al.*, 1979), and were extensively investigated by molecular dynamics calculations (Karplus & McCammon, 1980). This paper describes amide proton exchange measurements with the use of two-dimensional nuclear magnetic resonance. The results of these experiments bear on both the adaptation of the molecular conformation to changes of the protein environment and the time fluctuations of the protein about an equilibrium set of atomic co-ordinates.

In the above-mentioned, well-documented examples of conformational changes upon variation of the protein covalent structure or the protein environment, the initial and final states of the protein could both be studied by X-ray methods in single crystals (Perutz, 1970; Quiocho & Lipscomb, 1971; Huber, 1979). Different experimental techniques must be applied to investigate how the molecular conformation adapts to the change of environment when a protein is transferred from single crystals to a non-crystalline state. Such a technique should be capable of determining the polypeptide conformation with comparable detail to that of the single crystal X-ray structures. We have recently proposed that this could be achieved with ^{1}H n.m.r.† experiments (Wüthrich *et al.*, 1982), and as a first, fundamental step complete individual assignments were obtained for protein ^{1}H n.m.r. spectra (Billeter *et al.*, 1982; Wagner & Wüthrich, 1982; Wider *et al.*, 1982; Arseniev *et al.*, 1982). In this paper, 2D n.m.r. is used to investigate the exchange with the solvent of all the backbone amide protons in the basic pancreatic trypsin inhibitor. Combined with the previously obtained resonance assignments, these data serve to probe the molecular surface of BPTI in solution, which will then be correlated with data on the crystal structure.

† See footnote to p. 343.

Amide proton exchange measurements have long been applied for studies of internal fluctuations in proteins (for reviews, see e.g. Hvidt & Nielsen, 1966; Englander *et al.*, 1972), and already in 1958 Saunders & Wishnia demonstrated that proton exchange between proteins and solvent ^2H_2O can be observed by n.m.r. While other techniques for kinetic studies of specific groups of protons have been described (e.g. Rosa & Richards, 1979), the potentialities of n.m.r. for measurements of individual amide proton exchange rates appear to be quite unique (Wüthrich, 1976). With the experiments described in this paper, one obtains for the first time a complete data set, i.e. the individual exchange behaviour can be characterized for all the backbone amide protons in a protein. This should be of great value with regard to further clarification of the mechanistic aspects of amide proton exchange from globular proteins (see e.g. Wagner & Wüthrich, 1979; Hilton *et al.*, 1981), and one can hope to obtain a more complete view of the internal "breathing modes" (Hvidt & Nielsen, 1966) than has hitherto been possible.

2. Materials and Methods

Basic pancreatic trypsin inhibitor (Trasylol®, Bayer Leverkusen) was obtained from the Farbenfabriken Bayer AG. The solutions used for the n.m.r. recordings contained 0·02 M-BPTI and the pH was adjusted by the addition of minute amounts of HCl and NaOH, whereby in the ^2H_2O solution the pH meter readings were used without correction for isotope effects (Kalinichenko, 1976; Bundi & Wüthrich, 1979). For the experiment in Fig. 1, a mixed solvent of 90% H_2O and 10% ^2H_2O, pH 4·6, was used, so that all the backbone amide proton resonances were present in the spectrum. To observe a maximum number of amide protons in a protein solution in ^2H_2O, it is important that from the first contact between protein and ^2H_2O the p^2H is near the p^2H minimum for proton exchange (Wüthrich & Wagner, 1979). Therefore, to prepare ^2H_2O solutions of BPTI for the proton exchange measurements, the protein was first dissolved in H_2O and the pH adjusted to 3·5. Next, BPTI was lyophilized and dissolved at 24°C in ^2H_2O. The resulting p^2H of the BPTI solution was in all experiments in the range $3{\cdot}5 \pm 0{\cdot}2$, which was then adjusted to 3·5 with the use of a combination glass electrode.

The measurements in this paper were all done with homonuclear 2-dimensional (2D) correlated spectroscopy (COSY) (Aue *et al.*, 1976). COSY uses the pulse sequence (Aue *et al.*, 1976; Nagayama *et al.*, 1980; Wagner *et al.*, 1981):

$$[90^\circ - t_1 - 90^\circ - t_2]_n, \qquad (1)$$

where t_1 and t_2 are the evolution period and the observation period, respectively. To obtain a 2D n.m.r. spectrum, the measurement is repeated for a set of equidistant t_1 values. To improve the signal-to-noise ratio and to eliminate experimental artefacts, n groups of 16 recordings with different phases were added for each value of t_1 (Nagayama *et al.*, 1979, 1980). At the end of each recording, the system was allowed to reach equilibrium during a fixed delay of 1·2 s.

The spectra were recorded at 500 MHz on a Bruker WM 500 spectrometer. The spectrum in Fig. 1 was obtained from 512 measurements, with t_1 values from 0 to 47 ms and 2048 points in t_2. To reduce the observation time, the spectra in ^2H_2O were obtained from 256 measurements, with t_1 values from 0 to 24 ms and 1024 points in t_2. To end up with a 1024 × 1024 point data matrix in the frequency domain, which corresponds to a digital resolution of 5·3 Hz/point, the time domain matrix was in all experiments expanded to 2048 points in t_1 and 4096 points in t_2 by "zero-filling".

Quadrature detection was used for detection of the individual free induction decays, with the carrier frequency at the low-field end of the spectrum. For the measurement in H_2O, the solvent resonance was suppressed by selective, continuous irradiation at all times except during data acquisition (t_2: Anil Kumar *et al.*, 1980). Prior to Fourier transformation, the time domain data matrix was multiplied in the t_1 direction with a phase-shifted sine bell, $\sin(\pi(t+t_0)/t_s)$, and in the t_2 direction with a phase-shifted sine-squared bell, $\sin^2(\pi(t+t_0)/t_s)$. The length of the window functions, t_s, was adjusted for the bells to reach zero at the last experimental data point in the t_1 or t_2 direction, respectively. The phase shifts, t_0/t_s, were 1/32 and 1/64 in the t_1 and t_2 direction, respectively. The H_2O spectrum in Fig. 1 was further improved by symmetrization (Baumann *et al.*, 1981). The absolute value presentation was used in all the experiments.

Quantitative measurements of the amide proton exchange rates were made at 36°C and 68°C. The ^2H_2O solutions of BPTI used for these experiments were prepared at 24°C. Immediately after the protein had been dissolved and the p^2H adjusted to 3·5, the solution was heated to the desired exchange temperature for a certain length of time. Then it was cooled to 24°C, and the COSY spectrum was recorded during 12 h at 24°C. A new sample was prepared for each of the measurements with exchange times at 36°C of 0, 10, 240, 660 and 1720 min (Fig. 3). After the completion of the n.m.r. recording, the sample of the 1720 min experiment was further used for all the remaining measurements with longer exchange times (see Fig. 3), where the exchange during the n.m.r. recording at 24°C is negligibly small compared to the exchange during the waiting times at 36°C. Similarly, new samples were prepared for the individual measurements at 68°C, which were recorded after waiting times of 0, 5, 10 and 15 min. The same sample as for the 15 min experiment was further used for 2 additional measurements with exchange times of 60 and 120 min.

To measure resonance intensities, cross-sections through the COSY spectra were plotted (Nagayama *et al.*, 1978) and the height of the individual peaks was measured. For all the measurements, the spectra were recorded with identical instrument settings and the data were handled in identical ways. The peak heights in each spectrum were measured relative to those of the well-known and well-separated low-field $C^{\alpha}H$–$C^{\beta}H$ cross-peaks of Tyr21 and Cys30 (Wagner & Wüthrich, 1982). Since neither C^{α} nor C^{β} protons exchange with ^2H_2O, these two peaks have intensity 1 in each spectrum, independent of the exchange time and exchange temperature.

The rate constants in Table 1 were obtained from a non-linear least-squares fit of the experimental data to an exponential function. An error analysis was performed for all those protons, for which the time dependence of the peak intensity could be measured at 4 or more points.

For 19 amino acid residues in BPTI, the amide proton exchange is so fast that their NH–$C^{\alpha}H$ COSY cross-peaks were absent already from the spectrum recorded after the shortest exchange time used at 36°C. To obtain further information on these exchange reactions, new BPTI solutions in ^2H_2O were prepared at 4°C and the p^2H was again adjusted to 3·5. A COSY spectrum of one of these samples was recorded in 12 h at 10°C immediately after sample preparation. The second sample was heated to 36°C for 10 min before a COSY spectrum was recorded in 12 h at 10°C. For these experiments, resonance assignments at 10°C were obtained with the previously described techniques (Wagner & Wüthrich, 1982).

3. Results

In two-dimensional correlated spectroscopy, one obtains a square array of resonance peaks in the ω_1–ω_2 plane (Fig. 1). In the complete COSY spectrum, the so-called diagonal peaks with $\omega_1 = \omega_2$ correspond to the normal, one-dimensional spectrum (see e.g. Nagayama *et al.*, 1980; Wagner & Wüthrich, 1982) (the diagonal peaks are not shown in any of the Figures in this paper). Any two diagonal peaks at $\omega_1 = \omega_2 = \omega_A$ and $\omega_1 = \omega_2 = \omega_X$ that originate from protons linked by scalar spin–spin coupling (J-coupling), are connected by a pair of "cross-peaks" located in symmetrical positions with respect to the diagonal at ($\omega_1 = \omega_A$, $\omega_2 = \omega_X$) and

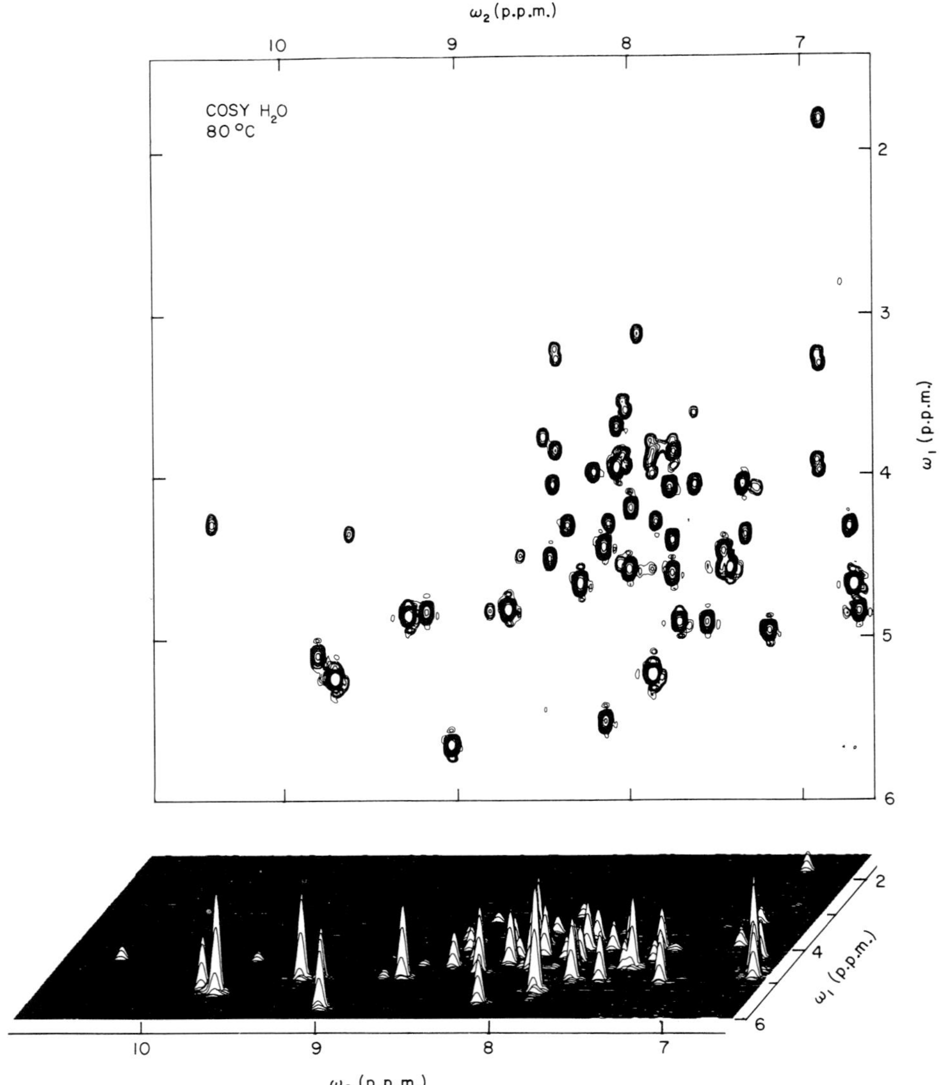

FIG. 1. Symmetrized (Baumann *et al.*, 1981), absolute value 500 MHz ^1H COSY spectrum of a 0·02 M solution of BPTI in a mixed solvent of 90% H_2O and 10% ^2H_2O, pH 4·6, at 80°C. The spectrum was recorded in approx. 24 h, the digital resolution is 5·3 Hz/point. The entire spectrum was presented previously (Wagner & Wüthrich, 1982). Here, the region (ω_1 = 1·5 to 6·0 p.p.m., ω_2 = 6·6 to 10·8 p.p.m.), is shown in a stacked plot representation in the lower part of the Figure, which affords a "3-dimensional" view of the spectrum, and as a contour plot in the upper part. The temperature of 80°C was chosen to avoid the appearance of cross-peaks with the labile protons of arginine and lysine side-chains, which show up at lower temperature at this pH value. The spectrum thus contains only NH–C$^{\alpha}$H cross-peaks, whereby all the amino acid residues can be observed except Arg1, Gly37 and the 4 proline residues (Wagner & Wüthrich, 1982).

($\omega_1 = \omega_X$, $\omega_2 = \omega_A$). When one works with a macromolecule such as a protein, a single COSY spectrum can provide information on all the proton–proton J-connectivities in the molecular structure. The present application of COSY for studies of amide proton exchange rates relies on measurements of the time dependence of the intensities of the polypeptide backbone amide proton–C^{α} proton cross-peaks, which we have previously called the "n.m.r. fingerprint of the protein structure" (Wagner & Wüthrich, 1982).

In a protein COSY spectrum recorded in H_2O at pH near 3·5, each amino acid residue gives rise to one NH–C^{α}H cross-peak. Exceptions are the glycine residues, which give in general two cross-peaks, the N-terminal residue, where the amino protons usually exchange too rapidly to be observed (Scheinblatt & Rahamin, 1976; Bundi & Wüthrich, 1979) and, of course, the proline residues (Wagner & Wüthrich, 1982). The NH–C^{α}H cross-peaks are usually all contained within a limited spectral region, which contains no or very few other cross-peaks, depending on the conditions of the experiment. When the amino acid sequence is known, inspection of this region of the COSY spectrum thus provides a rapid check of whether all or most of the amino acid residues can be observed and their resonances resolved in the 2D n.m.r. spectra. For BPTI, all NH–C^{α}H cross-peaks are located in the spectral region (ω_1 = 1·8 to 5·8 p.p.m., ω_2 = 6·7 to 10·5 p.p.m.), and at the pH and temperature of the experiment in Figure 1 this region contains no other cross-peaks (Wagner & Wüthrich, 1982)†. Figure 1 presents an almost complete fingerprint of the BPTI amino acid sequence. The NH–C^{α}H cross-peaks are resolved for 52 residues, i.e. all except Arg1, Gly37 and the four proline residues, and all the peaks were previously assigned to specific residues in the amino acid sequence (Wagner & Wüthrich, 1982).

A reduced fingerprint of the protein structure is afforded by the same region of a COSY spectrum recorded in ^2H_2O (Fig. 2). NH–C^{α}H cross-peaks are seen only for those residues for which the amide proton exchanges relatively slowly with ^{2}H of the solvent. Inspection of Figure 2 reveals that instead of the cross-peaks for 52 residues seen in Figure 1, only 33 residues are manifested. The peaks of these 33 residues are identified in Figure 3. Individual assignments for all the backbone amide protons of BPTI have been described at 68°C and pH 4·6 (Wagner & Wüthrich, 1982). For all the protons observed in ^2H_2O, the assignments at 24°C and p^{2}H 3·5 were independently established with the same techniques (some assignments at 24°C are documented by Wagner *et al.*, 1981). Compared to

† With regard to applications of the presently proposed experiments to other proteins, the following points should be considered. From the random coil chemical shifts (Bundi & Wüthrich, 1979) one might expect the NH–C^{α}H COSY cross-peaks to be contained in a more restricted region of the spectrum. Figure 1 shows that in BPTI approx. 80% of these cross-peaks are indeed within the area (ω_1 = 3·5 to 5·0 p.p.m., ω_2 = 6·6 to 9·3 p.p.m.). However, because of the extensive conformation-dependent chemical shifts (Wüthrich, 1976), at least the region shown in Fig. 1 should be searched to obtain a reliable fingerprint of the protein. At lower temperature and concomitantly slower exchange rates of the labile protons, this spectral region may further contain J cross-peaks with the amino protons of the lysine side-chains and with guanidinium NH groups of the arginine side-chains (Bundi & Wüthrich, 1979; Wagner & Wüthrich, 1982; Wider *et al.*, 1981). In principle, COSY peaks connecting $C^{\beta}H_2$ with C^{δ}H in aromatic side-chains might also occur in this spectral region. While these long-range J couplings were observed in the isolated amino acids (Nagayama *et al.*, 1978; Wider *et al.*, 1981), no such cross-peaks have been observed in COSY spectra of proteins.

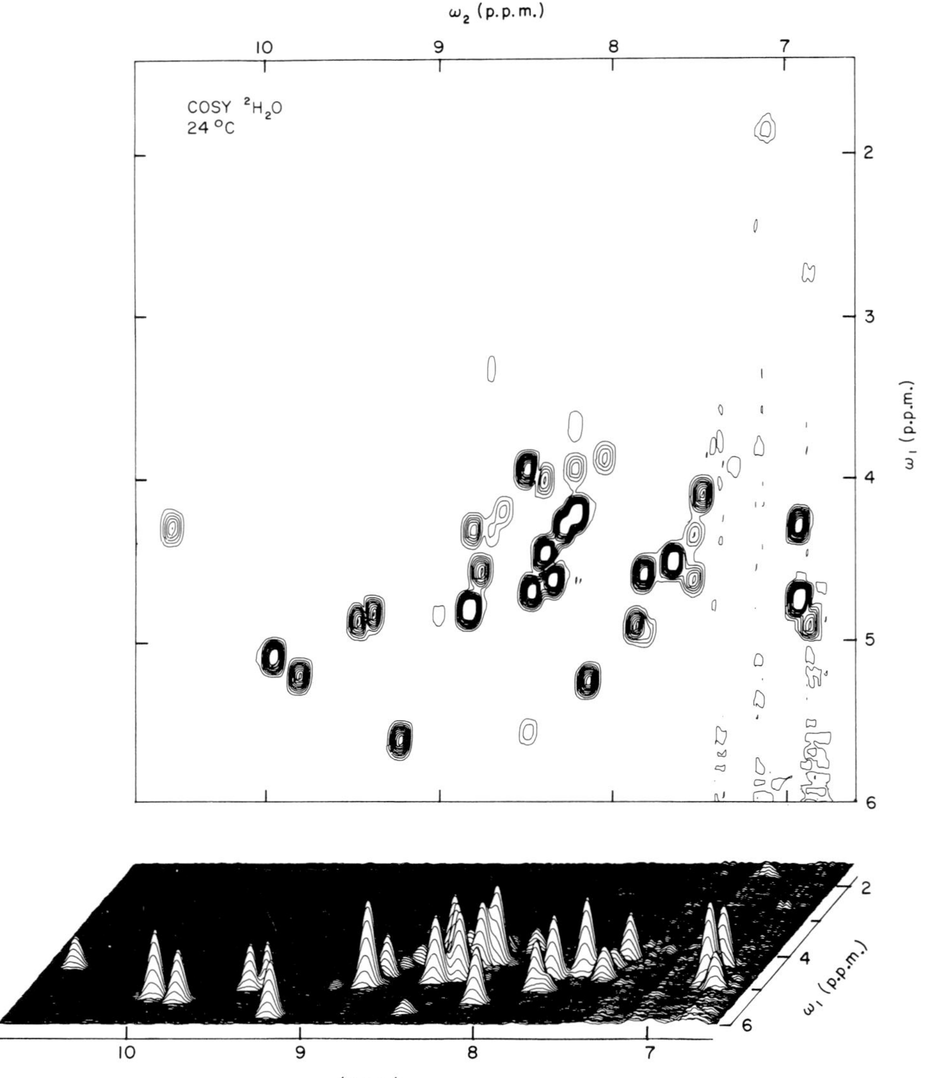

FIG. 2. Absolute value 500 MHz ^1H COSY spectrum of a 0·02 M solution of BPTI in ^2H_2O, p^{2}H 3·5, at 24°C. Immediately before the n.m.r. experiment, the protein was dissolved in ^2H_2O at 24°C. The solution was then kept at 36°C for 10 min and then cooled again to 24°C, at which temperature the spectrum was recorded in 12 h. The digital resolution is 5·3 Hz/point. The same spectral region is shown as in Fig. 1. Only the NH–C$^\alpha$H cross-peaks with the slowly exchanging amide protons (Wüthrich & Wagner, 1979; Wagner & Wüthrich, 1982) are observed. The vertical noise bands between 6·6 and 7·5 p.p.m. are "tails" of the strong aromatic signals in the adjacent spectral region.

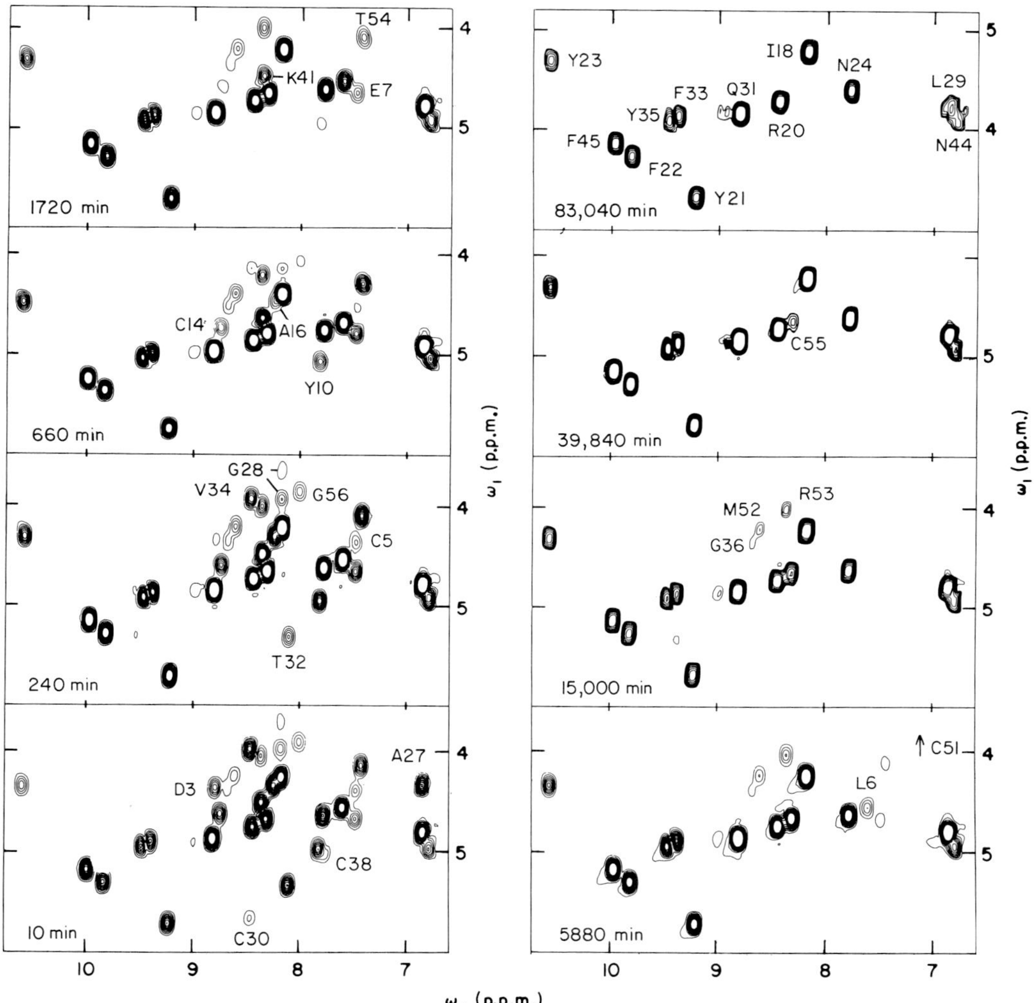

FIG. 3. Absolute value 500 MHz ^{1}H COSY spectra of 0·02 M solutions of BPTI in ^2H_2O recorded at different times after the protein was dissolved. The solutions were freshly prepared at 24°C and then kept at 36°C to allow exchange of protein amide protons with ^{2}H of the solvent. At the times indicated in the Figure, a particular solution was cooled to 24°C and a spectrum was recorded in 12 h. The digital resolution is 5·3 Hz/point. Compared to Fig. 2, the size of the plot was reduced to the region (ω_1 = 3·6 to 6·0 p.p.m., ω_2 = 6·6 to 10·8 p.p.m.), cutting off the cross-peak of Cys51 at (ω_1 = 1·85, ω_2 = 7·05 p.p.m.) and 1 of the 2 cross-peaks of Gly36 at (ω_1 = 3·4, ω_2 = 8·6 p.p.m.). Furthermore, the vertical noise bands between 6·6 and 7·5 p.p.m. have been covered with white paint. The peaks that disappeared in the course of the experiment are identified in the last spectrum, where they can be observed readily. Thus this Figure affords a qualitative survey of the exchange rates (for quantitative data, see Fig. 4 and Table 1). The peaks that did not disappear within 80,000 min are identified in the last spectrum. The disappearance of the C51 cross-peak, which is outside the spectral region shown, is indicated with an arrow in the spectrum taken after 5880 min.

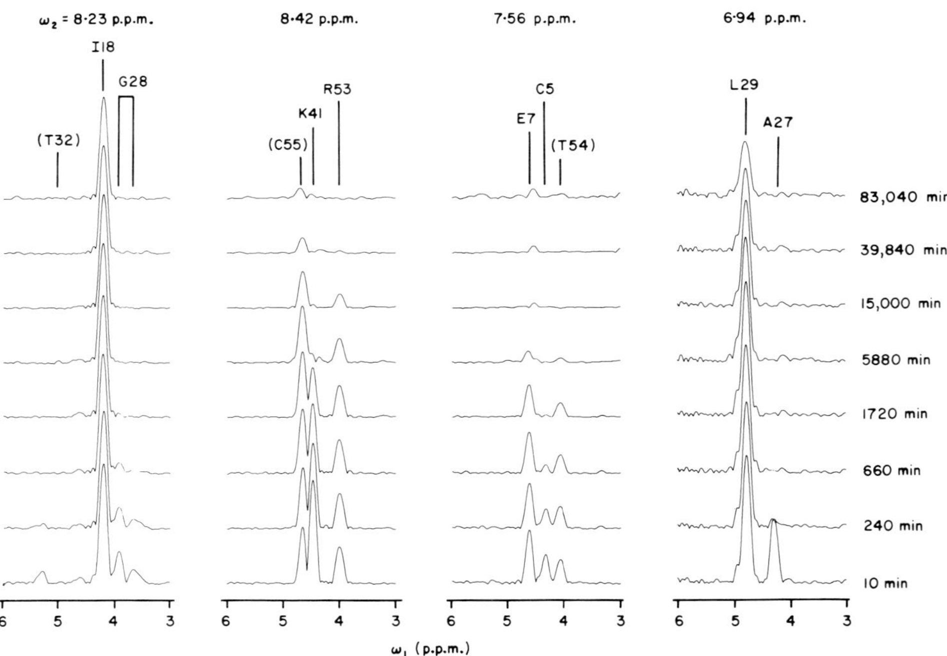

FIG. 4. Vertical cross-sections of the spectra shown in Fig. 3 taken at 4 different positions along the ω_2 axis. At the top of the Figure, those residues are identified for which ω_2 coincides with ω_2 of the cross-section. The cross-peaks of the residues indicated in parentheses are located so close to the cross-sections that tails of the peaks are observed in these presentations.

Figure 1, the cross-peaks in Figures 2 and 3 are considerably broader. This is a consequence of both the lower temperature and the smaller number of t_1 and t_2 values used to record the spectra (see Materials and Methods).

For the 33 residues that can be observed in Figure 2, the time-course of the cross-peak intensity was followed over a period of approximately two months (Fig. 3). For the quantitative measurements, cross-sections of the COSY spectra (Nagayama *et al.*, 1978) were used as illustrated in Figure 4. It seems worthwhile to point out the excellent signal-to-noise ratio achieved with the present experiments (Fig. 4) and to emphasize that the spectral analysis relied on measurements of the peak heights relative to peaks that are known not to vary with time (see Materials and Methods). For four residues, only upper limits for the exchange rates at 36°C were obtained (Table 1), since the peak intensities were essentially unchanged after 83040 minutes (Fig. 3). For 26 residues, the time change of the peak intensity could be followed in four or more of the spectra in Figure 3. For these, error limits for k_m (36°C) ranging from $\pm 3\%$ to $\pm 10\%$, depending on the proton, were computed (Table 1). For the remaining three residues, less than four data points were obtained, so that error limits for k_m could not be established reliably (Table 1).

TABLE 1

Rate constants k_m *(in* 10^{-3} min^{-1}, $\pm 10\%$*) for the exchange of individual backbone amide protons in BPTI at* p^2H *3·5 and 36°C and 68°C*

Amino acid residue	In solution			In the crystal structure		
	$k_m(36°C)$†	$k_m(68°C)$†	$k_{intr}(68°C)$‡	Accessible surface area§	Hydrogen bonding‖	Regular secondary structure¶
Arg1	n.o.	n.o.		27·5		
Asp3	10	~330	37,000	4·6		
Phe4	f	f	51,000	2·5		
Cys5	1·9	93	110,000	0·0		
Leu6	0·31	23	56,000	0·0		
Glu7	0·28	14	39,000	1·0		
Tyr10	1·2	93	24,000	0·0	+	
Thr11	f	f	39,000	3·2		
Gly12	~100	f	40,000	3·3		
Cys14	1·1	230	88,000	0·0	+	
Lys15	f	f	69,000	4·0		
Ala16	1·5	87	56,000	1·6	+	β
Arg17	f	f	35,000	4·8		
Ile18	0·0022	6·5	56,000	0·0	+	β
Ile19	f	f	30,000	3·6		
Arg20	<0·0004	7·1	35,000	0·0	+	β
Tyr21	<0·0004	5·6	44,000	0·0	+	β
Phe22	<0·0004	5·9	28,000	0·0	+	β
Tyr23	<0·0004	6·2	28,000	0·0	+	
Asn24	0·00049	12	88,000	0·0	+	β
Ala25	~100	f	78,000	1·8		
Lys26	f	f	35,000	0·5		
Ala27	~45	~320	56,000	0·0	+	β
Gly28	1·8	52	80,000	0·4	+	β
Leu29	0·011	9·6	59,000	0·0		
Cys30	~80	f	88,000	1·0		
Gln31	0·00048	11	78,000	0·0	+	β
Thr32	4·3	210	123,000	3·6		
Phe33	0·0010	6·8	44,000	0·0	+	β
Val34	3·1	140	36,000	4·6		
Tyr35	0·0044	10	24,000	0·0	+	β
Gly36	0·070	50	99,000	0·0	+	
Gly37	n.o.	n.o.	158,000	0·0		
Cys38	~80	f	175,000	0·0	+	
Arg39	f	f	69,000	7·2		
Ala40	f	f	56,000	2·6		
Lys41	0·51	180	35,000	0·2	+	
Arg42	f	f	70,000	1·8		
Asn43	~100	f	139,000	0·0		
Asn44	0·0075	37	196,000	0·0		
Phe45	0·0013	20	62,000	0·0	+	β
Lys46	f	f	37,000	2·4		
Ser47	~100	f	174,000	0·0		
Ala48	f	f	56,000	0·7		
Glu49	f	f	39,000	0·0		
Asp50	~100	f	72,000	0·0	+	
Cys51	0·22	~50	144,000	0·0	+	α
Met52	0·057	11	56,000	0·0	+	α

TABLE 1 *(continued*

Amino acid residue	In solution			In the crystal structure		
	k_m(36°C)†	k_m(68°C)†	k_{intr}(68°C)‡	Accessible surface area§	Hydrogen bonding‖	Regular secondary structure¶
Arg53	0·071	16	35,000	0·0	+	α
Thr54	0·49	33	174,000	0·0	+	α
Cys55	0·036	20	174,000	0·0	+	α
Gly56	2·1	79	156,000	0·0	+	α
Gly57	f	f	158,000	3·0		
Ala58	f	f	63,000	11·7		

For 68°C and p^{2}H 3·5 the intrinsic exchange rates, k_{intr} are also given. Further, the following properties of the individual amide protons in the crystal structure of BPTI are listed. Accessible surface area, hydrogen bonding and, where applicable, location in a polypeptide segment which forms regular secondary structure.

† The numbers give k_m in 10^{-3} min^{-1} ±10%. The sign ~ in front of a number indicates that less than 4 experimental points were measured and therefore no error analysis was warranted. n.o., not observed, indicates that the NH resonance of this residue could not be detected in the COSY spectrum. f indicates that the amide proton exchanged too rapidly to be seen in the COSY spectrum recorded at 10°C (see the text). The following limits for the exchange rates of these protons were estimated: k_m(36°C) > 0·1 min^{-1}; k_m(68°C) > 5 min^{-1} (see the text).

‡ Intrinsic rate constants at p^{2}H 3·5, which reflect inductive effects of neighbouring residues, were calculated according to the rules of Molday *et al.* (1972) starting from equation (2) of Englander *et al.* (1972). To obtain these data, pK_a values of 3·4, 3·8, 3·8, 3·0 and 2·9 were used for Asp3, Glu7, Glu49, Asp50 and the C-terminal Ala58, respectively (Wüthrich & Wagner, 1979). k_{intr}(68°C) in 10^{-3} min^{-1}.

§ Accessible surface area as defined by Lee & Richards (1971). The calculations for BPTI are described in Chothia & Janin (1975). The data used here are taken from this reference and from a complete listing of the solvent accessibilities for all the atoms in BPTI, which was kindly given to us by Dr C. Chothia.

‖ A + sign indicates that the amide proton of this residue was assigned to a hydrogen bond in the refined crystal structure (Deisenhofer & Steigemann, 1975).

¶ α and β, respectively, indicate that the amide proton was found to be in a hydrogen bond that is part of an α-helix or an antiparallel β-sheet in the refined crystal structure (Deisenhofer & Steigemann, 1975).

Corresponding experiments were done at 68°C. The exchange at this temperature could be followed for 31 amide protons (Table 1). For 28 residues, k_m(68°C) could be established with error limits of ±10%, and for three residues, approximate exchange rates without error limits were obtained.

Additional experiments were used to further discriminate among the 19 amide protons that exchange too rapidly to be seen in the first spectrum recorded in the experiment of Figure 3. A ^2H_2O solution of BPTI was prepared at 4°C and immediately thereafter a COSY spectrum was recorded at 10°C in 12 hours. Figure 5 shows that, compared to Figures 2 or 3, five additional residues could thus be observed. In a different experiment, where an identical sample was heated to 36°C for ten minutes before the COSY spectrum was recorded at 10°C, two of these five cross-peaks were too weak to be seen and the other three had lost much of their intensity. From these observations, we estimated that k_m(36°C) ~ 0·1 min^{-1} for the five protons identified in Figure 5. For the 14 protons not seen in the experiment at 10°C, we expect accordingly that k_m(36°C) > 0·1 min^{-1} (Table 1). (See Discussion for further details on how these values were obtained.)

4. Discussion

(a) *2D n.m.r. for studies of amide proton exchange rates*

The main advantage of the presently described use of COSY for studies of amide proton exchange in proteins is the improved spectral resolution, which allows observation of separate peaks for nearly all amide protons in BPTI. Compared to corresponding unidimensional n.m.r. experiments, the overlap of amide proton resonances with those of the aromatic protons is completely removed, and the resolution between different amide protons is greatly improved. With unidimensional experiments at 360 MHz, quantitative exchange data at temperatures near 36°C previously obtained for 16 individually assigned amide protons of BPTI (Wüthrich & Wagner, 1979; Richarz *et al.*, 1979). This already compared favourably with the amount of information on individual exchange rates that can be obtained from experiments other than n.m.r. (see e.g. Hvidt & Nielsen, 1966; Englander *et al.*, 1972; Rosa & Richards, 1979). With the use of COSY, quantitative exchange rates were obtained for 34 residues, and for four additional residues quantitative data could have been obtained by extension of the duration of the experiment (Table 1). For 14 additional residues, a lower limit for k_m(36°C) was established. Since this limiting rate is only approximately ten times slower than the average exchange rate for exposed amide protons in random coil model peptides (Englander *et al.*, 1972; and see below), we are quite confident that these 14 protons represent the exchange behaviour of solvent-accessible amide protons located near the molecular surface in the solution conformation of the protein. Overall, the COSY peaks of 52 of the total of 53 backbone amide protons had thus been individually assigned and the exchange behaviour of these protons was characterized.

In the present experiments, a quench method was applied; i.e. the sample temperature was lowered before the start of the n.m.r. measurement. Because of the relatively long recording times needed to obtain a COSY spectrum with workable signal-to-noise ratio, this will quite generally be needed. In a quench experiment, the observed resonance intensity, $I(t)$, where t is the time during which exchange took place before the reaction was quenched, can be written:

$$I(t) = \mathrm{I}_0\, \mathrm{e}^{-k_m t} \int_0^{t_{rec}} \frac{\mathrm{e}^{-k_{rec}t'}}{t_{rec}}\, \mathrm{d}t'. \qquad (2)$$

I_0 is the initial peak intensity, k_m the exchange rate constant of interest, t_{rec} the time used to record the spectrum and k_{rec} the exchange rate constant under the conditions that prevail during the recording of the spectrum. Equation (2) yields:

$$I(t) = \left[I_0 \frac{1}{-k_{rec}t_{rec}} (\mathrm{e}^{-k_{rec}t_{rec}} - 1) \right] \mathrm{e}^{-k_m t}. \qquad (3)$$

It is readily apparent from equation (3) that for the very slowly exchanging protons, with $k_{rec}t_{rec} \ll 1$, the exchange during the recording of the COSY spectrum can be neglected. For the more rapidly exchanging protons, straightforward quantitative evaluation of k_m is possible when fresh samples are prepared for each

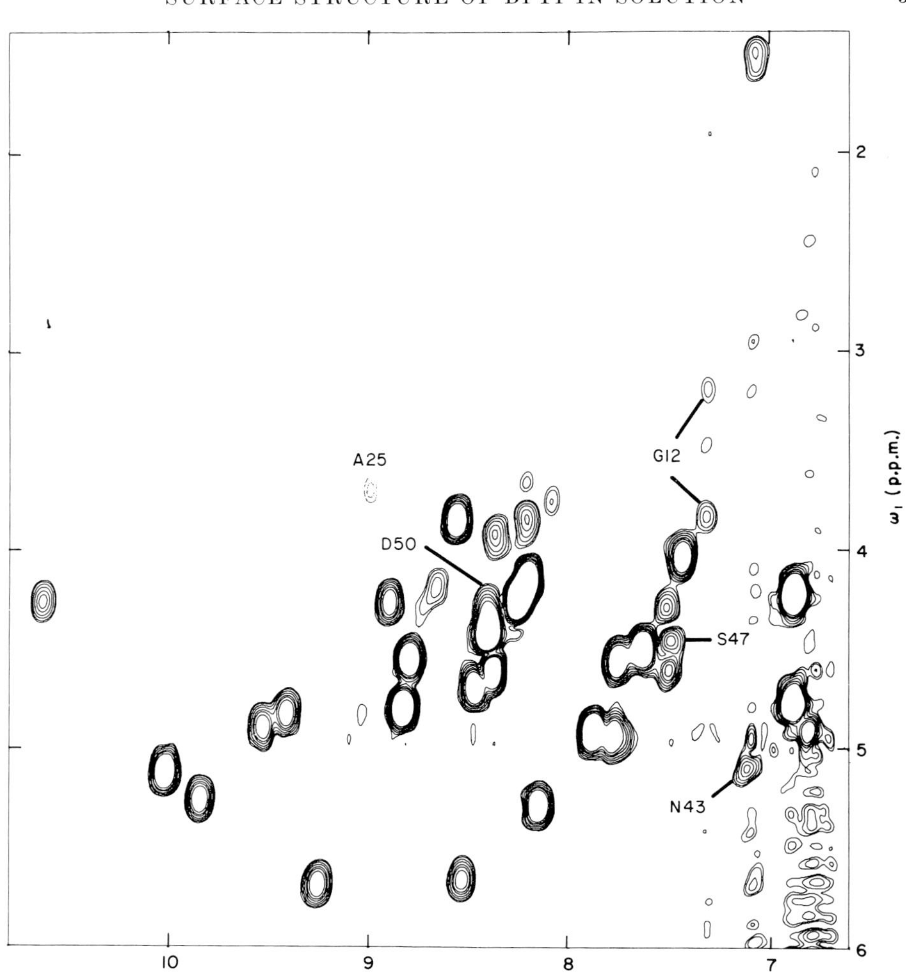

FIG. 5. Contour plot of the absolute value 500 MHz ^{1}H COSY spectrum of a 0·02 M solution of BPTI in ^2H_2O, p^{2}H 3·5, at 10°C. Immediately before the n.m.r. experiment, the protein was dissolved in ^2H_2O at 4°C. The spectrum was recorded in 12 h. The digital resolution is 5·3 Hz/point. The same spectral region is shown as in Figs 1 and 2. The assignments are indicated for the cross-peaks that could not be observed in the experiments of Figs 2 and 3. The vertical noise bands on the right of the spectrum are tails of the strong aromatic signals in the adjacent spectral region.

measurement of $I(t)$, and identical measuring times, t_{rec}, are used for all recordings of COSY spectra. In this case, the pre-exponential factor in equation (3) is the same for all measurements and the quantitative evaluation of k_m is not affected by the length of the recording time, provided that for each measurement $I(t)$ is $\gtrsim I_0/5$, which is the practical limit where the signal intensity can still be measured. With rapid quenching of both pH and temperature to the conditions of minimal exchange rate, and provided that the protein is not denatured under these conditions, the COSY experiment can thus be used for amide proton exchange studies over a wide range of experimental conditions. The spectrum of Figure 5 demonstrates that well-resolved COSY spectra can be obtained at conditions that are near those for minimal exchange rates (Englander *et al.*, 1972).

The fastest exchange between protons of the protein and deuterons of the solvent ^2H_2O that may be measured with the presently described use of COSY at 10°C (Fig. 5) is approximately 0·005 min^{-1}. One arrives at this value with the assumption that for quantitative studies the peak intensity after 12 hours at 10°C should be $\gtrsim I_0/5$. Following equation (2) of Englander *et al.* (1972), the average exchange rate for solvent-exposed amide protons in model peptides at 10°C and p^2H 3·5 is of the order of 0·055 min^{-1}. We thus know that solvent-exposed amide protons exchange too rapidly to be seen in COSY experiments in ^2H_2O, but the gap between the limiting observable value and the random coil exchange rate is only of the order of 10. The lower limits of the exchange rates at 36 and 68°C, respectively, for the protons that exchange too rapidly to be seen at 10°C, were obtained by extrapolation from 10°C to higher temperatures. This means that from a lower limit at 10°C, $k_m(10°C) > 0{\cdot}005\ min^{-1}$, lower limits at higher temperatures, $k_m(36°C) > 0{\cdot}1\ min^{-1}$ and $k_m(68°C) > 5\ min^{-1}$, were estimated with the use of equation (2) of Englander *et al.* (1972); i.e. it was assumed that the enthalpy of activation for these protons is not lower than that of solvent-exposed protons in model peptides.

(b) *Comparison of 2D n.m.r. with other techniques for proton exchange measurements*

Compared to the classical amide proton exchange measurements, which use radioactive tracers or infrared spectroscopy (Hvidt & Nielsen, 1966; Englander *et al.*, 1972; Rosa & Richards, 1979), n.m.r. has the advantage of being able to provide quantitative exchange rates for individual amide protons in specified locations along the polypeptide chain. As a consequence of the improved spectral resolution in COSY spectra as compared to conventional unidimensional n.m.r., much more extensive sets of individual exchange rates can be measured. In BPTI, a complete mapping of the amide proton exchange rates for the entire polypeptide chain was thus obtained.

An alternative, novel procedure to map amide proton exchange rates in protein uses neutron diffraction techniques (Kossiakoff, 1982). A map of qualitative exchange data was thus recently obtained for crystalline trypsin (Kossiakoff, 1982). Since neutron diffraction and 2D n.m.r. can provide nearly complete mappings of amide proton exchange rates in single crystals and in non-crystalline environments, respectively, the combined use of the two methods with the same protein should open an avenue for direct comparison of the molecular dynamics in the crystal and in solution. From the available information (Kossiakoff, 1982), the neutron diffraction technique should be more readily applicable for bigger proteins than 2D n.m.r., but it might not be practical for proteins of any size to obtain quantitative exchange rates from neutron diffraction experiments. COSY can provide quantitative exchange measurements under a variety of different experimental conditions, which will probably be of considerable importance for investigations of mechanistic aspects of proton exchange (see below). Considering that the use of both techniques for proton exchange studies is just being introduced, it would appear premature to evaluate further their respective potentialities, but already now there appears to be a good chance that they could in many ways provide complementary data on details of internal dynamics of proteins that were hitherto not amenable to experimental investigations.

(c) *Mechanism of amide proton exchange*

The availability of complete sets of individually assigned amide proton exchange rates measured with different conditions of pH and temperature should in the future help to clarify some uncertainties concerning the mechanisms by which the proton exchange reactions in BPTI proceed (Hilton & Woodward, 1978,1979; Wagner & Wüthrich, 1979; Wüthrich *et al.*, 1980*a*,*b*; Hilton *et al.*, 1981). Here, we would like to describe a preliminary observation that might have some bearing on our understanding of exchange mechanisms.

In Figure 6, the individual amide proton exchange rates for BPTI at 68°C and pH 3·5 are presented together with the intrinsic exchange rates. The latter are exchange rates for the random coil form of the polypeptide chain, which manifest the sequence effects on the exchange of solvent-accessible protons. Intrinsic rate constants for BPTI at 68°C and p^2H 3·5 were calculated with the rules of Molday *et al.* (1972). Figure 6 clearly shows that there is no simple relation between the two sets of rate constants and it can in most instances be excluded that the relative values of the exchange rates, k_m, of neighbouring residues in the amino acid sequence are determined by sequence effects. On the other hand, in an overall process dominated by exchange between denatured protein and the solvent one would intuitively expect the relative rates for individual neighboring protons to be

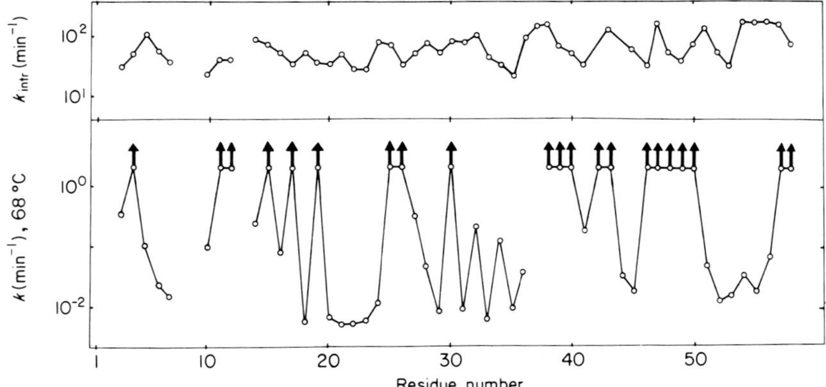

FIG. 6. Comparison of the experimental amide proton exchange rates in BPTI, k, at 68°C and p^2H 3·5 (lower trace, same as in Fig. 7) with the intrinsic exchange rates, k_{intr}, (upper trace) calculated with the rules of Molday *et al.* (1972) for the same conditions of p^2H and temperature. k_{intr} reflects the sequence effects on the exchange rates; i.e. mainly inductive effects from neighbouring residues. The empirical temperature dependence determined by Englander & Poulsen (1969) was used to obtain the data at 68°C (see the text). Rates are plotted on a logarithmic scale *versus* the amino acid sequence. No data are given for Arg1, the proline residues 2, 8, 9 and 13, and Gly37.

dominated by the intrinsic exchange rates. On this basis, the observations in Figure 6 might be forwarded as evidence against an exchange mechanism dominated by admixture of denatured protein molecules to the ensemble of BPTI conformers present in aqueous solution at 68°C (Hilton *et al.*, 1981).

(d) *Crystal structure of BPTI and amide proton exchange in solution*

Figure 7 presents the accessible surface areas (Lee & Richards, 1971) for the backbone amide groups in the BPTI crystal structure (Deisenhofer & Steigemann, 1975) and the proton exchange rates in solution at two different temperatures. Overall, the qualitative similarity of the patterns obtained in the three graphs representing the accessible surface areas, k_m(36°C) and k_m(68°C) along the sequence is quite striking. If one takes the patterns of exchange rates as an empirical manifestation of the three-dimensional protein structure, then the close similarity between the data obtained at 36°C and 68°C would indicate that the exchange at these two temperatures is from closely similar conformations. It would thus appear that Figure 7 provides additional evidence against the hypothesis that amide proton exchange at higher temperature is dominated by exchange from the denatured protein (Hilton *et al.*, 1981).

Some intriguing observations result from evaluation of the exchange rates in the light of the hydrogen-bonding network in the crystal structure. The conclusion that the exchange in the β-sheet is faster at both ends than in the central region, which resulted previously from a limited number of individually assigned exchange rates, is confirmed by the present, complete data set. The most slowly exchanging protons are those of residues 21 to 24, which form the central strand in a short region of triple-stranded β-sheet (Deisenhofer & Steigemann, 1975). Except for Ala16, the exchange rates at 36°C of all other hydrogen-bonded amide protons in the β-sheet tend to be slower than those for the hydrogen-bonded protons in the α-helix. The individual protons in the α-helix have also somewhat different rates, whereby the exchange is slowest for Cys55 and Met52. Quite possibly, this is an effect of the disulphide bonds formed by Cys51 and Cys55. Further, it seems worth noting that the amide protons of Tyr10, Cys14, Cys38 and Lys41, which in the crystal structure are hydrogen-bonded with internal water molecules (Fig. 7; Deisenhofer & Steigemann, 1975), do not have unusual exchange rates. As one might have anticipated from the vanishing or very small (Lys41) static accessible surface areas for these residues in the crystal structure, we observed exchange rates $k_m(36°C) < 0{\cdot}1\ \text{min}^{-1}$ (Table 1; Fig. 7).

The exchange studies of amide protons that are located near the protein surface in the single crystal structure should be a useful tool to probe the surface of the solution conformation. One might, in a preliminary evaluation for example, assume that all the amide groups that have a non-vanishing accessible surface area, and are hence in van der Waals' contact with the solvent (Lee & Richards, 1971), should exchange rapidly; i.e. with $k_m(36°C) > 0{\cdot}1\ \text{min}^{-1}$. k_m(36°C) values $\leq 0{\cdot}1\ \text{min}^{-1}$ would accordingly correspond to groups with zero accessible surface area. Figure 7 and Table 1 reveal that there is a single residue that has zero accessible surface area and $k_m(36°C) > 0{\cdot}1\ \text{min}^{-1}$; i.e. Glu49. If one assumes that

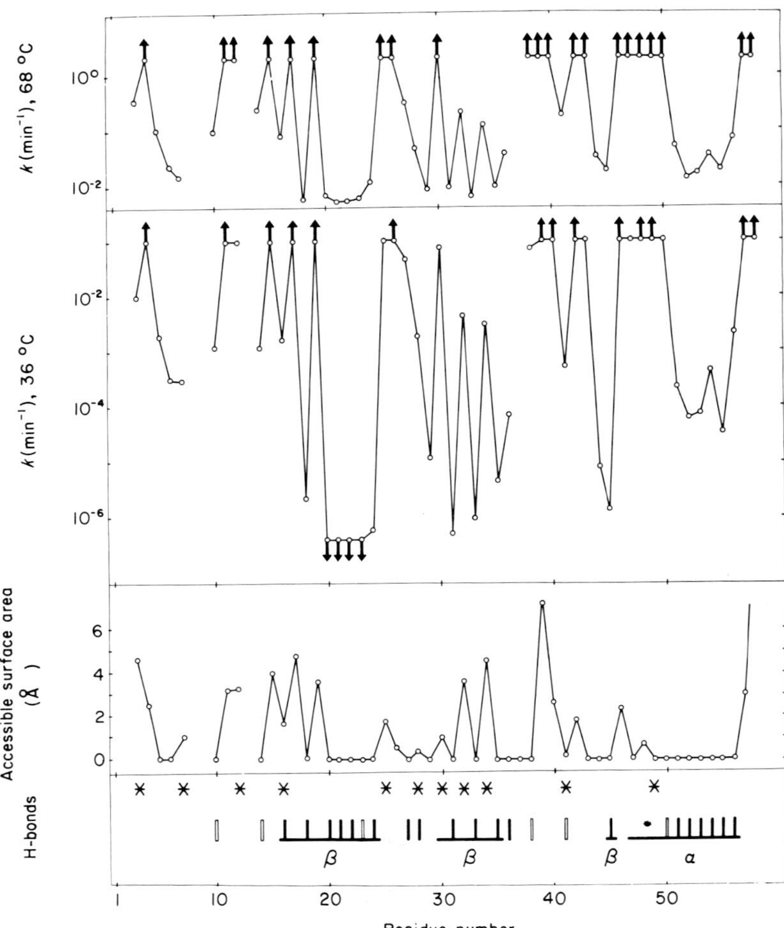

FIG. 7. Comparison of the amide protons exchange data for BPTI at p^2H 3·5 and 2 different temperatures (Table 1) with features of the crystal structure. The horizontal scale represents the amino acid sequence of BPTI. The 2 graphs at the top are logarithmic representations of the exchange rates at 68°C and 36°C. Arrows pointing upwards indicate that the exchange was too fast to be observed with COSY (see the text). Arrows pointing downwards show that the exchange was too slow to be measured quantitatively within the time of the experiment. This is the case for Arg20, Tyr21, Phe22 and Tyr23 at 36°C. No values are given for Arg1 or Gly37, which have not been observed, and of course for the 4 proline residues in positions 2, 8, 9 and 13. In the lower part of the Figure, the solvent-accessible surface areas (see Table 1 and the text) of the backbone peptide nitrogens are plotted. Furthermore, the intramolecular hydrogen bonds in the crystal structure of Deisenhofer & Steigemann (1975) are indicated. A filled vertical bar indicates that the amide protons of the residue in this location form intramolecular hydrogen bonds with main chain carbonyls, open vertical bars indicate hydrogen bonds to side-chains or internal water molecules. Residues that are part of regular α or β secondary structure are joined by horizontal bars. Finally, the stars identify the residues for which either the accessible surface area is zero and the proton exchange very fast, or which have a non-vanishing accessible surface area and a relatively slow exchange rate (see the text).

the protein surface is dynamic in the sense that the atoms undergo fluctuations about an equilibrium position, it is quite conceivable that the solvent exposure is increased as compared to the static accessibility in the crystal structure (Lee & Richards, 1971). On the other hand, there are ten residues that would in the crystal structure be in van der Waals' contact with the solvent but have $k_m(36°C) \leq 0{\cdot}1\ \text{min}^{-1}$ in solution. These are Asp3, Glu7, Gly12, Ala16, Ala25, Gly28, Cys30, Thr32, Val34 and Lys41. Here it appears intuitively rather unlikely that dynamic fluctuations of the protein structure could reduce the solvent accessibility as compared to the static accessibility in the crystal, and there is an indication that these residues are in different environments in the crystal and in solution.

The solvent-accessible surface area for the crystal structure (Table 1) was computed for an isolated BPTI molecule. Hence, lack of correlation with the amide proton exchange rates in solution could be due either to intermolecular aggregation at the high protein concentrations for which the data in Table 1 and Figure 7 were obtained, or to rearrangement of the protein surface structure between the crystal and the solution. Further studies of the apparent discrepancies between the conformations of BPTI in single crystals and in solution are in progress. These will include comparison of the proton exchange data in Table 1 with corresponding measurements of exchange rates in very dilute solutions of BPTI, to distinguish between influences of intermolecular aggregation and conformational changes in monomeric BPTI. We envisage that the observation of apparent discrepancies between the structures of the protein surfaces in the crystal and in solution by amide proton exchange studies could be a starting point for site-specific use of different, more laborious n.m.r. experiments capable of providing direct information on the surface conformation in these locations and/or on structural details of intermolecular aggregation in concentrated protein solutions.

We thank the Schweizerische Nationalfonds (project 3.538.79) for financial support, the Farbenfabriken Bayer AG for a generous gift of BPTI (Trasylol®), Dr C. Chothia for a complete listing of accessible surface areas in BPTI, and Mrs E. Huber and Mrs E. H. Hunziker for the careful preparation of the manuscript and the illustrations.

REFERENCES

Allerhand, A., Doddrell, D., Glushko, V., Cochran, D. W., Wenkert, E., Lawson, P. J. & Gurd, F. R. N. (1971). *J. Amer. Chem. Soc.* **93**, 544–546.

Anil Kumar, Wagner, G., Ernst, R. R. & Wüthrich, K. (1980). *Biochem. Biophys. Res. Commun.* **96**, 1156–1163.

Arseniev, A. S., Wider, G., Joubert, F. J. & Wüthrich, K. (1982). *J. Mol. Biol.* **159**, 323–352.

Aue, W. P., Bartholdi, E. & Ernst, R. R. (1976). *J. Chem. Phys.* **64**, 2229–2246.

Baumann, R., Wider, G., Ernst, R. R. & Wüthrich, K. (1981). *J. Magn. Reson.* **44**, 402–406.

Billeter, M., Braun, W. & Wüthrich, K. (1982). *J. Mol. Biol.* **155**, 321–346.

Bundi, A. & Wüthrich, K. (1979). *Biopolymers*, **18**, 285–298.

Chothia, C. & Janin, J. (1975). *Nature (London)*, **256**, 705–708.

Deisenhofer, J. & Steigemann, W. (1975). *Acta Crystallogr. sect. B*, **31**, 238–250.

Englander, S. W. & Poulsen, A. (1969). *Biopolymers*, **7**, 379–393.

Englander, S. W., Downer, N. W. & Teitelbaum, H. (1972). *Annu. Rev. Biochem.* **41**, 903–924.
Hilton, B. D. & Woodward, C. K. (1978). *Biochemistry*, **7**, 3325–3332.
Hilton, B. D. & Woodward, C. K. (1979). *Biochemistry*, **18**, 5834–5844.
Hilton, B. D., Trudeau, K. & Woodward, C. K. (1981). *Biochemistry*, **20**, 4679–4703.
Huber, R. (1979). *Trends Biochem. Sci.* **4**, 227–230.
Hvidt, A. & Nielsen, S. O. (1966). *Advan. Protein Chem.* **21**, 287–386.
Kalinichenko, P. (1976). *Stud. Biophys.* **58**, 235–240.
Karplus, M. & McCammon, J. A. (1981). *C.R.C. Crit. Rev. Biochem.* **9**, 293–349.
Kossiakoff, A. A. (1982). *Nature (London)*, **296**, 713–721.
Lee, B. & Richards, F. M. (1971). *J. Mol. Biol.* **55**, 379–400.
Molday, R. S., Englander, S. W. & Kallen, R. G. (1972). *Biochemistry*, **11**, 150–158.
Nagayama, K., Bachmann, P., Wüthrich, K. & Ernst, R. R. (1978). *J. Magn. Reson.* **31**, 133–148.
Nagayama, K., Wüthrich, K. & Ernst, R. R. (1979). *Biochem. Biophys. Res. Commun.* **90**, 305–311.
Nagayama, K., Anil Kumar, Wüthrich, K. & Ernst, R. R. (1980). *J. Magn. Reson.* **40**, 321–334.
Perutz, M. (1970). *Nature (London)*, **228**, 726–739.
Quiocho, F. A. & Lipscomb, W. N. (1971). *Advan. Protein Chem.* **25**, 1–59.
Ribeiro, A. A., King, R., Restivo, C. & Jardetzky, O. (1980). *J. Amer. Chem. Soc.* **102**, 4040–4051.
Richardson, J. S. (1981). *Advan. Protein Chem.* **34**, 167–339.
Richarz, R., Sehr, P., Wagner, G. & Wüthrich, K. (1979). *J. Mol. Biol.* **130**, 19–30.
Richarz, R., Nagayama, K. & Wüthrich, K. (1980). *Biochemistry*, **19**, 5189–5196.
Rosa, J. J. & Richards, F. M. (1979). *J. Mol. Biol.* **133**, 399–416.
Saunders, M. & Wishnia, A. (1958). *Ann. N.Y. Acad. Sci.* **70**, 870–874.
Scheinblatt, M. & Rahamin, Y. (1976). *Biopolymers*, **15**, 1643–1653.
Schulz, G. E. & Schirmer, R. H. (1979). *Principles of Protein Structure*, Springer, New York.
Wagner, G. & Wüthrich, K. (1979). *J. Mol. Biol.* **134**, 75–94.
Wagner, G. & Wüthrich, K. (1982). *J. Mol. Biol.* **155**, 347–366.
Wagner, G., De Marco, A. & Wüthrich, K. (1976). *Biophys. Struct. Mech.* **2**, 139–158.
Wagner, G., Anil Kumar & Wüthrich, K. (1981). *Eur. J. Biochem.* **114**, 375–384.
Wider, G., Baumann, R., Nagayama, K., Ernst, R. R. & Wüthrich, K. (1981). *J. Magn. Reson.* **42**, 73–87.
Wider, G., Lee, H. K. & Wüthrich, K. (1982). *J. Mol. Biol.* **155**, 367–388.
Wüthrich, K. (1976). *NMR in Biological Research: Peptides and Proteins*, North-Holland, Amsterdam.
Wüthrich, K. & Wagner, G. (1979). *J. Mol. Biol.* **130**, 1–18.
Wüthrich, K., Eugster, A. & Wagner, G. (1980*a*). *J. Mol. Biol.* **144**, 601–604.
Wüthrich, K., Wagner, G., Richarz, R. & Braun, W. (1980*b*). *Biophys. J.* **10**, 549–560.
Wüthrich, K., Wider, G., Wagner, G. & Braun, W. (1982). *J. Mol. Biol.* **155**, 311–319.

Edited by V. Luzzati

Reprinted from Biochemistry, 1985, *24*, 7396.
Copyright © 1985 by the American Chemical Society and reprinted by permission of the copyright owner.

Amide Proton Exchange in Proteins by EX_1 Kinetics: Studies of the Basic Pancreatic Trypsin Inhibitor at Variable p^2H and Temperature†

Heinrich Roder,‡ Gerhard Wagner,* and Kurt Wüthrich
Institut für Molekularbiologie und Biophysik, Eidgenössische Technische Hochschule, Zürich-Hönggerberg, CH-8093 Zürich, Switzerland

Received December 20, 1984

ABSTRACT: With the use of one-dimensional ^1H nuclear magnetic resonance, two-dimensional correlated spectroscopy, and two-dimensional nuclear Overhauser enhancement spectroscopy, the exchange mechanisms for numerous individual amide protons in the basic pancreatic trypsin inhibitor (BPTI) were investigated over a wide range of p^2H and temperature. Correlated exchange under an EX_1 regime was observed only for the most slowly exchanging protons in the central hydrogen bonds of the antiparallel β-sheet and only over a narrow range of temperature and p^2H, i.e., above ca. 55 °C and between p^2H 7 and 9, where the opening rates of the structure fluctuations which promote the exchange of these protons are of the order 0.1 min^{-1}. At p^2H below 7, the exchange of this most stable group of protons is uncorrelated and is governed by an EX_2 mechanism. At p^2H above 9, the exchange is also uncorrelated and occurs via either EX_2 or EX_1 processes promoted by strictly local structure fluctuations. For all other backbone amide protons in BPTI, the exchange was found to be uncorrelated and by an EX_2 mechanism under all conditions of p^2H and temperature where quantitative measurements could be obtained with the methods used, i.e., for $k_{ex} \lesssim 5$ min^{-1}. From these observations with BPTI it can be concluded that the amide proton exchange in globular proteins is quite generally via EX_2 processes, with rare exceptions for measurements with extremely stable protons at high temperature and basic p^2H. This emphasizes the need for further development of suitable concepts for the structural interpretation of EX_2 amide proton exchange [Wagner, G. (1983) *Q. Rev. Biophys. 16*, 1–57; Wagner, G., Stassinopoulou, C. I., & Wüthrich, K. (1984) *Eur. J. Biochem. 145*, 431–436] and for more detailed investigations of the intrinsic exchange rates for solvent-exposed amide protons in the "open" states of a protein [Roder, H., Wagner, G., & Wüthrich, K. (1985) *Biochemistry* (following paper in this issue)].

Among the numerous methods used to probe the dynamic nature of protein structures [e.g., see Gurd & Rothgeb (1979) and Wüthrich & Wagner (1984)], measurements of amide proton exchange rates have for a long time occupied a prominent position (Hvidt & Nielsen, 1966; Englander et al., 1972; Woodward & Hilton, 1979; Wagner & Wüthrich, 1979a; Barksdale & Rosenberg, 1982; Englander & Kallenbach, 1984). A major advance in experimental studies of proton exchange kinetics was achieved with the use of modern high-resolution nuclear magnetic resonance (NMR)[1] experiments. The power of NMR relies on the fact that separate resonance lines can be observed for the individual amide protons. For small proteins, essentially complete assignments for all polypeptide backbone amide protons can be obtained (Wagner & Wüthrich, 1982a; Strop et al., 1983), and therefore, specified amide proton exchange rates can be attributed to particular amino acid residues in the sequence. When the three-dimensional structure is known, a map of amide proton exchange rates across the protein can thus be obtained (Wagner & Wüthrich, 1982b; Wüthrich et al., 1984). BPTI is a small protein which was extensively studied with NMR (Richarz et al., 1979; Wagner & Wüthrich, 1979a; Hilton & Woodward, 1979) and for which a detailed map of amide proton exchange rates was obtained (Wagner & Wüthrich, 1982b).

Somewhat in contrast to the rapid progress made recently with the experimental measurements of amide proton exchange, the details of the mechanism of NH exchange in proteins and quantitative correlations between exchange kinetics and internal mobility are still controversial (Wüthrich et al., 1980; Hilton et al., 1981; Englander & Kallenbach, 1983). As a consequence, it is highly doubtful that the information contained in the experimental data has been fully exploited. In the present paper, we investigate mechanistic aspects of the proton exchange in BPTI on the basis of new experiments. These include studies of NOE's in partially exchanged BPTI samples, which provide direct evidence on the cooperativity of the exchange for neighboring amide protons in the spatial structure of the protein (Wagner, 1980, 1983). Furthermore, quench techniques are employed to extend the use of NMR measurements of individual proton exchange rates to the faster rates encountered at high temperatures, at high p^2H, and after the addition of denaturants.

Fundamental Considerations on Cooperative Proton Exchange

We interpret the exchange kinetics of interior amide protons in globular proteins within the framework of the "breathing model" (Hvidt & Nielsen, 1966; Englander et al., 1972; Wagner & Wüthrich, 1979a; Englander & Kallenbach, 1984). The exchange of an interior proton against a deuteron of the solvent can be described by the scheme:

† This work was supported by the Swiss National Science Foundation (Projects 3.528.79 and 3.284.82).

‡ Present address: Department of Biochemistry and Biophysics, University of Pennsylvania School of Medicine, Philadelphia, PA 19104.

[1] Abbreviations: BPTI, basic pancreatic trypsin inhibitor; TRAM-BPTI, chemical modification of BPTI obtained by transamination of the N-terminus; Gdn·HCl, guanidine hydrochloride; NMR, nuclear magnetic resonance; NOE, nuclear Overhauser effect; NOESY, two-dimensional nuclear Overhauser enhancement spectroscopy; COSY, two-dimensional correlated spectroscopy.

0006-2960/85/0424-7396$01.50/0 © 1985 American Chemical Society

$$\mathrm{N(H)} \underset{k_2}{\overset{k_1}{\rightleftharpoons}} \mathrm{O(H)} \xrightarrow[^2\mathrm{H_2O}]{k_3} \mathrm{O(D)} \underset{k_1}{\overset{k_2}{\rightleftharpoons}} \mathrm{N(D)} \quad (1)$$

N describes the ensemble of protein conformations in which the proton under consideration is protected from the solvent, and O stands for all conformations in which the proton is exposed to the solvent. In this general scheme, the N(H) to O(H) transition is characterized only to the extent that it leads to solvent exposure of the amide proton, which can be the result of, for example, local structure fluctuations in the folded protein or cooperative unfolding of larger portions of the molecule. The equilibrium between the two classes of states is described by an "opening rate" k_1 and a "closing rate" k_2. We assume that in the open states the proton exchanges with deuterium of the solvent ^2H_2O with the intrinsic exchange rate k_3. Values for individual k_3's can be computed with the rules by Molday et al. (1972) and have also been obtained directly by ^{1}H NMR exchange measurements in thermally unfolded BPTI (Roder et al., 1985). Under conditions favoring the folded states, i.e., $k_2 >> k_1$, the observed exchange rate, k_{ex}, is given by

$$k_{ex} = \frac{k_1 k_3}{k_2 + k_3} \quad (2)$$

Depending on the ratio of k_2 to k_3, one may encounter the following two limiting situations (Hvidt & Nielsen, 1966): (i) If $k_2 << k_3$ ("EX$_1$ limit"), each opening fluctuation leads to an isotope exchange, and eq 2 simplifies to

$$k_{ex} = k_1 \quad (3)$$

In this case, the exchange is limited by the opening rate k_1. Obviously, NH exchange measurements under EX$_1$ conditions provide readily interpretable data which can be correlated with a key feature of internal protein motility. (ii) If $k_2 >> k_3$ ("EX$_2$ limit"), the intrinsic exchange is the rate-limiting step, and eq 2 reduces to

$$k_{ex} = (k_1/k_2)k_3 = Kk_3 \quad (4)$$

$K = k_1/k_2$ is the equilibrium constant for the conformational transition between the states O and N. In contrast to measurements of k_{ex} under an EX$_1$ regime, k_{ex} resulting from an EX$_2$ process is not directly related to a kinetic process in the protein. However, provided that reliable values for k_3 are available (Roder et al., 1985), such experiments can be used to study the equilibria between O and N states for individual amide protons (Wagner et al., 1984).

Since in the EX$_2$ limit the protein must undergo many opening transitions before a particular proton is exchanged, the exchange of neighboring protons is expected to be uncorrelated, independent of the nature of the opening fluctuations. Information on the cooperativity of the dynamic processes can thus only be gained in the EX$_1$ limit, where the opening process is directly observed, and demonstration of correlated exchange for neighboring protons is a sufficient criterion for distinguishing exchange via an EX$_1$ mechanism from EX$_2$ exchange. Therefore, investigations on the cooperativity of proton exchange have a pivotal role in the present investigation. It must be added that while correlated exchange is a sufficient condition for identification of EX$_1$ exchange, it is not a necessary condition, since even in an EX$_1$ regime local protein fluctuations might expose different individual protons in an uncorrelated manner.

To illustrate the use of NOE experiments for distinguishing between correlated and uncorrelated NH exchange (Wagner, 1980), we consider a pair of protons located at a short distance in the protein structure, e.g., two amide protons in an antiparallel β-bridge (Figure 1) or in sequentially adjacent residues

N–H$_A$······O=C
C$^\alpha$ C$^\alpha$
C=O······H$_B$–N

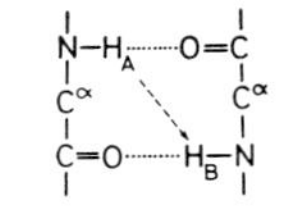

FIGURE 1: Schematic representation of a β-bridge in a regular, antiparallel β-sheet with the two amide protons H_A and H_B on opposite strands at a distance of ca. 2.6 Å (broken arrow).

of an α-helix. If proton A (Figure 1) is selectively saturated, magnetization is transferred to proton B through dipole–dipole interactions. The NOE, η_{AB}, is defined as the ratio of the magnetization change, ΔM_B, to the equilibrium magnetization, $M_B{}^0$:

$$\eta_{AB} = \Delta M_B / M_B{}^0 \quad (5)$$

η_{AB} depends on the inverse sixth power of the interproton distance and on the overall rotational tumbling and internal mobility of the protein. The influence due to other nearby protons can be reduced to a minimum by selecting sufficiently short buildup periods for the NOE's (Wagner & Wüthrich, 1979b; Anil Kumar et al., 1981).

In a partially deuterated sample, only molecules in which both positions , A and B in Figure 1, are protonated contribute to the NOE, and the magnetization transfer, $\Delta M_B{}^{ex}$, is reduced relative to the magnetization transfer in a fully protonated reference sample, $\Delta M_B{}^{ref}$, by the probability of residual protonation for both sites in the same molecule, p_{AB}:

$$\Delta M_B{}^{ex} = p_{AB} \Delta M_B{}^{ref} \quad (6)$$

The NOE observed in a partially exchanged protein preparation then becomes

$$\eta_{AB}{}^{ex} = \frac{\Delta M_B{}^{ex}}{M_B{}^{ex}} = \frac{p_{AB}\Delta M_B{}^{ref}}{p_B M_B{}^{ref}} = \frac{p_{AB}}{p_B}\eta_{AB}{}^{ref} \quad (7)$$

where p_A and p_B are the probabilities for independent protonation of sites A and B (Figure 1). p_A and p_B can be determined simply by comparing the intensities of the resonances A and B in the ^{1}H NMR spectra of the partially deuterated protein with those for nonlabile protons. In NOESY spectra, p_A and p_B can also be obtained from measurements of the intensites of cross-peaks manifesting NOE's between the labile protons and nonlabile protons in the partially deuterated protein. In practice, we expect to encounter the following three situations:

(*i*) *Uncorrelated Exchange.* If H_A and H_B are exchanged independently, we have

$$p_{AB} = p_A p_B \quad (8)$$

and eq 7 reduces to

$$\eta_{AB}{}^{ex} = p_A \eta_{AB}{}^{ref} \quad (9)$$

The NOE is reduced by the factor p_A, which describes the residual protonation of site A.

(*ii*) *Correlated Exchange.* If H_A and H_B can only be exchanged simultaneously, both positions are either protonated or deuterated. Therefore

$$p_{AB} = p_A = p_B \quad (10)$$

$$\eta_{AB}{}^{ex} = \eta_{AB}{}^{ref} \quad (11)$$

(*iii*) *Contributions from both Correlated and Uncorrelated Exchange.* If we consider H_A and H_B exchange by contributions from both correlated and uncorrelated exchange then

$$\eta_{AB}{}^{ex} = X\eta_{AB}{}^{ref} + (1 - X)p_A\eta_{AB}{}^{ref} \quad (12)$$

X is the degree of correlation, which describes the relative contribution from the correlated exchange processes:

$$X = (\eta_{AB}^{ex}/\eta_{AB}^{ref} - p_A)/(1 - p_A) \quad (13)$$

For completely correlated exchange, $X = 1$, and we have the limiting situation i. For completely uncorrelated exchange, $X = 0$, corresponding to situation ii.

Materials and Methods

BPTI (Trasylol) obtained from Bayer AG in Leverkusen, West Germany, was used without further purification. Two derivatives of BPTI with reduced thermal stability were used. TRAM-BPTI was prepared by transamination of the N-terminus according to the method of Dixon & Fields (1972) as described in Brown et al. (1978). The characterization of TRAM-BPTI with ^{1}H NMR and ^{13}C NMR by Brown et al. (1978) and Stassinopoulou et al. (1984) showed that after transamination the tertiary structure of BPTI was preserved, with the exception of the N-terminal region, where the modification inhibits the formation of a salt bridge between the C-terminus and the N-terminus.

For the ^{1}H NMR measurements of the NH exchange kinetics, 5–10 mM protein solutions were prepared by dissolving the lyophilized protein in buffered ^2H_2O. Depending on the p^{2}H region studied, 0.1 M phosphate buffer or 0.2 M glycine-d_5 was used. The BPTI solutions further contained 0.3 M NaCl. For a comparative study of BPTI and TRAM-BPTI, 3 M Gdn·HCl was added to the protein solutions. Before the NMR experiments, an approximate adjustment of the p^{2}H was achieved by adjusting the buffer solution without protein. Accurate p^{2}H values were obtained after the NMR experiments were completed by heating the solutions again to the temperatures at which the exchange was investigated (see below) and repeating the p^{2}H measurement with a combination glass electrode at this temperature.

Most of the measurements reported in this paper were obtained at high p^{2}H and elevated temperatures. In contrast to previously reported studies of BPTI [e.g., see Richarz et al. (1979)], the amide proton exchange was then so rapid that the NMR measurements could not be performed at the same conditions as those used for the exchange. Therefore, the following, simple quench procedure was used: 0.4 mL of a freshly prepared ^2H_2O solution of the protein was transferred to an NMR tube. This sample tube was then placed in a thermostat at the temperature where the exchange was to be measured, for example, at 70 °C. After a predetermined time span, the sample was rapidly cooled in ice–water. Subsequently, the ^{1}H NMR spectrum was recorded at 25 °C. This temperature provides a favorable compromise for good spectral resolution and negligible exchange of interior protons during the time needed to record a one-dimensional spectrum. The same protein solution was subjected to 5–10 subsequent exchange cycles, choosing suitable exchange intervals. The residual intensity of fully resolved NH resonances was determined with an accuracy of ca. 5% by simulating the spectra on the Aspect 2000 computer, using Lorentzian line shapes. The areas were normalized relative to the two-proton resonance of the nonlabile C$^\epsilon$ protons of Tyr-23 (Wagner & Wüthrich, 1982a). The kinetics were exponential within experimental error for all resolved NH resonances. Exchange rates for individual amide protons were calculated by exponential regression. With this simple procedure, reliable measurements of rates up to ca. 3 min^{-1} were possible, since the time required to equilibrate the sample in the thermostat, as measured with a fast-response thermocouple, was about 5 s. For some additional measurements of exchange rates which exceeded this limit, a flow apparatus was used which is described in the following paper (Roder et al., 1985).

^{1}H NMR spectra were recorded on Bruker HX 360 and WM 500 spectrometers, which are equipped with Aspect 2000 computers.

For studies of the cooperativity of the exchange in the central β-sheet region of BPTI, one-dimensional truncated-driven NOE difference spectra (Richarz & Wüthrich, 1978; Wagner & Wüthrich, 1979) were recorded with the following pulse sequence: $[-\tau_1(\omega_A)-90°$ observation pulse–RD–τ_1-$(\omega_{off})-90°$ observation pulse–RD–]. The proton resonance A at frequency ω_A was selectively saturated during the time τ_1, which was immediately followed by the nonselective observation pulse. After a relaxation delay (RD), during which the system was allowed to regain equilibrium, a reference spectrum was recorded with off-resonance irradiation in an empty region of the spectrum at a frequency ω_{off}. The free induction decays obtained with irradiation at ω_A and off-resonance, respectively, were accumulated in different memory locations. They were then subtracted and NOE difference spectra were obtained by Fourier transformation of the difference free induction decay. A value of 0.3 s was chosen for τ_1, which is sufficiently short to avoid falsification of the NOE data for BPTI by spin diffusion (Dubs et al., 1979). The delay time, RD, was 1.0 s. For the NOE experiments with native BPTI, protein concentrations of 20–30 mM were used, and 4000 scans were accumulated for each measurement. For TRAM-BPTI, the concentration was 10–15 mM, and between 8000 and 12 000 scans were added up. For the experiments with partially deuterated protein, the samples were exchanged to between 50% and 25% of the initial resonance intensity for the slowly exchanging amide protons. For improved ease of comparison, the NOE measurements for all experiments were obtained under identical conditions. For example, to investigate the cooperativity of the proton exchange at p^{2}H 8.5 and 68 °C, the partial exchange of the protein was obtained by exposing the sample to these conditions for a short time span. The exchange was then stopped by rapid cooling in ice–water, and subsequently the p^{2}H was adjusted to 4.5 by addition of ^{2}HCl. NOE spectra of this solution were then recorded at 25 °C. For the experiments where the exchange was studied in 3M Gdn·HCl, the Gdn·HCl was removed by ultrafiltration before the NOE measurements. To further ensure that the instrumental conditions were unchanged between different experiments, it was verified that corresponding NOE's between amide protons and nearby nonlabile protons, for example C$^\alpha$ protons, were quantitatively identical in the different experiments when measured relative to the residual amide proton resonance intensities in the partially exchanged protein.

Some results on exchange kinetics and cooperativity are also reported for the amide protons in the α-helix and the 3_{10}-helix of BPTI (Deisenhofer & Steigemann, 1975; Kabsch & Sander, 1983). Since these resonances are in a crowded region of the ^{1}H NMR spectrum, they could only be resolved in two-dimensional NMR spectra. Exchange rates were measured with COSY experiments, as described elsewhere in detail (Wagner & Wüthrich, 1982b; Wüthrich & Wagner, 1982). Comparison of the NOE's in the fully protonated and the partially deuterated protein was achieved by analysis of cross sections parallel to ω_1 in NOESY spectra recorded with a mixing time of 0.1 s (Anil Kumar et al., 1981). The procedures used for internal calibration of the relative peak intensities in the NOESY spectra of the fully protonated and partially deuterated protein are described under Results.

Results

p^{2}H Dependence of Exchange Rates. The pH dependence of the NH exchange rates in the range from 4 to 12 was

Table I: Activation Enthalpies for Exchange of Some Interior Amide Protons in BPTI

	$\Delta H^\ddagger$ (kcal M^{-1})						
p^2H, temp (°C)	Arg-20	Tyr-21	Phe-22	Tyr-23	Gln-31	Phe-45	kinetic mechanism
8.0, 73–88[a]	79	81	81	81	77	69	EX_1
7.7, 50–70[b]	48	65	65	57	≤30		EX_2–EX_1
8.0, 22–60[c]	29	40	38	42	20	34	EX_2
10.9, 69–81[a]		77	81	73	55		
3.5, 68–80[d]	89	84	88	88	87	81	EX_2

[a] Data from this work (0.3 M NaCl, 0.1 M $Na^2H_2PO_4$). [b] Data from Hilton & Woodward (1979) (0.3 M KCl). [c] Data from Richarz et al. (1979) (no salt). [d] Data from G. Wagner et al. (unpublished results) (no salt).

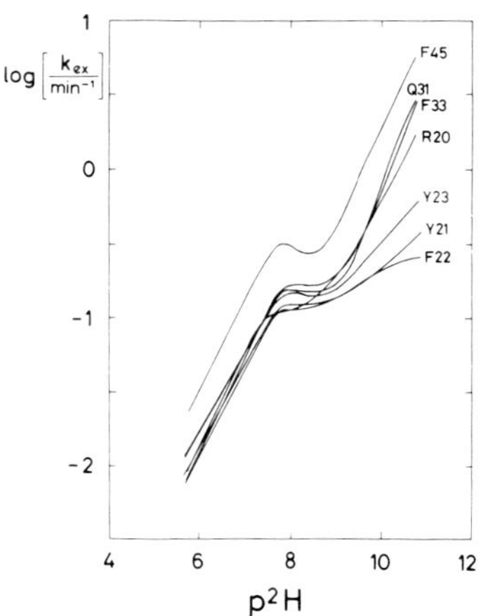

FIGURE 2: Logarithmic plots of the exchange rates vs. p^2H for seven slowly exchanging amide protons of BPTI at 68 °C in ^2H_2O containing 0.3 M NaCl and 0.1 M $Na^2H_2PO_4$ or 0.2 M glycine-d_5, depending on the p^2H.

studied for native BPTI at 68 °C in ^2H_2O, 0.3 M NaCl, and 0.1 M $Na^2H_2PO_4$ or 0.2 M glycine-d_5. Figure 2 presents logarithmic plots of the exchange rates vs. p^2H for those seven of the eight most slowly exchanging β-sheet amide protons in BPTI (Wagner & Wüthrich, 1982b) which have well-resolved lines in the one-dimensional ^1H NMR spectra, i.e., Arg-20, Tyr-21, Phe-22, Tyr-23, Glu-31, Phe-33, and Phe-45 (Wüthrich & Wagner, 1979). The locations of these protons in the antiparallel β-sheet are shown in Figure 3. The following qualitative features are readily apparent: For all seven protons, the exchange rate increases linearly with p^2H in the range from p^2H 5.0 to ca. 7.5. From p^2H 7.5 to ca. 9.0, there is a "plateau" where k_{ex} is nearly independent of p^2H. At p^2H above ca. 9.0, the exchange rates increase again with p^2H up to the highest values studied, which are near p^2H 12.0. For Phe-22, there is an indication that the plot of log k_{ex} vs. p^2H might level off again at p^2H above ca. 10.5. It can further be seen that closely similar exchange rates for six of these seven protons prevail in the p^2H range just below the plateau region and that there is a marked dispersion of the exchange rates in the region of strong p^2H dependence which follows the plateau. It should be added that all p^2H measurements were made at the exchange temperature of 68 °C, as described under Materials and Methods.

Additional measurements of the p^2H dependence of k_{ex} for the same seven amide protons were obtained for both BPTI and TRAM-BPTI at 55 °C in 3 M Gdn·HCl and 0.05 M NaH_2PO_4. In Figure 4, logarithmic plots of the p^2H dependence of k_{ex} in all three samples studied are superimposed. In all cases, the qualitative behavior is similar to that described above for BPTI at 68 °C. A nearly p^2H-independent plateau sets in between p^2H 7 and 8. Above p^2H 9 for BPTI and above

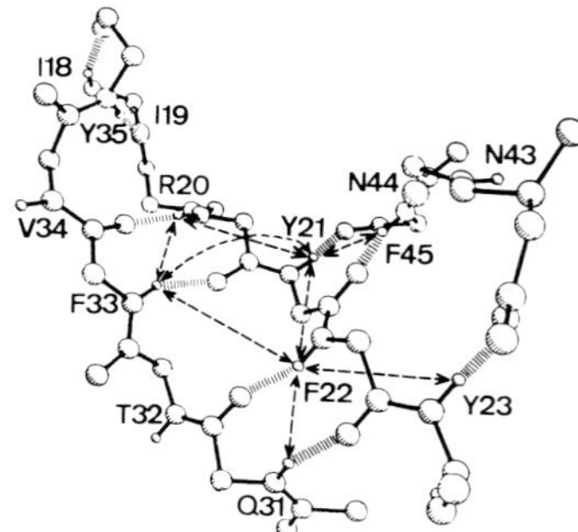

FIGURE 3: Computer drawing of the central part of the antiparallel β-pleated sheet in BPTI (Deisenhofer & Steigemann, 1975). Only the backbone atoms with NH's and carbonyl oxygens, but without $C^\alpha H$'s, are shown. The eight NH pairs studied in the NOE experiments are indicated by arrows.

p^2H 10 for TRAM-BPTI, the slope of log k_{ex} increases again.

For BPTI and TRAM-BPTI, the p^2H profiles below p^2H 7 were also recorded. They were found to be similar to those observed for the same protons in BPTI under somewhat different conditions (Richarz et al., 1979). The measurements for BPTI at 68 °C represent an extension of the data from Hilton & Woodward (1979) into the high p^2H range.

Temperature Dependence of Exchange Rates. The temperature dependence above 60 °C of the exchange rates for some interior amide protons in a solution of BPTI in 0.3 M NaCl and 0.1 M $Na^2H_2PO_4$ was measured at p^2H 8.0 and at p^2H 10.9. The rates were measured with the quench techniques described under Materials and Methods. The data are presented in Figure 5 in the Arrhenius representation. At p^2H 10.9, only the four most slowly exchanging protons had rates within the experimentally accessible range. Apparent activation enthalpies are listed in Table I. The points at the lowest temperatures studied (not shown in Figure 5) deviated systematically from linear behavior, and they were not included in the linear regression analysis.

NOE Studies of the Cooperativity of Exchange. One-dimensional truncated-driven nuclear Overhauser difference spectroscopy was used to study the cooperativity of exchange among the seven amide protons in the β-sheet of BPTI for which the p^2H dependence of the exchange was measured (Figures 2 and 4). The conclusions obtained relied on observation of NOE's between the eight pairs of protons which are connected by arrows in Figure 3.

Figure 6 presents typical experimental NOE results. The bottom trace of Figure 6A shows the region from 8 to 11 ppm of the ^1H NMR spectrum of a freshly prepared BPTI solution in ^2H_2O, which contains exclusively amide proton resonances (Masson & Wüthrich, 1973). The resonance assignments (Dubs et al., 1979) are also indicated. The upper trace in Figure 6A shows the NOE differenece spectrum obtained in this solution with preirradiation on the amide proton of Phe-22.

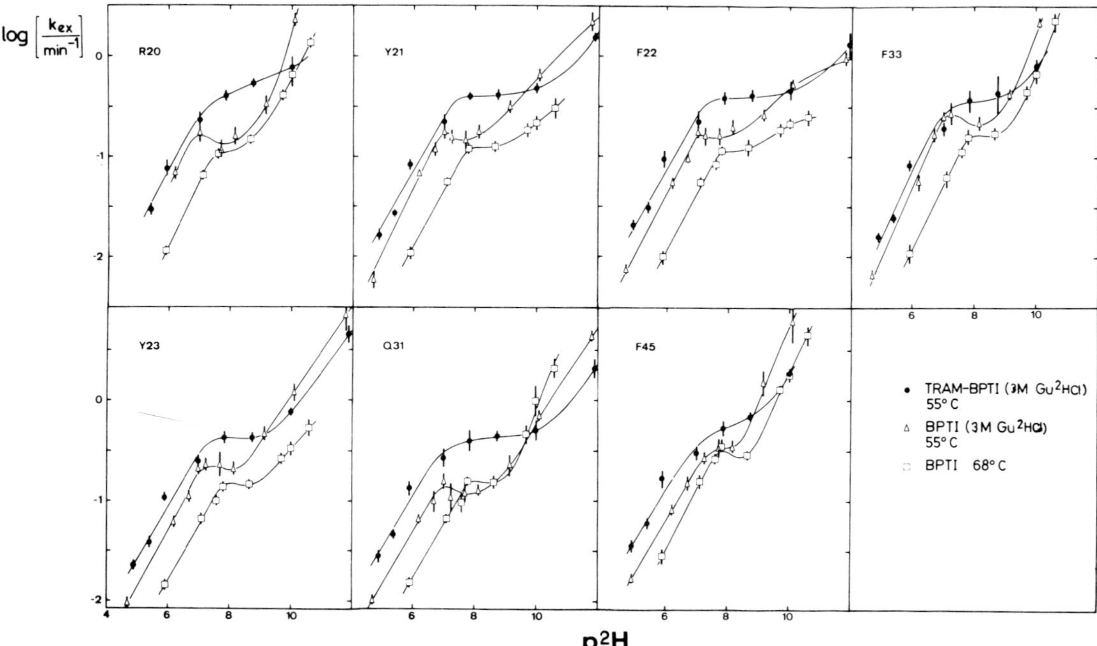

FIGURE 4: Logarithmic plots of the exchange rates as a function of p^2H for seven amide protons in the following systems: (□) BPTI exchanged at 68 °C in ^2H_2O containing 0.3 M NaCl and either 0.1 M $Na^2H_2PO_4$ or 0.2 M glycine-d_5, depending on the p^2H values. (△) BPTI exchanged at 55 °C in ^2H_2O containing 3 M Gdn·HCl. (●) TRAM-BPTI exchanged at 55 °C in ^2H_2O with 3 M Gdn·HCl and 0.05 M $Na^2H_2PO_4$. The vertical error bars indicate the standard deviation of the exponential regression used to calculate the rates. Curves are drawn by hand to guide the eye.

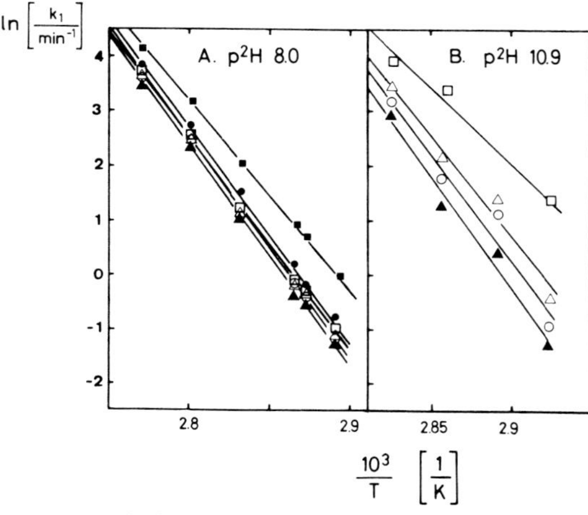

FIGURE 5: Arrhenius plots for selected amide proton exchange rates in BPTI at p^2H 8.0 and 10.9 in ^2H_2O containing 0.3 M NaCl and 0.1 M $Na^2H_2PO_4$: Arg-20 (●); Tyr-21 (○); Phe-22 (▲); Tyr-23 (△); Gln-31 (□); Phe-45 (■). Straight lines were calculated by linear regression using the rates obtained in the range $10^3/T < 2.90$ at p^2H 8.0 and $10^3/T < 2.95$ at p^2H 10.9.

Because of the limited selectivity of the preirradiation, the peaks at the positions of Phe-45 and Tyr-35 are due to direct saturation of these resonances. NOE's due to dipolar coupling with Phe-22 are observed for the amide protons of Tyr-23, Phe-33, Tyr-21, and Gln-31. Inspection of Figure 3 shows that these four amide protons are located nearest to that of Phe-22. Panels B and C of Figure 6 show the corresponding results obtained after ca. 50% of the amide protons were exchanged with ^2H at 68 °C and p^2H 6.0 or p^2H 8.0, respectively. After the exchange at p^2H 6.0 (Figure 6B), the NOE's on Tyr-23, Phe-33, Tyr-21, and Gln-31 had ca. 50% of the intensity observed before exchange, when the NOE's were calibrated against the intensities of the preirradiated line of Phe-22 in the two experiments. After the exchange at p^2H 8.0 (Figure 6C), NOE's essentially identical with those in Figure 6A were obtained. Following eq 9 and 11, we conclude that the exchange in the central β-sheet of BPTI is correlated at p^2H 8.0 and 68 °C and uncorrelated at p^2H 6.0 and 68 °C.

Experiments of the type of Figure 7 were carried out with preirradiation on each of the amide protons of residues 20–23, 31, 33, and 45 (Figure 3). For each pair of protons connected by an arrow, the average of the NOE's obtained with preirradiation on either proton was obtained. From these data, the degree of correlation, X, was then computed with eq 13. Figure 7 presents the p^2H dependence of X for BPTI in ^2H_2O solution at 45 and 68 °C. Figure 8 contains the corresponding data for BPTI in 3 M Gdn·HCl at 55 °C. The error bars in Figures 7 and 8 indicate the statistical errors estimated from the uncertainty of the intensity determination. The strong NOE's for the pairs Phe-22/Gln-31, Arg-20/Phe-33, and Tyr-21/Phe-45, which are all between residues in an antiparallel β-bridge (Figure 3), lead to the most reliable data. For the other pairs, the errors are bigger because of the intrinsically weaker NOE's due to the larger distance between the protons, and in some cases also because of the small separation of the resonances in the ^1H NMR spectrum (Figure 6), which limits the selectivity of irradiation. Figure 7 shows that for BPTI exchanged at 45 °C less than 20% contribution from correlated exchange was found between p^2H 6 and 10. Combined with the linear increase of log k_{ex} with p^2H observed by Richarz et al. (1979), this strongly indicates that under

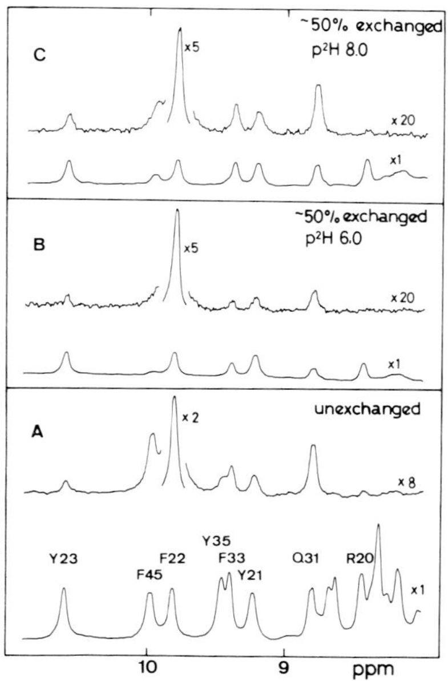

FIGURE 6: Truncated-driven NOE experiments in partially exchanged BPTI. The top trace in each panel shows the NH region of the NOE difference spectrum obtained by irradiating the NH resonance of Phe-22, and the lower trace shows the corresponding unperturbed spectrum. Spectra were recorded at 24 °C, p^{2}H 4.5. For a better presentation, the NOE difference spectra and the resonances of the irradiated protons were expanded vertically by the factors indicated in the figure. For reference, panel A shows the NOE spectrum of a fully protonated BPTI sample. Before the spectra in panels B and C were recorded, the samples were exchanged to ca. 50% of the NH intensity at 68 °C, p^{2}H 6.0 (B) and p^{2}H 8.0 (C), respectively.

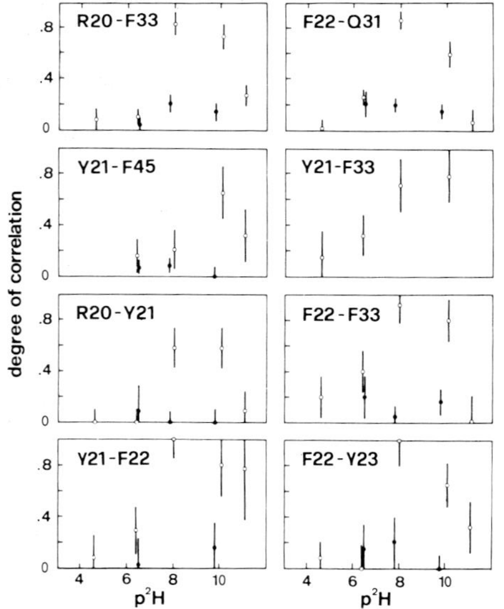

FIGURE 7: Plots of the degree of correlation, X, calculated with eq 13 as a function of p^{2}H for eight pairs of amide protons in BPTI after partial exchange at 68 (O) and 45 °C (●). The vertical bars indicate estimates of the error limits (see text).

these conditions the exchange is dominated by an EX_2 process, which is intrinsically uncorrelated. In BPTI exchanged at 68 °C, the degree of correlation increased for most NH pairs from

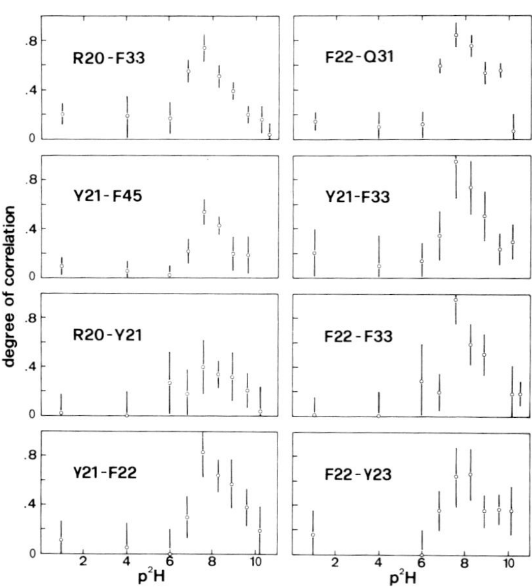

FIGURE 8: Degree of correlation as in Figure 8 for BPTI after partial exchange in ^2H_2O and 3 M deuterated Gdn·HCl at 55 °C.

values below 0.2 to peak values near 1 between p^{2}H 6 and 8. At p^{2}H above 8, the degree of correlation decreased again. Very similar behavior, with a pronounced peak in the degree of correlation near p^{2}H 8.0, was observed for BPTI which was dissolved in 3 M Gdn·HCl and partially exchanged at 55 °C (Figure 8).

To further investigate whether the unusual p^{2}H dependence of the exchange rates and the correlated exchange around p^{2}H 8.0 could be connected with the deprotonation of the N-terminal amino group, which has a pK_a value of 8.0 (Brown et al., 1978), NOE studies were also performed with TRAM-BPTI. For direct comparison with BPTI, partial exchange was achieved in 3 M Gdn·HCl at 55 °C. Results closely similar to those with BPTI were obtained. The transition from a low to a high degree of correlation occurred near p^{2}H 7.0. The regime of correlated exchange was found to be somewhat broader than for BPTI and to extend up to ca. p^{2}H 10.

Additional NOE measurements of the cooperativity of exchange were done for the amide protons in the α-helix from residues 47–54 and the 3_{10}-helix from residues 5–7 in BPTI (Deisenhofer & Steigemann, 1975; Kabsch & Sander, 1983). Compared to the central β-sheet, the exchange in the helical segments is considerably faster and was therefore only accessible to the NMR measurements when milder conditions of p^{2}H and temperature were used. Furthermore, the resonance lines for most of the helical amide protons are not sufficiently well resolved to be studied quantitatively with one-dimensional experiments. For these reasons, experiments of the type illustrated in Figure 9 were used.

Figure 9 shows a region of a NOESY spectrum of a freshly dissolved solution of BPTI in ^2H_2O, p^{2}H 3.5, which contains NOE cross-peaks with amide protons. Figure 10A shows a cross section along ω_1 at the ω_2-position of the amide proton of Met-52, which is located in the middle of the α-helix. This position is indicated in Figure 9 with an arrow and a broken line parallel to the ω_1 axis. In Figure 10A, the position of the diagonal amide proton line of Met-52 is indicated with a star. Three NOE cross-peaks to other amide protons are identified in Figure 10A, and some NOE's to nonlabile protons are

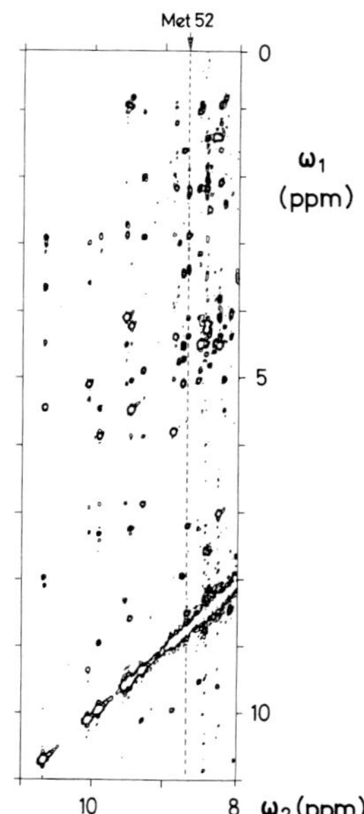

FIGURE 9: Low-field region of a 500-MHz NOESY spectrum of a 10 mM solution of BPTI recorded at p^2H 3.5, 24 °C, just after the protein was dissolved in ^2H_2O. The mixing time was 100 ms. The position of the cross section used for Figure 10 is indicated with an arrow and a broken line.

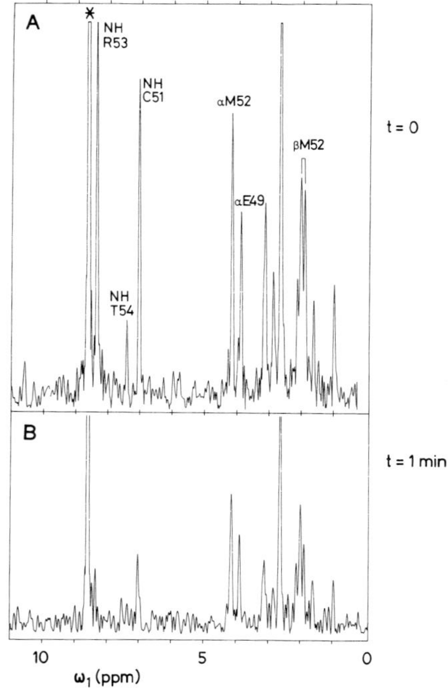

FIGURE 10: Cross sections from 500-MHz NOESY spectra of BPTI taken parallel to the ω_1 axis at the ω_2-position of the amide proton of Met-52. (A) Recorded in a freshly prepared ^2H_2O solution; (B) recorded at p^2H 3.5, 24 °C, after the sample had been kept for 1 min at 46 °C and p^2H 8.0. The partial exchange can be recognized from the decrease of the NOE cross-peaks to nonexchangeable protons. The degree of correlation can be obtained from cross-peaks to other labile protons. Some cross-peaks are identified in (A).

identified in the spectral region from ω_1 = 0–5 ppm in Figure 10A. A corresponding NOESY spectrum (Figure 10B) was recorded in a BPTI solution which had been partially exchanged by exposure for 1 min to 45 °C and p^2H 8.0 and then rapidly cooled and acidified to p^2H 3.5 by addition of ^2HCl. During 1 min at 45 °C and p^2H 8.0, the amide protons of the α-helix were exchanged to a residual protonation of 20–50% (Richarz et al., 1979). The NOESY spectrum was recorded with the same number of scans, and the same vertical expansion as in Figure 10A was used for the cross section in Figure 10B. Thus, the degree of protonation, $p_{\text{Met-52}} \approx 0.45$, could be obtained from the intensity ratios of the NOE cross-peaks to nonlabile protons in the two cross sections of Figure 10A,B. The probability for simultaneous protonation, p_{AB}, of two adjacent peptide groups can be obtained from the intensity ratios of the NOE cross-peaks between the amide protons. Thus, we found that $p_{\text{Met-52,Cys-51}} \approx 0.22$ and $p_{\text{Met-52,Arg-53}} \approx 0.15$. This shows that $p_{AB} = p_A p_B$, which indicates uncorrelated exchange at these conditions of temperature and p^2H. The same observation was made for all other amide protons of the α-helix and for all protons of the 3_{10}-helix.

A further NOESY experiment was carried out with a sample which had been kept for 2 min at 80 °C and p^2H 3.5, whereby the amide protons of the β-sheet were exchanged to about 50%. A comparison of the intensities of the NOE cross-peaks between the amide protons of Phe-22 and Glu-31, Ile-18 and Tyr-35, and Arg-20 and Phe-33 showed that the hydrogen–deuterium exchange was uncorrelated in the β-sheet under these experimental conditions.

Discussion

Considering the fundamental features of cooperative proton exchange expressed by eq 1–12, we arrive at the following qualitative conclusions on the exchange mechanisms which prevail for different individual amide protons in BPTI solutions at different conditions of p^2H and temperature. In the p^2H region from 1 to ca. 7.0, the data available (Richarz et al., 1979; Hilton & Woodward, 1978; Figures 2, 4, 7, and 8) indicate that the exchange is by an EX_2 mechanism at all experimentally accessible temperatures. At p^2H values from ca. 7 to 9 and at temperatures above ca. 55 °C, exchange by an EX_1 process becomes dominant. A plateau in the plots of log k_{ex} vs. p^2H (Figures 2 and 4) is reached in this p^2H range, since the efficiency of the EX_1 exchange mechanism changes little with further increase of p^2H above 7. [While EX_1 exchange appears to be common in nucleic acids (Englander & Kallenbach, 1984), we are not aware of previous experimental verifications of EX_1 exchange in proteins. In a broad discussion on the p^2H dependence of amide proton exchange in BPTI, Hilton & Woodward (1979) did, however, invoke the possibility that EX_1 exchange might also affect the experimental observations.] However, at p^2H above ca. 9, the rates increase again with p^2H (Figure 2). Simultaneously, the rates become different for the individual protons, and the exchange is predominantly uncorrelated again (Figures 7 and 8). This indicates that local, noncooperative structure fluctuations play a dominant role for hydrogen exchange in this p^2H range. Woodward & Hilton (1980) have discussed this phenomenon on the basis of a two-process model which is in several aspects consistent with our interpretation as discussed below. In the following, we evaluate the parameters which govern the exchange in the different regimes and investigate possible structural interpretations of the exchange kinetics.

Kinetic Parameters Governing EX_1 Exchange in BPTI. Exchange via EX_1 process is directly related to the opening

Table II: Representative Parameters Derived from NH Exchange in BPTI and TRAM-BPTI in the EX$_2$ Limit, in the EX$_1$ Limit, and at Basic p^{2}H

protein, temp (°C)	amino acid residue	p^{2}H$_{tr}$^{a}	k_2^g (min^{-1})^b	k_1^g (min^{-1}) at p^{2}H 8.6^c	k_{ex} (min^{-1}) at p^{2}H 10.6^d
BPTI, 68, ^2H_2O	R20	7.8	1.2×10^5	0.15	1.4
	Y21	7.8	1.5×10^5	0.13	0.31
	F22	7.8	9.4×10^4	0.12	0.25
	Y23	7.9	1.2×10^5	0.14	0.53
	Q31	7.8	2.6×10^5	0.15	2.1
	F33	7.9	1.9×10^5	0.17	2.2
	F45	7.7	1.7×10^5	0.29	4.4
BPTI, 55, 3 M Gdn·HCl	R20	7.3	1.6×10^4	0.12	2.3
	Y21	7.4	2.5×10^4	0.15	0.66
	F22	7.4	1.6×10^4	0.16	0.53
	Y23	7.3	1.2×10^4	0.23	1.2
	Q31	7.0	1.8×10^4	0.12	0.71
	F33	7.3	2.0×10^4	0.24	2.2
	F45	7.4	3.5×10^4	0.34	6.0
TRAM-BPTI, 55, 3 M Gdn·HCl	R20	6.9	6.2×10^3	0.40	10
	Y21	6.9	7.9×10^3	0.41	1.6
	F22	7.0	6.2×10^3	0.39	1.3
	Y23	7.1	7.9×10^3	0.43	4.6
	Q31	6.9	1.4×10^4	0.40	2.1
	F33	7.0	9.9×10^3	0.37	10
	F45	7.5	4.4×10^4	0.53	10

^{a}p^{2}H of transition from EX$_2$ to EX$_1$ mechanism at $k_2 = k_3$ as determined from Figure 4. ^{b}k_2^g, the closing rate for global fluctuations, was obtained from the transition from EX$_2$ to EX$_1$ exchange with $k_2^g = k_3$, which is the onset of the p^{2}H-independent exchange "plateau" in Figures 2 and 4. ^{c}k_1^g is the opening rate for global fluctuations. It was obtained from the exchange rate in the plateau region at p^{2}H 8.6 for BPTI at 68 °C, at p^{2}H 7.7 for BPTI in 3 M Gdn·HCl, and at p^{2}H 7.9 for TRAM-BPTI in 3 M Gdn·HCl. ^{d}At basic p^{2}H, the kinetic mechanism is still ambiguous, and only the experimental exchange rate, k_{ex}, is given.

rate, k_1 (eq 3), which can thus be evaluated from the data in the plateau regions above p^{2}H 7 (Figures 2 and 4). Table II shows that in the different systems studied, k_1 is between 0.12 and 0.53 min^{-1} at p^{2}H 8.6. The closing rate, k_2 (eq 1), can also be obtained by using $k_2 = k_3$ at the p^{2}H value where the EX$_2$ process changes to an EX$_1$ process. The values for k_2^g in Table II were thus evaluated on the basis of k_3 values computed with the rules of Molday et al. (1972) and activation enthalpies given by Englander & Poulsen (1969). The temperature dependence of the water dissociation constant was estimated after Covington et al. (1966) as described by Roder et al. (1985). The resulting k_2 values in the three systems studied cover the range from 6×10^3 to 2.6×10^5 min^{-1} (Table II).

Since individual resonance assignments are available for BPTI (Wagner & Wüthrich, 1982a) and TRAM-BPTI (Stassinopoulou et al., 1984), these fluctuation rates can be attributed to precisely defined locations in the amino acid sequence and the spatial structure (Deisenhofer & Steigemann, 1975). In BPTI at 68 °C, the residues in the central part of the β-sheet, Arg-20, Tyr-21, Phe-22, Tyr-23, Gln-31, and Phe-33 (Figure 3), have strikingly similar rates, k_1, extending over the narrow range from 0.12 to 0.17 min^{-1}. This is consistent with the observation that the exchange in the central β-sheet is correlated in the plateau region (Figures 7 and 8), and we conclude that these amide protons are exposed to the solvent by a cooperative opening of the whole β-sheet. The amide protons of the more peripheral β-sheet residues Ile-18, Tyr-35, and Phe-45 have considerably faster exchange rates (Richarz et al., 1979), which implies that additional, more strictly localized structure fluctuations contribute significantly to the exchange of these protons.

Limitations on Studies of EX$_1$ Exchange in Proteins. EX$_1$ exchange provides very direct access to the kinetic parameters needed for characterization of protein structure fluctuations. However, in BPTI, the EX$_1$ exchange prevails only over a narrow range of experimental conditions, i.e., at p^{2}H 7–9 and temperatures above 55 °C. More generally, the parameters in Table II imply that EX$_1$ exchange in proteins should only rarely be measurable with the presently available techniques. This can readily be rationalized, because the intrinsic exchange rates, k_3, are slow compared to the closing rates, k_2 (Table II), except possibly at high temperature and high p^{2}H. When such extreme conditions are approached, the amide proton exchange rates in most globular proteins become too fast to be measured by NMR, or the proteins even denature before EX$_1$ conditions are attained. An illustrative example is provided by the exchange in the α-helix of BPTI (Figures 9 and 10). Under the conditions where EX$_1$ exchange prevails for the central β-sheet in the same protein (Figures 2 and 7), the exchange in the helix is too rapid to be measured. Conversely, with all conditions which enable quantitative exchange measurements for the α-helix, uncorrelated exchange was observed (Figure 10).

Structural Interpretation of EX$_2$ Exchange. In contrast to EX$_1$ exchange, EX$_2$ exchange can for fundamental reasons (eq 4) not be directly related to the kinetics of protein structure fluctuations, and no unambiguous conclusions on the cooperativity of the exchange can be drawn from NH exchange studies in this regime. However, there are additional observations among the data collected for BPTI which indicate that in the p^{2}H range below p^{2}H 7 (Figures 2 and 4) the EX$_2$ exchange in the central part of the β-sheet is propagated by global fluctuations of this secondary structure element. The strongest inference comes from the observation in the EX$_1$ regime between p^{2}H 7 and 9 (Figures 2 and 4) that exchange for these amide protons is propagated by concerted fluctuations in the β-sheet. That the same type of internal mobility persists at lower p^{2}H is compatible with the small dispersion of the overall exchange rates for the amide protons of Arg-20, Tyr-21, Phe-22, Tyr-23, Gln-31, and Phe-33 in the p^{2}H range 5–7 (Figure 2) and with the data on the equilibria k_1/k_2 [eq 4; see below and Wagner et al. (1984)]. We therefore refer to the closing rate, k_2, in the transition region from EX$_2$ to EX$_1$ exchange near p^{2}H 7.0 as the closing rate for "global" fluctuations in the β-sheet, k_2^g (Table II).

For the practical purpose of studying protein internal motility with the use of amide proton exchange measurements, the most important conclusion from the present investigation is that such projects will usually have to rely on data from EX$_2$ exchange. It is therefore of imminent interest that while EX$_2$ exchange rates for amide protons cannot be directly correlated with the kinetics of intramolecular motions, EX$_2$ data combined with individual ^{1}H NMR assignments for the protons studied can be used to investigate the relative populations of closed and open states, N and O (eq 1), in a dynamic protein structure. Since this was recently discussed elsewhere in detail (Wagner, 1983; Wagner et al., 1984; Englander & Kallenbach, 1984), only a brief summary is included here.

The structural analysis of EX$_2$ exchange data relies on eq 4. Once sequence-specific resonance assignments for the amide protons have been obtained (Wagner & Wüthrich, 1982a), the intrinsic exchange rates, k_3, corresponding to the measured overall rates, k_{ex}, can be computed with the rules of Molday et al. (1972) [see also Roder et al. (1985)]. With eq 4, the equilibrium constant $K = k_1/k_2$ can then be evaluated. In logarithmic plots of k_{ex} vs. k_3

$$\log k_{ex} = \log k_3 + \log (k_1/k_2) \quad (14)$$

the data points for all those amide protons for which exchange is promoted by structure fluctuations with identical populations of the open states (eq 1) will lie on a straight line with slope 1. The intercepts of these lines with the log k_{ex} axis at log k_3 = 0 give log (k_1/k_2). From studies at variable temperature, the thermodynamic parameters governing the fluctuations between open and closed states of the protein conformation (eq 1) can then be further evaluated. On the basis of these considerations, it could be demonstrated for BPTI and a chemically modified form of BPTI that the fluctuations which promote the exchange from the central part of the β-sheet (Figure 3) have the same value of k_1/k_2 for all individual amide protons in this region. Different values for k_1/k_2 were found for peripheral locations in the β-sheet, and evidence for multiple fluctuation types promoting exchange within the C-terminal α-helix and within the N-terminal 3_{10}-helix was obtained (Wagner et al., 1984).

Amide Proton Exchange in BPTI at High p²H and High Temperature. While the available data are sufficient for an unambiguous characterization of the mechanisms which govern the exchange in BPTI solutions at p²H values up to ca. 9, the onset of uncorrelated exchange with sizable p²H dependence at p²H values above 9 (Figures 2, 4, 7, and 8) could a priori be caused by different exchange mechanisms. Considering that different exchange rates prevail for the individual amide protons and that the exchange is uncorrelated, local structure fluctuations must play an important role in this p²H range. It would then appear that there are two possible, fundamentally different mechanisms which may cause the transition from global to local exchange when these extreme conditions of temperature and p²H are approached. In the following, we refer to these as the "kinetic switch model" and the "deprotonation switch model". The kinetic switch model invokes that different types of fluctuations are present at all experimental conditions of p²H and that these are redistributed with temperature due to different enthalpies. Within the framework of this model, hydrogen exchange measurements at extreme conditions of temperature and p²H would provide interesting insights into protein fluctuations which are at different conditions masked by rare, but more efficient, global fluctuations. In contrast, the deprotonation switch model invokes fluctuations which are present only at high p²H. Following this model, the hydrogen exchange measurements at high p²H and high temperature would mainly provide an insight into the protein behavior on its way toward alkaline denaturation.

In the kinetic switch model, we assume that there are two classes of internal motions, which we call "global" and "local" with the kinetic parameters k_1^g and k_1^l, respectively. The global and local structure fluctuations must have sizably different closing rates, k_2, and different populations of open states, k_1/k_2, as follows. k_2^l must be much larger than k_2^g so that the transition from EX_2 to EX_1 exchange (at $k_2 = k_3$) occurs only at higher p²H for the local fluctuations. The equilibrium for the global fluctuations must be more strongly shifted to the open states, O, with $k_1^g/k_2^g > k_1^l/k_2^l$, so that the global fluctuations are dominant for the exchange as long as EX_2 exchange prevails for both the global and the local fluctuations. Note that this condition could be attained entirely as a consequence of k_2^l being large compared to k_2^g and that local openings might indeed be expected to close faster than global openings. With this model, the observed p²H profiles (Figures 2 and 4) and the loss of correlated exchange at high p²H (Figures 7 and 8) could be explained without any requirements for p²H-dependent changes of the relative populations of the global and local classes of fluctuations. If the exchange rates could be measured to even higher p²H than the data in Figures 2 and 4, the kinetic switch mechanism would predict another plateau for the p²H region where $k_3 \geq k_2^l$. Since this effect is not observed up to p²H 12 (Figure 4), k_2^l must be $\geq k_3$ (68 °C, p²H 12) $\approx 2.5 \times 10^7$ s^{-1}. For this estimate, the intrinsic exchange rate, k_3, was extrapolated from the model peptide data of Englander & Poulsen (1969) and Molday et al. (1972).[2]

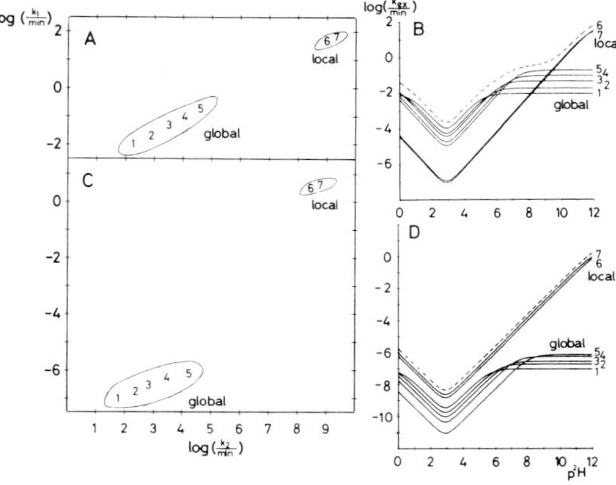

FIGURE 11: p²H and temperature dependence of the exchange of a single amide proton in BPTI resulting from a hypothetical distribution of global and local fluctuations. (A) Distribution of five global and two local fluctuations (characterized by parameters k_1^i and k_2^i; see text) at high temperature (68 °C) in a log k_1 vs. log k_2 plot. (B) Plot of log k_{ex} vs. p²H resulting from the structure fluctuations in (A). Solid lines, contributions from the individual fluctuations; dashed line, overall exchange rate. (C) Same as (A) for low temperature (36 °C). For k_1, a strong temperature dependence and for k_2 a small temperature dependence were assumed. (D) Same as (B) for the low-temperature structure fluctuations shown in (C).

A schematic survey of the manifestations of such a kinetic switch mechanism over the entire p²H range from 1 to 12 is provided by Figure 11. In this figure, we assume, in addition to the features discussed above, that we have two classes of intramolecular motions, global and local, respectively, with each of them containing a multitude of slightly different fluctuations, i, with discrete values of k_1 and k_2. Panels A and C of Figure 11 give hypothetical distributions of k_1^i and k_2^i, in a two-dimensional representation, at high temerature (68 °C) and low temperature (36 °C), respectively. Figure 11B,D shows the corresponding overall exchange rates in a log k_{ex} vs. p²H representation, where k_{ex} is the sum of the contributions from the different fluctuations:

$$k_{ex} = \sum_i \frac{k_1^i k_3}{k_2^i + k_3} \quad (15)$$

[2] The quantitative estimation of k_2 and k_1/k_2 from eq 2 and 4 depends crucially on the assumption that the intrinsic exchange rates, k_3, are identical with those in model peptides and can be extrapolated from the data of Molday et al. (1972) and Englander & Poulsen (1969). One could imagine, however, that for local fluctuations the resulting solvent accessibility of the amide protons is still limited when compared to global fluctuations (or, of course, small model peptides). In the model of eq 1, this might be accounted for by using a smaller apparent "intrinsic" exchange rate, k_3^l, for exchange from local openings. This would make local fluctuations less effective for promoting hydrogen exchange, and it would shift the EX_2 to EX_1 transition to higher p²H. Following these considerations, the present use of the transition condition $k_2 = k_3$ could lead to an overestimation of k_2.

k_1 and k_2 for the individual fluctuations were chosen so that they fit approximately the experimetnal data at 68 °C in the p^{2}H range 7–11. For the temperature dependence of the equilibrium constants, $k_1{}^i/k_2{}^i$, the enthalpy, ΔH, was taken to be 70–80 kcal/M for the global fluctuations and ~10 kcal/M for the local fluctuations. It was further assumed that the distribution of the individual $k_2{}^i$ is almost independent of temperature and that most of the temperature dependence of the equilibrium constant k_1/k_2 goes on account of k_1 (Pohl, 1976; Roder, 1981; Englander & Kallenbach, 1983). We thus have at high temperature that $k_1{}^g/k_2{}^g > k_1{}^l/k_2{}^l$, whereas at low temperature $k_1{}^g/k_2{}^g < k_1{}^l/k_2{}^l$ (Figure 11A,C). At high temperature (Figure 11B), only global fluctuations contribute to the hydrogen exchange up to p^{2}H ~9. The different global fluctuations reach the EX$_1$ limit at somewhat different p^{2}H values. As long as there remain some global fluctuations with $k_2{}^i > k_3$, predominant exchange by an EX$_2$ mechanism is observed overall. At p^{2}H $\gtrsim$7, all global fluctuations provide exchange via the EX$_1$ mechanism, and thus, overall this mechanism is dominant up to the onset of more efficient exchange by local fluctuations near p^{2}H 9. In the base-catalyzed regime between p^{2}H 4 and 7 at high temperature, this picture would predict a slope <1 in the plot of log k_{ex} vs. p^{2}H (see Figure 11B). At low temperature, local fluctuations would be dominant over the whole p^{2}H range, and a "normal" p^{2}H dependence should prevail (Figure 11D). These predictions of the model are consistent with experimental observations [compare, e.g., Hilton & Woodward (1979)].

In the deprotonation switch model, we assume that EX$_1$ exchange prevails over the entire p^{2}H range from the onset of the plateau near p^{2}H 7.5 to p^{2}H 12. The loss of correlated exchange at p^{2}H $\gtrsim$9 would be explained with a transition from global fluctuations to strictly local opening processes, where only single amide groups would be exposed to the solvent. In this model, the local character of the structure fluctuations at p^{2}H $\gtrsim$9 would be a consequence of the deprotonation of ionizable groups, which would lead to a loss of intramolecular salt bridges and thus reduce the cooperativity of the fluctuations. Since the onset of uncorrelated exchange lies at p^{2}H $\gtrsim$9, the titration of the N-terminus, the four tyrosines, and the four lysines would have to be responsible for this effect in BPTI. The increase of the exchange rates, $k_{ex} = k_1$, with p^{2}H would be due to increasing destabilization of the protein structure because of the titration of the same ionizable groups between p^{2}H 9 and 12 (Figure 2).

There are conditions where both exchange mechanisms appear to coexist, albeit in different regions of the molecule. Let us consider the pH, temperature, and correlation data measured in the absence of denaturant (Figures 2, 5, and 7). The amide protons of Tyr-21, Phe-22, and Tyr-23, which are located on the central strand of the β-sheet (Figure 3), have properties compatible with the deprotonation switch model: At both p^{2}H 8.0 and p^{2}H 10.9, their activation enthalpies are large (73–81 kcal M^{-1}; Table I); the rate increase toward basic p^{2}H is comparatively weak (Figure 2); Tyr-21 and Phe-22 exchange at similar rates (Figure 2), and the pairwise correlation at 68 °C remains high up to p^{2}H 11 (Figure 7). These properties are incompatible with the localized nature of the rapid fluctuations which would be required by the kinetic switch mechanism. With the deprotonation switch model, the rates are interpreted as opening rates, and the increase of k_1 above p^{2}H 8 reflects the destabilization of the native protein structure by the deprotonation of the basic residues (N-terminus, Tyr and Lys). The high-pH behavior of the remaining amide protons is, however, quite different and shows the characteristics of the kinetic switch mechanism. They are characterized by much stronger pH dependences (up to a factor of 8 rate increase per pH unit), which is indicative of an EX$_2$ mechanism, and smaller temperature dependences, as seen for Gln-31 in Figure 5B. The rate divergence with increasing pH, associated with decreased correlation, suggests that rapid, localized fluctuations dominate the exchange in all but the most stable regions of the molecule.

Additional insight can be gained by comparison of BPTI with TRAM-BPTI (Figure 4), which affords information on the influence of the deprotonation of the N-terminus on amide proton exchange. The transition from low-p^{2}H EX$_2$ exchange to EX$_1$ exchange is near the pK_a value for the N-terminal amino group in BPTI (Brown et al., 1978). Figure 4 and Table II show that in the measurements recorded at 55 °C and in 3 M Gdn·HCl, the transition is at a slightly lower p^{2}H for TRAM-BPTI than for BPTI. This indicates that the closing rate, k_2, is slightly decreased (by a factor $\leq$3) by the chemical modification of the N-terminus. Furthermore, the k_{ex} values due to EX$_1$ exchange in TRAM-BPTI are somewhat higher than in BPTI (Figure 4), indicating that the opening rates, k_1, are slightly higher (by factors of 1.5–3.3) in TRAM-BPTI (Table II). With regard to distinguishing between the two proposed mechanisms for the switch to noncooperative exchange at high p^{2}H, we note that the kinetic switch model would provide a straightforward explanation of the experimental data for TRAM-BPTI. Since k_1 is faster in TRAM-BPTI than in BPTI, higher p^{2}H would be needed for the local fluctuations to become dominant, according to $(k_1{}^l/k_2{}^l)k_3 > k_1{}^g$. Furthermore, since the transamination of the N-terminus in TRAM-BPTI does not affect the local fluctuations (Wagner et al., 1984), similar exchange rates at high p^{2}H would be expected for BPTI and TRAM-BPTI. Both of these predictions are consistent with the experimental data (Figure 4, Table II). On the other hand, the deprotonation switch model could also explain the TRAM-BPTI data. In this case, however, we would have to conclude that the deprotonation of groups (tyrosines, lysines) other than the N-terminus would cause the transition to uncorrelated exchange at high pH. Otherwise, one would expect a qualitatively different behavior for TRAM-BPTI: Since TRAM-BPTI at p^{2}H 7 and unmodified BPTI at p^{2}H 9 have identical charge distributions, the onset of uncorrelated exchange at the high-p^{2}H end of the EX$_1$ plateau in TRAM-BPTI would be expected at lower p^{2}H than in BPTI, or the EX$_1$ plateau might even be completely absent in TRAM-BPTI. This has not been observed in the experiments (Figure 4).

Activation Enthalpies for Amide Proton Exchange. Recent analyses of amide proton exchange measurements relied heavily on the activation enthalpies obtained from the temperature dependence of the exchange (Woodward & Hilton, 1980; Englander & Kallenbach, 1984). It therefore appeared of interest to investigate this parameter in the presently characterized, different exchange regimes.

At suitable p^{2}H values, the transition from the EX$_2$ to the EX$_1$ limit can be induced by increasing the temperature. For example, for BPTI at p^{2}H around 8, such a transition occurs between 45 and 68 °C (Figure 7). This may have two reasons: (i) The intrinsic exchange rate, k_3, might increase more strongly with temperature than the closing rate, k_2, so that EX$_1$ exchange would be reached at higher temperature. (ii) The EX$_2$ to EX$_1$ transition could be due to the dominance of different classes of fluctuations at low and high temperature, i.e., global fluctuations at higher temperature with a small closing rate ($k_2{}^g < k_3$) and local fluctuations at lower temperature with a fast closing rate ($k_2^{\text{I}} > k_3$). Probably both features are relevant for the EX_2 to EX_1 transition with temperature at $p^2H \sim 8$.

Because of the potential existence of different kinetic limits, and more generally the occurrence of different classes of fluctuations, much care should be exercised in the interpretation of temperature-dependent exchange rates in terms of activation enthalpies. The apparent activation enthalpies for the two kinetic limits are

$$\Delta H^{\ddagger}_{\text{app,EX}_1} = \Delta H^{\ddagger}_{k_1} \quad (16)$$

and

$$\Delta H^{\ddagger}_{\text{app,EX}_2} = \Delta H^{\ddagger}_{k_1} - \Delta H^{\ddagger}_{k_2} + \Delta H^{\ddagger}_{k_3} \quad (17)$$

From the available literature data, the closing rates, k_2, appear to have smaller temperature dependences than the intrinsic exchange rates, k_3 (Pohl, 1976; Roder, 1981; Englander & Kallenbach, 1983). Thus, for predetermined, single type of fluctuations, we expect that

$$\Delta H^{\ddagger}_{\text{app,EX}_2} > \Delta H^{\ddagger}_{\text{app,EX}_1} \quad (18)$$

This is consistent with the experimental high-temperature values of $\Delta H^{\ddagger}_{\text{app}}$ at p^2H 3.5 (EX_2) being larger than those at p^2H 8.0 (EX_1) (Table I), keeping in mind that the exchange at both p^2H values originates from global fluctuations but corresponds to different kinetic limits.

It is interesting to consider the experimental activation enthalpies in light of the equilibrium enthalpy difference, ΔH, for the denaturation of BPTI. Moses & Hinz (1980) have measured calorimetrically a ΔH value of 70 kcal M^{-1} at the denaturation temperature, and they found this value to be independent of pH. It deviates from the much larger value reported previously by Privalov (1979). After subtracting $\Delta H^{\ddagger}_{k_3}$ = 17 kcal M^{-1} (Englander & Poulsen, 1969) from $\Delta H^{\ddagger}_{\text{app}}$ at p^2H 3.5 in Table I, we obtain opening enthalpies that are very close to ΔH for global denaturation. This suggests that at elevated temperature exchange is promoted by global unfolding transitions. From analysis of the data for the EX_1 limit at p^2H 8.0 using ΔH = 70 kcal M^{-1}, we find that $\Delta H^{\ddagger}_{k_2} \approx$ 7–11 kcal M^{-1}. These parameters are consistent with typical values reported in the literature (Englander & Kallenbach, 1984; Roder, 1981; Englander et al., 1972).

Woodward and co-workers (Woodward & Hilton, 1979, 1980; Hilton et al., 1981; Simon et al., 1984) previously interpreted the amide proton exchange in BPTI with a model involving two distinct processes: (a) a high activation energy process associated with global unfolding (dominant at low pH and elevated temperature); (b) a low activation energy process which they associated with "exchange from the folded state". Process b was invoked to explain the exchange behavior at basic pH. The distinction of processes a and b was based on the temperature dependence of exchange (Woodward & Hilton, 1980) and total hydrogen–tritium exchange measurements (Woodward et al., 1981) which indicated that protons with low activation energy were not accelerated by addition of 8 M urea. With respect to the temperature dependence, our present measurements show that some amide protons exchange by a high activation enthalpy process even at p^2H 10.9 (Figure 5; Table I) where the two-process model would predict low activation enthalpy.

While our present interpretation of BPTI exchange data agrees with some aspects of the two-process model of the Woodward group (e.g., the global nature of the high actication energy process), it differs fundamentally in other aspects: We attribute the complicated pH dependence of BPTI exchange to the fact that different parts of the distribution of internal motion are sampled as the time window of the measurement, determined by the p^2H-dependent intrinsic exchange rates, is varied. Some protons in the core of the molecule are exchanged by major unfolding transitions under all conditions presently explored. Their exchange behavior at basic pH reflects the pH dependence of unfolding. In the view of Woodward et al. (1982), exchange in the low activation energy regime is attributed to an inherently different mechanism. Exchange is believed to occur within the more or less fully folded structure after penetration of the catalyst (H_3O^+ or OH^-). In contrast, Englander & Kallenbach (1984) stress that while access of the solvent is certainly necessary, it is not sufficient for exchange to occur. Existing internal hydrogen bonds have to be disrupted, which probably involves significant structural distortion. This problem and the conceptual difficulties associated with penetration of the charged and hydrated catalysts into the protein interior are avoided by the structural unfolding model.

ACKNOWLEDGMENTS

H.R. thanks Dr. S. W. Englander for stimulating discussions on the subject of this paper. We acknowledge the careful preparation of the figures and the manuscript by E. H. Hunziker, E. Huber, and R. Marani.

Registry No. BPTI, 9087-70-1; H_2, 1333-74-0.

REFERENCES

Anil Kumar, Wagner, G., Ernst, R. R., & Wüthrich, K. (1981) *J. Am. Chem. Soc. 103*, 3654–3658.

Barksdale, A. D., & Rosenberg, A. (1982) *Methods Biochem. Anal. 28*, 1–113.

Brown, L. R., DeMarco, A., Richarz, R., Wagner, G., & Wüthrich, K. (1978) *Eur. J. Biochem. 88*, 87–95.

Covington, A. K., Robinson, R. A., & Bates, R. G. (1966) *J. Phys. Chem. 70*, 3820–3824.

Deisenhofer, J., & Steigemann, W. (1975) *Acta Crystallogr., Sect. B: Struct. Crystallogr. Cryst. Chem. B31*, 238–250.

Dixon, H. B. F., & Fields, R. (1972) *Methods Enzymol. 25*, 409–419.

Dubs, A., Wagner, G., & Wüthrich, K. (1979) *Biochim. Biophys. Acta 577*, 177–194.

Englander, S. W., & Poulsen, A. (1969) *Biopolymers 7*, 379–393.

Englander, S. W., & Kallenbach, N. R. (1984) *Q. Rev. Biophys. 16*, 521–655.

Englander, S. W., Downer, N. W., & Teitelbaum, H. (1972) *Annu. Rev. Biochem. 41*, 903–924.

Englander, S. W., Calhoun, D. B., Englander, J. J., Kallenbach, N. R., Liem, R. K. H., Malin, E. L., Mandal, C., & Rogero, J. R. (1980) *Biophys. J. 32*, 577–589.

Gurd, F. R. N., & Rothgeb, T. M. (1979) *Adv. Protein Chem. 33*, 74–165.

Hilton, B. D., & Woodward, C. K. (1978) *Biochemistry 17*, 3325–3332.

Hilton, B. D., & Woodward, C. K. (1979) *Biochemistry 18*, 5834–5841.

Hilton, B. D., Trudeau, K., & Woodward, C. K. (1981) *Biochemistry 20*, 4697–4703.

Hvidt, A., & Nielsen, S. O. (1966) *Adv. Protein Chem. 21*, 287–386.

Kabsch, W., & Sander, C. (1983) *Biopolymers 22*, 2577–2637.

Masson, A., & Wüthrich, K. (1973) *FEBS Lett. 31*, 114–118.

Molday, R. S., Englander, S. W., & Kallen, R. G. (1972) *Biochemistry 11*, 150–158.

Moses, E., & Hinz, H.-J. (1983) *J. Mol. Biol. 170*, 765–776.

Pohl, F. M. (1976) *FEBS Lett. 65*, 293–296.

Privalov, P. L. (1979) *Adv. Protein Chem. 33*, 167–241.

Richarz, R., & Wüthrich, K. (1978) *J. Magn. Reson. 30*, 147–150.

Richarz, R., Sehr, P., Wagner, G., & Wüthrich, K. (1979) *J. Mol. Biol. 130*, 19–30.

Roder, H. (1981) Ph.D. Thesis, ETH Zürich.

Roder, H., Wagner, G., & Wüthrich, K. (1985) *Biochemistry* (following paper in this issue).

Simon, I., Tüchsen, E., & Woodward, C. (1984) *Biochemistry 23*, 2064–2068.

Stassinopoulou, C. I., Wagner, G., & Wüthrich, K. (1984) *Eur. J. Biochem. 145*, 423–430.

Štrop, P., Wider, G., & Wüthrich, K. (1983) *J. Mol. Biol. 166*, 641–667.

Wagner, G. (1980) *Biochem. Biophys. Res. Commun. 97*, 614–620.

Wagner, G. (1983) *Q. Rev. Biophys. 16*, 1–57.

Wagner, G., & Wüthrich, K. (1979a) *J. Mol. Biol. 134*, 75–94.

Wagner, G., & Wüthrich, K. (1979b) *J. Magn. Reson. 33*, 675–680.

Wagner, G., & Wüthrich, K. (1982a) *J. Mol. Biol. 155*, 347–366.

Wagner, G., & Wüthrich, K. (1982b) *J. Mol. Biol. 160*, 343–361.

Wagner, G., Tschesche, H., & Wüthrich, K. (1979) *Eur. J. Biochem. 95*, 239–248.

Wagner, G., Stassinopoulou, C. I., & Wüthrich, K. (1984) *Eur. J. Biochem. 145*, 431–436.

Woodward, C. K., & Hilton, B. D. (1979) *Annu. Rev. Biophys. Bioeng. 8*, 99–127.

Woodward, C. K., & Hilton, B. D. (1980) *Biophys. J. 32*, 561–575.

Woodward, C. K., Simon, I., & Tuchsen, E. (1982) *Mol. Cell. Biochem. 48*, 135–160.

Wüthrich, K., & Wagner, G. (1979) *J. Mol. Biol. 130*, 1–18.

Wüthrich, K., & Wagner, G. (1982) *Ciba Found. Symp. 93*, 310–328.

Wüthrich, K., & Wagner, G. (1984) *Trends Biochem. Sci. (Pers. Ed.) 9*, 152–154.

Wüthrich, K., Eugster, A., & Wagner, G. (1980) *J. Mol. Biol. 144*, 601–604.

Wüthrich, K., Štrop, P., Ebina, S., & Williamson, M. P. (1984) *Biochem. Biophys. Res. Commun. 122*, 1174–1178.

J. Mol. Biol. (1987) **196**, 227–231

Reinvestigation of the Aromatic Side-chains in the Basic Pancreatic Trypsin Inhibitor by Heteronuclear Two-dimensional Nuclear Magnetic Resonance

The ^{1}H nuclear magnetic resonance (n.m.r.) assignments for the aromatic spin systems of the four tyrosines and four phenylalanines in the basic pancreatic trypsin inhibitor (BPTI) were reinvestigated using novel ^{13}C–^{1}H heteronuclear two-dimensional experiments. Resonance lines which are degenerate in homonuclear ^{1}H n.m.r. spectra could thus be resolved. Based on this new evidence the previous assignments for Phe22 and Phe33 had to be corrected. This affects the earlier conclusions on aromatic ring flips in BPTI in that Phe22 is rotating rapidly on the n.m.r. time scale at 36 °C, rather than being immobilized up to 80 °C.

The basic pancreatic trypsin inhibitor (BPTI)† contains four tyrosine and four phenylalanine residues. These aromatic rings have long attracted interest since they undergo rotational motions about the C^{β}–C^{γ} axis, of which the frequencies as well as the activation energies and activation volumes could be measured quantitatively (Wagner, 1983; Wagner *et al.*, 1976; Wüthrich & Wagner, 1975). All these data were collected before the advent of two-dimensional n.m.r. and systematic sequential ^{1}H n.m.r. assignments (Wagner *et al.*, 1981; Wagner & Wüthrich, 1982; Wüthrich, 1986), whereby the identification and sequence-specific assignments of the aromatic spin systems relied on combined use of one-dimensional spin decoupling experiments (Wagner *et al.*, 1975, 1976), selective chemical modification (Snyder *et al.*, 1976), and comparison with homologous proteins (Wagner *et al.*, 1978). For six of the rings the resonance assignments were subsequently confirmed by sequential assignments using two-dimensional n.m.r. (Wagner & Wüthrich, 1982). However, we also had to conclude from the then available absolute value two-dimensional ^{1}H n.m.r. spectra that *for Phe22 and Phe33, cross-peaks are compatible with the assignments obtained from the comparison of homologous proteins, but the peaks are not sufficiently well-resolved to present clear-cut evidence for these assignments* (description of Fig. 9 of Wagner & Wüthrich, 1982). Recently, novel heteronuclear two-dimensional n.m.r. experiments were introduced, e.g. [^{13}C, ^{1}H]-COSY and [^{13}C, ^{1}H, ^{1}H]-RELAYED-COSY (Brühwiler & Wagner, 1986; Wagner & Brühwiler, 1986), which can yield a quite dramatic improvement of the resolution of the aromatic ^{1}H n.m.r. lines in proteins. Compared to homonuclear correlation experiments these techniques have the advantage that connectivities between nearly degenerate protons are manifested in heteronuclear cross-peaks which do not interfere with diagonal peaks, and that even connectivities between completely degenerate protons can be established. These new techniques have now been used to check on the resonance assignments and the data on rotational mobility for Phe22 and Phe33, which had been obtained with one-dimensional n.m.r. experiments (Wagner *et al.*, 1976).

Figure 1 shows the region of the cross-peaks for aromatic CH groups in a [^{13}C, ^{1}H]-COSY spectrum of BPTI. The pulse sequence described by Brühwiler & Wagner (1986) was used. The carbon spectrum is along ω_1, the proton spectrum along ω_2. The spectrum was recorded without refocusing delay prior to detection, and no broad-band carbon decoupling was applied during detection. For Figure 1 only positive cross-peaks were plotted so that only the right-hand components of the carbon doublets can be seen. Therefore, the spectrum has the same appearance as a broad-band carbon-decoupled spectrum. Since the heteronuclear one-bond ^{13}C–^{1}H coupling constants are nearly the same for all aromatic CH groups, the relative positions of the right-hand doublet components correspond to a good approximation to the relative chemical shifts. For convenience we shifted the ω_2 scale to the right by half the CH coupling constant, so that the observed cross-peak positions correspond directly to the proton chemical shifts.

Figure 2 shows the same spectral region from a [^{13}C, ^{1}H, ^{1}H]-RELAYED-COSY experiment. The spectrum was recorded with a refocusing delay prior to detection as described by Brühwiler & Wagner (1986), but no broad-band carbon decoupling was applied during detection. Thus, cross-peaks of direct connectivities appear as in-phase absorptive doublets, the components of which are connected with horizontal lines in Figure 2. Relayed connectivities appear at the same carbon position along ω_1 as the direct cross-peaks, with the ω_2 position corresponding to that of the remote proton. These cross-peaks are antiphase dispersive along both frequency directions, which gives rise to

†Abbreviations used: BPTI, basic pancreatic trypsin inhibitor; n.m.r., nuclear magnetic resonance; COSY, two-dimensional correlated spectroscopy; NOESY, two-dimensional nuclear Overhauser enhancement and exchange spectroscopy; p.p.m., parts per million.

0022–2836/87/130227–05 $03.00/0

© 1987 Academic Press Inc. (London) Ltd.

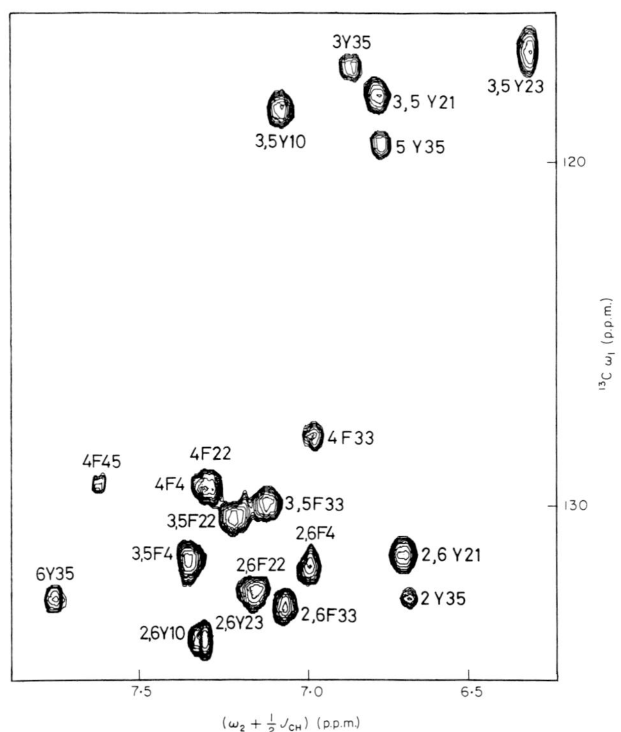

Figure 1. Aromatic region of a [^{13}C, ^{1}H]-COSY spectrum of a 20 mM-solution of BPTI in $^{2}H_2O$ at 36 °C and p^{2}H 4·6. The spectrum was recorded at a proton frequency of 360 MHz and a carbon frequency of 90 MHz. The pulse sequence of Fig. 1A of Brühwiler & Wagner (1986) was applied, i.e. no refocusing delay was used prior to detection. Only negative levels were plotted so that only the right-hand components of the ^{1}H–^{13}C doublets are seen. Since the ^{1}H–^{13}C coupling constants for these aromatic CH groups are to a good approximation all equal to 140 Hz, this spectrum has the same appearance as a [^{13}C, ^{1}H]-COSY spectrum with broad-band ^{13}C decoupling during the detection period. In the Figure the chemical shift axis was shifted by 70 Hz to the right so that the proton chemical shifts correspond to the positions of the single doublet component displayed. The cross-peaks are identified by the ring position of the CH group, the one-letter symbol for the amino acid and the sequence position. For Phe45 only the cross-peak between 4-C and 4-H is observed, since the other proton resonances are broadened due to rotational motions about the C^{β}–C^{γ} bond.

cross-like peak shapes with a positive center peak and four negative arms (the sign information is not shown in Fig. 2). This allows us to distinguish readily between remote and direct connectivities. For example, for Tyr21 the 3,5-carbons are at $\omega_1 = 118{\cdot}0$ p.p.m. and the 2,6-carbons at $\omega_1 = 131{\cdot}5$ p.p.m. The cross-shaped peak at ($\omega_1 = 118$ p.p.m, $\omega_2 = 6{\cdot}79$ p.p.m.) is a remote connectivity between the 3, 5-carbons and the 2,6-protons of this residue. Its ω_2 position coincides with the center of the doublet for the direct connectivity between 2,6-C and 2,6-H at $\omega_1 = 131{\cdot}5$ p.p.m. Such remote [^{13}C, ^{1}H, ^{1}H] connectivities contain the information of the corresponding cross-peak in the homonuclear ^{1}H COSY spectrum. In contrast to ^{1}H COSY, however, proton–proton connectivities can even be observed between protons with degenerate chemical shifts: since the direct cross-peak is split by the coupling with the directly bound ^{13}C, relayed cross-peaks with the remote proton can be observed in the empty space between the two doublet components of the direct connectivity (Wagner & Brühwiler, 1986).

Along ω_1 the cross-peaks for the tyrosine 3,5-CH groups are well-separated from the other cross-peaks. In Figure 1 the presence of five peaks originating from the 3,5-CH groups of the four tyrosine residues clearly confirms the earlier findings, that for Tyr35 at 36 °C the ring flips at a low frequency on the n.m.r. time scale. This interpretation received further support from a [^{13}C, ^{1}H]-COSY spectrum at 66 °C (not shown), which contains a single peak for the Tyr35 3,5-CH connectivity, manifesting the increased ring flip frequency at the elevated temperature (Wagner *et al.*, 1976). For all but one tyrosine residue, remote connectivities are observed between the 3,5-carbons and the 2,6-protons. The exception is Tyr23, where this cross-peak is missing because the 2,6-proton resonance is broadened by low frequency rotational motions of the ring (Wagner *et al.*, 1976). Reverse connectivities between 2,6-carbons and 3,5-protons

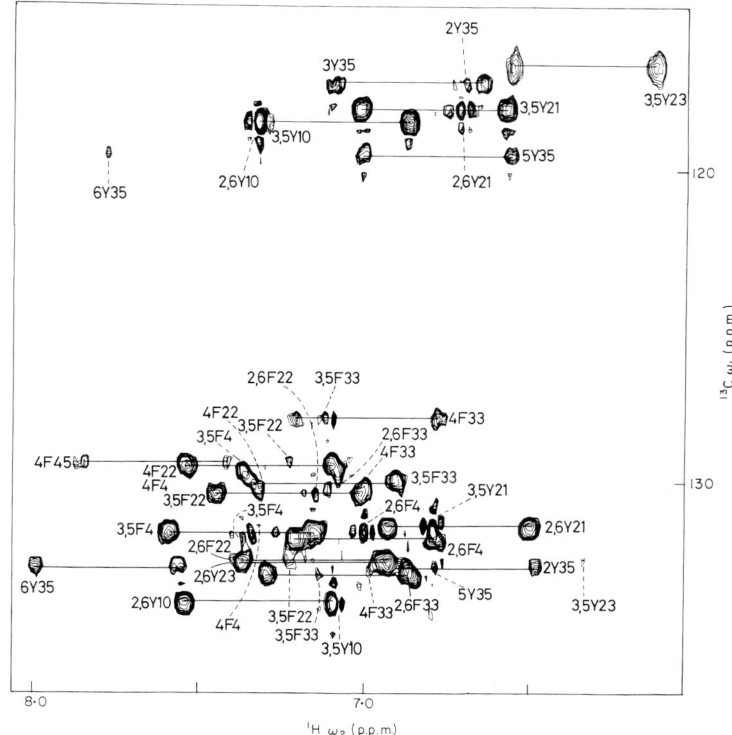

Figure 2. Aromatic region of a [^{13}C, ^{1}H, ^{1}H]-RELAYED-COSY spectrum of BPTI recorded with the same sample as Fig. 1. The pulse sequence of Fig. 1C of Brühwiler & Wagner (1986) was used. Cross-peaks for direct connectivities appear as in-phase absorptive doublets with a 140 Hz splitting along ω_2. In this transformation they are all positive. Remote connectivities appear as single peaks, which are in both frequency dimensions antiphase-dispersive with respect to the ^{1}H–^{1}H coupling constants responsible for the relayed coherence transfer. These peaks are thus cross-shaped, with a positive center peak and four negative arms. (In the black-and-white representation of the Figure, positive and negative components were plotted without distinction.) The in-phase doublet components of the direct connectivities are linked with straight horizontal lines and identified by the position of the ring CH fragment, the one-letter symbol of the amino acid and the sequence position. The cross-peaks for remote connectivities are indicated with broken lines and the peaks identified by the ring positions of the remote protons, the one-letter symbol and the sequence position of the amino acid residue.

are observed for all four tyrosine residues, whereby the relayed cross-peak to Tyr35 3-H is not resolved (Fig. 2). All these observations are fully compatible with the earlier results on the ^{1}H n.m.r. assignments and the dynamic properties of the four tyrosine residues (Wagner *et al.*, 1976).

For Phe4, Phe22 and Phe33 we observed all direct and remote connectivities. For each 2,6-carbon position there were direct connectivities to the 2,6-protons and remote connectivities to the 3,5-protons; for the 3,5-carbons there were direct connectivities to the 3,5-protons and remote connectivities to the 2,6- and the 4-protons; and for the 4-carbons there were direct connectivities to the 4-protons and remote connectivities to the 3,5-protons (Figs 1 and 2). The latter connectivity is crucial for distinction between immobilized and rotating phenylalanine rings. If there are two remote connectivities between the 4-carbon and the 3,5-protons, the side-chain must be immobilized. On the other hand, if there is only one such connectivity, the 3,5-protons are degenerate and it is likely that the ring rotation is rapid. Figures 1 and 2 provide no indication that either Phe22 or Phe33 is immobilized, since for both residues only one remote connectivity between the 4-carbon and the 3,5-protons is observed. This is at variance with the earlier conclusions from one-dimensional ^{1}H n.m.r. experiments that one of these rings is immobilized over the entire accessible temperature range (Wagner *et al.*, 1976). Finally, for Phe45 only the direct connectivity 4-C to 4-H was observed at 36°C (Figs 1 and 2), since the 2,6-H and 3,5-H resonances are broadened by the ring rotations. Just as for Tyr35 the missing resonance lines were observed at 66°C (not shown), which is fully compatible with the earlier results on ^{1}H n.m.r. assignments and dynamics of Phe45 (Wagner *et al.*, 1976).

In view of the apparent discrepancy between the

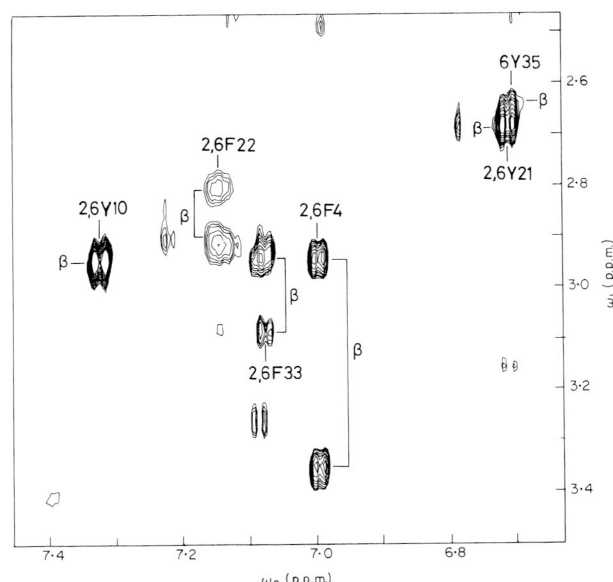

Figure 3. Phase-sensitive 500 MHz ^1H NOESY spectrum of BPTI in ^2H_2O solution at the same experimental conditions as for Figs 1 and 2. The mixing time was 100 ms. Prior to Fourier transformation the first free induction decay was scaled by a factor 0·5 to eliminate t_1-ridges (Otting *et al.*, 1986). The spectral region containing the intraresidue connectivities between β-CH_2 and the aromatic 2,6-protons is shown.

spin system identifications in Figures 1 and 2 and those previously published (Wagner *et al.*, 1976), we further reinvestigated the sequential assignments for the aromatic rings (Billeter *et al.*, 1982). Figure 3 shows the region of an absorption mode NOESY spectrum of BPTI in ^2H_2O, which contains the intraresidue cross-peaks between 2,6-protons (along ω_2) and β-protons (along ω_1) of the phenylalanine and tyrosine residues. The spectrum contains all such intraresidue connectivites except those for Tyr23

Table 1

Chemical shifts of the β-protons, the aromatic protons and the aromatic carbons of BPTI at 36°C, p^2H 4·6, and rates, activation energies, ΔG‡, and activation volumes, ΔV‡, for the rotational ring motions

Residue	Chemical shift†							Rate§	ΔG‡	ΔV‡
	CH_2	2,6-H	2,6-C	3,5-H	3,5-C	4-H	4-C	(s^{-1})	(kcal · M^{-1})	(Å^3)
Y10	2·96	7·32	133·5	7·09	118·5			**f**		
	2·96	7·32	133·5	7·09	118·5					
Y21	2·69	6·71	131·5	6·79	118·0			**f**		
	2·69	6·71	131·5	6·79	118·0					
Y23	3·46	7·20	132·0	6·33	116·5			300	14·7	
	2·72	7·20	132·0	6·33	116·5					
Y35	2·65	6·69	132·5	6·88	117·0			50	15·8	60
	2·50	7·77	132·5	6·78	119·5					
F4	3·35	7·00	131·5	7·37	131·0	7·31	129·0	**f**		
	2·94	7·00	131·5	7·37	131·0					
F22	2·90	7·15	132·0	7·23	130·0	7·32	129·0	**f**		
	2·80	7·15	132·0	7·23	130·0					
F33	3·07	7·07	132·5	7·13	130·0	6·99	128·0	**f**		
	2·95	7·07	132·5	7·13	130·0					
F45‡	3·40	(7·40)	(131·0)	(7·87)	(131·0)	6·63	129·0	1700	13·7	50
	2·78	(7·40)	(131·0)	(7·87)	(131·0)					

†In p.p.m. (±0·02) relative to sodium 3-trimethylsilyl-[2,2,3,3,-$^2H_4$] propionate.

‡Values in parentheses are at 66°C.

§ From Wagner *et al.* (1976), measured at 40°C and p^2H 7·8. **f** stands for high frequency of rotation on the n.m.r. time scale at 360 MHz.

‖From Wagner (1980).

and Phe45, where the 2,6-proton resonances are broadened at 36°C due to rotational motions. (These two missing connectivities are clearly manifested at different temperatures.) The spectrum of Figure 3 is much better resolved than the earlier absolute value spectra (Fig. 9 of Wagner & Wüthrich, 1982). It shows unambiguously that the AA'BB'C spin system of Phe, which had been attributed to Phe33 from comparison of BPTI with the homologous trypsin inhibitor from *Helix pomatia* (Wagner *et al.*, 1978), corresponds to position 22, and the newly found AA'BB'C system represents Phe33.

In conclusion, with the experiments described in this letter we finally had n.m.r. techniques at hand which produced sufficiently good spectral resolution for obtaining sequence-specific assignments and a characterization of the rotational dynamics of all eight aromatic rings in BPTI. The complete set of data for the aromatics is collected in Table 1. Compared to corresponding information derived from one-dimensional n.m.r. and comparison of homologous proteins (Wagner *et al.*, 1976, 1978), the chemical shifts for the aromatic protons of Phe22 and Phe33 are different, and Phe22 is found to be rotating rapidly rather than being immobilized up to 80°C. Furthermore, the 4-H chemical shift of Phe4 had to be revised. These revisions are fully compatible in light of the recent determination of the solution conformation of BPTI by n.m.r. (Wagner *et al.*, 1987). The earlier conclusions had been based on negative evidence rather than on direct observation of spectral features (Wagner *et al.*, 1976, 1978) and the results described in this letter make it clear that a proper documentation of the correct resonance assignments would have been impossible to achieve with one-dimensional n.m.r. experiments at 360 MHz. It should also be emphasized, however, that all those aromatic rings for which quantitative information on ring mobility was presented in 1976 had been correctly assigned, and that the conclusions on intramolecular motility of the BPTI structure and related proteins derived from these early investigations (e.g. see Wagner, 1983; Wagner & Wüthrich, 1978, 1979; Wüthrich & Wagner, 1978, 1984) retain their validity.

Financial support from the Schweizerischer Nationalfonds (projects 3.198.85 and 3.197.85) is gratefully acknowledged.

Gerhard Wagner
Daniel Brühwiler
Kurt Wüthrich

Institut für Molekularbiologie und Biophysik
Eidgenössische Technische Hochschule-Hönggerberg
CH-8093 Zürich, Switzerland
Switzerland

Received 2 January 1987

References

Billeter, M., Braun, W. & Wüthrich, K. (1982). *J. Mol. Biol.* **155**, 321–346.

Brühwiler, D. & Wagner, G. (1986). *J. Magn. Reson.* **69**, 546–551.

Otting, G., Widmer, H., Wagner, G. & Wüthrich, K. (1986). *J. Magn. Reson.* **66**, 187–193.

Snyder, G. H., Rowan, R., Karplus, S. & Sykes, B. D. (1976). *Biochemistry*, **14**, 3765–3777.

Wagner, G. (1980). *FEBS Letters* **112**, 280–284.

Wagner, G. (1983). *Quart. Rev. Biophys.* **16**, 1–57.

Wagner, G. & Brühwiler, D. (1986). *Biochemistry*, **25**, 5839–5843.

Wagner, G. & Wüthrich, K. (1978). *Nature (London)*, **275**, 247–248.

Wagner, G. & Wüthrich, K. (1979). *J. Mol. Biol.* **130**, 31–37.

Wagner, G. & Wüthrich, K. (1982). *J. Mol. Biol.* **155**, 347–366.

Wagner, G., DeMarco, A. & Wüthrich, K. (1975). *J. Magn. Reson.* **20**, 565–569.

Wagner, G., DeMarco, A. & Wüthrich, K. (1976). *Biophys. Struct. Mechan.* **2**, 139–158.

Wagner, G., Wüthrich, K. & Tschesche, H. (1978). *Eur. J. Biochem.* **89**, 367–377.

Wagner, G., Anil Kumar & Wüthrich, K. (1981). *Eur. J. Biochem.* **114**, 375–384.

Wagner, G., Braun, W., Havel, T. F., Schaumann, T., Gō, N. & Wüthrich, K. (1987). *J. Mol. Biol.*, in the press.

Wüthrich, K. (1986). *NMR of Proteins and Nucleic Acids*, Wiley, New York.

Wüthrich, K. & Wagner, G. (1975). *FEBS Letters*, **50**, 265–268.

Wüthrich, K. & Wagner, G. (1978). *Trends Biochem. Sci.* **3**, 227–230.

Wüthrich, K. & Wagner, G. (1984). *Trends Biochem. Sci.* **9**, 152–154.

Edited by M. F. Moody

Reprinted from Biochemistry, 1980, *19*, 5189
Copyright © 1980 by the American Chemical Society and reprinted by permission of the copyright owner.

Carbon-13 Nuclear Magnetic Resonance Relaxation Studies of Internal Mobility of the Polypeptide Chain in Basic Pancreatic Trypsin Inhibitor and a Selectively Reduced Analogue[†]

R. Richarz, K. Nagayama,[‡] and K. Wüthrich*

ABSTRACT: ^{13}C nuclear spin relaxation times and ^{13}C (^{1}H) nuclear Overhauser effects for the backbone α carbons, the protonated aromatic ring carbons, and the side-chain methyl carbons were measured in 25 mM solutions of the basic pancreatic trypsin inhibitor and a modified analogue obtained by reduction of the disulfide bond 14–38. The relaxation parameters for the methyl carbons could, on the basis of previous individual assignments, be correlated with specific locations in the molecular structure. Analysis in terms of a "wobbling in a cone" model, where isotropic overall rotational motion of the protein was assumed, showed that, in addition to the overall rotational motions of the molecule and the rotation of the methyl groups about the C–C bond, the relaxation data manifested librational motions of the polypeptide backbone and the amino acid side chains. The following parameters for the molecular mobility resulted from this analysis: for the overall rotational motions, $\tau_R = 4 \times 10^{-9}$ s; for the librational "wobbling" of the backbone α carbons, in a cone with $\theta_{max} = 20°$, $\tau_w = 1 \times 10^{-9}$ s; for the librational motions of individual aliphatic side chains in cones with θ_{max} varying between 30° and 60°, $\tau_w = 4 \times 10^{-10}$–$3 \times 10^{-9}$ s; for methyl rotation about the C–C bond, $\tau_F \lesssim 1 \times 10^{-11}$ s. From comparison of the two proteins, the molecular motions manifested in the ^{13}C relaxation parameters were found not to be correlated with the thermal stability of the globular conformation. This coincides with the behavior of aromatic ring flips and is different from that of the exchange rates for interior amide protons, which provides new information to further characterize the previously suggested hydrophobic cluster structure for globular proteins in solution.

The small globular protein BPTI,[1] which consists of a single polypeptide chain of 58 amino acid residues and has a molecular weight of 6500, has been used extensively for studies of fundamental aspects of protein conformation. A highly refined crystal structure at 1.5-Å resolution is available (Deisenhofer & Steigemann, 1975), and high-resolution NMR studies showed that the average spatial structure in aqueous solution corresponds very closely to that seen in single crystals (Richarz & Wüthrich, 1978; Wüthrich et al., 1978; Perkins & Wüthrich, 1978, 1979; Dubs et al., 1979). ^{1}H NMR was further used to investigate dynamic aspects of the globular form of BPTI and related proteins. These studies have so far mainly concentrated on measurements of the internal mobility of aromatic rings (Wagner et al., 1976), exchange of internal amide protons with the solvent (Wüthrich & Wagner, 1979; Richarz et al., 1979; Wagner & Wüthrich, 1979a), and thermal denaturation (Wagner & Wüthrich, 1978; Wüthrich et al., 1979a,b), i.e., relatively infrequent stochastic events, with characteristic times of $\gtrsim 1 \times 10^{-5}$ s, which consist of concerted motions involving sizable fractions of the protein structure (Hetzel et al., 1976; Wagner & Wüthrich, 1978, 1979b; Wüthrich & Wagner, 1978; Wüthrich et al., 1979b). In contrast, NMR relaxation studies (Doddrell et al., 1972; Wüthrich, 1976), X-ray techniques (Artymiuk et al., 1979; Frauenfelder et al., 1979; Huber, 1979), and molecular dynamics calculations (McCammon et al., 1977; Karplus & McCammon, 1979) have so far, for physical or practical reasons, provided exclusively information relating to much more frequent events, in the time range 1×10^{-8}–1×10^{-12} s. The present paper describes an attempt to correlate structural information obtained from measurements on largely different time scales. It reports on comparative ^{13}C NMR relaxation studies of BPTI and a chemically modified analogue, which were previously also investigated by the above-mentioned "slow" ^{1}H NMR experiments.

The strategy for the present investigation was influenced by previously recorded data on the ^{13}C NMR spectra of BPTI as well as by certain practical considerations. Since the relaxation parameters were markedly affected by higher BPTI concentrations (Wüthrich & Baumann, 1976), the present

† From the Institut für Molekularbiologie und Biophysik, Eidgenössiche Technische Hochschule, CH-8093 Zürich-Hönggerberg, Switzerland. *Received April 17, 1980.* This work was supported by the Swiss National Science Foundation (Project 3.0040.76).

‡ Present address: Department of Physics, University of Tokyo, Tokyo, Japan.

[1] Abbreviations used: BPTI, basic pancreatic trypsin inhibitor (Kunitz inhibitor, Trasylol, Bayer Wuppertal, Germany); RCAM-BPTI, basic pancreatic trypsin inhibitor obtained by reduction of the disulfide bond 14–38 with the cysteinyl residues protected by carboxamidomethylation; NMR, nuclear magnetic resonance; FT, Fourier transform; ppm, parts per million; Me_4Si, tetramethylsilane; TSP, 2,2,3,3-tetradeuterio-3-(trimethylsilyl)propionate; NOE, nuclear Overhauser enhancement.

experiments were done with a 25 mM protein concentration. So far, only one chemical modification of BPTI was prepared in sufficient quantity for ^{13}C NMR studies, i.e., RCAM-BPTI which denatures at 76 °C as compared to >95 °C for native BPTI in the same solvent (Wagner et al., 1979). Previously, individual assignments were obtained for nearly all the methyl carbon resonances in BPTI (Richarz & Wüthrich, 1977; 1978). Therefore, in addition to the α carbons of the polypeptide backbone and the protonated aromatic ring carbons, the relaxation studies concentrated mainly on the methyl carbons. For an adequate description of the complex motions of the peripheral methyl groups, a "wobbling in a cone" model (Kinosita et al., 1977) was adapted for the situations encountered with BPTI, as is described under Appendix.

Materials and Methods

The basic pancreatic trypsin inhibitor (BPTI, Trasylol, Bayer Leverkusen, Germany) was obtained from the Farbenfabriken Bayer AG. RCAM-BPTI was prepared by selective reduction of the disulfide bond 14–38 with borohydride as a reducing agent and iodoacetamide as a protecting agent (Kress et al., 1968). For the ^{13}C NMR experiments, 25 mM solutions of the proteins in D_2O were prepared. Oxygen was removed by bubbling nitrogen gas through the solutions. The pD was adjusted to 4.1 by the addition of minute amounts of 1 M DCl.

^{13}C NMR measurements at 25.1 MHz were carried out on a Varian XL-100 instrument and those at 90.5 MHz on a Bruker HX-360 spectrometer. The sample temperature in the probe of the XL-100 spectrometer was measured with a thermocouple which was inserted into the sample. On the HX-360 the temperature was measured with a maximum thermometer which was inserted into the sample. In both cases, the decoupling radio frequency field was applied during the temperature measurements. The accuracy of the temperature control was ±1 °C. Sample tubes with an outer diameter of 12 and 10 mm, respectively, were used at 25.1 and 90.5 MHz.

The longitudinal relaxation times T_1 at 25.1 MHz were measured with the inversion recovery method (Vold et al., 1968; Freemann & Hill, 1969). The pulse sequence $(180°-\tau-90°-T)_n$ was used, where τ is a variable delay time and T was at least 4 times the longest T_1 to be measured. Eleven spectra were recorded with delay times τ between 11 and 600 ms. Transients (40 000 per spectrum) were accumulated with T = 1.6 s. T_1's at 90.5 MHz were measured with the saturation recovery method (Markley et al., 1971; McDonald & Leigh, 1973) by using five 90° pulse in intervals of 1 ms between two pulses to saturate the spins. Ten spectra were recorded with delay times between 40 and 2400 ms. Transients (40 000 per spectrum) were accumulated and T_1 values were evaluated with a nonlinear least-squares fitting procedure (Gerhards & Dietrich, 1976) on a Hewlett-Packard 9830 computer. The integrated areas of the signals were used for these calculations. Nuclear Overhauser enhancements were measured by comparing the integrated intensities of the ^{13}C resonances in the proton noise-decoupled spectra with and without decoupling during a period of 2.4 s prior to the data acquisition.

Results

Figure 1 shows a ^{13}C NMR inversion recovery experiment with BPTI at 25.1 MHz. Corresponding data at 90.5 MHz were obtained with the saturation recovery method. Since the mutual overlap of lines in the spectral regions of the backbone α-carbon lines and the aromatic ring carbons makes measurements of T_1's for most of the individual resonances difficult, average values of T_1 for the groups of resonances "α" and "R" (Figure 1) were evaluated. The data for the α carbons in Table I correspond to the average for the resonances between 45 and 68 ppm (Figure 1), i.e., the glycine α carbons at around 43 ppm (Wüthrich, 1976) were not considered. The line widths are the average of the widths measured for the resolved lines, and the NOE's were obtained from integration of the resonance intensities over the spectral region α (Figure 1). Essentially identical data with those for the α carbons were obtained for the protonated aromatic ring carbons of phenylalanine and tyrosine. Figure 2 shows the high-field region of partially relaxed ^{13}C NMR spectra of BPTI with the previously established individual assignements of the methyl-carbon resonances (Richarz & Wüthrich, 1978). T_1 values and NOE's for the individual methyl carbons at 25.1 and 90.5 MHz are listed in Table II.

To investigate whether the ^{13}C relaxation parameters are correlated with the thermal stability of the globular protein,

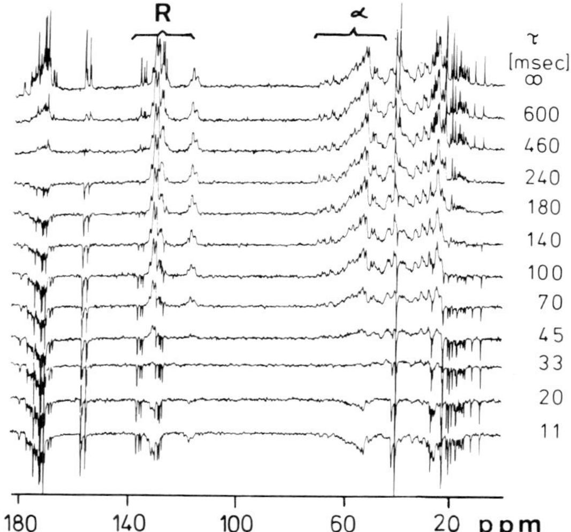

FIGURE 1: Partially relaxed proton noise-decoupled FT ^{13}C NMR spectra at 25.1 MHz of a 25 mM solution of BPTI in D_2O, at pD 4.1 and t = 39 °C. The spectra were obtained with a $(180°-\tau-90°-T)_n$ pulse sequence, with T = 1.6 s. The delay times τ are indicated on the right. The spectral regions where data for the T_1 values of the aromatic ring carbon atoms and the backbone α carbons were collected are indicated by R and α, respectively. In the R region the observed resonance intensity was corrected for the presence of the quaternary aromatic carbons.

Table I: ^{13}C NMR Relaxation Parameters for the α Carbon Resonances of BPTI[a]

resonance frequency (MHz)	parameter	exptl	calcd for best fit	
			isotropic rotation[c]	wobbling in a cone model[d]
25.1	T_1 (ms)	45 ± 3	40	47
	$\delta\nu_{1/2}$ (Hz)[b]	4 ± 1	5.2	4.4
	NOE	1.4 ± 0.2	1.4	1.4
90.5	T_1 (ms)	260 ± 70	210	240
	$\delta\nu_{1/2}$ (Hz)[b]	2 ± 1	3.1	2.6
	NOE	1.2 ± 0.2	1.2	1.2

[a] Measurements in a 25 mM solution of BPTI in D_2O, at pD 4.1 and t = 39 °C. [b] $\delta\nu_{1/2} = 1/(2\pi T_2)$. [c] The best fit with the assumption of isotropic rotation was obtained with $\tau_R = 3.9 \times 10^{-9}$ s. [d] The best fit with the wobbling in a cone model was obtained with the following parameters: $\tau_R = 4.0 \times 10^{-9}$ s; $\tau_W = 1.0 \times 10^{-9}$ s; $\theta_{max} = 20°$.

Table II: ^{13}C Relaxation Parameter for the Methyl-Carbon Resonances of BPTI

amino acid residue	resonance[a]	exptl and theoretical values[b] ν_0 = 25.1 MHz T_1 (ms)	ν_0 = 25.1 MHz NOE	ν_0 = 90.5 MHz T_1 (ms)	best fit parameters[c] τ_F (10^{-11} s)	τ_W (10^{-11} s)	θ_{max} (deg)
{Ala-16 / Ala-40}	{h[d] / i[d]}	160 (140)	(1.9)	260 (340)	1	100	30
Ala-25	g	155 (140)	1.9 (1.9)	320 (340)	1	100	30
Ala-27	k	150 (160)	2.3 (2.3)	210 (280)	1	300	50
Ala-48	e	150 (140)	2.0 (1.9)	310 (340)	1	100	30
Ala-58	j[d]	300 (300)	(2.5)	400 (430)	0.5	100	60
Ile-18γ	f	215 (200)	2.1 (2.3)	310 (330)	1	100	50
Ile-19γ	d	215 (230)	1.8 (2.2)	450 (410)	1	40	50
Ile-18δ	{a	385 (390)	2.1 (2.3)	640 (640)	0.5	40	60
Ile-19δ	b}	290 (290)	2.2 (2.0)	640 (620)	0.5	40	50
Leu-6	q	310 (290)	2.1 (2.0)	650 (620)	0.5	40	50
Met-52	c	325 (340)	2.1 (2.3)	560 (550)	0.5	70	60

[a] See Figure 2. [b] The first number in each column indicates the experimental value in a 25 mM solution of BPTI in D_2O, at pD 4.1 and t = 39 °C. The accuracy is ±15% for T_1 and ±0.2 for the NOE's. The numbers in parentheses indicate the corresponding theoretical values calculated with the wobbling in a cone model. [c] The theoretical values for T_1 and NOE were obtained as the best fits of the experimental data with the wobbling in a cone model with the parameters τ_F, τ_W, and θ_{max} given here. $\tau_R = 4 \times 10^{-9}$ s was used. [d] Because the resonances h, i, and j are mutually overlapping (Figure 2) the relaxation parameters for these lines are less accurate.

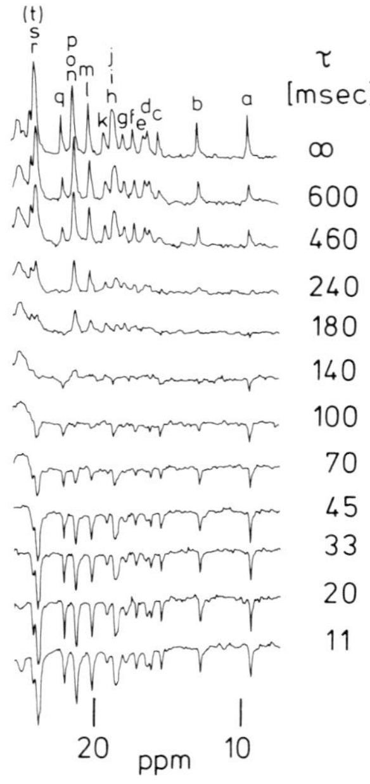

FIGURE 2: Expanded plot of the high-field region from 5 to 25 ppm of the partially relaxed ^{13}C NMR spectra of BPTI in Figure 1. The letters in the top trace indicate the previously established (Richarz & Wüthrich, 1978) assignments of the methyl-carbon lines: a and b, Ile-18δ and Ile-19δ; c, Met-52ϵ; d, Ile-19γ; e, Ala-48β; f, Ile-18γ; g, Ala-25β; (h and i), (Ala-16β and Ala-40β); j, Ala-58β; k, Ala-27β; m, Thr-11γ; n, Thr-54γ; o, Thr-32γ; q, Leu-6δ; (l, p, r, s, and t), (Leu-6δ, Leu-29δ, Leu-29δ, Val-34γ, and Val-34γ).

we compared corresponding data for BPTI and RCAM-BPTI. Previously, the denaturation temperatures at pD 5.0 were found to be >95 °C for BPTI and ~76 °C for RCAM-BPTI. Furthermore, the average spatial structures of the two proteins were found to be very similar, with local differences near the modification site in RCAM-BPTI, i.e., the reduced disulfide bond 14–38 (Wagner et al., 1979). The close similarity of the two proteins is also manifested in the ^{13}C NMR spectra. Figure 3 shows the high-field region of the spectrum of RCAM-BPTI. Comparison with Figure 2 shows that there is a 1:1 correspondence for all the methyl resonances except

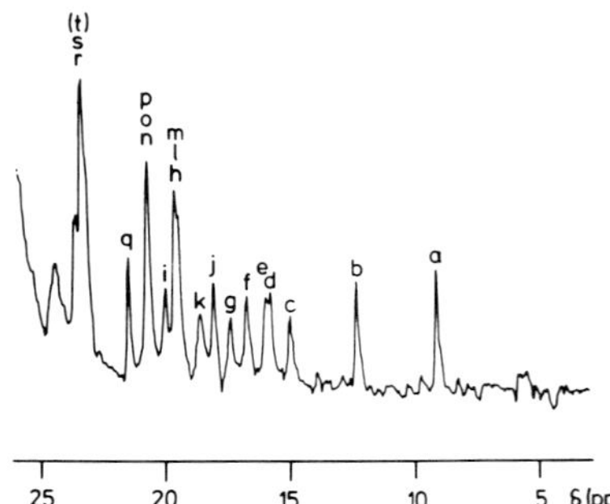

FIGURE 3: High-field region from 5 to 25 ppm of the proton noise-decoupled FT ^{13}C NMR spectrum at 25.1 MHz of a 25 mM solution of RCAM-BPTI at pD 4.1 and t = 39 °C. The letters indicate the assignments of the methyl-carbon resonances, with a 1:1 correspondence with the letters used for BPTI (see Figure 2).

Table III: Comparison of the ^{13}C Relaxation Times T_1 at 25.1 MHz in BPTI with those in RCAM-BPTI[a]

resonance[b]	assignment	T_1 (ms) BPTI	RCAM-BPTI
α carbons		45 ± 3	48 ± 3
Tyr C(2,6)		46 ± 3	45 ± 5
h	{Ala-16β	{160 ± 20}	180 ± 30
i	Ala-40β}		
g	Ala-25β	155	
k	Ala-27β	150 ± 20	160 ± 20
e	Ala-48β	150 ± 20	160 ± 20
j	Ala-58β	~300	300 ± 30
f	Ile-18γ	215 ± 30	170 ± 20
d	Ile-19γ	215 ± 30	200 ± 20
a}	{Ile-18δ	385 ± 30	360 ± 40
b}	Ile-19δ	290 ± 30	240 ± 30
q	Leu-6δ	310 ± 30	210 ± 40
c	Met-25ϵ	325 ± 30	330 ± 30

[a] The relaxation times were measured in 25 mM solutions of the proteins in D_2O, at pD 4.1 and t = 39 °C. [b] See Figures 2 and 3.

h and i, which have slightly different chemical shifts in the two proteins. Therefore the resonance assignments in Figure 3 were made analogous to those for BPTI. The two lines h and i were assigned to Ala-16 and Ala-40 which are immediately adjacent to the modification site. Values for the relaxation times T_1 for corresponding lines in the two proteins are listed in Table III.

Discussion

Overall Rotational Motions of BPTI. We adopt the generally used working hypothesis (Allerhand et al., 1971; Wilbur et al., 1976) that the relaxation parameters of the backbone α carbons are the most reliable manifestation of the overall rotational motions of a protein. With the simplifying assumptions that the rotational motions of the backbone atoms are strictly limited to the overall tumbling of the molecule and that the latter can be characterized by a single correlation time for isotropic rotational tumbling, τ_R, the experimental data of Table I correspond to $\tau_R = 3.9 \times 10^{-9}$ s. The experimental value of τ_R is thus 2–3 times longer than that estimated with the Stokes–Einstein relation (Abragam, 1962) for a spherical protein with the molecular weight of BPTI and with a hydration shell of a thickness of 2.5 Å in aqueous solution.

That the rotational correlation time is rather long, a phenomenon which was also reported from ^{13}C NMR studies of different proteins (Visscher & Gurd, 1974; Wilbur et al., 1976), appears to be primarily due to the high protein concentration. Support for this explanation comes from the observation that τ_R for BPTI in a 50 mM solution is considerably longer yet, i.e., 2×10^{-8} s (Wüthrich & Baumann, 1976). Part of the effect of high protein concentration can be directly assessed from the increased macroscopic viscosity, which is ~1.5 times that of pure D_2O in a 25 mM BPTI solution. As to other potential sources for the long τ_R, numerical calculations for an ellipsoidal shape of BPTI showed that the deviations of the molecular shape from a sphere should cause only a small increase of the effective τ_R, and, as is further discussed below, a more realistic protein model including internal segmental mobility yields nearly the same value for τ_R as the simple solid sphere approximation (Table I).

Analysis of Internal Motions by the Wobbling in a Cone Model. From basic principles it is obvious that in a protein molecule at ambient temperature all the atoms undergo intermolecular motions. In a mechanical description of the protein, one calculates a very large number of normal-mode vibrations to be assigned to bending or stretching of the individual covalent bonds. These coherent motions with frequencies $\gtrsim 1 \times 10^{12}$ s^{-1}, which are in principle accessible to observation by optical methods, are not noticeably manifested in the NMR relaxation. However, superimposed on these "elemental" thermal motions, there are further the relatively large time variations of dihedral angles about single bonds. Dampening of these rotational motions by the environment of the individual atoms in the hydrated protein structure leads to stochastic fluctuations of the peptide backbone and the amino acid side chains which may have frequency components in the range where the NMR spin relaxation can be affected.

Quite generally the rotational motions of a peptide fragment relative to the laboratory frame, which result from the combination of the overall rotation of the molecule and intramolecular motions, are very complex. Compromising between a realistic description of these complex movements and the requirement for a tractable mathematical model, we have chosen to analyze the relaxation data of BPTI in terms of a wobbling in a cone description (Kinosita et al., 1977) for stochastic internal motions. A detailed presentation of this model is given under Appendix. It can be seen that through a fit to the wobbling in a cone model the experimental relaxation parameters can be correlated with a combination of quite straightforward internal motions, such as rotation about single bonds and diffusion of a rotation axis inside a cone. In the following this treatment will be applied to different types of carbon atoms in BPTI.

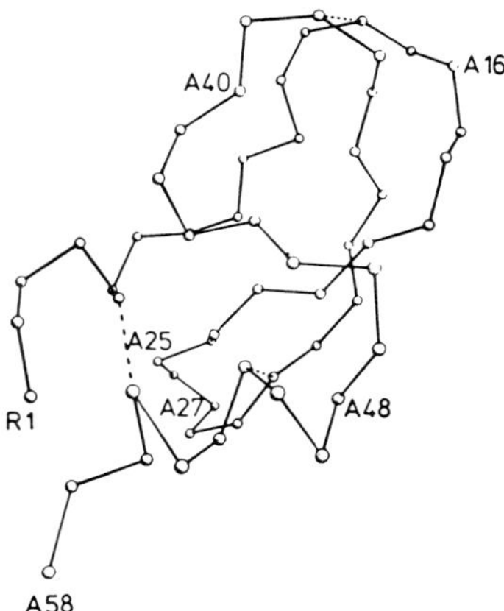

FIGURE 4: Computer drawing of the α-carbon positions in the crystal structure of BPTI (Deisenhofer & Steigemann, 1975). The locations of the three disulfide bonds 5–55, 14–38, and 30–51 are indicated by the broken lines, and the six alanines and the N terminus are identified by the IUB–IUPAC one-letter symbols and the position in the amino acid sequence.

Segmental Mobility of the Polypeptide Backbone. In a simplified form described under Appendix the wobbling in a cone model was used to analyze the relaxation parameters for the α-carbon atoms. While with the values for τ_R, τ_w, and θ_{max} given in Table I a somewhat better fit of the experimental data was obtained than with the rigid sphere approximation, the differences are too small to provide convincing evidence for backbone mobility.

More direct evidence for intramolecular motions of backbone fragments came from the data on the methyl carbons of the six alanines in BPTI. In a rigid molecular structure the internal mobility of the alanine methyl groups would be restricted to rotation about the C^α–C^β bond. This limiting situation was investigated by Woessner (1962) who showed that T_1 for the methyl carbon could be at most 3 times longer than that for the α-carbon atom. Since at 25.1 MHz the T_1's for the methyl carbons are consistently longer than this limiting value (Tables I and II), additional mobility besides methyl rotation had to be considered. Fitting the data with the wobbling in a cone model indicated that the C^α–C^β methyl rotation axes of Ala-16, -25, -40, and -48 undergo angular displacements on the order of 30° with a correlation time of ~1×10^{-9} s (Table II). A somewhat larger mobility is indicated for the backbone region near Ala-27, and the increased flexibility at the C-terminal Ala-58 is very clearly manifested.

It is interesting that the increased mobility of the polypeptide fragments near Ala-27 and near the C-terminal Ala-58 manifested in the ^{13}C NMR parameters coincides with predictions from molecular dynamics calculations on the BPTI structure (McCammon et al., 1977; Karplus & McCammon, 1979). These authors found that overall the time average of the structure of BPTI over a period of 1×10^{-10} s coincided closely with the crystal structure (Deisenhofer & Steigemann, 1975) yet that relatively large deviations were found for the polypeptide fragment 25–28, which links the two strands of the antiparallel β sheet (Figure 4), and for the two chain terminal fragments (Karplus & McCammon, 1979).

Internal Mobility of Amino Acid Side Chains. The internal mobility of the aromatic rings of phenylalanine and tyrosine in BPTI was previously investigated by ^{1}H NMR (Wagner et al., 1976). Precise information was obtained for five of the eight rings. For the remaining three rings the frequency of 180° flips about the C^{β}–C^{γ} bond (Hetzel et al., 1976) was rapid on the ^1H NMR time scale, and only an upper limit for the correlation time of the internal motions could be estimated, i.e., $\tau_{180°} < 2 \times 10^{-5}$ s. The present ^{13}C relaxation study showed an essentially identical behavior for the backbone α carbons and the protonated ring carbons of the aromatics (Figure 1 and Table III). From this it can be concluded, in agreement with a previous study (Wüthrich & Baumann, 1976), that the frequency of aromatic ring flips is small compared to the overall rotation of BPTI, i.e., $\tau_{180°} > 4 \times 10^{-9}$ s. Otherwise, the relaxation data would be compatible with small-amplitude librational motions of the aromatic rings on a time scale comparable to or faster than τ_R.

For the mobility of aliphatic side chains information was obtained from analysis of the methyl-carbon relaxation parameters with the wobbling in a cone model (Table II). There is a readily apparent trend to increased mobility for the peripheral carbons in longer side chains, which is manifested in both the increased frequency of the motions and the larger half-angle θ_{max} of the cone covered by the wobbling motions. This confirms earlier conclusions which were obtained without individually assigned ^{13}C NMR data on a variety of proteins [for a survey, see Allerhand (1979)].

Internal Mobility and Stability of Globular Protein Structure. Previously, ^1H NMR studies of a group of chemically modified analogues of BPTI were used to investigate correlations between internal mobility manifested in the amide proton exchange rates (Richarz et al., 1979) and the rotational flipping motions of the aromatic rings (Wagner et al., 1976) with the thermal stability of the globular form of the proteins. While the amide proton exchange rates were found to increase when the stability of the protein was lowered, the mobility of the aromatic rings was not affected (Wagner & Wüthrich, 1978). From these observations we have suggested a hydrophobic cluster architecture which might be found not only in BPTI but also quite generally in globular proteins (Wüthrich & Wagner, 1978; Wüthrich et al., 1979a; Wagner & Wüthrich, 1979b; Wüthrich et al., 1980). It was now interesting to see how the higher frequency internal motions manifested in the ^{13}C relaxation data were affected by different thermal stabilities of two otherwise nearly identical spatial protein structures.

Table III shows that nearly identical relaxation data were obtained for corresponding resonances in BPTI and RCAM-BPTI, which have denaturation temperatures of >95 °C and ~76 °C, respectively, under the conditions used for the ^{13}C NMR studies (Wagner et al., 1979). This holds for the backbone α carbons, the aromatic ring carbons, and the methyl carbons of aliphatic side chains, with the exception of Leu-6 δCH_3 where there is evidence for reduced mobility in RCAM-BPTI. There is also some indication that the mobility of the δ methyls of Ile-18 and Ile-19 might be somewhat more restricted in RCAM-BPTI. Overall, the observations for the ^{13}C relaxation parameters are thus very similar to those for the mobility of the aromatic rings, i.e., no global effect on the relaxation times for carbon atoms located in different regions of the molecule results from a localized modification of the covalent protein structure, such as the reduction of the covalent protein structure, such as the reduction of the disulfide bond 14–38 in RCAM-BPTI (Figure 4). This appears to further substantiate the hypothesis of a hydrophobic cluster architecture of globular proteins. The aliphatic side chains form an important part of the hydrophobic clusters surrounding the aromatic rings (Wüthrich et al., 1980). That their behavior in the two proteins with different stability parallels that seen previously for the aromatic rings supports that observation of the latter provides a reliable NMR probe for studies of internal fluctuations within these clusters.

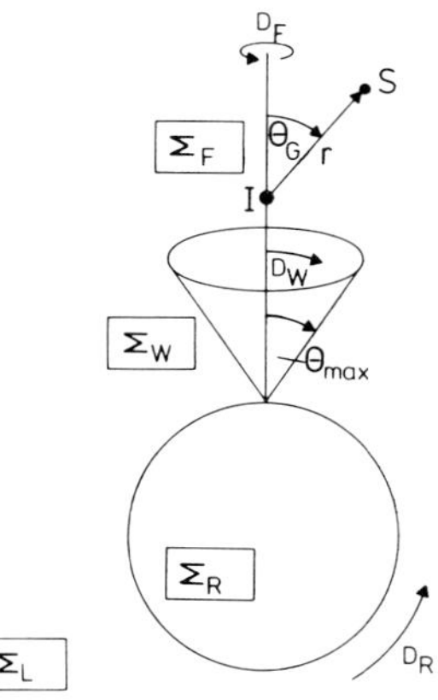

FIGURE 5: Wobbling in a cone model. We consider the relaxation of spin I by dipole–dipole coupling with spin S. The two spins are located in a spherical particle which undergoes isotropic rotational motions with a diffusion constant D_R relative to the laboratory frame Σ_L. The vector $\boldsymbol{r}$ which connects the two spins is attached at a fixed angle θ_G to an axis about which it rotates with a diffusion constant D_F. This rotation axis is further allowed to wobble, relative to the a coordinate system Σ_R fixed in the rotating molecule, with D_W inside a cone defined by the half-angle θ_{max}. For characterization of the resulting complex motion in space of the vector $\boldsymbol{r}$, the two coordinate systems Σ_W and Σ_F are used in addition to Σ_L and Σ_R. Σ_W is fixed with respect to the wobbling motion of the rotation axis inside the cone defined in the frame Σ_R. Σ_F is fixed with respect to the rotations about this axis.

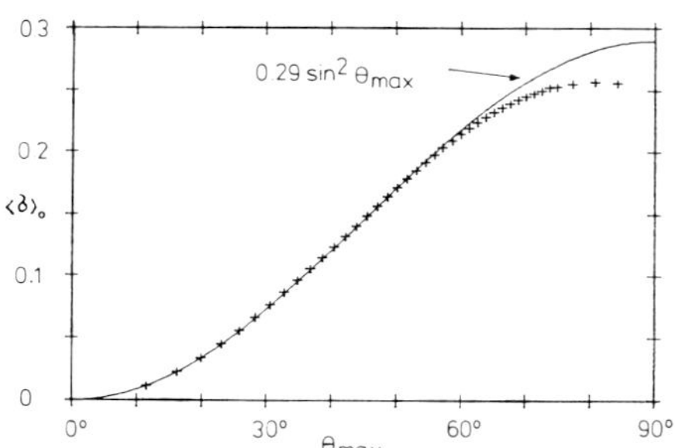

FIGURE 6: Wobbling in a cone model. Plot of $\langle\sigma\rangle_0$ vs. θ_{max}. The crosses represent the values calculated by Kinosita et al. (1977; personal communication), and the solid line represents the corresponding data obtained with the approximation $\langle\sigma\rangle_0 = 0.29 \sin^2 \theta_{max}$ (see text).

Outlook for Future Investigations. The wobbling in a cone model presents a quite realistic and tractable description of the complex intramolecular motions in a flexible protein structure (Karplus & McCammon, 1979; Wüthrich & Wagner, 1978). Overall, mainly as a result of the individual assignments for numerous ^{13}C NMR lines in BPTI (Richarz & Wüthrich, 1978), the data presented in this paper clearly provide evidence for wobbling-motion contributions to the relaxation parameters. In some instances, however, the manifestations of internal mobility were just barely discernible within the accuracy of the relaxation data. For future work it will therefore be highly desirable that more accurate relaxation parameters can be obtained. The use of the wobbling in a cone model should then also be suitable for a more

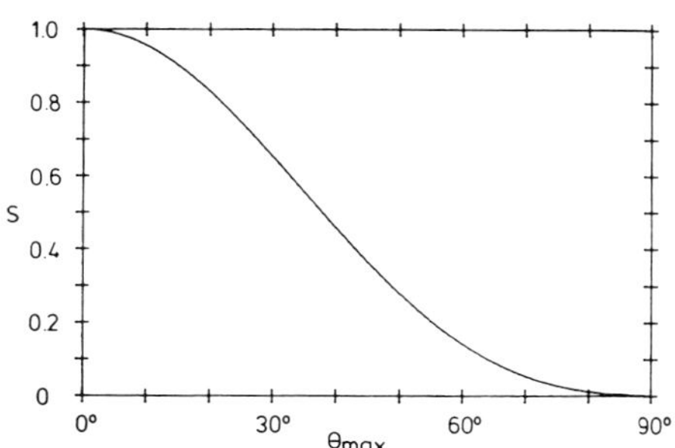

FIGURE 7: Wobbling in a cone model. Plot of S vs. θ_{max} (eq 14).

quantitative assessment of molecular regions with variable internal flexibility.

Acknowledgments

We thank Dr. R. Schmidt-Kastner, Farbenfabriken Bayer AG, for a generous gift of BPTI (Trasylol) and H. Roder for the preparation of the large amount of RCAM-BPTI needed for the ^{13}C NMR experiments. We are grateful to Dr. K. Kinosita for sending us his numerical results for the diffusion in a cone.

Appendix

The "Wobbling in a Cone" Model. Equations 1–3 describe the ^{13}C relaxation times T_1 and T_2 and the steady-state NOE which arise from dipole–dipole coupling between the observed ^{13}C nucleus and a single proton as functions of the spectral densities $J(\omega)$ (Abragam, 1962). γ_H and γ_C are the gyro-

$$T_1^{-1} = (1/3)\hbar^2\gamma_H^2\gamma_C^2 r^{-6}[J(\omega_C - \omega_H) + 3J(\omega_C) + 6J(\omega_C + \omega_H)] \quad (1)$$

$$T_2^{-1} = (1/6)\hbar^2\gamma_H^2\gamma_C^2 r^{-6}[4J(0) + J(\omega_C - \omega_H) + 3J(\omega_C) + 6J(\omega_H) + 6J(\omega_C + \omega_H)] \quad (2)$$

$$\text{NOE} = 1 + \frac{\gamma_H[6J(\omega_C + \omega_H) - J(\omega_C - \omega_H)]}{\gamma_C[J(\omega_C - \omega_H) + 3J(\omega_C) + 6J(\omega_C + \omega_H)]} \quad (3)$$

magnetic ratios for ^{1}H and ^{13}C, $\hbar$ is the Planck constant divided by 2π, r is the distance between the two nuclei, and ω_H and ω_C are the Larmor frequencies for ^{1}H and ^{13}C. The spectral densities $J(\omega)$ are the Fourier transforms of the autocorrelation functions G(t)

$$J(\omega) = \mathrm{F}[\mathrm{G}(t)] \quad (4)$$

For dipole–dipole coupling, G(t) has the form

$$\mathrm{G}^{(m)}(t) = 5\langle Y_{2m}[\Omega_{L\to F}(0)]Y_{2m}^*[\Omega_{L\to F}(t)]\rangle \quad (5)$$

where m is the magnetic quantum number, Y_{2m} are spherical harmonics of the second rank, and $\Omega_{L\to F}$ represents the Euler transformation between the laboratory and the final coordinate system (Figure 5). In the following a wobbling in a cone model is used for the description of the movements in space of the vector $\boldsymbol{r}$ linking ^{1}H and ^{13}C (Figure 5). The autocorrelation functions resulting from this type of spatial motion were computed with the formalism of Wallach (1967). In eq 6 the transformation between the laboratory system and the rotating coordinate system Σ_F (Figure 5) is separated into a series of steps, where the Euler angles α, β, and γ are used as arguments. In the derivation of eq 6 it was further assumed that the three types of diffusional motion characterized by the diffusion constants D_R, D_W, and D_F (Figure 5) are not correlated and that the angle θ_G is fixed. The functions

$$\mathrm{G}^{(m)}(t) = 5\sum_{a,b}\langle D_{ma}^{(2)}(\alpha_{L\to R},\beta_{L\to R},\gamma_{L\to R};0)D_{ma}^{(2)*} \times (\alpha_{L\to R},\beta_{L\to R},\gamma_{L\to R};t)\rangle\langle D_{ab}^{(2)}(\alpha_{R\to W},\beta_{R\to W},\gamma_{R\to W};0) \times D_{ab}^{(2)*}(\alpha_{R\to W},\beta_{R\to W},\gamma_{R\to W};t)\rangle|Y_{2b}(\beta_{W\to F};0)|^2 = \sum_{b=-2}^{2}\exp(6D_Rt)|Y_{2b}(\beta_{W\to F};0)|^2\sum_{a=-2}^{2}\langle D_{ab}^{(2)} \times (\alpha_{R\to W},\beta_{R\to W},\gamma_{R\to W};0)D_{ab}^{(2)*}(\alpha_{R\to W},\beta_{R\to W},\gamma_{R\to W};t)\rangle \quad (6)$$

$D_{mm'}^{(2)}(\alpha,\beta,\gamma)$ are elements of the Wigner rotation matrix (Wigner, 1959). For evaluation of the statistical average of the autocorrelation functions in eq 6, a Green's function for diffusion of a rigid body is generally required. The Green's function for diffusion of a unit vector in a cone was previously evaluated with the use of generalized Legendre functions $P_\nu^m(x)$, where ν takes noninteger values and m is an integer (Carslow & Jaeger, 1959). On the same theoretical basis Kinosita et al. (1977) calculated the correlation functions for the case $b = 0$. They found that the relaxation trend can be well approximated with the use of one average correlation time, while in reality the correlation function should be described by an infinite number of relaxation times. In analogy to the case $b = 0$, we used this result for the cases $b = 1$ and 2. Finally the sum of correlation functions in eq 6 is rewritten in the form

$$\sum_{a=-2}^{2}\langle D_{ab}^{(2)}(\alpha_{R\to W},\beta_{R\to W},\gamma_{R\to W};0)D_{ab}^{(2)*} \times (\alpha_{R\to W},\beta_{R\to W},\gamma_{R\to W};t)\rangle \cong [S_b + (1 - S_b)\exp(-D_Wt\langle\sigma\rangle^{-1})]\exp[-b^2(D_F - D_W)t] \quad (7)$$

where

$$S_b = (1/4)\cos^2\theta_{max}(1 + \cos\theta_{max})^2 \text{ for } b = 0 \quad (8a)$$

and

$$S_b = 0 \text{ for } b \neq 0 \quad (8b)$$

and where three individual correlation times are defined by

$$\tau_R = \frac{1}{6D_R} \qquad \tau_W = \frac{1}{D_W} \qquad \tau_F = \frac{1}{D_F} \quad (9)$$

The correlation time for diffusional wobbling in the cone was found to be defined by eq 10. In eq 10 $d_{ab}^{(2)}(\beta) = D_{ab}^{(2)}(0,\beta,0)$

$$\tau_W\langle\sigma\rangle_b(1 - S_b) = \sum_{a=-2}^{2}\int_0^\infty dt\int_0^{\theta_{max}}\int_0^{\theta_{max}}[d_{ab}^{(2)}(\beta)d_{ab}^{(2)}(\beta')p(\beta,\beta',t)p_0]\,d(\cos\beta)\,d(\cos\beta') \quad (10)$$

is an element of the reduced Wigner rotation matrix (Wigner, 1959). The running variable in the integration, β, corresponds to $\beta_{R\to W}$ in eq 6 and 7. θ_{max} is defined in Figure 5. p_0 is the initial distribution, and $p(\beta,\beta',t)$ is the transition probability for the time evolution of β which obeys diffusion eq 11 subject

$$-\frac{\partial}{\partial t}p(\beta,\beta',t) = D_W\left[\frac{\partial}{\partial\beta^2} + \cot\beta\frac{\partial}{\partial\beta} - \frac{1}{\sin^2\beta}(a^2 + b^2 - 2ab\cos\beta)\right]p(\beta,\beta',t) \quad (11)$$

to the initial condition $p(\beta,\beta',0) = \delta(\beta - \beta')$ and the boundary condition $(\partial p/\partial\beta)|_{\beta=\theta_{max}} = 0$. For the case $b = 0$, the summation in eq 10 is over all values of $|a| \leq 2$ except $a = 0$. From eq 10 one sees immediately that the average correlation time $\tau_W\langle\sigma\rangle_b$ depends on the angle θ_{max} even if one assumes that the diffusional wobbling in the cone can be described by a single diffusion constant, D_W.

The numerical calculation of $\langle\sigma\rangle_0$ has been first performed by Kinosita et al. (1977), and an analytical solution was recently proposed by Lipari & Szabo (1980). We find that in the range $0° < \theta_{max} < 70°$ it is well approximated within an error of 5% (see Figure 6) by the simple expression

$$\langle\sigma\rangle_0 = 0.29 \sin^2 \theta_{max} \quad (12)$$

The general expression for the spectral density is

$$J(\omega) = A\left[S\frac{\tau_A}{1 + (\tau_A\omega)^2} + (1 - S)\frac{\tau_A'}{1 + (\tau_A'\omega)^2}\right] + B\frac{\tau_B}{1 + (\tau_B\omega)^2} + C\frac{\tau_C}{1 + (\tau_C\omega)^2} \quad (13)$$

with

$$\begin{aligned}
\tau_A^{-1} &= \tau_R^{-1} \\
\tau_A'^{-1} &= \tau_R^{-1} + \tau_W^{-1}\langle\sigma\rangle_0^{-1} \\
\tau_B^{-1} &= \tau_R^{-1} + (1/6)\tau_F^{-1} + \tau_W^{-1}(\langle\sigma\rangle_1^{-1} - 1) \\
\tau_C^{-1} &= \tau_R^{-1} + (2/3)\tau_F^{-1} + \tau_W^{-1}(\langle\sigma\rangle_2^{-1} - 4) \\
S &= (1/4)\cos^2\theta_{max}(1 + \cos\theta_{max})^2 \quad (14) \\
A &= (1/4)(3\cos^2\theta_G - 1)^2 \\
B &= 3\sin^2\theta_G \cos^2\theta_G \\
C &= (3/4)\sin^4\theta_G \\
\langle\sigma\rangle_0 &= 0.29\sin^2\theta_{max} \quad (0° \leq \theta_{max} \leq 70°)
\end{aligned}$$

(See Figure 7.) For the situations of special interest in the present investigation, the terms in $\langle\sigma\rangle_1$ and $\langle\sigma\rangle_2$ vanish (see below). Hence these terms have not as yet been calculated.

Treatment of Relaxation of α Carbon in a Protein with the Wobbling in a Cone Model. The general case treated so far included wobbling of an axis in a cone combined with rotation about this axis (Figure 5). For the motion of the vector between an α carbon and the directly attached proton, no rotation about the wobbling axis must be considered. This case may be described by assuming that either $\theta_G = 0$ or $D_F = 0$ in eq 13 and 14, which leads to the expression for the spectral density

$$J(\omega) = S\frac{\tau_A}{1 + (\tau_A\omega)^2} + (1 - S)\frac{\tau_A'}{1 + (\tau_A'\omega)^2} \quad (15)$$

where

$$\begin{aligned}
\tau_A^{-1} &= \tau_R^{-1} \\
\tau_A'^{-1} &= \tau_R^{-1} + \tau_W^{-1}\langle\sigma\rangle_0^{-1} \\
S &= (1/4)\cos^2\theta_{max}(1 + \cos\theta_{max})^2 \\
\langle\sigma\rangle_0 &= 0.29\sin^2\theta_{max} \quad (0° \leq \theta_{max} \leq 70°)
\end{aligned}$$

Inserting eq 15 into eq 1–3 gives the final result for the relaxation times and the NOE of the α carbons in a protein for the situation where the flexibility of the backbone is described by a wobbling in a cone motion. Note that the well-known result for isotropic rotational motion is obtained either for $\tau_W \rightarrow \infty$ or for $\theta_{max} \rightarrow 0$.

Treatment of Relaxation of Methyl Carbons in a Protein with the Wobbling in a Cone Model. Here, the rotation of the methyl group about the C–C bond must be considered in addition to the wobble of the rotation axis (Figure 5). With the realistic assumption that the rotation of the methyl group described by τ_F is rapid compared to the wobbling rate, τ_B and τ_C in eq 13 and 14 can be reduced to

$$\begin{aligned}
\tau_B^{-1} &= \tau_R^{-1} + (1/6)\tau_F^{-1} \\
\tau_C^{-1} &= \tau_R^{-1} + (2/3)\tau_F^{-1} \quad (16)
\end{aligned}$$

for $\tau_F \ll \tau_W$. Note that one obtains the well-known formulas of Woessner (1962) for a rapidly rotating methyl group rigidly attached to a sphere when $\tau_W \rightarrow \infty$ or $\theta_{max} \rightarrow 0$.

References

Abragam, A. (1962) *The Principles of Nuclear Magnetism*, Clarendon Press, Oxford.

Allerhand, A. (1979) *Methods Enzymol. 61*, 458.

Allerhand, A., Doddrell, D., Glushko, V., Cochran, D. W., Wenkert, E., Lawson, P. J., & Gurd, F. R. N. (1971) *J. Am. Chem. Soc. 93*, 544.

Artymink, P. J., Blake, C. C. F., Grace, D. E. P., Oatley, S. J., Phillips, D. C., & Sternberg, M. J. E. (1979) *Nature (London) 280*, 563.

Carslaw, H. S., & Jaeger, J. C. (1959) *Conduction of Heat in Solids*, 2nd ed., Chapter 14, Clarendon Press, Oxford.

Deisenhofer, J., & Steigemann, W. (1975) *Acta Crystallogr., Sect. B 31*, 238.

Doddrell, D., Glushko, V., & Allerhand, A. (1972) *J. Chem. Phys. 56*, 3683.

Dubs, A., Wagner, G., & Wüthrich, K. (1979) *Biochim. Biophys. Acta 577*, 177.

Frauenfelder, H., Petsko, G. A., & Tsernoglou, D. (1979) *Nature (London) 280*, 558.

Freeman, R., & Hill, H. D. W. (1969) *J. Chem. Phys. 51*, 3140.

Gerhards, R., & Dietrich, W. (1976) *J. Magn. Reson. 23*, 21.

Hetzel, R., Wüthrich, K., Deisenhofer, J., & Huber, R. (1976) *Biophys. Struct. Mech. 2*, 159.

Huber, R. (1979) *Trends Biochem. Sci. (Pers. Ed.) 4*, 271.

Karplus, M., & McCammon, J. A. (1979) *Nature (London) 277*, 578.

Kinosita, K., Jr., Kawato, S., & Ikegami, A. (1977) *Biophys. J. 20*, 289.

Kress, L. F., Wilson, K. A., & Laskowski, M., Sr. (1968) *J. Biol. Chem. 243*, 1758.

Lipari, G., & Szabo, A. (1980) *Biophys. J.* (in press).

Markley, J. L., Horsley, W. J., & Klein, M. P. (1971) *J. Chem. Phys. 55*, 3604.

McCammon, J. A., Gelin, B. R., & Karplus, M. (1977) *Nature (London) 267*, 585.

McDonald, G. G., & Leigh, J. S. (1973) *J. Magn. Reson. 9*, 358.

Perkins, S. J., & Wüthrich, K. (1978) *Biochim. Biophys. Acta 536*, 406.

Perkins, S. J., & Wüthrich, K. (1979) *Biochim. Biophys. Acta 576*, 409.

Richarz, R., & Wüthrich, K. (1977) *FEBS Lett. 79*, 64.

Richarz, R., & Wüthrich, K. (1978) *Biochemistry 17*, 2263.

Richarz, R., Sehr, P., Wagner, G., & Wüthrich, K. (1979) *J. Mol. Biol. 130*, 19.

Visscher, R. B., & Gurd, F. R. N. (1975) *J. Biol. Chem. 250*, 2238.

Vold, R. L., Waugh, J. S., Klein, M. P., & Phelps, D. E. (1968) *J. Chem. Phys. 48*, 3831.

Wagner, G., & Wüthrich, K. (1978) *Nature (London) 275*, 247.

Wagner, G., & Wüthrich, K. (1979a) *J. Mol. Biol. 130*, 31.

Wagner, G., & Wüthrich, K. (1979b) *J. Mol. Biol. 134*, 75.

Wagner, G., DeMarco, A., & Wüthrich, K. (1976) *Biophys. Struct. Mech. 2*, 139.

Wagner, G., Tschesche, H., & Wüthrich, K. (1979) *Eur. J. Biochem. 95*, 239.

Wallach, D. (1967) *J. Chem. Phys. 47*, 5258.

Wigner, E. P. (1959) *Group Theory and Its Application to the Quantum Mechanics of Atomic Spectra*, Academic Press, New York.

Wilbur, D. J., Norton, R. S., Clouse, A. O., Addleman, R., & Allerhand, A. (1976) *J. Am. Chem. Soc. 98*, 8250.

Woessner, (1962) *J. Chem. Phys. 36*, 1.

Wüthrich, K. (1976) *NMR in Biological Research: Peptides and Proteins*, North-Holland Publishing Co., Amsterdam.

Wüthrich, K., & Baumann, R. (1976) *Org. Magn. Reson. 8*, 532.

Wüthrich, K., & Wagner, G. (1978) *Trends Biochem. Sci. (Pers. Ed.) 3*, 227.

Wüthrich, K., & Wagner, G. (1979) *J. Mol. Biol. 130*, 1.

Wüthrich, K., Wagner, G., Richarz, R., & Perkins, S. J. (1978) *Biochemistry 17*, 2253.

Wüthrich, K., Roder, H., & Wagner, G. (1979a) in *Protein Folding* (Balaban, A., & Jaenicke, R., Eds.) North-Holland Publishing Co. (in press).

Wüthrich, K., Wagner, G., & Richarz, R. (1979b) *Proc. FEBS Meet. 52*, 143.

Wüthrich, K., Wagner, G., Richarz, R., & Braun, W. (1980) *Biophys. J.* (in press).

Reprinted from Biochemistry, 1993, *32*. 3571
Copyright © 1993 by the American Chemical Society and reprinted by permission of the copyright owner.

Disulfide Bond Isomerization in BPTI and BPTI(G36S): An NMR Study of Correlated Mobility in Proteins†

Gottfried Otting, Edvards Liepinsh, and Kurt Wüthrich*

Institut für Molekularbiologie und Biophysik, Eidgenössische Technische Hochschule-Hönggerberg, CH-8093 Zürich, Switzerland

Received November 19, 1992; Revised Manuscript Received January 22, 1993

ABSTRACT: Two conformational isomers were observed in the ^{1}H nuclear magnetic resonance (NMR) spectra of the basic pancreatic trypsin inhibitor (BPTI) and of a mutant protein with Gly 36 replaced by Ser, BPTI(G36S). The less abundant isomer differs from the major conformation by different chirality of the Cys 14–Cys 38 disulfide bond. In BPTI, the population of the minor conformer increases from about 1.5% at 4 °C to 8% at 68 °C. In BPTI(G36S), the population of the minor conformation is about 15% of the total protein, so that a detailed structural study was technically feasible; a trend toward increasing population of the minor conformer at higher temperatures was observed also for this mutant protein. The activation parameters for the exchange between the two conformations were measured in the temperature range 4–68 °C, using uniformly ^{15}N-enriched protein samples. Below room temperature the exchange rate of the disulfide flip follows an Arrhenius-type temperature dependence, with negative activation entropy in both proteins. At higher temperatures the exchange rates are governed by a different set of activation parameters, which are similar to those for the ring flips of Tyr 35 about the C^β–C^γ bond. Although the equilibrium enthalpy and entropy were found to be largely temperature independent, the activation entropy changes sign and is positive at higher temperatures. These results suggest that, above room temperature, the disulfide flips are coupled to the same protein structure fluctuations as the ring flips of Tyr 35.

BPTI[1] is among the best characterized and most thoroughly investigated of all proteins. Three high-resolution crystal structures are available (Deisenhofer et al., 1975; Wlodawer et al., 1984, 1987), and a high-quality structure has been determined by NMR in aqueous solution (Berndt et al., 1992). Furthermore, the dynamic properties of BPTI in solution were extensively investigated by NMR (Wagner, 1980; Richarz et al., 1980; Wagner et al., 1984; Wagner & Nirmala, 1989), and there is a large body of data on structural effects of chemical modifications and site-directed mutagenesis (e.g., Kress et al., 1968; Wagner et al., 1984; Chazin et al., 1985; Vanmierlo et al., 1991). On the basis of this wealth of data, observations on subtle localized variations of structure and dynamics can now be made with BPTI, although similar effects might escape detection in other proteins. In the present study, investigations of the mutant BPTI(G36S) lead to the discovery of a slow dynamic equilibrium between two conformers with different chirality of the disulfide bridge formed by Cys 14 and Cys 38. The same two-state equilibrium was then also found in wild-type BPTI.

In BPTI(G36S), the hydroxyl-bearing side chain of Ser 36 replaces the internal water molecule that is present in a cavity formed between the peptide segments of residues 11–14 and 35–38 of the wild-type BPTI structure (Berndt et al., 1993). Amide proton exchange experiments and NOESY spectra recorded at 4 °C with BPTI(G36S) and BPTI indicated that the backbone conformation and the hydrogen-bonding network are virtually unaltered by the mutation, but more significant differences were observed at 36 °C. It had been noted earlier in wild-type BPTI that some proton resonances in the polypeptide segment 14–18 are broadened at 36 °C, and in the past only few NOEs with these resonances could be identified (Wagner & Wüthrich, 1982; Wagner et al., 1987; Berndt et al., 1992). Furthermore, increased transverse ^{13}C relaxation rates were observed for the α-carbon resonances of Cys 14 and Cys 38 (Wagner & Nirmala, 1989). These data, together with the observation of pronounced intensity changes of the COSY cross-peaks with temperature, have previously led to the conclusion that increased internal mobility prevailed in this part of the molecule (Wagner et al., 1987; Wagner & Nirmala, 1989). In BPTI(G36S), these dynamic effects could now be studied in more detail and characterized by an equilibrium between distinct, different molecular conformations.

In earlier work, slow local conformational equilibria in peptides and proteins were shown to involve cis–trans isomerization of Xxx–Pro peptide bonds (Grathwohl & Wüthrich, 1981) and 180° flips of aromatic rings about the C^β–C^γ bond (Wüthrich 1976, 1986). In the strongly constrained *cyclo*-L-cystine, slow chemical exchange has also been observed between two conformations that differ by the chirality of the disulfide bond (Donzel et al., 1972; Jung & Ottnad, 1974), and a similar exchange process was detected for the sterically constrained disulfide bridge in a cyclic octapeptide (Kopple et al., 1988). X-ray analyses of different crystal forms of proteins and peptides occasionally showed the occurrence of isoforms with different disulfide bond chiralities (Richardson, 1981). In addition to the experimental evidence for the presence of two BPTI conformations with different disulfide bond chirality, the present article presents a detailed investigation of the temperature dependence of the equilibrium constants and the interconversion rates between the two conformations in BPTI and BPTI(G36S). The kinetic analysis

† This work was supported by the Schweizerischer Nationalfonds (Project Nr. 31.32033.91).

[1] Abbreviations: BPTI, bovine pancreatic trypsin inhibitor; BPTI(G36S), BPTI with glycine in position 36 replaced by serine; NMR, nuclear magnetic resonance; COSY, two-dimensional correlated spectroscopy; 2QF, two-quantum filtered; TOCSY, two-dimensional total correlation spectroscopy; NOE; nuclear Overhauser enhancement; NOESY, two-dimensional NOE spectroscopy in the laboratory frame; ROESY, two-dimensional NOE spectroscopy in the rotating frame; TPPI, time-proportional phase incrementation.

0006-2960/93/0432-3571$04.00/0 © 1993 American Chemical Society

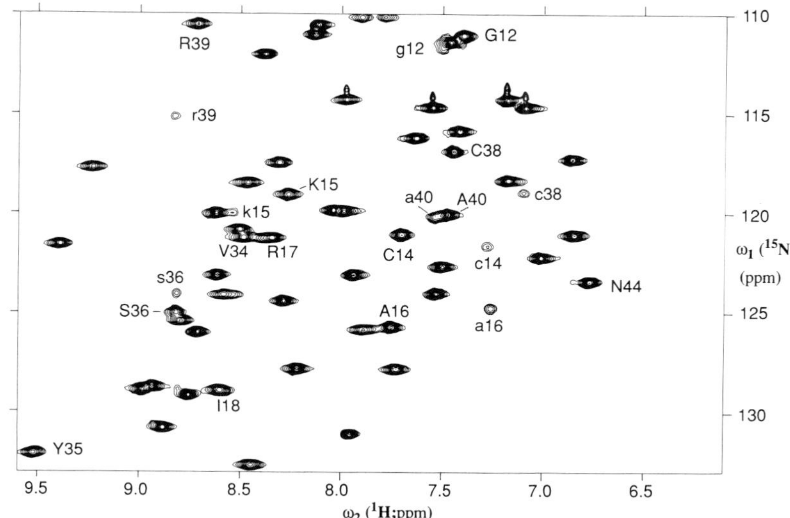

FIGURE 1: Region [ω_1(^{15}N) = 110–133 ppm, ω_2(^{1}H) = 6.2–9.6 ppm] of a [^{15}N,^{1}H]-COSY spectrum recorded with a 20 mM solution of uniformly ^{15}N-labeled BPTI(G36S) in 90% H_2O/10% ^2H_2O, pH 4.6, T = 10 °C. Spectral parameters used: t_{1max} = 72 ms, t_{2max} = 144 ms, recorded data size 800 × 2048 points, ^{1}H frequency = 600 MHz, and total recording time about 9 h. The cross-peaks of the amide groups with different chemical shifts in the major and minor conformations (see text) are identified with the one-letter amino acid symbol and the sequence number, where the cross-peaks of the minor conformation are labeled with lowercase letters. The minor peaks of Arg 17, Ile 18, Val 34, and Tyr 35 are clearly visible in the experiment of Figure 2.

of the disulfide bond isomerization gives new insights into mutual coupling of different modes of internal structure fluctuations in BPTI.

MATERIALS AND METHODS

Uniformly ^{15}N-enriched BPTI and BPTI(G36S) were obtained as a gift from Bayer A.G., Leverkusen, Germany. The preparation of these recombinant proteins was recently described elsewhere (Berndt et al., 1993). Due to the overexpression system used, both proteins contain an extra methionyl residue in position 0, i.e., amino-terminal of the regular amino acid sequence. Since in both the primary structure and the three-dimensional structure the amino terminus is well separated from the mutation site at residue 36, we refer to the otherwise unmutated BPTI sample from the overexpression system as "wild-type BPTI".

NMR spectra were recorded with 20 mM protein solutions at pH 4.6 either in 90% H_2O/10% ^2H_2O or in 100% ^2H_2O on Bruker AMX 500 and AMX 600 NMR spectrometers. [^{15}N,^{1}H]-COSY experiments were performed with the pulse scheme of Bodenhausen and Ruben (1980) using a spin-lock purge pulse for water suppression (Otting & Wüthrich, 1988; Messerle et al., 1989). A 2QF-COSY spectrum (Rance et al., 1983) and clean-TOCSY (Griesinger et al., 1988) and ROESY spectra (Bothner-By et al., 1984) with short mixing times were recorded with BPTI(G36S) in D_2O solution at 4 °C for the purpose of obtaining spectral assignments. Furthermore, NOESY and ROESY spectra of native BPTI recorded previously at 4, 36, 50, and 68 °C using a mixing time of 112 ms (Otting & Wüthrich, 1989) were used to collect some of the information reported here.

Quantitative exchange rates between the two protein conformations present in equilibrium were obtained with the [^{15}N,^{1}H]-two-spin-order-exchange difference experiment (Wider et al., 1991). The two data files recorded with the mixing time, τ_{mix}, inserted before or after the t_1-evolution period, respectively, were obtained in an interleaved manner and stored separately. For BPTI(G36S), each recording of a difference spectrum took between 1.7 and 3 h. To enable observation of the minor conformation, which is present at very low concentrations, the exchange experiments with BPTI were recorded using about 44 h per difference spectrum.

RESULTS

Experimental Evidence for the Presence of Two Protein Conformations in Equilibrium. Figure 1 shows the spectral region of the [^{15}N,^{1}H]-COSY spectrum of BPTI(G36S) at 10 °C that contains the ^{15}N–^{1}H cross-peaks of the backbone amide groups. In addition to a complete fingerprint (Wüthrich, 1986) of strong peaks, several signals of a less abundant protein species are observed. The exchange cross-peaks observed in the difference spectrum of the [^{15}N,^{1}H]-two-spin-order exchange experiment (Figure 2) (Wider et al., 1991) show that this minor species undergoes exchange with the main component: a characteristic rectangular cross-peak pattern is observed for each pair of exchanging resonances; the exchange cross-peak pattern is observed for each pair of exchanging resonances; the exchange cross-peaks are of opposite sign than the direct [^{15}N,^{1}H]-COSY cross-peaks, which were identified by comparison with the spectrum of Figure 1. Since the direct cross-peaks are largely suppressed in the difference spectrum (Wider et al., 1991), the rectangular exchange pattern can also be observed for some amino acid residues where the chemical shift differences between the minor and major conformers are too small for the peaks to be resolved in the conventional [^{15}N,^{1}H]-COSY spectrum (Figure 1). Examples are the exchange patterns of Arg 17, Ile 18, Val 34, and Tyr 35 (Figure 2). With only few exceptions, Figure 2 contains exchange cross-peaks for all the residues 12–18 and 35–40. The exceptions are Pro 13, which has no amide proton, Ser 36, for which the exchange cross-peaks and the direct cross-peaks coincide due to degenerate amide proton chemical shifts in the two conformers (Figure 1), and Gly 37, for which the exchange pattern was observed outside the spectral region shown in Figures 1 and 2.

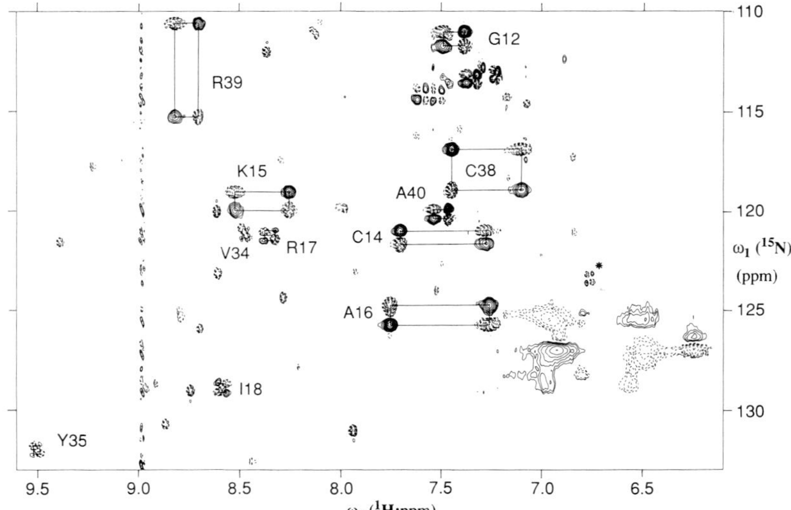

FIGURE 2: Same spectral region as in Figure 1 of the difference spectrum from a [^{15}N,^{1}H]-two-spin-exchange experiment (Wider et al., 1991) with BPTI(G36S). The same sample and same temperature are used as in Figure 1. Spectral parameters used: mixing time 25 ms, t_{1max} = 40 ms, t_{2max} = 144 ms, recorded data size 128 × 2048 points, ^{1}H frequency = 600 MHz, and total recording time about 4 h. States–TPPI quadrature detection with delayed acquisition was used in the ω_1 dimension (Marion et al., 1989; Bax et al., 1991), resulting in folding of the cross-peaks from the guanidium groups of the arginyl side chains into the spectral regions near ($\omega_1 \approx$ 127 ppm, $\omega_2 \approx$ 6.7 ppm) and near ($\omega_1 \approx$ 114 ppm, $\omega_2 \approx$ 7.4 ppm). Positive and negative levels were plotted with broken and solid lines, respectively, where the exchange cross-peaks connecting the direct ^{15}N–^{1}H cross-peaks of the major and minor conformers are positive. The rectangular patterns formed by two negative, direct COSY cross-peaks and by two positive exchange cross-peaks (Wider et al., 1991) are identified with solid lines, the amino acid one-letter symbol, and the sequence number of the exchanging residue. Additional positive signals that are not part of an exchange pattern arise from incomplete cancellation of [^{15}N,^{1}H]-COSY cross-peak intensity in the difference spectrum. The asterisk identifies an apparent exchange peak pattern at the position of Asn 44 (Figure 1).

Additional exchange cross-peaks for the NH_2 protons of the arginyl side chains arise from 180° flips about the N$^\epsilon$–C$^\zeta$ bond of the guanidinium groups. These cross-peaks are in principle outside the spectral region shown in Figure 2 but appear at about (ω_1 = 127 ppm, ω_2 = 6.7 ppm) because the spectrum has been folded in the ω_1 dimension to reduce the recording time. Similarly, residual intensity due to incomplete cancellation of the intense direct cross-peaks of the ϵNH groups of the arginyl side chains is folded into positions near (ω_1 = 114 ppm, ω_2 = 7.4 ppm) (Figure 2).

At 13 °C the less abundant conformation of BPTI amounts to only about 2% of the total protein concentration. Therefore, the intensities of the signals from the minor conformation in a [^{15}N,^{1}H]-COSY spectrum (not shown) are of similar intensity as those of small amounts of protein impurities present in the sample preparation used. The signals from the minor conformation could thus be unambiguously identified only in the difference spectrum of the [^{15}N,^{1}H]-two-spin-order-exchange experiment (Figure 3). Figure 3 shows that in BPTI the same amino acid residues give rise to discernible exchange cross-peaks as in BPTI(G36S). The only differences are observed for Val 34 and Tyr 35, which have only weak exchange patterns in BPTI(G36S) (Figure 2) and give no exchange peaks in BPTI, and for position 36. Gly 36 in BPTI gives rise to exchange cross-peaks (Figure 3), whereas no exchange peaks are observed for Ser 36 in BPTI(G36S) (Figure 2) because of the degeneracy of the amide proton chemical shifts in the two conformations (Figure 1).

Conformational Studies of the Less Abundant Form of the Proteins. Two factors interfered with an extensive collection of NOE distance constraints for a structure determination of the minor conformation. First, for both BPTI and BPTI-(G36S) the exchange with the more abundant conformation caused much stronger cross-peaks than those originating from NOEs between different protons of the minor conformation. For example, TOCSY and NOESY spectra recorded at 4 °C with mixing times of 60 ms showed strong exchange cross-peaks between corresponding diagonal peaks of the two conformers but only very weak intramolecular NOE cross-peaks of the minor conformation, because the magnetization of the latter relaxed rapidly into magnetization of the major conformation during the time period between evolution and detection. Effects of this relaxation pathway could only be suppressed by the use of very short mixing times. Second, for BPTI there was the additional difficulty that the abundance of the minor conformation is only of the order of 2% of the total protein concentration. As a consequence, we first concentrated on studies of BPTI(G36S) with experiments using very short mixing times and then transferred some of the results obtained with BPTI(G36S) to BPTI by systematic comparison of corresponding ^{15}N and ^{1}H chemical shifts, which are well known to respond sensitively to conformational differences (Wüthrich, 1976, 1986).

As a basis for collecting the desired local structural information, it was crucial to obtain stereospecific assignments for βCH_2 of Cys 14 and Cys 38. Figure 4 shows the region of a ROESY spectrum recorded in D_2O solution with a mixing time of 8 ms, which contains the β_2H–β_3H cross-peaks of Cys 14 and Cys 38 for both conformations. Even with this short mixing time, exchange cross-peaks are observed between the corresponding β_2H–β_3H cross-peaks of the major and minor conformations, where the exchange cross-peaks of Cys 14 have the same sign as the negative direct ROE cross-peaks, whereas positive exchange cross-peaks are observed for Cys 38 (see below). These exchange cross-peaks enabled the transfer of stereospecific resonance assignments between corresponding βCH_2 groups in the major and minor conformations.

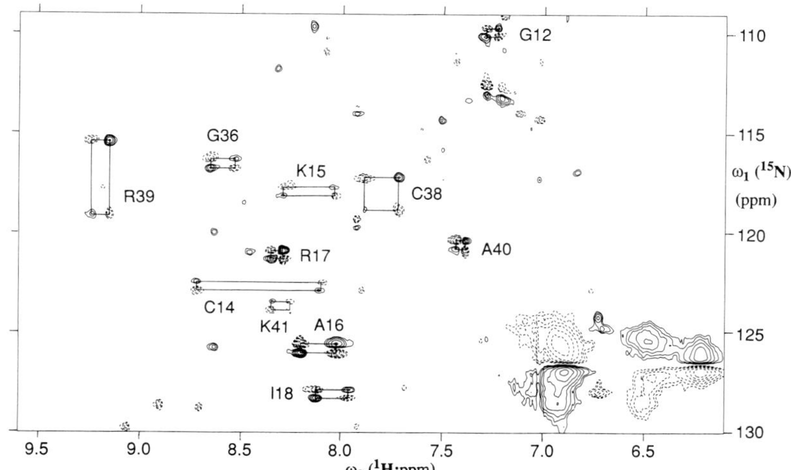

FIGURE 3: Spectral region [$\omega_1(^{15}N)$ = 109–130 ppm, $\omega_2(^1H)$ = 6.1–9.6 ppm] from the difference spectrum of a [$^{15}N,^1H$]-two-spin-exchange experiment with a 20 mM solution of uniformly ^{15}N-labeled BPTI in 90% H_2O/10% ^2H_2O at pH 4.6, T = 13 °C. Spectral parameters used: mixing time = 10 ms, t_{1max} = 40 ms, t_{2max} = 144 ms, recorded data size 256 × 2048 points, ^1H frequency = 600 MHz, and total recording time 44 h. The presentation is the same as in Figure 2. The spectrum also contains folded peaks of the Arg side chains in similar positions as in Figure 2.

At first, stereospecific resonance assignments were established in the major conformation, using the assumption that the major solution conformation around the Cys 14–Cys 38 disulfide bridge in BPTI(G36S) is identical to the crystal structures of native BPTI (Deisenhofer & Steigemann, 1975; Wlodawer et al., 1984, 1987). This assumption was based on the near identity of the structures of BPTI in solution and in crystals (Berndt et al., 1992) and on the close similarity of corresponding chemical shifts, NOE intensities, and spin–spin coupling constants in BPTI(G36S) and BPTI. Furthermore, the NOEs and spin–spin coupling constants involving hydrogen atoms of the residues near the Cys 14–Cys 38 disulfide bond in BPTI(G36S) are in full agreement with predictions based on the BPTI crystal structures: (i) In the crystal structure of BPTI, the β_2- and β_3-protons of Cys 14 are, respectively, trans and gauche with respect to the α-proton, with the amide proton of Lys 15 closer to the β_3-proton. Correspondingly, the β-proton resonances at high field and low field have αH–βH COSY cross-peak patterns indicative of a big and a small scalar coupling constant $^3J_{\alpha\beta}$, respectively. The low-field resonance shows a much stronger NOE with the α-proton of Cys 14 and with the amide proton signal of Lys 15 (not shown). The ensuing assignment of the high-field resonance to β_2 coincides with the stereospecific resonance assignments in wild-type BPTI (Berndt et al., 1992). (ii) For Cys 38, both β-protons are gauche with respect to the α-proton, with the amide proton of Arg 39 closer to the β_2-proton. As predicted from this local feature of the crystal structure, the αH–βH COSY cross-peak multiplets of the major conformation in BPTI(G36S) manifest small scalar coupling constants $^3J_{\alpha\beta}$ for both β-protons. The β-proton resonance at lower field shows a much stronger NOE with the amide proton of Arg 39 and was therefore assigned to the β_2-proton. The same result was obtained with wild-type BPTI, except that the sequential NOE between the lower field β-proton of Cys 38 and the amide proton of Arg 39 could be resolved only at temperatures above 50 °C. Therefore, no stereospecific resonance assignment of the β-protons of Cys 38 was used in the recent refinement of the NMR structure of BPTI, which was carried out at 36 °C (Berndt et al., 1992).

From the stereospecific assignments of the βCH_2 resonances of Cys 14 and Cys 38 in the major conformer the corresponding

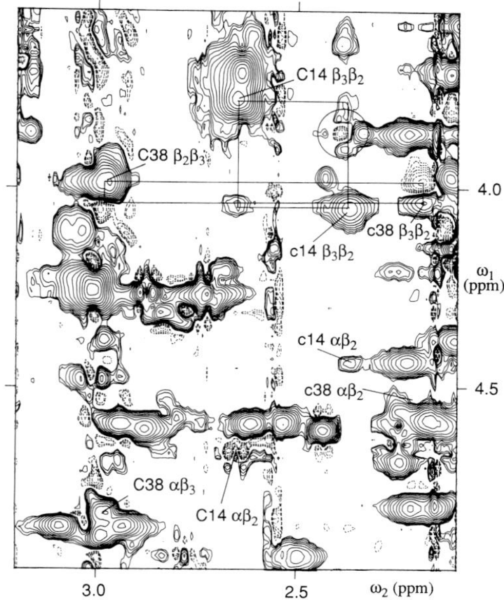

FIGURE 4: Region (ω_1 = 3.6–4.9 ppm, ω_2 = 2.1–3.2 ppm) of a [$^1H,^1H$]-ROESY spectrum recorded with a 20 mM solution of BPTI(G36S) in ^2H_2O, p^{2}H 4.6, T = 4 °C. Spectral parameters used: mixing time τ_m = 8 ms, t_{1max} = 50 ms, t_{2max} = 144 ms, recorded data size 700 × 2048 points, ^1H frequency = 600 MHz, and total recording time 52 h. Positive and negative levels were plotted with broken and solid lines, respectively. The positions of the intraresidual ROE cross-peaks of Cys 14 and Cys 38 are labeled with the same symbols as in Figure 1, where in addition the protons observed in the ω_1 and ω_2 dimensions are identified by the first and second Greek letter, respectively. Two rectangles connect ROE cross-peaks of the major and minor conformers of Cys 38 and Cys 14 with the respective exchange cross-peaks (see the text and Figure 5). A circle is drawn around a J cross-peak (Macura et al., 1982) that overlaps with an exchange cross-peak of Cys 14.

assignments for the minor conformer were obtained from the sign of the exchange cross-peaks in Figure 4. Figure 5 outlines schematically the cross-peak patterns observed for Cys 14 and Cys 38. There are three classes of cross-peaks present:

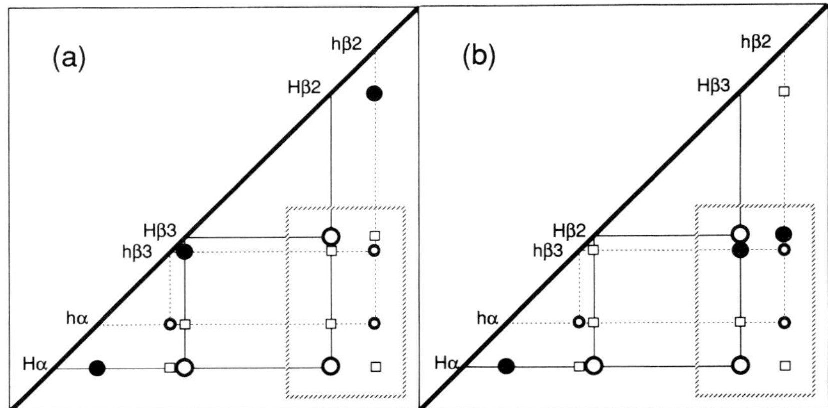

FIGURE 5: Schematic representation of the [^{1}H,^{1}H]-ROESY cross-peak patterns observed for (a) Cys 14 and (b) Cys 38. Upper- and lowercase characters along the diagonal identify the chemical shifts of the hydrogen atoms indicated with greek letters and numbers for the major and minor conformations, respectively. Open circles represent negative direct ROE cross-peaks, filled circles positive chemical exchange cross-peaks, and squares negative exchange-relayed ROE cross-peaks. Solid lines connect the diagonal peaks and cross-peaks observed for the major conformation, broken lines those of the minor conformation. The dashed rectangles correspond to the spectral region shown in Figure 4.

(i) Negative peaks representing direct ROEs (Bothner-By et al., 1984) (open circles); (ii) positive peaks representing chemical exchange (filled circles); (iii) negative peaks representing exchange-relayed ROE cross-peaks (open squares), which arise from magnetization transfer via a single ROE step followed by chemical exchange, or via chemical exchange followed by a single ROE step. For Cys 14 (Figure 5a), the chemical shift differences between corresponding resonances of the major and minor conformers are small, and therefore all positive chemical-exchange cross-peaks are close to the diagonal. The spectral region shown in Figure 4 contains exclusively negative ROE cross-peaks and exchange-relayed ROE cross-peaks of Cys 14. For Cys 38 (Figure 5b), the chemical shifts of the β_2H and β_3H resonances of the major and minor conformations are largely different. As a result, the positive chemical-exchange cross-peaks between corresponding β-protons are well separated from the diagonal. They are identified in Figure 4 by lines connecting them with the β_2H–β_3H cross-peak of Cys 38 in the major conformation and the corresponding β_3H–β_2H cross-peak in the minor conformation. The exchange-relayed ROE cross-peaks between αH and βCH$_2$ (Figure 5b) are not observed in Figure 4 because the αH–βH ROEs are much weaker than the ROEs between the geminal β-protons. The large differences between chemical shifts of corresponding hydrogen atoms in the two conformations point to direct involvement of the side chain of Cys 38 in the observed conformational rearrangement.

The aforementioned implications that the side chain orientations in the two conformations are similar for Cys 14 and different for Cys 38 are confirmed by a 2QF-COSY spectrum of BPTI(G36S) at 4 °C (Figure 6). The multiplet fine structure of the αH–β_2H cross-peaks of Cys 14 shows a large active coupling constant $^3J_{\alpha\beta_2}$ in both the minor and major conformations, which indicates that in both conformations the β_2-proton is in the trans orientation relative to the α-proton. The presence of a large and a small $^3J_{\alpha\beta}$ coupling constant for Cys 14 in both conformations is further inferred by the observation of strong and weak cross-peaks for, respectively, the αH–β_2H and αH–β_3H connectivities in a TOCSY spectrum recorded with a mixing time of 10 ms (not shown). A large $^3J_{\alpha\beta}$ coupling was then also indicated by the appearance at (ω_1 = 4.5 ppm, ω_2 = 2.2 ppm) in Figure 6 of the αH–β_2H COSY cross-peak of Cys 38 in the minor conformation. This cross-peak is of similar intensity as the

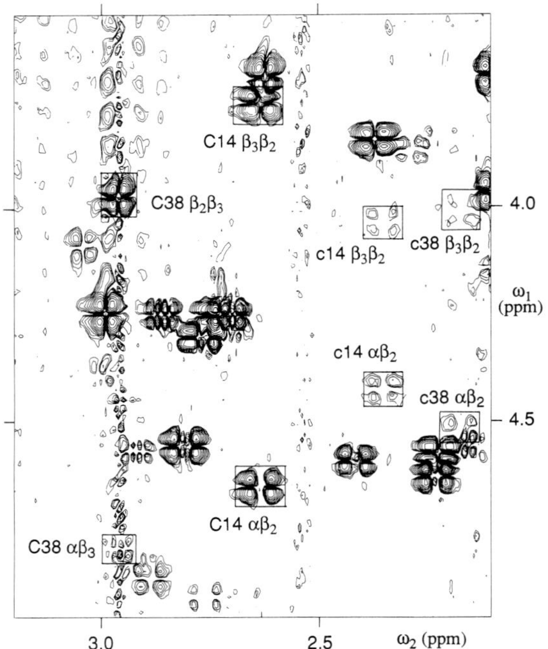

FIGURE 6: Homonuclear ^{1}H 2QF-COSY spectrum of a 10 mM solution of BPTI(G36S) in ^{2}H$_2$O, p^{2}H 4.6, T = 4 °C. Spectral parameters used: t_{1max} = 72 ms, t_{2max} = 144 ms, recorded data size 1024 × 2048 points, ^{1}H frequency = 6000 MHz, and total recording time 26 h. Positive and negative levels were plotted without distinction. The spectral region and symbols for cross-peak identification are the same as in Figure 4.

αH–β_3H cross-peak of Cys 38 in the major conformation, which is relatively weak due to the small value of $^3J_{\alpha\beta_3}$. These observations are again in line with the relative cross-peak intensities observed in the aforementioned TOCSY spectrum (not shown). The combined evidence from the chemical shift changes and the $^3J_{\alpha\beta}$ coupling constants shows that the local conformation of Cys 14 does not change significantly between the major and minor conformations, whereas a change of the χ^1 angle by about −120° is indicated for Cys 38, corresponding to a change from a gauche–gauche conformation in the major

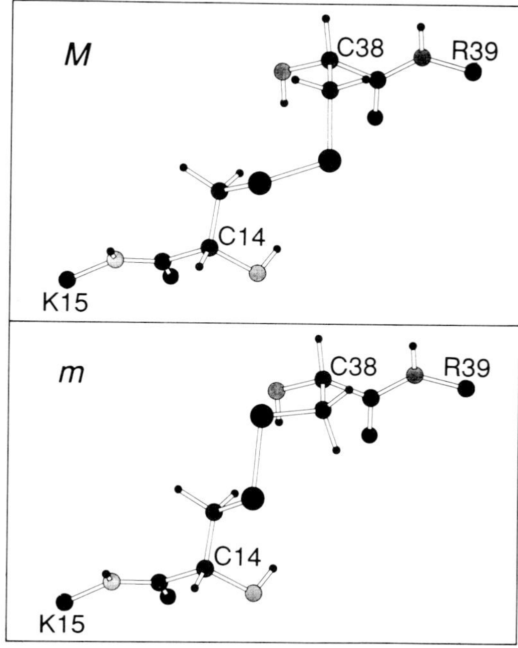

FIGURE 7: Model representations of the two different Cys 14–Cys 38 disulfide bond arrangements in the major and minor conformations of BPTI and BPTI(G36S). (M) More abundant species with right-handed chirality of the disulfide bridge. The atom coordinates were taken from the crystal structure of BPTI form II (Wlodawer et al., 1984). (m) Minor species with left-handed chirality of the disulfide bridge. The atom coordinates are the same as in M, except that relative to M the coordinates of the β-protons and the sulfur atom of Cys 38 were changed by a rotation of about −120° about the C^α–C^β bond. Atom code: carbon, black; oxygen, dark shaded; nitrogen, light-shaded; sulfur, big and dark-shaded; hydrogen, small and dark shaded. The α-carbons are labeled with the one-letter amino acid symbol and the sequence location.

conformer (Wagner et al., 1987) to a trans–gauche conformation in the less abundant conformer.

Figure 7 shows the two forms of the Cys14–Cys38 disulfide bridge which are implicated by the change of the χ^1 angle of Cys 38 from 60° in the major conformation to about −60° in the minor conformation. The disulfide bridge of the major conformation is the same as in the crystal structure and has a right-handed chirality (Figure 7a). In the minor conformation the disulfide bridge must assume a left-handed chirality, since the conformational change is restricted to the side chain of Cys 38 and the disulfide bond (Figure 7b). It is intriguing to note for Cys 38 that the chemical shift of the β_3-proton in the minor conformation is nearly the same as the chemical shift of the β_2-proton in the major conformation, since Figure 7 shows that these hydrogen atoms are in similar positions with respect to the framework provided by the parts of the protein that do not vary between the two conformations.

No evidence was obtained for further important conformational rearrangements. Table I lists the chemical shift differences between the major and minor conformations for all proton resonances in BPTI(G36S) for which resonance assignments are available in the minor conformation. It is seen that chemical shift changes larger than 0.2 ppm for carbon-bound protons occur exclusively in Cys 14 and Cys 38. The chemical shift change of the hydroxyl-proton of Ser 36 by 0.34 ppm could not be related to a significant conformational change of the side chain of Ser 36, since similar relative NOESY and TOCSY cross-peak intensities were

Table I: Chemical Shift Differences, $\Delta\delta = \delta_m - \delta_M$, between the ^1H Resonances of the Minor and Major Conformations of BPTI(G36S) at T = 4 °C and pH 4.6[a]

residue	$\Delta\delta$ (ppm) NH	αH	βH	others
Gly 12	0.10	−0.02, 0.19		
Pro 13		−0.10		
Cys 14	−0.43	−0.22	−0.18, 0.28	
Lys 15	0.26	0.01		
Ala 16	−0.50	−0.09	−0.02	
Arg 17	0.05			
Ile 18	−0.01			
Val 34	0.01			
Tyr 35	−0.01			
Ser 36	0.00	0.08	0.03, −0.05	γOH 0.34
Gly 37	−0.03	0.05, −0.15		
Cys 38	−0.34	−0.26	1.06, −1.81	
Arg 39	0.12	0.03	−0.07	
Ala 40	0.07			

[a] Only those residues are listed for which resonance assignments could be obtained for the less abundant conformation.

observed with all side chain proton resonances of Ser 36 in both conformations, including the NOEs with the side chain hydroxyl proton (in these measurements special care was exercised to avoid any bias from spin diffusion by using short mixing times of 25 ms for NOESY and 10 ms for TOCSY). Amide proton chemical shifts are generally very sensitive to even minor changes in the chemical environment (Wüthrich, 1976). In the presently investigated system the amide proton chemical shift changes observed between major and minor conformation (Table I) are comparable to those between BPTI and BPTI(G36S) (Figures 2 and 3). We therefore conclude that there is no indication of major conformational changes in the amide proton chemical shift data.

The low abundance of the minor conformation in solutions of wild-type BPTI makes it difficult to observe any ^1H–^1H cross-peaks of this minor form in homonuclear COSY or ROESY spectra (Figures 4 and 6), so that only the ^{15}N–^1H cross-peaks could be assigned for the minor conformation. However, close similarity between the minor forms of BPTI and BPTI(G36S) is inferred from the following chemical shift analysis. The amide proton and amide ^{15}N chemical shifts of the major conformations of BPTI and BPTI(G36S) differ by up to 1.1 and 5.0 ppm, respectively (Figures 2 and 3). However, all large chemical shift differences between the major forms of the two proteins are with amide groups near the mutation site. Most important, the differences in chemical shifts between the major and minor conformations of the same protein are very similar for all ^1H and ^{15}N spins of the amide groups in BPTI and BPTI(G36S) (Figure 8). The only exceptions are some amide proton chemical shift differences observed for the residues 35–40, which may be explained by the fact that the interior water molecule present near Cys 38 in BPTI is likely to respond differently to the conformational change of the side chain of Cys 38 than the covalently bound $-CH_2-OH$ side chain of Ser 36 in BPTI(G36S) (Berndt et al., 1993).

Further evidence for close similarity of the conformations of BPTI and BPTI(G36S) was obtained from the β-proton chemical shifts of Cys 14 and Cys 38. The β_2H and β_3H resonances of Cys 38 in the minor conformation in BPTI-(G36S) are shifted, respectively, by −1.8, and +1.1 ppm relative to the corresponding chemical shifts in the major conformation, whereas the β-proton chemical shifts of Cys 14 differ by less than 0.3 ppm between the two forms (Figures 5 and 6). The presence of similar chemical shift differences of the corresponding protons in BPTI is implicated by the temperature

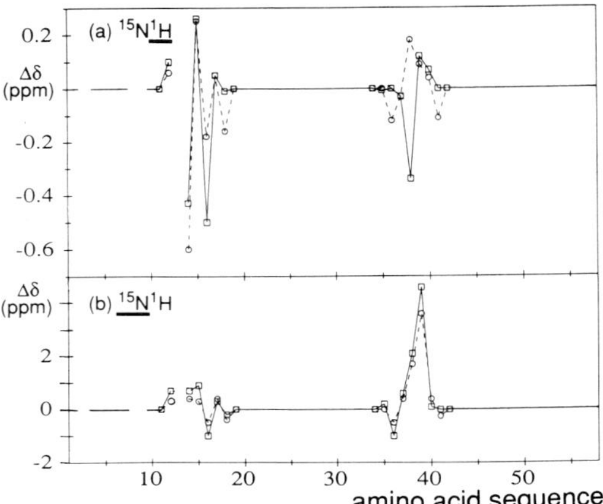

FIGURE 8: Plots of the chemical shift differences between the minor and major conformations of BPTI (squares connected by solid lines) and BPTI(G36S) (circles connected by dashed lines) at 10 °C, pH 4.6, versus the amino acid sequence. (a) Chemical shift differences of the amide protons. (b) Chemical shift differences of the amide ^{15}N spins.

dependence of the β-proton chemical shifts. Monitoring the β_2H–β_3H cross-peaks in NOESY spectra as a function of temperature, we found that the chemical shifts of the β_2H and β_3H resonances of Cys 38 changed by only about 0.01 ppm between 4 and 36 °C, whereas shifts of about –0.15 and +0.09 ppm, respectively, occurred between 36 and 68 °C. Furthermore, both β-proton resonances of Cys 38 are much broader between 50 and 68 °C than the β-proton signals of any of the other amino acid residues (see also the following section below). These observations can be explained by the facts that only the resonances of the major conformation of BPTI are seen between 4 and 36 °C and that there is coalescence with the signals of the minor form at higher temperatures. The chemical shifts of the coalesced lines correspond closely to those predicted from the chemical shifts of the major conformation of BPTI measured at 36 °C and those estimated for the minor conformation of BPTI with the assumption that they are similar to the corresponding shifts in BPTI(G36S). In contrast to the situation encountered with Cys 38, the line widths and chemical shifts of the β-proton resonances of Cys 14 in BPTI change only little with temperature. This again is what one would predict from the assumption that similar small chemical shift differences between the major and minor conformations prevail in BPTI and in BPTI(G36S).

Overall, the available data provide strong evidence that the minor conformation of BPTI comes about by a rearrangement of the disulfide bridge Cys14–Cys38, which would, in complete analogy with BPTI(G36S), include changes of the χ^1 angle of the Cys 38 and of the chirality of the disulfide bond. This conclusion is further corroborated by the fact that similar values were obtained for the activation enthalpies of the disulfide isomerization in the two proteins (see below).

Equilibrium Constants between the Major and Minor Conformations of BPTI and BPTI(G36S). All available data indicate that there is a two-state equilibrium between the major conformation, M, and the minor conformation, m (Figure 7):

$$\mathrm{M} \underset{k_\mathrm{m}}{\overset{k_\mathrm{M}}{\rightleftharpoons}} \mathrm{m} \tag{1}$$

k_M and k_m are the rate constants of the conformational exchange. In the temperature range of slow exchange, the relative populations of the two conformations could be measured from the cross-peak intensities observed in the [^{15}N,^{1}H]-COSY spectra (Figure 1). At 4 °C, the minor conformations of BPTI(G36S) and BPTI were thus found to be populated in the extent of 14.4 ± 2.0% and 1.5 ± 0.3%, respectively, of the total protein.

At temperatures above the coalescence temperature of the signals of the two isomeric forms, the relative populations were determined by an analysis of chemical shifts and line widths. In BPTI(G36S), the β_2- and β_3-proton resonances of Cys 38 in the major conformation broaden at 50 °C and become narrower again at 68 °C, indicating that in this temperature range these resonances merge with those of the minor conformation. Since for Cys 14 the chemical shift differences of βCH_2 between the major and minor conformations are much smaller, the coalescence temperature is lower than for Cys 38 and the signals are narrow at both 50 and 68 °C. On the basis of comparison of the chemical shifts of the averaged signals at 68 °C with those of the βCH_2 resonances of the major and minor conformations at 4 °C, the population of the minor conformation was estimated to be between 10 and 25% at 68 °C, and since the line broadening observed at 50 °C can only be explained when at least 15% of the protein assumes the minor conformation, this initial estimate of the population of m in BPTI(G36S) over the temperature range from 50 to 68 °C was adjusted to 20 ± 5%. This value is in agreement with the trend to increased population of the minor conformation at higher temperature evidenced by the exchange rate constants measured between 4 and 20 °C (see below).

In BPTI, the population of the minor conformer at 68 °C was estimated to be 8 ± 1%. This estimate was based on the β-proton chemical shifts of Cys 38 observed between 36 and 68 °C, and the assumptions that the chemical shift differences between major and minor conformations are temperature-independent and similar to those in BPTI(G36S). The fact that the population of the minor conformation increases with temperature is also supported by the extent of the line broadening observed at 50 and 68 °C, which is of the order of 20–40 Hz. This is much larger than what could be expected for any value of the exchange rate in the presence of only 2% of the minor conformation.

From the ratio of the populations of the minor and major conformations, [m]/[M], at 4 and 68 °C, equilibrium enthalpies, ΔH, and equilibrium entropies, ΔS, were estimated using the relation

$$-RT \ln \frac{[\mathrm{m}]}{[\mathrm{M}]} = \Delta H - T\Delta S \tag{2}$$

where R is the gas constant and T the absolute temperature. The results are listed in Table II. Equation 2 can also be used to estimate the populations at intermediate temperatures from the population ratios at 4 and 68 °C, whereby it is implicitly assumed that ΔH and ΔS are temperature-independent.

Measurements of the Frequency of Disulfide Bond Interconversions. The exchange rate constants k_M and k_m of the equilibrium of eq 1 were determined in the temperature range between 4 and 68 °C. In the temperature range 4–20 °C this was achieved with [^{15}N,^{1}H]-two-spin-order exchange experiments (Wider et al., 1991). The difference of the two recorded data files is devoid of most nonexchanging direct peaks and was used for the volume integration of the exchange cross-peaks. The experiment in which the mixing time precedes the t_1-evolution period (Wider et al., 1991) retains exclusively those [^{15}N,^{1}H]-cross-peaks that are also present in a conventional [^{15}N,^{1}H]-COSY spectrum. It was used to

Table II: Equilibrium Enthalpies, Equilibrium Entropies, Activation Enthalpies, and Activation Entropies for the Interconversion of the Chirality of the Disulfide Bond Cys 14–Cys 38 in BPTI and BPTI(G36S) Measured at Low Temperature[a]

parameter	BPTI	BPTI(G36S)
ΔH (kcal mol^{-1})	-0.5 ± 0.8^b	0.8 ± 1.3^c
ΔS (cal mol^{-1} K^{-1})	10.2 ± 3.2^b	0.4 ± 4.8^c
$\Delta H^\ddagger$ (kcal mol^{-1})		
M → m	13.0 ± 3.7^d	11.6 ± 2.3^d
M ← m	9.4 ± 5.9^d	9.3 ± 2.4^d
$\Delta S^\ddagger$ (cal mol^{-1} K^{-1})		
M → m	-13.9 ± 13.3^d	-13.3 ± 8.3^d
M ← m	-18.7 ± 20.8^d	-18.1 ± 8.5^d

[a] Measured at pH 4.6. The activation parameters were determined in the temperature range 4–20 °C. [b] From the equilibrium constants at 4 °C (population of the minor conformer 1.5 ± 0.3%) and at 68 °C (8 ± 1%). [c] From the equilibrium constants at 4 °C (population of the minor conformer (14.4 ± 2.0%) and at 68 °C (20 ± 5%). [d] The error ranges were determined from the steepest and least steep slopes compatible with the error bars indicated in Figure 10.

measure the volume integral, I, of the direct correlation peaks. For BPTI(G36S), spectra were recorded with mixing times, τ_{mix}, of 15, 30, 45, and 60 ms at 4 °C, τ_{mix} = 5, 15, 25, and 35 ms at 10 °C, and τ_{mix} = 5 and 10 ms at 7, 13, and 16 °C. The exchange rate constants k were determined from the initial buildup rates of the exchange cross-peaks in the difference spectrum, $\Delta I/\Delta\tau_{mix}$, and the intensities of the direct correlation peaks at zero mixing time, I_o:

$$k = \Delta I/(I_o \Delta\tau_{mix}) \qquad (3)$$

To avoid the recording of separate experiments with zero mixing time, I_o was obtained by extrapolation from the data recorded with short mixing times, assuming that the decay can be described by a single exponential.

In this procedure an important correction had to be made, since with increasing temperatures the exchange rates became sufficiently rapid for the chemical exchange during the refocusing delay, τ, between evolution and detection to cause the appearance of exchange peaks even in a conventional [^{15}N,^{1}H]-COSY spectrum. The refocusing delay τ used in the [^{15}N,^{1}H]-two-spin-order exchange experiments was about 5.4 ms. To account for the effective transverse relaxation caused by the chemical exchange during τ, which is different for the major and minor conformations, we corrected the peak intensities observed at the chemical shifts of the major and minor components by $\exp(k_m\tau)$ and $\exp(k_M\tau)$, respectively. Since initially only approximate values of the exchange rate constants k_m and k_M were available from the spectral analysis without correction for the exchange during τ, this correction was applied repeatedly, where in each subsequent cycle the new, improved rate constants were used until no further change in the k_m and k_M values was observed. The final, thus corrected rate constants differed by up to 26% from the rate constants obtained without this correction.

Because the low abundance of the minor conformation in wild-type BPTI required exceptionally long recording times for its detection, [^{15}N,^{1}H]-two-spin-order exchange experiments at 10 and 16 °C were recorded with only two different mixing times, 7.5 and 15 ms, and at 4, 7, and 13 °C only a single mixing time of 10 ms was used to evaluate the exchange rates in BPTI. The results obtained with mixing times of 7.5 and 15 ms showed that τ_{mix} = 10 ms was sufficiently short to approximate the initial-rate condition, where the volumes of the exchange cross-peaks are directly proportional to the exchange rates. Furthermore, with BPTI(G36S) it was observed that during the mixing time all direct correlation peaks relax to a similar degree at all temperatures. Therefore, it was assumed that the same amount of relaxation as in BPTI-(G36S) occurs also for the direct correlation peaks in BPTI, and a corresponding correction was applied to obtain the I_o value for those temperatures where only a single mixing time was used. For τ_{mix} = 10 ms, this correction factor was 1.12.

At low plot levels most of the direct cross-peaks of the minor conformations overlap with wings of other peaks in the [^{15}N,^{1}H]-COSY spectrum, which prohibits their accurate integration. In addition, different transverse relaxation rates during the defocusing and refocusing delays of the [^{15}N,^{1}H]-two-spin-exchange experiment resulted in different intensities of the exchange patterns observed for the different amino acid residues. Therefore, the rate constants were determined only from the cross-peaks of Cys 14 and Ala 16 in BPTI-(G36S) and from the relatively intense [^{15}N,^{1}H] cross-peaks of Ala 16 in BPTI. The measured exchange lifetimes ranged from about 16 ms for the minor conformation of BPTI(G36S) at 16 °C to 4.5 s for the major conformation of BPTI at 4 °C.

Above room temperature the evaluation of the exchange rate constants by this approach was made difficult by the aforementioned line broadening and the correspondingly reduced heights of the exchange peaks at the short mixing times needed to maintain the initial rate condition. In this limit, more accurate results for BPTI(G36S) were obtained by comparing the line widths of the minor and major conformations from spectra recorded with good signal-to-noise ratio. The line widths were obtained from the difference spectra and the reference spectra of [^{15}N,^{1}H]-two-spin-order exchange experiments recorded at 16 °C using a mixing time of 15 ms and a total recording time of 3 h and at 20 °C using a mixing time of 5 ms and a recording time of 9 h. Since the two conformations are present in different amounts, the chemical exchange leads to more severe line broadening in the minor form than in the major form. Assuming the same natural line width for both conformations, the difference of the full line widths at half-height of the minor and major conformations, $\Delta\nu_m - \Delta\nu_M$, is proportional to the difference in the rate constants k_M and k_m of the equilibrium of eq 1 and independent of the natural line width:

$$\Delta\nu_m - \Delta\nu_M = \frac{k_m - k_M}{\pi} \qquad (4)$$

Using

$$k_m = \frac{[M]}{[m]} k_M \qquad (5)$$

eq 4 can be rewritten as

$$k_M = \pi \frac{(\Delta\nu_m - \Delta\nu_M)}{\frac{[M]}{[m]} - 1} \qquad (6)$$

The exchange rate constants k_m and k_M were determined from eqs 5 and 6. A critical point in the data evaluation is the accurate determination of the ratio of the major versus the minor conformation, [M]/[m]. Since no signals from the minor conformation could be resolved in the one-dimensional ^{1}H or ^{15}N NMR spectra, this ratio was determined from the antiphase signals in two-dimensional [^{15}N,^{1}H]-COSY spectra recorded without refocusing of the heteronuclear antiphase magnetization, since any chemical exchange occurring during a refocusing delay would decrease [m] more than [M] (see above). At 16 °C the exchange rate constants measured from such line width comparisons agreed well with the rate constants determined from the buildup rates of the exchange cross-peaks. At lower temperatures and correspondingly slower

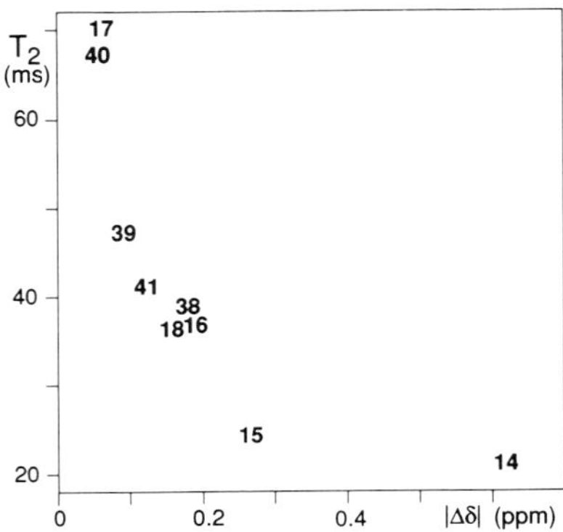

FIGURE 9: Plot of the T_2 relaxation times of the amide protons of residues 14–18 and 38–41 in BPTI versus the absolute ^{1}H chemical shift difference between the major and minor conformations (Table I). The T_2 values were obtained from a nonlinear fit of the cross-peak intensities observed in a set of J-modulated [^{15}N,^{1}H]-COSY experiments recorded at a ^{1}H frequency of 500 MHz with uniformly ^{15}N-enriched BPTI at 36 °C and pH 4.6. No data are given for glycyl residues, because from the experiments used no reliable T_2 values can be extracted for systems with two $^3J_{HN\alpha}$ coupling constants (Billeter et al., 1992).

exchange rates, the line width comparison yielded less accurate results, because the lines broadened too little compared with their natural line widths.

Because of the shorter lifetime, the resonances of the minor conformation broaden more rapidly with increasing temperature than those of the major conformation. Once the signals of the less abundant conformation are too broad to be detected, only the resonance lines of the major conformation can be monitored to extract the exchange rate constants. Similarly, above the coalescence temperature, only a single, average signal is observable. At 50 and 68 °C, where we had this situation, estimates of the exchange rate constants were obtained from the line widths of the $C^{\beta}H_2$ resonances of Cys 38 by comparison with values simulated for a range of exchange rates using the equilibrium constants determined at these temperatures. Since both β-protons change their chemical environment simultaneously during a disulfide flip, the predicted parameters could be checked against two different experimental line width measurements.

In BPTI(G36S) the β_2- and β_3-proton resonances of Cys 38 broaden by about 200 and 100 Hz at 50 °C, respectively. The same signals are only about half as broad at 68 °C. With a population of the minor conformation of 20 ± 5%, these line widths indicate exchange frequencies k_M of the disulfide flip of between 500 and 1000 s^{-1} at 50 °C and between 2000 and 6000 s^{-1} at 68 °C. The corresponding exchange line broadenings in BPTI were between 20 and 40 Hz at 68 °C. Using a population of the minor conformer of 8 ± 1% at 68 °C, the observed line broadenings correspond to a rate constant k_M of the order of 2000–3000 s^{-1} at 68 °C.

A further estimate of the exchange rate constant k_M for BPTI at 36 °C was obtained from the T_2 relaxation times of the amide protons involved in the exchange patterns of Figure 3. The T_2 relaxation times of the amide protons were obtained as a by-product of the determination of the scalar coupling constants $^3J_{HN\alpha}$ in uniformely ^{15}N-enriched BPTI from a series of J-modulated [^{15}N,^{1}H]-COSY spectra (Neri et al., 1990; Billeter et al., 1992). Strikingly, T_2 relaxation times shorter than 50 ms were observed exclusively for those residues which are affected by the disulfide flip in BPTI. Figure 9 shows that there exists an inverse correlation between the T_2 relaxation times of the rapidly relaxing amide protons and the ^{1}H chemical shift differences between the amide proton resonances of the major and minor conformations (Figure 3). A special case is presented by Tyr 35, which is not present in Figure 9. For the amide proton of Tyr 35, a T_2 relaxation time of 44 ms was determined. According to Figure 9, this value would correspond to a $\Delta\delta$ value of about 0.1 ppm, which should be sufficiently large for the observation of a well-resolved exchange pattern in the [^{15}N,^{1}H]-two-spin-exchange difference experiment (Figures 2 and 3). Although Tyr 35 gives rise to an exchange pattern in BPTI(G36S) (Figure 2), no such pattern was detected in the corresponding exchange spectrum of BPTI (Figure 3), which might be explained by degenerate ^{15}N chemical shifts in the minor and major conformations. Interpolation between the equilibrium populations at 4 and 68 °C based on eq 2 predicts the population of the minor form to be about 4% at 36 °C. Attributing the increased T_2 relaxation rates of the residues included in Figure 9 to the disulfide bond isomerization, the above population of the minor conformation, the chemical shifts of Figure 3 and a natural line width corresponding to T_2 = 70 ms were used to simulate the theoretical line widths for a range of different exchange rate constants using the analytical equation for a two-side exchange system with different populations at both sites (Gutowsky & Holm, 1956). The rate constant k_M obtained in this way was of the order of 40 ± 20 s^{-1}.

Activation parameters were determined in the temperature range 4–20 °C, where the exchange rates could be measured directly from [^{15}N,^{1}H]-two-spin-exchange experiments. The rate constants, k, determined at the different temperatures, T, were converted into ratios of free activation energies over the temperature, $\Delta G^{\ddagger}/T$, according to the Eyring theory:

$$\frac{\Delta G^{\ddagger}}{T} = -R\left(\ln k + \ln \frac{h}{\kappa k_B T}\right) \qquad (7)$$

In eq 7, h denotes the Planck constant, k_B the Boltzmann constant, and κ the transmission coefficient, which was assumed to be unity over the entire temperature range used. Figure 10 shows a plot of $\Delta G^{\ddagger}/T$ versus $1/T$. Activation enthalpies, $\Delta H^{\ddagger}$, and activation entropies, $\Delta S^{\ddagger}$, were determined from the slopes and the intersections with the ordinate according to

$$\frac{\Delta G^{\ddagger}}{T} = \frac{\Delta H^{\ddagger}}{T} - \Delta S^{\ddagger} \qquad (8)$$

Table II affords a survey over the activation parameters. The error ranges were obtained by fitting straight lines through the data points, assuming that the flattest and the steepest slopes compatible with the error bars in Figure 10 represent the uncertainties of the measurements. For BPTI(G36S), very similar exchange rate constants were determined from the cross-peaks of Cys 14 and Ala 16, as one would expect for a two-state equilibrium according to eq 1, where all protons change their chemical environment simultaneously.

Figure 10 shows that for both BPTI and BPTI(G36S) the differences between the free energies of activation of the forward reaction, M → m, and the backward reaction, m → M, in eq 1 decrease with increasing temperature. From the corresponding ratios k_m/k_M, the population of the minor conformation of BPTI was found to increase from 1.5% to 2.2% in the temperature range 4–16 °C. The corresponding increase in BPTI(G36S) was from 14% at 4 °C to 17% at 20 °C. The equilibrium constants determined from the ratios of the exchange rate constants agree well with the values determined independently from the intensity ratios of the direct [^{15}N,^{1}H]-COSY cross-peaks observed for the minor and major conformations. This result indicates that in the temperature

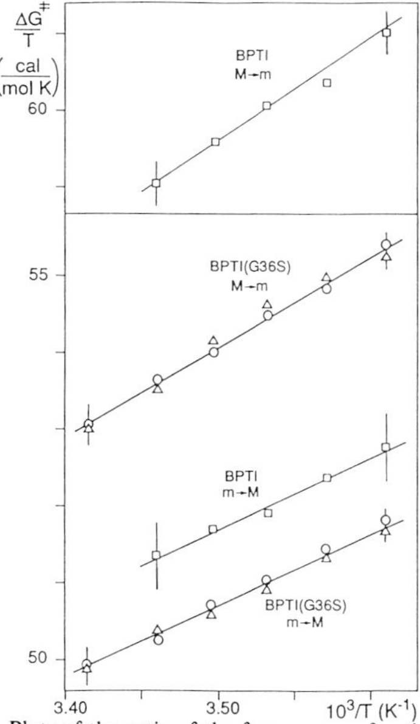

FIGURE 10: Plots of the ratio of the free energy of activation over the temperature, $\Delta G^{\ddagger}/T$, versus the inverse of the temperature, $1/T$, for the forward reaction of eq 1, M → m, and the backward reaction, M ← m, in BPTI and BPTI(G36S). The data were obtained from the $^{15}N-^{1}H$ cross-peaks of Ala 16 (squares and triangles) and from Cys 14 (circles). The error bars associated with the peripheral data points were obtained from error propagation, with the assumption of ±10% error in the volume integration or ±1.5 Hz error in the line width measurements. An error range of ±20% was attributed to the I_0 value of the minor conformation in BPTI in eq 3.

range between 4 and 20 °C, the exchange reaction proceeds via a single transition state, as expressed by eq 1.

DISCUSSION

Kinetic measurements with small organic model compounds have shown that the free energies of activation, $\Delta G^{\ddagger}$, for the exchange between the two different disulfide bond chiralities are of the order of 7–8 kcal/mol at temperatures around −100 °C (Fraser et al., 1971). This activation barrier is only about half of the free energy of activation for the disulfide flip Cys 14–Cys 38 in BPTI and BPTI(G36S). Correspondingly, no multiple conformers of free cystine in aqueous solution can be separately observed by NMR, and the particularly strongly constrained *cyclo*-L-cystine is the only low molecular weight compound for which the free energy of activation was found to be as high as 15.6 kcal/mol (Jung & Ottnad, 1974). These considerations show clearly that the observed slow exchange of the disulfide chirality in BPTI cannot be explained by the local covalent structure and must be due to steric constraints imposed by the protein environment.

The following considerations provide some indications why only one of the three disulfide bonds in BPTI shows slow interchange between two observably populated conformations. The disulfide flips of Cys 14–Cys 38 are coupled with minimal rearrangements of the rest of the protein molecule. Visual inspection of the other disulfide bonds in BPTI shows that a similar change of chirality would induce much larger changes of the average three-dimensional structure, so that the corresponding equilibria largely favor one of the two disulfide bond chiralities. The Cys 14–Cys 38 bond is also unique among the disulfide bonds in BPTI in that the χ^1 angle of 60° for Cys 38 in the major conformation places the sulfur atom in a sterically unfavorable position between the main chain atoms of Cys 38 (Figure 7; Richardson, 1981; Srinivasan et al., 1990). The change to a χ^1 angle of −60° in the minor conformation releases this local strain, which might contribute to the significant population of the minor conformational state.

Further insights into the special role of the Cys 14–Cys 38 bond might also derive from the fact that the minor conformation is more abundant in BPTI(G36S) than in BPTI. In the crystal structure of BPTI, an interior water molecule located near the disulfide bond Cys 14–Cys 38 is hydrogen-bonded to the carbonyl-oxygen of Cys 38. In BPTI(G36S) the hydroxyl group of Ser 36 replaces this water molecule and forms hydrogen bonds with the same partners (Berndt et al., 1993). However, because of steric constraints arising from the covalent attachment to the Ser side chain, it cannot approach the carbonyl oxygen of Cys 38 as closely as does the water molecule in BPTI. The ensuing change of the hydrogen bond to Cys 38 might explain that the proton chemical shift differences between the major and minor conformations of BPTI and BPTI(G36S) are significantly different near this residue (Figure 8a). A change in the hydrogen-bonding pattern would not be expected to be reflected in the ^{15}N chemical shifts (Figure 8b), since these are rather insensitive to hydrogen bonding (Glushka et al., 1989).

The proximity of the aromatic side chain of Tyr 35 to the disulfide bridge Cys 14–Cys 38 raises the question whether there might be some coupling between the presently described disulfide flips and the previously characterized 180° flips of this aromatic ring (Wagner et al., 1976). To enable a comparison between the different motional modes in BPTI and BPTI(G36S), we measured the ring flip frequencies of Tyr 35 in BPTI(G36S) and in BPTI with two-dimensional exchange experiments (Fejzo et al., 1990). These measurements yielded more accurate rate constants than could previously be obtained from the line-shape analysis in one-dimensional ^1H NMR spectra (Wagner et al., 1976). The ring flip rates of Tyr 35 in BPTI at 26, 40, and 50 °C were found to be $1.3 \pm 0.2\ s^{-1}$, $16 \pm 1\ s^{-1}$ and $80 \pm 20\ s^{-1}$, respectively, and in BPTI(G36S) the corresponding ring flip rates were about twice these values. From the temperature dependence of the ring flip rates, the activation parameters $\Delta H^{\ddagger}$ and $\Delta S^{\ddagger}$ in BPTI were determined to be about 33 kcal/mol and 53 cal/mol K, respectively. A survey of the kinetic results obtained for the aromatic ring flips of Tyr 35 and the disulfide flips in BPTI and BPTI(G36S) (Figure 11) indicates that the ring flips and the disulfide flips are not correlated at temperatures below room temperature. Their different nature is reflected in the significantly smaller free energies of activation for the disulfide flips (Figure 11). With increasing temperature, however, the disulfide flip rates increase much more rapidly than anticipated from the low temperature data of Table II. For example, on the basis of these activation parameters, rate constants k_M in the range of 10–70 s^{-1} and 150–600 s^{-1}, respectively, would be predicted for BPTI and BPTI(G36S) at 68 °C. The experimentally observed rate constants, however, are faster by one order of magnitude or more. This implies that the disulfide flip rates do not follow an Arrhenius-type temperature dependence over the entire range from 4 to 68 °C. Yet the temperature dependence of the equilibrium constants is still in line with the predictions from Figure 10. For example, using the curves drawn in Figure 10 and eqs 2 and 5, the population of the minor conformation in BPTI is estimated to be about 6% at 68 °C, which coincides closely with the value of 8% obtained from the averaged chemical shift of the $\beta_2H-\beta_3H$ cross-peaks of Cys 38. The situation of the accelerated reaction rates at the higher

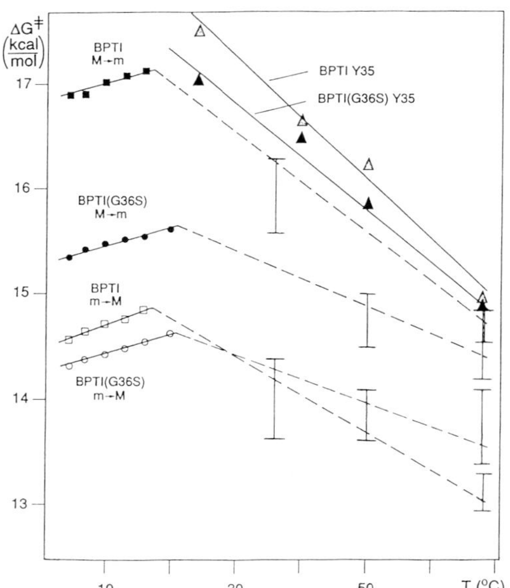

FIGURE 11: Plots of the free energies of activation, $\Delta G^{\ddagger}$, versus the temperature, T, for the forward reaction of eq 1, M → m, and the backward reaction, M ← m, in BPTI and BPTI(G36S). The data in the temperature range 4–20 °C are taken from Figure 10. Above 20 °C, vertical vars at 36, 50, and 68 °C, respectively, indicate the ranges of $\Delta G^{\ddagger}$ values that are compatible with the T_2 relaxation data of the amide protons of BPTI at 36 °C, the β-proton line widths of Cys 38 in BPTI(G36S) at 50 °C, and the β-proton line widths of Cys 38 in BPTI(G36S) and BPTI at 68 °C (see the text). To indicate the general trend, these large uncertainty ranges are connected by dashed lines. Also indicated with black and shaded triangles connected by solid lines are the $\Delta G^{\ddagger}$ values measured for the 180° ring flips of Tyr 35 in the two proteins (see the text).

temperatures with unchanged equilibrium parameters can be rationalized as intramolecular catalysis of the disulfide flips by local structure fluctuations similar to those manifested by the ring flips of Tyr 35.

It is intriguing to note that the disulfide flip rates k_M in BPTI and BPTI(G36S) are faster than the Tyr 35 ring flip rates over the entire temperature range from 4 to 68 °C. Using the activation parameters of the Tyr 35 ring flip, the flip rate at 4 °C is calculated to be 0.015 s^{-1}, which is too slow to be measured by ^{1}H NMR spectroscopy. The disulfide flip in BPTI occurs with much faster rates at this temperature, i.e., k_M = 0.27 s^{-1} and k_m = 16.8 s^{-1}, and even faster rates are observed in BPTI(G36S) (Figure 10). Thus, at low temperature the disulfide flips proceed much more easily than the Tyr 35 ring flips. On the basis of the activation parameters of the two different motions, the Tyr 35 ring flip rate would be expected to become faster than the disulfide flip rate at temperatures between 25 and 50 °C. The fact that the disulfide flip rates are actually faster over the entire temperature range implicates that above room temperature both NMR-detectable motions are triggered by similar structure fluctuations of the protein. A similar conclusion comes from the activation parameters $\Delta H^{\ddagger}$ and $\Delta S^{\ddagger}$ of the disulfide flip estimated from the rate constants k_M at 16 and 68 °C, which are more similar to the activation parameters of the Tyr 35 ring flip than to the parameters of Table II, with a positive value of the activation entropy, $\Delta S^{\ddagger}$.

In addition to the interest relative to protein structure and dynamics, the question arises whether the disulfide flips could interfere with the inhibitory binding of BPTI to trypsin. As it was observed earlier that selective reduction of the Cys 14–Cys 38 disulfide bond results in virtually unaltered binding affinity to trypsin (Kress et al., 1968), the strictly local changes coupled with the disulfide flips (Figure 7) make a significant influence on the functional properties rather unlikely.

In conclusion, the change in chirality of the disulfide bond Cys 14–Cys 38 proceeds with an ease that may be surprising in view of the conformational restraints imposed by the three-dimensional structure of a protein as rigid as BPTI. Therefore, flipping disulfide bonds may occur more frequently in protein solutions than what might be anticipated from the usually static model representations of protein structures. The present detailed analysis of structure, equilibrium, and kinetics of a disulfide flip in BPTI lead to the characterization of an example of intramolecular catalysis of conformational exchange in a protein. It seems plausible from this investigation that the activation parameters measured at high temperatures for different conformational exchange processes in proteins reflect quite generally the energy barrier of structural fluctuations involving entire protein subdomains rather than the energy barrier for specific, locally confined conformation changes.

ACKNOWLEDGMENTS

We thank Dr. L. P. M. Orbons for measuring the T_2 relaxation times of BPTI used in Figure 9, Bayer Leverkusen, Germany, for a generous gift of uniformly ^{15}N-labeled BPTI and BPTI(G36S), and Mr. R. Marani for the careful processing of the text.

SUPPLEMENTARY MATERIAL AVAILABLE

One table containing a survey of the data on the equilibrium and exchange rates between the two interconverting conformers observed in solutions of BPTI and BPTI(G36S) (2 pages). Ordering information is given on any current masthead page.

REFERENCES

Bax, A., Ikura, M., Kay, L. E., & Zhu, G. (1991) *J. Magn. Reson. 91*, 174–178.

Berndt, K. D., Güntert, P., Orbons, L. P. M., & Wüthrich, K. (1992) *J. Mol. Biol. 227*, 757–775.

Berndt, K. D., Beunink, J., Schröder, W., & Wüthrich, K. (1993) *Biochemistry* (in press).

Billeter, M., Neri, D., Otting, G., Qian, Y. Q., & Wüthrich, K. (1992) *J. Biomol. NMR 2*, 257–274.

Bodenhausen, G., & Ruben, D. J. (1980) *Chem. Phys. Lett. 69*, 185–189.

Bothner-By, A. A., Stephens, R. L., Lee, J., Warren, C. D., & Jeanloz, R. W. (1984) *J. Am. Chem. Soc. 106*, 811–813.

Chazin, W. J., Goldenberg, D. P., Creighton, T. E., & Wüthrich, K. (1985) *Eur. J. Biochem. 152*, 429–437.

Deisenhofer, J., & Steigemann, W. (1975) *Acta Crystallogr. B 31*, 238–250.

Donzel, B., Kamber, B., Wüthrich, K., & Schwyzer, R. (1972) *Helv. Chim. Acta 55*, 947–961.

Fejzo, S., Westler, W. M., Macura, S., & Markley, J. (1990) *J. Am. Chem. Soc. 112*, 2574–2577.

Fraser, R. R., Boussard, G., Saunders, J. K., Lambert, J. B., & Mixan, C. E. (1971) *J. Am. Chem. Soc. 93*, 3822–3823.

Glushka, J., Lee, M., Coffin, S., & Cowburn, D. (1989) *J. Am. Chem. Soc. 111*, 7716–7722.

Grathwohl, C., & Wüthrich, K. (1981) *Biopolymers 20*, 2623–2633.

Griesinger, C., Otting, G., Wüthrich, K., & Ernst, R. R. (1988) *J. Am. Chem. Soc. 110*, 7870–7872.

Gutowsky, H. S., & Holm, C. H. (1956) *J. Chem. Phys. 25*, 1228–1234.

Jung, G., & Ottnad, M. (1974) *Angew. Chem. 86*, 856–857.

Kopple, K. D., Wang, J., Cheng, A. G., & Bhandary, K. K. (1988) *J. Am. Chem. Soc. 110*, 4168–4176.

Kress, L. F., Wilson, K. A., & Laskowski, M., Sr. (1968) *J. Biol. Chem. 243*, 1758–1762.

Macura, S., Wüthrich, K., & Ernst, R. R., (1982) *J. Magn. Reson. 46*, 269–282.

Marion, D., Ikura, M., Tschudin, R., & Bax, A. (1989) *J. Magn. Reson. 85*, 393–399.

Messerle, B. A., Wider, G., Otting, G., Weber, C., & Wüthrich, K. (1989) *J. Magn. Reson. 85*, 608–613.

Neri, D., Otting, G., & Wüthrich, K. (1990) *J. Am. Chem. Soc. 112*, 3663–3665.

Otting, G., & Wüthrich, K. (1988) *J. Magn. Reson. 76*, 569–574.

Otting, G., & Wüthrich, K. (1989) *J. Am. Chem. Soc. 111*, 1871–1875.

Rance, M., Sørensen, O. W., Bodenhausen, G., Wagner, G., Ernst, R. R., & Wüthrich, K. (1983) *Biochem. Biophys. Res. Commun. 117*, 479–485.

Richardson, J. S. (1981) *Adv. Protein Chem. 34*, 167–330.

Richarz, R., Nagayama, K., & Wüthrich, K. (1980) *Biochemistry 19*, 5189–5196.

Srinivasan, N., Sowdhamini, R., Ramakrishnan, C., & Balaram, P. (1990) *Int. J. Pept. Protein Res. 36*, 147–155.

Vanmierlo, C. P. M., Darby, N. J., Neuhaus, D., & Creighton, T. E. (1991) *J. Mol. Biol. 222*, 373–390.

Wagner, G. (1980) *FEBS Lett. 112*, 280–284.

Wagner, G., & Wüthrich, K. (1982) *J. Mol. Biol. 155*, 347–366.

Wagner, G., & Nirmala, N. R. (1989) *Chem. Scr. 29A*, 27–30.

Wagner, G., DeMarco, A., & Wüthrich, K. (1976) *Biophys. Struct. Mech. 2*, 139–158.

Wagner, G., Stassinopoulou, C. I., & Wüthrich, K. (1984) *Eur. J. Biochem. 143*, 431–436.

Wagner, G., Braun, W., Havel, T. F., Schaumann, T., Gō, N., & Wüthrich, K. (1987) *J. Mol. Biol. 196*, 611–639.

Wider, G., Neri, D., & Wüthrich, K. (1991) *J. Biomol. NMR 1*, 93–98.

Wlodawer, A., Walter, J., Huber, R., & Sjölin, L. (1984) *J. Mol. Biol. 193*, 145–156.

Wlodawer, A., Nachman, J., Gilliland, G. L., Gallagher, W., & Woodward, C. (1987) *J. Mol. Biol. 198*, 469–480.

Wüthrich, K. (1976) *NMR in Biological Research: Peptides and Proteins*, North Holland, Amsterdam.

Wüthrich, K. (1986) *NMR of Proteins and Nucleic Acids*, Wiley, New York.

Journal of Biomolecular NMR, 3 (1993) 151–164
ESCOM

J-Bio NMR 095

Protein dynamics studied by rotating frame ^{15}N spin relaxation times

T. Szyperski, P. Luginbühl, G. Otting, P. Güntert and K. Wüthrich*

Institut für Molekularbiologie und Biophysik, Eidgenössische Technische Hochschule-Hönggerberg, CH-8093 Zürich, Switzerland

Received 2 October 1992
Accepted 3 December 1992

Keywords: Protein dynamics; Basic pancreatic trypsin inhibitor; Nuclear magnetic resonance spectroscopy; Rotating frame spin relaxation times

SUMMARY

Conformational rate processes in aqueous solutions of uniformly ^{15}N-labeled pancreatic trypsin inhibitor (BPTI) at 36 °C were investigated by measuring the rotating frame relaxation times of the backbone ^{15}N spins as a function of the spin-lock power. Two different intramolecular exchange processes were identified. A first local rate process involved the residues Cys^{38} and Arg^{39}, had a correlation time of about 1.3 ms, and was related to isomerization of the chirality of the disulfide bond Cys^{14}-Cys^{38}. A second, faster motional mode was superimposed on the disulfide bond isomerization and was tentatively attributed to local segmental motions in the polypeptide sequence - Cys^{14} - Ala^{15} - Lys^{16} -. The correlation time for the overall rotational tumbling of the protein was found to be 2 ns, using the assumption that relaxation is dominated by dipolar coupling and chemical shift anistropy modulated by isotropic molecular reorientation.

INTRODUCTION

Nitrogen-15 spin relaxation in peptide groups is mainly due to dipolar interactions with the amide proton and to chemical shift anisotropy interactions (Allerhand et al., 1971), so that relaxation times can be related to the spatial reorientation of the vector connecting the ^{15}N nucleus and the amide proton. Therefore, studies of ^{15}N relaxation times are an attractive approach for investigating global and local motional processes in the backbone of polypeptide chains. For example, the ^{15}N relaxation times of peptide bonds that are rigidly embedded in a

*To whom correspondence should be addressed.

Abbreviations: BPTI, basic pancreatic trypsin inhibitor; 2D, two-dimensional; COSY, 2D correlation spectroscopy; TOCSY, 2D total correlation spectroscopy; RF, radio frequency; CW, continuous wave; TPPI, time-proportional phase incrementation; CSA, chemical shift anisotropy; T_1, longitudinal relaxation time; T_2 transverse relaxation time; $T_{1\rho}$ relaxation time in the rotating frame τ_R, correlation time for overall rotational reorientation of the protein; τ^s_{ex}, τ^f_{ex}, correlation times for two conformational exchange processes (slow and fast).

0925-2738/$ 10.00 © 1993 ESCOM Science Publishers B.V.

regular secondary structure, such as a β-sheet or an α-helix, can be related to the correlation time for overall rotational tumbling of the molecule. Additional motional processes in other regions of the polypeptide chain can then be identified by comparing the observed relaxation times with those in the more rigid core formed by the regular secondary structures. For such studies with ^{15}N, a technical advantage arises because, even in uniformly ^{15}N-labeled proteins, measurements of relaxation times are not disturbed by scalar couplings to other heteronuclei. Two-dimensional heteronuclear NMR techniques efficiently measure longitudinal and transverse ^{15}N relaxation times (Kay et al., 1987, 1992; Sklenar et al., 1987; Nirmala and Wagner, 1988, 1989; Peng and Wagner, 1992). Kay et al. (1989) recently used relaxation times in conjunction with measurements of heteronuclear ^{1}H-^{15}N Overhauser effects to determine the correlation time for overall rotational tumbling and, following the model-free approach of Lipari and Szabo (1982), the order parameter, S, for a large number of backbone ^{15}N nuclei in staphylococcal nuclease. A similar approach has been used by Stone et al. (1992) for domain IIA of glucose permease from *Bacillus subtilis*, by Kördel et al. (1992) for calbindin D_{9K}, by Schneider et al. (1992) for human ubiquitin and, introducing a second-order parameter to describe fast intramolecular motions, by Clore et al. (1990) for interleukin-1β.

Conformational exchange can provide an adiabatic relaxation pathway for transverse magnetization (Kaplan and Fraenkel, 1980; Sandström, 1982). Deverell et al. (1970) showed that relaxation times in the rotating frame, $T_{1\rho}$, can be used to determine exchange rate constants, since the efficiency of relaxation of transverse magnetization locked along an effective magnetic field depends on the amplitude of the spin-lock field. Early applications of this principle with small molecules included determination of the rate constant for ring inversion in cyclohexane from measurements of $T_{1\rho}$ of the ring protons as a function of the spin-lock power, and evaluation of the rotational barriers in a series of urea derivatives on the basis of measurements of $T_{1\rho}$ of the ^{13}C nuclei (Stilbs and Moseley, 1978). In BPTI, it was previously noted that some of the proton resonances in the polypeptide segments of residues 14–18 and 36–41 are broadened at 36 °C, so that only few NOEs with these protons could be detected (Wagner and Wüthrich, 1982; Wagner et al., 1987; Berndt et al., 1992). Furthermore, measurements of ^{13}C spin relaxation times provided evidence for exchange between multiple conformations near the reactive site of BPTI (Wagner and Nirmala, 1989). There is a dynamic equilibrium between two conformational states with different chirality of the Cys^{14}–Cys^{38} disulfide bond (Otting, G., Liepinsh, E. and Wüthrich, K, unpublished results). While separate resonances for the two conformers could be resolved over the temperature range 4 °C to about 20 °C, the lines coalesced at 36 °C. In the present investigation we used nitrogen-15 $T_{1\rho}$ relaxation data of uniformly ^{15}N-labeled BPTI at 36 °C to determine the correlation time for overall rotational tumbling of BPTI, and to investigate the frequencies of the local motional processes near the residues Cys^{14} and Cys^{38}.

Analytical expressions for the determination of the overall rotational tumbling correlation time of a molecule, τ_R, from $T_{1\rho}$ relaxation times have been derived for dipolar relaxation (Blicharski, 1972; Jones, 1966; Peng et al., 1991a), quadrupolar relaxation, and spin-rotation relaxation (Blicharski, 1972). We have derived a corresponding expression for chemical shift anisotropy (CSA) relaxation, which is identical to the one recently published by Peng and Wagner (1992). Using this expression, we investigated the relative contribution of dipolar relaxation and of CSA relaxation to the overall relaxation times, which enabled a refined determination of the correlation time for overall rotational tumbling from ^{15}N $T_{1\rho}$ measurements.

THEORY

Influence of conformational exchange on $T_{1\rho}$

As discussed by Peng and Wagner (1992), the equation $T_{1\rho} = T_2$ in the on-resonance limit, where the effective magnetic field in the rotating frame is transverse, is valid only if there is no evolution of antiphase magnetization of the ^{15}N nuclei with respect to the amide protons, and if there are no rate processes leading to significant variations in the spectral density functions in the kHz frequency range. The contribution of such exchange processes with frequencies near the Larmor frequency of the applied RF spin-lock field, ω_1, to the spectral density function can be determined by carrying out a series of measurements with variable spin-lock power. The contribution of chemical exchange processes to $T_{1\rho}$ was evaluated earlier for a system of spins that exchange between two equally populated sites with chemical shifts $\Delta\Omega/2$ and $-\Delta\Omega/2$ relative to the ^{15}N carrier frequency, ω_0 (Eq. 5 in Deverell et al., 1970). Following Deverell et al. (1970) and using the general N-site jump model by Oppenheim et al. (1977), an analogous expression for $1/T_{1\rho}$ is obtained for the more general situation where the exchange process is between two sites, A and B, with arbitrary populations, p_A and p_B, and with rate constants, $k_{A\to B}$ and $k_{B\to A}$ (Brüschweiler, 1992):

$$\frac{1}{T_{1\rho}} = p_A\, p_B\, \Delta\Omega^2 \frac{\tau_{ex}}{1 + (\omega_1\, \tau_{ex})^2} + \frac{1}{T_{1\rho}^{\infty}} \qquad (1)$$

$\Delta\Omega = \Omega_A - \Omega_B$, where Ω_A and Ω_B are the chemical shifts of the spins in the sites A and B (in rad s^{-1}), respectively, relative to the ^{15}N carrier frequency. ω_1 is related to the applied RF spin-lock field, B_1, by $\omega_1 = -\gamma_N B_1$ and $T_{1\rho}^{\infty}$ is the relaxation time for an infinitely large spin-lock power. The equation for the detailed balance, $p_A\, k_{A\to B} = p_B\, k_{B\to A} = k/2$, connects the rate constants and the correlation time, τ_{ex}, of the exchange process, yielding

$$\tau_{ex} = \frac{2p_A p_B}{k} = \frac{1-p_A}{k_{A\to B}} \qquad (2)$$

As Eq. 1 is derived in the framework of time-dependent perturbation theory up to second- order, it is valid only for $\Delta\Omega\tau_{ex}<<1$ (see p.282 and p.517 of Abragam, 1961; Deverell et al., 1970). Another limitation of Eq. 1 arises because off-resonance effects are neglected. However, if both of these conditions are fulfilled, measurements of $T_{1\rho}$ as a function of ω_1 and subsequent fitting of the data according to Eq. 1 yields values for τ_{ex}, $1/T_{1\rho}^{\infty}$, and the product $p_A p_B \Delta\Omega^2$.

Once the rapid-exchange condition, $\Delta\Omega\tau_{ex}<<1$, is violated, the use of second-order pertubation theory is a priori no longer warranted, e.g., higher-order terms may have to be considered (see p.282 of Abragam, 1961). We therefore checked the results obtained from perturbation theory and the possible influence of off-resonance effects by solving the equation of motion in the rotating frame, where the radiofrequency field is static along the x-axis, for an ensemble of n non-interacting spins under the influence of the Hamiltonian in Eq. 3, where n is a large number and $\hbar=1$,

$$H(t) = \sum_{j=1}^{n} H_j(t) \quad \text{with} \quad H_j(t) = \omega_{1j} I_{jx} + \Omega_j(t) I_{jz}. \qquad (3)$$

For the spin j, I_{jx} and I_{jz} denote Cartesian spin operators. In order to account for spatial inhomogenities of the spin-lock field, we assumed that the ω_{1j} values (j = 1,...,n) are independent random variables which are uniformly distributed within the interval [0.75 ω_1, 1.25ω_1]. $\Omega_j(t)$ describes the Zeeman interaction, which is a function of time because of the exchange between the conformations A and B:

$$\Omega_j(t) = \begin{cases} \Omega_A, & \text{if spin j is in conformation A at time t;} \\ \Omega_B, & \text{if spin j is in conformation B at time t;} \end{cases} \tag{4}$$

We assumed that the exchange events of different individual spins are not correlated, and that the exchange reactions A→B and B→A occur with the rates $k_{A\to B}$ and $k_{B\to A}$, respectively. Therefore, the lifetimes, τ, of an individual spin in the conformations A and B are exponentially distributed, with probability densities $\rho_A(\tau)$ and $\rho_B(\tau)$, respectively:

$$\rho_A(\tau) = k_{A\to B} \exp(-\tau\, k_{A\to B}) \text{ and } \rho_B(\tau) = k_{B\to A} \exp(-\tau\, k_{B\to A}). \tag{5}$$

The expectation value for the macroscopic magnetization in the x-direction, M_x, is given by

$$M_x(t) = \sum_{j=1}^{n} \langle \psi_j(t) \,|\, I_{jx} \,|\, \psi_j(t) \rangle, \tag{6}$$

where $|\psi_j(t)\rangle$ denotes the state vector of spin j at time t. We assumed that all spins are aligned in the x-direction at t = 0, and that the probability that the spin j is initially in conformation A is given by $p_A = k_{B\to A}/(k_{A\to B} + k_{B\to A})$. Because the Hamiltonian in expression 3 is time independent between exchange events, $|\psi_j(t)\rangle$ can be calculated, for example, for the situation where the spin j is initially in conformation A and there are exchange events according to Eq. 4 at the times $t_1 < t_2 < \cdots < t_m$ (m odd), with $t_1 > 0$ and $t_m < t$:

$$|\psi_j(t)\rangle = e^{-iH_{jB}(t-t_m)}\, e^{-iH_{jA}(t_m-t_{m-1})} \cdots e^{-iH_{jB}(t_2-t_1)}\, e^{-iH_{jA}t_1}\, |\psi_j(0)\rangle, \tag{7}$$

where $H_{jA} = \omega_{1j}I_{jx} + \Omega_A I_{jz}$ and $H_{jB} = \omega_{1j}I_{jx} + \Omega_B I_{jz}$.

For the simulation for BPTI at 36 °C, we used the correlation time for the conformational exchange, τ_{ex}, and the product $p_A p_B \Delta\Omega^2$, which had been determined from fitting Eq. 1 to the experimental rotating frame relaxation times. Furthermore, use of the Ω_A and Ω_B values measured at temperatures below 20 °C (Otting, G., Liepinsh, E. and Wüthrich, K., unpublished results) enabled us to calculate p_A and p_B from the product $p_A p_B \Delta\Omega^2$ and the exchange rate constants $k_{A\to B}$ and $k_{B\to A}$ (Eq. 2). The simulation yielded the time course of M_x (Eqs. 6 and 7), from which the exchange contribution to the relaxation in the rotating frame, $T_{1\rho}^{sim}$, was determined by a least-squares fit of a single exponential function. The relaxation times were finally calculated for all experimental ω_1 values, using the relation

$$\frac{1}{T_{1\rho}^{calc}} = \frac{1}{T_{1\rho}^{sim}} + \frac{1}{T_{1\rho}^{\infty}}.$$

$T_{1\rho}^{\infty}$ denotes the relaxation time in the rotating frame for infinite spin-lock power as obtained from the experimental rotating frame relaxation times after fitting to Eq. 1.

NMR EXPERIMENTS

A 5 mM solution of uniformly ^{15}N-labeled BPTI in 90% H_2O/10% D_2O was used at a temperature of 36 °C and pH = 4.6. The NMR experiments were performed on Bruker AM500 and AM360 spectrometers. The ^{15}N resonance assignments of Glushka et al. (1989) were verified in a 500-MHz TOCSY-relayed-[^{15}N,^{1}H]-COSY experiment (Otting and Wüthrich, 1988).

Relaxation times were measured with the pulse schemes shown in Fig. 1, where ^{15}N broadband-decoupling during proton detection was achieved by using Waltz-16 (Shaka et al., 1983). Ten different relaxation delays (t_{rel} = 42.5, 82.5, 112.5, 202.5, 282.5, 382.5, 482.5, 582.5, 682.5 and 782.5 ms) were used to measure the longitudinal relaxation time, T_1. During the relaxation delay, the proton resonances were saturated with a train of 90° pulses at 5-ms intervals, as proposed by Nirmala and Wagner (1989) and Peng and Wagner (1992) to suppress the cross-correlation of dipolar and CSA relaxation (Goldman, 1984; Boyd et al., 1990; Kay et al., 1992; Palmer et al., 1992). Decoupling during t_1 was achieved with Waltz-16. Transverse relaxation times, T_2, were determined with the pulse scheme of Fig. 1B, using 12 different relaxation delays (t_{rel} = 9, 20, 30, 40, 50, 70, 80, 100, 120, 140, 200 and 280 ms). To suppress evolution of anti-phase magnetization of the ^{15}N-nuclei with respect to the amide protons, we applied Waltz-16 (Shaka et al., 1983) during the relaxation delay. Waltz-16 was also used to decouple the amide protons during the chemical shift evolution of the ^{15}N nuclei. Since the CW spin-lock decouples ^{15}N nuclei and amide protons during the relaxation delay when measuring $T_{1\rho}$ (ω_1) (Fig. 1C), Waltz-16 was applied only during t_1. $T_{1\rho}(\omega_1)$ was determined at ten different values of the CW spin-lock power (ω_1 = 1210, 1960, 3140, 4130, 5060, 5820, 6830, 8490, 11220 and 15240 rad s^{-1}). At each power level, eight different relaxation delays were used (t_{rel} = 20, 40, 60, 80, 100, 140, 180 and 260 ms). The relaxation measurements described so far were all performed at 500 MHz, but to assess the contributions from CSA, T_1, T_2 and $T_{1\rho}$ were also measured at 360 MHz. For T_1 the same relaxation delays were used at 360 MHz as at 500 MHz. T_2 values at 360 MHz were obtained from measurements with t_{rel} = 9, 20, 30, 40, 60, 80, 100, 140, 200 and 280 ms, and for $T_{1\rho}$ measurements at 360 MHz the maximum spin-lock field achievable on our instrument was used, i.e., ω_1 = 11640 rad s^{-1}, and t_{rel} = 30, 40, 50, 60, 80, 100, 140, 240 and 340 ms.

The 2D spectra were recorded as 256*1024 real matrices with 8 scans per t_1 value, resulting in a total recording time of 5 days for a total of 131 experiments. The water signal was suppressed with two spin-lock purge pulses of 500 μs and 2000 μs duration, respectively (Fig. 1 A–C) (Messerle et al., 1989). For the $T_{1\rho}$ measurements, the spin-lock purge pulses were applied *before* the relaxation delay, t_{rel}, (Fig. 1C), in order to avoid heteronuclear cross-polarization in the rotating frame (Maudsley et al., 1977). Quadrature detection in ω_1 was achieved by using TPPI (Marion and Wüthrich, 1983). Before Fourier transformation, the data were multiplied with a sine-bell window shifted by $\pi/2$ (DeMarco and Wüthrich, 1976). Baseline corrections with third-order polynomials were applied in each dimension, using standard Bruker software on a Bruker X32 workstation. The peak volumes were determined by using the EASY program (Eccles et al., 1991). The relaxation times were determined by fitting a single exponential function to the peak volumes by a least-squares fit routine. Finally, the rotating frame relaxation times of the backbone ^{15}N nuclei of Cys^{38} and Arg^{39} were fitted to Eq. 1, using the Levenberg–Marquardt algorithm (Press et al., 1986), yielding τ_{ex}, $1/T_{1\rho}^{\infty}$, and the product, $p_A p_B \Delta\Omega^2$.

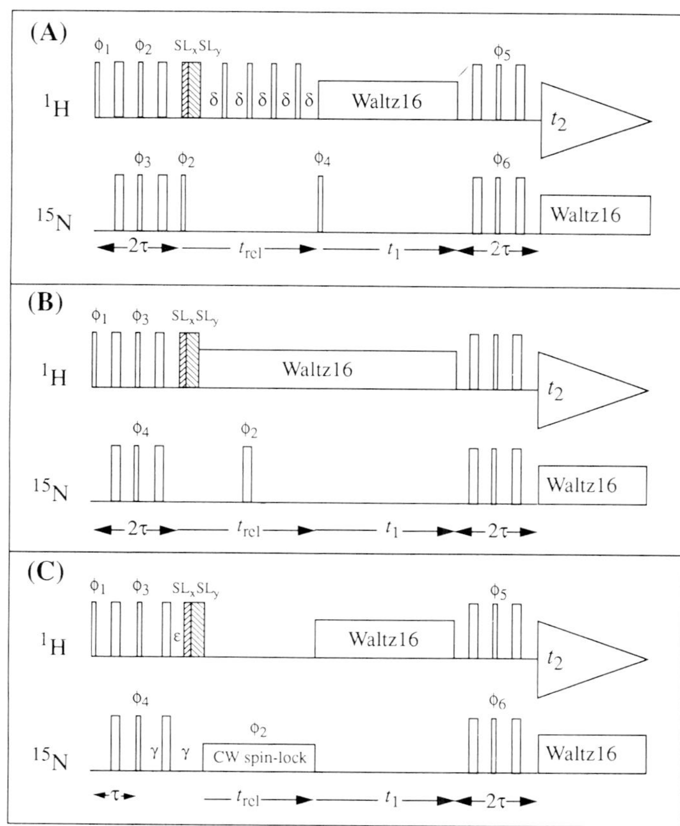

Fig. 1. Experimental schemes used for measurements of T_1 (A), T_2 (B) and $T_{1\rho}$ (C) of ^{15}N in polypeptide backbone amide groups. ^1H pulses are shown in the upper and ^{15}N pulses in the lower traces. 90° and 180° pulses are indicated by thin and thick vertical bars, respectively. Spin-lock purge pulses used to suppress the water signal are denoted as SL_x and SL_y, with pulse lenghts of SL_x = 500 μs and SL_y = 2000 μs. The delay τ was tuned to $1/2^1J(^1H,^{15}N)$. In (A) the delays δ were set to 5 ms. In (C) the delays γ and ε were tuned to $\gamma + \varepsilon = \tau$ and $\gamma - \varepsilon = SL_x + SL_y$. In all three schemes, a Waltz-16 composite pulse decoupling sequence was applied during the proton detection period. The relaxation delays are indicated as t_{rel}. Phase cycling for (A): ϕ_1 = 8(x), ϕ_2 = 4(y, −y), ϕ_3 = 4(x, −x), ϕ_4 = 2(x, x, −x, −x), ϕ_5 = 4(x), 4(−x), ϕ_6 = 4(y), 4(−y), receiver = 2(x −x, −x, x,); for (B) : ϕ_1 = 8(x), ϕ_2 = 2(x, −x, y, −y), ϕ_3 = 4(y), 4(−y), ϕ_4 = 4(x), 4(−x), receiver = 2(x −x, x, −x); for (C) : ϕ_1 = 8(x), ϕ_2 = 4(x, −x), ϕ_3 = 2(y, y, −y, −y), ϕ_4 = 2(x, x, −x, −x), ϕ_5 = ϕ_6 = 4(x), 4(−x), receiver = 8(x). The phases of all other pulses were set to x. TPPI (Marion and Wüthrich, 1983) may be achieved by simultaneous incrementation of the phases of all ^{15}N pulses before the evolution time, t_1.

RESULTS AND DISCUSSION

The analysis of the rotating frame relaxation times for the backbone ^{15}N nuclei of BPTI enabled the characterization of three different types of motion, namely the overall rotational tumbling of the molecule and two internal motional modes observed at residues 14–16 and at residues 38–39, respectively.

Determination of the correlation time for overall rotational tumbling of BPTI in aqueous solution

The correlation time, τ_R, for the overall rotational tumbling of BPTI at 36 °C and pH = 4.6 was calculated from the average of the relaxation times, T_1^{av} and $T_{1\rho}^{av}$, of the residues 18–24 and 29–36, which form a centrally located β-sheet (Deisenhofer and Steigemann, 1975; Berndt et al., 1992). The off-resonance angles, β, (Eq. 13 in Peng and Wagner, 1992) for these residues were all between 72° and 99° for the $T_{1\rho}$ measurements at the highest spin-lock power used (ω_1 = 15240 rad s^{-1}). Therefore, off-resonance effects were neglected (Peng et al., 1991a) and a value of β = 90° was used in all the calculations. In accordance with Kay et al. (1989), we also neglected possible contributions from fast internal motions, since residues in β-sheets are generally characterized by S ~ 1, where S is the order parameter (Lipari and Szabo, 1982; Kay et al., 1989; Clore et al., 1990; Stone et al., 1992). Assuming isotropic reorientation characterized by a Lorentzian spectral density function, J(ω) (Abragam, 1961), we calculated τ_R either for the (hypothetical) situation that the relaxation was entirely by dipolar interactions with the directly bonded amide proton (Peng et al., 1991), or with inclusion of CSA relaxation on the basis of Eq. 13 in Peng and Wagner (1992), thereby neglecting the cross-correlation of dipolar and CSA relaxation (Goldman, 1984). It has been shown that the ^{15}N chemical shift tensor is axially symmetric, and we set $\Delta\sigma = \sigma_{\parallel} - \sigma_{\perp}$ = −160 ppm (Hiyama et al., 1988) in Eq. 13 of Peng and Wagner (1992). $\sigma_{\parallel}$ and $\sigma_{\perp}$ are the parallel and perpendicular components of the ^{15}N chemical shift tensor. The bond length between the amide proton and the ^{15}N nucleus was set to r_{NH} = 1.02 Å (Keiter, 1986) in Eq. 13 of Peng and Wagner (1992).

In principle, two different τ_R values may correspond to the same value for the T_1 relaxation time (Eq. 88 on p.295 in Abragam, 1961). Comparison of the measurements at 360 MHz and 500 MHz enabled us to discriminate between the two possible solutions, because as the CSA interactions are dependent on B^2 (Eq. 141 on p.316 in Abragam, 1961), the CSA relaxation becomes more important at higher field strengths. At 500 MHz the CSA interactions are expected to account for approximately 30% of the T_1 and $T_{1\rho}$ relaxation (Table 1). Further inspection of Table 1 shows that the τ_R values derived from T_1 and $T_{1\rho}$ coincide within the experimental error, yielding τ_R = 2.0 ± 0.5 ns. Since we suppressed cross-correlation of dipolar and CSA relaxation for the T_1 measurements but not for the $T_{1\rho}$ measurements, we conclude that, in the framework of the present investigation, these cross-correlation effects are indeed negligible. In order to check for possible effects on the determination of the overall rotational correlation time which might arise from fast segmental motions characterized by the order parameters S, we also calculated τ_R from the ratio $T_1/T_{1\rho}$ (Kay et al., 1989). No significant differences were found relative to the τ_R values calculated from T_1 or $T_{1\rho}$ alone (Table 1).

Richarz et al. (1980) obtained a value of τ_R = 4 ns for BPTI at 36 °C from ^{13}C relaxation times. However, these earlier measurements were obtained at a protein concentration of 25 mM, where the viscosity is significantly higher than in the solutions used in the present studies. τ_R = 2.0 ns is also close to the value that has been estimated from the Stokes–Einstein relation for a spherical protein with a molecular weight similar to that of BPTI and with a hydration shell 2.5 Å thick in a solution with the viscosity of H_2O (Abragam, 1961; Richarz et al., 1980).

Investigation of intramolecular conformational exchange in BPTI

The longitudinal relaxation times, T_1, of the individual backbone ^{15}N nuclei of BPTI were longer than 380 ms throughout the amino acid sequence and, with the single exception of the

C-terminus, varied only within a narrow range for all residues (Fig. 2A). In contrast, the transverse relaxation times, T_2, for the residues 14–16, 36, and 38–40 were significantly shorter than those for the residues in the regular secondary structures (Fig. 2B), indicating that an additional rate process affects the spin relaxation of these residues. With the highest spin-lock power achievable on our instrument (ω_1 = 15240 rad s^{-1}), the $T_{1\rho}$ values of residues 36 and 38–40 were equal to those for the residues in the β-sheet (Fig. 2C). From Eq. 1 it follows that the correlation time of the exchange process manifested in the transverse ^{15}N relaxation of these residues must be longer than 0.1 ms. A different behavior was observed for residues 14–16, which had shorter $T_{1\rho}$ times, indicating that there must be two motional processes with different correlation times. The lower limit for the correlation time of the faster one of these processes, τ_{ex}^f, was indicated by the observation that the T_1 relaxation times of residues 14–16 were not affected (Fig. 2A), which shows that τ_{ex}^f must be long compared to the inverse of the ^{15}N Larmor frequency at 500 MHz, i.e., $\tau_{ex}^f >> 3$ ns. (Note that Gly[57] and Ala[58] undergo intramolecular motions that were fast compared to the ^{15}N Larmor frequency, so that T_1, T_2 and $T_{1\rho}$ were all significantly increased.) An upper limit for τ_{ex}^f may be derived from the observation that the reduction in the $T_{1\rho}$ values of residues

TABLE 1

CORRELATION TIMES FOR THE OVERALL ROTATIONAL TUMBLING OF BPTI IN AQUEOUS SOLUTION DETERMINED FROM MEASUREMENTS OF THE ^{15}N RELAXATION TIMES T_1, T_2, $T_{1\rho}$ AT 36 °C[a]

Measurement	T_{av}[b] [ms]	τ_R [ns]	
		dipolar[c]	dipolar + CSA[d]
T_1[11.7 T][e]	414 ± 22	2.5 ± 0.7 (4.0 ± 1.1)	1.5 ± 0.3 (6.3 ± 0.9)
T_1[8.4 T][e]	336 ± 25	2.5 ± 0.6 (7.5 ± 1.5)	2.0 ± 0.5 (9.0 ± 2.2)
$T_{1\rho}$[11.7 T][f]	274 ± 9	2.3 ± 0.3	1.8 ± 0.3
$T_{1\rho}$[8.4 T][g]	252 ± 9	2.3 ± 0.3	2.0 ± 0.3
$T_1/T_{1\rho}$[11.7 T]		2.2 ± 0.4	2.2 ± 0.4
$T_1/T_{1\rho}$[8.4 T]		1.8 ± 0.9	1.9 ± 0.8
T_2[11.7 T]	194 ± 12	3.7 ± 0.7	2.8 ± 0.6
T_2[8.4 T]	204 ± 10	3.0 ± 1.1	2.6 ± 0.6

[a] The protein concentration was 5 mM, pH = 4.6. Isotropic rotational tumbling was assumed in the calculation of τ_R. The error in the determination of τ_R was estimated from the longest and shortest measured relaxation times.

[b] Average of the relaxation times measured for the residues 18–24 and 29–35, which are all located in a β-sheet in the solution structure of BPTI (Wagner et al., 1987; Berndt et al., 1992). The errors represent the standard deviation of the relaxation times determined for these residues.

[c] Correlation time, τ_R, calculated on the assumption that the measured relaxation rate was due entirely to dipolar relaxation.

[d] Dipolar coupling and chemical shift anisotropy interactions were considered. The two relaxation pathways were assumed to be uncorrelated.

[e] The determination of τ_R from T_1 relaxation times yielded two solutions at any given field strength. The values that were excluded from comparison of the results obtained at different field strengths are given in parentheses.

[f] Relaxation times obtained with the highest spin-lock power used (ω_1 = 15240 rad s^{-1}; see text).

[g] ω_1 = 11640 rad s^{-1}.

14–16 relative to those observed for residues in the β-sheet was virtually identical to that of the corresponding T_2 values. Thus, the $T_{1\rho}$ values of residues 14–16 appeared to be independent of ω_1 for $0 \leqslant \omega_1 \leqslant 15240$ rad s^{-1}, so that 0.1 ms >> τ^f_{ex} >> 3 ns.

For a quantitative determination of the correlation time for the conformational exchange at residues 36 and 38–40, τ^s_{ex}, we measured the rotating frame relaxation times as a function of the spin-lock power. We set the carrier in the ^{15}N dimension at 115.5 ppm, which is equidistant from the ^{15}N chemical shifts of Cys^{38} (114.8 ppm) and Arg^{39} (116.2 ppm). Thus, off-resonance effects for observations with these two residues were minimized. (Figure 2B shows that the T_2 relaxation times of Cys^{38} and Arg^{39} were about 10 times shorter than those of rigidly embedded residues, which is the most pronounced manifestation of chemical exchange processes in the entire protein. The following experiments therefore focused on these two residues). At the lowest spin-lock power used (ω_1 = 1210 rad s^{-1}), the angle, β, was 80° for both residues. The dependence of the decay rate constants of the magnetization in the rotating frame, $1/T_{1\rho}$, on the Larmor frequency of the spin-lock field for Cys^{38} (Fig. 3A) and Arg^{39} (Fig. 3B) was fitted to Eq. 1. The resulting parameters are shown in Table 2. The correlation times obtained for the slower of the two conformational exchange processes, τ^s_{ex}, coincided within the accuracy of the measurements, i.e., they were 2.4 ± 1.8 ms for Cys^{38} and 1.3 ± 0.5 ms for Arg^{39}. The observation that the rotating frame relaxation times at infinite spin-lock power, $T^{\infty}_{1\rho}$, were in the same range as those for rigidly embedded residues (Fig. 2C) supports the notion that the short T_2 relaxation times for residues 38 and 39 (Fig. 2B) were solely due to a local conformational exchange process.

It has been shown (Otting, G., Liepinsh, E. and Wüthrich, K., unpublished results) that BPTI in solution exchanges between two conformations, A and B, which differ in the chirality of the Cys^{14}–Cys^{38} disulfide bond. The activation parameters of the conformational exchange in the temperature range 4°C to 16 °C were determined by longitudinal two-spin order exchange difference spectroscopy (Wider et al., 1991). In order to investigate whether the exchange-broadening of the ^{15}N resonances of Cys^{38} and Arg^{39} at 36 °C can be attributed to this disulfide bond isomerization, we compared the present results with those found by Otting, Liepinsh and Wüthrich (unpublished results) for the population of the major component, p_A, and the rate constant, $k_{A\rightarrow B}$, for the disulfide flip at 36 °C. Table 2 shows that this estimate of the parameters was close to the data measured, which suggests that disulfide bond isomerization is indeed the conformational exchange process that causes the short T_2 relaxation times observed for residues 38 and 39. The deviations between the estimated values and the measured data for the product $p_A p_B \Delta\Omega^2$ for Arg^{39} are possibly due to changes in the chemical shift difference, ΔΩ, when the temperature is increased from 16 °C to 36 °C. For residues 14–16, where one would also expect to observe the influence of disulfide bond isomerization, these effects were apparently masked by an additional, faster rate process.

Simulation of the exchange contribution to the rotating frame relaxation

Using the experimental correlation time, τ^s_{ex}, and chemical shift differences, ΔΩ, we found that $\Delta\Omega\tau^s_{ex}(Cys^{38}) = 0.7$ and $\Delta\Omega\tau^s_{ex}(Arg^{39}) = 1.5$. Since Eq. 1 was derived from second-order perturbation theory on the assumption that $\Delta\Omega\tau_{ex} << 1$, its validity for the system studied needed to be re-examined (see also p.282 of Abragam, 1961). Another approximation in the derivation of Eq. 1, which is not strictly valid in the present example, is that off-resonance effects were not considered. The observed average ^{15}N chemical shift, $\overline{\Omega}$, is related to the chemical shifts of the major and

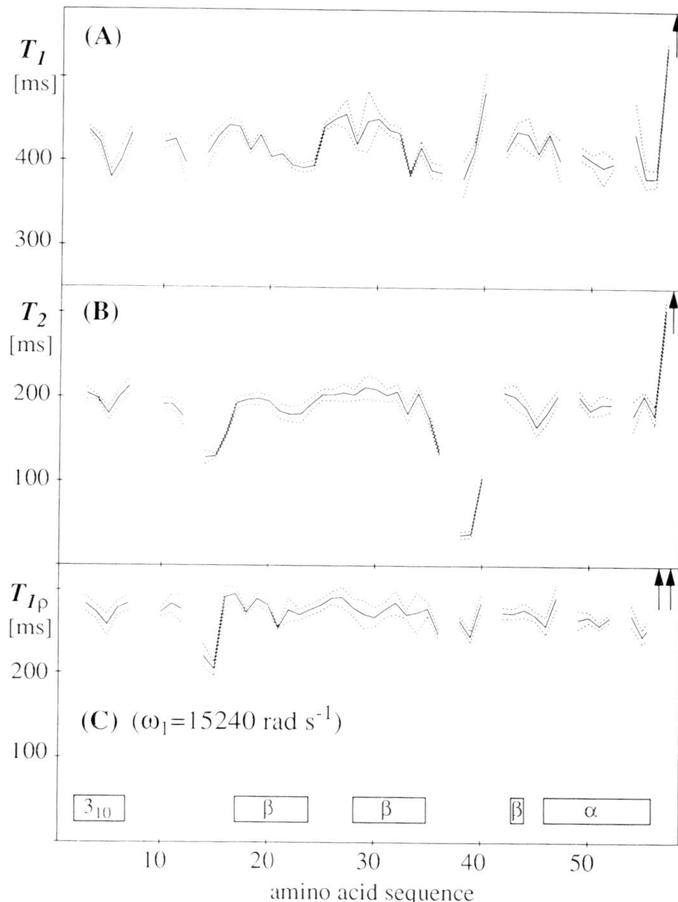

Fig. 2. Plots of the relaxation times T_1 (A), T_2 (B) and $T_{1\rho}$ (C) of ^{15}N in backbone amide groups of BPTI versus the amino acid sequence (solid line). Values that went off-scale are indicated by arrows. The dotted lines represent the 99% confidence interval corresponding to 2.5 σ, where σ is the standard deviation of the determination of the relaxation times. The measurements were made at 36 °C in a 5 mM aqueous solution of BPTI, pH = 4.6. At the bottom, the sequence locations of regular secondary structures in the solution conformation of BPTI are indicated, where 3_{10} and α identify the corresponding helix types, and β the individual strands of an antiparallel β-sheet.

minor conformers, A and B by $\overline{\Omega} = p_A\Omega_A + p_B\Omega_B$. Given that $\overline{\Omega}(\mathrm{Cys}^{38}) = -220$ rad s^{-1} and $\overline{\Omega}(\mathrm{Arg}^{39}) = 220$ rad s^{-1}, one obtains with $\Delta\Omega(\mathrm{Cys}^{38}) = 542$ rad s^{-1} and $\Delta\Omega(\mathrm{Arg}^{39}) = 1150$ rad s^{-1} that $\Omega_A(\mathrm{Cys}^{38}) = -133$ rad s^{-1}, $\Omega_B(\mathrm{Cys}^{38}) = -675$ rad s^{-1}, $\Omega_A(\mathrm{Arg}^{39}) = 300$ rad s^{-1}, and $\Omega_B(\mathrm{Arg}^{39}) = -850$ rad s^{-1}. The corresponding tip angles of the spin-lock axis for the major and minor conformers, β_A and β_B, were 76° and 55° for Arg^{39}, and 84° and 61° for Cys^{38}, respectively. Thus, although the off-resonance effect would be negligibly small for the *average* resonance frequency, under the experimental conditions chosen, the minor conformer experienced a significantly larger off-resonance effect for both residues.

To check the validity of Eq. 1 for the present system, we simulated the relaxation in the rotating

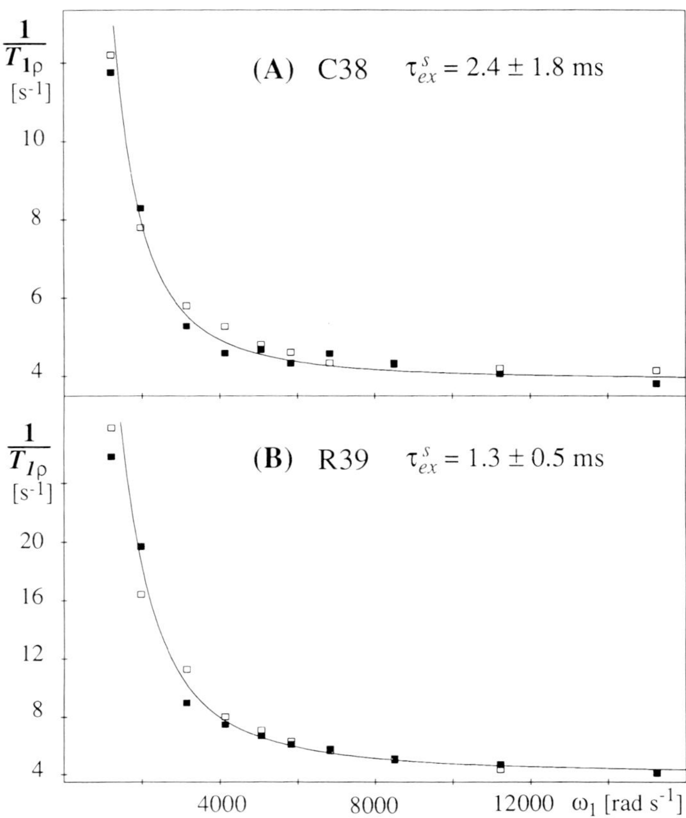

Fig. 3. Plots of the inverse of the measured ^{15}N rotating frame relaxation times, $T_{1\rho}$, (filled squares) versus the spin-lock power, ω_1, and comparison with the $T_{1\rho}$ model calculations described in the text (open squares). The solid curves represent fits of Eq. 1 to the experimental data; the correlation times for conformational exchange, τ^s_{ex}, obtained from this fitting procedure are also indicated.

frame arising from conformational exchange on the basis of a rigorous quantum-mechanical treatment (see Theory). We applied the parameters obtained using second-order perturbation theory (Table 2) and the chemical shift differences measured at 16 °C (Otting, G., Liepinsh, E. and Wüthrich, K., unpublished results). The calculated rotating frame relaxation times are shown in Fig. 3. Comparison with the experimental data (Fig. 3) demonstrates that the parameters obtained from Eq. 1 were correct, which implies that time-dependent perturbation theory is quite robust with respect to violations of the inequality $\Delta\Omega\tau_{ex} << 1$. Furthermore, calculation of the line shape for a two-site exchange, using the kinetic parameters and the chemical shift difference of Arg^{39} in Table 2 with the equations of Gutowsky and Saika (1953), showed that a single, coalesced resonance line was indeed expected for Arg^{39}. In contrast, two signals were expected when the product $\Delta\Omega\tau^s_{ex}$ was doubled, by increasing either τ_{ex} or $\Delta\Omega$ by a factor of 2, and the transverse magnetization displayed oscillatory behavior in simulations with $\Delta\Omega\tau^s_{ex} >> 1$. The present results thus suggest that the commonly used second-order perturbation theory is a good approximation whenever coalesced resonance lines are observed. On the basis of the close coincidence of the experimental and simulated data sets (Fig. 3), it appears that the influence of even rather large off-resonance effects (see above) is negligibly small.

Comparison of the absolute values for the relaxation times $T_{1\rho}$ and T_2

$T_{1\rho}$ relaxation times were in general longer than the corresponding T_2 relaxation times (Fig. 2B and C), although, in the absence of chemical exchange processes (Peng and Wagner, 1992) and evolution of antiphase magnetization (Peng et al., 1991b), they were expected to be equal. As discussed by Peng and Wagner (1992), cross-correlation between dipolar relaxation and CSA relaxation may lead to longer $T_{1\rho}$ values. However, the close agreement between the correlation times for overall rotational tumbling, τ_R, obtained from T_1 and $T_{1\rho}$ measurements, respectively (Table 1), indicates that such cross-correlation effects were negligible within the framework of the present investigation. Thus, the apparent discrepancy in the present measurements can probably be attributed to the use of the composite pulse decoupling scheme, Waltz-16 (Shaka et al., 1983), during the relaxation delay in the T_2 measurements (Fig. 1B). It has recently been reported by Kay et al. (1992) that the ^{15}N transverse magnetization decay rate is increased if composite pulse sequences are applied during the relaxation delay, since the scalar coupling between amide protons and α-protons diminishes the quality of the heteronuclear decoupling sequence and thus leads to incomplete heteronuclear decoupling (see also Shaka et al., 1988). The artifactually shortened T_2 relaxation times would correspond to longer apparent correlation times for the overall rotational tumbling, τ_R. Table 1 shows that the τ_R values derived from the T_2 relaxation

TABLE 2
PARAMETERS CHARACTERIZING THE CONFORMATIONAL EXCHANGE OBSERVED FOR RESIDUES CYS^{38} AND ARG^{39} IN BPTI[a]

Residue	τ^s_{ex} [ms][b]	$T^{\infty}_{1\rho}$[ms][c]	$p_A p_B \Delta\Omega^2$[rad^2 s^{-2}][d]	p_A[d,e]	$k_{A\to B}$[d][s^{-1}]
Cys^{38}[f]	**2.4 ± 1.8**	**256 ± 3**	**39000 ± 27000**	**0.84**	**66**
Estimate[g]	1.3 ± 0.7		11000 ± 3000	0.96 ± 0.01	40 ± 20
Arg^{39}[f]	**1.3 ± 0.5**	**245 ± 3**	**83000 ± 5000**	**0.93**	**53**
Estimate[g]	1.3 ± 0.7		51000 ± 13000	0.96 ± 0.01	40 ± 20

[a] Protein concentration 5 mM, T = 36 °C, pH = 4.6.

[b] Correlation time for the slow conformational exchange process (see text).

[c] Rotating frame relaxation time in the limit of infinite spin-lock power.

[d] Site A denotes the major conformer.

[e] The chemical shift difference ΔΩ was set to the value observed at 16 °C, i.e., for Cys^{38}, ΔΩ = 540 rad s^{-1}, and for Arg^{39}, ΔΩ = 1150 rad s^{-1} (Otting, G., Liepinsh, E. and Wüthrich, K., unpublished results).

[f] The experimental results are listed with bold numbers. They were determined from ^{15}N rotating frame relaxation time measurements, using a non-linear fit to Eq. 1.

[g] τ^s_{ex} was computed with Eq. 1 from the values of p_A and $k_{A\to B}$ derived from the amide proton line widths measured at 36 °C for the residues 14–18 and 38–41, which were all affected by the isomerization of the disulfide bond Cys^{14}–Cys^{38} (Fig. 2B), and using values of ΔΩ measured at 16 °C (Otting, G., Liepinsh, E. and Wüthrich, K., unpublished results). The interpretation of these proton line widths was based on the assumption that there was an exchange process between two isomeric states. The use of p_A = 0.96 at 36 °C was based on extrapolation from measurements in the temperature range 4 °C to 16 °C and at 68 °C (Otting, G., Liepinsh, E. and Wüthrich, K., unpublished results).

times obtained with the experimental scheme of Fig. 1B were indeed significantly longer than those derived from T_1 or $T_{1\rho}$.

CONCLUSIONS

The present investigation demonstrated that measurement of rotating frame relaxation times as a function of the spin-lock power allows the determination of correlation times for rate processes such as conformational exchange in the millisecond range. The determination of correlation times with this method is independent of knowledge of the chemical shift differences and the relative populations of the interchanging species. Furthermore, comparison with simulated data showed that second-order perturbation theory is adequate to evaluate $T_{1\rho}$ data within the limit $\Delta\Omega\tau_{ex}{\sim}1$, and that off-resonance effects become important only with relatively slow exchange rates that must be measured with small spin-lock frequencies, ω_1. On this basis, it appears that $T_{1\rho}$ measurements of ^{15}N might become an attractive tool for quantitative motional characterization of uniformly ^{15}N-enriched proteins.

ACKNOWLEDGEMENTS

Financial support was obtained from the Schweizerischer Nationalfonds (project 31.32033.91) and the Stipendien-Fonds im Verband der Chemischen Industrie (fellowship to Th.S.). We thank Bayer AG, Leverkusen, Germany, for a generous gift of ^{15}N-enriched BPTI, and Mr. R. Marani for the careful processing of the manuscript.

REFERENCES

Allerhand, A., Doddrell, D. and Komoroski, R. (1971) *J. Chem. Phys.*, **55**, 189–198.
Abragam, A. (1961) *The Principles of Nuclear Magnetic Relaxation*. Clarendon Press, Oxford.
Berndt, K.D., Güntert, P., Orbons, L.P.M. and Wüthrich, K. (1992) *J. Mol. Biol.*, **227**, 757–775.
Blicharski, J.S. (1972) *Acta Phys. Pol. A*, **41**, 223–236.
Boyd, J., Hommel, U. and Campbell, I.D. (1990) *Chem. Phys. Lett.*, **175**, 477–482.
Brüschweiler, R.P. (1992) Ph.D. thesis No. 9466, ETH Zürich.
Clore, G.M., Driscoll, P.C., Wingfield, P.T. and Gronenborn, A.M. (1990) *Biochemistry*, **29**, 7387–7401.
DeMarco, A. and Wüthrich, K. (1976) *J. Magn. Reson.*, **24**, 201–204.
Deisenhofer, J. and Steigemann, W. (1975) *Acta Cryst.*, **B31**, 238–250.
Deverell, C., Morgan, R.E. and Strange, J.H. (1970) *Mol. Phys.*, **18**, 553–559.
Eccles, C., Güntert, P., Billeter, M. and Wüthrich, K. (1991) *J. Biomol. NMR*, 1, 111–130.
Goldman, M. (1984) *J. Magn. Reson.*, **60**, 437–452.
Glushka, J., Lee, M., Coffin, S. and Cowburn, D. (1989) *J. Am. Chem. Soc.*, **111**, 7716–7722.
Gutowsky, H.S. and Saika, A. (1953) *J. Chem. Phys.*, **21**, 1688–1694.
Hiyama, Y., Niu, C., Silverton, J.V., Bavoso, A. and Torchia, D.A. (1988) *J. Am. Chem. Soc.*, **110**, 2378–2383.
Jones, G.P. (1966) *Phys. Rev.*, **148**, 332–335.
Kaplan, J.I. and Fraenkel, G. (1980) *NMR of chemically exchanging systems*. Academic Press, New York.
Kay, L.E., Jue, T., Bangerter, B. and Demou, P.C. (1987) *J. Magn. Reson.*, **73**, 558–564.
Kay, L.E., Torchia, D.A. and Bax, A. (1989) *Biochemistry*, **28**, 8972–8979.
Kay, L.E., Nicholson, L.K., Delaglio, F., Bax, A. and Torchia, D.A. (1992) *J. Magn. Reson.*, **97**, 359–375.
Keiter, E.A. (1986) Ph.D. Thesis, University of Illinois.
Kördel, J., Skelton, N.J., Akke, M., Palmer III, A.G. and Chazin, W.J. (1992) *Biochemistry*, **31**, 4856–4866.

Lipari, G. and Szabo, A. (1982) *J. Am. Chem. Soc.*, **104**, 4546–4559 and 4560–4570.

Marion, D. and Wüthrich, K. (1983) *Biochem Biophys. Res. Comm.*, **113**, 967–974.

Maudsley, A.A., Müller, L. and Ernst, R.R. (1977) *J. Magn. Reson.*, **28**, 463–469.

Messerle, B.A., Wider, G., Otting, G., Weber, C. and Wüthrich, K. (1989) *J. Magn. Reson.*, **85**, 608–613.

Nirmala, N.R. and Wagner, G. (1988) *J. Am. Chem. Soc.*, **110**, 7557–7558.

Nirmala, N.R. and Wagner, G. (1989) *J. Magn. Reson.*, **82**, 659–660.

Oppenheim, I., Shuler, K.E. and Weiss, G.H. (1977) *Stochastic Processes in Chemical Physics: The Master Equation*, MIT Press, Cambridge.

Otting, G. and Wüthrich, K. (1988) *J. Magn. Reson.*, **76**, 569–574.

Palmer III, A.G., Skelton, N.J., Chazin, W.J., Wright, P.E. and Rance, M. (1992) *Mol. Phys.*, **75**, 699–711.

Peng, J.W., Thanabal, V. and Wagner, G. (1991a) *J. Magn. Reson.*, **94**, 82–100.

Peng, J.W., Thanabal, V. and Wagner, G. (1991b) *J. Magn. Reson.*, **95**, 421–427.

Peng, J.W. and Wagner, G. (1992) *J. Magn. Reson.*, **98**, 308–332.

Press, W.H., Flannery, B.P., Teukolsky, S.A. and Vetterling, W.T. (1986) *Numerical Recipes*, Cambridge University Press, Cambridge.

Richarz, R., Nagayama, K. and Wüthrich, K. (1980) *Biochemistry*, **19**, 5189–5196.

Sandström, J. (1982) *Dynamic NMR spectroscopy*. Academic Press, New York.

Schneider, D.M., Dellwo, M.J. and Wand, A.J. (1992) *Biochemistry*, **31**, 3645–3652.

Sklenar, V., Torchia, D. and Bax, A. (1987) *J. Magn. Reson.*, **73**, 375–379.

Shaka, A.J., Keeler, J., Frenkiel, T. and Freeman, R. (1983) *J. Magn. Reson.*, **52**, 335–338.

Shaka, A.J., Lee, C.J. and Pines, A. (1988) *J. Magn. Reson.*, **77**, 274–293.

Stilbs, P. and Moseley, M.M. (1978) *J. Magn. Reson.*, **31**, 55–61.

Stone, M.J., Fairbrother, W.J., Palmer III, A.G., Reizer, J., Saier, M.H. and Wright, P.E. (1992) *Biochemistry*, **31**, 4394–4406.

Wagner, G. and Wüthrich, K. (1982) *J. Mol. Biol.*, **155**, 347–366.

Wagner, G. and Nirmala, N.R. (1989) *Chemica Scripta*, **29A**, 27–30.

Wagner, G., Braun, W., Havel, T.F., Schaumann, T., Gō, N. and Wüthrich, K. (1987) *J. Mol. Biol.*, **196**, 611–639.

Wider, G., Neri, D. and Wüthrich, K. (1991) *J. Biomol. NMR*, **1**, 93–98.

Reprinted from Biochemistry, 1994, *33*. 9303
Copyright © 1994 by the American Chemical Society and reprinted by permission of the copyright owner.

Transient Hydrogen Bonds Identified on the Surface of the NMR Solution Structure of Hirudin†

Thomas Szyperski,‡ Walfrido Antuch,‡ Martin Schick,‡ Andreas Betz,§ Stuart R. Stone,§ and Kurt Wüthrich*,‡

Institut für Molekularbiologie und Biophysik, Eidgenössische Technische Hochschule-Hönggerberg, CH-8093 Zürich, Switzerland, and Department of Haematology, University of Cambridge, MRC Centre, Hills Road, Cambridge CB2 2QH, U.K.

Received April 15, 1994®

ABSTRACT: Recombinant desulfatohirudin retains largely the thrombin-inhibitory activity of natural hirudin from *Hirudo medicinalis* and causes at most minimal immune response in humans. With regard to potential pharmaceutical applications it is of interest to further investigate the structural basis of hirudin functions. In this paper transient hydrogen bonds between backbone amide protons and side-chain carboxylates on the protein surface of desulfatohirudin (variant 1) have been identified using two-dimensional ^{1}H NMR experiments and site-directed mutagenesis. The analysis of pH titration curves measured with NMR enabled the determination of the p*K* values of all 13 carboxylates, and downfield shifts larger than 0.2 ppm arising from weak bonding interactions with carboxylates were observed for the amide protons of Gly 25, Ser 32, Glu 35, and Cys 39. For these backbone amide protons virtually identical titration parameters were observed in intact desulfatohirudin and the mutant, truncated hirudin(1–51), demonstrating that the hydrogen bond acceptors are located in the N-terminal polypeptide segment 1–51. The hydrogen bonds Gly 25 NH–Glu 43 δCOO^-, Ser 32 NH–Glu 35 δCOO^-, Glu 35 NH–Asp 33 γCOO^-, Glu 35 NH–Glu 35 δCOO^-, and Cys 39 NH–Glu 17 δCOO^- were identified by considering spatial proximity in the NMR solution structure of hirudin(1–51), and comparing the p*K* values for the amide protons and the carboxylates in desulfatohirudin and the mutants hirudin(E43Q), hirudin(E35Q), hirudin(D33N) and hirudin(E17A). Comparative structure calculations with and without distance constraints for these hydrogen bonds showed that although they are all compatible with the NMR solution structure, these hydrogen bonds are transient dynamic features of the protein surface which, with the sole exception of Cys 39 NH–Glu 17 δCOO^-, would not have been detected in a conventional NMR structure determination. Of special interest is the clear-cut information obtained on the fact that the lifetimes of the dynamic "bifurcated" hydrogen-bonding interactions of the amide proton of Glu 35 are in the millisecond time range or shorter.

Hirudin variant 1 (HV1;[1] Scharf et al., 1989) is a small protein of 65 amino acid residues occurring in the salivary glands of the leech *Hirudo medicinalis*. It is the most potent known inhibitor of the blood clotting enzyme thrombin, which is a serine protease that plays a central role in the pathology of thrombotic diseases (Johnson et al., 1989). HV1 contains a sulfated tyrosyl residue in position 63 (Badgy et al., 1976; Dodt et al., 1984), while recombinant desulfatohirudin lacks this posttranslational modification. Nevertheless, desulfatohirudin binds to thrombin nearly as tightly as natural hirudin (Stone & Hofsteenge, 1986) and causes no or only minimal immune response in humans (Close et al., 1994), which makes it an attractive polypetide for therapeutic applications in cardiovascular medicine (Lent, 1986; Märki & Wallis, 1990). Notwithstanding current limitations of our knowledge on structural features that determine immune response (Langone, 1989), future deeper insights will undoubtedly depend on the availability of detailed descriptions of the protein surfaces which constitute the epitopes that are recognized by the immune system. Further investigation of the protein surface of hirudin is thus clearly of direct interest, and in this paper we describe the characterization of transient hydrogen bonds on the surface of hirudin in aqueous solution by NMR techniques.

The NMR solution structure of desulfatohirudin contains a globular amino-terminal domain of residues 1–48 and a flexibly disordered carboxy-terminal tail of residues 49–65 (Folkers et al., 1989; Haruyama & Wüthrich, 1989). Desulfatohirudin contains 13 carboxylates (eight glutamates, four aspartates, and the C-terminal carboxylate), of which seven are located in the C-terminal segment 53–65. In earlier work (Haruyama et al., 1989), pH titration shifts of amide protons were detected which must be due to hydrogen-bonding interactions with some of these carboxylate groups (Bundi & Wüthrich, 1977, 1979). As was previously demonstrated in model peptides (Bundi & Wüthrich, 1977, 1979) as well as in proteins (Ebina & Wüthrich, 1984; O'Connell et al., 1993; Steinmetz et al., 1988), amide proton titration shifts may manifest transient hydrogen bonds with carboxylate groups that would not be identified in a convential NMR structure determination based on measurements of NOEs and spin–spin coupling constants. In desulfatohirudin, there is *a priori*

† This work was supported by the Schweizerischer Nationalfonds (Project 31.32033.91), MRC (S.R.S.), and by a Bundesstipendium to W.A.

‡ Eidgenössische Technische Hochschule-Hönggerberg.

§ University of Cambridge.

® Abstract published in *Advance ACS Abstracts*, July 1, 1994.

[1] Abbreviations: HV1, hirudin variant 1; desulfatohirudin, recombinant hirudin HV1; hirudin(1–51), N-terminal 51-residue polypeptide segment of desulfatohirudin; hirudin(E35Q), desulfatohirudin with Glu 35 replaced by Gln; hirudin(E17A), desulfatohirudin with Glu 17 replaced by Ala; hirudin(D33N), desulfatohirudin with Asp 33 replaced by Asn; hirudin-(E43Q), desulfatohirudin with Glu 43 replaced by Gln; NMR, nuclear magnetic resonance; 2D, two-dimensional; P. COSY, 2D purged correlation spectroscopy; TOCSY, 2D total correlation spectroscopy; NOE, nuclear Overhauser enhancement; NOESY, 2D nuclear Overhauser enhancement spectroscopy; TPPI, time-proportional phase incrementation; TSP, [2,2,3,3-$^2H_4$]trimethylsilyl)propionate; ppm, parts per million; δ_{HA}, chemical shift in ppm of the fully protonated state; δ_{A^-}, chemical shift in ppm of the fully deprotonated state; δ(pH), experimental chemical shift in ppm at a specified pH value; p*K*, negative logarithm of the acid dissociation constant.

0006-2960/94/0433-9303$04.50/0 © 1994 American Chemical Society

Table 1: Survey of 2D ^{1}H NMR Spectra Recorded (T = 22 °C)

experiment[a]	τ_m[b]	data matrix[c]	t_{1max}[d]	t_{2max}[d]	τ_{tot}[e]	resolution[f]
		6 mM desulfatohirudin/1.77, 2.14, 2.60, 3.30, 3.73, 4.10, 4.41, 4.77, 5.12, 5.75, 6.75[g]				
NOESY	80	230 × 2048	38	336	12	6.0/1.5
		6 mM desulfatohirudin/2.60, 2.80, 3.50, 3.95, 4.35, 5.30[g]				
clean-TOCSY	100	320 × 1024	57	182	8	5.5/2.7
		3 mM hirudin(E17A)/2.61, 3.21, 3.73, 4.13, 5.62, 6.48[g]				
NOESY	100	256 × 1024	42	168	9	6.0/3.0
		3 mM hirudin(E35Q)/2.02, 2.97, 3.55, 3.73, 4.23, 4.74, 5.06, 6.44[g]				
NOESY	100	195 × 1024	33	170	9	6.0/3.0
		3 mM hirudin(E43Q)/2.17, 3.86, 4.26, 4.62, 4.97, 6.43[g]				
NOESY	100	210 × 1024	37	180	9	5.6/2.7
		3 mM hirudin(D33N)/2.15, 2.97, 3.44, 3.99, 4.45, 5.02, 6.34[g]				
NOESY	100	500 × 1024	88	180	11.5	5.6/2.7
		2 mM hirudin(1–51)/2.60, 2.80, 3.50, 3.95, 4.35, 5.30[g]				
P.COSY		256 × 2048	42	336	13.5	6.0/1.5

[a] References for the experimental schemes used are as follows: P. COSY, Marion and Bax (1988); clean-TOCSY, Griesinger et al. (1988); NOESY, Anil-Kumar et al. (1980). [b] Mixing time in milliseconds. [c] Size of the acquired data matrices in complex points. [d] Maximal t_1 and t_2 values in milliseconds. [e] Approximate total measuring time per spectrum in hours. [f] Spectral resolution along ω_1/ω_2 after zero-filling, in Hz/point. [g] Sample/pH values.

the intriguing question to be answered whether such hydrogen-bonding contacts are exclusively between atom groups within the globular domain of residues 1–48 or also between amide protons in the structured domain and the numerous carboxylates in the C-terminal flexible tail. From the earlier studies with desulfatohirudin (Haruyama et al., 1989), some long-range interactions with residues near the C-terminus were implicated, but these conclusions had to be based on limited experimental data. In particular, the three-dimensional structure available for intact desulfatohirudin (Folkers et al., 1989; Haruyama & Wüthrich, 1989) was not of very good quality because of interference of NMR signals from the flexible tail with the data collection for the globular domain. Spectral overlap also prevented measurements of some of the 13 carboxylate pH titration curves over a sufficiently large range for a reliable determination of the pK values. In the meantime, a high-quality NMR structure was determined for the truncated hirudin(1–51) (Szyperski et al., 1992). In addition, as described in this paper, a sufficient selection of mutant proteins with single-residue replacements was prepared to resolve all ambiguities in the identification of amide proton–carboxylate hydrogen bonds, so that a complete description of this type of transient surface side-chain–backbone interactions can now be presented.

MATERIALS AND METHODS

Desulfatohirudin variant 1 (Scharf et al., 1989) was used as it was given to us by Ciba-Geigy AG, Basel, Switzerland (Meyhack et al., 1987; Grossenbacher et al., 1987). Hirudin-(1–51) and the mutants hirudin(E43Q), hirudin(E17A), hirudin(D33N), and hirudin(E35Q) were produced with recombinant DNA techniques as described by Dennis et al. (1990) and Braun et al. (1988).

NMR Spectroscopy. 2D ^{1}H NMR experiments were recorded on a Bruker AM500 spectrometer at 500 MHz and at 22 °C in the pure-phase absorption mode using TPPI (Marion & Wüthrich, 1983). The protein solutions were prepared in a mixed solvent of 90% H_2O and 10% ^2H_2O, and the pH value was adjusted by adding small amounts of HCl and NaOH. The proton chemical shifts were calibrated relative to TSP at 0.00 ppm. The spectra were processed on a Bruker X32 workstation using the program UXNMR. The residual water signal after preirradiation was further reduced using the convolution method of Marion et al. (1989). Before Fourier transformation, the time domain data were zero-filled and multiplied with shifted sine-bell windows (DeMarco & Wüthrich, 1976). The spectra were baseline-corrected using third-order polynomials. Further experimental details are given in Table 1.

Identification of Hydrogen Bonds with pH Titration Experiments. The protocol to identify hydrogen-bonding interactions between amide protons and carboxylates (Bundi & Wüthrich, 1977, 1979) is based on changing the pH from acidic to neutral. The sign of the pH variation in chemical shift indicates when a certain amide proton is involved in such a hydrogen bond; *i.e.*, upfield shifts are mediated via covalent bonds for the amide protons of Asp residues and the C-terminal amino acid, and downfield shifts are observed for amide protons that are hydrogen bonded to carboxylates. Obviously, the pH titration curves of the amide protons and the intrinsic titration curves of the interacting carboxylates, as observed on the γ-protons of Glu, the β-protons of Asp, or the α-proton of the C-terminal residue, must correspond to identical pK values. Provided that electrostatic interferences with other ionizable groups are negligible, the pH titration curves for interacting amide protons and carboxylate groups can be parametrized with a single pK value. In a globular protein, identity of the pK values for an amide proton and the corresponding carboxylate can be used as a necessary condition for identifying a hydrogen-bonding interaction, provided that the global conformation does not vary with pH over the range used. Having a sufficiently well resolved structural description of the protein at hand, it may then be possible to uniquely identify pairs of carboxylate groups and amide protons in spatial proximity that have identical pK values. Residual ambiguities due to, for example, degeneracy of two or multiple carboxylate pK values in the same protein can be resolved by site-directed mutagenesis. Any point mutation that replaces a carboxylate without changing the global protein structure leads to the disappearance of all amide proton downfield shifts arising from hydrogen bonding with this carboxylate and thus provides the desired identification of the acceptor residue.

Quantitative Data Analysis. Titration parameters were obtained by nonlinear least-squares fits of the one-proton titration curve (eq 1),

$$\delta\,(\mathrm{pH}) = \frac{\delta_{\mathrm{HA}} + \delta_{\mathrm{A^-}} 10^{\mathrm{pH}-\mathrm{p}K}}{1 + 10^{\mathrm{pH}-\mathrm{p}K}} \qquad (1)$$

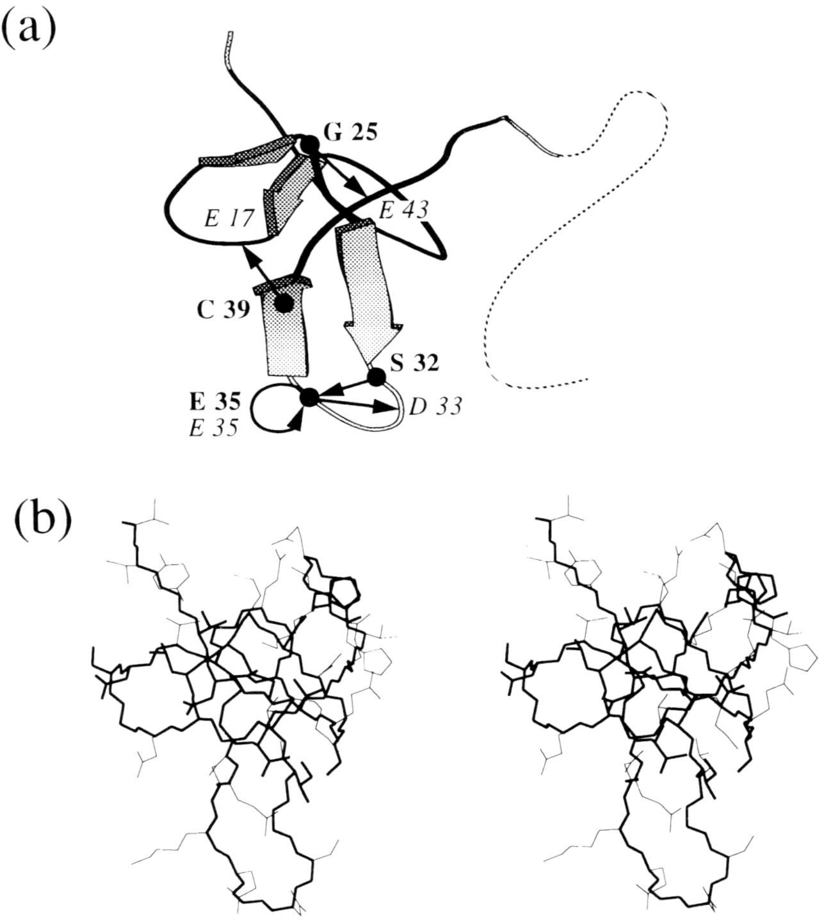

FIGURE 1: (a) Ribbon drawing of the NMR solution structure of desulfatohirudin variant 1 generated with the program MOLSCRIPT (Kraulis, 1991) using the conformer of hirudin(1–51) with the lowest DIANA target function value (Szyperski et al., 1992) and the knowledge that segment 52–65 is flexibly disordered in solution (Folkers et al., 1989; Haruyama & Wüthrich, 1989). The arrowed ribbons indicate position and direction of the β-sheet strands. A black rope represents the well-structured loop areas. Residues 1, 2, 32–36, and 49–51, which are only poorly constrained by the NMR data for hirudin(1–51), are indicated by a white rope, and residues 52–65, by a dotted line. The locations of backbone amide protons with $\Delta\delta(H^N) > 0.2$ ppm are identified by black circles and bold lettering, using the one-letter amino acid code and the sequence position. Arrows point to the C^α positions of the hydrogen bond acceptors identified by italic lettering. (b) Stereoview of an all-heavy-atom presentation of the hirudin(1–51) NMR conformer with the lowest DIANA target function value (Szyperski et al., 1992) in the same orientation as in (a). The backbone and the 19 best-defined side chains (amino acid residues 4–7, 9, 12, 14–16, 19, 20, 22, 26, 28, 29, 39, 40, 46, 48) are displayed with bold lines, and the other side chains, with thin lines.

to the experimental chemical shift, δ(pH), where δ_{HA} is the chemical shift in the acidic pH limit, while δ_{A^-} represents the chemical shift in the basic pH limit. Equation 1 can be rearranged to

$$\log\left[\frac{\delta\,(\mathrm{pH}) - \delta_{HA}}{\delta_{A^-} - \delta\,(\mathrm{pH})}\right] = \mathrm{pH} - \mathrm{p}K \qquad (2)$$

In order to obtain the standard deviation, σ, of the pK value, we introduced δ_{HA} and δ_{A^-} as obtained from the nonlinear fit of eq 1 into eq 2 and performed a linear fit of eq 2 to the experimental data. The overlap of the 99% confidence intervals of the pK values measured for the amide proton and the interacting carboxylate was used as a necessary condition for the formation of a hydrogen bond.

Calculations of the complete three-dimensional structure using the input data set described by Szyperski et al. (1992) supplemented by distance constraints for the presently identified hydrogen bonds were performed for hirudin(1–51) with the distance geometry program DIANA (Güntert et al., 1991). The distance between an amide proton and the oxygen atom of the interacting carboxylate was restricted to 1.8–2.0 Å, and the distance between the nitrogen atom of the amide group and the oxygen atom of the carboxylate was restricted to 2.7–3.0 Å (Williamson et al., 1985).

RESULTS

The molecular structure of desulfatohirudin (Figure 1), which represents the scaffold for the present investigation of transient features of the protein surface, includes a globular domain, with two antiparallel β-sheets (residues 14–16 and 20–22, and 27–31 and 36–40), three reverse turns (residues 8–11, type II; 17–20, type II′; and 23–26, type II), and a short stretch of a polyproline helix II (Sasisekharan, 1959) (residues 46–48), and the flexibly disordered segment 49–65. The NMR structure determination of hirudin(1–51) (Szyperski et al.,

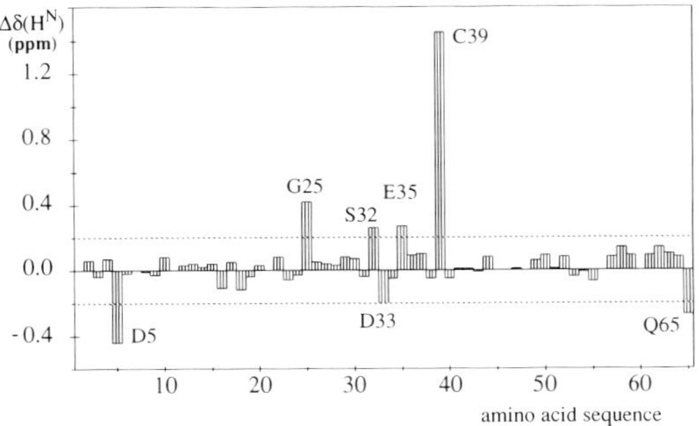

FIGURE 2: Plot of the titration shifts in ppm, $\Delta\delta(H^N)$, measured for the backbone amide protons of desulfatohirudin *versus* the amino acid sequence, which is VVYTD CTESG QNLCL CEGSN VCGQG NKCIL GSDGE KNQCV TGEGT PKPQS HNDGD FEEIP EEYLQ. $\Delta\delta(H^N) = \delta(H^N, pH = 1.77) - \delta(H^N, pH = 6.75)$, and amino acid residues with $\Delta\delta(H^N) > |0.2|$ ppm are labeled.

1992) revealed that removal of the carboxy-terminal polypeptide segment 52–65 does not noticeably affect the static and dynamic properties of the amino-terminal domain; the loop of residues 31–36 and the N-terminal dipeptide segment are flexibly disordered in both the intact and the truncated protein. For the present project, it is of particular relevancy that 12 of the 13 acidic groups in recombinant hirudin exhibit flexibly disordered conformations that allow for transient interactions with backbone amide protons, the only well-structured carboxylate being that of Asp 5.

NMR Experiments. The identification of the hydrogen-bonding partners described below depended importantly on detailed investigations of the pH dependence of their NMR parameters over the pH range 2–6 (Bundi & Wüthrich, 1977, 1979). An overview of all NMR experiments performed is afforded by Table 1. Phase-sensitive NOESY spectra (Anil-Kumar et al., 1980) at variable pH values were recorded for desulfatohirudin and for its mutants hirudin(E43Q), hirudin(E17A), hirudin(D33N), and hirudin(E35Q). Due to spectral overlap it was not possible to monitor the chemical shifts of the γ-protons of Glu 62 over a wide range of pH values in these spectra; therefore an additional series of clean-TOCSY spectra (Griesinger et al., 1988) were acquired for desulfatohirudin. The pH titration shifts of the amide proton resonances in hirudin(1–51) were followed by P.COSY spectra (Marion et al., 1988). In total, 50 2D NMR experiments were performed, with a total measurement time of about 21 days for the present investigation (Table 1). Systematic shifts of proton resonances that might occur due to variations in ionic strength or protein concentration were assessed from the observation of the proton resonances of the methyl groups of Val 21, Leu 15, and Leu 30. These chemical shifts remained constant within 0.01 ppm over the entire pH range from 2 to 6, so that global conformational rearrangements over this pH range could be excluded. This was further confirmed by the analysis of numerous medium- and long-range NOEs.

Identification of Amide Protons Acting as Hydrogen Bond Donors. Figure 2 presents a survey of the chemical shift variations of the backbone amide protons of recombinant hirudin, $\Delta\delta(H^N)$, when going from an acidic (pH = 1.77) to a neutral milieu (pH = 6.75). Large downfield shifts ($\Delta\delta(H^N) > 0.2$ ppm) indicative of through-space interactions with carboxylates are observed for the amide protons of Gly 25 ($\Delta\delta(H^N) = 0.42$ ppm), Ser 32 ($\Delta\delta(H^N) = 0.26$ ppm), Glu 35 ($\Delta\delta(H^N) = 0.27$ ppm), and Cys 39 ($\Delta\delta(H^N) = 1.45$ ppm). The corresponding titration parameters that were obtained

Table 2: Titration Parameters for Amide Protons with Titration Shifts Exceeding 0.2 ppm and for Side-Chain Protons of Glu and Asp in Desulfatohirudin and Hirudin Mutants at $T = 22$ °C

residue[a]	proton	protein[b]	δ_{HA}[c] (ppm)	$\Delta\delta$[c] (ppm)	p$K \pm 2.5\sigma$[d]
Asp 5	β^1		2.80	–0.32	4.25 ± 0.09
	β^2		2.91	–0.26	4.28 ± 0.13
	H^N		9.02	–0.44	4.32 ± 0.09
Glu 8	γ		2.48	–0.18	4.28 ± 0.17
Glu 17	γ^2		**2.48**	**–0.23**	**3.79 ± 0.17**
Gly 25	H^N		**8.72**	**0.42**	**4.25 ± 0.07**
	H^N	**hirudin(1–51)**	**8.69**	**0.47**	**4.29 ± 0.20**
	H^N	**hirudin(E43Q)**	**8.86**	**–0.01**	
Ser 32	H^N		**8.18**	**0.26**	**4.31 ± 0.07**
	H^N	**hirudin(1–51)**	**8.20**	**0.24**	**4.33 ± 0.20**
	H^N	**hirudin(E35Q)**	**8.18**	**–0.02**	
Asp 33	β^1		**2.85**	**–0.20**	**4.24 ± 0.17**
	β^2		**3.14**	**–0.29**	**4.12 ± 0.05**
	H^N		9.10	–0.20	3.93 ± 0.10
	β^2	**hirudin(E35Q)**	**3.16**	**–0.30**	**3.86 ± 0.10**
Glu 35	H^N		**7.64**	**0.27**	**4.00 ± 0.10**
	H^N	**hirudin(1–51)**	**7.65**	**0.27**	**3.87 ± 0.17**
	H^N	**hirudin(D33N)**	**7.63**	**0.14**	**4.32 ± 0.12**
Gln 35	H^N	**hirudin(E35Q)**	**7.66**	**0.15**	**3.91 ± 0.11**
Glu 35	γ^1		**2.60**	**–0.31**	**4.32 ± 0.07**
	γ^2		**2.65**	**–0.21**	**4.36 ± 0.07**
	γ^1	**hirudin(D33N)**	**2.58**	**–0.24**	**4.26 ± 0.10**
	γ^2	**hirudin(D33N)**	**2.63**	**–0.15**	**4.36 ± 0.12**
Cys 39	H^N		**8.80**	**1.45**	**3.76 ± 0.07**
	H^N	**hirudin(1–51)**	**8.79**	**1.41**	**3.78 ± 0.15**
	H^N	**hirudin(E17A)**	**9.10**	**–0.04**	
Glu 43	γ^1		**2.53**	**–0.29**	**4.25 ± 0.11**
	γ^2		**2.62**	**–0.23**	**4.24 ± 0.07**
Asp 53	β^1		2.90	–0.25	3.78 ± 0.15
	β^2		2.97	–0.26	3.79 ± 0.15
Asp 55	β^1		2.78	–0.25	4.07 ± 0.07
	β^2		2.86	–0.25	4.17 ± 0.06
Glu 57	γ		2.36	–0.17	4.63 ± 0.10
Glu 58	β^2		2.07	–0.09	4.69 ± 0.05
Glu 61	γ		2.48	–0.19	4.48 ± 0.15
Glu 62	γ^1		2.30	–0.13	4.53 ± 0.12
	γ^2		2.34	–0.14	4.51 ± 0.16
Gln 65	α		4.29	–0.16	4.01 ± 0.10
	H^N		8.16	–0.36	3.78 ± 0.15

[a] Bold letting identifies residues involved in NH–COO^- hydrogen bonds. [b] No entry means desulfatohirudin. [c] $\Delta\delta = \delta_{A^-} - \delta_{HA}$, with δ_{A^-} and δ_{HA} as obtained from a nonlinear least-squares fit of eq 1 to the experimental data. [d] σ is the standard deviation for the determination of the pK value (see text).

by nonlinear least-squares fits of eq 1 to the experimental data (Table 2) coincide closely with those reported by Haruyama et al. (1989). The spatial locations of the backbone amide protons in the polypeptide chain are indicated in Figure 1a. The downfield shifts of all side-chain amide protons were found to be smaller than 0.1 ppm, suggesting that none of the side-chain amide moieties is involved in a significantly populated hydrogen bond with a carboxylate. Two additional intriguing features in the data of Figure 2 are that there are numerous residues in the flexible segment 49–65 with $\Delta\delta(H^N) = 0.1$–0.15 ppm, indicating formation of significantly populated transient hydrogen bonds, and that only three residues (Asp 5, Asp 33, and Gln 65) show the expected large intrinsic high-field shifts (Bundi & Wüthrich, 1977), whereas for Asp 53 and Asp 55 these are presumably offset by downfield shifts arising from hydrogen-bonding interactions.

Identification of Carboxylates Acting as Hydrogen Bond Acceptors. The titration parameters of all 13 carboxylate groups in desulfatohirudin are given in Table 2. In addition, Figure 3 affords a visual display of the distribution of pK values and their 99% confidence intervals, showing that the pK values are well dispersed. The small standard deviations reflect that all titration curves exhibit sigmoidal shapes (see Figures 4–7). Thus, electrostatic interference between different carboxylates, which would typically lead to nonsigmoidal

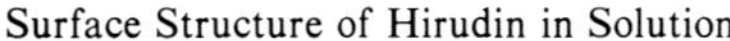

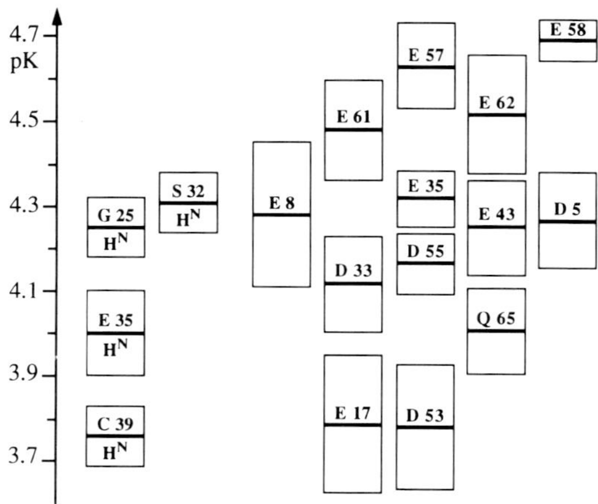

FIGURE 3: Graphical representation of the p*K* values of backbone amide protons with $\Delta\delta(H^N) > 0.2$ ppm (see Figure 2 and Table 2) and the p*K* values of the 13 carboxylates in desulfatohirudin (thick horizontal lines). For carboxylates where the p*K* value was determined by the observation of two protons, the mean value is displayed (see Table 2). The 99% confidence intervals (corresponding to 2.5σ, where σ is the standard deviation for the determination of the p*K* values) are represented by boxes.

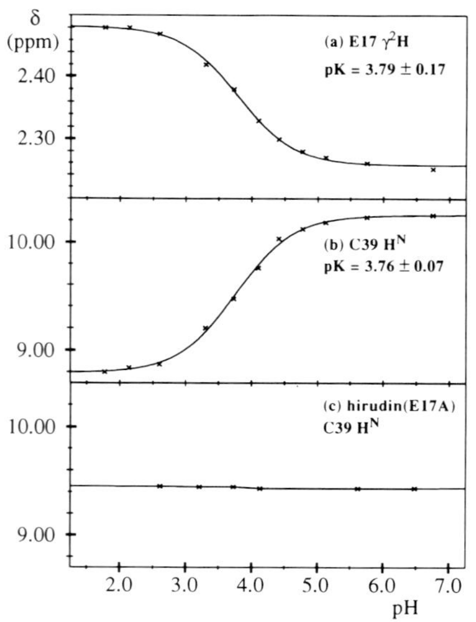

FIGURE 4: Plots of chemical shifts *versus* pH of (a) the γ^2 proton of Glu 17 in desulfatohirudin, (b) the amide proton of Cys 39 in desulfatohirudin, and (c) the amide proton of Cys 39 in hirudin-(E17A). The curves were determined by a nonlinear least-squares fit of eq 1 to the experimental data. The p*K* values are also given.

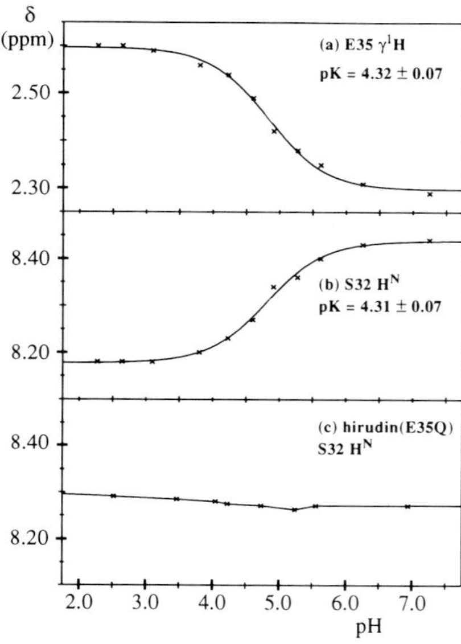

FIGURE 5: Same as Figure 4 for (a) the γ^1 proton of Glu 35 in desulfatohirudin, (b) the amide proton of Ser 32 in desulfatohirudin, and (c) the amide proton of Ser 32 in hirudin(E35Q).

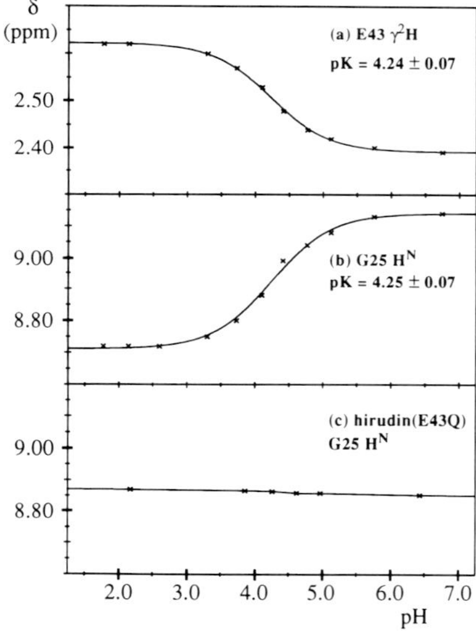

FIGURE 6: Same as Figure 4 for (a) the γ^2 proton of Glu 43 in desulfatohirudin, (b) the amide proton of Gly 25 in desulfatohirudin, and (c) the amide proton of Gly 25 in hirudin(E43Q).

titration curves, appears not to play a significant role, suggesting that the negative charges of the carboxylates are well screened due to the high dielectric constant of the aqueous solvent milieu with the elevated ionic strength arising from the high protein concentration used. Now that a complete data set is available, Figure 3 illustrates why unambiguous identification of the hydrogen bond acceptors cannot be based solely on the criterion that the 99% confidence intervals of the p*K* values of the interacting groups must overlap (Haruyama et al., 1989). Since the supplementary criterion of compatibility with the NMR structure of desulfatohirudin could not fully clarify the assignments, mainly because of the flexible nature of most carboxylate-bearing residues (see above and Figure 1b), we resorted to the use of mutant proteins tailored to resolve the remaining uncertainties.

A first important result was that the titration parameters of the backbone amide protons of Gly 25, Ser 32, Glu 35, and Cys 39 were virtually unchanged in hirudin(1–51) when compared with the values obtained for desulfatohirudin (Table 2), so that all carboxylates located in the flexible C-terminal tail comprising residues 49–65 can be excluded as potential hydrogen bond acceptors. Conclusive individual hydrogen bond assignments were then obtained by comparison of wild-type desulfatohirudin with mutants in which the most likely

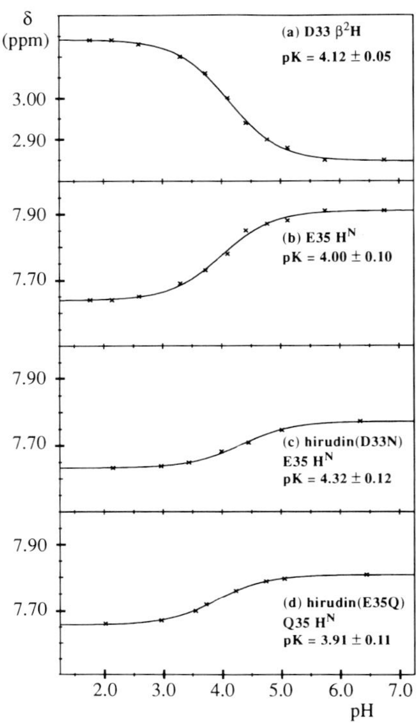

FIGURE 7: Same as Figure 4 for (a) the β^2 proton of Asp 33 in desulfatohirudin, (b) the amide proton of Glu 35 in desulfatohirudin, (c) the amide proton of Glu 35 in hirudin(D33N), and (d) the amide proton of Gln 35 in hirudin(E35Q).

acceptor groups were substituted by the corresponding amides or, in the case of Glu 17, by Ala. In all these mutants, chemical shifts of nonlabile protons and long-range NOEs were monitored over the pH range used to ascertain that observed variations in the titration parameters could not be attributed to global conformational rearrangements. Thus the interactions Cys 39 NH–Glu 17 δCOO$^-$ (Figure 4), Ser 32 NH–Glu 35 δCOO$^-$ (Figure 5), and Gly 25 NH–Glu 43 δCOO$^-$ (Figure 6) were clearly identified from the complete disappearence of the pH dependence of the chemical shifts in the mutant proteins (Table 2). A more complex situation is encountered for the backbone amide proton of Glu 35, since a residual downfield titration shift of $\Delta\delta(H^N) = 0.1$ ppm was observed in hirudin-(D33N) (Figure 7c), with a p*K* value close to that of the carboxylate of Glu 35 (Table 2). The downfield shift of the amide proton of Gln 35 in hirudin(E35Q) was also reduced to $\Delta\delta(H^N) = 0.14$ ppm (Figure 7d). Thus, the amide proton of Glu 35 interacts with both the carboxylate of Asp 33 and the carboxylate of its own side chain, while in turn the carboxylate of Glu 35 acts as a hydrogen bond acceptor for both the amide proton of Ser 32 and its own amide proton.

Structure Calculations with Hydrogen Bond Constraints. In order to test whether the presently identified hydrogen bonds are compatible with the input data set of 580 NOE upper distance constraints and 109 dihedral angle constraints used previously for the structure determination of hirudin-(1–51), we performed new distance geometry calculations using the program DIANA (Güntert et al., 1991) with supplementary distance constraints (Williamson et al., 1985) for the interresidual hydrogen bonds Gly 25 NH–Glu 43 δCOO$^-$, Ser 32 NH–Glu 35 δCOO$^-$, Glu 35 NH–Asp 33 γCOO$^-$, and Cys 39 NH–Glu 17 δCOO$^-$. Otherwise, the same protocol was used as reported by Szyperski et al. (1992). Compared to results obtained using the input without hydrogen bond constraints, the group of 20 conformers with the lowest residual DIANA target function values exhibited only slightly increased residual constraint violations, with target function values between 0.6 and 0.8 Å^2, as compared with 0.08–0.17 Å^2 (Szyperski et al., 1992). This clearly demonstrates that the hydrogen-bonding constraints are consistent with the constraints derived from ^{1}H–^{1}H NOEs and scalar spin–spin coupling constants. A visual impression of the hydrogen bonds is afforded by the stereo pictures in Figure 8, which were taken from the DIANA conformer with the lowest target function value (the structures of Figure 8 may be compared with the corresponding segments in the all-heavy-atom representation of hirudin(1–51) in Figure 1b, which was computed without supplementary hydrogen bond constraints). Clearly, the static view of Figure 8 is an oversimplified description of the protein surface, since we now know, for example, that the side chain of Glu 35 interacts with both the amide proton of Ser 32 and its own amide proton. Since only a single averaged resonance is seen for each of these two amide protons as well as for the Glu 35 side chain, these hydrogen bonds must be formed and cleaved in a rapid, dynamic equilibrium, with lifetimes of the order of milliseconds or shorter (Wüthrich, 1986).

DISCUSSION

Locations of the Amide Proton–Carboxylate Hydrogen Bonds. A survey of the locations of the hydrogen bonds in the desulfatohirudin structure is presented in Figure 1a. The hydrogen bond involving the carboxylate of Glu 17 and the backbone amide proton of Cys 39 connects the type II′ turn comprising residues 17–20 with the second strand of the β-sheet containing residues 27–31 and 36–40. These two secondary structure elements are locally and globally well defined in the NMR solution structure of hirudin(1–51) (see Figures 2 and 3 in Szyperski et al. (1992)), imply that the backbone atoms of Glu 17 and Cys 39 are in fixed relative orientations. Hence, the entropic expense for this hydrogen bond formation is restricted to the loss of the degrees of freedom about the side-chain torsion angles of Glu 17. The following three arguments indicate that this is a highly populated hydrogen bond. First, the downfield shift measured for the amide protons of Cys 39 (Figures 2 and 4) is more than 2 times larger than the largest previously observed values in peptides (Bundi & Wüthrich, 1977, 1979) or proteins (*e.g.*, Ebina and Wüthrich (1984), O'Connell et al. (1992), and Steinmetz et al. (1988)). Second, this hydrogen bond has been identified in 12 out of the 20 energy-refined DIANA conformers used to represent the NMR solution structure of hirudin(1–51) that was calculated without supplementary hydrogen bond constraints using the common distance and bond angle criteria for hydrogen bonds (Szyperski et al., 1992). Third, this hydrogen bond has also been observed in the X-ray crystal structure of the hirudin–thrombin complex (Rydel et al., 1990, 1991), where the carboxylate of Glu 17 forms both an intermolecular salt bridge to the guanidinium moiety of Arg 173 in thrombin and an intramolecular hydrogen bond with the amide proton of Cys 39.

The hydrogen bond involving the backbone amide proton of Gly 25 and the carboxylate of Glu 43 connects the type II turn comprising residues 23–26 with the polypeptide segment 41–45, which links the second strand of the β-sheet 27–31 and 36–40 with the short stretch of polyproline helix II containing

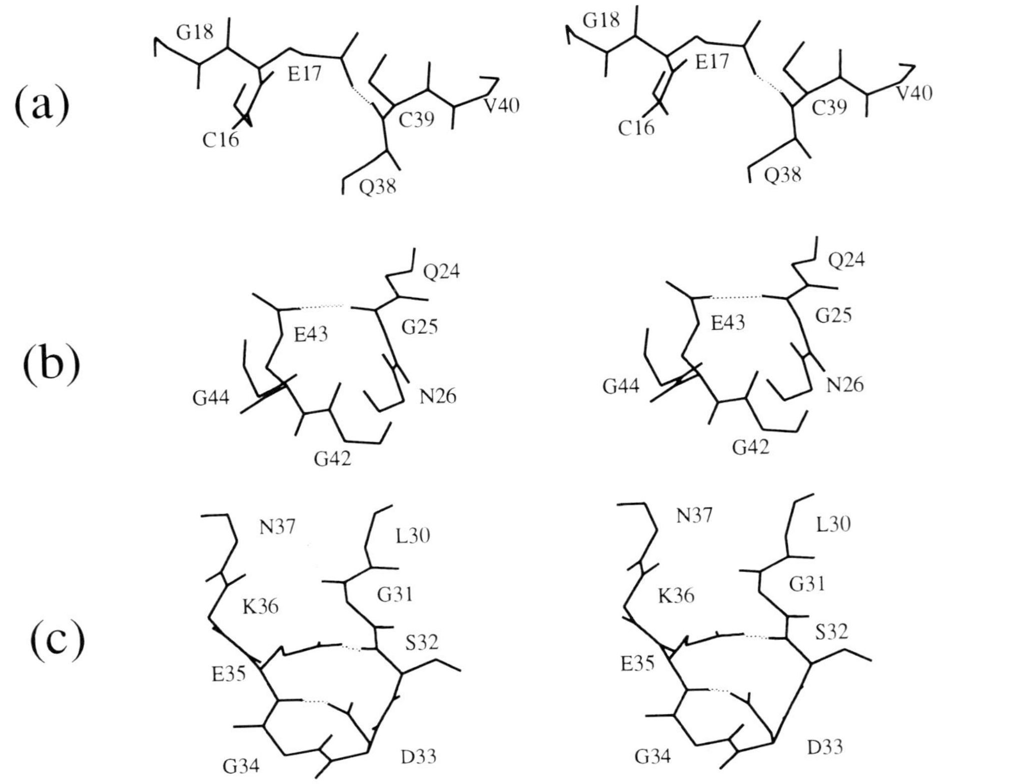

FIGURE 8: Stereo pictures of the hydrogen bonds (indicated as dotted lines) involving backbone amide protons and side-chain carboxylates in hirudin(1–51) as obtained from a structure calculation using hydrogen bond constraints (Williamson et al., 1985) in addition to the input of NOE upper distance constraints and dihedral angle constraints from spin-spin coupling constants (see text): (a) Cys 39 NH–Glu 17 δCOO⁻; (b) Gly 25 NH–Glu 43 δCOO⁻; (c) Ser 32 NH–Glu 35 δCOO⁻ and Glu 35 NH–Asp 33 γCOO⁻. The drawings were generated with the program XAM (Xia, 1992).

residues 46–48. While the type II turn is locally and globally well defined in the NMR solution structure of hirudin(1–51), the polypeptide segment 41–45 exhibits local conformational disorder (see Figures 2 and 3 in Szyperski et al. (1992)). The downfield shift of the backbone amide proton of Gly 25 is about 3 times smaller when compared with that of Cys 39, and the hydrogen bond was identified neither in the NMR solution structure of hirudin(1–51) (Szyperski et al., 1992) nor in the X-ray crystal structure of the hirudin–thrombin complex (Rydel et al., 1990, 1991), suggesting that the increased entropic expense for the fixation of the polypeptide segment 41–45 results in a reduced population of the hydrogen bond.

From entropy considerations, relatively low populations are expected also for the hydrogen bonds involving the backbone amide protons of Glu 35 and Ser 32 and the carboxylates of Glu 35 and Asp 33. All these residues are located in the flexibly disordered loop comprising residues 31–36. The amide proton downfield shifts are again smaller when compared with that of Gly 25, and none of these hydrogen bonds was identified in the NMR solution structure of hirudin(1–51) (Szyperski et al., 1992). In addition, the polypeptide segment 32–35 of hirudin in the X-ray crystal structure of the hirudin–thrombin complex exhibits no electron density, implicating structural disorder for these residues also in the crystals (Rydel et al., 1990, 1991).

It was previously suggested that the extent of the amide proton downfield titration shift might be used to estimate the population of the hydrogen bond with the carboxylate group (Bundi & Wüthrich, 1979). The resulting numbers must be used with care, however, since the shift will depend on the hydrogen bond geometry as well as on the population. If we assume that the shift of 1.45 ppm for Cys 39 reflects a population of 70–90%, we obtain the following approximate populations for the other interactions: Gly 25 NH–Glu 43 δCOO⁻, ~30%; Ser 32 NH–Glu 35 δCOO⁻, ~20%; Glu 35 NH–Asp 33 γCOO⁻, ~10%; Glu 35 NH–Glu 35 δCOO⁻, ~10%. The equilibrium populations of the hydrogen bonds appear to depend critically on the local mobilities of the interacting groups in the protein structure, since the enthalpic gain from hydrogen bond formation on the protein surface can apparently compensate the entropic expense for the fixation of a single side chain but not the entropic expense for immobilization of more extended polypeptide segments. This conclusion coincides with the observations of Dao-Pin et al. (1991) that engineered surface salt bridges between charged side chains do not contribute significantly to the protein stability, because the entropic cost of localizing a pair of solvent-exposed charged groups on the surface largely offsets the interaction energy expected from the formation of a defined salt bridge. Consistent with these general conclusions, no significantly populated hydrogen-bonding interactions between carboxylates and side-chain amide groups are implicated by the data on hirudin, suggesting that the enthalpic contribution of such an interaction is too small to act as a driving force for the formation of a defined hydrogen bond. It can thus be predicted that significantly populated hydrogen-bonding interactions between side-chain carboxylates and side-chain amide groups are scarce on protein surfaces in solution, whereas hydrogen bonds between side-chain carboxylates and backbone amide protons can be expected whenever the backbone atoms of the interacting residues are in a sterically favorable, fixed conformation. The hydrogen bond between the carboxylate of Glu 17 and the amide proton of Cys 39 appears to establish a paradigm for this latter situation which also indicates that in earlier work the populations of such hydrogen bonds were overestimated about 2-fold based on the extent of $\Delta\delta(H^N)$ (Bundi & Wüthrich, 1979).

Clearly, considerable effort was needed both for the preparation of NMR quantities of the mutant proteins and for the recording of the NMR data (Table 2). However, there is no alternative approach for obtaining the results described here. Although screening the NMR solution structure with the common bond length and bond angle criteria resulted in the identification of the secondary structure hydrogen bonds, the Cys 39 NH–Glu 17 δCOO^- bond was the only surface interaction thus found. The most important information is undoubtedly on the dynamic nature of the surface interactions. In all instances the spectra of the hydrogen-bonded and nonbonded forms of both the donor and acceptor groups are averaged on the chemical shift time scale (Wüthrich, 1986), which shows that the lifetimes of the hydrogen bonds are in the millisecond range or shorter. In particular, the amide proton of Glu 35 must exchange between two different hydrogen–bonded states with Glu 35 δCOO^- and Asp 33 γCOO^- and additional, nonbonded states, which presents a novel, dynamic picture of a "bifurcated" hydrogen bond. Although no detailed data are presently available, the sizeable downfield titration shifts of amide protons in the flexible C-terminal tail from residues 53–65 (Figure 2) would be compatible with the formation of possibly a multitude of rapidly interchanging hydrogen bonds with the numerous carboxylate groups in this polypeptide segment.

Finally, it may be added that the protocol of NMR pH titrations and comparison of single-residue mutations used here for desulfatohirudin and previously with model peptides (Bundi & Wüthrich, 1979) is generally applicable for studies of the surface structure in polypeptides and proteins that do not undergo global conformation changes over the pH range 2–6.

ACKNOWLEDGMENT

We thank Dr. P. Güntert for help with the structure calculations, Dr. M. Billeter for helpful discussions, Dr. H. Grossenbacher (Ciba-Geigy AG, Basel) for a generous gift of desulfatohirudin, and Mrs. E. Huber for the careful processing of the manuscript.

REFERENCES

Anil-Kumar, Ernst, R. R., & Wüthrich, K. (1980) *Biochem. Biophys. Res. Commun. 95*, 1–6.

Badgy, D., Barabas, E., Graf, L., Ellebaek, T., & Magnusson, S. (1976) *Methods Enzymol. 45*, 669–678.

Braun, P. J., Dennis, S., Hofsteenge, J., & Stone, S. R. (1988) *Biochemistry 27*, 6517–6522.

Bundi, A., & Wüthrich, K. (1977) *FEBS Lett. 77*, 11–14.

Bundi, A., & Wüthrich, K. (1979) *Biopolymers 18*, 299–311.

Close, P., Bichler, J., Kerry, R., et al. (1994) *Coronary Artery Disease* (in press).

Dao-Pin, S., Sauer, U., Nicholson, H., & Matthews, B. W. (1991) *Biochemistry 30*, 7142–7153.

DeMarco, A., & Wüthrich, K. (1976) *J. Magn. Reson. 24*, 201–204.

Dennis, S., Wallace, A., Hofsteenge, J., & Stone, S. R. (1990) *Eur. J. Biochem. 188*, 61–66.

Dodt, J., Müller, H. P., Seemüller, V., & Chang, J.-Y. (1984) *FEBS Lett. 165*, 180–184.

Ebina, S., & Wüthrich, K. (1984) *J. Mol. Biol. 179*, 283–288.

Folkers, P. J. M, Clore, G. M., Driscoll, P. C., Dodt, J., Köhler, S., & Gronenborn, A. M. (1989) *Biochemistry 28*, 2601–2617.

Griesinger, C., Otting, G., Wüthrich, K., & Ernst, R. R. (1988) *J. Am. Chem. Soc. 110*, 7870-7872.

Grossenbacher, H., Auden, J. A. L., Bill, K., Liersch, M., & Maerki, W. E. (1987) 11th Congress on Thrombosis and Haemostasis, Brussels, July 11, 1987, Abstract 34.

Güntert, P., Braun, W., & Wüthrich, K. (1991) *J. Mol. Biol. 217*, 517–530.

Haruyama, H., & Wüthrich, K. (1989) *Biochemistry 28*, 4301–4312.

Haruyama, H., Qian, Y.-Q., & Wüthrich, K. (1989) *Biochemistry 28*, 4312–4317.

Johnson, P. H., Sze, P., Winant, R., Payne, P. W., & Lazar, J. B. (1989) *Semin. Thromb. Hemostasis 15*, 302–315.

Kraulis, P. J. (1991) *J. Appl. Crystallogr. 24*, 946–950.

Langone, J. J., Ed. (1989) *Methods in Enzymology*, Vol. 178, Academic Press, San Diego.

Lent, C. (1986) *Nature 323*, 494.

Marion, D., & Wüthrich, K. (1983) *Biochem. Biophys. Res. Commun. 113*, 967–974.

Marion, D., & Bax, A. (1988) *J. Magn. Reson. 80*, 528–533.

Marion, D., Ikura, K., & Bax, A. (1989) *J. Magn. Reson. 84*, 425–430.

Märki, W. E., & Wallis, R. B. (1990) *Thromb. Haemostasis 64*, 344–348.

Meyhack, B., Heim, J., Rink, H., Zimmermann, W., & Maerki, W. E. (1987) 11th Congress on Thrombosis and Haemostasis, Brussels, July 11, 1987, Abstract 33.

O'Connell, J. F., Bougis, P. E., & Wüthrich, K. (1993) *Eur. J. Biochem. 213*, 891–900.

Rydel, T. J., Ravichandran, K. G., Tulinsky, A., Bode, W., Huber, R., Roitsch, C., & Fenton, J. W., II (1990) *Science 249*, 277–280.

Rydel, T. J., Tulinsky, A., Bode, W., & Huber, R. (1991) *J. Mol. Biol. 221*, 583–601.

Sasisekharan, V. (1959) *Acta Crystallogr. 12*, 897–903.

Scharf, M., Engels, M., & Tripier, D. (1989) *FEBS Lett. 255*, 105–110.

Steinmetz, W. E., Bougis, P. E., Rochat, H., Redwine, O. D., Braun, W., & Wüthrich, K. (1988) *Eur. J. Biochem. 172*, 101–116.

Stone, S. R., & Hofsteenge, J. (1986) *Biochemistry 25*, 4622–4624.

Szyperski, T., Güntert, P., Stone. S. R., & Wüthrich, K. (1992) *J. Mol. Biol. 228*, 1193–1205.

Williamson, M. P., Havel, T. F., & Wüthrich, K. (1985) *J. Mol. Biol. 182*, 295–315.

Wüthrich, K. (1986) *NMR of proteins and nucleic acids*, Wiley, New York.

Xia, T. (1992) Ph.D. Thesis 9831, ETH Zürich, Switzerland.

VII

HYDRATION OF PROTEINS AND NUCLEIC ACIDS IN SOLUTION

Introduction to papers 55 through 60

Having devoted almost three years of my research activity to investigations of the hydration of metal ions and metal complexes (see Part I, paper **2**), I was highly motivated to extend these studies to metal centers in proteins and quite generally to biological macromolecules. After two decades with a variety of futile attempts to contribute to the characterization of hydration water associated with proteins, NMR structure determination in solution could be extended in 1989 to include the location of individual solvent molecules. Today, high resolution NMR observation of hydration water is a fascinating aspect of the solution structures of proteins and nucleic acids, and the techniques used for this purpose are adaptable also for studies of the solvation of biopolymers with non-aqueous components in mixed solvents, for example, denaturants.[1]

BPTI contains four interior hydration water molecules, which are an integral part of the molecular architecture and are inaccessible to the solvent in a rigid model of the three-dimensional structure in crystals as well as in solution. On grounds of principle, one would have anticipated that the four individual water molecules have different chemical shifts, which should also be different from the bulk water chemical shift. However, it turned out that all water molecules in an aqueous solution of BPTI are observed at the same chemical shift. Selective observation of hydration water (paper **55**) was eventually achieved with the use of a novel approach for the suppression of the bulk water signal (paper **33** and ref. 2). In paper **56** an inorganic shift reagent was

[1] Wüthrich, K. (1994) in *Toward a Molecular Basis of Alcohol Use and Abuse* (B. Jansson, H. Jörnvall, U. Rydberg, L. Terenius, B.L. Vallee, eds.) pp. 261–268. Basel: Birkhäuser Verlag. NMR, alcohols, protein solvation and protein denaturation.

[2] Otting, G., Liepinsh, E., Farmer II, B.T. and Wüthrich, K. (1991) *J. Biomol. NMR 1,* 209–215. Protein hydration studied with homonuclear 3D ^{1}H NMR experiments.

used to demonstrate that the degeneracy of the chemical shifts of bound water and bulk water is indeed due to rapid exchange in and out of the protein molecule, and to establish a lower limit on the exchange rate constant.

In NMR studies the location of the hydration waters in proteins is determined by the observation of NOEs between water protons and hydrogen atoms of the polypeptide chain. Because of the dependence of the NOE on the inverse sixth power of the ^{1}H–^{1}H distance, only water molecules in a first hydration layer are observed. For the hydration studies it is further of crucial importance that the NOE intensity is also related to a correlation function describing the stochastic modulation of the dipole–dipole coupling between the interacting protons. This correlation function may be governed either by the Brownian rotational tumbling of the hydrated protein molecule, or by interruption of the dipolar interaction through translational diffusion of the interacting spins, whichever is faster. On this basis, the papers **57** and **58** showed that surface hydration of peptides and proteins is characterized by very short residence times of the water molecules in the hydration sites, in the range from about 20 to 300 picoseconds at 10° C, which is in excellent agreement with the results of long-time molecular dynamics simulations.[3]

The paper **59** describes NMR studies of the hydration of DNA duplexes in aqueous solution. These experiments showed that the formation of a "spine of hydration" in the minor groove, which had previously been characterized by X-ray diffraction in single crystals, is also manifested in the dynamic properties of the hydration water molecules. As paper **59** is the only reprint in this volume describing studies of a

[3] Brunne, R.M., Liepinsh, E., Otting, G., Wüthrich, K. and van Gunsteren, W.F. (1993) *J. Mol. Biol. 231,* 1040–1048. Hydration of proteins. A comparison of experimental residence times of water molecules solvating the bovine pancreatic trypsin inhibitor with theoretical model calculations.

nucleic acid free in solution, this seems the right moment to refer to some of our other projects with DNA, which include the development of a sequential assignment strategy based on heteronuclear ^{31}P-relayed ^{1}H–^{1}H correlation experiments,[4–6] work on assignment procedures for DNA duplexes that rely on observation of homonuclear NOEs,[7–10] and structure determinations of DNA duplexes[11,12] and drug–DNA com-

---

4 Pardi, A., Walker, R., Rapoport, H., Wider, G. and Wüthrich, K. (1983) *J. Am. Chem. Soc. 105,* 1652–1653. Sequential assignments for the ^{1}H and ^{31}P atoms in the backbone of oligonucleotides by two-dimensional nuclear magnetic resonance.

5 Neuhaus, D., Wider, G., Wagner, G. and Wüthrich, K. (1984) *J. Magn. Reson. 57,* 164–168. X-relayed ^{1}H–^{1}H correlated spectroscopy.

6 Frey, M.H., Leupin, W., Sørensen, O.W., Denny, W.A., Ernst, R.R. and Wüthrich, K. (1985) *Biopolymers 24,* 2371–2380. Sequence-specific assignment of the backbone ^{1}H- and ^{31}P-NMR lines in a short DNA duplex with homo- and heteronuclear correlated spectroscopy.

7 Chazin, W.J., Wüthrich, K., Hyberts, S., Rance, M., Denny, W.A. and Leupin, W. (1986) *J. Mol. Biol. 190,* 439–453. ^{1}H nuclear magnetic resonance assignments for d-$(GCATTAATGC)_2$ using experimental refinements of established procedures.

8 Wüthrich, K. (1986) *NMR of proteins and nucleic acids.* New York: Wiley, pp. 203–255.

9 Leupin, W., Wagner, G., Denny , W.A. and Wüthrich, K. (1987) *Nucl. Acids Res. 15,* 267–275. Assignment of the ^{13}C nuclear magnetic resonance spectrum of a short DNA-duplex with ^{1}H detected two-dimensional heteronuclear correlation spectroscopy.

10 Grütter, R., Otting, G., Wüthrich, K. and Leupin, W. (1988) *Eur. Biophys. J. 16,* 279–286. O_R3 operator of bacteriophage λ in a 23 base-pair DNA fragment: sequence-specific ^{1}H NMR assignments for the non-labile protons and comparison with the isolated 17 base-pair operator.

11 Celda, B., Widmer, H., Leupin, W., Chazin, W.J., Denny, W.A. and Wüthrich, K. (1989) *Biochemistry 28,* 1462–1471. Conformational studies of d-$(AAAAATTTTT)_2$ using constraints from nuclear Overhauser effects and from quantitative analysis of the cross-peak fine structures in two-dimensional ^{1}H nuclear magnetic resonance spectra.

12 Otting, G., Billeter, M., Wüthrich, K., Roth, H.J., Leumann, C. and Eschenmoser, A. (1993) *Helv. Chim. Acta 76,* 2701–2756. Warum Pentose- und nicht Hexose-Nucleinsäuren? 'Homo-DNS': ^{1}H-, ^{13}C-, ^{31}P- und ^{15}N-NMR-spektroskopische Untersuchung von ddGlc(A-A-A-A-A-T-T-T-T-T) in wässriger Lösung.

13 Leupin, W., Chazin, W.J., Hyberts, S., Denny, W.A. and Wüthrich, K. (1986) *Biochemistry 25,* 5902–5910. NMR studies of the complex between the decadeoxynucleotide d-$(GCATTAATGC)_2$ and a minor-groove-binding drug.

14 Leupin, W., Otting, G., Amacker, H. and Wüthrich, K. (1990) *FEBS Lett. 263,* 313–316. Application of $^{13}C(\omega_1)$-half-filtered [^{1}H,^{1}H]-NOESY for studies of a complex formed between DNA and a ^{13}C-labeled minor-groove-binding drug.

plexes.[13–15] The experience gained in these investigations was invaluable when we embarked on the aforementioned structure determination of a homeodomain–DNA complex (papers **41** and **42**). A detailed characterization of internal hydration of this complex showed that there is quite rapid exchange of water molecules between hydration sites in the protein–DNA interface and the bulk water, with lifetimes in the hydration sites in the range from milliseconds to nanoseconds. This observation presented novel insight into the nature of the intermolecular interactions that lead to specific DNA recognition.[16] The information obtained in this as well as in all other NMR studies of protein and nucleic acid hydration is genuinely different and complementary to that resulting from crystal diffraction experiments (paper **60**): As is by now well established, hydration water molecules continuously exchange in and out of particular hydration sites in proteins or nucleic acids in solution. NMR measures both the population of individual hydration sites and the life-times of the hydration water molecules in these sites with respect to exchange with the bulk water. Diffraction experiments also probe the total fraction of time that a particular hydration site is occupied by a water molecule, but in case there is exchange of hydration water molecules in and out of individual hydration sites in the crystals they will be insensitive to the residence time on any particular visit.

[15] Fede, A., Billeter, M., Leupin, W. and Wüthrich, K. (1993) *Structure 1,* 177–186. Determination of the NMR solution structure of the Hoechst 33258–d(GTGGAATTCCAC)$_2$ complex and comparison with the X-ray crystal structure.

[16] Qian, Y.Q., Otting, G. and Wüthrich, K. (1993) *J. Am. Chem. Soc. 115,* 1189–1190. NMR detection of hydration water in the intermolecular interface of a protein–DNA complex.

Reprinted from the Journal of the American Chemical Society, 1989, *111*, 1871.
Copyright © 1989 by the American Chemical Society and reprinted by permission of the copyright owner.

Studies of Protein Hydration in Aqueous Solution by Direct NMR Observation of Individual Protein-Bound Water Molecules†

Gottfried Otting and Kurt Wüthrich*

Contribution from the Institut für Molekularbiologie und Biophysik, Eidgenössische Technische Hochschule-Hönggerberg, CH-8093 Zürich, Switzerland. *Received June 24, 1988*

Abstract: Proton nuclear magnetic resonance was used to study individual molecules of hydration water bound to the protein basic pancreatic trypsin inhibitor (BPTI) in aqueous solution. The experimental observations are nuclear Overhauser effects (NOE) between protons of individual amino acid residues of the protein and those of sufficiently tightly bound water molecules. These NOEs were recorded by two-dimensional nuclear Overhauser enhancement spectroscopy (NOESY) in the laboratory frame, and by the corresponding experiment in the rotating frame (ROESY). The detection of NOEs with water protons was enabled by a solvent suppression technique which provides a uniform excitation profile in NOESY and ROESY, except at the ω_2 frequency of the water signal. From NOESY and ROESY spectra recorded at 5, 36, 50, and 68 °C, intermolecular $^1H-^1H$ NOEs between the protein and four water molecules buried in its interior were individually assigned, and additional NOEs between surface residues of the protein and labile protons with the chemical shift of the bulk water were identified. For the hydration waters that can be observed by NOEs at 36 °C, an upper limit for the proton-exchange rate with the bulk water is estimated to be 3×10^9 s^{-1}. These NOE-observable water molecules account only for a small percentage of the hydration waters seen in the crystal structure of BPTI. This observation supports the independently established picture of increased disorder near the molecular surface in protein structures in solution.

Three-dimensional protein structures can nowadays be determined either by X-ray diffraction in single crystals or by nuclear magnetic resonance (NMR)[1] in solution.[2] High-resolution crystal structures of globular proteins include typically numerous water molecules in defined hydration sites.[3,4] In aqueous solution, evidence for the presence of protein-bound hydration water could also be established, but the experiments used so far could not provide observations on individual molecules of hydration water. For example, measurements of the relaxation enhancement on the water signal in aqueous protein solutions led to the conclusion that the observed magnetization transfer is not only a consequence of chemical exchange between water protons and labile protons of the protein, but occurs also via cross relaxation at the water–protein interface.[5] In different experiments, pH-independent transfer of magnetization to slowly exchanging protein protons was observed upon irradiation of the water resonance in protein solutions.[6] In these experiments the observation of negative nuclear Overhauser effects (NOE)[6] showed that there must be hydration water bound with lifetimes that are comparable with the overall rotational correlation time of the hydrated protein, which is of the order of nanoseconds. Evidence for the presence, in certain proteins, of a small number of hydration waters that exchange with the bulk water on a time scale of seconds was deduced from ^{18}O tracer experiments.[7] None of these solution experiments could, however, provide information on the location of the hydration sites in the protein. The present paper describes a NMR technique that is capable of identifying individual molecules of hydration water and characterizing their binding sites on the protein molecule.

†This paper was presented at the 4th International Symposium on Biological and Artificial Intelligence Systems, Trento, Italy, September 18–22, 1988.

Our interest in detailed studies of protein hydration in solution was revived by recent observations with the protein Tendamistat, for which high-resolution structures were determined by NMR in solution and in single crystals.[8] Even though the global molecular architectures in the two states are very nearly identical, significant structural differences were identified near the protein surface. There is also evidence that compared to the core of the protein, the increase of structural disorder near the molecular surface is more pronounced in solution than in the crystals.[8] The experiments described in the present paper now demonstrate by observations with individual water molecules that in solution the hydration network on the protein surface is kinetically highly labile. Evidence for mobile hydration waters in water–protein systems

(1) Abbreviations and symbols used: NMR, nuclear magnetic resonance; NOE, nuclear Overhauser enhancement; ROE, rotating-frame Overhauser enhancement; 2D, two-dimensional; NOESY, two-dimensional nuclear Overhauser enhancement spectroscopy in the laboratory frame; ROESY, two-dimensional nuclear Overhauser enhancement spectroscopy in the rotating frame; ω_0, Larmor frequency; τ_c, rotational correlation time.

(2) Wüthrich, K. *NMR of Proteins and Nucleic Acids*; Wiley: New York, 1986.

(3) Deisenhofer, J.; Steigemann, W. *Acta Crystallogr., Sect. B* **1975**, *31*, 238.

(4) (a) Blundell, T. L.; Johnson, L. N. *Protein Crystallography*; Academic Press: New York, 1976. (b) Teeter, M. M. *Proc. Natl. Acad. Sci. U.S.A.* **1984**, *81*, 6014.

(5) Koenig, S. H.; Bryant, R. G.; Hallenga, K.; Jacob, G. S. *Biochemistry* **1978**, *17*, 4348.

(6) Stoesz, J. D.; Redfield, A. G.; Malinowksi, D. *FEBS. Lett.* **1978**, *91*, 320.

(7) Tüchsen, E.; Hayes, J. M.; Ramaprasad, S.; Copie, V.; Woodward, C. *Biochemistry* **1987**, *26*, 5163.

(8) (a) Kline, A. D.; Braun, W.; Wüthrich, K. *J. Mol. Biol.* **1986**, *189*, 377. (b) Kline, A. D.; Braun, W.; Wüthrich, K. *J. Mol. Biol.* **1988**, *204*, 675. (c) Pflugrath, J.; Wiegand, E.; Huber, R.; Vertĕsy, L. *J. Mol. Biol.* **1986**, *189*, 383.

0002-7863/89/1511-1871$01.50/0 © 1989 American Chemical Society

was previously also derived from measurements of NMR relaxation times and line widths,[9] which could, however, only provide a global view.

General Considerations on the Observation of Individual Molecules of Hydration Water by ^{1}H NMR

One can anticipate that the proton chemical shifts of protein-bound water molecules are influenced by the protein environment in a similar fashion as those of the individual amino acid residues upon incorporation into a globular protein structure.[2] On the basis of the resulting unique chemical shifts for different individual hydration waters, direct observation of distinct ^{1}H NMR lines corresponding to these water molecules should in principle be possible and thus provide direct evidence for their presence. In practice, however, it appears that as a rule the proton resonances of protein hydration water in aqueous solution are at the same chemical shift as the bulk water. From this one must conclude either that the effects of protein binding on the water chemical shifts are too small to be resolved, which would impose an upper limit of ca. 0.05 ppm on these shifts, or that exchange of protons from hydration water with the bulk water averages out the differences in chemical shift induced by the chemical nature of the hydration sites. This latter explanation applies for nearly all hydroxyl and carboxylate protons of the amino acid side chains of Ser, Thr, Tyr, Asp, and Glu in proteins, for which it is well-known that they have intrinsically different chemical shifts from that of H_2O.[2] Different time scales govern the averaging of the chemical shifts and the spin relaxation processes that lead to the appearance of NOEs.[2] In the situation where the resonance lines of both the OH groups of the protein and the hydration waters are merged with that of the bulk water, one may therefore still observe individual molecules of hydration water through specific intermolecular NOEs.

The degeneracy of all water proton and most hydroxyl proton chemical shifts requires that complete ^{1}H NMR assignments are available for the protein, and that the solution structure of the protein is known at high resolution, before unambiguous assignments of NOEs between protons of the protein and hydration water molecules can be obtained. In addition, one must rule out that the observed effects arise from chemical exchange of protons, and that the interactions could be with OH groups of the protein rather than with protein-bound water molecules.

Distinction between different possible mechanisms of intermolecular magnetization transfer can be based on differences in qualitative aspects of two-dimensional nuclear Overhauser experiments in the rotating frame (ROESY). A negative sign of a ROESY cross peak relative to the diagonal peaks shows that the magnetization transfer is by direct cross relaxation, since both transfer by first-order spin diffusion pathways or by chemical exchange lead to positive cross peaks.[10] On the other hand, ROESY spectra may contain extra peaks due to homonuclear Hartmann–Hahn transfer,[11] and the cross peak intensities have to be corrected for off-resonance effects in the ROESY spin lock.[12] As a control, comparisons with nuclear Overhauser enhancement experiments in the laboratory frame (NOESY) can therefore be helpful. In the present experiments we further used studies of the effects of temperature variation on the NOESY and ROESY spectra to identify cross peak intensity arising from chemical exchange. Once a cross peak between a protein proton and a proton at the bulk water chemical shift has thus been demonstrated to correspond to a direct NOE, distinction between intermolecular NOEs with hydration water and intramolecular NOEs with OH groups can only be achieved on the basis of the solution structure of the protein, from which the location of all the side-chain hydroxyl and carboxylate groups is known. Thereby we used as a criterion for the identification of a NOE with a hydration water molecule that the interacting protein proton must be at a distance of at least 4.0 Å from the nearest OH group; otherwise the distinction between hydration water and hydroxyl groups would remain undecided. Along similar lines, a distinction may be made between the two situations that either one molecule of hydration water interacts with two neighboring groups of protein protons or that two different water molecules interact with different protein protons, which are located far apart in the structure.

NMR experiments enable one to estimate limits on the lifetimes of the hydration water protons with respect to exchange in and out of the protein hydration sites. A lower limit can be obtained from the observation that the chemical shift is averaged with that of the bulk water by using an assumed value for the protein-induced chemical shift. For example, with the assumption that the protein-induced shift amounts to 0.02 ppm, observation of a single, averaged water resonance at a Larmor frequency of 600 MHz would indicate that the exchange rate is faster than 80 s^{-1}. An upper limit may be derived by using the fact that NOESY cross peaks manifesting direct NOEs are positive (i.e., the NOE has a negative sign) if the effective correlation time, τ_c, is longer than $1.12\omega_o^{-1}$, where ω_o is the Larmor frequency. Otherwise the NOESY cross peak would be negative. In practice, NOESY cross peaks manifesting NOEs (rather than proton exchange) will have significant intensity only for $\tau_c \gg 1.12\omega_o^{-1}$, whereas for $\tau_c \lesssim 1.12\omega_o^{-1}$ NOEs are more easily observed in ROESY experiments.[10]

NMR Experiments for Direct Observation of Individual Hydration Water Molecules

Commonly protein ^{1}H NMR spectra are recorded either in ^2H_2O solution or in H_2O solution with selective saturation to suppress the solvent resonance.[2] Such experiments do a priori preclude observation of hydration water. Alternative solvent suppression techniques that do not eliminate the water signal during the preparation period[13,14] have been used in studies of nucleic acids,[15-19] where the resonances of the labile protons would be suppressed by the saturation technique.[2] For work with proteins these solvent suppression schemes have rarely been used, since they have the disadvantage of nonuniform spectral excitation profiles along the ω_2 axis,[15-18] or along both ω_1 and ω_2[19] in 2D NMR experiments. However, with the use of recent advances in instrumentation, in particular digital phase shifters and fast switches for changing the radio-frequency power settings, selective pulses of controlled phase can now be applied to the water resonance, and it is possible to selectively eliminate the water signal within milliseconds immediately before the detection period.[20] The resulting spectra have a uniform excitation profile in both dimensions, except for a narrow spectral region parallel to the ω_1 axis at the ω_2 frequency of the water signal, which would be obscured by t_1 noise from the residual water signal also in spectra recorded with solvent presaturation.[2] For studies of hydration waters we employed a variant of an experiment proposed by Sklenář and Bax.[20] Figure 1 shows the pulse sequences applied for recording NOESY and ROESY spectra in H_2O, whereby the water signal is suppressed only at the end of the mixing time.

In NOESY experiments (Figure 1A) with $\tau_m \gtrsim 100$ ms, the water magnetization is, as a result of radiation damping,[18,21] mainly aligned along the z axis at the moment when the first selective pulse, $(\pi/2)_\theta$, is applied. For shorter mixing times a homospoil

(9) (a) Kuntz, I. D., Jr.; Kauzmann, W. *Adv. Protein Chem.* **1974**, *28*, 239. (b) Halle, B.; Andersson, T.; Forsén, S.; Lindman, B. *J. Am. Chem. Soc.* **1981**, *103*, 500. (c) Shirley, W. M.; Bryant, R. G. *J. Am. Chem. Soc.* **1982**, *104*, 2910. (d) Polnaszek, C. F.; Bryant, R. G. *J. Chem. Phys.* **1984**, *81*, 4038.
(10) Bothner-By, A. A.; Stephens, R. L.; Lee, J. *J. Am. Chem. Soc.* **1984**, *106*, 811.
(11) Neuhaus, D.; Keeler, J. *J. Magn. Reson.* **1986**, *68*, 568.
(12) Griesinger, C.; Ernst, R. R. *J. Magn. Reson.* **1987**, *75*, 261.
(13) Redfield, A. G.; Gupta, R. K. *J. Chem. Phys.* **1971**, *54*, 1418.
(14) Hore, P. J. *J. Magn. Reson.* **1983**, *55*, 283.
(15) Kearns, D. R.; Mirau, P. A.; Assa-Munt, N.; Behling, R. W. In *Nucleic Acids: The Vectors of Life*; Pullman, B., Jortner, J., Eds.; D. Reidel Publishing Co.: Dordrecht, The Netherlands, 1983; pp 113–125.
(16) Hilbers, C. W.; Heerschap, A.; Haasnoot, C. A. G.; Walters, J. A. L. I. *J. Biomol. Struct. Dyn.* **1983**, *1*, 183.
(17) Boelens, R.; Scheek, R. M.; Dijkstra, K.; Kaptein, R. *J. Magn. Reson.* **1985**, *62*, 378.
(18) Bax, A.; Sklenář, V.; Clore, G. M.; Gronenborn, A. M. *J. Am. Chem. Soc.* **1987**, *109*, 6511.
(19) Otting, G.; Grütter, R.; Leupin, W.; Minganti, C.; Ganesh, K. N.; Sproat, B. S.; Gait, M. J.; Wüthrich, K. *Eur. J. Biochem.* **1987**, *166*, 215.
(20) Sklenář, V.; Bax, A. *J. Magn. Reson.* **1987**, *75*, 378.
(21) Abragam, A. *The Principles of Nuclear Magnetism*; Oxford University Press: New York, 1961.

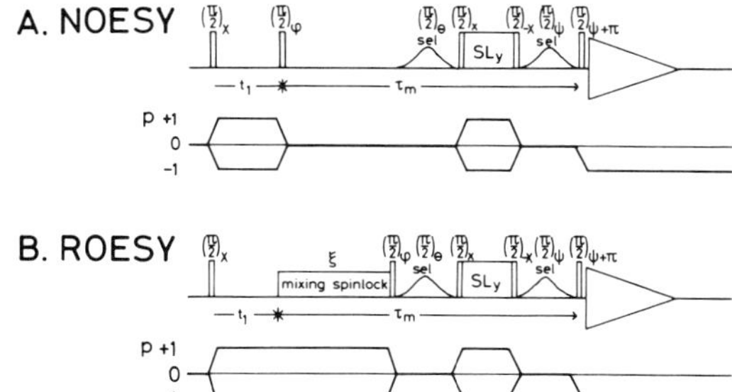

Figure 1. Pulse sequences used for the observation of intermolecular NOEs between water protons and protein protons by NOESY or ROESY, where SL denotes spin-lock pulses, t_1 and τ_m are the evolution period and the mixing time, and the observation period t_2 is indicated by a schematic free induction decay. The desired coherence transfer pathways shown below the pulse sequences are obtained by the following phase cycling: (A) NOESY: $\varphi = x, x, -x, -x, x, x, -x, -x$; $\theta = x, -x, x, -x, x, -x, x, -x$; $\psi = x, x, x, x, -x, -x, -x, -x$; receiver $= x, x, -x, -x, -x, -x, x, x$. Additional incrementation of the phases of all pulses and of the receiver in steps of 90° (CYCLOPS) results in a 32-step phase cycle. (B) ROESY: same as (A), except for the addition that, independent of all other phase, ξ is cycled between $-y$ and $+y$ after each package of eight scans, so that a 64-step phase cycle is obtained.

pulse can be used immediately before this first selective pulse to dephase all transverse magnetization present. The selective 90° pulse turns the longitudinal water magnetization vector into the transverse plane, from where it is immediately turned back to the z axis by the following nonselective $(\pi/2)_x$ pulse. The protein magnetization of interest, which is not excited by the selective pulse, is at that point aligned along the y axis. The following spin-lock pulse with phase y, SL_y, then locks the protein y magnetization and destroys with its radio-frequency inhomogeneity all coherences in the xz plane, which originate primarily from H_2O. The subsequent nonselective $(\pi/2)_{-x}$ pulse brings the spin-locked y magnetization, which contains the protein magnetization and residual contributions from H_2O, back to the z axis. The subsequent combination of a selective 90° pulse on the water resonance, $(\pi/2)_\psi$, and the nonselective pulse $(\pi/2)_{\psi+\pi}$ select again for protein z magnetization and convert it into observable coherence. In ROESY (Figure 1B), the mixing spin-lock period is followed by a $\pi/2$ pulse with its phase shifted by 90° relative to the phase of the spin lock. The spin-locked magnetization is thus turned to the z axis, and the same solvent suppression scheme as in the NOESY experiment of Figure 1A is then applied to the longitudinal magnetization. The following phase cycling is used with the pulse sequences for both NOESY and ROESY. Coherences of even order are selected during the mixing time by phase cycling of φ and ψ.[22] The spin-lock pulse, SL_y, further suppresses all multiple-quantum coherences. Cycling of the phase θ suppresses dispersive tails of the residual water signal. Axial peaks are shifted to the periphery of the spectrum by the use of time-proportional phase incrementation (TPPI).[22]

In the coherence transfer pathway of the ROESY experiment of Figure 1B, coherence order 0 is present during the water purging sequence following the mixing spin-lock. Therefore, longitudinal relaxation via NOE may occur during the selective pulses. However, since for macromolecules the buildup rate of rotating-frame Overhauser effects (ROE) is twice as fast as the NOE buildup, and ROEs have opposite sign to that of NOEs,[12] the ROE is expected to dominate the NOE for mixing spin-lock periods equal to or longer than a selective 90° pulse.

(22) Ernst, R. R.; Bodenhausen, G.; Wokaun, A. *Principles of Nuclear Magnetic Resonance in One and Two Dimensions*; Clarendon, Oxford, 1987.
(23) Kessler, H.; Griesinger, C.; Kerssebaum, R.; Wagner, K.; Ernst, R. R. *J. Am. Chem. Soc.* **1987**, *109*, 607.
(24) Morris, G. A.; Freeman, R. *J. Magn. Reson.* **1978**, *29*, 433.
(25) Bauer, C.; Freeman, R.; Frenkiel, T.; Keeler, J.; Shaka, A. J. *J. Magn. Reson.* **1984**, *58*, 442.
(26) Shinnar, M.; Leigh, J. S. *J. Magn. Reson.* **1987**, *75*, 502.

Finally, it should be pointed out that for instruments where homospoil pulses are available, the pulse schemes of Figure 1 can be modified so as to achieve the same favorable excitation profile without the spin-lock purge pulse SL_y. In practice, a 2-ms spin-lock was found to be sufficiently long for efficient purging of the solvent resonance and at the same time sufficiently short so that the extent of coherent transfer of magnetization during the spin-lock period is negligible. Such deleterious effects by the spin-lock are obviously completely avoided if the $(\pi/2)_x$–SL_y–$(\pi/2)_{-x}$ sequence is replaced by a homospoil pulse.

Results and Discussion

A series of NOESY and ROESY spectra recorded at 5, 36, 50, and 68 °C were used to assign the cross peaks between protein resonances and the water signal. Figures 2 and 3 are representative of the results obtained. Figure 2 shows contour plots of regions including the water resonance which were taken from NOESY spectra at 4 and 68 °C recorded with the pulse sequence of Figure 1A. At 4 °C the water resonance is at 5.01 ppm. Since there are no BPTI resonances at the same chemical shift,[27] all cross peaks seen at $\omega_1 = 5.01$ ppm must manifest interactions of protons from the protein with protons at the chemical shift of the water resonance. At 68 °C the water resonance is shifted to 4.45 ppm and overlaps with some protein lines, e.g., the α-protons of Lys-15 and Lys-46.[27] At this temperature some of the cross peaks at the ω_1 frequency of the water signal may therefore represent intramolecular NOEs with these α-protons rather than NOEs with water molecules or hydroxyl groups. Figure 3 shows that with few exceptions the same cross peaks at the ω_1 frequency of H_2O are observed by NOESY and ROESY in the ω_2 range from 6 to 10 ppm. The negative peak intensities seen in ROESY for the cross peaks between amide protons and the water line document that the magnetization transfer by ROE is more efficient than the chemical exchange of most of these protons in the temperature range 5–50 °C. At 5 °C positive ROESY cross peaks due to chemical exchange are observed for the side-chain amino protons of lysine residues and for a singlet resonance at 10.07 ppm, which was assigned to the side-chain hydroxyl proton of Tyr-35[28] by its NOEs to the 3,5 ring protons of Tyr-35, $C^\alpha H$ of Cys-38, and NH and β-CH_3 of Ala-40, which are all near neighbors of the Tyr-35 hydroxyl group in both the crystal structures of BPTI[3,29,30] and the solution structure determined by NMR.[31] Upon variation of the temperature from 5 to 68 °C, which is approximately 20 °C below the denaturation temperature of BPTI, the position and the intensity of the exchange peaks of the lysine amino protons change as they approach the coalescence temperature with the solvent resonance. Similarly, at 68 °C the OH resonance of Tyr-35 is broadened and shifted toward the water resonance. At 68 °C positive ROESY exchange cross peaks are seen also for several amide protons, and exchange cross peaks for the guanidino protons of arginine residues start to appear at 36 °C. The exchange peaks of the amide protons of Lys-46, Arg-39, Thr-11, Asp-3, and Glu-49 are identified in Figure 2. All the exchange peaks observed at 68 °C are from amide protons that were previously found to exchange too rapidly to be seen in a COSY spectrum recorded in ^2H_2O at 10 °C and pH 3.6 immediately after sample preparation[32,33] and are therefore among the most labile amide protons in BPTI.

(27) ^{1}H NMR assignments for BPTI are available at 36 °C (ref 31) and 68 °C (Wagner, G.; Wüthrich, K. *J. Mol. Biol.* **1982**, *155*, 347.) The presently used assignments at different temperatures and pH were established by comparison with these data.
(28) This resonance was not assigned previously, since it was not observed in the spectra recorded with water presaturation.
(29) Wlodawer, A.; Walter, S.; Huber, R.; Sjölin, L. *J. Mol. Biol.* **1984**, *180*, 301.
(30) Wlodawer, A.; Nachman, J.; Gilliland, G. L.; Gallagher, W.; Woodward, C. *J. Mol. Biol.* **1987**, *198*, 469.
(31) Wagner, G.; Braun, W.; Havel, T. F.; Schaumann, T.; Gō, N.; Wüthrich, K. *J. Mol. Biol.* **1987**, *196*, 611.
(32) Wagner, G.; Wüthrich, K. *J. Mol. Biol.* **1982**, *160*, 343.
(33) The assignments of Ile-19 and Asp-3 given in ref 32 must be interchanged: Tüchsen, E.; Woodward, C. *J. Mol. Biol.* **1985**, *185*, 405.

1874 J. Am. Chem. Soc., Vol. 111, No. 5, 1989

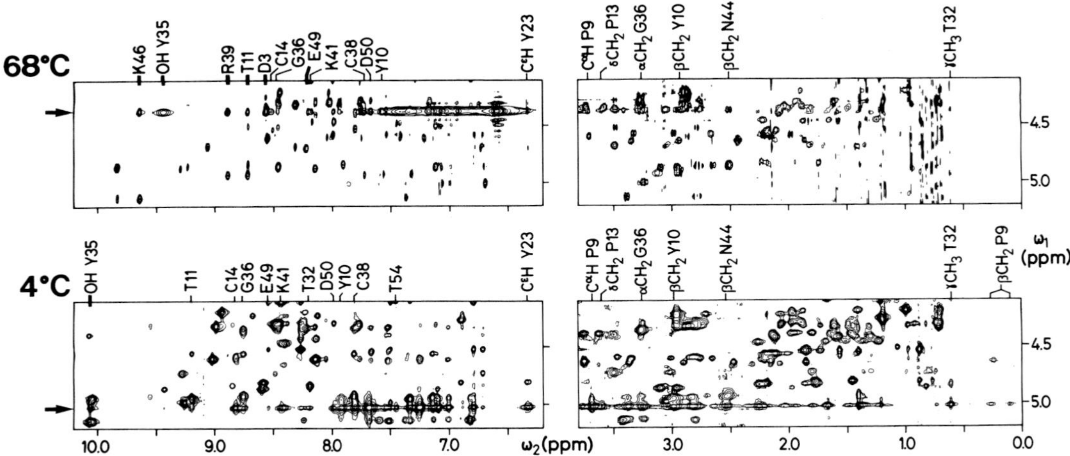

Figure 2. Spectral regions (ω_1 = 4.2–5.2 ppm, ω_2 = 0.0–3.8 ppm and 6.2–10.5 ppm) from contour plots of two NOESY spectra recorded at 600 MHz with a 15 mM solution of BPTI in 90% H_2O/10% ^2H_2O, 100 mM NaCl, pH 3.5, using the pulse sequence of Figure 1A with a mixing time of 112 ms. Arrows indicate the ω_1 frequency of the H_2O signal. NOE and exchange cross peaks with the water resonance are labeled above the spectra with the one-letter amino-acid symbol and the sequence position. Cross peaks due to chemical exchange are identified by the use of thicker lines, which are extended into the spectrum. The spectra were base-line corrected in both dimensions by using polynominals. Lower panel: spectrum at 4 °C, $t_{1_{max}}$ = 70 ms. Upper panel: spectrum at 68 °C, $t_{1_{max}}$ = 35 ms. Wiggles along ω_1 observed at the ω_2 position of the strongest peak arise from the truncation of the data set in t_1. At 68 °C the C$^\alpha$H–NH NOE cross peak of K46 overlaps with the exchange cross peak of K46 NH with the water (see text).

At 4 and 5 °C (Figures 2 and 3) the spectra contain a limited number of mostly well-separated cross peaks at the ω_1 frequency of the water. All NOESY cross peaks between protein resonances and the water signal that could be unambiguously assigned are labeled in Figure 2. Except for the cross peaks due to chemical exchange and the cross peaks with β-CH_2 of Pro-9, which probably arise from spin diffusion, all these peaks are manifested by negative ROESY cross peaks (Figure 3) and therefore come from direct NOE. Their intensity diminishes continuously with increasing temperature, but clearly there are still NOE cross peaks in the spectra at 68 °C. This is exemplified by the cross peak between the amide proton of Cys-14 and the water resonance, which is identified by an arrow in the ROESY spectra of Figure 3. In all, NOE cross peaks between protein protons and the water resonance position in NOESY and ROESY were identified for the amide protons of Tyr-10, Thr-11, Cys-14, Thr-32, Gly-36, Cys-38, Lys-41, Glu-49, Asp-50, and Thr-54, and for the nonexchangeable protons C$^\alpha$H of Pro-9, β-CH_2 of Tyr-10, δ-CH_2 of Pro-13, C$^\epsilon$H of Tyr-23, γ-CH_3 of Thr-32, α-CH_2 of Gly-36, and β-CH_2 of Asn-44.

As mentioned in the introduction, localization of oxygen atoms of hydration water molecules is usually part of a high-resolution protein crystal structure. For BPTI, structures have been determined at high resolution in three different crystal forms by X-ray crystallography and neutron diffraction,[3,29,30] and the structure determination by NMR has shown that the global molecular architecture in aqueous solution is the same as in these crystals.[31] In the three crystal forms, 60, 62, and 73 water molecules, respectively, have been located.[30] Sixteen water molecules were found to be in approximately the same locations in all three crystal forms, and 10 of these make the same hydrogen bonds. Four of these 10 fully conserved water molecules are located in the interior of the protein. They constitute an integral part of the protein structure and have no hydrogen bonds with hydration water on the protein surface. All other water molecules seen in the crystal structures are in the layer of hydration water on the protein surface.

In the crystal form II of BPTI, proton coordinates were determined by neutron diffraction.[29] Fifty-eight of the localized hydration water molecules have protons that are within a distance of ≤3.0 Å from protein protons, and there are about 240 water–proton to protein–proton distances within this range. A correspondingly large number of protein–H_2O NOEs would thus be expected. From the observations in Figures 2 and 3 it follows that of all the water molecules seen in the crystal structures, at most a small percentage gives rise to sizable negative NOEs with protein protons. In particular, all prominent NOE cross peaks with the water resonance, which could so far be assigned, can either be attributed to the four internal water molecules or could arise from side-chain hydroxyl protons as well as from hydration water protons. With the possible exception of a small number of discrete hydration sites, the first hydration sphere of BPTI in solution must therefore be kinetically highly labile.

The four internal water molecules in crystalline BPTI have been designated W111, W112, W113, and W122.[3,29,30] Among the protein protons seen in Figures 2 and 3 to have NOEs with the water resonance position, the amide protons of Cys-14, Gly-36, and Cys-38, and α-CH_2 of Gly-36 and δ-CH_2 of Pro-13 are all within 3.0 Å from the protons of the internal water molecule W122. Similarly, in the crystal structures the amide protons of Tyr-10 and Lys-41, C$^\alpha$H of Pro-9, β-CH_2 of Tyr-10, and β-CH_2 of Asn-44 make contacts with one or two of the water molecules W111, W112, and W113 in the interior of the protein. Since no side-chain hydroxyls or carboxylates are close to any of these protein protons (a single exception is the amide proton of Gly-36, which is at a distance of 3.7 Å from the OH of Tyr-10 in the crystal structure[29]) and since these are also far from the protein surface, the NMR data present convincing evidence for the presence of the four internal water molecules also in the solution conformation of BPTI. The additional individually assigned NOEs from the amide protons of Thr-11, Thr-32, Glu-49, Asp-50, and Thr-54, C$^\epsilon$H of Tyr-23, and γ-CH_3 of Thr-32 with the water resonance cannot be attributed unambiguously to intermolecular protein–water contacts, since all these protons are near neighbors of side-chain OH groups of either threonine, serine, or tyrosine. For all protons of hydration waters or OH groups of the protein, which are identified by positive NOESY cross peaks at 600 MHz, the lifetimes with respect to exchange with the bulk water must be longer than 3×10^{-10} s. This limit is obviously also valid for the lifetimes of the oxygen atoms of the same water molecules. Water molecules or OH groups of the protein that are not manifested by positive NOESY cross peaks have shorter proton-exchange lifetimes than 3×10^{-10} s (the oxygen atoms could of course have longer residence times).

In conclusion, NMR measurements using the experiments of Figure 1 at variable temperature over the range 4–68 °C provided

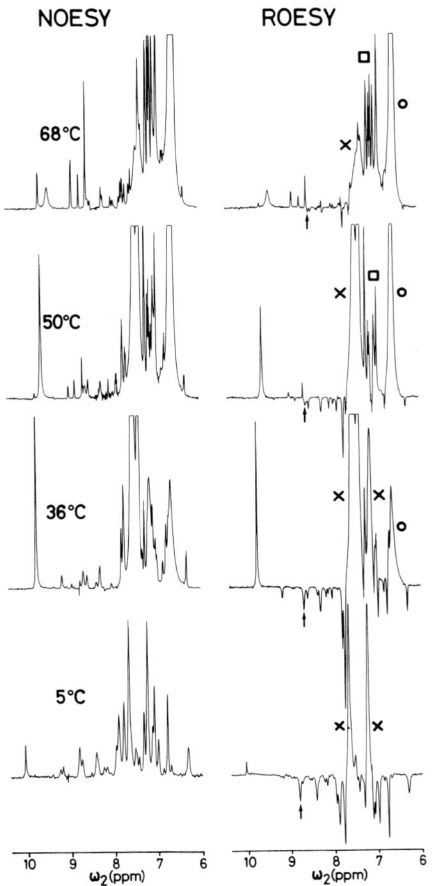

Figure 3. Corresponding cross sections parallel to the ω_2 axis taken at the ω_1 frequency of the water resonance in NOESY (left) and ROESY (right) spectra recorded at 5, 36, 50, and 68 °C. The NOESY spectra were recorded with a mixing time of 112 ms and the pulse sequence of Figure 1A. For the ROESY spectra the pulse sequence of Figure 1B was used with a mixing time of 87 ms. In the different ROESY cross sections, the NOE cross peak of the amide proton signal of Cys-14 with H_2O is identified with an arrow. Cross peaks with the water signal that manifest chemical exchange are identified in the ROESY cross sections with the following symbols: (×) Amino protons of lysine; (○, □) amino and imino protons, respectively, of the guanidinium group of arginine. With the exception of the NOESY cross section at 5 °C, where a reduced receiver gain setting was used, all cross sections are from spectra that were acquired and processed in an identical way.

unambiguous evidence that the four internal water molecules seen in the crystal structures of BPTI[3,29,30] are also an integral part of the solution conformation. Unambiguous evidence was further obtained for the presence of a discrete number of labile protons on the protein surface with lifetimes longer than 0.3 ns. It remains open if some or all of these protons correspond either to hydroxyl and carboxylate groups of the protein or to hydration waters. Compared to the corresponding crystal structures, the first hydration shell of BPTI in solution is clearly less well ordered. In particular, the present experiments show that the formation of a protein–water hydrogen-bonding network at the protein–water interface would, with the possible exception of a few unique sites, have to be very short-lived, with a lifetime shorter than 0.3 ns. This result is of special interest with respect to the indications obtained independently[8,31] that even when the core of a globular protein in aqueous solution is virtually identical with the corresponding crystal structure, the protein surface is generally more disordered in solution than in single crystals. Some of the questions raised by the present initial studies of protein hydration by high-resolution NMR may possibly be answered by rather straightforward further applications of the experiments in Figure 1. For example, measurements with other proteins than BPTI might reveal NOEs with protein surface groups that are sufficiently far apart from OH groups of the protein to produce unambiguous evidence for the presence of distinct molecules of hydration water on the protein surface. Also, studies of the influence of variable ionic strength, pH, etc. on the results obtained with these experiments might provide a wealth of additional insights relating to the characteristics of the interface between protein and solvent.

Experimental Section

Experiments were performed with the protein basic pancreatic trypsin inhibitor (BPTI). Material obtained from Bayer AG in Wuppertal, BRD (Trasylol) was used without further purification. For the NMR experiments we used a 15 mM solution of BPTI in a mixed solvent of 90% H_2O/10% ^2H_2O, containing 100 mM NaCl. A low pH value of 3.5 was chosen to be near the pH minimum for the rate of chemical exchange of labile protons from amide and guanidinium groups of the protein with the bulk water.[2]

The ^1H NMR experiments were recorded at 600 MHz on a Bruker AM 600 instrument. The following types of radio-frequency pulses were used in the experiments of Figure 1:

In ROESY a sequence of $\pi/4$ pulses was used as the mixing spin-lock.[23] To avoid TOCSY-type transfer of magnetization, the $\pi/4$ pulses were separated by delays 10 times longer than the width of the individual pulses.

The selective pulses applied to the water resonance during the mixing time in both NOESY and ROESY were Gaussian-shaped DANTE pulses consisting of a series of 32 pulses seperated by equal delays τ.[24-26] The shaping was achieved by increasing and decreasing the pulse lengths according to equidistant values on a Gaussian curve, starting at 1% of the maximum value. These Gaussian-shaped DANTE pulses are intrinsically very selective, but they exhibit additional excitation maxima at regular frequency intervals $1/\tau$. Therefore the delay τ must be chosen sufficiently short to place all excitation sidebands well outside of the spectrum. In practice, the pulse power was reduced by 27dB for the selective pulses relative to the nonselective pulses, and the shortest individual pulse used was longer than 0.2 μs. The exact length of these individual pulses was adjusted empirically to give the best water suppression. The optimum corresponded to a 70 or 80° pulse relative to a Gaussian pulse width corresponding to a 90° pulse. The 90° pulse width of the Gaussian pulse was determined by the maximum degree of water suppression in a two-pulse experiment, where a Gaussian pulse with phase x was immediately followed by a nonselective $(\pi/2)_{-x}$ pulse. The Gaussian-shaped DANTE pulses had a total length of 5 ms each and gave no significant excitation at 500 Hz away from the water signal. The suppression ratio was better than a factor 30 for a single scan at 4 °C, and appreciably higher suppression ratios resulted after completion of the phase cycle. The suppression efficiency was also increased at higher temperatures, where the water line is narrower, and could be further improved with the use of shorter, less selective Gaussian-shaped DANTE pulses with more pulse power.

Acknowledgment. We thank Bayer AG, Wuppertal, FRG, for a gift of BPTI (Trasylol), Dr. L. Orbons and Prof. G. Wagner for helpful discussions on the resonance assignment of BPTI, and R. Marani for the careful processing of the typescript. The pulse sequences with a homospoil pulse were tested at Spectrospin AG, Fällanden. Financial support by the Schweizerischer Nationalfonds (project 3.198.85) is gratefully acknowledged.

Reprinted from the Journal of the American Chemical Society, 1991, *113*.
Copyright © 1991 by the American Chemical Society and reprinted by permission of the copyright owner.

Proton Exchange with Internal Water Molecules in the Protein BPTI in Aqueous Solution

Gottfried Otting, Edvards Liepinsh, and Kurt Wüthrich*

Institut für Molekularbiologie und Biophysik
Eidgenössische Technische Hochschule-Hönggerberg
CH-8093 Zürich, Switzerland

Received January 14, 1991

A high-resolution protein crystal structure usually includes the positions of numerous hydration water molecules. The Brookhaven protein data bank[2] thus includes the locations of over 30 000 water oxygen atoms. The large majority of these water sites form a hydration shell that covers large parts of the molecular surface. In addition, a smaller number of water molecules can be located in the interior of a protein and represent an integral part of the molecular architecture. Using high-resolution NMR[1] experiments we have started investigations of protein hydration in aqueous solution.[3] In 2D [^{1}H,^{1}H]-NOESY and 3D ^{15}N-correlated [^{1}H,^{1}H]-NOESY experiments and the corresponding measurements in the rotating frame, NOE cross peaks between protein protons and protons of interior water molecules could be identified in several small proteins.[3-5] In contrast, the NOEs expected to arise from close proximity of surface hydration waters to the protein were, as a rule, not observed. Similar observations were recently reported for interleukin 1β.[6] These data imply that the NOEs between protein protons and surface hydration water are quenched because the effective correlation time for positional rearrangement of the water protons relative to the protein surface, which is determined either by chemical exchange or by independent rotational motions of the water molecules, is much shorter than the rotational correlation time of the protein. For interior water molecules a *lower limit* of 0.3 ns for the lifetime of the protons with respect to exchange with the bulk water was derived from the fact that the observed NOEs have a negative sign.[3] In the present note we report new experiments with paramagnetic shift reagents, which provide an *upper limit* for this lifetime, and thus contribute a further fundamental detail toward a precise characterization of protein hydration in aqueous solution.

In all the aforementioned experiments[3-6] it was observed that the chemical shifts of interior water protons manifested in the NOE cross peaks with polypeptide protons are identical with that of the bulk water. This could be rationalized by two limiting situations: (i) The water chemical shifts in the protein interior and the bulk water are indeed identical. (ii) The interior water protons experience *conformation-dependent shifts* similar to those observed for the protons of polypeptide chains in globular proteins;[7] the presence of separate ^{1}H NMR signals for the interior water molecules would then have to be concealed by proton exchange or exchange of intact water molecules. Because it is not known whether there is indeed a chemical shift difference between interior hydration water and bulk water, a proper distinction between the limiting situations i and ii was so far not possible. In the experiment described in this note a defined chemical shift difference between bulk water and "inaccessible" interior water molecules was therefore established by addition of an extrinsic paramagnetic shift reagent.

In the protein BPTI four internal waters were previously identified in three different crystal structures[8] and by NMR in solution.[3] After addition of $CoCl_2$ to an aqueous solution of BPTI, the equilibria 1 and 2 are of interest in the present context. H*

$$Co^{2+}(H_2O)_6 + H^*{}_2O \underset{k_{-1}}{\overset{k_1}{\rightleftharpoons}} Co^{2+}(H_2O)_5(H^*{}_2O) + H_2O \quad (1)$$

$$BPTI(H_2O)_4 + H'_2O \underset{k_{-m}}{\overset{k_m}{\rightleftharpoons}} BPTI(H_2O)_3(H'_2O) + H_2O \quad (2)$$

and H′ identify bulk water protons that are exchanged into binding sites of Co^{2+}, or interior sites in the protein, respectively. $H^*{}_2O$ and H'_2O are bound water molecules that contain one or two protons from the bulk water. The water molecules bound to Co^{2+} experience large chemical shifts and some broadening of the ^{1}H resonance lines due to the interactions with the unpaired electrons. It is known that the exchange k_1 is sufficiently fast for these paramagnetic effects to be averaged over the bulk water[9] and that a 30 mM concentration of Co^{2+} should cause a ^{1}H shift for the bulk water of approximately $\omega = 900\ s^{-1}$ at 600 MHz and 20 °C.[10]

Our measurements confirmed that the bulk water shift caused by addition of 30 mM $CoCl_2$ is approximately 0.25 ppm (arrows along ω_1 in Figure 1, parts A and B). Under these conditions a line broadening of 130 Hz for the bulk water is observed along ω_1 in Figure 1B, which is sufficiently small to allow observation of most of the previously identified NOEs between protons of the polypeptide chain of BPTI and the four interior water molecules (Figure 1B).[3] The NOESY cross peaks between different polypeptide protons of BPTI are virtually identical in parts A and B of Figure 1, which shows that the conformation of BPTI remained unchanged after addition of $CoCl_2$. The experiment of Figure

(1) Abbreviations used: NMR, nuclear magnetic resonance; 2D, two-dimensional; 3D, three-dimensional; NOESY, two-dimensional nuclear Overhauser enhancement spectroscopy; NOE, nuclear Overhauser enhancement; BPTI, basic pancreatic trypsin inhibitor.
(2) Bernstein, F. C.; Koetzle, T. F.; Williams, G. J. B.; Meyer, E. F., Jr.; Brice, M. D.; Rodgers, J. R.; Kennard, O.; Shimanouchi, T.; Tasumi, M. *J. Mol. Biol.* **1977**, *122*, 535–542.
(3) Otting, G.; Wüthrich, K. *J. Am. Chem. Soc.* **1989**, *111*, 1871–1875.
(4) Otting, G.; Wüthrich, K. In *Water and Ions in Biomolecular Systems*; Vasilescu, D., Jaz, J., Packer, L., Pullman, B., Eds.; Birkhäuser: Basel, 1990; pp 141–147.
(5) Wüthrich, K.; Otting, G. *Int. J. Quantum Chem.*, in press.
(6) Clore, G. M.; Bax, A.; Wingfield, P. T.; Gronenborn, A. *Biochemistry* **1990**, *29*, 5671–5676.
(7) Wüthrich, K. *NMR in Biological Research: Peptides and Proteins*; North-Holland: Amsterdam, 1976. Wüthrich, K. *NMR of Proteins and Nucleic Acids*; Wiley: New York, 1986.
(8) Deisenhofer, J.; Steigemann, W. *Acta Crystallogr.* **1975**, *B31*, 238–250. Wlodawer, A.; Walter, S.; Huber, R.; Sjölin, L. *J. Mol. Biol.* **1984**, *180*, 301–329. Wlodawer, A.; Nachman, J.; Gilliland, G. L.; Gallagher, W.; Woodward, C. *J. Mol. Biol.* **1987**, *198*, 469–480.
(9) Swift, T. J.; Connick, R. E. *J. Chem. Phys.* **1962**, *37*, 307–320.
(10) Luz, Z.; Shulman, R. G. *J. Chem. Phys.* **1965**, *43*, 3750–3756.

0002-7863/91/1513-4363$02.50/0 © 1991 American Chemical Society

4364

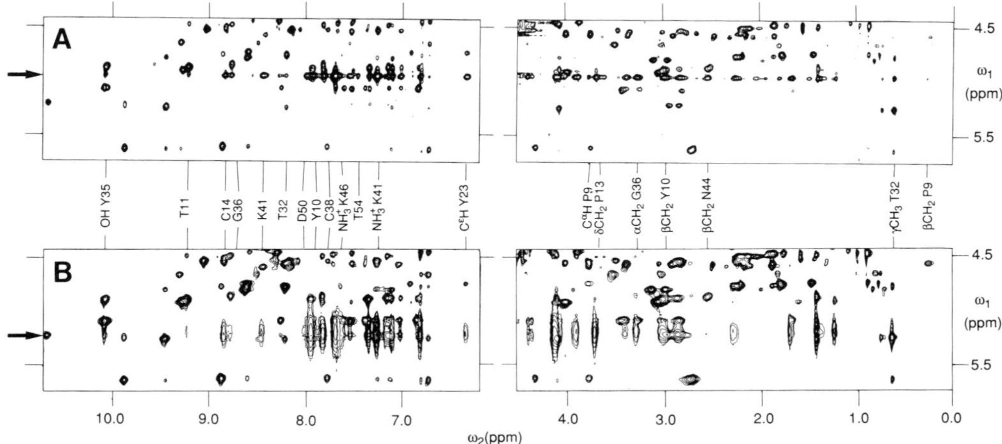

Figure 1. Spectral regions (ω_1 = 4.4–5.8 ppm, ω_2 = 0.0–4.5 and 6.2–10.2 ppm) of two homonuclear 2D [^1H,^1H]-NOESY spectra recorded at 4 °C with a 20 mM solution of BPTI in 90% H_2O/10% D_2O, pH 3.5, with and without addition of the shift reagent $CoCl_2$, using the experiment described in ref 3 (^1H frequency = 600 MHz, t_{1max} = 60 ms, t_{2max} = 150 ms, sweep width in ω_1 and ω_2 6944 Hz, total measuring time 12 h). The spectra were base-line-corrected in both dimensions, using polynominals. The arrows on the left indicate the ω_1 frequency of the water signal. The contour levels were plotted on an exponentially increasing scale, where each level is $2^{1/2}$ times higher than the preceding one. (A) Without $CoCl_2$, mixing time = 112 ms. (B) With 30 mM $CoCl_2$, mixing time = 90 ms. Lower contour levels are plotted than in A. In the center the cross peaks with the water resonance are identified with the one-letter amino acid symbol, the sequence position, and except for the backbone amide protons, the proton positions in the amino acid residues.[3] Some of the signals seen in A are below the lowest plot level in B because of the line broadening.

1 presents unambiguous evidence that the apparent ^1H chemical shift equivalence between bulk water and interior water molecules is related to the chemical exchange reaction described by eq 2. For the reaction rate a lower limit of $k_m > 50\ s^{-1}$ was established from the following considerations. Although the observations in Figure 1 would be compatible with the assumption that the resonance lines of the bulk water and the interior water are coalesced or that the resonances of the interior waters are separated from that of the bulk water but exchange-broadened beyond detection (this would be the case for $k_m > 10^3\ s^{-1}$), the above, more conservative estimate for k_m is obtained assuming that the resonance lines of the interior waters are resolved, but lie within 0.4 ppm of the bulk water resonance and thus are concealed by the water suppression technique used.[3] In this situation, an exchange rate of $k_m > 50\ s^{-1}$ would be sufficient to quench the cross peaks with the protein signals, since the magnetization would be rapidly transferred from the interior waters to the bulk water during the NOESY mixing time. The efficient relay of magnetization from the bulk water to the hydration sites would then also account for the cross peaks observed between bulk water and protein protons. Combined with the previously estimated lower limit,[3] we then have that, at 4 °C, 20 ms > τ_m > 0.3 ns. The available evidence[3–6] supports that these measurements with BPTI present a realistic guideline also for interior hydration of other globular proteins.

Because the experiment of Figure 1 provides direct information only on the upper limit for τ_m of the *protons* of interior water molecules, we performed similar experiments also with ^{17}O NMR, using 10% ^{17}O-enriched water as the solvent and $CoCl_2$ and $DyCl_3$ as shift reagents. In contrast to similar experiments with slowly exchanging water molecules bound to diamagnetic metal ions,[11] no separate ^{17}O line of BPTI-bound water could be observed. This result would be consistent with the view that the limiting value of τ_m < 20 ms for the lifetime of the interior hydration water in BPTI is valid not only for the water protons but also for the entire water molecules. It is not clear, however, that this conclusion is warranted, since there is the alternative explanation that the ^{17}O signal of protein-bound water could be broadened beyond detection due to the efficient nuclear quadrupole relaxation in the slow motional regime, which has not been properly excluded. In this context it should be recalled that, in earlier experiments with BPTI using gel filtration techniques with ^{18}O-enriched water, the ^{18}O exchange was complete within the deadtime of the experiment, which was 10 s.[12]

Acknowledgment. We thank Drs. R. E. Connick and R. G. Bryant for helpful suggestions and Mrs. E. Huber for the careful processing of the manuscript. Financial support was obtained from the Schweizerischer Nationalfonds (Project 31.25174.88).

(11) Jackson, J. A.; Lemons, J.; Taube, H. *J. Chem. Phys.* **1960**, *32*, 553–555. Connick, R. E.; Fiat, D. N. *J. Chem. Phys.* **1963**, *39*, 1349–1351.

(12) Tüchsen, E.; Hayes, J. M.; Ramaprasad, S.; Copie, V.; Woodward, C. *Biochemistry* **1987**, *26*, 5163–5172.

Reprinted with permission from *Science*, Vol. 254, pp. 974–980 (1991).
Copyright © 1991 American Association for the Advancement of Science.

Protein Hydration in Aqueous Solution

GOTTFRIED OTTING, EDVARDS LIEPINSH, KURT WÜTHRICH

High-resolution proton nuclear magnetic resonance studies of protein hydration in aqueous solution show that there are two qualitatively different types of hydration sites. A well-defined, small number of water molecules in the interior of the protein are in identical locations in the crystal structure and in solution, and their residence times are in the range from about 10^{-2} to 10^{-8} second. Hydration of the protein surface in solution is by water molecules with residence times in the subnanosecond range, even when they are located in hydration sites that contain well-ordered water in the x-ray structures of protein single crystals.

PROTEIN FOLDING, THAT IS, THE RELATIONS BETWEEN AMINO acid sequence, folding pathways, and kinetics, and the functional spatial arrangement of a polypeptide chain, is presently the least well understood step in a "central dogma" relating storage of genetic information with its expression by protein functions (*1*).

The authors are in the Institut für Molekularbiologie und Biophysik, Eidgenössische Technische Hochschule–Hönggerberg, CH-8093 Zürich, Switzerland.

New insights can be anticipated from structural characterization of both the unfolded and the functional folded polypeptide chain under the conditions of the folding milieu. Because water is excluded almost entirely from the interior of globular proteins (*1, 2*), different solvation of the polypeptide chain in the unfolded and folded forms must be an important factor. This article reports on investigations of the hydration of two polypeptides in aqueous solution. The hormone oxytocin has been chosen as a model for the highly solvated, unfolded state, since most atoms in this nonapeptide are solvent exposed in a predominantly flexible, nonglobular solution conformation (*3*). Bovine pancreatic trypsin inhibitor (BPTI) has been selected to represent globular proteins. For both molecules, high-resolution crystal structures are available that also include a range of hydration water molecules (*4–7*).

There are two classes of experiments capable of providing structural information at atomic resolution on protein molecules. One consists of x-ray diffraction and neutron diffraction studies with protein crystals (*8*). The coordinates of the oxygen atoms of numerous hydration water molecules are usually included in the description of a high-resolution protein crystal structure, suggesting that at least part of the hydration shell is well defined (*9*). Overall, the protein crystal structures deposited to date in the Brookhaven Protein Data Bank (*10*) include the coordinates of over 30,000 water oxygen atoms. The second experimental approach is nuclear magnetic resonance (NMR) spectroscopy with protein solutions (*11, 12*). Although the NMR method for protein structure determination has been available since 1985 (*13*), hydration water molecules proved to be evasive to detection in aqueous solution, and the observation of individual water molecules in a globular protein was reported only in 1989 (*14*). In the present study we used new NMR experiments (*15*) that enable studies of both the location and the residence time of individual hydration water molecules on the surface of flexible polypeptide chains or globular proteins in aqueous solution. Results obtained with this novel approach are presented and evaluated relative to the corresponding crystal structure data.

NMR and protein hydration. NMR experiments for studies of protein hydration rely primarily on phenomena related to nuclear spin relaxation (*16*). Beginning in the 1970's, measurements of relaxation dispersion in the bulk water signal of protein solutions have provided evidence that at least part of the water associated with proteins is highly mobile, with residence times in the hydration sites in the subnanosecond range [see, for example, (*17, 18*)]. In early one-dimensional (1D) high-resolution ^{1}H NMR experiments performed with selective water irradiation, nuclear Overhauser effects (NOE) between water protons and polypeptide protons were observed, but no further information on the kinetic stability of the hydration sites was obtained (*19*). Individual water molecules bound to hydration sites in a globular protein in solution were eventually observed with the use of 2D NOE spectroscopy (NOESY) (*14*) and heteronuclear 3D experiments (*20, 21*). However, in BPTI only four water molecules located in the interior of the protein were detected, and no NOEs were seen that would correspond to close contacts of protein protons with the surface hydration waters observed by x-ray diffraction in BPTI crystals (*5–7*). These observations were confirmed by NMR studies of interleukin-1β in aqueous solution, where similar NMR experiments detected exclusively hydration water molecules in the interior of the protein in identical locations as in the crystal structure (*21*).

The previous apparent absence of NOEs with water molecules in surface hydration sites is a consequence of the technical difficulties of the NMR experiments used. In all of the studies performed to date, the protein-bound hydration water molecules, including waters located in the interior, were found to exchange rapidly on the time scale of chemical shift differences, that is, with residence times in the hydration sites in the millisecond range or shorter (*22*). Therefore only a single signal is observed for the protons of the hydration water and the bulk water. For the same reason, the ^{1}H NMR lines of all -COOH groups, αNH_3^+, and most -OH groups of the polypeptide chain usually coincide with the bulk water resonance. As a consequence, NOEs can be assigned to different hydration water molecules only by reference to the individually assigned polypeptide ^{1}H NMR lines of the 3D protein structure (*14*). Furthermore, in 2D experiments the NOEs with the protons at the bulk water chemical shift are all located on the same 1D cross section, which is then quite crowded with lines. In the present study we obtained improved resolution with homonuclear 3D NMR experiments, and novel solvent suppression schemes were used for the suppression of the dominant bulk water signal after the NOE transfer of magnetization from the hydration waters to the protons of the polypeptide. These NMR technical details have been described in detail elsewhere (*15, 23*).

In the systems of interest here, ^{1}H spin relaxation and NOEs are dominated by time-dependent dipole-dipole coupling between nearby protons (*16*). In the simplest model the hydrated protein is represented by a rigid sphere (Fig. 1A). The intensity of the NOE between two hydrogen atoms i and j is then proportional to d_{ij}^{-6}, where d_{ij} is the distance between the two protons (Fig. 1A). It is further related to a correlation function describing the stochastic motions of the vector $\mathbf{d}_{ij}$ that connects the two protons. The strong distance dependence implies that NOEs can be observed only between spatially close protons, that is, in practice for $d_{ij} \lesssim 4.0$ Å. In the rigid model of Fig. 1A one would expect to observe more than 200 NOEs between protons of BPTI and the ~60 hydration waters reported in the crystal structures of this protein (*5–7*). However, these NOEs could be quenched by additional rapid motions of the water molecules relative to the protein surface, which would explain why they were not detected in the earlier studies (*14, 20, 21*).

The NOEs can be measured either by experiments in the laboratory frame of reference (NOESY) or in the rotating frame (ROESY) (*24*). If sufficiently short mixing times are used to minimize contributions from autorelaxation and spin diffusion (*11, 25*), the measured NOE intensities reflect directly the cross relaxation rates in the laboratory frame, σ^{NOE}, or in the rotating frame, σ^{ROE}, respectively. The two rates differ in their functional dependence on the spectral densities, $J(\omega)$ (*24*):

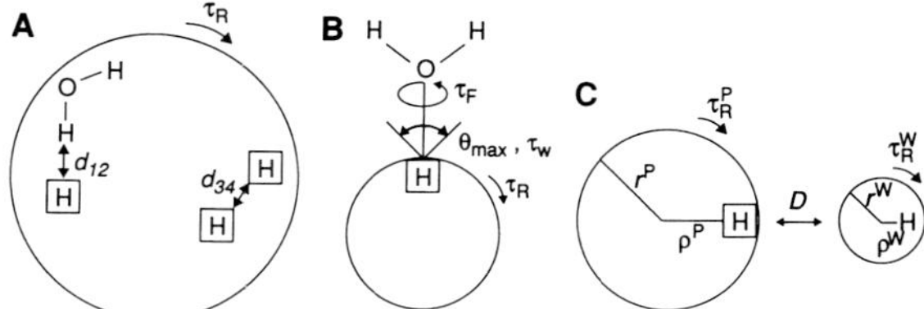

Fig. 1. Three models used in this article to calculate dipole-dipole cross relaxation rates σ^{NOE} and σ^{ROE}. (**A**) Isotropic rotational diffusion of a rigid sphere. Water molecules in the first hydration shell are considered to be part of the sphere representing the protein. Squares represent polypeptide protons. The vector d_{12} connects a polypeptide proton with a hydration water proton, and d_{34} connects two polypeptide protons. τ_R is the correlation time for the overall rotation of the sphere, which is in this model equal to the effective correlation time for the modulation of the dipolar interactions. (**B**) Wobbling in a cone model (*26, 27*). The residence time of the water molecules in the hydration sites is assumed to be long compared to the rotational correlation time τ_R, but the water rotates with a correlation time τ_F around a hydrogen bond to the protein, and this rotation axis is free to wobble with a correlation time τ_w within a cone with an opening angle $\theta_{max} = \pm 45°$. The overall rotational tumbling of the hydrated protein is again described by an equivalent sphere, with a correlation time τ_R. (**C**) Random, independent translation and rotation of the protein and the water molecule. The two molecules are represented by spheres of radius r^P and r^W, respectively, with the proton spin displaced from the center by ρ^P and ρ^W. The translational diffusion coefficient D characterizes the relative translational motions of the two molecules. The two spheres reorientate isotropically with correlation times τ_R^P and τ_R^W, respectively.

$$\sigma^{NOE} = 6J(2\omega_0) - J(0) \quad (1)$$

$$\sigma^{ROE} = 3J(\omega_0) + 2J(0) \quad (2)$$

where ω_0 is the Larmor frequency of the protons. Equations 1 and 2 show that σ^{ROE} is always positive because the spectral densities $J(\omega)$ have finite positive values at all frequencies (*16*). In contrast, the sign of σ^{NOE} depends on the explicit functional form of $J(2\omega_0)$ and $J(0)$, which in turn is related to the rate processes that govern the modulation of the dipole-dipole coupling. In the following section we investigate how sign and value of the ratio $\sigma^{NOE}/\sigma^{ROE}$ can be rationalized by different model representations of a hydrated protein.

Model representations of hydrated proteins. An important conclusion results from the simple model of Fig. 1A. In this rigid, spherical molecule the mobility of the vectors d_{ij} is governed by the overall rotational correlation time, τ_R, independent of whether these vectors connect different polypeptide protons or polypeptide protons with water protons (Fig. 1A). The results of computing σ^{NOE} and σ^{ROE} with the parameters given in the caption to Fig. 2 are displayed in Fig. 2A; $\sigma^{ROE} \geq \sigma^{NOE}$ over the entire range of τ_R values, and σ^{NOE} changes sign at $\omega_0\tau_R \approx 1.12$, corresponding to $\tau_R \approx 300$ ps at a Larmor frequency of 600 MHz. Under the experimental conditions used here (see captions to Figs. 3 and 4), measurements of ^{13}C relaxation times for α-carbon positions showed that the overall rotational tumbling of oxytocin can be characterized by a rotational correlation time for an equivalent sphere of $\tau_R \approx 2$ ns, and for BPTI $\tau_R \approx 8$ ns was extrapolated from earlier measurements at higher temperature (*26*). Correspondingly, for both molecules negative σ^{NOE} values were observed for all NOEs between different polypeptide protons. On the other hand, positive σ^{NOE} rates were detected for numerous intermolecular NOEs with water protons, showing that the polypeptide proton-water proton dipolar interactions must be modulated by additional, higher frequency rate processes.

The "wobbling in a cone" model (*26, 27*) assumes that the hydration water molecules are flexibly bound to a particular hydration site of the protein and have a long residence time compared to the effective rotational correlation time of the protein. Local motions of a water molecule in its hydration site are simulated by a combination of rotation about the hydrogen-bond axis (τ_F) and wobbling motions of this rotation axis inside a cone (τ_W). This model would not predict positive $\sigma^{NOE}/\sigma^{ROE}$ ratios for the protein proton–water proton NOE for any combination of τ_W and τ_F values (Fig. 2B). From studies with this and similar models we had to conclude that the observed positive values for $\sigma^{NOE}/\sigma^{ROE}$ cannot be rationalized by a description of the hydrated protein where the lifetime of the water protons in the hydration sites is long compared to τ_R.

Positive σ^{NOE} values were obtained with the assumption of short residence times of the hydration water, characterized by diffusion coefficients D greater than 3×10^{-6} cm²/s in the model of Fig. 1C (*28*). [Note that the self-diffusion coefficient of pure water at 6°C is about 12×10^{-6} cm²/s (*29*)]. The diffusion coefficients can be translated into residence times of the hydration water molecules using the Einstein-Smoluchowski relation

$$\tau = \overline{x^2}/D \quad (3)$$

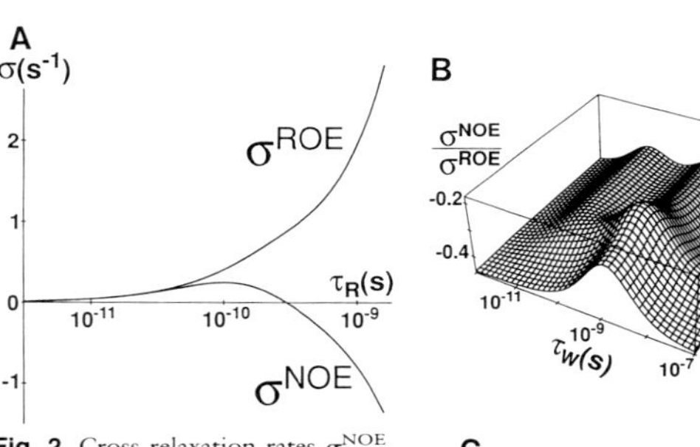

Fig. 2. Cross relaxation rates σ^{NOE} and σ^{ROE} for a pair of proton spins calculated on the basis of the three models of Fig. 1. (**A**) Model of Fig. 1A with a ^1H-^1H distance of 2.0 Å; ^1H frequency 600 MHz. (**B**) Wobbling in a cone model (Fig. 1B). H–O distance = 2.0 Å, $|\theta_{max}|$ = 45°, τ_R = 2 ns, ^1H frequency = 600 MHz. τ_R = 2 ns corresponds to the case of oxytocin. For BPTI, with τ_R = 8 ns, the maximum of $\sigma^{NOE}/\sigma^{ROE}$ over the complete range of the (τ_W,τ_F)-plane would be even lower, at about −0.4. (**C**) Diffusion model (Fig. 1C) with r^W = 2.0 Å, ρ^W = 1.0 Å, and τ_R^W = 4 ps. The solid curve was calculated with r^P = 12.0 Å, ρ^P = 11.0 Å, ^1H frequency = 500 MHz, and τ_R^P = 8 ns to simulate the intermolecular water proton–protein ^1H cross relaxation in the experiment with BPTI at 4°C (Fig. 4). The dashed curve was calculated with r^P = 4.0 Å, ρ^P = 3.0 Å, ^1H frequency = 600 MHz, and τ_R = 2 ns to approximate the situation for the oxytocin experiment at 6°C (Fig. 3). At the bottom the inverse of the translational diffusion coefficient, $1/D$, was converted into the corresponding lifetimes of the hydration water molecules with the Einstein-Smoluchowski relation (see text).

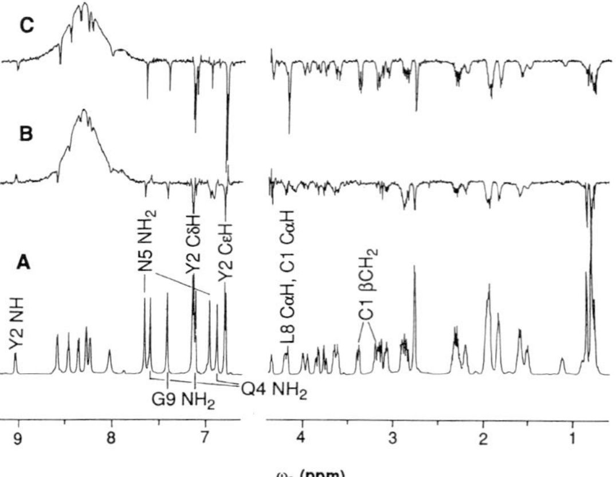

Fig. 3. Proton NMR spectra showing NOEs between oxytocin protons and water protons (oxytocin concentration 50 mM, solvent 90 percent H_2O–10 percent D_2O, T = 6°C, pH = 3.5, ^1H frequency = 600 MHz, experimental schemes of Fig. 1, A and B, in (*15*) with mixing time τ_m = 30 ms, spin locks $SL_{\phi4}$ = 0.5 ms and $SL_{\phi3} = SL_{\phi5}$ = 2 ms, and delay τ = 167 μs). All of the spectra were multiplied with the spectral excitation profile sin[0.63(δ − 4.9)], where δ is the chemical shift in ppm, which has excitation maxima at 2.45 and 7.35 ppm. The low-field region from 7 to 9 ppm was inverted for improved readability. (**A**) 1D spectrum. (**B**) Cross section through the NOESY spectrum along ω_2 at the ω_1 frequency of the water line (t_{1max} = 40 ms, t_{2max} = 328 ms, time domain data size 430 × 4096 points, and 32 scans per free induction decay); homospoil pulses of 0.5-ms duration were applied every 2 ms during the first 20 ms of the mixing time to prevent the decay of the water signal by radiation damping). The CεH multiplet fine structure of Tyr² is distorted by an artifact that was absent in similar spectra recorded with our AM500 spectrometer. (**C**) Cross section through the ROESY spectrum along ω_2 at the ω_1 frequency of the water line [processed and plotted with identical parameters as in (B)]. Resonance assignments for selected peaks are indicated with the one-letter amino acid symbol and the sequence position (*39*).

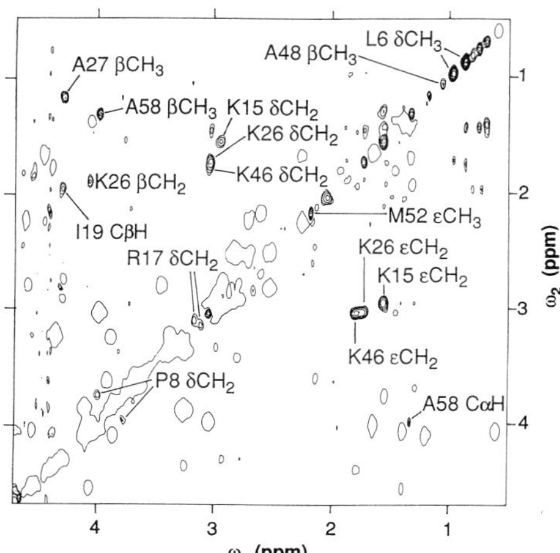

Fig. 4. ω_2-ω_3 cross plane through a homonuclear 3D ^{1}H NOESY-TOCSY spectrum taken at the ω_1 frequency of the water signal and showing NOE cross peaks between protons of BPTI and water protons [BPTI concentration = 20 mM, solvent 90 percent H_2O–10 percent D_2O, T = 4°C, pH = 3.5, ^{1}H frequency = 500 MHz, experimental scheme of Fig. 1C in (*15*), mixing time τ_m = 50 ms, and other parameters as in (*15*)]. For positive peaks only the lowest contour level is plotted. Selected negative peaks (corresponding to $\sigma^{NOE} > 0$) are identified by the assignment of the polypeptide proton that interacts with the water.

If we define an average displacement $(\overline{x^2})^{1/2}$ of 4.0 Å as the criterion for complete water proton exchange in and out of a hydration site, σ^{NOE} is positive for lifetimes shorter than ~500 ps (Fig. 2C). This is much shorter than the lifetime of a proton in a water molecule with respect to exchange by hydrolysis (*30*). We therefore conclude that the observation of positive σ^{NOE} rates indicates rapid exchange of complete water molecules between the bulk solvent and the protein hydration sites.

The NOE cross peaks resulting from positive σ^{NOE} should be small (Fig. 2C). This was confirmed by the NMR experiments (Figs. 3 and 4). At long mixing times, cross peaks arising from positive σ^{NOE} values may be canceled by spin-diffusion from cross peaks with negative σ^{NOE} rates, which are usually more intense.

NMR observations with oxytocin and BPTI. In Fig. 3 are shown the normal 1D ^{1}H NMR spectrum (Fig. 3A) of oxytocin and the cross sections through the 2D NOESY (Fig. 3B) and ROESY (Fig. 3C) spectra, which contain the water-polypeptide cross peaks. With few exceptions which are due to dominant effects from chemical exchange (*31*), the oxytocin signals show negative NOESY cross peaks with the water line, corresponding to positive σ^{NOE} values. In contrast, all intramolecular NOESY cross peaks between different protons of oxytocin were positive, as expected for a molecule in the slow tumbling regime of Fig. 2A (*32*). The rapid exchange of the hydration water molecules implicated by the different sign of intramolecular and intermolecular σ^{NOE} values is further substantiated by a comparison of Fig. 3, B and C, which shows that with few exceptions (*33*) the water-polypeptide cross peaks are two to three times less intense in NOESY than in ROESY. In the model of Fig. 1C this ratio corresponds to lifetimes of the water molecules in the hydration sites of 100 to 250 ps. Virtually all of the ^{1}H NMR lines in Fig. 3A are represented by NOEs in Fig. 3, B and C, showing that all parts of the oxytocin molecule are exposed to the solvent. However, no stably bound water molecules could be identified.

In contrast to oxytocin, the 1D cross section through the water resonance of a 2D ^{1}H NOESY spectrum of BPTI contains numerous overlapping groups of peaks (*14*). They can be virtually completely resolved in homonuclear 3D NOESY-TOCSY (TOCSY, total correlation spectroscopy) and 3D ROESY-TOCSY spectra (*15*). The high-field region from the 2D ω_2-ω_3 cross section taken at the ω_1 frequency of the water resonance in the 3D NOESY-TOCSY spectrum is shown in Fig. 4. The peaks on the diagonal come from the transfer of magnetization from the water line to the protein resonances during the NOESY mixing time. The fact that the ω_2 frequency equals the ω_3 frequency indicates that for these peaks there was no further magnetization transfer during the TOCSY mixing period (*34*). The off-diagonal peaks arise because magnetization precessing during t_2 is transferred to scalar-coupled protons during the TOCSY mixing period; they support the assignment of the diagonal NOE peaks (*15*). The previously reported cross peaks with the interior water molecules and some labile polypeptide protons (*14*) were again observed in these 3D NMR spectra, and many additional NOEs with the water signal could be assigned due to the presence of the off-diagonal peaks. Many of these newly found NOEs have positive σ^{NOE} values and come from solvent-accessible protons on the protein surface. They are much weaker than the positive NOE cross peaks with the interior water molecules, which correspond to negative σ^{NOE} rates (*14*). Except for the cross peaks with the interior waters, the $\sigma^{NOE}/\sigma^{ROE}$ ratios for the observed water-protein interactions are mostly in the range from 0.3 to 0.1. In the model of Fig. 1C these values indicate residence lifetimes in the range from 100 to 300 ps. Based on the experience with oxytocin, one might expect that most or all surface waters of BPTI should show positive σ^{NOE} rates with the water. We attribute the apparent absence of many of the expected peaks to the lower sensitivity (oxytocin was studied at higher concentration and has sharper lines than BPTI) because most of the negative cross peaks seen in Fig. 4 are with intense resonances of BPTI, such as those of methyl groups, or methylene groups with degenerate chemical shifts. This sensitivity criterion applies, however, only to weak NOEs with positive σ^{NOE} values from short-lived hydration waters. Long-lived surface hydration waters would produce much stronger NOEs with negative σ^{NOE} values, comparable to those seen for the internal waters. It is therefore very unlikely that NOEs with surface hydration water molecules bound with residence times >500 ps would have escaped detection by both NOESY and ROESY (Fig. 2C).

Observation of individual hydration water molecules in protein crystals and in aqueous solution. The NMR experiments in solution described above and x-ray diffraction experiments with protein single crystals are sensitive to different aspects of protein hydration. The intensity of the protein-water NOEs reflects primarily the residence times of the water molecules near the protein protons monitored by the NMR experiment (Fig. 2C). In contrast, the x-ray experiment probes the fraction of time that a water molecule is located at a particular point in space, but is largely insensitive to the residence time at that site on any particular visit (*35*). The main focus of the following discussion is to see how far the x-ray-observations of hydration water in single crystals can be correlated with the residence times of water molecules in corresponding locations observed by NMR in solution.

For oxytocin, where hydration of the entire molecular surface is observable by NMR (Fig. 3), there is no evidence that the sites of the seven and eight water molecules found, respectively, in the two crystal structures of deamino-oxytocin (*4*) are more stably hydrated in solution than the other surface areas. Since oxytocin is flexibly disordered in solution, a more detailed comparison does not seem to be warranted.

BPTI has the same molecular architecture in solution and in

crystals, which includes the four internal water molecules (*14, 36, 37*). In Fig. 5 the crystal structure atomic coordinates (*6*) were used to generate a molecular model visualizing selected surface properties of the protein and the x-ray and NMR observations on surface hydration. The "front view" shown is representative of the complete protein surface. With the possible exception of the protein-protein contact sites in the crystal lattice (yellow in Fig. 5A), the entire protein surface must be covered with hydration water molecules. Only part of the hydration water has so far been observed by either of the two methods considered here. In the BPTI crystal structure, ~40 percent of the protein surface is involved in protein-protein contacts, 25 percent is covered with x-ray–observable hydration water attributed to the central protein molecule, and for an additional 15 percent there are contacts with water molecules attributed to neighboring protein molecules (Fig. 5B). The remaining 20 percent of the protein surface must be in contact with water molecules that are not sufficiently well ordered to be seen by x-ray diffraction (see also Fig. 5A, where white color indicates water-accessible hydrogen atoms). More than 40 percent of the x-ray–observable water molecules are in contact with two protein molecules in the crystal lattice, and all of the observed waters are in contact with one protein molecule. Interestingly, most of the observed surface hydration water molecules are also accessible for contact with x-ray–unobservable water in the crystal. In aqueous solution all polypeptide hydrogen atoms giving rise to negative σ^{NOE} values with the water resonance (brown in Fig. 5C) are located near the amino terminus or a carboxyl or hydroxyl proton (magenta in Fig. 5C) or near one of the four interior water molecules (green in Fig. 5C). On the basis of additional NMR measurements and observations on the locations of the x-ray–observable hydration water molecules in BPTI crystals (see below), we arrive at two important conclusions on surface hydration in solution: (i) Nearly all NOEs with negative σ^{NOE} values between the water resonance and resonances of protons on the

A

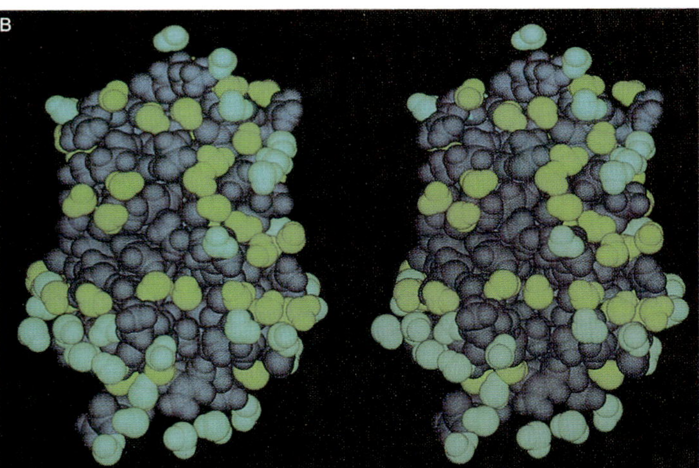

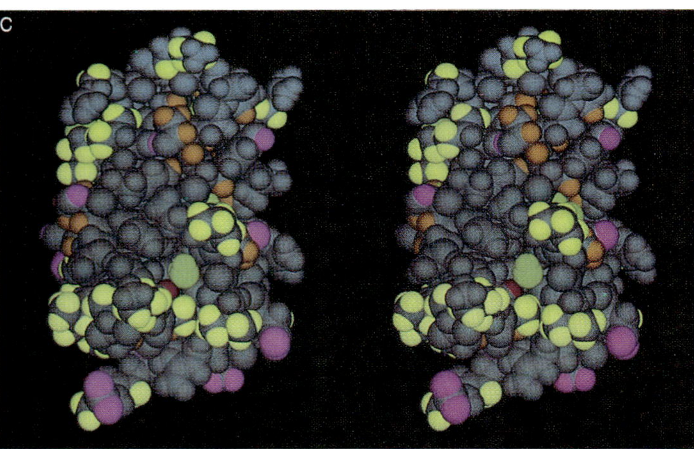

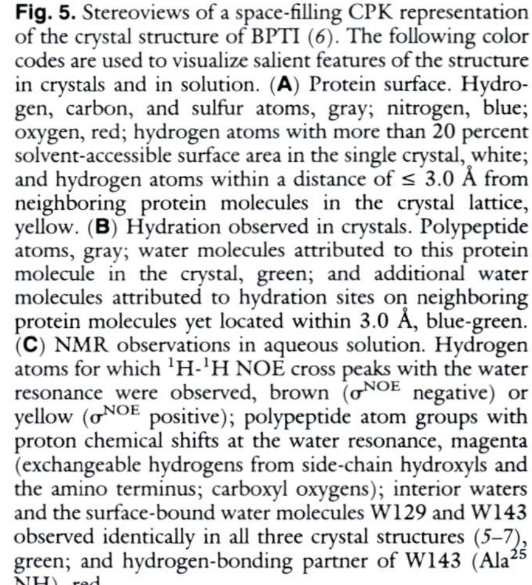

Fig. 5. Stereoviews of a space-filling CPK representation of the crystal structure of BPTI (*6*). The following color codes are used to visualize salient features of the structure in crystals and in solution. (**A**) Protein surface. Hydrogen, carbon, and sulfur atoms, gray; nitrogen, blue; oxygen, red; hydrogen atoms with more than 20 percent solvent-accessible surface area in the single crystal, white; and hydrogen atoms within a distance of ≤ 3.0 Å from neighboring protein molecules in the crystal lattice, yellow. (**B**) Hydration observed in crystals. Polypeptide atoms, gray; water molecules attributed to this protein molecule in the crystal, green; and additional water molecules attributed to hydration sites on neighboring protein molecules yet located within 3.0 Å, blue-green. (**C**) NMR observations in aqueous solution. Hydrogen atoms for which ^1H-^1H NOE cross peaks with the water resonance were observed, brown (σ^{NOE} negative) or yellow (σ^{NOE} positive); polypeptide atom groups with proton chemical shifts at the water resonance, magenta (exchangeable hydrogens from side-chain hydroxyls and the amino terminus; carboxyl oxygens); interior waters and the surface-bound water molecules W129 and W143 observed identically in all three crystal structures (*5–7*), green; and hydrogen-bonding partner of W143 (Ala^{25} NH), red.

protein surface are due to the hydroxyl protons of Ser, Thr, and Tyr, rather than to stably bound water molecules; and (ii) the surface hydration sites with well-ordered, x-ray–observable water molecules in the crystal structure have similar residence times for bound water in solution as other surface areas for which no hydration water is seen by x-ray diffraction (Fig. 5).

The conclusion (i) is supported by NMR observations made at 4°C in the pH range 5.0 to 6.5, where the chemical exchange of the hydroxyl protons of Ser, Thr, and Tyr becomes sufficiently slow [figure 2.3 in (*11*)] for separate signals to be seen away from the water line. Under these conditions the NOEs between nonlabile polypeptide protons and the hydroxyl proton resonances were observed as strong cross peaks with negative σ^{NOE} values that are well separated from the water chemical shift. These same NOEs are observed in the cross plane of Fig. 4 at the water frequency, since at pH 3.5 the hydroxyl protons exchange more rapidly and their resonances are coalesced with the water signal.

The conclusion (ii) results from two quite independent observations. First, out of a total of approximately 60 x-ray–observable surface hydration waters there are only six water molecules that have conserved hydrogen-bonding partners in all three single-crystal structures of BPTI (*5*–*7*). Of those, the hydration sites W143 and W129 (Fig. 5C; W, water) could so far be characterized by the NMR data. In the crystal structures, W143 is in hydrogen-bonding distance to the amide proton of Ala^{25}, which has vanishing cross peak intensity with the water line in NOESY and a weak NOE cross peak in the ROESY spectrum. W129 is within hydrogen-bonding distance of the amide proton of Ile^{19}, which interacts with a positive σ^{NOE} rate with the water. From these data, upper limits for the residence times can be established as <500 ps for W143 and <300 ps for W129, that is, the same as for other surface hydration waters. Second, in the crystal structure 5PTI (*6*) more than 50 percent of the x-ray–observable water molecules are within 3.0 Å of a backbone carbonyl oxygen, 40 percent are near a charged group, and only ~10 percent (7 out of a total of 63 water molecules) are in contact with the uncharged -OH groups of the eight residues of Ser, Thr, and Tyr. If the x-ray–observable surface hydration sites were characterized by outstandingly long residence times of the hydration water molecules in solution and correspondingly large negative σ^{NOE} values (Fig. 2C), there would be clear-cut discrepancies with the results of the NMR experiments. These experiments showed strong negative NOEs with the water resonance for protons near the -OH groups of Ser, Thr, and Tyr, which have already been shown to correspond to direct NOEs with these protons. Only a small number of additional NOEs with negative σ^{NOE} values were observed between the water resonance and polypeptide protons located on the protein surface, of which all but one could be assigned to proximity to the NH_2-terminal amino group. There were no strong NOE cross peaks with negative σ^{NOE} values left that could be attributed to the preferred binding sites for ordered hydration water, which are near the carbonyl oxygens and charged groups in the crystal structure. It can therefore be excluded that water molecules in these sites in the solution structure have significantly longer residence times than the other surface hydration waters.

Implications for protein hydration. X-ray data on protein crystals and NMR observation of individual hydration water molecules in solution agree in one aspect of protein hydration: Interior waters, which are part of the protein molecular architecture, are observed in identical locations of the protein molecule in solution and in crystals (*14, 21*). A general structural characterization of hydration sites that give rise to "interiorlike" behavior of the bound water molecules is therefore an interesting project for future research. A crucial difference between the results obtained with the two methods is that although there is a priori no x-ray evidence to distinguish between the properties of the interior waters and highly occupied surface hydration sites, these two types of hydration sites are clearly distinguished by the residence times of the water molecules manifested in the sign of σ^{NOE} in solution. For the interior waters the residence time is in the range of about 10^{-2} to 10^{-8} s (*22*), whereas for surface hydration waters it is in the subnanosecond range even at 4°C. The presently available evidence as described in this article implies that the extent to which the surface hydration water molecules are ordered, and hence observable by x-ray diffraction in protein crystals, cannot be correlated with significantly longer residence times of water molecules in the corresponding sites in solution. The experience gained with the NMR experiments shows that at 4°C, different residence times in the range <500 ps can be found for different individual surface hydration waters (yellow in Fig. 5C). Although a quantification of these differences would at present be premature, future improvements of the sensitivity of the NOE experiments should enable further refinements of the description of protein surface hydration in solution.

With regard to protein structure and function in aqueous milieus the surface hydration water provides a flexible matrix enabling the polypeptide chain to respond efficiently to environmental changes during processes such as protein folding, protein-protein complexation, and enzyme-substrate interaction. The permanent rearrangement of the hydration network on the protein surface must be an important factor for rapid approach to a near-global energy minimum in these processes, which also shows that any intermolecular recognition of a protein surface includes the characteristics of an induced fit [see, for example, (*38*)].

REFERENCES AND NOTES

1. F. M. Richards, *Sci. Am.* **264**, 34 (January 1991).
2. G. E. Schulz and R. H. Schirmer, *Principles of Protein Structure* (Springer-Verlag, New York, 1979).
3. J. D. Glickson, in *Peptides: Chemistry, Structure, Biology*, R. Walter and J. Meienhofer, Eds. (Ann Arbor Science, Ann Arbor, MI, 1975), pp. 787–802; V. J. Hruby, in *Topics in Molecular Pharmacology*, A. S. V. Burgen and G. C. K. Roberts, Eds. (North-Holland, Amsterdam, 1981), pp. 100–125.
4. S. P. Wood *et al.*, *Science* **232**, 633 (1986).
5. J. Deisenhofer and W. Steigemann, *Acta Crystallogr.* **B31**, 238 (1975).
6. A. Wlodawer, J. Walter, R. Huber, L. Sjölin, *J. Mol. Biol.* **180**, 301 (1984).
7. A. Wlodawer, J. Nachman, G. L. Gilliland, W. Gallagher, C. Woodward, *ibid.* **198**, 469 (1987).
8. T. L. Blundell and L. N. Johnson, *Protein Crystallography* (Academic Press, New York, 1976).
9. In one case, crambin, ring structures formed by mutually hydrogen-bonded water molecules on the protein surface were reported: M. M. Teeter, *Proc. Natl. Acad. Sci. U.S.A.* **81**, 6014 (1984).
10. F. C. Bernstein *et al.*, *J. Mol. Biol.* **122**, 535 (1977).
11. K. Wüthrich, *NMR of Proteins and Nucleic Acids* (Wiley, New York, 1986).
12. ———, *Science* **243**, 45 (1989).
13. ———, *Acc. Chem. Res.* **22**, 36 (1989).
14. G. Otting and K. Wüthrich, *J. Am. Chem. Soc.* **111**, 1871 (1989).
15. G. Otting, E. Liepinsh, B. T. Farmer II, K. Wüthrich, *J. Biomol. NMR* **1**, 209 (1991).
16. F. Solomon, *Phys. Rev.* **99**, 559 (1955); A. Abragam, *Principles of Nuclear Magnetism* (Clarendon, Oxford, 1961).
17. R. G. Bryant and W. S. Shirley, *Biophys. J.* **32**, 3 (1980); B. Halle, T. Andersson, S. Forsén, B. Lindman, *J. Am. Chem. Soc.* **103**, 500 (1981); S. H. Koenig, *Biophys. J.* **53**, 91 (1988).
18. C. F. Polnaszek and R. G. Bryant, *J. Chem. Phys.* **81**, 4038 (1984).
19. T. P. Pitner, J. D. Glickson, J. Dadok, G. R. Marshall, *Nature* **250**, 582 (1974); J. D. Glickson *et al.*, *Biochemistry* **15**, 1111 (1976); J. D. Stoesz, A. G. Redfield, D. Malinowski, *FEBS Lett.* **91**, 320 (1978).
20. G. Otting and K. Wüthrich, in *Water and Ions in Biomolecular Systems*, D. Vasilescu, J. Jaz, L. Packer, B. Pullman, Eds. (Birkhäuser, Basel, 1990), pp. 141–147.
21. G. M. Clore, A. Bax, P. T. Wingfield, A. Gronenborn, *Biochemistry* **29**, 567 (1990).
22. G. Otting, E. Liepinsh, K. Wüthrich, *J. Am. Chem. Soc.* **113**, 4363 (1991).
23. B. A. Messerle, G. Wider, G. Otting, C. Weber, K. Wüthrich, *J. Magn. Reson.* **85**, 608 (1989).
24. A. A. Bothner-By, R. L. Stephens, J. Lee, C. D. Warren, R. W. Jeanloz, *J. Am. Chem. Soc.* **106**, 811 (1985).
25. A. Kalk and H. J. C. Berendsen, *J. Magn. Reson.* **24**, 343 (1976).
26. R. Richarz, K. Nagayama, K. Wüthrich, *Biochemistry* **19**, 5189 (1980).

27. T. Fujiwara and K. Nagayama, *J. Chem. Phys.* **83**, 3110 (1985).
28. Y. Ayant, E. Belorizky, P. Fries, J. Rosset, *J. Phys.* (*Paris*) **38**, 325 (1977).
29. K. T. Gillen, D. C. Douglass, M. J. R. Hoch, *J. Chem. Phys.* **57**, 5117 (1972).
30. S. Meiboom, *ibid.* **34**, 375 (1961).
31. In Fig. 3 such exceptions with positive NOESY cross peaks are the amide proton of Tyr^2 and the side-chain amide protons of Gln^4, where the chemical exchange with the water dominates over the NOE in the NOESY spectrum. The broad positive peak at 8.35 ppm is due to chemical exchange of the amino group of Cys^1.
32. The ratio $\sigma^{NOE}/\sigma^{ROE}$ is −0.5 for the intensities of the CαH-NH cross peaks between sequentially neighboring amino acid residues, in agreement with $\tau_R = 2$ ns.
33. Exceptions with significantly different $\sigma^{NOE}/\sigma^{ROE}$ values are the CαH and CβH resonances of Cys^1, which show strong cross peaks with the water signal in the ROESY spectrum (Fig. 3C) but vanishing cross peak intensity in the corresponding NOESY trace (Fig. 3B). This is due to a superposition of NOEs with hydration water and NOEs with the protons of the αNH_3^+ group of Cys^1. Both NOEs have the same sign in ROESY but opposite sign in NOESY. Similar cancellation may be effective in the cross peak between the water signal and the CεH resonance of Tyr^2, where the σ^{NOE} rate with the labile side chain hydroxyl proton of Tyr^2 would be of opposite sign than the σ^{NOE} rate due to the interaction with hydration water.
34. The diagonal in Fig. 4 corresponds to the 1D cross section through the water line in a 2D NMR spectrum as shown in Fig. 3, except that because of the lesser digital resolution of about 0.1 ppm per point in the ω_1 dimension, the water signal in the 3D NMR spectrum is not completely resolved from all α-proton resonances of the protein, that is, the α-proton resonance of Cys^{38} overlaps with the water signal, and therefore its intraresidual NOEs with βCH_2 appear on the diagonal. See also (*15*).
35. W. Saenger, *Annu. Rev. Biophys. Biophys. Chem.* **16**, 93 (1987).
36. G. Wagner *et al.*, *J. Mol. Biol.* **196**, 611 (1987).
37. K. Berndt, P. Güntert, L. Orbons, K. Wüthrich, unpublished results.
38. D. E. Koshland, G. Neméthy, D. Filmer, *Biochemistry* **5**, 365 (1966).
39. Abbreviations for the amino acid residues are: A, Ala; C, Cys; G, Gly; I, Ile; K, Lys; L, Leu; M, Met; N, Asn; P, Pro; Q, Gln; R, Arg; and Y, Tyr.
40. Supported by the Schweizerischer Nationalfonds (project no. 31.25174.88). We thank T. J. Richmond for helpful discussions, Xia Tai-He for help with the solvent accessibility computation using the program ANAREA [T. J. Richmond, *J. Mol. Biol.* **178**, 63 (1984)], and R. Marani for the careful processing of the manuscript.

9 August 1991; accepted 16 October 1991

Reprinted from the Journal of the American Chemical Society, 1992, *114*.
Copyright © 1992 by the American Chemical Society and reprinted by permission of the copyright owner.
7093

Polypeptide Hydration in Mixed Solvents at Low Temperatures

Gottfried Otting, Edvards Liepinsh, and Kurt Wüthrich*

Contribution from the Institut für Molekularbiologie und Biophysik, Eidgenössische Technische Hochschule-Hönggerberg, CH-8093 Zürich, Switzerland. *Received November 5, 1991*

Abstract: The hydration of the nonapeptide oxytocin dissolved in a mixed solvent of 60% H_2O and 40% [$^2H_6$]acetone was investigated by nuclear magnetic resonance (NMR) spectroscopy. In the temperature range between 0 and −15 °C the nuclear Overhauser effects (NOEs) between the individual proton resonances of the peptide and the water signal were found to change their sign. At the temperature of the sign change the residence time of the hydration water molecules on the peptide surface is about 500 ps. The temperature-dependent transition from positive to negative NOEs provides a sensitive probe for assessing differences in the residence times of the hydration water molecules bound to different atom groups of a polypeptide chain.

The characterization of three-dimensional protein structures in solution at atomic resolution by nuclear magnetic resonance (NMR)[1] spectroscopy[2] has recently been extended to include a description of surface hydration.[3] A view at individual hydration sites in solution is afforded by the observation of nuclear Overhauser effects (NOEs) between the individual protons of the polypeptide and nearby water molecules.[4] Quite generally, hydration sites on polypeptide and protein surfaces are characterized by hydration lifetimes shorter than about 500 ps, as indicated by the positive sign of the cross-relaxation rate, σ^{NOE}, measured in NOESY cross peaks between polypeptide protons and water protons.[3] Water molecules bound with significantly longer residence times were so far observed only in interior cavities of globular proteins; their NOEs with the protein protons are characterized by large negative σ^{NOE} rates indicating residence times much longer than 500 ps for these waters.

In this paper we further investigate the residence times of water molecules near different atom groups of a polypeptide chain using the nonapeptide oxytocin

$$\text{H-}\overline{\text{Cys-Tyr-Ile-Gln-Asn-Cys}}\text{-Pro-Leu-Gly-NH}_2$$

as a model system. Oxytocin is flexibly disordered in aqueous solution[5] and is devoid of interior cavities that could contain bound water molecules. Correspondingly, all NOEs between oxytocin protons and water protons were found to have positive σ^{NOE} rates at 8 °C, which is characteristic of hydration lifetimes shorter than 500 ps.[3] In the following we show that as the temperature is lowered in the range between 0 and −15 °C the sign of the σ^{NOE} rates becomes negative for all oxytocin–water NOEs. The temperature-dependent transitions from positive to negative values of σ^{NOE} enable a distinction of different hydration sites by their hydration lifetimes.

Results and Discussion

To investigate the water proton–polypeptide proton NOEs at temperatures below 0 °C, oxytocin was dissolved in a mixed solvent of 60% H_2O and 40% [$^2H_6$]acetone. This solution has a freezing point of approximately −28 °C. Two-dimensional NOE spectra in the laboratory frame (NOESY) and the rotating frame (ROESY) were recorded with the experimental schemes of Figure 1, as well as of Figure 1B of ref 8. Figures 2 and 3 show respectively cross sections through the NOESY and ROESY spectra taken at the ω_1 chemical shift of the water signal. Positive cross peaks in the ROESY spectrum indicate chemical exchange with the water, whereas all NOE cross peaks are negative (Figure 3).[9] Over the entire temperature range from 10 to −25 °C chemical exchange peaks are observed for the labile protons of the N-terminal α-amino group and the –OH group of Tyr 2. For these protons the exchange with the water is sufficiently rapid to result in efficient bleaching of the cross-peak intensities at 10 °C. At all four temperatures studied, the intrapeptide NOEs with these two resonances are therefore transferred to the water resonance, so that the intensities of the cross peaks with αCH and βCH_2 of Cys 1 and εCH of Tyr 2 (Figures 2 and 3) are partly or entirely due to such transfer effects rather than to direct NOE cross peaks with the water line. The only other exchange peak in Figure 3 is with the amide proton of Tyr 2 at +10 °C. Since this proton exchanges much more slowly than the aforementioned examples, the NOE magnetization transfer from the water is more effective than chemical exchange at temperatures below 0 °C (Figure 3). All other peaks are negative throughout and represent direct NOE interactions with hydration water at all temperatures studied. With only a few exceptions due to overlap with intrapeptide cross peaks (circles and crosses in Figures 2 and 3), the NOE cross peaks with the water line could readily be assigned to individual protons of oxytocin.

Since oxytocin is flexibly disordered at all temperatures studied,[5] only intraresidual and sequential NOEs[2] were observed. The NOESY cross peaks between different peptide resonances were positive throughout, indicating that the peptide is in the slow motional regime. In contrast, the sign of the NOESY cross peaks between polypeptide protons and the water changed with temperature (Figure 2), although the corresponding peaks were always negative in the ROESY spectra (Figure 3). This behavior is a consequence of the different dependence of the cross relaxation rates in NOESY and ROESY on the spectral density function, $J(\omega)$, where ω_0 is the angular Larmor frequency:

$$\sigma^{NOE} = 6J(2\omega_0) - J(0) \quad (1)$$

$$\sigma^{ROE} = 3J(\omega_0) + 2J(0) \quad (2)$$

The spectral density function $J(\omega)$ depends further on the effective correlation time for the vector connecting the water proton with the proton of the peptide. σ^{NOE} is respectively positive and negative for rapid and slow changes in orientation and length of this interproton vector, where positive σ^{NOE} values are manifested by negative NOESY cross peaks and vice versa. Quite independent of the motional model used to calculate $J(\omega)$,[10,11] the sign change

(1) Abbreviations and symbols used: NMR, nuclear magnetic resonance; NOE, nuclear Overhauser enhancement; σ^{NOE}, cross-relaxation rate in the laboratory frame; σ^{ROE}, cross-relaxation rate in the rotating frame; 2D, two-dimensional; NOESY, two-dimensional nuclear Overhauser enhancement spectroscopy; ROESY, two-dimensional nuclear Overhauser enhancement spectroscopy in the rotating frame.

(2) (a) Wüthrich, K. *NMR of Proteins and Nucleic Acids*; Wiley: New York, 1986. (b) Wüthrich, K. *Science* **1989**, *243*, 45–50. (c) Wüthrich, K. *J. Biol. Chem.* **1990**, *265*, 22059–22062.

(3) Otting, G.; Liepinsh, E.; Wüthrich, K. *Science* **1991**, *254*, 974–980.

(4) Otting, G.; Wüthrich, K. *J. Am. Chem. Soc.* **1989**, *111*, 1871–1875.

(5) (a) Glickson, J. D. *Peptides: Chemistry, Structure, Biology*; Walter, R., Meierhofer, J., Eds.; Ann Arbor Science: Ann Arbor, 1975; pp 787–802. (b) Hruby, V. J. *Topics in Molecular Pharmacology*; Burgen, A. S. V., Roberts, G. C. K., Eds.; North-Holland: Amsterdam, 1981; pp 100–125.

(6) Leroy, J. L.; Broseta, D.; Guéron, M. *J. Mol. Biol.* **1985**, *184*, 165–178.

(7) Hoult, D. I.; Richards, R. E. *Proc. R. Soc. London, Ser. A* **1975**, *344*, 311–340.

(8) Otting, G.; Liepinsh, E.; Farmer, B. T., II; Wüthrich, K. *J. Biomol. NMR* **1991**, *1*, 209–215.

(9) Bothner-By, A. A.; Stephens, R. L.; Lee, J.; Warren, C. D.; Jeanloz, R. W. *J. Am. Chem. Soc.* **1984**, *106*, 811–813.

(10) (a) Richarz, R.; Nagayama, K.; Wüthrich, K. *Biochemistry* **1980**, *19*, 5189–5196. (b) Fujiwara, T.; Nagayama, K. *J. Chem. Phys.* **1985**, *83*, 3110–3117.

(11) Ayant, Y.; Belorizky, E.; Fries, P.; Rosset, J. *J. Phys. Paris* **1977**, *38*, 325–337.

0002-7863/92/1514-7093$03.00/0 © 1992 American Chemical Society

7094 J. Am. Chem. Soc., Vol. 114, No. 18, 1992

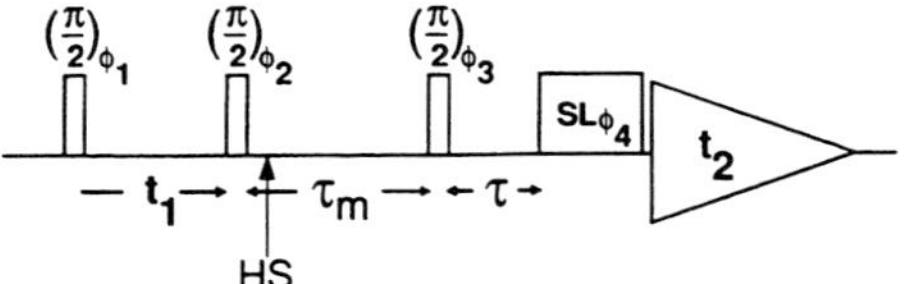

Figure 1. Experimental scheme used to record the NOESY spectra of Figure 2. One or several homospoil pulses, HS, at the beginning of the mixing time, τ_m, defocus transverse magnetization and thus prevent radiation damping.[6] The carrier is set at the water frequency and the water signal is suppressed using a spin lock pulse of 1–2 ms duration, $SL_{\phi 4}$, which is separated from the mixing time by a delay τ of about 1/SW, where SW is the spectral width in Hz. The excitation profile along the ω_2 frequency axis is given by sin ($\Omega\tau$), where Ω is the angular frequency relative to the carrier frequency. The phase cycle is $\phi_1 = 8(x)8(-x)$, $\phi_2 = 8(x,-x)$, $\phi_3 = 4(x,x,-x,-x)$, $\phi_4 = 2(4(x),4(-x))$, $\phi_5 = 2(x,-x,-x,x)2(-x,x,x,-x)$, which can be extended to 64 steps by CYCLOPS.[7] The experiment is also well suited for the measurement of water proton–peptide proton NOEs with mixing times shorter than 10 ms. (In a previously published NOESY pulse sequence (Figure 1A of ref 8) an additional spin lock pulse was applied immediately after the $\pi/2$ pulse following τ_m, instead of using a homospoil pulse during the mixing time. The advantage of the present experiment lies in the fact that the additional spin lock pulse used in ref 8 could contribute significant amounts of ROE mixing when used with short mixing times τ_m.)

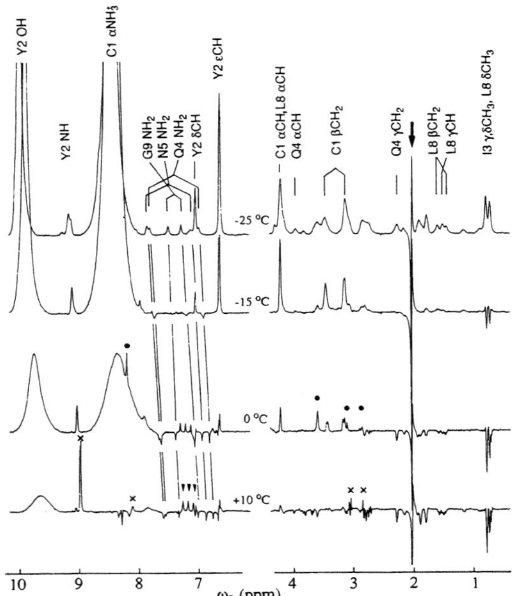

Figure 2. Cross sections along ω_2 taken in NOESY spectra of oxytocin at the ω_1 chemical shift of the water resonance; the peaks represent either NOEs or chemical exchange between oxytocin protons and water protons (oxytocin concentration 50 mM, solvent 60% H_2O/40% [$^2H_6$]acetone, $pH_{app} = 3.5$, ^1H frequency 600 MHz, time domain data size 450 × 2048 points; the experimental scheme of Figure 1 was used with $t_{1max} = 36$ ms, $t_{2max} = 156$ ms, 30 ms mixing time, $\tau = 135$ µs, and a 1 ms spin lock pulse). Since the excitation profile leads to sign inversion of all signals on both sides of the water line, the low-field region of each cross section was inverted in this figure for improved readability. The spectra were baseline corrected in both dimensions using polynomials. Selected peaks are labeled with the one-letter amino acid symbol and the sequence position. The temperature-dependent resonance positions of the amide NH_2 protons are identified by lines between the spectra. The cross peak of acetone at 2.05 ppm (arrow) was used as a chemical shift reference. Chemical exchange peaks from NH_4^+ ions present as an impurity are labeled with triangles, and intrapeptide cross peaks overlapping with the water line are identified with filled circles in the spectrum at 0 °C and with crosses at 10 °C.

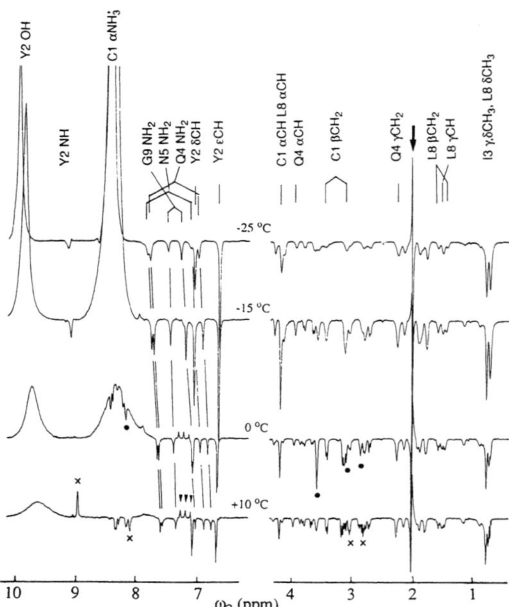

Figure 3. Cross sections along ω_2 taken in ROESY spectra of oxytocin at the ω_1 chemical shift of the water resonance. The spectra were recorded with an experimental scheme that corresponds to the NOESY scheme of Figure 1 and is described in detail in ref 8. See Figure 2 for sample conditions and presentation of the spectra.

in σ^{NOE} is expected to occur when the effective correlation time is of the order of the inverse of the angular Larmor frequency, i.e., about 500 ps at 600 MHz.[3] This time span is much shorter than the lifetime of a water proton with respect to hydrolysis,[12] so that these NOE observations must represent exchange of all the water molecules rather than proton exchange by hydrolysis. At 10 °C the negative NOESY cross peaks (Figure 2) are due to the translational diffusion of water molecules relative to the peptide protons, with diffusion coefficients that are only about four times smaller than the self-diffusion coefficient of pure water at the same temperature.[3] The positive cross peaks seen at −25 °C show that at this temperature the hydration water molecules bind to the peptide surface with residence times that are longer than 500 ps.

With regard to future applications of these principles for detailed studies of protein hydration, it is of special interest that individual polypeptide protons in oxytocin show different behavior in the experiment of Figure 2. Thus, although the NOE cross peaks with the amide protons (spectral range 8.1–8.5 ppm) and the α-protons (spectral range 3.7–5.0 ppm) vanish in the NOESY spectrum at 0 °C (Figure 2) (at the same temperature the corresponding cross peaks have nonvanishing intensity in the ROESY spectra of Figure 3), most NOESY cross peaks with side chain protons invert their sign at lower temperatures, i.e., between 0 and −15 °C, and negative NOESY cross peaks for the side chain methyl groups of Ile 3 and Leu 8 and the side chain amide group of Gln 4 are observed even at −15 °C. It remains to be seen to what extent the decreased effective correlation time for the water proton–polypeptide proton interactions is due to increased segmental mobility of the side chains or to shorter water residence times on the side chains than on the backbone, respectively.

Increased side chain mobility is indicated by the smaller $\sigma^{NOE}/\sigma^{ROE}$ ratio observed for the NOEs between backbone protons and side chain protons than between protons of the backbone. A quantitative evaluation of the polypeptide–water NOEs would

(12) Meiboom, S. *J. Chem. Phys.* **1961**, *34*, 375–388.

therefore require the calculation of the spectral density function $J(\omega)$ in the presence of side chain motions. To date, no analytical formula has been derived for a model that would describe that situation. Assuming similar mobilities for all side chains, it is interesting to note that the sign inversion of σ^{NOE} occurs at a similar temperature for the side chain amide protons as for the methyl resonances of Ile 3 and Leu 8. This seems to indicate that the water does not bind with significantly increased residence times to the polar amide groups when compared to the apolar side chains.

Figure 3 shows that the intensities of the ROESY cross peaks between the water signal and the polypeptide resonances increase with decreasing temperature between +10 and −15 °C and are again decreased at −25 °C. This behavior is a consequence of the fact that the cross-relaxation rates increase with decreasing molecular motions. With the mixing time of 30 ms used to record the spectra of Figure 2 and 3, only the spectra at 10 °C fulfill the initial rate condition, where the cross-peak intensities are directly proportional to σ^{ROE}. At lower temperatures the increased cross-relaxation rates lead to more pronounced relaxation and spin diffusion, and therefore to a decrease of the overall intensities in the ROESY spectrum at −25 °C. Additional NOESY and ROESY spectra recorded with mixing times as short as 3 ms confirmed that both σ^{NOE} and σ^{ROE} increase steadily with lower temperature (data not shown).

While the sign inversion of σ^{NOE} provides a sensitive tool to probe the residence times of the hydration waters around individual atom groups of a polypeptide chain, the *size* of σ^{ROE} is far less sensitive to the hydration lifetime in the temperature range studied, and increases only about 3-fold between 0 and −15 °C. Therefore, the size of σ^{ROE} is mainly indicative of the average spatial proximity of the water protons to the polypeptide proton and of the number of water protons involved in the interaction. Variations in the size of σ^{ROE} between different polypeptide protons are most easily assessed by comparison of the one-dimensional ^{1}H NMR spectrum of the solute with the one-dimensional cross section along ω_2 taken at the ω_1 chemical shift of the water resonance in a 2D ROESY spectrum recorded with a short mixing time. Interestingly, no significant variations in σ^{ROE} were observed for the oxytocin–water NOEs in aqueous solution at 8 °C (Figure 3, A and B, of ref 3). This observation supports the view that as a consequence of the motional and orientational disorder of polypeptide hydration the average number, orientation, and residence time of the hydration water molecules are similar for all types of protons in a polypeptide chain. Similar behavior was observed with the oxytocin solution in 60% H_2O and 40% [$^2H_6$]acetone at even lower temperatures.

To assess the role of acetone in the peptide environment, we recorded NOESY and ROESY spectra at 10 and −15 °C with a solution of oxytocin in 60% ^2H_2O/40% acetone. The same experimental schemes were used as for the spectra of Figures 2 and 3, and the carrier was set to the frequency of the acetone resonance for solvent signal suppression. At either temperature only a few very weak cross peaks with the acetone signal were observable, and all these cross peaks had positive σ^{NOE} values. This shows clearly that in this mixed solvent oxytocin is more stably solvated by water than by acetone.

Conclusions

Quite generally, the ratio $\sigma^{NOE}/\sigma^{ROE}$ may be used to characterize the hydration lifetimes of different hydration sites on a polypeptide or a protein surface. However, the quantitative determination of the cross-relaxation rates is difficult for weak NOEs, since it would require that NOESY and ROESY spectra are recorded with sufficiently short mixing times to fulfill the initial rate condition, where the cross peaks are weak. As a favorable alternative, the temperature-dependent transitions from positive to negative values of σ^{NOE} can be reliably observed even in experiments with longer mixing times, which enables an accurate distinction of different hydration sites by their residence times in a much simpler way than by quantitative measurements of $\sigma^{NOE}/\sigma^{ROE}$ ratios at a single temperature.

The experiments of Figures 2 and 3 present a clearcut corroboration of the previous findings[3] that the residence times for water molecules on peptide or protein surfaces are in the subnanosecond time range at temperatures above 5 °C. These measurements clearly have the potential to discriminate between different individual hydration sites on the basis of the different effective correlation times (Figure 2). If effects from segmental mobility can be assessed, it will be possible to obtain a quantitative description of possible variations of the hydration water residence times in different sites on the protein surface. This information is not otherwise available with present methods; in particular, it cannot be derived from observations on protein hydration made by X-ray diffraction in single crystals.[3] Experiments of the type shown in Figures 2 and 3 can thus contribute to a more detailed characterization of the hydration of polypeptide chains in solution, which is of fundamental interest for a better understanding of protein folding, protein stability, and intermolecular interactions with proteins. As is indicated by the aforementioned experiments with oxytocin in 60% ^2H_2O/40% acetone, similar experimental approaches can be employed for investigations on preferred solvation in solutions with mixed solvents. This could be particularly attractive for studies of the solvation of unfolded forms of proteins resulting from the addition of chemical denaturants.

Acknowledgment. We acknowledge financial support by the Schweizerischer Nationalfonds (Project 31-24174.88) and thank Mr. R. Marani for the careful processing of the manuscript.

© 1992 Oxford University Press Nucleic Acids Research, 1992, Vol. 20, No. 24 6549–6553

NMR observation of individual molecules of hydration water bound to DNA duplexes: direct evidence for a spine of hydration water present in aqueous solution

Edvards Liepinsh, Gottfried Otting and Kurt Wüthrich*
Institut für Molekularbiologie und Biophysik, ETH-Hönggerberg, CH-8093 Zürich, Switzerland

Received September 19, 1992; Accepted October 21, 1992

ABSTRACT

The residence times of individual hydration water molecules in the major and minor grooves of DNA were measured by nuclear magnetic resonance (NMR) spectroscopy in aqueous solutions of d-(CGCGAATT-CGCG)$_2$ and d-(AAAAATTTTT)$_2$. The experimental observations were nuclear Overhauser effects (NOE) between water protons and the protons of the DNA. The positive sign of NOEs with the thymine methyl groups shows that the residence times of the hydration water molecules near these protons in the major groove of the DNA must be shorter than about 500 ps, which coincides with the behavior of surface hydration water in peptides and proteins. Negative NOEs were observed with the hydrogen atoms in position 2 of adenine in both duplexes studied. This indicates that a 'spine of hydration' in the minor groove, as observed by X-ray diffraction in DNA crystals, is present also in solution, with residence times significantly longer than 1 ns. Such residence times are reminiscent of 'interior' hydration water molecules in globular proteins, which are an integral part of the molecular architecture both in solution and in crystals.

INTRODUCTION

The hydration of DNA duplexes has been the subject of great interest in context with attempts to rationalize the sequence-specific recognition of DNA by proteins and other compounds (*e.g.*, 1, 2), and with the possible role of water in stabilizing sequence-dependent conformational variations in double-helical DNA (3–9). Dickerson and coworkers noted the importance of hydration water in the crystal structure of the self-complimentary dodecamer duplex d-(CGCGAATTCGCG)$_2$, where the minor groove of the central base pairs is filled with an ordered zig-zag array of water molecules, the 'spine of hydration' (3, 4). Subsequent conformational energy calculations suggested that the presence of the spine of hydration is a prime reason for the significant narrowing of the minor groove of poly(dA)·poly(dT) tracts in B-type DNA conformations (7), while other theoretical studies indicated that a similar spine of hydration may also be present in the minor groove of G·C rich DNA sequences (8, 9). Studies of biological macromolecules in aqueous solution have recently added an important facet to rationalizing the important structural role of the hydration water, by demonstrating conclusively that the residence times at the hydration sites are usually very short. Thus, the residence times of surface hydration water molecules in proteins were shown by nuclear magnetic resonance (NMR) experiments to be shorter than about 500 ps, while the residence times observed for hydration water molecules in the protein interior, where they represent an integral part of the protein architecture, were found to be in the range of about 10^{-3} to 10^{-8} s (10). The present paper describes NMR evidence that the water molecules of the spine of hydration in DNA duplexes have residence times longer than about one nanosecond, which is comparable to the behavior of interior waters in globular proteins. This emphasises the important role of these water molecules in the molecular architecture of B-DNA.

MATERIAL AND METHODS

NMR detection of hydration water

Hydration water near DNA protons can be detected by the observation of nuclear Overhauser effects (NOE) between the protons of the DNA and the protons of the water. Because of water exchange between the hydration water sites and the bulk water, all water protons appear at the chemical shift of the dominant signal of the bulk water (11). Therefore, in two-dimensional (2D) [^{1}H, ^{1}H]-NMR experiments the water–DNA NOEs are detected in a single cross section along ω_2 taken at the ω_1 chemical shift of the water line (12).

In assigning water–DNA NOE cross peaks, care has to be taken to discriminate between these chemical exchange peaks, and NOEs of non-labile protons of the DNA with labile protons of the DNA which exchange sufficiently rapidly with the water to appear at the bulk water chemical shift (10–13). In the present work the assignment of direct water–DNA NOE cross peaks was based on the following two criteria: (i) the DNA protons involved in the cross peaks do not exchange rapidly with the water; (ii) the DNA protons with NOEs to the water resonance are spatially well separated from potentially labile and rapidly exchanging DNA protons in the B-DNA type conformations. Rapidly exchanging, labile DNA protons were identified by their

* To whom correspondence should be addressed

Reprinted by permission of Oxford University Press

chemical exchange cross peaks with the water signal, which have positive sign in NOESY (NOE spectroscopy in the laboratory frame of reference) *and* ROESY (NOE spectroscopy in the rotating frame of reference) spectra, while NOE cross peaks are negative in ROESY (12, 14). The aforementioned criterion (ii) is applied to eliminate potential errors in the assignment of direct water−DNA NOEs for water molecules in the slow motional regime (*i.e.*, water molecules bound with residence times exceeding 1 ns), which cannot *a priori* be distinguished from NOEs between non-labile and labile DNA protons with magnetization transfer to the bulk water signal by rapid chemical exchange.

NMR sample preparation

The self-complementary DNA sequences d-$(CGCGAATTCGCG)_2$ (*dodecamer*) and d-$(A_5T_5)_2$ (*decamer*) were investigated. To slow down the chemical exchange of the labile protons in the DNA duplexes with the solvent, the samples were desalted by extensive ultrafiltration to remove exchange catalysts such as phosphate ions (*e.g.*, 15). The lyophilized DNA samples were dissolved in a mixture of 90% H_2O/10% D_2O and the pH adjusted by the addition of minute amounts of HCl or NaOH. The final sample concentrations were about 1.0 mM and 1.2 mM in duplex for the *decamer* and the *dodecamer*, respectively, with a pH value of 6.0 for the *decamer* and 7.0 for the *dodecamer*.

NMR measurements

All NMR measurements were performed at low temperatures, *i.e.*, 10°C and 4°C, to slow down the exchange of the imino and amino protons of the DNA. Two-dimensional homonuclear ^{1}H NOESY and ROESY spectra were recorded under identical conditions on a Bruker AMX 600 NMR spectrometer. Adequate water suppression was achieved with the use of the SL_x-τ-SL_y element before the acquisition period, where SL_x and SL_y denote spin-lock pulses of 0.5 and 2 ms duration, respectively (16), or with a modified scheme where the first spin-lock pulse was replaced by a homospoil pulse applied at the beginning of the NOESY mixing period (13). The delay τ was set to 156 μs, which results in optimum spectral excitation near 2.3, 7.7 and 13.1 ppm (16). After Fourier transformation, all spectra were baseline corrected in both dimensions to avoid interference of baseline artefacts with the one-dimensional cross section through the water line.

RESULTS

Intermolecular NOEs of d-$(CGCGAATTCGCG)_2$, with hydration water molecules

Figure 1 shows the one-dimensional ^{1}H NMR spectrum and cross sections through the two-dimensional NOESY and ROESY spectra of d-$(CGCGAATTCGCG)_2$ taken along ω_2 at the ω_1 chemical shift of the water resonance. Intermolecular water-DNA NOEs are observed for the protons in position 2 of A5 and A6 and for the methyl groups of T7 and T8. All these protons show negative cross peaks in ROESY (Figure 1B). Since they are far from any rapidly exchanging labile protons of the DNA, they must represent direct NOEs between the DNA and hydration water molecules. Most important, the cross peaks of A5 2H and A6 2H are positive in the NOESY cross section, with the cross peak of A5 2H appearing as a shoulder of the more intense cross peak of C 1 4NH^b (Figure 1C). In the same spectrum, negative cross peaks are observed with the methyl groups of T7 and T8, the protons in position 8 of guanine, the proton in position 3 of C3, and some of the deoxyribose 2′ protons. All these protons are located in the major groove of the DNA. As shown previously (10), negative NOESY cross peaks indicate rapid modulation of the internuclear vector connecting the protons of the macromolecule with those of the hydration water molecules, showing that the hydration water residence times are shorter than about 500 ps (10). (Note that in the presentation of Figure 1, negative NOESY cross peaks correspond to positive cross relaxation rates, σ^{NOE}, and *vice versa*.) In this way, the different signs observed for the NOEs with adenine 2H and with different protons in the major groove present a direct experimental criterion to distinguish highly mobile hydration water molecules in the major groove of the DNA, which have residence times shorter than about 0.5 ns, from water molecules near A5 2H and A6 2H in the minor groove of the DNA, which must be bound with residence times longer than 1 ns.

The identification of the aforementioned DNA−H_20 NOEs was dependent on a detailed analysis of the origin of the other

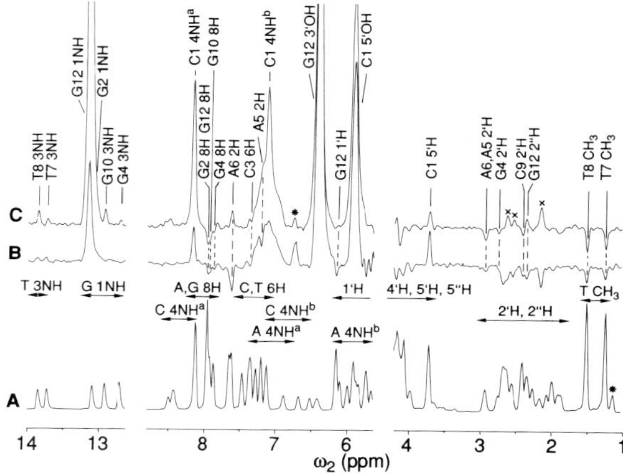

Figure 1. Proton NMR spectra showing NOEs between protons of d-$(CGCGAATTCGCG)_2$ and water protons (*dodecamer* concentration = 1.2 mM in duplex, solvent 90% H_20/10% D_2O, T = 10°C, pH = 7.0, ^{1}H frequency = 600 MHz; the experimental schemes of Figure 1, A and B, in ref. (16) were used, with mixing times τ_m = 60 ms for NOESY and τ_m = 30 ms for ROESY, $SL_{\phi4}$ = 0 and $SL_{\phi5}$ = 2 ms; a homospoil pulse of 4 ms duration was applied at the beginning of the mixing time in the NOESY experiment (13); time domain data size = 330×2048 points, t_{1max} = 33 ms and t_{2max} = 84 ms, total experimental time about 20 h per spectrum; spectral excitation profile sin[0.59(δ-5.0)], where δ is the chemical shift relative to TSP ([2,2,3,3-D_4]-trimethyl-silylpropionate) in ppm, with excitation maxima near 7.7 and 2.3 ppm; the spectral region between 5.8 and 8.7 ppm has been inverted for improved readability). (A) Conventional one-dimensional ^{1}H NMR spectrum obtained by projecting the NOESY spectrum along ω_1 onto the ω_2 frequency axis. (B) Cross section through the ROESY spectrum along ω_2 at the ω_1 frequency of the water line. (C) Cross section through the NOESY spectrum along ω_2 at the ω_1 frequency of the water line. In (A) the double arrows indicate the chemical shift ranges for the different hydrogen positions in the *dodecamer*. In (C), resonance assignments for individual peaks are indicated with the one-letter symbol for the nucleotide and the sequence position in d-$(C_1G_2C_3G_4A_5A_6T_7T_8C_9G_{10}C_{11}G_{12})$, where the same numeration is used for both strands to account for the two-fold symmetry in the NMR spectrum. For amino groups the two protons are distinguished by the superscripts a and b. Crosses (×) in (C) identify cross peaks corresponding to intramolecular NOEs with those 3′ deoxyribose protons that have their chemical shifts at or near the water frequency, and asterisks in (A) and (C) identify peaks arising from impurities.

peaks in Figure 1. The largest signals in both the NOESY and ROESY cross sections are the positive exchange cross-peaks from the imino proton of the terminal base pair, G12 1NH, at 13.1 ppm, and the hydroxyl protons G12 3′OH at 6.4 ppm and C1 5′OH at 5.95 ppm. Note that with the single exceptions of G12 1NH and possibly G2 1NH, all imino protons exchange too slowly with the solvent to lead to strong exchange cross peaks at 10°C. Intense exchange-relayed cross peaks (17) are observed with the $4NH_2$ group of C1, which arise from the following sequence of magnetization transfers during the mixing time: $H_20 \rightarrow$ G12 1NH by chemical exchange, G12 1NH$\rightarrow$C1 $4NH^a$ by NOE, and C1 $4NH^a \rightarrow$ C1 $4NH^b$ by exchange due to the rotation of the $4NH_2$ group about the 4C-4N bond. As one would expect from this cascade of transfers, the cross peak intensities with the water resonance decrease in the same order (Figure 1, B and C). (The corresponding cross peaks in the ROESY cross section (Figure 1B) have much smaller intensities than in NOESY, because the carner frequency of the spin-lock was placed at the water frequency. This caused strong off-resonance effects for the imino proton region in the ROESY experiment (18), which hindered efficient magnetization transfer from the water resonance to the imino protons, and from the imino protons to the C1 $4NH_2$ signals.) The broad exchange cross peak undemeath the C1 $4NH^b$ cross peak was attributed to the NH_2 group of G12, which is broadened by the rotation about the 2C-2N bond. The weak cross peaks of G12 2′′H and G12 1′H arise from intramolecular NOEs with the labile chain-terminal hydroxyl proton G12 3′OH, which appear at the water frequency due to chemical exchange of this hydroxyl proton with the water. Similarly, the cross peak with C1 5′H at about 3.7 ppm arises from interaction with C1 5′OH. The fact that the cross

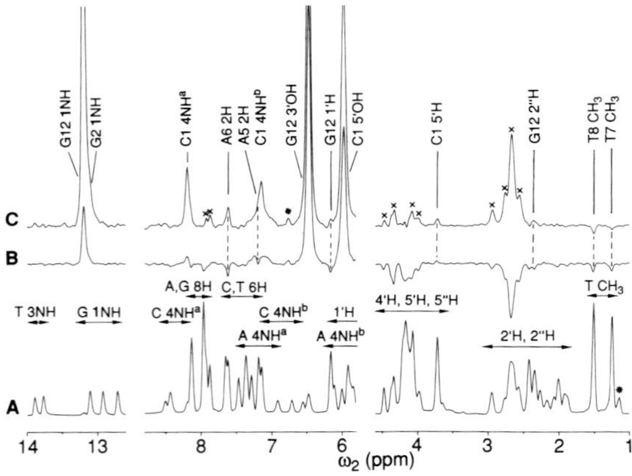

Figure 2. Proton NMR spectra showing NOEs between protons of d-(CGCG-AATTCGCG)$_2$ and water protons (*dodecamer* concentration = 1.2 mM in duplex, solvent 90% H_2O/10% D_2O, T = 4°C, pH 7.0, ^1H frequency = 600 MHz; same experimental schemes as in Figure 1 with identical parameters, except that the homospoil pulse was of 5 ms duration, the time domain data size was 580×2048 points, t_{1max} was 37 ms and the total experimental time about 45 h per spectrum). (A) One-dimensional ^1H NMR spectrum obtained by projecting the NOESY spectrum along ω_1 onto the ω_2 frequency axis. (B) Cross section through the ROESY spectrum along ω_2 at the ω_1 frequency of the water line. (C) Cross section through the NOESY spectrum along ω_2 at the ω_1 frequency of the water line. Peak identification as in Figure 1.

peak is positive in the ROESY cross section (Figure 1B) is explained by a homonuclear Hartmann–Hahn effect with the hydroxyl proton. While this magnetization transfer mechanism between scalar coupled protons may occur in ROESY experiments (19), it is not present in NOESY, where the corresponding cross peak represents a NOE with the rapidly exchanging proton of the C1 5′OH group. Further intense cross peaks are from intramolecular NOEs with the 3′-protons of G2, T8 and G10, which have virtually the same chemical shift as the water resonance (crosses in Figure 1).

A further interesting aspect is revealed by a comparison of the cross peak intensities in the NOESY and ROESY cross sections of Figure 1. Because a two-fold longer mixing time was used in the NOESY experiment and because of the off-resonance effect in ROESY which decreases the signal intensities for the resonance frequencies far from the water frequency (see above), the exchange cross peaks and NOE cross peaks are more intense in NOESY than in ROESY. An important exception is presented by the water–DNA NOE of A6 2H, which is almost twice as intense in ROESY. Considering the two-fold longer mixing time used in the NOESY experiment of Figure 1C, the cross relaxation rate, σ, between this proton and the protons of the hydration water molecules must be about four times faster in ROESY (σ^{ROE}) than in NOESY (σ^{NOE}), although σ^{ROE} would be expected to be at most two times larger than σ^{NOE} for hydration water molecules that are stably bound with a lifetime $\geq$ 1 ns. Therefore, the reduced NOE intensity in NOESY indicates either local reorientation of the hydration water during the residence time, or exchange with the bulk water on a time scale shorter than 1 ns (compare Figure 2 of ref. 10).

The aforementioned result is supported by corresponding experiments recorded at 4°C. Figure 2, B and C, shows the cross sections taken at the ω_1 chemical shift of the water line through the NOESY and ROESY experiments. Most notably, the NOE cross peak between the water signal and the A6 2H resonance is more intense in NOESY (Figure 2C) than in ROESY (Figure 2B). This shows that at this temperature σ^{NOE} has reached its maximum value attainable in the slow motional regime, where σ^{NOE} is half as big as σ^{ROE} (14). The more intense NOESY cross peak in Figure 2C is then explained by the two-fold longer mixing time used in the NOESY experiment and by the faster auto-relaxation rate during the ROESY mixing time. While the hydration water near A6 2H is thus shown to be immobile at 4°C on a time scale of about 1 ns, the NOESY cross peaks between the water and the thymine methyl goups in the major groove are still negative, which indicates that the residence times of the hydration water molecules near these groups are still shorter than about 500 ps at 4°C. From the spectrum of Figure 2C it cannot be decided whether the other negative NOESY cross peaks observed at 10°C (Figure 1C) are also present at 4°C, because the water resonance is shifted to lower field by about 0.07 ppm when going from 10°C to 4°C, so that there is more pronounced overlap with some intramolecular NOEs with the 3′H resonances of the DNA (crosses in Figure 2).

Intermolecular NOEs of d-$(A_5T_5)_2$ with hydration water molecules

Hydration water molecules that are stably bound in the DNA minor groove are also evidenced by experiments corresponding to those of Figures 1 and 2 performed with d-$(A_5T_5)_2$ (Figure 3). A detailed NMR investigation of this decamer was previously reported (20), and here we limit the discussion to the spectral

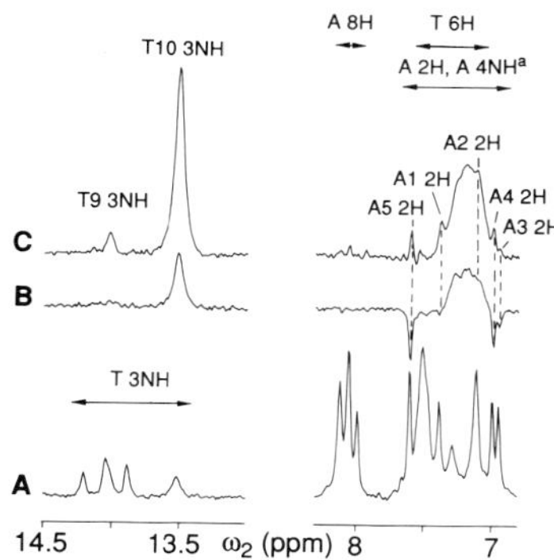

Figure 3. Spectral regions (6.8–8.3 ppm and 13.0–14.5 ppm) from NMR spectra showing NOEs between protons of d-$(A_5T_5)_2$ and water protons (*decamer* concentration 1.0 mM in duplex, solvent 90% H_2O/10% D_2O, T = 4°C, pH = 6.0, ^1H frequency = 600 MHz; same experimental schemes as in Figure 1, except that $SL_{\phi 4}$ = 0.5 ms and no homospoil pulse was used; time domain data size = 440×2048 points, t_{1max} = 30 ms, $t2_{2max}$ = 84 ms τ_m = 30 ms in both NOESY and ROESY, total experimental time about 46 h; same spectral excitation profile as in Figure 1; the spectral region between 6.8 and 8.3 ppm was inverted before plotting). (A) One-dimensional ^1H NMR spectrum obtained by projecting the NOESY spectrum along ω_1 onto the ω_2 frequency axis. (B) Cross section through the ROESY spectrum along ω_2 at the ω_1 frequency of the water line. (C) Cross section through the NOESY spectrum along ω_2 at the ω_1 frequency of the water line. Peak identification as in Figure 1.

regions containing the resonances of the imino protons, amino protons and base protons of A and T. The most important observation is that positive NOESY cross peaks are observed with the protons in position 2 of A3, A4 and A5 (Figure 3C). The cross peak with A2 2H is very weak and the cross peak with A1 2H is probably due to an exchange-relayed NOE with the labile imino proton of T10 rather than a direct NOE with hydration water. We conclude that in the non-terminal base pairs of the DNA duplex, the hydration water near the A 2H protons is characterized by residence times longer than 1 ns, while the weaker NOE cross peak intensities observed towards the chain ends in the duplex indicate that the fraying of the ends also reduces the residence lifetimes of the hydration water molecules. No evidence was obtained for stably bound water molecules in the major groove. The weak peaks observed for the protons in position 8 of adenine (Figure 3C) come from A 3′H – A 8H cross peaks which have their maximal intensities in neighboring cross sections. Relatively broad lineshapes were observed for the methyl resonances of T (not shown), so that no cross peaks could be discerned between these signals and the water line in the NOESY spectrum. However, hydration with residence times longer than 1 ns near these methyl goups, near the protons in position 8 of adenine, or near position 6 in thymine would be expected to result in strong, positive NOE cross peaks, which were definitely not observed.

As with the *dodecamer* (Figures 1 and 2), these conclusions are based on tracing the origins of all other peaks in Figure 3, B and C. Comparison of Figure 3, A and C, shows that only the imino protons of T10 and T9 give rise to observable chemical exchange cross peaks. The exchange cross peak at 7.2 ppm was assigned to the adenine $4NH_2$ protons of the terminal base pairs, with a magnetization transfer pathway similar to that described above for the cytosine $4NH_2$ resonances of the *dodecamer*. In Figure 2 the same mixing time was used for NOESY and ROESY, which results in comparable size of the exchange cross peaks in the two spectra, except for the imino proton region where off-resonance effects (18) are dominant in the ROESY experiment (see above), while the NOE cross peaks are about two times more intense in the ROESY cross section (Figure 2B) than in the NOESY cross section (Figure 2C), as expected for hydration water molecules bound with residence times exceeding 1 ns.

DISCUSSION

The key implication from the present work is that hydration water molecules in the minor groove of A_nT_n tracts in DNA duplexes have residence times exceeding 1 ns, although these hydration sites are accessible to the bulk solvent (3–6). This observation supports the notion that these water molecules have an important structural role in the duplex architecture. Unlike interior water in protein structures, which is typically completely inaccessible to the solvent, all hydration water molecules form hydrogen bonds to the bulk water in both B- and A-type DNA, and are in this sense 'surface hydration waters'. Since surface hydration water in proteins has very short residence times in the hydration sites (10), the long residence times of water molecules in the minor groove of the DNA are a rather unexpected result. Two stably bound water molecules with access to the bulk water have recently also been reported for the active site of the *Lactobacillus casei* dihydrofolate reductase–methotrexate–NADPH complex (21). However, these water molecules are located at the end of a channel in the complex and the extent of exposure to the bulk solvent is through hydrogen bonds with a single water molecule at a time (22), and is thus much more restricted than for the hydration water of the DNA.

The hydration water molecules detected by the NOEs with the adenine 2H signals must be part of the 'spine of hydration' that has first been observed in the X-ray crystal structure of the *dodecamer* (3, 4). In this crystal structure, the spine of hydration contains two different types of hydration sites. The innermost hydration water molecules form hydrogen bonds with the thymine 2O and adenine 3N atoms of adjacent base pairs on opposite strands of the DNA duplex, while the second type of hydration water is located further away from the bases and connects two adjacent water molecules of the inner hydration layer. Based on the crystal coordinates of the *dodecamer* (3–6), only the inner type of hydration water molecules would be expected to give observable NOEs with the adenine 2H protons. The water molecules which are hydrogen-bonded to A6 3N and T8 2O of the opposite strand have a proton–proton distance of about 2.5 Å to A6 2H and of about 3.0 Å to A5 2H. The NOEs with A6 2H are expected to be further enhanced by the water molecule at the central A–T step of the *dodecamer*, which is hydrogen bonded to the thymine 2O atoms of base pairs 6 and 7, with its protons at a distance of about 3.0 Å from A6 2H. No conclusive experimental evidence could be obtained for the outer water molecules of the spine of hydration. Their protons are about 4.0 Å away from the adenine 2H protons and even farther from any other non-labile proton of the DNA that could be resolved in the present experiments.

It should be noted, however, that the detailed locations of the water molecules representing the spine of hydration in the single crystal structure of the *dodecamer* may be different in aqueous solution. A recent single crystal X-ray analysis of d-(CGC-AAATTTGCG)$_2$ showed a somewhat different arrangement of the water molecules in the spine of hydration (23). In this structure, fewer water molecules were found that bridge the two DNA strands. Because there was no clear distinction between an inner and an outer hydration layer, the authors referred to the hydration water in the minor groove of this crystal structure as a 'ribbon of hydration' (23). Interestingly, the X-ray analysis resulted in a much better definition of the water molecules in the minor groove than of those in the major groove. The NMR studies of d-$(CGCGAATTGCGC)_2$ and d-$(A_5T_5)_2$ now show that the increased order of the hydration water in the minor groove of the DNA in the single crystal is also reflected by significantly longer residence times in aqueous solution.

Quite generally, the detection of hydration water molecules in DNA is limited by the small number of DNA proton resonances that can be used as reporter signals. For example, most of the NOEs between the water signal and deoxyribose protons are obscured by overlap with intraresidual NOEs with 3′H signals near the water frequency (Figure 1), and the 1′H region is further obscured by its proximity to the water resonance and the strong exchange cross peaks of the chain terminal 3′ and 5′ hydroxyl protons. Furthermore, the rotation of NH_2 groups about the C-N bond broadens the signals of these groups and thus interferes with the observation of their NOEs with the water signal. In particular, the NH_2 resonances of G are broadened beyond detection by this exchange process, so that no conclusive experimental evidence could presently be obtained relating to the possible presence of a stable spine of hydration also in the minor groove of DNA segments with G·C base pairs (8, 9). In contrast to the minor groove, the hydration water in the major groove of the DNA is highly mobile, with residence times shorter than 500 ps evidenced by negative NOESY cross peaks with the methyl goups, several of the 8H and 6H resonances of the bases, and some 2′H signals of the sugar moieties. Additional details on DNA hydration, including studies of possible long-lived hydration water in the minor groove of G_nC_n tracts, may in the future emerge from similar NMR measurements with more concentrated DNA samples, which will allow the use of three-dimensional experiments to assign those water–DNA NOEs which are obscured by overlap with intraresidual cross peaks in the cross sections of Figures 1–3 (16), or by applying heteronuclear NMR experiments with ^{13}C-labeled DNA.

ACKNOWLEDGEMENTS

We thank Dr W.Leupin for the sample of d-$(A_5T_5)_2$. Financial support by the Schweizerischer Nationalfonds (project no. 31.32033.91) is gratefully acknowleged.

REFERENCES

1. Zhang,R.G., Joachimiak,A., Lawson,C.L., Schevitz,R.W., Otwinowski,Z. and Sigler,P.B.(1987) *Nature* **327**, 591–597.
2. Aggarwal,A.K., Rodgers,D.W., Drottar,M., Ptashne,M. and Harrison,S.C. (1988) *Science* **242**, 899–907.
3. Drew,H.R. and Dickerson,R.E. (1981) *J. Mol. Biol.* **151**, 535–556.
4. Kopka,M.L., Fratini,A.V., Drew,H.R. and Dickerson,R.E. (1983) *J. Mol. Biol.* **163**, 129–146.
5. Westhof,E., Prange,T., Chevrier,B. and Moras,D. (1985) *Biochimie* **67**, 811–817.
6. Westhof,E. (1987) *J. Biomol. Struct. Dyn.* **5**, 581–600.
7. Chuprina,V.P. (1987) *Nucleic Acids Res.* **15**, 293–311.
8. Subramanian,P.S. and Beveridge,D.L. (1989) *J. Biomol. Struct. Dyn.* **6**, 1093–1122.
9. Subramanian,P.S., Swaminathan,S. and Beveridge,D.L. (1990) *J. Biomol. Struct. Dyn.* **7**, 1161–1165.
10. Otting,G., Liepinsh,E. and Wüthrich,K. (1991) *Science* **254**, 974–980.
11. Otting,G., Liepinsh,E. and Wüthrich,K. (1991) *J. Am. Chem. Soc.* **113**, 4363–4364.
12. Otting,G. and Wüthrich,K. (1989) *J. Am. Chem. Soc.* **111**, 1871–1875.
13. Otting,G., Liepinsh,E. and Wüthrich,K. (1992) *J. Am. Chem. Soc.* **114**, 7093–7095.
14. Bothner-By,A.A., Stephens,R.L., Lee,J., Warren,C.D. and Jeanloz,R.W. (1984) *J. Am. Chem. Soc.* **106**, 811–813.
15. Guéron,M., Kochoyan,M. and Leroy,J. (1987) *Nature* **328**, 89–92.
16. Otting,G., Liepinsh,E., Farmer,B.T.II and Wüthrich,K. (1991) *J. Biomol. NMR* **1**, 209–215.
17. Van de Ven,F.J.M., Janssen,H.G.J.M., Gräslund,A. and Hilbers,C.W. (1988) *J. Magn. Reson.* **79**, 221–235.
18. Griesinger,C. and Ernst,R.R. (1987) *J. Magn. Reson.* **75**, 261–271.
19. Neuhaus,D. and Keeler,J. (1986) *J. Magn. Reson.* **68**, 568–574.
20. Celda,B., Widmer,H., Leupin,W., Chazin,W.J., Denny,W.A. and Wüthrich,K. (1989) *Biochemistry* **28**, 1462–1471.
21. Gerothanassis,I.P., Birdsall,B., Bauer,C.J., Frenkiel,T.A. and Feeney,J. (1992) *J. Mol. Biol.* **226**, 549–554.
22. Bolin,J.T., Filman,D.J., Matthews,D.A., Hamlin,R.C. and Kraut,J. (1982) *J. Biol. Chem.* **257**, 13650–13662.
23. Edwards,K.J., Brown,D.G., Spink,N., Skelly,J.V. and Neidle,S. (1992) *J. Mol. Biol.* **226**, 1161–1173.

Hydration of Biological Macromolecules in Solution: Surface Structure and Molecular Recognition

K. WÜTHRICH
Institut für Molekularbiologie und Biophysik, Eidgenössische Technische Hochschule-Hönggerberg, CH-8093 Zürich, Switzerland

The presently available three-dimensional protein and nucleic acid structures at atomic resolution have, with few exceptions, been determined either by X-ray diffraction in protein crystals (Blundell and Johnson 1976) or by nuclear magnetic resonance (NMR) spectroscopy in protein solutions (Wüthrich 1986). These two techniques produce a rapidly growing pool of data (Hendrickson and Wüthrich 1991, 1992, 1993), which represent today's structural basis for detailed investigations on the functionality of biological macromolecules.

NMR spectroscopy and X-ray diffraction can provide complementary information on the same molecule, which results primarily from the facts that the time scales of the two types of measurements are widely different, and that the two techniques use, respectively, proteins in solution and protein single crystals (see, e.g., Wüthrich 1990, 1991). Since protein structure determinations by NMR or by X-ray diffraction can be performed independently (Blundell and Johnson 1976; Wüthrich 1986), meaningful comparisons of corresponding structures in single crystals and in noncrystalline states can be obtained. This is highly relevant, since the solution conditions for NMR studies may coincide closely with the physiological fluids, and comparative studies of corresponding crystal and solution structures promise to result in more relevant ways of analyzing crystal data with regard to protein functions in physiological milieus. Providing this kind of fundamental information may well turn out to be the major impact of NMR in structural biology, considering that X-ray crystallography continues to provide a dominant fraction of the new macromolecular structures (Hendrickson and Wüthrich 1993). Although NMR structure determination in solution will foreseeably be limited to proteins with molecular weights below about 30,000–40,000 (see, e.g., Wüthrich 1990), many conclusions from comparative studies with relatively small proteins should also be applicable to crystal structures of bigger molecules.

In comparisons of corresponding crystal and solution structures of proteins, both global conformational rearrangements and extensive conservation between the two states have been observed. Major rearrangements are usually seen in nonglobular polypeptides (see, e.g., Braun et al. 1983) and on the surface of globular proteins, whereas close coincidence is commonly encountered for the core of globular proteins (Billeter 1992). However, even for proteins with virtually identical molecular architecture in crystals and in solution, the two techniques provide different information on the internal mobility of the molecular core. NMR can provide direct, quantitative measurements of the frequencies of certain high-activation-energy motional processes in the interior of globular proteins, and at least semiquantitative information on additional, higher-frequency processes. The corresponding information from X-ray structure determinations commonly consists of an outline of the conformation space covered by the combination of static disorder and high-frequency structure fluctuations. The protein surface is most likely to be influenced by crystal packing, or by solvent interactions, respectively, and it is thus of special interest to investigate these effects by combined use of NMR and crystallography. On the one hand, a detailed description of the protein surface in near-physiological solution represents a proper reference for investigating mechanistic aspects of intermolecular interactions with proteins, and novel NMR techniques enabling the observation of hydration water on the molecular surface (Otting et al. 1991a) have greatly added to the characterization of protein surfaces in solution. On the other hand, the structural basis of intermolecular recognition between different macromolecules can be investigated in a large sample of X-ray crystal structures of proteins in binary complexes or multimolecular assemblies, and even the influence of crystal-packing effects on the protein surface may be indicative of the types of protein–protein interactions that are important in physiological recognition processes.

In this paper, I survey novel insights into protein structures in solution that result from NMR investigations of protein hydration. Special emphasis is on protein surface hydration in aqueous solution and comparison with corresponding data from diffraction experiments with protein crystals. As an illustration, hydration data on a homeodomain–DNA complex (Billeter et al. 1993; Qian et al. 1993a) are discussed.

METHODS

NMR structure determinations. The solution structure of the protein basic pancreatic trypsin inhibitor (BPTI) was solved with the use of two-dimensional (2-D) homonuclear ^{1}H NMR experiments (Berndt et al. 1992). Sequence-specific resonance assignments were obtained with the conventional method of establishing sequential relations via sequential ^{1}H–^{1}H nuclear Overhauser effects (NOE) (Billeter et al. 1982; Wüthrich 1986). The input for the structure calculation consisted of more than 10 NOE upper distance constraints per residue, on average, and numerous dihedral angle constraints derived from measurements of vicinal ^{1}H–^{1}H spin–spin coupling constants. The structure calculations were performed with the program DIANA (Güntert et al. 1991), followed by energy minimization in vacuo using the AMBER force field (Singh et al. 1986). The solution structure of BPTI is represented by a group of 20 conformers.

The structure determination of the complex formed between the mutant *Antp(C39S)* homeodomain and the DNA duplex with d-(GAAAGCCATTAGAG) made use of isotope labeling of the protein with ^{15}N or ^{13}C, combined with heteronuclear half-filter experiments and heteronuclear 3-D NMR (Qian et al. 1993b). The input for the structure calculation consisted of more than 10 intramolecular NOE upper distance constraints per residue in the protein, more than 10 DNA–DNA NOE upper distance constraints per base pair, and 39 intermolecular protein–DNA NOE distance constraints. The structure calculation was started by docking the independently determined structure of the DNA-bound homeodomain (Qian et al. 1993b) against the DNA duplex in a rigid, regular B-form, using the REPEL option of the program X-PLOR (Brünger 1990). Of a group of 20 conformers of the complex obtained by applying the docking procedure to 20 combinations of the protein and the DNA with randomly generated relative starting positions, the 16 best solutions were refined by simulated annealing with the program AMBER (Singh et al. 1986). For the refinement, the DNA structure was released, so that all atoms of the complex were allowed to move during the calculations, and the experimental DNA–DNA NOE distance constraints were used in addition to the intramolecular protein–protein NOEs and the intermolecular protein–DNA NOEs. Furthermore, upper distance constraints of 2.0 Å between donor protons and acceptor atoms for Watson-Crick hydrogen bonds were added whenever a NOE between the donor proton and a proton located on the other base of the base pair was observed. The refined conformers of the *Antp(C39S)* homeodomain–DNA complex resulting from the simulated annealing were then surrounded by a water shell of 5 Å minimal thickness, which corresponds to about 1000 water molecules. The conformers in the water bath were first energy-minimized, then a molecular dynamics trajectory of 10 picoseconds duration at 300 K was recorded, and finally the conformers were minimized again. During these AMBER calculations, all the aforementioned NOE and hydrogen bond distance constraints were active, with a calibration such that violations of 0.2 Å corresponded to an energy of kT/2. During the molecular dynamics runs, one-third of the water molecules moved by less than 5 Å, and three-quarters of the water molecules were displaced by less than 10 Å. For the purpose of modeling the hydration of the complex, the solution structure is represented by 15 of the resulting 16 refined conformers (Billeter et al. 1993).

NMR studies of protein hydration in aqueous solution. Identification of individual molecules of hydration water by NMR in solution relies on the observation of NOEs between hydrogen atoms of the polypeptide chain and water protons. These NOEs are due to time-dependent dipole–dipole coupling between nearby protons. The intensity of the NOE between two hydrogen atoms i and j is proportional to d_{ij}^{-6}, where d_{ij} is the distance between the two protons. Because of the strong distance dependence, NOEs can be observed only between spatially close protons, i.e., for $d_{ij} \lesssim 4.0$ Å. The NOE intensity is further related to a correlation function describing the stochastic modulation of the dipole–dipole coupling between the two protons. For studies of hydration, it is important that this correlation function may be governed either by the Brownian rotational tumbling of the hydrated protein molecule or by interruption of the dipolar interaction through translational diffusion of the interacting spins, whichever is faster (Otting et al. 1991a).

The NOEs can be measured either by experiments in the laboratory frame of reference (NOESY) or in the "rotating frame" (ROESY) (Bothner-By et al. 1984). In suitably executed experiments, the measured NOE intensities reflect directly the cross-relaxation rates in the laboratory frame, σ^{NOE}, or in the rotating frame, σ^{ROE}, respectively. The two rates differ in their functional dependence on the spectral densities in such a way that studies of the sign and value of the ratio $\sigma^{NOE}/\sigma^{ROE}$ can be used for investigations of the rate processes that determine the effective correlation time for modulation of the dipole–dipole coupling. As mentioned above, such rate processes include the Brownian rotational tumbling of the molecules considered and translational diffusion of the two interacting spins, for example, upon rapid dissociation of a bimolecular complex. As an illustration, the plots of σ^{NOE} and σ^{ROE} versus the effective rotational correlation time (Fig. 1) shows that positive values of $\sigma^{NOE}/\sigma^{ROE}$ are expected only for very short correlation times, i.e., shorter than about 0.3 nanoseconds at a proton frequency of 600 MHz (for modulation by translational diffusion, the sign change occurs at about 0.5 nanoseconds), and that the NOE and ROE intensities will increase rapidly for longer correlation times. NOEs with hydration water have been observed using homonuclear 2-D NOESY and 2-D ROESY (Otting and Wüthrich 1989), homonuclear 3-D TOCSY-relayed NOESY and ROESY (Otting et al. 1991c), and 3-D ^{15}N- or ^{13}C-correlated NOESY and ROESY (Qian et al. 1993a). These NMR experiments have to be performed in H_2O solution, and the intense solvent resonance can be suppressed only at the very end of the NOESY or ROESY pulse sequences. As an illustration, Figure 2 shows cross-sections from 2-D NOESY

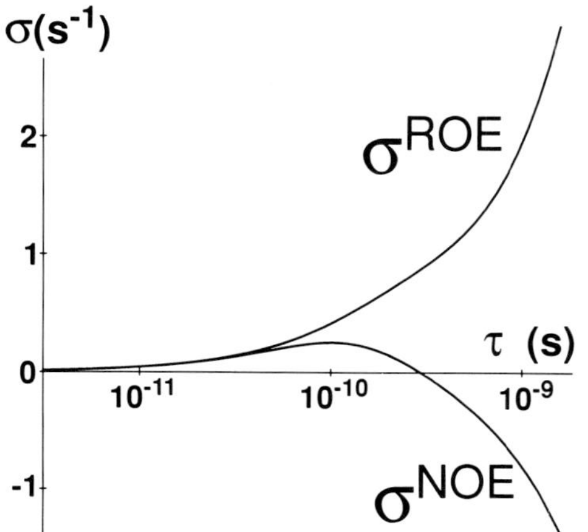

Figure 1. Plot of the cross-relaxation rates σ^{NOE} and σ^{ROE} between two protons versus the effective correlation time τ for modulation of the dipole-dipole coupling by Brownian rotational tumbling. The curves were calculated for a ^{1}H frequency of 600 MHz and a ^{1}H–^{1}H distance of 2.0 Å. (Adapted from Otting et al. 1991a.)

and 2-D ROESY experiments, which were taken at the chemical shift of the solvent water and contain NOEs between the polypeptide chain and hydration water molecules. Nearly all the resonances have opposite sign in the two experiments, showing that the system is in the "slow motion regime," with effective correlation times much greater than 0.5 nanoseconds (Fig. 1).

Observation of hydration water by NMR in solution and by diffraction experiments with protein crystals. In view of the discussions in the following sections of this paper, it is useful to recall that NMR experiments in solution and diffraction experiments with protein single crystals are sensitive to different aspects of protein hydration and that the information collected with the two techniques is therefore largely complementary. On the basis of the dependence of the cross-relaxation rates on the effective correlation time (Fig. 1), it has been demonstrated that the sign and intensity of the protein–water NOEs observed by NMR reflect primarily the residence times of the water molecules in the hydration sites near the protein protons. In other words, individual hydration water molecules continuously exchange in and out of particular hydration sites on the protein, and NMR measures the average duration of these "visits." In contrast, diffraction experiments probe the total fraction of time during which a particular hydration site is occupied by a water molecule, but they are largely insensitive to the residence time at that site on any particular visit (Saenger 1987). Typically, hydration water molecules in protein crystals are observed in discrete sites covering 30–60% of the molecular surface, depending on the protein and the crystal form (Fig. 3). On this basis, we have started to investigate how the presence or absence of ordered, diffraction-observable hydration water in protein crystals can be correlated with the residence times of water molecules in corresponding hydration sites as observed by NMR in solution.

RESULTS

This section first summarizes some key data on protein hydration in aqueous solution, which were collected with peptides and small proteins, and then surveys corresponding data on the *Antp(C39S)* homeodomain–DNA complex.

Considering the high water concentration in dilute

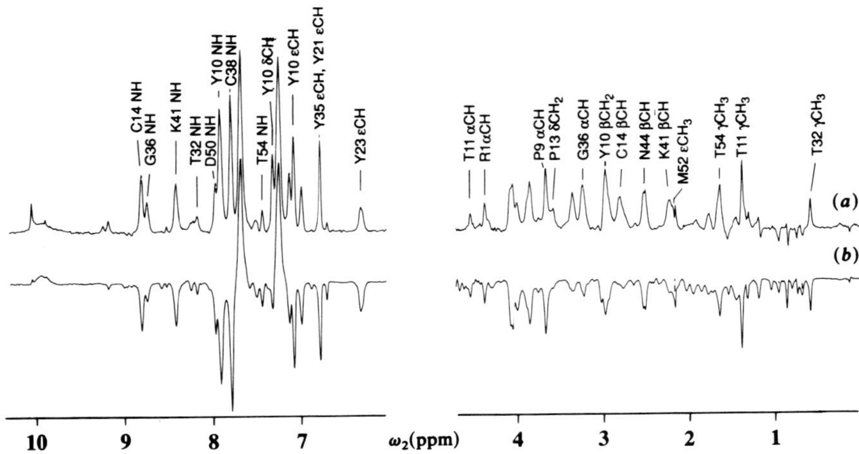

Figure 2. ^{1}H-NMR spectra containing NOE cross-peaks between polypeptide hydrogen atoms of the protein BPTI and hydration water protons. Cross-sections along ω_2 are shown, which were taken at the ω_1 chemical shift of the water resonance through a 2-D NOESY (*a*) and a 2-D ROESY (*b*) spectrum ("soft-NOESY" and "soft-ROESY," respectively; ^{1}H frequency = 600 MHz; mixing time = 50 msec; BPTI concentration = 20 mM; solvent = 90% H_2O/10% D_2O; pH = 3.5; T = 4°C). The peaks which were identified as NOEs with the H_2O resonance are identified above the NOESY cross-section with the one-letter amino acid symbol, the sequence number in the polypeptide chain, and the proton type. (Adapted from Wüthrich et al. 1992.)

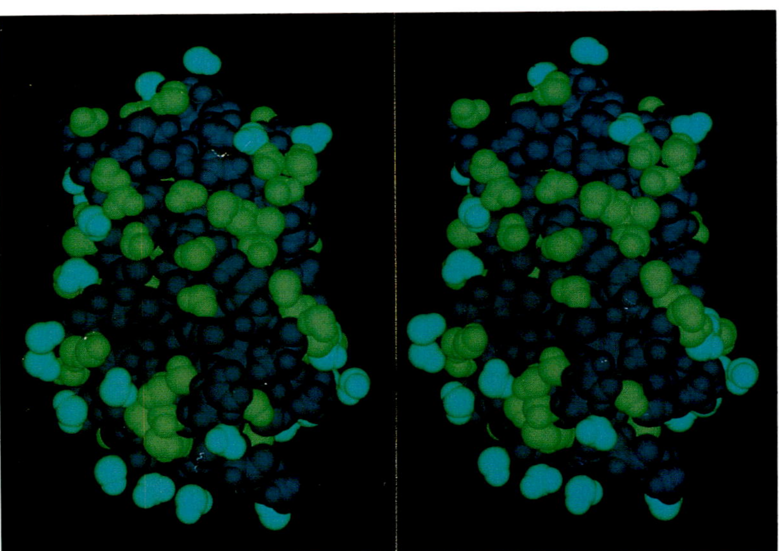

Figure 3. Stereo view of a space-filling representation of the protein BPTI showing the surface hydration in the crystal structure 5PTI (Wlodawer et al. 1984). All polypeptide atoms are gray, the hydration water molecules attributed to this protein molecule in the crystal are green, and additional water molecules attributed to hydration sites on neighboring protein molecules yet located within 3.0 Å are blue. (Reprinted, with permission, from Wüthrich et al. 1992.)

aqueous protein solutions, one anticipates intuitively that the entire surface of the solute is covered with water molecules. In experiments with the nonapeptide oxytocin, all polypeptide hydrogen atoms did indeed show NOEs with the water, and the residence times for all hydration water molecules were found to be very short, in the approximate range of 20–200 picoseconds at 10°C (Otting et al. 1991a, 1992).

The protein BPTI contains four interior hydration water molecules, which are inaccessible to the solvent in a rigid model of the 3-D protein structure. Nonetheless, these waters exchange with the bulk solvent at rates corresponding to residence times in the protein hydration sites which are shorter than 20 milliseconds (Otting et al. 1991b). The effective correlation time for NOEs with these water molecules is of the order of several nanoseconds, as manifested by the different sign for the NOESY and ROESY cross peaks (Fig. 2). Because of overlap with these intense peaks, only part of the much weaker NOEs with surface hydration waters (see Fig. 1) could so far be identified in BPTI, which have residence times in the range of approximately 20–200 picoseconds at 4°C (Otting et al. 1991a).

In the NMR experiments with the *Antp(C39S)* homeodomain–DNA complex, it was clear a priori that the NOEs with short-lived surface hydration waters would be too weak to be detected, since only solutions with low concentrations of the complex could be prepared. On the other hand, using 3-D ^{13}C- and ^{15}N-correlated [^{1}H,^{1}H]-NOESY and [^{1}H,^{1}H]-ROESY with complexes containing the homeodomain in ^{15}N- and ^{13}C-labeled form, NOEs with longer-lived hydration waters located in a cavity in the protein–DNA interface could be identified (Qian et al. 1993a). These experiments also showed that, similar to the aforementioned interior waters in BPTI, the interfacial water molecules in the *Antp(C39S)* homeodomain–DNA complex (Fig. 4) exchange with the bulk water with lifetimes somewhere between milliseconds and a nanosecond. Further insights into the behavior of these internal hydration sites were obtained from the molecular dynamics refinement of the complex in a water bath, as described in Methods. These calculations show that in the individual ones of the 15 conformers used to represent the solution structure of the complex, the cavity contains from 1 to 5 water molecules. In concert with several flexibly disordered amino acid side chains, in particular the functionally important Gln-50 and Asn-51, these waters occupy different positions within the cavity in the different conformers, so that their locations are not precisely defined (Fig. 5) (Billeter et al. 1993). In 15 internally hydrated conformers of the complex, a total of 39 internal hydration sites were identified. The envelope of these 39 interior water molecules from the 15 conformers (Fig. 4) illustrates the maximum confines of the internal cavity. The surface of this envelope covers a large part of the recognition helix of the homeodomain: More precisely, it is in contact with the residues 43–44, 47–48, 50–52, and 54 of the homeodomain; the nucleotides C 7, A 8, and T 9 of one DNA strand; and the nucleotides A 9 and A 10 of the other DNA strand.

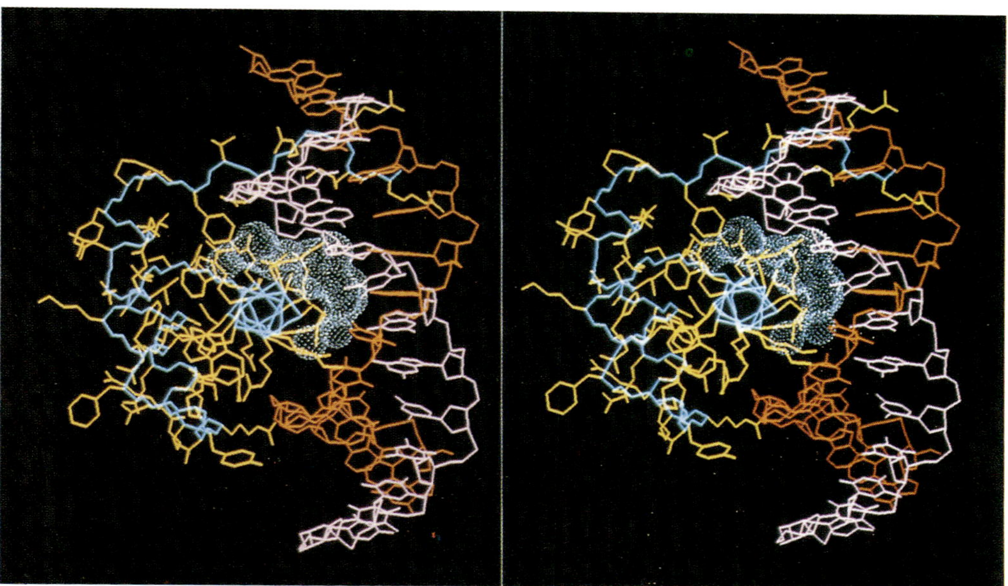

Figure 4. Stereo view of the *Antp(C39S)* homeodomain–DNA complex. For improved clarity, only the residues 3–56 of the homeodomain are shown, and only the heavy atoms have been drawn. Color code: DNA strands, brown and magenta; polypeptide backbone, blue; amino acid side chains, yellow. Blue dots identify an envelope surrounding possible locations of hydration water molecules in the protein–DNA interface (see text and Fig. 5). Drawing prepared using the atomic coordinates from Billeter et al. (1993).

DISCUSSION

Hydration and Internal Mobility in Globular Proteins

The locations of the four interior hydration water molecules determined by NMR in the solution structure of BPTI coincide with four water positions observed in three different crystal forms of BPTI (Deisenhofer and Steigemann 1975; Wlodawer et al. 1984, 1987). The exchange of these water molecules with the bulk water demonstrates that there is significant internal mobility ("breathing") in the protein structure. These motions have higher frequencies than 50 sec^{-1}, and they must have an amplitude of about 1.5 Å in order to provide a passage for the water molecules. Interestingly, these numbers coincide nearly identically with the frequencies and amplitudes of the internal structure fluctuations implicated by the "ring flips" of the phenylalanine and tyrosine side chains (180° rotations of the aromatic rings about the C^{β}–C^{γ} bond) in BPTI (Wüthrich and Wagner 1975; Wüthrich 1986). Internal large-amplitude structure fluctuations with activation energies, ΔG‡, in the range from about 10 to 20 kcal M^{-1}, as observed in BPTI, appear to be a common feature of globular proteins. Ring flips have been demonstrated in a wide variety of different proteins, and water exchange from interior hydration sites has also been evidenced for several different proteins. The emerging dynamic image of the core of globular proteins is also in line with earlier qualitative indications of internal breathing of protein molecules, based primarily on amide proton exchange data (see, e.g., Hvidt and Nielsen 1966; Englander and Kallenbach 1984) and fluorescence quenching experiments with internal tryptophan residues (Lakowicz and Weber 1973).

Protein Surface Structure in Solution and in Crystals

Figure 6 visualizes the result of a high-quality NMR structure determination of BPTI (Berndt et al. 1992) in the form of a superposition of 20 conformers. This is a commonly used presentation of NMR structures (Wüthrich 1986): All conformers were calculated from the same NMR data, using different boundary conditions; close coincidence of the different conformers indicates well-defined regions of the molecular structure, whereas divergence indicates "structural disorder" (Wüthrich 1989, 1991). The protein core of BPTI consisting of the polypeptide backbone of residues 2–56 and the majority of the interior side chains (blue in Fig. 6) is well defined and is nearly identical to the crystal structure (Deisenhofer and Steigemann 1975). In contrast, most of the surface side chains (red in Fig. 6) are disordered. Technically speaking, the surface disorder implicated by the result of the NMR structure determination (Fig. 6) arises primarily from the scarcity of both steric constraints on surface side chains and NOE distance constraints involving peripheral side-chain hydrogen atoms. On the basis of these data alone, it is therefore difficult to assess the significance

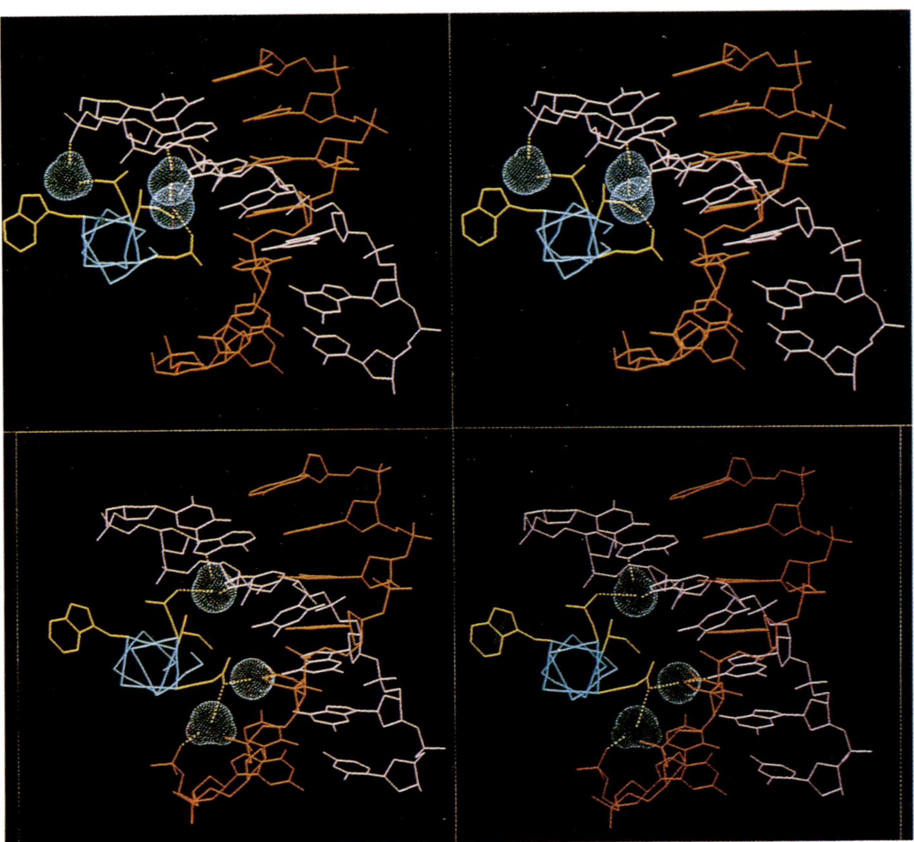

Figure 5. Stereo views of 2 out of 15 conformers of the *Antp(C39S)* homeodomain–DNA complex which were selected to represent the hydrated solution structure (see text). Same color code as in Fig. 4, but for the homeodomain, only the recognition helix with the side chains of Ile-47, Trp-48, Gln-50, and Asn-51 is shown. The dotted yellow lines indicate likely water-mediated intramolecular hydrogen bonds. The envelope in Fig. 4 surrounds all the water positions found in all 15 conformers, each of which contained from 1 to 5 water molecules in the interfacial cavity. Drawing prepared using the atomic coordinates from Billeter et al. (1993).

of the indications of surface disorder. However, there is additional evidence indicating that the scarcity of NOE distance constraints for molecular regions that appear disordered in the final result of the structure determination is in part also a consequence of locally increased mobility. For example, there is a striking similarity between the local displacements among the NMR conformers and the local crystallographic temperature factors along the polypeptide backbone (Wüthrich 1991; Billeter 1992). Spin relaxation measurements provide direct evidence that the disordered regions in NMR structures tend to have increased mobility. Finally, and of particular interest in the context of the main theme of this paper, surface side chains may be more intimately coupled with the solvent environment than with the protein core, as is quite apparent from Figure 6. The NMR observation of flexibly disordered states for most of the surface side chains of proteins in solution can thus be tentatively rationalized by the interactions with the rapidly exchanging solvent matrix, which, in contrast to the discrete hydration sites in crystal structures (Fig. 3), covers the entire protein surface. Overall, it is typical for high-quality NMR structures of proteins in solution that the same molecule contains a well-structured, albeit "breathing" core and a flexibly disordered interface with the solvent, which may include individual amino acid side chains, the ends of the polypeptide chain (see, e.g., Folkers et al. 1989; Haruyama and Wüthrich 1989), or nonterminal polypeptide segments that form solvent-exposed loops (see, e.g., Widmer et al. 1989; Szyperski et al. 1992).

For the biologist or biochemist who makes use of three-dimensional protein structures in the analysis of his experimental data or in the development of new research projects, knowledge on the hydrated molecular surface under solution conditions similar to the physiological milieu is of special interest. For example, a solution structure with a flexibly disordered molecular surface (Fig. 6) is probably the most meaningful repre-

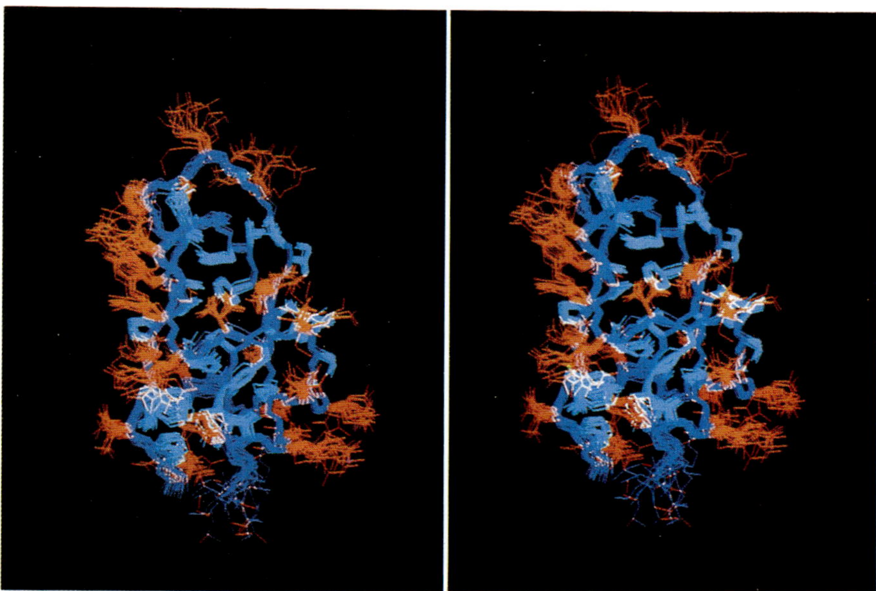

Figure 6. Stereo view of the NMR solution structure of the protein BPTI represented as a superposition of 20 conformers. All heavy atoms are displayed. Color code: polypeptide backbone and core side chains, blue; surface side chains, red. A close fit among the 20 conformers indicates regions of the molecule where the structure is well defined, whereas a large dispersion implicates structural disorder. Drawing prepared using the atom coordinates from Berndt et al. (1992).

sentation of a "free" protein to be used as a reference state for studies of the structural basis of specific intermolecular interactions with the protein. Knowledge that surface hydration waters have residence times in the subnanosecond range shows that polypeptide hydration is not a rate-limiting event in processes such as protein folding or complexation of proteins with other macromolecules. With regard to specific protein functions, it will be of special interest for the future to look for discrete, longer-lived surface hydration sites, as they have recently been observed in the minor groove of DNA duplexes (Kubinec and Wemmer 1992; Liepinsh et al. 1992).

Although surface hydration waters can also be detected by diffraction experiments with protein crystals, it appears that future, systematic studies of the surface hydration of proteins and nucleic acids in solution will have to be performed directly by NMR. As was discussed in detail elsewhere (Otting et al. 1991a; Wüthrich et al. 1992), it is unlikely that data on protein surfaces, and in particular surface hydration, can be extrapolated from the corresponding crystal structures (Fig. 3), even if one disregards the trivial case of those surface areas that are not hydrated in the crystals because of van der Waals contacts with neighboring protein molecules in the crystal lattice. Overall, crystallographic surface hydration (Fig. 3) appears to be a property of the protein crystals rather than the molecular structures of the proteins. For example, similar residence times of surface hydration waters in solution have been found for crystallographic hydration sites as well as for surface areas that are not covered with water in the crystal structure (Fig. 3) (Otting et al. 1991a), and this experimental observation is also supported by theoretical simulations of protein hydration (Brunne et al. 1993).

Hydration Water and Homeodomain–DNA Recognition

The refinement of the NMR structure of the *Antp(C39S)* homeodomain–DNA complex showed that the complex is well stabilized by 14 salt bridges between arginine and lysine side chains of the protein and the DNA backbone, and by a large number of additional direct protein–DNA contacts (Billeter et al. 1993). Two specific contacts to DNA bases are indicated with Arg-3 and Arg-5 in the minor groove of the DNA, but most of the specific contacts are with the recognition helix of the homeodomain, in particular, the side chains of Ile-47, Gln-50, Asn-51, and Met-54. The functional importance of position 50 has been amply documented (Hanes and Brent 1989; Treisman et al. 1989; Schier and Gehring 1992), and a similarly important role has been implicated for Asn-51 by its strict conservation in all known homeodomain sequences (Wüthrich and Gehring 1992). It is then interesting to note that the steric situation for the formation of hydrogen bonds between Gln-50 or Asn-51 and the DNA bases across the cavity (Fig. 4) is unfavorable, unless the hydrogen bonds are mediated by water molecules (Fig. 5). Involvement of hydration water in protein–DNA recognition has previously been proposed, in particular, by Sigler and his colleagues from their work on a *trp* repressor–DNA complex (Otwinowski et al. 1988; Sigler 1992). However, a novel facet of DNA recognition evolving from the presently discussed NMR observations is the indication that at least part of the specific protein–DNA interactions are short-lived. As mentioned earlier, the residence times of the cavity hydration waters with respect to exchange with the bulk water are shorter than 20 milliseconds. This time scale overlaps with that for the independently evidenced slow-motional averaging for the side chain of Asn-51 (Qian et al. 1993a; Billeter et al. 1993), which indicates that this side chain jumps forth and back between two or multiple contact sites on the DNA on a time scale of milliseconds to microseconds. Overall, the essential interactions of the DNA with the highly conserved Gln-50 and the invariant Asn-51 are thus best rationalized as a fluctuating network of short-lived, weak-bonding interactions involving the DNA bases A 9 and A 10, as well as interfacial hydration water molecules (Fig. 5).

Already on the basis of the initial, "low-resolution" NMR structure of the *Antp(C39S)* homeodomain–DNA complex (Otting et al. 1990), it was readily apparent that the global features of this NMR structure are very similar to those of subsequently determined X-ray crystal structures of DNA complexes with homologous homeodomains (Kissinger et al. 1990; Wolberger et al. 1991), including the particular location of the recognition helix in the major groove of the DNA and the contacts of the amino-terminal peptide segment with the minor groove. Using an input data set derived from the atomic coordinates of the crystal structure of the *engrailed* homeodomain–DNA complex of Kissinger et al. (1990), we calculated a "simulated NMR structure" of this complex, using a molecular dynamics treatment in a water bath identical to that used for the complex with *Antp(C39S)* (see Methods). The result shows the presence of a hydrated interfacial cavity similar to that in the complex with *Antp(C39S)* (Fig. 4) (Billeter and Wüthrich 1993), indicating that the DNA recognition by short-lived, water-mediated specific intermolecular interactions could also readily be accommodated within the framework of the crystallographic data on homologous homeodomain–DNA complexes. (The hydration water in the interfacial cavity was not observed by the X-ray diffraction experiments, presumably because of the limited resolution of 2.8 Å, or possibly because this water is "disordered" in the crystal.)

ACKNOWLEDGMENTS

NMR studies of protein hydration were pursued in collaboration with Drs. G. Otting and E. Liepinsh in my laboratory, those of the *Antp* system with Drs. M. Billeter, G. Otting, and Y.Q. Qian in my laboratory and the group of Dr. W. Gehring at the Biozentrum of the University of Basel. I particularly acknowledge help in the preparation of the figures by Dr. G. Otting (Fig. 3), Drs. M. Billeter and Y.Q. Qian (Figs. 4 and 5), and Dr. K. Berndt (Fig. 6). I thank R. Marani for the careful processing of the manuscript and the Schweizerischer Nationalfonds for financial support (project 91.32033.91).

REFERENCES

Berndt, K.D., P. Güntert, L.P.M. Orbons, and K. Wüthrich. 1992. Determination of a high-quality nuclear magnetic resonance solution structure of the bovine pancreatic trypsin inhibitor and comparison with three crystal structures. *J. Mol. Biol.* **227:** 757.

Billeter, M. 1992. Comparison of protein structures determined by NMR in solution and by X-ray diffraction in single crystals. *Q. Rev. Biophys.* **25:** 325.

Billeter, M. and K. Wüthrich. 1993. Model studies relating nuclear magnetic resonance data with the three-dimensional structure of protein-DNA complexes. *J. Mol. Biol.* (in press).

Billeter, M., W. Braun, and K. Wüthrich. 1982. Sequential resonance assignments in protein ^{1}H nuclear magnetic resonance spectra: Computation of sterically allowed proton–proton distances and statistical analysis of proton–proton distances in single crystal protein conformations. *J. Mol. Biol.* **155:** 321.

Billeter, M., Y.Q. Qian, G. Otting, M. Müller, W.J. Gehring, and K. Wüthrich. 1993. Determination of the NMR solution structure of an *Antennapedia* homeodomain-DNA complex. *J. Mol. Biol.* (in press).

Blundell, T.L. and L.N. Johnson. 1976. *Protein crystallography*. Academic Press, New York.

Bothner-By, A.A., R.L. Stephens, J. Lee, C.D. Warren, and R.W. Jeanloz. 1984. Structure determination of a tetrasaccharide: Transient nuclear Overhauser effects in the rotating frame. *J. Am. Chem. Soc.* **106:** 811.

Braun, W., G. Wider, K.H. Lee, and K. Wüthrich. 1983. Conformation of glucagon in a lipid-water interphase by ^{1}H nuclear magnetic resonance. *J. Mol. Biol.* **169:** 921.

Brünger, A.T. 1990. *X-PLOR version 2.1*. Yale University, New Haven, Connecticut.

Brunne, R.M., E. Liepinsh, G. Otting, K. Wüthrich, and W.F. van Gunsteren. 1993. Hydration of proteins. A comparison of experimental residence times of water molecules solvating the bovine pancreatic trypsin inhibitor with theoretical model calculations. *J. Mol. Biol.* **231:** 1040.

Deisenhofer, J. and W. Steigemann. 1975. Crystallographic refinement of the structure of bovine pancreatic trypsin inhibitor at 1.5 Å resolution. *Acta Crystallogr. B.* **31:** 238.

Englander, S.W. and N.R. Kallenbach. 1984. Hydrogen exchange and structural dynamics of proteins and nucleic acids. *Q. Rev. Biophys.* **16:** 521.

Folkers, P.J.M., G.M. Clore, P.C. Driscoll, J. Dodt, S. Köhler, and A.M. Gronenborn. 1989. Solution structure of recombinant hirudin and the Lys47→Glu mutant: A nuclear magnetic resonance and hybrid distance geometry-dynamical simulated annealing study. *Biochemistry* **28:** 2601.

Güntert, P., W. Braun, and K. Wüthrich. 1991. Efficient computation of three-dimensional protein structures in solution from nuclear magnetic resonance data using the program DIANA and the supporting programs CALIBA, HABAS and GLOMSA. *J. Mol. Biol.* **217:** 517.

Hanes, S.D. and R. Brent. 1989. DNA specificity of the bicoid activator protein is determined by homeodomain recognition helix residue 9. *Cell* **57:** 1275.

Haruyama, H. and K. Wüthrich. 1989. Conformation of recombinant desulfatohirudin in aqueous solution determined by nuclear magnetic resonance. *Biochemistry* **28:** 4301.

Hendrickson, W.A. and K. Wüthrich, eds. 1991. *Macromolecular structures 1991*. Current Biology, London.

———. 1992. *Macromolecular structures 1992*. Current Biology, London.

———. 1993. *Macromolecular structures 1993*. Current Biology, London.

Hvidt, A. and S.O. Nielsen. 1966. Hydrogen exchange in proteins. *Adv. Protein Chem.* **21:** 287.

Kissinger, C.R., B. Liu, E. Martin-Blanco, T.B. Kornberg, and C.O. Pabo. 1990. Crystal structure of an *engrailed* homeodomain-DNA complex at 2.8 Å resolution: A framework for understanding homeodomain-DNA interactions. *Cell* **63:** 579.

Kubinec, M.G. and D.E. Wemmer. 1992. NMR evidence for DNA bound water in solution. *J. Am. Chem. Soc.* **114:** 8739.

Lakowicz, J. R. and G. Weber. 1973. Quenching of protein fluorescence by oxygen. Detection of structural fluctuations in proteins on the nanosecond time scale. *Biochemistry* **12:** 4171.

Liepinsh, E., G. Otting, and K. Wüthrich. 1992. NMR observation of individual molecules of hydration water bound to DNA duplexes: Direct evidence for a spine of hydration water present in aqueous solution. *Nucleic Acids Res.* **20:** 6549.

Otting, G. and K. Wüthrich. 1989. Studies of protein hydration in aqueous solution by direct NMR observation of individual protein-bound water molecules. *J. Am. Chem. Soc.* **111:** 1871.

Otting, G., E. Liepinsh, and K. Wüthrich. 1991a. Protein hydration in aqueous solution. *Science* **254:** 974.

———. 1991b. Proton exchange with internal water molecules in the protein BPTI in aqueous solution. *J. Am. Chem. Soc.* **113:** 4363.

———. 1992. Polypeptide hydration in mixed solvents at low temperatures. *J. Am. Chem. Soc.* **114:** 7093.

Otting, G., E. Liepinsh, B.T. Farmer II, and K. Wüthrich. 1991c. Protein hydration studied with homonuclear 3D ^{1}H NMR experiments. *J. Biomol. NMR* **1:** 209.

Otting, G., Y.Q. Qian, M. Billeter, M. Müller, M. Affolter, W.J. Gehring, and K. Wüthrich. 1990. Protein-DNA contacts in the structure of a homeodomain-DNA complex determined by nuclear magnetic resonance spectroscopy in solution. *EMBO J.* **9:** 3085.

Otwinowski, Z., R.W. Schevitz, R.G. Zhang, C.L. Lawson, A. Joachimiak, R.Q. Marmorstein, B.F. Luisi, and B.P. Sigler. 1988. Crystal structure of *trp* repressor/operator complex at atomic resolution. *Nature* **335:** 321.

Qian, Y.Q., G. Otting, and K. Wüthrich. 1993a. NMR detection of hydration water in the intermolecular interface of a protein-DNA complex. *J. Am. Chem. Soc.* **115:** 1189.

Qian, Y.Q., G. Otting, M. Billeter, M. Müller, W. Gehring, and K. Wüthrich. 1993b. NMR spectroscopy of a DNA complex with the uniformly ^{13}C-labeled Antennapedia homeodomain and structure determination of the DNA-bound homeodomain. *J. Mol. Biol.* (in press).

Saenger, W. 1987. Structure and dynamics of water surrounding biomolecules. *Annu. Rev. Biophys. Biophys. Chem.* **16:** 93.

Schier, A.F. and W.J. Gehring. 1992. Direct homeodomain-DNA interaction in the autoregulation of the *fushi tarazu* gene. *Nature* **356:** 804.

Sigler, P.B. 1992. The molecular mechanism of *trp* repression. In *Transcriptional regulation* (ed. S.L. McKnight and K.R. Yamamoto), p. 475. Cold Spring Harbor Laboratory Press, Cold Spring Harbor, New York.

Singh, U.C., P.K. Weiner, J.W. Caldwell, and P.A. Kollman. 1986. *AMBER 3.0*. University of California, San Francisco.

Szyperski, T., P. Güntert, S.R. Stone, and K. Wüthrich. 1992. Nuclear magnetic resonance solution structure of hirudin (1–51) and comparison with corresponding three-dimensional structures determined using the complete 65-residue hirudin polypeptide chain. *J. Mol. Biol.* **228:** 1193.

Treisman, J., P. Gönczy, M. Vashishtha, E. Harris, and C. Desplan. 1989. A single amino acid can determine the DNA binding specificity of homeodomain proteins. *Cell* **59:** 553.

Widmer, H., M. Billeter, and K. Wüthrich. 1989. Three-dimensional structure of the neurotoxin ATX Ia from *Anemonia sulcata* in aqueous solution determined by nuclear magnetic resonance spectroscopy. *Proteins* **6:** 357.

Wlodawer, A., J. Walter, R. Huber, and L. Sjölin. 1984. Structure of bovine pancreatic trypsin inhibitor: Results of joint neutron and X-ray refinement of crystal form II. *J. Mol. Biol.* **180:** 301.

Wlodawer, A., J. Nachman, G.L. Gilliland, W. Gallagher, and C. Woodward. 1987. Structure of form III crystals of bovine pancreatic trypsin inhibitor. *J. Mol. Biol.* **198:** 469.

Wolberger, C., A.K. Vershon, B. Liu, A.D. Johnson, and C.O. Pabo. 1991. Crystal structure of a *MATα2* homeodomain-operator complex suggests a general model for homeodomain-DNA interactions. *Cell* **67:** 517.

Wüthrich, K. 1986. *NMR of proteins and nucleic acids*. Wiley, New York.

———. 1989. Protein structure determination in solution by nuclear magnetic resonance spectroscopy. *Science* **243:** 45.

———. 1990. Protein structure determination in solution by NMR spectroscopy. *J. Biol. Chem.* **265:** 22059.

———. 1991. Six years of protein structure determination by NMR spectroscopy: What have we learned? *Ciba Found. Symp.* **161:** 136.

Wüthrich, K. and W.J. Gehring. 1992. Transcriptional regulation by homeodomain proteins: Structural, functional, and genetic aspects. In *Transcriptional regulation* (ed. S.L. McKnight and K.R. Yamamoto), p. 535. Cold Spring Harbor Laboratory Press, Cold Spring Harbor, New York.

Wüthrich, K. and G. Wagner. 1975. NMR investigations of the dynamics of the aromatic amino acid residues in the basic pancreatic trypsin inhibitor. *FEBS Lett.* **50:** 265.

Wüthrich, K., G. Otting, and E. Liepinsh. 1992. Protein hydration in aqueous solution. *Faraday Discuss. Chem. Soc.* **93:** 35.

VIII

PROTEIN FOLDING

Introduction to papers 61 through 66

The "protein folding problem" is an area where the outlook for successful use of NMR techniques during the next years is particularly promising, be it for studies of folding kinetics or for structural characterization of unfolded states and folding intermediates. So far our activities relating to protein folding have, for a variety of reasons, mainly focused on characterization of non-globular states of polypeptides. The six papers collected in this part are representative of projects of this type in my group.

In the 1970s, we were involved in studies of polypeptide hormones such as calcitonin,[1] human parathyroid hormone[2] and glucagon,[3] and synthetic fragments thereof. As we know today, these peptides adopt predominantly flexibly extended solution conformations. NMR approaches used for studies of globular proteins were not suitable even for qualitative structural characterization of these compounds, and there was keen interest in novel, more informative experiments. The paper **61** describes a proposal to use the *cis–trans* equilibrium of Xxx–Pro peptide bonds as an NMR probe for the detection of non-random conformations in non-globular polypeptides,[4] which was subsequently complemented with studies of the exchange rates between the *cis* and *trans* forms of Xxx–Pro in a series of linear and cyclic oligopeptides.[5] In connection with the projects on non-globular polypeptides, we also spent considerable effort on

---

[1] Wüthrich, K. (1976) *NMR in Biological Research: Peptides and Proteins.* Amsterdam: North Holland.

[2] Bundi, A., Andreatta, R.H. and Wüthrich, K. (1978) *Eur. J. Biochem. 91,* 201–208. Characterisation of a local structure in the synthetic parathyroid hormone fragment 1–34 by ^{1}H nuclear magnetic resonance techniques.

[3] Bösch, C., Bundi, A., Oppliger, M. and Wüthrich, K. (1978) *Eur. J. Biochem. 91,* 209–214. ^{1}H nuclear magnetic resonance studies of the molecular conformation of monomeric glucagon in aqueous solution.

[4] Grathwohl, C. and Wüthrich, K. (1976) *Biopolymers 15,* 2043–2057. NMR studies of the molecular conformations in the linear oligopeptides H–(L-Ala)$_n$–L-Pro–OH.

[5] Grathwohl, C. and Wüthrich, K. (1981) *Biopolymers 20,* 2623–2633. NMR studies of the rates of proline *cis–trans* isomerization in oligopeptides.

measurements of "random coil" chemical shifts in model peptides,[6] and paper **62** describes amide proton titration experiments for the identification of amide proton–carboxylate hydrogen bonds that were developed in this context. The approaches described in the papers **61** and **62** both have the potential to detect non-random traits in unfolded, "denatured" polypeptides, and amide proton pH titration shifts have also been used for detailed studies of the surface structure in globular proteins (paper **54**).

The paper **63** introduced the use of 2D NMR for spatio-temporal resolution of amide proton trap experiments in studies of early events in protein folding. This project was part of Heinrich Roder's Ph.D. thesis in 1981 (see Appendix **A2**), but the work was published only several years later, as was also the case for paper **49**. Amide proton trap experiments became very popular in the late 1980s and are now one of the standard techniques in laboratories working on the protein folding problem. We recently used a variant of this approach for investigations on the conformational states of chaperone-bound polypeptides (paper **64**).

In analogy to work with folded proteins, complete sequence-specific NMR assignments are needed as a basis for detailed structural characterization of unfolded states of globular proteins (paper **65**). Recent progress in achieving this goal have been surveyed in paper **66**.

[6] Bundi, A. and Wüthrich, K. (1979) *Biopolymers 18,* 285–297. ^{1}H NMR parameters of the common amino acid residues measured in aqueous solutions of the linear tetrapeptides H–Gly–Gly–X–L–Ala–OH.

BIOPOLYMERS VOL. 15, 2025–2041 (1976)

The X-Pro Peptide Bond as an Nmr Probe for Conformational Studies of Flexible Linear Peptides

CHRISTOPH GRATHWOHL and KURT WÜTHRICH, *Institut für Molekularbiologie und Biophysik, Eidgenössische Technische Hochschule, 8093 Zürich-Hönggerberg, Switzerland*

Synopsis

The equilibrium between the *cis* and *trans* forms of X-Pro peptide bonds can readily be measured in the ^{13}C nmr spectra. In the present paper we investigate how observation of this equilibrium could be used as an nmr probe for conformational studies of flexible polypeptide chains. The experiments include studies by ^{13}C nmr of a series of linear oligopeptides containing different X-L-Pro peptide bonds, with X = Gly, L-Ala, L-Leu, L-Phe, D-Ala, D-Leu, and D-Phe. Overall the study confirms that X-Pro peptide bonds can generally be useful as ^{13}C nmr probes reporting the formation of nonrandom conformations in flexible polypeptide chains. It was found that the *cis–trans* equilibrium of X-Pro is greatly affected by the side chain of X and the configuration of the α-carbon atom of X. On the basis of these observations some general rules are suggested for practical applications of the X-Pro nmr probes in conformational studies of polypeptide chains.

INTRODUCTION

The development during the last two decades of our knowledge on the molecular conformations in polypeptide chains was characterized by numerous investigations of molecules with relatively rigid spatial structures, such as cyclic peptides and globular proteins. Both X-ray methods in single crystals and spectroscopic techniques in solutions have been extensively used in studies of these classes of oligopeptides and polypeptides. On the other hand it proved exceedingly more difficult to obtain information on the spatial solution structures of polypeptides which occur predominantly in flexible extended forms; these include, e.g., a variety of polypeptide hormones.[1–3] Since it appears rather unlikely for this class of molecules that the solution conformations would be readily inferable from single crystal X-ray results, if they became available, it is all the more important to develop experiments which are applicable for conformational studies in solution. In this paper we present experimental evidence for a ^{13}C nmr probe which should be quite generally suitable for investigations of flexible proline-containing polypeptide chains in solution.

An adequate description of the molecular conformations in flexible linear

© 1976 by John Wiley & Sons, Inc.

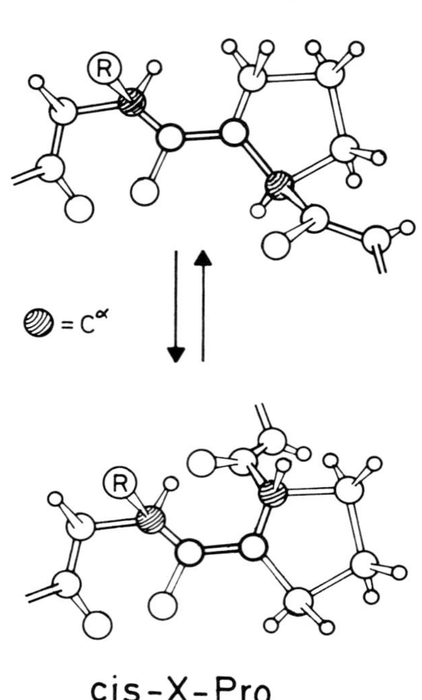

Fig. 1. *Cis* and *trans* forms of the X-L-Pro peptide bond.

polypeptide chains can be obtained with the assumption that the latter consist of planar standard peptide groups[4] linked together by single bonds via the α-carbon atoms. Except for the N-substituted peptide groups, where the *cis* and *trans* forms are taken into consideration (Fig. 1), the peptide bonds are assumed to be in the *trans* form. For a given primary structure, different backbone conformations can be generated by variation of the torsion angles ϕ_i and ψ_i about the single bonds between the peptide groups and the α-carbon atoms, ω_i about X-Pro and possibly other N-substituted peptide bonds, and χ_i^j about the bonds between nonhydrogen atoms of the amino-acid side chains.[4,5] The contours of conformational energy maps for flexible linear polypeptides will quite generally be characterized by wide regions of very similar energies encompassing a multitude of molecular species which differ in the torsion angles ϕ_i, ψ_i, and χ_i^j. Because of the low barriers for rotational motions about single bonds, there will be rapid interconversion among these different species. The random-coil form of a flexible polypeptide chain may then be defined as an ensemble of rapidly interconverting species containing all possible combinations of the sterically allowed torsion angles in the individual dipeptide fragments.[4] Energetically preferred nonrandom flexible conformations would be characterized by thermal populations of certain rotation states being different from those in the random-coil form.

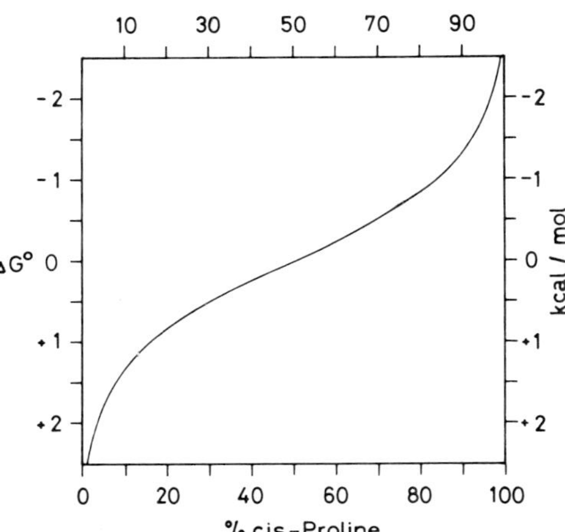

Fig. 2. Correspondence between $\Delta G°$ for the equilibrium between the *cis* and *trans* forms of X-Pro peptide bonds and the relative concentrations of the *cis* and *trans* forms.

A principal difficulty for studies of flexible peptides arises since on the time scale of the common spectroscopic techniques suited for studies in solution, e.g., nmr and ORD/CD, the observed spectra correspond to the average of all the rapidly interconverting species. Manifestations of different thermal populations of individual rotation states in the average spectral parameters are in most cases too small to be detected with the presently available spectral resolution. Thus it is quite common that "typical random-coil spectra" are reported for flexible linear polypeptide chains.[1–3] A more favorable situation for the application of spectroscopic techniques is encountered when different polypeptide conformations differ by *cis–trans* isomerization of a peptide group (Fig. 1). Since there is an energy barrier of $\Delta G^{\ddagger} \approx 20$ kcal/mol for *cis–trans* interconversion of peptide groups,[6–8] this process is usually slow on the time scale of most relevant experimental techniques. Thus the two species can be observed separately if a sufficient spectral resolution is available. The slow rate of the *cis–trans* isomerization of X-Pro peptide bonds is an essential factor in the experiments discussed below.

The *cis–trans* equilibrium of an X-Pro peptide group can be described either by the equilibrium constant K or by the free enthalpy $\Delta G°$

$$trans \rightleftarrows cis;\ \Delta G° = -RT \ln K;\ K = \frac{[cis]}{[trans]} \tag{1}$$

As was shown previously[9–12] and will be evidenced with some new examples in this paper, $\Delta G°$ at ambient temperature is of the order −2.0–2.0 kcal/mol in a variety of linear oligopeptides. The *cis* and *trans* forms are then simultaneously present (Fig. 2) and $\Delta G°$ can be determined from the relative concentrations of the two species. In an earlier paper[11] it was suggested that the values of $\Delta G°$ thus obtained could be decomposed into two terms

$$\Delta G^\circ = \Delta G^\circ_{XP} + \Delta G^\circ_{conf} \qquad (2)$$

where ΔG°_{XP} would be characteristic for the covalent structure of the fragment X-L-Pro-, and ΔG°_{conf} would account for effects on the equilibrium (1) arising from the occurrence of energetically preferred conformations in the polypeptide chain containing the X-L-Pro- peptide bond. ΔG°_{conf} will only be different from 0 if the populations of the rotational states described by ϕ_i, ψ_i, and χ_i^j are different in the molecular species containing, respectively, *cis* and *trans* proline. The difference between the two quantities ΔG° and ΔG°_{XP} can thus be used as a naturally built-in probe indicating the occurrence of nonrandom preferred conformations in proline-containing peptides. The present paper investigates some practical aspects of the use of different peptide groups X-L-Pro for studies of peptide conformation.

On the basis of Eq. (2), any fragment X-L-Pro can *a priori* be used to detect conformational changes in a given polypeptide chain and to estimate the relative conformational energies of the observed species, e.g., when the molecule is subjected to different solvent media or interacting with other biopolymers. A more quantitative use of the X-L-Pro probes for measurements of the conformational energies of the observed molecular species relative to the random-coil polypeptide chain, however, requires that ΔG°_{XP} [Eq. (2)] is known. The experiments described in what follows concentrated mainly on the relations between ΔG° and the structure of X, and on the experimental determination of ΔG°_{XP}. As in the previous communications[9,11], only ^{13}C nmr will be considered for the experimental measurements of the *cis–trans* equilibrium of X-L-Pro.

MATERIALS AND METHODS

The peptides H-Gly-L-Pro-OH, H-Gly-L-Pro-L-Ala-OH, H-(L-Ala)$_m$-L-Pro-(L-Ala)$_n$-OH (m = 1,2,3; n = 0,1,2), Z-L-Ala-L-Pro-OH, Z-L-Ala-L-Pro-L-Ala-OH, Z-L-Ala-L-Pro-L-Ala-OMe, H-L-Ala-L-Phe-L-Pro-OH, H-L-Ala-L-Phe-L-Pro-L-Ala-OH, and CF_3CO-Gly-Gly-L-Pro-L-Ala-OMe were obtained from Bachem AG, Liestal, Switzerland; H-D-Ala-L-Pro-OH, H-L-Leu-L-Pro-OH, and H-D-Leu-L-Pro-OH were obtained from Chemspec, Bühler, Switzerland; H-L-Phe-L-Pro-OH, H-D-Phe-L-Pro-OH, and Z-L-Ala-L-Pro-NH_2 were obtained as a gift from Dr. M. Brugger and Dr. W. Rittel, Ciba-Geigy AG, Basel. The protecting groups are denoted by Z = carbobenzoxy, CF_3CO = trifluoroacetyl, Me = methyl. H-Gly-Gly-L-Pro-L-Ala-OH was prepared by treating CF_3CO-Gly-Gly-L-Pro-L-Ala-OMe with aqueous NaOH in methanol. H-L-Ala-L-Pro-NH_2 was prepared through hydrogenation of Z-L-Ala-L-Pro-NH_2 in methanol in the presence of Pd-active coal.

For the nmr measurements approximately 0.1 *M* solutions of the peptides in D_2O, CD_3OD, and d_6-Me_2SO were prepared. In D_2O solutions dioxane was used as an internal reference; in CD_3OD and d_6-Me_2SO solutions, Me_4Si was added for this purpose. To obtain the different ionization states of the peptides in aqueous solutions, the pH was adjusted to 1.5, 6.5, or 10.5 by the addition of minute amounts of 4 *N* NaOH or HCl. Solutions in CD_3OD and d_6-Me_2SO of the cationic and zwitterionic forms of the unprotected peptides, the uncharged and anionic forms of the peptides with protected terminal amino groups, and the cationic form of H-L-Ala-L-Pro-NH_2 were obtained by lyophilizing the peptides from H_2O solutions with pH 1.5 and 6.5, respectively, and then dissolving the lyophilized materials in the desired solvents. Because the peptides were decomposed during lyophilization from basic aqueous solutions, the anionic forms of the unprotected peptides could not be studied in methanol and Me_2SO.

^{13}C nmr spectra were recorded at 25.16 MHz on a Varian XL-100 spectrometer, using 12-mm sample tubes. The relative concentrations of the *cis* and *trans* forms of the X-Pro peptide bonds were obtained from measurements of the peak heights of corresponding resonances in the two species.

RESULTS

Since in natural polypeptide chains any of the common amino acids can precede proline, the present investigation should ideally have included all the different X-Pro peptide bonds. For practical reasons we have limited the study to the four common amino acids Gly, L-Ala, L-Leu, and L-Phe. Since they carry no functional groups, it appeared reasonable as a first approximation that different behavior of these four amino-acid residues could be interpreted in terms of a single structural parameter, i.e., the bulkiness of the hydrophobic side chains. Variation of a second structural parameter, i.e., the stereospecificity at the α-carbon atom, was then included in that D-Ala, D-Leu, and D-Phe were also examined.

A selection of representative ^{13}C nmr spectra of the peptide fragments X-L-Pro are shown in Figure 3. It is seen that both the C^{β} and C^{γ} proline ring carbon resonances for the *cis* and *trans* forms of the X-L-Pro peptide bond are resolved in the dipeptides H-Gly-L-Pro-OH, H-Ala-L-Pro-OH, and H-Phe-L-Pro-OH,[13–16] whereas only the C^{β} lines are resolved in H-Leu-L-Pro-OH. It is readily apparent in the figure that the *cis–trans* ratio of the X-L-Pro peptide bond is strongly influenced by X. The relative abundance of *cis* proline increases when one goes from X = Gly to L-Ala to L-Leu to L-Phe, and it is also markedly dependent on the configuration at the α-carbon atom of X.

The *cis–trans* ratios measured for the accessible states of protonation of the seven dipeptides in Figure 3 in the three solvents D_2O, CD_3OD, and d_6-Me_2SO are collected in Table I. The table confirms the observation in Figure 3 that the *cis–trans* ratio depends on the bulkiness of the side chain of X and on the configuration at the α-carbon atom. The dipeptides with D-amino-acid residues contain less *cis* proline than the corresponding peptides with L-amino-acid residues. In the L-L dipeptides, the concentration of *cis* proline increases with increasing bulkiness of the side chain of X, whereas the opposite behavior is observed for the D-L dipeptides.

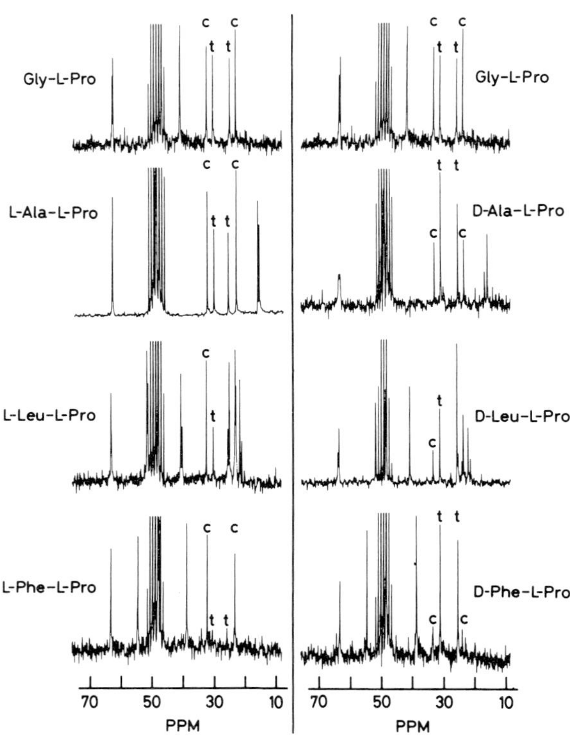

Fig. 3. Spectral region from 10 to 70 ppm of the ^{13}C nmr spectra of various dipeptides H-X-L-Pro-OH in d_4-methanol. The peptides were dissolved in the zwitterionic state $^+H_2$-X-L-Pro-O^-. The ring carbon resonances C^β and C^γ of *cis* and *trans* proline are indicated by *c* and *t*, respectively. The spectra were recorded at 25.16 MHz with broad-band ^1H decoupling. Internal Me_4Si is at 0 ppm, the solvent resonance is at 49.0 ppm, t = 25°C.

Table I also shows that the *cis–trans* equilibrium of X-Pro depends on the protonation state of the dipeptides and on the solvent. Quite generally the fully protonated cationic forms of the dipeptides contain relatively little *cis* proline, and the influence of the structural parameters of X on the *cis–trans* equilibrium is not very pronounced. For the zwitterionic dipeptides, which contain in all cases more *cis* proline than the corresponding cations, the general trend is that the concentration of *cis* proline increases when one goes from the solutions in D_2O to CD_3OD to d_6-Me_2SO. In contrast to the solutions in D_2O and CD_3OD, where the above-mentioned dependence on the structure of X is clearly manifested, the *cis–trans* equilibrium for the zwitterionic L-L dipeptides in d_6-Me_2SO is essentially independent of X. In the anionic peptides in basic aqueous solution, the concentration of *cis* proline is somewhat higher or equal to that observed for the zwitterionic peptides in D_2O.

The dependence of the *cis–trans* equilibrium of the X-Pro bond on the protonation of the peptide and on the solvent is qualitatively very similar for all the peptides in Table I. In analogy to H-L-Ala-L-Pro-OH, which is discussed in detail in Ref. 17, we conclude that a conformational energy term ΔG°_{conf} [Eq. (2)] is operative for all these peptides, i.e., electrostatic interactions with the negatively charged carboxylic acid group of proline stabilize molecular conformations which include *cis* proline. Table I and earlier observations in different peptides[11,12] thus illustrate that all the seven dipeptide segments X-L-Pro in Figure 3 function as natural probes indicating changes of the peptide conformations.

Provided that ΔG°_{XP} had independently been determined, ΔG°_{conf} can in principle be quantitatively evaluated as the difference between ΔG° observed for the molecular species in question and ΔG°_{XP} [Eq. (2)]. As an empirical approach for the evaluation of ΔG°_{XP}, we have chosen to measure the *cis–trans* equilibrium of the X-L-Pro peptide bonds in a series of different peptides which one might expect from the amino-acid sequence to be preferentially in a random-coil conformation. Thus a number of peptides H-(L-Ala)$_m$-L-Pro-(L-Ala)$_n$-OH, with m = 1,2,3 and n = 0,1,2, and

TABLE I

Cis–Trans Equilibrium of the X-L-Pro Peptide Bond in the Peptides of Figure 3 in Different States of Protonation and in Three Different Solvents at 25°C[a]

Peptide and Solvent	Cation	Zwitterion	Anion[b]
D_2O			
H-Gly-L-Pro-OH	15	37	45
H-L-Ala-L-Pro-OH	10	40	54
H-L-Leu-L-Pro-OH	5	49	64
H-L-Phe-L-Pro-OH	29	76	76
H-D-Ala-L-Pro-OH	7.5	22	38
H-D-Leu-L Pro-OH	10	20	43
H-D-Phe-L-Pro-OH	⩽3	10	20
CD_3OD			
H-Gly-L-Pro-OH	24	55	
H-L-Ala-L-Pro-OH	15	57	
H-L-Leu-L-Pro-OH	10	68	
H-L-Phe-L-Pro-OH	10	88	
H-D-Ala-L-Pro-OH	14	33	
H-D-Leu-L-Pro-OH	20	30	
H-D-Phe-L-Pro-OH	10	15	
d_6-Me_2SO			
H-Gly-L-Pro-OH	25	80	
H-L-Ala-L-Pro-OH	10	85	
H-L-Leu-L-Pro-OH	⩽3	85	
H-L-Phe-L-Pro-OH	10	78	
H-D-Ala-L-Pro-OH	20	66	
H-D-Leu-L-Pro-OH	25	50	
H-D-Phe-L-Pro-OH	10	26	

[a] The numbers indicate the concentration of *cis* proline in % of the total peptide concentration.

[b] The anionic forms of the peptides could not be studied in CD_3OD and d_6-Me_2SO because diketopiperazine was formed during lyophilization from the basic aqueous solutions.

several analogs with protected end groups were examined to determine ΔG°_{AP}. The results are presented in Table II.

Table II shows that most of the compounds studied contain an equilib-

TABLE II
Cis–Trans Equilibrium of the L-Ala-L-Pro Peptide Bond in Various Linear Oligopeptides in Three Different Solvents at 25°C[a]

Peptide[b] and Solvent	Without Charge[c]	Cation	Zwitterion	Anion
D_2O				
H-Ala-Pro-OH		10	40	54
Z-Ala-Pro-OH	10			>25
H-Ala-Pro-NH_2	15	13		
Z-Ala-Pro-NH_2	d			
H-Ala-Pro-Ala-OH[f]		10	11	15
Z-Ala-Pro-Ala-OH	5			5
Z-Ala-Pro-Ala-OMe	d			
H-Ala_2-Pro-Ala-OH		5	10	10
H-Ala_3-Pro-Ala-OH		10	10	4
H-Ala_2-Pro-Ala_2-OH		5	5	5
H-Ala_3-Pro-Ala_2-OH		8	5	8
CD_3OD				
H-Ala-Pro-OH		15	57	e
Z-Ala-Pro-OH	10			35
H-Ala-Pro-NH_2	—	15		
Z-Ala-Pro-NH_2	11			
Z-Ala-Pro-Ala-OH	13			5
Z-Ala-Pro-Ala-OMe	13			
H-Ala_2-Pro-Ala-OH		15	d	e
H-Ala_3-Pro-Ala-OH		10	10	e
H-Ala_2-Pro-Ala_2-OH		19	20	e
H-Ala_3-Pro-Ala_2-OH		10	d	e
d_6-Me_2SO				
H-Ala-Pro-OH		10	85	e
Z-Ala-Pro-OH	10			58
H-Ala-Pro-NH_2	—	16		
Z-Ala-Pro-NH_2	12			
Z-Ala-Pro-Ala-OH	13			12
Z-Ala-Pro-Ala-OMe	11			
H-Ala_2-Pro-Ala-OH		10	d	e
H-Ala_3-Pro-Ala-OH		13	d	e
H-Ala_2-Pro-Ala_2-OH		15	17	e
H-Ala_3-Pro-Ala_2-OH		11	d	e

[a] The numbers indicate the concentration of *cis* proline in % of the total peptide concentration.

[b] All the amino-acid residues are in the L-configuration. Z = carbobenzoxy protecting group, C_6H_5—CH_2—O—CO—, Me = —CH_3.

[c] Fully protected peptides, protonated form of peptides with protected amino end and deprotonated form of peptides with protected carboxylic acid group.

[d] Not measured because of poor solubility.

[e] Not measured because the peptide was decomposed during lyophilization from basic aqueous solutions.

[f] H-Ala-Pro-Ala-OH was examined only in aqueous solution.

rium concentration of *cis* proline of approximately 10%. The simplest one of these molecules is the dipeptide H-L-Ala-L-Pro-OH in the fully protonated cationic form. The *cis–trans* ratio in this dipeptide was only little affected when either the amino terminus or the carboxylic acid terminus or both terminal groups were protected with one of the common protecting groups. In the tripeptide H-L-Ala-L-Pro-L-Ala-OH and its partly or fully protected analogs the same *cis–trans* ratio of proline was observed as in the cationic form of H-L-Ala-L-Pro-OH. This equilibrium state was also maintained when additional alanyl residues were attached either at the amino end or the carboxylic acid end of the peptide chain. In aqueous solution and also in other solvents, in as far as they were examined, the *cis–trans* ratio in the peptides with terminal alanine was found to be essentially independent of the state of protonation. These observations are compatible with the assumption that for the peptides in Table II, an equilibrium concentration of approximately 10% *cis* proline is indicative of a random-coil molecular conformation. With $\Delta G^\circ_{conf} = 0$, the *cis–trans* equilibria in Table II can be used to determine ΔG°_{AP}, which is thus found to be essentially independent of the covalent molecular structure outside the fragment -L-Ala-L-Pro-.

The only species in Table II for which the *cis–trans* ratio of the L-Ala-L-Pro bond deviates markedly from the random-coil value of ca. 10% *cis* proline, are the deprotonated forms of the peptides containing proline at the carboxylic acid terminus. These compounds form nonrandom conformations containing *cis* proline, as is discussed in detail elsewhere.[17]

In view of the different solvation in different solvents, one might quite generally expect ΔG°_{XP} to depend on the solvent. Table II shows that only a small solvent effect was observed for ΔG°_{AP}. In aqueous solution, the relative concentration of *cis*-proline in all the peptides studied was within the limits 10% ± 5%; in methanol and dimethyl sulfoxide solutions it was 15% ± 5%. From Figure 2 one then finds that the relative concentrations of *cis* proline observed in the different model peptides (Table II) correspond to a range of values for ΔG°_{AP} of 1.7 kcal/mol $\geqslant \Delta G^\circ_{AP} \geqslant$ 1.0 kcal/mol in water, and 1.3 kcal/mol $\geqslant \Delta G^\circ_{AP} \geqslant$ 0.82 kcal/mol in methanol and dimethylsulfoxide solution.

One conclusion of considerable practical interest which can be drawn from Table II is that a reliable value for ΔG°_{AP} could be obtained from the fully protonated form of the dipeptide H-L-Ala-L-Pro-OH. To investigate whether this procedure could also be applied for different residues X, we examined some model peptides containing a Gly-L-Pro peptide bond, or a L-Phe-L-Pro peptide bond, respectively. The results for the Gly-L-Pro peptides are presented in Table III. In solutions in D_2O and methanol, analogous results to those for L-Ala-L-Pro peptides (Table II) were obtained, i.e., the *cis–trans* ratio was found to be the same in the protonated dipeptide H-Gly-L-Pro-OH and in the three tripeptides and tetrapeptides of Table III. The equilibrium concentration of *cis* proline for the different peptides studied was 17% ± 3% in water, and 22% ± 2% in methanol, corresponding to a range of values for ΔG°_{GP} of 1.1 kcal/mol $\geqslant \Delta G^\circ_{GP} \geqslant$ 0.82

kcal/mol in water, and 0.82 kcal/mol $\geqslant \Delta G^{\circ}_{GP} >$ 0.68 kcal/mol in methanol (Fig. 2). In dimethylsulfoxide solution, the dipeptide H-Gly-L-Pro-OH yields a value of ΔG°_{GP} which agrees closely with that observed in methanol solution. In the tripeptides and tetrapeptides examined, the *cis–trans* equilibrium is shifted to higher concentrations of *cis* proline. From observations in the ^{1}H nmr spectra of these peptides we concluded that the most reliable value for ΔG°_{GP} in Me_2SO is that derived from the experiments with the dipeptide, i.e., $\Delta G^{\circ}_{GP} \approx$ 0.68 kcal/mol; for the other peptides in Table III, the ^{1}H nmr data indicated the occurrence of nonrandom conformations in this solvent.[12] As was mentioned previously (Table I), the remarkably high concentration of *cis* proline in the deprotonated forms of H-Gly-L-Pro-OH arises as a consequence of the formation of nonrandom conformations,[17] and is therefore not included in the present discussion of ΔG°_{GP}.

The data on the *cis–trans* equilibrium of the L-Phe-L-Pro bond in three linear peptides are presented in Table IV. Comparison of the protonated dipeptide H-L-Phe-L-Pro-OH in the three solvents D_2O, CD_3OD, and d_6-Me_2SO would seem to indicate that ΔG° is more extensively affected

TABLE III

Cis-Trans Equilibrium of the Gly-L-Pro Peptide Bond in Various Linear Oligopeptides in Different States of Protonation and in Three Different Solvents at 25°C[a]

Peptide and Solvent[b]	Without Charge[c]	Cation	Zwitterion	Anion[d]
D_2O				
H-Gly-L-Pro-OH		15	37	45
H-Gly-L-Pro-L-Ala-OH		16	15	18
H-Gly-Gly-L-Pro-L-Ala-OH		13	15	20
CF_3CO-Gly-Gly-L-Pro-L-Ala-OMe	16			
CD_3OD				
H-Gly-L-Pro-OH		24	55	
H-Gly-L-Pro-L-Ala-OH		22	e	
H-Gly-Gly-L-Pro-L-Ala-OH		22	21	
CF_3CO-Gly-Gly-L-Pro-L-Ala-OMe	23			
d_6-Me_2SO				
H-Gly-L-Pro-OH		25	80	
H-Gly-L-Pro-L-Ala-OH		29	e	
H-Gly-Gly-L-Pro-L-Ala-OH		35	e	
CF_3CO-Gly-Gly-L-Pro-L-Ala-OMe	35			

[a] The numbers indicate the concentration of *cis* proline in % of the total peptide concentration.

[b] CF_3CO = trifluoroacetyl protecting group; Me = $-CH_3$.

[c] Peptide with protected end groups.

[d] The anionic forms of the peptides could not be studied in CD_3OD and d_6-Me_2SO because the peptides were decomposed during lyophilization from the basic aqueous solutions.

[e] Not measured because of poor solubility.

by the solvent than in the corresponding dipeptides with X = L-Ala (Table II) or Gly (Table III). This is probably a consequence of the hydrophobic character of the side chain of L-Phe. According to the criteria used above for the evaluation of ΔG°_{XP} in the L-Ala-L-Pro peptides (Table II), the value $\Delta G^{\circ}_{FP} \approx 0.53$ kcal/mol obtained from the protonated dipeptide H-L-Phe-L-Pro-OH appears to be quite representative for the L-Phe-L-Pro peptide bond in random-coil peptides in aqueous solution. In the nonaqueous solvents, on the other hand, there are quite marked differences between the *cis–trans* equilibria in the cationic dipeptide and in the longer peptide molecules. Even though the numbers in Table IV may well serve as reference values for work with L-Phe-L-Pro peptide bonds, it seems premature in view of the small number of compounds studied to assign any of the values observed in the different model peptides to ΔG°_{FP} in the strict sense of the definition introduced with Eq. (2). As for the proline terminal peptides in Tables I–III, the markedly increased *cis* proline concentration in the deprotonated forms of the -L-Phe-L-Pro-OH peptides is a consequence of the occurrence of nonrandom molecular species,[17] and is therefore not further discussed in the present context.

DISCUSSION

This discussion will focus mainly on one aspect of the data presented in Tables I–IV, i.e., the practical aspects of using X-L-Pro peptide bonds as ^{13}C nmr probes for conformational studies of polypeptide chains. Addi-

TABLE IV

Cis–Trans Equilibrium of the L-Phe-L-Pro Peptide Bond in Various Linear Oligopeptides in Different States of Protonation and in Three Different Solvents at 25°C[a]

Peptide and Solvent	Cation	Zwitterion	Anion[b]
D_2O			
H-L-Phe-L-Pro-OH	29	76	76
H-L-Ala-L-Phe-L-Pro-OH	22	57	43
H-L-Ala-L-Phe-L-Pro-L-Ala-OH	29	35	27
CD_3OD			
H-L-Phe-L-Pro-OH	10	88	
H-L-Ala-L-Phe-L-Pro-OH	19	64	
H-L-Ala-L-Phe-L-Pro-L-Ala-OH	30	42	
d_6-Me_2SO			
H-L-Phe-L-Pro-OH	10	78	
H-L-Ala-L-Phe-L-Pro-OH	12	c	
H-L-Ala-L-Phe-L-Pro-L-Ala-OH	22	22	

[a] The numbers indicate the concentration of *cis* proline in % of the total peptide concentration.

[b] The peptides could not be studied in the anionic form in CD_3OD and d_6-Me_2SO because they were decomposed during lyophilization from the basic aqueous solutions.

[c] Not measured because of poor solubility.

tional points of interest relating to the occurrence of particular molecular conformations in individual ones of the model peptides (Tables I–IV) will be briefly mentioned at the end of this section.

On the basis of Eqs. (1) and (2), observation of the *cis–trans* equilibrium of X-L-Pro peptide bonds can be used to measure the relative energies of different conformations of a given polypeptide chain. Because of the limited sensitivity of ^{13}C nmr, applications of this probe are in practice confined to the range of free energies between approximately 1.5 kcal/mol $\geqslant \Delta G° \geqslant -1.5$ kcal/mol. It is one of the attractive features of the proposed technique that within this range small differences in $\Delta G°$ of the order of several tenth of a kcal/mol are readily discernible (Fig. 2). Tables I–IV list more than 20 oligopeptides for which $|\Delta G°| \lesssim 1.5$ kcal/mol, and recent publications show that many additional oligopeptides have similar characteristics.[9–32] It thus appears that the X-L-Pro probe can in principle be applied to a wide variety of synthetic and natural proline-containing peptides. Potential uses may, for example, include studies of denaturation by heat and by chemical agents, or detection of conformational rearrangements during interactions of peptides with other macromolecules, membranes, and perhaps even with intact tissue.

Demonstration of the occurrence of thermodynamically preferred peptide conformations can thus, in suitable systems, be achieved with comparatively modest effort, and on this basis applications of more laborious techniques for studies of structural details may then be directed more efficiently.

The practical application of X-L-Pro ^{13}C nmr probes will in each individual peptide depend on the spectral resolution and the sensitivity for measurements of the relative concentrations of *cis* and *trans* proline. These experimental aspects will now be briefly considered.

The *cis–trans* equilibrium of the X-L-Pro bonds is most readily measured by the relative intensities of the C^{β} and C^{γ} ring carbon resonances of proline. For these two carbon atoms the relative chemical shifts between *cis* and *trans* proline are of the order of 2 ppm, and hence the resonances of the two isomeric forms of proline are well separated (Fig. 3).[13–16] Furthermore, the chemical shifts for C^{β} and C^{γ} are remarkably similar in different peptides, so that these resonance lines are usually readily identified.[10,12,33] In small peptides with up to approximately 10 amino-acid residues, the C^{β} and C^{γ} resonances of proline can be recorded at natural abundance of ^{13}C (Fig. 3), unless the amino-acid composition is such that the spectral regions from 23–26 ppm and from 30–33 ppm are crowded with other lines. Amino-acid residues likely to give lines overlapping with C^{β} of proline are Val, Glu, Gln, Lys, Arg, and Met, while resonances of Leu, Ile, Lys, and Arg occur near the C^{γ} resonances of proline.[25,33] When there is overlap of both the C^{β} and C^{γ} resonances with other lines of the peptide, a reliable measurement of the *cis–trans* ratio may still be achieved by suitable ^{13}C enrichment of proline.[33] Preliminary experiments with the polypeptide hormones calcitonin M and ACTH indicated that one will, in general, also have to recur to specific ^{13}C enrichment techniques to obtain reliable data for polypeptide chains which are longer than approximately 10 amino-acid residues.

Experience so far indicated that measurements of the relative peak heights of the *cis* and *trans* C^{β} and C^{γ} resonances of proline gave the most reliable values for the concentrations of the two forms of the peptide bond. Since studies of the ^{13}C spin relaxation in small proline-containing peptides[23,32] showed that T_1 values in molecules with *cis* X-L-Pro bonds are very similar to those in corresponding molecules with *trans* proline, errors in the measured *cis–trans* intensity ratios which might arise from different relaxation rates in the isomeric species should be small. Errors might result from peak height measurements when variations of the line widths would arise from the exchange between *cis* and *trans* proline.[34] So far, however, we found no evidence in the systems studied for a low energy of activation and concomitant rapid rate of the *cis–trans* isomerization of the X-L-Pro peptide bonds.

Provided that reliable values for ΔG°_{XP} can be obtained it should in principle be possible with the X-L-Pro probe to evaluate the average conformational energy of the preferentially populated species relative to the random-coil form of the same peptide molecule. The data in Tables I–IV indicate that ΔG°_{XP} is in the range from 0.8 to 1.8 kcal/mol for the different residues X which were investigated; measurements of ΔG°_{XP} in random-coil peptides should thus quite generally be possible. For practical purposes it is of interest that for X = L-Ala and Gly, the *cis–trans* equilibria in the fully protonated dipeptides H-X-L-Pro-OH yielded values for ΔG°_{XP} which are valid for the peptide fragments -X-L-Pro- also in longer polypeptide chains. However, the present study also indicates certain limitations for the general application of such values of ΔG°_{XP} for quantitative studies. Even for X = L-Ala or Gly, where ΔG°_{XP} is to a good approximation independent of the covalent peptide structure outside the fragment -X-L-Pro-, the variation of the ΔG°_{XP} values obtained with the different model peptides is of the order of 0.6 kcal/mol (Tables II and III). These rather large uncertainties arise mainly because the absolute values of ΔG°_{AP} and ΔG°_{GP} are at around 1.3 kcal/mol, where the accuracy of the measured *cis–trans* ratios is particularly critical (Fig. 2). This has also the additional consequence for -Gly-L-Pro- and -L-Ala-L-Pro- peptides that negative values of ΔG°_{conf}, i.e., conformation induced increase of the concentration of *cis* proline, can be measured with greater accuracy than positive values of ΔG°_{conf}.

Whereas the values obtained for ΔG°_{XP} of Gly and L-Ala appear to be quite useful, the results with a small selection of L-Phe-L-Pro peptides (Table IV) indicate that the determination of ΔG°_{XP} is, in general, more critical for amino-acid residues X with bulkier side chains. The ^{13}C chemical shifts of the phenylalanyl residues in the different peptides of Table IV indicate a likely explanation for the apparently somewhat erratic variations of ΔG°.

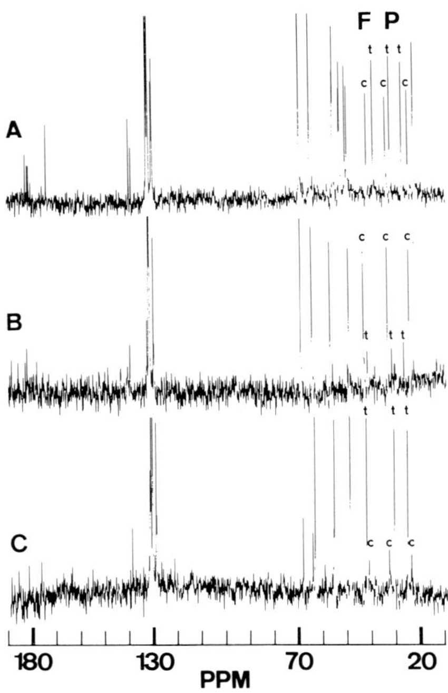

Fig. 4. ^{13}C nmr spectra. (A) H-L-Ala-L-Phe-L-Pro-OH. (B) H-L-Phe-L-Pro-OH. (C) H-D-Phe-L-Pro-OH in D_2O in the anionic state (pD = 10.3). The resonances of C^β of Phe (F) and C^β and C^γ of Pro (P) are indicated by *c* and *t* for the *cis* and *trans* forms of the Phe–Pro bond, respectively. The spectra were recorded at 25.16 MHz with broad-band ^{1}H decoupling, t = 25°C. The chemical shifts are relative to external Me_4Si where the internal standard dioxane was taken to be at 67.8 ppm.

Figure 4 shows that an outstandingly large relative chemical shift of 2.7 ppm for C^β of phenylalanine is observed between the two species of the tripeptide H-L-Ala-L-Phe-L-Pro-OH which contain the L-Phe-L-Pro peptide bond in the *cis* and *trans* form, respectively. The corresponding chemical shifts for C^β of phenylalanine in the dipeptides H-L-Phe-L-Pro-OH and H-D-Phe-L-Pro-OH are much smaller, i.e., of the order 1 ppm. ^{13}C chemical shifts are known to be sensitive to changes in the hybridization of the atom considered, which may be caused by preferred population of certain rotation states about interatomic bonds or by steric strain.[35] Thus a possible explanation of the data in Figure 4 would seem to be that the side chain of phenylalanine is, because of its bulkiness, confined to relatively narrow ranges of rotation states about χ^1 by the amino-acid residue Y in the fragment -Y-Phe-L-Pro-. This appears to be borne out independently by model considerations. CPK molecular models imply that the populations of different rotation states of the side chain of phenylalanine are greatly affected by the residue Y in the sequence -Y-Phe-L-Pro-, and that different conformations of the side chain are preferred in the species containing, respectively, the *cis* and *trans* forms of the Phe-L-Pro peptide bond. On this basis it would not be surprising that different values

$\Delta G^{\circ}_{FP)apparent}$ result from studies of different "random-coil" model peptides, in particular when dipeptides H-Phe-L-Pro-OH are compared with longer peptide chains. An additional comment on the data in Figure 4 is of practical interest. If in the species containing *cis* and *trans* proline, respectively, there are sizeable chemical shift differences in carbon positions other than C^{β} and C^{γ} of proline, as, e.g., in the tripeptide H-L-Ala-L-Phe-L-Pro-OH, such resonances may be used to measure the *cis–trans* equilibrium in molecules where the proline resonances cannot readily be resolved in the ^{13}C nmr spectrum.

Overall the present results and model considerations on ΔG°_{XP} appear to indicate that the empirical values for ΔG°_{AP} and ΔG°_{GP} obtained from

TABLE V

^{13}C nmr Chemical Shifts of the Dipeptide H-D-Ala-L-Pro-OH in the Different States of Ionization Which Were Accessible in D_2O, CD_3OD, and d_6-Me_2SO[a]

Resonance Assignment		Cation		Zwitterion		Anion	
Solvent D_2O		*cis*	*trans*	*cis*	*trans*	*cis*	*trans*
Pro	C^{α}	61.6	61.0	63.7	63.4	63.7	63.2
	C^{β}	32.5	30.1	32.8	30.7	32.7	30.8
	C^{γ}	22.9	25.5	23.2	25.4	23.6	25.5
	C^{δ}		48.6	48.7	48.6	48.5	48.5
Ala	C^{α}		49.4	49.2	49.5	48.8	48.8
	C^{β}	17.0	16.1	16.7	16.1	20.3	19.7
Pro	C′		177.1				
Ala	C′		170.3				
Solvent CD_3OD							
Pro	C^{α}		60.7	63.8	63.4		
	C^{β}	32.2	30.2	33.0	31.0		
	C^{γ}	23.1	25.6	23.6	25.5		
	C^{δ}		48.2				
Ala	C^{α}		49.3				
	C^{β}		16.1	17.0	16.1		
Pro	C′		175.3				
Ala	C′						
Solvent d_6-Me_2SO							
Pro	C^{α}		58.9	61.7	61.2		
	C^{β}	30.7	28.6	31.2	29.1		
	C^{γ}	21.7	24.1	22.2	24.0		
	C^{δ}		46.3	46.2			
Ala	C^{α}		47.1				
	C^{β}	16.0	15.7	17.1			
Pro	C′		172.8	175.1			
Ala	C′		167.7	169.3			

[a]Where no numbers are given, the resonances had not been unambiguously assigned. The chemical shifts are with respect to the following reference compounds.

In D_2O: external Me_4Si, where internal dioxane was taken to be at 67.8 ppm.

In CD_3OD: internal Me_4Si, where the solvent resonance was taken to be at 49.0 ppm.

In d_6-Me_2SO: internal Me_4Si, where the solvent resonance was taken to be at 39.4 ppm.

Tables II and III are quite reliable numbers which are essentially independent of the amino-acid sequence outside the fragment -X-L-Pro-. On the other hand, for amino-acid residues X with bulky side chains or with side chains including functional groups, somewhat different values of ΔG°_{XP} are to be expected depending on the amino-acid residue Y preceding X in the sequence -Y-X-L-Pro-.

A quite striking feature in Tables I–IV is the dependence of ΔG° on the ionization state of the peptides containing proline at the carboxylic acid terminus. These data were completed by additional experiments with different proline peptides and are discussed in detail in the following paper.[17] It was found that the increased concentration of *cis* proline in the deprotonated species is a consequence of the formation of nonrandom molecular conformations which are stabilized mainly by electrostatic interactions with the terminal carboxylate group.[17] This then called our attention to an additional observation in Table I, i.e., the dependence of ΔG°_{conf} on the amino-acid residue X, which is even more pronounced than the influence of X on ΔG°_{XP}. An interpretation of this phenomenon on the basis of semi-empirical energy calculations will be presented in a forthcoming paper (manuscript in preparation).

APPENDIX

On request by one of the referees we present in Table V the ^{13}C nmr chemical shifts for the dipeptide H-D-Ala-L-Pro-OH in the different states of ionization which were accessible in the three solvents D_2O, CD_3OD, and d_6-Me_2SO. The chemical shifts of the proline resonances in this dipeptide and the pH dependence of the chemical shifts are quite representative for the compounds discussed in the present and the following[17] paper. A complete listing of the ^{13}C nmr chemical shifts of the molecular species discussed in these two papers is included in the thesis of Ch. Grathwohl.[12]

For the purpose of using the *cis–trans* equilibrium of the X-Pro peptide bonds as an nmr probe for conformational studies, the pH dependence of the C^{β} and C^{γ} resonances of proline is of particular interest. Table V shows that while there are some variations of the C^{β} and C^{γ} chemical shifts with pH, these are only of the order of several tenths of a ppm. The assignment of the C^{β} and C^{γ} resonance lines of *cis* and *trans* proline is therefore quite straightforward at all pH values. Otherwise, the pH dependence of the ^{13}C chemical shifts corresponds essentially to the expectations on the basis of earlier systematic studies of amino acids and terminal amino-acid residues in peptides.[33]

The authors would like to thank H. Grogg of Bachem AG and Dr. J. Haegele of Chemspec for their cooperation in the synthesis of the model peptides used in this study, and Dr. M. Brugger and Dr. W. Rittel for providing some peptides as a gift. Financial support by the Schweizerischer Nationalfonds (project 3.1510.73) and the Stiftung der Ciba für Naturwissenschaftliche, Medizinische und Technische Forschung (predoctoral fellowship for Ch. G.) is gratefully acknowledged.

References

1. Patel, D. J. (1970) *Macromolecules* **3,** 448–449.
2. Patel, D. J. (1971) *Macromolecules* **4,** 251–254.
3. Masson, A. (1974) Ph.D. thesis, Eidgenössische Technische Hochschule, Zürich.

4. Ramachandran, G. N. & Sasisekharan, V. (1968) *Advan. Protein Chem.* **23,** 283–437.

5. Dickerson, R. E. & Geis, I. (1969) *The Structure and Action of Proteins,* Harper and Row, New York.

6. La Planche, L. A. & Rogers, M. T. (1964) *J. Amer. Chem. Soc.* **86,** 337–341.

7. Maia, H. L., Orrell, K. G. & Rydon, H. N. (1971) *Chem. Commun.,* 1209–1210.

8. Love, A. L., Alger, T. D. & Olson, R. K. (1972) *J. Phys. Chem.* **76,** 853–855.

9. Wüthrich, K., Grathwohl, Ch. & Schwyzer, R. (1974) in *Peptides, Polypeptides and Proteins,* Blout, E. R., Bovey, F. A., Goodman, M. & Lotan, N., Eds., Wiley-Interscience, New York, pp. 300–307.

10. Dorman, D. E. & Bovey, F. A. (1973) *J. Org. Chem.* **38,** 2379–2383.

11. Wüthrich, K. & Grathwohl Ch. (1974) *FEBS Letters* **43,** 337–340.

12. Grathwohl, Ch. (1975) Ph.D. thesis Nr 5640, Eidgenössische Technische Hochschule, Zürich.

13. Wüthrich, K., Tun-Kyi, A. & Schwyzer, R. (1972) *FEBS Letters* **25,** 104–108.

14. Thomas, W. A. & Williams, M. K. (1972) *Chem. Commun.,* 994.

15. Bovey, F. A. (1972) in *Chemistry and Biology of Peptides,* Meienhofer, J., Ed., Ann Arbor Science Publishers, Ann Arbor, Mich., pp. 3–10.

16. Smith, I. C. P., Deslauriers, R. & Walter, R. (1972) in *Chemistry and Biology of Peptides,* Meienhofer, J., Ed., Ann Arbor Science Publishers, Ann Arbor, Mich., pp. 29–34.

17. Grathwohl, Ch. & Wüthrich, K. (1976) *Biopolymers* **15,** 2043–2057.

18. Deslauriers, R., Walter, R. & Smith, I. C. P. (1972) *Biochem. Biophys. Res. Commun.* **48,** 854–859.

19. Deslauriers, R., Walter, R. & Smith, I. C. P. (1973) *Biochem. Biophys. Res. Commun.* **53,** 244–250.

20. Deslauriers, R., Garrigou-Lagrange, C., Bellocq, A. & Smith, I. C. P. (1973) *FEBS Letters* **31,** 59–66.

21. Dorman, D. E. & Bovey, F. A. (1973) *J. Org. Chem.* **38,** 2379–2383.

22. Bedford, G. R. & Sadler, P. J. (1974) *Biochim. Biophys. Acta* **343,** 656–662.

23. Deslauriers, R., Smith, I. C. P. & Walter, R. (1974) *J. Biol. Chem.* **249,** 7006–7010.

24. Evans, Ch. A. & Rabenstein, D. L. (1974) *J. Amer. Chem. Soc.* **96,** 7312–7317.

25. Grathwohl, Ch. & Wüthrich, K. (1974) *J. Magn. Res.* **13,** 217–225.

26. Keim, P., Vigna, R. A., Nigen, A. M., Morrow, J. S. & Gurd, F. R. N. (1974) *J. Biol. Chem.* **249,** 4149–4156.

27. Torchia, D. A. & Lyerla, J. R. (1974) *Biopolymers* **13,** 97–114.

28. Torchia, D. A. Lyerla, J. R. & Deber, C. M. (1974) *J. Amer. Chem. Soc.* **96,** 5009–5011.

29. Voelter, W., Oster, O. & Zech, K. (1974) *Angew. Chem.* **86,** 46–48.

30. Bellocq, A. M., Dubieu, M. & Dupart, E. (1975) *Biochem. Biophys. Res. Commun.* **4,** 1393–1399.

31. Fermandjian, S., Tran-Dinh, S., Švarda, J., Sala, E., Mermet-Bouvier, R., Bricas, E. & Fromageot, P. (1975) *Biochim. Biophys. Acta* **399,** 313–337.

32. Fossel, E. T., Easwaran, K. R. K. & Blout, E. R. (1975) *Biopolymers* **14,** 927–935.

33. Wüthrich, K. (1976) *NMR in Biological Research: Peptides and Proteins,* ASP Biological and Medical Press B. V., Amsterdam.

34. Young, P. E. & Deber, Ch. M. (1975) *Biopolymers* **14,** 1547–1549.

35. Stothers, J. B. (1972) *Carbon-13 NMR Spectroscopy,* Academic, New York.

Received January 23, 1976
Accepted April 8, 1976

Use of Amide ¹H-NMR Titration Shifts for Studies of Polypeptide Conformation

ARNO BUNDI and KURT WÜTHRICH, *Institut für Molekularbiologie und Biophysik, Eidgenössische Technische Hochschule, CH-8093 Zürich-Hönggerberg, Switzerland*

Synopsis

This paper shows that backbone amide proton titration shifts in polypeptide chains are a very sensitive manifestation of intramolecular hydrogen bonding between carboxylate groups and backbone amide protons. The population of specific hydrogen-bonded structures in the ensemble of species that constitutes the conformation of a flexible nonglobular linear peptide can be determined from the extent of the titration shifts. As an illustration, an investigation of the molecular conformation of the linear peptide H-Gly-Gly-L-Glu-L-Ala-OH is described. The proposed use of amide proton titration shifts for investigating polypeptide conformation is based on 360-MHz ¹H-nmr studies of selected linear oligopeptides in H_2O solutions. It was found that only a very limited number of amide protons in a polypeptide chain show sizable intrinsic titration shifts arising from through-bond interactions with ionizable groups. These are the amide proton of the C-terminal amino acid residue, the amide protons of Asp and the residues following Asp, and possibly the amide proton of the residue next to the N-terminus. Since the intrinsic titration shifts are upfield, the downfield titration shifts arising from conformation-dependent through-space interactions, in particular hydrogen bonding between the amide protons and carboxylate groups, can readily be identified.

INTRODUCTION

The amide protons in polypeptide chains have played a prominent role in numerous nmr studies of peptide and protein conformation, whereby the chemical shift and its dependence on temperature and solvent, the spin–spin coupling constants $^3J_{H\alpha NH}$, and the rate of amide proton exchange with the solvent were the principal sources of information concerning the molecular structures.[1–9] The present paper deals with an additional feature of amide protons, i.e., the pH titration shifts in the ¹H-nmr spectra recorded in aqueous solution.

The pH-dependent amide proton resonances of the small globular proteins, basic pancreatic trypsin inhibitor,[9] snake neurotoxin,[10,11] and snake cardiotoxin,[10,11] were observed in H_2O solutions, and to some extent in D_2O solutions also. Both high- and low-field titration shifts were observed.[9,11] To relate the pH-dependent amide proton chemical shifts with polypeptide conformations, the corresponding intrinsic titration shifts in flexible extended "random-coil" polypeptide chains had to be determined. Preliminary experiments suggested that probably only a very limited number of amide protons in a linear peptide chain experienced observable intrinsic

Biopolymers, Vol. 18, 299–311 (1979)
© 1979 John Wiley & Sons, Inc.
0006-3525/79/0018-0299$01.00

titration shifts, i.e., the amide protons of the C-terminal residue, the Asp residues, and the residues following Asp in the amino acid sequence.[12] These results were now completed by studies of the influence of the amino acid sequence on the C-terminal amide proton titration shift, measurements of the intrinsic titration shift arising from ionizable side chains other than Asp, and experiments with peptides containing multiple ionizable groups. On the basis of these data, the potential of amide proton titration shifts for studies of polypeptide conformation is assessed and their use illustrated with investigations of some oligopeptides.

MATERIALS AND METHODS

The protected analogs of the peptides listed in Tables I and IA were purchased from Bachem AG, Liestal. The unprotected peptides were prepared by hydrolysis of the protected analogs with $Ba(OH)_2$. The purity and composition of the compounds were checked by TLC with n-butanol/pyridine/acetic acid/water 50:12:12:25 and by ^{13}C-nmr.[13,14] For the ^1H-nmr studies, 0.05M solutions of the peptides in a mixed solvent of 90% H_2O/10% D_2O were prepared and the pH value adjusted by the addition of HCl or NaOH.

The ^1H-nmr spectra were recorded with the Fourier transform technique on a Bruker HXS 360 spectrometer. The system was locked on the internal D_2O. The intensity of the H_2O resonance was reduced by double resonance irradiation, i.e., the water line was irradiated for approximately 1 sec prior to the observation pulse. Chemical shifts are relative to internal TSP (sodium-2,2,3,3-tetradeutero-3-trimethylsilyl propionate). The amide proton titration curves were corrected for the small pH dependence of the reference TSP by[15]

$$\delta_{\text{corr}}[\text{ppm}] = \delta_{\text{meas}} - 0.019 \cdot (1 + 10^{5.0-\text{pH}})^{-1} \qquad (1)$$

RESULTS

Measurements of Amide Proton NMR Titration Shifts

The linear oligopeptides used in this study are listed in Tables I and IA. It is seen that most of the compounds are of the type H-Gly-Gly-X-L-Ala-OH, where X stands for one of the common amino acid residues. Table I presents the amide proton titration parameters δ_{HA}, [that is, the chemical shift of the fully protonated peptide, $\Delta\delta = (\delta_{A^-} - \delta_{HA})$, i.e., the titration shift] and pK_a for the peptides with X = Gly, Val, Thr, Met, Gln, Glu, and Asp-Asp. For the tetrapeptides with X = Ile, Leu, Ser, Tyr, and Trp, the titration shifts $\Delta\delta$ for the amide proton of Ala 4 are listed in Table IA. In addition, Table I contains the amide proton titration parameters of some protected oligopeptides that were used to study the influence of deprotonation of the ionizable side chains of Asp, Asp-Asp, and Glu. C-terminal Asp was investigated in H-Gly-Ala-Asp-OH.

TABLE I
The pH Titration Parameters of the Amide Protons in Selected Linear Oligopeptides at 25°C

Peptide	αNH of Residue	δ_{HA}[a]	$\Delta\delta$[b]	pK_a[c]
H-Gly(1)-Gly(2)-Gly(3)-Ala(4)-OH	2	8.620	0.037	3.00 ± 0.15
	3	8.405	0.025	2.8 ± 0.5
	4	8.320	−0.423	3.33 ± 0.05
H-Gly-Gly-Val-Ala-OH	2	8.546	±0.009[d]	—
	3	8.186	0.000	—
	4	8.572	−0.548	3.38 ± 0.10
H-Gly-Gly-Thr-Ala-OH	2	8.580	0.024	3.01 ± 0.15
	3	8.238	−0.017	3.81 ± 0.15
	4	8.523	−0.460	3.33 ± 0.10
H-Gly-Gly-Met-Ala-OH	2	8.555	0.028	2.80 ± 0.15
	3	8.398	0.020	3.00 ± 0.15
	4	8.531	−0.518	3.31 ± 0.10
CF_3CO-Gly-Gly-Asp-Ala-OCH_3	2	8.544	0.000	—
	3	8.410	−0.036	4.20 ± 0.15
	4	8.343	−0.062	4.20 ± 0.10
H-Gly-Gly-Gln-Ala-OH	2	8.568	0.021	3.55 ± 0.15
	3	8.408	−0.006	3.38 ± 0.20
	4	8.543	−0.451	3.43 ± 0.10
H-Gly-Gly-Glu-Ala-OH	2	8.556	0.062	3.81 ± 0.15
	3	8.371	0.172	4.12 ± 0.10
	4	8.528	−0.513	3.52 ± 0.10
H-Gly-Gly-Glu-Ala-OCH_3	2	8.561	0.077	4.00 ± 0.15
	3	8.374	0.228	4.04 ± 0.10
	4	8.542	0.000	—
H-Gly-Ala-Asp-OH	2	8.580	0.022	3.39 ± 0.15
	3	8.572	−0.499	3.09 ± 0.10
H-Gly-Gly-Asp-Asp-Ala-OH	2	8.588	0.071	3.51 ± 0.15
	3	8.559	−0.087	3.96 ± 0.15
	4	8.498	−0.271	3.86 ± 0.10
	5	8.207	−0.366	3.48 ± 0.10
H-Gly-Gly-Asp-Asp-Ala-OCH_3	2	8.572	0.078	3.65 ± 0.15
	3	8.544	−0.089	3.95 ± 0.15
	4	8.494	−0.244	3.82 ± 0.10
	5	8.275	±0.015[d]	

[a] δ_{HA} is the chemical shift in ppm of the amide proton in the protonated peptide.

[b] $\Delta\delta = (\delta_{A^-} - \delta_{HA})$ is the ^{1}H-nmr titration shift in ppm of the amide proton. Negative numbers indicate upfield shifts upon deprotonation.

[c] pK_a is the acidity constant obtained by a least-squares fit of the amide proton chemical shifts with Eq. (2).

[d] No simple titration. The chemical shift varied within the limits indicated about δ_{HA}.

The experimental procedures are illustrated with H-Gly-Gly-Glu-Ala-OH. Figure 1 shows the amide proton resonances in this peptide at various pH values. The resonance of Gly 2 was assigned from its triplet fine structure; the doublet resonances of Glu 3 and Ala 4 were individually assigned by spin decoupling. Independently, the three resonances were also identified from the different exchange rates with water protons, which

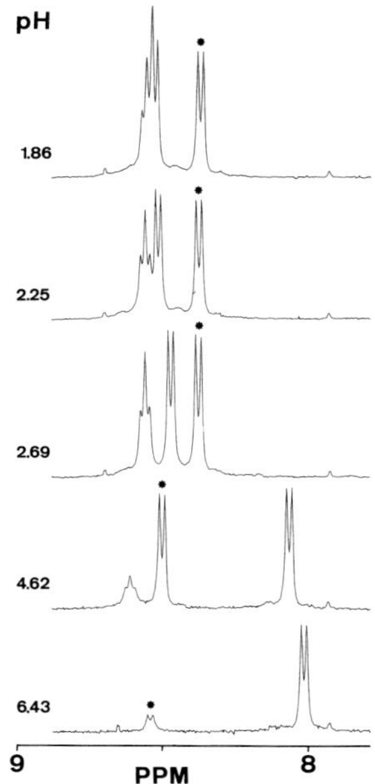

Fig. 1. Region of the amide proton resonances between 8 and 9 ppm in the 360-MHz ^{1}H-nmr spectra of a 0.05M solution of H-Gly-Gly-Glu-Ala-OH in a mixed solvent of 90% H_2O/10% D_2O at different pH values, t = 25°C. The resonances were assigned by spin decoupling. The triplet comes from Gly 2, the doublet marked with * from Glu 3 and the other doublet from Ala 4. The intensity of the water resonance was reduced by saturation through double resonance irradiation. The reduced intensities of the amide proton lines of Gly 2 and Glu 3 in the spectra at pH 4.62 and 6.43 arise from saturation transfer and are related to the proton exchange rates with the solvent (see text).

decrease when one goes from the N- to the C-terminus.[6] In Fig. 1, the different exchange rates are manifested at pH 4.62 by the reduced intensity of the Gly 2 resonance and at pH 6.43 by the disappearance of the Gly 2 resonance and the reduced intensity of the Glu 3 doublet. These intensity changes are a consequence of the saturation of the H_2O line in these experiments. Proton exchange reactions with the solvent mediate saturation transfer to the solute molecules.[1] Since the rate of the amide proton exchange increases with pH, it thus gradually leads to saturation of the individual amide proton resonances. Because of too rapid exchange with H_2O, the NH resonance of Gly 1 was not observed, neither in the protected nor in the free peptides.[6] In Figure 2, the chemical shifts of the three backbone amide protons and the γ-CH_2 protons of Glu in H-Gly-Gly-Glu-Ala-OH are plotted vs pH. The titration parameters in Table I were obtained by nonlinear least-squares fits of the experimental data to one-proton titration curves:

$$\delta(\text{pH}) = \frac{\delta_{HA} + \delta_{A^-} \cdot 10^{(\text{pH}-\text{p}K_a)}}{1 + 10^{(\text{pH}-\text{p}K_a)}} \tag{2}$$

Here, δ (pH) is the experimental chemical shift in ppm, δ_{HA} is the chemical

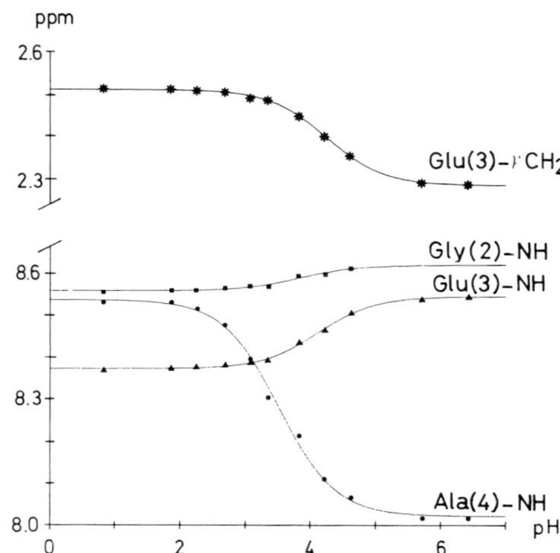

Fig. 2. The pH titration shifts at 25°C of the amide protons and the γ-CH_2 protons of Glu in the tetrapeptide H-Gly-Gly-Glu-Ala-OH. The experimental points were obtained from experiments of the type shown in Fig. 1. The solid lines correspond to nonlinear least-squares fits of one-proton titration curves [Eq. (2)] to the experimental data. The titration parameters are listed in Table I.

shift of the fully protonated peptide, and $\delta_{A^-} = (\delta_{HA} + \Delta\delta)$, the chemical shift of the fully deprotonated peptide.

Conformational Studies of Glutamic Acid-Containing Peptides

Among the data in Fig. 2, the upfield titration shift for Ala 4 is comparable to the corresponding shifts in the other peptides. The downfield shifts of the amide protons 2 and 3, however, were rather unexpected. From earlier experience,[12] it could be excluded that they resulted from "intrinsic" through-bond interactions with either of the two carboxylic acid groups in the molecule, indicating that the downfield titration shifts of amide protons 2 and 3 manifested a particular conformation of the peptide. This was further evidenced by the following experiments.

To ascertain that the titration curves of the amide protons 2 and 3 in Fig. 2 were due to intramolecular interactions, the dependence on peptide concentration was studied over the range from 0.001 to 0.1*M*. Since the titration parameters were essentially independent of concentration, it was excluded that they could be related to interactions between different peptide molecules.

The pK_a values (Table I) indicated that the pH dependence of amide proton 3 in H-Gly-Gly-Glu-Ala-OH was related essentially entirely with the deprotonation of the γ-carboxyl group of Glu 3, whereas the amide proton 2 seemed to be affected by both the C-terminal and the Glu side-chain carboxyl. This was confirmed by studying the influence of the two carboxyl titrations separately in the two analog peptides H-Gly-Gly-Glu-Ala-OCH_3 and H-Gly-Gly-Gln-Ala-OH (Fig. 3 and Table I). Interaction with the γ-carboxyl group of Glu in the protected peptide caused a down-

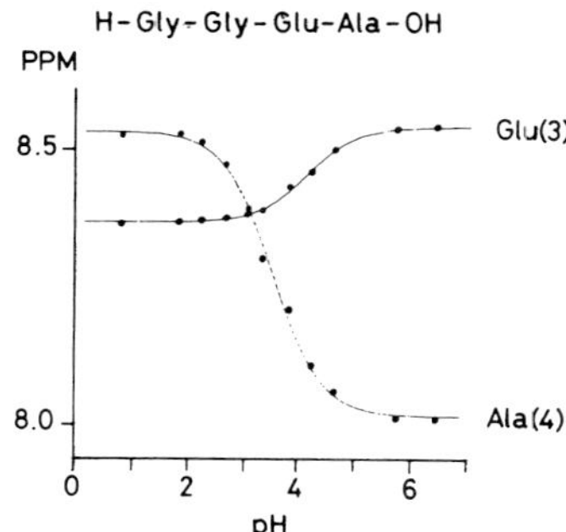

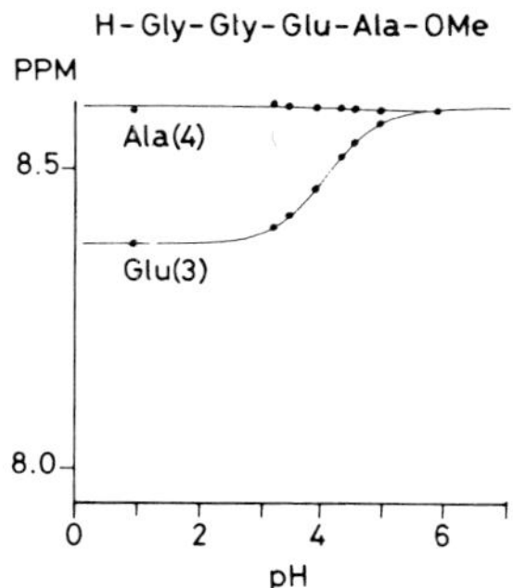

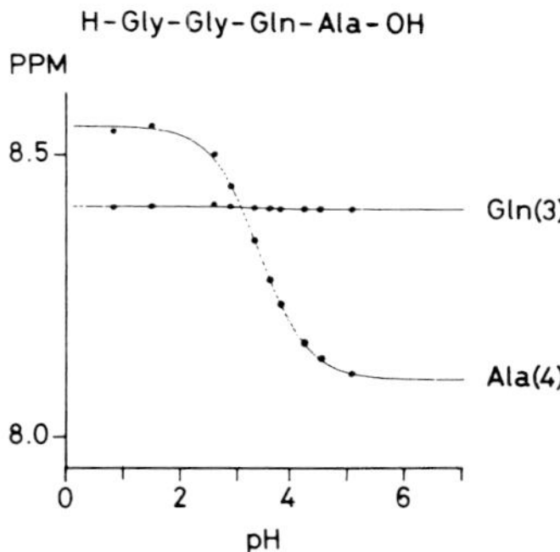

Fig. 3. The pH titration shifts at 25°C of the amide protons in the three tetrapeptides H-Gly-Gly-Glu-Ala-OH, H-Gly-Gly-Glu-Ala-OCH_3, and H-Gly-Gly-Gln-Ala-OH. The experimental points were obtained from experiments of the type shown in Fig. 1. The solid lines correspond to nonlinear least-squares fits of one-proton titration curves [Eq. (2)] to the experimental data. The titration parameters are listed in Table I.

field titration of 0.077 ppm for amide proton 2 and 0.228 ppm for amide proton 3. The C-terminal carboxyl group in the Gln peptide gave rise to a titration shift of 0.021 ppm for residue 2 and −0.006 ppm for residue 3, which is comparable to the corresponding values in the other tetrapeptides (Table I).

A possible explanation for the unusual amide proton titration shifts arising from interactions with the Glu side chain seemed to be that the solution conformation of H-Gly-Gly-Glu-Ala-OH included admixtures of species where the γ carboxyl group of Glu would be hydrogen bonded to the backbone amide protons. This was supported by the temperature dependence of the amide proton chemical shifts in the protected peptide H-Gly-Gly-Glu-Ala-OCH_3. At pH 3.24, 3.90, and 5.80, the temperature coefficients in 10^3 ppm/deg were 6.94, 6.60, and 6.45 for Glu 3, and 8.15, 8.17, and 8.58 for Ala 4. Hence, there is a small but significant decrease of the temperature coefficient for Glu 3 as compared to Ala 4, indicating the amide proton of Glu 3 is involved in intramolecular hydrogen bonding.[1]

The difference between the temperature coefficients for Glu 3 and Ala 4 in H-Gly-Gly-Glu-Ala-OH increased with pH between 3.24 and 5.80. This appeared to confirm the intuitive expectation that the deprotonated Glu side chain would form more stable hydrogen bonds with backbone amide protons than the protonated carboxyl group. The downfield amide proton titration shifts arising from the presence of the glutamic acid residue would hence be caused primarily by the charged deprotonated form of the Glu side chain. This was supported by two additional observations. One was that in 4*M* NaCl the downfield titration shifts of Gly 2 and Glu 3 in the two Glu-containing tetrapeptides (Table I) were reduced to approximately 50% of the shifts observed at low ionic strengths, whereas the "intrinsic" upfield titration shift of Ala 4 was only little affected by increasing the ionic strength. Second, whereas the two γCH_2 protons of Glu 3 are chemical shift equivalent[1] at acidic pH, their chemical shifts differ by 0.03 ppm at neutral pH.[16] This is compatible with unhindered rotational freedom about the C^{β}—C^{γ} bond in the non-hydrogen-bonded protonated Glu side chain at low pH, and restricted mobility in the hydrogen-bonded deprotonated side chain.

DISCUSSION

Intrinsic Amide Proton Titration Shifts in Random-Coil Polypeptide Chains

The term "intrinsic titration shift" denotes pH-dependent variations of amide proton chemical shifts that arise from interactions with an ionizable group mediated exclusively via covalent bonds. Intrinsic titration shifts can manifest different polypeptide conformations when these affect the freedom of rotation about the single bonds linking the ionizable group and the amide proton.

Intrinsic nmr titration shifts of amide protons in polypeptide chains can be caused by the deprotonation of the chain termini and the ionizable side chains. For practical reasons, only the intrinsic titration shifts due to the carboxyl groups of the C-terminus and the side chains of Asp and Glu can readily be studied in H_2O solutions of oligopeptides. At neutral or basic pH, where deprotonation occurs for the N-terminus and the side chains of His, Lys, Tyr, and Arg, the proton exchange with the solvent is too fast for the amide proton resonances to be observed.[1] Hence, the peptides used for the present study (Tables I and IA) were selected so that the effects of deprotonation of the C-terminus or the side chains of Asp and Glu on the amide proton chemical shifts could be measured separately in different amino acid sequences, or that combined effects could be studied in peptides containing more than one carboxyl group. Since the present data should provide a basis for using amide proton titration shifts in studies of polypeptide conformation, it was also hoped that the peptides would predominantly adopt an extended flexible random-coil form in aqueous solution. This will be discussed in more detail below.

Intrinsic titration shifts are to higher field upon deprotonation. By far the largest titration shifts, on the order of −0.5 ppm, were observed for the C-terminal amino acid residue, where the pK_a for the amide proton titration was in all cases identical to the pK_a of the α-carboxyl group determined by ^{13}C-nmr.[14] The influence of the residue X on the extent of the amide proton titration shift of Ala 4 in the peptides H-Gly-Gly-X-Ala-OH will be discussed below. The effect of the C-terminal carboxyl group on the amide proton of the penultimate residue is negligibly small. The side-chain carboxyl titration of Asp 3 in the peptide CF_3CO-Gly-Gly-Asp-Ala-OCH_3 caused upfield shifts of −0.036 and −0.062 ppm for the amide protons 3 and 4, respectively. Inasmuch as this is evidenced by the data in Table I, the effect of Asp side-chain deprotonation is negligibly small for all the amide protons in a linear peptide, except that of Asp and the residue following Asp. Intrinsic titration shifts arising from deprotonation of Glu could not be measured because of the dominant through-space interactions. They were estimated to be smaller than −0.01 ppm for all the amide protons in a linear peptide chain.

The observations with peptides containing a single ionizable group imply that, for an amide proton to experience combined intrinsic shifts from two ionizable groups, it has to be located at the C-terminus with either the ultimate or the penultimate residue carrying an ionizable side chain, or else it must link two nonterminal ionizable amino acid residues. One of these three situations was investigated in H-Gly-Ala-Asp-OH. The titration shift of −0.499 ppm for the C-terminal Asp was well within the range observed in peptides with inert side chains (Table I), and hence no conclusive information was obtained on the combined effects of end group and Asp side-chain deprotonation. It could be excluded, however, that the effects of the two carboxyl titrations were additive, since $\Delta\delta$ was −0.337 ppm for a C-terminal Asn[17] and −0.036 ppm for a nonterminal Asp (Table I). In H-Gly-Gly-Asp-Asp-Ala-OCH_3, upfield titration shifts of −0.089 and −0.244 ppm were observed for the amide protons 3 and 4, respectively.

Both these shifts have the direction of intrinsic titrations shifts, and from the pK_a values they have to be attributed to the deprotonation of the Asp side chains. However, neither these large effects for residues 3 and 4 nor the lack of a titration shift for the amide proton 5 (Table I) can be interpreted as a superposition of the effects for two isolated Asp residues. A possible explanation appears to be that this pentapeptide adopts a particular conformation that also affects the intrinsic titration shifts, e.g., by limiting the rotational motions about certain single bonds and in which the intrinsic titration shift of the amide proton 5 would be cancelled by a hydrogen-bonding interaction with one of the Asp side chains (see below). Evidence for a non-random conformation of this peptide seems to come also from the small titration shift of the C-terminal Ala 5 amide proton in the unprotected analogue (Table I) and the sizable downfield titration of the Gly 2 amide proton (see below).

Overall, the data in Table I show that only a very limited number of amide protons in a linear polypeptide chain experience sizable intrinsic pH titration shifts. In the pH range from 1 to 5, these are the amide protons of the C-terminal residue, the Asp residues, and the residues following Asp in the amino acid sequence. For the C-terminus, intrinsic titration shifts are from ca. −0.3 to −0.6 ppm; for Asp, ca. −0.05 ppm; and for the residue following Asp, ca. −0.05 ppm (Table I). Furthermore, there is an indication that rather large upfield shifts of approximately −0.25 ppm may occur for amide groups linking two Asp residues. Since at pH values above 5 the amide proton exchange with H_2O in linear oligopeptides is too fast for the resonances to be readily observed, intrinsic titration shifts that might possibly arise from deprotonation of the N-terminus or the side chains of His, Lys, Tyr, and Arg are probably of little practical importance for studies of small peptides. In globular proteins, however, resonances of interior amide protons have been observed over the entire pH range from 1 to 12.5.[3,8,9] Judging from the large titration shift for the carbonyl ^{13}C-resonance of the N-terminal residue in polypeptide chains,[1] a sizable intrinsic titration shift might be anticipated for the amide proton of the residue following the N-terminus. On the other hand, since the ionizable group of Asp is closer to the backbone than those of any of the other titratable side chains (i.e., it is separated by four and five covalent bonds, respectively, from the two nearest amide protons, whereas the corresponding distances in the other residues are five and six bonds or more), the present data on Asp and Glu imply that no intrinsic titration shifts are to be expected from deprotonation of the basic amino acid side chains.

Molecular Conformations of the Glu-Containing Peptides

The experiments described in Results showed that hydrogen bonding with the deprotonated Glu side chain caused a large downfield titration shift of the amide proton resonance of Glu in H-Gly-Gly-Glu-Ala-OH. The downfield titration with the pK_a of Glu observed for the Gly 2 amide proton in the protected analog peptide (Table I) implied hydrogen-bond formation occurred also between the Glu side chain and amide proton 2.

Downfield titration shifts for an amide proton interacting with a carboxyl group may arise because of the more pronounced tendency of the negatively charged deprotonated carboxylic acid to form a hydrogen bond with a donor group. In the hydrogen bond with the carboxylate group, the N—H bond is strongly polarized and hence the amide proton is deshielded and its resonance shifts to lower field:

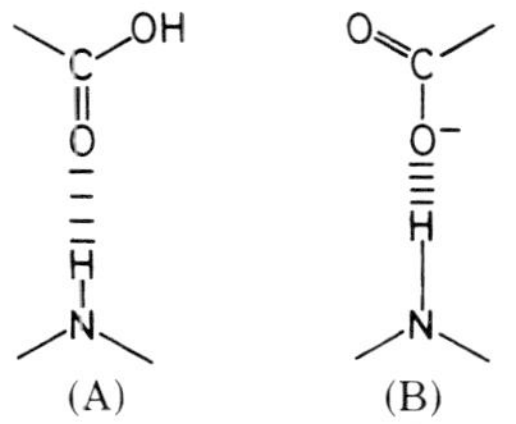

(A) (B)

(A) Acidic pH. Weak hydrogen bonding with little polarization of the NH bond. (B) Neutral or basic pH. Strong hydrogen bonding with pronounced polarization of the NH bond.

The solution conformations of H-Gly-Gly-Glu-Ala-OH and the protected analog peptide have to be visualized as dynamic ensembles of rapidly interconverting species. The protonated peptide in acidic solution appears to be predominantly in extended forms with full flexibility of the peptide backbone and the Glu side chain. The deprotonated form in neutral solution includes species with hydrogen bonds between the Glu side chain and the backbone amide proton 2 or 3, respectively. As will be discussed in the last section of this discussion, we suggest that the size of the downfield titration shifts may be used to obtain an estimate of the absolute populations of individual species. In H-Gly-Gly-Glu-Ala-OCH_3, hydrogen-bonded forms involving the amide protons 2 and 3 would accordingly account for ~20 and ~50%, respectively, of the total number of molecules. Direct evidence for a rather high population of the hydrogen-bonded species was independently obtained from the observation that the restricted mobility of the Glu side chain in these structures is clearly manifested in the nonequivalence of the γ-CH_2 protons at neutral pH.[16] Interestingly, predominant orientation towards the N-terminal end of linear peptides as found here for glutamic acid was independently evidenced for the aromatic amino acid side chains also,[18,19] and may quite likely prevail for bulky side chains in general.[18]

Molecular Conformations of the Peptides H-Gly-Gly-*X*-Ala-OH with Inert Side Chains of *X*

Studies of the dependence on the peptide concentration had shown that the small downfield titration shifts of the amide protons 2 and 3 in peptides H-Gly-Gly-*X*-Ala-OH with nonionizable side chains of *X* (Table I) were not due to intermolecular interactions. These shifts thus had to be attributed to intramolecular interactions with the terminal carboxylate group. The pK_a values obtained from the Gly 2 amide proton titration shifts were indeed quite similar to those of the terminal carboxyl group (Table I). Analogous to the above-discussed situation in the Glu peptides, these low-field titration shifts thus appear to come from hydrogen bonding with a carboxylate group. For hydrogen bonding between the amide proton 2 and the terminal carboxylate of Ala 4, this would be compatible with the formation of a γ-turn[20] in these peptides:

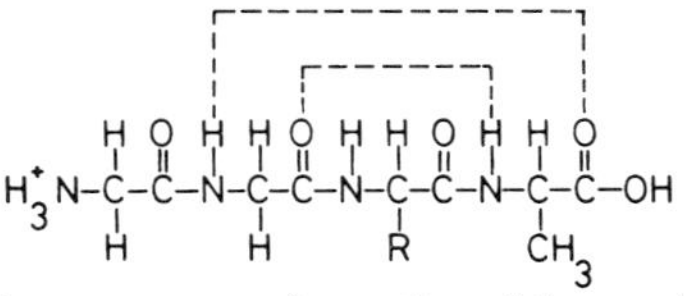

```
             H O H H O H H O H H O
             | ‖ | | ‖ | | ‖ | | ‖
          H3+N-C-C-N-C-C-N-C-C-N-C-C-OH
               |     |     |     |
               H     H     R     CH3
```

Hydrogen bonding scheme in a γ-turn conformation of the peptides H-Gly-Gly-*X*-Ala-OH (Ref. 20).

Since the downfield shifts $\Delta\delta$ for Gly 2 are small compared with the shifts observed in the Glu peptides, it would appear that the γ-turn is only little populated in aqueous solution, i.e., probably less than 5% of the total number of molecules. Therefore, it is not surprising that the occurrence of γ-turns was not manifested in the spin–spin coupling constants $^3J_{H\alpha NH}$, which were previously found to be compatible with a flexible extended random-coil form of the peptides.[16–18] The low-field shifts of the amide protons 3 (Table I) would then further imply that small percentages of other hydrogen-bonded species might also be contained in the ensembles of rapidly interchanging species that constitute the solution conformations of these peptides.

Amide Proton Titration Shifts for ^{1}H-NMR Studies of Polypeptide Conformation

The present experiments illustrate that measurements of downfield amide proton titration shifts are a very sensitive technique for studies of hydrogen-bonding interactions between carboxylate groups and amide protons. They are a much more sensitive manifestation of this type of hydrogen bond than the temperature coefficients of the amide proton chemical shifts,[1] as evidenced by the studies of H-Gly-Gly-Glu-Ala-OCH_3. Admixtures of hydrogen-bonded γ-turn-type structures in the dynamic ensemble of species constituting the solution conformation of the peptides H-Gly-Gly-*X*-Ala-OH were clearly indicated in the amide proton titration shifts, even though the population of these species was so small that the spin–spin coupling constants $^3J_{H\alpha NH}$ were well within the range of values expected for a flexible extended random-coil form of the peptide backbone.[1]

On the following basis, we propose that amide proton titration shifts can be used to estimate the population of specific hydrogen-bonded structures in the ensemble of species constituting the solution conformation of flexible polypeptide chains. In the crystal structure of the globular protein basic pancreatic trypsin inhibitor (BPTI), which was obtained from crystals grown at pH 10.5, the side chain of Glu 49 is hydrogen bonded to its amide proton.[21] The pK_a value for Glu 49 in BPTI was measured by ^{13}C-nmr to be in the range of 3.6–3.8.[22] In H_2O solution of BPTI, an amide proton titrated 0.408 ppm downfield with a pK_a value of 3.85, and no other amide proton resonance, showed a titration shift with a pK_a between 3.5 and 7.9.[9] These data combined with the observations on Glu peptides in this paper imply that the titration shifts of 0.408 ppm have to be assigned to Glu 49. Assuming that the hydrogen bond linking the side-chain carboxylate with the amide proton of Glu 49 is nearly 100% populated in the globular protein and that all the hydrogen bonds in question have similar geometries (or else that the titration shift is not markedly dependent on the hydrogen-bond geometry), we have that a 10% population of a carboxylate-amide proton hydrogen bond in a dynamic ensemble of molecular species should give rise to a downfield titration shift of 0.04 ppm. The populations of hydrogen-bonded species thus estimated for the peptides H-Gly-Gly-X-Ala-OH, including X = Glu (see above), are compatible with all the other nmr parameters of these molecules.

The aforementioned interpretation of downfield amide proton titration shifts depended on the knowledge of the intrinsic titration shifts in a polypeptide chain. Whether the sizable variations of the intrinsic titration shifts for the C-terminal residue (Tables I and IA) can be used to characterize the conformation of the chain terminus remains to be seen (see also the Appendix).

APPENDIX

When searching for structural interpretations of the data in Table I, we discovered a quite surprising correlation between the amide proton titration shifts $\Delta\delta$ of Ala 4 in the peptides H-Gly-Gly-X-Ala-OH and the β-sheet conformational parameters[23] P_β of the residues X. Here, we present this correlation without further comments.

Table IA lists the Ala 4 amide proton titration shifts $\Delta\delta$ and the β-sheet conformational parameters P_β for 10 of the 11 residues X that we studied. Glu was not considered because of the dominant effects of the side-chain

TABLE IA
Amide Proton Titration Shifts $\Delta\delta$ of Ala 4 in the Peptides H-Gly-Gly-X-Ala-OH and β-Sheet Conformational Parameters of the Residues X

Residue X	$\Delta\delta$ (ppm) of Ala 4	P_β of X[a]
Val	−0.543	H_β
Ile	−0.531	H_β
Met	−0.518	H_β
Leu	−0.507	h_β
Tyr	−0.472	h_β
Thr	−0.460	h_β
Gln	−0.451	h_β
Gly	−0.423	i_β
Trp	−0.412	h_β
Ser	−0.364	b_β

[a] P_β are the β-sheet conformational parameters of Chou and Fasman (Ref. 23): H_β, strong β former; h_β, β former; i_β, indifferent; b_β, β breaker.

conformation. It is seen that there is an almost perfect correlation between the decrease of $\Delta\delta$ for Ala 4 and the decrease of the β-sheet formation tendency of X. The only exception concerns the peptides with X = Gly and Trp, where the order is reversed. The probability that this correlation is entirely fortuitous is approximately 4×10^{-4}, and the probability that there is accidentally only one error, as in Table IA, is $<10^{-2}$.[17]

We thank Dr. G. Wagner for communicating to us the amide proton titration shifts in the basic pancreatic trypsin inhibitor. Financial support by the Swiss National Science Foundation (Project No. 3.0040.76) is gratefully acknowledged.

References

1. Wüthrich, K. (1976) *NMR in Biological Research: Peptides and Proteins,* North-Holland, Amsterdam.
2. Ohnishi, M. & Urry, D. W. (1969) *Biochem. Biophys. Res. Commun.* **36,** 194–202.
3. Masson, A. & Wüthrich, K. (1973) *FEBS Lett.* **31,** 114–118.
4. Karplus, S., Snyder, G. H. & Sykes, B. D. (1973) *Biochemistry* **12,** 1323–1329.
5. Llinás, M. & Klein, M. P. (1975) *J. Am. Chem. Soc.* **97,** 4731–4737.
6. Scheinblatt, M. & Rahamin, Y. (1976) *Biopolymers* **15,** 1643–1653.
7. Kopple, K. D. & Go, A. (1976) *Biopolymers* **15,** 1701–1715.
8. Wagner, G., De Marco, A. & Wüthrich, K. (1976) *Biophys. Struct. Mech.* **2,** 139–153.
9. Wagner, G. (1977) Ph.D. thesis, No. 5992, ETH, Zürich.
10. Lauterwein, J., Wüthrich, K., Schweitz, H., Vincent, J. P. & Lazdunski, M. (1977) *Biochem. Biophys. Res. Commun.* **76,** 1071–1078.
11. Lauterwein, J., Wüthrich, K., Schweitz, H., Vincent, J. P. & Lazdunski, M. (1977) Communication at the 11th FEBS Meeting, Copenhagen, Abstract B5-1 363.
12. Bundi, A. & Wüthrich, K. (1977) *FEBS Lett.* **77,** 11–14.
13. Grathwohl, Ch. & Wüthrich, K. (1974) *J. Magn. Reson.* **13,** 217–225.
14. Richarz, R. & Wüthrich, K. (1978) *Biopolymers* **17,** 2133–2141.
15. De Marco, A. (1977) *J. Magn. Reson.* **26,** 527–528.
16. Bundi, A. & Wüthrich, K. (1979) *Biopolymers* **18,** 285–297.
17. Bundi, A. (1977) Ph.D. thesis, No. 6036, ETH, Zürich.
18. Bundi, A., Grathwohl, Ch., Hochmann, J., Keller, R., Wagner, G. & Wüthrich, K. (1975) *J. Magn. Reson.* **18,** 191–198.
19. Wüthrich, K. & De Marco, A. (1976) *Helv. Chim. Acta* **59,** 2228–2235.
20. Némethy, G. & Printz, M. P. (1972) *Macromolecules* **5,** 755–758.
21. Deisenhofer, J. & Steigemann, W. (1975) *Acta Crystallogr., Sect. B* **31,** 238–250.
22. Richarz, R. & Wüthrich, K. (1978) *Biochemistry,* in press.
23. Chou, P. Y. & Fasman, G. D. (1974) *Biochemistry* **13,** 222–245.

Received March 16, 1978
Accepted April 19, 1978

PROTEINS: Structure, Function, and Genetics 1:34–42 (1986)

Reprinted by permission of John Wiley & Sons, Inc.

Protein Folding Kinetics by Combined Use of Rapid Mixing Techniques and NMR Observation of Individual Amide Protons

Heinrich Roder and Kurt Wüthrich
Institut für Molekularbiologie und Biophysik, Eidgenössische Technische Hochschule-Hönggerberg, CH-8093 Zürich, Switzerland

ABSTRACT **A method to be used for experimental studies of protein folding introduced by Schmid and Baldwin (*J. Mol. Biol.* 135: 199–215, 1979), which is based on the competition between amide hydrogen exchange and protein refolding, was extended by using rapid mixing techniques and ^{1}H NMR to provide site-resolved kinetic information on the early phases of protein structure acquisition. In this method, a protonated solution of the unfolded protein is rapidly mixed with a deuterated buffer solution at conditions assuring protein refolding in the mixture. This simultaneously initiates the exchange of unprotected amide protons with solvent deuterium and the refolding of protein segments which can protect amide groups from further exchange. After variable reaction times the amide proton exchange is quenched while folding to the native form continues to completion. By using ^{1}H NMR, the extent of exchange at individual amide sites is then measured in the refolded protein. Competition experiments at variable reaction times or variable pH indicate the time at which each amide group is protected in the refolding process. This technique was applied to the basic pancreatic trypsin inhibitor, for which sequence-specific assignments of the amide proton NMR lines had previously been obtained. For eight individual amide protons located in the β-sheet and the C-terminal α-helix of this protein, apparent refolding rates in the range from 15 s^{-1} to 60 s^{-1} were observed. These rates are on the time scale of the fast folding phase observed with optical probes.**

Key words: hydrogen exchange, BPTI, folding pathway, protein dynamics

INTRODUCTION

Considerable effort has been directed at determining the folding pathways of globular proteins. Although the existence of folding intermediates has been demonstrated in several cases (reviewed by Kim and Baldwin[1]), specific structural information is still lacking, especially with respect to the early events of structure condensation. A particularly promising approach introduced by Baldwin and co-workers[2–4] and applied to characterize the slow folding phase of RNase A relies on investigations of the competition between amide proton exchange from the unfolded protein and protein refolding, using tritium labeling of amide groups. The present paper describes an extension of this competition method, which can also be applied to proteins where a fast phase dominates refolding.

After inducing refolding of the protein by rapid mixing into a deuterated solvent medium, amide proton exchange is quenched after a predetermined time, and the extent of exchange for individual amide protons is determined by ^{1}H NMR experiments with the refolded protein. Sequence-specific assignments in the native protein can thus be directly applied for the interpretation of these experiments. The procedure is illustrated with studies on the refolding of BPTI. Rapid mixing is required since refolding of BPTI is dominated by a process in the 10-ms time range.[5]

METHODS

The mixing procedure used for the competition experiments is schematically illustrated in Figure 1. An H_2O solution of the unfolded protein (syringe S1) is rapidly mixed in the mixing chamber (M) with a larger volume of D_2O buffer (S2). Refolding and hydrogen-deuterium exchange are allowed to proceed during the time τ before the exchange reaction is quenched by rapid cooling (Q).

The kinetic analysis of the experiment is straightforward if we assume that the formation of folded structure affecting the exchange of a particular amide proton occurs irreversibly in a single step (in the discussion we will deal with more general kinetic situations). Equation (1) describes the competition between refolding and exchange for an amide proton that is protected against exchange upon refolding.

Received March 18, 1986; accepted May 21, 1986.

Address reprint requests to Dr. Heinrich Roder, Dept. of Biochemistry and Biophysics, University of Pennsylvania, Philadelphia, PA 19104.

Abbreviations: BPTI, basic pancreatic trypsin inhibitor; RCAM-BPTI, 14–38 reduced and carboxamidomethylated BPTI; RNase A, bovine pancreatic ribonuclease A; NMR, nuclear magnetic resonance; CD, circular dichroism.

© 1986 ALAN R. LISS, INC.

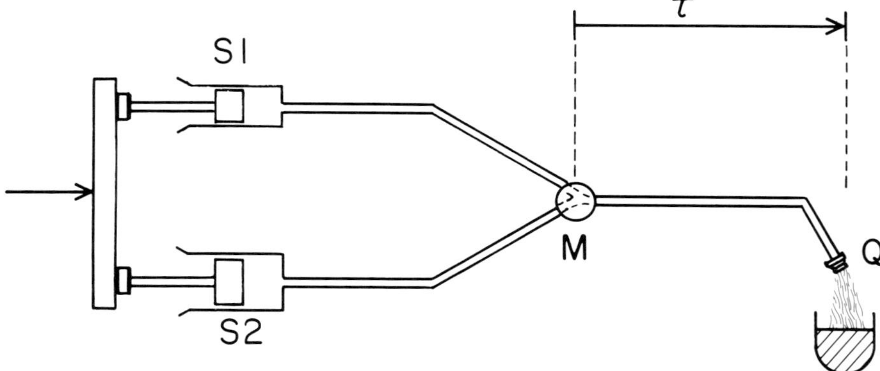

Fig. 1. Schematic illustration of the mixing apparatus. A mechanical actuator advances the syringe plungers by 13.9 mm in 2 revolution of a flywheel (maximum speed 700 rpm). The solutions contained in syringes S1 and S2 (8 and 12 mm diameter) are mixed in the four-jet mixing chamber M. At the quencher Q, consisting of a nozzle with five 0.1-mm holes, the solution is sprayed into cold liquid (ice water or isopentane at dry ice temperature). Syringes, mixing chamber, and tubing up to Q are thermostated at 70°C (± 0.5°). At a given flow rate, dV/dt, the volume V, between M and Q determines the competition time, τ = V (dV/dt), during which exchange and refolding are allowed to proceed at 70°C.

$$\begin{array}{ccc} U^H & \xrightarrow{k_f} & F^H \\ k_c \downarrow & & \\ U^D & \xrightarrow{k_f} & F^D \end{array} \qquad (1)$$

Initially, all amide groups are protonated in the unfolded form, U^H. After mixing with deuterated buffer solution, the proton can either exchange at the intrinsic rate k_c, producing U^D, or it can be trapped within the structure formed during refolding (F^H). With the initial condition at $\tau=0$ (Fig. 1) as $U^H(0)=1$, $U^D(0)=F^H(0)=F^D(0)=0$, we obtain the following functions of the competition time τ.

$$U^H(\tau) = \exp\,[-(k_f+k_c)\tau] \qquad (2)$$

$$F^H(\tau) = \frac{k_f}{k_f+k_c} \cdot \{1 - \exp[-(k_f+k_c)\tau]\} \qquad (3)$$

When exchange is stopped by rapid cooling, the U → F transition goes to completion. Since the protons observed in this study have exchange times on the order of months in the native protein at room temperature and neutral or slightly acidic pH,[6] the experimentally observable quantity is the total remaining proton population ($U^H + F^H$) at the time of quenching. The measured proton occupancy, P, as a function of the competition time τ is then given by

$$\begin{aligned} P(\tau) &= U^H(\tau)+F^H(\tau) \\ &= P_\infty+(1-P_\infty)\exp[-(k_f+k_c)\tau] \end{aligned} \qquad (4)$$

$$P_\infty = k_f/(k_f+k_c) \qquad (5)$$

P_∞ is the final proton occupancy after a long competition time ($\tau >> 1/(k_f+k_c)$). The background due to the fraction f_H of H_2O present in the reaction mixture is taken into account by normalizing the measured NH intensity, I_m, as follows:

$$P(\tau) = [(I_m(\tau)/I_o) - f_H]/(1 - f_H) \qquad (6)$$

where I_o is the resonance intensity of the fully protonated amide group.

Equations (4) and (5) suggest two different methods for measuring folding rates with observation of individual amide protons:

1. Measurement of $P(\tau)$ at constant pH and temperature. According to Eq. (4) the proton occupancy decays exponentially from unity to a final level $P_\infty = k_f/(k_f+k_c)$, with a rate constant k_f+k_o. Under conditions where k_f and k_c are similar in magnitude, both rates can be determined from an exponential fit of the normalized NH resonance intensity as a function of the competition time.

2. Measurement of the pH dependence of P_∞. In the base-catalysed regime the intrinsic exchange rate, k_c, increases by a factor of 10 per pH unit. Variation of pH can therefore be used to adjust k_c until it equals the folding rate, k_f. According to Eqs. (5) and (6), we then have that $P_\infty = 0.5$ (provided that the folded protein is stable under these conditions). At sufficiently low pH, where $k_c << k_f$, the proton will be fully trapped, so that $P_\infty = 1$; at high pH, where $k_c >> k_f$, the amide proton will be completely exchanged before folding is complete, and $P_\infty = 0$. Since a reliable calibration for the intrinsic amide proton exchange rates at variable pH and temperature is available from model peptide studies[7,8] and from measurements in thermally unfolded BPTI,[9] folding rates can be obtained from the pH location of the sigmoidal transition in plots of P_∞ vs. pH, as described by Eq. (5).

Intrinsic exchange rates, k_c, for the present study with BPTI were calculated with the equation

$$k_c(T,pH) = \tfrac{1}{2} \cdot k_c(T_o,pH_{min}) \cdot e^{-E_a\left(\frac{1}{T}-\frac{1}{T_o}\right)/R} \cdot [10^{(pH-pH_{min})} + 10^{(pH_{min}-pH)}] \quad (7)$$

using the values for pH_{min} and k_c (T_o, pH_{min}) measured for the same protons in thermally unfolded RCAM-BPTI at T_o=86°C.[9] An activation energy E_a = 17 kcal mol^{-1} was used for the base-catalyzed exchange reaction.[7] In addition, a correction was applied to account for the effect of solvent additives on NH exchange.[10,11] In 13% n-propanol, we find the intrinsic NH exchange to be slowed by a factor 1.3 (measured with poly-DL-alanine at pH 4.0, 25°C).

EXPERIMENTAL

BPTI (Trasylol) was a gift from Bayer AG, Leverkusen Germany. A home-built rapid mixing apparatus, kindly provided by Dr. H. Dutler, was modified for the present experiment (Fig. 1). An H_2O solution of the fully denatured protein at 70°C in 40% (v/v) n-propanol at pH 2.0 was mixed with 2 volumes of 0.2 M phosphate buffer in D_2O at a range of pD values between about 4.5 and 7.5. This mixture produces refolding conditions at pH values above 4.0 in a 1:2 H_2O/D_2O mixture containing 13% n-propanol. The initial protein concentration in the H_2O solution was 3 mM. pH was measured at 70°C in aliquots of the refolding mixture and is reported without correction for isotope effects or the presence of n-propanol. Syringes, mixer, and reaction tubing were thermostated at 70±0.5°C. For quenching, the mixture was injected through a nozzle into isopentane at −70°C or ice water. The nozzle was placed a few millimeters above the liquid. The reaction time was determined from the calibrated flow rate and the precisely measured reaction volume, i.e., the volume between mixer and quencher. The shortest reaction time achieved was 30 ms. Longer times were set by reduction of the flow rate and/or increased length of the reaction tubing (thermostated at 70°C).

The quenched protein samples were stored at dry ice temperature pending further processing. When isopentane was used for quenching, the frozen solution particles were collected by evaporating the organic solvent and immediately thereafter lyophilized. In experiments that involved injection into ice water, samples were concentrated by ultrafiltration at 4°C and lyophilized. Prior to the NMR measurements the lyophilized protein was redissolved in D_2O and adjusted to pD 4.0 by addition of DCl.

At 360 MHz on a Bruker HX 360 spectrometer ^1H NMR spectra were recorded to measure the extent of H-D exchange for resolved, individually assigned NH resonances.[12,13] Resonance intensities (I_m in Eq. 6) were determined by comparison with simulated spectra. The accuracy thus achieved was about ±5%. The intensity of a resolved line corresponding to a nonlabile proton was used for I_o. Since the determination of the proton occupancy, P, involves two intensity measurements (Eq. 6), its error is estimated to be about ±10%. (f_H was determined more accurately).

The exchange kinetics of poly-DL-alanine (Miles-Yeda Ltd., Revohot) in D_2O with and without addition of 13% (v/v) n-propanol (25°C, pH 4.0) was studied by the spectrophotometric method of Englander et al.[14] by using a Cary 118 spectrophotometer.

Circular dichroism experiments were performed on a Jasco 500 C spectopolarimeter by using thermostated quartz cuvettes with 1-mm pathlength.

RESULTS

In a search for suitable experimental conditions we performed CD measurements at 375 nm to monitor the thermal unfolding transition of BPTI. The high stability of BPTI requires addition of denaturants for complete unfolding at experimentally accessible temperatures. At pH 2.2 in the presence of 40% n-propanol, thermal unfolding was found to be fully reversible and had the characteristics of a two-state transition with a melting temperature of 60.5°C, an enthalpy change of 61 kcal mol^{-1}, and a heat capacity change of about 1 kcal mol^{-1} K^{-1}. The temperature chosen for the competition experiments, 70°C, lies at the top end of the transition (95% unfolded). In a 13% n-propanol solution at pH 4.5, the melting temperature was found to be above 90°C. Thus, our refolding conditions (70°C, 13% n-propanol, pH > 4.0) correspond to an equilibrium situation with at least 98% refolded protein. NMR spectra recorded after refolding were indistinguishable from those measured with a fresh solution.

Measurements of $P(\tau)$ (method 1) were performed at pH 6.7 (70°C, 13% propanol, 0.13 M phosphate buffer). The incubation time τ of the mixture of 70°C was varied between 30 and 400 ms by adjusting the flow rate (Fig. 1). The exchange was quenched by injection of the solution into isopentane at −70°C (see Fig. 1). These procedures were repeated for eight values of τ with a common stock of fresh protein solution. NMR samples were prepared as described in the experimental section.

The fraction of amide protons retained in the refolded protein was measured in the ^1H NMR spectrum for six resolved lines and one pair of overlapped NH resonances. The proton occupancy, $P(\tau)$, for the NH of Phe 22, normalized according to Eq 6, is plotted in Figure 2 as a function of the competition time. Similar results were obtained for the other amide protons studied. The calculated curve in Figure 2 represents a least-squares fit of Eq. (4) with the parameters k_f+k_c = 40 s^{-1} and P_∞ = 0.17. With Eq. (5) we obtain the individual rate constants k_f = 7 s^{-1} and k_c = 33 s^{-1}. For the exponential fit it was necessary to shift the zero time by 30 ms (Fig. 2), which probably reflects the mixing dead time. This adds considerably to the uncertainty of the analysis. Only

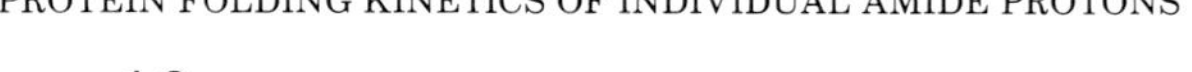

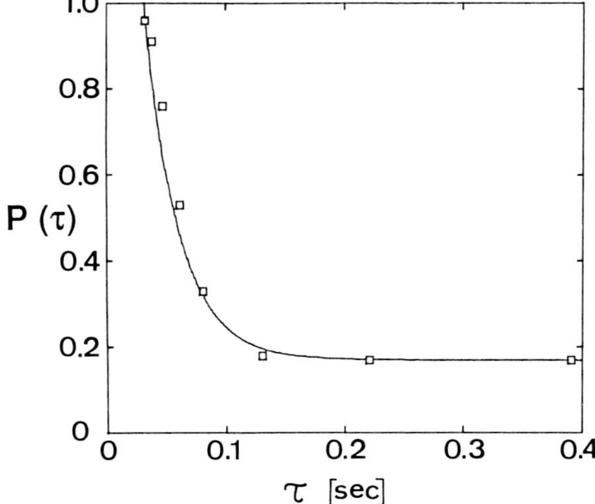

Fig. 2. Plot of the proton occupancy, $P(\tau)$, calculated with Eq. (6), vs. the competition time, τ, for the Phe 22 NH in BPTI at 70°C and pH 6.7 (0.13 M phosphate, 13% n-propanol). The data resulted from NMR analysis of eight different protein samples subjected to the mixing procedures outlined in the Methods section (method 1). The theoretical curve is a fit of Eq. (4) with $k_f + k_c = 40\ s^{-1}$ and $P_\infty = 0.17$. In this fitting procedure the zero point of the time axis was shifted by 30 ms (see text).

qualitative conclusions can be drawn from these results. Quantitiative results could be obtained from method 2, which is technically less demanding.

The pH dependence of P_∞ (method 2) was measured in the same refolding medium at a constant competition time of 0.4 s. Experiments were performed by using different D_2O phosphate buffer solutions producing pH values between 4.0 and 7.5 during the competition period. At this longer competition time, injection into ice water was used for quenching. The intensity of the resolved amide proton resonances in the ^{1}H-NMR spectrum was measured and Eq. (6) was used to calculate the residual protonation for eight individual amide groups in the protein. Results are displayed in Figure 3. In each case the trapped proton intensity decreases with increasing pH in a sigmoidal manner, with a midpoint near pH 6.0. The curves in Figure 3 were calculated with Eq. (5) and fitted to the data by variation of k_f. The values for k_c (pH) were calculated with Eq. (7). In independent experiments using method 1 it was found that $P(0.4\ s) = P_\infty$ at all pH values used for this study (Fig. 2).

The apparent folding rates resulting from this procedure are reported in Table I. The quality of the fits in Figure 3 indicates statistical errors of about 20%. Additional errors are introduced by the calibration of k_c[9]. The extrapolation of k_c and solvent effects may cause systematic errors affecting the absolute magnitude of the rates, but not the relative rates among different amide protons.

The competition data recorded with method 1 (Fig. 2) provide some important experimental controls. First, we note that the proton occupancy lies close to 1 for the points at the shortest competition times used. In these samples that were exposed to the D_2O refolding medium at 70°C only for a short time, the total amide proton intensity was recovered after the exchange was quenched by rapid cooling. This demonstrates that quenching was efficient and that no further exchange took place during the subsequent steps of sample handling and the NMR measurement.

The evaluation of the data obtained by measurement of the pH dependence of P_∞ (Fig. 3) relies on the condition that the long-time limit is reached before the end of the competition period used, i.e., the condition $\tau >> 1/(k_f+k_c)$ must be fulfilled. For the present experiments with BPTI the time course of $P(\tau)$ in Figure 2 shows that at pH 6.7 the proton level remains constant for competition times longer than ca. 0.15 s. The value of $\tau=0.4$ s chosen for method 2 therefore ensures stationary conditions at pH 6.7, and the magnitude of the k_f values reported in Table I (15 s^{-1} or larger for all NH studied) shows that this is true at all other pH values of this study. In similar ways, suitable experimental conditions would have to be established for work with other proteins.

As a final control, we have to verify that the structure formed during competition experiments with BPTI at 70°C is able to protect the interior amide protons against exchange. In previous exchange studies of BPTI at elevated temperature[15] the amide protons studied here were found to exchange at rates between 10^{-3} and $2 \cdot 10^{-2}\ s^{-1}$ at 70°C and pH 8.0. Exchange from molecules refolded into nativelike conformations is thus expected to be negligible on the time scale of the mixing experiments, which were

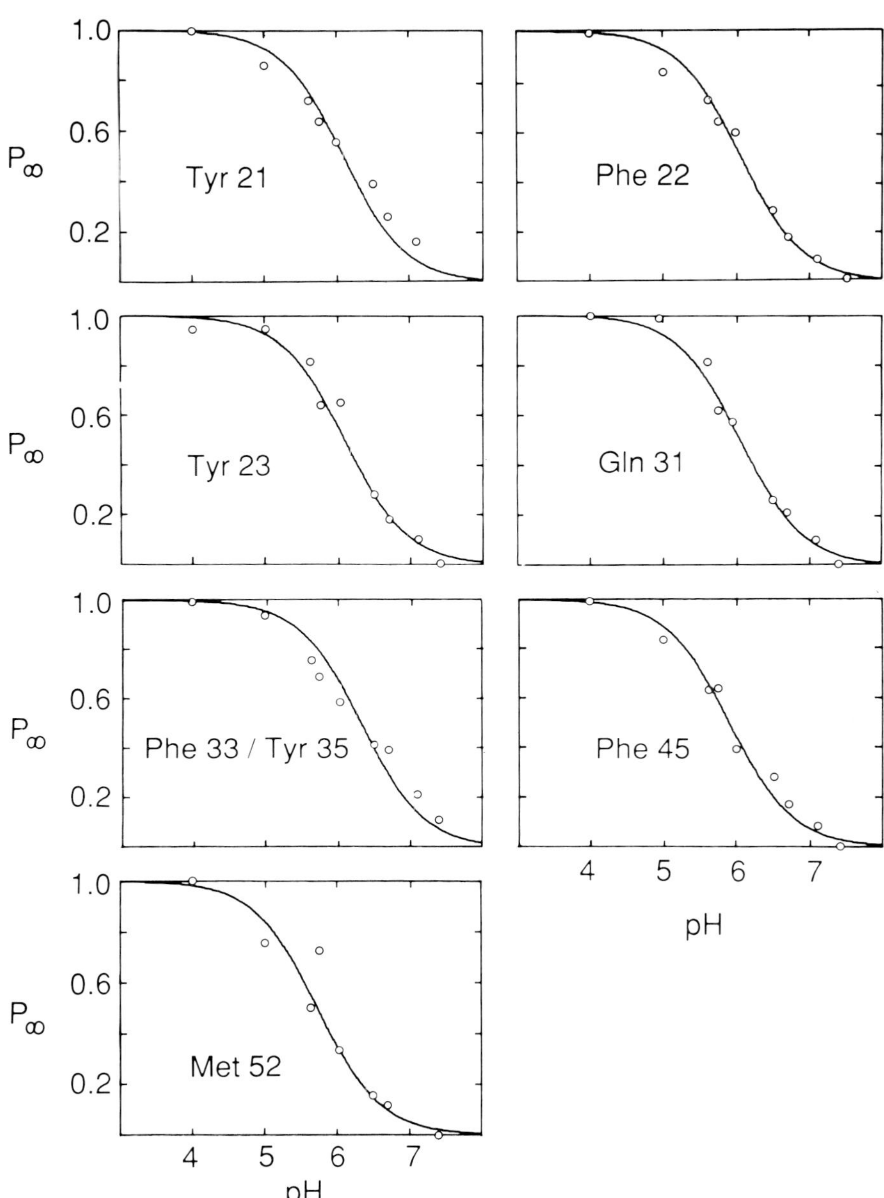

Fig. 3. Plot of the limiting proton occupancy, P_∞, vs. pH for individual amide groups in BPTI at 70°C (method 2). The theoretical curves are fits of Eq. (5) with adjustment of k_f (k_c was computed using Eq. (7)). Apparent folding rates, k_f, resulting from these fits are reported in Table I.

conducted at pH values below 7.3. This is further demonstrated by Figure 2, which shows that after the initial exponential phase, $P(\tau)$ does not change further at times $\tau \geq 0.15$ s. In contrast, if exchange from refolded BPTI were appreciable, we would have observed a second phase with the proton occupancy decaying toward the solvent proton background level ($P=0$).

DISCUSSION

The potential of experimental techniques using amide protons as probes of structure formation in protein folding studies has been recognized for some time.[2–4,16–18] Most such studies so far used tritium labeling and manual mixing and were therefore limited to slow folding steps. The present experiments

TABLE I. Apparent Folding Rates for Individual Amide Protons in BPTI From the pH Dependence of the Competition Between Refolding and NH Exchange*

	Tyr 21	Phe 22	Tyr 23	Gln 31	Phe 33/ Tyr 35†	Phe 45	Met 52
$pH_{½}$‡	6.1	6.0	6.1	6.0	6.3	6.0	5.6
k_f [s^{-1}]**	30	25	30	60	50	50	15

*Refolding conditions were 70°C, 2:1 D_2O:H_2O, 13% (v/v) n-propanol, 0.15 M phosphate.
†The NH resonances of Phe 33 and Tyr 35 were overlapped.
‡pH where 50% of the amide protons in this position is trapped in the refolded protein (Fig. 3).
**Apparent folding rates, k_f, determined by fitting Eq. (5) to the data in Figure 3 (method 2). The average standard deviation of the fit is 20%. Intrinsic NH exchange rates, k_c, were obtained by extrapolation of the data from Roder et al.[9], using Eq. (7) with E_a = 17 kcal mol^{-1}.

combine rapid mixing techniques with NMR observation of individual amide protons at known sequence positions.[12,13] This enables extension of folding studies to faster processes and leads to the structural resolution of folding rates for individual amide proton sites.

The approach described here relies on measurements of the competition between two rate processes, i.e., between the exchange of solvent deuterons for amide protons in unfolded parts of the protein, and the formation of folded structure which protects the amide protons against solvent exchange. This competition method makes it possible to label short-lived folding intermediates in a nonperturbing way, so that the intrinsically slow NMR technique can be used subsequently to obtain a high degree of structural resolution. The detailed interpretation of competition data depends on the structural features responsible for slowing hydrogen exchange, among which intramolecular hydrogen bonding appears to be the most important factor.[19] As a general observation, in previously studied proteins one notes that the most slowly exchanging amide protons are involved in the hydrogen bonds of regular secondary structures.[20,21–23] The time resolution of the competition method is determined by the intrinsic amide proton exchange rate, which can be varied over a wide range by the choice of pH and temperature, and by the speed of mixing, which can be controlled by the choice of the experimental setup (Fig. 1).

The kinetic interpretation of competition data is straight forward if the simple two-state situation described by Eq. (1) applies; i.e., a particular amide proton is protected against exchange in a single step without transient formation of intermediate states that would be sufficiently long-lived to observably retard exchange. In this case the apparent folding rate, k_f, obtained in the competition experiment (method 1 or 2) can be attributed to the rate of local structure formation near the observed amide proton. It should be stressed that a two-state situation for individual amide protons does not imply that the overall folding process has to be a two-state reaction. Even if individual segments of secondary structure are formed in a concerted way, different parts of the protein may fold at different times. The competition method can thus in principle resolve the sequence of folding events throughout the structure.

In general, however, more complicated kinetic situations than Eq. (1) have to be considered. While it is usually possible to ensure essentially irreversible formation of the final folded state by choosing strongly native refolding conditions, unstable folding intermediates are likely to be encountered in the early phases of refolding. Such a situation is described by the following generalized scheme (cf.[2]):

$$\begin{array}{l} U^H \underset{k_U}{\overset{k_I}{\rightleftharpoons}} I^H \overset{k_N}{\rightarrow} N^H \\ k_c\downarrow \\ U^D \underset{k_U}{\overset{k_I}{\rightleftharpoons}} I^D \overset{k_N}{\rightarrow} N^D \end{array} \quad (8)$$

where k_N is the rate-limiting step ($k_N << k_I$). The intermediate state, I, affects the competition behavior of a particular amide proton only if the structure formed in I is able to protect that proton against exchange. If this is the case, then three limiting situations can be distinguished:

1. If I is stable on the timescale of the exchange reaction and formed at a rate k_I comparable to k_c ($k_c \sim k_I >> k_U$, k_N), the model reduces to the scheme of Eq. (1) with k_I substituted for k_f in Eqs. (4) and (5); i.e., formation of I is now the kinetic step competing with exchange.
2. If I is only marginally stable ($k_I \sim k_U$) and $k_I >> k_c$, the folding rate k_f in Eqs. (4) and (5) can be interpreted as an effective rate of protection, given by

$$k_f = k_N \cdot (1 + k_I/k_U) \quad (9)$$

i.e., the rate k_f determined on the basis of the simple scheme of Eq. (1) would overestimate the rate constant k_N. (Protons are lost via the backreaction to the unfolded form).

3. If I is unstable ($k_U >> k_I$), it is not detected by the competition experiment. Eq. (9) reduces to $k_f = k_N$ and we have again the simple situation of Eq. (1).

Different limiting cases in folding processes with nonnegligible transient accumulation of structured intermediate states cannot a priori be distinguished by the competition method, so that in general a unique interpretation in terms of elementary rates in Eq. (8) is not possible.

The competition experiment can, however, still provide structural characterization of complex folding pathways by comparing the apparent folding rates measured in different parts of the protein. Obtaining structural resolution relies on the following. First, the amide proton resonances in the folded protein must be assigned to specific residues in the amino acid sequence.[13] Second, the spatial protein structure in solution must be known, either from direct determination by NMR[24–26] or from experiments (primarily NMR) enabling detailed comparison with the crystal structure. Resolution in the spatial protein structure is then achieved, since each data point measured with the competition technique can be attributed to a specified location. This spatial resolution can be obtained also in the absence of knowledge on the individual elementary rates in the scheme of Eq. (8). This is in the following illustrated with the data on BPTI in Figures 2 and 3 and Table I.

An important initial observation for BPTI is that the close agreement between the data in Figure 3 and the curve shape predicted by the simple kinetic model of Eq. (1) demonstrates that the assumption of pH-independent folding rates is reasonable. This was to be expected, since BPTI has no titrating groups in the range between about pH 4.5 and 7.5.[27]

The groups observed in this initial competition study with BPTI include seven amide protons located in the core of the β-sheet and one amide proton in the C-terminal α-helix.[28] The variation in apparent folding rates for these sites is relatively small (Table I), indicating that the observed folding transition is quite cooperative. On a finer level we can distinguish three groups of protons on the basis of the different apparent folding rates. The smallest rate is observed for Met 52 located in the C-terminal α-helix; its pH transition is displaced by ca. 0.5 pH units to lower pH relative to the other amide protons (Fig. 3, Table I). The amide protons of Tyr 21, Phe 22, and Tyr 23 on the central strand of the β-sheet form a group with intermediate rates. The amide protons on the two peripheral strands, i.e., Gln 31, Phe 33, Tyr 35, and Phe 45, are protected at somewhat higher rates. While the lower rate of protection for Met 52 can be rationalized by the fact that this proton is in a different structural element, and also in terms of the more peripheral location of the C-terminal helix, inspection of the native BPTI structure reveals no obvious reason for the distinction among the different β-sheet protons.

Jullien and Baldwin[5] studied the folding kinetics of reduced BPTI (RCAM-BPTI) by stopped-flow measurements monitored by tyrosine absorption. The structure of this derivative, which is obtained by selective cleavage of the 14–38 disulfide bond, has been shown to resemble closely that of the native protein.[29] In 1.9 M GuHCl at 25°C and pH 6.8, three distinct folding phases were observed, with the major fraction (75% of the amplitude) formed at a rate of 25 s^{-1} and two minor fractions with rates of 0.07 and 0.005 s^{-1}. At 60°C and pH 6.1, a fast phase with a rate of 70 s^{-1} and a single slow process at 0.3 s^{-1} were observed with about equal amplitude. Although different experimental conditions were chosen for our competition experiments, it is apparent that the proton trapping manifested in Figures 2 and 3 occurs on the time scale of the fastest optically detected process. Furthermore, our observation of complete proton trapping on the acidic side of the pH-dependent competition data (Fig. 3) indicates that structure capable of protecting the central β-sheet amide protons is formed rapidly not only in the fast refolding form, but also in the slowly refolding forms of BPTI[5]. (A significant fraction (> 10%) of the protein remaining fully unfolded over a period of seconds would lead to a lower protonation level than that observed at the low pH end of the curves in Fig. 3).

The narrow range of folding rates for the β-sheet protons suggests that an early intermediate is formed with nativelike secondary structure in the core of the molecule. The limited evidence from the single amide proton in the C-terminal helix suggests that this more peripheral secondary structure is folded in an independent step. Optical probes[5] apparently monitor much slower, minor structure rearrangements which set in after the main features of the folded structure have been formed. These probably also include local structural rearrangements associated with proline isomerization (cf.[1]).

Further information on local folding rates resulted from NH exchange studies of the core protons in the BPTI β-sheet under destabilizing conditions,[15] where the transition predicted by the structural opening model[30] between the exchange-limited (EX_2) and the opening-limited (EX_1) region of the exchange kinetics was observed. Structurally resolved folding rates estimated from the location of the transition on the pH scale ranged between $1.5 \cdot 10^3$ and $4 \cdot 10^3$ s^{-1} at 68°C, and between 200 and 600 s^{-1} at 55°C in the presence of 3 M GuHCl. Additional observations suggested that these rates reflect refolding from only partially unfolded conformations. While the absolute rates should therefore not be directly compared, it is interesting that the pattern of relative rates for the different β-sheet protons ($k_2{}^g$ in Table II in[15]) is comparable with the results of the competition experiments reported in Table I.

The present results demonstrate that early events in protein folding can to a certain extent be structurally and kinetically characterized by ^{1}H-NMR observation of the competition between hydrogen exchange and structure condensation. A number of improvements and extensions of the method can be considered. Two-dimensional (2D) correlated spectroscopy (COSY)[31,32] or 2D nuclear Overhauser enhancement spectroscopy (NOESY)[33] could be used to resolve a more complete set of amide protons throughout the protein structure (cf[20]). The mixing technique could be improved by increasing the D_2O:H_2O ratio during the competition period. This would reduce the H_2O background present in the mixture and permit more extensive dilution of the denaturant; the resulting more strongly native conditions favor the population of folding intermediates.[1] Complementary information might be gained by using the pulse labeling method developed by the Baldwin group.[3,4] This variation of the competition method would offer improved flexibility and the possibility to characterize intermediates at any stage of folding, though this is achieved at the cost of further reduced time resolution compared to the methods used in this study. Finally and probably most important in practice, the number of proteins for which sequence-specific resonance assignments are available is growing rapidly, and it can be forseen that we shall soon be able to work with a number of proteins for which the solution conformation has been determined by NMR. On this basis we expect the presently proposed competition techniques to become instrumental in future efforts to unravel the protein folding puzzle.

ACKNOWLEDGMENTS

Helpful discussions with Drs. Robert L. Baldwin and S. Walter Englander are gratefully acknowledged. This work was taken from the Ph.D. thesis of H.R., which was completed in 1981. The project was supported by the Swiss National Science Foundation (project 3.528.79).

REFERENCES

1. Kim, P.S., Baldwin, R.L. Specific intermediates in the folding reactions of small proteins and the mechanism of folding. Annu. Rev. Biochem. 51:459–489, 1982.
2. Schmid, F.X., Baldwin, R.L. Detection of an early intermediate in the folding of ribonuclease A by protection of amide protons against exchange. J. Mol. Biol. 135:199-–215, 1979.
3. Kim, P.S., Baldwin, R.L. Structural intermediates trapped during the folding of ribonuclease A by amide proton exchange. Biochemistry 19:6124–6129, 1980.
4. Brems, D.N., Baldwin, R.L. Protection of amide protons in folding intermediates of ribonuclease A measured by pH-pulse exchange curves. Biochemistry 24:1689–1693, 1985.
5. Jullien, M., Baldwin, R.L. The role of proline residues in the folding kinetics of the bovine pancreatic trypsin inhibitor derivative RCAM (14–38). J. Mol. Biol. 145:265–280, 1981.
6. Richarz, R., Sehr, P., Wagner, G., Wüthrich, K. Kinetics of the exchange of individual amide protons in the basic pancreatic trypsin inhibitor. J. Mol. Biol. 130:19–30, 1979.
7. Englander, S.W., Poulsen, A. Hydrogen-tritium exchange of the random chain polypeptide. Biopolymers 7:379–393, 1969.
8. Molday, R.S., Englander, S.W., Kallen, R.G. Primary structure effects on peptide group hydrogen exchange. Biochemistry 11:150–159, 1972.
9. Roder, H., Wagner, G. & Wüthrich, K. Individual amide proton exchange rates in thermally unfolded basic pancreatic trypsin inhibitor. Biochemistry 24:7407–7411, 1985.
10. Woodward, C.K., Ellis, L., Rosenberg, A. Solvent accessibility in folded proteins: studies of hydrogen exchange in trypsin. J. Biol. Chem. 250:432–444, 1975.
11. Englander, J.J., Rogero, J.R., Englander, S.W. Protein hydrogen exchange studied by the fragment separation method. Anal. Biochem. 147:234–244, 1985.
12. Dubs., A., Wagner, G., Wüthrich, K. Individual assignments of amide proton resonances in the proton NMR spectrum of the basic pancreatic trypsin inhibitor. Biochim. Biophys. Acta 577:177–194, 1979.
13. Wagner, G., Wüthrich, K. Sequential resonance assignments in protein ^{1}H NMR spectra: basic pancreatic trypsin inhibitor. J. Mol. Biol. 155:347–366, 1982.
14. Englander, J.J., Calhoun, D.B., Englander, S.W. Measurements and calibration of peptide group hydrogen-deuterium exchange by ultraviolet spectrophotometry. Anal. Biochem. 92:517–524, 1979.
15. Roder, H., Wagner, G., Wüthrich, K. Amide proton exchange in proteins by EX_1 kinetics: studies of the basic pancreatic trypsin inhibitor at variable p^{2}H and temperature. Biochemistry 24:7396–7407, 1985.
16. McPhie, P. Swine pepsin folding intermediates are highly structured, motile molecules. Biochemistry 21:5509–5515, 1982.
17. Kuwajima, K., Kim, P.S., Baldwin, R.L. Strategy for trapping intermediates in folding of ribonuclease A and for using ^{1}H-NMR to determine their structure. Biopolymers 22:59–67, 1983.
18. Beasty, A.M., Matthews, C.R. Characterization of an early intermediate in the folding of the α subunit of tryptophan synthase by a hydrogen exchange measurement. Biochemistry 24:3547–3553, 1985.
19. Englander, S.W., Kallenbach, N.R. Hydrogen exchange and structural dynamics of proteins and nucleic acids. Q. Rev. Biophys. 16:521–655, 1984.
20. Wagner, G., Wüthrich, K. Amide proton exchange and surface conformation of the basic pancreatic trypsin inhibitor in solution. J. Mol. Biol. 160:343–361, 1982.
21. Wüthrich, K., Strop, P., Ebina, S., Williamson, M.P. A globular protein with slower amide proton exchange from an α helix than from antiparallel β sheets. Biochem. Biophys. Res. Commun. 122:1174–1178, 1984.
22. Kline, A.D., Wüthrich, K. Secondary structure of the α-amylase polypeptide inhibitor tendamistat from streptomyces tendae determined in solution by ^{1}H NMR. J. Mol. Biol. 183:503–507, 1985.
23. Wand, A.J., Roder, H., Englander, S.W. Two-dimensional ^{1}H NMR studies of cytochrome c: hydrogen exchange in the N-terminal helix. Biochemistry 25:1107–1114, 1986.
24. Williamson, M.P., Havel, T.F., Wüthrich, K. Solution conformation of proteinase inhibitor IIA from bull seminal plasma by ^{1}H nuclear magnetic resonance and distance geometry. J. Mol. Biol. 182:295–315, 1985.
25. Kline, A.D., Braun, W., Wüthrich, K. Studies by ^{1}H nuclear magnetic resonance and distance geometry of the solution conformation of the α-amylase inhibitor tendamistat. J. Mol. Biol. 189:377–382, 1986.
26. Wüthrich, K. "MNR of Proteins and Nucleic Acids." New York: J. Wiley, in press, 1986.
27. Wagner, G., Wüthrich, K. Correlation between the amide proton exchange rates and the denaturation temperature in globular proteins related to the basic pancreatic trypsin inhibitor. J. Mol. Biol. 130:31–37, 1979.
28. Deisenhofer, J., Steigemann, W. Crystallographic refinement of the structure of bovine pancreatic trypsin inhibitor

at 1.5 Å resolution. Acta Crystallogr. B31:238–250, 1975.

29. Stassinopoulou, C.I., Wagner, G., Wüthrich, K. Two-dimensional ^{1}H NMR of two chemically modified analogs of the basic pancreatic trypsin inhibitor: sequence specific resonance assignments and sequence location of conformation changes relative to the native protein. Eur. J. Biochem. 145:423–430, 1984.
30. Hvidt, A., Nielsen, S.O. Hydrogen exchange in proteins. Adv. Protein Chem. 21:287–386, 1966.
31. Aue, W.P., Bartholdi, E., Ernst., R.R. Two-dimensional spectroscopy: application to nuclear magnetic resonance. J. Chem. Phys. 64:2229–2246, 1976.
32. Nagayama, K., Anil-Kumar, Wüthrich, K., Ernst, R.R. Experimental techniques of two-dimensional correlated spectroscopy. J. Magn. Reson. 40:321–334, 1980.
33. Anil-Kumar, Ernst, R.R., Wüthrich, K. A 2D NOE experiment for the elucidation of complete proton-proton cross-relaxation networks in biological macromolecules. Biochem. Biophys. Res. Commun. 95:1–6, 1980.

Reprinted with permission from *Nature* **368**, 261–265 (1994)
Copyright © 1994 MacMillan Magazines Limited.

LETTERS TO NATURE

Destabilization of the complete protein secondary structure on binding to the chaperone GroEL

Ralph Zahn*, Claus Spitzfaden†, Marcel Ottiger†, Kurt Wüthrich† & Andreas Plückthun*‡

* Max-Planck-Institut für Biochemie, Protein Engineering Group, Am Klopferspitz, D-82152 Martinsried, Germany
† Institut für Molekularbiologie und Biophysik, ETH-Hönggerberg, CH-8093 Zürich, Switzerland

PROTEIN folding *in vivo* is mediated by helper proteins, the molecular chaperones[1–3], of which Hsp60 and its *Escherichia coli* variant GroEL are some of the best characterized. GroEL is an oligomeric protein with 14 subunits each of M_r 60K[4–6], which possesses weak, co-operative ATPase activity[7–9] and high plasticity[10]. GroEL seems to interact with non-native proteins, binding one or two molecules per 14-mer[11–19] in a 'central cavity'[20], but little is known about the conformational state of the bound polypeptides. Here we use nuclear magnetic resonance techniques to show that the interaction of the small protein cyclophilin[21,22] with GroEL is reversible by temperature changes, and all amide protons in GroEL-bound cyclophilin are exchanged with the solvent, although this exchange does not occur in free cyclophilin. The complete secondary structure of cyclophilin must be disrupted when bound to GroEL.

Exchange of cyclophilin amide protons (Fig. 1) in the presence of GroEL was studied under conditions in which cyclophilin could bind to GroEL, that is, at 30 °C and pH 6.0 (see below). An equimolar mixture of ^{15}N-labelled cyclophilin and GroEL was heated in D_2O to 30 °C for 8 h to induce cyclophilin binding, followed by cooling to 6 °C for 14 h, to induce protein–chaperone dissociation (Fig. 2*a*). This cycle was repeated three times to ensure that most cyclophilin molecules were bound at least once even if the turnover of bound cyclophilin was slow (at pH 6.0 and 30 °C, only 50% of cyclophilin is bound to GroEL at equilibrium). Before the nuclear magnetic resonance (NMR) experiments, cyclophilin and GroEL were separated by cation-exchange chromatography in D_2O at 6 °C. The resulting cyclophilin fractions were pooled and concentrated to a protein concentration of 0.4 mM. A two-dimensional [^{15}N, ^{1}H]-correlation ([^{15}N, ^{1}H]-COSY) spectrum of this solution was completely empty (not shown), demonstrating that all amide protons had been exchanged with deuterium.

To show that the cyclophilin recovered in D_2O from the GroEL–cyclophilin complex is in the native folded form, it was

‡ To whom correspondence should be addressed at: Biochem. Institut, Universität Zürich, Winterthurerstr. 190, CH-8057 Zürich, Switzerland.

LETTERS TO NATURE

re-exchanged into H_2O and kept at 26 °C for 2 weeks. A [^{15}N, ^{1}H]-COSY spectrum of this sample (Fig. 2*d*) had the same appearance as a spectrum of native cyclophilin (Fig. 2*b*), except that the signals corresponding to the 39 most slowly exchanging amide protons (Fig. 1) were very weak. A [^{15}N, ^{1}H]-COSY spectrum of cyclophilin treated as before (Fig. 2*a*) in the absence of GroEL showed that the 39 backbone amide protons which exchange most slowly in free cyclophilin[23] gave strong peaks. Residual peak intensities were observed for some residues with medium-fast exchange (Figs 1 and 2*c*), corresponding to the behaviour of free cyclophilin in D_2O solution without the special treatment used in Fig. 2*a*. Experiments with cross-sections[24] support these findings. The cross-section Fig. 2*b'* contains peaks from a rapidly and a slowly exchanging amide proton. In cross-section Fig. 2*c'* only the peak of the slowly exchanging amide proton is left. In cross-section Fig. 2*d'* the rapidly exchanging proton has been fully re-exchanged, whereas less than 10% of the slowly exchanging protons have been re-exchanged. A complete survey of the backbone amide proton exchange in free and GroEL-bound cyclophilin is given in Fig. 2*e*, *f*, showing that residues that were strongly protected against exchange in a solution of free cyclophilin in D_2O re-exchange only to a limited extent in the experiment shown in Fig. 2*d*.

To investigate the kinetic stability of the complex formed between cyclophilin and GroEL, a cyclophilin solution containing only 0.1 equivalents of GroEL (but otherwise identical to the solutions used in Fig. 2*c* and *d*) was subjected to the treatment of Fig. 2*a*. If the GroEL–cyclophilin complex is relatively stable under these conditions, one would predict a 14% reduction in the intensity of the cross peaks corresponding to slowly exchanging amide protons, since in each cycle of Fig. 2*a* only about half of GroEL would bind cyclophilin (Fig. 3*d*), leading to exchange. A much bigger intensity loss could result if cyclophilin exchanged in and out of the GroEL binding site. The [^{15}N, ^{1}H]-COSY spectrum from this experiment (not shown) was very similar to that in Fig. 2*c*. Only a small exchange enhancement can thus be attributed to cyclophilin turnover, and the GroEL–cyclophilin complex at 30 °C and pH 6.0 must be kinetically rather stable.

When 0.005 stoichiometric equivalents of GroEL were present during the NMR observation of cyclophilin (much more would not be soluble at these high molar concentrations), the exchange of the slowly exchanging protons was not measurably affected (see above). There was significant line broadening in the NMR spectra however, which persisted after eightfold dilution of the protein sample, but disappeared on addition of 6 mM Mg–ATP

TABLE 1 Kinetics of cyclophilin folding

	pH	T-shift (°C)	[Cyclophilin] (μM)	[GroEL] (μM)	Rate constant
Aggregation	7.0	25→54	25	—	$1.3\times10^{3}\ s^{-1}\ M^{-1}$
Unfolding	7.0	25→48	25	25	$2.3\times10^{-3}\ s^{-1}$
	7.0	25→46	25	25	$1.9\times10^{-3}\ s^{-1}$
	6.0	6→30	1	5	$7.6\times10^{-5}\ s^{-1}$
Refolding	7.0	46→1	25	25	$2.9\times10^{-4}\ s^{-1}$
	7.0	46→5	25	25	$5.5\times10^{-4}\ s^{-1}$
	6.0	30→6	1	5	$1.8\times10^{-5}\ s^{-1}$

The folding kinetics were measured after shifting the temperature of a cyclophilin-containing solution to the value indicated and measuring the enzymatic activity as a function of time. The rate constants were determined by a three-parameter fit (starting activity, final activity, rate constant) to first-order kinetics (for GroEL-assisted unfolding and refolding) or second-order kinetics (for irreversible aggregation in the absence of GroEL).

(Fig. 2*g*). Under the conditions of these experiments (pH 6.0, 26 °C), the GroEL present contains some stably bound cyclophilin but is probably not saturated. Cyclophilin must thus interact transiently with either the free GroEL or the GroEL–cyclophilin complex, or both. While they are bound, these cyclophilin molecules would be restricted to the slow rotational tumbling of the GroEL–cyclophilin complex, causing the observed line broadening. The broadening in turn indicates that the cyclophilin stays bound for more than about 1 ns. Mg–ATP suppresses this interaction, presumably by causing a change in GroEL conformation[9]. The fact that this type of binding induces no exchange of the most stable amide protons (Fig. 1) indicates that it does not involve unfolding or destabilization of native cyclophilin, setting it apart from that seen in Fig. 2*d*.

The conditions for these experiments were selected on the basis of biochemical measurements of the kinetics and equilibria of cyclophilin folding in the presence of GroEL. Gel chromatography at pH 7.0 (data not shown) showed that micromolar concentrations of cyclophilin do not give rise to a stable complex with equimolar amounts of GroEL at room temperature, but that a significant proportion of cyclophilin co-elutes with GroEL once the temperature of the column is raised to 58 °C. If the solution is subsequently cooled to 25 °C, the complex of GroEL and cyclophilin dissociates again.

Measurements of enzymatic activity (Fig. 3*a*) show that there is concentration-dependent, irreversible unfolding of cyclophilin in the absence of GroEL, the kinetics of which are consistent with a second-order reaction (Table 1) as is typical of aggregation processes. In the presence of equivalent concentrations of GroEL, most of the cyclophilin is protected from irreversible

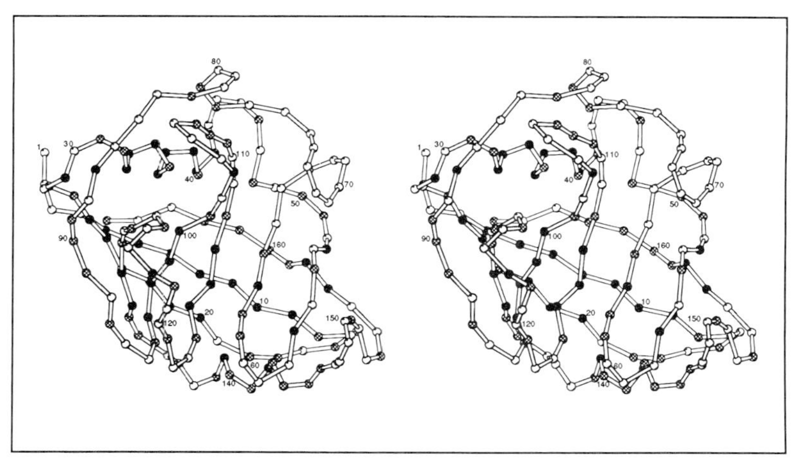

FIG. 1 Stereo view of the polypeptide backbone in the three-dimensional NMR solution structure of cyclophilin[28] with identification of residues with slowly exchanging amide protons. Backbone amide nitrogen positions represented by circles are connected by virtual bonds; every tenth residue is numbered. The black circles identify the previously identified[23] locations of the 39 residues for which complete amide proton exchange could not be achieved in free cyclophilin without irreversible denaturation of the protein. Shaded and open circles identify residues with medium and fast amide proton exchange, respectively[23]. The plot was made with the program MOLSCRIPT[29].

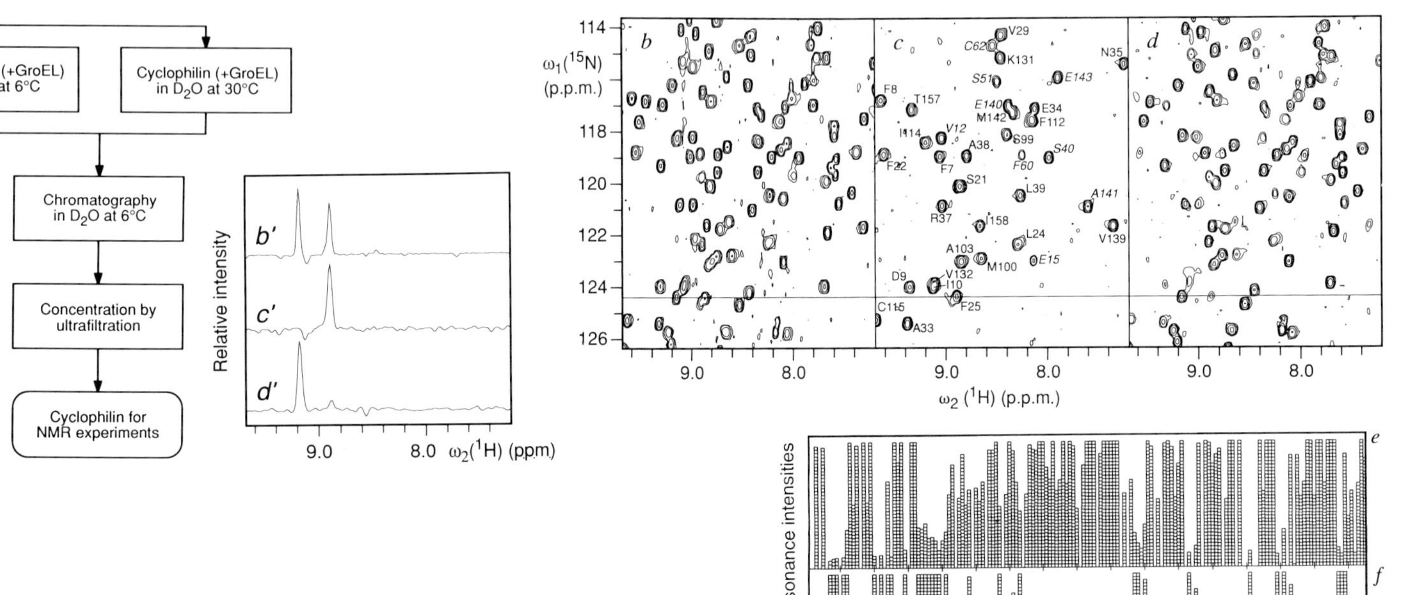

FIG. 2 *a*, Diagrammatic representation of the preparation of the protein solutions used to study GroEL-promoted amide proton exchange in cyclophilin by NMR. *b–d*, NMR spectra used to study the amide proton exchange of cyclophilin in the presence and absence of GroEL. The region ($\omega_1(^{15}N)$ = 114–126 p.p.m., $\omega_2(^1H)$ = 7.2–9.7 p.p.m.) of [^{15}N, ^1H]-COSY spectra (^1H frequency = 600 MHz, 26 °C, pH or pD 6.0) is shown. *b*, Cyclophilin (4 mM) in 90% H_2O/10% D_2O. *c*, A 0.5 mM D_2O solution of cyclophilin that had been subjected to the procedure outlined in *a* with no GroEL added. *d*, A 0.3 mM H_2O solution of cyclophilin that had been subjected to the procedure in *a* in the presence of a stoichiometric amount of GroEL, after subsequent re-equilibration in 90% H_2O/10% D_2O for 2 weeks at 26 °C. The cross peaks which correspond to very slowly exchanging amide protons[23] (Fig. 1) are identified in *c* with the amino acid one-letter symbol and the sequence position. Some additional peaks corresponding to amide protons with medium–slow exchange are similarly identified using italics. *b′–d′*, Cross-sections along $\omega_2(^1H)$ taken at the position of the horizontal line $\omega_1(^{15}N)$ = 124.4 p.p.m. in the spectra *b*, *c* and *d*. *e*, *f*, Plots against the amino acid sequence of cyclophilin of the [^{15}N, ^1H]-COSY cross-peak volumes for the individual amino acid residues normalized by the corresponding peak intensities in the reference spectrum *b*, where *e* shows data from the experiment shown in *d* and *f* those for cyclophilin after amide proton exchange in D_2O for 37 h at 26 °C. The two data sets can be compared only qualitatively because of the different exchange periods of 336 h for *e* and 37 h for *f*. *g*, High-field region from −0.8 to 0.6 p.p.m. from 1D ^1H NMR spectra: (*i*), 0.3 mM cyclophilin; (*ii*), 0.3 mM cyclophilin +1.5 μM GroEL; (*iii*), same as (*ii*), 1.5 h after addition of 6 mM Mg–ATP (26 °C, ^1H frequency 500 MHz, solvent 90% H_2O/10% D_2O).

METHODS. Recombinant ^{15}N-labelled cyclophilin was over-expressed in *E. coli* and purified as described previously[30]. GroEL was purified from lysates of cells harbouring the multicopy plasmid pOF39, as described previously[19]. The concentration of GroEL is always given for the 14-subunit oligomeric form. For the NMR experiments *c* and *d*, a 20 μM solution of cyclophilin in D_2O (99.8% D_2O, 20 mM 2-[N-morpholino] ethane sulphonate (MES) pD 6.0, 2 mM DTT, 2 mM EDTA, 0.02% NaN_3) was subjected to the procedure in *c* without and in *d* with addition of 20 μM of GroEL. The two samples of 25 ml volume were subjected to three cycles of heating to 30 °C for 8 h and cooling to 6 °C for 14 h (see *a*). After standing at 6 °C for 40 h after the last temperature shift, cyclophilin was separated from GroEL on a HiLoad S-Sepharose column (Pharmacia) equilibrated with the same MES buffer in D_2O at 6 °C. The cyclophilin fractions were concentrated to about 0.4 mM by ultrafiltration at 6 °C.

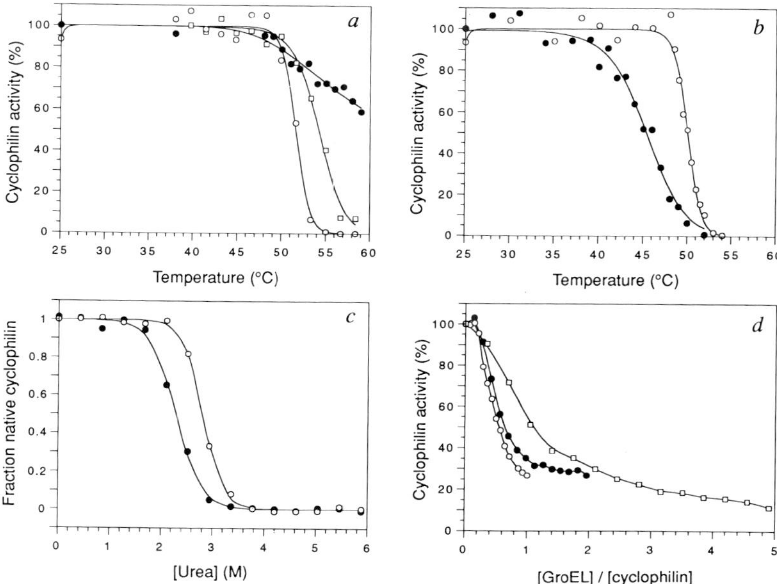

FIG. 3 Folding of cyclophilin in the absence and presence of GroEL. *a*, Irreversibly unfolded cyclophilin after heating at pH 7.0. □, low cyclophilin concentration (0.9 μM) in the absence of GroEL; ● and ○, high cyclophilin concentration (9 μM) in the presence or absence of GroEL (10 μM), respectively. *b*, Portion of total (reversibly and irreversibly) inactivated cyclophilin after heating at pH 7.0 in the presence (●) or absence (○) of GroEL. *c*, Urea denaturation curves of cyclophilin at 30 °C and pH 6.0 (●) or pH 7.0 (○). *d*, Titration of cyclophilin with GroEL at 46 °C and pH 7.0 (●, ○), and at 30 °C and pH 6.0 (□). The concentration of cyclophilin was 1 μM (●, □) or 10 μM (○).

METHODS. *a*, Cyclophilin was incubated for 20 min at the temperatures indicated in a buffer containing 100 mM K_2HPO_4/KH_2PO_4 pH 7.0 and 10 mM DTT. After cooling the samples to 10 °C and a subsequent waiting period of 1 h at 25 °C (to allow unfolded enzyme to refold), the enzymatic peptidyl-prolyl *cis-trans*-isomerase activity of cyclophilin was measured at 10 °C as described[31]. This experiment determines the sum of residual folded and refoldable cyclophilin. *b*, Solutions of 25 μM cyclophilin in 100 mM K_2HPO_4/KH_2PO_4 at pH 7.0 and 10 mM DTT were incubated at various temperatures in the presence of 25 μM GroEL or in the absence of GroEL. After 20 min the solutions were rapidly cooled to 1 °C by 100-fold dilution in the same buffer (to freeze the folding state of cyclophilin). The cyclophilin activity was immediately measured at 4.5 °C. In this experiment, only the amount of residual enzymatically active cyclophilin present in the incubation mixture is measured as a function of the incubation temperature. The remainder is reversibly or irreversibly inactivated. *c*, The enzymatic activity of a 1 μM cyclophilin solution, incubated for 22 h at 30 °C at different urea concentrations, was measured at 4.5 °C. The solution was either at pH 6.0 (100 mM MES, 10 mM DTT) or at pH 7.0 (100 mM K_2HPO_4/KH_2PO_4, 10 mM DTT). From these data, the fraction of native cyclophilin was calculated[32] (note that these are not true equilibrium curves, however, because cyclophilin cannot be refolded after denaturation in 8 M urea under these conditions). *d*, Solutions of cyclophilin containing different concentrations of GroEL were incubated for 20 min at 46 °C in a buffer containing 100 mM K_2HPO_4/KH_2PO_4, 10 mM DTT, pH 7.0, or for 19 h at 30 °C in a buffer containing 90 mM MES/38 mM K_2HPO_4/KH_2PO_4, 10 mM DTT, pH 6.0. Immediately after heating, the enzymatic activity of cyclophilin, which is given relative to that of a GroEL-free cyclophilin solution, was measured at 4.5 °C.

denaturation (Fig. 3*a*), probably by forming a stable reversible complex with GroEL, as shown by the gel filtration experiments. In the absence of GroEL, the total unfolded portion of cyclophilin after heating at pH 7.0 (Fig. 3*b*) is similar to that in Fig. 3*a*, indicating that once cyclophilin is unfolded by heat, it cannot be folded back under these conditions. In the presence of GroEL, however, a reduction of cyclophilin activity occurs far below the denaturation temperature for free cyclophilin at a rate consistent with first-order kinetics (Table 1.) When the temperature was subsequently shifted back from 46 °C to 1 °C, cyclophilin recovered enzymatic activity with slow first-order kinetics (Table 1).

Urea denaturation curves show that cyclophilin is less stable at pH 6.0 than at pH 7.0 (Fig. 3*c*). The titration curves of cyclophilin with GroEL at 30 °C show that the net interaction between the chaperone and the substrate is stronger at pH 6.0 than at pH 7.0 (Fig. 3*d*), probably because of the lower thermodynamic stability of cyclophilin at pH 6.0, where it can be unfolded at 30 °C (Table 1). The close similarity of the titration curves obtained at pH 7.0 with different absolute concentrations of GroEL and cyclophilin indicates that most of the cyclophilin molecules may already be transiently bound to GroEL at the concentrations used; the observed transition in enzymatic activity would thus occur in the complex and be independent of concentration. Even at the highest GroEL concentrations used, the residual cyclophilin activity exceeded 20%.

These data can be interpreted in terms of the model:

$$EL + N \underset{\text{fast}}{\overset{\text{fast}}{\rightleftharpoons}} EL\text{–}N \underset{\text{slow}}{\overset{\text{medium}}{\rightleftharpoons}} EL\text{–}U \underset{\text{fast}}{\overset{\text{slow}}{\rightleftharpoons}} EL + U$$

where EL is GroEL, N is folded, enzymatically active cyclophilin with 39 exchange-protected amide protons, U is unfolded, enzymatically inactive cyclophilin with no exchange-protected amide protons, EL–N is the complex of EL and N, and EL–U is the complex of EL and U. The rates shown apply to the unfolding reaction observed at high temperature.

The exchange experiment of Fig. 2*d*, which started with native cyclophilin and led to tightly GroEL-bound cyclophilin, would thus start on the left, go to EL–U, and return to the left after the temperature had been lowered. In these experiments, cyclophilin was destabilized by lowering the pH, allowing unfolding to occur even at 30 °C (Table 1). At pH 7.0, cyclophilin remains stable at this temperature, but can be reversibly unfolded by GroEL at around 45 °C (Fig. 3*a*, *b*). By contrast, in the experiment with transiently bound cyclophilin (Fig. 2*g*) only the state EL–N is transiently occupied in the absence of ATP. The transient binding of native cyclophilin may be analogous to the previously reported stabilization of helical structure in a GroEL-bound peptide with a short residence time[25,26]. *In vivo*, the reaction would presumably start on the right-hand side with U and EL, and as the EL–N↔EL–U equilibrium in the cell will normally be on the left, the reaction would proceed all the way from right to left, the last step being aided by ATP.

As cyclophilin is not bound instantaneously and quantitatively by GroEL under the conditions used (Fig. 3*d*), yet all protons are quantitatively lost, the protection factor of the amide protons labelled black in Fig. 1 must be reduced from $>10^7$ in native cyclophilin to $<10^3$ in GroEL-bound cyclophilin. By contrast, the protection factors of α-lactalbumin are 10^4 in the native state and are lowered to 10^2 in the 'molten globule' state[27]. Cyclophilin bound to GroEL thus seems to be fully unfolded, or in a low-energy molten globule state. The apparent absence of any stable secondary structure may be essential for the observed substrate promiscuity of GroEL[11-19]. The chaperone may interact with interior side-chains to shift the equilibrium towards an unfolded state. By disrupting all native structure at least transiently, it may thus direct folding towards the native state by favouring stable structures and thereby avoiding aggregation, regardless of the final topology of the substrate. □

Received 22 September 1993; accepted 4 January 1994.

1. Ellis, R. J. & van der Vies, S. M. *A. Rev. Biochem.* **60**, 321–347 (1991).
2. Gething, M.-J. & Sambrook, J. *Nature* **355**, 33–45 (1992).
3. Jaenicke, R. *Curr. Opin. struct. Biol.* **3**, 104–112 (1993).
4. Hendrix, R. W. *J. molec. Biol.* **129**, 375–392 (1979).
5. Hohn, T., Hohn, B., Engel, A., Wurtz, M. & Smith, P. R. *J. molec. Biol.* **129**, 359–373 (1979).
6. Hemmingsen, S. M. *et al. Nature* **333**, 330–334 (1988).
7. Gray, T. E. & Fersht, A. R. *FEBS Lett.* **292**, 254–258 (1991).
8. Bochkareva, E. S., Lissin, N. M., Flynn, G. C., Rothman, J. E. & Girshovich, A. S. *J. biol. Chem.* **267**, 6796–6800 (1992).
9. Jackson, G. S. *et al. Biochemistry* **32**, 2554–2563 (1993).
10. Zahn, R., Harris, J. R., Pfeifer, G., Plückthun, A. & Baumeister, W. *J. molec. Biol.* **229**, 579–584 (1993).
11. Goloubinoff, P., Christeller, J. T., Gatenby, A. A. & Lorimer, G. H. *Nature* **342**, 884–889 (1989).
12. Laminet, A. A., Ziegelhoffer, T., Georgopoulos, C. & Plückthun, A. *EMBO J.* **9**, 2315–2319 (1990).
13. Badcoe, G. *et al. Biochemistry* **30**, 9195–9200 (1991).
14. Buchner, J. *et al. Biochemistry* **30**, 1586–1591 (1991).
15. Höll-Neugebauer, B., Rudolph, R., Schmidt, M. & Buchner, J. *Biochemistry* **30**, 11609–11614 (1991).
16. Martin, J. *et al. Nature* **352**, 36–42 (1991).
17. Mendoza, J. A., Rogers, E., Lorimer, G. H. & Horowitz, P. M. *J. biol. Chem.* **266**, 13044–13049 (1991).
18. Viitanen, P. V., Donaldson, G. K., Lorimer, G. H., Lubben, T. H. & Gatenby, A. A. *Biochemistry* **30**, 9716–9723 (1991).
19. Zahn, R. & Plückthun, A. *Biochemistry* **31**, 3249–3255 (1992).
20. Langer, T., Pfeifer, G., Martin, J., Baumeister, W. & Hartl, F.-U. *EMBO J.* **11**, 4757–4765 (1992).
21. Handschumacher, R. E., Harding, M. W., Rice, J., Drugge, R. J. & Speicher, D. W. *Science* **226**, 544–547 (1984).
22. Schreiber, S. L. *Science* **251**, 283–287 (1991).
23. Wüthrich, K., Spitzfaden, C., Memmert, K., Widmer, H. & Wider, G. *FEBS Lett.* **285**, 237–247 (1991).
24. Wüthrich, K. *NMR of Proteins and Nucleic Acids* (Wiley, New York, 1986).
25. Landry, S. J. & Gierasch, L. M. *Biochemistry* **30**, 7359–7362 (1991).
26. Landry, S. J. Jordan, R., McMacken, R. & Gierasch, L. M. *Nature* **355**, 455–457 (1992).
27. Chyan, C.-L., Wormald, C., Dobson, C. M., Evans, P. A. & Baum, J. *Biochemistry* **32**, 5681–5691 (1993).
28. Spitzfaden, C., Wider, G., Widmer, H. & Wüthrich, K. *Abstr. XV Int. Conf. Magnetic Resonance in Biological Systems*, Jerusalem, 192 (1992).
29. Kraulis, P. J. *J. appl. Crystallogr.* **24**, 946–950 (1991).
30. Weber, C. *et al. Biochemistry* **30**, 6563–6574 (1991).
31. Fischer, G., Bang, H. & Mech, C. *Biomed. biochim. Acta* **43**, 1101–1111 (1984).
32. Pace, C. N. *Trends Biotechnol.* **8**, 93–98 (1990).

ACKNOWLEDGEMENTS. We thank K. Memmert (Sandoz) for the cyclophilin expression plasmid and discussions about the production of ^{15}N-labelled protein. The works was supported by the Deutsche Forschungsgemeinschaft, the Schweizerischer Nationalfonds and Sandoz, and a predoctoral fellowship from the Fonds der chemischen Industrie (to R.Z.).

Reprinted with permission from *Science*, Vol. 257, pp. 1559–1563 (1992).
Copyright © 1992 American Association for the Advancement of Science.

NMR Determination of Residual Structure in a Urea-Denatured Protein, the 434-Repressor

Dario Neri,* Martin Billeter, Gerhard Wider, Kurt Wüthrich†

A nuclear magnetic resonance (NMR) structure determination is reported for the polypeptide chain of a globular protein in strongly denaturing solution. Nuclear Overhauser effect (NOE) measurements with a 7 molar urea solution of the amino-terminal 63-residue domain of the 434-repressor and distance geometry calculations showed that the polypeptide segment 54 to 59 forms a hydrophobic cluster containing the side chains of Val^{54}, Val^{56}, Trp^{58}, and Leu^{59}. This residual structure in the urea-unfolded protein is related to the corresponding region of the native, folded protein by simple rearrangements of the residues 58 to 60. Based on these observations a model for the early phase of refolding of the 434-repressor(1-63) is proposed.

Investigations of proteins in strongly denaturing solution are of general interest relative to the protein folding problem (*1, 2*). For example, strongly denaturing solutions are often used as the "random-coil" reference state in refolding studies, and residual nonrandom structure found under such conditions might be indicative of nucleation sites for the refolding process. Structure determinations of partially or fully unfolded polypeptides are intrinsically difficult, and structural data on such species are scarce. Here we use the NMR method that is now in widespread use for studies of native, folded proteins (*3, 4*) for a structure determination of the urea-unfolded form of the amino-terminal 63-residue domain of the 434-repressor.

For a polypeptide consisting of the amino-terminal residues 1 to 69 of the 434-repressor, which includes the DNA-binding domain (*5*), the three-dimensional structure has been determined at high resolution by x-ray diffraction in single crystals (*6*) and by NMR in solution (*7*). The same molecular architecture with five α helices was observed in the crystals and in solution, and in both states the carboxyl-terminal peptide of residues 64 to 69 was found to be unstructured. Initial studies of urea denaturation showed that the native, folded form and an unfolded form coexist over a wide range of urea concentrations at pH 4.8 and 18°C (*8*). In the absence of urea, only the NMR spectrum of the folded protein is seen. In 4.2 M urea, the two forms are present in equal concentrations and have an exchange life time of ~1 s. In 7 M urea.

Institut für Molekularbiologie und Biophysik, ETH-Hönggerberg, CH-8093 Zürich, Switzerland.

*Present address: Cambridge Centre for Protein Engineering, Hills Road, Cambridge CB2 2QII, United Kingdom.
†To whom correspondence should be addressed.

only the NMR spectrum of the unfolded form is seen. Sequence-specific NMR assignments for the urea-unfolded 434 repressor(1-69) showed significant deviations of some ^{1}H NMR chemical shifts from the random-coil values (9), indicating the presence of residual nonrandom structure (8). Combination of the chemical shift data with observations on slowed backbone amide proton exchange with the solvent then led to the conclusion that the nonrandom structure was located in the polypeptide segment 45 to 60, whereas for the remainder of the polypeptide chain a nonglobular flexible form was implicated.

For further characterization of the residual nonrandom structure in the urea-unfolded protein, we cloned a new polypeptide consisting of the amino-N-terminal 63 residues of the 434-repressor, and overexpressed it with uniform ^{15}N labeling and with selective ^{13}C labeling of the methyl groups of Val and Leu. By proteolytic cleavage of the 434-repressor(1-63), we further prepared the identically isotope-labeled polypeptide 434-repressor(44-63). The previously described (*10*) sequence-specific NMR assignments showed that the ^{1}H chemical shift deviations from the random coil values (9) observed in 434-repressor(1-69) are also present in both of these shorter polypeptides. Moreover, identical behavior at variable urea concentrations as described above for the 434-repressor(1-69) was observed for the 434-repressor(1-63). We used heteronuclear editing of ^{1}H nuclear Overhauser exchange spectroscopy (NOESY) spectra recorded with the ^{15}N- and ^{13}C-labeled polypeptides (*11–15*) to collect a set of NOE distance constraints (Fig. 1) and found a sufficiently high density of constraints for the polypeptide segment 53 to 60 to warrant a structure calculation (Fig. 2).

The input data used for the structure calculation (Fig. 1) are detailed in Table 1. An upper limit of 5.0 Å was attributed to all observable NOEs (Table 1) except for two intraresidual NOEs in Leu^{59} and Leu^{60}, for which a limit of 3.0 Å was determined from comparison with NOEs between vicinal protons and between the two methyl groups in the same residues. This procedure is in line with our long-standing practice of estimating NOE distance constraints involving peripheral amino acid side chain protons by using the uniform averaging model to account for time variations of the proton-proton distances due to internal mobility of the protein (*3*, *16*). A more quantitative evaluation of the distance constraints was not warranted, considering that the data were collected from different spectra with different labeling of the heavy atoms (see legend to Fig. 1 for details). The usual correction of 1.0 Å was added to the upper distance limits when pseudoatoms were used to represent a methyl or methylene group (*17*). Five hundred calculations with the program DIANA (*18*) were started with random polypeptide conformations, and the 20 resulting conformers with the lowest residual target function values were further refined with the program AMBER (*19*) by using restrained energy minimization (*20*). These 20 energy-minimized DIANA conformers are used to represent the solution structure (Fig. 2). They contain no violations of NOE upper distance limits by more than 0.32 Å or of van der Waals constraints by more than 0.19 Å.

The result of the above structure determination with 434-repressor(1-63) shows that the polypeptide backbone of residues 54 to 59 and the orientations of the side chains of Val^{54}, Val^{56}, Trp^{58}, and Leu^{59} are well defined, with the isopropyl groups of all these Val and Leu residues in contact with the indole ring of Trp^{58} (Fig. 2). On the surface of the hydrophobic cluster formed by the side chains of residues 54, 56, 58, and 59, the hydrophilic side chains of Ser^{55} and Asp^{57} point toward the solvent and are disordered, and the backbone carbonyl oxygens also show a tendency of pointing outward toward the solvent. This visual impression of a well-defined local structure is confirmed by the root-mean-square deviations (RMSDs) of 0.74 Å for the backbone atoms of residues 54 to 59 and 1.07 Å for the backbone and the four hydrophobic side chains (Table 2). Moreover, the notion that Fig. 2 represents a meaningful structure determination for the urea-unfolded protein is supported by comparison with a corresponding group of 20 conformers calculated with identical techniques but without use of any NOE upper distance constraints in the input. These conformers have much larger RMSD values of 2.08 Å for the backbone atoms which increases to 3.39 Å when the hydrophobic side chains are included (Table 2). As an additional reference, Table 2 shows also that the same polypeptide segment has significantly smaller RMSD values in the NMR structure of the native 434 repres-

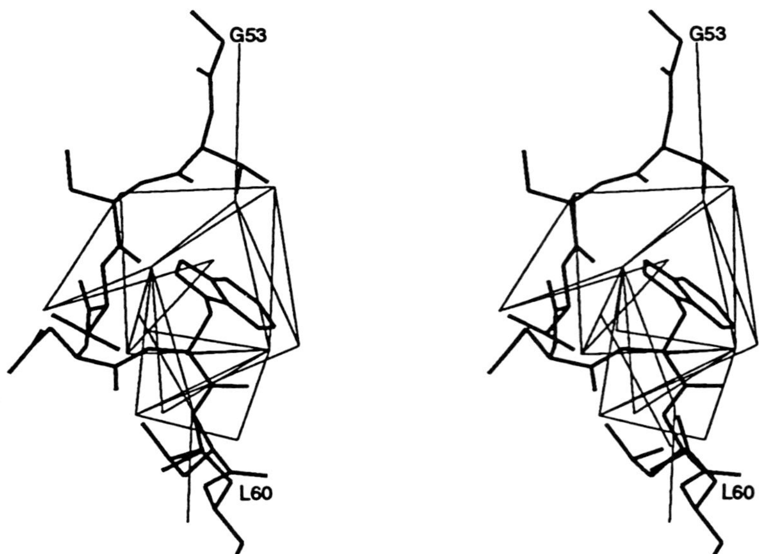

Fig. 1. Survey of the NMR constraints used as input for the presently reported structure determination. The figure shows an all–heavy-atom stereo view of one of the conformers of the polypeptide segment 53 to 60 from 434-repressor shown in Fig. 2, which were calculated from NMR data measured in 7 M aqueous urea solutions of 434-repressor(1-63) and 434-repressor(44-63) at pH 4.8 and 18°C (*25*). Thin lines connect those pairs of hydrogen atom positions [or pseudoatom positions representing the methyl groups (*17*)] for which a NOE distance constraint was measured and used in the input for the structure calculation. The NOEs were collected from five two-dimensional NMR spectra: heteronuclear-resolved ^{13}C(ω_2)-half-filter ^{1}H,^{1}H-NOESY (*15*) with the two protein fragments enriched with ^{13}C at the Val and Leu methyl positions; NOE-relayed ^{15}N,^{1}H-COSY (*26*) with the two uniformly ^{15}N-labeled polypeptides; and ^{1}H,^{1}H-NOESY of the unlabeled 434-repressor(44-63). A mixing time of 300 ms was used, which was sufficiently short to avoid significant spin diffusion, as evidenced by the following controls: (i) for the indole NH proton of Trp^{58}, Nϵ^1H, intraring NOEs were observed only to δ^1H and ζ^2H; (ii) the NOE buildup curves were to a good approximation linear for mixing times in the range from 0 to 450 ms; and (iii) in the Leu residues, largely different NOE intensities were observed between the α proton and the δ^1 or δ^2 methyl group, respectively.

sor(1-69), which was obtained at high quality from a carefully quantitated input of NOE distance constraints and dihedral-angle constraints (*7*).

The approach used for the structure determination in Fig. 2 was based on the fact that the NMR spectrum at 7 M urea contains only one resonance line for each proton (*8, 10*), which shows that all protein conformations present in this solution are in rapid exchange on the chemical-shift time scale (*3*). Because a quantitative evaluation of the NOE intensities was not warranted (see above) and in particular no absolute NOE intensities were obtained, the NOE data provide no direct evidence for the extent of the population of the structure shown in Fig. 2, except that the NOESY cross peaks would be undetectably weak if the population of the folded conformation were less than ~10%. However, the observation that the data of Table 1 can be satisfied by a single, quite well-defined structure (Fig. 2 and Table 2) implies that whatever the population of this structure is, all other conformations present in the solution do not significantly contribute to the NOEs in Table 1. In other words, because the NOE intensity decreases with the inverse sixth power of the distance between two protons (*3*), the molecular conformations in which a given ^1H-^1H distance exceeds 5.0 Å do not contribute to the overall intensity of the corresponding NOE. This averaging is in contrast to the observed chemical shifts, which represent an average of the chemical shifts in all conformations present and the contribution from each molecular species is simply weighted by its population in the solution studied. This different averaging of NOE intensities and chemical shifts (*21*) is manifest in the result of the present structure determination: Although the previously reported deviations of the ^1H chemical shifts in the polypeptide segment 53 to 60 from the random coil values (*8, 10*) can be qualitatively rationalized by the close approach of the aliphatic side chains to the indole ring of Trp^{58} (Fig. 2), ring current calculations for the methyl protons in the structure of Fig. 2 predict larger deviations from the random-coil shifts (9) than those observed (*8, 10*), with the largest discrepancies for γ^2CH_3 of Val^{54}, γ^1CH_3 of Val^{56}, and δ^2CH_3 of Leu^{59}. This result indicates that the observed chemical shifts include important contributions from fully unfolded polypeptides with random coil shifts. On this basis, the chemical shift measurements provide presently the most direct indication that in 7 M urea solution the structure of Fig. 2 is present in an equilibrium with conformers that do not contain a compact structure involving the residues 54 to 59.

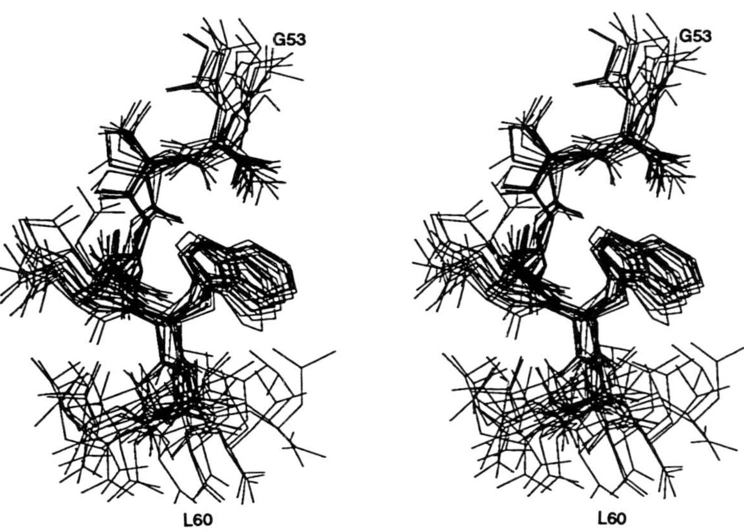

Fig. 2. Result of the structure calculations from the input data of Table 1 for the polypeptide segment Gly^{53}-Val-Ser-Val-Asp-Trp-Leu-Leu^{60} of the 434-repressor(1-63) unfolded in 7 M urea at pH 4.8 and 18°C. The figure shows an all–heavy-atom stereo view of the 20 conformers selected to represent the solution structure. The conformers 2 through 20 were superimposed with conformer 1 for minimal RMSD of the backbone atoms of residues 54 to 59.

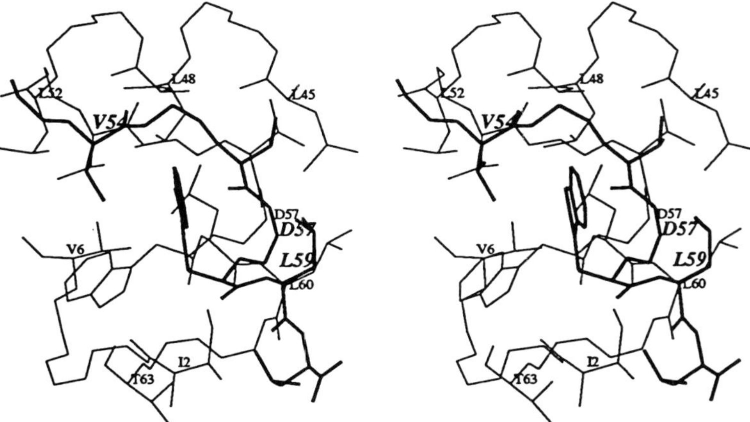

Fig. 3. Stereo view of the all–heavy-atom presentation of one of the conformers from Fig. 2 of the segment Gly^{53}-Val-Ser-Val-Asp-Trp-Leu-Leu^{60} in the urea-unfolded 434-repressor(1-63) (thick lines, bold labels) superimposed with fragments of the native solution structure of the 434-repressor(1-69) (*7*) (thin lines, small labels). The two structures were superimposed for optimal fit of the backbone atoms of residues 54 to 57. For the native structure, the backbone is drawn for the residues 2 to 6 and 45 to 63 and the side chains are included for Ile^2, Val^6, Leu^{45}, Leu^{48}, Ala^{49}, Leu^{52}, residues 53 to 60, and Thr^{63}.

The structure of residues 54 to 59 in urea-unfolded 434-repressor are compared with the corresponding peptide segment in the folded protein (*6, 7*) in Fig. 3. In the two species, a similar loop is formed by the backbone atoms of residues 54 to 57 and the side chains of Val^{54} and Val^{56} are in nearly identical locations. Because of the formation of helix 5 in the native structure, the indole ring of Trp^{58} in the hydrophobic cluster formed with residues 54, 56, and 59 in the unfolded form is replaced by the side chain of Leu^{59}, the space of which is in turn occupied by Leu^{60}. In this way the hydrophobic cluster observed in the urea-unfolded form is essentially preserved in the folded protein. Similarity of the folded and urea-unfolded forms of the 434-repressor in the region of residues 54 to 59 is also evidenced by the fact that the RMSD values between these two structures are less than those between the urea-unfolded protein and a randomized form, especially when the hydrophobic side chains are included (Table 2). Figure 3 shows further that the polypeptide fold in the globular form of the 434-repressor(1-69) (6, 7) brings the side chains of Val^{54}, Val^{56}, Leu^{59}, and Leu^{60} in contact with the apolar residues Ile^{2}, Val^{6}, Leu^{45}, Leu^{48}, Ala^{49}, and Leu^{52}. The local hydrophobic cluster relating to that seen in the urea-unfolded form of the protein is thus expanded to a global hydrophobic core by interactions of residues that are further apart in the amino acid sequence. Such longer range interactions also ensure the formation of a predominantly hydrophobic environment for Trp^{58} in the folded protein.

From the experience gained with the presently described project, two conclusions can be drawn relating to NMR experiments that are frequently used with proteins. First, the importance of having obtained individual, stereospecific assignments for all methyl resonances of Val and Leu (8, *10*) cannot be overemphasized. Otherwise, the hydrophobic cluster formed by residues 54, 56, 58, and 59 could hardly have been characterized. For unfolded forms of proteins these stereospecific assignments can in practice only be obtained by using the method of biosynthetically directed fractional ^{13}C labeling (*22*). Second, considering that the 434-repressor(1-63) is a protein devoid of disulfide bonds or prosthetic groups that might stabilize residual local structure (5), the observation of nonrandom structure in 7 M urea implies that it may be difficult with certain proteins to prepare a fully random reference state, for example, for the analysis of amide proton exchange measurements during refolding (*23*). Extreme care should be exercised to distinguish between experimental manifestations of such residual local structure in the denatured reference state and manifestations of the formation of folding intermediates after initiation of the refolding process.

Finally, it is tempting to speculate on possible folding pathways starting from the polypeptide conformation observed in 7 M urea, that is, an ensemble of nonglobular flexible forms of the polypeptide chain including a sizeable population of conformers containing a local nonrandom hydrophobic cluster. Figure 4 illustrates that the polypeptide chain in the folded, globular state of the 434-repressor(1-63) (6, 7) forms two subdomains containing, respectively, the helices 1 to 3 (residues 1 to 36) and the helices 4 and 5 (residues 45 to 60). The two subdomains are covalently linked by a somewhat flexible loop (7). Formation of this structure could be initiated by the strictly localized "nucleation cluster" with the side chains of residues 54, 56, 58, and 59 (Fig. 2). In a subsequent step, the helix 57 to 60 would be formed, resulting in the replacements of Trp^{58} by Leu^{59} and of Leu^{59} by Leu^{60} in the nucleation cluster (Fig. 3). As another event following the

Table 1. Input for the calculation of the residual structure of the fragment 53 to 60 in the 434-repressor(1-63) dissolved in water containing 7 M urea at pH = 4.8 and 18°C.

NOE distance constraints		
Atom 1*	Atom 2*	Upper limit [Å]†
Gly^{53} αCH$_2$	Val^{54} γ^2CH$_3$	7.0
Val^{54} γ^1CH$_3$	Ser^{55} αH	6.0
Val^{54} γ^1CH$_3$	Trp^{58} ζ^3H	6.0
Val^{54} γ^1CH$_3$	Trp^{58} ϵ^3H	6.0
Val^{54} γ^1CH$_3$	Trp^{58} δ^1H	6.0
Val^{54} γ^2CH$_3$	Trp^{58} ζ^3H	6.0
Val^{54} γ^2CH$_3$	Trp^{58} ϵ^3H	6.0
Val^{54} γ^2CH$_3$	Trp^{58} δ^1H	6.0
Ser^{55} αH	Val^{56} γ^1CH$_3$	6.0
Ser^{55} αH	Val^{56} γ^2CH$_3$	6.0
Val^{56} γ^1CH$_3$	Trp^{58} ζ^2H	6.0
Val^{56} γ^1CH$_3$	Trp^{58} ϵ^3H	6.0
Val^{56} γ^1CH$_3$	Trp^{58} δ^1H	6.0
Val^{56} γ^2CH$_3$	Trp^{58} ζ^2H	6.0
Val^{56} γ^2CH$_3$	Trp^{58} δ^1H	6.0
Val^{56} αH	Leu^{59} δ^1CH$_3$	6.0
Trp^{58} HN	Trp^{58} ϵ^3H	5.0
Trp^{58} HN	Trp^{58} δ^1H	5.0
Trp^{58} αH	Trp^{58} ϵ^3H	5.0
Trp^{58} δ^1H	Leu^{59} δ^2CH$_3$	6.0
Trp^{58} ϵ^3H	Leu^{59} HN	5.0
Trp^{58} δ^1H	Leu^{59} HN	5.0
Trp^{58} ϵ^3H	Leu^{59} αH	5.0
Trp^{58} ζ^3H	Leu^{59} δ^2CH$_3$	6.0
Trp^{58} ϵ^3H	Leu^{59} δ^2CH$_3$	6.0
Trp^{58} δ^1H	Leu^{60} δ^2CH$_3$	6.0
Leu^{59} αH	Leu^{59} δ^2CH$_3$	4.0
Leu^{60} αH	Leu^{60} δ^2CH$_3$	4.0

*These two columns identify the pairs of protons for which an NOE has been observed. The protons are identified by the three-letter amino acid code, the sequence position, and the location in the amino acid residue in the standard IUB-IUPAC nomenclature. †Upper limit to the proton-proton distance after correction for use of pseudoatoms to represent methylene and methyl groups (*17*). The lower limit was set equal to the sum of the van der Waals radii.

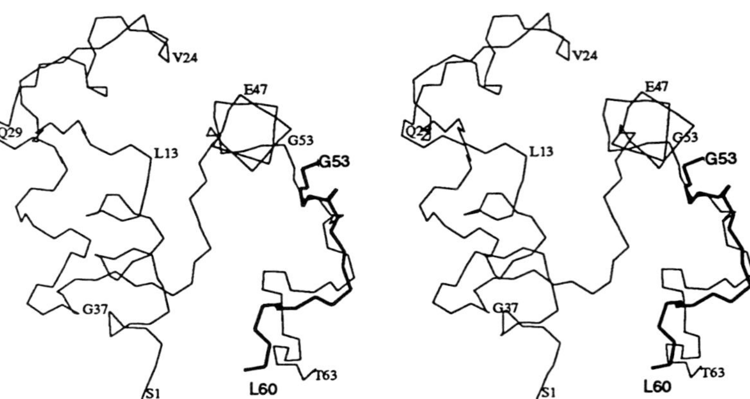

Fig. 4. Stereo view of the polypeptide backbone for the segment 1 to 63 in the NMR solution structure of native 434-repressor(1-69) (thin line, small labels) (*7*). Also drawn with a thick line and bold labels is the backbone of the segment 53 to 60 of the urea-unfolded 434-repressor(1-63). The two structures were superimposed for optimal fit of the backbone atoms of residues 54 to 57.

Table 2. Comparisons of the polypeptide segment 53 to 60 in the urea-unfolded 434-repressor(1-63) with the corresponding segment in native 434-repressor(1-69) and in a hypothetical random-coil form.

Atoms compared*	RMSD (Å)				
	⟨Urea⟩†	⟨Random⟩‡	⟨Native⟩§	⟨Urea⟩-⟨native⟩‖	⟨Urea⟩-⟨random⟩¶
Backbone of residues 54 to 59	0.74	2.08	0.17	1.67	2.01
Same + side chains of 54, 56, 58, and 59	1.07	3.39	0.31	2.80	4.13

*The backbone atoms used for calculating the RMSDs are N, Cα, and C′. †Group of 20 energy-minimized DIANA conformers used to represent the solution structure in 7 M urea. This column lists the average of the pairwise RMSDs in this group of conformers. ‡Group of 20 energy-minimized DIANA conformers calculated with identical procedures as ⟨urea⟩ and ⟨native⟩, except that no NOE upper distance constraints were used. This column lists the average of the pairwise RMSDs in this group of conformers. §Group of 20 energy-minimized DIANA conformers representing the native solution structure of 434-repressor(1-69) (*7*). This column lists the average of the pairwise RMSDs in this group of conformers. ‖Average of the pairwise RMSDs for all combinations of conformers in ⟨urea⟩ with those in ⟨native⟩. ¶Average of the pairwise RMSDs for all combinations of conformers in ⟨urea⟩ with those in ⟨random⟩.

folding of the nucleation cluster, formation of the helix 4 would induce a "hydrophobic collapse" (*24*) in the subdomain on the right of Fig. 4, resulting in a more extensive hydrophobic core around the nucleation cluster (Fig. 3). In further steps, hydrophobic contacts with the first subdomain mediated by the apolar side chains of Leu^{48}, Leu^{52}, and Leu^{59} would lead to further growth of the hydrophobic core and proper spatial positioning of the two subdomains (Fig. 4). Although this selection of distinct folding events derived from combined inspection of the structures of folded and urea-unfolded 434-repressor(1-63) is largely hypothetical, it may provide a platform for additional experiments to investigate spatial and temporal patterns of the order in which distinct, individual folding events take place.

REFERENCES AND NOTES

1. F. M. Richards, *Sci. Am.* **264**, 54 (January 1991).
2. C. M. Dobson, *Curr. Opin. Struct. Biol.* **1**, 22 (1991).
3. K. Wüthrich, *NMR of Proteins and Nucleic Acids* (Wiley, New York, 1986).
4. ______, *Science* **243**, 45 (1989).
5. J. Anderson, M. Ptashne, S. C. Harrison, *Proc. Natl. Acad. Sci. U.S.A.* **81**, 1307 (1984).
6. A. Mondragon, S. Subbiah, S. C. Almo, M. Drottar, S. C. Harrison, *J. Mol. Biol.* **205**, 189 (1989).
7. D. Neri, M. Billeter, K. Wüthrich, *ibid.* **223**, 743 (1992).
8. D. Neri, G. Wider, K. Wüthrich, *Proc. Natl. Acad. Sci. U.S.A.* **89**, 4397 (1992).
9. A. Bundi and K. Wüthrich, *Biopolymers* **18**, 285 (1979).
10. D. Neri, G. Wider, K. Wüthrich, *FEBS Lett.* **303**, 129 (1992).
11. G. Otting, H. Senn, G. Wagner, K. Wüthrich, *J. Magn. Reson.* **70**, 500 (1986).
12. S. W. Fesik, *Nature* **332**, 865 (1988).
13. R. H. Griffey and A. G. Redfield, *Q. Rev. Biophys.* **19**, 51 (1987).
14. G. Otting and K. Wüthrich, *ibid.* **23**, 39 (1990).
15. G. Wider, C. Weber, K. Wüthrich, *J. Am. Chem. Soc.* **113**, 4676 (1991).
16. W. Braun, Ch. Bösch, L. R. Brown, N. Gō, K. Wüthrich, *Biochim. Biophys. Acta* **667**, 377 (1981).
17. K. Wüthrich, M. Billeter, W. Braun, *J. Mol. Biol.* **169**, 949 (1983).
18. P. Güntert, W. Braun, K. Wüthrich, *ibid.* **217**, 517 (1991).
19. U. C. Singh, P. K. Weiner, J. W. Caldwell, P. A. Kollman, *AMBER 3.0* (University of California, San Francisco, 1986).
20. M. Billeter, Th. Schaumann, W. Braun, K. Wüthrich, *Biopolymers* **29**, 695 (1990).
21. In this context, one should also recall that only a short lifetime of ~1 ns is needed for a protein conformation to give rise to negative NOEs [G. Otting, E. Liepinish, K. Wüthrich, *Science* **254**, 974 (1991)], as observed in the urea-unfolded 434-repressor(1-63) and 434-repressor(44-64). In contrast, chemical shift averaging is typically observed between different conformers even when they have a life time of the order of 1 ms.
22. H. Senn *et al.*, *FEBS Lett.* **249**, 113 (1989); D. Neri, Th. Szyperski, G. Otting, H. Senn, K. Wüthrich, *Biochemistry* **28**, 7510 (1989).
23. H. Roder and K. Wüthrich, *Proteins* **1**, 34 (1986); H. Roder, G. A. Elove, S. W. Englander, *Nature* **335**, 700 (1988); J. B. Udgaonkar and R. L. Baldwin, *ibid.*, p. 694.
24. H. S. Chan and K. A. Dill, *Proc. Natl. Acad. Sci. U.S.A.* **87**, 6388 (1990).
25. Abbreviations for the amino acid residues are: A, Ala; C, Cys; D, Asp; E, Glu; F, Phe; G, Gly; H, His; I, Ile; K, Lys; L, Leu; M, Met; N, Asn; P, Pro; Q, Gln; R, Arg; S, Ser; T, Thr; V, Val; W, Trp; and Y, Tyr.
26. G. Otting and K. Wüthrich, *J. Magn. Reson.* **76**, 569 (1988).

27. Supported by the Schweizerischer Nationalfonds (project no. 31.25174.88). We thank R. Marani for the careful processing of the manuscript.

1 May 1992; accepted 15 July 1992

NMR assignments as a basis for structural characterization of denatured states of globular proteins

Kurt Wüthrich

ETH, Zürich, Switzerland

NMR has unique potential for detailed structural characterization of denatured states of proteins, provided that workable spectral resolution and sequence-specific assignments can be obtained in spite of the lack of conformation-dependent dispersion of the ^{1}H chemical shifts. Structures can be determined in the presence of dynamic conformational polymorphism. This has recently been achieved for urea-unfolded bacteriophage 434 repressor, and NMR assignments were determined for urea-unfolded FK 506-binding protein.

Current Opinion in Structural Biology 1994, **4**:93–99

Introduction

NMR spectroscopy is well established as a method for protein and nucleic acid structure determination [1]. Although most powerful for use with small globular proteins, the fact that NMR structures are determined in solution makes this technique attractive for studies of proteins in denaturing solvents. This review focuses on structural characterization of unfolded states of proteins based on complete NMR assignments. Such data have recently been reported for urea-unfolded forms of bacteriophage 434 repressor [2•–4•] and FK 506-binding protein (FKBP) [5•]. As the material for review is thus still scarce, I have included some historical background to place these studies in perspective with other NMR investigations.

Interest in structural characterization of proteins in strongly denaturing environments is related to the use of such protein solutions as the 'random-coil' reference in folding studies. Meaningful analysis of the results of such investigations should be based on knowledge of the unfolded structure so that, for example, one is able to distinguish between manifestations of residual spatial structure in the denatured state and those of 'early folding events'. Furthermore, residual non-random structure found under strongly denaturing solution conditions might be indicative of nucleation sites for the refolding process. As an illustration Fig. 1 and 2 show a hydrophobic cluster of 434 repressor in 7 M urea, and its relation to the conformation of the corresponding polypeptide segment in the native protein [2•]. The key observation is that the transition from the denatured to the native conformation can be rationalized by the rearrangement of the backbone dihedral angles ϕ and ψ of residues 57–60, which is required to form helical secondary structure as it was observed in the native protein (Fig. 2). Starting from the structure in 7 M urea (Fig. 1), the generation of this helical backbone structure moves Trp58 out of the hydrophobic cluster, where it is replaced by Leu59. The position occupied by Leu59 in the urea-unfolded form is in turn assumed by Leu60, which is not part of the hydrophobic cluster in 7M urea (Figs 1 and 2). Considering that in the native protein, Trp58 forms the core of a hydrophobic cluster involving sequentially distant residues (Fig. 2), it is tempting to speculate that a 'hydrophobic collapse' leading to the local structure of Fig. 1 and subsequent rearrangement due to the formation of the helix 56–60 are indeed early folding steps.

In general, although any succession of distinct folding events proposed on the basis of combined inspection of folded and unfolded forms of a protein is largely hypothetical, it provides a platform for devising additional experiments to investigate the order of spatial and temporal patterns in which distinct, individual folding events take place.

Historical perspective

In 1967, McDonald and Phillips [6] demonstrated the phenomenon of ^{1}H chemical shift dispersion in globular proteins. This dispersion ensures that multiple

Abbreviations

2D—two dimensional; **3D**—three dimensional; **COSY**—correlation spectroscopy; **FKBP**—FK 506-binding protein; **NOE**—nuclear Overhauser enhancement; **NOESY**—nuclear Overhauser enhancement spectroscopy; **RMSD**—root mean square deviation; **TOCSY**—total correlation spectroscopy.

© Current Biology Ltd ISSN 0959-440X

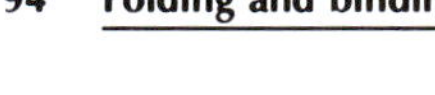

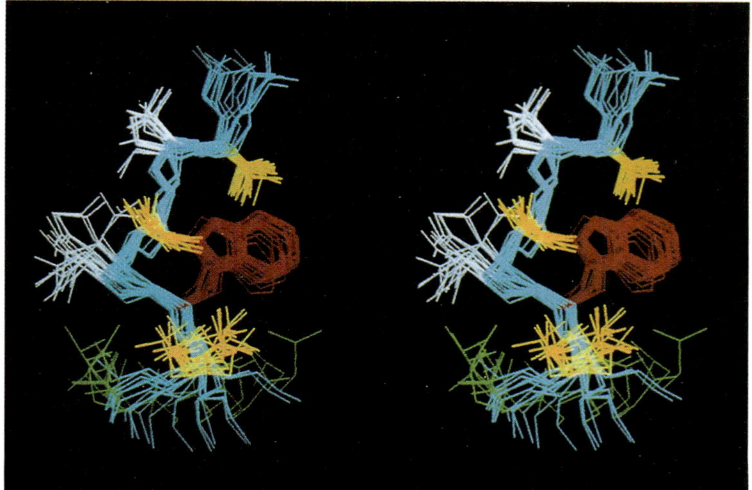

Fig. 1. Stereo view of an all heavy atom presentation of the 20 conformers selected to represent the solution structure of the polypeptide segment Gly53-Val-Ser-Val-Asp-Trp-Leu–Leu60 of the 434 repressor unfolded in 7 M urea at pH 4.8 and 18°C. The conformers 2 through 20 were superimposed with conformer 1 for minimal RMSD of the backbone atoms of residues 54 to 59. Color code: blue, polypeptide backbone; red, Trp58; yellow, Val54, Val56 and Leu59; green, Leu60; white, Ser55 and Asp57.

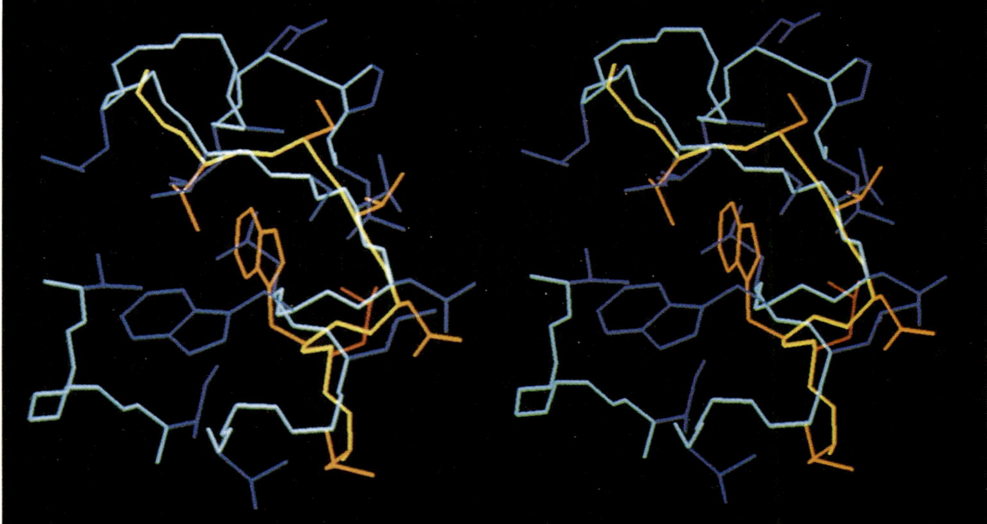

Fig. 2. Stereo view of a superposition of the all heavy atom presentations of one of the conformers from Fig. 1 of the segment Gly53-Val-Ser-Val-Asp-Trp-Leu-Leu60 in the urea-unfolded 434 repressor with fragments of the native solution structure of the protein [17]. The two structures were superimposed for optimal fit of the backbone atoms of residues 54 to 57. For the native structure, the backbone is drawn for the residues 2–6 and 45–63, and side chains are included for Ile2, Val6, Leu45, Leu48, Ala49, Leu52, residues 53–60, and Thr63. Colour code: yellow, polypeptide backbone of residues 53–60 in the unfolded state; light blue, backbone of all residues shown for the native state; orange, side chains in the unfolded state; dark blue, side chains in the native state.

copies of the same amino acid residue in a polypeptide chain can be observed as separate resonance lines, since the chemical shifts for each copy deviate differently from the random coil shifts [1]. On this basis, NMR structure determinations of folded proteins with molecular weights up to about 10 kDa can be performed using homonuclear ^{1}H NMR [1]. With increasing protein size the incidence of accidental degeneracy of two or more ^{1}H NMR lines becomes more frequent, and these degeneracies are nowadays routinely resolved by ^{13}C and/or ^{15}N editing of the ^{1}H NMR spectra of recombinant, uniformly isotope-labeled proteins [7].

When working with unfolded states of proteins one has at best very small ^{1}H chemical shift dispersion. As a consequence, the analysis of the ^{1}H NMR spectra for relatively small unfolded polypeptides may be as limited as it is for much larger globular proteins. These inherent difficulties have been mastered in work with short linear oligopeptides, for example, with model studies on 'nascent secondary structures' [8], or with polypeptide fragments from folded proteins with or without addition of structure-inducing solvents [9,10]. For the systems of central interest in this review, however, one has to resort to heteronuclear NMR techniques of the type used for larger, folded proteins [7]. This direct approach of resonance assignment by heteronuclear scalar correlations was first applied to the nonglobular natural polypeptide Apamin by Bystrov and colleagues [11]. Alternatively, sequence-specific assignments for the unfolded polypeptide chain may result from magnetization transfer with the folded protein under solution conditions where both states coexist in an equilibrium. This approach was originally introduced for transferring resonance assignments between paramagnetic states (with large chemical shift dispersion) and diamagnetic states of metalloproteins [12]. Before the advent of sequential resonance assignments it was used for the identification of resonance lines in folded proteins by reference to the random coil shifts in an unfolded state [13].

NMR assignment of unfolded proteins

I shall first summarize the two different approaches described in references [5•], and [3•,4•], respectively, and then discuss the relative merits of the two strategies.

In the 'direct' assignment approach, FKBP labeled uniformly with ^{15}N and ^{13}C was dissolved in 6.3 M urea at pH 6.3, where FKBP is 'fully denatured' [5•]. The assignments were then obtained entirely *via* heteronuclear scalar correlation experiments — that is, all side chain and backbone proton and carbon resonances were correlated to the amide ^{15}N and ^{1}H frequencies, which have the highest spectral dispersion in denatured proteins (see Fig. 4). Initially, the individual amino acid ^{1}H spin systems were identified in a 3D ^{15}N-correlated [^{1}H,^{1}H]-total-correlation-spectroscopy (TOCSY) experiment [14]. Thanks to the favorable relaxation times in denatured proteins, nearly all non-labile proteins in each spin system can usually be correlated with the amide resonances (see Fig. 4). The individual spin systems were then sequentially linked with a 3D H(C)(CO)NH-TOCSY experiment, in which the entire spin system of the residue is correlated with the amide ^{1}H and ^{15}N resonances, and finally, the ^{13}C chemical shift assignments were obtained from 3D (H)C(CO)NH-TOCSY [15,16].

The investigation of the unfolded 434 repressor started from complete assignments for the native state (Fig. 3a) [17], using uniformly ^{15}N-labeled protein. A complete ^{15}N–^{1}H fingerprint [1] of the urea-unfolded state was observed in 7M urea at pH 4.8 and 18°C (Fig. 3c). At 4.2 M urea, the fingerprints of native (Fig. 3a) and unfolded (Fig. 3c) protein were present with approximately equal intensities (Fig. 3b). In a first step, using the protein solution of Fig. 3b, the amide ^{15}N and ^{1}H assignments of the native protein were transferred to the unfolded form by exchange of heteronuclear longitudinal two-spin order in a 2D difference [^{15}N,^{1}H]-correlation spectroscopy (COSY) experiment [18]. In the second step, using the unfolded protein of Fig. 3c, the sequence-specific assignments of the amide ^{15}N and ^{1}H resonances were extended to the complete ^{1}H spin systems of the individual amino acid residues with a TOCSY-relayed [^{15}N,^{1}H]-COSY experiment (Fig. 4) [19]. Fig. 4 nicely illustrates that the complete amino acid spin systems can usually be observed in ^{1}H TOCSY experiments with unfolded proteins, and that NMR studies with unfolded proteins rely critically on the spectral resolution of the amide ^{15}N chemical

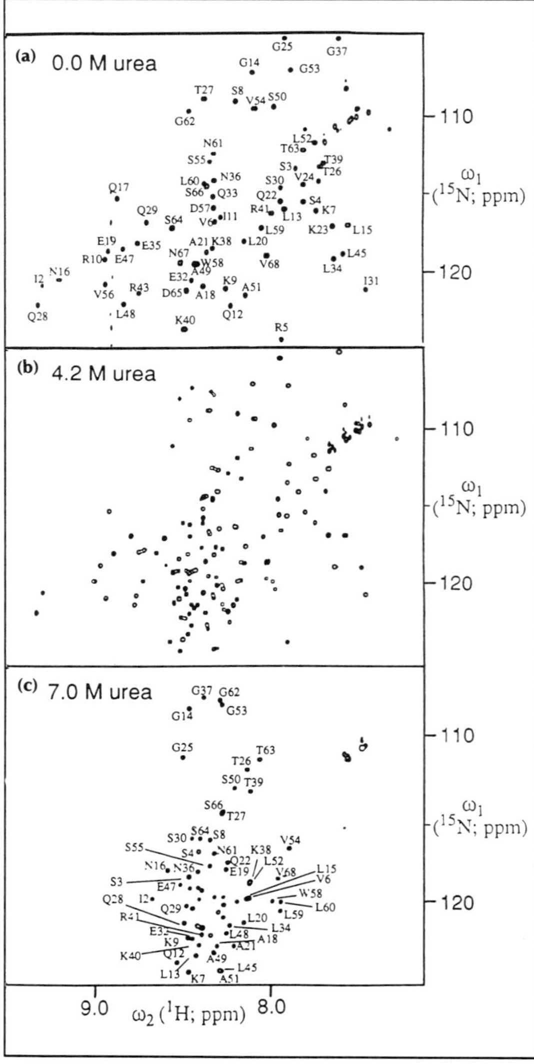

Fig. 3. (a) Two-dimensional [^{15}N,^{1}H]-COSY spectrum of the uniformly ^{15}N-labeled 434 repressor (^{1}H frequency = 600 MHz, 3 mM protein concentration, mixed solvent of 90 % H_2O/10 % $^{2}H_2O$ containing 20 mM $NaClO_4$ at pH 4.8, T = 18°C). The assignments are given by the one-letter amino acid code and the sequence positions. Phe44 and Arg69 lie outside of the plotted region. (b) Same as in (a), but the sample contained 4.2 M urea. (c) Same as in (a) but the sample contained 7.0 M urea, so that only the urea-unfolded form is present. Some sequence-specific resonance assignments are indicated as in (a); for the other peaks, the assignments can be found in ref. [3•]. (Reproduced with permission from [3•]).

shifts. Note, for example, that Fig. 4 contains 10 spin systems of Glu and Gln, which all have nearly identical ^1H chemical shifts but are well separated along the ^{15}N shift axis ω_1. For parts of the polypeptide chain, the sequential NOE connectivities [1] could be identified in a 2D NOESY-relayed [^{15}N,^1H]-COSY experiment [19] recorded in 7M urea (Fig. 3c), which also resulted in sequence-specific assignments of the Pro residues.

For both assignment procedures, applications to unfolded proteins rely critically on the sequence-dependent dispersion of the amide ^{15}N chemical shifts [20] (D Braun, G Wider and K Wüthrich, unpublished data), and hence the two approaches should have similar limitations relative to the size-range of the proteins to be studied. The direct approach has the conceptual advantage that it does not require knowledge of the NMR assignments for the native state, or depend on suitable interconversion rates between the native and unfolded forms (see below). When starting from the assignments for the native protein, the unfolded form can be assigned with less expense of isotope labeling and NMR measurement time, but its use will probably be limited to two-state systems and depends on finding conditions where the interconversion rate between folded and unfolded states is sufficiently slow for observation of separate spectra (Fig. 3b), yet fast enough to ensure that magnetization is not lost by relaxation during the transfer. These conditions are usually met when the interconversion lifetimes are comparable to the nuclear spin relaxation times. The strongest argument in favour of the latter procedure, besides the fact that it is technically quite straightforward, is that for many globular proteins the available literature indicates that it is possible to find combinations of temperature, denaturant concentration and pH at which the native and denatured forms coexist, and the interconversion rates are suitable for exchange studies by NMR [13,21,22]. These systems are also potentially of interest for structural characterization and other physico-chemical measurements based on sequence-specific NMR assignments.

Three-dimensional structure determination of unfolded proteins from NMR data

In determining the three-dimensional structure of unfolded proteins one generally faces the situation that the NMR spectrum of the unfolded state represents an average over an ensemble of molecular conformations that interconvert rapidly relative to the chemical shift time scale [1] (Fig. 5). As a result, the observed chemical shifts are the population-weighted arithmetic average of the shifts in the individual interconverting conformations **S** and **U**$^{\mathbf{i}}$, the spin-spin couplings are the population-weighted average calculated using a Karplus-type dependence on the dihedral angles, and the nuclear Overhauser enhancement (NOE) intensities are averaged with $1/r_{ij}^6$, where r_{ij} is the distance between two protons i and j in a given conformation, again weighted by the populations and possibly further by different correlation functions [1]. Procedures using time-averaged NOEs or spin-spin coupling constants

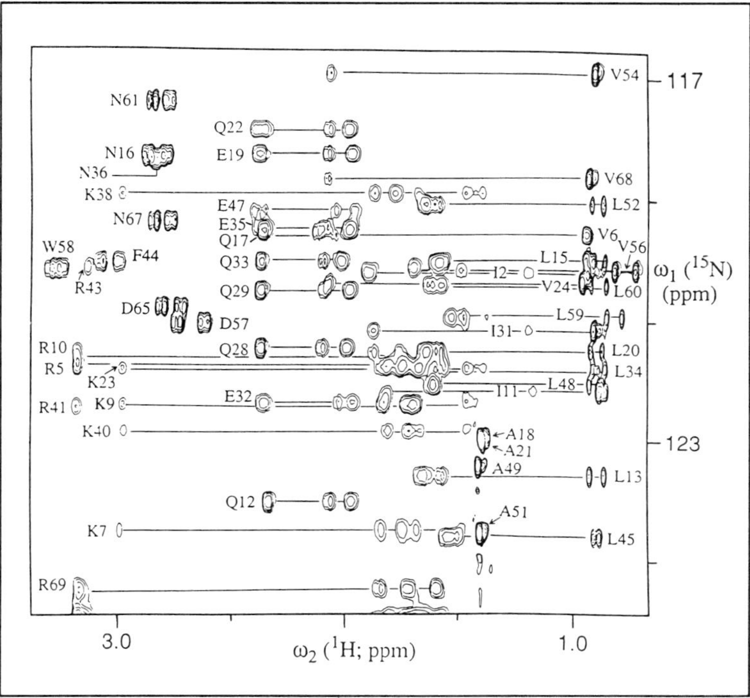

Fig. 4. Spectral region (ω_1 = 116.8–125.9 ppm; ω_2 = 0.7–3.4 ppm) of a TOCSY-relayed [^{15}N,^1H]-COSY spectrum recorded with uniformly ^{15}N-labeled 434 repressor (same conditions as in Fig. 3c), containing the amino acid side chain ^1H-TOCSY peaks along ω_2 at the amide ^{15}N chemical shifts along ω_1. Relayed peaks corresponding to protons that belong to the same spin system are connected by horizontal lines, and the amino acid residue is indicated with the one-letter amino acid code and the sequence position. To provide an impression of the different typical peak patterns present in the spectrum, the labels for the different amino acid types are placed near the relayed peaks with the most characteristic proton chemical shift; for example, the β-CH_3 resonance of Ala, the γ-CH_2 resonance of Glu and Gln, or the ε-CH_2 resonance of Lys. (Reproduced with permission from [3•].)

for NMR structure determinations have been discussed [23,24], but these consider averaging within the conformation space occupied by a 'structured' polypeptide chain (Fig. 1) and would hardly be practical for the treatment of systems with multiple, largely different interconverting conformations. In practice, for example in the identification of 'nascent secondary structures' in short linear polypeptides by NOE pattern recognition [1], the interpretation of the NMR data was based on the assumption that contributions to NOE intensities from 'unstructured' molecules ($\mathbf{U}^i$ in Fig. 5) was negligibly small [9]. Measures taken to use a similar approximation for the determination of the structure in Fig. 1 are described below.

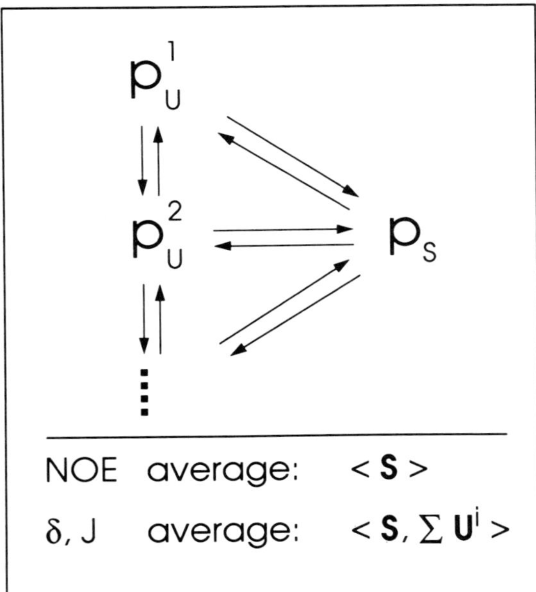

Fig. 5. Upper part: Conformational equilibria in a solution of an 'unfolded' protein. $\mathbf{p}_S$ is the population of conformers containing a localized non-random structure, such as the one shown in Fig. 1. $\mathbf{p}_U^i$ are the populations of different conformers i which do not contain the local non-random structure **S**. Lower part: Averaging of the parameters NOE, chemical shift, δ, and spin-spin coupling constants, J, in the limiting situation where interconversion among all conformations is rapid on the chemical shift time scale (life times in any of the conformation states ≤ 1 ms) so that a single NMR spectrum is observed for the ensemble of all conformers **S** and **U**. It is further assumed that all ^{1}H-^{1}H distances related to NOEs in the conformer **S** are so long in all conformers $\mathbf{U}^i$ that only the conformer **S** contributes to the observed NOE intensities (see text). The brackets, < >, indicate rapid averaging within the conformation space attributed to the structure **S** (for the polypeptide segment 53–60 of the 434 repressor in 7 M urea this is represented approximately by the conformation space spanned by the 20 conformers in Fig. 1), or **S** and all conformers $\mathbf{U}^i$, respectively.

Deviations of ^{1}H chemical shifts from the random coil values [1] and slowed amide proton exchange in ^2H_2O indicated the presence of non-random structure in the polypeptide segment Val54–Ser–Val–Asp–Trp–Leu–Leu60 of the 434 repressor. For the structure determination we wanted to collect NOE distance constraints between peripheral side chain hydrogen atoms which could be short in a structure **S** but would be too long for NOE observation, that is >5.0 Å, in the flexibly extended forms of the polypeptide chain, $\mathbf{U}^i$ (Fig. 5). Considering the sequence 54–60, ^{13}C-labeling of individual methyl groups in all residues of Val and Leu was an obvious choice for identifying a significant proportion of the desired long-range NOEs [1]. In addition, these methyl resonances were stereospecifically assigned [4•], and NOEs with indole ring protons of Trp58 could be identified from the unique ^{1}H chemical shifts, since residue 58 is the only Trp in the protein. A data set consisting exclusively of NOE upper distance constraints involving indole ring protons of Trp and isopropyl methyls of Val and Leu was collected. Since no absolute NOE intensities were obtained, these data provided no direct evidence for the population $\mathbf{p}_S$ (Fig. 5), except that the nuclear Overhauser enhancement spectroscopy (NOESY) cross peaks would be undetectably weak if the folded conformation ammounted to less than ~10 % of the total protein. However, the fact that the input data could be satisfied by a single, quite well-defined structure (Fig. 1) showed that whatever the population of this structure, all other conformations present in the solution did not significantly contribute to the NOEs in the input for the structure calculation. The aforementioned different averaging of NOE intensities and chemical shifts is clearly manifested in the structure of Fig. 1. Although deviations of the ^{1}H chemical shifts from the random coil values could be qualitatively rationalized by the close approach of the aliphatic side chains to the indole ring of Trp58, ring current calculations for the methyl protons of the 20 conformers in Fig. 1 predicted larger deviations from the random coil shifts than those actually observed, indicating that the observed chemical shifts include contributions from fully unfolded polypeptides with random coil shifts. The chemical shift measurements thus provide the most direct indication that in 7 M urea solution the structure in Fig. 1 is present in an equilibrium of the type shown in Fig. 5, where the conformers $\mathbf{U}^i$ do not contain a compact structure involving the residues 54–59.

It may be instructive at this point to add two comments. First, most of the sequential NOEs between H$^\alpha$ and the amide proton of the following residue, $d_{\alpha N}$ [1], for the polypeptide segment 53–60 of the 434 repressor had also been identified, as had other NOEs between pairs of protons located near each other in the covalent structure. Since $d_{\alpha N}$ varies only between 2.2 and 3.6 Å for all possible conformations [1], the observed NOE intensities are a complex average over non-vanishing contributions from all conformations **S** and $\mathbf{U}^i$ (Fig. 5), and they were therefore not used in the calculation of the structure **S**. Second, non-random structure in polypeptide chains that do not fold back on themselves must be stabilized by short-range interactions, which provides an additional criterion for checks on structure determinations in denatured proteins — that is, a local non-random structure should be conserved in shorter polypeptide fragments. Indeed, the structure of polypeptide segment 53–60 of 434 repressor (Fig. 1) is retained in the fragment 1–63 as well as in the shorter fragment 44–63 [2•]. A different and particularly nice illustration is provided by some old work with the 29-residue polypeptide hormone, glucagon, and the synthetic fragment 1–34 of human parathyroid hormone. In dilute aqueous solution these polypeptides are predominantly in extended flexible forms, with local non-random structure manifested by ^{1}H chemical shifts. Using synthetic polypeptide fragments, this local structure was found to involve the tetrapeptide segment 22–25 in glucagon [25], and the tetrapeptide segment 21–24 in parathyroid hormone [26]. The spatial structure was conserved in the pentapeptide fragments 22–26 and 20–24, respectively, where it could be reversibly denatured by the addition of dimethylsulfoxide [27].

Conclusion

In contrast to globular proteins, the polypeptide chain in denatured states does not fold back on itself, and there is little or no dispersion of the ^{1}H chemical shifts. In this situation, sequence-specific NMR assignments can nonetheless be obtained using isotope-labeling and heteronuclear NMR techniques similar to those applied to bigger, folded proteins. Within the next few years, complete NMR assignments should be obtainable for the unfolded states of a variety of small proteins (M < 20kDa) in use for folding studies, providing a more reliable platform for the analysis of future physico-chemical investigations. Structure determinations in denatured proteins will intrinsically be limited by the presence of conformational polymorphism (Fig. 5). As shown in the example of the 434 repressor, long-range NOE distance constraints with methyl groups will generally play a dominant role, since these are easiest to observe. Stereospecific assignments of the diastereotopic pairs of methyl groups in Val and Leu will be indispensible, and these can be obtained directly for denatured proteins using biosynthetically-directed fractional ^{13}C-labeling [28].

Conceptually, the conformation **S** in Fig. 5 could be considered to represent a 'folding intermediate'. In the absence of non-random structure in the denatured state, which would then consist exclusively of conformations $\mathbf{U}^i$ (Fig. 5), the approaches discussed here could be applied to structure determinations of true folding intermediates, provided these can be accumulated to more than about 10% of the total protein concentration. It should be possible in the near future to determine the presence, or absence of non-random structure in denatured states of at least those few proteins that are most frequently used in physico-chemical studies relating to the protein folding problem. Clearly, we look forward to important contributions by NMR techniques that will establish more reliable foundations for a wide array of investigations in the folding field.

Acknowledgements

NMR studies of denatured proteins in my laboratory involve projects of M Billeter, V Dötsch, D Neri, G Siegal and G Wider. Special thanks go to M Billeter for preparing Fig. 1 and 2, and R Marani for processing the manuscript. Financial support by the Schweizerischer Nationalfonds (project 31.32033.91) is gratefully acknowledged.

References and recommended reading

Papers of particular interest, published within the annual period of review, have been highlighted as:

- • of special interest
- •• of outstanding interest

1. WÜTHRICH K: *NMR of Proteins and Nucleic Acids*. New York: Wiley; 1986.

2. • NERI D, BILLETER M, WIDER G, WÜTHRICH K: **NMR Determination of Residual Structure in a Urea-Denatured Protein, the 434 Repressor.** *Science* 1992, **257**:1559–1563.
Describes the determination of the three-dimensional structure of a hydrophobic cluster in the DNA-binding domain of the 434 repressor in 7M urea.

3. • NERI D, WIDER G, WÜTHRICH K: **Complete ^{15}N and ^{1}H NMR Assignments for the Amino-Terminal Domain of the Phage 434 Repressor in the Urea-Unfolded Form.** *Proc Natl Acad Sci USA* 1992, 89:4397–4401.
Sequence-specific ^{1}H and ^{15}N NMR assignments were obtained for the 69-residue polypeptide chain of the DNA-binding domain of the 434 repressor in 7M urea. Using uniform ^{15}N-labelling, complete ^{15}N and amide proton assignments for the polypeptide backbone were obtained by saturation transfer between the native and the unfolded states of the protein in 4M urea, and ^{1}H assignments for the amino acid side chains resulted from ^{15}N-resolved 2D scalar ^{1}H correlation experiments in 7M urea.

4. • NERI D, WIDER G, WÜTHRICH K; **^{1}H, ^{15}N and ^{13}C NMR Assignments of the 434 Repressor Fragments 1-63 and 44-64 Unfolded in 7 M Urea.** *FEBS Lett* 1992, **303**:129–135.
Sequence-specific ^{1}H and ^{15}N NMR assignments were obtained for the fragments 1–63 and 44–63 of the uniformly ^{15}N labeled DNA-binding domain of the 434 repressor in 7 M urea. In a different protein preparation, the methyl groups of Val and Leu in these two polypeptides were labeled with ^{13}C and stereospecifically assigned, which provided the basis for the structure determination described in [2] above.

5. • LOGAN TM, OLEJNICZAK ET, XU RX, FESIK SW; **A General Method for Assigning NMR Spectra of Denatured Proteins Using 3D HC(CO)NH-TOCSY Triple Resonance Experiments.** *J Biomol NMR* 1993, 3:225–231.
Using the uniformly ^{13}C, ^{15}N-doubly-labeled protein, sequence-specific ^{1}H, ^{13}C, and ^{15}N assignments were obtained for the 107-residue protein FKBP in 6.3 M urea. The resonance assignments relied on 3D heteronuclear scalar correlation spectroscopy. This direct approach is called for whenever suitable conditions cannot be found for magnetization transfer with the native protein, or when the native form has not been assigned.

6. MCDONALD CC, PHILLIPS WD; **Manifestations of the Tertiary Structures of Proteins in High-Frequency Nuclear Magnetic Resonance.** *J Amer Chem Soc* 1967, 89:6332–6341.

7. BAX A, GRZESIEK S: **Methodological Advances in Protein NMR.** *Acc Chem Res* 1993, 26:131–138.

8. DYSON HJ, MERUTKA G, WALTHO JP, LERNER RA, WRIGHT PE: Folding of Peptide Fragments Comprising the Complete Sequence of Proteins. Models for Initiation of Protein Folding I. Myohemerytrin. *J Mol Biol* 1992, 226:795–817.

9. WALTHO JP, FEHER VA, MERUTKA G, DYSON HJ, WRIGHT P: Peptide Models of Protein Folding Initiation Sites. 1. Secondary Structure Formation by Peptides Corresponding to the G- and H-Helices of Myoglobin. *Biochemistry* 1993, 32:6337–6347.

10. SANCHO J, NEIRA JL, FERSHT AR: An N-Terminal Fragment of Barnase has Residual Helical Structure Similar to that in a Refolding Intermediate. *J Mol Biol* 1992, 224:749–758.

11. BYSTROV VF, ARSENIEV AS, GAVRILOV YD; NMR Spectroscopy of Large Peptides and Small Proteins. *J Magn Reson* 1978, 30:151–184.

12. REDFIELD AG, GUPTA RK: Pulsed NMR Study of the Structure of Cytochrome *c*. *Cold Spring Harbor Symposia on Quantitative Biology* 1971, 36:405–411.

13. WÜTHRICH K, WAGNER G, RICHARZ R, PERKINS SJ: Individual Assignments of the Methyl Resonances in the ^{1}H Nuclear Magnetic Resonance Spectrum of the Basic Pancreatic Trypsin Inhibitor. *Biochemistry* 1978, 17:2253–2263.

14. MARION D, DRISCOLL PC, KAY LE, WINGFIELD PT, BAX A, GRONENBORN AM, CLORE GM: Overcoming the Overlap Problem in the Assignment of ^{1}H NMR Spectra of Larger Proteins by Use of Three-Dimensional Heteronuclear ^{1}H-^{15}N Hartmann-Hahn-Multiple Quantum Coherence and Nuclear Overhauser-Multiple Quantum Coherence Spectroscopy: Application to Interleukin 1β. *Biochemistry* 1989, 28:6150–6156.

15. LOGAN TM, OLEJNICZAK ET, XU RX, FESIK SW: Side Chain and Backbone Assignments in Isotopically Labeled Proteins from Two Heteronuclear Triple-Resonance Experiments. *FEBS Lett* 1992, 314:413–418.

16. MONTELIONE GT, LYONS BA, EMERSON SD, TASHIRO M: An Efficient Triple-Resonance Experiment Using Carbon-13 Isotropic Mixing for Determining Sequence-Specific Resonance Assignments of Isotopically Enriched Proteins. *J Amer Chem Soc* 1992, 114:10974–10975.

17. NERI D, BILLETER M, WÜTHRICH K: Determination of the NMR Solution Structure of the DNA-Binding Domain 1-69 of the 434 Repressor and Comparison with the X-Ray Crystal Structure. *J Mol Biol* 1992, 223:743–767.

18. WIDER G, NERI D, WÜTHRICH K: Studies of Slow Conformational Equilibria in Macromolecules by Exchange of Heteronuclear Longitudinal 2-Spin-Order in a 2D Difference Correlation Experiment. *J Biomol NMR* 1991, 1:93–98.

19. OTTING G, WÜTHRICH K: Efficient Purging Scheme for Proton-Detected Heteronuclear Two-Dimensional NMR. *J Magn Reson* 1988, 76:569–574.

20. GLUSHKA J, LEE M, COFFIN S, COWBURN D: ^{15}N Chemical Shifts of Backbone Amides in Bovine Pancreatic Trypsin Inhibitor and Apamin. *J Amer Chem Soc* 1989, 111:7716–7722; 12:2843.

21. BAUM J, DOBSON CM, EVANS PA, HANLEY C: Characterization of a Partly Folded Protein by NMR Methods: Studies on the Molten Globule State of Guinea Pig α-Lactalbumin. *Biochemistry* 1989, 28:7–13.

22. EVANS PA, TOPPING KD, WOOLFSON DN, DOBSON CM: Hydrophobic Clustering in Nonnative States of a Protein: Interpretation of Chemical Shifts in NMR Spectra of Denatured States of Lysozyme. *Proteins* 1991, 9:248–266.

23. TORDA AE, SCHEEK RM, VAN GUNSTEREN WF: Time-Averaged Nuclear Overhauser Effect Distance Restraint Applied to Tendamistat. *J Mol Biol* 1990, 214:223–235.

24. TORDA AE, BRUNNE RM, HUBER T, KESSLER H, VAN GUNSTEREN WF: Structure Refinement Using Time-Averaged J-Coupling Constant Restraints. *J Biomol NMR* 1993, 3:55–66.

25. BÖSCH C, BUNDI A, OPPLIGER M, WÜTHRICH K: ^{1}H Nuclear Magnetic Resonance Studies of the Molecular Conformation of Monomeric Glucagon in Aqueous Solution. *Eur J Biochem* 1978, 91:209–214.

26. BUNDI A, ANDREATTA RH, WÜTHRICH K: Characterization of a Local Structure in the Synthetic Parathyroid Hormone Fragment 1-34 by ^{1}H Nuclear Magnetic Resonance Techniques. *Eur J Biochem* 1978, 91:201–208.

27. BUNDI A, ANDREATTA R, RITTEL W, WÜTHRICH K: Conformational Studies on the Synthetic Fragment 1-34 of Human Parathyroid Hormone by NMR Techniques. *FEBS Lett* 1976, 64:126–129.

28. NERI D, SZYPERSKI T, OTTING G, SENN H, WÜTHRICH K: Sterospecific Nuclear Magnetic Resonance Assignments of the Methyl Groups of Valine and Leucine in the DNA-Binding Domain of the 434 Repressor by Biosynthetically-Directed Fractional ^{13}C Labeling. *Biochemistry* 1989, 28:7510–7516.

K Wüthrich, Institut für Molekularbiologie und Biophysik, Eidgenössische Technische Hochschule-Hönggerberg, CH-8093 Zürich, Switzerland.

APPENDICES

AND

INDICES

A1. Academic and staff appointments

The top part is an alphabethical list of the scientists that were appointed as Assistants or Senior Assistants in my research group at the ETH Zürich from 1970–94. The list includes, in parentheses, the duration of the appointment, the area of the University degrees, and the position held at the end of the employment (PD stands for Privatdozent). The lower part lists the appointments in technical and secretarial staff positions.

Billeter, Martin (1987– , Theoretical physics, Oberassistent/PD)

Braun, Werner A. (1984– , Theoretical physics, Oberassistent/PD)

Brown, Larry R. (1976–81, Chemistry, Assistent)

Brunisholz, René (1994 – , Biochemistry, Oberassistent/PD)

Bundi, Arno (1977–79, Physics, Assistent)

Frank, Gerhard (1994– , Biochemistry, Oberassistent)

Grathwohl, Christoph (1978–80, Physics, Assistent)

Güntert, Peter (1994 – , Theoretical physics, Assistent)

Keller, Regula M. (1977–82, Physics, Oberassistentin)

Leiting, Barbara (1992–93, Molecular Biology, Assistentin)

Leupin, Werner (1983–88, Chemistry, Oberassistent)

Otting, Gottfried (1987–92, Chemistry, Oberassistent)

Senn, Hans (1980–86, Microbiology, Assistent)

Sidler, Walter (1994 – , Microbiology, Oberassistent/PD)

Szyperski, Thomas (1993– , Chemistry, Assistent)

Wagner, Gerhard (1977–87, Physics, Oberassistent/PD)

Wider, Gerhard (1988– , Physics, Oberassistent)

Baumann, Rudolf (1971– , Chemical engineer)

Eugster, Albert (1978–88, Chemical engineer)

Frey, Alexandra S. (1975–77, Chemical engineer)

Hunziker-Kwik, Eng-hiang (1979–88, Chemical engineer)

Braun, Daniel (1988– , Chemical engineer)

Huber, Edith (1979–94, Secretary)

Hug, Rosmarie (1994 – , Secretary)

Marani, Renato (1983 – , Secretary)

Rutz, Lotti (1977–79, Secretary)

Schumacher, Antoinette (1975–77, Secretary)

A2. Ph.D. students

Alphabethical list of my Ph.D. students at the ETH Zürich from 1970–94. Each entry includes the title of the thesis and, in parentheses, the time period spent as a graduate student, the ETH identification number of the thesis, and the area of specialization in which the Ph.D. degree was awarded. Entries without identification number refer to currently active graduate students, with indication of the research themes.

Altmann, Serge: *NMR studies of nucleic acids and nucleic acid–polypeptide complexes.* (1993– , Biochemistry).

Antuch-Garcia, Walfrido: *NMR structure determination of proteinase inhibitors and polypeptide toxins.* (1992– , Chemistry).

Bartels, Christian: *Methoden der Zuordnung mehrdimensionaler magnetischer Kernspinresonanzspektren zur Strukturbestimmung von Makromolekülen.* (1990–94, ETH Nr. 10966, Biology).

Billeter, Martin: *Strukturermittlung von Polypeptiden und kleinen Proteinen auf der Basis von ^{1}H-NMR Daten mittels Computer-Graphik und Optimierung.* (1980–85, ETH Nr. 7723, Theoretical physics).

Bösch, Chris: *Konformationsstudien am Polypeptidhormon Glucagon mittels hochauflösender Kernresonanzspektroskopie in wässriger Lösung und an einer hydrophob/hydrophilen Grenzfläche.* (1976–79, ETH Nr. 6488, Physics).

Bundi, Arno: *Konformationsstudien an kleinen Modellpeptiden und am menschlichen Parathyroidhormon-Fragment 1–34 mittels hochauflösender Kernresonanz.* (1973–77, ETH Nr. 6036, Physics).

Dötsch, Volker: *Charakterisierung der Wechselwirkungen zwischen Proteinen und Lösungsmittelmolekülen.* (1991–94, ETH Nr. 10856, Chemistry).

Fede, Agostino: *NMR studies and solution structure determination of 1:1 complexes formed between Hoechst 33258 and two dodecamer DNA duplexes.* (1989–92, ETH Nr. 9886, Biochemistry).

Fernandez-Estrabao, César: *NMR structure determination of proteins in solution.* (1993– , Chemistry).

Freyberg-Eisenberg, Berthold, Freiherr von: *Studium der Mobilität von Polypeptid- und Proteinstrukturen mittels Monte Carlo-Verfahren und unter Berücksichtigung von Solvatationseffekten.* (1989–93, ETH Nr. 10031, Theoretical physics).

Grathwohl, Christoph: *^{13}C-Kernresonanz-Studien an kleinen Peptiden.* (1971–75, ETH Nr. 5640, Physics).

Grütter, Rolf: *Methodische Aspekte der in vivo ^{31}Phosphor-Kernspinresonanz-Spektroskopie in der pädiatrischen Diagnostik.* (1986–90, ETH Nr. 9138, Physics).

Güntert, Peter: *Neue Rechenverfahren für die Proteinstrukturbestimmung mit Hilfe der magnetischen Kernspinresonanz.* (1988–93, ETH Nr. 10135, Theoretical physics).

Hetzel, Robert: *Untersuchung der räumlichen Strukturen einiger X–Pro-Peptide mit empirischen Energierechnungen.* (1972–78, ETH Nr. 6125, Theoretical physics).

Hochmann, Jiri: *Physikalisch-chemische Studien an Hämoglobinen.* (1971–75, ETH Nr. 5772, Physics).

Hideo, Iwai: *Cloning, expression and NMR studies of stability and hydration of recombinant mutants of 434 repressor.* (1994– , Biology).

Hyberts, Sven G.: *NMR solution structure of the proteinase inhibitor eglin c. Development of methods and ensemble statistics.* (1985–92, ETH Nr. 10006, Physics).

Koradi, Reto: *Computer support of protein structure determination by NMR.* (1992– , Computer sciences).

Luginbühl, Peter: *Theoretical foundations and experimental implementation of NMR studies on protein dynamics in denaturing solvents.*(1991– , Theoretical physics).

Masson, André: *Konformationsstudien an Calcitonin M und am Trypsin-Inhibitor BPTI mittels hochauflösender Kernresonanz.* (1970–73, ETH Nr. 5229, Physics).

Méraldi, Jean-Paul: *Etude des conformations moléculaires de cinq pentapeptides cycliques par la résonance magnétique nucléaire des protons et par la méthode des potentiels semi-empiriques d'énergie.* (1970–74, ETH Nr. 5283, Physics).

Mumenthaler, Christian: *Theoretical investigations on protein structure, protein dynamics and protein folding.* (1994– , Physics).

Ottiger, Marcel: *NMR structure determination of human cyclophilin and studies of chaperone-bound states of cyclophilin.* (1992– , Biochemistry).

Otting, Gottfried H.: *Strukturermittlung an kleinen Proteinen mit NMR: Neue Methoden und Anwendungen am Beispiel der Cardiotoxine von Naja Mossambica Mossambica und des P22 c2 Repressors.* (1984–87, ETH Nr. 8314, Chemistry).

Pellecchia, Maurizio: *NMR structure determination of the heat shock factor Dna J and studies of the molecular dynamics.* (1993– , Chemistry).

Qian, Yan-Qiu: *Determination of the NMR solution structures of the Antennapedia homeodomain and an Antennapedia homeodomain–DNA complex.* (1989–92, ETH Nr. 9972, Physics).

Richarz, René: *^{13}C Kernresonanzstudien an kleinen Modellpeptiden sowie am basischen pankreatischen Trypsin-Inhibitor und seinem Komplex mit Proteasen.* (1976–79, ETH Nr. 6495, Physics).

Roder, Heinrich: *Interne Mobilität in Proteinen unter nativen und denaturierenden Bedingungen: Untersuchung von Trypsin-Inhibitoren mit spektroskopischen Methoden.* (1978–81, ETH Nr. 6932, Physics).

Schaumann, Thomas: *Strukturverfeinerung von Polypeptiden und Proteinen mit Newton-Raphson-Energieminimierung.* (1984–88, ETH Nr. 8489, Theoretical physics).

Schott, Oliver: *NMR structure determination of the DNA-binding domain of the transcription factor LFB 1 and its complex with the operator DNA.* (1993– , Physics).

Schultze, Peter A. : *Numerische und Computergraphische Modellstudien an DNA und Metallothionein.* (1983–88, ETH Nr. 8547, Chemistry).

Senn, Hans: *Zusammenhänge zwischen Aminosäuresequenz, Häm-Eisen-Koordinations-geometrie und funktionellen Eigenschaften in Cytochromen c: ^{1}H-NMR-Studien.* (1979–83, ETH Nr. 7314, Microbiology).

Spitzfaden, Claus: *Strukturbestimmung des Cyclophilin–Cyclosporin-Komplexes mittels NMR-Spektroskopie.* (1989–93, ETH Nr. 10406, Chemistry).

Szyperski, Thomas: *Charakterisierung der strukturellen Eigenschaften von Hirudin mit NMR in Lösung sowie Entwicklung und Anwendung neuer NMR-Methoden zur Untersuchung von Proteinen in Lösung.* (1989–92, ETH Nr. 9981, Chemistry).

Wagner, Gerhard: *Konformation und Dynamik von Protease-Inhibitoren: ^{1}H NMR-Studien.* (1973–77, ETH Nr. 5992, Physics).

Weber, Christoph B.: *Ermittlung der räumlichen Struktur von an Cyclophilin gebundenem Cyclosporin A mittels isotopengefilterter Protonenresonanzspektroskopie in wässriger Lösung.* (1987–90, ETH Nr. 9343, Biochemistry).

Wider, Gerhard: *Zweidimensionale Kernresonanz-Spektroskopie von Polypeptiden und Proteinen. Anwendung für Konformationsstudien von an volldeuterierte Lipid-Micellen gebundenem Glucagon.* (1978–82, ETH Nr. 7040, Physics).

Widmer, Hans: *Ermittlung der räumlichen Struktur des Seeanemonentoxins ATX Ia mittels Kernresonanzspektroskopie in Lösung und Entwicklung eines Simulationsprogramms zur Analyse der Kreuzpeakfeinstrukturen in zweidimensionalen Korrelationsspektren.* (1983–87, ETH Nr. 8369, Chemistry).

Xia, Tai-he: *Software for Determination and Visual Display of NMR Structures of Proteins: the Distance Geometry Program DGPLAY and the Computer Graphics Programs CONFOR and XAM.* (1987–92, ETH Nr. 9831, Theoretical physics).

A3. Undergraduate students

Alphabethical list of the undergraduate students at the ETH Zürich from 1970–94 who worked in my laboratory for about 6 months on their diploma thesis. The list includes, in parentheses, the year of graduation and the subject of the diploma degree.

Amacker, Hugo (1988, Physics)

Arnet, Cyrill (1995, Biochemistry)

Billeter, Martin (1980, Theoretical physics)

Bösch, Christoph (1976, Physics)

Brühwiler, Daniel (1986, Physics)

Grilc, Matej (1993, Physics)

Dubs, Andreas (1978, Physics)

Denk, Winfried (1984, Physics)

Elber, Kaspar (1972, Chemistry)

Fede, Agostino (1989, Biochemistry)

Gremlich, Hans-Ulrich (1975, Chemistry)

Grütter, Rolf (1986, Physics)

Güntert, Peter (1987, Theoretical physics)

Hänggi, Gabriel (1992, Physics)

Hässler, Martin (1983, Physics)

Hyberts, Sven (1984, Physics)

Kaissl, Wolfgang (1988, Physics)

Luginbühl, Peter (1991, Theoretical physics)

Meier, Willi (1975, Chemistry)

Moonen, Chrit (1979, Biology)

Mumenthaler, Christian (1994, Physics)

Phuntsok, Ernst (1984, Chemistry)

Picot, Daniel (1978, Physics)

Richarz, René (1975, Physics)

Riek, Roland (1995, Physics)

Riva, Francesco (1986, Chemistry)

Roder, Heinrich (1978, Physics)

Schaefer, Nikolaus (1992, Mathematics)

Schaffner, Johannes (1979, Physics)

Schick, Martin (1991, Physics)

Schmidiger, Markus (1979, Physics)

Schott, Oliver (1992, Physics)

Sehr, Peter (1974, Chemistry)

Teleman, Olle (1980, Physics)

Vranesic, Ivan (1988, Physics)

A4. Postdoctoral fellows and visiting scientists

Alphabetical list of the visitors that joined my research group at the ETH Zürich from 1970–94 either as visiting faculty, postdoctoral fellows, or as postgraduate exchange students. The list includes, in parentheses, identification of visiting faculty by the letter S, for "sabbatical", the period of the stay in Zürich, and the area of the University degrees.

Anil-Kumar (1979–80/S84, Physics)
Arseniev, Alexandre (1981/S86, Chemistry)
Avilés, Francesco (S1987, Biochemistry)
Bachmann, Peter (1976–79, Chemistry)
Berndt, Kurt (1989–92, Chemistry)
Braun, Werner (1978–82, Theoretical physics)
Brown, Larry (1975–76/S91–92, Chemistry)
Bushweller, John H. (1990–93, Chemistry)
Celda, Bernardo (1985–86, Chemistry)
Chary, Kandala V. R. (1988–90, Physics)
Chazin, Walter J. (1983–85, Chemistry)
Chreszczyk, Adela (1978–80, Chemistry)
Coll, Miguel S. (1990, Biochemistry)
Darbon, Hervé (1987–88, Chemistry)
DeMarco, Antonio (1974–76, Chemistry)
Donzel, Bernard (1971–72, Chemistry)
Eccles, Craig D. (1987–90, Physics)
Farmer, Bennett Th. (1987, Chemistry)
Frey, Michael H. (1983–84, Chemistry)
Gilboa, Aaron J. (S1974–75, Biochemistry)
Gō, Mitiko (S1979, Theoretical physics)
Gō, Nobuhiro (S1979, Theoretical physics)
Gordon, Sidney L. (S1977–78, Chemistry)
Guerlesquin, Françoise (1982–83, Chemistry)
Haruyma, Hideyuki (1986–88, Chemistry)
Havel, Timothy (1982–84, Biology)
Hosur, Ramakrishna (1981–83, Chemistry)
Hua, Qingxin (1987, Biophysics)
Kline, Allen D. (1984–87, Chemistry)
Kovacs, Helena (1992–93, Chemistry)
Johansson, Jan (1992–93, Medicine)
Labhardt, Alex M. (1984–85, Biophysics)
Lauterwein, Jürgen (1976–78, Physics)
Lee, Kong H. (1980–83, Chemistry)
Leiting, Barbara (1990–92, Biochemistry)
Liepinsh, Edvards (1990–92, Chemistry)
Llinás, Miguel (1974–76, Biophysics)
Loth, Klaus (1976–78, Chemistry)
Macura, Slobodan (1980–81, Physics)
Maeda, Tadakazu (S1990, Physics)
Marion, Dominique (1982, Chemistry)
Messerle, Barbara (1987–89, Chemistry)
Montelione, Gaetano(1984–86, Biochemistry)
Moore, Jonathan M. (1989–90, Chemistry)
Mronga, Siggi (1991–93, Chemistry)
Müller, Norbert (1984–85, Chemistry)
Nagayama, Kuniaki (1976–79, Physics)
Neuhaus, David (1982–83, Chemistry)
Ohlenschläger, Oliver (1994, Chemistry)
Orbons, Leonard P. M. (1987–90, Chemistry)
O'Connell, John F. (1990–92, Chemistry)
Pardi, Arthur (1982–83, Chemistry)
Perkins, Stephen J. (1976–78, Biochemistry)
Pervushin, Konstantine (1994– , Physics)
Prêcheur, Bénédicte (1994, Chemistry)
Qian, Yan-Qiu (1987–89, Physics)

Ramaprasad, Subbaraya (1979–80, Physics)
Redwine, Oscar (1986–87, Chemistry)
Stassinopoulou, Cariclia (1976/S82, Chemistry)
Steinmetz, Wayne (S1979–80/86, Chemistry)
Strop, Petr (1981–82, Chemistry)
Satoshi, Ebina (1983–84, Chemistry)
Sevilla, Maria Paz (1988–89, Chemistry)
Siegal, Gregory (1992–94, Biochemistry)
Sodano, Patrick (1988–91, Chemistry)
Targonski, Jolan (1972–73, Mathematics)
Vendrel, Josep (1989–90, Biochemistry)
Viti, Vincenza (1976, Physics)
Wang, Qiwen (S1984–85, Chemistry)
Watnick, Paula Y. (1982–83, Chemistry)
Williamson, Michael P. (1982–83, Chemistry)
Wörgötter, Erich (1984–86, Chemistry)
Wynants, Chantal (1984, Chemistry)
Yabuki, Sadato H. (S1979, Physics)
Zerbe, Oliver (1994– , Chemistry)
Zuiderweg, Erik R. P. (1982–83, Chemistry)

A5. List of outside collaborations

Alphabethical list of scientists working outside of my research group with whom we co-authored publications during the period 1970–94. The list includes the affiliation at the time of the collaboration.

Affolter, M., Biozentrum der Universität, Basel, Switzerland.

Alagón, A.C., Instituto de Investigaciones Biomédicas, Universidad Nacional Autónoma de México, Mexico.

Altschuh, D., Institut de Biologie Moléculaire et Cellulaire du C.N.R.S., Strasbourg, France.

Andreatta, R., Forschungslaboratorium Pharmazeutika, Ciba-Geigy AG, Basel, Switzerland.

Arens, A., Bayer AG, Wuppertal, Germany.

Aslund, F., Department of Chemistry, Karolinska Institute, Stockholm, Sweden.

Assmann, G., Institut für Klinische Chemie und Laboratoriumsmedizin, Westfälische Wilhelms-Universität, Münster, Germany.

Aue, W.P., Laboratorium für Physikalische Chemie, ETH Zürich, Switzerland.

Avilés, F.X., Departament de Bioqímica i Biologia Molecular, Universitat Autónoma de Barcelona, Spain.

Aviram, I., Department of Biochemistry, Tel-Aviv University, Israel.

Bachmann, P., Laboratorium für Physikalische Chemie, ETH Zürich, Switzerland.

Bender, R., Hoechst AG, Frankfurt, Germany.

Betz, A., Department of Haematology, MRC Centre, University of Cambridge, UK.

Beunink, J., Verfahrensentwicklung Biochemie, Bayer AG, Wuppertal, Germany.

Björnberg, O., Department of Chemistry, Karolinska Institute, Stockholm, Sweden.

Bode, W., Max-Planck-Institut für Biochemie, Martinsried, Germany.

Bodenhausen, G., Laboratorium für Physikalische Chemie, ETH Zürich, Switzerland.

Boelens, R., Department of Physical Chemistry, University of Groningen, The Netherlands.

Böhme, H., Fakultät für Biologie, Universität Konstanz, Germany.

Bougis, P.E., Laboratoire de Biochimie, Institut National de la Recherche Scientifique, Marseille, France.

Bradshaw, R.A., Department of Biological Chemistry, University of California, Irvine, USA.

Brunne, R.M., Laboratorium für Physikalische Chemie, ETH Zürich, Switzerland.

Brunner, P., Spectrospin AG, Fällanden, Switzerland.

Brunori, M., Dipartimento di Scienze Biochimiche, Università degli Studi di Roma, Italy.

Bruschi, M., Laboratoire de Chimie Bactérienne, C.N.R.S., Marseille, France.

Bur, D., Präklinische Forschung, F. Hoffmann-La Roche AG, Basel, Switzerland.

Burgess, A.W., Ludwig Institute for Cancer Research, Melbourne, Australia.

Carlsson, F.H.H., National Chemical Research Laboratory, Pretoria, South Africa.

Cechová, D., Institute of Molecular Genetics, Czechoslovak Academy of Sciences, Prague, CSSR.

Chávez, M.A., Facultad de Biología, Universidad de la Habana, Cuba.

Coll, M., Unitat de Química Macromolecular , Enginyers Industrials, Barcelona, Spain.

Cortese, R., Istituto di Ricerche di Biologia Molecolare P. Angeletti, Roma, Italy.

Creighton, T.E., MRC Laboratory of Molecular Biology, Cambridge, UK.

Curstedt, T., Department of Clinical Chemistry, Karolinska Institute at Danderyd Hospital, Danderyd, Sweden.

Cusanovich, M.A., Department of Biochemistry, University of Arizona, Tucson, AZ, USA.

Debrunner, P.G., Department of Physics, University of Illinois, Urbana, IL, USA.

Delfín, J., Facultad de Biología, Universidad de la Habana, Cuba.

Dorn, A., Präklinische Forschung, F. Hoffmann-La Roche AG, Basel, Switzerland.

Duc, G., Universitäts-Frauenklinik, Zürich, Switzerland.

De Francesco, R., Istituto di Ricerche di Biologia Molecolare P. Angeletti, Roma, Italy.

Deisenhofer, J., Max-Planck-Institut für Biochemie, Martinsried, Germany.

Denny, W.A., Cancer Research Laboratory, University of Auckland, Auckland, New Zealand.

Dent M.A.R., Instituto de Investigaciones Biomédicas, Universidad Nacional Autónoma de México, Mexico.

Eschenmoser, A., Laboratorium für Organische Chemie, ETH Zürich, Switzerland.

Engeli, M., FIDES Treuhandgesellschaft, Zürich, Switzerland.

Engels, J.W., Institut für Organische Chemie, J.W. Goethe Universität, Frankfurt, Germany.

Epp, O., Max-Planck-Institut für Biochemie, Martinsried, Germany.

Ernst, R.R., Laboratorium für Physikalische Chemie, ETH Zürich, Switzerland.

Fanconi, A., Universitäts-Kinderklinik, Zürich, Switzerland.

Furukubo-Tokunaga, K., Biozentrum der Universität, Basel, Switzerland.

Fuchs, J.A., Department of Biochemistry, University of Minnesota, St. Paul, MN, USA.

Gait, M.J., MRC Laboratory of Molecular Biology, Cambridge, UK.

Ganesh, K.N., MRC Laboratory of Molecular Biology, Cambridge, UK.

Gehring, W.J., Biozentrum der Universität, Basel, Switzerland.

Georgopoulos, C., Département de Biochimie Médicale, Université de Genève, Switzerland.

Giacometti, G., Dipartimento di Scienze Biochimiche, Università degli Studi di Roma, Italy.

Gibson, K.D., Department of Chemistry, Cornell University, Ithaca, NY, USA.

Goldenberg, D.P., MRC Laboratory of Molecular Biology, Cambridge, UK.

Griesinger, C., Laboratorium für Physikalische Chemie, ETH Zürich, Switzerland.

Grossenbacher, H., Abteilung Biotechnologie, Ciba-Geigy AG, Basel, Switzerland.

Groudinsky, O., Centre de Génétique Moléculaire du CNRS, Gif-sur-Yvette, France.

Guasch, A., Departament de Bioqímica i Biologia Molecular, Universitat Autónoma de Barcelona, Spain.

Formanek, H., Max-Planck-Institut für Biochemie, Martinsried, Germany.

Hawthorne, T., Abteilung Biotechnologie , Ciba-Geigy AG, Basel, Switzerland.

Höhener, A., Laboratorium für Physikalische Chemie, ETH Zürich, Switzerland.

Holmgren, A., Department of Chemistry, Karolinska Institute, Stockholm, Sweden.

Huber, R., Max-Planck-Institut für Biochemie, Martinsried, Germany.

Jeener, J., Pool de Physique, Université Libre de Bruxelles, Belgium.

Joubert, F.J., National Chemical Research Laboratory, Pretoria, South Africa.

Kägi, J.H.R., Biochemisches Institut, Universität Zürich, Switzerland.

Kallen, J., Präklinische Forschung, Sandoz Pharma AG, Basel

Kamber, B., Forschungslaboratorium Pharmazeutika, Ciba-Geigy AG, Basel, Switzerland.

Kaptein, R., Department of Physical Chemistry, University of Groningen, The Netherlands.

Klein, M.P., Chemical Biodynamics Laboratory, University of California, Berkeley, CA, USA.

Kogler, H., Laboratorium für Physikalische Chemie, ETH Zürich, Switzerland.

Koller, K.P., Hoechst AG, Frankfurt, Germany.

Kren, B., Department of Biochemistry, University of Minnesota, St. Paul, MN, USA.

Leumann, C., Laboratorium für Organische Chemie, ETH Zürich, Switzerland.

Leupin, W., Präklinische Forschung, F. Hoffmann-La Roche AG, Basel, Switzerland.

Levitt, M.H., Laboratorium für Physikalische Chemie, ETH Zürich, Switzerland.

Lazdunski, M., Faculté des Sciences, Université de Nice, France.

Linder, M., Laboratorium für Physikalische Chemie, ETH Zürich, Switzerland.

Luporini, P., Dipartimento di Biologia Molecolare, Università degli Studi di Camerino, Italy.

Marchot, P., Laboratoire de Biochimie, Institut National de la Recherche Scientifique, Marseille, France.

Martin, E., Universitäts-Kinderklinik, Zürich, Switzerland.

Meier, B.U., Laboratorium für Physikalische Chemie, ETH Zürich, Switzerland.

Memmert, K., Präklinische Forschung, Sandoz Pharma AG, Basel, Switzerland.

Mertens, M.L., Biochemisches Institut, Universität Zürich, Switzerland.

Minganti, C., MRC Laboratory of Molecular Biology, Cambridge, UK.

Möschler, H.J., Institut für Molekularbiologie und Biophysik, ETH Zürich, Switzerland.

Müller, L., Laboratorium für Physikalische Chemie, ETH Zürich, Switzerland.

Müller M., Biozentrum der Universität, Basel, Switzerland.

Müri, M., Spectrospin AG, Fällanden, Switzerland.

Nice, E.C., Ludwig Institute for Cancer Research, Melbourne, Australia.

Ortenzi, C., Dipartimento di Biologia Molecolare, Università degli Studi di Camerino, Italy.

Pecht, I., Department of Chemical Immunology, Weizmann Institute of Science, Rehovot, Israel.

Percival-Smith, A., Biozentrum der Universität Basel, Switzerland.

Pettigrew, G.W., Department of Molecular Biology and Biochemistry, University of Edinburgh, Scotland.

Pfändler, P., Laboratorium für Physikalische Chemie, ETH Zürich, Switzerland.

Plückthun, A., Max-Planck-Institut für Biochemie, Martinsried, Germany.

Possani, L., Instituto de Investigaciones Biomédicas, Universidad Nacional Autónoma de México, Mexico.

Rapoport, H., Department of Chemistry, University of California, Berkeley, CA, USA.

Rauenbusch, E., Bayer AG, Wuppertal, Germany.

Resendez-Perez, D., Biozentrum der Universität, Basel, Switzerland.

Rink, H., Forschungslaboratorium Pharmazeutika, Ciba-Geigy AG, Basel, Switzerland.

Rittel, W., Forschungslaboratorium Pharmazeutika, Ciba-Geigy AG, Basel, Switzerland.

Robbins, A.H., Miles Research Center, West Haven, CT, USA.

Rochat, H., Laboratoire de Biochimie, Institut National de la Recherche Scientifique, Marseille, France.

Roth, H.J., Laboratorium für Organische Chemie, ETH Zürich, Switzerland.

Schäffer, A., Biochemisches Institut der Universität, Zürich, Switzerland.

Scharf, M., Institut für Organische Chemie, J. W. Goethe-Universität, Frankfurt, Germany.

Scheek, R.M., Department of Physical Chemistry, University of Groningen, The Netherlands.

Schejter, A., Department of Biochemistry, Tel-Aviv University, Israel.

Scheraga, H.A., Department of Chemistry, Cornell University, Ithaca, NY, USA.

Schiffer, C.A., Laboratorium für Physikalische Chemie, ETH Zürich, Switzerland.

Schröder, W., Bayer AG, Wuppertal, Germany.

Schutt, H., Bayer AG, Wuppertal, Germany.

Schweitz, H., Faculté des Sciences, Université de Nice, France.

Schwotzer, W., Institut für Organische Chemie, Universität Zürich, Switzerland.

Schwyzer, R., Institut für Molekularbiologie und Biophysik, ETH Zürich, Switzerland

Seedorf, U., Institut für Arterioskleroseforschung, Westfälische Wilhelms-Universität, Münster, Germany.

Senn, H., Präklinische Forschung, Sandoz Pharma AG, Basel, Switzerland.

Siekmann, J., Lehrstuhl für Biochemie,Universität Bielefeld, Germany.

Sørensen, O.W., Laboratorium für Physikalische Chemie, ETH Zürich, Switzerland.

Sproat, B.S., MRC Laboratory of Molecular Biology, Cambridge, UK.

Stone, S.R., University of Cambridge, Department of Haematology, MRC Centre, Cambridge, UK.

Stout, C.D., The Scripps Research Institute, La Jolla, CA, USA.

Thierry, C., Institut de Biologie Moléculaire et Cellulaire du C.N.R.S., Strasbourg, France.

Tomei, L., Istituto di Ricerche di Biologia Molecolare P. Angeletti, Roma, Italy.

Traber R., Präklinische Forschung, Sandoz Pharma AG, Basel, Switzerland.

Truscheit, E., Bayer AG, Wuppertal, Germany.

Tschesche, H., Lehrstuhl für Biochemie, Universität Bielefeld, Germany.

Tulinsky, A., Department of Chemistry, Michigan State University, East Lansing, MI, USA.

Tun-Kyi, A., Institut für Molekularbiologie und Biophysik, ETH Zürich, Switzerland.

Vasak, M., Biochemisches Institut, Universität Zürich, Switzerland.

Vendrell, J., Departament de Bioquímica i Biologia Molecular, Universidad Autónoma de Barcelona, Spain.

Villegas, V., Departament de Bioquímica i Biologia Molecular, Universidad Autónoma de Barcelona, Spain.

van Gunsteren, W.F., Laboratorium für Physikalische Chemie, ETH Zürich, Switzerland.

Vincent, J.P., Faculté des Sciences, Université de Nice, France.

Visser, L., National Chemical Research Laboratory, Pretoria, South Africa.

Vix, O., Institut de Biologie Moléculaire et Cellulaire du C.N.R.S., Strasbourg, France.

Vogel, R., Institut für Molekularbiologie und Biophysik, ETH Zürich, Switzerland.

von Philipsborn, W., Institut für Organische Chemie, Universität Zürich, Switzerland.

Walker, R., Department of Chemistry, University of California, Berkeley, CA, USA.

Walkinshaw, M.D., Präklinische Forschung, Sandoz Pharma AG, Basel, Switzerland.

Wall, D., Department of Molecular Biology, University of Utah, Salt Lake City, UT, USA.

Weber, H.P., Präklinische Forschung, Sandoz Pharma AG, Basel, Switzerland.

Wenzel, H.R., Lehrstuhl für Biochemie, Universität Bielefeld, Germany

Wider, G., Spectrospin AG, Fällanden, Switzerland.

Widmer, H., Präklinische Forschung, Sandoz Pharma AG, Basel, Switzerland.

Winterhalter, K.H., Friedrich Miescher-Institut, Basel, Switzerland.

Wokaun, A., Laboratorium für Physikalische Chemie, ETH Zürich, Switzerland.

Wagner, G., Department of Biological Chemistry and Molecular Pharmacology, Harvard Medical School, Boston, MA, USA.

Zahn, R., Max-Planck-Institut für Biochemie, Martinsried, Germany.

Zurini, M.G.M., Präklinische Forschung, Sandoz Pharma AG, Basel, Switzerland.

AUTHOR INDEX

The first pages of the articles contributed by Wüthrich's coauthors are listed.

SUBJECT INDEX

For each keyword the first pages of the most relevant articles are listed.